Introduction to Health and Safety in Construction

nebosh

Endorsed by NEBOSH

This publication is endorsed by NEBOSH as offering high quality support for the delivery of NEBOSH qualifications. NEBOSH endorsement does not imply that this publication is essential to achieve a NEBOSH qualification, nor does it mean that this is the only suitable publication available to support NEBOSH qualifications. No endorsed material will be used verbatim in setting any NEBOSH examination and all responsibility for the content remains with the publisher. Copies of official specifications for all NEBOSH qualifications may be found on the NEBOSH website – **www.nebosh.org.uk**

Fifth Edition

Introduction to Health and Safety in Construction

For the NEBOSH
National Certificate in Construction Health and
Safety

Phil Hughes MBE, MSc, CFIOSH

Chairman NEBOSH 1995–2001. President of IOSH 1990–1991

Ed Ferrett PhD, BSc (Hons Eng), CEng, MIMechE, MIET, CMIOSH

Vice Chairman NEBOSH 1999–2008

Endorsed by NEBOSH

nebosh
Endorsed by NEBOSH

Routledge
Taylor & Francis Group

LONDON AND NEW YORK

Fifth edition published 2016
by Routledge
2 Park Square, Milton Park, Abingdon, Oxon OX14 4RN

and by Routledge
711 Third Avenue, New York, NY 10017

Routledge is an imprint of the Taylor & Francis Group, an informa business

First edition published by Butterworth Heinemann in 2005
Third edition published by Butterworth Heinemann in 2008
Fourth edition published by Routledge 2011

Notice
No responsibility is assumed by the publisher for any injury and/or damage to persons or property
as a matter of products liability, negligence or otherwise, or from any use or operation of any
methods, products, instructions or ideas contained in the material herein. Because of rapid
advances in the medical sciences, in particular, independent verification of diagnoses and drug
dosages should be made.

Trademark notice: Product or corporate names may be trademarks or registered trademarks, and
are used only for identification and explanation without intent to infringe.

British Library Cataloguing-in-Publication Data
A catalogue record for this book is available from the British Library

Library of Congress Cataloging in Publication Data
Hughes, Phil (Phillip W.), 1946- , author.
Introduction to health and safety in construction : for the NEBOSH national certificate in
construction health and safety / Phil Hughes and Ed Ferrett. -- Fifth edition.
p. ; cm.
Includes bibliographical references and index.
I. Ferrett, Ed., author. II. Title.
[DNLM: 1. National Examination Board in Occupational Safety and Health. 2. Occupational Health–
Great Britain–Examination Questions. 3. Accidents, Occupational–prevention & control–Great
Britain–Examination Questions. 4. Safety Management–Great Britain–Examination Questions.
5. Workplace–legislation & jurisprudence–Great Britain–Examination Questions. WA 18.2]
RC965.C75
362.110684–dc23
2015011445

ISBN: 978-0-415-82436-1 (pbk)
ISBN: 978-1-315-85870-8 (ebk)

Typeset in Univers Lt by
Servis Filmsetting Ltd, Stockport, Cheshire

Printed by Bell and Bain Ltd, Glasgow

MIX
Paper from
responsible sources
FSC® C007785

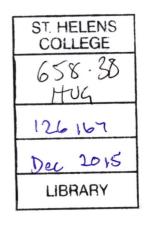

Contents

Contents

vii

Contents

List of illustrations

Tables

Boxes

Preface to the fifth edition

The *Introduction to Health and Safety in Construction* has quickly established itself as the standard text for students taking the NEBOSH National Certificate in Construction Health and Safety, and for those taking other courses in building or construction. It is also of great value to those working in the construction industry at all levels – particularly construction site managers and foremen. As it has become a significant work of reference for managers with health and safety responsibilities, it is a matter of prime importance that it should be kept up to date, as far as is possible, with new legislation and recent developments.

There has been concern over a number of years at the poor record of health and safety in the construction industry. The legal health and safety requirements for all places of work are numerous and complex; it is the intention of the authors to offer an introduction to the subject for all those who have the maintenance of good health and safety standards as part of their employment duties or those who are considering the possibility of a career as a health and safety professional. Health and safety is well recognised as an important component of the activities of any organisation, not only because of the importance of protecting people from harm but also because of the growth in the direct and indirect costs of accidents. These costs have increased higher than the rate of retail price inflation by a considerable amount in the last few years as the number of civil claims and awards have risen each year. It is very important that basic health and safety legal requirements are clearly understood by all organisations, whether public or private, large or small. A good health and safety performance is normally only achieved when health and safety is effectively managed so that significant risks are identified and reduced by adopting appropriate high quality control measures.

The NEBOSH National Certificate in Construction Health and Safety is established as a leading health and safety qualification for the construction industry, with over 15,000 successful candidates. It is designed for supervisors and managers within the construction industry and to provide a sound breadth of underpinning knowledge that enables them to discharge more effectively their duties with respect to health and safety in construction activities. Many larger construction organisations choose the NEBOSH National Construction Certificate as a key part of their supervisors' or management development programme. By ensuring that line managers have a sound understanding of the principles of risk management they build an effective safety culture in the company. Smaller construction organisations often choose the NEBOSH National Construction Certificate as the appropriate qualification for the manager taking the lead on health and safety issues.

The course is divided into three distinct units, each of which is assessed separately. The three units are: NGC1 – Management of health and safety, NCC1 – Managing and controlling hazards in construction activities and NCC2 – Construction health and safety practical application. This development offers the opportunity for additional and more flexible course formats and students may now study parallel courses (in, say, general health and safety and fire) without repeating the management unit. Students who decide to take individual units will, on passing, receive a Unit Certificate. However, it has necessitated the need for an additional chapter (Chapter 7) on construction law and management to deal with those construction topics that were in the original management syllabus.

This fifth edition has been produced to include all the recent syllabus changes and to update the health and safety legislation contained within it, with particular regard to the following changes in legislation:

- ▶ The Enterprise and Regulatory Reform Act 2013 (Section 69)
- ▶ The Health and Safety at Work etc. Act 1974 (Civil Liability) (Exceptions) Regulations 2013
- ▶ The Report of Injuries, Diseases and Dangerous Occurrences Regulations 2013 (RIDDOR)
- ▶ The Control of Asbestos Regulations 2012
- ▶ Classification, Labelling and Packaging of Substances and Mixtures Regulation (European) adopting into EU UN Globally Harmonised System of Classification and Labelling of Chemicals (GHS);
- ▶ Construction (Design and Management) (CDM) Regulations 2015
- ▶ The revocation of The Notification of Conventional Tower Cranes Regulations 2010

▶ The revocation of The Construction (Head Protection) Regulations 1969

▶ The revocation of The Site Waste Management Plans Regulations 2008.

Other changes made by the Health and Safety Executive (HSE) have also been reflected in this edition. These include:

▶ Fees for intervention

▶ The withdrawal of the Approved Code of Practice (L21) for the Management of Health and Safety at Work Regulations

▶ The revised health and safety management system outlined in HSG65 – Managing for health and safety.

The publication of the amended HSG65 – Managing for health and safety, recommends a new model for health and safety management based on the 'Plan, Do, Check, Act' principle replacing the 'Policy, Organising, Planning, Measuring performance, Auditing and Review (POPMAR)' model. This has produced a significant change to NGC1 – the management unit. This change produced a very large Do element which we have split into two chapters – Do1 that covers 'organising' and Do2 that covers 'risk assessment and controls' resulting in an extra chapter in this edition. The other major change to NGC1 is the Construction (Design and Management) (CDM) Regulations 2015.

In a similar way, the hazards unit NCC1 syllabus has been amended to reflect changes to the Construction (Design and Management) (CDM) Regulations 2015 and other relevant revoked legislation. The tutor references for all elements have been updated.

Since the first edition of this book was published, NEBOSH has allowed us to use past NEBOSH examination questions at the end of each chapter. Over the last few years, it has become evident that a small number of candidates have memorised these questions and the contents of the accompanying examiner reports. As a result of this problem, NEBOSH has withdrawn permission to use past examination questions and changed the format of examiner reports. We have, therefore, included our own questions at the end of each chapter using the NEBOSH format. Candidates that can successfully answer these questions unaided should have no problems in the examinations.

NEBOSH is anxious to dispel the myths surrounding their examinations and have provided training courses for course providers to introduce changes to the syllabuses and to answer any queries so that their students get the best possible preparation for the assessment tasks. The NEBOSH website is also a very useful channel of communication with course providers and students.

As mentioned in previous editions, it is the policy of NEBOSH to examine new relevant legislation 6 months after its introduction whether it is specifically mentioned in the syllabus document or not.

This book has been written specifically for students who are studying for the NEBOSH National Certificate in Construction Safety and Health but will also be useful for those studying a variety of building and construction courses, such as the Higher National Certificate and Diploma. It was felt appropriate to produce a textbook which mirrored the Construction Certificate syllabus in its revised unitised form and in a single volume to the required breadth and depth. The syllabus, which follows the general pattern for health and safety management set by the Health and Safety Executive in their guidance HSG65, is risk and management based so it does not start from the assumption that health and safety is best managed by looking first at the causes of failures. Fortunately, failures such as accidents and ill-health are relatively rare and random events in most workplaces.

The book is also intended as a useful reference guide for managers and directors with health and safety responsibilities and for safety representatives. Chapter 19 summarises all the most commonly used Acts and Regulations. It was written to provide an easily accessible reference source for students during and after the course and many others in industry and commerce such as managers, supervisors and safety representatives.

Finally, since one of the objectives of the book is to provide a handbook for the use of any person who has health and safety as part of their responsibilities, we thought that it would be helpful to add a few useful topics which are outside the syllabus. These include fast-track settlement of compensation claims following the Woolfe reforms (Chapter 5) and demolition using explosives (Chapter 18). We have also added a chapter on the international and environmental aspects of construction health and safety that are not included in the NCC syllabus. This will be useful for those who need to travel and work overseas.

We hope that you find this new edition to be useful.

Phil Hughes
Ed Ferrett

Acknowledgements

Throughout the book, definitions used by the relevant legislation and the Health and Safety Executive and advice published in Approved Codes of Practice or various Health and Safety Commission/Executive publications have been utilised. Most of the references produced at the end of each Act or Regulation summary in Chapter 19 are drawn from the HSE Books range of publications.

The authors' grateful thanks go to Liz Hughes and Jill Ferrett for proof reading and patience and their administrative help during the preparation of this edition. The authors are particularly grateful to Liz for the excellent study guide that she has written for all NEBOSH students, which is included at the end of this book, for the section on report writing in Chapter 5 and the sections on home safety and cycle safety in Chapter 20. Liz gained an honours degree in psychology at the University of Warwick, later going on to complete a Master's degree at the same university. She taught psychology in further and higher education, where most of her students were either returning to education after a gap of many years, or were taking a course to augment their existing professional skills. She went on to qualify as a social worker specialising in mental health, and later moved into the voluntary sector where she managed development for a number of years. Liz then helped to set up and manage training for the National Schizophrenia Fellowship (now called Rethink) in the Midlands.

We would also like to acknowledge the additional contribution made by Jill Ferrett for the help that she gave during the research for the book and with some of the word processing. Given her background in economics and higher education, her advice on certain legal and economic issues has been particularly valuable.

We would like to thank Teresa Budworth, the Chief Executive of NEBOSH, for her support during this fifth edition and various NEBOSH and HSE staff for their generous help and advice. Finally we would like to thank Stephen Vickers, the immediate past Chief Executive of NEBOSH for his encouragement at the beginning of the project and Sadé Lee and all the production team at Routledge who have worked hard to translate our dream into reality.

About the authors

Phil Hughes MBE is a well-known UK safety professional with over 40 years worldwide experience as Head of Environment, Health and Safety at two large multinationals: Courtaulds and Fisons. Phil started work in health and safety in the Factory Inspectorate at the Derby District UK in 1969 and moved to Courtaulds in 1974. He joined IOSH in that year and became Chairman of the Midland Branch, then National Treasurer and was President in 1990–1991. Phil was very active on the NEBOSH Board for over 10 years and served as Chairman from 1995 to 2001. He was also a Professional Member of the American Society of Safety Engineers for many years and has lectured widely throughout the world. Phil received the RoSPA Distinguished Service Award in May 2001 and was a Director and Trustee of RoSPA from 2003 to 2010. He received an MBE in the New Year Honours List 2005 for services to Health and Safety. Phil is a Chartered Fellow of IOSH.

Ed Ferrett is an experienced health and safety consultant who has practised for over 25 years. With a PhD and an honours degree in Mechanical Engineering from Nottingham University, Ed spent 30 years in Higher and Further education, retiring as the Head of the Faculty of Technology of Cornwall College in 1993. Since then he has been an independent consultant to several public and private sector organisations including construction businesses, the Regional Health and Safety Adviser for the Government Office (West Midlands), and was Chair of West of Cornwall Primary Care NHS Trust for 6 years until 2006.

Ed was a member of the NEBOSH Board from 1995 until 2010 and Vice Chair from 1999 to 2008. He has delivered many health and safety courses and has been a lecturer in NEBOSH courses for various course providers. He has been an External Examiner for an MSc course and BSc course in Health and Safety at two UK Universities, a Reporting Inspector for Independent Further and Higher Education with the British Accreditation Council and a NEBOSH Ambassador. Ed is a Chartered Engineer and a Chartered Member of IOSH.

How to use this book and what it covers

Introduction to Health and Safety in Construction Fifth Edition is basically designed to:

1. cover the syllabus of the NEBOSH National Certificate 2015 editions in Construction Health and Safety (NGC1, NCC1 and NCC2);

2. go beyond the NEBOSH syllabus in covering some construction, environmental, home safety and international aspects;

3. provide a good basis in OSH for students who wish to progress to the NEBOSH Diploma or a University first or second degree;

4. provide a text which more than covers the IOSH Managing Safely syllabus or other similar awards;

5. give summaries of UK OSH legislation relevant to the NGC1 and NCC1;

6. help students study, revise and sit the examinations;

7. provide brief guidance to students who carry out the practical assessment;

8. provide brief guidance for searching the internet;

9. supply a range of significant OSH websites;

10. provide a good updated reference text for managers with OSH responsibilities and OSH practitioners in industry and commerce;

11. provide numerous templates for typical fire and safety forms.

We expect the book to be used as a basis for training, and as further reference when students are back in their own workplaces. We believe that all questions can be answered from the material in the book but we would also urge students to study some of the documents given as reference sources at the end of each chapter. It would be helpful to visit some of the websites where further detailed guidance is available. The websites featured in the text were found to be correct at the time of writing in April 2015.

There is a companion website (http://www.routledge.com/cw/hughes/) where animated versions of the workplace inspection exercises in Chapter 5 can be accessed. They show the hazards and then a corrected version appropriately labelled. Copies of the forms in Chapter 23 can be found in Word; many of the illustrations are also available for downloading and use in training materials; and there is a range of multiple answer quizzes for revision purposes.

Poor

Corrected

Poor

Corrected

Animated inspection exercises

Usable images

Figure X1 shows an overview of occupational health and safety and how it fits with the NEBOSH National Certificate in Construction Health and Safety syllabus.

For more detail see the NEBOSH syllabus guide at **www.nebosh.org.uk**

The extra chapters in Figure X2 are designed to help the student understand UK OSH legislation. There is information on how to study, the standard for NEBOSH answers, how to research the internet and essential websites for OSH information, plus a range of form templates which can be freely used by readers.

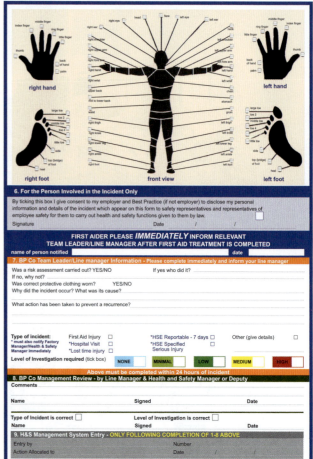

Forms in Word that can be downloaded for use at work (see Chapter 23 for full set)

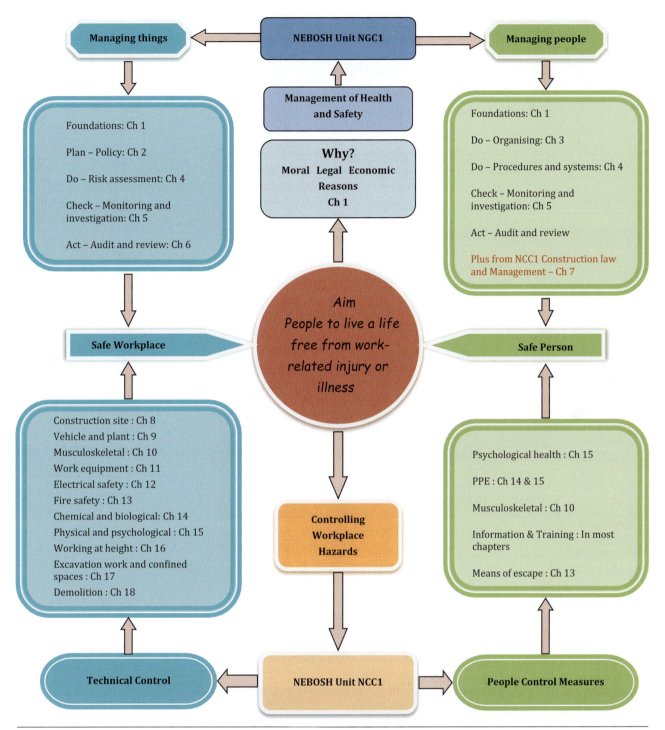

Figure X1 Health and safety overview and link to NEBOSH NGC1 and NCC1 syllabus

Table X.1		Syllabus for the NEBOSH NGC1 and NCC1&2	
Element No	Chapter	Title	Recommended Study Hours
Unit NGC1 : Management of health and safety			
1	1	Foundations in health & safety	7
2	2	Health and safety management systems – Plan	3
3	3	Health and safety management systems – Organising – Do 1	7
3	4	Health and safety management systems – Risk assessment and controls – Do 2	10
4	5	Health and safety management systems – Monitoring investigation and recording – Check	5
5	6	Health and safety management systems – Audit and review – Act	4
		Minimum total tuition time for Unit NGC1	**36**
		Recommended private study time for NGC1	**23**
Unit NCC1 : Managing and controlling hazards in construction activities			
1	7	Construction law and management	5
2	8	Construction site hazards and controls	7
3	9	Vehicle and plant movement – hazards and control	5
4	10	Musculoskeletal hazards and control	7
5	11	Work equipment – hazards and risk control	6
6	12	Electrical safety	4
7	13	Fire safety	6
8	14	Chemical and biological health – hazards and risk control	8
9	15	Physical and psychological health – hazards and risk control	5
10	16	Working at height – hazards and risk control	7
11	17	Excavation work and confined spaces – hazards and risk control	5
12	18	Demolition and deconstruction hazards and risk control	3
		Minimum total tuition time for Unit NCC1	**68**
		Recommended private study time for NCC1	**29**
Unit NCC2 : Construction health and safety practical application			
1	21	**Construction health and safety practical application**	**2**
		Minimum unit tuition time	**2**
		Recommended private study time	**6**
Minimum total tuition time			**106**
Recommended private study time			**58**
Total overall hours			**164**

The NEBOSH NCC syllabus is divided into three units. Each of the first two units NGC1 and NCC1 is further divided into a number of elements

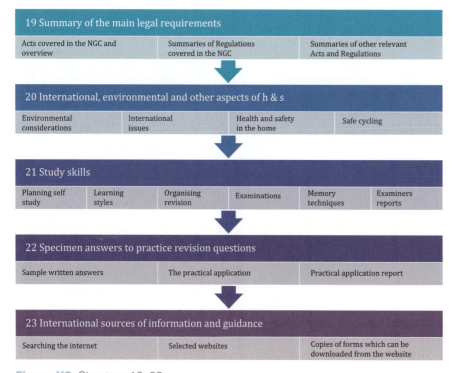

Figure X2 Chapters 19–23

List of principal abbreviations

Most abbreviations are defined within the text. Abbreviations are not always used if it is not appropriate within the particular context of the sentence. The most commonly used ones are as follows:

ACGIH American Conference of Governmental Industrial Hygienists
ACL Approved Carriage List
ACM Asbestos Containing Material
ACoP Approved Code of Practice
ADR Accord dangereux routier (European agreement concerning the international carriage of dangerous goods by road)
AFNOR French Standards Association
AFSSET French Agency for Environmental and Occupational Health Safety
AIB Asbestos Insulation Board
ALARP As low as reasonably practicable
AND European provisions concerning the international carriage of dangerous goods by inland waterways
ANSI American National Standards Institute
ASCC Australian Safety and Compensation Council
ASEAN Association of Southeast Asian Nations
ASSE American Society of Safety Engineers
ASTM American Society for Testing and Materials (now ASTM International)
ATSDR Agency for Toxic Substances and Diseases Registry (USA)
ATEX Atmosphere Explosive (used in the context of two European Directives, 94/9/EC and 1999/92/EC)
BA Breathing apparatus
BAT Best available techniques
BEBOH British Examining Board in Occupational Hygiene
BIOH British Institute of Occupational Hygiene
BLR Blue light radiation
BPM Best practicable means
BRE Building Research Establishment
BSI British Standards Institution
CAR Control of Asbestos Regulations (UK)
CAS Chemical Abstracts Service (USA)
CBI Confederation of British Industry
CD Consultative document
CDM Construction (Design and Management) Regulations (UK)

CEN Comité Européen de Normalisation
CENELEC Comité Européen de Normalisation Électrotechnique
CIB Chartered Institute of Building
CIRC Centre International de Recherche sur le Cancer (France)
CIS International Occupational Safety and Health Information Centre
CISDOC International Labour Organisation database available on OSHROM
CISPR Comité International Spécial des Perturbations Radioelectriques
CLAW Control of Lead at Work Regulations
CLP Classification Packaging and Labelling
CNS Comité de Normalisation de la Soudure (France)
CO Carbon monoxide
COMAH Control of Major Accident Hazards Regulations (UK)
CONIAC Construction Industry Advisory Committee
COSHH Control of Substances Hazardous to Health Regulations (UK)
CPR Cardiopulmonary resuscitation
CSA Canadian Standards Association
CTS Carpal tunnel syndrome
CVD Cardiovascular disease
dB Decibels
DB Dry bulb
dB(A) Decibel (A-weighted)
dB(C) Decibel (C-weighted)
DSE Display Screen Equipment
DSEAR Dangerous Substances and Explosive Atmospheres Regulations (UK)
E&W England and Wales
EA Environment Agency
EAV Exposure Action Value
EC European Community
EEF Engineering Employers Federation
ELV Exposure Limit Value
EMAS Employment Medical Advisory Service
EPA Environmental Protection Act 1990 (UK)
EU European Union
EU-OSH European Agency for Safety and Health at Work
FAO Food and Agriculture Organisation of the United Nations

FOPS	Falling-Object Protective Structure(s)
FPO	Fire Prevention Officer
GATT	General Agreement on Tariffs and Trade
GHGB	Good Health is Good Business
GHS	Globally Harmonised System of Classification and Labelling of Chemicals
GTAW	Gas tungsten arc welding
HACCP	Hazard analysis critical control point
HAI	Hospital acquired infections
HAM	Hand-held monitor
HASAC	Health and Safety Advice Centre
HAV	Hand–Arm Vibration
HGV	Heavy Goods Vehicle
HOPE	Healthcare, Occupational and Primary for Employees
HSCER	Health and Safety (Consultation with Employees) Regulations (UK)
HSE	Health and Safety Executive
HSG	Health and Safety Guidance Booklet
HSW Act	Health and Safety at Work etc. Act 1974 (UK)
HWL	Healthy Working Lives
IAC	Industry Advisory Committee
IChemE	Institution of Chemical Engineers
IEA	International Ergonomics Association
IEC	International Electrotechnical Commission
IEE	Institution of Electrical Engineers
IET	Institution of Engineering and Technology
ILO	International Labour Organisation
INDG	Industry Guidance
IOH	Institution of Occupational Hygienists
IOSH	Institution of Occupational Safety and Health
IPCS	International Programme on Chemical Safety
IPMS	Institution of Professionals, Managers and Specialists
IPPC	Integrated pollution prevention and control
IPPR	Institute for Public Policy Research
IPR	Integrated pollution regulation
IRM	Institute of Risk Management
IRSM	International Institute of Risk and Safety Management
ISBN	International Standard Book Number(ing)
ISO	International Organisation for Standardisation
LD50	Lethal dose fifty
LDLo	Lethal dose low
LEA	Local Enterprise Agency
LEAL	Lower Exposure Action Level
LEL	Lower explosive limit
Leq	Equivalent continuous sound level
Leq(8)hr	Equivalent continuous sound level (normalised to 8 hours)
LEV	Local exhaust ventilation
LNG	Liquefied natural gas
LOLER	Lifting Operations and Lifting Equipment Regulations (UK)
LPG	Liquefied petroleum gas
MEL	Maximum exposure limit
mg/m³	Milligrams per cubic metre
MHOR	Manual Handling Operations Regulations (UK)
MHSW	Management of Health and Safety at Work Regulations (UK)
MORR	Management of Occupational Road Risk
MOT	Ministry of Transport (still used for vehicle tests in UK)
MSD	Musculoskeletal disorder
MSDS	Material Safety Data Sheet(s)
NEBOSH	National Examination Board in Occupational Safety and Health
NIOSH	National Institute for Occupational Safety and Health (NIOSH), USA
NVQ	National Vocational Qualification
OECD	Organisation for Economic Cooperation and Development
OEL	Occupational exposure limit
OES	Occupational exposure standard
OHS	Occupational Health Service
OHSAS	Occupational Health and Safety Assessment Series
OHSLB	Occupational Health and Safety Lead Body
OIAC	Oil Industry Advisory Committee
OSH	Occupational Safety and Health or Occupational Health and Safety
OSHA	Occupational Safety and Health Administration (USA)
PPE	Personal Protective Equipment
ppm	Parts per million
PTFE	Polytetrafluoroethylene
PUWER	Provision and Use of Work Equipment Regulations (UK)
PVC	Polyvinyl chloride
RCD	Residual current device
REACH	Registration Evaluation and Authorisation and Restriction of Chemicals
RIDDOR	Reporting of Injuries, Diseases and Dangerous Occurrences Regulations (UK)
ROES	Representative(s) of Employee Safety
ROPS	Rollover Protective Structure(s)
RoSPA	Royal Society for the Prevention of Accidents
RPE	Respiratory protective equipment
RRFSO	Regulatory Reform Fire Safety Order (UK)
RTA	Road traffic accident
SaHW	Safe and Healthy Working
SFAIRP	So far as is reasonably practicable
SMEs	Small and medium-sized enterprises
SPL	Sound Pressure Level
STEL	Short-term Exposure Limit
SWL	Safe working load
SWP	Safe working pressure
TLV	Threshold limit value
TUC	Trades Union Congress
TWA	Time-Weighted Average
UEAL	Upper Exposure Action Level
UK	United Kingdom

ULD	Upper Limb Disorder
UNEP	United Nations Environment Programme
UNESCO	United Nations Educational, Scientific and Cultural Organisation
VAWR	Vibration at Work Regulations (UK)
WAHR	Work at Height Regulations (UK)
WBV	Whole-Body Vibration
WEL	Workplace Exposure Limit

WHO	World Health Organisation
WRULD	Work-related Upper Limb Disorder

See ILO for more information on abbreviations and acronyms at: http://www.ilo.org/legacy/english/protection/safework/cis/products/safetytm/acronym.htm

Safety signs

PROHIBITION SIGNS

No smoking | No open flame; Fire, open ignition source and smoking prohibited | No access for unauthorised persons | No access for pedestrians | Not drinking water | No access for fork lift trucks and other industrial vehicles | Do not extinguish with water | Do not touch

WARNING SIGNS

General danger | Explosive material | Radioactive material | Laser beam | Non-ionising radiation | Magnetic field | Obstacles | Drop (fall) | Biological hazard

Low temperature | Electricity | Fork lift trucks and other industrial vehicles | Overhead load | Toxic material | Risk of fire/flammable materials | Corrosive substance | Oxidizing substance

MEANS OF ESCAPE AND EMERGENCY EQUIPMENT (SAFE CONDITION) SIGNS

Emergency exit (left hand) | Emergency exit (right hand) | First aid | Emergency telephone | Eyewash station | Safety shower | Stretcher

FIRE SAFETY SIGNS

Fire extinguisher | Fire hose reel | Ladder | Fire alarm call point | Emergency fire telephone

MANDATORY ACTION SIGNS

General mandatory action sign | Wear safety footwear

Wear protective gloves | Wear protective clothing (overalls symbol) | Wear ear protection | Wear eye protection (not opaque image) | Wear face shield | Wear head protection | Wear respiratory protection | Wear safety harness | Pedestrians must use this route

SIGNS FOR GLOBALLY HARMONISED SYSTEM (GHS) OF CLASSIFICATION AND LABELLING OF CHEMICALS

Warning | Fatal if swallowed/inhaled/in contact with skin | May cause damage to organs/genetic defects/cancer/damage to fertility or the unborn child | Flammable | May cause or intensify fire; oxidizer (oxidising image) | Contains gas under pressure | Heating may cause explosion | May be corrosive to metals | Hazardous to the environment

HSE coded hand signals

A. General signals

START
Both arms are extended horizontally with the palms facing forwards

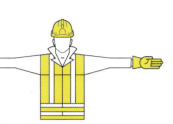

STOP
End of movement the right arm points upwards with the palm facing forwards

END
Of the operation both hands are clasped at chest height

B. Vertical movements

RAISE
The right arm points upwards with the palm facing forward and slowly makes a circle

LOWER
The right arm points downwards with the palm facing inwards and slowly makes a circle

VERTICAL DISTANCE
The hands indicate the relevant distance

C. Horizontal movements

MOVE FORWARDS
Both arms are bent with the palms facing upwards, and the forearms make slow movements towards the body

MOVE BACKWARDS
Both arms are bent with the palms facing downwards, and the forearms make slow movements away from the body

RIGHT (of the signalman)
The right arm is extended more or less horizontally with the palm facing downwards and slowly makes small movements to the right

LEFT (of the signalman)
The left arm is extended more or less horizontally with the palm facing downwards and slowly makes small movements to the left

HORIZONTAL DISTANCE
The hands indicate the relevant distance

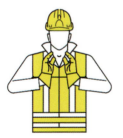

D. Danger

DANGER
Emergency stop – both arms point upwards with the palms facing forwards

Artwork from www.hse.gov.uk/pubns/priced/164.pdf

Foundations in health and safety

> **This chapter covers the following NEBOSH learning objectives:**
> 1 Outline the scope and nature of occupational health and safety
> 2 Explain the moral, legal and financial reasons for promoting good standards of health and safety in the workplace
> 3 Explain the legal framework for the regulation of health and safety including sources and types of law
> 4 Explain the scope, duties and offences of employers, managers, employees and others under the Health and Safety at Work etc. Act 1974
> 5 Explain the scope, duties and offences of employers, managers, employees and others under the Management of Health and Safety at Work Regulations
> 6 Outline the legal and organisational health and safety roles and responsibilities of clients and their contractors
> 7 Outline the principles of assessing and managing contractors

1.1 The scope and nature of occupational health and safety

1.1.1 Introduction

Occupational health and safety is relevant to all branches of industry, business and commerce including traditional industries, information technology companies, the National Health Service, care homes, schools, universities, leisure facilities and offices.

The purpose of this chapter is to introduce the foundations on which appropriate health and safety management systems may be built. Occupational health and safety affects all aspects of work. In a low hazard organisation, health and safety may be supervised by a single competent manager. In a high hazard manufacturing plant, many different specialists, such as engineers (electrical, mechanical and civil), lawyers, medical doctors and nurses, trainers, work planners and supervisors may be required to assist the professional health and safety practitioner in ensuring that there are satisfactory health and safety standards within the organisation.

Construction is the largest industry in the UK and accounts for 8% of its gross domestic product. It employs 10% of the working population and has an annual turnover of over £250 billion. The construction industry has a world reputation for the quality of its work but it remains one of the most dangerous in the UK. In 2004/05, the fatal injury rate (per 100,000 workers) was 3.4 while the industrial average was 0.8. In 2006/07, there was a 28% increase in fatalities in the industry, which accounted for 32% of all notifiable fatal injuries. In August 2010, the Health and Safety Executive (HSE) reported that in 2009, the industry saw an 11.5% drop in output (compared to 5% for the economy as a whole) followed by a slow return to growth over the last few years. However, the construction industry still represents 8.3% of the whole UK economy and there are over 300,000 construction enterprises employing in excess of 2.5m workers.

In an attempt to reduce the rate of fatal and major injury to its workers, the construction industry set itself a target to reduce these injuries significantly over a five-year period and some progress has been made. The construction client who commissions the work is a very important agent in the drive for improved health and safety standards. The client should insist on evidence of a good health and safety record and performance of a contractor at the tendering stage, and ensure that health and safety standards are being met on site. He/she should also require that all the people working on the site are properly trained for their particular job.

There are many barriers to the achievement of good standards. The pressure of production or performance targets, financial constraints and the complexity of the organisation are typical examples of such obstacles. In difficult economic times, organisations need to reduce their costs and this often impacts on the management of health and safety such as the maintenance of complex machinery and the training of workers in safe systems of work. However, there are some powerful incentives for organisations to strive for high health and safety standards. These incentives are moral, legal and financial.

Corporate responsibility, a term used extensively in the 21st century world of work, covers a wide range of issues. It includes the effects that an organisation's business has on the environment, human rights and Third World poverty. Health and safety in the workplace is an important corporate responsibility issue.

Corporate responsibility has various definitions. However, broadly speaking, it covers the ways in which organisations manage their core business to add social, environmental and economic value in order to produce a positive, sustainable impact on both society and the business itself. Terms such as 'corporate social responsibility', 'socially responsible business' and 'corporate citizenship' all refer to this concept.

The Health and Safety Executive's (HSE) mission is to ensure that the risks to health and safety of workers are properly controlled. In terms of corporate responsibility, it is working to encourage organisations to:

▶ improve health and safety management systems to reduce injuries and ill-health;

▶ demonstrate the importance of health and safety issues at board level;

▶ report publicly on health and safety issues within their organisation, including their performance against targets.

The HSE believes that effective management of health and safety:

▶ is vital to employee well-being;

▶ has a role to play in enhancing the reputation of businesses and helping them achieve high-performance teams;

▶ is financially beneficial to business.

This chapter covers the legal responsibilities that exist between people who control premises and those who use them, and between contractors and those who hire them; and the duties of suppliers, manufacturers and designers of articles and substances for use at work.

The chapter also describes the legal responsibilities that exist between duty-holders under the Construction (Design and Management) Regulations (CDM) to ensure that health and safety is fully integrated into the management of any construction project and to encourage everyone involved with the project (the client, designer and principal contractor) to work together effectively.

1.1.2 Some basic definitions

Before a detailed discussion of health and safety issues can take place, some basic occupational health and safety definitions are required.

▶ **Health** – The protection of the bodies and minds of people from illness resulting from the materials, processes or procedures used in the workplace.

▶ **Safety** – The protection of people from physical injury. The borderline between health and safety is ill-defined and the two words are normally used together to indicate concern for the physical and mental well-being of the individual at the place of work.

▶ **Welfare** – The provision of facilities to maintain the health and well-being of individuals at the workplace. Welfare facilities include washing and sanitation arrangements, the provision of drinking water, heating, lighting, accommodation for clothing, seating (when required by the work activity or for rest), eating and rest rooms. First-aid arrangements are also considered as welfare facilities.

▶ **Occupational or work-related ill-health** – This is concerned with those illnesses or physical and mental disorders that are either caused or triggered by workplace activities. Such conditions may be induced by the particular work activity of the individual, or by activities of others in the workplace. The time interval between exposure and the onset of the illness may be short (e.g. asthma attacks) or long (e.g. deafness or cancer).

Figure 1.1 At work in Southampton 2015 – site operated well into the night

▶ **Environmental protection** – These are the arrangements to cover those activities in the workplace which affect the environment (in the form of flora, fauna, water, air and soil) and, possibly, the health and safety of employees and others. Such activities include waste and effluent disposal and atmospheric pollution.

▶ **Accident** – This is defined by the Health and Safety Executive (HSE) as 'any unplanned event that results in injury or ill-health of people, or damage or loss to property, plant, materials or the environment or a loss of a business opportunity'. Other authorities define an accident more narrowly by excluding events that do not involve injury or ill-health. This book will always use the HSE definition.

▶ **Near miss** – This is any incident that could have resulted in an accident. Knowledge of near misses is very important as research has shown that, approximately, for every 10 'near miss' events at a particular location in the workplace, a minor accident will occur.

▶ **Dangerous occurrence** – This is a 'near miss' which could have led to serious injury or loss of life. Dangerous occurrences are defined in the Reporting of Injuries, Diseases and Dangerous Occurrences Regulations (often known as RIDDOR) and are always reportable to the enforcement authorities.

Examples include the collapse of a scaffold or a crane or the failure of any passenger-carrying equipment.

▶ **Hazard and risk** –
  ▷ A **hazard** is something with the *potential* to cause harm (this can include articles, substances, plant or machines, methods of working, the working environment and other aspects of work organisation). Hazards take many forms including, for example, chemicals, electricity and noise. A hazard can be ranked relative to other hazards or to a possible level of danger.
  ▷ A **risk** is the *likelihood* of potential harm from that hazard being realised. Risk (or strictly the level of risk) is also linked to the severity of its consequences. A risk can be reduced and the hazard controlled by good management.

It is very important to distinguish between a *hazard* and a *risk* – the two terms are often confused and activities such as construction work are frequently called high risk when they are high hazard. Although the hazard will continue to be high, the risks will be reduced as controls are implemented. The level of risk remaining when controls have been adopted is known as the **residual risk**. There should only be high residual risk where there is poor health and safety management and inadequate control measures.

1.2 Moral, legal and financial reasons for promoting good standards of health and safety

The first concern of most managers when they start work at a new organisation is to understand the implications of their new role and to form good relationships with other members of the team. Concerns about health and safety are often not a first or even second consideration. So why bother about health and safety?

There are three basic drivers for good health and safety management; these are moral, legal and financial reasons. The moral reasons are centred on the need to protect people from injury and disease while they are at work. The legal reasons are embodied in the criminal and civil law, and the financial reasons come as a consequence of infringements of health and safety law with the consequent fines, compensation payments, associated financial costs and even, in extreme cases, imprisonment. Each of these reasons will now be discussed in turn.

1.2.1 Moral reasons

The moral reasons are supported by the occupational accident and disease rates.

Accidents

Accidents at work can lead to serious injury and even death. Over recent years, between 130 and 190 workers have been fatally injured at their place of work and over 20,000 others suffered major injuries such as amputations, fractures and burns. Statistics are collected on all people who are injured at workplaces – not just employees – and over 350 members of the public each year have been fatally injured at places of work. Since 1995 suicides and trespassers on the railways have been included in the HSE figures – this has led to a significant increase in the overall fatality figures.

Table 1.1 shows a typical annual accident breakdown between employees, self-employed and members of the public. These figures give an indication of the scale of the problem even though the actual figures for any given year may be slightly higher or lower. The industries with the highest fatality and major accident rates (per 100,000 employees) are agriculture, construction, transport, waste and recycling and manufacturing and the most common causes are slipping or tripping (41%), and falls from a height (16%). A further large number of injuries to employees caused an absence from work of over seven days. Of these less serious injuries, the most common causes were handling, lifting or carrying (36%), and slipping or tripping (24%).

Accident statistics are published each year by the HSE and indicate that there is a need for health and safety awareness even in occupations which many would consider very low hazard, such as the health services and hotels. In fact over 70% of all deaths occur in the service sector and manufacturing is considerably safer than construction and agriculture. These latter two industries account for almost half of all fatal injuries to workers. Some of the most common causes of deaths and serious injuries in the agricultural sector include handling livestock and using tractors, quad bikes and chainsaws. Finally, a further significant proportion of work-related accidents occur while travelling on roads and not at the workplace.

These injury figures show that there is clearly a very strong moral case for improvement in health and safety performance.

Table 1.1 Annual accidents for different groups of people

	Fatalities	Major
Total	520	46,000
Employees	130	20,000
Self-employed	40	1,000
Members of the public	350	25,000

Disease

Work-related ill-health and occupational disease can lead to absence from work and, in some cases, to death. Such occurrences may also lead to costs to the State (the Industrial Injuries Scheme) and to individual employers (sick pay and, possibly, compensation payments). Each year thousands of people die from work-related diseases mainly due to past working conditions. The industry sectors having ill-health rates that consistently have been higher than the rate for all industries are health and social work, public administration and education.

Table 1.2 Approximate proportions (%) of cases of work-related ill-health reported by General Practitioners in any year

Type of illness	Percentage
Musculoskeletal disorders	53
Mental ill-health (stress, anxiety)	36
Dermatitis and other skin disorders	5
Other diagnosis including infections	4
Respiratory disease	2
Hearing loss	0.1

Stress and musculoskeletal disorders are the largest causes of work-related ill-health. There are on average over 400,000 workers suffering from stress-related ill-health each year, of which 50% have suffered for one year or longer causing over 10 million working days lost each year. Over 400,000 workers suffer from musculoskeletal disorders (mainly back pain and upper limb disorders) and 60% of these people suffer for one year or longer causing over 7 million working days lost each year. Data from 300 General Practitioners (GPs) (Table 1.2) confirms that musculoskeletal disorders are the most common type of work-related illness, but mental ill-health (usually caused by stress) accounts for more working days lost (Table 1.4).

Occupational asthma is the UK's fastest growing workplace disease and affects between 1,500 and 3,000 people each year. Every year in the UK, 7,000 people are thought to contract occupational asthma that is either caused by their work or is made worse by it. In some cases people are left disabled and unable

to work. Other work-related respiratory diseases include chronic obstructive pulmonary disease (COPD), pneumoconiosis and silicosis. According to GPs, there are 8,500 new cases of work-related respiratory diseases each year.

Work-related cancer is another serious problem and results in 8,000 deaths and 13,500 new cases each year. The leading cause of these deaths is occupational exposure to asbestos which accounts for at least 4,500 deaths each year although this figure is expected to increase in the future. The most common forms of such cancers are lung cancer and mesothelioma.

Recent research has shown that one in five people who are on sickness leave from work for 6 weeks will stay off work and leave paid employment.

1.2.2 Legal reasons

The legal reasons concerning the employer's duty of care in criminal and civil law will be covered later in this chapter.

Some statistics on legal enforcement indicate the legal consequences resulting from breaches in health and safety law. There have been some very high compensation awards for health and safety cases in the Civil Courts and fines in excess of £100,000 in the Criminal Courts. Table 1.3 shows the typical number of enforcement notices served each year in Great Britain. Most notices are served in the manufacturing, construction and agricultural sectors. Local authorities serve 40% of all improvement notices and 20% of all prohibition notices. A small number of enforcement notices are also issued by the Office of Rail Regulation (ORR).

Table 1.3 also indicates the typical number of prosecutions by the HSE and local authorities each year. The HSE (together with the Procurator Fiscal in Scotland) present 80% of the prosecutions and the remainder are presented by Local Authority Environmental Health Officers. These prosecutions result in approximately £15 million in fines each year. Most of these prosecutions were for infringements of various Construction Regulations (including the Work at Height Regulations) and the Provision and Use of Work Equipment Regulations.

There are clear legal reasons for effective health and safety management systems.

Table 1.3 Typical recent annual health and safety enforcement activity in Great Britain

	Improvement notices	Deferred prohibition	Immediate prohibition	Offences prosecuted
HSE	6,664	25	3,430	1,000
Local Authorities	2,412	24	1,235	200
Total	9,076	49	4,665	1,200

Note

* Includes 42 prosecutions by the Procurator Fiscal in Scotland.

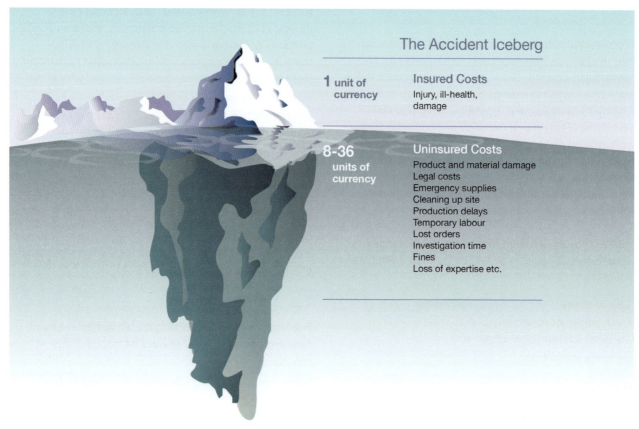

Figure 1.2 Insured and uninsured costs

1.2.3 **Financial reasons**

Costs of accidents

Any accident or incidence of ill-health will cause both direct and indirect costs and incur an insured and an uninsured cost. It is important that all of these costs are taken into account when the full cost of an accident is calculated. In a study undertaken by the HSE, it was shown that indirect costs or hidden costs could be 36 times greater than direct costs of an accident. In other words, the direct costs of an accident or disease represent the tip of the iceberg when compared to the overall costs (Figure 1.2). Annually over 27 million days are lost in the UK due to injuries and work-related ill-health and Table 1.4 shows the causes of these lost days. The total cost of illness and injury to UK industry has been estimated as £14 billion – 60% on ill-health and about 40% on injuries and fatalities.

There is clearly a strong moral, financial and legal case to do more to reduce this unacceptable level of injury and ill-health in the workplace

Table 1.4 Causes of working days lost in the UK

Cause	Percentage
Stress and anxiety	40
Musculoskeletal disorders	28
Injury	17
Other illnesses	15

Source: HSE.

1.2.4 **Societal expectations of good standards of health and safety**

Societal expectations are not static and tend to rise over time, particularly in a wealthy nation like the UK. For example, the standards of safety accepted in a motor car 50 years ago would be considered to be totally inadequate at the beginning of the 21st century. People expect safe, quiet, comfortable cars that do not break down and which retain their appearance for many thousands of miles. Industry should strive to deliver these same high standards for the health and safety of employees or service providers. The question is whether societal expectations are as great an influence on workplace safety standards as they are on product safety standards. Society can influence standards through:

▶ people only working for good employers. This is effective in times of low unemployment;
▶ national and local news media highlighting good and bad employment practices;
▶ schools teaching good standards of health and safety;
▶ the purchase of fashionable and desirable safety equipment, such as trendy crash helmets for mountain bikes;
▶ buying products only from responsible companies. The difficulty of defining what is responsible has been partly overcome through ethical investment criteria but this is possibly

not widely enough understood to be a major influence;

▶ watching TV and other programmes which improve safety knowledge and encourage safe behaviour from an early age.

1.2.5 The business case for health and safety

The business case for health and safety is centred on the potential costs of poor standards of health and safety. Fines in excess of £250,000 and even higher levels of compensation payments are not uncommon. As mentioned earlier, the costs may be direct or indirect and insured or uninsured. Some examples of these follow.

Direct costs

These are costs which are directly related to the accident and may be insured or uninsured.

Insured direct costs normally include:

▶ claims on employers and public liability insurance;
▶ damage to buildings, equipment or vehicles;
▶ any attributable production and/or general business loss;
▶ the absence of employees.

Uninsured direct costs include:

▶ fines resulting from prosecution by the enforcement authority;
▶ sick pay;
▶ some damage to product, equipment, vehicles or process not directly attributable to the accident (e.g. caused by replacement staff);
▶ increases in insurance premiums resulting from the accident;
▶ any compensation not covered by the insurance policy due to an excess agreed between the employer and the insurance company;
▶ legal representation following any compensation claim.

Indirect costs

These are costs which may not be directly attributable to the accident but may result from a series of accidents. Again these may be insured or uninsured.

Insured indirect costs include:

▶ a cumulative business loss;
▶ product or process liability claims;
▶ recruitment of replacement staff.

Uninsured indirect costs include:

▶ loss of goodwill and a poor corporate image;
▶ accident investigation time and any subsequent remedial action required;
▶ production delays;
▶ extra overtime payments;

▶ lost time for other employees, such as a first-aider, who tend to the needs of the injured person;
▶ the recruitment and training of replacement staff;
▶ additional administration time incurred;
▶ first-aid provision and training;
▶ lower employee morale possibly leading to reduced productivity.

Some of these items, such as business loss, may be uninsurable or too prohibitively expensive to insure. Therefore, insurance policies can never cover all of the costs of an accident or disease because either some items are not covered by the policy or the insurance excess is greater than the particular item cost.

1.2.6 Employers' Liability Compulsory Insurance

The Employers' Liability (Compulsory Insurance) Act makes it a legal requirement for all employers to have employers' liability insurance. This ensures that any employee, who successfully sues his/her employer following an accident, is assured of receiving compensation irrespective of the financial position of the employer.

For employers, the insurance covers the cost of legal fees and compensation in the event of a claim by a worker. Only very few businesses are not required to have employers' liability insurance.

There is a maximum penalty of up to £2,500 for every day without appropriate cover for employers who do not have such insurance. In addition, one or more copies of the current certificate must be displayed at each place of business and be 'reasonably protected' from being defaced or damaged. Recently, the rules requiring an employer to display the certificate have changed, so that the requirement will be satisfied if the certificate is made available in electronic format and is reasonably accessible to relevant employees.

1.3 The legal framework for the regulation of health and safety including sources and types of law

1.3.1 Sub-divisions of law

There are two sub-divisions of the law that apply to health and safety issues: criminal law and civil law.

Criminal law

Criminal law consists of rules of behaviour laid down by the Government or the State and, normally, enacted by Parliament through Acts of Parliament. These rules or Acts are imposed on the people for the protection of the people. Criminal law is enforced by several different Government Agencies who may prosecute individuals and organisations for contravening criminal laws. It is important to note that, except for very rare cases, only these agencies are able to decide whether to prosecute an individual or not.

An individual or organisation who breaks criminal law is deemed to have committed an offence or crime and, if he/she is prosecuted, the court will determine whether he/she is guilty or not. If the individual is found guilty, the court could sentence him/her to a fine or imprisonment. Owing to this possible loss of liberty, the level of proof required by a Criminal Court is very high and is known as proof 'beyond reasonable doubt', which is as near certainty as possible. Although the prime object of a Criminal Court is the allocation of punishment, the court can award compensation to the victim or injured party. One example of criminal law is the Road Traffic Act, which is enforced by the police. However, the police are not the only criminal law enforcement agency. The Health and Safety at Work (HSW) etc. Act is another example of criminal law and this is enforced either by the HSE or Local Authority Environmental Health Officers (EHOs). Other agencies which enforce criminal law include the Fire Authority, the Environment Agency, Trading Standards and Customs and Excise.

There is one important difference between procedures for criminal cases in general and criminal cases involving health and safety. The prosecution in a criminal case has to prove the guilt of the accused beyond reasonable doubt. Although this obligation is not totally removed in health and safety cases, Section 40 of the HSW Act 1974 transferred, where there is a duty to do something 'so far as is reasonably practicable' or 'so far as is practicable' or to 'use the best practicable means', the onus of proof to the accused to show that there was no better way to discharge his/her duty under the Act. However, when this burden of proof is placed on the accused, they only need to satisfy the court on the balance of probabilities that what they are trying to prove has been done.

Civil law

Civil law concerns disputes between individuals or individuals and companies. An individual sues another individual or company to address a civil wrong or tort (or delict in Scotland). The individual who brings the complaint to court is known as the claimant or plaintiff (pursuer in Scotland), and the individual or company who is being sued is known as the defendant (defender in Scotland).

The Civil Court is concerned with liability and the extent of that liability, rather than guilt or non-guilt. Therefore, the level of proof required is based on the 'balance of probability', which is a lower level of certainty than that of 'beyond reasonable doubt' as required by the Criminal Court. If a defendant is found to be liable, the court would normally order him/her to

pay compensation and possibly costs to the plaintiff. However, the lower the balance of probability, the lower the level of compensation awarded. In extreme cases, where the balance of probability is just over 50%, the plaintiff may 'win' the case but lose financially because costs may not be awarded and the level of compensation is low. The level of compensation may also be reduced through the defence of **contributory negligence**, which is discussed later under 1.3.6. For cases involving health and safety, civil disputes usually follow accidents or illnesses and concern negligence or a breach of statutory duty. The vast majority of cases are settled 'out of court'. Although actions are often between individuals, where the defendant is an employee who was acting in the course of his/her employment during the alleged incident, the defence of the action is transferred to his/her employer – this is known as **vicarious liability**. The civil action then becomes one between the individual and the employer.

1.3.2 The legal system in England and Wales

The description that follows applies to England and Wales (and with a few minor differences to Northern Ireland). Only the court functions concerning health and safety are mentioned. Figure 1.3 shows the court hierarchy in schematic form.

Criminal law

Magistrates Courts

Most criminal cases begin and end in the Magistrates Courts. Health and safety cases are brought before the court by enforcement officers (Health and Safety Executive or Local Authority Environmental Health Officers) and they are tried by a bench of three lay magistrates (known as Justices of the Peace) or a single district judge. The lay magistrates are members of the public, usually with little previous experience of the law, whereas the district judge is legally qualified.

Since March 2015, the Magistrates Court may impose an unlimited fine for health and safety offences. Magistrates are also able to imprison for up to 12 months. The vast majority of health and safety criminal cases are dealt with in the Magistrates Court. See Table 19.1 in Chapter 19 for details of penalties under the Health and Safety Offences Act 2008.

Crown Court

The Crown Court hears the more serious cases (indictable), which are passed to them from the Magistrates Court – normally because the sentences available to the magistrates are felt to be too lenient. Cases are heard by a judge and jury, although some cases are heard by a judge alone. The penalties available

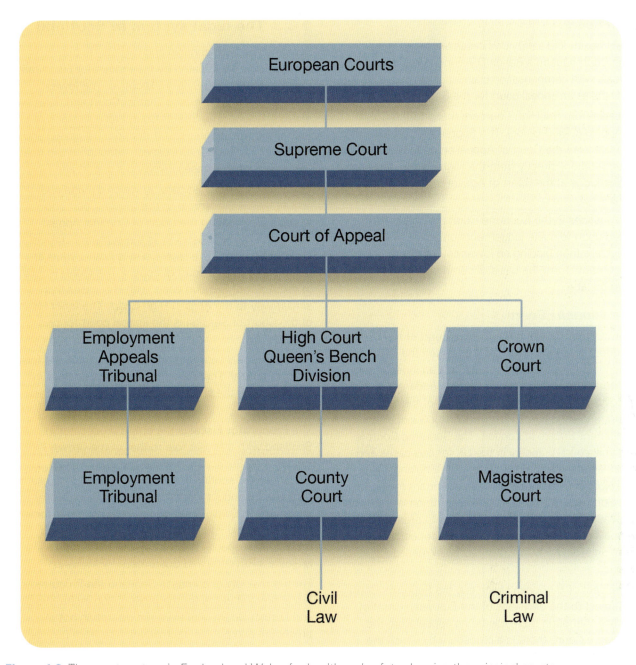

Figure 1.3 The court system in England and Wales for health and safety showing the principal courts

to the Crown Court are an unlimited fine and up to 2 years' imprisonment for breaches of enforcement notices. The Crown Court also hears appeals from the Magistrates Court.

Appeals from the Crown Court are made to the Court of Appeal (Criminal Division), who may then give leave to appeal to the most senior court in the country – the Supreme Court. The most senior judge at the Court of Appeal is the Lord Chief Justice.

In England and Wales, the **Crown Prosecution Service (CPS)** is the main prosecuting authority for criminal cases prepared by the police and other investigators. CPS prosecutors present cases in both the Magistrates' Courts and the higher courts. The head of the CPS is the Director of Public Prosecutions (DPP). In Northern Ireland, the main prosecuting authority is the **Public**

Prosecution Service for Northern Ireland (PPSNI) and has a similar role to the CPS. Both organisations are independent of the police and decide whether a prosecution is to proceed to court. For health and safety offences, the CPS decides on manslaughter and corporate manslaughter cases (see 1.4).

Civil law

County Court

The lowest court in civil law is the County Court, which only deals with minor cases (for compensation claims of up to £50,000 if the High Court agrees). Cases are normally heard by a judge sitting alone. For personal injury claims of less than £5,000, a small claims court is also available.

High Court

Many health and safety civil cases are heard in the High Court (Queen's Bench Division) before a judge only. It deals with compensation claims in excess of £50,000 and acts as an appeal court for the County Court.

Appeals from the High Court are made to the Court of Appeal (Civil Division). The Supreme Court receives appeals from the Court of Appeal or, on matters of law interpretation, directly from the High Court. The most senior judge at the Court of Appeal is the Master of the Rolls.

The Justices in the Supreme Court are sometimes called upon to make judgements on points of law, which are then binding on lower courts. Such judgements form the basis of common law, which is covered later.

The Supreme Court

From October 2009, the Supreme Court of the United Kingdom was established and assumed the judicial functions of the House of Lords. It is an independent institution, presided over by 12 independently appointed judges, known as Justices of the Supreme Court, and is housed in the Middlesex Guildhall on Parliament Square in London. The Supreme Court is the final court of appeal in the UK for civil cases and hears appeals in criminal cases from England, Wales and Northern Ireland. It plays an important role in the development of United Kingdom common law as was the case previously with the House of Lords. As an appeal court, the Supreme Court cannot consider a case unless a relevant order has been made in a lower court.

In respect of European Union law the Supreme Court is subject to the decisions of the European Court of Justice. Although there can be no appeal from the Supreme Court, there is a procedure by which the Supreme Court may refer to the European Court questions of European law which arise in cases before it, and obtain a definitive ruling before the Supreme Court gives its judgement.

Other courts – Employment Tribunals

These were established in 1964 and primarily deal with employment and conditions of service issues, such as unfair dismissal. However, they also deal with appeals over health and safety enforcement notices, disputes between recognised safety representatives and their employers and cases of unfair dismissal involving health and safety issues. There are usually three members who sit on a tribunal. These members are appointed and are often not legally qualified. Appeals from the tribunal may be made to the Employment Appeal Tribunal or, in the case of enforcement notices, to the High Court. Appeals from tribunals can only deal with the clarification of points of law.

1.3.3 The legal system in Scotland

Scotland has both criminal and civil courts but prosecutions are initiated by the Procurator Fiscal rather than the Health and Safety Executive. The lowest Criminal Court is called the District Court and deals with minor offences. The Sheriff Court has a similar role to that of the Magistrates Court (for criminal cases) and the County Court (for civil cases), although it can deal with more serious cases involving a sheriff and jury.

The High Court of Justiciary, in which a judge and jury sit, has a similar role to the Crown Court and appeals are made to the Court of Criminal Appeal. The High Court of Justiciary is the supreme criminal court of Scotland. The High Court is both a court of first instance and a court of appeal. The Outer House of the Court of Session deals with civil cases in a similar way to the English High Court. The Inner House of the Court of Session is the appeal court for civil cases.

For both appeal courts, the Supreme Court is the final court of appeal. There are Employment Tribunals in Scotland with the same role as those in England and Wales.

The **Procurator Fiscal** is the public prosecutor in Scotland with a similar role to that of the Crown Prosecution Service in England and Wales. They investigate all sudden and suspicious deaths (similar to a coroner), conduct Fatal Accident Inquiries (similar to an inquest in England and Wales) and handle criminal complaints against the police. They also receive reports from specialist reporting agencies such as Her Majesty's Revenue and Customs.

For the majority of crimes in Scotland the Procurators Fiscal present cases for the prosecution in the Sheriff and District Courts and have the discretion not to prosecute and pursue alternatives free from political interference. However, they are always subject to the directions of Crown Office and the Lord Advocate.

Procurators Fiscal make preliminary investigations into criminal cases, take written statements from witnesses (known as precognition) and are responsible for the investigation and prosecution of crime. This includes the power to direct the police in their investigation, but except for serious crimes such as murder the police normally complete their enquiries before involving the Procurator Fiscal. They have the power to offer fixed penalties instead of prosecution (a fiscal fine) and to give warnings, compensation orders, work orders and road traffic fixed penalties. They can also divert offenders from prosecution to social workers, psychological counselling or psychiatric treatment.

All suspicious, sudden and accidental deaths must be reported to the Procurator Fiscal, who prosecutes on behalf of the HSE. They have a responsibility to identify if any criminal action has occurred and, where appropriate, prosecute. Where a criminal offence is suspected to have occurred, the Procurator Fiscal will instruct the local police to investigate. The Procurator Fiscal also investigates and, where appropriate, prosecutes cases of homicide and corporate homicide involving health and safety issues. However, if the police decide that there is not a case of homicide or a corporate homicide, the investigation will be handed over to the HSE to investigate. As in the rest of the UK, the HSE (or local authorities) in Scotland will investigate non-fatal work-related incidents. Finally it is important to note that no offence in Scottish law can be proved unless there is evidence from more than one source – this is known as corroboration.

1.3.4 European Courts

There are two European Courts – the European Court of Justice and the European Court of Human Rights.

The European Court of Justice, based in Luxembourg, is the highest court in the European Union (EU). It deals primarily with community law and its interpretation. It is normally concerned with breaches of community law by member states and cases may be brought by other member states or institutions. Its decisions are binding on all member states. There is currently no right of appeal.

The European Court of Human Rights, based in Strasbourg, is not directly related to the EU – it covers most of the countries in Europe including the EU member states. As its title suggests, it deals with human rights and fundamental freedoms. With the introduction of the Human Rights Act 1998 in October 2000, many of the human rights cases are now heard in the UK.

1.3.5 Sources of law (England and Wales)

There are two sources of law – common law and statute law.

Common law

Common law dates from the 11th century when William I set up Royal Courts to apply a uniform (common) system of law across the whole of England. Prior to that time, there was a variation in law, or the interpretation of the same law, from one town or community to another. Common law is based on judgements made by courts (or strictly judges in courts). In general, courts are bound by earlier judgements on any particular point of law – this is known as 'judicial precedent'. Lower courts must follow the judgements of higher courts. Hence judgements made by the Justices in the Supreme Court form the basis of most of the common law currently in use.

In health and safety, the legal definition of negligence, duty of care and terms such as 'practicable' and 'as far as is reasonably practicable' are all based on legal judgements and form part of the common law. Common law also provides the foundation for most civil claims made on health and safety issues.

Statute law

Statute law is a law which has been laid down by Parliament as Acts of Parliament and other legislation. In health and safety, an Act of Parliament, the HSW Act 1974, lays down a general legal framework. Specific health and safety duties are, however, covered by Regulations or Statutory Instruments – these are also examples of statute law. If there is a conflict between statute and common law, statute law takes precedence. However, as with common law, judges interpret statute law usually when it is new or ambiguous. Although for health and safety, statute law is primarily the basis of criminal law, there is a tort of breach of statutory duty. However, the application of this tort has been considerably reduced by the Enterprise and Regulatory Reform Act 2013 (see later under 1.3.6).

The relationship between the sub-divisions and sources of law

The two sub-divisions of law may use either of the two sources of law (as shown in Figure 1.4). For example, murder is a common law crime. In terms of health and safety, however, criminal law is only based on statute law, whereas civil law may be based on either common law or statute law.

In summary, criminal law seeks to protect everyone in a society whereas civil law seeks to protect and recompense any individual citizen or organisation.

1.3.6 Common law torts and duties

Negligence

The only tort (civil wrong) of real significance in health and safety is negligence. Negligence is the lack of reasonable care or conduct which results in injury, damage (or financial loss) of or to another. Whether the act or omission was reasonable is usually decided as a result of a court action or out-of-court settlement.

There have been two important judgements that have defined the legal meaning of negligence. In 1856, negligence was judged to involve actions or omissions and the need for reasonable and prudent behaviour. In 1932, Lord Atkin said the following:

'You must take reasonable care to avoid acts or omissions which you reasonably foresee would be likely to injure your neighbour. Who then, in law is my neighbour? The answer seems to be persons who are so closely and directly affected by my act that I ought reasonably to have them in contemplation as being so affected when I am directing my mind to the acts or omissions which are called in question.'

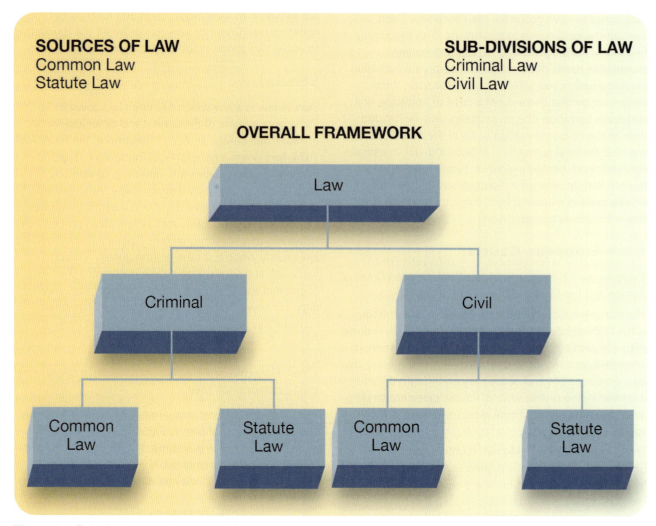

SOURCES OF LAW
Common Law
Statute Law

SUB-DIVISIONS OF LAW
Criminal Law
Civil Law

OVERALL FRAMEWORK

Figure 1.4 Sub-divisions and sources of law

This judgement followed the case of Donoghue v. Stevenson, in which Mrs Donoghue, the claimant, consumed part of a bottle of ginger beer that contained a decomposed snail. The snail was not visible, as the bottle was opaque. Neither her friend, who bought it for her, nor the shopkeeper who sold it was aware of its presence. Mrs Donoghue became ill and sued the manufacturer, Mr Stevenson, for negligence. The case worked its way through the court system up to the House of Lords where the judges agreed that Mrs Donoghue had a valid claim, but disagreed on the remedy. One judge thought that it was a product liability case but Lord Atkin argued that the law should recognise a unifying principle that everyone owes a duty of reasonable care to their neighbour. He quoted the Bible in support of his argument, specifically the general biblical principle that 'Thou shalt love thy neighbour.' In framing this landmark judgement, he created the doctrine that nobody should harm their neighbour.

It can be seen, therefore, that for negligence to be established, it must be reasonable and foreseeable that the injury could result from the act or omission. In practice, the Court may need to decide whether the injured party is the neighbour of the perpetrator.

A collapsing scaffold could easily injure a member of the public, who could be considered a neighbour to the scaffold erector. A recent court case has ruled that hindsight after an incident cannot be used to establish negligence. The Supreme Court has decided in recent cases that foreseeability must play a part in determining whether a place is or was safe. The law does not aim to create an environment that is entirely risk-free. It is directed at situations where there is a material risk to health and safety, which any reasonable person would appreciate and take steps to guard against. If a risk cannot be reasonably foreseen, then it is not material.

An employee who is suing his/her employer for negligence needs to establish the following three criteria:

1. A duty was owed to the employee by the employer and the incident took place during the course of his/her employment.
2. There was a breach of that duty because the event was foreseeable and all reasonable precautions had not been taken.
3. The breach resulted in the specific injury, disease, damage and/or loss suffered.

These tests should also be used by anyone affected by the employer's undertaking (such as contractors and members of the public), who is suing the employer for negligence.

If the employer is unable to defend against the three criteria, two further partial defences are available. It could be argued that the employee was fully aware of the risks that were taken by not complying with safety instructions (known as *volenti non fit injuria* or 'the risk was willingly accepted'). This defence is unlikely to be totally successful because courts have ruled that employees have not accepted the risk voluntarily as economic necessity forces them to work.

The second possible defence is that of **contributory negligence** where the employee is deemed to have contributed to the negligent act. This partial defence, if successful, can significantly reduce the level of compensation (by up to 80% in some cases).

There are two other defences that are not often used – Act of God and a disagreement on the facts of the case. An Act of God is a legal term that describes an accident that could not have been prevented by human foresight but occurred due to some form of unpredictable natural disaster. Any negligence claim must be made within a set time period (currently 3 years).

Duty of care

Several judgements have established that employers owe a duty of care to each of their employees. This duty cannot be assigned to others, even if a consultant is employed to advise on health and safety matters, or if the employees are sub-contracted to work with another employer. This duty may be sub-divided into five groups. Employers must:

1. provide a safe place of work, including access and egress;
2. provide safe plant and equipment;
3. provide a safe system of work;
4. provide safe and competent fellow employees;
5. provide adequate levels of supervision, information, instruction and training.

Employer duties under common law are often mirrored in statute as in the HSW Act 1974. This, in effect, makes them both common law and statutory duties.

These duties apply even if the employee is working at a third-party premises or if he/she has been hired by the employer to work for another employer.

Breach of statutory duty

Although for health and safety, statute law is primarily the basis of criminal law, there is a tort of breach of statutory duty which can be used when a person is seeking compensation following an accident or illness.

In the past when there was a breach of statutory duty, such as the duties laid down by many health and safety regulations, an individual, who was affected by the breach, could, subject to certain conditions, pursue a civil action to obtain compensation.

However, the Enterprise and Regulatory Reform Act 2013 in Section 69 has amended Section 47(2) of the HSW Act 1974 as follows:

> '(2) Breach of a duty imposed by a statutory instrument containing (whether alone or with other provision) health and safety regulations shall not be actionable except to the extent that regulations under this section so provide.'

This means that breaches of statutory duty, as described by regulations made under the HSW Act, no longer give rise to civil liability unless there is a specific provision to the contrary. One specific provision is given in the Health and Safety at Work etc. Act 1974 (Civil Liability) (Exceptions) Regulations 2013 which allows workers who are pregnant, or new or expectant mothers (including agency workers), to continue to bring claims for breach of statutory duty in relation to breaches of Regulations 16–17A of the Management of Health and Safety at Work Regulations to the extent that the breach causes damage. These regulations relate to requirements to carry out risk assessments and make particular arrangements for new and expectant mothers to protect their health and safety. These employees may sue their employer both for negligence and the breach of a statutory duty. Such an action is termed a **double-barrelled action**. However, it is important to stress that double damages would not result because the compensation is awarded for the actual harm suffered.

The requirement to provide competent fellow employees includes the provision of adequate supervision, instruction and training. As mentioned earlier, employers are responsible for the actions of their employees (**vicarious liability**) provided that the action in question took place during the normal course of their employment.

In response to the growing 'compensation culture', the Compensation Act became law in 2006. It contains provisions relating to the law of negligence and breach of statutory duty. It expects courts not to prevent or impede a desirable activity from taking place when they consider the particular steps to be taken to meet a reasonable standard of care following a claim for negligence or breach of statutory duty. The Act attempts to ensure that normal activities are not prevented because of a fear of litigation and excessive risk aversion. The emphasis needs to be on risk management rather than risk aversion. It states that in claims for negligence or breach of statutory duty, an apology, offer of treatment or other redress shall not of itself amount to an admission of liability.

The Compensation Act also contains provisions on the damages for mesothelioma and the regulation of claims management services.

Levels of statutory duty

There are three principal levels of statutory duty which form a hierarchy of duties. These levels are used extensively in health and safety statutory (criminal) law, but have been defined by judges under common law. The three levels of duty are absolute, practicable and reasonably practicable.

Absolute

This is the highest level of duty and often occurs when the risk of injury is so high that injury is inevitable unless safety precautions are taken. It is a rare requirement regarding physical safeguards, although it was more common before 1992 when certain sections of the Factories Act 1961 were still in force. The duty is absolute and the employer has no choice but to undertake the duty. The verbs used in the Regulations are 'must' and 'shall'.

An example of this is Regulation 11(1a) of the Provision and Use of Work Equipment Regulations which requires that every employer shall ensure that effective measures are taken to prevent access to any dangerous part of machinery or rotating stock bar.

For certain health and safety regulations, such as The Electricity at Work Regulations and The Control of Substances Hazardous to Health Regulations, an absolute duty may be defended using the argument that 'all reasonable precautions and all due diligence' were taken to comply with the law. To rely on this defence, proof is needed that a control system for the risks outlined in the particular Regulation is in place. Proof of due diligence requires that this control system has suitable safeguards and control measures and that they operate effectively.

Many of the health and safety management requirements contained in health and safety law place an absolute duty on the employer. Examples of this include the need for written safety policies and risk assessments when employee numbers rise above a basic threshold.

Practicable

This level of duty is more often used than the absolute duty as far as the provision of safeguards is concerned and, in many ways, has the same effect.

A duty that 'the employer ensures, so far as is practicable, that any control measure is maintained in an efficient state' means that if the duty is technically possible or feasible then it must be done irrespective of any difficulty, inconvenience or cost. Examples of this duty may be found in the Provision and Use of Work Equipment Regulations [Regulation 11(2) (a and b)].

Reasonably practicable

This is the most common level of duty in health and safety law and was defined by Judge Asquith in Edwards v. the National Coal Board (1949) as follows:

> ' "Reasonably practicable" is a narrower term than "physically possible", and seems to me to imply that computation must be made by the owner in which the quantum of risk is placed on one scale and the sacrifice involved in the measures necessary for averting the risk (whether in time, money or trouble) is placed in the other, and that, if it be shown that there is a gross disproportion between them – the risk being insignificant in relation to the sacrifice – the defendants discharge the onus on them.'

In other words, if the risk of injury is very small compared with the cost, time and effort required to reduce it, no action is necessary. It is important to note that money, time and trouble must 'grossly outweigh', not balance (Figure 1.5), the risk. This duty requires judgement on the part of the employer (or his/her adviser) and clearly needs a risk assessment to be undertaken with conclusions noted. Continual monitoring is also required to ensure that risks do not increase. There are numerous examples of this level of duty, including the HSW Act [e.g. Section 2(3)], the Manual Handling Operations Regulations (see Chapter 19) and the Control of Substances Hazardous to Health Regulations (see Chapter 19). If organisations follow 'good practice', defined by the HSE as 'those standards which satisfy the law', then they are likely to satisfy the 'reasonably practicable' test. Approved Codes of Practice and Guidance published by the HSE describe and define many of these standards.

The term 'suitable and sufficient' is used to define the scope and extent required for health and safety risk assessment and may be interpreted in a similar way to 'reasonably practicable'. More information is given on this definition in Chapter 4.

Figure 1.5 Diagrammatic view of 'reasonably practicable'

1.3.7 The influence of the European Union on health and safety

As Britain is part of the European Union, much of the health and safety law originates in Europe. Proposals from the European Commission may be agreed by member states. The member states are then responsible for making them part of their national law.

In Britain itself and in much of the EU, health and safety law is based on the principle of risk assessment described in Chapter 4. The main role of the EU in health and safety is to harmonise workplace and legal standards and remove barriers to trade across member states. A Directive from the EU is legally binding on each member state and must be incorporated into the national law of each member state. Directives set out specific minimum aims which must be covered within the national law. Some member states incorporate Directives more speedily than others.

Directives are proposed by the European Commission, comprising commissioners, who are citizens of each of the member states. The proposed Directives are sent to the European Parliament, which is directly elected from the member states. The European Parliament may accept, amend or reject the proposed Directives. The proposed Directives are then passed to the Council of Ministers, who may accept the proposals on a qualified majority vote, unless the European Parliament has rejected them, in which case they can only be accepted on a unanimous vote of the Council. The Council of Ministers consists of one senior government minister from each of the member states.

The powers of the EU in health and safety law are derived from the Treaty of Rome 1957 and the Single European Act 1986. For health and safety, the Single European Act added two additional Articles – 100A and 118A. The 1997 Treaty of Amsterdam renumbered them as 95A and 138A, respectively.

Article 95A is concerned with health and safety standards of equipment and plant and its Directives are implemented in the UK by the Department for Business, Innovation and Skills.

Article 138A is concerned with minimum standards of health and safety in employment and its Directives are implemented by the Health and Safety Executive.

The objective of the Single European Act 1986 is to produce a 'level playing field' for all member states so that goods and services can move freely around the EU without any one state having an unfair advantage over another. The harmonisation of health and safety requirements across the EU is one example of the 'level playing field'.

The first introduction of an EU Directive into UK health and safety law occurred on 1 January 1993 when a Framework Directive on Health and Safety Management and five daughter Directives were introduced using powers contained in the HSW Act 1974. These Directives, known as the European Six Pack, covered the following areas:

- management of health and safety at work;
- workplace;
- provision and use of work equipment;
- manual handling;
- personal protective equipment;
- display screen equipment.

Since 1993, several other EU Directives have been introduced into UK law. Summaries of the more common UK Regulations are given in Chapter 19.

1.4 The scope, duties and offences of employers, managers, employees and others under the Health and Safety at Work etc. Act 1974

1.4.1 The Health and Safety at Work Act 1974 (HSW Act)

Background to the Act

The HSW Act resulted from the findings of the Robens Report, published in 1972. Earlier legislation had tended to relate to specific industries or workplaces. This resulted in over 5 million workers being unprotected by any health and safety legislation. Contractors and members of the public were generally ignored. The law was more concerned with the requirement for plant and equipment to be safe rather than the development of parallel arrangements for raising the health and safety awareness of employees.

A further serious problem was the difficulty that legislation had in keeping pace with developments in technology. For example, following a court ruling in 1955 which, in effect, banned the use of grinding wheels throughout industry, it then took 15 years to produce the Abrasive Wheels Regulations 1970 to address the problem raised by the 1955 court judgement (John Summers and Sons Ltd. v. Frost). In summary, health and safety legislation before 1974 tended to be reactive rather than proactive.

Lord Robens was asked, in 1970, to review the provisions made for the health and safety of people at work. His report produced conclusions and recommendations upon which the HSW Act 1974 was based. The principal recommendations were as follows:

- There should be a single Act that covers all workers and that Act should contain general duties which should 'influence attitudes'.
- The Act should cover all those affected by the employer's undertaking such as contractors, visitors, students and members of the public.
- There should be an emphasis on health and safety management and the development of safe systems of work. This would involve the encouragement of

employee participation in accident prevention. (This was developed many years later into the concept of the health and safety culture.)

▶ Enforcement should be targeted at 'self-regulation' by the employer rather than reliance on prosecution in the courts.

These recommendations led directly to the introduction of the HSW Act in 1974.

Figure 1.6 HSW Act

An overview of the Act

Health and Safety Executive (HSE)

The HSW Act established the Health and Safety Commission and gave it the responsibility to draft new Regulations and to enforce them either through its executive arm, known as the HSE, or through the Local Authority Environmental Health Officer (EHO).

In April 2008, the Health and Safety Commission and HSE merged into a single unified body called the HSE. The merger increased the outside, or non-executive, input to its work. This improved communication, accountability and the oversight of longer term strategy. The merger means that the size of the Board of the Executive is no more than 12 members. The Board of the Executive has assumed responsibility for running all aspects of the organisation, including setting the overall strategic direction, financial and performance management and prioritisation of resources.

The HSE retains its independence, reflecting the interests of employers, employees and local authorities and is committed to maintaining its service delivery.

Although occupational health and safety in Scotland is a reserved issue (it was not devolved to the Scottish Executive), there are some differences in some areas of activity. A summary of the concordat between the HSE and the Scottish Executive is given in Chapter 19 and covers several of these differences.

In Northern Ireland, the **Health and Safety Executive of Northern Ireland (HSENI)** has a very similar role to that of the HSE.

The prime function of the HSE remains the same – to monitor, review and enforce health and safety law and to produce codes of practice and guidance. However, the HSE also undertakes many other activities, such as the compilations of health and safety statistics, leading national health and safety campaigns, investigations into accidents or complaints, visiting and advising employers and the production of a very useful website.

Regulations

The HSW Act is an **Enabling Act** which allows the Secretary of State to make further laws (known as Regulations) without the need to pass another Act of Parliament. Regulations are laws, approved by Parliament. These are usually made under the HSW Act, following proposals from the HSE. This applies to Regulations based on EU Directives as well as 'home-grown' ones. It is a criminal offence to breach a Regulation and any breaches may result in enforcement action as explained later in this chapter.

The HSW Act, and general duties in the Management Regulations, aim to help employers to set goals, but leave them free to decide how to control hazards and risks which they identify. Guidance and Approved Codes of Practice give advice, but employers are free to take other routes to achieving their health and safety goals, so long as they do what is reasonably practicable. But some hazards are so great, or the proper control measures so expensive, that employers cannot be given discretion in deciding what to do about them. Regulations identify these hazards and risks and set out specific action that must be taken. Often these requirements are absolute – employers have no choice but to follow them and there is no qualifying phrase of 'reasonably practicable' included.

Some Regulations apply across all organisations – the Manual Handling Operations Regulations would be an example. These apply wherever things are moved by hand or bodily force. Equally, the Health and Safety (Display Screen Equipment) Regulations apply wherever visual display units are used at work. Other Regulations apply to hazards unique to specific industries, such as mining or construction.

Wherever possible, the HSE will set out the Regulations as goals and describe what must be achieved, but not how it must be done.

Sometimes it is necessary to be *prescriptive*, and to spell out what needs to be done in detail, as some standards are absolute. For example, all mines should have two exits; contact with live electrical conductors should be avoided. Sometimes European law requires prescription.

Some activities or substances are so dangerous that they have to be licensed, for example explosives and asbestos. Large, complex installations or operations require 'safety cases', which are large-scale risk assessments subject to scrutiny by the Regulator. An

1

example would be the privatised railway companies who are required to produce safety cases for their operations.

Approved Code of Practice (ACoP)

An ACoP is produced for most sets of Regulations by the HSE and attempts to give more details on the requirements of the Regulations. It also attempts to give the level of compliance needed to satisfy the Regulations. ACoPs are aimed at what the HSE calls 'experienced health and safety audiences' to explain the law and enable the control of more complex risks. It is important to note that although ACoPs are not law, they do have a special legal status (sometimes referred to as quasi-legal), which means if the advice in an ACoP is followed properly a duty-holder can be reasonably assured they are complying with the law.

The relationship of an ACoP to a Regulation is similar to the relationship of the Highway Code to the Road Traffic Act. A person is never prosecuted for contravening the Highway Code but can be prosecuted for contravening the Road Traffic Act. If a company is prosecuted for a breach of health and safety law and it is proved that it has not followed the relevant provisions of the ACoP, a court can find it at fault, unless the company can show that it has complied with the law in some other way.

As most health and safety prosecutions take place in a Magistrates Court, it is likely that the lay magistrates will consult the relevant ACoP as well as the Regulations when dealing with a particular case. Therefore, in practice, an employer must have a good reason for not adhering to an ACoP.

Approved Codes of Practice generally are only directly legally binding if:

▶ the Regulations or Act indicates that they are so, for example The Safety Signs and Signals Regulations Schedule 2 specify British Standard Codes of Practice for alternative hand signals;
▶ they are referred to in an Enforcement Notice.

Guidance

Guidance, which has no formal legal standing, comes in two forms – legal and best practice. The Legal Guidance series of booklets is issued by the HSE to cover the technical aspects of Health and Safety Regulations. These booklets generally include the Regulations and the ACoP, where one has been produced.

Best practice guidance is normally published in the HSG series of publications by the HSE. Examples of best practice guidance books include *Health and Safety in Construction* HSG150 and *Lighting at Work* HSG38.

An example of the relationship between these forms of requirement/advice can be shown using a common problem found throughout industry and commerce – minimum temperatures in the workplace.

Regulation 7 of the Workplace (Health, Safety and Welfare) Regulations states '(1) During working hours, the temperature in all workplaces inside buildings shall be reasonable'.

The ACoP states 'The temperature in workrooms should normally be at least 16 degrees Celsius unless much of the work involves severe physical effort in which case the temperature should be at least 13 degrees Celsius'.

It would, therefore, be expected that employers would not allow their workforce to work at temperatures below those given in the ACoP unless it was reasonable for them to do so (if, for example, the workplace was a refrigerated storage unit).

Best practice guidance to cover this example is given in HSG194 *Thermal Comfort in the Workplace* in which possible solutions to the maintenance of employee welfare in low-temperature environments are given.

General duties and key sections of the Act

A summary of the HSW Act is given in Chapter 19 – only an outline is given here.

Section 2 Duties of employers to employees

To ensure, so far as is reasonably practicable, the health, safety and welfare of all employees. In particular:

▶ safe plant and systems of work;
▶ safe use, handling, transport and storage of substances and articles;
▶ provision of information, instruction, training and supervision;
▶ safe place of work, access and egress;
▶ safe working environment with adequate welfare facilities;
▶ a written safety policy together with organisational and other arrangements (if there are five or more employees);
▶ consultation with safety representatives and formation of safety committees where there are recognised trade unions.

Section 3 Duties of employers to others affected by their undertaking

A duty to safeguard those not in their employment but affected by the undertaking. This includes members of the public, contractors, patients, customers, visitors and students.

Section 4 Duties of landlords, owners and those in control of premises

This includes a duty to ensure that means of access and egress are safe for those using the premises. Those in control of non-domestic premises also have a duty to ensure, so far as is reasonably practicable, that the premises, the means of access and exit, and any plant (such as boilers and air conditioning units) or substances are safe and without risks to health.

Section 6 Duties of suppliers

Persons who design, manufacture, import or supply any article or substance for use at work must ensure, so far as is reasonably practicable, that it is safe and without risk to health. Articles must be safe when they are set, cleaned, used and maintained. Substances must be without risk to health when they are used, handled, stored or transported. This requires that information must be supplied on the safe use of the articles and substances. There may be a need to guarantee the required level of safety by undertaking tests and examinations.

Section 7 Duties of employees

Two main duties are:

▶ to take reasonable care for the health and safety of themselves and others affected by their acts or omissions;
▶ to cooperate with the employer and others to enable them to fulfil their legal obligations.

Section 8

No person is to misuse or interfere with safety provisions (sometimes known as the 'horseplay section').

Section 9

Employees cannot be charged for health and safety requirements such as personal protective equipment.

Section 36 Offences due to the fault of another person

Where an offence is committed by a person that was caused by the act of a second person, the second person may also be charged with the first person. This situation may arise during the trial of the first person when it becomes apparent that the action was the fault of a second person. The HSE have the power under this section to prosecute either both persons or just one. The first person is usually the employer and the second person could be, for example, a manufacturer of a faulty product or a self-employed contractor.

Section 37 Personal liability of directors

Where an offence is committed by a corporate body with the consent or connivance of, or is attributable to any neglect of, a director or other senior officer of the body, both the corporate body and the person are liable to prosecution.

HSE figures show that there has been an increase over the past few years in prosecutions of senior managers and directors under Section 37. Recent case law has confirmed that directors cannot avoid a charge of neglect under Section 37 by arranging their organisation's business so as to leave them ignorant of circumstances which would trigger their obligation to address health and safety breaches.

Enforcement of the Act

Fee for Intervention (FFI)

The Health and Safety (Fees) Regulations has introduced the Fee for Intervention (FFI) scheme which places a duty on the HSE to recover its costs for carrying out its regulatory functions from those found to be in **material breach** of health and safety law. A **material breach** is when, in the opinion of the HSE inspector, there is or has been a contravention of health and safety law that requires them to issue a notice in writing of that opinion to the duty-holder.

Written notification is given by the inspector in either a simple notification of the particular contravention or an improvement or prohibition notice, or a prosecution. The notification will include the following information:

▶ the law that the inspector's opinion relates to;
▶ the reasons for their opinion; and
▶ notification that a fee is payable to the HSE.

Any organisation that is served such a notification is liable for fines to cover HSE-related costs, including call-outs, inspections and investigations and the taking of enforcement action. Breaches of health and safety law ranging from slips, trips and falls to not providing enough toilets or washing facilities have resulted in HSE fines. The two sectors that received the most fines in the first year of FFI were manufacturing (38%) and construction (36%).

Figure 1.7 Employees at work taking reasonable care of themselves

Powers of inspectors

Inspectors under the Act work either for the HSE or the Local Authority. Local authorities are responsible for retail and service outlets and the HSE is responsible for all other work premises including the local authorities themselves. Both groups of inspectors have the same powers.

Table 1.5 shows the breakdown of responsibilities between the HSE and the Local Authorities.

Table 1.5 Premises inspected by HSE and Local Authorities

HSE inspect:	Local Authorities inspect:
factories	offices (except government offices)
farms	shops
building sites	hotels
nuclear installations	restaurants
mines	leisure premises
schools and colleges	nurseries and playgroups
fairgrounds	pubs and clubs
gas, electricity and water systems	museums (privately owned)
hospitals and nursing homes	places of worship
central and local government premises	sheltered accommodation
offshore installations	care homes

The detailed powers of inspectors are given in Chapter 19. In summary, an inspector has the right to:

▶ enter premises at any reasonable time, accompanied by a police officer, if necessary;
▶ examine, investigate and require the premises to be left undisturbed;
▶ take samples, photographs and, if necessary, dismantle and remove equipment or substances;
▶ require the production of books or other relevant documents and information;
▶ seize, destroy or render harmless any substance or article;
▶ issue enforcement notices and initiate prosecutions.

An inspector may issue a formal caution when an offence has been committed, but it is deemed that the public interest does not require a prosecution. Formal cautions are not normally considered if the offender has already had a formal caution.

Amongst many other duties the **Office of Rail Regulation (ORR)** has responsibility for the independent health and safety regulation of the railway industry in the UK. It is a statutory body and is the combined economic and safety regulatory authority for the UK's railway network.

Figure 1.8 The inspector inspects

Its role in health and safety on the railways was originally undertaken by the HM Railway Inspectorate (HMRI). HMRI, which began in 1840, was responsible for overseeing safety on Britain's railways and tramways. From 1990 to 2006, it became part of the Health and Safety Executive and then was transferred to the Office of Rail Regulation.

In 2009, a single rail regulatory body covering both safety and economic issues, the Safety Directorate, was created within the ORR but the individual inspectors continue to be known as Her Majesty's Railway Inspectors.

Enforcement notices

There are two types of enforcement notice.

Improvement Notice – This identifies a specific breach of the law and specifies a date by which the situation is to be remedied. An appeal may be made to the Employment Tribunal and must be made within 21 days. The notice is then suspended until the appeal is either heard or withdrawn. There are five main grounds for an appeal:

1. The inspector interpreted the law incorrectly.
2. The inspector exceeded his/her powers.
3. The breach of the law is admitted but the proposed solution is not practicable or reasonably practicable.
4. The time allowed to comply is too short.
5. The breach of the law is admitted but the breach is so insignificant that the notice should be cancelled.

Prohibition Notice – This is used to halt an activity which the inspector feels could lead to a serious personal injury. The notice will identify which legal requirement is being or is likely to be contravened. The notice takes effect as soon as it is issued. As with the improvement notice, an appeal may be made to the Employment Tribunal but, in this case, the notice remains in place during the appeal process.

There are two forms of prohibition notice:

▶ an immediate prohibition notice – this stops the work activity immediately until the specified risk is reduced;
▶ a deferred prohibition notice – this stops the work activity within a specified time limit.

Penalties

In 2008, the Health and Safety (Offences) Act raised the maximum penalties available to the courts in respect of certain health and safety offences by amending Section 33 of the HSW Act. Further, under the Legal Aid, Sentencing and Punishment of Offenders Act 2015 Section 85(1) from 12 March 2015, Magistrates Courts can issue unlimited fines and imprison for up to 12 months.

Magistrates Court (or lower court)

The Magistrates Court deals with summary offences.

It can impose an unlimited fine and/or up to 12 months imprisonment. The sentencing option of 12 months applies in Scotland but only applies in England and Wales when Section 154(1) of the Criminal Justice Act 2003 is enacted.

Crown Court (or higher court)

The Crown Court deals with indictable (solemn in Scotland) offences and some summary cases referred to it by the lower court which may feel that its powers are too limited.

Fines are unlimited in the Crown Court for all health and safety offences. Possible imprisonment for up to 2 years may also be imposed by the court.

The courts also have the power to disqualify directors who are convicted of health and safety offences for up to 5 years at the lower court and 15 years at the higher court.

Imprisonment for health and safety offences has been rare. Over a 10-year period from 1996, only five people were sent to prison for health and safety offences. With the changes introduced in the Health and Safety Offences Act, this number may increase. Imprisonment is a possible penalty for most health and safety offences in both the Magistrates Court and the Crown Court and a possible penalty for an offence committed by a corporate body. Additionally, certain offences, which were triable only in the lower courts, such as the failure to comply with an improvement notice, are now triable in either court. This could lead to the offender receiving a higher sentence if their case was referred to the higher court. Courts (or the Crown Prosecution Service in England and Wales or the Procurator Fiscal's Office in Scotland) may refer cases to the next highest court if the case is complex or the powers of the lower court are thought to be too limited.

Indictable and summary offences

Most minor offences, including the majority of health and safety offences, are called summary offences and are tried without a jury only in a Magistrates Court (or Sheriff Court in Scotland). The most serious offences are called indictable offences and are tried only by the Crown Court (or the High Court of Justiciary in Scotland). If the defendant is to be tried on indictment, the Magistrates Court (or lower court) will conduct committal proceedings.

Summary of the actions available to an inspector

Following a visit by an inspector to premises, the following actions are available to an inspector:

- take no action;
- give verbal advice;
- give written advice;
- formal caution;
- serve an improvement notice;
- serve a prohibition notice;
- prosecute.

In any particular situation, more than one of these actions may be taken.

The decision to prosecute will be made either when there is sufficient evidence to provide a realistic possibility of conviction or when prosecution is in the public interest.

Work-related deaths

Work-related deaths are investigated by the police jointly with the HSE or Local Authority to ascertain whether a charge of manslaughter (culpable homicide in Scotland) or corporate manslaughter is appropriate. An inquest by a coroner may be made into any workplace death whether due to an accident or industrial disease. Before the inquest, the coroner will decide whether to order a post-mortem. At the inquest, the coroner sits with a jury and decides which witnesses will be called to give evidence. Decisions are usually made on the balance of probability rather than 'beyond reasonable doubt'. The appropriate enforcement agency, such as the HSE, may attend the inquest either to give evidence or to decide whether the evidence presented warrants a prosecution.

If the police decide not to pursue a manslaughter charge, the HSE or Local Authority will continue the investigation under health and safety law. Recent procedural changes now allow prosecution to take place before the inquest where manslaughter or homicide charges are not relevant – if it is considered appropriate and in the interest of justice.

The Corporate Manslaughter and Corporate Homicide Act 2007

The Act created a new statutory offence of corporate manslaughter which replaced the common law offence of manslaughter by gross negligence where corporations and similar entities were concerned. In Scotland, the new offence is called 'corporate homicide'. The Act introduces a new offence but does not introduce any new health and safety duties. An organisation will have committed the offence if:

- it owes a duty of care to another person in defined circumstances;
- there is a management failure by its senior managers; and
- it is judged that its actions or inaction amount to a gross breach of that duty resulting in a person's death.

The health and safety duties relevant to the Act are:

- employer and occupier duties including the provision of safe systems of work and training on any equipment used;

duties connected with:
  ▷ the supply of goods and services to customers;
  ▷ the operation of any activity on a commercial basis;
  ▷ any construction and maintenance work;
  ▷ the use or storage of plant, vehicles or any other item.

A breach of duty will be gross if the management of health and safety at the organisation falls far below that which would reasonably be expected. It is, therefore, a judgement on the health and safety culture of the organisation (as described in Chapter 3) at the time of the incident. The emphasis of the Act is on the responsibility of the senior management of the organisation rather than a single individual. A conviction is only likely if the management and organisation of health and safety by senior management is a substantial element in the breach of the duty of care.

On conviction, the offence is punishable by an unlimited fine and the courts are able to make:

▶ a remedial order requiring the organisation to take steps to remedy the management failure concerned;
▶ a publicity order requiring the organisation to publicise details of its conviction and fine.

If a court makes a publicity order, it will stipulate in which medium a public announcement should be made and will consider requiring a statement to be displayed on the company's website. The prosecution will provide the court (and serve on the defendant) a draft copy of the suggested order, which the judge will then personally endorse.

The publicity order will probably require an organisation convicted of corporate manslaughter to publicise:

▶ the fact that it has been convicted;
▶ the particulars of the offence;
▶ the amount of any fine;
▶ the terms of any remedial order.

The convicted organisation will normally be ordered to pay the prosecution's costs. Normally orders for compensation are considered by the civil courts, but it is possible that occasional cases may arise where a criminal court considers a compensation order in respect of bereavement and funeral expenses.

It is important to note that the Act does not create a new individual liability. Individuals may still be charged with the existing offence of manslaughter by gross negligence and/or a breach of Section 37 of the HSW Act. Crown immunity will not apply to the offence, although a number of public bodies and functions will be exempt from it (such as military operations and the response of emergency services).

The police will investigate the death with the technical support of the relevant enforcement agency (often but not always the HSE). However, the Crown Prosecution Service (in Scotland the Procurator Fiscal) decides whether to initiate a prosecution after the police investigation.

Sentencing guidelines have been published to help courts deal with bodies found guilty of corporate manslaughter. Such bodies can be given up to a seven-figure fine and seldom less than £500,000. However, although the level of the fine should not be linked to the turnover or profit of the body, the financial circumstances of the body and the effect on those innocent parties affected by the undertaking must be considered.

1.4.2 Employers' duties

Employers have duties under both criminal and civil law. The general duties of employers under the HSW Act relate to:

▶ the health, safety and welfare at work of employees and others whether part-time, casual, temporary or homeworkers, on work experience, government training schemes or on site as contractors – that is anyone working under their control or direction;
▶ the health and safety of anyone who visits or uses the workplace and anyone who is allowed to use the organisation's equipment;
▶ the health and safety of those affected by the work activity, for example neighbours, and the general public.

Employers must protect the health and safety of employees and others who might be affected by their work activities. Health and safety is about sensible, proportionate actions that protect people – not unnecessary bureaucracy and paperwork.

This 10-point list shows some of the key duties required by the HSW Act and other legislation that apply to nearly every organisation:

1. Obtain Employers' Liability Compulsory Insurance and display the certificate.
2. Ensure that the organisation has competent health and safety advice available. This does not have to be an external consultant.
3. Develop a health and safety policy that outlines the health and safety management system.
4. Undertake risk assessments on the business activities that include details of the required control measures.
5. Ensure that relevant actions are taken following the risk assessments to prevent accidents and ill-health.
6. Provide basic welfare facilities, such as toilets, washing facilities and drinking water.
7. Provide free health and safety training for workers.
8. Consult workers on health and safety.
9. Display the health and safety law poster or give workers a leaflet with the information.
10. Report certain work-related accidents, diseases and dangerous occurrences.

Failure to comply with these duties can have serious consequences – for both organisations and individuals. As has already been stated, sanctions include fines, imprisonment and disqualification.

Employers are required to do a risk assessment of the work activities carried out by homeworkers. It may be necessary for employers to visit the home of the worker to carry out a risk assessment, although homeworkers can also help in identifying the hazards for their employers themselves.

The Working Time Regulations set a limit on the time that a worker can be required to work, which averages at 48 hours a week (over a 17-week rolling reference period), although workers can voluntarily opt out to work longer. Young workers must not work more than eight hours per day and 40 hours per week. Night workers are limited to an average of eight hours' work in a 24-hour period and entitled to a free health assessment. Records and monitoring of working times, rest breaks and holiday entitlement are also covered by the regulations.

Visitors and general public

Employers have a duty to ensure the health and safety of the public while on their premises, even if the individuals concerned, like children, are not supposed to be there. Two cases reported in 2008 have highlighted how far this liability can extend.

> ### CASE STUDIES
>
> **Two companies pleaded guilty to charges of breaching the HSW Act 1974 following the drowning of a 9-year-old girl, who was playing with other children in the company car park when they strayed onto nearby reservoirs. At the time of the accident, in 2004, the main gates to the factory were off their hinges because work was being carried out on the site and a second gate, which led to the reservoir, was only secured with a nylon rope.**
>
> **In a separate case, a construction company has been found guilty of failing to prevent unauthorised persons, including children, from gaining access to an area where construction material and equipment were stored. A child was seriously injured by falling paving stones while playing on a partly built housing estate where materials were being stored during construction work.**

Misunderstandings over who is responsible for monitoring and protecting contractors and sub-contractors – leading to the failure to carry out responsibilities – is often at the heart of high-profile health and safety cases. But situations such as these could be avoided with a clearer understanding of employers' and contractors' responsibilities.

Visitors to a site whether authorised or not are often more at risk than employees because:

▶ they are unfamiliar with the workplace processes, the hazards and associated risks they present;
▶ they may not have the appropriate personal protective equipment (PPE);
▶ they will have a lack of knowledge of the site or premises layout;
▶ walkways are often inadequate, unsigned or poorly lit;
▶ they are not familiar with the emergency procedures or means of escape;
▶ they may be particularly vulnerable if they suffer from a disability or are very young.

Many of these problems with visitors can be overcome by, for example:

▶ visitors signing in and being provided with a site escort;
▶ providing appropriate PPE and identity badges;
▶ providing simple induction procedures with a short video and information on site rules, hazards and emergency procedures;
▶ clear marking of walkways and areas where unauthorised people are not permitted.

Night working

Employers should ascertain whether they employ people who would be classified as night workers. If so, they should check:

▶ the number of hours that the night workers normally work;
▶ if night workers work more than eight hours per day on average, whether the number of hours can be reduced and if any exceptions apply;
▶ that a health assessment is made on each night worker;
▶ that a regular programme of health checks is made on each night worker;
▶ that proper records of night workers are maintained, including details of health assessments; and
▶ that night workers are not involved in work which is particularly hazardous.

It is important to note that the Working Time Regulations state that night-shifts must not exceed eight hours in every 24. There are several risks to which night workers are exposed due to fatigue, lack of sleep, insufficient rest breaks and an increase in possible accidents due to tiredness. Research is currently being undertaken whether there is a link between night working and the development of breast cancer in women.

Temporary workers Directive adopted by EU Parliament

The European Parliament has adopted a Directive on temporary agency work which enables temporary workers to be treated equally with those of the employer company. However, following an agreement reached in 2008, agency workers should receive the same pay and conditions as permanent staff after being employed for 12 weeks.

The agreement has been made between the UK Government, the CBI and the TUC, and the following points have been agreed:

- After 12 weeks in a given job, there will be entitlement to equal treatment.
- Equal treatment will be defined to mean 'at least the basic working and employment conditions that would apply to the workers concerned if they had been recruited directly by that undertaking to occupy the same job'. This does not cover occupational social security schemes.
- The Government will consult regarding the implementation of the Directive, in particular how disputes will be resolved regarding the definition of equal treatment and how both sides of the industry can reach appropriate agreements on the treatment of agency workers.
- The new arrangements will be reviewed at an appropriate point 'in the light of experience'.

1.4.3 Duties of managers and supervisors

Managers and supervisors are essential communication links between the Board and the workforce. They should ensure that the agreed health and safety standards are communicated to the workers and adherence to those standards is monitored effectively. They have a responsibility to deliver the duty of care of the employer to the workforce.

Therefore managers should:

- familiarise themselves with the health and safety management system of the organisation;
- ensure that there is an adequate and appropriate level of supervision for all workers;
- ensure that supervisors are aware of:
  - the health and safety standards of the organisation;
  - the specific hazards within their area of supervision;
  - the need to set a good example on health and safety issues;
  - the need to monitor the health and safety performance of their workforce; and
  - the training needs, including induction training, of their workforce;
- ensure that sufficient resources are available to allow tasks to be completed safely and without risks to health; and
- communicate to the Chief Executive and the Board the adherence or otherwise of the health and safety standards agreed by the Board.

Effective supervision of the workforce is one of the important keys to successful health and safety management. It will enable managers to monitor the effectiveness of the training that workers have received, and whether they have the necessary capacity and competence to do the job assigned to them. Supervisors should lead by example in health and safety matters (for example in the wearing of PPE).

Hence supervisors should ensure that:

- workers in their department understand the risks associated with their workplace and the measures available to control them;
- the risk control measures are up to date and are being properly used, maintained and monitored;
- particular attention is paid to new, inexperienced or young people and those whose first language is not English;
- workers are encouraged to raise concerns over any shortcomings in health and safety provision;
- arrangements are in place to supervise the work of contractors.

1.4.4 Duties of employees and others

Employees and agency workers

They have specific duties under the HSW Act, which are:

- to take reasonable care for the health and safety of themselves and of other persons who may be affected by their acts or omissions at work. This involves the same wide group that the employer has to cover, not just the people on the next desk or bench;
- to cooperate with employers in assisting them to fulfil their statutory duties;
- not to interfere with deliberately or misuse anything provided, in accordance with health and safety legislation, to further health and safety at work. See also Chapter 19.

Self-employed

There are 4.2 million self-employed workers in the UK of whom 3 million do not employ anyone. Approximately half of the self-employed use their home as a workplace.

The duties of the self-employed under the HSW Act are fairly limited. Under the HSW Act the self-employed are:

- responsible for their own health and safety;
- responsible to ensure that others who may be affected are not exposed to risks to their health and safety.

These responsibilities are extended by the Management of Health and Safety at Work Regulations (see 1.5), which requires self-employed people to:

- What can the manufacturer or supplier do while working on the product or service to reduce the chance of anything going wrong later?
- What can be done at the point of handover to limit the chances of anything going wrong?
- What steps should customers take to reduce the chances of something going wrong? What precisely would they need to know?

Depending on what is being provided to customers, the customer information may need to comply with the following legislation:

- Supply of Machinery (Safety) Regulations;
- Provision and Use of Work Equipment Regulations;
- Control of Substances Hazardous to Health Regulations; and
- CLP Regulation (Classification, labelling and packaging of substances).

This list is not exhaustive. See Chapter 19 for summaries.

Buying problems

Examples of problems that may arise when purchasing include:

- second-hand equipment which does not conform to current safety standards such as an office chair which does not provide adequate back support or have five feet/castors;
- starting to use new substances which do not have safety data sheets;
- machinery which, while well-guarded for operators, may pose risks for a maintenance engineer.

A risk assessment should be done on any new product, taking into account the likely life expectancy (delivery, installation, use, cleaning, maintenance, disposal, etc.). The supplier should be able to provide the information needed to do this. This will help the purchaser make an

informed decision on the total costs because the risks will have been identified as will the precautions needed to control those risks. A risk assessment will still be needed for a CE-marked product. The CE marking signifies the manufacturer's declaration that the product conforms to relevant European Directives. Declarations from reputable manufacturers will normally be reliable. However, purchasers should be alert to fake or inadequate declarations and technical standards which may affect the health and safety of the product despite the CE marking. The risk assessment is still necessary to consider how and where the product will be used, what effect it might have on existing operations and what training will be required.

Employers have some key duties when buying plant and equipment:

- They must ensure that work equipment is safe and suitable for its purpose and complies with the relevant legislation. This applies equally to equipment which is adapted to be used in ways for which it was not originally designed.
- When selecting work equipment, they must consider existing working conditions and health and safety issues.
- They must provide adequate health and safety information, instructions, training and supervision for operators. Manufacturers and suppliers are required by law to provide information that will enable safe use of the equipment, substances, etc. and without risk to health.

Some of the issues that will need to be considered when buying in product or plant include:

- ergonomics – risk of work-related upper limb disorders (WRULDs);
- manual handling needs;
- access/egress;
- storage, for example of chemicals;
- risk to contractors when decommissioning old plant or installing new plant;
- hazardous materials – provision of extraction equipment or personal protective equipment;
- waste disposal;
- safe systems of work;
- training;
- machinery guarding;
- emissions from equipment/plant, such as noise, heat or vibration.

Supply chains in construction

Studies by the Building Research Establishment and The Logistics Business Ltd have shown that there is a long way to go before the construction sector matches the best of manufacturing and retail businesses in their management of supply chains.

This is partly due to the 'one-off' approach applied to building design and the British and others' love of

Figure 1.11 Inadequate chair – it should have five feet and an adjustable backrest – take care when buying second-hand

traditional buildings using small components glued or fixed together on site both for the structure (e.g. brickwork) and services (e.g. electrics and plumbing).

Understanding customer needs

The studies found a number of problems in construction which included the following:

▶ There is a failure to see the chain of customer– supplier relationships.

▶ The customer needs and service levels are not defined and monitored and, hence, improved.

Supplier partnerships

▶ There are generally too many suppliers.

▶ The relationships are adversarial rather than managed through service level agreements.

▶ Supplier selection is driven by price, rather than by value.

▶ There is a tendency to secrecy rather than sharing of good quality information.

Control of materials and information

▶ No effort is made to reduce material waste, reduce double handling or improve efficiency generally through an understanding of where goods are and what has to be done to them.

▶ Sources of wasted time are generally not well understood.

▶ Waste is not monitored and no targets are set for improvement.

▶ Care must be taken with second-hand equipment which may not conform to current standards.

Influencing change

The attitude of the end client or principal contractor is extremely important in determining the way in which the supply chain acts. For example, supermarkets demand absolute compliance with legislation and codes of practice in matters of food hygiene, because this is paramount to their business. They have rigorous systems of monitoring and inspection to ensure that agreed procedures are being followed. Unfortunately similar regimes do not exist for the health and safety of people at work. The HSE are concerned to encourage larger organisations to influence the standards adopted by smaller sub-contractors.

Companies want to ensure that they engage safe contractors. This involves a close attention to detail from the pre-contract stage to the end of the project. Large firms often have their own approved list of contractors and suppliers. To get on this list will usually involve a demonstration that the supplier has all the necessary policies and procedures in place and has a good health and safety track record. Failure to meet the client's requirements because of accidents or

enforcement action may mean a loss of business and even exclusion from the approved list. Some clients or principal contractors may want to audit a supplier to check compliance.

Firms applying to tender may not always understand that large firms will usually make enquiries of other firms with experience of their health and safety performance. Many large firms regard health and safety performance as a good indicator of the general competence of a smaller firm, precisely because it is an aspect that is often neglected. An adverse report will, at the very least, mean that a bidder must do even better on other aspects to succeed and, in these circumstances, companies generally make special arrangements to ensure that their safety standards are met by the contractors.

Appendix 1.1 gives a useful checklist compiled by the HSE for supply chain health and safety management.

1.4.6 Role and functions of external agencies

The HSE and the Local Authorities (a term used to cover county, district and unitary councils) are all external agencies that have a direct role in the monitoring and enforcement of health and safety standards. There are, however, three other external agencies that have a regulatory influence on health and safety standards in the workplace.

Fire and Rescue Authority

The Fire and Rescue Authority is situated within a single or group of local authorities and is normally associated with rescue, fire-fighting and the offering of general advice. It has also been given powers to enforce fire precautions within places of work under fire safety law. The powers of the Authority are very similar to those of the HSE on health and safety matters. The Authority issues Alteration Notices to workplaces and conducts routine and random fire inspections (often to examine fire risk assessments). It should be consulted during the planning stage of proposed building alterations when such alterations may affect the fire safety of the building (e.g. means of escape).

The Fire and Rescue Authority can issue both Enforcement and Prohibition Notices. The Authority can prosecute for offences against fire safety law.

The Environment Agency (Scottish Environment Protection Agency)

The Environment Agency was established in 1995 and was given the duty to protect and improve the environment. It is the regulatory body for environmental matters and has an influence on health and safety issues.

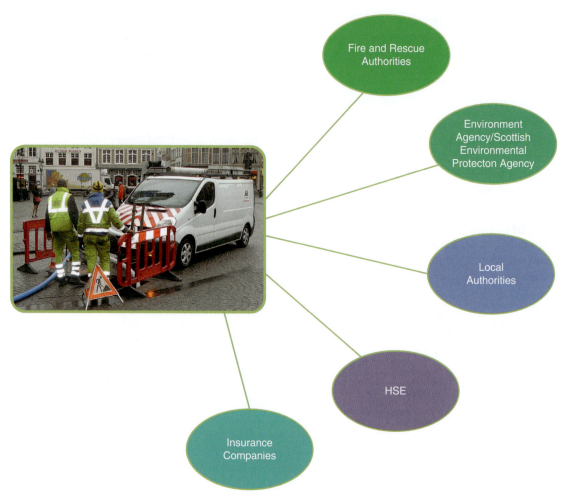

Figure 1.12 Diagram showing the main external agencies that impact on the workplace

It is responsible for authorising and regulating emissions from industry.

Other duties and functions of the Agency include:

▶ ensuring effective controls of the most polluting industries;
▶ monitoring radioactive releases from nuclear sites;
▶ ensuring that discharges to controlled waters are at acceptable levels;
▶ setting standards and issuing permits for collecting, transporting, processing and disposal of waste (including radioactive waste);
▶ enforcement of the Producer Responsibility Obligations (Packaging Waste) Regulations. These resulted from an EU Directive which seeks to minimise packaging and recycle at least 50% of it;
▶ enforcement of the Waste Electrical and Electronic Equipment (WEEE) Directive and its associated Directives.

The Agency may prosecute in the Criminal Courts for the infringement of environmental law – in one case a fine of £4 million was imposed.

Insurance companies

Insurance companies play an important role in the improvement of health and safety standards. Since 1969, it has been a legal requirement for employers to insure their employees against liability for injury or disease arising out of their employment. This is called employers' liability insurance and was covered earlier in this chapter. Certain public sector organisations are exempted from this requirement because any compensation is paid from public funds. Other forms of insurance include fire insurance and public liability insurance (to protect members of the public).

Figure 1.13 Good standards prevent harm and save money

Premiums for all these types of insurance are related to level of risks which is related to standards of health and safety. In recent years, there has been a considerable increase in the number and size of compensation claims and this has placed further pressure on insurance companies.

Insurance companies are becoming effective health and safety Regulators by weighing the premium offered to an organisation according to its safety and/or fire precaution record.

1.5 The scope, duties and offences of employers, managers, employees and others under the Management of Health and Safety at Work Regulations

As mentioned earlier, on 1 January 1993, following an EC Directive, the Management of Health and Safety at Work Regulations became law in the UK. These Regulations were updated in 1999 and are described in detail in Chapter 19. In many ways, the Regulations were not introducing new concepts or replacing the 1974 Act – they simply reinforced or amended the requirements of the HSW Act. Some of the duties of employers and employees were re-defined. The duties of employers, employees and others are summarised below but the influence of these duties on health and safety management is discussed in detail in chapters 3 and 4.

1.5.1 Employers' duties

Employers must:

▶ undertake suitable and sufficient written risk assessments when there are five or more employees;
▶ undertake preventative and protective measures on the basis of the principles of prevention specified in Schedule 1 to the regulations;
▶ put in place effective arrangements for the planning, organisation, control, monitoring and review of health and safety measures in the workplace (including health surveillance). Such arrangements should be recorded if there are five or more employees;
▶ employ (to be preferred) or contract competent persons to help them comply with health and safety duties;
▶ develop suitable emergency procedures. Ensure that employees and others are aware of these procedures and can apply them;
▶ provide health and safety information to employees and others, such as other employers, the self-employed and their employees who are sharing the same workplace and parents of child employees or those on work experience;
▶ cooperate in health and safety matters with other employers who share the same workplace;

▶ provide employees with adequate and relevant health and safety training;
▶ provide temporary workers and their contract agency with appropriate health and safety information;
▶ protect new and expectant mothers and young persons from particular risks;
▶ under certain circumstances, as outlined in Regulation 6, provide health surveillance for employees.

The information that should be supplied by employers under the Regulations is:

▶ risks identified by any risk assessments including those notified to the employer by other employers sharing the same workplace;
▶ the preventative and protective measures that are in place;
▶ the emergency arrangements and procedures and the names of those responsible for the implementation of the procedures.

Regulation 19 requires employers to protect young persons from risks resulting from their immaturity.

The protection of young persons (under the age of 18) from risks associated with their immaturity is covered in detail in Chapter 4. This immaturity is often associated with the inability to perceive hazards and may be intellectual and/or physical. The influence of peer pressure is also an important factor amongst young people.

Finally, it is important to stress, as noted above, that Schedule 1 of the Regulations outline the principles of prevention which employers and the self-employed need to apply so that health and safety risks are addressed and controlled. These principles are discussed in detail in Chapter 4.

1.5.2 Employees' duties

Employees must:

▶ use any equipment or substance in accordance with any training or instruction given by the employer;
▶ report to the employer any serious or imminent danger;
▶ report any shortcomings in the employer's protective health and safety arrangements.

1.5.3 Joint occupation of premises

The Management of Health and Safety at Work Regulations specifically state that where two or more employers share a workplace – whether on a temporary or a permanent basis – each employer shall:

▶ cooperate with other employers;
▶ take reasonable steps to coordinate between other employers to comply with legal requirements;
▶ take reasonable steps to inform other employers where there are risks to health and safety.

All employers and self-employed people involved should satisfy themselves that the arrangements adopted are adequate. Where a particular employer controls the premises, the other employers should help to assess the shared risks and coordinate any necessary control procedures. Where there is no controlling employer, the organisations present should agree joint arrangements to meet regulatory obligations, such as appointing a health and safety coordinator.

1.6 The legal and organisational health and safety roles and responsibilities of clients and their contractors

1.6.1 Introduction

A contractor is anyone who is brought in to work and is not an employee. People on work experience or on labour only contracts or temporary staff are considered to be employees. Contractors are used for maintenance, repairs, installation, construction, demolition, computer work, cleaning, security, health and safety and many other tasks. Sometimes there are several contractors on site at any one time. Clients need to think about how their work may affect each other's and how they interact with the normal site occupier.

1.6.2 Legal considerations

The HSW Act applies to all work activities. It requires employers to ensure, so far as is reasonably practicable, the health and safety of:

▶ their employees;
▶ other people at work on their site, including contractors;
▶ members of the public who may be affected by their work.

All parties to a contract have specific responsibilities under health and safety law, and these cannot be passed on to someone else:

▶ Employers are responsible for protecting people from harm caused by work activities. This includes the responsibility not to harm contractors and sub-contractors on site.
▶ Employees and contractors have to take care not to endanger themselves, their colleagues or others affected by their work.
▶ Contractors also have to comply with the HSW Act and other health and safety legislation. Clearly, when contractors are engaged, the activities of different employers do interact. So cooperation and communication are needed to make sure all parties can meet their obligations.
▶ Employees have to cooperate with their employer on health and safety matters, and not do anything that puts them or others at risk.

▶ Employees must be trained and clearly instructed in their duties.
▶ Self-employed people must not put themselves in danger, or others who may be affected by what they do.
▶ Suppliers of chemicals, machinery and equipment have to make sure their products or imports are safe, and provide information on this.

The Management of Health and Safety at Work Regulations apply to everyone at work and encourage employers to take a more systematic approach to dealing with health and safety by:

▶ assessing the risks which affect employees and anyone who might be affected by the site occupier's work, including contractors;
▶ setting up emergency procedures;
▶ providing training;
▶ cooperating with others on health and safety matters, for example contractors who share the site with an occupier;
▶ providing temporary workers, such as contractors, with health and safety information.

Most contracted work is governed by a legally binding contract and it is, therefore, very important that the contract covers all parts of the work – for example, competent workers, safe access when working at height, fire precautions and safe waste disposal. The principles of cooperation, coordination and communication between organisations underpin the Management of Health and Safety at Work Regulations and the CDM Regulations.

1.6.3 Construction (Design and Management) (CDM 2015) Regulations

Businesses often engage contractors for construction projects at one time or another to build, convert or extend premises and demolish buildings. The CDM 2015 Regulations apply to all construction projects. Larger projects which are notifiable (see details in Chapter 19) have more extensive requirements.

All projects require the following:

▶ clients to check the competence of all their appointees; ensure there are suitable management arrangements for the project; allow sufficient time and resources for all stages and provide pre-construction information to designers and contractors. There are special arrangements under the regulations for domestic clients (see later in this section);
▶ designers to eliminate, reduce or control hazards and reduce risks during the design and construction stage and during the maintenance and use of a building once it is built. They should also provide information about remaining risks;

▶ contractors to plan, manage and monitor their own work and that of employees; check the competence of all their appointees and employees; train their own employees; provide information to their employees; comply with the requirements for health and safety on site detailed in Part 4 of the Regulations and other Regulations such as the Work at Height Regulations; and ensure there are adequate welfare facilities for their employees;

▶ everyone, including workers, to: assure their own competence; cooperate with others and coordinate work so as to ensure the health and safety of construction workers and others who may be affected by the work; report obvious risks; take account of the general principles of prevention in planning or carrying out construction work; and comply with the requirements in Part 4 of CDM 2015 and other relevant regulations for any work under their control.

For even small projects, clients should ensure that contractors provide:

▶ information regarding the contractor's health and safety policy;

▶ information on the contractor's health and safety organisation detailing the responsibilities of individuals;

▶ information on the contractor's procedures and standards of safe working;

▶ the method statements for the project in hand;

▶ details on how the contractor will audit and implement its health and safety procedures;

▶ procedures for investigating incidents and learning the lessons from them.

Smaller contractors may need some guidance to help them produce suitable method statements. While they do not need to be lengthy, they should set out those features essential to safe working, for example access arrangements, PPE, control of chemical risks and fire precautions.

Copies of relevant risk assessments for the work to be undertaken should be requested. These need not be very detailed but should indicate the risk and the control methods to be used.

The client, principal designer, designer, principal contractor and other contractors all have specific roles under CDM 2015 Regulations.

1.6.4 Explanation of terms used and duties in CDM 2015

(a) **The Client** is an organisation or individual for whom a construction project is undertaken. This can include, for example, local authorities, school governors, insurance companies and project originators for private finance initiative (PFI) projects. If it is likely that more than one contractor will be working on a project at any one time, the client must appoint a principal designer and a principal contractor. The client has the following duties:

▶ make arrangements for managing the project including the allocation of sufficient time and other resources and to ensure that construction work is carried out as safely so far as is reasonably practicable;

▶ ensure that the welfare requirements in Schedule 2 of the regulations are followed;

▶ appoint in writing a designer with control over the pre-construction phase as principal designer and a contractor as principal contractor. If the client does not appoint a principal designer, the client must fulfil the duties of the principal designer;

▶ provide pre-construction information to each designer and each contractor engaged on the project;

▶ take reasonable steps to ensure that the health and safety arrangements are maintained and reviewed throughout the project;

▶ ensure that the principal designer complies with their duties under the regulations;

▶ ensure that the principal contractor complies with their duties under the regulations;

▶ ensure before the construction phase begins that the principal contractor (or contractor if only one is used) draws up a construction phase plan;

▶ ensure that the principal designer prepares an appropriate health and safety file for the project which:
  ▷ includes information on asbestos as required under the Control of Asbestos Regulations;
  ▷ is revised from time to time;
  ▷ is kept available for inspection; and
  ▷ is passed on if interest in the project passes to another client.

Domestic clients are a special case and do not have specific duties under CDM 2015 (see the note at the end of this section).

(b) **The Principal Designer** must plan, manage, monitor and coordinate the pre-construction phase of a project, taking into account the general principles of prevention to ensure:

▶ the project is carried out safely so far as is reasonably practicable;

▶ assistance is given to the client for the provision of pre-construction information;

▶ health and safety risks are identified, eliminated, controlled or reduced;

▶ the cooperation of everyone working on the project;

▶ the preparation and revision as necessary of the health and safety file;

▶ the provision of pre-construction information to every designer and contractor; and

▶ there is liaison with the principal contractor for the duration of the project, particularly over the preparation of the construction phase plan.

(c) **Designers** are those who have a trade or a business which involves them in:

▶ preparing designs for construction work, including variations. This includes preparing drawings, design details, specifications, bills of quantities and the specification (or prohibition) of articles and substances, as well as all the related analysis, calculations and preparatory work; and

▶ arranging for their employees or other people under their control to prepare designs relating to a structure or part of a structure.

It does not matter whether the design is recorded (e.g. on paper or a computer) or not (e.g. it is communicated only orally).

Designers may include:

▶ architects, civil and structural engineers, building surveyors, landscape architects, other consultants, manufacturers and design practices (of whatever discipline) contributing to, or having overall responsibility for, any part of the design, for example drainage engineers;

▶ anyone who specifies or alters a design, or who specifies the use of a particular method of work or material, such as a design manager, quantity surveyor who insists on specific material, or a client who stipulates a particular layout for a new building;

▶ building service designers, engineering practices or others designing plant which forms part of the permanent structure (including lifts, heating, ventilation and electrical systems), for example a specialist provider of permanent fire-extinguishing installations;

▶ those purchasing materials where the choice has been left open, for example those purchasing building blocks and so deciding the weights that a bricklayer must handle;

▶ contractors carrying out design work as part of their contribution to a project, such as an engineering contractor providing design, procurement and construction management services;

▶ temporary works engineers, including those designing auxiliary structures, such as formwork, falsework, façade retention schemes, scaffolding and sheet piling;

▶ interior designers, including shop-fitters who also develop the design;

▶ heritage organisations who specify how work is to be done in detail, for example providing detailed requirements to stabilise existing structures; and

▶ those determining how buildings and structures are altered, for example during refurbishment, where this has the potential for partial or complete collapse.

In summary, therefore, under CDM, a designer includes architects, consulting engineers, quantity surveyors,

building service engineers and temporary works engineers.

The duties of the designer are:

▶ before starting work be satisfied that the client is aware of their duties under CDM 2015;

▶ to take into account the general principles of prevention to eliminate risks for anyone:
  ▷ working on the project;
  ▷ maintaining or cleaning a structure; or
  ▷ using a structure designed as a workplace;

▶ if elimination is not possible the designer must so far as is reasonably practicable:
  ▷ take steps to reduce and control health and safety risks through the design process;
  ▷ provide information about health and safety risks to the principal designer;
  ▷ ensure appropriate information is contained in the health and safety file; and
  ▷ provide with the design sufficient information to help clients, other designers and contractors to fulfil their duties under CDM.

(d) **The Principal Contractor** is the contractor appointed by the client. The principal contractor can be an organisation or an individual, and is the main or managing contractor. The principal contractor must consult and engage with workers on the project. A principal contractor's key duties are to coordinate and manage the construction phase of the project and to ensure the health and safety of everybody involved with the construction work, or who are affected by it. Other duties are to ensure that:

▶ construction work is carried out safely so far as is reasonably practicable;

▶ a construction phase plan is prepared prior to setting up a site and reviewed and revised as necessary;

▶ there is coordination of the implementation of legal requirements and the construction phase plan for all people on site;

▶ necessary site health and safety rules are drawn up;

▶ suitable site induction is provided;

▶ access by unauthorised persons is controlled preventing unauthorised access;

▶ the welfare requirements in Schedule 2 of the regulations are provided throughout the construction phase;

▶ there is liaison with the principal designer throughout the project, particularly regarding information for the health and safety file and management of the construction phase plan; and

▶ the health and safety file is appropriately updated, reviewed and revised to take account of changes.

(e) **The Contractor** is any person who, in the course or furtherance of a business, undertakes or manages

1

construction work. A contractor has the following duties:

- not to undertake construction work on a project unless satisfied that the client is aware of their duties under CDM 2015;
- to plan, manage and monitor the way construction work is done to ensure it is safe so far as is reasonably practicable;
- where there is no principal contractor, to ensure that a construction phase plan is drawn up prior to setting up the site;
- to provide any employees or persons under their control, any information and instruction to ensure the safety of the project including:
  - ▷ suitable site induction;
  - ▷ emergency procedures;
  - ▷ information on risks to their health and safety either identified by their risk assessments or from another contractor's operations;
- to provide any other health and safety training required by the Management of Health and Safety at Work Regulations;
- to not start work until access to unauthorised persons is controlled;
- to ensure that the welfare requirements in Schedule 2 of the regulations are provided for employees and others under their control;
- to comply with any directions given by the principal designer or principal contractor and any site rules; and
- to consult with their workers.

(f) **Pre-construction health and safety information**
The client must provide designers and contractors and with specific health and safety information needed to identify the hazards and risks associated with the design and construction work. The information must be identified, assembled and supplied in good time, so that those who need it during the preparation of a bid for construction work or the planning of such work may estimate the resources required to enable the design, planning and construction work to be properly organised and carried out. The topics which should be addressed in the pre-construction health and safety information are given in Appendix 1.2.

(g) **Construction phase health and safety plan**
The principal contractor must define the way in which the construction phase will be managed and the key health and safety issues for a particular project must be recorded. The health and safety plan should outline the organisation and arrangements required to manage risks and coordinate the work on site. It should not be a repository for detailed generic risk assessments, records of how decisions were reached or detailed method statements; but it may, for example, set out when such documents will need to be prepared.

It should be well focused, clear and easy for contractors and others to understand – emphasising key points and avoiding irrelevant material. It is crucial that all relevant parties are involved and cooperate in the development and implementation of the plan as work progresses.

The plan must be tailored to the particular project and relevant photographs and sketches will simplify and shorten explanations. The plan should be organised so that relevant sections can easily be made available to designers and contractors.

The construction phase health and safety plan for the initial phase of the construction work must be prepared before any work begins. Further details may be added as completed designs become available and construction work proceeds. Appendix 1.3 gives details of what should be included in the health and safety plan.

(h) **Health and safety file**
This is a record of health and safety information for the client to retain. It alerts those who are responsible for the structure and equipment within it to the significant health and safety risks that will need to be addressed during subsequent use, construction, maintenance, cleaning, repair, alterations, refurbishment and demolition work. Appendix 1.4 gives details of what should be included in the health and safety file.

(i) **Method statement**
This is a formal document that describes the sequences of operations for safe working that will ensure health and safety during the performance of a task. It results from the risk assessment carried out for the task or operation and the identified control measures. If the risk is low, a verbal statement may normally suffice.

(j) **Notifiable work**
Construction work is notifiable to the HSE if it lasts longer than 30 working days and has more than 20

Figure 1.14 Domestic client: CDM applies but not notifiable

Figure 1.15 Large-scale contract: CDM applies and it is notifiable

workers simultaneously or will involve more than 500 person days of work (e.g. 50 people working for 10 days). Holidays and weekends are not counted if there is no construction work on those days. The client is responsible for making the notification as soon as possible after their appointment to the particular project. The notice must be displayed where it can be read by people working on the site and must be updated as necessary. (See Chapter 19 for the details required.) Notification does not give rise to any additional duties.

(k) **Domestic clients**

Unless there is a written agreement between the domestic client and the principal designer, the duties of the client must be carried out by the contractor or principal contractor where there is more than one.

Where there is more than one contractor, the domestic client must appoint a principal designer and principal contractor in writing. Where no appointment is made, the first designer and contractor appointed are deemed to be the principal one in each case.

1.7 The principles of assessing and managing contractors

1.7.1 Scale of contractor use

The use of contractors is increasing as many companies turn to outside resources to supplement their own staff and expertise. Contractors are often used for information technology projects, maintenance, repairs, installation, construction, demolition and many other jobs. Organisations have increasingly used contractors over recent years rather than employ more full-time employees. This may be to supplement their staff particularly for specialist tasks, or to undertake non-routine activities. There may also be other reasons for the increased use of contractors:

▶ Demand for products or services is uncertain.
▶ Contractors can be used when more flexibility is required.
▶ Contractors usually supply their own tools and equipment associated with the contract.
▶ There are no permanent staff available to perform the work.
▶ The financial overheads and legal employment obligations are lower.
▶ Most of the costs associated with increasing and reducing employee numbers as product demand varies do not relate to contractors.
▶ Permanent staff can concentrate on the core business of the organisation.

The advantage of the flexibility of contractors in uncertain economic times is the main reason for the increased use of them over recent years. There are, however, some disadvantages in the use of independent contractors. These include:

▶ Contractors/sub-contractors may cost more than the equivalent daily rate for employing a worker.
▶ By relying on contractors, the skills of permanent staff are not developed.
▶ There is less control over contractors than permanently employed staff. This can be a significant problem if the contractor sub-contracts some of the work.
▶ The control of the contractor and the quality of the work is crucially dependent on the terms of the contract.

1.7.2 Contractor selection

The selection of the right contractor for a particular job is probably the most important element in ensuring that the risks to the health and safety of everybody involved in the activity and people in the vicinity are reduced as far as possible. Ideally, selection should be made from a list of approved contractors who have demonstrated that they are able to meet the client's requirements.

The selection of a contractor has to be a balanced judgement with a number of factors being taken into account. Fortunately, a contractor who works well and meets the client's requirements in terms of the quality and timeliness of the work is also likely to have a better than average health and safety performance. Cost, of course, will have to be part of the judgement but may not provide any indication of which contractor is likely to give the best performance in health and safety terms. In deciding which contractor should be chosen for a task, the following should be considered:

▶ Do they have an adequate health and safety policy?
▶ Can they demonstrate that the person responsible for the work is competent?
▶ Can they demonstrate that competent safety advice will be available?

- Do they monitor the level of accidents at their work site?
- Do they have a system to assess the hazards of a job and implement appropriate control measures?
- Will they produce a method statement which sets out how they will deal with all significant risks?
- Do they have guidance on health and safety arrangements and procedures to be followed?
- Do they have effective monitoring arrangements?
- Do they use trained and skilled staff who are qualified where appropriate? (Judgement will be required, as many construction workers have had little or no training except training on the job.) Can the company demonstrate that the employees or other workers used for the job have had the appropriate training and are properly experienced and, where appropriate, qualified?
- Can they produce good references indicating satisfactory performance?

Figure 1.16 Contractors at work unloading steel beams

1.7.3 Management and authorisation of contractors

The management of contractors is essential since accidents tend to happen more easily when the work of contractors is not properly supervised or the hazards of their job have not been identified and suitable controls introduced. It is important that contractors are made aware of:

- the health and safety procedures and rules of the organisation;
- the hazards on your site, particularly those associated with the project;
- any special equipment or personal protective equipment that they need to use;
- the emergency procedures and the sound of the alarm; and
- the safe disposal of waste.

Good communication and supervision of contractors is essential with a named point of contact from the organisation known to the contractors. Regular liaison

is important so that the progress of the contract to a satisfactory completion may be monitored.

It is important that the activities of the contractors within the organisation of the client are effectively planned and coordinated so there is minimal interference with the normal activities of the organisation and the health and safety of its staff is not put at risk. This means that staff should be made aware of the contractor activity and any particular hazards associated with it, such as dust and noise.

Control of construction contractors

On being selected, contractors should be expected to:

- familiarise themselves with those parts of the health and safety plan which affect them and their employees and/or sub-contractors;
- cooperate with the principal contractor in his/her health and safety duties to contractors;
- comply with their legal health and safety duties.

On arrival at the site, sub-contractors should ensure that:

- they report to the site office and the site manager;
- they abide by any site rules, particularly in respect of personal protective equipment;
- the performance of their work does not place others at risk;
- they are familiar with the first-aid and accident reporting arrangements of the principal contractor;
- they are familiar with all emergency procedures on the site;
- any materials brought onto the site are safely handled, stored and disposed of in compliance, where appropriate, with the current Control of Substances Hazardous to Health Regulations;
- they adopt adequate fire precaution and prevention measures when using equipment which could cause fires;
- they minimise noise and vibration produced by their equipment and activities;
- any ladders, scaffolds and other means of access are erected in conformity with good working practice and any statutory requirements;
- any welding or burning equipment brought onto the site is in safe operating condition and used safely with a suitable fire extinguisher to hand;
- any lifting equipment brought onto the site complies with the current Lifting Operation and Lifting Equipment Regulations;
- all electrical equipment complies with the current Electricity at Work Regulations;
- connection to the electricity supply is from a point specified by the principal contractor and is by proper cables and connectors. For outside construction work, only 110V equipment should be used;
- any restricted access to areas on the site is observed;

- welfare facilities provided on site are treated with respect;
- any vehicles brought onto the site observe any speed, condition or parking restriction.

The control of sub-contractors can be exercised by monitoring them against the criteria listed above and by regular site inspections. On completion of the contract, the work should be checked to ensure that the agreed standard has been reached and that any waste material has been removed from the site.

SITE SAFETY

Under the Health & Safety at Work Act 1974 all persons entering this site must comply with all regulations under this Act.
All visitors must report to the site office and obtain permission to proceed on to the site or any work area.
Safety signs and procedures must be observed and personal protection and safety equipment must be used at all times.

 Construction work in progress. Parents are advised to warn children of the dangers of entering this site.

 Safety helmets must be worn

 Unauthorised entry to this site is strictly forbidden.

Figure 1.17 Site safety rules

Contractor authorisation

Contractors, their employees, and sub-contractors and their employees, should not be allowed to commence work on any client's site without authorisation signed by the company contact. The authorisation should clearly define the range of work that the contractor can carry out and set down any special requirements, for example protective clothing, fire exits to be left clear, and isolation arrangements.

Permits will be required for operations such as hot work. All contractors should keep a copy of their authorisation at the place of work. A second copy of the authorisation should be kept at the site and be available for inspection.

The company contact signing the authorisation will be responsible for all aspects of the work of the contractor. The contact will need to check as a minimum the following:

- that the correct contractor for the work has been selected;
- that the contractor has made appropriate arrangements for supervision of staff;
- that the contractor has received and signed for a copy of the contractor's safety rules;
- that the contractor is clear what is required, the limits of the work and any special precautions that need to be taken;

- that the contractor's personnel are properly qualified for the work to be undertaken.

The company contact should check whether sub-contractors will be used. They will also require authorisation, if deemed acceptable. It will be the responsibility of the company contact to ensure that sub-contractors are properly supervised.

Appropriate supervision will depend on a number of factors, including the risk associated with the job, experience of the contractor and the amount of supervision the contractor will provide. The responsibility for ensuring there is proper supervision lies with the person signing the contractor's authorisation.

The company contact will be responsible for ensuring that there is adequate and clear communication between different contractors and company personnel where this is appropriate.

1.7.4 Safety rules for contractors

In the conditions of contract, there should be a stipulation that the contractor and all of their employees adhere to the contractor's safety rules. Contractors' safety rules should contain as a minimum the following points:

- **Health and safety** – that the contractor operates to at least the minimum legal standard and conforms to accepted industry good practice;
- **Supervision** – that the contractor provides a good standard of supervision of their own employees;
- **Sub-contractors** – that they may not use sub-contractors without prior written agreement from the organisation;
- **Authorisation** – that each employee must carry an authorisation card issued by the organisation at all times while on site.

Figure 1.18 Rules at site entrance with viewing panel to see inside the site

1.7.5 Example of rules for contractors

Contractors engaged by the organisation to carry out work on its premises will:

▶ familiarise themselves with so much of the organisation's health and safety policy as affects them and will ensure that appropriate parts of the policy are communicated to their employees, and any sub-contractors and employees of sub-contractors who will do work on the premises;

▶ cooperate with the organisation in its fulfilment of its health and safety duties to contractors and take the necessary steps to ensure the like cooperation of their employees;

▶ comply with their legal and moral health, safety and food hygiene duties;

▶ ensure the carrying out of their work on the organisation's premises in such a manner as not to put either themselves or any other persons on or about the premises at risk;

▶ ensure that where they wish to avail themselves of the organisation's first-aid arrangements/facilities while on the premises, written agreement to this effect is obtained prior to first commencement of work on the premises;

▶ supply a copy of their statement of policy, organisation and arrangements for health and safety written for the purposes of compliance with The Management of Health and Safety at Work Regulations and Section 2(3) of the HSW Act where applicable and requested by the organisation;

▶ abide by all relevant provisions of the organisation's safety policy, including compliance with health and safety rules and CDM 2015;

▶ ensure that on arrival at the premises, they and any other persons who are to do work under the contract report to reception or their designated organisation contact.

Without prejudice to the requirements stated above, contractors, sub-contractors and employees of contractors and sub-contractors will, to the extent that such matters are within their control, ensure:

▶ the safe handling, storage and disposal of materials brought onto the premises;

▶ that the organisation is informed of any hazardous substances brought onto the premises and that the relevant parts of the Control of Substances Hazardous to Health Regulations in relation thereto are complied with;

▶ that fire prevention and fire precaution measures are taken in the use of equipment which could cause fires;

▶ that steps are taken to minimise noise and vibration produced by their equipment and activities;

▶ that scaffolds, ladders and other such means of access, where required, are erected and used in accordance with Work at Height Regulations and good working practice;

▶ that any welding or burning equipment brought onto the premises is in safe operating condition and used in accordance with all safety requirements;

▶ that any lifting equipment brought onto the premises is adequate for the task and has been properly tested/certified;

▶ that any plant and equipment brought onto the premises is in safe condition and used/operated by competent persons;

▶ that for vehicles brought onto the premises, any speed, condition or parking restrictions are observed;

▶ that compliance is made with the relevant requirements of the Electricity at Work Regulations;

▶ that connection(s) to the organisation's electricity supply is from a point specified by its management and is by proper connectors and cables;

▶ that they are familiar with emergency procedures existing on the premises;

▶ that welfare facilities provided by the organisation are treated with care and respect;

▶ that access to restricted parts of the premises is observed and the requirements of food safety legislation are complied with;

▶ that any major or lost-time accident or dangerous occurrence on the organisation's premises is reported as soon as possible to their site contact;

▶ that where any doubt exists regarding health and safety requirements, advice is sought from the site contact.

The foregoing requirements do not exempt contractors from their statutory duties in relation to health and safety, but are intended to assist them in attaining a high standard of compliance with those duties.

1.8 Further information

The Health and Safety at Work etc. Act 1974

The Enterprise and Regulatory Reform Act 2013 (Section 69)

The Health and Safety at Work etc. Act 1974 (Civil Liabilities) (Exceptions) Regulations 2013

The Interpretation Act 1978

The Management of Health and Safety at Work Regulations 1999 (as amended)

The Construction (Design and Management) Regulations 2015

Corporate Manslaughter and Corporate Homicide Act 2007

Health and Safety (Fees) Regulations 2012 (Regulations 23–25)

Guidance on the application of Fees for Intervention (FFI), HSE47, HSE Books http://www.hse.gov.uk/pubns/hse47.pdf

Managing for health and safety, HSG65 third edition 2013, HSE Books, ISBN 978-0-7176-6456-6 http://www.hse.gov.uk/pubns/priced/hsg65.pdf

HSE, 'Managing for health and safety' http://www.hse.gov.uk/managing/

HSE, The Health and Safety Toolbox: how to control risks at work http://www.hse.gov.uk/toolbox/

Leading health and safety at work (INDG 417), HSE Books, ISBN 978 0 7176 6267 8 www.hse.gov.uk/pubs/indg417.htm

Managing Health and Safety in Construction (Guidance) (L153), HSE Books, ISBN 978 0 7176 6223 4 http://www.hse.gov.uk/pubns/books/l153.htm

Managing Health and Safety in Construction, ACoP, (L144), HSE Books, ISBN 978-0-7176-6223-4 (to be replaced by L153 during 2015)

Health and Safety in Construction (Guidance) (HSG150rev), HSE Books, ISBN 978 0 7176 6182 4 http://www.hse.gov.uk/pubns/books/hsg150.htm

The Construction (Design and Management) Regulations 2015, Industry guidance for Clients, produced by CONIAC, ISBN 978-1-85751-389-9 http://www.citb.co.uk/documents/cdm%20regs/industry-guidance-clients.pdf

The Construction (Design and Management) Regulations 2015, Industry guidance for Contractors, produced by CONIAC, ISBN 978-1-85751-391-2 http://www.citb.co.uk/documents/cdm%20regs/industry-guidance-contractors.pdf

The Construction (Design and Management) Regulations 2015, Industry guidance for Designers, produced by CONIAC, ISBN 978-1-85751-392-9 http://www.citb.co.uk/documents/cdm%20regs/industry-guidance-designers.pdf

The Construction (Design and Management) Regulations 2015, Industry guidance for Principal contractors, produced by CONIAC, ISBN 978-1-85751-393-6 http://www.citb.co.uk/documents/cdm%20regs/industry-guidance-principal-contractors.pdf

The Construction (Design and Management) Regulations 2015, Industry guidance for Principal Designers, produced by CONIAC, ISBN 978-1-85751-390-5 http://www.citb.co.uk/documents/cdm%20regs/industry-guidance-principal-designer.pdf

The Construction (Design and Management) Regulations 2015, Industry guidance for Workers, produced by CONIAC, ISBN 978-1-85751-394-3 http://www.citb.co.uk/documents/cdm%20regs/industry-guidance-workers.pdf

Occupational Health and Safety Assessment Series (OHSAS 18000): Occupational Health and Safety Management Systems OHSAS 18001:2007 ISBN 978-0-5805-9404-5 OHSAS18002:2008 ISBN 978-0-5806-2686-9

HSE Statistics http://www.hse.gov.uk/statistics/index.htm

1.9 Practice revision questions

1. In relation to occupational health and safety, **explain**, using an example in **EACH** case, the meaning of the following terms:
   (a) hazard
   (b) risk
   (c) welfare
   (d) work-related ill-health
   (e) near miss.

2. (a) **Explain** the reasons that organisations are required to have employer liability insurance.
   (b) **Outline SIX** possible direct **AND SIX** possible indirect costs to an organisation following a serious accident in the workplace. Identify those costs that are insurable and those that are not.

3. (a) **Outline** the key differences between civil law and criminal law.
   (b) **Identify** the main differences between common law and statute law and **explain** by giving **ONE** example how these two sources of law are applied in occupational health and safety cases.

4. In relation to health and safety, **outline** the role of the following:
   (a) employment tribunals;
   (b) Civil Courts;
   (c) Criminal Courts.

5. (a) **Identify** the courts in which summary and indictable health and safety offences are heard.
   (b) Giving an example in each case, **outline** the common law duties of care that employers have towards their employees for their health, safety and welfare.

6. (a) **Explain** the meaning of the term 'negligence' in health and safety matters.
   (b) **Outline** the three absolute defences and two partial defences available to an employer in a case of alleged negligence brought by an employee.
   (d) **Explain** the significance of vicarious liability for an employer following an injury to an employee resulting from the negligence of another employee.

7. (a) **Explain**, giving an example in each case, the differences between the levels of statutory duty – absolute, practicable and reasonably practicable.

(b) **Outline** the meaning of the terms statute-barred breaches of statutory duty and double-barrelled actions.

(c) **Identify** the general duties placed on employees by:
   (i) the Health and Safety at Work etc. Act 1974
   (ii) the Management of Health and Safety at Work Regulations 1999.

8. (a) **Outline** the general duties placed on employers by the Health and Safety at Work etc. Act 1974 in Sections 2, 3 and 6.

(b) **Identify** the possible defences that an employer may give following a prosecution under the Health and Safety at Work etc. Act 1974.

(c) **Identify FOUR** enforcement agencies that have the power to prosecute companies for non-compliance with health and safety legislation.

(d) **Explain**, using relevant examples, the circumstances under which a health and safety inspector may serve an improvement notice or a prohibition notice on an employer. Briefly outline the appeal procedure in each case.

9. While visiting a company, a health and safety enforcement officer observes a contractor operating a dangerous part of a machine that is unguarded.

(a) **Describe** the powers available to the officer under the Health and Safety at Work etc. Act 1974.

(b) **Identify** the sections of the Health and Safety at Work etc. Act 1974 that may have been breached.

(c) **Identify** the actions available and the likely one that was taken by the officer.

10. Giving an example in each case, **explain** the meaning, legal status and roles of:

(a) health and safety regulations
(b) HSE Approved Codes of Practice
(c) HSE guidance.

11. **Describe** the possible penalties that a court could impose on a company that is convicted of an offence under the Corporate Manslaughter and Corporate Homicide Act 2007.

12. (a) **Outline** the measures that directors of corporations can take to promote good health and safety standards within their organisations.

(b) **Outline** ways by which the health and safety of visitors to the workplace may be protected.

(c) **Describe** the issues that should be addressed when assessing risks to night workers.

(d) **Describe** the important factors that will help to achieve cooperation and coordination in a jointly occupied workplace.

13. **Describe** the possible health and safety issues that may arise in the management of a supply chain.

14. **Outline FIVE** duties of employers to their employees under the Management of Health and Safety at Work Regulations 1999.

15. Following an investigation of an accident to a contractor while on an employer's premises, it was found that the employer had failed to provide relevant health and safety information to the contractor. The employer was also unable to demonstrate any source of competent health and safety advice.

(a) **Outline** the duties that the employer owes to the contractor under the Health and Safety at Work etc. Act 1974.

(b) **Outline** the duties that the employer owes under the Management of Health and Safety at Work Regulations 1999 that may not have been discharged.

(c) **Outline** the health and safety information that should have been provided to and from the contractor before work commenced.

16. **Outline FOUR** duties of each of the following persons under the Construction (Design and Management) Regulations:

(a) the client
(b) the principal designer
(c) the principal contractor.

17. Several construction contractors are to be employed by a printing company to refurbish its building.

(a) **Outline** the health and safety duties of the self-employed.

(b) **Outline** the main issues to be considered during the selection of the contractors by the printing company.

(c) **Outline** the matters to be addressed by the company while supervising the contractors and the company employees.

APPENDIX 1.1 Checklist for supply chain health and safety management

This checklist is a reminder of the topics that might need to be discussed with people with whom individual contractors may be working.

It is not intended to be exhaustive and not all questions will apply at any one time, but it should help people to get started.

1. Responsibilities

▶ What are the hazards of the job?

▶ Who is to assess particular risks?

▶ Who will coordinate action?

▶ Who will monitor progress?

2. The job

▶ Where is it to be done?

▶ Who with?

▶ Who is in charge?

▶ How is the job to be done?

▶ What other work will be going on at the same time?

▶ How long will it take?

▶ What time of day or night?

▶ Do you need any permit to do the work?

3. The hazards and risk assessments

Site and location

Consider the means of getting into and out of the site and the particular place of work – are they safe? – and:

▶ Will any risks arise from environmental conditions?

▶ Will you be remote from facilities and assistance?

▶ What about physical/structural conditions?

▶ What arrangements are there for security?

Substances

▶ What suppliers' information is available?

▶ Is there likely to be any microbiological risk?

▶ What are the storage arrangements?

▶ What are the physical conditions at the point of use? Check ventilation, temperature, electrical installations, etc.

▶ Will you encounter substances that are not supplied, but produced in the work, for example fumes from hot work during dismantling plant? Check how much, how often, for how long, method of work, etc.

▶ What are the control measures? For example, consider preventing exposure, providing engineering controls, using personal protection (in that order of choice).

▶ Is any monitoring required?

▶ Is health surveillance necessary, for example for work with sensitisers? (Refer to health and safety data sheet.)

Plant and equipment

▶ What are the supplier/hirer/manufacturer's instructions?

▶ Are any certificates of examination and test needed?

▶ What arrangements have been made for inspection and maintenance?

▶ What arrangements are there for shared use?

▶ Are the electrics safe to use? Check the condition of power sockets, plugs, leads and equipment. (Don't use damaged items until they have been repaired.)

▶ What assessments have been made of noise levels?

4. People

▶ Is information, instruction and training given, as appropriate?

▶ What are the supervision arrangements?

▶ Are members of the public/inexperienced people involved?

▶ Have any disabilities/medical conditions been considered?

5. Emergencies

▶ What arrangements are there for warning systems in case of fire and other emergencies?

▶ What arrangements have been made for fire/emergency drills?

▶ What provision has been made for first-aid and fire-fighting equipment?

▶ Do you know where your nearest fire exits are?

▶ What are the accident reporting arrangements?

▶ Are the necessary arrangements in place for availability of rescue equipment and rescuers?

6. Welfare

Who will provide:

▶ shelter

▶ food and drinks

▶ washing facilities

▶ toilets (male and female)

▶ clothes changing/drying facilities?

There may be other pressing requirements which make it essential to re-think health and safety as the work progresses.

APPENDIX 1.2 Pre-construction information

This is taken from the industry guidance produced by members of CONIAC (Construction Industry Advisory Committee) – The Construction (Design and Management) Regulations 2015 Industry guidance for Principal Contractors. The ACoP for CDM 2015 was not available when this book was published but it is unlikely that the content of the Pre-construction information will change significantly.

When drawing up the pre-construction information, each of the following topics should be considered. Information should be included where the topics are relevant to the work proposed. The pre-construction information provides information for those bidding for or planning work, and for the development of the construction phase plan. The level of detail in the information should be proportionate to the risks involved in the project.

Pre-construction information

1. Description of project

(a) Project description and programme details including:
  ▶ key dates (including planned start and finish of the construction phase), and
  ▶ the minimum time to be allowed between appointment of the principal contractor and instruction to commence work on site.

(b) Details of client, principal designer, designers, and other consultants

(c) Whether or not the structure will be used as a workplace – if so, the finished design will need to take account of the relevant requirements of the Workplace (Health, Safety and Welfare) Regulations

(d) Extent and location of existing records and plans.

2. Client's considerations and management requirements

(a) Arrangements for:
  ▶ planning for and managing the construction work, including any health and safety goals for the project
  ▶ communications and liaison between client and others
  ▶ security of the site
  ▶ welfare provisions.

(b) Requirements relating to the health and safety of the client's employees or customers or those involved in the project such as:
  ▶ site hoarding requirements
  ▶ site transport arrangements or vehicle movement restrictions

  ▶ client permit-to-work systems
  ▶ fire precautions
  ▶ emergency procedures and means of escape
  ▶ 'no-go' areas or other authorisation requirements for those involved in the project
  ▶ any areas the client has designated as confined spaces
  ▶ smoking and parking restrictions.

3. Environmental restrictions and existing on-site risks

(a) Safety hazards, including:
  ▶ boundaries and access, including temporary access, e.g. narrow streets, lack of parking, turning or storage
  ▶ any restrictions on deliveries or waste collection or storage
  ▶ adjacent land uses, e.g. schools, railway lines or busy roads
  ▶ existing storage of hazardous materials
  ▶ location of existing services particularly those that are concealed – water, electricity, gas, etc.
  ▶ ground conditions, underground structures or water courses where these might affect the safe use of plant, e.g. cranes, or the safety of ground work
  ▶ information about existing structures – stability, structural form, fragile or hazardous materials, anchorage points for fall arrest systems (particularly where demolition is involved)
  ▶ previous structural modifications, including weakening or strengthening of the structure (particularly where demolition is involved)
  ▶ fire damage, ground shrinkage, movement or poor maintenance which may have adversely affected the structure
  ▶ any difficulties relating to plant and equipment in the premises, such as overhead gantries whose height restricts access
  ▶ health and safety information contained in earlier design, construction or 'as-built' drawings, such as details of pre-stressed or post-tensioned structures.

(b) Health hazards, including:
  ▶ asbestos, including results of surveys (particularly where demolition is involved)
  ▶ existing storage of hazardous material
  ▶ contaminated land, including results of surveys
  ▶ existing structures containing hazardous materials
  ▶ health risks arising from client's activities.

4. Significant design and construction hazards

(a) Significant design assumptions and suggested work methods, sequences or other control measures

(b) Arrangements for coordination of ongoing design work and handling design changes

(c) Information on significant risks identified during design

(d) Materials requiring particular precautions.

5. The health and safety file

Description of its format and any conditions relating to its content.

APPENDIX 1.3 Construction phase plan

This is taken from the industry guidance produced by members of CONIAC (Construction Industry Advisory Committee) – The Construction (Design and Management) Regulations 2015 Industry guidance for Principal Contractors. The ACoP for CDM 2015 was not available when this book was published but it is unlikely that the content of the Construction phase plan will change significantly.

When drawing up the construction phase plan, employers should consider each of the following topics: information should be included in the plan where the topic is relevant to the work proposed; the plan sets out how health and safety is to be managed during the construction phase. The level of detail should be proportionate to the risks involved in the project.

Construction phase plan

1. Description of project

(a) Project description and programme details including key dates

(b) Details of client, principal designer, designers, principal contractor and other consultants

(c) Extent and location of existing records and plans that are relevant to health and safety on site, including information about existing structures when appropriate.

2. Management of the work

(a) Management structure and responsibilities

(b) Health and safety goals for the project and arrangements for monitoring and review of health and safety performance

(c) Arrangements for:

- ▶ regular liaison between parties on site
- ▶ consultation with the workforce
- ▶ exchange of design information between the client, principal designer, designers and contractors on site
- ▶ handling design changes during the project
- ▶ the selection and control of contractors
- ▶ the exchange of health and safety information between contractors

- ▶ site security
- ▶ site induction
- ▶ on-site training
- ▶ welfare facilities and first-aid
- ▶ the reporting and investigation of accidents and incidents including near misses
- ▶ the production and approval of risk assessments and written systems of work
- ▶ site rules
- ▶ fire and emergency arrangements.

3. Arrangements for controlling significant site risks

(a) Safety risks, including:

- ▶ delivery and removal of materials (including waste) and work equipment taking account of any risks to the public, e.g. during access to and egress from the site
- ▶ dealing with services – water, electricity and gas, including overhead power lines and temporary electrical installations
- ▶ accommodating adjacent land use
- ▶ stability of structures whilst carrying out construction work, including temporary structures and existing unstable structures
- ▶ preventing falls
- ▶ work with or near fragile materials
- ▶ control of lifting operations
- ▶ the maintenance of plant and equipment
- ▶ work on excavations and work where there are poor ground conditions
- ▶ work on wells, underground earthworks and tunnels
- ▶ work on or near water where there is a risk of drowning
- ▶ work involving diving
- ▶ work in a caisson or compressed air working
- ▶ work involving explosives
- ▶ traffic routes and segregation of vehicles and pedestrians
- ▶ storage of materials (particularly hazardous materials) and work equipment
- ▶ any other significant safety risks.

(b) health risks, including:

- the removal of asbestos
- dealing with contaminated land
- manual handling
- use of hazardous substances, particularly where there is a need for health monitoring
- reducing noise and vibration

- working with ionising radiation; any other significant health risks.

4. The health and safety file

(a) layout and format
(b) arrangements for the collation and gathering of information
(c) storage of information.

APPENDIX 1.4 The health and safety file

This is taken from the industry guidance produced by members of CONIAC (Construction Industry Advisory Committee) - The Construction (Design and Management) Regulations 2015 Industry guidance for Principal Contractors. The ACoP for CDM 2015 was not available when this book was published but it is unlikely that the content of the health and safety file will change significantly.

- a brief description of the work carried out;
- any residual hazards which remain and how they have been dealt with (e.g. surveys or other information concerning asbestos, contaminated land, water-bearing strata and buried services);
- key structural principles (e.g. bracing, sources of substantial stored energy – including pre- or post-tensioned members) and safe working loads (SWL) for floors and roofs, particularly where these may

preclude placing scaffolding or heavy machinery there;
- hazardous material used (e.g. lead paint, pesticides and special coatings which should not be burnt off);
- information regarding the removal or dismantling of installed plant and equipment (e.g. any special arrangements for lifting, order of or other special instructions for dismantling);
- health and safety information about equipment provided for cleaning or maintaining the structure;
- the nature, location and markings of significant services, including underground cables, gas supply equipment and fire-fighting services;
- information and as-built drawings of the structure, its plant and equipment (e.g. the means of safe access to and from service voids, fire doors and compartmentalisation).

CHAPTER 2

Health and safety management systems – PLAN

2.1 Key elements of a health and safety management system

2.1.1 Introduction

Every organisation should have a clear policy for the systematic management of health and safety so that health and safety risks may be effectively addressed and controlled and for those with five or more employees, it is a legal requirement that this policy must be in writing. A good health and safety policy will indicate the goals that the adopted health and safety management system hopes to achieve. The health and safety policy and management system will also complement those policies in areas such as quality, the environment and human resources. As for those areas, for the health and safety policy and its associated management system to be successful, it must have realistic goals and the active support and involvement of all levels of management within the organisation.

Planning to implement the health and safety policy is the key element of any health and safety management system. An effective planning system for health and safety requires organisations to set up, operate and maintain a management system that can detect, eliminate and control hazards and risks. This process is particularly important when there are long latency periods before a health problem becomes apparent, such as asbestosis or lung cancer caused by asbestos fibre inhalation.

Since the main objective of any health and safety plan is to prevent people being harmed at work, workplace precautions must be provided and maintained at the point of risk. Risks are created in the workplace as resources and information are used to create products and services in the manufacturing process (see Figure 2.1). This business activity model also applies to other industries including construction, education, leisure, hospitals and local authorities. Workplace precautions to match the hazards and risks are needed at each stage of the business activity. They can include machine guards, electrical safeguards, flammable liquid storage, dust controls, safety instructions and systems of work.

This chapter is concerned with the planning for managing occupational health and safety within an organisation. The preparation of an effective health and safety policy is an essential first step in the formulation of the health and safety management system. The policy will only remain as words on paper, however good the intentions, until there is an effective organisation set up to implement and monitor

its requirements. Some policies are written so that most of the wording concerns strict requirements laid on employees and only a few vague words cover managers' responsibilities. Generally, such policies do not meet the requirements of the HSW Act or the Management of Health and Safety at Work Regulations, which require an effective policy with a robust organisation and arrangements to be set up. For a policy to be effective, it must be honoured in the spirit as well as the letter. A good health and safety policy together with an effective occupational health and safety management system will also enhance the performance of the organisation in areas other than health and safety, help with the personal development of the workforce and reduce financial losses.

2.1.2 The key elements of a health and safety management system

Most of the key elements required for effective health and safety management are very similar to those required for good quality, finance and general business management. Commercially successful organisations usually have good health and safety management systems in place. The principles of good and effective management provide a sound basis for the improvement of health and safety performance.

HSG65

The HSE document HSG65, Managing for health and safety, describes the occupational health and safety management system used extensively in the UK. This is the principal one required for the NEBOSH National General Certificate course. In 2013, the HSE's guidance in HSG65 moved from using the POPMAR model (Policy, Organising, Planning, Measuring performance, Auditing and Review) to a 'Plan, Do, Check, Act' approach. Figure 2.1 illustrates this model.

The HSE have argued that the move towards Plan, Do, Check, Act achieves a better balance between the systems and behavioural aspects of management. It also treats health and safety management as an integral part of good management generally, rather than as a stand-alone system. A description of this four-step approach to occupational health and safety management is:

PLAN – establish standards for health and safety management based on risk assessment and legal requirements.

DO – implement plans to achieve objectives and standards.

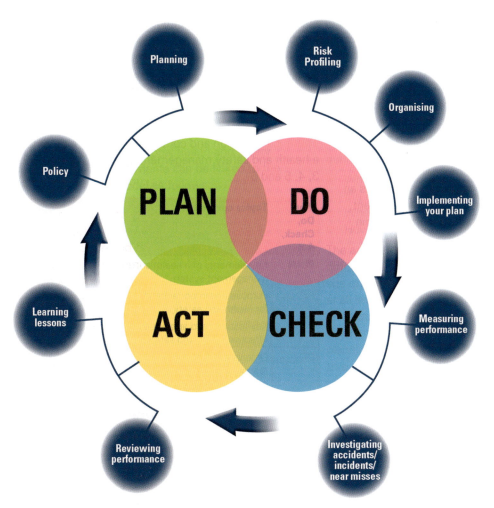

Figure 2.1 The Plan, Do, Check, Act cycle

CHECK – measure progress with plans and compliance with standards.

ACT – review against objectives and standards and take appropriate action.

The **Plan, Do, Check, Act** model for occupational health and safety management also forms the basis of the two occupational health and safety management systems: OHSAS 18001 and ILO-OSH 2001.

All recognised occupational health and safety management systems, including HSG65, have some basic and common elements. These are:

▶ a planning phase;
▶ a performance phase;
▶ a performance assessment phase; and
▶ a performance improvement phase.

The planning phase – PLAN

The planning phase always includes a **policy statement** which outlines the health and safety aims, objectives and commitment of the organisation and lines of responsibility. **Hazard identification and risk assessment** takes place during this phase and the significant hazards may well be included in the policy statement. It is important to note that in some reference texts, in particular those in languages other than English, the whole process of hazard identification, risk determination and the selection of risk reduction or control measures is termed 'risk assessment'. However, all three occupational health and safety management systems described in this chapter refer to the individual elements of the process separately and use the term 'risk assessment' for the determination of risk only.

At the **planning stage**, emergency procedures should be developed and relevant health and safety legal requirements and other standards identified together with appropriate benchmarks from similar industries. An **organisational structure** must be defined so that health and safety responsibilities are allocated at all levels of the organisation and issues such as competent persons and health and safety training are addressed. Realistic targets should be agreed within the organisation and be published as part of the policy.

The performance phase – DO

The performance phase (Do) will only be successful if there is good **communication** at and between all levels of the organisation. This implies employee participation as both **worker representatives** and on **safety committees**. Effective communication with

values and beliefs, at all levels of the organisation, is an essential component of a positive health and safety culture. For a positive health and safety culture to be achieved, an organisation must have clearly defined health and safety responsibilities so that there is always management control of health and safety throughout the organisation. The formal organisational structure should be such that the promotion of health and safety becomes a collaborative activity between the workforce, safety representatives and the managers. An effective organisation is essential for the **DO** stage of the management system and will be noted for good staff involvement and participation, high-quality communications, the promotion of competency and the empowerment of all employees to make informed contributions to the work of the organisation.

3. **A clear health and safety plan** – This involves the setting and implementation of performance standards and procedures using an effective occupational health and safety management system. The plan is based on risk assessment methods to decide on priorities and set objectives for controlling or eliminating hazards and reducing risks. Measuring success requires the establishing of performance standards against which achievements can be identified.

4. **The measurement of health and safety performance** – This includes both active (sometimes called proactive) and reactive monitoring to see how effectively the occupational health and safety management system is working. It forms part of the **CHECK** stage of the management system. Active monitoring involves looking at the premises, plant and substances plus the people, procedures and systems. Reactive monitoring discovers through investigation of accidents and incidents why controls have failed. It is also important to measure the organisation against its own long-term goals and objectives.

5. **Reviewing performance** – The results of monitoring and independent audits should be systematically reviewed to evaluate the performance of the management system against the objectives and targets established by the health and safety policy. It is at the review or **ACT** stage of the management system that the objectives and targets set in the health and safety policy may be changed. Changes in the health and safety environment in the organisation, such as an accident, should also trigger a performance review – this is discussed in more detail later in this chapter. Performance reviews are not only required by the HSW Act but are part of any organisation's commitment to continuous improvement. Comparisons should be made with internal performance indicators and the external performance indicators of similar

organisations with exemplary practices and high standards.

6. **Auditing** – An independent and structured audit of all parts of the health and safety management system reinforces the review process. Such audits may be internal or external (the differences are discussed in Chapter 6). The audit assesses compliance with the health and safety management arrangements and procedures. If the audit is to be really effective, it must assess both the compliance with stated procedures and the performance in the workplace. It will identify weaknesses in the health and safety policy and procedures and identify unrealistic or inadequate standards and targets.

2.1.5 Other aspects of health and safety management systems

Any occupational health and safety management system will fail unless there is a positive health and safety culture within the organisation and the active involvement of internal and external stakeholders.

A structured and well-organised occupational health and safety management system is essential for the maintenance of high health and safety standards within all organisations and countries. Some systems, such as OHSAS 18001, offer the opportunity for integration with quality and environmental management systems. This enables a sharing of resources although it is important that technical activities, such as health and safety risk assessment, are only undertaken by persons trained and competent in that area.

For an occupational health and safety management system to be successful, it must address workplace risks and be 'owned' by the workforce. It is, therefore, essential that the audit process examines shop floor health and safety behaviour to check that it mirrors that required by the health and safety management system.

Finally, whichever system is adopted, there must be continual improvement in health and safety performance if the application of the occupational health and safety management system is to succeed in the long term.

2.1.6 Key characteristics of a health and safety management system

The four basic elements common to all occupational health and safety management systems, as described earlier in this chapter, contain the different activities of the system together with the detailed arrangements and activities required to deliver those activities. However, there are four key characteristics of a **successful** occupational health and safety management system:

▶ a positive health and safety culture;
▶ the involvement of all stakeholders;
▶ an effective audit; and
▶ continual improvement.

A positive health and safety culture

In Chapter 3, the essential elements for a successful health and safety culture are detailed and discussed. In summary, they are:

▶ leadership and commitment to health and safety throughout the organisation;

▶ an acceptance that high standards of health and safety are achievable;

▶ the identification of all significant hazards facing the workforce and others;

▶ a detailed assessment of health and safety risks in the organisation and the development of appropriate control and monitoring systems;

▶ a health and safety policy statement outlining short- and long-term health and safety objectives. Such a policy should also include national codes of practice and health and safety standards;

▶ relevant communication and consultation procedures and training programmes for employees at all levels of the organisation;

▶ systems for monitoring equipment, processes and procedures and the prompt rectification of any defects found;

▶ the prompt investigation of all incidents and accidents and reports made detailing any necessary remedial actions.

Some of these essential elements form part of the health and safety management system but unless all are present within the organisation, it is unlikely that occupational health and safety will be managed successfully no matter which system is introduced. The chosen management system must be effective in reducing risks in the workplace else it will be nothing more than a paper exercise.

The involvement of stakeholders

There are a number of internal and external stakeholders of the organisation who will have an interest and influence on the introduction and development of the occupational health and safety management system.

The internal stakeholders include:

▶ **Directors and trustees of the organisation** – Following several national and international reports on corporate governance in recent years (such as the Combined Code of Corporate Governance 2003), the measurement of occupational health and safety performance and the attainment of health and safety targets have been recognised as being as important as other measures of business performance and targets. A report on health and safety performance should be presented at each board meeting and be a periodic agenda item for sub-committees of the Board such as audit and risk management.

▶ **The workforce** – Without the full cooperation of the workforce, including contractors and temporary employees, the management of health and safety will not be successful. The workforce is best qualified to ensure and provide evidence that health and safety procedures and arrangements are actually being implemented at the workplace. So often this is not the case even though the occupational health and safety management system is well designed and documented. Worker representatives can provide useful evidence on the effectiveness of the management system at shop floor level and they can also provide a useful channel of communication between the senior management and the workforce. One useful measure of the health and safety culture is the enthusiasm with which workers volunteer to become and continue to be occupational health and safety representatives. For a representative to be really effective, some training is essential. Organisations which have a recognised trade union structure will have fewer problems with finding and training worker representatives. More information on the duties of worker representatives is given in Chapter 3.

▶ **Health and safety professionals** – Such professionals will often be appointed by the organisation to manage the occupational health and safety management system and monitor its implementation. They will, therefore, have a particular interest in the development of the system, the design of objectives and the definition of targets or goals. Unless they have a direct-line management responsibility, which is not very common, they can only act as advisers, but still have a positive influence on the health and safety culture of the organisation. The appointed health and safety professional(s) also liaises with other associated competent persons, such as for electrical appliance and local exhaust ventilation testing.

The **external stakeholders** include:

▶ **Insurance companies** – As compensation claims increase, insurance companies are requiring more and more evidence that health and safety is being effectively managed. There is increasing evidence that insurance companies are becoming less prepared to offer cover to organisations with a poor health and safety record and/or occupational health and safety management system.

▶ **Investors** – Health and safety risks will, with other risks, have an effect on investment decisions. Increasingly, investment organisations require evidence that these risks are being addressed before investment decisions are made.

▶ **Regulators** – In many parts of the world, particularly in the Far East, national regulators and legislation require certification to a recognised international occupational health and safety management system standard. In many countries, regulators use such a standard to measure the awareness of a particular organisation of health and safety issues.

▶ **Customers** – Customers and others within the supply chain are increasingly insisting on some form of formal occupational health and safety management system to exist within the organisation. The construction industry is a good example of this trend. Much of this demand is linked to the need for corporate social responsibility and its associated guidelines on global best practice.

▶ **Neighbours** – The extent of the interest of neighbours will depend on the nature of the activities of the organisation and the effect that these activities have on them. The control of noise, and dust and other atmospheric contaminants are examples of common problem areas which can only be addressed on a continuing basis using a health, safety and environmental management system.

▶ **International organisations** – The United Nations, the ILO, the International Monetary Fund, the World Bank and the World Health Organisation are all examples of international bodies which have shown a direct or indirect interest in the management of occupational health and safety. In particular, the ILO is keen to see minimum standards of health and safety established around the world. The ILO works to ensure for everyone the right to work in freedom, dignity and security – which includes the right to a safe and healthy working environment. More than 70 ILO Conventions and Recommendations relate to questions of health and safety. In addition the ILO has issued more than 30 Codes of Practice on Occupational Health and Safety. For more information see their website www.ilo.org/safework.

There is a concern of many of these international organisations that as production costs are reduced by relocating operations from one country to another, there is also a lowering in occupational health and safety standards. The introduction of internationally recognised occupational health and safety management systems will help to alleviate such fears.

An effective audit

An effective audit is the final step in the occupational health and safety management system control cycle. The use of audits enables the reduction of risk levels and the effectiveness of the occupational health and safety management system to be improved. The auditing process is discussed in detail in Chapter 6 as is the important choice that must be made between the use of internal or external auditors.

Continual improvement

Continual improvement is recognised as a vital element of all occupational health and safety management systems if they are to remain effective and efficient as internal and external changes affect the organisation.

Internal changes may be caused by business reorganisation, such as a merger, new branches and changes in products, or by new technologies, employees, suppliers or contractors. External changes could include new or revised legislation, guidance or industrial standards, new information regarding hazards or campaigns by regulators.

Continual improvement is part of the **ACT** stage of the management system and need not necessarily be done at high cost or add to the complexity of the management system. Continual improvement is discussed in detail in Chapter 6.

2.1.7 The benefits and problems associated with occupational health and safety management systems

Occupational health and safety management systems have many benefits, of which the principal ones are:

▶ It is much easier to achieve and demonstrate legal compliance. Enforcement authorities have more confidence in organisations which have a health and safety management system in place.

▶ They ensure that health and safety is given the same emphasis as other business objectives, such as quality and finance. They will also aid integration, where appropriate, with other management systems.

▶ They enable significant health and safety risks to be addressed in a systematic manner.

▶ They can be used to show legal compliance with terms such as 'practicable' and 'so far as is reasonably practicable'.

▶ They indicate that the organisation is prepared for an emergency.

▶ They illustrate that there is a genuine commitment to health and safety throughout the organisation.

There are, however, several problems associated with occupational health and safety management systems, although most of them are soluble because they are caused by poor implementation of the system. The main problems are:

▶ The arrangements and procedures are not apparent at the workplace level and the audit process is only concerned with a desktop review of procedures.

▶ The documentation is excessive and not totally related to the organisation due to the use of generic procedures.

▶ Other business objectives, such as production targets, lead to ad hoc changes in procedures.

▶ Integration, which should really be a benefit, can lead to a reduction in the resources and effort applied to health and safety.

▶ A lack of understanding by supervisors and the workforce leads to poor system implementation.

▶ The performance review is not implemented seriously thus causing cynicism throughout the organisation.

2.2 Purpose and importance of setting a policy for health and safety

Every organisation should have a clear policy for the management of health and safety so that everybody associated with the organisation is aware of its health and safety aims and objectives and how they are to be achieved. For a policy to be effective, it must be honoured in the spirit as well as the letter. A good health and safety policy will also enhance the performance of the organisation in areas other than health and safety, help with the personal development of the workforce and reduce financial losses. A clear health and safety policy contributes to business efficiency and continuous improvement throughout the operation. The demonstration of senior management involvement provides evidence to all stakeholders that responsibilities to people and the environment are taken seriously. The policy should state the intentions of the business in terms of clear aims, objectives, organisation, arrangements and targets for all health and safety issues.

2.2.1 Legal duties concerning health and safety policy

Section 2(3) of the Health and Safety at Work (HSW) Act and the Employers' Health and Safety Policy Statements (Exception) Regulations 1975 require employers, with **five or more** employees, to prepare and review on a regular basis a written health and safety policy together with the necessary organisation and arrangements to carry it out and to bring the policy and any revision of it to the notice of their employees.

This does not mean that organisations with four or fewer employees do not need to have a health and safety policy – it simply means that it does not have to be written down. The number of employees is the maximum number at any one time whether they are full time, part time or seasonal.

This obligation on employers was introduced for the first time by the HSW Act and is related to the employers' reliance in the Act on self-regulation to improve health and safety standards rather than on enforcement alone. A good health and safety policy involves the development, monitoring and review of the standards needed to address and reduce the risks to health and safety produced by the organisation.

The law requires that the written health and safety policy should include the following three sections:

▶ a health and safety policy statement of intent which includes the health and safety aims and objectives of the organisation;
▶ the health and safety organisation detailing the people with specific health and safety responsibilities and their duties;
▶ the health and safety arrangements in place in terms of systems and procedures.

The Management of Health and Safety at Work Regulations also require the employer to 'make and give effect to such arrangements as are appropriate, having regard to the nature of their activities and the size of their undertaking, for the effective planning, organisation, control, monitoring and review of the preventative and protective measures'. They further require that these arrangements must be recorded when there are five or more employees.

When an inspector from the Health and Safety Executive (HSE) or Local Authority visits an establishment, it is very likely that they will wish to see the health and safety policy as an initial indication of the management attitude and commitment to health and safety. There have been instances of prosecutions being made due to the absence of a written health and safety policy. (Such cases are, however, usually brought before the courts because of additional concerns.)

Figure 2.3 Well-presented policy documents

2.3 Key features and appropriate content of an effective health and safety policy

2.3.1 Policy statement of intent

The health and safety policy statement of intent is often referred to as the health and safety policy statement or sometimes, incorrectly, as the health and safety policy. It should contain the aims (which are not measurable) and objectives (which are measurable) of the organisation or company. Aims will probably remain unchanged during policy revisions whereas objectives will be reviewed and modified or changed every year. The statement should be written in clear and simple language so that it is easily understandable. It should also be fairly brief and broken down into a series of smaller statements or bullet points.

The statement should be signed and dated by the most senior person in the organisation. This will demonstrate management commitment to health and safety and give authority to the policy. It will indicate where ultimate responsibility lies and the frequency with which the policy statement is reviewed.

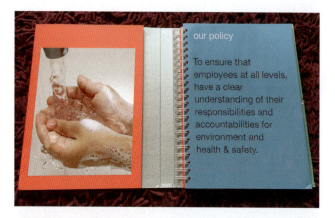

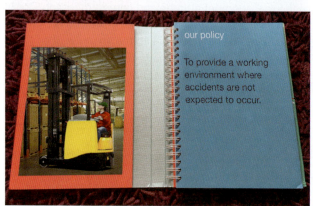

Figure 2.4 (a) and (b) Part of a policy commitment

The most senior manager is normally the Chief Executive Officer (CEO) or the Managing Director. It is the responsibility of the CEO, under the HSW Act, to ensure that the health and safety policy is developed and communicated to all employees in the organisation. They will need to ensure the following:

▶ key functions of health and safety management, such as monitoring and audit, accident investigation and training, are included in the organisational structure;
▶ adequate resources are available to manage health and safety effectively;
▶ the production of various health and safety arrangements in terms of rules and procedures;
▶ arrangements for the welfare of employees;
▶ the regular review and, if necessary, updating of the health and safety policy.

The policy statement should be written by the organisation and not by external consultants, as it needs to address the specific health and safety issues and hazards within the organisation. In large organisations,

it may be necessary to have health and safety policies for each department and/or site with an overarching general policy incorporating the individual policies. Such an approach is often used by local authorities and multinational companies.

The following points should be included or considered when a health and safety policy statement is being drafted:

▶ the aims, which should cover health and safety, welfare and relevant environmental issues;
▶ the position of the senior person in the organisation or company who is responsible for health and safety (normally the chief executive);
▶ the names of the Health and Safety Adviser and any safety representatives or other competent health and safety persons;
▶ a commitment to the basic requirements of the HSW Act (access, egress, risk assessments, safe plant and systems of work, use, handling, transport and handling of articles and substances, information, training and supervision);
▶ a commitment to the additional requirements of the Management of Health and Safety at Work Regulations (risk assessment, emergency procedures, health surveillance and employment of competent persons);
▶ duties towards the wider general public and others (contractors, customers, students, etc.);
▶ the principal hazards in the organisation;
▶ specific policies of the organisation (e.g. smoking policy, violence to staff, etc.);
▶ a commitment to employee consultation possibly using a safety committee or staff council;
▶ duties of employees (particularly those defined in the Management of Health and Safety at Work Regulations);
▶ specific health and safety performance targets for the immediate and long-term future;
▶ a commitment to provide the necessary resources to achieve the objectives outlined in the policy statement.

Health and safety **performance targets** are an important part of the statement of intent because:

▶ they indicate that there is management commitment to improve health and safety performance;
▶ they motivate the workforce with tangible goals resulting, perhaps, in individual or collective rewards;
▶ they offer evidence during the monitoring, review and audit phases of the management system.

The type of target chosen depends very much on the areas which need the greatest improvement in the organisation. The following list, which is not exhaustive, shows common health and safety performance targets:

▶ a specific reduction in the number of accidents, incidents (not involving injury) and cases of work-related ill-health (perhaps to zero);

- a reduction in the level of sickness absence;
- a specific increase in the number of employees trained in health and safety;
- an increase in the reporting of minor accidents and 'near miss' incidents;
- a reduction in the number of civil claims;
- no enforcement notices from the HSE or Local Authority;
- a specific improvement in health and safety audit scores;
- the achievement of a nationally recognised health and safety management standard such as HSG65.

The policy statement of intent should be posted on prominent notice boards throughout the workplace and brought to the attention of all employees at induction and refresher training sessions. It can also be communicated to the workforce during team briefing sessions, at 'toolbox' talks which are conducted at the workplace or directly by email, intranet, newsletters or booklets. It should be a permanent item on the agenda for health and safety committee meetings where it and its related targets should be reviewed at each meeting.

2.3.2 Setting health and safety objectives

A health and safety plan is necessary to guide the organisation – setting out the objectives for a specified time period. The speed with which the objectives will be achieved will depend on the resources available, the current state of health and safety compliance and the policy aims of the organisation. If the policy is to achieve excellence, the objectives will need to be tougher than those wishing simply to comply with minimum legal standards. It is essential to decide on priorities, with the emphasis on effective and adequate workplace precautions that meet legal requirements. High hazard/risk activities should receive priority. In some cases short-term measures may be needed quickly to minimise risks while longer-term solutions are designed and implemented.

There are two basic methods of setting objectives within an organisation – from the top or from the bottom – and either has its merits. Many policies are set from the boardroom and it is here that the resources for health and safety and the associated standards can be agreed. This is the top-down approach.

However, at times directors may be somewhat remote from the workplaces where most accidents occur. In some world-leading businesses, objectives are set at the workplace by those who are exposed to the significant risks. These are then approved at higher levels and coordinated with other workplaces within the organisation. The overall aims may be set centrally and then the details organised at site or workplace levels. This is the 'bottom-up' approach and helps to achieve commitment from people at the workplace.

Whatever approach is used, it is vitally important to ensure that safety representatives and/or representatives of employee safety are fully involved through consultation when setting the objectives and during their implementation. The objectives should be properly documented and may differ in detail at each functional level.

Health and safety objectives need to be specific, measurable, achievable, agreed with those who deliver them, realistic and set against a suitable timescale (SMART). Both short- and long-term objectives should be set and prioritised against business needs. Objectives at different levels or within different parts of an organisation should be aligned so they support the overall policy objectives. Personal targets can also be agreed with individuals to secure the attainment of objectives of the organisation and an example of such objectives is given in Box 2.1.

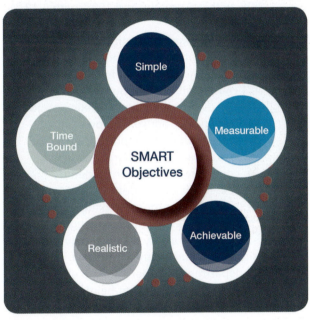

Figure 2.5 SMART performance standards or objectives

There are three complementary outputs from the planning process:

- health and safety plans with objectives for developing, maintaining and improving the health and safety management system, such as:
  - ▷ requiring each site of a multi-site organisation to have an annual health and safety plan and an accident and incident investigation system;
  - ▷ establishing a reliable risk assessment process for COSHH;
  - ▷ involving employees in preparing workplace precautions;
  - ▷ completing all manual handling assessments by the end of the current year;
  - ▷ providing a new guard for a particular machine;

Box 2.1 Example of objectives

An organisation might wish to minimise the risk to workers of working at height and aim to prevent all serious injury from falls when working at height. The objectives at each functional level may be as follows:

1. Board room – objective is to eliminate all serious injuries from working at height within one year. The Board may set one of its members as the champion to make this happen and provide advice from in-house expertise or external consultants. The budget to be approved within the first month.

2. The Board champion may set the following objectives/plan to achieve the Board's aim in consultation with the organisation's central health and safety steering committee:
   a) All sites to reduce serious accidents by 50% within 6 months and 100% within 12 months.
   b) All site managers to audit existing work at height operations; determine what can be done to avoid working at height; and report in 1 month with a draft budget.
   c) Safety adviser to establish existing accident performance; produce best practice guidance with safety representatives for those work-at-height operations which are considered unavoidable; identify suitable levels of competency and training requirements for those working at height.

3. Site managers may set the following in consultation with safety representatives and the site health and safety committee:
   a) All departmental supervisors to ensure immediately that work-at-height operations are inspected and risk assessed before work starts, all equipment to be assessed for suitability and checked for condition.
   b) All people working at height to be assessed for competence and specific training given as needed, for example working in mobile elevating work platforms.
   c) Toolbox talks on general work-at-height awareness to be provided.

▶ specifications for management arrangements, risk control and workplace precautions; and
▶ performance standards for implementing the health and safety management system, identifying the contribution of individuals to implementing the system (this is essential to building a positive health and safety culture).

Health and safety targets will need to be set so that the agreed performance standards are approached

and met. Since there may well be many standards, it will be necessary to prioritise important standards and associated targets. These prioritised targets are known as **Key Performance Indicators (KPIs)**. The targets must be realistic and a procedure known as benchmarking will help to ascertain whether this is the case.

Benchmarking

Benchmarking compares the performance of the organisation with that of similar organisations. A small construction company will benchmark itself against the performance of the UK construction industry perhaps using statistics supplied by the HSE. Such benchmarks, or examples of good practice, are defined by comparison with the health and safety performance of other parts of the organisation or the national performance of the occupational group of the organisation. The HSE publish an annual report, statistics and a bulletin, all of which may be used for this purpose. Typical benchmarks include accident rates per employee and accident or disease causation. It is important that the correct benchmarks are chosen and the organisation is assured of the accuracy of the data. The advantages of benchmarking are that:

▶ the key performance indicators for a particular organisation may be easily identified;
▶ it helps with continuous improvement;
▶ it focusses attention on weaker performance areas;
▶ it gives confidence to various stakeholders; and
▶ it is useful feedback for Boards, Chief Executives and managers.

The relationship of health and safety performance to other business activities

The best health and safety policies do not separate health and safety and human resources management since people are a key resource. Organisations want a fit, competent and committed workforce rather than simply preventing accidents and ill-health. This integrated approach will extend to people outside the organisation with policies for the control of environmental pollution and product safety. Organisations often fail to manage health and safety effectively because they see it as something distinct from other managerial tasks. The principles and approach to managing health and safety are exactly the same as those required for managing quality or the environment.

Well-managed health and safety is important in business terms because stakeholders (such as shareholders, customers and the general public) are influenced by any bad publicity resulting from poor health and safety standards. The identification, assessment and control of health and safety and other risks is a managerial responsibility and of equal importance to production and quality. There is good evidence that good health and safety management also produces good quality products at a reasonable cost. However, the practical implications of health and safety must be carefully thought through to avoid conflict between the demands of policy and other operational requirements. Insufficient attention to health and safety can lead to problems with production – for example, work schedules that fail to take into account the problems of fatigue or inadequate resources allocated to training.

The considerable advances in information technology have been very helpful in the development of health and safety management systems. It has enabled the rapid identification of data that is critical to the management of health and safety (e.g. accident and ill-health trends, health and safety performance variations between departments within the organisation). It has also simplified the collection and analysis of essential data.

Research by the HSE found that 'in many organisations the visible leadership and emphasis on continual improvement with respect to health and safety lagged behind that for quality of a product or service'.

2.3.3 Organisation of health and safety

This section of the policy defines the names, positions and responsibilities of those within the organisation or company who have a specific responsibility for health and safety. Therefore, it identifies those health and safety responsibilities and the reporting lines through the management structure. This section will include the following groups together with their associated responsibilities:

▶ directors and senior managers (responsible for setting policy, objectives and targets);
▶ supervisors (responsible for checking day-to-day compliance with the policy);
▶ health and safety advisers (responsible for giving advice during accident investigations and on compliance issues);
▶ other specialists, such as an occupational nurse, chemical analyst and an electrician, responsible for giving specialist advice on particular health and safety issues;
▶ health and safety representatives (responsible for representing employees during consultation meetings on health and safety issues with the employer);
▶ employees (responsible for taking reasonable care of the health and safety of themselves and others who may be affected by their acts or omissions);
▶ fire marshals (responsible for the safe evacuation of the building in an emergency);
▶ first-aiders (responsible for administering first-aid to injured persons).

For smaller organisations, some of the specialists mentioned above may well be employed on a consultancy basis.

For the health and safety organisation to work successfully it must be supported from the top (preferably at Board level) and some financial resource made available.

It is also important that certain key functions are included in the organisation structure. These include:

▶ accident investigation and reporting;
▶ health and safety training and information;
▶ health and safety monitoring and audit;
▶ health surveillance;
▶ monitoring of plant and equipment, their maintenance and risk assessment;
▶ liaison with external agencies;
▶ management and/or employee safety committees – the management committee will monitor day-to-day problems and any concerns of the employee health and safety committee.

The role of the health and safety adviser is to provide specialist information to managers in the organisation and to monitor the effectiveness of health and safety procedures. The adviser is not 'responsible' for health and safety or its implementation; that is the role of the line managers.

Finally the job descriptions, which define the duties of each person in the health and safety organisational structure, must not contain responsibility overlaps or

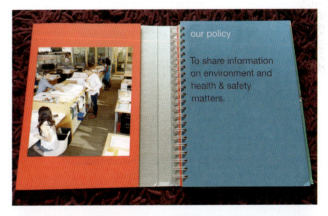

Figure 2.6 (a) and (b) Good information, training and working with employees is essential

Figure 2.7 Providing guidance and training is essential

blur chains of command. Each individual must be clear about his/her responsibilities and the limits of those responsibilities.

2.3.4 Arrangements for health and safety

The arrangements section of the health and safety policy gives details of the specific systems and procedures used to assist in the implementation of the policy statement. This will include health and safety rules and procedures and the provision of facilities such as a first-aid room and washrooms. It is common for risk assessments (including COSHH, manual handling and personal protective equipment (PPE) assessments) to be included in the arrangements section, particularly for those hazards referred to in the policy statement. It is important that arrangements for fire and other emergencies and for information, instruction, training and supervision are also covered. Local codes of practice (e.g. for fork-lift truck drivers) should be included.

The following list covers the more common items normally included in the arrangements section of the health and safety policy:

▶ employee health and safety code of practice;
▶ employee consultation procedures
▶ accident and illness reporting and investigation procedures;
▶ emergency procedures, first-aid;
▶ fire drill procedure;
▶ procedures for undertaking risk assessments;
▶ control of exposure to specific hazards (noise, vibration, radiation, manual handling, hazardous substances, etc.);
▶ machinery safety (including safe systems of work, lifting and pressure equipment);
▶ electrical equipment (maintenance and testing);
▶ maintenance procedures;
▶ permits to work procedures;
▶ use of PPE;
▶ monitoring procedures including health and safety inspections and audits;

▶ procedures for the control and safety of contractors and visitors;
▶ provision of welfare facilities;
▶ training procedures and arrangements;
▶ catering and food hygiene procedures;
▶ arrangements for consultation with employees;
▶ terms of reference and constitution of the safety committee;
▶ procedures and arrangements for waste disposal.

The three sections of the health and safety policy are usually kept together in a health and safety manual and copies distributed around the organisation.

2.3.5 Review of health and safety policy

It is important that the health and safety policy is monitored and reviewed on a regular basis. For this to be successful, a series of benchmarks needs to be established.

Figure 2.8 The policy might be good but is it put into practice – unsafe use of a ladder

There are several reasons for a review of the health and safety policy. The more important reasons are that:

▶ significant organisational changes may have taken place;
▶ there have been changes in key personnel;

- there have been changes in legislation and/or guidance;
- new work methods have been introduced;
- there have been alterations to working arrangements and/or processes;
- there have been changes following consultation with employees;
- the monitoring of risk assessments or accident/ incident investigations indicates that the health and safety policy is no longer totally effective;
- information from manufacturers has been received;
- advice from an insurance company has been received;
- the findings of an external health and safety audit have been received;
- enforcement action has been taken by the HSE or Local Authority;
- a sufficient period of time has elapsed since the previous review.

A positive promotion of health and safety performance will achieve far more than simply preventing accidents and ill-health. It will:

- support the overall development of personnel;
- improve communication and consultation throughout the organisation;
- minimise financial losses due to accidents and ill-health and other incidents;
- directly involve senior managers in all levels of the organisation;
- improve supervision, particularly for young persons and those on occupational training courses;
- improve production processes;
- improve the public image of the organisation or company.

It is apparent, however, that some health and safety policies appear to be less than successful. There are many reasons for this. The most common are:

- the statements in the policy and the health and safety priorities not understood by or properly communicated to the workforce;
- minimal resources made available for the implementation of the policy;
- too much emphasis on rules for employees and too little on management policy;
- a lack of parity with other activities of the organisation (such as finance and quality control) due to mistaken concerns about the costs of health and safety and the effect of those costs on overall

performance. This attitude produces a poor health and safety culture;
- lack of senior management involvement in health and safety, particularly at board level;
- employee concerns that their health and safety issues are not being addressed or that they are not receiving adequate health and safety information. This can lead to low morale among the workforce and, possibly, high absenteeism;
- high labour turnover;
- inadequate or no PPE;
- unsafe and poorly maintained machinery and equipment;
- a lack of health and safety monitoring procedures.

In summary, a successful health and safety policy is likely to lead to a successful organisation or company. A checklist for assessing any health and safety policy has been produced by the HSE and has been reproduced in Appendix 2.1.

2.4 Further information

The Health and Safety at Work etc. Act 1974

Employers' Health and Safety Policy Statements (Exception) Regulations 1975

The Management of Health and Safety at Work Regulations 1999 (as amended)

Managing for health and safety, HSG65 third edition 2013, HSE Books, ISBN 978-0-7176-6456-6 http://www.hse.gov.uk/pubns/priced/hsg65.pdf

HSE, 'Managing for health and safety' http://www.hse.gov.uk/managing/

HSE, The Health and Safety Toolbox: how to control risks at work http://www.hse.gov.uk/toolbox/

Health and safety made simple, INDG449, HSE Books, http://www.hse.gov.uk/pubns/indg449.htm

Plan, Do, Check, Act – An introduction to managing for health and safety, INDG275 (rev 1), HSE books http://www.hse.gov.uk/pubns/indg275.htm

Occupational Health and Safety Assessment Series (OHSAS 18000): Occupational Health and Safety Management Systems OHSAS 18001:2007 ISBN 978-0-5805-9404-5 OHSAS18002:2008 ISBN 978-0-5806-2686-9

Figure 2.9 Emergency procedures

Figure 2.10 Ladders and scaffold maintained in good condition and frequently inspected

Emergencies

▶ Ensuring that fire exits are marked, unlocked and free from obstruction;
▶ maintenance and testing of fire-fighting equipment, fire drills and evacuation procedures;

▶ first-aid, including name and location of person responsible for first-aid and deputy, and location of first-aid box.

Communication

▶ Giving employees information about the general duties under the HSW Act and specific legal requirements relating to their work;
▶ giving employees necessary information about substances, plant, machinery and equipment with which they come into contact;
▶ discussing with contractors, before they come on site, how they plan to do their job, whether they need any equipment from your organisation to help

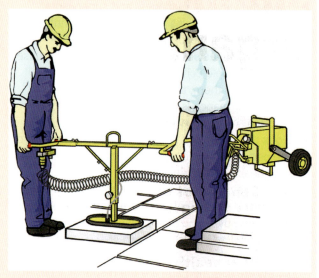

Figure 2.11 Vacuum-operated paving stone placer

them, whether they can operate in a segregated area or only when part of the plant is shut down and, if not, what hazards they may create for your employees and vice versa.

Training

▶ Training employees, supervisors and managers to enable them to work safely and to carry out their health and safety responsibilities efficiently.

Supervising

▶ Supervising employees so far as necessary for their safety – especially young workers, new employees and employees carrying out unfamiliar tasks.

Keeping check

▶ Regular inspections and checks of the workplace, machinery appliances and working methods.

CHAPTER 3

Health and safety management systems – Organising – DO 1

This chapter covers the following NEBOSH learning objectives:

1. Outline the organisational health and safety roles and responsibilities of employers, directors, managers and supervisors
2. Explain the concept of health and safety culture and its significance in the management of health and safety in an organisation
3. Outline the human factors which influence behaviour at work in a way that can affect health and safety
4. Explain how health and safety behaviour at work can be improved

Introduction

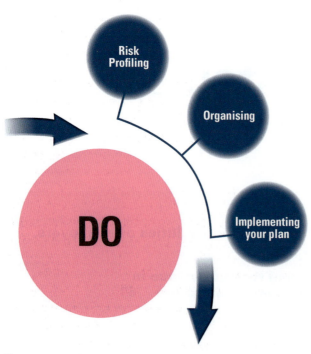

Figure 3.1 DO part of the management cycle involves Risk Profiling (Chapter 4), Organising and Implementing plans

This chapter is about managers in construction organisations setting out clear responsibilities and lines of communications for everyone in the enterprise. The policy sets the direction for health and safety within the enterprise and forms the written intentions of the principals or directors of the business. The organisation needs to be clearly communicated and people need to know what they are responsible for in the day-to-day operations. A vague statement that 'everyone is responsible for health and safety' is misleading and fudges the real issues. Everyone is responsible (Figure 3.2), but management in particular. There is no equality of responsibility under law between those who provide direction and create policy and those who are employed to follow. Principals, or employers in terms of the Health and Safety at Work (HSW) Act, have substantially more responsibility than employees.

In 1972, the Robens report recognised that the introduction of health and safety management systems was essential if the ideal of self-regulation of health and safety by industry was to be realised. It

further recognised that a more active involvement of the workforce in such systems was essential if self-regulation was to work. Self-regulation and the implicit need for health and safety management systems and employee involvement were incorporated into the Health and Safety at Work (HSW) Act.

Since the introduction of the HSW Act, health and safety standards have improved considerably but there have been some catastrophic failures. One of the worst was the fire on the offshore oil platform, Piper Alpha, in 1988, when 167 people died. At the subsequent enquiry, the concept of a safety culture was defined by the Director General of the Health and Safety Executive (HSE) at that time, J. R. Rimington. This definition has remained as one of the key points for a successful health and safety management system.

3.1 Organisational health and safety roles and responsibilities of employers, directors, managers and supervisors

3.1.1 Control of the health and safety organisation

Like all management functions, establishing control and maintaining it day in day out is crucial to effective health and safety management. Managers, particularly at senior levels, must take proactive responsibility for controlling issues that could lead to ill-health, injury or loss. A nominated senior manager at the top of the organisation needs to oversee policy implementation and monitoring. The nominated person will need to report regularly to the most senior management team and will be a director or principal of the organisation.

Health and safety responsibilities will need to be assigned to line managers and expertise must be made available, either inside or outside the enterprise, to help them achieve the requirements of the HSW Act and the Regulations made under the Act. The purpose of the health and safety organisation is to harness the collective enthusiasm, skills and effort of the entire workforce with managers taking key responsibility and providing clear direction. The prevention of accidents and ill-health through management systems of control becomes the focus rather than looking for individuals to blame after the incident occurs.

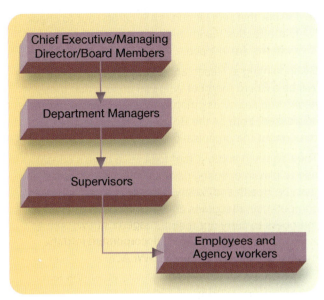

Figure 3.2 Everyone from senior manager down has health and safety responsibilities

The control arrangements should be part of the written health and safety policy. Performance standards will need to be agreed and objectives set which link the outputs required to specific tasks and activities for which individuals are responsible. For example, the objective could be to carry out a workplace inspection once a week to an agreed checklist and rectify faults within three working days. The periodic, say annual, audit would check to see if this was being achieved and, if not, what the reasons for non-compliance with the objective were.

People should be held accountable for achieving the agreed objectives through existing or normal procedures such as:

- job descriptions, which include health and safety responsibilities;
- performance appraisal systems, which look at individual contributions;
- arrangements for dealing with poor performance;
- where justified, the use of disciplinary procedures.

Such arrangements are only effective if health and safety issues achieve the same degree of importance as other key management concerns, and a good performance is considered to be an essential part of career and personal development.

3.1.2 Organisational health and safety responsibilities – employers

The organisational health and safety responsibilities of employers are closely linked to their statutory duties which are covered in detail in chapters 1 and 19.

3.1.3 Organisational health and safety responsibilities – directors

There is an absence of explicit legal duties that deal with directors' responsibilities for health and safety,

but if the organisation fails to fulfil the employers' duties, directors can be held personally responsible. There is voluntary guidance to inform directors of the requirements for good management of health and safety within their organisations. The HSE and the Institute of Directors have published a guide *INDG417 Leading health and safety at work*. This guide is based on a plan that will deliver, monitor and review the management concept and the following information is closely based on this guide.

The following are quotations from health and safety leaders in the public and private sectors taken from the guide:

'Health and safety is integral to success. Board members who do not show leadership in this area are failing in their duty as directors and their moral duty, and are damaging their organisation.'

'An organisation will never be able to achieve the highest standards of health and safety management without the active involvement of directors. External stakeholders viewing the organisation will observe the lack of direction.'

For many organisations, health and safety is a corporate governance issue. The Board should integrate health and safety into its main governance structures, including Board sub-committees, such as for risk, quality, remuneration and audit. The Turnbull guidance on the Combined Code on Corporate Governance requires listed companies to have robust systems of internal control, covering not just 'narrow' financial risks but also risks relating to the environment, business reputation and health and safety.

Effective health and safety performance comes from the top; members of the Board have both collective and individual responsibility for health and safety. Directors and boards need to examine their own behaviours, both individually and collectively, against the guidance given by the HSE. If they fall short of the standards it sets them, then they must change so that they become effective leaders in health and safety.

Directors and board members need to take action so that:

- the health and safety of employees and others, such as members of the public, may be protected;
- risk management includes health and safety risks and includes health and safety as a key business risk in board decisions; and
- health and safety duties imposed by legislation are followed.

In INDG417, the Institute of Directors and the HSE state that:

'Protecting the health and safety of employees or members of the public who may be affected by the activities of an organisation is an essential part of risk management and must be led by the Board.'

The starting points are the following essential principles. These principles are intended to underpin the actions in this guidance and so lead to good health and safety performance.

▶ Strong and active leadership from the top:
  ▷ visible, active commitment from the Board;
  ▷ establishing effective 'downward' communication systems and management structures;
  ▷ integration of good health and safety management with business decisions.
▶ Worker involvement:
  ▷ engaging the workforce in the promotion and achievement of safe and healthy conditions;
  ▷ effective 'upward' communication;
  ▷ providing high quality training.
▶ Assessment and review:
  ▷ identifying and managing health and safety risks;
  ▷ accessing (and following) competent advice;
  ▷ monitoring, reporting and reviewing performance.

The HSE in its guidance on Directors and Board responsibility for health and safety recommends the following five action points:

▶ The Board needs to accept formally and publicly its collective role in providing health and safety leadership in its organisation.
▶ Each member of the Board needs to accept their individual role in providing health and safety leadership for their organisation.
▶ The Board needs to ensure that all board decisions reflect its health and safety intentions, as articulated in the health and safety policy statement.
▶ The Board needs to recognise its role in engaging the active participation of workers in improving health and safety.
▶ The Board needs to ensure that it is kept informed of, and alert to, relevant health and safety risk management issues. The Health and Safety Executive recommends that Boards appoint one of their number to be the 'Health and Safety Director'.

Directors need to ensure that the Board's health and safety responsibilities are properly discharged. To achieve this, the Board will need to:

▶ carry out an annual review of health and safety performance;
▶ keep the health and safety policy statement up to date with current board priorities and review the policy at least every year;
▶ ensure that there are effective management systems for monitoring and reporting on the organisation's health and safety performance;
▶ ensure that any significant health and safety failures and their investigation are communicated to Board members;
▶ ensure that when decisions are made the health and safety implications are fully considered;

▶ ensure that regular audits are carried out to check that effective health and safety risk management systems are in place.

By appointing a 'Health and Safety Director' there will be a Board member who can ensure that these health and safety risk management issues are properly addressed, both by the Board and more widely throughout the organisation.

The Chairman and/or Chief Executive have a critical role to play in ensuring risks are properly managed and that the Health and Safety Director has the necessary competence, resources and support of other Board members to carry out their functions. Indeed, some Boards may prefer to see all the health and safety functions assigned to their Chairman and/or Chief Executive. As long as there is clarity about the health and safety responsibilities and functions, and the Board properly addresses the issues, this is acceptable.

The health and safety responsibilities of all Board members should be clearly articulated in the organisation's statement of health and safety policy and arrangements. It is important that the role of the Health and Safety Director should not detract either from the responsibilities of other directors for specific areas of health and safety risk management or from the health and safety responsibilities of the Board as a whole. Some form of health and safety training will probably be required for directors.

INDG417 states that there are four elements that boards need to incorporate into their management of health and safety. These are:

▶ planning the direction of health and safety;
▶ delivering the plan for health and safety;
▶ monitoring health and safety performance; and
▶ reviewing health and safety performance.

Details of these four elements are given in Appendix 3.1.

3.1.4 Typical managers' organisational responsibilities

In addition to the legal responsibilities on management, there are many specific responsibilities imposed by each organisation's health and safety policy. A summary of the organisational responsibilities for health and safety of typical line managers and their accountability of each level of the line organisation is given below. Many organisations will not fit this exact structure but most will have those who direct, those who manage or supervise and those who have no line responsibility, but have responsibilities to themselves and fellow workers. A more detailed list of the responsibilities is given in Appendix 3.2.

Managing Directors/Chief Executives

Managing Directors/Chief Executives are responsible for the health, safety and welfare of all those who work or visit the organisation. In particular, they:

1. are responsible and accountable for health and safety performance within the organisation;
2. must ensure that adequate resources are available for the health and safety requirements within the organisation including the appointment of a senior member of the senior management with specific responsibility for health and safety;
3. appoint one or more competent persons and adequate resources to provide assistance in meeting the organisation's health and safety obligations including specialist help where appropriate;
4. establish, implement and maintain a formal, written health and safety programme for the organisation that encompasses all areas of significant health and safety risk;
5. approve, introduce and monitor all site health and safety policies, rules and procedures; and
6. review annually the effectiveness, and, if necessary, require revision, of the health and safety programme.

Departmental managers

The principal departmental managers may report to the Site Manager, Managing Director or Chief Executive. In particular, they:

1. are responsible and accountable for the health and safety performance of their department;
2. are responsible for the engagement and management of contractors and that they are properly supervised;
3. must ensure that any machinery, equipment or vehicles used within the department are maintained, correctly guarded and meet agreed health and safety standards. Copies of records of all maintenance, statutory and insurance inspections must be kept by the Departmental Manager;
4. develop a training plan that includes specific job instructions for new or transferred employees and follow up on the training by supervisors. Copies of records of all training must be kept by the Departmental Manager; and
5. personally investigate all lost workday cases and dangerous occurrences and report to their line manager. Progress any required corrective action.

Supervisors

The supervisors are responsible to and report to their Departmental Manager. In particular, they:

1. are responsible and accountable for their team's health and safety performance;
2. enforce all safe systems of work procedures that have been issued by the Departmental Manager;
3. instruct employees in relevant health and safety rules, make records of this instruction and enforce all health and safety rules and procedures;
4. supervise any contractors that are working within their area of supervision; and

5. enforce personal protective equipment requirements, make spot checks to determine that protective equipment is being used and periodically appraise condition of equipment. Record any infringements of the personal protective equipment policy.

3.1.5 Role and functions of health and safety practitioners and other advisers

Competent person

One or more competent persons must be appointed to help managers comply with their duties under health and safety law. The essential point is that managers should have access to expertise to help them fulfil their legal requirements. However, they will always remain as advisers and will not assume responsibility in law for health and safety matters. This responsibility always remains with line managers and cannot be delegated to an adviser whether inside or outside the organisation. The appointee could be:

▶ the employer themselves if they are competent. This may be appropriate in a small, low-hazard business;
▶ one or more employees, provided that they have sufficient time and other resources to undertake the task properly;
▶ a person(s) from outside the organisation who has sufficient expertise to help.

The HSE have produced a leaflet entitled *Getting specialist help with health and safety* INDG420. This gives simple guidance for those looking for specialist health and safety help.

When there is an employee with the ability to do the job, it is better for them to be appointed than to use outside specialists. Many health and safety issues can be tackled by people with an understanding of current best practice and the ability to judge and solve problems. Some specialist help is needed on a permanent basis whilst other help may only be required for a short period. There is a wide range of specialists available for different types of health and safety problem, these include:

▶ engineers for specialist ventilation or chemical processes;
▶ occupational hygienists for assessment and practical advice on exposure to chemical (dust, gases, fumes, etc.), biological (viruses, fungi, etc.) and physical (noise, vibration, etc.) agents;
▶ occupational health professionals for medical examinations and diagnosis of work-related disease, pre-employment and sickness advice, health education;
▶ ergonomists for advice on suitability of equipment, comfort, physical work environment, work organisation;
▶ physiotherapists for prevention and treatment of musculoskeletal disorders;

3

- radiation protection advisers for advice on compliance with the Ionising Radiations Regulations;
- health and safety practitioners for general advice on implementation of legislation, health and safety management, risk assessment, control measures and monitoring performance.

Health and safety practitioner

Status and **competence** are essential for the role of health and safety practitioners and other advisers. They must be able to advise management and employees or their representatives with authority and independence. They need to be able to advise on:

- creating and developing health and safety policies. These will be for existing activities in addition to new acquisitions or processes;
- the promotion of a positive health and safety culture. This includes helping managers to ensure that an effective health and safety policy is implemented;
- health and safety planning. This will include goal-setting, deciding priorities and establishing adequate systems and performance standards. Short- and long-term objectives need to be realistic;
- day-to-day implementation and monitoring of policy and plans. This will include accident and incident investigation, reporting and analysis;
- performance reviews and audit of the whole health and safety management system.

To do this properly, health and safety practitioners need to:

- have proper training and be suitably qualified – for example NEBOSH Diploma or competence-based IOSH membership, relevant degree and, where appropriate, a Chartered Safety and Health Practitioner or a NEBOSH Certificate in small to medium-sized low-hazard premises, like offices, call centres, warehouses and retail stores;
- keep up-to-date information systems on such topics as civil and criminal law, health and safety management and technical advances;
- know how to interpret the law as it applies to their own organisation;
- actively participate in the establishment of organisational arrangements, systems and risk control standards relating to hardware and human performance. Health and safety practitioners will need to work with management on matters such as legal and technical standards;
- undertake the development and maintenance of procedures for reporting, investigating, recording and analysing accidents and incidents;
- develop and maintain procedures to ensure that senior managers get a true picture of how well health and safety is being managed (where a benchmarking role may be especially valuable). This will include monitoring, review and auditing; and

Figure 3.3 Safety practitioner at the front line

- be able to present their advice independently and effectively.

A national register of occupational safety consultants has been established by the HSE to help employers access good quality, proportionate advice. The Occupational Safety Consultants Register (OS-CR) provides firms with details of consultants who have met the highest qualification standard of recognised professional bodies and who are bound by a code of conduct that requires them to only give advice which is sensible and proportionate.

Relationships within the organisation

Health and safety practitioners:

- support the provision of authoritative and independent advice;
- report directly to directors or senior managers on matters of policy and have the authority to stop work if it contravenes agreed standards and puts people at risk of injury;
- are responsible for professional standards and systems.

They may also have line management responsibility for other health and safety practitioners, in a large group of companies or on a large and/or high-hazard site.

Relationships outside the organisation

Health and safety practitioners also have a function outside their own organisation. They provide the point of liaison with a number of other agencies including the following:

- environmental health officers and licensing officials;
- architects and consultants;
- HSE and the fire and rescue authorities;
- the police;
- HM coroner or the procurator fiscal;
- local authorities;
- insurance companies;
- contractors;
- clients and customers;
- the public;

- equipment suppliers;
- the media;
- general practitioners;
- IOSH and occupational health specialists and services.

Many safety advisers may think it unlikely they will ever be investigated and if they are, they will be covered by insurance. This may not necessarily be the case. If a safety adviser is being investigated, the normal procedure is to claim under the insurance cover that is in place, either the professional indemnity (PI) if self-employed, or under the employer's policy if employed in-house. It is not unusual to find that cover for advice and representation in criminal investigations and proceedings is excluded from these policies. Therefore, health and safety advisers, particularly those that are self-employed, should always check their insurance cover.

3.2 Concept of health and safety culture and its significance in the management of health and safety in an organisation

3.2.1 Definition of a health and safety culture

The health and safety culture of an organisation may be described as the development stage of the organisation in health and safety management at a particular time. HSG65 gives the following definition of a health and safety culture:

'The safety culture of an organisation is the product of individual and group values, attitudes, perceptions, competencies and patterns of behaviour that determine the commitment to, and the style and proficiency of, an organisation's health and safety management.'

'Organisations with a positive safety culture are characterised by communications founded on mutual trust, by shared perceptions of the importance of safety and by confidence in the efficacy of preventive measures.'

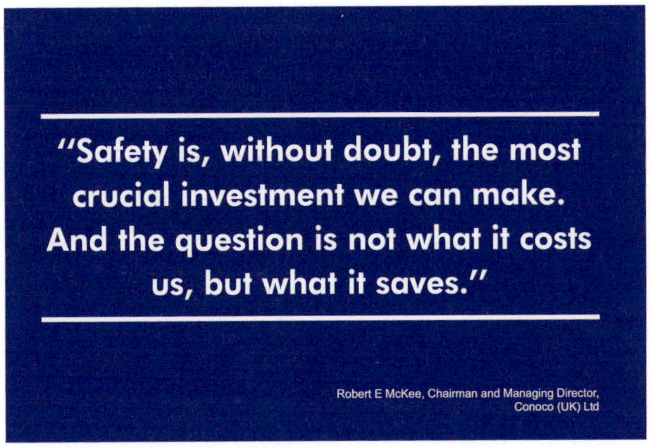

> **"Safety is, without doubt, the most crucial investment we can make. And the question is not what it costs us, but what it saves."**
>
> Robert E McKee, Chairman and Managing Director, Conoco (UK) Ltd

Figure 3.4 Safety investment

There is concern among some health and safety professionals that many health and safety cultures are developed and driven by senior managers with very little input from the workforce. Others argue that this arrangement is sensible because the legal duties are placed on the employer. A positive health and safety culture needs the involvement of the whole workforce just as a successful quality system does. There must be a joint commitment in terms of attitudes and values. The workforce must believe that the safety measures put in place will be effective and followed even when financial and performance targets may be affected.

3.2.2 Safety culture and safety performance

The relationship between health and safety culture and health and safety performance

The following elements are the important components of a **positive** health and safety culture:

- leadership and commitment to health and safety throughout the organisation, which is demonstrated in a genuine and visible way;
- health and safety treated as seriously as other corporate aims and adequately resourced;
- health and safety must be a line-management responsibility;
- acceptance that high standards of health and safety are achievable as part of a long-term strategy formulated by the organisation requiring sustained effort and interest;
- a detailed assessment of health and safety risks in the organisation and the development of appropriate control and monitoring systems;
- a health and safety policy statement that conveys a sense of optimism and outlines short- and long-term health and safety objectives. Such a policy should also include codes of practice and required health and safety standards;
- realistic and achievable targets and performance measured against them;
- relevant employee training programmes, communication and consultation procedures to ensure ownership and participation in health and safety throughout the organisation;
- systems for monitoring equipment, processes, behaviour and procedures and the prompt rectification of any defects; and
- the prompt investigation of all incidents and accidents and reports made detailing any necessary remedial actions.

If the organisation adheres to these elements, then a basis for a good performance in health and safety will have been established. However, to achieve this level of performance, sufficient financial and human resources must be made available for the health and safety function at all levels of the organisation.

Managers at all levels of the organisation must show their commitment to health and safety by leading through example and ensuring that adequate resources in terms of time, money and people are allocated to health and safety management. Good managers appear regularly on the shop floor, talk about health and safety and visibly demonstrate their commitment by their actions – such as stopping production to resolve issues. If employees perceive that managers are not seriously committed to high health and safety standards, they will assume that they are expected to put commercial interests first, and safety initiatives or programmes will be undermined as a result.

All managers, supervisors and members of the governing body (e.g. directors) should receive training in health and safety and be made familiar during training sessions with the health and safety targets of the organisation. The depth of training undertaken will depend on the level of competence required of the particular manager. Managers should be accountable for health and safety within their departments and be rewarded for significant improvements in health and safety performance. They should also be expected to discipline employees within their departments who infringe health and safety policies or procedures.

Important indicators of a health and safety culture

As the safety culture of an organisation develops, increasing use should be made of specific leading indicators and the setting and monitoring of specific goals. The main indicators for the development of health and safety must:

▶ be objective and easy to measure and collect;
▶ be relevant to the organisation;
▶ provide immediate and reliable indication of the performance level;
▶ be cost-effective in terms of equipment, personnel and additional technology required to gather the information; and
▶ be understood and owned by the organisation.

The goals must be relevant and improve safety performance over a reasonable time period and be specific to the key risks and activities of the organisation.

There are several outputs or indicators of the state of the health and safety culture of an organisation. The most important are the numbers of accidents, near misses and occupational ill-health cases occurring within the organisation.

Although the number of accidents may give a general indication of the health and safety culture, a more detailed examination of accidents and accident statistics is normally required. A calculation

of the rate of accidents enables health and safety performance to be compared between years and organisations.

The simplest measure of accident rate is called the incidence rate and is defined as:

$$\frac{\text{Total number of accidents}}{\text{Number of persons employed}} \times 1{,}000$$

or the total number of accidents per 1,000 employees.

A similar measure (per 100,000) is used by the HSE in its annual report on national accident statistics and enables comparisons to be made within an organisation between time periods when employee numbers may change. It also allows comparisons to be made with the national occupational or industrial group relevant to the organisation.

There are four main problems with this measure which must be borne in mind when it is used. These are:

▶ there may be a considerable variation over a time period in the ratio of part-time to full-time employees;
▶ the measure does not differentiate between major and minor accidents and takes no account of other incidents, such as those involving damage but no injury (although it is possible to calculate an incidence rate for a particular type or cause of accident);
▶ there may be significant variations in work activity during the periods being compared;
▶ under-reporting of accidents will affect the accuracy of the data.

Some industries prefer the accident frequency rate per million hours worked. This method, by counting hours worked rather than the number of employees, avoids distortions by including part-time as well as full-time employees and overtime worked.

The calculation is:

$$\frac{\text{Number of accidents in the period}}{\text{Total hours worked in the period}} \times 1{,}000{,}000$$

There is more information on these two measures in Chapter 5.

Indications of a poor health and safety culture

Subject to the above limitations, an organisation with a high accident incidence or frequency rate is likely to have a negative or poor health and safety culture.

There are other indications of a poor health and safety culture or climate. These include:

▶ a high sickness, ill-health and absentee rate among the workforce;
▶ the perception of a blame culture;
▶ high staff turnover leading to a loss of momentum in making health and safety improvements;

- no resources (in terms of budget, people or facilities) made available for the effective management of health and safety;
- management decisions that consistently put production or cost before health and safety considerations;
- a lack of compliance with relevant health and safety law and the safety rules and procedures of the organisation;
- regular procedural violations;
- poor selection procedures and management of contractors;
- poor levels of communication, cooperation and control;
- a weak health and safety management structure;
- either a lack, or poor levels, of health and safety competence; and
- high insurance premiums.

A positive health and safety culture is not enforceable through legislation. However, there can be enforcement that addresses the outcomes of a poor culture. For example if a company has unsuccessfully relied on procedural controls to avoid major accidents, there could be enforcement action to ensure compliance with relevant health and safety legislation or for the provision of appropriate safeguards.

In summary, a poor health and safety performance within an organisation is an indication of a negative health and safety culture.

Factors affecting a health and safety culture

The most important factor affecting the culture is the commitment to health and safety from the top of an organisation. This commitment may be shown in many different ways. It needs to have a formal aspect in terms of an organisational structure, job descriptions and a health and safety policy, but it also needs to be apparent during crises or other stressful times. The health and safety procedures may be circumvented or simply forgotten when production or other performance targets are threatened.

Structural reorganisation or changes in market conditions will produce feelings of uncertainty among the workforce which, in turn, will affect the health and safety culture.

Poor levels of supervision, health and safety information and training are very significant factors in reducing health and safety awareness and, therefore, the culture.

Finally, the degree of consultation and involvement with the workforce in health and safety matters is crucial for a positive health and safety culture. Most of these factors may be summed up as human factors.

A safety culture questionnaire is given in Appendix 3.3.

3.3 Human factors which influence behaviour at work

3.3.1 Human factors

Over the years, there have been several studies undertaken to examine the link between various accident types, graded in terms of their severity, and near misses. One of the most interesting was conducted in the USA by H. W. Heinrich in 1950. He looked at over 300 accidents/incidents and produced the ratios illustrated in Figure 3.5.

Figure 3.5 Heinrich's accidents/incidents ratios

This study indicated that for every 10 near misses, there will be an accident. Although the accuracy of this study may be debated and other studies have produced different ratios, it is clear that if near misses are continually ignored, an accident will result. Further, the HSE Accident Prevention Unit has suggested that 90% of all accidents are due to human error and 70% of all accidents could have been avoided by earlier (proactive) action by management. It is clear from many research projects that the major factors in most accidents are human factors.

While Heinrich's Triangle is useful in many ways, it does not suggest that minor and major incidents have the same causes as has been suggested by some technical papers over the years. Neither does it imply that the causes of incident frequency are the same as the causes of severe (major) injuries. The causes of severe injuries are often different to those of the causes of minor injuries. It is not reasonable, therefore, to assume that the majority of major injuries or incidents will occur where the majority of incidents take place. The hazards in a construction company's main office will not present the same major hazards as those at the company's various construction sites.

The HSE has defined human factors as, 'environmental, organisational and job factors, and human and individual characteristics which influence behaviour at work in a way which can affect health and safety'.

In simple terms, in addition to the environment, the health and safety of people at work are influenced by:

▶ the organisation;
▶ the job; and
▶ individual (personal) factors.

These are known as human factors as they each have a human involvement. The individual factors which differentiate one person from another are only one part of those factors – and not always the most important.

Each of these elements will be considered in turn.

The organisation

The organisation is the company or corporate body and has the major influence on health and safety. It must have its own positive health and safety culture and produce an environment in which it:

▶ manages health and safety throughout the organisation, including the setting and publication of a health and safety policy and the establishment of a health and safety organisational structure;
▶ measures the health and safety performance of the organisation at all levels and in all departments. The performance of individuals should also be measured. There should be clear health and safety targets and standards and an effective reporting procedure for accidents and other incidents so that remedial actions may be taken;
▶ motivates managers within the organisation to improve health and safety performance in the workplace in a proactive rather than reactive manner.

The HSE has recommended that an organisation needs to provide the following elements within its management system:

▶ a clear and evident commitment from the most senior manager downwards, which provides a climate for safety in which management's objectives and the need for appropriate standards are communicated and in which constructive exchange of information at all levels is positively encouraged;
▶ an analytical and imaginative approach identifying possible routes to human factor failure. This may well require access to specialist advice;
▶ procedures and standards for all aspects of critical work and mechanisms for reviewing them;
▶ effective monitoring systems to check the implementation of the procedures and standards;
▶ incident investigation and the effective use of information drawn from such investigations;
▶ adequate and effective supervision with the power to remedy deficiencies when found.

It is important to recognise that there are often reasons for these elements not being present, resulting in weak management of health and safety. The most common reason is that individuals within the management organisation do not understand their roles – or their roles have never been properly explained to them. The

higher a person is within the structure the less likely it is that he/she has received any health and safety training. Such training at board level is rare.

Objectives and priorities may vary across and between different levels in the structure, leading to disputes which affect attitudes to health and safety. For example, a warehouse manager may be pressured to block walkways so that a large order can be stored prior to dispatch.

Motivations can also vary across the organisation, which may cause health and safety to be compromised. The production controller will require that components of a product are produced as near simultaneously as possible so that their final assembly is performed as quickly as possible. However, the health and safety adviser will not want to see safe systems of work compromised.

In an attempt to address some of these problems, the HSE has produced guidance on the safety duties of company directors. Each director and the Board, acting collectively, will be expected to provide health and safety leadership in the organisation. The Board will need to ensure that all its decisions reflect its health and safety intentions and that it engages the workforce actively in the improvement of health and safety. The Board will also be expected to keep itself informed of changes in health and safety risks. (See 3.1.3 earlier in this chapter for more details on directors' responsibilities.)

The following simple checklist may be used to check any organisational health and safety management structure.

Does the structure have:

▶ an effective health and safety management system?
▶ a positive health and safety culture?
▶ arrangements for the setting and monitoring of standards?
▶ adequate supervision?
▶ effective incident reporting and analysis?
▶ learning from experience?
▶ clearly visible health and safety leadership?
▶ suitable team structures?
▶ efficient communication systems and practices?
▶ adequate staffing levels?
▶ suitable work patterns?

HSG48 *'Reducing Error and Influencing Behaviour'* gives the following causes for failures in organisational and management structures:

▶ poor work planning leading to high work pressure
▶ lack of safety systems and barriers
▶ inadequate responses to previous incidents
▶ management based on one-way communications
▶ deficient coordination and responsibilities
▶ poor management of health and safety
▶ poor health and safety culture.

Organisational factors play a significant role in the health and safety of the workplace. However, this role is often forgotten when health and safety is being reviewed after an accident or when a new process or piece of equipment is introduced.

The job

Jobs may be highly dangerous or present only negligible risk of injury. Health and safety is an important element during the design stage of the job and any equipment, machinery or procedures associated with the job. Method study helps to design the job in the most cost-effective way and ergonomics helps to design the job with health and safety in mind. Ergonomics is the science of matching equipment, machines and processes to people rather than the other way round. An ergonomically designed machine will ensure that control levers, dials, meters and switches are sited in a convenient and comfortable position for the machine operator. Similarly, an ergonomically designed workstation will be designed for the comfort and health of the operator. Chairs, for example, will be designed to support the back properly throughout the working day (Figure 3.6).

All dimensions in millimetres

Workstation where workers can sit or stand

Figure 3.6 Well-designed workstation for sitting or standing

Physically matching the job and any associated equipment to the person will ensure that the possibility of human error is minimised. It is also important to ensure that there is cognitive matching of the person's information and decision-making requirements. A person must be capable, either through past experience or through specific training, of performing the job with the minimum potential for human error.

The major considerations in the design of the job, which would be undertaken by a specialist, have been listed by the HSE as follows:

▶ the identification and detailed analysis of the critical tasks expected of individuals and the appraisal of any likely errors associated with those tasks;

▶ evaluation of the required operator decision making and the optimum (best) balance between the human and automatic contributions to safety actions (with the emphasis on automatic whenever possible);

▶ application of ergonomic principles to the design of man–machine interfaces, including displays of plant and process information, control devices and panel layout;

▶ design and presentation of procedures and operating instructions in the simplest terms possible;

▶ organisation and control of the working environment, including the workspace, access for maintenance, lighting, noise and heating conditions;

▶ provision of the correct tools and equipment;

▶ scheduling of work patterns, including shift organisation, control of fatigue and stress and arrangements for emergency operations;

▶ efficient communications, both immediate and over a period of time.

For some jobs, particularly those with a high risk of injury, a job safety analysis should be undertaken to check that all necessary safeguards are in place. All jobs should carry a job description and a safe system of work for the particular job. The operator should have sight of the job description and be trained in the safe system of work before commencing the job. More information on both these latter items is given in Chapter 4.

The following simple checklist may be used to check that the principal health and safety considerations of the job have been taken into account:

▶ Have the critical parts of the job been identified and analysed?

▶ Have the employee's decision-making needs been evaluated?

▶ Has the best balance between human and automatic systems been evaluated?

▶ Have ergonomic principles been applied to the design of equipment displays, including displays of plant and process information, control information and panel layouts?

▶ Has the design and presentation of procedures and instructions been considered?

▶ Has the guidance available for the design and control of the working environment, including the workspace, access for maintenance, lighting, noise and heating conditions, been considered?

▶ Have the correct tools and equipment been provided?

▶ Have the work patterns and shift organisation been scheduled to minimise their impact on health and safety?

▶ Has consideration been given to the achievement of efficient communications and shift handover?

HSG48 gives the following causes for failures in job health and safety:

▶ illogical design of equipment and instruments;
▶ constant disturbances and interruptions;
▶ missing or unclear instructions;
▶ poorly maintained equipment;
▶ high workload;
▶ noisy and unpleasant working conditions.

It is important that health and safety monitoring of the job is a continuous process. Some problems do not become apparent until the job is started. Other problems do not surface until there is a change of operator or a change in some aspect of the job.

It is very important to gain feedback from the operator on any difficulties experienced because there could be a health and safety issue requiring further investigation.

Individual (personal) factors

Individual or personal factors, which affect health and safety, may be defined as any condition or characteristic of an individual which could cause or influence him/her to act in an unsafe manner. They may be physical, cognitive or psychological in nature. Individual factors, therefore, include issues such as attitude, motivation, training and human error and their interaction with the physical, mental and perceptual capability of the individual. Typical individual factors could include:

▶ self-interest in the case of incentive or bonus schemes to maximise earnings;
▶ loss of/or poor memory;
▶ loss of/or poor hearing;
▶ language issues and communication problems;
▶ physical health;
▶ experience and competence;
▶ personality and attitude;
▶ age;
▶ the position in the team; and
▶ training and competence.

Language and communication issues may be particular problems for some migrant workers. The law requires that employers provide workers with comprehensible and relevant information about risks and about the procedures they need to follow to ensure they can work safely and without risk to health. This does not have to be in English. The employer may make special arrangements, which could include translation, using interpreters or replacing written notices with clearly understood symbols or diagrams. Any health and safety training provided must take into account the worker's capabilities, including language skills.

The following **negative** factors can also affect the individual:

▶ low skill and competence levels;
▶ tired colleagues;
▶ bored or disinterested colleagues;
▶ individual medical or mental problems;

Figure 3.7 Most construction rubbish can burn. Make sure that it is swept up and removed from the site as soon as possible

▶ complacency from repetitive tasks and lack of awareness training;
▶ inexperience, especially if the employee is new or a young person; and
▶ peer pressure to conform to the 'group' or an individual's perception of how a task should be completed.

These factors have a significant effect on health and safety. Some of them, normally involving the personality of the individual, are unchangeable but others, involving skills, attitude, perception and motivation, can be changed, modified or improved by suitable training or other measures. In summary, the person needs to be matched to the job.

Studies have shown that the most common individual factors which contribute to accidents are low skill and competence levels, tiredness, boredom, low morale and individual medical problems. The setting of realistic timescales and work task schedules will lead to considerable improvements in these factors. However, the key issue is effective and relevant training. This can help to improve attitudes and the understanding of risks and safe working. It will also enhance or improve skills.

It is difficult to separate all the physical, mental or psychological factors because they are interlinked. However, the three most common factors are psychological factors – attitude, motivation and perception.

Attitude is the tendency to behave in a particular way in a certain situation. One of the principal aims of a good safety culture is to change behaviour. But, if this change is to become permanent, then attitude will have to change. However, good intentions are not enough and attitudes and behaviour do not always coincide. Attitudes are influenced by the prevailing health and safety culture within the organisation, the commitment of the management, the experience of the individual and the influence of the peer group. Peer group pressure is a particularly important factor among young people, and health and safety training must be designed with this in mind by using examples or case studies that are relevant to them. Behaviour may be changed by training, the formulation and enforcement of safety rules and meaningful consultation – attitude change often follows.

Motivation is the driving force behind the way a person acts or the way in which people are stimulated to act. Involvement in the decision-making process in a meaningful way will improve motivation as will the use of incentive schemes. However, there are other important influences on motivation such as recognition and promotion opportunities, job security and job satisfaction. Self-interest, in all its forms, is a significant motivator and individual factor.

Perception is the way in which people interpret the environment or the way in which a person believes or understands a situation (Figure 3.9). In health and safety, the perception of hazards is an important concern. Many accidents occur because people do not perceive that there is a risk. There are many common examples of this, including the use of personal protective equipment (such as hard hats) and guards on drilling machines and the washing of hands before meals. It is important to understand that when perception leads to an increased health and safety risk, it is not always caused by a conscious decision of the individual concerned. The stroboscopic effect caused by the rotation of a drill at certain speeds under fluorescent lighting will make the drill appear stationary. It is a well-known phenomenon, especially among illusionists, that people will often see what they expect to see rather than reality. Routine or repetitive tasks will reduce attention levels, leading to the possibility of accidents.

Other individual factors which can affect health and safety include physical stature, age, experience, health, hearing, intelligence, language, skills, level of competence and qualifications.

Memory is an important individual factor, as it is influenced by training and experience. The efficiency of memory varies considerably between people and during the lifetime of an individual. The overall health of a person can affect memory as can personal crises. Owing to these possible problems with memory, important safety instructions should be available in written as well as verbal form.

Finally, it must be recognised that some employees do not follow safety procedures either due to peer pressure or a wilful disregard of those procedures.

The following checklist is modelled on HSG48 and may

Setting goals leads to vigorous activity if:

The goal is realistic.

A serious commitment is made, especially if it is made publicly.

Feedback is received.

Figure 3.8 Motivation and activity

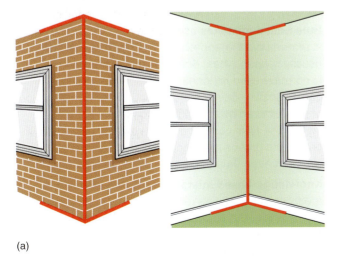  

(a) (b) (c)

Figure 3.9 Visual perceptions: (a) Are the lines of the same length? (b) Faces or vase? (c) Face or saxophone player?

be used to check that the relevant individual factors have been covered:

▶ Has the job specification been drawn up and included age, physique, skill, qualifications, experience, aptitude, knowledge, intelligence and personality?

▶ Have the skills and aptitudes been matched to the job requirements?

▶ Have the personnel selection policies and procedures been set up to select appropriate individuals?

▶ Has an effective training system been implemented?

▶ Have the needs of special groups of employees been considered?

▶ Have the monitoring procedures been developed for the personal safety performance of safety critical staff?

▶ Have fitness for work and health surveillance been provided where it is needed?

▶ Have counselling and support for ill-health and stress been provided?

Individual factors are the attributes that employees bring to their jobs and may be strengths or weaknesses. Negative individual factors cannot always be neutralised by improved job design. It is, therefore, important to ensure that personnel selection procedures should match people to the job. This will reduce the possibility of accidents or other incidents.

3.3.2 Human errors and violations

In 1985, a report by the HSE found that senior management failures resulted in 61% of all accidents, and 47% of those accidents had human causes. Human failures in health and safety are classified either as errors or violations. An error is an unintentional deviation from an accepted standard, whereas a violation is a deliberate deviation from the standard (Figure 3.10).

Human errors

Human errors fall into three groups – slips, lapses and mistakes, which can be further sub-divided into rule-based and knowledge-based mistakes.

Slips and lapses

These are very similar in that they are caused by a momentary memory loss often due to lack of attention or loss of concentration. They are not related to levels of training, experience or motivation and they can usually be reduced by re-designing the job or equipment or minimising distractions.

Slips are failures to carry out the correct actions of a task. Examples include the use of the incorrect switch, reading the wrong dial or selecting the incorrect component for an assembly. A slip also describes an action taken too early or too late within a given working procedure.

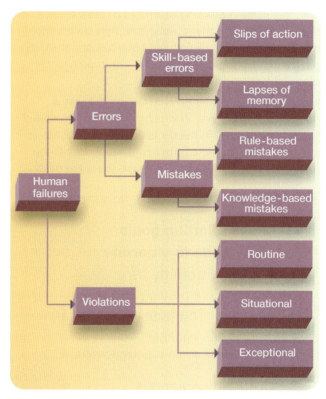

Figure 3.10 Types of human failure

Lapses are failures to carry out particular actions which may form part of a working procedure. A fork-lift truck driver leaving the keys in the ignition lock of his truck is an example of a lapse as is the failure to replace the petrol cap on a car after filling it with petrol. Lapses may be reduced by re-designing equipment so that, for example, an audible horn indicates the omission of a task. They may also be reduced significantly by the use of detailed checklists.

Mistakes

Mistakes occur when an incorrect action takes place but the person involved believes the action to be correct. A mistake involves an incorrect judgement. There are two types of mistake – rule-based and knowledge-based.

Rule-based mistakes occur when a rule or procedure is remembered or applied incorrectly. These mistakes usually happen when, due to an error, the rule that is normally used no longer applies. For example, a particular job requires the counting of items into groups of ten followed by the adding together of the groups so that the total number of items may be calculated. If one of the groups is miscounted, the final total will be incorrect even though the rule has been followed.

Knowledge-based mistakes occur when well-tried methods or calculation rules are used inappropriately. For example, the depth of the foundations required for a particular building was calculated using a formula. The formula, which assumed a clay soil, was used to calculate the foundation depth in a sandy soil. The resultant building was unsafe.

The HSE has suggested the following points to consider when the potential source of human errors is to be identified:

▶ What human errors can occur with each task? There are formal methods available to help with this exercise.
▶ What influences are there on performance? Typical influences include time pressure, design of controls, displays and procedures, training and experience, fatigue and levels of supervision.
▶ What are the consequences of the identified errors?
▶ What are the significant errors?
▶ Are there opportunities for detecting each error and recovering it?
▶ Are there any relationships between the identified errors? Could the same error be made on more than one item of equipment due, for example, to the incorrect calibration of an instrument?

Errors and mistakes can be reduced by the use of instruction, training and relevant information. However, communication can also be a problem, particularly at shift handover times. Environmental and organisational factors involving workplace stress will also affect error levels.

The following steps are suggested to reduce the likelihood of human error:

▶ Examine and reduce the workplace stressors (e.g. noise, poor lighting) which increase the frequency of errors.
▶ Examine and reduce any social or organisational stressors (e.g. insufficient staffing levels, peer pressure).
▶ Design plant and equipment to reduce error possibilities – poorly designed displays, ambiguous instructions.
▶ Ensure that there are effective training arrangements.
▶ Simplify any complicated or complex procedures.
▶ Ensure that there is adequate supervision, particularly for inexperienced or young trainees.
▶ Check that job procedures, instructions and manuals are kept up to date and are clear.
▶ Include the possibility of human error when undertaking the risk assessment.
▶ Isolate the human error element of any accident or incident and introduce measures to reduce the risk of a repeat.
▶ Monitor the effectiveness of any measures taken to reduce errors.

Violations

There are three categories of violation – routine, situational and exceptional.

Routine violation occurs when the breaking of a safety rule or procedure is the normal way of working. It becomes routine not to use the recommended procedures for tasks. An example of this is the regular speeding of fork-lift trucks in a warehouse so that orders can be fulfilled on time.

There are many reasons given for routine violations; for example:

▶ taking short cuts to save time and energy;
▶ a belief that the rules are unworkable or too restrictive;
▶ lack of knowledge of the procedures;
▶ perception that the rules are no longer applied;
▶ poor supervision and a lack of enforcement of the rules;
▶ new workers thinking that routine violations are the norm and not realising that this was not the safe way of working.

Finally, it must be recognised that there are some situations where peer pressure or simply a wilful disregard for procedures or other people's safety may result in routine violations. Routine violations can be reduced by regular monitoring, ensuring that the rules are actually necessary, or re-designing the job.

The following features are very common in many workplaces and often lead to routine violations:

▶ poor working posture due to poor ergonomic design of the workstation or equipment;
▶ equipment difficult to use and/or slow in response;
▶ equipment difficult to maintain or pressure on time available for maintenance;
▶ procedures unduly complicated and difficult to understand;
▶ unreliable instrumentation and/or warning systems;
▶ high levels of noise and other poor aspects to the environment (fumes, dusts, humidity);
▶ associated PPE either inappropriate, difficult and uncomfortable to wear or ineffective due to lack of maintenance.

Situational violations occur when particular job pressures at particular times make rule compliance difficult. They may happen when the correct equipment is not available or weather conditions are adverse. A common example is the use of a ladder rather than a scaffold for working at height to replace window frames in a building. Situational violations may be reduced by improving job design, the working environment and supervision.

Exceptional violations rarely happen and usually occur when a safety rule is broken to perform a new task. Good examples of this are the violations which can occur during the operations of emergency procedures such as for fires or explosions. These violations should be addressed in risk assessments and during training sessions for emergencies (e.g. fire training).

Everybody is capable of making errors. It is one of the objectives of a positive health and safety culture to reduce them and their consequences as much as possible.

3.4 How health and safety behaviour at work can be improved

Safe behaviour should be encouraged and strengthened through positive reinforcement by good example and supervision. Workers should be encouraged to set their own safety targets. Positive feedback on behaviour can sometimes reverse normal peer pressure so that the safe working becomes the accepted social norm. However, safe methods of working need to be identified and communicated to workers. This requires a significant amount of time and resources, but will result in an improvement in health and safety performance. The success of behavioural change is very dependent on the commitment of management.

No single section or department of an organisation can develop a positive health and safety culture on its own. In addition to the commitment by the management, there needs to be the promotion of health and safety standards, effective communication within the organisation, cooperation from and with the workforce and an effective and developing training programme. Each of these topics will be examined in turn to show their effect on improving the health and safety culture in the organisation.

3.4.1 Commitment of management

As mentioned earlier, there needs to be a commitment from the very top of the organisation and this commitment will, in turn, produce higher levels of motivation and commitment throughout the organisation. Probably the best indication of this concern for health and safety is shown by the status given to health and safety and the amount of resources (money, time and people) allocated to health and safety. The management of health and safety should form an essential part of a manager's responsibility and they should be held to account for their performance on health and safety issues. Specialist expertise should be made available when required (e.g. for noise assessment), either from within the workforce or by the employment of external contractors or consultants. Health and safety should be discussed on a regular basis at management meetings at all levels of the organisation. If the organisation employs sufficient people to make direct consultation with all employees difficult, there should be a health and safety committee at which there is employee representation. In addition, there should be recognised routes for anybody within the organisation to receive health and safety information or have their health and safety concerns addressed.

The health and safety culture is enhanced considerably when senior managers appear regularly at all levels of an organisation whether it be the shop floor, the hospital ward or the general office and are willing to discuss health and safety issues with staff. A visible management is very important for a positive health and safety culture.

Finally, the positive results of a management commitment to health and safety will be the active involvement of all employees in health and safety, the continuing improvement in health and safety standards and the subsequent reduction in accident and occupational ill-health rates. This will lead, ultimately, to a reduction in the number and size of compensation claims.

HSG48 makes some interesting suggestions to managers on the improvements that may be made to health and safety which will be seen by the workforce as a clear indication of their commitment. The suggestions are:

▶ review the status of the health and safety committees and health and safety practitioners. Ensure that any recommendations are acted upon or implemented;
▶ ensure that senior managers receive regular reports on health and safety performance and act on them;
▶ ensure that any appropriate health and safety actions are taken quickly and are seen to have been taken;
▶ any action plans should be developed in consultation with employees, based on a shared perception of hazards and risks, be workable and continually reviewed.

3.4.2 The promotion of health and safety standards

For a positive health and safety culture to be developed, everyone within the organisation needs to understand the standards of health and safety expected by the organisation and the role of the individual in achieving and maintaining those standards. Such standards are required to control and minimise health and safety risks.

Standards should clearly identify the actions required of people to promote health and safety. They should also specify the competencies needed by employees and should form the basis for measuring health and safety performance.

Health and safety standards cover all aspects of the organisation. Typical examples include:

▶ the design and selection of premises;
▶ the design and selection of plant and substances (including those used on site by contractors);
▶ the recruitment of employees and contractors;
▶ the control of work activities, including issues such as risk assessment;
▶ competence, maintenance and supervision;
▶ emergency planning and training;
▶ the transportation of the product and its subsequent maintenance and servicing.

Having established relevant health and safety standards, it is important that they are actively promoted within the organisation by all levels of management. The most effective method of promotion is by leadership and example. There are many ways to do this such as:

▶ the involvement of managers in workplace inspections and accident investigations;
▶ the use of PPE (e.g. goggles and hard hats) by all managers and their visitors in designated areas;
▶ ensuring that employees attend specialist refresher training courses when required (e.g. first-aid and fork-lift truck driving);
▶ full cooperation with fire drills and other emergency training exercises;
▶ comprehensive accident reporting and prompt follow-up on recommended remedial actions.

The benefit of good standards of health and safety will be shown directly in less lost production, accidents and compensation claims, and lower insurance premiums. It may also be shown in higher product quality and better resource allocation.

An important and central necessity for the promotion of high health and safety standards is health and safety competence. What is meant by 'competence'?

Competence

The word 'competence' is often used in health and safety literature. One definition, made during a civil case in 1962, stated that a competent person is:

> 'a person with practical and theoretical knowledge as well as sufficient experience of the particular machinery, plant or procedure involved to enable them to identify defects or weaknesses during plant and machinery examinations, and to assess their importance in relation to the strength and function of that plant and machinery.'

This definition concentrates on a manufacturing rather than service industry requirement of a competent person.

Regulation 7 of the Management of Health and Safety at Work Regulations requires that:

> 'every employer shall employ one or more competent persons to assist him in undertaking the measures he needs to take to comply with the requirements and prohibitions imposed upon him by or under the relevant statutory provisions.'

In other words, competent persons are required to assist the employer in meeting his/her obligations under health and safety law. This may mean a health and safety adviser in addition to, say, an electrical engineer, an occupational nurse and a noise assessment specialist. The number and range of competent persons will depend on the nature of the business of the organisation.

It is recommended that competent employees are used for advice on health and safety matters rather than external specialists (consultants). It is recognised, however, that if employees competent in health and safety are not available in the organisation, then an external service may be enlisted to help. The key is that management and employees need access to health and safety expertise.

The Regulations do not define 'competence' but do offer advice to employers when appointing health and safety advisers who should have:

▶ a knowledge and understanding of the work involved, the principles of risk assessment and prevention and current health and safety applications;
▶ the capacity to apply this to the task required by the employer in the form of problem and solution identification, monitoring and evaluating the effectiveness of solutions and the promotion and communication of health and safety and welfare advances and practices.

Such competence does not necessarily depend on the possession of particular skills or qualifications. It may only be required to understand relevant current best practice, be aware of one's own limitations in terms of experience and knowledge and be willing to supplement existing experience and knowledge. However, in more complex or technical situations, membership of a relevant professional body and/or the possession of an appropriate qualification in health and safety may be necessary. It is important that any competent person employed to help with health and safety has evidence of relevant knowledge, skills and experience for the tasks involved. The appointment of a competent person as an adviser to an employer does not absolve the employer from his/her responsibilities under the HSW Act and other relevant statutory provision.

Finally, it is worth noting that the requirement to employ competent workers is not restricted to those having a health and safety function but covers the whole workforce.

Competent workers must have sufficient training, experience, knowledge and other qualities to enable them to properly undertake the duties assigned to them.

3.4.3 Identifying and keeping up to date with legal requirements

The comparison with legal standards and requirements is an essential part of the setting of health and safety objectives. The organisation should have systems to identify legal requirements and keep up to date with changes to the law. This might be achieved by:

▶ regular checking of the HSE website looking at the 'news' section. This website gives detailed guidance on compliance – many published guides are now free to be downloaded as pdf files;

▶ free subscription to HSE bulletins on the internet;

▶ reading various health and safety periodicals;

▶ members of trade associations being kept up to date through their journals or websites;

▶ the text of legislation which can be obtained free of charge online.

There is more information on OSH websites and guidance on searching the internet in Chapter 23. Today there are so many ways of keeping up to date through the internet that the comment 'ignorance of the law is no excuse' is more poignant than ever.

3.4.4 Types of communication

Many problems in health and safety arise due to poor communication. It is not just a problem between management and workforce – it is often a problem the other way or indeed at the same level within an organisation. It arises from ambiguities or, even, accidental distortion of a message.

There are three basic methods of communication in health and safety – verbal, written and graphic.

Verbal communication is the most common. It is communication by speech or word of mouth. Verbal communication should only be used for relatively simple pieces of information or instruction. It is most commonly used in the workplace, during training sessions or at meetings.

There are several potential problems associated with verbal communication. The speaker needs to prepare the communication carefully so that there is no confusion about the message. It is very important that the recipient is encouraged to indicate their understanding of the communication. There have been many cases of accidents occurring because a verbal instruction has not been understood. There are several barriers to this understanding from the point of view of the recipient, including language and dialect, the use of technical language and abbreviations, background noise and distractions, hearing problems, ambiguities in the message, mental weaknesses and learning disabilities, and lack of interest and attention.

Having described some of the limitations of verbal communication, it does have some merits. It is less formal, enables an exchange of information to take place quickly and the message to be conveyed as near to the workplace as possible. Training or instructions that are delivered in this way are called toolbox talks and can be very effective.

Written communication takes many forms from the simple memo to the detailed report.

A memo should contain one simple message and be written in straightforward and clear language. The title should accurately describe the contents of the memo.

In recent years, emails have largely replaced memos, as it has become a much quicker method to ensure that the message gets to all concerned (although a recent report has suggested that many people are becoming overwhelmed by the number of emails which they receive!). The advantage of memos and emails is that there is a record of the message after it has been delivered. The disadvantage is that they can be ambiguous or difficult to understand or, indeed, lost within the system.

Reports are more substantial documents and cover a topic in greater detail. The report should contain a detailed account of the topic and any conclusions or recommendations. The main problem with reports is that they are often not read properly due to the time constraints on managers. It is important that all reports have a summary attached so that the reader can decide whether it needs to be read in detail (see Chapter 5).

The most common way in which written communication is used in the workplace is the **notice board**. For a notice board to be effective, it needs to be well positioned within the workplace and there needs to be a regular review of the notices to ensure that they are up to date and relevant. The use of notice boards as a means of communicating health and safety information to employees has some limitations that include:

▶ the information may not be read;

▶ the notice boards may not be accessible;

▶ the information may become outdated or defaced;

▶ some employees may not be able to read while others may not understand what they have read;

▶ there may be language barriers;

▶ the information is mixed in with other non-health and safety information; and

▶ there is no opportunity offered for feedback.

The following alternative methods could be used for the communication of essential health and safety information:

▶ memos, emails and company intranet;

▶ toolbox talks and team briefings;

▶ induction training and any further back-up training sessions;

▶ newsletters, bulletins and payslips;

▶ digital video media including DVDs;

▶ a staff handbook; and

▶ through safety committees, safety representatives, and representatives of employee safety.

Graphic communication is communication by the use of drawings, photographs or DVDs. It is used to impart either health and safety information (e.g. fire exits) or health and safety propaganda. The most common forms of health and safety propaganda are the poster and the DVD. Both can be used very effectively as training aids, as they can retain interest and impart a simple message. Their main limitation is that they can become out of date fairly quickly or, in the case of posters, become largely ignored.

There are many sources of health and safety information which may need to be consulted before an accurate communication can be made. These include regulations, judgements, Approved Codes of Practice, guidance, British and European standards, periodicals, case studies and HSE publications.

3.4.5 The health and safety poster

The Health and Safety (Information for Employees) Regulations require that the approved poster entitled *'Health and Safety Law – what you should know'* is displayed or the approved leaflet is distributed. This information tells employees in general terms about the requirements of health and safety law. Employers previously had to inform employees of the local address of the enforcing authority (either the HSE or the Local Authority) and the Employment Medical Advisory Service (EMAS), marked on the poster or supplied with the leaflet.

Research by the HSE showed that the earlier versions of the law poster and law leaflet were visually unappealing and rarely read. They were re-designed to be more readable and engaging. The poster and leaflet are available in a range of formats as part of the commitment to make health and safety information more accessible.

The Health and Safety Information (Amendment) Regulations allows the HSE to approve and publish new posters and leaflets which do not need organisations to update or add enforcing authority and Employment Medical Advisory Service (EMAS) contact information. However, the poster will still need to be displayed and provide employees with basic health and safety information. The HSE is concerned that the poster needs to be understood by employees:

▶ who have visual and/or learning difficulties;
▶ who have poor English reading skills;
▶ who work in an environment where the risk of being denied employment rights is high.

The poster (Figure 3.11) is a simplified version of the previous one and contains a single Incident Contact Centre number for the HSE helpline. The leaflets that employers could give to workers have been replaced with pocket cards.

In addition to the health and safety poster, the following types of health and safety information could be displayed on a workplace notice board:

▶ a copy of the Employer's Liability Insurance Certificate;
▶ details of first-aid arrangements;
▶ emergency evacuation and fire procedures;
▶ minutes of the last health and safety committee meeting;

Figure 3.11 Health and Safety Law poster – must be displayed or brochure given to employees

▶ details of health and safety targets and performance against them;
▶ health and safety posters and campaign details.

There are many other examples of written communications in health and safety, such as employee handbooks, company codes of practice, minutes of safety committee meetings and health and safety procedures.

3.4.6 Consultation with the workforce

General

It is important to gain the cooperation of all employees if a successful health and safety culture is to become established. This cooperation is best achieved by consultation. Joint consultation can help businesses be more efficient and effective by reducing the number of accidents and work-related ill-health and also to motivate staff by making them aware of health and safety problems.

There are two pieces of legislation that cover health and safety consultation with employees and both are summarised in Chapter 19. They are the Safety Representatives and Safety Committees Regulations 1977 and the Health and Safety (Consultation with Employees) Regulations 1996. These Regulations require employers to consult with their employees

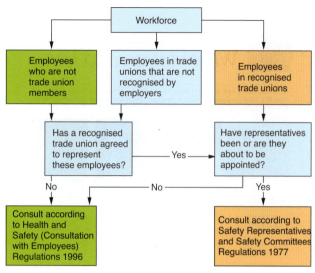

Figure 3.12 The law on consulting employees about health and safety in your workplace. References to the Regulations are colour-coded to help find the parts that are most relevant to a particular organisation: for workplaces where the Safety Representatives and Safety Committees Regulations 1977 apply; for workplaces where the Health and Safety (Consultation with Employees) Regulations 1996 apply.

on health and safety matters and make provision for both trade-union-appointed safety representatives and **representatives of employee safety (ROES)** elected by the workforce.

Employers must consult their workforce on a range of health and safety issues, including:

▶ any measures at the workplace which may substantially affect their health and safety, for example changes in systems of work, types of equipment or chemicals being used;

▶ arrangements for getting competent people to help them comply with health and safety requirements;

▶ the information that must be given to employees about risks to health and safety and any preventative measures;

▶ planning and organising health and safety training;

▶ the health and safety consequences of using new technology or substances.

Also, in workplaces in which the employer recognises a trade union, the employer must consult with trade-union-appointed safety representatives on health and safety matters affecting the employees they represent.

Trade-union-appointed safety representatives may:

▶ investigate possible dangers at work, the causes of accidents there and general complaints by employees on health and safety issues and take these up with the employer;

▶ carry out inspections of the workplace;

▶ represent employees in discussions with the HSE inspectors and receive information from them;

▶ attend safety committee meetings.

In workplaces in which trade unions are not recognised:

▶ employees must be consulted on health and safety, either directly or through their elected representatives.

Elected ROES may:

▶ take up with employers concerns about possible risks and dangerous events in the workplace that may affect the employees they represent;

▶ take up with employers general matters affecting the health and safety of the employees they represent;

▶ represent employees who elected them in consultations with health and safety inspectors.

Finally, an employer should ensure that all representatives receive reasonable training in health and safety at the employer's expense, are allowed time during working hours to perform their duties, and provide other facilities and assistance that the representatives should reasonably require.

The Safety Representatives and Safety Committees Regulations

These Regulations apply only to those organisations that have recognised trade unions for collective bargaining purposes. The recognised trade union may appoint safety representatives from among the employees and notify the employer in writing. In some workplaces, safety representatives have agreed to represent the entire workforce.

The Regulations give safety representatives several functions not duties (a failure to carry out a function is not a breach of law as it would be if they were legal duties). Functions include:

(a) representing employees in consultation with the employer;

(b) investigating potential hazards and dangerous occurrences;

(c) being involved with risk assessment procedures;

(d) investigating the causes of accidents, cases of work-related diseases or ill-health and dangerous occurrences;

(e) investigating employee complaints relating to health, safety and welfare;

(f) making representations to the employer on health, safety and welfare matters;

(g) carrying out inspections of the workplace. They must also be allowed to inspect the workplace at least once a quarter or sooner if there has been a substantial change in the conditions of work;

(h) representing employees at the workplace in consultation with enforcing inspectors;

(i) receiving information from health and safety inspectors;

(j) attending safety committee meetings.

Representatives must be allowed time off with pay to fulfil these functions and to undergo health and safety training. They must have access to suitable facilities and assistance to carry out their functions.

Safety committees

If two or more safety representatives have requested in writing that a safety committee be set up, the employer has three months to comply.

In medium to large organisations the easiest and often the most effective method of consultation is the health and safety committee. It will realise its full potential if its recommendations are seen to be implemented and both management and employee concerns are freely discussed. It will not be so successful if it is seen as a talking shop.

The committee should have stated objectives which mirror the objectives in the organisation's health and safety policy statement, and its own terms of reference. Terms of reference should include the following:

- the study of accident and notifiable disease statistics to enable reports to be made of recommended remedial actions;
- the examination of health and safety audits and statutory inspection reports;
- the consideration of reports from the external enforcement agency;
- the review of new legislation, Approved Codes of Practice and guidance and their effect on the organisation;
- the monitoring and review of all health and safety training and instruction activities in the organisation;
- the monitoring and review of health and safety publicity and communication throughout the organisation;
- development of safe systems of work and safety procedures;
- reviewing risk assessments;
- considering reports from safety representatives;
- continuous monitoring of arrangements for health and safety and revising them whenever necessary.

There are no fixed rules on the composition of the health and safety committee except that it should be representative of the whole organisation. It should have representation from the workforce and the management including at least one senior manager (other than the Health and Safety Adviser). Managers and representatives should agree who should chair the meetings, how often meetings should be held and what they hope to achieve.

Accident and ill-health investigations

Properly investigated accidents and ill-health can reveal weaknesses which need to be remedied. A joint investigation with safety representatives is more likely to inspire confidence with workers so that they cooperate fully with the investigation, as in many cases those involved may be concerned about being blamed for the accident.

Safety representatives are entitled to contact enforcing authority inspectors. If this is just for information, they can be contacted directly. If it is a formal complaint against the employer, the inspector will need to know if the employer has been informed. The inspectors can be contacted anonymously. They will keep the person's identity secret in such circumstances.

Training, facilities and assistance

Safety representatives are legally entitled to paid time off for training, which is usually freely available from their trade union or the TUC.

Training course topics often include:

- the role and functions of the safety representative;
- health and safety legislation;
- how to identify and minimise hazards;
- how to carry out a workplace inspection and accident investigation;
- employer's health and safety arrangements, including emergency procedures, risk assessments and health and safety policies;
- further information on training courses.

The employer is also required to provide facilities and assistance for safety representatives. Depending on the circumstances, these could include:

- notice board;
- telephone;
- lockable filing cabinet;
- access to an office to meet workers in private;
- camera;
- key health and safety information;
- access to specialist assistance and support in understanding technical issues.

The Health and Safety (Consultation with Employees) Regulations

In non-unionised workplaces where there are no safety representatives, or in a workplace that has trade union recognition but either the trade union has not appointed a safety representative or they do not represent the whole workforce, the Health and Safety (Consultation with Employees) Regulations will apply.

Employers must consult either on an individual basis (e.g. in very small companies) or through ROES elected by the workforce. See Table 3.1 for a comparison of functions with union-appointed safety representatives.

The guidance to these Regulations emphasises the difference between informing and consulting. Consultation involves listening to the opinion of employees on a particular issue and taking it into account before a decision is made. Informing employees means providing information on health and safety issues such as risks, control systems and safe systems of work.

Table 3.1 A comparison of the functions of health and safety representatives

Safety Representatives and Safety Committees Regulations 1977	Health and Safety (Consultation with Employees) Regulations 1996
Representatives	
Appointed in writing by a trade union recognised for collective bargaining purposes.	Elected by the workforce, where the employer has decided not to consult directly.
Known as	
Safety representatives.	Representatives of employee safety.
Functions	
Investigate potential hazards and dangerous occurrences at the workplace, and complaints by an employee relating to health, safety and welfare at work, and examine causes of workplace accidents.	
Representation to the employer on the above investigations, and on general matters affecting the health and safety of employees they represent.	Representation to the employer on: • potential hazards and dangerous occurrences • general matters affecting the health and safety of employees they represent • specific matters on which the employer must consult.
Inspect the workplace	
Represent employees in dealings with health and safety inspectors.	Represent employees in dealings with health and safety inspectors.
Receive certain information from inspectors.	
Attend health and safety committee meetings.	

Source: HSE INDG232(rev) page 4 HSE Web.

The functions of these ROES are to:

▶ represent the interest of workers on health and safety matters to the employer;

▶ approach the employer regarding potential hazards and dangerous occurrences at the workplace;

▶ approach the employer regarding general matters affecting the health and safety of the people they represent;

▶ to speak for the people they represent in consultation with inspectors.

The employer must consult ROES on the following:

▶ the introduction of any measure or change which may substantially affect employees' health and safety;

▶ the arrangements for the appointment of competent persons to assist in following health and safety law;

▶ any information resulting from risk assessments or their resultant control measures which could affect the health, safety and welfare of employees;

▶ the planning and organisation of any health and safety training required by legislation;

▶ the health and safety consequences to employees of the introduction of new technologies into the workplace.

However, the employer is not expected to disclose information if:

▶ it violates a legal prohibition;

▶ it could endanger national security;

▶ it relates specifically to an individual without his/her consent;

▶ it could harm substantially the business of the employer or infringe commercial security;

▶ it was obtained in connection with legal proceedings.

There are several benefits that accrue to organisations that consult with their employees. These include:

▶ healthier and safer workplaces because employees can help to identify hazards and develop relevant procedures to eliminate, reduce or control risks;

▶ a stronger commitment to implementing procedures and decisions because employees have been involved proactively in the decision-making process;

▶ the development of better work practices;

▶ a reduction in workplace accidents; and

▶ greater cooperation and trust between employers and employees so that the overall performance of the organisation improves.

Consultation does not remove the employer's right and duty to manage since the employer will always make the final decision following the consultation. But talking to employees and involving them in the decision-making process is a vital part of successfully managing health and safety.

3.4.7 Health and safety training

The provision of information and training for employees will develop their awareness and understanding of the specific hazards and risks associated with their jobs and working environment. It will inform them of the control measures that are in place and any related safe procedures that must be followed. Apart from satisfying legal obligations, several benefits will accrue to the employer by the provision of sound information and training to employees. These benefits include:

▶ a reduction in accident severity and frequency;

▶ a reduction in injury and ill-health related absence;

▶ a reduction in compensation claims and, possibly, insurance premiums;

▶ an improvement in the health and safety culture of the organisation;

▶ improved staff morale and retention.

Health and safety training is a very important part of the health and safety culture and it is also a legal requirement, under the Management of Health and Safety at Work Regulations and other regulations (see Appendix 3.4), for an employer to provide such training. Training is required on recruitment, at induction or on being exposed to new or increased risks due to:

▶ being transferred to another job or given a change in responsibilities;
▶ the introduction of new work equipment or a change of use in existing work equipment;
▶ the introduction of new technology;
▶ the introduction of a new system of work or the revision of an existing system of work;
▶ an increase in the employment of more vulnerable employees (young or disabled persons);
▶ particular training required by the organisation's insurance company (e.g. specific fire and emergency training).

Additional training may well be needed following a single or series of accidents or near misses, the introduction of new legislation, the issuing of an enforcement notice or as a result of a risk assessment or safety audit.

Figure 3.13 Health and safety training needs and opportunities

It is important during the development of a training course that the target audience is taken into account. If the audience are young persons, the chosen approach must be capable of retaining their interest and any illustrative examples used must be within their experience. The trainer must also be aware of external influences, such as peer pressures, and use them to advantage. For example, if everybody wears PPE then it will be seen as the thing to do. Levels of literacy and numeracy are other important factors.

The way in which the training session is presented by the use of DVDs, power point slides, case studies, lectures or small discussion groups needs to be related to the material to be covered and the backgrounds of the trainees. Supplementary information in the form of copies of slides and additional background reading is often useful. The environment used for the training sessions is also important in terms of room layout and size, lighting and heating.

Attempts should be made to measure the effectiveness of the training by course evaluation forms issued at the time of the session, by a subsequent refresher session and by checking for improvements in health and safety performance (such as a reduction in specific accidents).

There are several different types of training: these include induction, job specific, supervisory and management, and specialist. Informal sessions held at the place of work are known as 'toolbox talks'. Such sessions should only be used to cover a limited number of issues. They can become a useful route for employee consultation.

Induction training

Induction training should always be provided to new employees, trainees and contractors. While such training covers items such as pay, conditions and quality, it must also include health and safety. It is useful if the employee signs a record to the effect that training has been received. This record may be required as evidence should there be a subsequent legal claim against the organisation.

New workers are most vulnerable to accidents during their first six months at a new workplace. In the construction industry, eight out of 16 fatal accidents happened during the first ten days on site, half of them on the very first day. The causes of these accidents were:

▶ lack of experience of working in a new industry or workplace;
▶ lack of familiarity with the job and the work environment;
▶ reluctance to raise concerns (or not knowing how to); and
▶ eagerness to impress workmates and managers.

They may not recognise hazards as a potential source of danger, possibly due to a lack of familiarity with the workplace, and may ignore warning signs and rules, or cut corners.

A key requirement for all new workers is adequate supervision particularly during the first few weeks. They must know how and with whom they can raise any concerns about any health and safety issues. It is important that they have understood the information, instruction and training needed to work safely. They must also have understood any emergency arrangements or procedures. Young persons and migrant workers are two groups of potentially vulnerable workers who may require particular attention at induction.

Most induction training programmes would include the following topics:

▶ the health and safety policy of the organisation including a summary of the organisation and arrangements including employee consultation;

▶ a brief summary of the health and safety management system including the name of the employee's direct supervisor, safety representative and source of health and safety information;

▶ the employee responsibility for health and safety including any general health and safety rules (e.g. smoking prohibitions);

▶ the accident reporting procedure of the organisation, the location of the accident book and the location of the nearest first-aider;

▶ the fire and other emergency procedures including the location of the assembly point;

▶ the hazards that are specific to the workplace;

▶ a summary of any relevant risk assessments and safe systems of work;

▶ the location of welfare, canteen facilities and rest rooms;

▶ procedures for reporting defects or possible hazards and the name of the responsible person to whom the report should be made;

▶ details of the possible disciplinary measures that may be enacted for non-compliance with health and safety rules and procedures.

Additional items which are specific to the organisation may need to be included such as:

▶ internal transport routes and pedestrian walkways (e.g. fork-lift truck operations);

▶ the correct use of PPE and maintenance procedures;

▶ manual handling techniques and procedures;

▶ details of any hazardous substances in use and any procedures relating to them (e.g. health surveillance).

There should be some form of follow-up with each new employee after three months to check that the important messages have been retained. This is sometimes called a refresher course, although it is often better done on a one-to-one basis.

It is very important to stress that the content of the induction course should be subject to constant review and updated following an accident investigation, new legislation, changes in the findings of a risk assessment or the introduction of new plant or processes. A well-designed induction programme may help to encourage a positive health and safety culture and reduce the effect of negative peer pressure.

Job-specific training

Job-specific training ensures that employees undertake their job in a safe manner. Such training, therefore, is a form of skill training and is often best done 'on the job' – sometimes known as 'toolbox training'. Details of the safe system of work or, in more hazardous jobs, a

permit-to-work system, should be covered. In addition to normal safety procedures, emergency procedures and the correct use of PPE also need to be included. The results of risk assessments are very useful in the development of this type of training. It is important that any common causes of human errors (e.g. discovered as a result of an accident investigation), any standard safety checks or maintenance requirements, are addressed.

It is common for this type of training to follow an operational procedure in the form of a checklist which the employee can sign on completion of the training. The new employee will still need close supervision for some time after the training has been completed.

Supervisory and management training

Supervisory and management health and safety training follows similar topics to those contained in an induction training course but will be covered in more depth. There will also be a more detailed treatment of health and safety law. There has been considerable research over the years into the failures of managers that have resulted in accidents and other dangerous incidents. These failures have included:

▶ lack of health and safety awareness, enforcement and promotion (in some cases, there has been encouragement to circumvent health and safety rules);

▶ lack of consistent supervision of and communication with employees;

▶ lack of understanding of the extent of the responsibility of the supervisor.

It is important that all levels of management, including the Board, receive health and safety training. This will not only keep everybody informed of health and safety legal requirements, accident prevention techniques and changes in the law, but also encourage everybody to monitor health and safety standards during visits or tours of the organisation.

Specialist training

Specialist health and safety training is normally needed for activities that are not related to a specific job but more to an activity. Examples include first-aid, fire prevention, fork-lift truck driving, overhead crane operation, scaffold inspection and statutory health and safety inspections. These training courses are often provided by specialist organisations and successful participants are awarded certificates. Details of two of these courses will be given here by way of illustration.

Fire prevention training courses include the causes of fire and fire spread, fire and smoke alarm systems, emergency lighting, the selection and use of fire extinguishers and sprinkler systems, evacuation procedures, high-risk operations and good housekeeping principles.

A fork-lift truck drivers' course would include the general use of the controls, loading and unloading procedures, driving up or down an incline, speed limits, pedestrian awareness (particularly in areas where pedestrians and vehicles are not segregated), security of the vehicle when not in use, daily safety checks and defect reporting, refuelling and/or battery charging and emergency procedures.

Details of other types of specialist training appear elsewhere in the text.

Training is a vital part of any health and safety programme and needs to be constantly reviewed and updated. Many health and safety regulations require specific training (e.g. manual handling, PPE and display screens). Additional training courses may be needed when there is a major reorganisation, a series of similar accidents or incidents, or a change in equipment or a process. Finally, the methods used to deliver training must be continually monitored to ensure that they are effective.

A list of typical legislation that requires some form of health and safety training is given in Appendix 3.4. The list has been compiled to cover the needs of most small- to medium-sized organisations.

3.4.8 Internal influences on health and safety performance

There are many influences on health and safety standards, some are positive and others negative (Figure 3.14). No business, particularly small businesses, is totally divorced from its suppliers, customers and neighbours. This section considers the internal influences on a business, including management commitment, production demands, communication, competence and employee relations.

Figure 3.14 Internal influences on safety culture

Management commitment

Managers, particularly senior managers, can give powerful messages to the workforce by what they do for health and safety. Managers can achieve the level of health and safety performance that they demonstrate they want to achieve. Employees soon get the negative message if directors disregard safety rules and ignore written policies to get urgent production to the customer or to avoid personal inconvenience. It is what they do that counts not what they say. Earlier in this chapter, managers' organisational roles were listed showing the ideal level of involvement for senior managers. Depending on the size and geography of the organisation, senior managers should be personally involved in:

▶ health and safety inspections or audits;
▶ meetings of the central health and safety or joint consultation committees;
▶ the investigation of accidents, ill-health and incidents. The more serious the incident the more senior the manager who takes an active part in the investigation.

Production/service demands

Managers need to balance the demands placed by customers with the action required to protect the health and safety of their employees. How this is achieved has a strong influence on the standards adopted by the organisation. The delivery driver operating to near impossible delivery schedules and the manager agreeing to the strident demands of a large customer regardless of the risks to employees involved are typical dilemmas facing workers and managers alike. The only way to deal with these issues is to be well organised and agree with the client/ employees ahead of the crisis as to how they should be prioritised.

Rules and procedures should be intelligible, sensible and reasonable. They should be designed to be followed under normal production or service delivery conditions. If following the rules involves very long delays or impossible production schedules, they should be revised rather than ignored by workers and managers alike until it is too late and an accident occurs. Sometimes the safety rules are simply used in a court of law, in an attempt to defend the company concerned, following an accident.

Managers who plan the impossible schedule or ignore a safety rule to achieve production or service demands are held responsible for the outcomes. It is acceptable to balance the cost of the action against the level of risk being addressed but it is never acceptable to ignore safety rules or standards simply to get the work done. Courts are not impressed by managers who put profit considerations ahead of safety requirements.

Communication

Communications were covered in depth earlier in the chapter and clearly will have significant influence on health and safety issues. This will include:

▶ poorly communicated procedures that will not be understood or followed;

▶ poor verbal communications which will be misunderstood and will demonstrate a lack of interest by senior managers;

▶ missing or incorrect signs which may cause accidents rather than prevent them;

▶ managers who are nervous about face-to-face discussions with the workforce on health and safety issues, which will have a negative effect.

Managers and supervisors should plan to have regular discussions to learn about the problems faced by employees and discuss possible solutions. Some meetings, like that of the safety committee, are specifically planned for safety matters, but this should be reinforced by discussing health and safety issues at all routine management meetings. Regular one-to-one talks should also take place in the workplace, preferably to a planned theme or safety topic, to get specific messages across and get feedback from employees.

Competence

Competent people, who know what they are doing and have the necessary skills to do the task correctly and safely, will make the organisation more effective. Competence can be bought in through recruitment or consultancy but it is often much more effective to develop it among employees. It demonstrates commitment to health and safety and a sense of security for the workforce. The loyalty that it creates in the workforce can be a significant benefit to safety standards. More details on how to achieve competence were given earlier in this chapter.

Employee representation

Given the resources and freedom to fulfil their function effectively, enthusiastic, competent employee safety representatives can make a major contribution to good health and safety standards. They can provide the essential bridge between managers and employees. People are more willing to accept the restrictions that some precautions bring if they are consulted and feel involved, either directly in small workforces or through their safety representatives.

3.4.9 External influences on health and safety performance

The role and function of some external organisations are set out in Chapter 1. Here the influence they can have is briefly discussed, including societal expectations, legislation and enforcement, insurance companies,

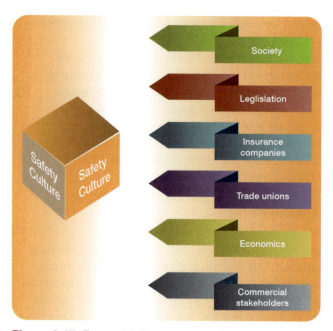

Figure 3.15 External influences on safety culture

trade unions, economics and commercial stakeholders (Figure 3.15).

Societal expectations

Societal expectations have been covered in Chapter 1. In general society assumes that health and safety considerations have been considered in all aspects of their lives and it becomes concerned when this does not seem to be the case.

Legislation and enforcement

Good comprehensible legislation should have a positive effect on health and safety standards. Taken together, legislation and enforcement can affect standards by:

▶ providing a level to which every employer has to conform;

▶ insisting on minimum standards which also enhances peoples' ability to operate and perform well;

▶ providing a tough, visible threat of getting shut down or a heavy fine;

▶ stifling development by being too prescriptive; for example, woodworking machines did not develop quickly in the 20th century, partly because the Regulations were so detailed in their requirements that new designs were not feasible;

▶ providing well-presented and easily read guidance for specific industries at reasonable cost or free.

On the other hand, a weak enforcement regime can have a powerful negative effect on standards.

Insurance companies

Insurance companies can influence health and safety standards mainly through financial incentives.

Employers' liability insurance is a legal requirement in the UK and therefore all employers have to obtain this type of insurance cover. Insurance companies can influence standards through:

▶ discounting premiums to those in the safest sectors or best individual companies;
▶ insisting on risk reduction improvements to remain insured. This is not very effective where competition for business is fierce;
▶ encouraging risk reduction improvements by bundling services into the insurance premium;
▶ providing guidance on standards at reasonable cost or free.

Trade unions

Trade unions can influence standards by:

▶ providing training and education for members;
▶ providing guidance and advice cheaply or free to members;
▶ influencing governments to regulate, enhance enforcement activities and provide guidance;
▶ influencing employers to provide high standards for their members. This is sometimes confused with financial improvements with health and safety getting a lower priority;
▶ encouraging members to work for safer employers;
▶ helping members to get proper compensation for injury and ill-health if it is caused through their work.

Economics

Economics can play a major role in influencing health and safety standards. The following ways are the most common:

▶ lack of orders and/or money can cause employers to try to ignore health and safety requirements;
▶ if employers were really aware of the actual and potential cost of accidents and fires, they would be more concerned about prevention. The HSE believes that the ratio between insured and uninsured costs of accidents is between 1:8 and 1:36;
▶ perversely, when the economy is booming activity increases and, particularly in the building industry, accidents can sharply increase. The pressures to perform and deliver for customers can be safety averse;
▶ businesses that are only managed on short-term performance indicators seldom see the advantage of the long-term gains that are possible with a happy, safe and fit workforce.

The cost of accidents and ill-health, in both human and financial terms, needs to be visible throughout the organisation so that all levels of employee are encouraged to take preventative measures.

Commercial stakeholders

A lot can be done by commercial stakeholders to influence standards. This includes:

▶ insisting on proper arrangements for health and safety management at supplier companies before they tender for work or contracts;
▶ checking on suppliers to see if the workplace standards are satisfactory;
▶ encouraging ethical investments;
▶ considering ethical as well as financial standards when banks provide funding;
▶ providing high-quality information for customers;
▶ insisting on high standards to obtain detailed planning permission (where this is possible);
▶ providing low-cost guidance and advice.

3.5 Further information

The Health and Safety at Work etc. Act 1974,

The Management of Health and Safety at Work Regulations 1999 (as amended)

The Safety Representatives and Safety Committees Regulations 1977

The Health and Safety Information for Employees Regulations 1989

The Health and Safety (Consultation with Employees) Regulations 1996

Corporate Manslaughter and Corporate Homicide Act 2007

Occupational Health and Safety Assessment Series (OHSAS 18000): Occupational Health and Safety Management Systems OHSAS 18001:2007 ISBN 978-0-5805-9404-5; OHSAS18002:2008 ISBN 978-0-5806-2686-9

Reducing Error and Influencing Behaviour (HSG48), HSE Books, ISBN 978-0-7176-2452-2 http://www.hse.gov.uk/pubns/priced/hsg48.pdf

First-aid at Work, The Health and Safety (First-Aid) Regulations 1981 (L74, third edition 2013), HSE Books, ISBN 978-0-7176-6560-0 http://www.hse.gov.uk/pubns/priced/l74.pdf

Leading health and safety at work (INDG417 rev 1 2013), HSE Books www.hse.gov.uk/pubns/indg417.htm

Human factors and ergonomics http://www.hse.gov.uk/humanfactors/

Consulting workers on health and safety, Safety Representatives and Safety Committee Regulations 1977 (as amended) and Health and Safety (Consultation with Employees) Regulations 1996 (as amended), ACoP, L146, HSE Books, ISBN 978-0-7176-6461-0 http://www.hse.gov.uk/pubns/priced/l146.pdf

3

3.6 Practice revision questions

1. (a) **Describe** the four elements that boards of directors should take into account when overseeing the management of health and safety within their organisation.
   (b) **Outline** the responsibilities that departmental managers have to ensure the health, safety and welfare of those working in their department.
   (c) **Outline** the factors that will determine the level of supervision that a new employee should receive during their initial period of employment within an organisation.

2. (a) **Explain** the meaning of the term 'competent person' when related to health and safety.
   (b) **Outline** the factors that the employer should consider when selecting an individual for the role of health and safety competent person within the organisation.
   (c) **Outline** the organisational factors that may cause a person to work unsafely even though they are competent.
   (d) **Identify FOUR** types of specialist help that may be required to support the health and safety function.

3. (a) **Explain** the meaning of the term *'health and safety culture'*.
   (b) **Describe FIVE** requirements of a positive health and safety culture.
   (c) **Identify** ways in which the health and safety culture of an organisation might be improved.
   (d) **Outline EIGHT** possible indications of a poor health and safety culture in an organisation.

4. (a) The number of accidents is a measure of the health and safety culture in a company. **State** the definition of 'accident incidence rate'.
   (b) **Describe** other measures of health and safety culture. **Outline** how information on accidents could be used to promote health and safety in the workplace.

5. (a) **Explain**, using an example, the meaning of the term 'attitude'.
   (b) **Outline THREE** influences on the attitude towards health and safety of employees within an organisation.

6. (a) **Explain** the meaning of the term 'motivation'.
   (b) Other than lack of motivation, **outline FOUR** reasons why employees may fail to comply with safety procedures at work.
   (c) **Outline** ways in which employers may motivate their employees to comply with health and safety procedures.

7. (a) **Explain** the meaning of the term 'perception'.
   (b) **Outline** the factors relating to the individual that may influence a person's perception of an occupational risk.
   (c) **Outline** ways in which employees' perceptions of hazards in the workplace might be improved.

8. The human factors which influence behaviour in the workplace are the organisation, the job and personal factors.
   (a) **Describe** the essential elements in the organisation and the job that will ensure good health and safety behaviour.
   (b) **Outline** personal or individual factors that may contribute to human errors occurring at work.

9. (a) **Explain** the meaning of the term 'human error'.
   (b) **Describe** the difference between human errors and violations.
   (c) **Describe**, using practical examples, **THREE** types of human error and **THREE** types of violation that can lead to accidents in the workplace.

10. (a) **Identify FIVE** topics for which there would be health and safety standards in a company.
    (b) **Outline** the possible actions that a chief executive and senior managers could take to improve the health and safety standards in the workplace.

11. Following a significant increase in accidents, a health and safety campaign is to be launched within a manufacturing company to encourage safer working by employees.
    (a) **Outline** how the company might ensure that the nature of the campaign is effectively communicated to, and understood by, employees.
    (b) **Identify** a range of methods that an employer can use to provide health and safety information directly to individual employees.
    (c) **Outline** reasons why the safety procedures may not have been followed by all employees.

12. **Outline** the advantages and disadvantages of communicating health and safety messages to employees:
    (a) Verbally
    (b) In writing
    (c) Graphically.

13. Notice boards are often used to display health and safety information in the workplace.

   (a) **Outline** the limitations of relying on this method to communicate to employees and how these limitations may be addressed.

   (b) **Identify FOUR** types of health and safety information that might usefully be displayed on a notice board.

   (c) **Identify FOUR** other methods which the employer could use to communicate essential health and safety information to their employees.

14. (a) **Explain** the circumstances in which an employer must form a health and safety committee under the Safety Representatives and Safety Committees Regulations.

   (b) **Outline SIX** reasons why a health and safety committee may prove to be ineffective in practice.

15. With reference to the Health and Safety (Consultation with Employees) Regulations 1996:

   (a) **explain** the difference between consulting and informing and the ways in which an employer can consult with the workforce;

   (b) **outline FOUR** health and safety matters on which employers must consult their employees;

   (c) **outline** the rights and functions of representatives of employee safety elected under the Regulations;

   (d) **identify** four types of information that an employer is not obliged to disclose to an employee representative.

16. (a) **Outline** the importance of induction training for new employees in reducing the number of accidents in a workplace.

   (b) **Outline** reasons for additional refresher health and safety training at a later stage of employment within a workplace.

   (c) **Identify** the reasons for a review of the health and safety training programme within an organisation.

17. A building contractor is to undertake building maintenance work in the roof space of a busy warehouse. **Outline** the issues that should be covered in an induction programme for the contractor's employees.

18. Following an accident at a manufacturing company, an investigation concluded that there had been a poor record of specialist and supervisory training.

   (a) **Identify THREE** types of specialist training that may have been relevant to the company.

   (b) **Outline** the possible supervisory failures that could have caused an accident.

19. **Describe TWO** internal and **TWO** external influences on the health and safety culture of an organisation.

APPENDIX 3.1 Leadership actions for directors and board members

This guidance, taken from INDG417, sets out an agenda for the effective leadership of health and safety by all directors, governors, trustees, officers and their equivalents in the private, public and third sectors. It applies to organisations of all sizes.

There are four elements that boards need to incorporate into their management of health and safety. These are:

▶ planning the direction of health and safety;
▶ delivering the plan for health and safety;
▶ monitoring health and safety performance; and
▶ reviewing health and safety performance.

1. Planning the direction of health and safety

The Board should set the direction for effective health and safety management. Board members need to establish a health and safety policy that is much more than a document – it should be an integral part of the organisation's culture, its values and performance standards.

All board members should take the lead in ensuring the communication of health and safety duties and benefits throughout the organisation. Executive directors must develop the policy since they have responsibility for the day-to-day operation of the organisation and must respond quickly where difficulties arise or new risks are introduced; non-executive directors must make sure that health and safety is properly addressed.

Essential actions

For a policy to be effective, all board members must be aware of the significant risks faced by their organisation. The policy should include the Board's role and that of individual board members in leading the management of health and safety in the organisation. The Board needs to 'own' and understand the key issues involved and decide on the best methods to communicate, promote and champion health and safety.

The health and safety policy is a 'living' document and it should evolve over time; for example, during major organisational changes such as restructuring or a significant acquisition.

Good practice guide

▶ Health and safety should appear regularly on the agenda for board meetings.
▶ The Chief Executive should give the clear leadership, but it may also be useful to name one of their members as the health and safety 'champion'.
▶ The presence on the Board of a Health and Safety Director gives a strong signal that the issue is being

taken seriously and that its strategic importance is understood.
▶ The setting of targets helps the Board to measure performance.
▶ A non-executive director can act as a scrutiniser to ensure that robust processes are in place to support boards that face significant health and safety risks.

2. Delivering the plan for health and safety

Delivery depends on an effective management system to ensure, so far as is reasonably practicable, the health and safety of employees, customers and members of the public. Organisations should aim to protect people by introducing management systems and practices that ensure risks are dealt with sensibly, responsibly and proportionately.

Essential actions

To take responsibility and 'ownership' of health and safety, members of the Board must ensure that:

▶ health and safety arrangements are adequately resourced;
▶ they obtain competent health and safety advice;
▶ risk assessments are carried out;
▶ employees or their representatives are involved in decisions that affect their health and safety.

The Board should consider the health and safety implications of introducing new processes, working practices or personnel and dedicate adequate resources to the task and seek competent advice when necessary. All board decisions, particularly when implementing change, must be made in the context of the organisation's health and safety policy.

Good practice guide

▶ Leadership is more effective if visible – board members can reinforce health and safety policy by being seen on the 'shop floor', following all safety measures themselves and addressing any breaches immediately.
▶ Health and safety should be considered when senior management appointments are made.
▶ Having good procurement standards for goods, equipment and services should prevent the introduction of expensive health and safety hazards.
▶ The health and safety arrangements of partners, key suppliers and contractors should be assessed regularly since their performance could adversely affect a director's own performance.
▶ A separate risk management or health and safety sub-committee of the Board, chaired by a senior executive, should ensure that key issues are addressed and guard against time and effort

being wasted on trivial risks and unnecessary bureaucracy.

▶ Health and safety training to some or all board members will promote understanding and knowledge of the key health and safety issues in the organisation.

▶ Worker involvement in health and safety, above the legal duty to consult worker representatives, will improve participation and indicate the commitment of senior management.

3. Monitoring health and safety performance

Monitoring and reporting are vital parts of a health and safety system. Management systems must allow the Board to receive both specific (e.g. incident-led) and routine reports on the performance of health and safety policy. Much day-to-day health and safety information need be reported only at the time of a formal review. But only a strong system of monitoring can ensure that the formal review will proceed as planned – and that relevant events in the interim are brought to the Board's attention.

Core actions

The Board should ensure that:

▶ appropriate weight is given to reporting both preventative information (such as progress of training and maintenance programmes) and incident data (such as accident and sickness absence rates);

▶ periodic audits of the effectiveness of management structures and risk controls for health and safety are carried out;

▶ the impact of changes such as the introduction of new procedures, work processes or products, or any major health and safety failure, is reported as soon as possible to the Board;

▶ there are procedures to implement new and changed legal requirements and to consider other external developments and events.

Good practice guide

▶ Effective monitoring of sickness absence and workplace health will alert the Board to underlying problems that could seriously damage performance or result in accidents and long-term illness.

▶ The collection of workplace health and safety data will allow the Board to benchmark the organisation's performance against others in its sector.

▶ Appraisals of senior managers should include an assessment of their contribution to health and safety performance.

▶ Boards should receive regular reports on the health and safety performance and actions of contractors.

▶ Organisations will find that they win greater support for health and safety from workers by involving them in the monitoring process.

4. Reviewing health and safety performance

A formal boardroom review of health and safety performance is essential. It allows the Board to establish whether the essential health and safety principles – strong and active leadership, worker involvement, and assessment and review – have been embedded in the organisation. It tells senior managers whether their system is effective in managing risk and protecting people.

Core actions

The Board should review health and safety performance at least once a year. The review process should:

▶ examine whether the health and safety policy reflects the organisation's current priorities, plans and targets;

▶ examine whether risk management and other health and safety systems have been effectively reported to the Board;

▶ report health and safety shortcomings, and the effect of all relevant board and management decisions;

▶ decide actions to address any weaknesses and a system to monitor their implementation;

▶ instigate immediate reviews in the light of major shortcomings or events.

Good practice guide

▶ Performance on health and safety and well-being should be reported in the organisations' annual report to investors and stakeholders.

▶ Board members should make extra 'shop floor' visits to gather information for the formal review.

▶ Good health and safety performance should be celebrated at central and local level.

APPENDIX 3.2 Detailed health and safety responsibilities

Managing Directors/Chief Executives

1. Are responsible and accountable for health and safety performance at the organisation;

2. Develop a strong, positive health and safety culture throughout the company and communicate it to all managers. This should ensure that all managers have a clear understanding of their health and safety responsibilities;

3. Provide guidance and leadership on health and safety matters to their management team;

4. Establish minimum acceptable health and safety standards within the organisation;

5. Ensure that adequate resources are available for the health and safety requirements within the organisation and authorise any necessary major health and safety expenditures;

6. Evaluate, approve and authorise health and safety related projects developed by the organisation's health and safety advisers;

7. Review and approve health and safety policies, procedures and programmes developed by the organisation's managers;

8. Ensure that a working knowledge of the areas of health and safety that are regulated by various governmental agencies, particularly the HSE, Local Authorities and the Environment Agency, are maintained;

9. Ensure that health and safety is included as an agenda topic at all formal senior management meetings;

10. Review and act upon major recommendations submitted by outside loss prevention consultants and insurance companies;

11. Ensure that health and safety is included in any tours such as fire inspections of the organisation's sites and note any observed acts or conditions that fall short of or exceed agreed health and safety standards;

12. Ensure that all fatalities, major property losses, serious lost workday injuries and dangerous occurrences are investigated;

13. Establish, implement and maintain a formal, written health and safety programme for the organisation that encompasses all areas of significant health and safety risk;

14. Establish controls to ensure uniform adherence to the health and safety programme across the organisation;

15. Attend the health and safety committee meetings at the organisation;

16. Review, on a regular basis, all health and safety activity reports and performance statistics;

17. Review health and safety reports submitted by outside agencies and determine that any agreed actions have been taken;

18. Review annually the effectiveness of, and, if necessary, require revision of, the site health and safety programme;

19. Appraise the performance of the health and safety advisers and provide guidance or training where necessary;

20. Monitor the progress of managers and others towards achieving their individual health and safety objectives.

Departmental Managers

1. Are responsible and accountable for the health and safety performance of their department;

2. Contact each supervisor frequently (daily) to monitor the health and safety standards in the department;

3. Hold departmental health and safety meetings for supervisors and employee representatives at least once a month;

4. Ensure that any machinery, equipment or vehicles used within the department are maintained and correctly guarded and meet agreed health and safety standards. Copies of records of all maintenance, statutory and insurance inspections must be kept by the Departmental Manager;

5. Ensure that all fire and other emergency equipment is properly maintained on a regular basis with all faults rectified promptly and that all departmental staff are aware of fire and emergency procedures;

6. Ensure that there is adequate first-aid cover on all shifts and all first-aid boxes are adequately stocked;

7. Ensure that safe systems of work procedures are in place for all jobs and that copies of all procedures are submitted to the site Managing Director for approval;

8. Review all job procedures on a regular basis and require each supervisor to check that the procedures are being used correctly;

9. Approve and review, annually, all departmental health and safety risk assessments, rules and procedures, maintain strict enforcement and develop plans to ensure employee instruction and re-instruction;

10. Ensure that all health and safety documents (such as the organisation's health and safety manual, risk assessments, rules and procedures) are easily accessible to all departmental staff;

11. Establish acceptable housekeeping standards, defining specific areas of responsibility, and assign areas to supervisors; make a weekly spot check across the department, hold a formal inspection with supervisors at least monthly and submit written reports of the inspections to the health and safety adviser with deadlines for any required actions;

12. Authorise purchases of tools and equipment necessary to attain compliance with the organisation's specifications and relevant statutory Regulations;

13. Develop a training plan that includes specific job instructions for new or transferred employees and follow up on the training by supervisors. Copies of records of all training must be kept by the Departmental Manager;

14. Review the health and safety performance of their department each quarter and submit a report to the Managing Director/Chief Executive;

15. Personally investigate all lost workday cases and dangerous occurrences and report to the Managing Director/Chief Executive. Progress any required corrective action;

16. Adopt standards for assigning PPE to employees, insist on strict enforcement and make spot field checks to determine compliance;

17. Evaluate the health and safety performances of supervisors;

18. Develop in each supervisor strong health and safety attitudes and a clear understanding of their specific duties and responsibilities;

19. Instil, by action, example and training, a positive health and safety culture among all departmental staff;

20. Instruct supervisors in site procedures for the care and treatment of sick or injured employees;

21. Ensure that the names of any absentees, written warnings and all accident reports are submitted to the Human Resources Manager.

Supervisors

1. Are responsible and accountable for their team's health and safety performance;

2. Conduct informal health and safety meetings with their employees at least monthly;

3. Enforce all safe systems of work procedures that have been issued by the Departmental Manager;

4. Report to the Departmental Manager any weaknesses in the safe system of work procedures or any actions taken to revise such procedures. These weaknesses may be revealed by either health and safety risk assessments or observations;

5. Report any jobs that are not covered by safe systems of work procedures to the Departmental Manager;

6. Review any unsafe acts and conditions and either eliminate them or report them to the Departmental Manager;

7. Instruct employees in relevant health and safety rules, make records of this instruction and enforce all health and safety rules and procedures;

8. Make daily inspections of assigned work areas and take immediate steps to correct any unsafe or unsatisfactory conditions, report to the Departmental Manager those conditions that cannot be immediately corrected and instruct employees on housekeeping standards;

9. Instruct employees that tools/equipment are to be inspected before each use and make spot checks of tools'/equipment's condition;

10. Instruct each new employee personally on job health and safety requirements in assigned work areas;

11. Provide on-the-job instruction on safe and efficient performance of assigned jobs for all employees in the work area;

12. Report any apparent employee health problems to the Departmental Manager;

13. Enforce PPE requirements; make spot checks to determine that protective equipment is being used and periodically appraise condition of equipment. Record any infringements of the PPE policy;

14. In the case of a serious injury, ensure that the injured employee receives prompt medical attention, isolate the area and/or the equipment as necessary and immediately report the incident to the Departmental Manager. In case of a dangerous occurrence, the Supervisor should take immediate steps to correct any unsafe condition and, if necessary, isolate the area and/or the equipment. As soon as possible, details of the incident and any action taken should be reported to the Departmental Manager;

15. Investigate all accidents, serious incidents and cases of ill-health involving employees in assigned work areas. Immediately after an accident, complete an accident report form and submit it to the Departmental Manager for onward submission to the health and safety adviser. A preliminary investigation report and any recommendations for preventing a recurrence should be included on the accident report form;

16. Check for any changes in operating practices, procedures and other conditions at the start of each shift/day and before relieving the 'on duty' Supervisor (if applicable). A note should

be made of any health and safety related incidents that have occurred since their last working period;

17. At the start of each shift/day, make an immediate check to determine any absentees. Report any absentees to the Departmental Manager;

18. Make daily spot checks and take necessary corrective action regarding housekeeping, unsafe acts or practices, unsafe conditions, job procedures and adherence to health and safety rules;

19. Attend all scheduled and assigned health and safety training meetings;

20. Act on all employee health and safety complaints and suggestions;

21. Maintain, in their assigned area, health and safety signs and notice boards in a clean and legible condition.

Employees and agency workers

1. Are responsible for their own health and safety;

2. Ensure that their actions will not jeopardise the safety or health of other employees;

3. Obey any safety rules, particularly regarding the use of PPE or other safety equipment;

4. Learn and follow the operating procedures and health and safety rules and procedures for the safe performance of the assigned job;

5. Must correct, or report to their Supervisor, any observed unsafe practices and conditions;

6. Maintain a healthy and safe place to work and cooperate with managers in the implementation of health and safety matters;

7. Make suggestions to improve any aspect of health and safety;

8. Maintain an active interest in health and safety;

9. Follow the established procedures if accidents occur by reporting any accident to the Supervisor;

10. Report any absence from the company caused by illness or an accident.

APPENDIX 3.3 Safety culture questionnaire

The HSE have suggested that the following questionnaire can be used to inspect the safety culture of an organisation.

1. Management commitment

Where is safety perceived to be in management's priorities (senior/middle/first line)?
How do they show this?
How often are they seen in the workplace?
Do they talk about safety when in the workplace and is this visible to the workforce?
Do they 'walk the talk'?
Do they deal quickly and effectively with safety issues raised?
What balance do their actions show between safety and production?
Are management trusted over safety?

2. Communication

Is there effective two-way communication about safety?
How often are safety issues discussed:
  ▷ with line manager/subordinate?
  ▷ with colleagues?
What is communicated about the safety programme of the company?
How open are people about safety?

3. Employee involvement

How are people (all levels, especially operators) involved in safety?

How often are individual employees asked for their input on safety issues?
How often do operators report unsafe conditions or near misses, etc?
Is there active, structured operator involvement, e.g. workshops, projects, safety circles?
Is there a continuous improvement/total quality approach?
Whose responsibility is safety regarded to be?
Is there genuine cooperation over safety – a joint effort between all in the company?

4. Training/information

Do employees feel confident that they have all the training that they need?
How accurate are employees' perceptions of hazards and risks?
How effective is safety training in meeting needs (including managers!)?
How are needs identified?
How easily available is safety information?

5. Motivation

Do managers give feedback on safety performance (and how)?
Are they likely to notice unsafe acts?
Do managers (all levels – senior/middle/first line) always confront unsafe acts?
How do they deal with them?
Do employees feel they can report unsafe acts?

How is discipline applied to safety?
What do people believe are the expectations of managers?
Do people feel that this is a good place to work (why/why not)?
Are they proud of their company?

6. Compliance with procedures

What are written procedures used for?
What decides whether a particular task will be captured in a written procedure?
Are they read?
Are they helpful?
What other rules are there?
Are there too many procedures and rules?

How well are people trained in them?
Are they audited effectively?
Are they written by users?
Are they linked to risks?

7. Learning organisation

Does the company really learn from accident history, incident reporting, etc?
Do employees feel confident in reporting incidents or unsafe conditions?
Do they report them?
Do reports get acted upon?
Do they get feedback?

APPENDIX 3.4 List of typical legislation requiring health and safety training

	Legislation
1.	Management of Health and Safety at Work Regulations
2.	Regulatory Reform (Fire Safety) Order 2005
3.	Provision and Use of Work Equipment Regulations
4.	Control of Substances Hazardous to Health Regulations
5.	Health and Safety (First-aid) Regulations
6.	Health and Safety (Display Screen Equipment) Regulations
7.	Manual Handling Operations Regulations
8.	Control of Noise at Work Regulations
9.	Personal Protective Equipment at Work Regulations
10.	Health and Safety (Safety Signs and Signals) Regulations
11.	Control of Asbestos Regulations
12.	Confined Spaces Regulations
13.	Health and Safety (Consultation with Employees) Regulations
14.	Control of Vibration at Work Regulations

CHAPTER 4

Health and safety management systems – Risk assessment and controls – DO 2

Introduction

Risk assessment and minimum levels of risk prevention or control are regulated by health and safety legal requirements. Risk assessment is an essential part of the planning stage of any health and safety management system (see Figure 4.1). Risk assessment methods are used to decide on priorities and to set objectives for eliminating hazards and reducing risks. Wherever possible, risks should be eliminated through the selection and design of facilities, equipment and processes. If risks cannot be eliminated, they should be minimised by the use of physical controls or, as a last resort, through systems of work and personal protective equipment. The control of risks is essential to secure and maintain a healthy and safe workplace, which complies with the relevant legal requirements. In this chapter, hazard identification and risk assessment are covered together with appropriate risk control measures. A hierarchy of control methods is discussed that gives a preferred order of approach to risk control.

This chapter also concerns the principles of prevention that should be adopted when deciding on suitable measures to eliminate or control both acute and immediate and long-term risks to the health and safety of people at work. The principles of control can be applied to both health risks and safety risks although health risks have some distinctive features that require a special approach. The chapter concludes with emergency procedures and the treatment of first-aid.

4.1 Principles and practice of risk assessment

4.1.1 Introduction

There are several descriptions of the meaning of 'risk assessment'. NEBOSH defines risk assessment as:

> *'the identification of preventative and protective measures by the evaluation of the risk(s) arising from a hazard(s), taking into account the adequacy of any existing controls, and deciding whether or not the risk(s) is acceptable.'*

4.1.2 Legal aspects of risk assessment

The general duties of employers to their employees in section 2 of the HSW Act 1974 imply the need for risk assessment. This duty was also extended by Section 3 of the Act to anybody else affected by activities of the employer – contractors, visitors, customers or members of the public. However, the Management of Health and Safety at Work Regulations are much more specific concerning the need for risk assessment. The following requirements are laid down in those Regulations:

▶ the risk assessment shall be 'suitable and sufficient' and cover both employees and non-employees affected by the employer's undertaking (e.g. contractors, members of the public, students, patients, customers); every self-employed person shall make a 'suitable and sufficient' assessment of the risks to which they or those affected by the undertaking may be exposed;

▶ any risk assessment shall be reviewed if there is reason to suspect that it is no longer valid or if a significant change has taken place;

▶ where there are five or more employees, the significant findings of the assessment shall be recorded and any specially at risk group of

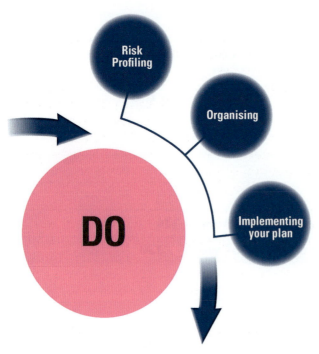

Figure 4.1 Risk assessment or profiling is covered by the DO part of the management cycle.

employees identified. (This does not mean that employers with four or less employees need not undertake risk assessments.)

The term 'suitable and sufficient' is important as it defines the limits to the risk assessment process. A suitable and sufficient risk assessment should:

▶ identify the significant risks and ignore the trivial ones;
▶ identify and prioritise the measures required to comply with any relevant statutory provisions;
▶ remain appropriate to the nature of the work and valid over a reasonable period of time;
▶ identify the risk arising from or in connection with the work. The level of detail should be proportionate to the risk.

The significant findings that should be recorded include a detailed statement of the hazards and risks, the preventative, protective or control measures in place and any further measures required to reduce the risks present.

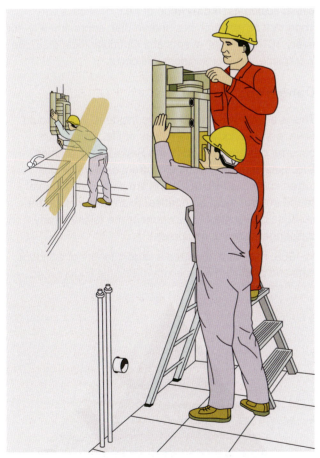

Figure 4.2 Reducing the risk – finding an alternative safer method when fitting a wall-mounted boiler

When assessing risks under the Management of Health and Safety at Work Regulations, reference to other Regulations may be necessary even if there is no specific requirement for a risk assessment in

those Regulations. For example, reference to the legal requirements of the Provision and Use of Work Equipment Regulations will be necessary when risks from the operation of machinery are being considered. However, there is no need to repeat a risk assessment if it is already covered by other Regulations (e.g. a risk assessment considering personal protective equipment is required under the COSHH Regulations so there is no need to undertake a separate risk assessment under the Personal Protective Equipment Regulations).

Apart from the duty under the Management of Health and Safety at Work Regulations to undertake a health and safety assessment of the risks to any person (employees, contractors or members of the public), who may be affected by the activities of the organisation, the following Regulations require a specific risk assessment to be made:

▶ Ionising Radiations Regulations;
▶ Control of Asbestos Regulations;
▶ Control of Noise at Work Regulations;
▶ Manual Handling Operations Regulations;
▶ Health and Safety (Display Screen Equipment) Regulations;
▶ Personal Protective Equipment at Work Regulations;
▶ Confined Spaces Regulations;
▶ Work at Height Regulations;
▶ Regulatory Reform (Fire Safety) Order (not under HSW Act);
▶ Control of Vibration at Work Regulations;
▶ Control of Lead at Work Regulations;
▶ Control of Substances Hazardous to Health Regulations.

See http://www.hse.gov.uk/risk/ the HSE's risk management site.

4.1.3 Forms of risk assessment

There are two basic forms of risk assessment.

A **quantitative** risk assessment attempts to measure the risk by relating the probability of the risk occurring to the possible severity of the outcome and then giving the risk a numerical value. This method of risk assessment is used in situations where a malfunction could be very serious (e.g. aircraft design and maintenance or the petrochemical industry).

The more common form of risk assessment is the **qualitative** assessment, which is based purely on personal judgement and is normally defined as high, medium or low. Qualitative risk assessments are usually satisfactory as the definition (high, medium or low) is normally used to determine the time frame over which further action is to be taken.

The term **'generic'** risk assessment is sometimes used, and describes a risk assessment which covers similar activities or work equipment in different departments, sites or companies. Such assessments are often produced by specialist bodies, such as trade associations. If used, they must be appropriate to the particular job and they will need to be extended to cover additional hazards or risks.

4.1.4 Some definitions

Some basic definitions were introduced in Chapter 1 and those relevant to risk assessment are reproduced here.

Hazard and risk

A hazard is something with the **potential** to cause harm (this can include articles, substances, plant or machines, methods of working, the working environment and other aspects of work organisation). Hazards take many forms including, for example, chemicals, electricity or noise. A hazard can be ranked relative to other hazards or to a possible level of danger.

A risk is the **likelihood** of potential harm from that hazard being realised. Risk (or strictly the level of risk) is also linked to the severity of its consequences. A risk can be reduced and the hazard controlled by good management.

It is very important to distinguish between a hazard and a risk – the two terms are often confused and activities often called high risk are in fact high hazard. There should only be high residual risk where there is poor health and safety management and inadequate control measures.

Electricity is an example of a high hazard as it has the potential to kill a person. The risk associated with electricity – the likelihood of being killed on coming into contact with an electrical device – is, hopefully, low.

Occupational or work-related ill-health

This is concerned with those acute and chronic illnesses or physical and mental disorders that are either caused or triggered by workplace activities. Such conditions may be induced by the particular work activity of the individual or by activities of others in the workplace. The time interval between exposure and the onset of the illness may be short (e.g. acute asthma attacks) or long (e.g. chronic deafness or cancer).

Accident

This is defined by the Health and Safety Executive (HSE) as *'any unplanned event that results in injury or ill-health of people, or damage or loss to property, plant, materials or the environment or a loss of a business opportunity'*. Other authorities define an accident more narrowly by excluding events that do not involve injury or ill-health.

Figure 4.3 Accident at work

Incident and near miss

HSE states that an *'Incident includes all undesired circumstances and "near misses" which could cause accidents'*. Knowledge of near misses is very important as research has shown that, approximately, for every 10 'near miss' events at a particular location in the workplace, a minor accident will occur.

Dangerous occurrence

This is a 'near miss' or 'Damage Incident' which could have led to serious injury or loss of life. Dangerous occurrences are defined in the Reporting of Injuries, Diseases and Dangerous Occurrences Regulations (often known as RIDDOR) and are always reportable to the enforcement authorities. Examples include the collapse of a scaffold or a crane or the failure of any passenger-carrying equipment.

In 1969, F. E. Bird collected a large quantity of accident data and produced a well-known triangle (Figure 4.4).

Figure 4.4 Bird's well-known accident triangle

It can be seen that damage and near miss accidents occur much more frequently than injury accidents and are, therefore, a good indicator of risks. The study also shows that most accidents are predictable and avoidable.

4.1.5 The objectives of risk assessment

The *main* objective of risk assessment is to determine the measures required by the organisation to comply with relevant health and safety legislation and, thereby, reduce the level of occupational injuries and ill-health. The purpose is to help the employer or self-employed person to determine the measures required to comply with their legal statutory duty under the HSW Act 1974 or its associated Regulations. The risk assessment will need to cover all those who may be at risk, such as customers, contractors and members of the public. In the case of shared workplaces, an overall risk assessment may be needed in partnership with other employers.

In Chapter 1, the moral, legal and financial arguments for health and safety management were discussed in detail. The important distinction between the direct and indirect costs of accidents is reiterated here.

Any accident or incidence of ill-health will cause both direct and indirect costs and incur an insured and an uninsured cost. It is important that all of these costs are taken into account when the full cost of an accident is calculated. In a study undertaken by the HSE, it was shown that indirect or hidden costs could be 36 times greater than direct costs of an accident. In other words, the direct costs of an accident or disease represent the tip of the iceberg when compared with the overall costs.

Direct costs are costs that are directly related to the accident. They may be insured (claims on employers' and public liability insurance, damage to buildings, equipment or vehicles) or uninsured (fines, sick pay, damage to product, equipment or process).

Indirect costs may be insured (business loss, product or process liability) or uninsured (loss of goodwill, extra overtime payments, accident investigation time, production delays).

There are many reasons for the seriousness of a hazard not to be obvious to the person exposed to it. It may be that the hazard is not visible (radiation, certain gases and biological agents) or have no short-term effect (work-related upper limb disorders). Some common causes of accidents include lack of attention, lack of experience, not wearing appropriate PPE, sensory impairment and inadequate information, instruction and training.

4.1.6 Accident categories

There are several categories of accident, all of which will be dealt with in more detail in later chapters. The principal categories are as follows:

- contact with moving machinery or material being machined;
- struck by a moving, flying or falling object;
- hit by a moving vehicle;
- struck against something fixed or stationary;
- injured while handling, lifting or carrying;
- slips, trips and falls on the same level;
- falls from a height;
- trapped by something collapsing;
- drowned or asphyxiated;
- exposed to, or in contact with, a harmful substance;
- exposed to fire;
- exposed to an explosion;
- contact with electricity or an electrical discharge;
- injured by an animal;
- physically assaulted by a person;
- other kind of accident.

4.1.7 Health risks

Risk assessment is not only concerned with injuries in the workplace but also needs to consider the possibility of occupational ill-health. Health risks fall into the following four categories:

1. chemical (e.g. paint solvents, exhaust fumes);
2. biological (e.g. bacteria, pathogens);
3. physical (e.g. noise, vibrations);
4. psychological (e.g. occupational stress).

There are two possible health effects of occupational ill-health.

They may be **acute**, which means that they occur soon after the exposure and are often of short duration, although in some cases emergency admission to hospital may be required.

They may be **chronic**, which means that the health effects develop with time. It may take several years for the associated disease to develop and the effects may be slight (mild asthma) or severe (cancer).

Health risks are discussed in more detail in chapters 14 and 15.

4.1.8 The management of risk assessment

Risk assessment is part of the planning and performance stages of the health and safety management system recommended by the HSE in its publication HSG65. All aspects of the organisation, including health and safety management, need to be covered by the risk assessment process. This will involve the assessment of risk in areas such as maintenance procedures, training programmes and supervisory arrangements. A general risk assessment of the organisation should reveal the significant hazards present and the general control measures that are in place. Such a risk assessment should be completed first and then followed by more specific risk assessments that examine individual work activities.

Risk assessors

It is important that the risk assessment team is selected on the basis of its competence to assess risks in the particular areas under examination in the organisation. The Team Leader or Manager should have health and safety experience and relevant training in risk assessment. It is sensible to involve the appropriate line manager, who has responsibility for the area or activity being assessed, as a team member. Other members of the team will be selected on the basis of their experience, their technical and/or design knowledge and any relevant standards or Regulations relating to the activity or process. At least one team member must have communication and report writing skills. A positive attitude and commitment to the risk assessment task are also important factors. It is likely that team members will require some basic training in risk assessment.

4.1.9 The practice of risk assessment

The HSE has produced a free leaflet entitled 'Risk assessment – A brief guide to controlling risks in the workplace', INDG163 (Figure 4.5). It gives practical advice on assessing risks and recording the findings, and is aimed at small and medium-sized companies in the service and manufacturing sectors. There are five steps in risk assessment. These are:

1. look for the hazards;
2. decide who might be harmed, and how;
3. evaluate the risks and decide whether existing precautions are adequate or more should be done;
4. record the significant findings;
5. review the assessment and revise it if necessary.

Each of these steps will be examined in turn in the following sections

Figure 4.5 Five steps to risk assessment

Step 1 – Look for the hazards

The essential first step in risk assessment is to seek out and identify hazards. Relevant sources of information include:

► legislation and supporting Approved Codes of Practice which give practical guidance and include basic minimum requirements;
► process information;
► product information provided under Section 6 of the HSW Act;
► relevant British, European and international standards;
► industry or trade association guidance;
► the personal knowledge and experience of managers and employees;
► accident, ill-health and incident data from within the organisation, from other organisations or from central sources;
► expert advice and opinion and relevant research.

See also Section 3.4.3 and Chapter 19.

There should be a critical appraisal of all routine and non-routine business activities. People exposed may include not just employees, but also others such as members of the public, contractors and users of the products and services. Employees and safety representatives can make a useful contribution in identifying hazards.

In the simplest cases, hazards can be identified by observation and by comparing the circumstances with the relevant information (e.g. single-storey premises will not present any hazards associated with stairs). In more complex cases, measurements such as air sampling or examining the methods of machine operation may be necessary to identify the presence of hazards from chemicals or machinery. In the most complex or high-risk cases (for example, in the chemical or nuclear industry) special techniques and systems may be needed such as hazard and operability studies (HAZOPS) and hazard analysis techniques such as event or fault-tree analysis. Specialist advice may be needed to choose and apply the most appropriate method.

Only significant hazards, which could result in serious harm to people, should be identified. Trivial hazards are a lower priority.

A tour of the area under consideration by the risk assessment team is an essential part of hazard identification as is consultation with the relevant section of the workforce.

A review of accident, incident and ill-health records will also help with the identification. Other sources of information include safety inspection, survey and audit reports, job or task analysis reports, manufacturers' handbooks or data sheets and Approved Codes of Practice and other forms of guidance.

Hazards will vary from workplace to workplace but the checklist in Appendix 4.1 shows the common hazards that are significant in many workplaces. Many questions in the NEBOSH examinations involve several common hazards found in most workplaces.

It is important that unsafe conditions are not confused with hazards, during hazard identification. Unsafe conditions should be rectified as soon as possible after observation. Examples of unsafe conditions include missing machine guards, faulty warning systems and oil spillage on the workplace floor.

Step 2 – Decide who might be harmed and how

Employees and contractors who work full time at the workplace are the most obvious groups at risk and it will be a necessary check that they are competent to perform their particular tasks. However, there may be other groups who spend time in or around the workplace. These include young workers, trainees, new and expectant mothers, cleaners, contractor and maintenance workers and members of the public. Members of the public will include visitors, patients, students or customers as well as passers-by.

The risk assessment must include any additional controls required due to the vulnerability of any of these groups, perhaps caused by inexperience or disability. It must also give an indication of the numbers of people from the different groups who come into contact with the hazard and the frequency of these contacts.

Step 3 – Evaluating the risks and the adequacy of current controls

This step is really two steps – evaluating the risks and evaluating the adequacy of current controls.

Evaluating the risks

During most risk assessments it will be noted that some of the risks posed by the hazard have already been addressed or controlled. The purpose of the risk assessment, therefore, is to reduce the remaining risk. This is called the **residual risk**.
The goal of risk assessment is to reduce all residual risks to as low a level as reasonably practicable. In a relatively complex workplace, this will take time so that **a system of ranking risk** is required – the higher the risk level the sooner it must be addressed and controlled.

For most situations, a **qualitative** risk assessment will be perfectly adequate. (This is certainly the case for NEBOSH Certificate candidates and is suitable for use during the practical assessment.) During the risk assessment, a judgement is made as to whether the risk level is high, medium or low in terms of the risk of somebody being injured. This designation defines a timetable for remedial actions to be taken thereby reducing the risk. High-risk activities should normally be addressed in days, medium risks in weeks and low risks in months or in some cases no action will be required.

It will usually be necessary for risk assessors to receive some training in risk level designation.

A **quantitative** risk assessment attempts to quantify the risk level in terms of the likelihood of an incident and its subsequent severity. Clearly the higher the likelihood and severity, the higher the risk will be. The likelihood depends on such factors as the control measures in place, the frequency of exposure to the hazard and the category of person exposed to the hazard. The severity will depend on the magnitude of the hazard (voltage, toxicity, etc.).

Likelihood of occurrence	Likelihood level
Harm is certain or near certain to occur	High 3
Harm will often occur	Medium 2
Harm will seldom occur	Low 1
Severity of harm	**Severity level**
Death or major injury (as defined by RIDDOR)	Major 3
7-day injury or illness (as defined by RIDDOR)	Serious 2
All other injuries or illnesses	Slight 1
Risk = Severity x Likelihood	

The HSE has suggested a simple 3 x 3 matrix to determine risk levels.

Likelihood	Severity		
	Slight 1	**Serious 2**	**Major 3**
Low 1	Low 1	Low 2	Medium 3
Medium 2	Low 2	Medium 4	High 6
High 3	Medium 3	High 6	High 9

It is possible to apply such methods to organisational risk or to the risk that the management system for health and safety will not deliver in the way in which it was expected or required. Such risks will add to the activity or occupational risk level. In simple terms, poor supervision of an activity will increase its overall level of risk. A risk management matrix has been developed which combines these two risk levels, as shown below.

RISK MANAGEMENT MATRIX	Occupational risk levels		
	Low	Medium	High
Organisational risk level — Low	Low	Low	Medium
Organisational risk level — Medium	Medium	Medium	High
Organisational risk level — High	High	High	Unsatisfactory

Whichever type of risk evaluation method is used, the level of risk simply enables a timetable of risk reduction to an acceptable and tolerable level to be formulated. The legal duty requires that all risks should be reduced to as low as is reasonably practicable.

In established workplaces, some control of risk will be in place already. The effectiveness of these controls needs to be assessed so that an estimate of the residual risk may be made. Many hazards have had specific Acts, Regulations or other recognised standards developed to reduce associated risks. Examples of such hazards are fire, electricity, lead and asbestos. The relevant legislation and any accompanying Approved Codes of Practice or guidance should be consulted first and any recommendations implemented. Advice on control measures may also be available from trade associations, trade unions or employers' organisations.

Where there are existing preventative measures in place, it is important to check that they are working properly and that everybody affected has a clear understanding of the measures. It may be necessary to strengthen existing procedures, for example by the introduction of a permit to work system. More details on the principles of control are given below.

Evaluating the controls

1. Hierarchy of risk control

When assessing the adequacy of existing controls or introducing new controls, a hierarchy of risk controls should be considered. The health and safety management system ISO 45001 (to replace OHSAS 18001 in 2016) states that the organisation shall establish a process for achieving risk reduction based upon the following hierarchy:

(a) eliminate the hazard;
(b) substitute with less hazardous materials, processes, operations or equipment;
(c) use engineering controls;
(d) use safety signs, markings and warning devices and administrative controls;
(e) use personal protective equipment.

The organisation shall ensure that the Occupational Health and Safety risks and determined controls are taken into account when establishing, implementing and maintaining its Occupational Health and Safety management system.

The hierarchy reflects that risk elimination and risk control by the use of physical engineering controls and safeguards can be more reliably maintained than those which rely solely on people. These concepts are now written into the Control of Substances Hazardous to Health (COSHH) Regulations and the Management of Health and Safety at Work (MHSW) Regulations.

Where a range of control measures are available, it will be necessary to weigh up the relative costs of each against the degree of control each provides, both in the short and long term. Some control measures, such as eliminating a risk by choosing a safer alternative substance or machine, provide a high degree of control and are reliable. Physical safeguards such as guarding a machine or enclosing a hazardous process need to be maintained. In making decisions about risk control, it will therefore be necessary to consider the degree of control and the reliability of the control measures along with the costs of both providing and maintaining the measure. The ISO 45001 hierarchy will now be discussed in detail.

(a) Elimination of the hazard

The best and most effective way of reducing risks is by avoiding a hazard and its associated risks. For example avoid working at height by installing a permanent working platform with stair access; avoid entry into a confined space by, for example, using a sump pump in a pit which is removed by a lanyard for maintenance; eliminate the fire risks from tar boilers by using bitumen which can be applied cold.

(b) Substitution

Substitution describes the use of a less hazardous form of a substance or process. There are many examples of substitution such as the use of water-based rather than oil-based paints; the use of asbestos substitutes and the use of compressed air as a power source rather than electricity to reduce both electrical and fire risks; and the use of mechanical excavators instead of hand digging.

In some cases it is possible to change the method of working so that risks are reduced. For example use rods to clear drains instead of strong chemicals; use a long-handled water hose brush to clean windows instead of ladders. Sometimes the pattern of work can be changed so that people can do things in a more natural way, for example when placing components for packing consider whether people are right- or left-handed; encourage people in offices to take breaks from computer screens by getting up to photocopy, fetch files or print documents.

Care must be taken to consider any additional hazards which may be involved, and thereby introduce additional risks, as a result of a substitution.

(c) Engineering controls

This describes the control of risks by means of engineering design rather than a reliance on preventative actions by the employee. There are several ways of achieving such controls:

1. Control the risks at the source (e.g. the use of more efficient dust filters or the purchase of less noisy equipment).
2. Control the risk of exposure by:

   ▶ **isolating** the equipment by the use of an enclosure, a barrier or guard;

   ▶ **insulating** any electrical or temperature hazard;

   ▶ **ventilating** away any hazardous fumes or gases either naturally or by the use of extractor fans and hoods (Figure 4.6).

(d) Safety signs, markings and warning devices and administrative controls

Safety signs, markings and warning devices

A summary of the Safety Signs and Signals Regulations is given in Chapter 19. All general health and safety signs used in the workplace must include a pictorial

Figure 4.6 Proper control of gases and vapours in a laboratory

symbol categorised by shape, colour and graphic image (Figure 4.7).

All workplaces need to display safety signs of some kind, but deciding what is required can be confusing. Presented here are the basic requirements for the majority of small premises or sites, such as small construction sites, canteens, shops, small workshop units and offices. This section does not cover any signs which food hygiene law may require.

Most requirements are covered by the Health and Safety (Safety Signs and Signals) Regulations. These require signs wherever a risk has not been controlled by other means. For example if a wet area of floor is cordoned off, a warning sign will not be needed, because the barrier will keep people out of the danger area. Signs are not needed where the sign would not reduce the risk or the risk is insignificant.

The following signs are typical of some of the ones most likely to be needed in these premises. Others may be necessary, depending on the hazards and risks present.

(i) Overhead obstacles, construction site and prohibition notices (Figures 4.8 and 4.9)

Safety Helmet Must be worn

Overhead load

Safety harness Must be worn

Not Drinkable

No access for Pedestrians

Figure 4.8 Examples of warning, mandatory and prohibition signs

Danger Falling objects

SITE SAFETY

Under the Health & Safety at Work Act 1974 all persons entering this site must comply with all regulations under this Act.
All visitors must report to the site office and obtain permission to proceed on to the site or any work area.
Safety signs and procedures must be observed and personal protection and safety equipment must be used at all times.

Construction work in progress.
Parents are advised to warn children of the dangers of entering this site.

Safety helmets must be worn

Unauthorised entry to this site is strictly forbidden.

Figure 4.9 Falling object and construction site entrance signs

Prohibition
A red circular band with diagonal crossbar on a white background, the symbol within the circle to be black denoting a safety sign that indicates that a certain behaviour is prohibited.

Warning
A yellow triangle with black border and symbol within the yellow area denoting a safety sign that gives warning of a hazard.

Mandatory
A blue circle with white symbol denoting a sign that indicates that a specific course of action must be taken.

Safe condition
A green oblong or square with symbol or text in white denoting a safety sign providing information about safe conditions.

Fire equipment
A red oblong or square with symbol in white denoting a safety sign that indicates the location of fire-fighting equipment.

Figure 4.7 Colour categories and shapes of signs

(ii) Wet floors

These need to be used wherever a slippery area is not cordoned off. Lightweight stands holding double-sided signs are readily available (Figure 4.10).

Figure 4.10 Wet floor signs

(iii) Chemical storage

Where hazardous cleaning chemicals are stored, apart from keeping the store locked, a suitable warning notice should be posted if it is considered this would help to reduce the risk of injury (Figure 4.11).

Figure 4.11 Examples of chemical warning signs

(iv) Fire safety signs

The Regulations apply in relation to general fire precautions. The guidance under the Fire Safety Order requires signs to comply with BS5499-4 and 5 and the Safety Signs and Signals Regulations.

Since 1998 the older, text-only 'fire exit' signs should have been supplemented or replaced with pictogram signs. Fire safety signs complying with BS5499-4 and 5 already contain a pictogram and do not require changing (Figure 4.12).

Figure 4.12 Examples of fire safety signs

(v) Fire action signs

These and other fire safety signs, such as fire extinguisher location signs, will be needed (Figure 4.13).

Figure 4.13 Examples of fire action signs

(vi) First-aid

Signs showing the location of first-aid facilities will be needed. Advice on the action to take in the case of electric shock is no longer a legal requirement but is recommended (Figure 4.14).

Figure 4.14 Examples of first-aid signs

(vii) Gas pipes and LPG cylinder stores

LPG cylinder stores should have the sign shown in Figure 4.15.

Figure 4.15 LPG sign

(viii) No smoking

Areas substantially enclosed should have the sign shown in Figure 4.16 under smoke-free legislation.

Figure 4.16 Smoke-free – no smoking sign

(ix) Fragile roofs

Signs should be erected at roof access points and at the top of outside walls where ladders may be placed (Figure 4.17).

Figure 4.17 Fragile roof signs

(x) Obstacles or dangerous locations
Signs that warn of obstacles or dangerous locations such as low head height or slippery floors should have alternating yellow and black stripes.

(xi) Other signs and posters

▶ Health and Safety Law – What you should know (there is a legal requirement to display this poster or distribute equivalent leaflet).

▶ Certificate of Employer's Liability Insurance (there is a legal requirement to display this).

▶ Scalds and burns are common in kitchens. A poster showing recommended action is advisable, for example 'First-aid for Burns'.

(xii) Sign checklist
Existing signs should be checked to ensure that:

▶ they are correct and up to date;
▶ they carry the correct warning symbol where appropriate;
▶ they are relevant to the hazard;
▶ they are easily understood;
▶ they are suitably located and not obscured;
▶ they are clean, durable and weatherproof where necessary;
▶ illuminated signs have regular lamp checks;
▶ they are used when required (e.g. 'Caution wet floor' signs);
▶ they are obeyed and effective.

Administrative controls

Reduced time exposure

This involves reducing the time during the working day that the employee is exposed to the hazard, by giving the employee either other work or rest periods.

It is normally only suitable for the control of health hazards associated with, for example, noise, vibration, excessive heat or cold, display screens and hazardous substances. However, it is important to note that for many hazards, there are short-term exposure limits as well as normal occupational exposure limits (OELs) over an 8-hour period (see Chapter 14). Short-term limits must not be exceeded during the reduced time exposure intervals.

It cannot be argued that a short time of exposure to a dangerous part of a machine is acceptable. However, it is possible to consider short bouts of intensive work with rest periods when employees are engaged in heavy labour, such as manual digging when machines are not permitted due to the confines of the space or buried services.

Isolation/segregation

Controlling risks by isolating them or segregating people and the hazard is an effective control measure and is used in many instances; for example separating vehicles and pedestrians on factory sites, providing separate walkways for the public on road repairs, providing warm rooms on sites or noise refuges in noisy processes.

The principle of isolation is usually followed with the storage of highly flammable liquids or gases which are put into open, air-ventilated compounds away from other hazards such as sources of ignition, or away from people who may be at risk from fire or explosion.

Safe systems of work

Operating procedures or safe systems of work are probably the most common form of control measure used in industry today and may be the most economical and, in some cases, the only practical way of managing a particular risk. They should allow for methodical execution of tasks. The development of safe operating procedures should address the hazards that have been identified in the risk assessment. The system of work describes the safe method of performing the job or activity. A safe system of work is a requirement of the Health and Safety at Work Act and is dealt with in detail later.

If the risks involved in the task are high or medium, the details of the system should be in writing and should be communicated to the employee formally in a training session. Details of systems for low-risk activities may be conveyed verbally. There should be records that the employee (or contractor) has been trained or instructed in the safe system of work and that they understand it and will abide by it.

Training

Training helps people acquire the skills, knowledge and attitudes to make them competent in the health and safety aspects of their work. There are generally two types of safety training:

► Specific safety training (or on the job training) which aims at tasks where training is needed due to the specific nature of such tasks. This is usually a job for supervisors, who by virtue of their authority and close daily contact, are in a position to convert safety generalities to the everyday safe practice procedures that apply to individual tasks, machines, tools and processes;
► Planned training, such as general safety training, induction training, management training, skill training or refresher courses that are planned by the organisation, and relate to managing risk through policy, legislative or organisational requirements that are common to all employees.

Before any employee can work safely, they must be shown safe procedures for completing their tasks. The purpose of safety training should be to improve the safety awareness of employees and show them how to perform their jobs while employing acceptable, safe behaviour.

See Chapter 3 for more detail on health and safety training.

Information

Organisations need to ensure that they have effective arrangements for identifying and receiving relevant health and safety information from outside the organisation including:

► ensuring that pertinent health and safety information is communicated to all people in the organisation who need it;
► ensuring that relevant information is communicated to people outside the organisation who require it;
► encouraging feedback and suggestions from employees on health and safety matters.

Anyone who is affected by what is happening in the workplace will need to be given safety information. This does not only apply to staff. It can also apply to visitors, members of the public and contractors.

Information to be provided for people in a workplace includes:

► who is at risk and why;
► how to carry out specific tasks safely;
► correct operation of equipment;
► emergency action;
► accident and hazard reporting procedures;
► the safety responsibilities of individual people.

Information can be provided in a variety of ways. These include safety signs, posters, newsletters, memos, emails, personal briefings, meetings, toolbox talks, formal training, written safe systems of work and written health and safety arrangements.

Welfare

Welfare facilities include general workplace ventilation, lighting and heating and the provision of drinking water, sanitation and washing facilities. There is also a requirement to provide eating and rest rooms. Risk control may be enhanced by the provision of eye washing and shower facilities for use after certain accidents (Figure 4.18).

Figure 4.18 Welfare washing facilities: washbasin should be large enough for people to wash their forearms

Good housekeeping is a very cheap and effective means of controlling risks. It involves keeping the workplace clean and tidy at all times and maintaining good storage systems for hazardous substances and other potentially dangerous items. The risks most likely to be influenced by good housekeeping are fire and slips, trips and falls.

See Chapter 8 for more information on the work environment.

Monitoring and supervision

All risk control measures, whether they rely on engineered or human behavioural controls, must be monitored for their effectiveness, with supervision to ensure that they have been applied correctly. Competent people who have a sound knowledge of the equipment or process should undertake this monitoring. Checklists are useful to ensure that no significant factor is forgotten. Any statutory inspection or insurance company reports should be checked to see whether any areas of concern were highlighted and if any recommendations were implemented. Details of any accidents, illnesses or other incidents will give an indication on the effectiveness of the risk control measures. Any emergency arrangements should be tested during the monitoring phase including first-aid provision.

It is crucial that the operator should be monitored to ascertain that all relevant procedures have been understood and followed. The operator may also be able to suggest improvements to the equipment or system of work. The supervisor is an important source of information during the monitoring process.

Where the organisation is involved with shift work, it is essential that the risk controls are monitored on all shifts to ensure the uniformity of application.

The effectiveness and relevance of any training or instruction given should be monitored.

Periodically the risk control measures should be reviewed. Monitoring and other reports are crucial for the review to be useful. Reviews often take place at safety committee and/or at management meetings. A serious accident or incident should lead to an immediate review of the risk control measures in place.

(e) Personal protective equipment

Personal protective equipment (PPE) should only be used as a last resort. There are many reasons for this. The most important limitations are that PPE:

- only protects the person wearing the equipment, not others nearby;
- relies on people wearing the equipment at all times;
- must be used properly;
- must be replaced when it no longer offers the correct level of protection. This last point is particularly relevant when respiratory protection is used.

The benefits of PPE are:

- it gives immediate protection to allow a job to continue while engineering controls are put in place;
- in an emergency it can be the only practicable way of effecting rescue or shutting down plant in hazardous atmospheres;
- it can be used to carry out work in confined spaces where alternatives are impracticable. But it should never be used to allow people to work in dangerous atmospheres, which are, for example, enriched with oxygen or potentially explosive.

See Chapter 14 for more details on PPE.

2. Other hierarchies of risk control

There are several other similar hierarchies of risk control, which have been used over many years. A typical example is as follows:

- elimination;
- substitution;
- changing work methods/patterns;
- reduced or limited time exposure;
- engineering controls (e.g. isolation, insulation and ventilation);
- good housekeeping;
- safe systems of work;
- training and information;
- personal protective equipment;
- welfare;
- monitoring and supervision; and
- review.

Figure 4.19 Good dust control for a chasing operation. A dust mask is still required for complete protection

Figure 4.20 Respiratory protection and disposable overalls are needed when working in high levels of asbestos dust

Prioritisation of risk control

The prioritisation of the implementation of risk control measures will depend on the risk rating (high, medium and low) but the timescale in which the measures are introduced will not always follow the ratings. It may be convenient to deal with a low-level risk at the same time as a high-level risk or before a medium-level risk. It may also be that work on a high-risk control system is delayed due to a late delivery of an essential component – this should not halt the overall risk reduction work. It is important to maintain a continuous programme of risk improvement rather than slavishly following a predetermined priority list.

Step 4 – Recording significant findings

It is very useful to keep a written record of the risk assessment even if there are less than five employees in the organisation. For an assessment to be 'suitable and sufficient', only the significant hazards and conclusions need be recorded. The record should also include details of the groups of people affected by the hazards and the existing control measures and their effectiveness. The conclusions should identify any new controls required and a review date. The HSE booklet *Five Steps to Risk Assessment* provides a very useful guide and examples of the detail required for most risk assessments.

There are many possible layouts which can be used for the risk assessment record. Examples are given in Appendix 4.2 and Chapter 23 C2 and C3. It should be noted that in Appendix 4.2, the initial qualitative risk level at the time of the risk assessment is given – the residual risk level when all the additional controls have been implemented will be 'low'. This should mean that an annual review will be sufficient.

The written record provides excellent evidence to a health and safety inspector of compliance with the law. It is also useful evidence if the organisation should become involved in a civil action.

The record should be accessible to employees and a copy kept with the safety manual containing the safety policy and arrangements.

Step 5 – Monitoring and review

As mentioned earlier, the risk controls should be reviewed periodically. This is equally true for the risk assessment as a whole. Review and revision may be necessary when conditions change as a result of the introduction of new machinery, processes or hazards. There may be new information on hazardous substances or new legislation. There could also be changes in the workforce, for example the introduction of trainees. The risk assessment needs to be revised only if significant changes have taken place since the last assessment was done. An accident or incident or a series of minor ones provides a good reason for a review of the risk assessment. This is known as the post-accident risk assessment.

Examples of completed risk assessments are given in Appendix 4.2 and Chapter 23 C2 and C3. They show how small and medium-sized businesses have approached risk assessment.

4.1.10 Cost-benefit analysis

In recent years, risk assessment has been accompanied by a cost-benefit analysis that attempts to evaluate the costs and benefits of risk control and reduction. The costs could include capital investment, maintenance and training and produce benefits such as reduced insurance premiums, higher productivity and better product quality. The pay-back period for most risk reduction projects (other than the most simple) has been shown to be between two and five years. Although the benefits are often difficult to quantify, cost-benefit analysis does help to justify the level of expenditure on a risk reduction project.

4.1.11 Special cases

There are several groups of persons who require an additional risk assessment due to their being more 'at risk' than other groups. Four such groups will be considered – young persons, expectant and nursing mothers, workers with a disability and lone workers.

Young persons

There are about 20 fatalities of young people at work each year. Therefore, any risk assessment involving young people needs to consider the particular vulnerability of young persons in the workplace. Young workers clearly have a lack of experience and awareness of risks in the workplace, a tendency to be subject to peer pressure and a willingness to work hard. Many young workers will be trainees or on unpaid work experience. Young people are not fully developed, have poor perception of risk and are more vulnerable to physical, biological and chemical hazards than adults.

An amendment to the HSW Act enables trainees on government-sponsored training schemes to be treated as employees as far as health and safety is concerned. The Management of Health and Safety at Work Regulations defines a young person as anybody under the age of 18 years and stipulates that a special risk assessment must be completed which takes into account their immaturity and inexperience. The assessment must be completed before the young person starts work. If the young person is of school age (16 years or less), the parents or guardian of the child should be notified of the outcome of the risk assessment and details of any safeguards which will be used to protect the health and safety of the child.

The following key elements should be covered by the risk assessment:

▶ details of the work activity, including any equipment or hazardous substances;
▶ details of any prohibited equipment or processes;
▶ details of health and safety training to be provided;
▶ details of supervision arrangements.

The extent of the risks identified in the risk assessment will determine whether employers should restrict the work of the people they employ. Except in special circumstances, young people should not be employed to do work which:

▶ is beyond their physical or psychological capacity;
▶ exposes them to substances chronically harmful to human health, for example toxic or carcinogenic substances, or effects likely to be passed on genetically or likely to harm the unborn child;
▶ exposes them to radiation;
▶ involves a risk of accidents which they are unlikely to recognise because of, for example, their lack of experience, training or insufficient attention to safety;
▶ involves a risk to their health from extreme heat, noise or vibration.

These restrictions will not apply in **special circumstances** where young people over the minimum school leaving age are doing work necessary for their training, under proper supervision by a competent person, and providing the risks are reduced to the lowest level, so far as is reasonably practicable. Under no circumstances can children of compulsory school age do work involving these risks, whether they are employed or under training such as work experience.

Induction training is important for young workers and such training should include site rules, restricted areas, prohibited machines and processes, fire precautions, emergency procedures, welfare arrangements and details of any further training related to their particular job. At induction, they should be introduced to their mentor and given close supervision, particularly during the first few weeks of their employment.

If work-experience trainees are not at school on any days during the placement, then they should not work more than 8 hours a day or 40 hours a week (Working Time Directive) and no night work or overtime. They should also have a 20-minute rest within a 4.5-hour working period.

A guide is available to employers who organise site visits for young people. The guide 'Guidance for employers: Getting involved in work-experience and work-related learning' is available from LANTRA (www.lantra.co.uk).

Expectant and nursing mothers

The Management of Health and Safety at Work Regulations 1999 incorporates the Pregnant Workers Directive from the EU. Pregnant women and nursing mothers are more at risk from certain types of hazard. In most of these instances the hazard presents a risk not only to the woman but also to the baby. Many of these hazards can cause miscarriage, birth defects or ill-health in the baby. If any type of work could present a particular risk to expectant or nursing mothers, the risk assessment must include an assessment of such risks. Should these risks be unavoidable, then the woman's working conditions or hours must be altered to avoid the risks. The alternatives for her are to be offered other work or be suspended from work on full pay. The woman must notify the employer in writing that she is pregnant, or has given birth within the previous six months and/or is breastfeeding.

Pregnant workers should not be exposed to chemicals, such as pesticides and lead, or to biological hazards, such as hepatitis. Female agricultural workers, veterinaries or farmers' wives who are pregnant should not assist with lambing, so that any possible contact with ovine chlamydia is avoided.

Other work activities that may present a particular risk to pregnant women at work are radiography, involving possible exposure to ionising radiation, and shop work when long periods of standing are required during shelf filling or stock-taking operations.

Other hazards which might affect such women are:

▶ manual handling especially later in pregnancy;
▶ chemical or biological agents (e.g. lead and the rubella virus);
▶ passive smoking;
▶ lack of rest room facilities;
▶ temperature variations;
▶ ergonomic issues related to prolonged standing, sitting or the need for awkward body movement;
▶ whole-body vibration;
▶ issues associated with the use and wearing of personal protective equipment;
▶ working excessive hours;
▶ night working; and
▶ stress and violence to staff.

Detailed guidance is available on the HSE website at www.hse.gov.uk/mothers.

Workers with a disability

Organisations have been encouraged for many years to employ workers with disabilities and to ensure that their premises provide suitable access for such people. From a health and safety point of view, it is important that workers with a disability are covered by special risk assessments so that appropriate controls are in place to protect them. Workers with disabilities may be at greater risk from particular hazards depending on the nature and extent of their disability. For example, employees with a hearing problem will need to be warned when the fire alarm sounds or a fork-lift truck

approaches. Special vibrating signals or flashing lights may be used. Similarly workers in wheelchairs will require a clear, wheelchair-friendly route to a fire exit and onwards to the assembly point. Safe systems of work and welfare facilities need to be suitable for any workers with disabilities.

The special risk assessment should identify:

▶ the jobs with particular health and fitness requirements;
▶ the types of disability that would make certain jobs unsuitable;
▶ the staff whose disabilities would exclude them from undertaking those jobs safely;
▶ and screen staff against these criteria – this may have the effect of excluding those with a certain disability from doing these jobs (e.g. fork-lift truck driving).

It is important to recognise that there is other employment and anti-discrimination legislation that should be considered before the findings of the risk assessment are finalised. The Equality Act places duties on service providers where physical features make access to their services impossible or unreasonable for people with a disability. The Act requires equal opportunities for employment and access to workplaces to be extended to all people with disabilities.

Lone workers

People who work alone, like those in small workshops, remote areas of a large site, social workers, sales personnel or mobile maintenance staff, should not be at more risk than other employees (Figure 4.21). Lone workers are a group of workers who are especially vulnerable in certain situations. If, for example, the worker is in contact with members of the public, they may be at risk of assault from people who could be violent. People who work alone in confined spaces could also be at risk if they have an accident or become ill. There are, however, no absolute restrictions on working alone; it will depend on the findings of a risk assessment.

When risk assessment shows that it is not possible for the work to be done safely by a lone worker, arrangements for providing help or back-up should be put in place. Where a lone worker is working at the premises of another employer, that employer should inform the employer of the lone worker of any risks and the precautions that should be taken. A risk assessment is, therefore, essential for all instances of lone working.

Initially, it is important to consider whether the risks of the job can be properly controlled by one person. Other considerations in the risk assessment include:

▶ does the particular workplace present a special risk to someone working alone?
▶ is there safe entrance and exit from the workplace?

Figure 4.21 A lone worker – special arrangements required. Sand or shot blasting inside a tank with an air-fed helmet and vest

▶ can all the equipment and substances be safely handled by one person?
▶ is violence from others a risk?
▶ would women and young persons be specially at risk?
▶ is the worker medically fit and suitable for working alone?
▶ are special training and supervision required?
▶ has the worker access to first-aid?

The risk assessment should help decide the right level of supervision.

The controls resulting from the risk assessment for lone workers and the safe systems of work are covered in 4.4.9.

4.2 General principles of prevention in relation to risk reduction measures

4.2.1 Introduction

The control of risks is essential to secure and maintain a healthy and safe construction site or workplace which complies with the relevant legal requirements. Hazard identification and risk assessment are covered earlier in this chapter and these together with appropriate risk control measures form the core of the HSG65 'PLAN' section of the management model.

In industry today, including the construction sector, safety is controlled through a combination of **engineered measures** such as the provision of safety protection (e.g. guarding and warning systems), and **operational measures** in training, safe work practices, operating procedures and method statements, along with **management supervision**.

These measures (collectively) are commonly known in health and safety terms as **control measures**. Some of these more common measures will be explained in more detail later.

This section concerns the principles that should be adopted when deciding on suitable measures to eliminate or control both acute and chronic risks to the health and safety of people on construction operations. The principles of control can be applied to both health risks and safety risks, although health risks have some distinctive features that require a special approach.

Chapters 8–18 deal with specific workplace hazards and controls, subject by subject. The principles of prevention now enshrined in the Management of Health and Safety at Work (MHSW) Regulations need to be used jointly with the hierarchy of control methods which give the preferred order of approach to risk control.

When risks have been analysed and assessed, decisions can be made about workplace precautions.

All final decisions about risk control methods must take into account the relevant legal requirements, which establish minimum levels of risk prevention or control. Some of the duties imposed by the HSW Act and the relevant statutory provisions are **absolute** and must be complied with. Many requirements are, however, qualified by the words **so far as is reasonably practicable**, or **so far as is practicable**. These require an assessment of cost, along with information about relative costs, effectiveness and reliability of different control measures. Further guidance on the meaning of these three expressions is provided in Chapter 1.

Figure 4.22 When controls break down

4.2.2 General principles of prevention

The MHSW Regulations Schedule 1 specifies the general principles of prevention which are set out in Article 6(2) of the European Council Directive 89/391/EEC. For the first time the principles have been enshrined directly in Regulations which state, at Regulation 4, that '*Where an employer implements any preventative measures he shall do so on the basis of the principles specified in Schedule 1*'. These principles are:

1. **Avoiding risks**

   This means, for example, trying to stop doing the task or using different processes or doing the work in a different, safer way.

2. **Evaluating the risks which cannot be avoided**

   This requires a risk assessment to be carried out.

3. **Combating the risks at source**

   This means that risks, such as a dusty work atmosphere, are controlled by removing the cause of the dust rather than providing special protection; or that slippery floors are treated or replaced rather than putting up a sign.

4. **Adapting the work to the individual**

   This involves the design of the workplace, the choice of work equipment and the choice of working and production methods, with a view, in particular, to alleviating monotonous work and work at a predetermined work rate and to reducing their effect on health.

   This will involve consulting those who will be affected when workplaces, methods of work and safety procedures are designed. The control individuals have over their work should be increased, and time spent working at predetermined speeds and in monotonous work should be reduced where it is reasonable to do so.

5. **Adapting to technical progress**

   It is important to take advantage of technological and technical progress, which often gives designers and employers the chance to improve both safety and working methods. With the Internet and other international information sources available, very wide knowledge, going beyond what is happening in the UK or Europe, will be expected by the enforcing authorities and the courts.

6. **Replacing the dangerous by the non-dangerous or the less dangerous**

   This involves substituting, for example, equipment or substances with non-hazardous or less hazardous substances.

7. **Developing a coherent overall prevention policy**

   This covers technology, organisation of work, working conditions, social relationships and the influence of factors relating to the working environment. Health and safety policies should be prepared and applied by reference to these principles.

8. **Giving collective protective measures priority over individual protective measures**

   This means giving priority to control measures which make the workplace safe for everyone working there so giving the greatest benefit, for example removing hazardous dust by exhaust ventilation rather than providing a filtering respirator to an individual worker. This is sometimes known as a 'Safe Place' approach to controlling risks.

9. **Giving appropriate instruction to employees**

   This involves making sure that employees are fully aware of company policy, safety procedures, good practice, official guidance, any test results and legal requirements. This is sometimes known as a 'Safe Person' approach to controlling risks where the focus is on individuals. A properly set-up health and safety management system should cover and balance both a Safe Place and Safe Person approach.

4.3 Sources of health and safety information

When anybody, whether a health and safety professional, a manager or an employee, is confronted with a health and safety problem, they will need to consult various items of published information to ascertain the scale of the problem and its possible remedies. The sources of this information may be internal to the organisation and/or external to it.

4.3.1 Internal sources which should be available within the organisation include

▶ accident and ill-health records and investigation reports;
▶ absentee records;
▶ inspection and audit reports undertaken by the organisation and by external organisations such as the HSE;
▶ maintenance, risk assessment (including COSHH) and training records;
▶ documents which provide information to workers;
▶ any equipment examination or test reports.

4.3.2 External sources, which are available outside the organisation, are numerous and include

▶ health and safety legislation;

- HSE publications, such as Approved Codes of Practice, guidance documents, leaflets, journals, books and their website;
- International (e.g. International Labour Organisation, ILO), European and British Standards;
- health and safety magazines and journals;
- information published by trade associations, employer organisations and trade unions;
- specialist technical and legal publications;
- information and data from manufacturers and suppliers;
- the internet and encyclopedias.

Figure 4.23 Checking the label for health risks

See Chapter 19 for more information and website addresses. Many of these sources of information will be referred to throughout this book.

4.4 Factors that should be considered when developing and implementing a safe system of work for general work activities

4.4.1 What is a safe system of work?

A safe system of work has been defined as:

> 'The integration of personnel, articles and substances in a laid out and considered method of working which takes proper account of the risks to employees and others who may be affected, such as visitors and contractors, and provides a formal framework to ensure that all of the steps necessary for safe working have been anticipated and implemented.'

In simple terms, a safe system of work is a defined method for doing a job in a safe way. It takes account of all foreseeable hazards to health and safety and seeks to eliminate or minimise these. Safe systems of work are normally formal and documented, for example in written operating procedures but, in some cases, they may be verbal.

The particular importance of safe systems of work stems from the recognition that most accidents are caused by a combination of factors (plant, substances, lack of training and/or supervision, etc.). Hence prevention must be based on an integral approach and not one which only deals with each factor in isolation. The adoption of a safe system of work provides this integral approach because an effective safe system:

- is based on looking at the job as a whole;
- starts from an analysis of all foreseeable hazards, for example physical, chemical, health;
- brings together all the necessary precautions, including design, physical precautions, training, monitoring, procedures and PPE.

It follows from this that the use of safe systems of work is in no way a replacement for other precautions, such as good equipment design, safe construction and the use of physical safeguards. However, there are many situations where these will not give adequate protection in themselves, and then a carefully thought-out and properly implemented safe system of work is especially important. The best example is maintenance and repair work, which will often involve, as a first-stage, dismantling the guard or breaking through the containment, which exists for the protection of the ordinary process operator. In some of these operations, a permit-to-work procedure will be the most appropriate type of safe system of work.

The operations covered may be simple or complex, routine or unusual.

Whether the system is verbal or written, and whether the operation it covers is simple or complex, routine or unusual, the essential features are forethought and planning – to ensure that all foreseeable hazards are identified and controlled. In particular, this will involve scrutiny of:

- the sequence of operations to be carried out;
- the equipment, plant, machinery and tools involved;
- chemicals and other substances to which people might be exposed in the course of the work;
- the people doing the work – their skill and experience;
- foreseeable hazards (health, safety, environment), whether to the people doing the work or to others who might be affected by it;
- practical precautions which, when adopted, will eliminate or minimise these hazards (Figure 4.24);
- the training needs of those who will manage and operate under the procedure;
- monitoring systems to ensure that the defined precautions are implemented effectively.

Figure 4.24 Multi-padlocked hasp for locking off an isolation valve – each worker puts on their own padlock

4.4.2 Legal requirements

The HSW Act Section 2 requires employers to provide safe plant and systems of work. In addition, many Regulations made under the Act, such as the Provision and Use of Work Equipment Regulations 1998, require information and instruction to be provided to employees and others. In effect, this is also a more specific requirement to provide safe systems of work. Many of these safe systems, information and instructions will need to be in writing. There is also a need for employers to provide a safe system of work to fulfil their common law duty of care.

4.4.3 Assessment of what safe systems of work are required

Requirement

It is the responsibility of the management in each organisation to ensure that its operations are assessed to determine where safe systems of work need to be developed.

This assessment must, at the same time, decide the most appropriate form for the safe system; that is:

▶ Is a written procedure required?
▶ Should the operation only be carried out under permit to work?
▶ Is an informal system sufficient?

Factors to be considered

It is recognised that each organisation must have the freedom to devise systems that match the risk potential of their operations and which are practicable in their situation. However, they should take account of the following factors in making their decision:

▶ types of risk involved in the operation;
▶ magnitude of the risk, including consideration of the worst foreseeable loss;
▶ complexity of the operation;

▶ past accident and loss experience;
▶ requirements and recommendations of the relevant health and safety authorities;
▶ the type of documentation needed;
▶ resources required to implement the safe system of work (including training and monitoring).

4.4.4 Development of safe systems

Role of competent person

The competent person appointed under the MHSW Regulations and/or appointed under the CDM Regulations should assist managers to draw up guidelines for safe systems of work. This will include, where necessary, particularly in construction work, method statements. The competent person should prepare suitable forms and should advise management on the adequacy of the safe systems produced.

Role of managers

Primarily management is responsible for providing safe systems of work, as they will know the detailed way in which the task should be carried out.

Management is responsible to ensure that employees are adequately trained in a specific safe system of work and are competent to carry out the work safely. Managers need to provide sufficient supervision to ensure that the system of work is followed and the work is carried out safely. The level of supervision will depend on the experience of the particular employees concerned and the complexity and risks of the task.

When construction work is involved, principal contractors will need to monitor sub-contractors to check that they are providing suitable safe systems of work, have trained their employees and are carrying out the tasks in accordance with the safe systems.

Role of employees/consultation

Many people operating a piece of machinery or a manufacturing process are in the best position to help with the preparation of safe systems of work. Consultation with those employees who will be exposed to the risks, either directly or through their representatives, is also a legal requirement. The importance of discussing the proposed system with those who will have to work under it, and those who will have to supervise its operation, cannot be emphasised enough.

Employees have a responsibility to follow the safe system of work.

Analysis

The safe system of work should be based on a thorough analysis of the job or operation to be covered by the system. The way this analysis is done will depend on the nature of the job/operation.

If the operation being considered is a new one involving high loss potential, the use of formal hazard analysis techniques such as hazard and operability (HAZOP) study, fault tree analysis (FTA) or failure modes and effects analysis should be considered.

However, where the potential for loss is lower, a more simple approach, such as job safety analysis (JSA), will be sufficient. This will involve three key stages:

▶ identification of the key steps in the job/operation – What activities will the work involve?

▶ analysis and assessment of the risks associated with each stage – What could go wrong?

▶ definition of the precautions or controls to be taken – What steps need to be taken to ensure the operation proceeds without danger, either to the people doing the work, or to anyone else?

The results of this analysis are then used to draw up the safe operating procedure or method statement. (See Chapter 23 form M7 for a suitable form.)

Introducing controls

There are a variety of controls that can be adopted in safe systems of work. They can be split into the following three basic categories:

(i) Technical – these are engineering or process type controls which engineer out or contain the hazard so that the risks are acceptable. For example exhaust ventilation, a machine guard, dust respirator.

(ii) Procedural – these are ways of doing things to ensure that the work is done according to the procedure, legislation or cultural requirements of the organisation. For example a supervisor must be involved, the induction course must be taken before the work commences, a particular type of form or a person's signature must be obtained before proceeding, the names of the workforce must be recorded.

(iii) Behavioural – these are controls which require a certain standard of behaviour from individuals or groups of individuals. For example no smoking is permitted during the task, hard hats must be worn, all lifts are to be in tandem between two workers.

4.4.5 Preparation of safe systems

A checklist for use in the preparation of safe systems of work is set out as follows:

▶ What is the work to be done?
▶ What are the potential hazards?
▶ Is the work covered by any existing instructions or procedures? If so, to what extent (if any) do these need to be modified?
▶ Who is to do the work?

▶ What are their skills and abilities – is any special training needed?
▶ Under whose control and supervision will the work be done?
▶ Will any special tools, protective clothing or equipment (e.g. breathing apparatus) be needed? Are they ready and available for use?
▶ Are the people who are to do the work adequately trained to use the above?
▶ What isolations and locking-off will be needed for the work to be done safely?
▶ Is a permit to work required for any aspect of the work?
▶ Will the work interfere with other activities? Will other activities create a hazard to the people doing the work?
▶ Have other departments been informed about the work to be done, where appropriate?
▶ How will the people doing the work communicate with each other?
▶ Have possible emergencies and the action to be taken been considered?
▶ Should the emergency services be notified?
▶ What are the arrangements for handover of the plant/equipment at the end of the work? (For maintenance/project work, etc.)
▶ Do the planned precautions take account of all foreseeable hazards?
▶ Who needs to be informed about or receive copies of the safe system of work?
▶ What arrangements will there be to see that the agreed system is followed and that it works in practice?
▶ What mechanism is there to ensure that the safe system of work stays relevant and up to date?

4.4.6 Documentation

Safe systems of work should be properly documented. Wherever possible, they should be incorporated into normal process operating procedures. This is so that:

▶ health and safety are seen as an integral part of, and not an add-on to, normal production procedures;
▶ the need for operators and supervisors to refer to separate manuals is minimised.

Whatever method is used, all written systems of work should be signed by the relevant managers to indicate approval or authorisation. Version numbers should be included so that it can quickly be verified that the most up-to-date version is in use. Records should be kept of copies of the documentation, so that all sets are amended when updates and other revisions are issued.

As far as possible, systems should be written in a non-technical style and should specifically be designed to be as intelligible and user-friendly as possible. It may be necessary to produce simple summary sheets which contain all the key points in an easy-to-read format.

4.4.7 Communication and training

People doing work or supervising work must be made fully aware of the laid-down safe systems that apply. The preparation of safe systems will often identify a training need that must be met before the system can be implemented effectively.

In addition, people should receive training in how the system is to operate. This applies not only to those directly involved in doing the work but also to supervisors/managers who are to oversee it.

In particular, the training might include:

- why a safe system is needed;
- what is involved in the work;
- the hazards which have been identified;
- the precautions which have been decided and, in particular:
  - the isolations and locking-off required, and how this is to be done;
  - details of the permit-to-work system, if applicable;
  - any monitoring (e.g. air testing) which is to be done during the work, or before it starts;
  - how to use any necessary PPE;
  - emergency procedures.

4.4.8 Monitoring safe systems

Safe systems of work should be monitored to ensure that they are effective in practice. This will involve:

- reviewing and revising the systems themselves, to ensure they stay up to date;
- inspecting to identify how fully they are being implemented.

In practice, these two things go together, as it is likely that a system that is out of date will not be fully implemented by the people who are intended to operate it.

All organisations are responsible for ensuring that their safe systems of work are reviewed and revised as appropriate. Monitoring of implementation is part of all line managers' normal operating responsibilities, and should also take place during health and safety audits.

4.4.9 Definition of and specific examples of safe systems of work

Confined spaces

A confined space is defined in **Confined Spaces Regulations** as: '*any place, including any chamber, tank, vat, silo, pit, trench, pipe, sewer, flue, well or other similar space in which, by virtue of its enclosed nature, there arises a reasonably foreseeable specified risk*'.

These specified risks mean a risk to a worker of:

- serious injury arising from a fire or explosion;

- loss of consciousness arising from an increase in body temperature;
- loss of consciousness or asphyxiation arising from gas, fume, vapour or the lack of oxygen;
- drowning arising from an increase in the level of liquid;
- asphyxiation arising from a free-flowing solid or because of entrapment by it.

Therefore, confined spaces include chambers, tanks (sealed and open-top), vessels, furnaces, ducts, sewers, manholes, pits, flues, excavations, boilers, reactors and ovens.

The principal hazards associated with a confined space are the difficult access and egress, which can make escape and rescue more difficult. Other hazards associated with confined spaces include:

- asphyxiation due to oxygen depletion;
- poisoning by toxic substance or fumes;
- explosions due to gases, vapours and dust;
- fire due to flammable liquids;
- fall of materials leading to possible head injuries;
- free-flowing solid such as grain in a silo;
- electrocution from unsuitable equipment;
- difficulties of rescuing injured personnel;
- drowning due to flooding; and
- fumes from plant or processes entering confined spaces.

Since work in confined spaces is a high risk work activity, a risk assessment is essential. The following items should be included in the assessment:

- the task;
- the working environment;
- working materials and tools;
- the suitability of those carrying out the task; and
- arrangements for emergency rescue.

If the assessment identifies risks of serious injury from work in the confined space, then the following key duties specified in the Confined Spaces Regulations apply. These duties are:

- if possible, avoid entry to a confined space by doing the work outside the space (e.g. remote cameras can be used for internal inspection of vessels);
- if entry to a confined space is unavoidable, follow a safe system of work; and
- ensure that adequate emergency arrangements are in place before the work starts.

If entry into a confined space cannot be avoided, then a safe system of work for working inside the space must be developed that uses the results of the risk assessment to identify the controls needed to reduce the risk of injury. These controls will depend on the nature of the confined space, the associated risk and the work involved. It is important that the agreed safe system of work is fully implemented and everyone who is to work in the confined space must be trained and instructed on the controls required.

The following topics need to be addressed in a safe system of work for a confined space:

▶ the appointment of a supervisor;

▶ the competence and experience required of the workers in the confined space;

▶ the isolation of mechanical and electrical equipment in an emergency;

▶ the size of the entrance must enable rapid access and exit in an emergency;

▶ the provision of adequate ventilation;

▶ the testing of the air inside the space to ensure that it is fit to breathe. If the air inside the space is not fit to breathe, then breathing apparatus will be essential;

▶ the provision of special tools and lighting, such as extra low voltage equipment, non-sparking tools and specially protected lighting;

▶ the emergency arrangements to cover the necessary equipment, training, practice drills and the raising of the alarm; and

▶ adequate communications arrangements to enable communication between people inside and outside the confined space and to summon help in an emergency.

The provision of suitable rescue and resuscitation equipment will depend on the likely emergencies identified. Where such equipment is provided for rescuers to use, training in the correct operation of the equipment is essential. Rescuers should also be trained in all aspects of the emergency procedures and relevant first-aid procedures.

For particularly hazardous confined space working a permit to work may be required; this is discussed later in 4.5.4.

Lone working

Section 4.1.11 covered the need and contents of a risk assessment for lone workers. In this section the controls required to protect lone workers is discussed.

People who work by themselves without close or direct supervision are found in many work situations. In some cases they are the sole occupant of small workshops or warehouses; they may work in remote sections of a large site; they may work out of normal hours, like cleaners or security personnel; they may be working away from their main base as installers, or maintenance people; they could be people giving a service, like domiciliary care workers, drivers and estate agents.

There is no general legal reason why people should not work alone, but there may be special risks which require two or more people to be present; for example during entry into a confined space in order to effect a rescue. It is important to ensure that a lone worker is not put at any higher risk than other workers. This is achieved by carrying out a specific risk assessment and introducing special protection arrangements for their safety.

The overall health and suitability of a person to work alone should be taken into account. It is important to ascertain whether the work should be performed alone particularly where there is a possibility of a serious risk, such as violence, being confronted by the worker.

Typical control procedures may include:

▶ documented records of the location or itineraries of the lone workers;

▶ periodic visits from the supervisor to observe what is happening;

▶ regular voice contact, using mobile phones or radios, between the lone worker and the supervisor;

▶ automatic warning devices to alert others if a specific signal is not received from the lone worker;

▶ other devices to raise the alarm, which are activated by the absence of some specific action;

▶ checks that the lone worker has returned safely home or to their base;

▶ special arrangements for first-aid to deal with minor injuries – this may include mobile first-aid kits;

▶ arrangements for emergencies – these should be established and employees trained.

One of the largest increases in lone working has been in work-related driving. Just under half of such workers face road-rage incidents at least once a year and 11% are assaulted.

All lone workers at risk of violence should receive training to help them recognise and anticipate violence and difficult situations. Such training should develop communication skills to help the workers remain diplomatic and non-confrontational when facing potentially violent circumstances.

The Institution of Occupational Safety and Health (IOSH) have produced a very useful document on lone working that contains an audit checklist for remote working.

4.5 Role and function of a permit-to-work system

4.5.1 Introduction

Safe systems of work are crucial in work such as the maintenance of chemical plant where the potential risks are high and the careful coordination of activities and precautions is essential to safe working. In this situation and others of similar risk potential, the safe system of work is likely to take the form of a permit-to-work procedure.

The role of the permit-to-work procedure is to provide a specialised type of safe system of work for ensuring that potentially very dangerous work (e.g. entry into process plant and other confined spaces) is done safely.

Although this procedure has been developed and refined by the chemical industry, the principles of the permit-to-work procedure are equally applicable to the management of complex risks in other industries.

The function of the permit-to-work procedure is to ensure that certain defined operations are prohibited without the specific permission of a responsible manager, this permission being only granted once stringent checks have been made to ensure that all necessary precautions have been taken and that it is safe for work to go ahead.

The people doing the work take on responsibility for following and maintaining the safeguards set out in the permit, which will define the work to be done (no other work being permitted) and the timescale in which it must be carried out.

To be effective, the permit system requires the training needs of those involved to be identified and met, and the monitoring procedures must ensure that the system is operating as intended.

4.5.2 The principles that apply to permits to work

Permit systems must adhere to the following eight principles:

1. Wherever possible, and especially with routine jobs, hazards should be eliminated so that the work can be done safely without requiring a permit to work.
2. Although the Site Manager may delegate the responsibility for the operation of the permit system, the overall responsibility for ensuring safe operation rests with him/her.
3. The permit must be recognised as the master instruction which, until it is cancelled, overrides all other instructions.
4. The permit applies to everyone on site, including contractors.
5. Information given in a permit must be detailed and accurate. It must state:
   (a) which plant/equipment has been made safe and the steps by which this has been achieved;
   (b) work may be done;
   (c) the time at which the permit comes into effect.
6. The permit remains in force until the work has been completed and the permit is cancelled by the person who issued it or by the person nominated by management to take over the responsibility (e.g. at the end of a shift or during absence).
7. No work other than that specified is authorised. If it is found that the planned work has to be changed, the existing permit should be cancelled and a new one issued.
8. Responsibility for the plant must be clearly defined at all stages.

4.5.3 Permit-to-work procedures

The permit-to-work procedure is a specialised type of safe system of work, under which certain categories of high risk-potential work may only be done with the specific permission of an authorised manager. This permission (in the form of the permit to work) will be given only if the laid down precautions are in force and have been checked.

The permit document should specify the following key items of information:

- the date, time and duration of the permit;
- a description and assessment of the task to be performed and its location;
- the plant/equipment involved, and how it is identified;
- the persons authorised to do the work;
- the steps which have already been taken to make the plant safe;
- potential hazards which remain, or which may arise as the work proceeds;
- the precautions to be taken against these hazards;
- the action to be taken prior to the task being started, such as:
  - ▷ the isolation of sources of energy and outlets;
  - ▷ emergency procedures and equipment;
  - ▷ ensuring the competency of those involved;
  - ▷ communication arrangements; and
  - ▷ reference to any other relevant documents;
- the equipment to be released to those who are to carry out the work.

In accepting the permit, the person in charge of doing the authorised work normally undertakes to take/maintain whatever precautions are outlined in the permit, such as:

- isolation of the area;
- carrying out atmospheric monitoring;
- the provision and use of personal protective equipment;
- the provision of suitable equipment including lighting and tools;
- ensuring an adequate level of supervision; and
- arrangements for any extension to or handover of the permit.

The permit will also include spaces for:

- signature certifying that the work is complete; and
- signature confirming re-acceptance of the plant/equipment.

See Chapter 22 for an example of a general permit to work.

4.5.4 Work requiring a permit

The nature of permit-to-work procedures will vary in their scope depending on the job and the risks involved. However, a permit-to-work system is unlikely to be needed where, for example:

(a) the assessed risks are low and can be controlled easily;
(b) the system of work is very simple;

(c) other work being done nearby cannot affect the work concerned in say a confined space entry, or a welding operation.

However, where there are high risks and the system of work is complex and other operations may interfere, a formal permit to work should be used.

The main types of permit and the work covered by each are identified below. Chapter 23 form M8 illustrates the essential elements of a permit-to-work form with supporting notes on its operation.

General permit

The general permit should be used for work such as:

▶ alterations to or overhaul of plant or machinery where mechanical, toxic or electrical hazards may arise. This is particularly important for:
  ▷ large machines where visual contact between workers is difficult
  ▷ where work has to be done near dangerous parts of the machine
  ▷ where there are multiple isolations for energy sources and/or dangerous substances, and
  ▷ where dangerous substances are being used in confined areas with poor ventilation;
▶ work on or near overhead crane tracks;
▶ work on pipelines with hazardous contents;
▶ repairs to railway tracks, tippers, conveyors;
▶ work with asbestos-based materials;
▶ work involving ionising radiation; and
▶ excavations to avoid underground services.

Typical work tasks that might require a permit to work

Hot work

Hot work is potentially hazardous because:
▶ It may act as a source of ignition in any plant in which flammable materials are handled;
▶ It may act as a cause of fires in all processes, regard-less of whether flammable materials are present.

Hot work includes cutting, welding, brazing, soldering and any process involving the application of a naked flame. Drilling and grinding should also be included where a flammable atmosphere is potentially present (Figure 4.25).

Hot work should therefore be done under the terms of a hot work permit, the only exception being where hot work is done in a designated maintenance area suitable for the purpose. Typical controls include:

▶ a suitable fire extinguisher nearby;
▶ prompt removal of flammable waste material; and
▶ the damping down of nearby wooden structures such as floors.

Figure 4.25 A hot work permit is usually essential for welding, cutting and burning except in designated areas like a welding shop

Work on high-voltage apparatus (including testing)

Work on high-voltage apparatus (over about 600V) is potentially high risk. Hazards include:

▶ possibly fatal electric shock/burns to the people doing the work;
▶ electrical fires/explosions;
▶ consequential danger from disruption of power supply to safety-critical plant and equipment.

In view of the risk, this work must only be done by suitably trained and competent people acting under the terms of a high-voltage permit. The most important control is to ensure that the necessary isolation is provided. Other controls for this type of work are covered in Chapter 12.

Confined spaces

The safe systems of work required for confined spaces were discussed under 4.4.9. If the work and/or confined space is very hazardous, then a permit to work will be required. This will ensure that a formal check is undertaken to make sure all the elements of a safe system of work are in place before people are allowed to enter or work in the confined space. It will also be a means of communication between site management, supervisors, and those carrying out the hazardous work. The essential features of such a permit to work are:

▶ the clear identification of who may authorise particular jobs (and any limits to their authority) and who, including any contractors, are responsible for specifying the necessary precautions (e.g. isolation, air testing and emergency arrangements);
▶ any particular training and instruction that may be required; and
▶ monitoring and auditing of the permit-to-work system.

Many fatal accidents have occurred where inadequate precautions were taken before and during work involving entry into confined spaces (Figure 4.26). Two hazards are the potential presence of toxic or other dangerous substances, and the absence of adequate oxygen. In addition, there may be mechanical hazards (entanglement on agitators), ingress of fluids, risk of engulfment in a free-flowing solid like grain or sugar, and raised temperatures. The work to be carried out may itself be especially hazardous when done in a confined space, for example cleaning using solvents, or cutting/welding work. Should the person working in a confined space get into difficulties for whatever reason, getting help in and getting the individual out may prove difficult and dangerous.

Stringent preparation, isolation, air testing and other precautions are therefore essential, and experience shows that the use of a confined space entry permit is essential to confirm that all the appropriate precautions have been taken.

The Confined Spaces Regulations are summarised in Chapter 19. They detail the specific controls that are necessary when people enter confined spaces.

Figure 4.26 Entering a confined space with breathing apparatus, rescue tripod and rescue watcher

Machinery maintenance

The Provision and Use of Work Equipment Regulations require that: all work equipment should be maintained in an efficient state and in good repair. Any maintenance operations on the equipment should be carried out safely. The frequency and nature of maintenance should be determined through risk assessment, taking full account of:

▶ the manufacturer's recommendations;
▶ the intensity and frequency of use;
▶ operating environment (e.g. the effect of temperature, corrosion, weathering);
▶ user knowledge and experience; and

▶ the risk to health and safety from any foreseeable failure or malfunction.

Maintenance work should only be undertaken by those who are competent to do the work, who have been provided with sufficient information, instruction and training. For high-risk equipment, positive means of disconnecting the equipment from the energy source may be required (e.g. isolation), along with means to prevent inadvertent reconnection (e.g. by locking off). Such procedures are best controlled using permits to work. Where possible, equipment should normally be shut down and any residual/stored energy safely released (e.g. pneumatic pressure dumped, parts with gravitational/rotational energy stopped or brought to a safe position). In some cases, it may not be possible to avoid particular significant hazards during the maintenance of work equipment so appropriate measures should be taken to protect people and minimise the risk. These may include:

▶ physical measures, such as temporary guarding, slow speed hold-to-run control devices, safe means of access and personal protective equipment;
▶ management issues, including safe systems of work, supervision and monitoring; and
▶ personnel competence (training, skill, awareness and knowledge of risk).

Work at height

A permit to work may be required for some hazardous work at height such as roof work to ensure that a fall arrest strategy is in place. This is particularly important where there are:

▶ no permanent work platforms with fixed handrails on flat roofs;
▶ on sloping or fragile roofs;
▶ where specialist access equipment, like rope hung cradles, is required; and
▶ where access is difficult.

4.5.5 Responsibilities

The effective operation of the permit system requires the involvement of many people. The following specific responsibilities can be identified:

(Note: all appointments, definitions of work requiring a permit, etc. must be in writing. All the categories of people identified below should receive training in the operation of the permit system as it affects them.)

Site Manager

▶ has overall responsibility for the operation and management of the permit system;
▶ appoints a senior manager (normally the Chief Engineer) to act as a senior authorised person.

Senior authorised person

▶ is responsible to the Site Manager for the operation of the permit system;

▶ defines the work on the site which requires a permit;

▶ ensures that people responsible for this work are aware that it must only be done under the terms of a valid permit;

▶ appoints all necessary authorised persons;

▶ appoints a deputy to act in his/her absence.

Authorised persons

▶ issue permits to competent persons and retain copies;

▶ personally inspect the site to ensure that the conditions and proposed precautions are adequate and that it is safe for the work to proceed;

▶ accompany the competent person to the site to ensure that the plant/equipment is correctly identified and that the competent person understands the permit;

▶ cancel the permit on satisfactory completion of the work.

Competent persons

▶ receive permits from authorised persons;

▶ read the permit and make sure they fully understand the work to be done and the precautions to be taken;

▶ signify their acceptance of the permit by signing both copies;

▶ comply with the permit and make sure those under their supervision similarly understand and implement the required precautions;

▶ on completion of the work, return the permit to the authorised person who issued it.

Operatives

▶ read the permit and comply with its requirements, under the supervision of the competent person.

Specialists

A number of permits require the advice/skills of specialists in order to operate effectively. Such specialists may include chemists, electrical engineers, health and safety advisers and fire officers. Their role may involve:

▶ isolations within his/her discipline – for example electrical work;

▶ using suitable techniques and equipment to monitor the working environment for toxic or flammable materials, or for lack of oxygen;

▶ giving advice to managers on safe methods of working.

Specialists must not assume responsibility for the permit system. This lies with the Site Manager and the senior authorised person.

Engineers (and others responsible for work covered by permits)

▶ ensure that permits are raised as required.

Contractors

The permit system should be applied to contractors in the same way as to direct employees.

The contractor must be given adequate information and training on the permit system, the restrictions it imposes and the precautions it requires.

4.6 Emergency procedures and arrangements for contacting the emergency services

4.6.1 Introduction

Emergency procedures are a critically important part of the health and safety organisation. They are concerned about the control procedures and equipment required to limit the damage to people and property caused by an incident. Local fire and rescue authorities will often be involved and are normally prepared to give advice to employers.

Under Regulation 8 of the Management of Health and Safety at Work Regulations, procedures must be established and set in motion when necessary to deal with serious and imminent danger to persons at work. Necessary links must be maintained with local authorities, particularly with regard to first-aid, emergency medical care and rescue work.

Although fire is the most common emergency likely to be faced, there are many other possibilities which should be considered, including:

▶ gas explosion (Figure 4.27);

▶ electrical burn or electrocution;

▶ escape of toxic gases or fumes;

Figure 4.27 Emergency services at work

▶ discovery of dangerous dusts like asbestos in the atmosphere;

▶ terrorist threat;

▶ large vehicle crashing into the premises;

▶ aircraft crash if near a flight path;

▶ spread of highly infectious disease;

▶ severe weather with high winds and flooding.

For fire emergencies, see Chapter 13.

4.6.2 Points to include in emergency procedures

Emergency procedures form an essential part of any health and safety management system. A quick and effective response to an emergency may well help to ease the situation and reduce the consequences. In any emergency, people are more likely to respond quickly and effectively if they:

▶ are well trained and competent;

▶ take part in regular and realistic practice exercises; and

▶ have clearly agreed, recorded and rehearsed plans, actions and responsibilities.

The following points should be considered when developing an emergency procedure:

▶ decide on the possible nature of any emergency that might occur and the potential consequences to the organisation;

▶ consider how the alarm will be raised. This may be particularly important when the organisation is closed at weekends or holiday times;

▶ decide on how the emergency services will be contacted – again including times when the organisation is closed (if there are 25 tonnes or more of dangerous substances on the premises, the fire and rescue service must be notified and warning signs erected);

▶ decide on an emergency assembly point for staff;

▶ ensure there are enough emergency exits for everyone to escape quickly, and keep emergency doors and escape routes unobstructed and clearly marked;

▶ nominate competent people to take control in the event of an emergency;

▶ decide which other key people are needed, such as activators of emergency equipment, a nominated incident controller, and first-aiders;

▶ plan essential actions such as emergency plant shutdown, isolation or making processes safe. Clearly identify important items such as shut-off valves and electrical isolators;

▶ ensure that everyone is included in emergency procedures including people with disabilities and vulnerable workers; and finally

▶ decide on the arrangements for liaising with the media during and after the event.

4.6.3 Supervisory duties

A member of the site staff should be nominated to supervise and coordinate all emergency arrangements. This person should be in a senior position or at least have direct access to a senior manager. Senior members of the staff should be appointed as departmental fire/emergency procedure wardens, with deputies for every occasion of absence, however brief. They should ensure that the following precautions are taken:

▶ Everyone on site can be alerted to an emergency.

▶ Everyone on site knows what signal will be given for an emergency and knows what to do.

▶ Someone who has been trained in what to do is on site and ready to coordinate activities.

▶ Emergency routes are kept clear, signed and adequately lit.

▶ There are arrangements for calling the fire and rescue services and to give them special information about high-hazard work, for example, in tunnels or confined spaces.

▶ There is adequate access to the site for the emergency services and this is always kept clear.

▶ Suitable arrangements for treating and recovering injured people are set up.

▶ Someone is posted to the site entrance to receive and direct the emergency services.

4.6.4 Assembly and roll-call

Assembly points should be established for use in the event of evacuation. They should be in positions, preferably undercover, which are unlikely to be affected at the time of emergency. In some cases, it may be necessary to make mutual arrangements with the client or occupiers of nearby premises.

In the case of small sites, a complete list of the names of all staff should be maintained so that a roll-call can be made if evacuation becomes necessary.

In those premises where the number of staff would make a single roll-call difficult, each area warden should maintain a list of the names of employees and contractors in their area. Roll-call lists must be updated regularly.

4.6.5 Contacting the emergency services

Emergency services should be called as soon as possible after the start of an emergency. The emergency services prefer to be called early as this gives them the best chance to contain the emergency and save lives. It is important that the organisation has a well-defined policy to call the emergency services in good time and a person nominated to make the call. Many fires have become established because everyone assumed that somebody else had called the fire service. A rota may need to be organised so that there is a nominated person available at all times even when the premises of the organisation are unoccupied. Some businesses use a specialist company to monitor the premises when it is unoccupied.

An emergency telephone number call may be answered by either a telephone operator or an emergency service dispatcher. The nature of the emergency (police, fire or medical) is determined first. If the call has been answered by a telephone operator, the call is transferred to the appropriate emergency service, which dispatches the appropriate help. If multiple services are required, the most urgent need must be determined and other services called in as needed.

Emergency dispatchers are trained to control the calling order to provide help in an appropriate manner. The dispatcher may find it necessary to give urgent advice in life-threatening situations. Some dispatchers have special training in telling people how to perform first-aid or **cardiopulmonary resuscitation (CPR)** in an effort to return life to a person in cardiac arrest. The emergency dispatcher will need the following information:

- the caller's telephone number;
- the exact location of the incident (e.g. the road name and any important details about approaching and accessing the site);
- the type and seriousness of the incident; and
- details of any hazards, (e.g. gas leak or fire).

The emergency service may be able to identify the telephone number of the caller. This is normally done using the system that the telephone company uses to bill calls, making the number visible even for users who have unlisted numbers or who block caller identity. For an individual fixed landline telephone, the caller's number is often associated with the caller's address and therefore their location. However, with mobile phones and business telephones, the address may be a mailing address rather than the caller's location.

4.6.6 Testing and training for emergencies

It is important that the emergency procedures are covered during the induction training. Separate training courses should also be given to staff on a regular basis on the following topics:

- use of fire-fighting equipment;
- regular refresher first-aid training;
- suitable training (and competency assessment) for all those allocated particular roles in an emergency; and
- regular timed fire drills.

Emergency lighting should be regularly tested and fire-fighting equipment examined. In high hazard plants, there should be an occasional full scale rehearsal of the emergency procedure in action.

4.7 Requirements for, and effective provision of, first-aid in the workplace

4.7.1 Introduction

People at work can suffer injuries or fall ill. It does not matter whether the injury or the illness is caused by the work they do. What is important is that they receive immediate attention and that an ambulance is called in serious cases. First-aid at Work (FAW) covers the arrangements employers must make to ensure this happens. It can save lives and prevent minor injuries becoming major ones.

The Health and Safety (First-Aid) Regulations 1981 require employers to provide adequate and appropriate equipment, facilities and personnel to enable first-aid to be given to employees if they are injured or become ill at work.

What is adequate and appropriate will depend on the circumstances in a particular workplace.

The minimum first-aid provision on any work site is:

- a suitably stocked first-aid box;
- an appointed person to take charge of first-aid arrangements.

It is also important to remember that accidents can happen at any time. First-aid provision needs to be available at all times to people who are at work.

Many small firms will only need to make the minimum first-aid provision. However, there are factors which might make greater provision necessary. The following checklist covers the points that should be considered.

4.7.2 Aspects to consider

The risk assessments carried out under the Management of Health and Safety at Work and COSHH Regulations should show whether there are any specific risks in the workplace. The following should be considered:

- Are there hazardous substances, dangerous tools and equipment, dangerous manual handling tasks, electrical shock risks, dangers from neighbours or animals?
- Are there different levels of risk in parts of the premises or site?
- What is the history of accident and ill-health, and type and location of incidents?
- What is the total number of persons likely to be on site?
- Are there young people, pregnant or nursing mothers on site and employees with disabilities or special health problems?
- Are the facilities widely dispersed with several buildings or compact in a multi-storey building?

▶ What is the pattern of working hours? Does it involve night work?

▶ Is the site remote from emergency medical services?

▶ Do employees travel or work alone for a considerable amount of time?

▶ Do any employees work at sites occupied by other employers?

▶ Are members of the public regularly on site?

4.7.3 Impact on first-aid provision if risks are significant

Qualified first-aiders may need to be appointed if risks are significant. This will involve a number of factors which must be considered, including:

▶ training for first-aiders;

▶ additional first-aid equipment and the contents of the first-aid box;

▶ siting of first-aid equipment to meet the various demands in the premises. For example provision of equipment in each building or on several floors. There needs to be first-aid provision at all times during working hours;

▶ informing local medical services of the site and its risks;

▶ any special arrangements that may be needed with the local emergency services.

Any first-aid room provided under these Regulations must be easily accessible to stretchers and to other equipment needed to convey patients to and from the room. They must be sign posted according to the Safety Signs and Signals Regulations (Figure 4.28).

If employees travel away from the site, the employer needs to consider:

▶ issuing personal first-aid kits and providing training;

▶ issuing mobile phones to employees;

▶ making arrangements with employers on other sites.

Although there are no legal responsibilities for non-employees, the HSE strongly recommends that they are included in any first-aid provision.

Figure 4.28 (a) First-aid and stretcher sign; (b) first-aid sign

4.7.4 Contents of the first-aid box

There is no standard list of items to put in a first-aid box. It depends on what the employer assesses the needs to be. Where there is no special risk in the workplace, a minimum stock of first-aid items is listed (see Table 4.1).

Tablets or medicines should not be kept in the first-aid box. Table 4.1 shows a suggested contents list only; equivalent but different items will be considered acceptable.

Table 4.1 Typical contents of first-aid box – low risk

Stock for up to 50 persons:	
A leaflet giving general guidance on first-aid, for example the HSE leaflet *Basic advice on first-aid at work*	
Medical adhesive plaster	40
Sterile eye pads	4
Individually wrapped triangular bandages	6
Safety pins	6
Individually wrapped medium sterile unmedicated wound dressings	8
Individually wrapped large sterile unmedicated wound dressings	4
Individually wrapped wipes	10
Paramedic shears	1
Pairs of latex gloves	2
Sterile eyewash if no clean running water	2

4.7.5 Appointed persons

An appointed person is someone who is appointed by management to:

▶ take charge when someone is injured or falls ill. This includes calling an ambulance if required;

▶ look after the first-aid equipment, for example keeping the first-aid box replenished;

▶ keeping records of treatment given.

Appointed persons should never attempt to give first-aid for which they are not competent. Short emergency first-aid training courses are available. An appointed person should be available at all times when people are working at the work site – this may mean appointing more than one. The training should be repeated every three years to keep up to date.

4.7.6 A first-aider

A first-aider is someone who has undergone an HSE-approved training course in administering FAW and holds a current FAW certificate. Lists of local training organisations are available from the local environmental officer or HSE offices. The training should be repeated every three years to maintain a valid certificate and keep the first-aider up to date.

It is not possible to give definitive rules on when or how many first-aiders or appointed persons might be needed. This will depend on the circumstances of each particular organisation or work site. Table 4.2 offers suggestions on how many first-aiders or appointed persons might be needed in relation to categories of risk and number of employees. The details in the table are suggestions only; they are not definitive, nor are they a legal requirement.

Employees must be informed of the first-aid arrangements. Putting up notices telling staff who and where the first-aiders or appointed persons are and where the first-aid box is will usually be enough. Special arrangements will be needed for employees with reading or language difficulties.

To ensure cover at all times when people are at work and where there are special circumstances, such as remoteness from emergency medical services, shift work or sites with several separate buildings, there may need to be more first-aid personnel than set out in Table 4.2.

4.7.7 Implementation of changes to first-aid training and approval arrangements

The HSE published a report in 2003 following a consultation of the Health and Safety (First-aid) at Work Regulations. Since these Regulations were introduced in 1981, there has been a significant transformation of the economy from manufacturing to service based industries. The report identified the following points:

▶ small organisations found it difficult to release employees to attend a four-day first-aid training course;

▶ confusion over the role and training requirement of an appointed person;

▶ a significant depletion in the retained information of

a trained first-aider after a short period of time after the training course;

Training guidelines have been issued by the HSE to address these issues. These guidelines are:

▶ a short first-aid course – Emergency First-aid at Work (EFAW). This lasts for one day and will enable the first-aider to give emergency first-aid following an accident or ill-health problem. This is suitable for an organisation where the risk assessment has indicated that it is suitable for the level of risks present;

▶ a reduction to three days for the First-aid at Work (FAW) course;

▶ all first-aiders should undertake a half-day refresher course annually during the three-year certification period of the FAW and EFAW. This should address the depletion of information problem.

The first-aid training requirements have been amended so that the four-day course has been replaced by a First-aid at Work (FAW) course taught over three days. The syllabuses of the two courses are very similar with some topics being brought up to date. For example, the protocols used for resuscitation have become much easier to understand. Many medical professionals recommend that there should be an annual refresher course to accompany the course. The one-day Emergency First-aid at Work (EFAW) course is designed for low risk organisations and appointed persons.

Employers are able to send suitable employees on either a 6-hour (minimum) emergency first-aid at work (EFAW) or an 18-hour (minimum) FAW course, based on the findings of their first-aid needs assessment (see Figure 4.29). After three years, first-aiders need to complete another course (either a 6-hour EFAW or 12-hour FAW requalification course, as appropriate) to

Table 4.2 Number of first-aid personnel

Category of risk	Numbers employed at any location	Suggested number of first-aid personnel
Lower risk		
For example shops and offices, libraries	Fewer than 50	At least one appointed person
	50–100	At least one first-aider
	More than 100	One additional first-aider for every 100 employees
Medium risk		
For example light engineering and assembly work, food processing, warehousing	Fewer than 20	At least one appointed person
	20–100	At least one first-aider for every 50 employed (or part thereof)
	More than 100	One additional first-aider for every 100 employees
Higher risk		
For example most construction, slaughterhouses, chemical manufacture, extensive work with dangerous machinery or sharp instruments	Fewer than 5	At least one appointed person
	5–50	At least one first-aider
	More than 50	One additional first-aider for every 50 employees

Source: HSE.

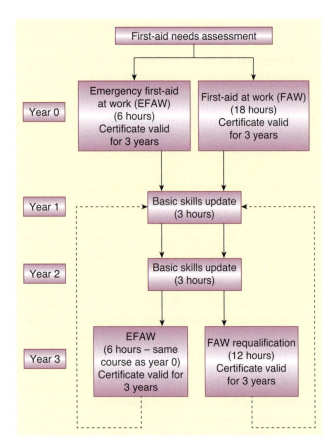

Figure 4.29 Flow chart showing courses to be completed over a 3-year certification period for EFAW and FAW. The dotted line indicates the route to be taken in subsequent years after completion of the relevant course at year 3

obtain a new certificate. Within any 3-year certification period, first-aiders should complete two annual refresher courses, covering basic life support/skills updates, which will each last for at least three hours.

Detailed guidance for employers is available as a revision to the document produced by the HSE: *First-aid at Work – The Health and Safety (First-Aid) Regulations 1981 – Approved Code of Practice and Guidance (L74)*. Within this, it is only the guidance that has been revised – the Regulations and Approved Code of Practice will remain the same. The HSE has revised its guidance for small- and medium-sized enterprises in *First-aid at Work – Your Questions Answered* (INDG214).

Detailed guidance for first-aid training organisations is available on the HSE website: First-aid training and qualifications for the purposes of the Health and Safety (First-Aid) Regulations 1981 www.hse.gov.uk/pubns/web41.pdf.

4.8 Further information

The Health and Safety at Work etc. Act 1974

The Management of Health and Safety at Work Regulations 1999 (as amended)

The Health and Safety (First-aid) Regulations 1981 as amended

Confined Spaces Regulations 1997

Emergency Procedures, HSE website guidance, http://www.hse.gov.uk/toolbox/managing/emergency.htm

Managing for health and safety, HSG65 third edition 2013, HSE Books, ISBN 978-0-7176-6456-6 http://www.hse.gov.uk/pubns/priced/hsg65.pdf

Occupational Health and Safety Assessment Series (OHSAS 18000): Occupational Health and Safety Management Systems OHSAS 18001:2007 ISBN 978-0-5805-9404-5; OHSAS18002:2008 ISBN 978-0-5806-2686-9

HSE, 'Managing for health and safety', Parts 3 and 4 http://www.hse.gov.uk/managing/

Five Steps to Risk Assessment (INDG163, rev 3), HSE Books, ISBN 978-0-7176-6440-5 http://www.hse.gov.uk/pubns/indg163.pdf

Young people and work experience, A brief guide to health and safety for employers, INDG364 (rev 1), HSE books, http://www.hse.gov.uk/pubns/indg364.pdf

New and expectant mothers http://www.hse.gov.uk/mothers/

Safe work in confined spaces; Confined Spaces Regulations 1997, ACoP regulations and guidance, (L101), second edition, HSE Books, ISBN 978-0-7176-6233-3 http://www.hse.gov.uk/pubns/priced/l101.pdf

Guidance on permit-to-work systems. A guide for the petroleum, chemical and allied industries (HSG250) HSE Books, ISBN 978-0-7176-2943-5 http://www.hse.gov.uk/pubns/priced/hsg250.pdf

Bird F E, 1974. *Management guide to Loss Control*, Institute Press, Atlanta, Georgia, USA

First-aid at Work, The Health and Safety (First-Aid) Regulations 1981 (L74, third edition 2013), HSE Books, ISBN 978-0-7176-6560-0 http://www.hse.gov.uk/pubns/priced/l74.pdf

4.9 Practice revision questions

1. General risk assessments on activities within a workplace are required under the Management of Health and Safety at Work Regulations:

   (a) **Explain** the criteria that must be met for the assessment to be deemed 'suitable and sufficient'.

   (b) **Out**line the FIVE key stages of the risk assessment process, identifying the issues that would need to be considered at **EACH** stage.

   (c) **State** the legal requirements for recording workplace risk assessments.

2. A plastics manufacturing company has moved to a new larger site and a programme of general risk assessments needs to be undertaken.

   (a) **Outline** the factors that should be considered when planning the risk assessment programme.

   (b) **Identify** the important characteristics required of the risk assessment team to ensure that it has the necessary competence.

   (c) **Outline** the content of a training course for the risk assessment team.

3. (a) **Outline** the hazards that might be encountered in a busy hotel kitchen.

   (b) **Identify SIX** hazards that might be considered when assessing the risk to health and safety of a multi-storey car park attendant.

   (c) **Outline** the hazards that might be encountered by a gardener employed by a local authority parks department.

4. (a) **Define** the meaning of the term 'young person' as used in health and safety legislation.

   (b) **Identify FOUR** 'personal' factors that may place young persons at a greater risk from workplace hazards.

   (c) **Outline** the factors to be taken into account when undertaking a risk assessment on young persons who are to be employed in the workplace.

5. (a) **Identify** work activities that may present a particular risk to pregnant women at work giving an example of **EACH** type of activity.

   (b) **Outline** the factors that the employer should consider when undertaking a specific risk assessment in relation to a pregnant employee.

6. **Outline** the issues to be considered to ensure the health and safety of disabled workers in the workplace.

7. **Outline** the issues that should be considered to ensure the health and safety of cleaners employed in a school out of normal working hours.

8. (a) **Identify** the personal factors that may increase the risks to an individual who is required to work alone away from their workplace.

   (b) **Describe** the controls that employers could implement to help minimise the risk to lone workers.

9. (a) **Identify SIX** of the General Principles of Prevention given in Schedule 1 of the Management of Health and Safety at Work Regulations that could be used to reduce the risk of injury from hazards found in a workplace.

   (b) **Outline**, with examples, the general hierarchy of control that should be applied when controlling health and safety risks in the workplace.

10. **Outline** the possible consequences of poor standards of housekeeping on health and safety in the workplace.

11. **Identify** the shape and colour, and give a relevant example, of each of the following types of safety sign:

    (a) prohibition
    (b) warning
    (c) mandatory
    (d) safe condition.

12. **Explain** why personal protective equipment (PPE) should be considered as a last resort in the control of occupational health hazards.

13. **Outline FOUR** sources of internal and **FOUR** sources of external information that may be consulted when dealing with a health and safety issue in the workplace.

14. (a) **Explain** the meaning of the term 'safe system of work'.

    (b) **State** the legal requirements for an employer to provide a safe system of work to an employee.

    (c) **Outline** the factors that should be considered when developing a safe system of work.

    (d) **Outline SIX** issues that should be addressed during the preparation of a safe system of work.

15. (a) **Define** the term 'permit-to-work system'.

    (b) **Outline FIVE** types of work situation that

131

may require a permit-to-work system, giving reasons in **EACH** case for the requirement.

(c) **Outline** the details that a permit to work should specify.

16. With reference to the Confined Space Regulations:

(a) **Define** the meaning of 'confined space', giving **FOUR** workplace examples.

(b) **Outline** specific hazards associated with working in confined spaces.

(c) **Outline** the controls that should be in place to ensure the safety of employees undertaking maintenance work in an underground storage vessel.

17. (a) **Identify FOUR** types of emergency procedure that an organisation might need to have in place.

(b) **Outline** reasons why workplace emergency procedures should be developed and practised.

(c) **Explain** why visitors to a workplace should be informed of the emergency procedures.

18. (a) **Identify** the two main functions of first-aid treatment.

(b) **Outline** the factors to consider when making a risk assessment of first-aid provision in a workplace.

APPENDIX 4.1 Hazard checklist

The following checklist may be helpful.

1. Equipment/mechanical

entanglement

friction/abrasion

cutting

shearing

stabbing/puncturing

impact

crushing

drawing-in

air or high-pressure fluid injection

ejection of parts

pressure/vacuum

display screen equipment

hand tools

2. Transport

works vehicles

mechanical handling

people/vehicle interface

3. Access

slips, trips and falls

falling or moving objects

obstruction or projection

working at height

confined spaces

excavations

4. Handling/lifting

manual handling

mechanical handling

5. Electricity

fixed installation

portable tools and equipment

6. Chemicals

dust/fume/gas

toxic

irritant

sensitising

corrosive

carcinogenic

nuisance

7. Fire and explosion

flammable materials/gases/liquids

explosion

means of escape/alarms/detection

8. Particles and dust

inhalation

ingestion

abrasion of skin or eye

9. Radiation

ionising

non-ionising

10. Biological

bacterial

viral

fungal

11. Environmental

noise

vibration

light

humidity

ventilation

temperature

overcrowding

12. The individual

individual not suited to work

long hours

high work rate

violence to staff

unsafe behaviour of individual

stress

pregnant/nursing women

young people

13. Other factors to consider

people with a disability

lone workers

poor maintenance

lack of supervision

lack of training

lack of information

inadequate instruction

unsafe systems

APPENDIX 4.2 Risk assessment example: office cleaning

Name of Company: His and Hers Hairdressers
Name of Assessor: A. R. Smith

Date of Assessment: 11 January 2014
Date of Review: 12 July 2014

Hazards	Persons affected	Risks	Initial risk level	Existing controls	Additional controls	Action by whom?	Action by when?	Done
Hairdressing products and chemicals Various bleaching and cleansing products in particular: Lightening (bleach) product Hydrogen peroxide Oxidative colourants	Staff and customers	Eye and/or skin irritation Possible allergic reaction	Medium	• COSHH assessment completed • Non-latex gloves are provided for staff when using products • Customers are protected with single use towels • Only non-dusty bleaches used • Staff report any allergies at induction • Store-room and salon well ventilated • Products stored as per manufacturer's recommendations • Staff are specifically trained in the correct use of products • Staff check whether customers have allergies to any products	• Needs to be reviewed • Eye baths to be purchased for treatment of eye splashes • Repeat allergy checks every 3 months and records kept • Storage in salon kept to 1 day requirement • Refresher training every 3 months and records kept • Records kept of customer allergies and simple patch tests introduced	Manager Owner Manager Manager Manager Staff Manager	12/2/14 12/2/14 14/2/14 then every 3 months 18/1/14 11/4/14 25/4/14	8/2/14 24/1/14 12/2/14 16/1/14 8/4/14 7/5/14
Sharp instruments	Staff and customers	Cuts, grazes and blood-borne infections	Medium	• All sharp instruments sterilised after use • Sterilising liquid changed daily • Sharps box available for disposable blades • First-aid box available	• Regular recorded checks that sterilisation procedures correctly followed • Contents of first-aid box checked weekly	Manager Manager	18/1/14 then monthly 18/1/14 then weekly	18/1/14 18/1/14
Fire	Staff and customers	Smoke inhalation and burns	Low	• Fire risk assessment completed	• No flammable products will be displayed in the windows or near heater	Manager	18/1/14	18/1/14

(Continued)

Hazards	Persons affected	Risks	Initial risk level	Existing controls	Additional controls	Action by whom?	Action by when?	Done
Electricity	Staff and customers	Electrical shocks and burns. Also a risk of fire	Low	• Any damaged cables, plugs or electrical equipment is reported to the manager	• An appropriate fire extinguisher (carbon dioxide) should be available	Owner	25/1/14	23/1/14
				• All portable electrical equipment and thermostats checked every 6 months by a competent person	• Make a visual check of electrical equipment, cables and sockets every month and record findings	Manager	25/1/14 then every month	25/1/14
				• All electrical equipment purchased from a reliable supplier • Staff shown how to use and store hairdryers, etc. safely and isolate electrical supply	• All internal wiring should be checked by a qualified electrician	Owner	25/4/14	3/4/14
Standing for long periods	Staff	Back pain and pain in neck, shoulders, legs and feet – musculoskeletal injuries	Medium	• Staff given regular breaks • Stools available for staff for use while trimming hair • Customer chairs are adjustable in height	• Develop a formal rota system for breaks	Manager	25/1/14	21/1/14
Wet hand work	Staff	Skin sensitisation, dry skin, dermatitis	Medium	• Staff are trained to wash and dry hands thoroughly between hair washes • Non-latex gloves are provided for staff • Moisturising hand cream is provided for staff	• Ensure that a range of glove sizes are available for staff • Staff will always wear gloves for all wet work	Manager Manager	25/1/14 25/1/14	31/1/14 1/2/14
Slips and trips	Staff and customers	Bruising, lacerations and possible fractures	Low	• Cut hair is swept up regularly • Staff must wear slip-resistant footwear • No trailing leads on floor • Any spills are cleaned immediately • Door mat provided at shop entrance	• Organise repair of worn floor covering	Manager	8/2/14	5/2/14

APPENDIX 4.3 Asbestos examples of safe systems of work

Standards required

All work involving asbestos in any form will be carried out in accordance with the current Control of Asbestos Regulations and Approved Code of Practice. Any asbestos removal must be done in accordance with the Asbestos Licensing Regulations. The use of new asbestos containing materials is prohibited.

Planning procedures

All work will be tendered for or negotiated in accordance with the approved standards.

The Contracts Manager will ascertain at an early stage whether asbestos in any form is likely to be present or used on the site. If details provided by the client are inconclusive, then an occupational hygiene specialist will be asked to take and analyse samples.

Method statements will be prepared by the Contracts Manager in conjunction with an occupational hygiene specialist, and, where necessary, a licensed asbestos removal contractor will be selected to carry out the work.

The Contracts Manager will ensure that any requirement to give notice of the work to the Health and Safety Executive is complied with.

Where any work involving asbestos materials not subject to the licensing requirements is to be carried out by employees, the working methods, precautions, safety equipment, protective clothing, special tools, etc. will be arranged by the Contracts Manager.

Supervision

Before work starts, all information on working methods and precautions agreed will be issued to site supervision by the Contracts Manager in conjunction with the Safety Adviser/Officer.

The Site Supervisor in conjunction with management will ensure that the licensed contractor contracted to carry out the removal work has set up operations in accordance with the agreed method statement and that the precautions required are fully maintained throughout the operation so that others not involved are not exposed to risk.

Where necessary, smoke testing of the enclosure and monitoring of airborne asbestos fibre concentrations outside the removal enclosure will be carried out by an occupational hygiene specialist.

The Site Supervisor will ensure that when removal operations have been completed, no unauthorised person enters the asbestos removal area until clearance samples have been taken by an occupational hygiene specialist and confirmation received that the results are satisfactory.

Where employees are required to use or handle materials containing asbestos not subject to the Licensing Regulations, the Site Supervisor will ensure that the appropriate safety equipment and protective clothing is provided and that the agreed safe working procedures are understood by employees and complied with.

All warning labels will be left in place on any asbestos materials used on site.

Examples

1. Painting undamaged asbestos insulating boards

Description

This task guidance sheet can be used where undamaged asbestos insulating boards need to be painted. This may be to protect them, or for aesthetic reasons.

It is not appropriate where the material is damaged. Use a specialist contractor licensed by HSE.

Carry out this work only if you are properly trained.

PPE

▶ disposable overalls fitted with a hood;
▶ boots without laces (laced boots can be difficult to decontaminate);
▶ disposable particulate respirator (FFP3).

Equipment

▶ 500 gauge polythene sheeting and duct tape;
▶ warning tape and notices;
▶ type H vacuum cleaner to BS5415 (if dust needs to be removed from the asbestos insulating board);
▶ paint conforming to the original specification, for example fire resistant. Select one low in hazardous constituents, for example solvents;
▶ low-pressure spray or roller/brush;
▶ bucket of water and rags;
▶ suitable asbestos waste container, for example a labelled polythene sack;
▶ appropriate lighting.

Preparing the work area

This work may be carried out at height; if so, the appropriate precautions MUST be taken.

▶ Carry out the work with the minimum number of people present.
▶ Restrict access, for example close the door and/or use warning tape and notices.
▶ Use polythene sheeting, secured with duct tape, to cover surfaces within the segregated area, which could become contaminated.
▶ Ensure adequate lighting.

Painting

- Never prepare surfaces by sanding.
- Before starting, check there is no damage.
- Repair any minor damage.
- If dust needs to be removed, use a Type H vacuum cleaner or rags.
- Preferably use the spray to apply the paint.
- Spray using a sweeping motion.
- Do not concentrate on one area as this could cause damage.
- Alternatively, apply the brush/roller lightly to avoid abrasion/damage.

Cleaning

- Use wet rags to clean the equipment.
- Use wet rags to clean the segregated area.
- Place debris, used rags, polythene sheeting and other waste in the waste container.

Personal decontamination

- Use suitable personal decontamination procedure.

Clearance procedure

- Visually inspect the area to make sure that it has been properly cleaned.
- Clearance air sampling is not normally required.

2. Removal of asbestos cement sheets, gutters, etc.

Description

This task guidance sheet can be used where asbestos cement sheets, gutters, drains and ridge caps, etc. need to be removed.

For the large-scale removal of asbestos cement, for example demolition, read *Working with Asbestos Cement* HSG189/2, HSE Books, 1999, ISBN 978 0 7176 1667 1.

It is not appropriate for the removal of asbestos insulating board.

Carry out this work only if you are properly trained.

PPE

- Use disposable overalls fitted with a hood.
- Waterproof clothing may be required outside.
- Boots without laces (laced boots can be difficult to decontaminate).
- Use disposable particulate respirator (FFP3).

Equipment

- 500 and 1000 gauge polythene sheeting and duct tape;
- warning tape and notices;
- bolt cutters;
- bucket of water, garden type spray and rags;
- suitable asbestos waste container, for example a labelled polythene sack;
- lockable skip for larger quantities of asbestos cement;
- asbestos warning stickers;
- appropriate lighting.

Preparing the work area

This work may be carried out at height; if so, the appropriate precautions to prevent the risk of fails MUST be taken.

- Carry out the work with the minimum number of people present.
- Restrict access, for example close the door and/or use warning tape and notices.
- Use 500-gauge polythene sheeting, secured with duct tape, to cover any surface within the segregated area, which could become contaminated.
- It is dangerous to seal over exhaust vents from heating units in use.
- Ensure adequate lighting.

Overlaying

- Instead of removing asbestos cement roofs, consider overlaying with a non-asbestos material.
- Attach sheets to existing purlings but avoid drilling through the asbestos cement.
- Note the presence of the asbestos cement so that it can be managed.

Removal

- Avoid breaking the asbestos cement products.
- If the sheets are held in place with fasteners, dampen and remove – take care not to create a risk of slips.
- If the sheets are bolted in place, use bolt cutters avoiding contact with the asbestos cement. Remove bolts carefully.
- Unbolt or use bolt cutters to release gutters, drain pipes and ridge caps, avoiding contact with the asbestos cement.
- Lower the asbestos cement to the ground. Do not use rubble chutes.
- Check for debris in fasteners or bolt holes. Clean with wet rags.
- Single asbestos cement products can be double wrapped in 1000-**gauge** polythene sheeting (or placed in waste containers if small enough). Attach asbestos warning stickers.
- Where there are several asbestos cement sheets and other large items, place in a lockable skip.

Cleaning

- Use wet rags to clean the equipment.
- Use wet rags to clean segregated area.
- Place debris, used rags, polythene sheeting and other waste in the waste container.

Personal decontamination

- Use a suitable personal decontamination system.

Clearance procedure

▶ Visually inspect the area to make sure that it has been properly cleaned.

▶ Clearance air sampling is not normally required.

3. Personal decontamination system

Description

This guidance sheet explains how you should decontaminate yourself after working with asbestos materials.

If you do not decontaminate yourself properly, you may take asbestos fibres home on your clothing. You or your family and friends could be exposed to them if they were disturbed and became airborne.

It is important that you follow the procedures given in the task guidance sheets and wear PPE such as overalls correctly; this will make cleaning easier.

Removing and decontaminating PPE

▶ Remove your respirator last.

▶ Clean your boots with wet rags.

▶ Where available, use a Type H vacuum cleaner to clean your overalls.

▶ Otherwise use a wet rag – use a 'patting' action – rubbing can disturb fibres.

▶ Where two or more workers are involved they can help each other by 'buddy' cleaning.

▶ Remove overalls by turning inside out – place in a suitable asbestos waste container.

▶ Use wet rags to clean waterproof clothing.

▶ Disposable respirators can then be removed and placed in a suitable asbestos waste container.

Personal decontamination

▶ Site-washing facilities can be used but restrict access during asbestos work.

▶ Wash each time you leave the work area.

▶ Use wet rags to clean washing facilities at the end of the job.

▶ Clean facilities daily if the job lasts more than a day.

▶ Visually inspect the facilities once the job is finished.

▶ Clearance air sampling is not normally required.

Further Information

These examples are taken from Asbestos Essentials Task Manual HSG210 (now revised), HSE Books, 2008, ISBN 978 0 7176 6263 0. Many more examples are contained in the publication including equipment and method guidance sheets. Obtainable from HSE Books.

CHAPTER 5

Health and safety management systems – Monitoring, investigation and recording – CHECK

> **This chapter covers the following NEBOSH learning objectives:**
> 1. Outline the principles, purpose and role of active and reactive monitoring
> 2. Explain the purpose of, and procedures for, investigating incidents (accidents, cases of work-related ill-health and other occurrences)
> 3. Describe the legal and organisational requirements for recording and reporting incidents

5.1 Active and reactive monitoring

5.1.1 Introduction

This chapter concerns the checking or monitoring of health and safety performance, including both positive measures such as inspections, and negative measures like injury statistics. It is about reviewing progress to see if something better can be done and auditing to ensure that what has been planned is being implemented. It is concerned with the recording of incidents and accidents at work, their investigation, the legal reporting requirements and simple analysis of incidents to help managers benefit from the investigation and recording process.

Measurement is a key step in any management process and forms the basis of continuous improvement. If measurement is not carried out correctly, the effectiveness of the health and safety management system is undermined and there is no reliable information to show managers how well the health and safety risks are controlled.

Managers should check by asking key questions to ensure that arrangements for health and safety risk control are in place, comply with the law as a minimum, and operate effectively.

There are two basic types of checking or monitoring:

Active monitoring, by taking the initiative before things go wrong, involves routine inspections and checks to make sure that standards and policies are being implemented and that controls are working.

Reactive monitoring, after things go wrong, involves looking at historical events to learn from mistakes and see what can be put right to prevent a recurrence.

The UK Health and Safety Executive's (HSE's) experience is that organisations find health and safety performance measurement a difficult subject. They struggle to develop health and safety performance measures which are not based solely on injury and ill-health statistics.

5.1.2 The traditional approach to measuring health and safety performance

Senior managers often measure company performance by using, for example, percentage profit, or return on investment or market share. A common feature of these measures would be that they are generally positive in nature, which demonstrates achievement, rather than negative, which demonstrates failure.

Yet, if senior managers are asked how they measure their companies' health and safety performance, it is likely that the only measure would be accident or injury statistics. Although the general business performance of an organisation is subject to a range of positive measures, for health and safety it too often comes down to one negative measure of failure.

Health and safety differs from many areas measured by managers because:

▶ improvement in performance means **fewer** outcomes from the measure (injuries or ill-health) rather than more;
▶ a low injury or ill-health rate trend over years is still no guarantee that risks are being controlled and that incidents will not happen in the future. This is particularly true in organisations when major hazards are present where, fortunately, there is a low probability of an incident occurring, but it would be catastrophic if it happened.

There is no single reliable measure of health and safety performance. What is required is a 'basket' of measures, providing information on a range of health and safety issues.

There are some significant problems with the use of injury/ill-health statistics in isolation:

▶ there may be under-reporting – focusing on injury and ill-health rates as a measure, especially if a reward system is involved, can lead to non-reporting to keep up performance;
▶ it is often a matter of chance whether a particular incident causes an injury, and it may not show whether or not a hazard is under control. Luck, or a reduction in the number of people exposed, may produce a low injury/accident rate rather than good health and safety management;
▶ an injury is the particular consequence of an incident and often does not reflect the potential severity. For example an unguarded machine could result in a cut finger or an amputation;
▶ people can be absent from work for reasons which are not related to the severity of the incident;
▶ there is evidence to show that there is little relationship between 'occupational' injury statistics (e.g. slips, trips and falls) and the reasons for the lack of control of major accident hazards (e.g. loss of containment of flammable or toxic material);

- a small number of accidents may lead to complacency;
- injury statistics demonstrate outcomes not causes.

Because of the potential shortcomings related to the use of accident/injury and ill-health data as a single measure of performance, more proactive (active) or 'upstream' measures are required. These require a systematic approach to deriving positive measures and how they link to the overall risk control process, rather than a quick-fix based on things that can easily be counted, such as the numbers of training courses or numbers of inspections, which has limited value. The resultant data provide no information on how the figure was arrived at, whether it is 'acceptable' (i.e. good/bad) or the quality and effectiveness of the activity. A more disciplined approach to health and safety performance standards measurement is required. This needs to develop as the health and safety management system develops.

5.1.3 Why measure performance?

'You can't manage what you can't measure.'
– Drucker

Measurement is an accepted part of the 'plan, do, check, act' management process discussed in Chapter 2. Measuring performance is as much part of a health and safety management system as financial, production or service delivery management. The HSG65 framework for managing health and safety, illustrated in Figure 5.1, shows where measuring performance standards fits within the overall health and safety management system shown in Figure 2.1.

The main purpose of measuring health and safety performance is to provide information on the progress and current status of the strategies, processes and activities employed to control health and safety risks. Effective measurement not only provides information on what the levels are but also why they are at this level, so that corrective action can be taken.

Answering questions

Active checking or monitoring of health and safety performance standards should seek to answer such questions as the following:

- Where is the position relative to the overall health and safety aims and objectives?
- Where is the position relative to the control of hazards and risks?
- How does the organisation compare with others?
- What is the reason for the current position?
- Is the organisation getting better or worse over time?
- Is the management of health and safety doing the right things?
- Is the management of health and safety doing things right consistently?
- Is the management of health and safety proportionate to the hazards and risks?
- Is the management of health and safety efficient?
- Is an effective health and safety management system in place across all parts of the organisation?
- Is the culture supportive of health and safety, particularly in the face of competing demands?

These questions should be asked at all management levels throughout the organisation. The aim of checking or monitoring performance standards should be to provide a complete picture of how well an organisation's health and safety risks are being controlled.

Decision making

The checking or monitoring of performance standards helps in deciding:

- where the organisation is in relation to where it wants to be;
- what progress is necessary and reasonable in the circumstances;
- how that progress might be achieved against particular restraints (e.g. resources or time);
- priorities – what should be done first and what is most important;
- effective use of resources.

Addressing different information needs

Information from the monitoring of performance standards is needed by a variety of people. These will include directors, senior managers, line managers, supervisors, health and safety professionals and employees/safety representatives. They each need information appropriate to their position and responsibilities within the health and safety management system.

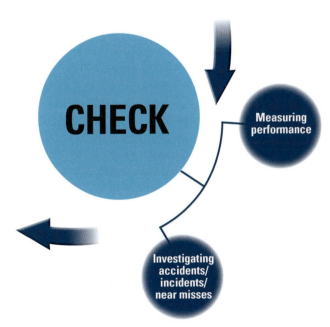

Figure 5.1 CHECK involves measuring performance and investigating incidents

For example what the Chief Executive Officer of a large organisation needs to know from the performance standards measurement system will differ in detail and nature from the information needs of the manager of a particular location.

A coordinated approach is required so that individual measuring activities fit within the general performance standards measurement framework.

Although the primary focus for performance standards measurement is to meet the internal needs of an organisation, there is an increasing need to demonstrate to external stakeholders (regulators, insurance companies, shareholders, suppliers, contractors, members of the public, etc.) that arrangements to control health and safety risks are in place, operating correctly and effectively.

5.1.4 What to measure

In order to achieve an outcome of no injuries or work-related ill-health, and to satisfy stakeholders, health and safety risks need to be controlled. Effective risk control is founded on an effective health and safety management system. This is illustrated in Figure 5.2.

▶ **Input: monitoring** the scale, nature and distribution of hazards created by the organisation's activities – measures of the hazard burden.

▶ **Process: active monitoring** of the adequacy, development, implementation and deployment of the health and safety management system and the activities to promote a positive health and safety culture – measures of success.

▶ **Outcomes: reactive monitoring** of adverse outcomes resulting in injuries, ill-health, loss and accidents with the potential to cause injuries, ill-health or loss – measures of failure.

5.1.5 Active monitoring – how to measure performance

Introduction and types of inspection

Active monitoring is a means of verifying the adequacy of, and the degree of compliance with, the risk control systems that have been established. It is intended to identify deficiencies in the safety management system for subsequent remedial action, and thus prevent accidents by eradicating their potential causes. It also enables a company to gain a picture of its health

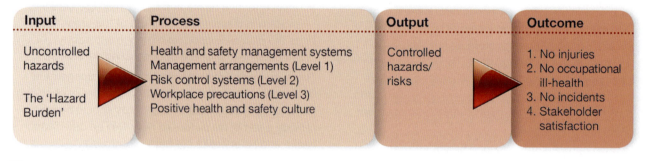

Input	Process	Output	Outcome
Uncontrolled hazards The 'Hazard Burden'	Health and safety management systems Management arrangements (Level 1) Risk control systems (Level 2) Workplace precautions (Level 3) Positive health and safety culture	Controlled hazards/ risks	1. No injuries 2. No occupational ill-health 3. No incidents 4. Stakeholder satisfaction

Figure 5.2 Effective risk control

Effective risk control

The health and safety management system comprises three levels of control (see Figure 5.2):

▶ Level 3 – effective workplace precautions provided and maintained to prevent harm to people who are exposed to the risks;

▶ Level 2 – risk control systems (RCSs): the basis for ensuring that adequate workplace precautions are provided and maintained;

▶ Level 1 – the key elements of the health and safety management system: the management arrangements (including plans and objectives) necessary to organise, plan, control and monitor the design and implementation of RCSs.

The health and safety culture must be positive to support each level.

Performance measurement should cover all elements of Figure 5.2 and be based on a balanced approach which combines:

and safety performance and chart the progress of improvement.

Active monitoring employs several complementary methods which address differing aspects and areas of the organisation. These methods may be usefully categorised as follows:

(i) The **Safety Management System (SMS) Audit** see chapter 6 for more information on auditing.

(ii) The **Safety Survey** is usually a detailed assessment of one aspect of an organisation's SMS, e.g. the organisation's training arrangements.

(iii) The **Safety Inspection** consists of a formal assessment of workplace safety, and the identification of hazardous conditions or practices, for subsequent remedial action. It is normally carried out by first line managers referring to, and completing, a checklist.

(iv) The **Safety Tour** addresses the 'people' aspects of workplace safety, and by discussions with a range of staff establishes their familiarity with safety procedures and requirements. It is normally carried out by middle and senior management, as one means of demonstrating their commitment to safety. A questionnaire is frequently used.

(v) **Safety sampling** is a useful technique that helps organisations to concentrate on one particular area or subject at a time. A specific area is chosen which can be inspected in about 30 minutes. A checklist is drawn up to facilitate the inspection looking at specific issues. These may be different types of hazard; they may be unsafe acts or conditions noted; they may be proactive, good behaviour or practices noted. The inspection team or person then carries out the sampling at the same time each day or week in the specified period. The results are recorded and analysed to see if the changes are good or bad over time. Of course, defects noted must be brought to the notice of the appropriate person for action on each occasion.

Management roles

Performance should be measured at each management level from directors downwards. It is not sufficient to monitor by exception, where, unless problems are raised, it is assumed to be satisfactory. Senior managers must satisfy themselves that the correct arrangements are in place and working properly. Responsibilities for both active and reactive monitoring must be laid down and managers need to be personally involved in making sure that plans and objectives are met and compliance with standards is achieved. Although systems may be set up with the guidance of safety professionals, managers should be personally involved and given sufficient training to be competent to make informed judgements about monitoring performance.

Other people, such as safety representatives, will also have the right to inspect the workplace. Each employee should be encouraged to inspect their own workplace frequently to check for obvious problems and rectify them if possible or report hazards to their supervisors.

Specific statutory (or thorough) examinations, for example of lifting equipment or pressure vessels, have to be carried out at intervals laid down in written schemes by competent persons – usually specially trained and experienced inspection/insurance company personnel.

This frequency of monitoring and inspections will depend on the level of risk and any statutory inspection requirement. Directors may be expected to examine the premises formally at an annual audit, whereas departmental supervisors may be expected to carry out inspection every week. Senior managers should regularly monitor the health and safety plan to ensure that objectives are being met and to make any changes to the plan as necessary.

Data from reactive monitoring should be considered by senior managers at least once a month. In most

organisations serious events would be closely monitored as they happen.

Depending on the size and complexity of the organisation there should be arrangements for active monitoring and comparing the results of such monitoring, checking that appropriate actions are being taken. It is at this point that decisions may be required on resources (e.g. investment in a new machine which does not spill oil) and priorities, and that consultation with employees and/or their representatives is desirable (see also section on Consultation). It is important that all places and activities are included because it is often at the margins of responsibility that hazards are overlooked.

The measurement process can gather information through:

▶ direct observation of conditions and of people's behaviour (sometimes referred to as unsafe acts and unsafe conditions monitoring);

▶ talking to people to elicit facts and their experiences as well as gauging their views and opinions;

▶ examining written reports, documents and records.

These information sources can be used independently or in combination. Direct observation includes inspection activities and the monitoring of the work environment (e.g. temperature, dust levels, solvent levels, noise levels) and people's behaviour. Each risk control system (RCS) should have a built-in monitoring element that will define the frequency of monitoring; these can be combined to form a common inspection system.

5.1.6 Safety inspection programme

Within any active monitoring programme there should be a system for workplace safety inspections. It often forms part of the preventative maintenance scheme for plant and equipment. However, they may also be covered by legal examination and inspection requirements. Equipment in this category includes steam boilers, other pressure vessels, hoists, lifts, cranes, chains, ropes, lifting tackle, power presses, scaffolds, trench supports and local exhaust ventilation. Workplace inspections should include other precautions, such as those covering the use of premises, welfare facilities, behaviour of people and systems of work.

Inspections should be done by people who are competent to identify the relevant hazards and risks and who can assess the conditions found. Preferably they should mainly be local staff who understand the problems and can immediately present solutions. It is essential that they do not, in any way, put themselves or anyone else at risk during the inspection. Particular care must be taken with regard to safe access. In carrying out inspections, the safety of people's actions should be considered, in addition to the safety of the conditions they are working in – a ladder might be in perfect condition but it has to be used properly too.

Figure 5.3 Poor conditions: (a) inspection needed; (b) inspection in progress

A properly thought-out approach to inspection will include:

▶ well-designed inspection forms to help plan and initiate remedial action by requiring those doing the inspection to rank any deficiencies in order of importance;
▶ summary lists of remedial action with names and deadlines to track progress on implementing improvements;
▶ periodic analysis of inspection forms to identify common features or trends which might reveal underlying weaknesses in the system;
▶ information to aid judgements about any changes required in the frequency or nature of the inspection programme.

The inspection programme should be properly targeted whilst considering all risks in the premises. For example, low risks could be covered by monthly or bi-monthly general inspections involving a wide range of workplace safety issues such as the general condition of premises, floors, passages, stairs, heating, ventilation, lighting, welfare facilities and first-aid. Higher risks need more frequent and detailed inspections; perhaps weekly or even, in extreme cases, daily or before using a piece of equipment like an injection moulding machine or mobile plant.

The daily, weekly or monthly inspections will be aimed at checking conditions in a specified area against a fixed checklist drawn up by local management. It will cover specific items, such as the guards at particular machines, whether access/agreed routes are clear, whether fire extinguishers are in place, etc. The checks should be carried out primarily by competent staff of the department who should sign off the checklist. They should not last more than half an hour, perhaps less. This is not a specific hazard-spotting operation, but there should be a space on the checklist for the inspectors to note down any particular problems encountered.

In order to get maximum value from inspection checklists, they should be designed so that they require objective rather than subjective judgements of conditions. For example asking the people undertaking a general inspection of the workplace to rate housekeeping as good or bad, begs questions as to what does good and bad mean, and what criteria should be used to judge this. If good housekeeping means there is no rubbish left on the floor, all waste bins are regularly emptied and not overflowing, floors are swept each day and cleaned once a week, decorations should be in good condition with no peeling paint, then this should be stated. Adequate expected standards should be provided in separate notes so that those inspecting know the standards that are required.

The checklist or inspection form should facilitate:

▶ the planning and initiation of remedial action, by requiring those doing the inspection to rank deficiencies in priority order (those actions which are most important rather than those which can easily be done quickly);
▶ identifying those responsible for taking remedial actions, with sensible timescales to track progress on implementation;
▶ periodic monitoring to identify common themes which might reveal underlying problems in the system;
▶ management information on the frequency or nature of the monitoring arrangements.

The checklist could be structured using the 'four Ps' (note that the examples are not a definitive list) (see Chapter 23 for an example):

Premises, including:

▶ work at height
▶ access
▶ working environment
▶ welfare
▶ services
▶ fire precautions

Plant and substances, including:

▶ work equipment
▶ manual and mechanical handling
▶ vehicles
▶ dangerous/flammable substance
▶ hazardous substance

Procedures, including:

- risk assessments
- safe systems of work
- permits to work
- personal protective equipment
- contractors
- notices, signs and posters

People, including:

- health surveillance
- people's behaviour
- training and supervision
- appropriate authorised persons
- violence
- those especially at risk.

Appendix 5.3 shows a detailed construction site inspection checklist which follows a subject-based approach rather than the four Ps.

Key points in becoming a good observer

To improve health and safety performance, managers and supervisors must eliminate unsafe acts by observing them, taking immediate corrective action and following up to prevent recurrence. To become a good observer, they must improve their observation skills and must learn how to observe effectively. Effective observation includes the following key points:

- be selective
- know what to look for
- practice
- keep an open mind
- guard against habit and familiarity
- do not be satisfied with general impressions
- record observations systematically.

Observation techniques

In addition, to become a good observer, a person must:

- stop for 10–30 seconds before entering a new area to ascertain where employees are working;
- be alert for unsafe practices that are corrected as soon as you enter an area;
- observe activity – do not avoid the action;
- remember ABBI – look above, below, behind, inside;
- develop a questioning attitude to determine what injuries might occur if the unexpected happened and how the job might be accomplished more safely. Ask 'why?' and 'what could happen if … ?';
- use all senses: sight, hearing, smell, touch;
- maintain a balanced approach. Observe all phases of the job;
- be inquisitive;
- observe for ideas – not just to determine problems;
- recognise good performance.

Some commonly observed poor work practices include:

- using machinery or tools without authority
- operating at unsafe speeds or in other violation of safe work practice
- removing guards or other safety devices, or rendering them ineffective
- using defective tools or equipment or using tools or equipment in unsafe ways
- using hands or body instead of tools or push sticks
- overloading, crowding, or failing to balance materials or handling materials in other unsafe ways, including improper lifting
- repairing or adjusting equipment that is in motion, under pressure, or electrically charged
- failing to use or maintain, or improperly using, personal protective equipment or safety devices
- creating unsafe, unsanitary, or unhealthy conditions by improper personal hygiene, by using compressed air for cleaning clothes, by poor housekeeping, or by smoking in unauthorised areas
- standing or working under suspended loads, scaffolds, shafts, or open hatches.

Additional principles for conducting inspections:

- Draw attention to the presence of any immediate danger – other items can await the final report.
- Shut down and 'lock out' any hazardous items that cannot be brought to a safe operating standard until repaired.
- Do not operate equipment. Ask the operator for a demonstration. If the operator of any piece of equipment does not know what dangers may be present, this is cause for concern. Items should never be ignored because the observer did not have the knowledge to make an accurate assessment of safety.
- Clearly describe each hazard and its exact location in rough notes. Allow 'on-the-spot' recording of all findings before they are forgotten. Record what have or have not been examined in case the inspection is interrupted.
- Ask questions, but do not unnecessarily disrupt work activities. This may interfere with efficient assessment of the job function and may also create a potentially hazardous situation.
- Consider the static (stop position) and dynamic (in motion) conditions of the item you are inspecting. If a machine is shut down, consider postponing the inspection until it is functioning again.
- Discuss as a group, 'Can any problem, hazard or accident generate from this situation when looking at the equipment, the process or the environment?' Determine what corrections or controls are appropriate.
- Do not try to detect all hazards simply by relying on human senses or by looking at them during the inspection. Monitoring equipment may have to be used to measure the levels of exposure to chemicals, noise, radiation or biological agents.

5

▶ Take photographs to help with any description of a particular situation. Make sure it is safe to operate a camera in the workplace concerned.

Reports from inspections

Some of the items arising from safety inspections will have been dealt with immediately; other items will require action by specified people. Where there is some doubt about the problem, and what exactly is required, advice should be sought from the safety adviser or external expert. A brief report of the inspection and any resulting action list should be submitted to the safety committee. While the committee may not have the time available to consider all reports in detail, it will want to be satisfied that appropriate action is taken to resolve all matters; it will be necessary for the committee to follow up the reports until all matters are resolved.

Essential elements of a report are:

▶ identification of the organisation, workplace, inspector and date of inspection;
▶ list of observations;
▶ priority or risk level;
▶ actions to be taken;
▶ timescale for completion of the actions.

Appendix 5.1 gives examples of poor workplaces which can be used for practice exercises. Appendix 5.3 shows a detailed construction site inspection checklist. Chapter 23 shows a specimen workplace inspection form which can be adapted for use at many workplaces. There is more information on report writing in general in Section 5.1.7.

Chapter 23 also provides a workplace inspection checklist which can be used to assist in drawing up a specific checklist for any general workplace or as an aide-memoire for the workplace inspection report form.

Figure 5.4 The use of a checklist

Advantages of using a checklist

The strengths of using a checklist for inspection of a workplace are:

▶ the opportunity given for prior preparation and planning in order to ensure that the inspection is both structured and systematic;
▶ the degree of consistency obtained first by those carrying out the inspection and second in the areas and issues to be covered;
▶ it can be easily adapted or customised for different areas;
▶ it results in an immediate record of findings;
▶ it provides an easy method of comparison and audit.

Checklists can be used to improve safety performance by:

▶ checking adherence to standards;
▶ allowing an analysis of trends to be made;
▶ identifying problems which if quickly rectified can boost the morale of the workforce;
▶ demonstrating the commitment of management; and
▶ the involvement of employees which will increase the ownership of health and safety within the organisation.

There are a few downsides with checklists including:

▶ the lack of comments so that reasons for passing or failing are not given;
▶ the tick-box mentality when it is all too easy to merely tick a box than provide any original thinking;
▶ people may shut their eyes to items not on the list.

These can be overcome by requiring inspectors to fill in the comments section with each item and providing space for other issues with perhaps a few prompts of additional hazards which could be present.

5.1.7 Effective report writing

There are three main aims to the writing of reports and they are all about communication. A report should aim to:

▶ get a message through to the reader;
▶ make the message and the arguments clear and easy to understand;
▶ make the arguments and conclusions persuasive.

Communication starts with trying to get into the mind of the reader, imagining what would most effectively catch the attention, what would be most likely to convince, and what will make this report stand out among others.

A vital part of this is presentation; so while a handwritten report is better than nothing if time is short, a well-organised, typed report is very much clearer. To the reader of the report, who may well be very busy with a great deal of written information to wade through, a clear, well-presented report will produce a positive attitude from the outset, with instant benefit to the writer.

Five factors which help to make reports effective are:

▶ structure;
▶ presentation of arguments;

- style;
- presentation of data;
- how the report itself is presented.

Structure

The structure of a report is the key to its professionalism. Good structuring will:

- help the reader to understand the information and follow the arguments contained in the report;
- increase the writer's credibility;
- ensure that the material contained in the report is organised to the best advantage.

The following list shows a frequently used method of producing a report, but always bear in mind that different organisations use different formats:

1. title page
2. summary
3. contents list
4. introduction
5. main body of the report
6. conclusions
7. recommendations
8. appendices
9. references.

It is important to check with the organisation requesting the report in case their in-house format differs.

1. Title page

This will contain:

- a title and often a subtitle;
- the name of the person or organisation to whom the report is addressed;
- the name of the writer(s) and their organisation;
- the date on which the report was submitted.

As report writing is about communication, it is a good idea to choose a title that is eye-catching and memorable as well as being informative, if this is appropriate to the subject.

2. Summary

Limit the summary to between 150 and 500 words. Do not include any evidence or data. This should be kept for the main report. Include the main conclusions and principal recommendations and place the summary near the front of the report.

3. Contents list

Put the contents list near the beginning of the report. Short reports do not need a list but if there are several headings, it does help the reader to grasp the overall content of the report in a short time.

4. Introduction

The introduction should contain the following:

- information about who commissioned the report and when;
- the reason for the report;
- objectives of the report;
- terms of reference;
- preparation of the report (type of data, research undertaken, subjects interviewed, etc.);
- methodology used in any analysis;
- problems and the methods used to tackle them;
- details about consultation with clients, employees, etc.

There may be other items that are specific to the report.

5. The main body of the report

This part of the report should describe, in detail, what was discovered (the facts), and the significance of these discoveries (analysis) and their importance (evaluation). Graphs, tables and charts are often used at this stage in the report. These should have the function of summarising information rather than giving large amounts of detail. The more detailed graphics should be made into appendices.

To make it more digestible, this part of the report should be divided into sections, using numbered headings and sub-headings. Very long and complex reports will need to be broken down into chapters.

When writing an inspection report, the recommendations should be made in this section (according to the examiners' report for IGC1 in 2011). Priorities should be noted and recommendations made clearly for each action, with a proposed timescale. There should be a comparison made with previous inspections to indicate whether standards have improved or deteriorated. The report should acknowledge strengths, for example where a good standard of compliance has been found.

6. Conclusions

The concluding part of the report should be a reasonably detailed 'summing up'. It should give the conclusions arrived at by the writer and explain why the writer has reached these conclusions.

7. Recommendations

The use of this section depends on the requirements of the person commissioning the report. If recommendations are required, provide as few as possible, to retain a clear focus. Report writers are often asked only to provide the facts. In the case of an inspection report, the recommendations should be made in the main body of the report (see no. 5 above).

8. Appendices

This part of the report should contain sections that may be useful to a reader who requires more detail. Examples would be the detailed charts, graphs and tables, any questionnaires used in constructing the evidence mentioned in no. 5, forms, case studies and so on. The appendices are the background material of the report.

9. References

If any books, papers or journal articles have been used as source material, this should be acknowledged in a reference section. There are a number of accepted referencing methods used by academics.

Because the reader is likely to be a person with some degree of expertise in the subject, a report must be reliable, credible, relevant and thorough. It is therefore important to avoid emotional language, opinions presented as facts and arguments that have no supporting evidence. To make a report more persuasive, the writer needs to:

▶ present the information clearly;
▶ provide reliable evidence;
▶ present arguments logically;
▶ avoid falsifying, tampering with or concealing facts.

In conclusion

Expertise in an area of knowledge means that distortions, errors and omissions will quickly be spotted by the discerning reader and the presence of any of these will cast doubt on the credibility of the whole report.

Reports are usually used as part of a decision-making process. If this is the case, clear, unembellished facts are needed. Exceptions to this would be where the report is a proposal document or where a recommendation is specifically requested. Unless this is the case, it is better not to make recommendations.

A report should play a key role in organising information for the use of decision makers. It should review a complex and/or extensive body of information and make a summary of all the important issues.

It is relatively straightforward to produce a report, as long as the writer keeps to a clear format. Using the format described here, it should be possible to tell the reader as clearly as possible:

▶ what happened and why;
▶ who was involved;
▶ what it cost if appropriate;
▶ what the result was.

There may be a request for a special report and this is likely to be longer and more difficult to produce. Often it will relate to a 'critical incident' and the decision makers will be looking for information to help them:

▶ decide whether this is a problem or an opportunity;
▶ decide whether to take action;
▶ decide what action, if any, to take.

Finally, report writing should be kept simple. Nothing is gained in the use of long, complicated sentences, jargon and official-sounding language. When the report

is finished, it is helpful to run through it with the express intention of simplifying the language and making sure that it says what was intended in a clear and unambiguous way.

5.1.8 Measuring failure – reactive monitoring

So far, this chapter has concentrated on measuring activities designed to prevent the occurrence of injuries and work-related ill-health (active monitoring). Failures in risk control also need to be measured (reactive monitoring), to provide opportunities to check performance, learn from failures and improve the health and safety management system.

Reactive monitoring arrangements include systems to identify and report:

▶ injuries and work-related ill-health (details of the incidence rate calculation are given in Chapter 3);
▶ other losses such as damage to property;
▶ incidents, including those with the potential to cause injury, ill-health or loss (near misses);
▶ hazards and faults;
▶ weaknesses or omissions in performance standards and systems, including complaints from employees and enforcement action by the authorities.

Each of the above provides opportunities for an organisation to check performance, learn from mistakes, and improve the health and safety management system and risk control systems. In some cases, the organisation must send a report of the circumstances and causes of incidents, such as major injuries, to the appropriate enforcing authority. Statutorily appointed safety representatives and representatives of employee safety are entitled to investigate.

The 'corporate memory' is enhanced by these events and information obtained by investigations can be used to reinforce essential health and safety messages. Safety committees and safety representatives should be involved in discussing common features or trends. This will help the workforce to identify jobs or activities which cause the most serious or the greatest number of injuries where remedial action may be most beneficial. Investigations may also provide valuable information if an insurance claim is made or legal action taken.

Most organisations should not find it difficult to collect data on serious injuries and ill-health. However, learning about minor injuries, other losses, incidents and hazards can prove more challenging. There is value in collecting information on all actual and potential losses to learn how to prevent more serious events. Accurate reporting can be promoted by:

▶ training which clarifies the underlying objectives and reasons for identifying such events;
▶ a culture which emphasises an observant and responsible approach and the importance of

having systems of control in place before harm occurs;

▶ open, honest communication in a just environment, rather than a tendency merely to allocate 'blame';

▶ cross-referencing and checking first-aid treatments, health records, maintenance or fire reports and insurance claims to identify any otherwise unreported events.

Guidance on recording, investigating and analysing these incidents is given later in this chapter.

5.2 Investigating incidents

5.2.1 Function of incident investigation

Incidents and accidents rarely result from a single cause and many turn out to be complex. Most incidents involve multiple, interrelated causal factors. They can occur whenever significant deficiencies, oversights, errors, omissions or unexpected changes occur. Any one of these can be the precursor of an accident or incident. There is a value in collecting data on all incidents and potential losses as it helps to prevent more serious events.

Incidents and accidents, whether they cause damage to property or more serious injury and/or ill-health to people, should be properly and thoroughly investigated (Figure 5.5). Good investigation is a key element to making improvements in health and safety performance.

Figure 5.5 Dangerous occurrence: aftermath of a fire

Incident investigation is considered to be part of a reactive monitoring system because it is triggered after an event.

Incident/accident investigation is based on the logic that:

▶ all incidents/accidents have causes – eliminate the cause and eliminate future incidents;

▶ the direct and indirect causes of an incident/accident can be discovered through investigation;

▶ corrective action indicated by the causation can be taken to eliminate future incidents/accidents.

Investigation is not intended to be a mechanism for apportioning blame. There are often strong emotions associated with injury or significant losses. It is all too easy to look for someone to blame without considering the reasons why a person behaved in a particular way. Often short cuts to working procedures that may have contributed to the accident give no personal advantage to the person injured. The short cut may have been taken out of loyalty to the organisation or ignorance of a safer method.

Valuable information and understanding can be gained from carrying out accident/incident investigations. These include:

▶ an understanding of how and why problems arose which caused the accident/incident;

▶ an understanding of the ways people are exposed to substances or situations which can cause them harm;

▶ a snapshot of what really happens, for example why people take short cuts or ignore safety rules;

▶ identifying deficiencies in the control of risks in the organisation.

The legal reasons for conducting an investigation are:

▶ to ensure that the organisation is acting in accordance with legal requirements;

▶ that it forms an essential part of the MHSW Regulation 5 requirements to plan, organise, control, monitor and review health and safety arrangements;

▶ to comply with the Woolf Report on civil action which changed the way cases are run. Full disclosure of the circumstances of an accident/incident has to be made to the injured parties considering legal action. The fact that a thorough investigation was carried out and remedial action taken would demonstrate to a court that a company has a positive attitude to health and safety. The investigation will also provide essential information for insurers in the event of an employer's liability or other claim.

There are many **benefits** from investigating accidents/incidents. These include:

▶ the prevention of similar events occurring again. Where the outcomes are serious injuries the enforcing authorities are likely to take a tough stance if previous warnings have been ignored;

▶ the prevention of business losses due to disruption immediately after the event, loss of production, loss of business through a lowering of reputation or inability to deliver, and the costs of criminal and legal actions;

▶ improvement in employee morale and general attitudes to health and safety particularly if they have been involved in the investigations;

▶ improving management skills to improve health and safety performance throughout the organisation.

The case for investigating near misses and undesired circumstances may not be so obvious but it is just as useful and much easier as there are no injured people to deal with. There are no demoralised people at work or distressed families and seldom a legal action to answer. Witnesses will be more willing to speak the truth and help with the investigation.

Managers need to:

▶ communicate the type of accident and incident that needs to be reported;

▶ provide a system for reporting and recording;

▶ check that proper reports are being made;

▶ make appropriate records of accidents and incidents;

▶ investigate incidents and accidents reported;

▶ analyse the events routinely to check for trends in performance and the prevalence of types of incident or injury;

▶ monitor the system to make sure that it is working satisfactorily.

5.2.2 Types of incident or adverse events

In HSG245 *Investigating Injury and Accidents at Work* 2004, the HSE define different types of incident as follows:

Figure 5.6 Accident at work – reconstruction of a ladder accident showing where the deceased person was found under the ladder which had toppled over while he was attempting to adjust the height of the extending ladder

'Adverse event' includes:

1. **Accident:** is defined by the HSE as 'any unplanned event that results in injury or ill-health of people, or damage or loss to property, plant, materials or the environment or a loss of a business opportunity'. Other authorities define an accident more narrowly by excluding events that do not involve injury or ill-health. This book will always use the HSE definition.

2. **Incident:**
   – near miss: an event that, while not causing harm, has the potential to cause injury or ill-health. (In this guidance, the term near miss will be taken to include dangerous occurrences);
   – undesired circumstance: a set of conditions or circumstances that have the potential to cause injury or ill-health, e.g. untrained nurses handling heavy patients.

3. **Dangerous occurrence:** one of a number of specific, reportable adverse events, as defined in the Reporting of Injuries, Diseases and Dangerous Occurrences Regulations 2013 (RIDDOR).

4. **Immediate cause:** the most obvious reason why an adverse event happens, e.g. the guard is missing; the employee slips, etc. There may be several immediate causes identified in any one adverse event.

5. **Root cause:** an initiating event or failing from which all other causes or failings spring. Root causes are generally management, planning or organisational failings.

6. **Underlying cause:** the less obvious 'system' or 'organisational' reason for an adverse event happening, e.g. pre-start-up machinery checks are not carried out by supervisors; the hazard has not been adequately considered via a suitable and sufficient risk assessment; production pressures are too great.

Consequence:

1. **Fatal:** work-related death.

2. **Specified serious injury/ill-health:** (as defined in RIDDOR 2013, Regulation 4; see Chapter 19), including a
   ▷ fracture, other than to fingers, thumbs and toes;
   ▷ amputation of an arm, hand, finger, thumb, leg, foot or toe;
   ▷ permanent loss of sight or reduction of sight;
   ▷ crush injuries leading to internal organ damage;
   ▷ serious burns (covering more than 10% of the body, or damaging the eyes, respiratory system or other vital organs);
   ▷ scalpings (separation of skin from the head) which require hospital treatment;
   ▷ unconsciousness caused by head injury or asphyxia;
   ▷ any other injury arising from working in an enclosed space, which leads to hypothermia, heat-induced illness or requires resuscitation or admittance to hospital for more than 24 hours.

3. Over 7-day **injury/ill-health:** where the person affected is unfit to carry out his or her normal work for more than seven consecutive days.

4. **Minor injury:** all other injuries, where the injured person is unfit for his or her normal work for less than seven days.

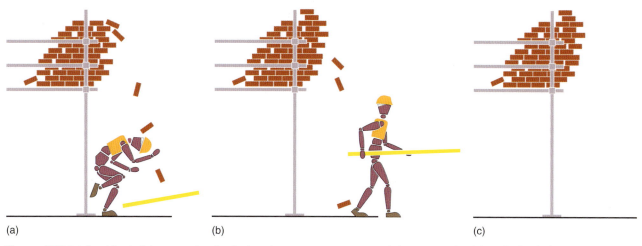

(a) (b) (c)

Figure 5.7 (a) Accident; (b) near miss (includes dangerous occurrence) damage only; (c) undesired circumstances

5. Damage only: damage to property, equipment, the environment or production losses.

Figure 5.7 demonstrates the difference between an accident, near miss and undesired circumstances.

Each type of event/incident gives the opportunity to:

▶ check performance;
▶ identify underlying deficiencies in management systems and procedures;
▶ learn from mistakes and add to the corporate memory;
▶ reinforce key health and safety messages;
▶ identify trends and priorities for prevention;
▶ provide valuable information if there is a claim for compensation;
▶ help meet legal requirements for reporting certain incidents to the authorities.

5.2.3 Accident triangles and their limitation

In 1969, F. E. Bird collected a large quantity of accident data and produced a well-known triangle (Figure 5.8). It can be seen that damage and near miss incidents occur much more frequently than injury accidents and,

therefore, may be a good indicator of risks. The study also shows that most accidents are predictable and avoidable.

There are other accident ratio triangles which have been produced, for example by Heinrich (see Chapter 3) and the HSE. They are all very similar despite different actual numbers but the concept has limitations as recent research in the USA by Behavioral Science Technology (BST) has shown.

The safety triangle, or pyramid, states that reducing minor injuries or near misses (the large number at the bottom) leads to a proportionate reduction in severe injuries and deaths (the small number at the top). However, statistics over the past 10 years or so have shown that minor injuries have steadily declined while the number of serious injuries and deaths has not changed. The reasons for this are suggested as:

▶ Similar injuries may have completely different causes;
▶ Some incidents have the potential for serious injury yet may only result in a minor one. For example falling off a step ladder may only cause a bruise but could be very serious;
▶ Serious injuries have different underlying causes to minor injuries. For example missing controls, poor procedures and badly designed equipment create high-risk situations that may lead to a major incident;
▶ Many injuries are musculoskeletal sprains and strains which could not have ended up as fatal injuries. Whereas falling from height often results in serious injury or death;
▶ Insurance companies may concentrate more on frequent minor injuries because serious injuries, although individually expensive, are very rare;
▶ The solutions to major injury risks may be expensive and difficult to design or effect, for example changes to aircraft controls to prevent an accident like Virgin Galactic.

It is clear that a good safety record based solely on few lost time injuries or minor accidents does not guarantee

Figure 5.8 F. E. Bird's well-known accident triangle

Serious or disabling injuries — 1
Minor injuries (first aid injuries) — 10
Damage accidents — 30
Near miss accidents — 600

a safe workplace. Companies must systematically examine their procedures, designs, supervision and the standards implemented in the workplace to identify the potential for serious injuries and major incidents, as well as paying attention to the causes of minor injuries.

5.2.4 Which incidents/accidents should be investigated?

The types of incident which may need to be investigated and the depth of the investigation are usually determined by their outcomes or consequences. Should every accident be investigated or only those that lead to serious injury? In fact the main determinant is the potential of the accident to cause harm rather than the actual harm resulting. For example, a slip can result in an embarrassing flailing of arms or, just as easily, a broken leg. The frequency of occurrence of the accident type is also important – a stream of minor cuts from paper needs looking into.

As it is not possible to determine the potential for harm simply from the resulting injury, the only really sensible solution is to investigate all accidents. The amount of time and effort spent on the investigation should, however, vary depending on the level of risk (severity of potential harm, frequency of occurrence). The most effort should be focused on significant events involving serious injury, ill-health or losses and events which have the potential for multiple or serious harm to people or substantial losses. These factors should become clear during the accident investigation and be used to guide how much time should be taken.

Figure 5.9 has been developed by the Health and Safety Executive (HSE) to help to determine the level of investigation which is appropriate. The potential worst injury consequences in any particular situation should be considered when using the figure. A particular incident like a scaffold collapse may not have caused an injury but had the potential to cause major or fatal injuries.

In a **minimal-level** investigation, the relevant supervisor will look into the circumstances of the accident/incident and try to learn any lessons which will prevent future incidents.

A **low-level** investigation will involve a short investigation by the relevant supervisor or line manager into the circumstances and immediate underlying and root causes of the accident/incident, to try to prevent a recurrence and to learn any general lessons.

A **medium-level** investigation will involve a more detailed investigation by the relevant supervisor or line manager, the health and safety adviser and employee representatives and will look for the immediate, underlying and root causes.

A **high-level** investigation will involve a team-based investigation, involving supervisors or line managers, health and safety advisers and employee representatives. It will be carried out under the supervision of senior management or directors and will look for the immediate, underlying and root causes.

Likelihood of recurrence	Potential worst injury consequences of accident/incident			
	Minor	**Serious**	**Major**	**Fatal**
Certain				
Likely				
Possible				
Unlikely				
Rare				

Risk	Minimal	Low	Medium	High
Investigation level	**Minimal Level**	**Low Level**	**Medium Level**	**High Level**

Figure 5.9 Appropriate levels of investigation

5.2.5 Basic incident investigation procedures

Investigations should be led by supervisors, line managers or other people with sufficient status and knowledge to make recommendations that will be respected by the organisation. The person to lead many investigations will be the Department Manager or Supervisor of the person/area involved because they:

▶ know about the situation;
▶ know most about the employees;
▶ have a personal interest in preventing further incidents/accidents affecting 'their' people, equipment, area, materials;
▶ can take immediate action to prevent a similar incident;
▶ can communicate most effectively with the other employees concerned;
▶ can demonstrate practical concern for employees and control over the immediate work situation.

The investigation should be carried out as soon as possible after the incident to allow the maximum amount of information to be obtained. There may be difficulties which should be considered in setting up the investigation quickly – if, for example, the victim is removed from the site of the accident, or if there is a lack of a particular expert. An immediate investigation is advantageous because:

▶ factors are fresh in the minds of witnesses;
▶ witnesses have had less time to talk (there is an almost automatic tendency for people to adjust their story of the events to bring it into line with a consensus view);
▶ physical conditions have had less time to change;
▶ more people are likely to be available, for example delivery drivers, contractors and visitors, who will quickly disperse following an incident, making contact very difficult;
▶ there will probably be the opportunity to take immediate action to prevent a recurrence and to demonstrate management commitment to improvement;
▶ immediate information from the person suffering the accident often proves to be most useful.

Consideration should be given to asking the person to return to site for the accident investigation if they are physically able, rather than wait for them to return to work. A second option, although not as valuable, would be to visit the injured person at home or even in hospital (with their permission) to discuss the accident.

A number of people will potentially be involved in accident or incident investigation. For most of these people this will only be necessary on very few occasions. Training, guidance and help will therefore be required. Training can be provided in accident/incident investigation in courses run on site and also in numerous off-site venues. Computer-based training courses are also available. These are intended to provide refresher training on an individual basis or complete training at office sites, for example, where it may not be feasible to provide practical training.

Initial action

There are a lot of things that have to be done when an incident occurs. The success of an investigation comes in the first few moments. A line manager's initial action varies for every event. The person on the scene must be the judge of what is critical. These steps are guidelines to apply as appropriate.

▶ Take control at the scene – line managers need to take charge, directing and approving everything that is done.
▶ Ensure first-aid is provided and call for emergency services.
▶ Control potential secondary events – these events such as explosions and fire are usually more serious. Positive actions need to be taken quickly after careful thought of the consequences.
▶ Identify sources of evidence at the scene.
▶ Preserve evidence from alteration or removal.
▶ Notify appropriate site management.

Investigation method

There are four basic elements to a sound investigation:

1. Collect facts about what has occurred.
2. Assemble and analyse the information obtained.
3. Compare the information with acceptable industry and company standards and legal requirements to draw conclusions.
4. Implement the findings and monitor progress.

Information should be gathered from all available sources, for example witnesses, supervisors, physical conditions, hazard data sheets, written systems of work, training records. Photographs are invaluable aids to investigation, but of course with digital photography they can be easily altered. It is a good idea to print on the time and date of the picture and be prepared to verify that it is accurate. Printing out the picture as soon as possible captures it for the records. Plans and simple sketches of the incident site are also valuable.

The amount of time spent should not, however, be disproportionate to the risk. The aim of the investigation should be to explore the situation for possible underlying factors, in addition to the immediately obvious causes of the accident. For example, in a machinery accident it would not be sufficient to conclude that an accident occurred because a machine was inadequately guarded. It is necessary to look into the possible underlying system failure that may have occurred.

Investigations have three facets, which are particularly valuable and can be used to check against each other:

▶ direct observation of the scene, premises, workplace, relationship of components, materials and substances being used, possible reconstruction of events and injuries or condition of the person concerned;

▶ documents including written instructions, training records, procedures, safe operating systems, risk assessments, policies, records of inspections or tests and examinations carried out;

▶ interviews (including written statements) with persons injured, witnesses, people who have carried out similar functions or examinations and tests on the equipment involved and people with specialist knowledge.

Investigation interview techniques

It must be made clear at the outset and during the course of the interview that the aim is not to apportion blame but to discover the facts and use them to prevent similar accidents or incidents in the future.

A witness should be given the opportunity to explain what happened in their own way without too much interruption and suggestion. Questions should then be asked to elicit more information. These should be of the open type, which do not suggest the answer. Questions starting with the words in Figure 5.10 are useful.

Figure 5.10 Questions to be asked in an investigation

'Why' should not be used at this stage. The facts should be gathered first, with notes being taken at the end of the explanation. The investigator should then read them or give a summary back to the witness, indicating clearly that they are prepared to alter the notes, if the witness is not content with them.

If possible, indication should be given to the witness about immediate actions that will be taken to prevent a similar occurrence and that there could be further improvements depending on the outcome of the investigation.

Seeing people injured can often be very upsetting for witnesses, which should be borne in mind. This does not mean they will not be prepared to talk about what has happened. They may in fact wish to help, but questions should be sensitive; upsetting the witness further should be avoided.

5.2.6 Incident causes and analysis

Immediate causes – unsafe acts and conditions

An investigation should look at the following factors as they can provide useful information about **immediate causes** that have been manifested in the incident/accident and result directly from unsafe acts and conditions in the workplace.

Personal factors:

▶ behaviour of the people involved (unsafe acts)
▶ suitability of people doing the work
▶ training and competence.

Task factors:

▶ workplace conditions and precautions or controls (unsafe conditions)
▶ actual method of work adopted at the time
▶ ergonomic factor
▶ normal working practice either written or customary.

Underlying and root causes – management system failures

A thorough investigation should also look at the following factors as they can provide useful information about **underlying and root causes** that have been manifested in the incident/accident:

Underlying causes are the less obvious system or organisational reasons for an accident or incident such as:

▶ pre-start-up machinery checks were not made by supervisors;
▶ the hazard had not been considered in the risk assessment;
▶ there was no suitable method statement;
▶ pressures of production had been more important;
▶ the employee was under a lot of personal pressure at the time;
▶ have there been previous similar incidents?
▶ was there adequate supervision, control and coordination of the work involved?

Root causes involve an initiating event or failing from which all other causes or failings arise. Root causes are generally management, planning or organisational failings including:

▶ quality of the health and safety policy and procedures;
▶ quality of consultation and cooperation of employees;
▶ the adequacy and quality of communications and information;
▶ deficiencies in risk assessments, plans and control systems;
▶ deficiencies in monitoring and measurement of work activities;
▶ quality and frequency of reviews and audits.

Comparison with relevant standards

There are usually suitable and relevant standards which may come from the HSE, industry or the organisation itself. These should be carefully considered to see if:

▶ suitable standards are available to cover legal standards and the controls required by the risk assessments;
▶ the standards are sufficient and available to the organisation;
▶ the standards were implemented in practice;
▶ the standards were implemented, why there was a failure;
▶ changes should be made to the standards.

Simple root cause analysis – the 'five whys'

This type of simple root cause analysis is ideal for a minimal or low level of investigation. More complex methods like the Tree of Causes are suitable for a higher level of investigation and are not covered here. Now is the time after obtaining all the initial information to ask 'why'.

▶ Basic Question – Keep asking 'What caused or allowed this condition/practice to occur?' until you get to root causes.
▶ The 'five whys' is one of the simplest of the root cause analysis methods. It is a question-asking method used to explore the cause/effect relationships underlying a particular problem. Ultimately, the goal of applying the five whys method is to determine a **root cause** of a defect or problem.

The following example demonstrates the basic process:

Car will not start. (the problem)
1. *Why?* – The battery is dead. (first 'why')
2. *Why?* – The alternator is not functioning. (second 'why')
3. *Why?* – The alternator belt has broken. (third 'why')
4. *Why?* – The alternator belt was well beyond its useful service life and has never been replaced. (fourth 'why')
5. *Why?* – I have not been maintaining my car according to the recommended service schedule. (fifth 'why' and the root cause)

Benefits of asking the five whys:

▶ **Simplicity.** It is easy to use and requires no advanced mathematics or tools.
▶ **Effectiveness.** It truly helps to quickly separate symptoms from causes and identify the root case of a problem.
▶ **Comprehensiveness.** It aids in determining the relationships between various problem causes.
▶ **Flexibility.** It works well alone and when combined with other quality improvement and trouble shooting techniques.
▶ **Engaging.** By its very nature, it fosters and produces teamwork and teaming within and without the organisation.
▶ **Inexpensive.** It is a guided, team focused exercise. There are no additional costs.

Often the answer to the one 'why' uncovers another reason and generates another 'why'. It often takes 'five whys' to arrive at the root cause of the problem. Investigators will probably find that they ask more or less than 'five whys' in practice.

5.2.7 Remedial actions

The investigation should have highlighted both immediate causes and underlying causes. Recommendations with priorities, both for immediate action and for longer term improvements, should come out of this. It may be necessary to ensure that the report goes further up the management chain if the improvements recommended require authorisation, which cannot be given by the investigating team.

It is essential that a follow-up is made to check on the implementation of the recommendations. It is also necessary to review the effect of the recommendations to check whether they have achieved the desired result and whether they have had unforeseen 'knock-on' effects, creating additional risks and problems.

The accident or incident investigation should not only be used to generate recommendations but should also be used to generate safety awareness. The investigation report or a summary should therefore be circulated locally to relevant people and, when appropriate, summaries circulated throughout the organisation. The accident or incident does not need to have resulted in a 7-day lost time injury for this system to be used.

Investigation form

Headings which could be used to compile an accident/incident investigation form are given below:

▶ date and location of accident/incident;
▶ circumstances of accident/incident;
▶ immediate cause of accident/incident;

- underlying cause of accident/incident;
- immediate action taken;
- recommendation for further improvement;
- report circulation list;
- date of investigation;
- signature of investigating team leader;
- names of investigating team.

Follow-up

- were the recommendations implemented?
- were the recommendations effective?

Key data for medium level of investigation

The HSE has suggested that the key data included in Box 5.1 should be covered in an investigation report. This level of data is more appropriate to a medium level of investigation.

5.3 Recording and reporting incidents

5.3.1 Typical examples

This section covers the legal and organisational requirements for recording and reporting accidents and other incidents. It also covers simple analysis of the data to help employers and employees benefit from the investigation and recording process.

Typical examples covered by this section are:

- a work-related death;
- a Specified Injury, such as a fractured leg, amputation of an arm or hand, loss of an eye or a serious burn. Typical construction accidents which are likely to result in a specified serious injury are:
  - ▷ falls while working at height on scaffolds, structure, roofs or ladders
  - ▷ falling through holes in uncompleted structures or when lifting covers over holes
  - ▷ being hit by falling objects from scaffolds or partly completed buildings
  - ▷ dropping heavy items like bricks on unprotected feet
  - ▷ hit by moving pieces of plant or inadequately guarded elevators
  - ▷ crushed by overturning vehicle
  - ▷ contact with live overhead power lines
  - ▷ overcome in a confined space like a sewer;

Box 5.1 Key data for medium level of investigation

Key data to be covered in accident, ill-health and incident reports.

The event

- Details of any injured person, including age, gender, experience, training, etc.
- A description of the circumstances, including the place, time of day and conditions.
- Details of the event, including:
  - ▷ any actions which led directly to the event;
  - ▷ the direct causes of any injuries, ill-health or other loss;
  - ▷ the immediate causes of the event;
  - ▷ the underlying causes – for example failures in workplace precautions, risk control systems or management arrangements.
- Details of the outcomes, including in particular:
  - ▷ the nature of the outcome – for example injuries or ill-health to employees or members of the public, damage to property, process disruption, emissions to the environment, creation of hazards;
  - ▷ the severity of the harm caused, including injuries, ill-health and losses.
- The immediate management response to the situation and its adequacy:
  - ▷ Was it dealt with promptly?
  - ▷ Were continuing risks dealt with promptly and adequately?
  - ▷ Was the first-aid response adequate?
  - ▷ Were emergency procedures followed?
  - ▷ Whether the event was preventable and if so how?

The potential consequences

- What was the worst that could have happened?
- What prevented the worst from happening?
- How often could such an event occur (the *'recurrence potential'*)?
- What was the worst injury or damage which could have resulted (the *'severity potential'*)?
- How many people could the event have affected (the *'population potential'*)?

Recommendations

- Prioritised actions with responsibilities and targets for completion.

▶ a Reportable Disease, such as asbestosis from contact with asbestos-containing material, anthrax from contact with an infected animal, lead poisoning from breathing lead fumes while burning off old paint, severe occupational dermatitis from wet cement, occupational asthma from adhesives and solvents, severe back injury from lifting or manual handling of loads; and

▶ a Dangerous Occurrence, such as the failure of a pressure vessel, collapse of a crane or a scaffold from overloading, incorrect design or high winds or a serious fire from welding, cutting or tar boilers.

More detail is given later in this chapter and a full summary of RIDDOR is contained in Chapter 19.

5.3.2 Statutory requirements for recording and reporting incidents

Accident book

Under the Social Security (Claims and Payments) Regulations 1979, Regulation 25, employers must keep a record of accidents at premises where more than 10 people are employed. Anyone injured at work is required to inform the employer and record information on the accident in an accident book, including a statement on how the accident happened.

The employer is required to investigate the cause and enter this in the accident book if they discover anything that differs from the entry made by the employee. The purpose of this record is to ensure that information is available if a claim is made for compensation.

The HSE Accident Book BI 510 (see Figure 5.11) Second Edition 2012 has notes on these Regulations, and the Reporting of Injuries, Diseases and Dangerous Occurrences Regulations 2013 (RIDDOR). It also complies with the Data Protection Act 1998 by requiring injured people to give their consent for the information to be passed on to Safety Representatives or Representatives of Employee Safety. The accident/incident report form (see Chapter 23) can be used as an initial record and as a substitute for form BI 510.

RIDDOR 2013

RIDDOR 2013 requires employers, the self-employed and those in control of premises, to report certain more serious accidents and incidents to the HSE or other enforcing authority and to keep a record. There are no exemptions for small organisations. A full summary of the Regulations is given in Chapter 19. The reporting and recording requirements are as follows:

Death or serious injury

If an accident occurs at work and:

▶ an employee or self-employed person working on the premise is killed or suffers a specified injury (including the effects of physical violence) (see

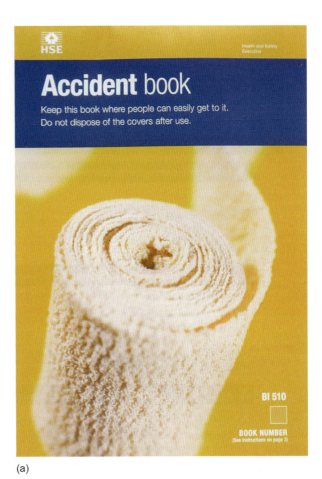

(a)

Figure 5.11 (a) The Accident Book BI 510 (Second Edition) ISBN 97807176640580

Chapter 19 under RIDDOR for definition of 'specified injury'). Specified Injuries include:

▷ a fracture, other than to fingers, thumbs and toes;

▷ amputation of an arm, hand, finger, thumb, leg, foot or toe;

▷ permanent loss of sight or reduction of sight;

▷ crush injuries leading to internal organ damage;

▷ serious burns (covering more than 10% of the body, or damaging the eyes, respiratory system or other vital organs);

▷ scalpings (separation of skin from the head) which require hospital treatment;

▷ unconsciousness caused by head injury or asphyxia;

▷ any other injury arising from working in an enclosed space, which leads to hypothermia, heat-induced illness or requires resuscitation or admittance to hospital for more than 24 hours;

▶ a member of the public or person not at work is killed or taken to hospital from the scene of the accident. If the accident occurred at a hospital, the report only needs to be made if the injury is a 'specified injury'.

The responsible person must notify the enforcing authority without delay by the quickest practicable

Report Number

Accident record

1 About the person who had the accident

Name ....

Address ....

.... Postcode ....

Occupation ....

2 About you, the person filling in this record

▼ If you did not have the accident write your address and occupation.

Name ....

Address ....

.... Postcode ....

Occupation ....

3 About the accident   *Continue on the back of this form if you need to*

▼ Say when it happened.   Date   /   /   Time

▼ Say where it happened. State which room or place.

....

▼ Say how the accident happened. Give the cause if you can.

....

....

....

....

▼ If the person who had the accident suffered an injury, say what it was.

....

▼ Please sign the record and date it.

Signature ....   Date .... / .... / ....

4 For the employee only

▼ By ticking this box I give my consent to my employer to disclose my personal information and details of the accident which appear on this form to safety representatives and representatives of employee safety for them to carry out the health and safety functions given to them by law.   ☐

Signature ....   Date   /   /

5 For the employer only

▼ Complete this box if the accident is reportable under the Reporting of Injuries, Diseases and Dangerous Occurrences Regulations (RIDDOR). To report, go to page 4 of this book or go to www.hse.gov.uk/riddor/report.htm2.

How was it reported? ....

Date reported   /   /   Signature ....

(b)

Figure 5.11 (b) Record form from BI 510

means, for example telephone to the HSE ICC (0845 300 9923).

Over 7-day lost time injury

If there is an accident connected with work (including physical violence), and an employee or self-employed person working on the premises suffers an injury and is away from work or not doing his/her normal duties for more than 7 days (including weekends, rest days or holidays but not counting the day of the accident), the responsible person must report to the enforcing authority within 15 days. The best and quickest way is via telephone or the internet to the ICC.

Reportable occupational disease

If a doctor notifies the responsible person that an employee suffers from a reportable work-related disease they must report to the enforcing authority within 10 days and a completed disease report form, F2508A, must be sent to the enforcing authority. A summary is included in Chapter 19 and a full list is included with a pad of report forms. The best and quickest way is via telephone or the internet to the ICC. These diseases include:

▶ carpal tunnel syndrome;
▶ severe cramp of the hand or forearm;
▶ occupational dermatitis;
▶ hand–arm vibration syndrome;
▶ occupational asthma;
▶ tendonitis or tenosynovitis of the hand or forearm;
▶ any occupational cancer;
▶ any disease attributed to an occupational exposure to a biological agent.

The HSE Incident Contact Centre can be contacted to check whether a particular disease is reportable.

Reportable dangerous occurrences

Dangerous occurrences are certain, specified near-miss events. Not all such events require reporting. There are 27 categories of dangerous occurrence that are relevant to most workplaces. For example:

▶ the collapse, overturning or failure of load-bearing parts of lifts and lifting equipment;
▶ plant or equipment coming into contact with overhead power lines;
▶ the accidental release of any substance which could cause injury to any person.

See Chapter 17 for a summary of dangerous occurrences. All dangerous occurrences must be reported immediately to the authorities. The best and quickest way is via telephone to the ICC (on 0845 300 9923). The HSE Incident Contact Centre can be contacted to check whether a dangerous occurrence is reportable.

Reportable gas incidents

If a distributor, filler, importer or supplier of flammable gas learns, either directly or indirectly, that someone has died, lost consciousness, or been taken to hospital for treatment to an injury arising in connection with the gas they distributed, filled, imported or supplied, this can be reported online.

A gas engineer registered with the Gas Safe Register must provide details of any gas appliances or fittings that they consider to be dangerous to the extent that people could die, lose consciousness or require hospital treatment. This may be due to the design, construction, installation, modification or servicing, and could result in:

▶ an accidental leakage of gas;
▶ inadequate combustion of gas; or
▶ inadequate removal of products of the combustion of gas.

Records

Records of incidents covered by RIDDOR are also important. They ensure that sufficient information to allow properly managed health and safety risks is collected. This information is a valuable management tool that can be used as an aid to risk assessment, helping to develop solutions to potential risks. In this way, records also help to prevent injuries and ill-health, and control costs from accidental loss.

Records must be kept of:

▶ any **accident, occupational disease or dangerous occurrence** which requires reporting under RIDDOR; and
▶ any other occupational accident causing injuries that result in a worker being away from work or **incapacitated for more than three consecutive days** (not counting the day of the accident but including any weekends or other rest days). Over-three-day injuries do not have to be reported, unless the incapacitation period goes on to exceed seven days.

Organisations can keep the record in any form they wish, for example, they may choose to keep records by:

▶ keeping copies of report forms in a file;
▶ recording the details on a computer;
▶ using the Accident Book entry;
▶ maintaining a written log.

RIDDOR records must be produced when asked for by the HSE, local authority or Office of Rail Regulator inspectors.

Whom to report to

Online

Go to www.hse.gov.uk/riddor and complete the appropriate online report form. The form will then be submitted directly to the RIDDOR database. A copy will be received for records.

Telephone

All incidents can be reported online but a telephone service remains for reporting **fatal and specified injuries only**. Incident Contact Centre can be called on 0845 300 9923 (opening hours Monday to Friday 8.30 a.m. to 5 p.m.).

Reporting out of hours

The HSE has an out-of-hours duty officer. Circumstances where the HSE may need to respond out of hours include:

▶ a work-related death or situation where there is a strong likelihood of death following an incident at, or connected with, work;

▶ a serious accident at a workplace so that the HSE can gather details of physical evidence that would be lost with time;

▶ following a major incident at a workplace where the severity of the incident, or the degree of public concern, requires an immediate public statement from either the HSE or government ministers.

Information supplied to the HSE in a RIDDOR report is **not** passed on to the organisation's insurance company. If the relevant insurer needs to know about a work related accident, injury, or case of ill-health they must be contacted separately – insurers have told the HSE that reporting injuries and illnesses at work to them quickly could save time and money.

5.3.3 Internal systems for collecting and analysing incident data

Managers need effective internal systems to know whether the organisation is getting better or worse, to know what is happening and why, and to assess whether objectives are being achieved. Earlier in the chapter, Section 5.1 deals with monitoring generally; but here, the basic requirements of a collection and analysis system for incidents are discussed.

The incident report form (discussed earlier) is the basic starting point for any internal system. Each organisation needs to lay down what the system involves and who is responsible to do each part of the procedure. This will involve:

▶ what type of incidents should be reported;

▶ who completes the incident report form – normally the manager responsible for the investigation;

▶ how copies should be circulated in the organisation;

▶ who is responsible to provide management measurement data;

▶ how the incident data should be analysed and at what intervals;

▶ the arrangements to ensure that action is taken on the data provided.

The data should seek to answer the following questions:

▶ are failure incidents occurring, including injuries, ill-health and other loss incidents?

▶ where are they occurring?

▶ what is the nature of the failures?

▶ how serious are they?

▶ what are the potential consequences?

▶ what are the reasons for the failures?

▶ how much has it cost?

▶ what improvements in controls and the management system are required?

▶ how do these issues vary with time?

▶ is the organisation getting better or worse?

Type of accident/incident

Most organisations will want to collect data on:

▶ all injury accidents;

▶ cases of ill-health;

▶ sickness absence;

▶ dangerous occurrences;

▶ damage to property, the environment, personal effects and work in progress;

▶ incidents with the potential to cause serious injury, ill-health or damage (undesired circumstances).

Not all of these are required by law, but this should not deter the organisation that wishes to control risks effectively.

Analysis

All the information, whether in accident books or report forms, will need to be analysed so that useful management data can be prepared. Many organisations look at the analysis every month and annually. However, where there are very few accidents/incidents, quarterly may be sufficient. The health and safety information should be used alongside other business measures and should receive equal status.

There are several ways in which data can be analysed and presented. The most common ways are:

▶ by causation using the classification used on the RIDDOR form F2508. This has been used on the example accident/incident report form (see Box 5.2 and Chapter 23);

▶ by the nature of the injury, such as cuts, abrasions, asphyxiation and amputations;

▶ by the part of the body affected, such as hands, arms, feet, lower leg, upper leg, head, eyes, back and so on. Sub-divisions of these categories could be useful if there were sufficient incidents;

▶ by age and experience at the job;

▶ by time of day;

▶ by occupation or location of the job;

▶ by type of equipment used.

There are a number of up-to-date computer recording programs which can be used to manipulate the data if significant numbers are involved. The trends can be shown against monthly, quarterly and annual past performance of, preferably, the same organisation. If indices are calculated, such as Incident Rate, comparisons can be made nationally with HSE figures and with other similar organisations or businesses in the same industrial group. This is really of major value only to larger organisations with significant numbers of events.

Although the number of accidents may give a general indication of the health and safety performance, a more detailed examination of accidents and accident statistics is normally required. A calculation of the rate of accidents enables health and safety performance to be compared between years and organisations.

The simplest measure of accident rate is called the incidence rate.

The HSE's formula for calculating an annual injury incidence rate is:

$$\frac{\textit{Number of reportable injuries in financial year}}{\textit{Average number employed during the year}} \times 100{,}000$$

This gives the rate per 100,000 employees. The formula makes no allowances for variations in part-time employment or overtime. It is an annual calculation and the figures need to be adjusted pro rata if they cover a shorter period. Such shorter-term rates should be compared only with rates for exactly similar periods – not the national annual rates.

While the HSE and industry calculate injury **incidence rates** per 100,000 or 1,000 employees, some parts of industry prefer to calculate injury **frequency rates**, usually per million hours worked. This method, by counting hours worked rather than the number of employees, avoids distortions which may be caused in the incidence rate calculations by part- and full-time employees and by overtime working. Frequency rates can be calculated for any time period.

The calculation is:

$$\frac{\textit{Number of injuries in the period}}{\textit{Total hours worked during the period}} \times 1{,}000{,}000$$

The HSE produces annual bulletins of national performance along with a detailed statistical report, which can be used for comparisons (see: http://www. hse.gov.uk/statistics/index.htm). There are difficulties in comparisons across Europe and, say, with the USA, where the definitions of accidents or time lost vary.

Reports should be prepared with simple tables and graphs showing trends and comparisons. Line graphs, bar charts and pie charts are all used quite extensively with good effect. All analysis reports should be made available to employees as well as managers. This can often be done through the Health and Safety Committee and safety representatives, where they exist, or directly to all employees in small organisations. Other routine meetings, team briefings and notice boards can all be used to communicate the message.

It is particularly important to make sure that any actions recommended or highlighted by the reports are taken quickly and employees kept informed.

The report form shown in Chapter 23 uses these immediate causes which can be used for analysis purposes. The categories can easily be changed to suit local needs.

Box 5.2 The following categories of immediate causes of accident are used in F2508:

1. contact with moving machinery or material being machined;
2. struck by moving, including flying or falling, object;
3. struck by moving vehicle;
4. struck against something fixed or stationary;
5. injured whilst handling, lifting or carrying;
6. slip, trip or fall on same level;
7. fall from height; indicate approximate distance of fall in metres;
8. trapped by something collapsing or overturning;
9. drowning or asphyxiation;
10. exposure to or contact with harmful substances;
11. exposure to fire;
12. exposure to an explosion;
13. contact with electricity or an electrical discharge;
14. injured by an animal;
15. violence;
16. other kind of accident.

5.3.4 Collection of information, compensation and insurance issues

Accidents/incidents arising out of the organisation's activities resulting in injuries to people and incidents resulting in damage to property can lead to compensation claims. The second objective of an investigation should be to collect and record relevant information for the purposes of dealing with any claim. It must be remembered that, in the longer term, prevention is the best way to reduce claims and must be the first objective in the investigation. An overzealous approach to gathering information concentrating on the compensation aspect can, in fact, prompt a claim from the injured party where there was no particular intention to take this route before the investigation. Nevertheless, relevant information should be collected. Sticking to the collection of facts is usually the best approach.

As mentioned earlier the legal system in England and Wales changed dramatically with the introduction of the Woolf reforms in 1999. These reforms apply to injury claims. This date was feared by many because of the uncertainty and the fact that the pre-action protocols were very demanding.

The essence of the pre-legal action protocols is as follows:

1. 'letter of claim' to be acknowledged within 21 days;
2. ninety days from date of acknowledgement to either accept liability or deny. If liability is denied then full reasons must be given;
3. agreement to be reached on using a single expert.

The overriding message is that to comply with the protocols quick action is necessary. It is also vitally important that records are accurately kept and accessible.

Lord Woolf made it clear in his instructions to the judiciary that there should be very little leeway given to claimants and defendants who did not comply.

What has been the effect of these reforms on day-to-day activity?

Initially there was a 25% drop in the number of cases moving to litigation. Whilst this was the primary objective of the reforms, the actual drop is far more significant than anyone anticipated. By 2007 cases began to rise significantly again and in some cases judges have been very reluctant to strike out cases because they are out of time, except in extreme circumstances.

However, some of the positive effects have been:

▶ the elimination of speculative actions due to the requirement to fully outline the claimant's case in the letter of claim;
▶ earlier and more comprehensive details of the claim allowing a more focused investigation and response;
▶ 'Part 36 offers' (payments into court) seeming to be having greater effect in deterring claimants from pursuing litigation;
▶ overall faster settlement.

The negative effects have principally arisen from failure to comply with the timescales, particularly relating to the gathering of evidence and records and having no time therefore to construct a proper defence.

Section 47 of the Health and Safety at Work Act 1974, by virtue of Section 6 of the Enterprise and Regulatory Reform Act 2013, removes the standard of strict liability from certain health and safety regulations. No civil claim may be brought for breach of statutory duty unless a regulation expressly provides for it; this effectively reverses the previous position.

In almost all cases it will be for the injured employee to rely on common law negligence and prove that their injuries were caused by the employer's negligence. Negligence is generally a higher hurdle for employees than a breach of statutory duty – particularly those imposing strict liability – as the standard required of employers is higher; greater emphasis is placed on the 'reasonable practicability' defence which balances the expense of potential preventative measures against the scale of the risk.

The removal of strict liability for health and safety breaches is one of many changes designed to combat the perceived 'compensation culture' and reflects a growing trend against punishing employers for injuries that they took all reasonable steps to prevent. Under the previous regime businesses have paid out significant sums in compensation with around 78,000 civil liability claims brought each year for injuries sustained at work.

Overburdened employers are likely to welcome this cutting of red tape and the freedom to fulfil their obligations under health and safety law without worrying of potential, unforeseen claims against them. This in turn will increase employers' confidence to not only protect their employees but also to develop and expand their business.

Employees on the other hand might not welcome the changes so readily. They are now faced with the prospect of discharging a heavier legal and evidential burden to establish liability. In fact the Trades Union Congress argues that the changes do nothing to remove unfairness but merely shift it to another place.

Appendix 5.2 provides a checklist of headings, which may assist in the collection of information. It is not expected that all accidents and incidents will be investigated in depth and a dossier with full information prepared. Judgement has to be applied as to which incidents might give rise to a claim and when a full record of information is required. All accident/incident report forms should include the names of all witnesses as a minimum. Where the injury is likely to give rise to lost time, a photograph(s) of the situation should be taken.

5.3.5 Lessons learnt from an incident

After an appropriate investigation there should be an action plan for the implementation of additional risk control measures The action plan should have SMART objectives, i.e. Specific, Measurable, Agreed, and Realistic, with Timescales.

A good knowledge of the organisation and the way it carries out its work is essential to know where improvements are needed. Management, safety professionals, employees and their representatives should all contribute to a constructive discussion on what should be in the action plan in order to make the proposals SMART.

Not every recommendation for further risk controls will be implemented, but the ones accorded the highest priority should be implemented immediately. Organisations need to ask 'What is essential to securing the health and safety of the workforce today? What cannot be left until another day? How high is the risk to employees if this risk control measure is not implemented immediately?'

Despite financial constraints, failing to put in place measures to control serious and imminent risks is totally unacceptable – either reduce the risks to an acceptable level, or stop the work. Each risk control measure should be assigned a priority, and a timescale with a designated person to carry out the recommendation. It is crucial that the action plan as a whole is properly monitored with a specific person, preferably a director, partner or senior manager, made responsible for its implementation.

Progress on the action plan should be regularly monitored and significant departures should be explained and risk control measures rescheduled, if necessary. There should be regular consultation with employees and their representatives to keep them fully informed of progress with implementation of the action plan.

Relevant safety instructions, safe working procedures and risk assessments should be reviewed after an incident. It is important to ask what the findings of the investigation indicate about risk assessments and procedures in general, to see if they really are suitable and sufficient.

It is also useful to estimate the cost of incidents to fully appreciate the true cost of accidents and ill-health to the organisation. To find out more about the costs of accidents and incidents visit the HSE's website cost calculator at: www.hse.gov.uk/costs

5.4 Further information

Investigating Incidents and Accidents at Work HSG245, HSE Books, 2004, ISBN 978 0 7176 2827 8. http://www.hse.gov.uk/pubns/books/hsg245.htm

Managing for health and safety, HSG65 third edition 2013, HSE Books, ISBN 978-0-7176-6456-6 http://www.hse.gov.uk/pubns/priced/hsg65.pdf

OHSAS 18001:2007 Occupational health and safety management systems – Requirements ISBN 978 0 580 50802 8 http://www.bsigroup.com/en/Assessment-and-certification-services/management-systems/Standards-and-Schemes/BSOHSAS-18001/

Reporting accidents and incidents at work, A brief guide to the Reporting of Injuries, Diseases and Dangerous Occurrences Regulations 2013 (RIDDOR), INDG453 (rev 1), October 2013 http://www.hse.gov.uk/pubns/indg453.pdf

The Management of Health and Safety at Work Regulations 1999, see HSE management site at: http://www.hse.gov.uk/managing/index.htm

The Reporting of Injuries, Disease and Dangerous Occurrences Regulations 2013, see HSE reporting website: http://www.hse.gov.uk/riddor/index.htm

5

5.5 Practice revision questions

1. **Identify FIVE** active (or proactive) and **FIVE** reactive measures that can be used to monitor an organisation's health and safety performance.

2. A large company is planning to introduce a programme of regular inspections of the workplace.
   (a) **Outline** the factors that should be considered when planning such inspections.
   (b) **Outline** factors that determine the frequency with which health and safety inspections should be undertaken.

3. (a) **Define** the terms 'safety survey', 'safety tour' and 'safety sampling'.
   (b) **Outline** the issues which should be considered when a safety survey of a workplace is to be undertaken.

4. (a) **Outline** the strengths **AND** weaknesses of using a checklist to undertake a health and safety inspection of a workplace.
   (b) **Identify** the questions that might be included on a checklist to gather information following an accident involving slips, trips and falls.

5. An employee of a company is to be given duties to undertake a safety inspection.

   (a) **Outline** the competencies needed to carry out the duties.
   (b) **Identify** the principal issues that should be included in a safety inspection report so that managers can make decisions on any required remedial actions.
   (c) **Explain** how the report should be structured and presented so as to increase the likelihood of action being taken by managers.

6. (a) **Give FOUR** reasons why an organisation should have a system for the internal reporting of accidents.
   (b) **Identify** the issues that should be included in a typical workplace accident reporting procedure.
   (c) **Outline** factors that may discourage employees from reporting accidents at work.

7. An employee has been seriously injured after being struck by a fork-lift truck in a warehouse.
   (a) **Give FOUR** reasons why the accident should be investigated by the person's employer.
   (b) **Outline** the information that should be included in the investigation report.
   (c) **Outline FOUR** possible immediate causes and **FOUR** possible underlying (root) causes of the accident.

(d) Giving reasons in **EACH** case, **identify FOUR** people who may be considered useful members of the accident investigation team.

8. **Outline** the issues to be considered to ensure an effective witness interview following a workplace accident.

9. (a) **Outline**, using a workplace example, the meaning of the terms:
    (i) near miss;
    (ii) dangerous occurrence.
   (b) **Explain** the purpose and benefits of collecting 'near-miss' incident data.
   (c) **Outline** how an 'accident ratio study' can contribute to an understanding of accident prevention.

10. An employee sustained a serious injury while using an unguarded drilling machine and was admitted to hospital where he remained for several days. The machine had been unguarded for several days before the accident.

    (a) **Outline** the legal requirements for reporting the accident to the enforcing authority.

   (b) **Identify** the possible immediate **AND** root causes of the accident.
   (c) **Outline** the actions that should be taken by the management to improve health and safety standards in the workplace following the accident.

11. With reference to the Reporting of Injuries, Diseases and Dangerous Occurrences Regulations 2013:

    (a) **Identify** the legal requirements for reporting a fatality resulting from an accident at work to an enforcing authority.
    (b) **Identify FOUR** types of specified (serious) injury and **THREE** reportable work-related illnesses.

12. **Outline** reasons why employers should keep records of occupational ill-health amongst employees.

13. An employee is claiming compensation for injuries received during a fall down a flight of stairs at the place of work. **Identify** the documented information required when preparing a possible defence against the claim.

APPENDIX 5.1 Workplace inspection exercises

Figures 5.12–5.15 show workplaces with numerous inadequately controlled hazards. They can be used to practise workplace inspections and risk assessments.

To see the safe versions of these scenes with the corrected faults listed, visit the book's companion site at: www.routledge.com/cw/hughes/

Figure 5.12 Construction site

Figure 5.13 Road repair

Figure 5.14 Workshop

Figure 5.15 Roof repair and unloading flammable liquids

APPENDIX 5.2 Information for insurance/compensation claims

Factual information needs to be collected where there is the likelihood of some form of claim either against the organisation or by the organisation (e.g. damage to equipment). This aspect should be considered as a second objective in accident/incident investigation, the first being to learn from the accident/incident to reduce the possibility of accidents/incidents occurring in the future.

Workplace claims

- accident book entry;
- first-aider report;
- surgery record;
- foreman/supervisor accident report;
- safety representatives accident report;
- RIDDOR report to HSE;
- other communications between defendants and HSE;
- minutes of Health and Safety Committee meeting(s) where accident/incident considered;
- report to Department of Work and Pensions;
- documents listed above relative to any previous accident/incident identified by the claimant and relied upon as proof of negligence.

Documents produced to comply with requirements of the Management of Health and Safety at Work Regulations 1999:

- pre-accident/incident risk assessment;
- post-accident/incident re-assessment;
- Accident/incident Investigation Report prepared in implementing the requirements; Health Surveillance Records in appropriate cases; information provided to employees;
- documents relating to the employee's health and safety training.

Workplace claims – examples of disclosure where specific Regulations apply

Section A – Workplace (Health, Safety and Welfare) Regulations 1992

- repair and maintenance records;
- housekeeping records;
- hazard warning signs or notices (traffic routes).

Section B – Provision and Use of Work Equipment Regulations 1998

- manufacturers' specifications and instructions in respect of relevant work equipment establishing its suitability to comply with Regulation 5;
- maintenance log/maintenance records required;

documents providing information and instructions to employees; documents provided to the employee in respect of training for use;

- any notice, sign or document relied upon as a defence against alleged breaches dealing with controls and control systems.

Section C – Personal Protective Equipment at Work Regulations 1992

- documents relating to the assessment of the PPE;
- documents relating to the maintenance and replacement of PPE;
- record of maintenance procedures for PPE;
- records of tests and examinations of PPE;
- documents providing information, instruction and training in relation to the PPE;
- instructions for use of PPE to include the manufacturers' instructions.

Section D – Manual Handling Operations Regulations 1992

- manual handling risk assessment carried out;
- re-assessment carried out post-accident;
- documents showing the information provided to the employee to give general indications related to the load and precise indications on the weight of the load and the heaviest side of the load if the centre of gravity was not positioned centrally;
- documents relating to training in respect of manual handling operations and training records.

Section E – Health and Safety (Display Screen Equipment) Regulations 1992

- analysis of workstations to assess and reduce risks;
- re-assessment of analysis of workstations to assess and reduce risks following development of symptoms by the claimant;
- documents detailing the provision of training including training records;
- documents providing information to employees.

Section F – Control of Substances Hazardous to Health Regulations 2002

- risk assessments and any reviews;
- copy labels from containers used for storage handling and disposal of carcinogens;
- warning signs identifying designation of areas and installations which may be contaminated by carcinogens;
- documents relating to the assessment of the PPE;
- documents relating to the maintenance and replacement of PPE;

- record of maintenance procedures for PPE;
- records of tests and examinations of PPE;
- documents providing information, instruction and training in relation to the PPE;
- instructions for use of PPE to include the manufacturers' instructions;
- air monitoring records for substances assigned a workplace exposure limit;
- maintenance examination and test of control measures records;
- monitoring records;
- health surveillance records;
- documents detailing information, instruction and training including training records for employees;
- labels and health and safety data sheets supplied to the employers.

Section G – Construction (Design and Management) Regulations 2015

- Documents relating to the appointment of principal designer and/or contractor;
- Documents relating to the pre-construction phase health and safety information;
- Documents relating to the construction phase health and safety plan;
- Documents relating to the health and safety file;
- Documents covering appropriate risk assessments method statements;
- Information on sub contractors working on site who are relevant to the incident concerned;
- Any relevant inspection records of the work site, scaffolds, excavations or pieces of plant.

APPENDIX 5.3 Checklist of items to be covered in a construction site inspection

This checklist identifies some of the hazards most commonly found on construction sites. The questions it asks are intended to help you decide whether your site is a safe and healthy place to work. **It is not an exhaustive list**. More detailed information can be found in HSG150 *Health and safety in construction* and other HSE publications.

Access on site

- Can everyone reach their place of work safely; for example, are roads, gangways, passageways, passenger hoists, staircases, ladders and scaffolds in good condition?
- Are access routes free from obstructions and clearly signposted?
- Are holes and openings securely guard-railed, provided with an equivalent standard of edge protection or provided with fixed, clearly marked covers to prevent falls?
- Are temporary structures stable, adequately braced and not overloaded?
- Will permanent structures remain stable during any refurbishment or demolition work?
- Is the site tidy, and are materials stored safely?
- Are there proper arrangements for collecting and disposing of waste materials?
- Is the work area adequately lit?
- Is lighting adequate especially when work is carried on after dark outside or inside buildings?

Welfare

- Have suitable and sufficient numbers of toilets been provided and are they kept clean and properly lit?
- Are there clean wash basins, hot and cold (or warm) water, soap and towels or means of drying?

- Are washbasins large enough to wash up to the elbow?
- Is suitable clothing provided for those who have to work in wet, dirty or otherwise adverse conditions?
- Are there facilities for changing, drying and storing clothes?
- Are drinking water and cups provided?
- Is there a site hut or other accommodation where workers can sit, make hot drinks and prepare food?
- Are welfare facilities easily and safely accessible to all who need to use them?

Scaffolds

- Are scaffolds erected, altered and dismantled by competent persons?
- Is there safe access to the scaffold platform?
- Are all uprights provided with base plates (and, where necessary, timber sole plates) or prevented in some other way from slipping or sinking?
- Are all the uprights, ledgers, braces and struts in position?
- Is the scaffold secured to the building or structure in enough places to prevent collapse?
- Are there double guard rails and toe boards or an equivalent standard of protection at every edge to prevent falling?
- Where guard rails and toe boards or similar are used:
  - ▷ are the toe boards at least 150 mm in height?
  - ▷ are upper guard rails positioned at a height of at least 910 mm above the work area?
  - ▷ are additional precautions, for example intermediate guard rails or brick guards, in place to ensure that there is no unprotected gap of more than 470 mm between the toe board and upper guard rail?

▶ Are the working platforms fully boarded and are the boards arranged to avoid tipping or tripping?

▶ Are there effective barriers or warning notices in place to stop people using an incomplete scaffold (e.g. where working platforms are not fully boarded)?

▶ Has the scaffold been designed and constructed to cope with the materials stored on it and are these distributed evenly?

▶ Does a competent person inspect the scaffold regularly, for example at least once a week, if the working platform is 2 m or above in height or at suitable intervals if less than 2 m, and always after it has been substantially altered or damaged and following extreme weather?

▶ Are the results of inspections recorded and kept?

▶ Have proprietary tower scaffolds been inspected and are they erected and used in accordance with suppliers' instructions?

▶ Are the wheels of tower scaffolds locked and outriggers deployed when in use and are the platforms empty when they are moved?

Powered access equipment

▶ Has the equipment been erected/installed by a competent person?

▶ Is fixed equipment (e.g. mast climbers) rigidly connected to the structure against which it is operating?

▶ Does the working platform have adequate secure guard rails and toe boards or other barriers to prevent people and materials failing off?

▶ Have precautions been taken to prevent people being struck by:
  ▷ the moving platform,
  ▷ projections from the building or
  ▷ falling materials,
  for example, by a barrier or fence around the base?

▶ Are the operators trained and competent?

▶ Is the safe working load clearly marked?

▶ Is the equipment inspected by a competent person?

▶ Is the power supply isolated and the equipment secured at the end of the working day?

Ladders

▶ Does the risk assessment conclude that ladders are the right way to do the job? Don't work from a ladder if there is a safer way using more suitable equipment.

▶ Are all ladders in good condition?

▶ Are they correctly inclined at a ratio of 1 out to 4 up?

▶ Are they secured to prevent them slipping sideways or outwards?

▶ Do ladders rise a sufficient height above their landing place (about 5 rungs or 1 m)? If not, are there other handholds available?

▶ Are the ladders positioned so that users don't have to overstretch or climb over obstacles to work?

▶ Does the ladder rest against a solid surface and not on fragile or insecure materials?

Roof work

▶ Are there enough barriers and is there other edge protection to stop people or materials falling from roofs?

▶ Do the roof battens provide safe hand and foot holds? If not, are crawling ladders or boards provided and used?

▶ During industrial roofing, are precautions taken (e.g. nets) to stop people falling from the leading edge of the roof or from fragile or partially fixed sheets which could give way?

▶ Where nets are used have they been rigged safely by a competent person?

▶ Have fragile surfaces been identified (e.g. fibre, cement sheets or roof lights)?

▶ Are suitable barriers, guard rails or covers, etc. provided where people pass or work near fragile material such as asbestos cement sheets and roof lights?

▶ Are crawling boards provided where work on fragile materials cannot be avoided?

▶ Are people excluded from the area below the roof work? If this is not possible, have additional precautions been taken to stop debris falling onto them?

Traffic, vehicles and plant

▶ Are vehicles and pedestrians kept separate wherever possible? If not are:
  ▷ they separated as much as possible and then barriers used?
  ▷ people told about the problem and what to do about it?
  ▷ warning signs displayed?

▶ Have one-way systems or turning points been provided to minimise the need for reversing?

▶ Where vehicles have to reverse, are they controlled by properly trained signallers?

▶ Are vehicles maintained; do the steering, handbrake and footbrake work properly?

▶ Have drivers received proper training and are they competent for the vehicles and plant they are operating?

▶ Are loads properly secured?

▶ Are passengers prevented from riding in dangerous positions and only on vehicles designed to carry them?

▶ Are vehicles and plant prevented from operating on dangerous slopes?

5

169

▶ Can zero tail swing excavators be used or is there adequate clearance around slewing vehicles?

Hoists

▶ Has the hoist and its equipment been installed by a competent person?

▶ Is the hoist protected by a substantial enclosure to prevent someone from being struck by any moving part of the hoist or falling down the hoistway?

▶ Are gates provided at all landings, including ground level?

▶ Are the gates kept shut except when the platform is at the landing?

▶ Are the controls arranged so that the hoist can be operated from one position only?

▶ Is the hoist operator trained and competent?

▶ Is the hoist's safe working load or rated capacity clearly marked?

▶ If the hoist is for materials only, is there a warning notice on the platform or cage to stop people riding on it?

▶ Does the hoist have a current report of thorough examination and a record of inspection by a competent person?

▶ Are the results of inspection recorded?

Cranes and lifting appliances

▶ Is the crane suitable for the job?

▶ Has the lift been properly planned by an 'Appointed person'?

▶ Is the crane on a firm level base? Are the riggers properly set?

▶ Are the safe working loads and corresponding radii known and considered before any lifting begins?

▶ Who is the appointed 'crane supervisor' responsible for controlling the lifting operation on site?

▶ Is the load secure?

▶ Are all operators trained and competent?

▶ Have arrangements been made to make sure that the driver can see the load or has a signaller been provided to help?

▶ Has the signaller/slinger been trained to give signals and to attach loads correctly?

▶ Do the operator and slinger find out the weight and centre of gravity of the load before trying to lift it?

▶ Does the crane have a current report of thorough examination and a record of inspection by a competent person?

▶ Are people stopped from walking beneath a raised load?

Excavations

▶ Is an adequate supply of timber, trench sheets, props or other supporting material made available

before excavation work begins, or has the excavation been sloped or battered back to a safe angle?

▶ If the sides of the excavation are sloped back or battered, is the angle of batter sufficient to prevent collapse?

▶ Is this material strong enough to support the sides?

▶ Is a safe method used for putting in the support, i.e. one that does not rely on people working within an unsupported trench?

▶ Is there safe access to the excavation (e.g. by a sufficiently long, secured ladder)?

▶ Are there guard rails or other equivalent protection to stop people falling in?

▶ Are properly secured stop blocks or other safeguards provided to prevent vehicles falling in?

▶ Does the excavation affect the stability of neighbouring structures or services?

▶ Are materials, spoil or plant stored away from the edge of the excavation in order to reduce the likelihood of a collapse of the side?

▶ Is the excavation regularly inspected by a competent person; for example, at the start of every shift, and after any accidental collapse or event likely to have affected its stability?

Manual handling

▶ Has the risk of manual handling injuries been assessed?

▶ Can lighter or smaller sizes of material be used, for example cement in 25 kg bags, and so avoid the repetitive laying of building blocks weighing more than 20 kg?

▶ Are hoists, telehandlers, wheelbarrows and other plant or equipment used so that manual lifting and handling of heavy objects is kept to a minimum?

▶ Have people been instructed and trained how to lift safely?

Tools and machinery

▶ Are the right tools and machinery being used for the job?

▶ Are all dangerous parts guarded (e.g. exposed gears, chain drives, projecting engine shafts)?

▶ Are guards secured and in good repair?

▶ Are tools and machinery maintained in good repair and are all safety devices operating correctly?

▶ Are all operators trained and competent?

Fire and emergencies

General

▶ Have emergency procedures been developed, for example, evacuating the site in case of fire or rescue from a confined space?

▶ Are people on site aware of the procedures?

▶ Is there a means of raising the alarm and does it work?

▶ Is there a way of contacting emergency services from site?

▶ Are there adequate escape routes and are these kept clear?

▶ Is there adequate first-aid provision?

Fire

▶ Is the quantity of flammable materials, liquids and gases on site kept to a minimum?

▶ Are there proper storage areas for flammable materials, flammable liquids and gases?

▶ Are containers and cylinders returned to ventilated stores at the end of the shift?

▶ Are suitable containers used for flammable liquids?

▶ If liquids are transferred from their original containers are the new containers suitable for flammable materials?

▶ Is smoking banned in areas where gases or flammable liquids are stored and used? Are other ignition sources also prohibited?

▶ Are gas cylinders, associated hoses and equipment properly maintained and in good condition?

▶ When gas cylinders are not in use, are the valves fully closed?

▶ Are adequate bins or skips provided for storing flammable and combustible waste?

▶ Is flammable and combustible waste removed regularly?

▶ Are the right number and type of fire extinguishers available and accessible?

Hazardous substances

▶ Have all harmful materials (e.g. asbestos, lead, solvents, paints, cement and dust) been identified?

▶ Have the risks to everyone who might be exposed to these substances been assessed?

▶ Has a check been made to see whether a licensed contractor is needed to deal with asbestos on site (most work with asbestos needs a licence, although some limited work can be done with asbestos without a licence)?

▶ Have precautions to prevent or control exposure to hazardous substances been identified and put into place by:
  ▷ doing the work in a different way, to remove the risk entirely;
  ▷ using less hazardous materials;
  ▷ using tools fitted with dust suppression (water lubrication) or extraction;
  ▷ providing suitable personal protective equipment?

▶ Have workers received information and training so they know what the risks are from the hazardous substances used and produced on site, and what they need to do to avoid the risks?

▶ Are there procedures to prevent contact with wet cement?

▶ Has health surveillance been arranged for people using certain hazardous substances like lead?

Noise

▶ Have workers received information and training so they know what the risks are from noise on site, and what they need to do to avoid the risks?

▶ Has workers' exposure to noise been assessed?

▶ Can the noise be reduced by using different working methods or selecting quieter plant, for example, by fitting breakers and other plant or machinery with silencers?

▶ Are barriers erected to reduce the spread of noise?

▶ Is work sequenced to minimise the number of people exposed to noise?

▶ Are others not involved in the work kept away from the source of noise?

▶ Is suitable hearing protection provided and worn in noisy areas?

▶ Have hearing protection zones been marked?

▶ Has health surveillance been arranged for people exposed to high levels of noise?

Hand–arm vibration

▶ Have workers received information and training so they know what the risks are from hand–arm vibration (HAV) on site, and what they need to do to avoid the risks?

▶ Have risks to workers from prolonged use of vibrating tools such as breakers, angle grinders, hammer drills and chainsaws been identified and assessed?

▶ Has exposure to HAV been reduced as much as possible by selecting suitable work methods and equipment?

▶ Have reduced vibration tools been used whenever possible?

▶ Are vibrating tools properly maintained?

▶ Have arrangements for health surveillance for people exposed to high levels of HAV (especially when exposed for long periods) been put in place?

Electricity and other services

▶ Have all necessary services been provided on site before work begins and have existing services present on site been identified (e.g. electricity cables, gas and water mains) and effective steps taken to prevent danger from them?

▶ Is the supply voltage for tools and equipment the lowest necessary for the job (could battery operated tools and reduced voltage systems, e.g. 110V, or even lower in wet conditions, be used)?

▶ Where mains voltage has to be used, are trip

devices (e.g. residual current devices (RCDs)) provided for all equipment?

▶ Are RCDs protected from damage, dust and dampness and checked daily by users?

▶ Are cables and leads protected from damage by sheathing, protective enclosures or positioning away from causes of damage?

▶ Are all connections to the system properly made and are suitable plugs used?

▶ Is there an appropriate system of user checks, formal visual examinations by site managers and combined inspection and test by competent persons for all tools and equipment?

▶ Are scaffolders, roofers, etc., or cranes or other plant, working near or under overhead lines? Has the electricity supply been turned off, or have other precautions, such as 'goal posts' or taped markers been provided to prevent them contacting the lines?

▶ Have underground electricity cables and other services been located (with a locator and plans) and marked, and precautions for safe digging and other work been taken?

Confined spaces

▶ Is work done in confined spaces (including tanks, pits, sewers and manholes) where there may be an inadequate supply of oxygen or the presence of dangerous or flammable gases, dusts or fumes?

▶ If so have all necessary precautions been taken, including operating a formal permit-to-work system?

Protective clothing

▶ Has adequate personal protective equipment (e.g. hard hats, safety boots, gloves, goggles and dust masks) been provided?

▶ Is the equipment in good condition and worn by all who need it?

Protecting the public

▶ Are the public fenced off or otherwise protected from the work?

▶ Are roadworks barriered off and lit, and is a safe alternative route provided?

▶ Are the public protected from falling material?

▶ Has a safe route been provided through roadworks for people with prams, wheelchair users and visually impaired people?

▶ When work has stopped for the day:
  ▷ are the gates secured?
  ▷ is the perimeter fencing secure and undamaged?
  ▷ are all ladders removed or their rungs boarded so that they cannot be used?
  ▷ are excavations and openings securely covered or fenced off?
  ▷ is all plant immobilised to prevent unauthorised use?
  ▷ are bricks and materials safely stacked?
  ▷ are flammable or dangerous substances locked away in secure storage places?

CHAPTER 6

Health and safety management systems – Audit and review – ACT

6.1 Health and safety auditing

6.1.1 Audits – definition, scope and purpose

The final ACT steps in the health and safety management control cycle are auditing, performance review and taking action on lessons learned. Organisations need to be able to reinforce, maintain and develop the ability to reduce risks and control hazards in the workplace. This is not a once and for all procedure and should form part of a continual improvement programme.

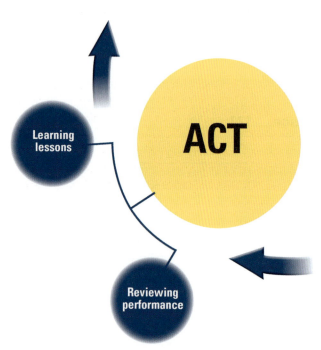

Figure 6.1 ACT part of the health and safety management system

Audit is a business discipline which is frequently used, for example in finance, environmental matters and quality. It can equally well be applied to health and safety.

The term is often used to mean inspection or other monitoring activity. Here, the following definition is used, which follows HSG65:

'The structured process of collecting independent information on the efficiency, effectiveness and reliability of the total health and safety management system and drawing up plans for corrective action.'

The **Safety Management System (SMS) Audit** is a major exercise, typically carried out every 2–4 years, as a means of assessing the adequacy of the whole organisation's SMS. It addresses all aspects of safety management in a structured manner, using written

questions. Answers will be confirmed by a review of records, staff interviews, and observation of workplaces and operations. The SMS Audit may benefit from being undertaken by a person or persons independent of the organisation and should, where practicable, be carried out in real time.

Over time, it is inevitable that control systems will decay and may even become obsolete as things change. Auditing is a way of supporting monitoring by providing managers with information. It will show how effectively plans and the components of health and safety management systems are being implemented. In addition, it will provide a check on the adequacy and effectiveness of the management arrangements and risk control systems (RCS).

Auditing is critical to a health and safety management system, but it is not a substitute for other essential parts of the system like workplace inspections (see Chapter 5). Companies need systems in place to manage cash flow and pay the bills – this cannot be managed through an annual or less frequent audit. In the same way, health and safety needs to be managed on a day-to-day basis and, for this, organisations need to have systems in place. A periodic audit will not achieve this.

The main difference between audits and other forms of monitoring (safety surveys, inspections, tours and sampling) is primarily the breadth and depth of an **audit**. **Surveys** look at only one aspect of the safety management system, inspections are frequent regular local monitoring normally carried out by line managers, **tours** concentrate on management commitment, **sampling** looks at only one area or subject over a short limited time.

The aims of auditing should be to establish that the three major components of a safety management system are in place and operating effectively. It should show that:

▶ appropriate management arrangements are in place;
▶ adequate Risk Control Systems (RCS) exist, are implemented and are consistent with the hazard profile of the organisation;
▶ appropriate workplace precautions are in place.

Where the organisation is spread over a number of sites, the management arrangements linking the centre with the business units and sites should be covered by the audit.

There are a number of ways in which this can be achieved and some parts of the system do not need auditing as often as others. For example, an audit to verify the implementation of RCSs would be made more frequently than a more overall audit of the capability of the organisation or of the management arrangements for health and safety. Critical RCSs, which control the principal hazards of the business, would need to be audited more frequently. Where there are complex workplace precautions, it may be necessary to undertake technical audits. An example would be chemical process plant integrity and control systems.

A well-structured auditing programme will give a comprehensive picture of the effectiveness of the health and safety management system in controlling risks. Such a programme will indicate when and how each component part will be audited. Managers, safety representatives and employees, working as a team, will effectively widen involvement and cooperation needed to put together the programme and implement it.

The process of auditing involves:

▶ gathering information from all levels of an organisation about the health and safety management system;
▶ making informed judgements about its adequacy and performance.

6.1.2 Pre-audit preparations

Decisions will need to be made about the level and detail of the audit before starting to gather information about the health and safety management of an organisation. Auditing involves sampling; so initially it is necessary to decide how much sampling is needed for the assessment to be reliable. The type of audit and its complexity will relate to its objectives and scope, to the size and complexity of the organisation and to the length of time that the existing health and safety management system has been in operation.

Information sources including interviewing people, looking at documents and checking physical conditions are usually approached in the following order and form the audit process (see Figure 6.2):

a) Preparatory work
▶ meet with relevant managers and employee representatives to discuss and agree the objectives and scope of the audit;
▶ prepare and agree the audit procedure with managers;
▶ gather and consider documentation.

b) On site
▶ interviewing;
▶ review and assessment of additional documents;
▶ observation of physical conditions and work activities.

c) Conclusion
▶ assemble the evidence;
▶ evaluate the evidence;
▶ write an audit report;
▶ presentation of findings to management and workforce representatives where appropriate.

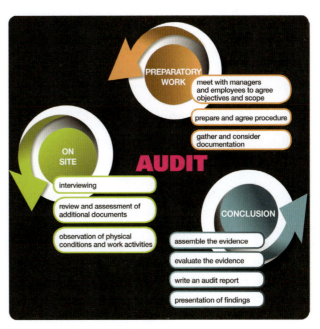

Figure 6.2 The Audit Process

It is essential to start with a relevant standard or benchmark against which the adequacy of a health and safety management system can be judged. If standards are not clear, assessment cannot be reliable. Audit judgements should be informed by legal standards, HSE guidance and applicable industry standards. HSG65 sets out benchmarks for management arrangements and for the design of Risk Control Systems (RCSs). This book follows the same concepts.

Auditing should not be seen as a fault-finding activity. It should make a valuable contribution to the health and safety management system and to learning. It should recognise achievement as well as highlight areas where more needs to be done.

Scoring systems can be used in auditing along with judgements and recommendations. This can be seen as a useful way to compare sites or monitor progress over time. However, there is no evidence that quantified results produce a more effective response than the use of qualitative evidence. Indeed, the introduction of a scoring system can, the HSE believes, have a negative effect, as it may encourage managers to place more emphasis on high-scoring questions which may not be as relevant to the development of an effective health and safety management system.

An organisation can use its own auditing system or one of the proprietary systems on the market or, as it is unlikely that any ready-made system will provide a perfect fit, a combination of both. With any scheme, cost and benefits have to be taken into account. Common problems include:

▶ systems can be too general in their approach. These may need considerable work to make them fit the needs and risks of the organisation;
▶ systems can be too cumbersome for the size and culture of the organisation;

- scoring systems may conceal problems in underlying detail;
- organisations may design their management system to gain maximum points rather than using one which suits the needs and hazard profile of the business.

The HSE encourages organisations to assess their health and safety management systems using in-house or proprietary schemes but without endorsing any particular one.

People selected for interviews need reasonable notice and should agree to be interviewed. They should be assured that the audit is not to apportion blame but to improve health and safety management in the organisation.

To achieve the best results, auditors should be competent people who are independent of the area and of the activities being audited. External consultants or staff from other areas of the organisation can be used. It is essential that auditors are experienced in meeting and interviewing people as well as knowledgeable in the subject being audited. In limited audits and smaller companies one person may carry out the audit but in larger sites and organisations it is often necessary to have a team approach with a variety of disciplines. Sufficient time must be allocated to achieve the objectives of the audit and it is essential that local management organise themselves to give adequate time to the audit process including pre-audit preparation, interviews, accompanying auditors on plant inspection and post-audit remedial work.

Figure 6.3 Using the audit questions for interviews and collecting information

6.1.3 Responsibility for audits – external v. internal audits

Directors and senior managers have a primary duty to ensure that they establish adequate systems for managing health and safety issues. Equally (as set out in 'Leading health and safety at work', HSE UK INDG417(rev1) and the ILO-OSH 2001) they have a duty to ensure that:

- appropriate weight is given to reporting both preventative information (such as progress of training and maintenance programmes) and incident data (such as accident and sickness absence rates);
- periodic audits of the effectiveness of management structures and risk controls for health and safety are carried out;
- the impact of changes such as the introduction of new procedures, work processes or products, or any major health and safety failure, is reported as soon as possible to the Board;
- there are procedures to implement new and changed legal requirements and to consider other external developments and events.

The summary of audit results and the fact that they are carried out should be part of directors' annual reporting procedures just as they are required to have financial audited reports. This gives confidence to workers, investors and customers that the company is operating proper health and safety management systems and considers the welfare of all those affected by its business operations. A failure to do this adequately has severely damaged the reputation of many large organisations.

Auditing may be carried out by internal or external consultants or, as is often the case, a combination of the two. Internal auditors:

- are familiar with the workplace, systems, processes and the organisation;
- are likely to be aware of what is practicable for the industry;
- have the ability to see improvements or a deterioration from the last audit;
- are familiar with the workforce and an individual's qualities and attitude; and
- may be less costly and easier to arrange than an external auditor.

But on the other hand internal auditors:

- may miss or gloss over some issues because of their familiarity;
- may not get honest views from the workforce for fear of the consequences;
- may not be in possession of recognised auditing skills;
- may not be up to date with legal requirements;
- may be less likely to be aware of best practice in other organisations;
- may be subject to pressure from management and the workforce; and
- have time constraints imposed upon them.

External auditors:

► come with a new perspective, a fresh pair of eyes;
► need to ask more questions to understand the systems in operation which can elicit underlying problems;
► may have solutions learnt elsewhere that would benefit the organisation;
► can be more impartial in their presentation of the audit results;
► are more likely to have the necessary auditing skills;
► will not be inhibited from criticising members of management or the workforce;
► are more likely to be up to date with legal requirements and best practice in other companies.

The disadvantage of external auditors is that they:

► are unlikely to be familiar with the workplace, tasks and processes;
► will not be familiar with the workforce and their individual attitudes to health and safety and will have difficulty in obtaining their full cooperation;
► may be unfamiliar with the industry and seek unrealistic standards; and
► may well be more costly than an internal member of staff.

Multi-site organisations often use a hybrid arrangement with cross-site audits where people from other locations inside the organisation will audit a particular site, perhaps with one independent external expert on the team.

6.1.4 Actions taken after the audit

It is essential for the organisation including, and especially, the board of directors to take the appropriate action after the audit report is received. It is often a very good idea for the lead auditor to make a presentation to senior managers on the findings of the audit and the recommendations. It may be necessary to ensure that some or all board members have some health and safety training to promote their understanding of the importance and recommendations from audit and other health and safety monitoring or investigation reports.

An implementation programme will probably be needed to ensure that adequate capital finance is made available. This will need to be closely monitored by a responsible manager allocated to the task. This should be periodically reviewed by senior managers to ensure that implementation is on track. A follow-up independent audit to check progress might be useful.

The audit report and recommendations should be a live document which helps to improve the health and safety management system and produce a new impetus to improve performance and enhance risk control systems. The results should be fed into formal internal reviews, covered in Section 6.2, to see whether policy or other changes are required.

Figure 6.4 The audit report should be reviewed by senior managers with an action plan and follow-up

6.2 Review of health and safety performance

6.2.1 Purpose of reviewing performance

When performance is reviewed, judgements are made about its adequacy and decisions are taken about how and when to rectify problems. The ACT part in the health and safety management system is needed by organisations so that they can see whether the health and safety management system is working as intended. This feedback on both successes and failures is an essential element to keep employees at all levels motivated to improve health and safety performance. Many successful organisations encourage positive reviews and concentrate on those indicators that demonstrate risk control improvements.

The information for review of performance comes from audits of RCSs (Risk Control Systems) and workplace precautions, and from the measurement of activities. There may be other influences, both internal and external, such as reorganisation, new legislation or changes in current good practice. These may result in the necessity to redesign or change parts of the health and safety management system or to alter its direction or objectives.

SMART (see Figure 2.5) performance standards need to be established which will identify the systems requiring change, responsibilities, and completion dates. It is essential to feed back the information about success and failure so that employees are motivated to maintain and improve performance.

In a review, the following areas will need to be examined:

► the operation and maintenance of the existing system;

▶ how the safety management system is designed, developed and installed to accommodate changing circumstances.

6.2.2 People involved and planned intervals

Reviewing is a continuous process. It should be undertaken at different levels within the organisation. Responses will be needed as follows:

▶ by first-line supervisors or other managers to remedy failures to implement workplace precautions which they observe in the course of routine activities;
▶ to remedy sub-standard performance identified by active and reactive monitoring;
▶ to the assessment of plans at individual, departmental, site, group or organisational level;
▶ to the results of audits.

Figure 6.5 Review of performance

Senior managers, departmental managers and health and safety professionals will be involved in many reviews where these are relevant to their responsibilities. Review plans may include:

▶ monthly reviews of individuals, supervisors or sections;
▶ three-monthly reviews of departments;
▶ annual reviews of sites or of the organisation as a whole.

The frequency of review at each level should be decided upon by the organisation and reviewing activities should be devised which will suit the measuring and auditing activities. The review will identify specific remedial actions which establish who is responsible for implementation and set deadlines for completion. Reviews should always be looking for opportunities to improve performance and be prepared to recommend changes to policies and the organisation to achieve

the necessary change. The output from reviews must be consistent with the organisation's policy, performance, resources and objectives. They must be properly documented since they will be the base line from which continual improvement will be measured.

6.2.3 Items to be considered in reviews

Reviews will be wide ranging and may cover one specific subject or a range of subjects for an area of the organisation. They should aim to include:

▶ evaluation of compliance with legal and organisational requirements;
▶ incident data, recommendations and action plans from investigations;
▶ inspections, surveys, tours and sampling;
▶ absences and sickness records and their analysis;
▶ any reports on quality assurance or environmental protection;
▶ audit results and implementation;
▶ monitoring of data, reports and records;
▶ communications from enforcing authorities and insurers;
▶ any developments in legal requirements or best practice within the industry;
▶ changed circumstances or processes;
▶ benchmarking with other similar organisations;
▶ complaints from neighbours, customers and the public;
▶ effectiveness of consultation and internal communications;
▶ whether health and safety objectives have been met;
▶ whether actions from previous reviews have been completed.

6.2.4 Role of directors and senior managers

(a) Reporting on performance

The management systems must allow the Board and senior managers to receive both specific (e.g. incident-led) and routine reports on the performance of health and safety policy. Much day-to-day health and safety information need be reported only at the time of a formal review. But only a strong system of monitoring can ensure that the formal review can proceed as planned – and that relevant events in the interim are brought to the Board's attention.

The Board should ensure that:

▶ appropriate weight is given to reporting both preventative information (results of active monitoring), such as progress of training and maintenance programmes, and incident data (results of reactive monitoring), such as accident and sickness absence rate;

- periodic audits of the effectiveness of management structures and risk controls for health and safety are carried out;
- the impact of changes such as the introduction of new procedures, work processes or products, or any major health and safety failure, is reported as soon as possible to the Board;
- there are procedures to implement new and changed legal requirements and to consider other external developments and events.

Good practice involves:

- effective monitoring of sickness absence and workplace health. This can alert the Board to underlying problems that could seriously damage performance or result in accidents and long-term illness;
- the collection of workplace health and safety data. This can allow the Board to benchmark the organisation's performance against others in its sector;
- appraisals of senior managers which includes an assessment of their contribution to health and safety performance;
- boards receiving regular reports on the health and safety performance and actions of contractors;
- winning greater support for health and safety by involving workers in monitoring.

(b) Reviewing health and safety

Reviewing performance should be supported at the highest level in the organisation and built into the safety management system. A formal boardroom review of health and safety performance is essential. It allows the Board to establish whether the essential health and safety principles – strong and active leadership, worker involvement, and assessment and review – have been embedded in the organisation. It tells senior managers whether their system is effective in managing risk and protecting people. Directors will need to know whether they are provided with sufficient information to make sound health and safety judgements about the organisation.

Performance on health and safety and well-being is increasingly being recorded in organisations' annual reports to investors and stakeholders. Board members can make extra 'shop floor' visits to gather information for the formal review. Good health and safety performance can be celebrated at central and local level both inside and outside the organisation (for example at RoSPA or IOSH annual awards).

The Board should review health and safety performance at least once a year. The review process should:

- examine whether the health and safety policy reflects the organisation's current priorities, plans and targets;

- examine whether risk management and other health and safety systems have been effectively reported to the board;
- report health and safety shortcomings, and the effect of all relevant board and management decisions;
- decide actions to address any weaknesses and a system to monitor their implementation;
- consider immediate reviews in the light of major shortcomings or events.

Setting up a separate risk management or health and safety committee as a subset of the board, chaired by a senior executive, can make sure the key issues are addressed and guard against time and effort being wasted on trivial risks and unnecessary bureaucracy. This is referred to in Chapter 3 in more detail. The results of these reviews needs to be properly recorded and fed into action and development plans for the coming year or so. This committee can ensure ongoing monitoring and review of performance and provide an annual report for the formal board review.

6.2.5 Continual improvement

Many organisations have a policy on continual improvement which should be applied to health

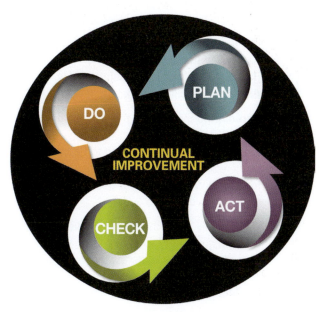

Figure 6.6 Continual improvement part of the health and safety management process

and safety management in the same way as other management issues. The health and safety commitment on continuous improvement might include statements like: 'The organisation will continuously seek to improve its safety performance.' Continual improvement of safety performance will be achieved through:

- active and reactive evaluations of facilities, equipment, documentation and procedures through safety audits and surveys;

▶ active evaluation of each individual's performance to verify the fulfilment of their safety responsibilities; and

▶ a reactive evaluation in order to verify the effectiveness of the system for control and mitigation of risk.

The organisation will also continuously seek to improve its safety management processes.

Measures that could be used to improve safety management include:

▶ more succinct procedures;

▶ improved safety reviews, studies and audits;

▶ improved reporting and analysis tools;

▶ improved hazards identification and risk assessment processes and improved awareness of risks in the organisation;

▶ improved relations with the sub-contractors, suppliers and customers regarding safety;

▶ improved communication processes, including feedback from the personnel.

6.3 Further information

HSE, 'Managing for health and safety' http://www.hse.gov.uk/managing/

HSE, The Health and Safety Toolbox: how to control risks at work http://www.hse.gov.uk/toolbox/

Managing for health and safety, HSG65 third edition 2013, HSE Books, ISBN 978-0-7176-6456-6 http://www.hse.gov.uk/pubns/priced/hsg65.pdf

Occupational Health and Safety Assessment Series (OHSAS 18000): OHSAS 18001:2007 Occupational health and safety management systems – Requirements ISBN 978-0-5805-9404-5, OHSAS18002:2008 ISBN 978-0-5806-2686-9 http://www.bsigroup.com/en/Assessment-and-certification-services/management-systems/Standards-and-Schemes/BSOHSAS-18001/

The Management of Health and Safety at Work Regulations 1999, see HSE management site at: http://www.hse.gov.uk/managing/index.htm

6.4 Practice revision questions

1. **Outline** how the following techniques may be used to improve health and safety performance within an organisation and the differences between them.
   (a) Safety inspections.
   (b) Externally led health and safety audits.
   (c) Analysis of accident statistics.

2. (a) **Explain** the meaning of the term health and safety *'audit'*.
   (b) **Outline** the issues that need to be considered at the planning stage of the audit.
   (c) **Identify TWO** methods of gathering information during an audit.

3. (a) **Outline** key areas that may be covered within a health and safety audit.
   (b) **Identify** the documents that are likely to be examined during a health and safety audit.
   (c) **Explain** how the findings of a health and safety audit can be used to improve health and safety performance.

4. **Outline FOUR** advantages **AND FOUR** disadvantages of undertaking a health and safety audit of an organisation's activities by:
   (a) an internal auditor; or
   (b) an external auditor.

5. A health and safety audit of a manufacturing company has identified a lack of compliance with many of its health and safety procedures.
   (a) **Describe** the possible reasons for procedures not being followed.
   (b) **Outline** the practical measures that could be taken to motivate employees to comply with health and safety procedures.

6. (a) **Outline** ways in which an organisation can monitor its health and safety performance.
   (b) **Identify EIGHT** measures that could be used by an organisation in order to monitor its health and safety performance.
   (c) **Outline** the reasons why an organisation should review and monitor its health and safety performance.

7. The Board of Directors of a large company decides to review its health and safety performance over the past year.
   (a) **Outline** the role of the Board in such a review.
   (b) **List** the topics that should be covered in the review.
   (c) **Outline** the possible review periods for the various parts of the company.

CHAPTER 7

Construction law and management

This chapter covers the following NEBOSH learning objectives:

1. Identify the scope, definition and particular issues relating to construction activities
2. Outline the legal, moral and financial consequences of failing to manage health and safety within the construction industry
3. Outline the scope and application of the Construction (Design and Management) Regulations 2015
4. Identify the nature and main sources of external construction health and safety information

---

7.1 The scope, definition and particular issues relating to construction activities

7.1.1 Introduction

Management of health and safety in the construction industry is more important now than it has ever been. Approximately 2.2 million people work in Britain's construction industry, making it the country's biggest industry. The number of construction projects has grown over the last decade in terms of both new build and refurbishment. The labour force is also changing as more and more migrant workers join the industry. At the same time, the fatalities and injuries to the workforce, even when measured as an incidence rate, are increasing. Although the record of the UK construction industry compares well with other countries, the fatal accident rate remains over four times the average for all other UK industries and remains one of the largest contributors to worker fatalities for any sector. The fragmented and peripatetic nature of the sector, the shortage of skills and casualisation present significant challenges to both the industry and the regulator.

This chapter covers the scope and nature of construction activities and the moral, legal and financial consequences of failing to manage health and safety in the construction industry. Details of the Construction (Design and Management) (CDM) Regulations are given. Since these regulations now incorporate the Construction (Health, Safety and Welfare) Regulations (CHSW), they form the legal foundations for the management of health and safety in construction activities.

It is important to stress that this chapter must be read in conjunction with Chapter 1 which dealt with the foundations of health and safety law.

7.1.2 The scope and definition of construction

The scope of the construction industry is very wide. The most common activity is general building work, which is domestic, commercial or industrial in nature. This work may be new building work, such as a building extension or, more commonly, the refurbishment, renovation, alteration, maintenance or repair of existing buildings. The buildings may be occupied or unoccupied. Such projects may begin with a partial or total demolition of a structure which is a particularly hazardous operation.

Larger civil engineering projects involving road and bridge building, water supply and sewage schemes, and river and canal work all come within the scope of construction.

Most construction projects cover a range of activities such as site clearance, the demolition or dismantling of building structures or plant and equipment, the felling of trees and the safe disposal of waste materials. The work could involve hazardous operations, such as roof work or excavation, or contact with hazardous materials, such as asbestos or lead. The site activities will include the loading, unloading and storage of materials and site movements of vehicles and pedestrians. Finally, the construction processes themselves are often hazardous.

Figure 7.1 Building site entrance

These processes include fabrication, decoration, cleaning, installation and the removal and maintenance of services (electricity, water, gas and telecommunications). Construction also includes the use of woodworking workshops together with woodworking machines and their associated hazards, painting and decorating and the use of heavy machinery. It will often require work to take place in confined spaces, such as excavations and underground chambers. At the end of most projects, the site is landscaped, which will introduce a new set of hazards.

7.1.3 Particular issues relating to the construction industry

Nearly all construction sites are temporary in nature and, during the construction process, are constantly changing. The workforce, itself, is transitory in nature and, at any given time, there are often many young people receiving training on site in the various construction trades. These trainees need supervision and structured training programmes. The level of numeracy and literacy among some of the workforce may make the communication of health and safety information difficult.

The progress of work on the site can be impeded by poor weather conditions leading to the possibility of contract penalties. Clients are usually keen to see the work finished as soon as possible and put additional pressure on the contractor. All these issues always lead to the temptation to compromise on health and safety issues, such as the provision of adequate welfare facilities or the safe re-routing of site traffic.

In recent years, some additional management issues have arisen with the increasing employment of migrant workers.

Figure 7.2 Demolition and ground clearance

7.1.4 Migrant workers in the construction industry

A recent survey has shown that up to 8% of the national construction workforce are migrant workers – in London, 40% of the construction workforce were migrants. Approximately 75% of this national migrant workforce lives in the South East of England. Overseas workers are valued in the UK construction industry, as they provide skills and fill labour shortages. Many are experienced tradespeople and some have entered the workforce through the Highly Skilled Migrants Programme (now closed for new applicants).

There is no direct evidence that these workers are more vulnerable to site accidents. However, they tend to be more vulnerable on small sites where health and safety management is poor. This increased vulnerability is caused by language difficulties, exploitation and lack of health and safety training. Between 2005 and 2008, 25 foreign workers were killed in the construction industry

in Britain. The biggest single cause of fatal injury was falls from height, accounting for 11 cases. In five further cases workers were overcome by fumes after running either a petrol- or diesel-fuelled generator or similar machine in a confined space.

Research published as part of the Secretary of State for Work and Pensions Inquiry into the 'Underlying Causes of Accidents in Construction' revealed that HSE Inspectors repeatedly identify four reasons why foreign or migrant workers may be more at risk than British-born workers. These are:

▶ language skills;
▶ unscrupulous managers or employers;
▶ inexperience or lack of understanding of UK Health and Safety standards;
▶ cultural differences.

Workers with limited English language skills can obtain a Construction Skills Certification Scheme (CSCS) card. The Construction Skills touchscreen health and safety test can be taken in a variety of languages or with the help of an interpreter. This test is a robust and independent assurance of a basic level of understanding of health and safety knowledge.

Construction Skills, the Sector Skills Council for the construction industry, has developed web-based materials to help integrate overseas workers into the industry. It provides construction employers with information and support tools to help ensure that any overseas workers they employ are properly qualified, competent and safe. Increasingly employers in construction are using multi-lingual supervisors, and translating instructions and guidance into the first language for their workers. Inductions, toolbox talks and training materials are becoming available in a range of languages.

Research by the HSE showed that many migrants were ignorant of their rights at work and had no knowledge of the HSE. The HSE rarely received complaints or phone calls from migrant workers. As a result of these concerns, the HSE established an outreach team whose brief was to:

▶ raise awareness of the HSE and UK health and safety standards within the migrant communities;
▶ assist HSE inspectors and other staff; and
▶ improve the HSE's knowledge of the targeted communities.

Over 110,000 pocket information cards, containing basic health and safety information, have been distributed to migrants and their families. Websites, migrant newspapers, radio and specialist help lines have also been used to communicate health and safety messages.

It is not easy to define where the responsibility for health and safety lies – with the gangmaster or the construction employer. It is recommended that there should be liaison between the two so that all relevant issues, such as personal protective equipment (PPE), are covered. The HSE have held various events to familiarise migrants with the important aspects of the health and safety regime in the United Kingdom and provided several translations of health and safety publications including the practical checklist in *Health and Safety in Construction HSG150*. As mentioned above, Construction Skills has developed web-based material to provide construction employers with information so that they can ensure that any migrants they employ are properly qualified, competent and safe. For many migrants there are cultural as well as language differences that need to be taken into account. Sometimes the cultural differences are more important than language differences.

Figure 7.3 Recent migrant workers, whose standards may not match those in Europe, are employed in the UK and the EU in general. In this instance, language was a problem, hard hats and gloves would have helped, boots were good protection but not steel-toed. Concrete delivery equipment was very up to date

Before migrants begin work, the following checks should be made by the employer:

▶ Are any special vocational skills required for the job? If so, ensure that any worker supplied has those qualifications or skills.
▶ Have they a reasonable command of written and spoken English?
▶ What information, instruction and training will need to be provided as a result of the risk assessment?

A special risk assessment is important in assessing the risks posed to migrant workers when on site. Particular issues include:

▶ the ability to speak and write English to a reasonable standard so that basic communication on site is possible;
▶ the required basic competencies in terms of literacy, numeracy, skills and past work experience;
▶ any physical and health attributes needed for the job;

▶ the compatibility or equivalence of foreign vocational qualifications with those required in the UK;
▶ the provision of necessary information, instruction, training and supervision;
▶ the provision of safe equipment, substances and PPE; and
▶ the provision of any special welfare and emergency arrangements.

The risk assessment will need to be reviewed on a very regular basis to reflect changes in migrant nationalities and changes to work processes and procedures.

It is important to give migrant workers relevant induction and general health and safety training and to ensure that such training is fully understood. The key points are to ensure that:

▶ essential induction information is given;
▶ any necessary job-related and/or vocational training is given;
▶ consideration is given to the needs of workers with a poor command of English; the use of accredited translation services will probably be needed together with visual non-verbal material, such as pictures, signs, videos and DVDs;
▶ any safety critical documents, such as safe systems of work, permits to work, method statements and emergency procedures, are translated by a professional interpreter;
▶ the workers know the identity of their supervisor and with whom they can raise any concerns about health and safety issues; and
▶ the workers have clearly understood all the information, instruction and training that they have received.

Finally, while it is not the responsibility of construction employers to provide transport to and from the site, it is reasonable for them to check that some form of safe transport is available to the migrant workers.

7.2 The legal, moral and financial consequences of failing to manage health and safety within the construction industry

7.2.1 Moral reasons

The moral reasons are reflected by the occupational accident and disease rates.

Accident rates

Accidents at work can lead to serious injury and even death. A major accident is a serious accident typically involving a fracture of a limb or a 24-hour stay in a hospital. An 'over 7-day accident' is an accident that leads to more than seven days off work. Statistics are collected on all people who are injured at construction workplaces, not just employees.

Figure 7.4 A serious accident waiting to happen on a small building site: no top guard on the circular saw – a very common safety fault

The main causes of fatal accidents in the construction industry are:

▶ falling through roof lights and fragile roofs;
▶ falling from ladders, scaffolds and other workplaces;
▶ being struck by excavators, lift trucks or dumpers;
▶ being struck by falling loads and equipment; and
▶ being crushed by collapsing structures.

There were 42 fatal injuries to workers in construction in 2013/14, 14 of these fatalities were to the self-employed. This compares with an average of 46 over the previous five years, including an average of 17 to the self-employed. The rate of fatal injury per 100,000 construction workers was 2.0 in 2013/14 compared with a five-year average of 2.1. There have been significant reductions in the numbers and rates of injury over the last 20 years or more. Nevertheless, construction remains a high-risk industry. Although it accounts for only about 5% of the employees in Britain it accounted for 31% of fatal injuries to employees, 10% of reported major/specified injuries and 6% of over 7-day injuries to employees. The types of fatalities in 2013/14 are shown in Table 7.1.

Table 7.1 Types of fatal injury in construction

Types of injury	Fatal injuries 2013/14	% of total fatal injuries to workers in construction
Fall from height	19	45%
Contact with machinery	3	7%
Struck by object	3	7%
Struck by moving vehicle	3	7%
Contact with electricity	3	7%
Struck against something fixed or stationary	1	2%
Slip, trip, fall on same level	1	2%
Trapped by something collapsing	1	2%

Source: HSE.

More than 2,000 construction workers suffer major injuries each year, many of which change lives forever, and each year around 50 people are killed on construction sites in Britain. With more than five incidents a day recorded in the construction sector, one of the most common causes of injury or fatality is falling from a height. Such incidents are usually preventable and are often the result of poor management. Table 7.2 shows the common causes of major injuries in the construction industry.

Table 7.2 Causes of major injuries in construction

Causes of major injuries	
Falls from height	31%
Slips, trips and falls on level	27%
Struck by moving/falling object	13%
Handling objects	9%

Source: HSE.

The construction industry has one of the highest rates of fatal injuries, making it one of the most dangerous industries in which to work. Table 7.3 shows the number of accidents in various industries over a three-year period. These figures indicate that there is a need for health and safety awareness even in occupations which many would consider very low hazard, such as schools and health and social work. In fact, 78% of all deaths occur in the service sector and manufacturing appears safer than construction.

Table 7.3 Accidents to all people in various employment sectors over a three-year period

Sector	Deaths	Major*
Service industries	420 (78%)	18,898
Manufacturing	26 (5%)	4,087
Construction	47 (9%)	3,120
Agriculture	45 (8%)	640
Total	538	26,745

Source: HSE.

Note:

* Major accident figures for members of the public are not available.

There is an unproven concern that the increase in the number of migrant workers in the industry may be one of the causes of this level of accidents. The highest number of fatalities occurred in the repair, refurbishment and new-build housing sector of the industry. Table 7.4 shows the proportion of fatalities in various construction activities over a 10-year period and it can be seen that the largest proportion of fatalities occurred in refurbishment work, particularly in the domestic sector.

7

Table 7.4 Proportion (%) of fatalities in various construction activities

Construction activity	Percentage
Refurbishment/repair/maintenance (domestic)	34
Refurbishment/repair/maintenance (non-domestic)	34
New build – housing	12
New build – commercial	5
New build – industrial	2.5
Roadworks	9
Civil engineering	0
Other/demolition	3

Source: HSE.

The occupations where the proportion of major injuries due to falls from height was higher included:

▶ painters and decorators (62%);
▶ bricklayers and masons (38%);
▶ scaffolders, stagers and steeplejacks (36%); and
▶ electrical fitters (36%).

Injuries due to handling, lifting and carrying and being hit by moving/falling objects, each accounted for 16% of the major injuries in the construction industry.

Disease rates

Work-related ill-health and occupational diseases can lead to long absences from work and even death. In the construction industry, recent figures indicate that average prevalence rates for all illnesses are similar to those found in other industries with some notable exceptions.

Table 7.5 shows the annual cases seen by specialist doctors in the construction and all industries over a three-year period.

The incidence rates show that asbestos-related diseases, musculoskeletal disorders and the effects of exposure to vibration are far more prevalent in the construction industry. Diffuse pleural thickening of the lung lining is thought to be principally due to exposure to asbestos but can be caused by other diseases, such as tuberculosis, or certain drugs. One positive fact is that the construction industry has one of the lowest rates for stress.

Over a three-year period an estimated 70,000 construction workers in the UK suffered ill-health as a result of their work. Vibration white finger, carpal tunnel syndrome, occupational deafness, musculoskeletal disorders and dermatitis are the most common cases of non-lung diseases in the construction industry. There has been a significant decline in the incidence of allergic contact dermatitis since 2005 when the marketing and use of cement containing more than two parts per million of chromate was restricted in the industry.

Airborne materials from spray painting, welding or cutting/grinding metals, and dusts from stone, cement, brick or concrete were all implicated as significant

Table 7.5 Annual cases and incidence rates for work-related ill-health seen by the Health and Occupational Reporting Network Disease Specialists over a three-year period

	Construction		All industries	
	Average annual cases	Rate per 100,000	Average annual cases	Rate per 100,000
Diffuse pleural thickening	506	23.0	735	3.0
Mesothelioma	363	17.5	490	2.0
All musculoskeletal disorders	160	7.2	1,350	5.5
Asbestosis	66	3.0	122	0.5
Dermatitis	110	5.0	980	4.0
Vibration white finger	44	2.0	73.5	0.3
Stress	66	3.0	1,102	4.5
Asthma	11	0.5	245	1.0

causes of 'breathing or lung problems'. Analysis by industry has shown that the construction industry has the largest burden of occupational cancer amongst the industrial sectors; over 40% of the occupational cancer deaths and cancer registrations were from construction. Most of them were caused by past exposures to asbestos and silica. Solar radiation, coal tars and pitches were responsible for an additional 1,300 cancer registrations.

The most significant carcinogen is still past exposure to asbestos (69%), followed by silica (17%), painting and diesel engine exhaust (6–7% each).

An estimated 1.7 million working days were lost due to ill-health in 2013/14 – 0.8 days per worker.

7.2.2 Legal reasons

In 2002, the Construction Division of the HSE undertook four construction blitzes across the country. They concentrated on falls from height and gave advance warning of their visits. Work was stopped on nearly 50% of sites visited. A number of prosecutions followed involving large as well as small companies. Thus there were some clear legal reasons for implementing sound health and safety management systems.

Since then there have been several similar large-scale unannounced site inspections across the UK. During the inspections, HSE inspectors considered whether:

▶ jobs that involve working at height have been identified and properly planned to ensure that appropriate precautions, such as proper support of structures, are in place;
▶ equipment is correctly installed/assembled, inspected, maintained and used properly;

proper monitoring and control arrangements to prevent unnecessary exposure to harmful dusts are in place;

▶ sites are well organised, to avoid trips and falls, walkways and stairs are free from obstructions; and

▶ work areas are clear of unnecessary materials and waste and welfare facilities are adequate.

Almost one in five of the construction sites visited in one recent group of inspections required enforcement action to be taken. A total of 631 enforcement notices was served across 433 sites for dangerous work practices, with 451 notices ordering that work cease immediately until problems were rectified. The inspectors also encountered numerous examples of poor practice – from lack of edge protection on stairwells and scaffolding, to unsafe storage of flammable materials and inadequate personal protective equipment. Most of the enforcement action taken was against dangerous work at height. The HSE continue to focus on monitoring health and safety in refurbishment work as well as new-build sites.

Over a three-year period, local authorities, through their Environmental Health Officers, served 40% of all improvement notices and 20% of all prohibition notices – the remainder were served by the HSE. Although construction received only 15% of all improvement notices, it received 59% of all prohibition notices. The HSE present 80% of all prosecutions and the remainder are presented by Local Authority Environmental Health Officers. For construction, most of these prosecutions are for infringements of the various Construction Regulations and the Provision and Use of Work Equipment Regulations.

7.2.3 Financial reasons

Costs of accidents

Any accident or incidence of ill-health causes both direct and indirect costs and incurs an insured and an uninsured cost. This is particularly the case in the construction industry where the cost of non-injury accidents is considerably higher than the costs of ill-health and injury accidents. A non-injury accident is an accident that causes damage or loss to property, plant, materials, the environment or a business opportunity. An estimated 2.3 million working days were lost in 2013/14, 1.7 million due to ill-health and 592,000 due to workplace injury, making a total of 1.1 days lost per worker. It has been estimated by the HSE that these injuries and new cases of ill-health resulting largely from current working conditions of workers in construction cost society over £1.1 billion a year. The costs to the construction industry are also high and non-injury incident costs are 80% of the total costs and between 3 and 11 times the injury accident costs.

Direct costs and indirect costs

Direct costs are directly related to the accident. They may be insured (claims on employers' and public liability insurance, damage to buildings, equipment or vehicles) or uninsured (fines, sick pay, damage to product, equipment or process). **Indirect costs** may also be insured (business loss, product or process liability) or uninsured (loss of goodwill, extra overtime payments, accident investigation time, production delays).

It is important that all of these costs are taken into account when the full cost of an accident is calculated.

Chapter 1 covered this topic in detail and most of the costs mentioned are also relevant to construction. Probably one of the most significant costs for the construction industry is contract penalty costs.

7.2.4 The role of the regulator in the improvement of health and safety in the construction industry

The health and safety problem in the construction industry is its poor record when compared with the other parts of British industry. This performance deteriorated in 2000 and certain actions were taken by the HSE. A new construction division was launched in April 2002 and a new intervention strategy was developed. Clients and developers as well as construction sites were targeted.

Several initiatives have been undertaken by the government and HSE including the use of spot site inspections mentioned earlier. By 2010, the number of fatalities was 42, which was a reduction from the 1999/2000 figures of 39 (48%) and the number of major accidents was 3,120 which was a reduction of 1,629 (34%).

The Secretary of State for Work and Pensions convened a Construction Forum which was attended by the government, the HSE, employers, trade unions, trade associations, contractors and suppliers. A Framework for Action was agreed to reduce the number of future fatalities and major accidents. Key areas singled out for urgent action were:

▶ sharing best practice – particularly in house building, repair and refurbishment projects;

▶ ensuring that all construction workers receive induction training before starting work on a site;

▶ raising competence levels of all site workers in the house-building sector, enabling them to carry a Construction Skills Certification Scheme (CSCS) card, or be able to demonstrate their competence;

▶ encouraging worker involvement, by ensuring that all projects include trade union and worker safety representatives;

▶ tackling the problem of the informal economy and the use of casual labour.

In contrast to the situation in the United Kingdom, the Republic of Ireland almost halved its construction fatalities during the construction boom years and increased its site inspections by 13%. The Irish Health and Safety Authority also introduced safe system of work plans for house building, ground works and demolition. These plans, in the form of pictograms, must be completed and signed off by the site manager and shared with the rest of the site team.

The HSE has published on their website a list of topics that they will be checking during site visits. These include:

► competence and control of contractors and employees;
► risk assessments, particularly for work at height;
► communication of risk controls to the workforce;
► selection, use, inspection and maintenance of equipment; and
► welfare facilities and site tidiness.

7.3 Scope and application of the Construction (Design and Management) Regulations 2015

7.3.1 Introduction

The Construction (Design and Management) Regulations 2015 (CDM 2015) replaced those of CDM 2007 in April 2015 – there were transitional arrangements for projects that had already started before that date. Under CDM 2015, it is the duty of the client to inform the HSE for a construction project which lasts 30 days with more than 20 workers occupying the site simultaneously or exceeds 500 person days. The role of the client has received greater emphasis under CDM 2015. The client must maintain and review the CDM arrangements to ensure they remain relevant throughout the life of the project. The CDM coordinator (created under CDM 2007) is replaced with the principal designer who is appointed by the client. The need to appoint a principal designer occurs when a project involves more than one contractor on site. The HSE hope that the introduction of the principal designer, who has responsibilities for health and safety in the design team, will ensure that health and safety may be integrated within the project from the outset. Domestic clients also have duties for domestic projects, which can be transferred to the principal designer and/or principal contractor.

The Regulations remove the need to assess competency and have been replaced with a legal obligation on the client to ensure that the parties they appoint are able to demonstrate appropriate information, instruction, training and supervision. The construction phase coordination duties remain with the principal contractor, but the current proposals do not make any provision for an independent role, as was provided by

the CDM coordinator, to protect the client. The HSE has produced detailed guidance on the responsibilities of the five duty-holders under CDM and one for workers, and the actions that are required of them to deliver a safe and healthy construction project. This guidance is available on the HSE website. The technical standards set out in Part 4 of the new Regulations will remain essentially unchanged from those in guidance related to CDM 2007. The HSE's targeting and enforcement policy, as a proportionate and modern regulator, also remains unchanged.

In Chapter 1, a comprehensive description of CDM 2015 was given and this covered the following topics:

► The duties of the client, principal designer and the designers, principal contractor and the contractors, workers and domestic clients.
► The criteria for the notification of a construction project to the HSE.
► The meaning and contents of:
  ▷ pre-construction health and safety information;
  ▷ the construction phase health and safety plan; and
  ▷ the health and safety file.

However, the role of the principal designer will be covered in a little more detail since the role is new to CDM.

Figure 7.5 Design and management of construction work

7.3.2 The principal designer

The principal designer is the designer with control over the pre-construction phase of the project and is appointed in writing by the client. The principal designer is usually an organisation (or, on smaller projects, an individual) that has:

(a) technical knowledge of the construction industry relevant to the project; and
(b) the understanding and skills to manage and coordinate the pre-construction phase, including any design work carried out after construction begins.

Where the principal designer is an organisation it should have the organisational capability to undertake the role in addition to the necessary skills, knowledge and experience required of an individual designer. Working with the client and principal contractor, the principal designer has an important role in influencing how the risks to health and safety should be managed and incorporated into the wider management of a project. Decisions about the design taken during the pre-construction phase can have a significant effect on whether the project is delivered in a way that secures health and safety. The role of the principal designer involves coordinating the work of others in the project team to ensure that significant and foreseeable risks are managed throughout the design process.

The principal designer must liaise with the principal contractor for the duration of their appointment. During the pre-construction phase this must cover sharing information that may affect the planning, management, monitoring and coordination of the construction phase – in particular, the information needed by the principal contractor to prepare the construction phase plan. This liaison should extend into the construction phase to deal with ongoing design and to obtain information for the health and safety file. The principal designer should also arrange for the handover of the health and safety file to the principal contractor and alert them to any issues that may need to be taken into account in reviewing, updating and completing it.

7.3.3 The appointment and required competence of relevant parties

Before making an appointment, those responsible for appointing people, such as the principal designer, principal contractor, designers and contractors, to work on a construction project must ensure that they have the skills, knowledge and experience to undertake the work in a way that secures health and safety. If an organisation is to be appointed, it must also have the appropriate organisational capability.

Clients must appoint principal designers and principal contractors as soon as practicable and before the start of the construction phase, so they have enough time to plan and manage both the pre-construction and construction phases.

The principal designer should be appointed as early as possible in the design process, if practicable at the concept stage. Appointing the principal designer early will provide the client with help in matters such as assembling the pre-construction information. The duration of the principal designer's appointment should take into account any design work which may continue into the construction phase or any issues that may arise during construction involving the need to make suitable modifications to the designs.

The principal contractor should be appointed early enough in the pre-construction phase to help the client to drawn up the construction phase plan before the

construction phase begins. An early appointment will also enable the principal contractor to liaise with the principal designer. If a client fails to appoint either the principal designer or principal contractor, then the client must carry out their duties.

Where a principal designer's appointment finishes before the end of the project, the client should ensure that the principal designer passes the health and safety file to the principal contractor so that it can be revised during the remainder of the project if necessary.

The required competence of relevant parties is split into its component parts of skills, knowledge, training and experience, and – if it relates to an organisation – organisational capability. This will provide clarity and help the industry to both assess and demonstrate that construction project teams have the right attributes to deliver a healthy and safe project. When a contractor employs or appoints an individual to work on a construction site, they should make enquiries that the individual:

(a) has the necessary skills, knowledge, training and experience to undertake the advertised job in a way that secures health and safety for anyone working on the site; or

(b) are in the process of obtaining them.

Reliance should not be placed on an industry certification card or similar as evidence that a worker has the necessary qualities. Nationally recognised qualifications (such as NVQs and SVQs) will provide contractors with assurance that the holder has the necessary skills, knowledge, training and experience to carry out a task in such a way. Contractors should recognise that training on its own is not enough. Newly trained individuals need to be supervised and given the opportunity to gain positive experience of working in a range of conditions.

Those making appointments will find the questions that are incorporated in *Publicly Available Specification 91, 2013: Construction related procurement – Prequalification Questionnaires* (PAS 91)3 a useful aid. PAS 91 sets out standardised pre-qualification questions that include health and safety questions along with other question sets on matters such as financial information. This is one way of assessing organisational capability.

In addition to making pre-qualification checks on organisations, those responsible for making appointments should also check that the designer or contractor has sufficient experience and a good track record in managing the health and safety risks involved in the project. These checks are ideally carried out at the final stage after pre-qualification checks have been completed and before appointments are made.

When considering the requirements for designers, architects and other construction professionals, membership of an established professional institution or body relevant to the advertised post should be expected. However, questions should still be asked of individuals to ensure that they have sufficient skills, knowledge and experience required for the post and have kept those capabilities up to date through continuing professional development.

Companies that are members of the Safety Schemes in Procurement (SSIP) Forum, an umbrella body with binding agreements in place to ensure member schemes recognise each other's pre-qualification assessments, can provide pre-qualification assessment services. Further information about SSIP is available from their website (www.ssip.org.uk/). SSIP assessment is one way in which a designer or contractor can demonstrate organisational capability at the pre-qualification stage of the appointment process, but it is not the only way.

7.3.4 Notification of projects

A project is notifiable if the construction work on a construction site is scheduled to:

(a) last longer than 30 working days and have more than 20 workers working simultaneously at any point in the project; or

(b) exceed 500 person days.

Where a project is notifiable, the client must give notice in writing to the relevant enforcing authority (HSE, Office of Rail Regulation (ORR) or Office for Nuclear Regulation (ONR)) as soon as is practicable before the construction phase begins. All days on which construction work is likely to take place (including weekends and bank holidays) count towards the period of construction work. If a construction project is not notifiable at first, but there are subsequent changes to its scope so that it becomes notifiable, the client must notify the work to the relevant enforcing authority as soon as possible. The notice should be submitted as soon as practicable before the construction phase

Figure 7.6 (a) Domestic client: CDM applies but only a short duration contract, no notification required – would be notifiable if a large project

Figure 7.6 (b) Large site (over 30 days, more than 20 workers simultaneously or exceeds 500 person days): CDM applies and client must notify the relevant enforcing authority

begins and any modifications or updates to the notification should be sent making clear that they relate to an earlier notification. The client must ensure that an up-to-date copy of the notice is displayed in the construction site office so that it is accessible to anyone working on the site and in a form that can be easily understood.

The easiest way to notify any project is to use the electronic F10 notification form on the HSE's website. Further information on how to notify construction work is on the HSE's construction web pages (www.hse.gov.uk/construction).

7.3.5 Preparation of pre-construction information

Pre-construction information provides the health and safety information needed by:

▶ designers and contractors who are bidding for work on the project, or who have already been appointed, to enable them to carry out their duties;

▶ principal designers and principal contractors in planning, managing, monitoring and coordinating the work of the project.

Pre-construction information is information that is already in the possession of the client (such as any existing health and safety file, an asbestos survey and structural drawings) or can be obtained by reasonable enquiry. The information must be relevant to the project, have an appropriate level of detail and be proportionate given the nature of the health or safety risks.

The client has the main duty for the provision of pre-construction information. This must be provided as soon as practicable to each designer (including the principal designer) and contractor (including the principal contractor) who is bidding for work on the project or has already been appointed. For projects involving more than one contractor, the client should expect the principal designer to help in bringing the pre-construction information together and provide it to the designers and contractors appointed, or being considered for appointment. The information should be in a convenient form and be clear, concise and easily understandable to help other duty-holders involved in the project with their duties. It will evolve as the project progresses towards the construction phase. The extent of the help required will depend on the nature of the project, the risks involved and the client's level of knowledge and experience of construction work.

The client and the principal designer must also decide when designers and contractors will need pre-construction information to enable them to carry out their duties so that:

▶ designers or contractors who are seeking appointment for work on the project should have sufficient information made available to them at a time which allows them to put together a bid based on a clear understanding of the nature of the work involved;

▶ designers already appointed should be provided with sufficient information (when it becomes available) at a stage early enough to enable them to judge whether it is reasonably practicable to eliminate any foreseeable health and safety risks in the design process and, where it is not, the steps that they should take to reduce or control the remaining risks; and

▶ contractors already appointed should be provided with the information that they will need to plan, manage and monitor their work.

Neither the principal contractor nor the contractor have specific duties in relation to pre-construction information. However, they must liaise with the principal designer for the duration of the principal designer's appointment and share any information relevant to the planning, management, monitoring or coordination of the pre-construction phase.

More information on pre-construction information is given in Appendix 1.2 of Chapter 1.

7.3.6 The construction phase plan

The client must ensure that a construction phase plan for the project is prepared before the construction phase begins. The plan outlines the health and safety arrangements, site rules and specific measures concerning any work involving the particular risks listed in Schedule 3 of CDM 2015 (such as hazardous chemical or biological substances and work near high-voltage power lines). For single-contractor projects, the contractor must ensure the plan is prepared, and for projects involving more than one contractor, it is the duty of the principal contractor.

The client must ensure that the principal contractor (or, where relevant, the contractor) is provided with all the available relevant information, such as the pre-construction information, that they need to draw up the plan. The client must also ensure that:

(a) when it is drawn up, the plan adequately addresses the arrangements for managing the risks; and

(b) the principal contractor (or contractor) regularly reviews and revises the plan to ensure that the plan takes account of any changes that occur as construction progresses and that it continues to be fit for purpose.

Figure 7.7 Protection of the public in main shopping area

The information included in the plan should be relevant to the project and have sufficient detail to enable the arrangements, site rules and special measures needed to manage the construction phase to be clear to all project personnel. However, it should be proportionate to the scale and complexity of the project and the risks involved. It should not contain irrelevant documents such as generic risk assessments or generic method statements. The following topics should be included:

1. A description of the project
2. The management of the work
3. The arrangements for controlling the significant site risks.

More detail on these topics is given in Appendix 1.3 of Chapter 1.

The principal designer must help the principal contractor to prepare the construction phase plan by providing any relevant information that they hold. Before the start of the construction phase, the principal designer should regularly check that the principal contractor has the necessary information to prepare the plan. They must continue to liaise as the construction phase progresses to share any information that is relevant to the planning and management of the construction phase. The plan must be drawn up or arrangements made for it to be drawn up during the pre-construction phase before the construction site is set up. For projects involving more than one contractor, the principal contractor must take the lead in preparing, reviewing, updating and revising the construction phase plan. The principal contractor must also liaise with the contractors to ensure that the plan takes into account their views on the arrangements for managing the construction phase and ensure that the construction phase plan is appropriately reviewed, updated and revised from time to time. The planning process must consider the risks to all those affected – workers, members of the public and the client's employees if working in an occupied premises. It must cover:

(a) the risks likely to arise during construction work;
(b) the measures needed to protect those affected by planning to provide:
 ▷ and maintain the right plant and equipment;
 ▷ the necessary information, instruction and training; and
 ▷ the right level of supervision;
(c) the resources needed to organise and deliver the work including its management, monitoring and coordination.

Where the plan includes site rules, the rules should cover (but not be limited to) topics such as personal protective equipment (PPE), parking, use of radios and mobile phones, smoking, restricted areas, hot works and emergency arrangements. The rules should be clear and easily understandable and should be brought to the attention of everyone on site who should be expected to follow them. Further guidance, including a template for a construction phase plan, is provided for contractors working on small-scale, routine and domestic projects on the HSE website.

7.3.7 The level of supervision, information and instructions

The level of supervision, instructions and information required will depend on the health and safety risks involved in the project and the level of skills, knowledge, training and experience of the workforce. Contractors (including principal contractors) must ensure that supervision is effective and that suitable site inductions are provided together with information that will be needed – such as the procedures to be followed in the event of serious and imminent danger to health and safety.

Supervision

Supervisors are a vital part of effective management arrangements. Effective supervisors are those who have the skills, knowledge, training, experience and leadership qualities appropriate to the construction project. Good communication and people management skills on site are important qualities for supervisors. Where site workers are promoted to a supervisory role, they should be provided with nationally recognised site supervisor training which includes components covering leadership and communication skills.

The role of the supervisor may include team leading, briefing and carrying out toolbox talks. It may also include coaching and encouragement of individual workers and supporting other formal and informal means of engaging with workers. The supervisor has a particularly important part to play as a front-line decision maker in emergencies or when workers on site face immediate risks that may require work to stop.

A contractor who employs workers or manages workers under their control must ensure that appropriate supervision is provided. The level of supervision provided will depend on the risks to health and safety involved, and the skills, knowledge, training and experience of the workers concerned. Workers will require closer supervision if they are young, inexperienced, or starting a new work activity. In these circumstances, arrangements should be put in place for supervision to continue even when the supervisor is not present. Other factors that should be considered when assessing the level of supervision needed include the level of individuals' safety awareness, education, physical agility, literacy and attitude. Even experienced workers may need an appropriate level of supervision if they do not have some or all of the skills, knowledge, training and experience required for the job and the risks involved.

Figure 7.8 Contractors at work

Information and instruction

Any duty-holder under CDM 2015 who is required to provide health and safety information or instruction to others must ensure that it is easy to understand.

Information about hazards is essential to all project workers and managers to ensure they understand the risks involved with the work. Instructions are those agreed actions that must be followed to prevent or minimise those risks.

Any information or instruction provided should be in simple, clear English (and/or other languages where appropriate) and it should also be set out in logical order (possibly with illustrations). The use of photographs or diagrams in written communication can also be useful. The amount of detail should be proportionate to the scale and complexity of the project, the health and safety risks, and the nature and purpose of the messages that need to be communicated. Examples include the:

(a) pre-construction information the client is required to provide to designers and contractors;

(b) health and safety information about the design that designers are required to provide to other duty-holders;

(c) information that the principal designer must provide to enable preparation of the construction phase plan;

(d) site rules that are part of the construction phase plan; and

(e) information that principal contractors must provide to workers (or worker representatives).

Figure 7.9 Barriers to prevent unauthorised entry – also advertising involvement with the Considerate Constructors Scheme

Site inductions

The principal contractor must ensure a suitable site induction is provided to every site worker. The induction should be site specific and highlight any particular risks (including those listed in Schedule 3) and the associated control measures. The following issues should be considered for inclusion in a site induction:

▶ the commitment of senior management to health and safety;

▶ outline of the project;

▶ management of the project;

▶ first-aid arrangements;

▶ accident and incident reporting arrangements;

▶ arrangements for briefing workers on an ongoing basis, such as toolbox talks;

▶ arrangements for consulting the workforce on health and safety matters;

▶ individual responsibility of workers for health and safety.

Site inductions should also be provided to those who do not regularly work on the site, but who visit it on an occasional (e.g. architects) or once-only basis (e.g. students). The inductions should be proportionate to the nature of the visit. Inductions provided to escorted visitors need not have the detail that unescorted visitors should have. Escorted visitors only need to be made aware of the main hazards and control measures.

Consultation

Proper consultation with the workforce is to be encouraged. Consultation about health and safety is a two-way process. It involves giving information to workers, listening to them and taking into account their views before making decisions. For example, meetings between managers and workers could be held before work starts to discuss the work planned that day, identify all the health and safety risks and agree the appropriate control measures.

7.3.8 The preparation of the health and safety file

A health and safety file is only required for projects involving more than one contractor. The client must ensure that the principal designer prepares a health and safety file for their project. Its purpose is to ensure that, at the end of the project, the client has information about health and safety risks that will be required by anyone carrying out subsequent construction work on the building. To ensure that an appropriate health and safety file is produced at the end of the project, the client must:

▶ provide the principal designer with any existing file produced as part of an earlier project to enable the information it contains to be used to plan the pre-construction phase of the current project;

▶ ensure the principal designer prepares a new file (or revises any existing one); and

▶ as the project progresses, the client must ensure that the principal designer regularly updates, reviews and revises the health and safety file to take account of the work and any changes that have occurred.

The principal designer has primary responsibility for preparing the file, and reviewing, updating and revising it as the project progresses. If their appointment continues to the end of the project they must also pass the completed file to the client to retain. If the principal designer's appointment finishes before the end of the project, the file must be passed to the principal contractor for the remainder of the project. The principal contractor must then take on the responsibility for reviewing, updating and revising it and passing it to the client when the project finishes. The file must contain information about the current project that is likely to be needed to ensure health and safety during any subsequent work such as maintenance, cleaning, refurbishment or demolition. The file should not include things that will be of no help when planning future construction work such as pre-construction information, the construction phase plan, contractual documents and method statements. Information must be in a convenient form, clear, concise and easily understandable.

The principal designer is responsible for preparing the health and safety file. They are accountable to the client and should liaise closely to agree the structure and content of the file as soon as practicable after appointment. In preparing the file, the principal designer should expect the client to provide any health and safety file that may exist from an earlier project.

The principal designer must also cooperate with the rest of the project team and should expect their cooperation in return. Cooperation with the principal contractor is particularly important in agreeing the structure and content of the information included in the file. Liaison with designers and other contractors is also important. Where it is not possible to eliminate health and safety risks when preparing or modifying designs, designers must ensure appropriate information is included in the health and safety file about the reasonably practicable steps they have taken to reduce or control those risks.

For the duration of the principal designer's appointment, the principal contractor plays a secondary role in ensuring the health and safety file is fit for purpose. They must provide the principal designer with any relevant information that needs to be included in the health and safety file. In doing this, they should ensure the client understands the structure and content of the file and its significance for any subsequent project.

The client must retain the file and ensure that it is available to anyone who may need it for as long as it is relevant – normally the lifetime of the building – to enable them to comply with health and safety requirements during any subsequent project. It can be kept electronically, on paper, on film, or any other durable form. If a client disposes of their interest in the building, they must provide the file to the individual or organisation who takes on the client duties and ensure that the new client is aware of the nature and purpose of the file. If they sell part of a building, any relevant information in the file must be passed or copied to the new owner. If the client leases out all or part of the building, arrangements should be made for the file to be made available to leaseholders. If the leaseholder

acts as a client for a future construction project, the leaseholder and the original client must arrange for the file to be made available to the new principal designer.

More information on the health and safety file is given in Appendix 1.4 of Chapter 1.

7.3.9 The duties of domestic clients

A domestic client is someone who has construction work done on their own home, or the home of a family member, which is not done in connection with a business. If the work is in connection with a business attached to domestic premises, such as a shop, the client is not a domestic client. A domestic client is not required to carry out the duties placed on commercial clients.

Where the project involves:

(a) only one contractor, the client duties must instead be carried out by the contractor. In practice, this should involve doing little more than manage the work to ensure health and safety;

(b) more than one contractor, the client duties must instead be carried out by the principal contractor as well as the duties they already have as principal contractor. If the domestic client has not appointed a principal contractor then the duties of the client will be carried out by the contractor in control of the construction work.

In many situations, domestic clients wishing to extend, refurbish or demolish parts of their own property will, in the first instance, engage an architect or other designer to produce possible designs for them. It is also recognised that construction work does not always follow immediately after design work is completed. If they so wish, a domestic client has the flexibility of agreeing (in writing) with their designer that the designer coordinates and manages the project, rather than this role automatically passing to the principal contractor. Where no such agreement is made, then the principal contractor will automatically take over the project management responsibilities.

7.3.10 Health, safety and welfare requirements

Part 4 of the Regulations relates to health and safety on construction sites. Schedule 2 covers welfare arrangements. These parts of CDM 2015 effectively specify the requirements contained previously in the now revoked Construction (Health, Safety and Welfare) Regulations 1997. Chapter 19 covers these requirements in more detail in the summary of the Regulations and the issues arising from these requirements are covered in chapters 8–18.

The requirements described are:

1. safe place of construction work
2. good order and site security

3. stability of structures
4. demolition and dismantling
5. explosives
6. excavations
7. cofferdams and caissons
8. reports of inspections
9. energy distribution installations
10. prevention of drowning
11. traffic routes
12. vehicles
13. prevention of risk from fire
14. emergency procedures
15. emergency routes and exits
16. fire detection and fire-fighting
17. fresh air
18. temperature and weather protection
19. lighting
20. sanitary conveniences
21. washing facilities
22. drinking water
23. changing rooms and lockers
24. facilities for rest.

7.4 Sources of external construction health and safety information

7.4.1 General sources of information

All the sources of information outlined in Chapter 4 are relevant to the construction industry in addition to the following:

▶ construction legislation (particularly CDM 2015 – see Chapter 19);
▶ specific construction publications and reports from the HSE. See the HSE's micros site for Construction URL: http://www.hse.gov.uk/construction/index.htm.
▶ Also see their infonet and news bulletin sign-up page URL: http://www.hse.gov.uk/construction/infonet.htm to receive regular free information on construction health and safety;
▶ reports from trade organisations and construction forums, such as CITB (Construction Industry Training Board), CIRIA (Construction Industry Research and Information Association, see 7.4.3), CECA (Civil Engineering Contractors Association, see 7.4.4) and UKCG (the UK Contractors Group, see 7.4.5);
▶ CIOB (Chartered Institute of Building): the Health and Safety pages of the CIOB's website provide all the latest construction health and safety news, as well as several valuable resources for those working in the industry. http://www.ciob.org
▶ CSCS (Construction Skills Certification Scheme) was established in 1995 with the objective of helping the construction industry to get quality up, accidents down and cowboy builders out. Since then a great deal of progress has been made. In May 2007 they

presented the millionth CSCS card. That's a million people who have formally proved their competency and their awareness of health and safety, a positive move for them, their colleagues, employers and clients as well as the productivity and well-being of the industry. See their website: http://www.cscs.uk.com

The Construction Skills Certification Scheme (CSCS) has recently been amalgamated with the occupational health initiative Constructing Better Health (CBH). CBH is a not-for-profit organisation that aims to set standards for work-related health provisions and occupational health provision; collect health data; and promote fit-for-task information so employers can manage health risks at site level.

▶ International Labour Organisation: http://www.ilo.org/global/about-the-ilo/lang--en/index.htm
▶ CEN, the European Committee for Standardisation, was founded in 1961 by the national standards bodies in the European Economic Community and EFTA countries. Now CEN is contributing to the objectives of the European Union and European Economic Area with voluntary technical standards which promote free trade, the safety of workers and consumers, interoperability of networks, environmental protection, exploitation of research and development programmes, and public procurement. CEN is a non-profit-making technical organisation set up under Belgian law. https://www.cen.eu/Pages/default.aspx
▶ BSI, the British Standards Institution, publish more than 2,000 new and revised standards every year, for all industry sectors. Standards are available from British Standards Online as either a hard copy or electronic format or from their Customer Services department. http://www.bsigroup.com/
▶ the internet, a truly global resource for those in the construction industry. Only a very few links have been given here but each of the websites has other links to sources of information. A quick search will also provide numerous sources of specific information.

In addition to vocational qualifications, CSCS recognises the following construction related academic qualifications:

▶ HNC and HND
▶ Degrees
▶ NEBOSH Construction Certificates and National Diplomas
▶ CIOB Certificates.

Successful card applicants holding these academic qualifications are issued with an AQP (Academically Qualified Person) card which has a five-year life and is currently non-renewable. However, this is under review. Holders of the AQP card are expected to work toward achieving a relevant vocational qualification or achieving membership of a recognised

7

professional body (a full listing is available on the CSCS website).

7.4.2 Construction health and safety forums

There are several construction health and safety forums to help people discuss and share best practice in the industry. This will assist in the review of standards to determine whether an organisation is performing well.

Two of the principal organisations are:

1. CIRIA – Construction Industry Research and Information Association
2. CECA – Civil Engineering Contractors Association.

7.4.3 Construction Industry Research and Information Association (CIRIA)

The Construction Industry Research and Information Association was founded in 1960. In recent times the organisation's name was shortened to CIRIA. Today, CIRIA is known solely by its abbreviated name. CIRIA's mission is to improve the performance of those in the construction and related industries.

CIRIA works with the construction industry, government and academia to provide performance improvement products and services in the construction and related industries and currently engages with around 700 subscribing organisations. Activities include collaborative projects, networking, publishing, workshops, seminars and conferences. Each year CIRIA runs about 40 projects, holds over 90 events and publishes 25 best practice guides.

CIRIA does not answer technical enquiries. CIRIA provides consensus-driven best practice guidance through publications, seminars, training and multimedia solutions. Many of CIRIA's members and funders may be able to provide assistance.

Website: http://www.ciria.org.uk

7.4.4 Civil Engineering Contractors Association (CECA)

The Civil Engineering Contractors Association (CECA) was established in November 1996 at the request of contractors to represent the interests of civil engineering contractors registered in the UK as well as to provide a full range of services to members.

CECA promotes the positive contribution that the civil engineering industry makes to the nation. The industry is an integral part of the economy and provides the transport and other infrastructure necessary for civilised life.

CECA's current membership is in excess of 200 civil engineering companies, which range in size from large and well-known national names to the medium- and smaller-sized company which may operate at a more regional level.

CECA's main objectives are:

▶ to ensure that CECA is, and is recognised to be, fully representative of British civil engineering contractors;
▶ to promote and lobby effectively on behalf of civil engineering contractors;
▶ to provide an effective range of services to CECA regional companies and CECA members.

Website: http://www.ceca.co.uk

7.4.5 UK Contractors Group (UKCG)

The UKCG represents the UK's large construction companies. UKCG members carry out over £20 billion worth of construction work each year.

UKCG represents the interests of major contractors to government and other decision makers.

The UKCG Health and Safety Charter for member companies is as follows:

▶ Sharing common goals, the member companies of the UKCG are committed to operating construction sites that provide a working environment which is both safe and free from health hazards for all stakeholders within the construction industry and for members of the public.

UKCG member companies working with their supply chains will commit to:

▶ leading behavioural change on all our sites to eliminate accidents and incidence of ill-health;
▶ a fully qualified workforce;
▶ an effective site-specific induction process before anyone is allowed to work on site;
▶ all workers being consulted on health and safety matters in a way that engages them in improving health and safety;
▶ exchanging best practice and lessons learned in order to establish the root cause of incidents;
▶ raising awareness and insisting on the highest standards of PPE;
▶ publishing an annual report of progress made against the commitments of this Charter;
▶ reducing the incidence rate of work-related ill-health in the construction industry by health surveillance, education, rehabilitation and reducing exposure. Companies will monitor progress by measuring the total number of days lost due to sickness absence.

Website: www.ukcg.org.uk

7.5 Further information

The Construction (Design and Management) Regulations 2007

The Health and Safety at Work etc. Act 1974 – Sections 2, 3, 20–25, 33 and 39–40

The Management of Health and Safety at Work Regulations 1999 – Regulation 5

Managing for health and safety, HSE Books, ISBN 978-0-7176-6456-6. http://www.hse.gov.uk/pubns/priced/hsg65.pdf

HSE's The health and safety toolbox: How to control risks at work. http://www.hse.gov.uk/toolbox/

Managing Health and Safety in Construction CDM ACoP, L144, ISBN 9780717662234 (to be replaced by L153 during 2015)

Health and Safety in Construction (Guidance) (HSG150rev), HSE Books, ISBN 978 0 7176 6182 4. http://www.hse.gov.uk/pubns/books/hsg150.htm

The Construction (Design and Management) Regulations 2015, Industry guidance for Clients, produced by CONIAC, ISBN 978-1-85751-389-9. http://www.citb.co.uk/documents/cdm%20regs/industry-guidance-clients.pdf

The Construction (Design and Management) Regulations 2015, Industry guidance for Contractors, produced by CONIAC, ISBN 978-1-85751-391-2. http://www.citb.co.uk/documents/cdm%20regs/industry-guidance-contractors.pdf

The Construction (Design and Management) Regulations 2015, Industry guidance for Designers, produced by CONIAC, ISBN 978-1-85751-392-9. http://www.citb.co.uk/documents/cdm%20regs/industry-guidance-designers.pdf

The Construction (Design and Management) Regulations 2015, Industry guidance for Principal contractors, produced by CONIAC, ISBN 978-1-85751-393-6. http://www.citb.co.uk/documents/cdm%20regs/industry-guidance-principal-contractors.pdf

The Construction (Design and Management) Regulations 2015, Industry guidance for Principal Designers, produced by CONIAC, ISBN 978-1-85751-390-5. http://www.citb.co.uk/documents/cdm%20regs/industry-guidance-principal-designer.pdf

The Construction (Design and Management) Regulations 2015, Industry guidance for Workers, produced by CONIAC, ISBN 978-1-85751-394-3. http://www.citb.co.uk/documents/cdm%20regs/industry-guidance-workers.pdf

7.6 Practice revision questions

1. A principal contractor needs to employ sub-contractors, some of whom may not have English as their first language.

    (a) **Outline** how the principal contractor could ensure that all site operatives fully understand site rules with regard to health and safety.

    (b) **Outline** some of the other pressures on principal contractors that could lead to accidents on construction sites.

2. (a) **Outline** the role of the Health and Safety Executive in improving the health and safety performance of the construction industry.

    (b) **Outline** the moral, legal and financial benefits to a construction company of maintaining high standards of health and safety.

3. The Construction (Design and Management) Regulations require that pre-construction information is provided to all designers and contractors.

    (a) **Outline** the purpose of providing this information.

    (b) **Identify** the information that needs to be provided in the following sections:
    (i) Description of site;
    (ii) Client considerations and management requirements;
    (iii) Environmental restrictions and on-site hazards;
    (iv) Significant design and construction hazards; and
    (v) The health and safety file.

4. A former factory is to be demolished to make way for a new block of offices on a trading estate.

    (a) **Outline** the criteria for notifying the project to the HSE.

    (b) **Identify** the information that should be included in the pre-construction information for prospective principal contractors for this project.

    (c) **Identify** the information that would be required from the various parties involved in the project on completion of the works so that the Health and Safety File could be completed.

5. The principal contractor on a major new-build project requires a number of sub-contractors to work for him/her.

    (a) **Outline** what health and safety information should be requested from prospective contractors in order to aid the principal contractor to decide which sub-contractors he/she should appoint.

    (b) **Describe**, with relevant examples, the information that the principal contractor should provide to sub-contractors so that he/she fulfils his/her duties under the

Construction (Design and Management) Regulations.

6. (a) **Outline** the duties of the principal designer under the Construction (Design and Management) Regulations 2015.

   (b) **Outline**, with practical examples where appropriate, how failures on the part of designers to comply with their duties under the Construction (Design and Management) Regulations can lead to health and safety problems during the construction phase of a project.

7. The extension and refurbishment of an existing office block is planned and this will also involve building a new car park. The client will continue to occupy the building while the work is in progress.

   (a) **Identify** the pre-construction information that should be provided to potential principal contractors in relation to the following topics:
   (i) description of project;
   (ii) the client's management requirements and provisions;

   (iii) environmental restrictions and existing site risks; and
   (iv) design and construction hazards.

   (b) **Outline** the precautions that should be in place to protect the health and safety of the occupants of the building during the construction work.

8. (a) **Outline** the main duties of a principal contractor under the Construction (Design and Management) Regulations.

   (b) **Identify FOUR** items of information in the health and safety file for an existing building that might be needed by a contractor carrying out refurbishment work.

   (c) **Outline** the meaning of a 'notifiable project' and the particulars that must be provided to the enforcing authority when a project is required to be notified.

9. **Outline** the sources of published information that may be consulted when dealing with any health and safety problems on a construction site.

CHAPTER 8

Construction site issues – hazards and risk control

This chapter covers the following NEBOSH learning objectives:

1. Explain the factors which should be considered when carrying out an initial assessment of a site to identify significant hazards and their risks
2. Explain the appropriate general site control measures needed in setting up and organising a site
3. Identify the health, welfare and work environment requirements on construction sites
4. Explain the hazards and appropriate control measures for violence at work
5. Explain the hazards and appropriate control measures for substance misuse at work
6. Explain the hazards associated with the movement of people on construction sites and the control measures for pedestrians

---

8.1 Initial site assessment

8.1.1 Introduction

The construction industry covers a wide range of activities ranging from large-scale civil engineering projects to very small house extensions. It has approximately 200,000 firms of which only 12,000 employ more than seven people – many of these firms are much smaller. The use of sub-contractors is very common at all levels of the industry.

Figure 8.1 Concrete being pumped to upper floors on a large construction site

Over many years, the construction industry has had a poor but improving health and safety record. In 1966, there were 292 fatalities in the industry but by 2014 this figure had reduced to 42. Although most of these fatalities (over 70%) were caused by falls from height, there are other significant causes. The Health and Safety Executive (HSE) has listed the main causes of fatalities as:

▶ falling through fragile roofs and roof lights;
▶ falling from ladders, scaffolds and other workplaces;
▶ being struck by excavators, fork-lift trucks or dumper trucks;
▶ overturning vehicles;
▶ being crushed by collapsing structures.

During a concentrated month-long period of inspections in 2014, the HSE demanded improvements at many sites and in some cases put an immediate stop to work activities, where they fell short of expected standards.

In addition to safety issues, inspectors focused on significant health risk issues, such as respiratory risks from dusts containing silica materials, exposure to other hazardous substances such as cement and lead paint, manual handling, noise and vibration. Conditions were so poor in some situations that the work had to be stopped on at least 13 occasions. A total of 1,748 repair and refurbishment sites were visited and unacceptable and dangerous practices were observed at nearly half of the sites. At 20% of the sites conditions were so poor that enforcement action was taken – 13 Prohibition Notices, 107 Improvement Notices and a total of 239 health-related Notices of Contravention were served.

In a separate month-long inspection initiative in the South West of England, almost a third of the construction sites visited failed health and safety checks. A total of 61 of the 215 sites inspected were found to have significant failings and 40 enforcement notices were issued as a result. Specific work activities on some of the sites were deemed so dangerous that 27 Prohibition Notices were served by inspectors, immediately halting further work until standards had been improved.

The following problems were found during all these spot checks:

▶ failing to protect workers working at height;
▶ exposure to harmful dust particularly respirable silica;
▶ inadequate welfare facilities;
▶ structural instability of access equipment, such as scaffolds;
▶ asbestos issues; and
▶ untidy sites.

Over recent years, the HSE has undertaken several spot checks of construction sites and during these checks HSE inspectors checked that:

▶ all jobs involving working at height had been identified and properly planned so that appropriate controls were in place;
▶ all equipment was correctly installed, maintained and operated;
▶ construction sites were well organised, avoiding trips and falls;
▶ stairs and walkways were free from obstructions;

- work areas were clear of unnecessary materials and waste;
- risks to health from exposure to dust such as silica were controlled;
- other health risks, such as exposure to noise and vibration, manual handling and hazardous substances were properly managed; and
- the workforce were aware of the risk control measures.

Due to the fragmented nature of the industry and its accident and ill-health record, the recent construction industry legal framework has concentrated on hazards associated with the industry, welfare issues and the need for management and control at all stages of a construction project.

Finally, it is important to note that the construction site may be inside a building (refurbishment work) as well as an open-air site (Figure 8.2).

Figure 8.2 Keeping corridors clear during refurbishment

8.1.2 General hazards and controls

The Construction (Design and Management) (CDM) Regulations deal with the main hazards likely to be found on a construction site. In addition to these specific hazards, there will be the more general hazards (manual handling, electricity, noise, etc.), which are discussed in more detail in other chapters. The hazards and controls identified in CDM 2015 are as follows.

Safe place of work

Safe access to and egress from the site and the individual places of work on the site are fundamental to a good health and safety environment. This clearly requires that all ladders, scaffolds, gangways, stairways

and passenger hoists are safe for use. It further requires that all excavations are fenced, the site is tidy and proper arrangements are in place for the storage of materials and the disposal of waste. The site needs to be adequately lit and secured against intruders, particularly children, when it is unoccupied. Such security will include:

- secure and locked gates with appropriate notices posted;
- a secure and undamaged perimeter fence with appropriate notices posted;
- all ladders either stored securely or boarded across their rungs;
- all excavations covered;
- all mobile plant immobilised, fuel removed, where practicable, and services isolated;
- secure storage of all inflammable and hazardous substances;
- a programme of visits to local schools to explain the dangers present on a construction site. This has been shown to reduce the number of child trespassers;
- security patrols and closed circuit television (CCTV) – may need to be considered if unauthorised entry persists.

Figure 8.3 Secure site access gate with added protection to prevent vehicles entering at night or on Sundays

Protection against falls

Falls are the most common cause of serious injury or death in the construction industry. These falls are normally from a height but may also be on the same level. It is important that trip hazards, such as building and waste materials, are not left on site on walkways or roadways.

Fragile roofs

Roof work, particularly work on pitched roofs, is hazardous and requires a specific risk assessment and method statement prior to the commencement of work. Further details are given in Chapter 16.

Protection against falling objects

Construction workers and members of the public need to be protected from the hazards associated with falling objects. Both groups should be protected by the use of covered walkways or suitable netting to catch falling debris. Waste material should be brought to ground level by the use of chutes or hoists. Waste should not be thrown from scaffolding and only minimal quantities of building material should be stored on working platforms. The Construction (Head Protection) Regulations 1989 virtually mandated employers to supply head protection (hard hats) to employees whenever there was a risk of head injury from falling objects. (Sikhs wearing turbans could be exempted from this requirement.) The employer was also responsible for ensuring that hard hats were properly maintained and replaced when they were damaged in any way. Self-employed workers had to supply and maintain their own head protection. Visitors to construction sites had to be supplied with head protection and mandatory head protection signs displayed around the site.

Following the recommendations of the Lofstedt report in 2011, the Construction (Head Protection) Regulations were revoked since it was felt that they largely replicated the Personal Protective Equipment at Work Regulations which could be relied on to regulate the use of head protection on construction sites. Sikhs wearing turbans are still exempted from the use of head protection.

Under the Personal Protective Equipment at Work Regulations there will still be a requirement for employers to provide, and for workers to wear, head protection where there is a risk of head injury. The HSE will review and update existing guidance (either published or on its website) on compliance with the Regulations to ensure it adequately covers the provision and use of head protection on construction sites.

Injuries due to falling materials are not limited to materials falling from height. Many fatalities have been caused by the collapse of walls, particularly retaining walls, and by the collapse of excavations.

Demolition

Demolition is one of the most hazardous construction operations and is responsible for more deaths and major injuries than any other activity. The management of demolition work is controlled by the CDM Regulations and requires a principal designer and a principal contractor and a health and safety plan (as covered in Chapter 7). A more detailed discussion of demolition is given in Chapter 18.

Excavations

This topic is covered in more detail in Chapter 17. Excavations must be constructed so that they are safe environments for construction work to take place. They must also be fenced and suitable notices posted so that neither people nor vehicles fall into them.

Prevention of drowning

Where construction work takes place over water, steps should be taken to prevent people falling into the water and rescue equipment should be available at all times.

Figure 8.4 Prevention of drowning. Rescue and safety equipment must always be easily available and in good condition

Vehicles and traffic routes

All vehicles used on site should be regularly maintained and records kept. Only trained drivers should be allowed to drive vehicles and the training should be relevant to the particular vehicle (fork-lift truck, dumper truck, etc.). Vehicles should be fitted with reverse warning systems. HSE investigations have shown that in over 30% of dumper truck accidents, the drivers had little experience and no training. Common forms of accident include driving into excavations, overturning whilst driving up steep inclines and runaway vehicles which have been left unattended with the engine running. Many vehicles such as mobile cranes require regular inspection and test certificates.

Traffic routes and loading and storage areas need to be well designed with enforced speed limits (a maximum of 10mph is the limit on most sites), good visibility and the separation of vehicles and pedestrians being considered. The use of one-way systems and separate site access gates for vehicles and pedestrians may be required. Finally, the safety of members of the public must be considered particularly where vehicles cross public footpaths.

Fire and other emergencies

Emergency procedures relevant to the site should be in place to prevent or reduce injury arising from fire, explosion, flooding or structural collapse. These procedures should include the location of fire points and assembly points, extinguisher provision, site evacuation, contact with the emergency services, accident reporting and investigation and rescue from excavations and confined spaces. There also needs to be training in these procedures at the induction of new workers and ongoing training for all workers.

Electricity

Electrical hazards are covered in detail in Chapter 12, and all the control measures mentioned apply on a construction site. However, due to the possibility of wet conditions, it is recommended that only 110V equipment is used on site. If mains electricity is used (perhaps during the final fitting out of the building), residual current devices should be used with all electrical equipment. Where workers or tall vehicles are working near or under overhead power lines, either the power should be turned off or 'goal posts' or taped markers should be used to prevent contact with the lines. Similarly, underground supply lines should be located and marked before digging takes place.

Noise

Noisy machinery should be fitted with silencers. When machinery is used in a workshop (such as woodworking machines), a noise survey should be undertaken and, if the noise levels exceed the second action level, the use of ear defenders becomes mandatory. If it is necessary to shout when talking to a nearby colleague, then the noise level is excessive and ear defenders should be worn. A cement mixer can be a particularly noisy piece of machinery on a construction site. The levels of noise exposure to workers on the site can be reduced by fixing silencers to diesel-powered mixers, ensuring that the mixer is regularly maintained, minimising the exposure time of the workers by job rotation and providing ear defenders to those working near to the mixer. A better solution would be to use a mixer with lower noise emissions.

Excessive noise hazards may also be present during demolition work and by the use of pneumatic drills, compressors and vehicles used on site.

Health hazards

Health hazards are present on a construction site. These hazards include vibration, dust (including asbestos), cement, solvents and paints, and cleaners. A Control of Substances Hazardous to Health Regulations (COSHH)

assessment is essential before work starts with regular updates as new substances are introduced. Copies of the assessment and the related safety data sheets should be kept in the site office for reference after accidents or fires. They will also be required to check that the correct personal protective equipment (PPE) is available. A manual handling assessment should also be made to ensure that the lifting and handling of heavy objects is kept to a minimum. Additional health hazards may also be present when working near water whether the water is stagnant or not.

8.1.3 Initial site assessment process

Before work commences, an initial site assessment should be undertaken. This assessment should not only consider hazards associated with the site itself but also the hazards which may well be present during the construction phase.

The previous and current use of the site will form the basis of the assessment. A greenfield site is likely to present fewer problems than a brownfield site, particularly when the site has been previously contaminated by earlier activities. Chemical contamination is common when the site formerly housed industrial plant. It is not uncommon in such situations for oils or other pollutants to be stored in underground storage chambers; and although normally the bulk of such substances will have been removed, residues will remain and present both environmental and cost problems to the construction developer who needs to understand the various remedial measures that may be used to prevent pollution of nearby streams. Remedial measures may involve the capping of the chamber or its excavation and the complete disposal of the pollutants (possibly including soil washing and thermal treatment). In many parts of the country, there are old mine workings, shafts and wells and in many cases there is no record of these on available public documents. A detailed site survey may well discover that such old workings exist and indicate that sufficient funds must be allocated to ensure that they are safely capped.

Knowledge of the history of the site is, therefore, important during the initial site assessment. Details will also be needed of any other activities, including non-construction ones, which are taking place at or near the site.

If the project involves **demolition and refurbishment**, then information on the occupancy of the premises or nearby premises needs to be ascertained. A structural survey should be made of the building to investigate the types of structural defect that may be apparent during a visual inspection and explain the causes of those defects. Typical defects include corrosion of steelwork, cracking due to fatigue in structural steel and welds, cracked brickwork and concrete, concrete degradation (flaking), dropped lintels and distorted doors or window frames, wood rot, evidence of timber infestation, dampness above the damp-proof course, cracks or dips in floors and ceilings, dips in the roof line, deformed structural members and bulging walls. Possible causes could include the quality of the initial design, the age of the building, lack of maintenance, poor quality materials, poor construction standards, solar radiation, damage from extreme weather conditions, vibration from machinery or traffic, chemical attack, overloading, accidental impact, ground settlement or subsidence, leaking drains, unauthorised modifications, fire damage, vandalism, the action of tree roots and pest infestation.

The provision of temporary supports where required, suitable access equipment and barriers or covers to prevent falls through floor openings and the need to carry out an assessment of the manual handling operations are additional factors to be considered. Asbestos, vermin infestation and exposure to lead from paintwork and pipes are possible health hazards and evidence of any of these must also be included in any assessment.

Also an emergency evacuation plan and the use of hot work permits may need to be considered at the initial site assessment stage.

A **survey of the topography** of the site, including the ground conditions, ground stability and type of ground or soil (e.g. gravel, sand or clay), is essential prior to the commencement of any construction work.

Access to the site, particularly for vehicles, is an important consideration especially if access is from a busy road or close to a school or hospital.

Finally **the nature of the surroundings** of the site should be included in the assessment and should include the following considerations:

▶ the proximity of roads, footpaths, railways, rivers and other waterways;

▶ the details of any fencing or hedging and general vegetation;

▶ the position and description of any trees which need to be felled and those that must be untouched (particularly if they are subject to a tree preservation order);

▶ the proximity of residential, commercial or industrial properties and concerns;

▶ the proximity of schools, colleges, hospitals and playing fields;

▶ the presence of overhead power lines and details of the responsible authority;

▶ the presence of buried services including sewers and disused wells;

▶ the availability of services, such as electricity, telephone lines and water, and drainage systems;

▶ any evidence of fly tipping or other unauthorised activity;

▶ measures required to prevent unauthorised access by vehicles.

8.2 Appropriate general site control measures

Many of the site controls have been covered earlier in this chapter in the previous section. Site controls can be conveniently sub-divided under five headings – site planning, site preparation, site security, environmental considerations and the arrangements with the client and/or occupier of the premises.

8.2.1 Site planning

An essential element of site planning is the layout of the site. Due to a lack of forethought or planning, many sites rapidly degenerate into a muddled or confused state. This leads to a corresponding reduction in health and safety standards and an increase in accidents.

During the site planning stage, arrangements will be needed to ensure that there is safe access to and around the site for both pedestrians and vehicles. Care must be taken during the planning of traffic routes so that pedestrians and vehicles are kept apart as much as possible. It will also be necessary to provide suitable loading and unloading areas, which may need to be moved as the construction work progresses, and recognised parking areas.

Method statements (see Chapter 4) should be developed at this stage. They will detail how a specific job or process is to be undertaken together with the associated hazards and the necessary control measures that should be in place to ensure health and safety. If contractors are to be used, a pre-tender questionnaire should be submitted by them so that their competence may be determined. The method statement should be checked and approved by the employer or a responsible person, such as the site manager or supervisor, before the work starts. When the work has started, the employer is obliged, under Section 3 of the Health and Safety at Work Act, to ensure adherence to the method statement and that a safe system of work is being used while the work is being undertaken.

Adequate lighting must be provided in all parts of the site especially along vehicle routes, although natural light should be made available whenever possible. When lighting failure could lead to serious risks of accidents, such as working on a tower scaffold, secondary lighting should be provided, perhaps in the form of torches. Suitable signs will be required on the site – these may be health and safety, warning or directional signs. Areas on the site will also need to be designated for site offices, materials and on-site machinery (fixed and mobile) storage and welfare facilities for the workforce. Arrangements for the day-to-day management and control of the site, monitoring, review and inspection of procedures, and equipment and workforce supervision must be made during the planning stage.

If demolition or refurbishment is to be undertaken, then hazards associated with structural collapse, fire, working at height, lifting and carrying and electricity must be considered together with health hazards from lead, asbestos, vermin and insects. All alteration, demolition and dismantling work must be carefully planned and undertaken by competent people to avoid unplanned structural collapse. Under CDM 2015, the client must provide the principal contractor with relevant information about the building structure, including its stability and structural form and any significant design assumptions, suggested work methods and sequences. The contractor must then use that information to plan and carry out the work safely.

As mentioned in Chapter 18, a structural survey before work commences should reduce the risk of unintentional structural collapse. Such a survey should be undertaken by a competent person before any potentially load-bearing parts of a structure are altered. The structural survey should consider:

▶ the age of the structure;
▶ previous use;
▶ type of construction; and
▶ any nearby buildings or structures.

A competent person should decide the method and design of temporary supports. Temporary supports provided must be designed, installed and maintained to withstand foreseeable loads and structures should never be overloaded.

The issue of hot work permits, provision of fire-fighting equipment and suitable storage facilities for flammable substances should control the fire risk. Health risks can be controlled, for example, by the use of licensed contractors for the removal of asbestos, specialist contractors to eradicate vermin and the provision of suitable personal protective equipment.

Emergency arrangements need to be in place before work begins. These arrangements should deal with fire precautions, accident reporting and investigation and the provision of an accident book together with RIDDOR reporting forms.

Whatever the type of construction project, the disposal of any waste produced must be planned. There are four requirements for the transfer of waste from the site to final treatment or disposal:

1. The waste must be clearly identified. This will involve the determination of the chemical and physical properties of the material, the identification of any special problems associated with the waste, and ensuring that it is properly labelled with a written description.

2. The waste must be securely stored with procedures to deal with spillages and prevent access, particularly by scavengers or children.

3. The waste must be transferred only to an authorised person. Classes of authorised person include registered waste carriers, holders of waste management licences or exempt parties such as charities or waste collection authorities. Waste transfer notes must be transferred from one duty-holder to the next along the waste chain. Checks are advised to ensure that prospective waste contractors are competent to deal with waste.

4. The adequate records of waste consignments must be kept for at least 2 years in the case of controlled waste, and 3 years for any hazardous waste. (These categories of waste are defined in Chapter 14 which includes additional possible legal requirements for waste disposal on construction sites.)

Waste skip selection should be made during the site planning process. The selected skip must be suitable for the particular job. The following points should be considered:

▶ sufficient strength to cope with its load;
▶ stability while being filled;
▶ a reasonable uniform load distribution within the skip at all times;
▶ the immediate removal of any damaged skip from service and the skip inspected after repair before it is used again;
▶ sufficient space around the skip to work safely at all times;
▶ the skip should be resting on firm level ground;
▶ the skip should never be overloaded or overfilled; and
▶ there must be sufficient headroom for the safe removal of the skip when it is filled.

There are hazards present during the movement of a loaded skip from the ground to the back of a skip loader vehicle. Entanglement with the vehicle lifting mechanisms, such as the hydraulic arms and lifting chains, is a major hazard. Other hazards include contact by the skip with overhead obstructions, movement of the skip contents and skip overload leading to mechanical or structural failure. Slip hazards may be present due to spillages from the skip and the skip contents could be contaminated with biological material, asbestos or syringes. Passing traffic during the loading operation may also present a hazard.

The control measures for these hazards include the use of outriggers to increase the stability of the loader vehicle and the provision of steps for the driver to alight from the cab or the vehicle flatbed. The contents of the skip should be secured using netting or tarpaulin. Adhering to the safe working loads of the skip and lifting equipment and the use of a banksman during the lifting

8

process are additional controls. The area around the vehicle may need to be cordoned off to protect passing pedestrians and road traffic. All workers concerned with the operation should wear suitable personal protective equipment, such as high visibility jackets, gloves and suitable footwear. All lifting equipment, including chains and shackles, must be subject to a periodic statutory examination.

Finally the HSE should be notified before work begins (using F10 (rev) or the details required by F10) when the project construction phase involves more than 30 days of work and has more than 20 workers working simultaneously or involves more than 500 person days of construction work. The CDM client is responsible for this notification to the HSE. See Chapter 19 for details of the information required.

8.2.2 Site preparation

It is quite possible that specialist activities, such as lifting, piling or steelwork, will be carried out on site during the construction phase and controls will need to be in place to address associated risks. Only workers who are competent in the specialist activity should be employed or contracted. One of the ways of assessing competence is to check on relevant method statements supplied by the workers. When the activity involves working at height, then the safeguards outlined in Chapter 16 must be made available. Care must also be taken to ensure that other major hazards are addressed, such as noise and vibration hazards associated with piling work and crane hazards during lifting operations. Site rules for site workers should be formulated during the planning stage and a specimen set of such rules is given in Appendix 8.1.

The preparation of the site may well involve the removal or de-branching of trees using chainsaws and a wheeled excavator to remove the roots. These are particularly hazardous operations when there are overhead or buried power lines at the site.

Chainsaws must only be used by trained and competent operators using a safe system of work which fully describes the hazards involved and the measures required to reduce the risks. Such measures include the necessary personal protective equipment (PPE), for example goggles, hearing protection, and ballistic trousers or overalls; the provision of safe and secure access to higher branches such as ladders or cherry pickers; and the provision of fall arrest equipment such as harnesses. There must be a high level of supervision at all times. There is much more information on chainsaws and the necessary precautions when used in Chapter 11.

When site preparation requires the removal of contaminated topsoil, it is important to ensure that appropriate PPE, including respirators, is issued to the workforce and the suitable welfare facilities,

including a decontamination unit, are available. A decontamination unit will have a dirty chamber where contaminated clothing can be left, a shower facility and a clean chamber where clean clothes will be kept. Arrangements will also be needed to ensure that eating, rest and first-aid areas are free from contamination and that emergency decontamination facilities are available.

Finally, high-voltage overhead power lines should always be made dead to prevent contact with the energised cable. If this is not possible, other control measures should be used such as height restrictions for site vehicles, signs, goalpost barriers and the use of banksmen (Figure 12.12a&b). It is also necessary to locate and identify any underground services. It is essential that the electrical supply company is informed of the construction work and that sound emergency procedures are in place.

8.2.3 Site security

Site security is required to ensure that construction materials, equipment and accessories are protected and this is usually achieved by the erection of a strong perimeter fence and a lockable gate. However, it is also important to protect members of the public including trespass by children. This is particularly relevant for a refurbishment project when strong hoardings or secure fencing must be provided. There is also a need to ensure that the site itself is safe should children manage to trespass onto the site. This involves the removal of access ladders, providing trench supports, covering holes where possible, immobilising plant, safe stacking of materials, locking away dangerous chemicals, isolating electrical supplies and locking off fuel storage tanks. Many of these problems may be avoided by visiting local schools to inform children of construction site dangers. Accidents to children on construction sites are a national concern and a widespread problem everywhere. A brickwork company has been fined after being convicted of failing to provide and maintain a suitable barrier around the boundary of a quarry where a 13-year-old schoolgirl died.

Generally members of the public must not be put at risk by the activities on the construction site. The client's pre-construction information should include:

▶ project boundaries;
▶ adjacent land use;
▶ access; and
▶ measures to exclude unauthorised people.

The determination of the site boundary is an important aspect of managing public risk and requires the:

▶ planning of the form that the perimeter will take;
▶ provision of the fencing; and
▶ maintenance of the fencing.

Figure 8.5 Well organised site with internal storage compounds and site accommodation behind with means of escape staircase in case of fire (also from the UK, in background on Southampton water)

Typically, in populated areas, this will mean a 2 m high small mesh fence or hoarding around the site. In addition the principal contractor must take reasonable steps to prevent unauthorised people accessing the site. The elderly, children and people with certain disabilities may need special attention. Work in premises such as schools and hospitals will need careful thought and planning.

The hazards that face members of the public and visitors include:

▶ Falling objects – objects must not fall outside the site boundary.

▶ Delivery and other site vehicles – pedestrians may be at risk from vehicles entering or leaving the site.

▶ Scaffolding and other access equipment – people outside the boundary may be struck during the erection or dismantling of access equipment.

▶ Storing and stacking materials – materials should be stored within the site perimeter, preferably in secure compounds or away from the perimeter fencing.

▶ Openings and excavations – barriers or covers should be used to prevent people from falling into excavations, manholes, stairwells or from open floor edges.

▶ Slips, trips and falls within pedestrian areas.

▶ Plant machinery and equipment.

▶ Hazardous substances.

▶ Electricity and other energy sources.

▶ Dust, noise and vibration.

▶ Road works.

The provision of warning signs at the perimeter fence and suitable viewing points for the public to observe the construction work safely has been shown to reduce the incidents of site trespass.

Whilst storage is an important factor for site security, it is also needed for health and safety reasons. Cement, which is a hazardous substance, must be securely stored in a dry store. Gas cylinders, fuel and other flammable substances must be stored in a separate flameproof store (more details of these requirements are given in Chapter 13).

8.2.4 Environmental considerations

The public must also be protected from environmental hazards resulting from the work, such as excessive dust, noise and mud on the public highway. Excessive dust may be controlled by damping down haul roads regularly, the sheeting of lorry loads and imposing speed limits on roads close to the site. During road repair operations, dust should be prevented from entering surface drains and precautions taken to prevent or restrict the effects of fuel spillage.

Noise levels may be reduced by the regular maintenance of equipment and the erection of noise barriers. Other controls include the use of quieter equipment and restricting the use of noisy machinery to late morning or early afternoon.

The principal contractor should endeavour to ensure that mud cannot get into public areas and should arrange for the road/s to be cleaned if it does. Where possible, work should be arranged so that road-going vehicles do not drive onto mud on the site. This may be achieved by laying site roads and using site-based plant to move materials to holding areas at the edge of a site road. Either routinely, or as a back-up plan, the site should have arrangements in place to call in a road-sweeping machine at short notice. Site management should be monitoring conditions outside the site regularly during periods of material movement.

Regular road sweeping will reduce mud on surrounding roads. The problem will also be alleviated by washing vehicle wheels before they leave the site.

For any of these measures to be successful, the site workforce must be informed of the procedures at induction and by posters and signs around the site.

8.2.5 Arrangements with the client and/or occupier of the premises

Such arrangements will include site rules (see Appendix 8.1 for an example) for the protection of members of the public, visitors and other employees. They should also cover issues such as shared facilities (e.g. first-aid arrangements and the use of catering facilities) and the need for full cooperation from the occupier and his/her staff in the observation of health and safety rules.

When the construction work involves a public highway, measures should be introduced to ensure the safe passage of vehicles and pedestrians. This will involve the erection of warning signs, the use of cones with lead-in and exit tapers and an indication of a safety zone between the cones and the work area. Pedestrian safety will require the use of barriers and lighting and traffic may need to be controlled using traffic lights or stop/go boards. Certain neighbours, such as schools, hospitals or factories, may also need to be informed of the timing of some aspects of the work.

8.2.6 Arrangements for site induction

Properly organised and prepared site inductions for all site workers are very important. As mentioned in Chapter 7, site induction training is seen by the HSE as an essential element in the drive to reduce accidents on construction sites. Induction training in general was covered in Chapter 3 and the issues raised there should be included in an induction event on a construction site. However, the following additional topics should also be covered:

- senior management commitment to health and safety;
- site health and safety policy;
- the outline of the project;
- the managerial and consultation arrangements on site including the identity of their supervisor;
- site health and safety plans;
- site welfare facilities;
- any site-specific health and safety risks, for example in relation to access, transport, site contamination, hazardous substances and manual handling;
- those site method statements which are relevant to the workers at the induction;
- the findings of any relevant risk assessments;

- details of any hazardous operations, such as work at height, site vehicular movement and confined space working;
- control measures on the site, including:
  - ▷ any site rules
  - ▷ any permit-to-work systems
  - ▷ traffic routes
  - ▷ security arrangements
  - ▷ hearing protection zones
  - ▷ arrangements for PPE, including what is needed, where to find it and how to use it
  - ▷ manual handling procedures and precautions
  - ▷ arrangements for housekeeping and materials storage
  - ▷ facilities available, including welfare facilities
  - ▷ emergency procedures, including fire precautions, the action to take in the event of a fire, escape routes, assembly points, responsible people and the safe use of any firefighting equipment;
- arrangements for first-aid;
- arrangements for reporting accidents and other incidents;
- details of any planned training, such as 'toolbox' talks;
- arrangements for consulting and involving workers in health and safety, including the identity and role of any:
  - ▷ appointed trade union representatives
  - ▷ representatives of employee safety
  - ▷ safety committees; and
- information about the individual's responsibilities for health and safety.

The quality of the information presented at the site meeting will determine the effectiveness of the induction training. Some use should be made of videos or DVDs in the presentation and the opportunity for interactive discussion given. A copy of the site rules, health and safety policy and a simple organisational chart should be issued to each worker at the induction meeting.

8.2.7 Working in occupied premises

When construction work is being undertaken on occupied premises, such as a retail store or an office block, certain important health and safety matters would need to be discussed and agreed with store or office management. These matters would include the location and isolation of the construction working areas, supervision of the work, procedures for the evacuation of the premises in the event of an emergency, the use of welfare and canteen facilities, the protection of the occupier's employees and procedures for handing back areas of the store when the work has been completed. The construction contractor must provide protected access into the building, ensuring that the area to be occupied by him/her is provided with adequate fences or barriers and that all waste materials are safely removed.

Other issues which need to be addressed include the selection of equipment (preferably using reduced voltage electrical equipment) and personal protective equipment and the erection of appropriate warning signs. Other precautions and controls include the provision of:

▶ safe access for workers and materials;
▶ a safe working platform;
▶ clean-up procedures and good housekeeping;
▶ fall arrest equipment;
▶ protection for all employees of the occupier of the premises;
▶ a permit system for hot work where this is necessary;
▶ fire extinguishers and procedures for emergency evacuation of the building.

Finally, some or all of the existing services may need to be isolated at certain times, and these times should be agreed with the occupier; adequate levels of lighting must be provided, where necessary, to carry out the work safely.

The impact on workplaces from hazards associated with works of a temporary nature

Works of a temporary nature include such projects as building maintenance, renovation and refurbishment. These works usually take place in a host organisation by contractors and often include building demolition and excavation work required to supply new underground services. The projects must be properly planned, supervised and monitored, particularly when the project is to take place within or near an occupied building. There are many other possible hazards associated with work of a temporary nature. Some of them are:

▶ slips, trips and falls are a frequent source of injury to members of the host organisation and the public;
▶ unfenced excavations;
▶ inadequate protection of holes, uneven surfaces, poor reinstatement, trailing leads and cables (especially on stairways), spillage of oils and gravel;
▶ poor lighting and ventilation in and around the temporary works;
▶ noise and vibration;
▶ hazardous substances including dust and asbestos fibres;
▶ falling or flying objects;
▶ the accumulation of waste materials;
▶ falling or moving objects;
▶ poor storage of materials and equipment and other obstructions in public areas, including inadequate control of waste materials, are common causes of accidents;

▶ fire and/or explosions;
▶ lack of project management and supervision; and
▶ hazards created outside the site perimeter (for example, unloading materials from a delivery lorry outside the perimeter).

Two relatively common problems with building projects concern asbestos and demolition. An up-to-date asbestos survey together with relevant plans and drawings should, therefore, be available before any renovation work is commenced.

Most of the health and safety risks in demolition activities are related to an unplanned or premature collapse of a structure or of a part of it. Various accident investigation reports following such incidents have found:

▶ the absence of temporary structures to support unstable elements;
▶ the lack of risk assessment at design stage, neglecting CDM requirements;
▶ the lack of any preliminary structural survey or site investigation;
▶ poor planning of demolition sequences;
▶ the lack of demolition method statements; and
▶ the lack of supervision while undertaking demolition activities.

Several of these conclusions are valid for a wide range of large and small building projects, namely the lack of good communications, supervision, planning and risk assessment and the neglect of legislative requirements.

The majority of temporary work projects involve some form of construction work. Details of some of the basic hazards and controls for construction work are given later in this section.

Main control measures relating to the management of works of a temporary nature

The important controls for work of a temporary nature are communication and the cooperation of both contractors and host employees, a comprehensive risk assessment for the project and the management and supervision of the project to ensure that all the agreed controls are implemented.

Communication and cooperation

The organisation of liaison and consultation arrangements is essential to ensure good communication with both the host employees and the contractors. This should provide the cooperation and coordination of all those responsible for the project to ensure the health and safety of everyone in the workplace and enable any concerns to be addressed. Good communication between clients, contractors and residents is an important means of controlling the risks and minimising discomfort. It can also be essential

8

when arranging alternative escape routes during work. Advice should be sought from the local fire authority where it is necessary to block or alter the permanent escape routes.

Risk assessment

A detailed risk assessment should be made to cover the temporary works. This should include the identification of:

▶ all aspects of the works including the health and safety implications to the host employees and the contractors;

▶ the people at risk from the work including, where appropriate, members of the public;

▶ any risks to the contractors from the business of the host organisation; and

▶ any additional controls and procedures required to protect everyone affected by the project including:

▷ any equipment that should or should not be worked on or used;

▷ the personal protective equipment that should be used and who will provide it;

▷ all the working procedures, including any permits to work and method statements that are to be used;

▷ details of any site access restrictions;

▷ any amendment required to existing emergency procedures and access to the host building; and

▷ details of any safety signs and/or barriers required.

Often work must be undertaken in areas which need to remain occupied. The risk assessment should indicate the nature of the perimeter of the project and how it will be maintained. It might be possible for the work to take place outside normal hours – if not then the work areas must be segregated or access controlled by physical barriers or warning signs.

Management and supervision of the project

The first two important tasks for the management and supervision of the project are the formulation of a detailed specification of the project and the appointment of competent managers and contractors to complete the project (see Chapter 1).

A Temporary Works Coordinator should be appointed for all temporary works to oversee the project and to sign it off on completion. Such an appointment is particularly important for refurbishment projects involving complex activities that require information related to the existing structure and interrelated health and safety issues.

The coordinator would be responsible for:

▶ the appointment of competent contractors to undertake the project;

▶ the provision of information, instruction and training to the employees of the host organisation

on any hazards associated with the temporary works;

▶ the provision of information to the temporary works contractors on any hazards from the activities of the host organisation and the controls in place to address them;

▶ the management and supervision of the work of the contractors and agreement on the nature of any temporary controls required before work begins; and

▶ the formulation of a detailed job specification for the work required which should include:

▷ safe access to both the temporary work site and to the rest of the building;

▷ safe method statements to cover all aspects of the work; and

▷ the arrangements for the examination of any temporary access equipment.

The provision of welfare facilities including first-aid arrangements must be discussed and agreed prior to the commencement of the project. It may be that project contractors can have use of the same facilities as the host employees. If this is not possible, then separate arrangements should be included in the project specification.

Other control measures

Where construction work is carried out in, or near, occupied premises, such as houses, hospitals, factories and shops, it may be necessary to evacuate part or all of these premises, either for the full duration of the work, or for a limited period during hazardous operations. Certain organisations such as housing associations and care homes often have an established policy which identifies the circumstances when it is appropriate to evacuate premises.

Other control measures that will be relevant to many projects include the:

▶ use of temporary flooring material, such as plywood or steel plates, to cover uneven ground or potholes;

▶ covering or fixing of any trailing cables which need to cross pedestrian areas;

▶ provision of lighting at night and in dark areas;

▶ prompt removal of all waste and rubbish as it arises;

▶ storage of all hazardous substances in suitable containers or in secure compounds when not in use;

▶ provision of site storage compounds to accommodate all the plant, equipment and materials outside working hours. Strict control over the amount and timing of deliveries will help keep storage to a minimum outside this compound area;

▶ safe storing and stacking of materials and a limit to the height of all material stacks;

▶ use of suitably protected electrical cordless tools or reduced voltage equipment;

Figure 8.6 Working in or close to occupied premises

▶ use of gas cylinders and similar appliances fitted with valves which require special tools to turn on the supply;

▶ isolation of gas cylinders when not in use and locking them in a secure cage out of hours;

▶ protection against hazards that are created within the project site as the work develops, such as around deep excavations which will need to be covered or fenced;

▶ protection of vulnerable groups such as the elderly, children and people with certain disabilities – work in certain premises such as schools and hospitals requires careful planning,

▶ protection of wheelchair users and partially sighted people where construction work affects pedestrian routes;

▶ provision of suitable protection:
  ▷ from dust, asbestos fibres, noise and vibration (see chapters 14 and 15):
  ▷ to contractors, host employees and members of the public from being struck by falling or ejected materials (see 8.6.2): and
  ▷ from delivery and other moving site vehicles (see Chapter 9). Consideration should be given to ensuring that deliveries are scheduled at times outside of large movements of people such as rush hours or the journeys to and from school.

▶ posting of warning signs around the site and at all entrances to the main building.

Whenever possible, school projects which present high risks to staff and pupils should be carried out during school holidays, at weekends or out of school hours. If this is not possible, the work should be programmed to avoid busy periods such as school start, finish and break times. Deliveries should also be arranged to avoid these times.

Certain premises have their particular hazards that will need to be addressed. Residents in care homes may need to be evacuated when hazardous substances are being used (such as timber treatment or damp course work). Where work takes place within occupied factories, offices, shops and other premises, the host organisation and the contractor will need to manage the risks created by their own work, and cooperate closely to manage those risks created by sharing the workplace.

In healthcare premises, the vulnerability of those who are within the premises must be considered – children and outpatients with restricted mobility or with partial sight. Some patients, such as those who have had major surgery, will be more open to infection than healthy people. Refurbishment works may disturb fungal spores and other organisms which could be a serious hazard to patients. Site vehicles and pedestrians must be separated from the hospital traffic as much as possible.

8.3 Health, welfare and work environment requirements

8.3.1 Introduction

People are most often involved in accidents as they walk around the construction site or when they come into contact with vehicles in or around the site. It is therefore important to understand the various common accident causes and the control strategies that can be employed to reduce them. Slips, trips and falls account for the majority of accidents to pedestrians and the more serious accidents between pedestrians and vehicles can often be traced back to excessive speed or other unsafe vehicle practices, such as lack of driver training. Many of the risks associated with these hazards can be significantly reduced by an effective management system.

Finally, there are welfare and work environment requirements to ensure a healthy and safe workplace.

8.3.2 Health and welfare provision

The HSE have been concerned for some time at the poor standard of welfare facilities on many construction sites. The pre-construction information prepared by the client should include the arrangements required for welfare provision. The construction phase must not start unless the client is satisfied that there are arrangements for welfare facilities in place. The facilities must be maintained by the principal contractor throughout the life of the project.

Welfare arrangements include the provision of sanitary conveniences and washing facilities, drinking water, accommodation for clothing, facilities for changing clothing and facilities for rest and eating meals. First-aid provision is also a welfare issue and is covered in Chapter 4. However, the scale of first-aid provision will be related to the size and complexity of the project, the number of workers on site at any given time and the proximity of emergency hospital facilities.

Sanitary and washing facilities (including showers if necessary) with an adequate supply of drinking water should be provided for everybody working on the site. Accommodation will be required for the changing and storage of clothes and rest facilities for break times. There should be adequate first-aid provision (an accident book) and protective clothing against adverse weather conditions. Welfare issues are covered by the CDM 2015 and its guidance document L153.

Detailed guidance is also available in HSG150 *Health and Safety in Construction*. Welfare requirements are also covered in the Workplace (Health, Safety and Welfare) Regulations and the ACoP.

Sanitary conveniences and washing facilities must be provided together and in proportion to the size of the workforce. The Approved Code of Practice of the Workplace (Health, Safety and Welfare) Regulations provides two tables offering guidance on the requisite number of water closets, wash stations and urinals for varying sizes of workforce (approximately one of each for every 25 employees). Special provision should be made for disabled workers and there should normally be separate facilities for men and women. There should be adequate protection from the weather and only as a last resort should public conveniences be used. A good supply of warm water, soap and towels must be provided as close to the sanitary facilities as possible. Hand dryers are permitted but there are concerns about their effectiveness in drying hands completely and thus removing all bacteria. In the case of temporary or remote work sites, sufficient chemical closets and sufficient washing water in containers must be provided.

All such facilities should be well ventilated and lit, and cleaned regularly.

Drinking water must be readily accessible to all the workforce. The supply of drinking water must be adequate and wholesome. Normally mains water is used and should be marked as 'drinking water' if water not fit for drinking is also available. On some sites, particularly during the early stages of construction, it may be necessary to provide bottled water or water in tanks. In this event, the water should be changed regularly and protected from contamination. Cups or mugs should be available near the tap unless a drinking fountain is provided.

Accommodation for clothing and facilities for changing clothing must be provided which is clean, warm, dry, well ventilated and secure. Where workers are required to wear special or protective clothing, arrangements should be such that the workers' own clothing is not contaminated by any hazardous substances. On smaller sites, the site office may be suitable for the storage of clothing, provided that it can be kept secure and used as a rest facility.

Facilities for rest and eating meals must be provided so that workers may sit down during break times in areas where they do not need to wear PPE. It should be possible for workers to make hot drinks and prepare food. Separate areas should be provided for smokers and non-smokers. Facilities should also be provided for pregnant women and nursing mothers to rest. Arrangements must be in place to ensure that food is not contaminated by hazardous substances. Many fires have been caused by placing clothes on an electric heater. Damp clothes should not be positioned in contact with the heater. There should be adequate ventilation around the heater. If possible, the heater should be fitted with a high temperature cut-out device.

The location of the welfare facilities on the site is important. They should be located adjacent to each other and as close to the main working area as possible. It is likely that, as the work progresses, the welfare facilities will need to be moved. It may be appropriate to use facilities in neighbouring premises. However, all such arrangements must be agreed in advance with the relevant neighbour. Reliance on public conveniences should be avoided due to their unreliability and lack of hot water for washing (normally).

Welfare arrangements for transient sites will be different to those for fixed sites. Transient sites include sites where the construction work is of short duration (up to one week in many cases), such as road works and emergency work. In these cases, welfare facilities (toilets, hand-washing basins, rest, eating and first-aid arrangements) may be provided by a suitably equipped vehicle or by siting fixed installations at intervals along the length of the job.

The HSE have produced two useful information leaflets on the provision of welfare facilities at fixed construction sites (CIS No. 18 (rev 1)) and transient construction sites (CIS No. 46).

8.3.3 Work environment issues

Work environment issues are covered by the CDM 2015 Regulations together with an Approved Code of Practice.

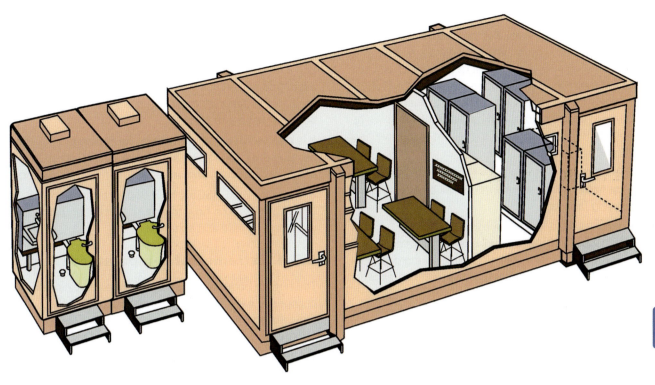

Figure 8.7 A wide range of portable welfare facilities like these are available. It may be possible when refurbishing buildings to use the facilities already on site

The issues governing the workplace environment are ventilation, heating and temperature, lighting, workstations and seating.

Ventilation

Ventilation of the workplace should be effective and sufficient and free of any impurity and air inlets should be sited clear of any potential contaminant (e.g. a chimney flue). Care needs to be taken to ensure that workers are not subject to uncomfortable draughts. The ventilation plant should have an effective visual or audible warning device fitted to indicate any failure of the plant. The plant should be properly maintained and records kept. The supply of fresh air should not normally fall below 5 to 8 litres per second per occupant.

Smoking is now forbidden in enclosed or substantially enclosed areas. See Appendix 8.2.

Heating and temperature

During working hours, the temperature in all workplaces inside buildings shall be reasonable (not uncomfortably high or low). 'Reasonable' is defined in the Approved Code of Practice as at least 16°C, unless much of the work involves severe physical effort, in which case the temperature should be at least 13°C. These temperatures refer to readings taken close to the workstation at working height and away from windows. The Approved Code of Practice recognises that these minimum temperatures cannot be maintained where rooms open to the outside or where food or other products have to be kept cold. A heating or

cooling method must not be used in the workplace which produces fumes, or is injurious or offensive to any person. Such equipment needs to be regularly maintained.

A sufficient number of thermometers should be provided and maintained to enable workers to determine the temperature in any workplace inside a building (but need not be provided in every workroom).

Where, despite the provision of local heating or cooling, the temperatures are still unreasonable, suitable protective clothing and rest facilities should be provided.

Lighting

Every workplace shall have suitable and sufficient lighting and this shall be natural lighting so far as

Figure 8.8 A large building site well lit at night

is reasonably practicable. Suitable and sufficient emergency lighting must also be provided and maintained in any room where workers are particularly exposed to danger in the event of a failure of artificial lighting (normally due to a power cut and/or a fire). Windows and skylights should be kept clean and free from obstruction, but the shading of windows or skylights may be necessary to prevent excessive heat or glare.

When deciding on the suitability of a lighting system, the general lighting requirements will be affected by the following factors:

▶ the availability of natural light;
▶ the specific areas and processes, in particular any colour rendition aspects or concerns over stroboscopic effects (associated with fluorescent lights);
▶ the type of equipment to be used and the need for specific local lighting;
▶ the lighting characteristics required (type of lighting, its colour and intensity);
▶ the location of visual display units and any problems of glare;
▶ structural aspects of the workroom and the reduction of shadows;
▶ the presence of atmospheric dust;
▶ the heating effects of the lighting;
▶ lamp and window cleaning and repair (and disposal issues);
▶ the need and required quantity of emergency lighting.

Light levels are measured in illuminance, having units of lux (lx), using a light meter. A general guide to lighting levels in different workplaces is given in Table 8.1.

Poor lighting levels will increase the risk of accidents such as slips, trips and falls. More information is available on lighting from *Lighting at Work*, HSG38, HSE Books.

Table 8.1 Typical workplace lighting levels

Workplace or type of work	Illuminance (lx)
Warehouses and stores	150
General factories or workshops	300
Offices	500
Drawing offices (detailed work)	700
Fine working (ceramics or textiles)	1000
Very fine work (watch repairs or engraving)	1400

Workstations and seating

Workstations should be arranged so that work may be done safely and comfortably. The worker should be at a suitable height relative to the work surface and there should be no need for undue bending and stretching. Workers must not be expected to stand for long periods of time, particularly on solid floors. A suitable seat

should be provided when a substantial part of the task can or must be done sitting. The seat should, where possible, provide adequate support for the lower back and a footrest provided for any worker whose feet cannot be placed flat on the floor. It should be made of materials suitable for the environment, be stable and, possibly, have arm rests.

It is also worth noting that sitting for prolonged periods can present health risks, such as blood circulation and pressure problems, and vertebral and muscular damage. *Seating at Work*, HSG57, HSE Books, provides useful guidance on how to ensure that seating in the workplace is safe and suitable.

8.3.4 Extremes of temperature

The human body is very sensitive to relatively small changes in external temperatures. Food not only provides energy and the build-up of fat reserves, but also generates heat, which needs to be dissipated to the surrounding environment. The body also receives heat from its surroundings. The body temperature is normally around 37°C and the body will attempt to maintain this temperature irrespective of the temperature of the surroundings. Therefore, if the surroundings are hot, sweating will allow heat loss to take place by evaporation caused by air movement over the skin. On the other hand, if the surroundings are cold, shivering causes internal muscular activity, which generates body heat.

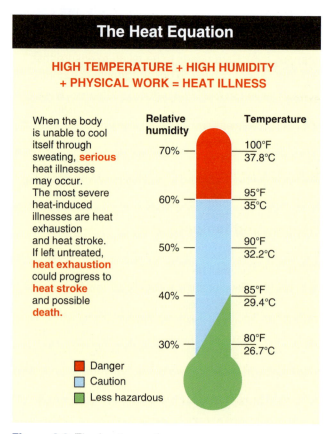

Figure 8.9 The heat equation

At high temperatures, the body has more and more difficulty in maintaining its natural temperature unless sweating can take place and therefore water must be replaced by drinking. If the surrounding air has high humidity, evaporation of the sweat cannot take place and the body begins to overheat. This leads to heart strain and, in extreme cases, heat stroke. It follows that when working is required at high temperatures, a good supply of drinking water should be available and, further, if the humidity is high, a good supply of ventilation air is also needed. On construction sites, weather is a relevant factor and is dealt with in detail in Chapter 15. In general, construction workers must be dressed to suit the prevailing weather conditions, such as warm clothing in cold weather and waterproof clothing during wet weather. Temperature and heat exhaustion are very relevant to working in a confined space, as described in Chapter 17.

At low temperatures, the body will lose heat too rapidly and the extremities of the body will become very cold leading to frostbite and possibly the loss of limbs. Under these conditions, thick, warm (thermal) clothing, the provision of hot drinks and external heating will be required. For those who work in sub-zero temperatures, such as cold store workers, additional precautions will be needed. The store doors must be capable of being unlocked from the inside and an emergency alarm system should be installed. Appropriate equipment selection and a good preventative maintenance system is very important as well as a regular health surveillance programme for the workers, who should be provided with information and training on the hazards associated with working in very low temperatures.

In summary, extremes of temperature require special measures, particularly if accompanied by extremes of humidity. Frequent rest periods will be necessary to allow the body to acclimatise to the conditions.

8.4 Violence at work

Violence at work, particularly from dissatisfied customers, clients, claimants or patients, can cause stress and in some cases injury. This is not only physical violence as people may face verbal and mental abuse, discrimination, harassment and bullying. Fortunately, physical violence is still rare, but violence of all types has risen significantly in recent years. Violence at work is known to cause pain, suffering, anxiety and stress, leading to financial costs due to absenteeism and higher insurance premiums to cover increased civil claims. It can be very costly to ignore the problem. Violence from members of the public is a higher risk with several occupations, e.g. the health and social services, police and fire-fighters, various types of enforcement officers, education, benefit services, various service industries and debt collectors. In one year, there was a total of 54,758 physical assaults reported on NHS staff working in all care sectors.

The risk of being a victim of actual or threatened violence at work over recent years was similar to the last few years with an estimated 1.1% of working adults the victims of one or more violent incidents at work. Strangers were the offenders in 60% of cases of workplace violence. Victims of actual or threatened violence at work said that the offender was under the influence of alcohol in 38% of incidents, and that the offender was under the influence of drugs in 26% of incidents. A survey found 51% of assaults at work resulted in injury, with minor bruising or a black eye accounting for the majority of the injuries recorded.

In 1999 the Home Office and the HSE published a comprehensive report entitled *Violence at Work: Findings from the British Crime Survey*. This report is updated annually and the HSE publishes an annual report – *Violence at work 2013/14: Findings from the Crime Survey for England and Wales*. It shows the extent of violence at work and how it has changed during the period 2006–2014 (Table 8.2).

Table 8.2 Trend in physical assaults and threats at work, 2006–2014 (based on working adults of working age)

Number of incidents (000s)	2006	2008	2010	2012	2014
All violence	752	554	683	628	583
Assaults	338	202	300	310	269
Threats	414	352	363	318	314
Number of victims (000s)					
All violence	350	302	328	318	257
Assaults	200	110	148	150	125
Threats	150	192	180	168	152

Source: British Crime Survey 2014.

Such incidents, whilst they are still high, have halved since 1999. Over this period many of the protections outlined in this chapter and advocated by the HSE have been in place. For example, several hospitals now employ their own police officers to police accident and emergency departments and there has been much more use of CCTV.

Workers who are most at risk from violence are those that:

▶ handle money;
▶ provide a service to the public (such as shop workers, teachers and nurses);
▶ are lone workers;
▶ represent authority (such as police, traffic wardens and even school crossing patrols).

Many people resort to violence due to frustration. Common causes of such frustration are the following:

▶ dissatisfaction with a product or service, including the cost;

▶ a perception of being unreasonably penalised over an incident such as car parking; and

▶ a general lack of information following a problem, such as aircraft delays or long delays at hospitals.

The report defines violence at work as:

'All assaults or threats which occurred while the victim was working and were perpetrated by members of the public.'

Physical assaults include the offences of common assault, wounding, robbery and snatch theft. Threats include both verbal threats, made to or against the victim, and non-verbal intimidation. These are mainly threats to assault the victim and, in some cases, to damage property.

Excluded from the survey are violent incidents where there was a relationship between the victim and the offender and also where the offender was a work colleague. The latter category was excluded because of the different nature of such incidents.

The HSE Survey reported approximately 314,000 threats of violence and 269,000 physical assaults by members of the public on workers in the UK during the year.

Figure 8.10 Security access and surveillance CCTV camera

Approximately 257,000 workers had experienced at least one incident of violence at work. This resulted in one fatality, 866 major or specified injuries and 4,069 over 7-day injuries. Those workers particularly affected in 2008/09 were:

▶ social welfare workers (2.6%);

▶ healthcare professionals (3.8%); and

▶ police officers (9%).

In over a third of the incidents, the victim believed the offender was under the influence of alcohol, and in nearly a fifth of incidents under the influence of drugs.

It is interesting that almost half of the assaults and a third of the threats happened after 1800 hrs, which suggests that the risks are higher if people work at night or in the late evening. About 16% of the assaults involved offenders under the age of 16 and were mainly against teachers or other education workers.

Violence at work is defined by the HSE as:

'any incident in which an employee is abused, threatened or assaulted in circumstances relating to their work.'

In recognition of this, the HSE has produced a useful guide to employers which includes a four-stage action plan and some advice on precautionary measures (*Violence at Work: a Guide for Employers*, INDG69 (rev)). The employer is just as responsible, under health and safety legislation, for protecting employees from violence as they are for any other aspects of their safety.

The Health and Safety at Work Act puts broad general duties on employers and others to protect the health and safety of staff. In particular, Section 2 of the HSW Act gives employers a duty to safeguard, so far as is reasonably practicable, the health, safety and welfare at work of their staff.

Employers also have a common law general duty of care towards their staff, which extends to the risk of violence at work. Legal precedents (see West Bromwich Building Society v Townsend (1983) IRLR 147 and Charlton v Forrest Printing Ink Company Limited (1980) IRLR 331) show that employers have a duty to take reasonable care to see that their staff are not exposed to unnecessary risks at work, including the risk of injury by criminals. In carrying out their duty to provide a safe system of work and a safe working place, employers should, therefore, have regard to, and safeguard their staff against, the risk of injury from violent criminals.

The HSE recommends the following four-point action plan:

1. find out if there is a problem;
2. decide on what action to take;
3. take the appropriate action;
4. check that the action is effective.

8.4.1 Find out if there is a problem

This involves a risk assessment to determine what the real hazards are. It is essential to ask people at the workplace and, in some cases, a short questionnaire may be useful. Record all incidents to get a picture of what is happening over time, making sure that all relevant detail is recorded. The records should include:

▶ a description of what happened;
▶ details of who was attacked, the attacker and any witnesses;
▶ the outcome, including how people were affected and how much time was lost;
▶ information on the location of the event.

Owing to the sensitive nature of some aggressive or violent actions, employees may need to be encouraged to report incidents and be protected from future aggression.

All incidents should be classified so that an analysis of the trends can be examined.

Consider the following:

▶ fatalities;
▶ major injury;
▶ less severe injury or shock which requires first-aid treatment, outpatient treatment, time off work or expert counselling;
▶ threat or feeling of being at risk or in a worried or distressed state.

8.4.2 Decide on what action to take

It is important to evaluate the risks and decide who may be harmed and how this is likely to occur. The threats may be from the public or co-workers at the workplace or may be as a result of visiting the homes of customers. Consultation with employees or other people at risk will improve their commitment to control measures and will make the precautions much more effective. The level of training and information provided, together with the general working environment and the design of the job, all have a significant influence on the level of risk.

Those people at risk could include those working in:

▶ reception or customer service points;
▶ enforcement and inspection;
▶ lone working situations and community-based activities;
▶ front-line service delivery;
▶ education and welfare;
▶ catering and hospitality;
▶ retail petrol and late-night shopping operations;
▶ leisure facilities, especially if alcohol is sold;
▶ healthcare and voluntary roles;
▶ policing and security;

▶ mental health units or in contact with disturbed people;
▶ cash handling or control of high value goods.

Consider the following issues:

▶ quality of service provided;
▶ design of the operating environment;
▶ type of equipment used;
▶ designing the job.

Some violence may be deterred if measures are taken which suggest that any violence may be recorded. Many public bodies use the following measures:

▶ informing telephone callers that their calls will be recorded;
▶ displaying prominent notices that violent behaviour may lead to the withdrawal of services and prosecution;
▶ using CCTV or security personnel.

Four in ten of all employment enquiries to one employment law firm in 2009 were related to bullying. Of those who reported suffering bullying, 80% said it had affected their physical and mental health and a third had taken time off work or left their jobs as a result.

Quality of service provided

The type and quality of service provision has a significant effect on the likelihood of violence occurring in the workplace. Frustrated people whose expectations have not been met and who are treated in an unprofessional way may believe they have the justification to cause trouble.

Sometimes circumstances are beyond the control of the staff member and potentially violent situations need to be defused. The use of correct skills can turn a dissatisfied customer into a confirmed supporter simply by careful response to their concerns. The perceived lack of or incorrect information can cause significant frustrations.

Design of the operating environment

Personal safety and service delivery are very closely connected and have been widely researched in recent years. This has resulted in many organisations altering their facilities to reduce customer frustration and enhance sales. It is interesting that most service points experience less violence when they remove barriers or screens, but the transition needs to be carefully planned in consultation with staff and other measures adopted to reduce the risks and improve their protection.

The layout, ambience, colours, lighting, type of background music, furnishings including their comfort, information, things to do while waiting and even smell all have a major impact. Queue-jumping causes a lot of anger and frustration and needs effective signs and proper queue management, which can help to reduce the potential for conflict.

8

Wider desks, raised floors and access for special needs, escape arrangements for staff, carefully arranged furniture and screening for staff areas can all be utilised.

Type of security equipment used

There is a large amount of equipment available and expert advice is necessary to ensure that it is suitable and sufficient for the task. Some measures that could be considered include the following:

- **Access control** to protect people and property. There are many variations from staffed and friendly receptions, barriers with swipe-cards and simple coded security locks. The building layout and design may well partly dictate what is chosen. People inside the premises need access passes so they can be identified easily.
- **Closed circuit television** is one of the most effective security arrangements to deter crime and violence. Because of the high cost of the equipment, it is essential to ensure that proper independent advice is obtained on the type and the extent of the system required.
- **Alarms** – there are three main types:
  - ▷ intruder alarms fitted in buildings to protect against unlawful entry, particularly after working hours;
  - ▷ panic alarms used in areas such as receptions and interview rooms covertly located so that they can be operated by the staff member threatened
  - ▷ personal alarms carried by an individual to attract; attention and to temporarily distract the attacker.
- **Radios and pagers** can be a great asset to lone workers in particular, but special training is necessary as good radio discipline with a special language and codes are required.
- **Mobile phones** are an effective means of communicating and keeping colleagues informed of people's movements and problems such as travel delays. Key numbers should be inserted for rapid use in an emergency.

Job design

Many things can be done to improve the way in which the job is carried out to improve security and avoid violence. These include:

- using cashless payment methods;
- keeping money on the premises to a minimum;
- careful checking of customer or client's credentials;
- careful planning of meetings away from the workplace;
- team work where suspected aggressors may be involved;
- regular contact with workers away from their base. There are special services available to provide contact arrangements;

- avoidance of lone working as far as is reasonably practicable;
- thinking about how staff who have to work shifts or late hours will get home. Safe transport and/or parking areas may be required;
- setting up support services to help victims of violence and, if necessary, other staff who could be affected. They may need debriefing, legal assistance, time off work to recover or counselling by experts.

A busy accident and emergency department of a general hospital, for example, has to balance the protection of staff from violent attack with the need to offer patients a calm and open environment. Protection could be given to staff by the installation of wide counters, coded locks on doors, CCTV systems, panic buttons and alarm systems. The employment of security staff and strict security procedures for the storage and issuing of drugs are two further precautions taken by such departments. Awareness training for staff so that they can recognise early signs of aggressive behaviour and an effective counselling service for those who have suffered from violent behaviour should be provided.

8.4.3 Take the appropriate action

The arrangements for dealing with violence should be included in the safety policy and managed like any other aspect of the health and safety procedures. Action plans should be drawn up and followed through using the consultation arrangements as appropriate. The police should also be consulted to ensure that they are happy with the plan and are prepared to play their part in providing back-up and the like.

8.4.4 Check that the action is effective

Ensure that the records are being maintained and any reported incidents are investigated and suitable action taken. The procedures should be regularly audited and changes made if they are not working properly.

Victims should be provided with help and assistance to overcome their distress, through debriefing, counselling, time off to recover, legal advice and support from colleagues.

8.5 Substance misuse at work

Alcohol and drug abuse damages health and causes absenteeism and reduced productivity. Both forms of abuse can lead to serious accidents in the construction industry particularly when working occurs at height. The HSE is keen to see employers address the problem and offers advice in two separate booklets.

Alcohol abuse is a considerable problem when vehicle driving is part of the job, especially if driving is required on public roads. Misuse of alcohol can reduce productivity, increase absenteeism, increase accidents at work and, in some cases, endanger the public. The HSE has estimated that between 3% and 5% of all absences from work are due to alcohol and result in approximately 14 million working days lost each year. Employers need to adopt an alcohol policy following employee consultation. The following matters need to be considered:

▶ how the organisation expects employees to restrict their level of drinking;
▶ how drinking problems can be recognised and help offered; and
▶ at what point and under what circumstances will an employee's drinking be treated as a disciplinary rather than a health problem.

Prevention of the problem is better than remedial action after a problem has occurred. During working hours, there should be no drinking, and drinking during break periods should be discouraged. Induction training should stress this policy and managers should set it. Posters can also help to communicate the message. However, it is important to recognise possible symptoms of an alcohol problem, such as lateness and absenteeism, poor work standards, impaired concentration, memory and judgement, and deteriorating relations with colleagues. It is always better to offer counselling rather than dismissal. The policy should be monitored to check on its effectiveness.

Drug and solvent abuse presents similar problems to those found with alcohol abuse – absenteeism, reduced productivity and an increase in the risk of accidents. The drugs that are involved in drug abuse may be legal drugs, such as pain killers of various types, or illegal drugs such as heroin. A study undertaken by Cardiff University found that 'although drug use was lower among workers than the unemployed, one in four workers under the age of 30 years reported having used drugs in the previous year. There are well documented links between drug use and impairments in cognition, perception

and motor skills, both at the acute and chronic levels. Associations may therefore exist between drug use and work performance'. The following conclusions were also drawn from the study:

▶ about 13% of working respondents reported drug use in the previous year. The rate varied with age, from 3% of those over 50 years to 29% of those under 30 years;
▶ drug use is strongly linked to smoking and heavy drinking in that order;
▶ there is an association between drug use and minor injuries among those who are also experiencing other minor injury risk factors;
▶ the project has shown that recreational drug use may reduce performance efficiency and safety at work.

A successful drug misuse policy will benefit the organisation and employees by reducing absenteeism, poor productivity and the risk of accidents. There is no simple guide to the detection of drug abuse, but the HSE has suggested the following signals:

▶ sudden mood changes;
▶ unusual irritability or aggression;
▶ a tendency to become confused;
▶ abnormal fluctuations in concentration and energy;
▶ impaired job performance;
▶ poor time-keeping;
▶ increased short-term sickness leave;
▶ a deterioration in relationships with colleagues, customers or management;
▶ dishonesty and theft (arising from the need to maintain an expensive habit).

A policy on drug abuse can be established by:

1. Investigation of the size of the problem

Examination of sicknesses, behavioural and productivity changes and accident and disciplinary records is a good starting point.

2. Planning actions

Develop an awareness programme for all staff and a special training programme for managers and supervisors. Employees with a drug problem should be encouraged to seek help in a confidential setting.

3. Taking action

Produce a written policy that includes everyone in the organisation and names the person responsible for implementing the policy. It should include details of the safeguards to employees and the confidentiality given to anyone with a drug problem. It should also clearly outline the circumstances in which disciplinary and/or reported action will be taken (the refusal of help, gross misconduct and possession of/dealing in drugs).

4. Monitoring the policy

The policy can be monitored by checking for positive changes in the measures made during the initial investigation (improvements in the rates of sickness and accidents). Drug screening and testing is a

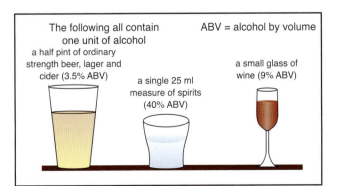

Figure 8.11 It takes a healthy liver about one hour to break down and remove one unit of alcohol. A unit is equivalent to 8 mg or 10 ml (1 cl) of pure alcohol

sensitive issue and should only be considered with the agreement of the workforce (except in the case of pre-employment testing). Screening will only be acceptable if it is seen as part of the health policy of the organisation and its purpose is to reduce risks to the misusers and others. Laboratories are accredited by the United Kingdom Accreditation Service (UKAS) and are able to undertake reliable testing.

It is important to stress that some drugs are prescribed and controlled. The side effects of these can also affect performance and pose risks to colleagues. Employers should encourage employees to inform them of any possible side effects from prescribed medication and be prepared to alter work programmes accordingly.

8.6 Safe movement of people on construction sites

8.6.1 Hazards in the workplace

The most common hazards to pedestrians at work are slips, trips and falls on the same level, falls from height, collisions with moving vehicles, being struck by moving, falling or flying objects and striking against fixed or stationary objects. Each of these will be considered in turn, including the conditions and environment in which the particular hazard may arise.

Slips, trips and falls on the same level

Slipping and tripping is the single most common cause of injury in the workplace. These are the most common of the hazards faced by pedestrians and account for 38% of all the major accidents every year and 24% of minor injuries. Every 25 minutes someone breaks or fractures a bone due to slipping, tripping or falling at work. In addition to 40 fatalities each year, there are over 15,000 major injuries to workers, as well as more than 30,000 workers having to take over three days off work. HSE figures show that slips and trips are the most common cause of major workplace injury in Britain.

It has been estimated that the annual cost of these accidents to the nation is over £650 million and a direct cost to employers of £300 million. Civil compensation claims are becoming more common and costly to employers. The highest reported injuries are reported in the food and related industries. Older workers, especially women, are the most severely injured group from falls resulting in fractures of the hips and/or femur. Civil compensation claims are now being made by members of the public who have tripped on uneven paving slabs on pavements or in shopping centres. A worker was awarded compensation after slipping on a wet wheelchair ramp. The accident could have been avoided if the ramp had been covered in an anti-slip material.

The HSE has been so concerned at the large number of such accidents that it has identified slips, trips and falls

on the same level as a key risk area. The costs of slips, trips and falls on the same level are high to the injured employee (lost income and pain), the employer (direct and indirect costs including lost production) and to society as a whole in terms of health and social security costs.

Slip hazards are caused by:

► wet or dusty floors;
► the spillage of wet or dry substances – oil, water, flour dust and plastic pellets used in plastic manufacture;
► loose mats on slippery floors;
► wet and/or icy weather conditions;
► unsuitable footwear or floor coverings or sloping floors.

Figure 8.12 Tripping hazards on untidy site

Trip hazards are caused by:

► loose floorboards or carpets;
► obstructions, low walls, low fixtures on the floor;
► cables or trailing leads across walkways or uneven surfaces; leads to portable electrical hand tools and other electrical appliances (vacuum cleaners and overhead projectors);
► raised telephone and electrical sockets – also a serious trip hazard (this can be a significant problem when the display screen workstations are re-orientated in an office);
► rugs and mats – particularly when worn or placed on a polished surface;
► poor housekeeping – obstacles left on walkways, rubbish not removed regularly;
► poor lighting levels – particularly near steps or other changes in level;
► sloping or uneven floors – particularly where there is poor lighting or no handrails;
► unsuitable footwear – shoes with a slippery sole or lack of ankle support.

Over half of all trip accidents are caused by poor housekeeping. Many of these housekeeping problems can easily be solved by:

- ensuring that all walkways are suitable for purpose and clear of obstructions (e.g. trailing leads);
- ensuring that workers wear suitable footwear;
- training workers in the maintenance of trip free working areas;
- regular inspections of work areas by supervisors.

Slip and trip accidents increase during the Autumn and Winter season for a number of reasons: there is less daylight, leaves fall onto paths and become wet and slippery and cold weather spells cause ice and snow to build up on paths.

The selection of suitable footwear for the workplace should involve consideration of the types of hazard present, the general environment and ergonomic issues.

Figure 8.13 Cleaning must be done carefully to prevent slipping or falling using bosun's chair and rope support with trained worker

Falls from work at height

More workplace deaths are triggered by falls from height than any other cause. These are the most common cause of serious injury or death in the construction industry. These accidents are often concerned with falls of greater than about 2 m and often result in fractured bones, serious head injuries, loss of consciousness and death. Twenty-five per cent of all deaths at work and 19% of all major accidents are due to falls from a height. Falls down staircases and stairways, through fragile surfaces, off landings

and stepladders and from vehicles, all come into this category. Injury, sometimes serious, can also result from falls below 2 m, for example using swivel chairs for access to high shelves.

Collisions with moving vehicles

These can occur within the workplace premises or on the access roads around the building. It is a particular problem where there is no separation between pedestrians and vehicles or where vehicles are speeding. Poor lighting, blind corners, the lack of warning signs and barriers at road crossing points also increase the risk of this type of accident. Eighteen per cent of fatalities at work are caused by collisions between pedestrians and moving vehicles with the greatest number occurring in the service sector (primarily in retail and warehouse activities). Collisions are also a significant cause of accidents in the construction industry.

Figure 8.14 Falling from a height – tower scaffold with inadequate handrail (too low) and no middle rail. Access ladder should be internal and it should never be moved with people on the scaffold

Being struck by moving, falling or flying objects

This causes 18% of fatalities at work and is the second highest cause of fatality in the construction industry. It also causes 15% of all major and 14% of over 3-day accidents. Moving objects include articles being moved, moving parts of machinery or conveyor belt systems, and flying objects are often generated by the disintegration of a moving part or a failure of a system under pressure. Falling objects are a major problem in construction (due to careless working at height) and in warehouse work (due to careless stacking of pallets on racking). The head is particularly vulnerable to these hazards. Items falling off high shelves and moving loads are also significant hazards in many sectors of industry.

Striking against fixed or stationary objects

This accounts for over 1,000 major accidents every year. Injuries are caused to a person either by colliding with a fixed part of the building structure, work in progress, a machine member or a stationary vehicle or by falling against such objects. The head appears to be the most vulnerable part of the body to this particular hazard and this is invariably caused by the misjudgement of the height of an obstacle. Concussion in a mild form is the most common outcome and a medical check-up is normally recommended. It is a very common injury during maintenance operations when there is, perhaps, less familiarity with particular space restrictions around a machine. Effective solutions to all these hazards need not be expensive, time consuming or complicated. Employee awareness and common sense combined with a good housekeeping regime will solve many of the problems.

8.6.2 Control measures for the safe movement of people in the workplace

Slips, trips and falls on the same level

These may be prevented or, at least, reduced by several control strategies. These and all the other pedestrian hazards discussed should be included in the workplace risk assessments required under the Management of Health and Safety at Work Regulations by identifying slip or trip hazards, such as poor or uneven floor/pavement surfaces, badly lit stairways and puddles from leaking roofs. There is also a legal requirement in the Workplace (Health, Safety and Welfare) Regulations for all floors to be suitable, in good condition and free from obstructions. Traffic routes must be organised to enable people to move around the workplace safely.

The key elements of a health and safety management system are as relevant to these as to any other hazards:

Planning – remove or minimise the risks by using appropriate control measures and defined working practices (e.g. covering all trailing leads).

Organisation – involve employees and supervisors in the planning process by defining responsibility for keeping given areas tidy and free from trip hazards.

Control – record all cleaning and maintenance work. Ensure that anti-slip covers and cappings are placed on stairs, ladders, catwalks, kitchen floors and smooth walkways. Use warning signs when floor surfaces have recently been washed.

Monitoring and review – carry out regular safety audits of cleaning and housekeeping procedures and include trip hazards in safety surveys. Check on accident records to see whether there has been an improvement or if an accident black spot can be identified.

Slip and trip accidents are a major problem for large retail stores for both customers and employees. The provision of non-slip flooring, a good standard of lighting and minimising the need to block aisles during the restocking of merchandise are typical measures that many stores use to reduce such accidents. Other measures include the wearing of suitable footwear by employees, adequate handrails on stairways, the highlighting of any floor level changes and procedures to ensure a quick and effective response to any reports of floor damage or spillages. Good housekeeping procedures are essential. The design of the store layout and any associated warehouse can also ensure a reduction in all types of accident. Many of these measures are valid for a range of workplaces.

Cleaning arrangements should be chosen to suit both the type of floor and the users. Floors with surface roughness are not difficult to clean despite the popular belief that they are. Stairs are a particularly hazardous part of the building and become even more so when being cleaned.

The prevention of access during cleaning or drying needs to be effective. Signs that warn people that floors are wet are only partially effective. The HSE make the following suggestions:

▶ Physically exclude people from wet cleaning areas by the use of barriers or locking off an area.
▶ Clean during quiet hours.
▶ Clean in sections so there is always a dry path through the area.

Floor cleaners must be provided with appropriate slip-resistant footwear which can help reduce the risk of slipping.

Stairs can also present significant potential for harm to their users, with falls on stairs often leading to serious injury or even death. Around 20% of all major injuries reported to the HSE each year from slips and trips occurred on stairs.

Figure 8.15 Good stairs with handrail leading from site accommodation

Falls from work at height

The Work at Height Regulations give a legal framework for the protection of workers from falls when working at height. The principal means of preventing falls of people or materials includes the use of fencing, guard rails, toe boards, working platforms, access boards and ladder hoops, safety nets and safety harnesses. Safety harnesses arrest the fall by restricting the fall to a given distance due to the fixing of the harness to a point on an adjacent rigid structure. Safety nets and safety harnesses should only be used when all other possibilities are not reasonably practicable. The use of banisters on open sides of stairways and handrails fitted on adjacent walls will also help to prevent people from falling. Holes in floors and pits should always be fenced or adequately covered. Precautions should be taken when working on fragile surfaces. More detail on the requirements of the Work at Height Regulations is given in Chapter 16.

Staircases are also a source of accidents included within this category of falling from a height and the following design and safety features will help to reduce the risk of such accidents:

▶ adequate width of the stairway, depth of the tread and provision of landings and banisters or handrails and intermediate rails; the treads and risers should always be of uniform size throughout the staircase and designed to meet the Building Regulations requirements for angle of incline (i.e. steepness of staircase);
▶ provision of non-slip surfaces and reflective edging;
▶ adequate lighting;
▶ adequate maintenance;
▶ special or alternative provision for disabled people (e.g. personnel elevator at the side of the staircase).

Great care should be used when people are loading or unloading vehicles; as far as possible people should avoid climbing onto vehicles or their loads. For example, sheeting of lorries should be carried out in designated places using properly designed access equipment.

Collisions with moving vehicles

These are best prevented by completely separating pedestrians and vehicles, providing well-marked, protected and laid-out pedestrian walkways. People should cross roads by designated and clearly marked pedestrian crossings. Suitable guard rails and barriers should be erected at entrances and exits from buildings and at 'blind' corners at the end of racking in warehouses. Particular care must be taken in areas where lorries are being loaded or unloaded. It is important that separate doorways are provided for pedestrians and vehicles and all such doorways should be provided with a vision panel and an indication of the safe clearance height, if used by vehicles. Finally, the enforcement of a sensible speed limit, coupled, where practicable, with speed governing devices, is another effective control measure.

Being struck by moving, falling or flying objects

These hazards may be prevented by guarding or fencing the moving part (as discussed in Chapter 11) or by adopting the measures outlined for construction work at height (Chapter 16). Both construction workers and members of the public need to be protected from the hazards associated with falling objects. Both groups should be protected by the use of covered walkways or suitable netting to catch falling debris where this is a significant hazard. Waste material should be brought to ground level by the use of chutes or hoists. Waste should not be thrown from a height and only minimal quantities of building materials should be stored on working platforms. Appropriate personal protective equipment, such as hard hats or safety glasses, should be worn at all times when construction operations are taking place.

It is often possible to remove high-level storage in offices and provide driver protection on lift truck cabs in warehouses. Storage racking is particularly vulnerable and should be strong and stable enough for the loads it has to carry. Damage from vehicles in a warehouse can easily weaken the structure and cause collapse. Uprights need protection, particularly at corners.

The following action can be taken to keep racking serviceable:

▶ Inspect them regularly and encourage workers to report any problems.
▶ Post notices with maximum permissible loads and never exceed the loading.
▶ Use good pallets and safe stacking methods.
▶ Band, box or wrap articles to prevent items falling.
▶ Set limits on the height of stacks and regularly inspect to make sure that limits are being followed.
▶ Provide instruction and training for staff and special procedures for difficult objects.

Striking against fixed or stationary objects

This hazard can only be effectively controlled by:

▶ having good standards of lighting and housekeeping by ensuring that all waste debris and construction materials, particularly timber, are safely stored and/or removed from the site;
▶ defining walkways and making sure they are used;
▶ the use of awareness measures, such as training and information in the form of signs or distinctive colouring;
▶ the use of appropriate personal protective equipment, such as head protection, as discussed previously.

8

General preventative measures for pedestrian hazards

Minimising pedestrian hazards and promoting good work practices requires a mixture of sensible planning, good housekeeping and common sense. A few of the required measures are costly or difficult to introduce and, although they are mainly applicable to slips, trips and falls on the same level and collisions with moving vehicles, they can be adapted to all types of pedestrian hazard. Typical measures include the following:

▶ Develop a safe workplace as early as possible and ensure that suitable floor surfaces and lighting are selected and vehicle and pedestrian routes are carefully planned. Lighting should not dazzle approaching vehicles nor should pedestrians be obscured by stored products. Lighting is very important where there are changes of level or stairways. Any physical hazards, such as low beams, vehicular movements or pedestrian crossings, should be clearly marked. Staircases need particular attention to ensure that they are slip resistant and the edges of the stairs marked to indicate a trip hazard.

▶ Consider pedestrian safety when re-orientating the workplace layout (e.g. the need to reposition lighting and emergency lighting).

▶ Adopt and mark designated walkways.

▶ Apply good housekeeping principles by keeping all areas, particularly walkways, as tidy as possible and ensure that any spillages are quickly removed.

▶ Ensure that all workers are suitably trained in the correct use of any safety devices (such as machine guarding or personal protective equipment) or cleaning equipment provided by the employer.

▶ Only use cleaning materials and substances that are effective and compatible with the surfaces being cleaned, so that additional slip hazards are not created.

▶ Ensure that a suitable system of maintenance, cleaning, fault reporting and repair are in place and working effectively. Areas that are being cleaned must be fenced and warning signs erected. Care must also be taken with trailing electrical leads used with the cleaning equipment. Records of cleaning, repairs and maintenance should be kept.

▶ Ensure that all workers are wearing high visibility clothing and appropriate footwear with the correct type of slip-resistant soles suitable for the type of flooring.

▶ Consider whether there are significant pedestrian hazards present in the area when any workplace risk assessments are being undertaken.

It can be seen, therefore, that floors and traffic routes should be of sound construction. If there are frequent, possibly transient, slip hazards, the provision of slip-resistant coating and/or mats should be considered and warning notices posted. Any damaged areas must be cordoned off until repairs are completed. Risk assessments should review past accidents and near misses to enable relevant controls, such as suitable footwear, to be introduced. Employees can often indicate problem areas, so employee consultation is important.

Electrically powered gates

The HSE has recently issued a warning on the risks to pedestrians from crushing zones on electrically powered gates, such as those commonly used at car parks. Two young children suffered fatal injuries after being trapped by automatic sliding gates between the closing edge of the gate and the gate post at the end of the gates' travel. They were trapped because:

▶ Their presence in the vicinity of the closing edge was not detected; and

▶ The closing force of the gate when they obstructed it was not limited to the values specified in Annex A of BS EN 12453:2001.

The HSE Safety Notice reinforced and updated previous information to organisations and individuals involved in the design, construction, installation and commissioning of electrically powered gates and organisations in control of their use and/or maintenance. It is also relevant to companies carrying out ongoing maintenance of these types of gates. The British Standard (BS EN 12453:2001) recommends a minimum

Figure 8.16 Typical pedestrian/vehicle crossing area

Figure 8.17 A designated waste collection area

level of safeguarding against the crushing hazard at the closing edge of the gate depending on the type of environment in which the gate is operating. For electrically powered gates that operate in an area accessible to the public, the advised level of safeguarding is to:

▶ Limit forces according to Annex A of the Standard using force limitation devices or sensitive protective equipment AND provide a means for the detection of the presence of a person or an obstacle standing on the floor at one side of the gate; or

▶ Provide a means for detection of the presence a person, which is designed in a way that in no circumstances can that person be touched by the moving gate leaf.

The HSE recommend the following actions:

▶ All designers and installers of electrically powered gates should ensure that the forces generated by a gate when meeting a person or an obstacle are limited and that they do not exceed the values specified in Annex A of BS EN 12453:2001.

▶ Forces should be periodically re-measured and checked as part of the planned preventative maintenance schedule for the gates.

▶ In addition to force limitation, additional safeguards, such as pressure sensitive strips on the closing edge and photoelectric sensing devices, should be fitted where the risk assessment identifies the gate as high risk, in that it is operating automatically in a public place where children and other members of the public may be present.

▶ Persons or organisations in control of powered gates should periodically review their risk assessments to ensure that they identify any changes to the environment or operating conditions and that they have taken appropriate steps to address them. This is particularly important when the responsibility for management of the gate passes from one person or organisation to another.

▶ All safety devices and features should be checked on a regular basis and in accordance with the manufacturer's instructions to ensure they continue to function as designed to ensure that safety is maintained. This should be specified in a planned preventative maintenance schedule agreed by persons responsible for the gate's management and their appointed maintenance company.

Maintenance activities

Maintenance-related accidents are a serious cause of concern. For example, analysis of data from recent years indicates that 25–30% of manufacturing industry fatalities in Great Britain were related to maintenance activity. Maintenance work is often more hazardous than other types of work because it is often performed under pressure – possibly due to a plant failure during a production run – or the areas requiring access are difficult to reach.

The majority of recent accidents to workers engaged in maintenance work occurred during the following activities:

▶ roofwork;
▶ machinery and plant maintenance;
▶ building maintenance including painting work;
▶ maintenance of services (gas, electricity, telecommunications and water);
▶ vehicle repair work; and
▶ window and building cleaning.

The main hazards associated with maintenance work are due to:

▶ electricity – electric shock;
▶ mechanical including entanglement, ejection and unexpected start-up;
▶ hazardous substances – gases, dusts and fume;
▶ physical – noise and vibration;
▶ obstructions of various types;
▶ access – confined spaces and work at height; and
▶ unexpected releases of pressure or other forms of energy.

Inspections and even thorough examinations are not substitutes for properly maintaining equipment. The information gained in the maintenance work, inspections and thorough technical examinations should inform one another. A maintenance log should be kept and be up to date. The whole maintenance system will require proper management systems. The frequency will depend on the equipment, the conditions in which it is used and the manufacturers' instructions.

8.6.3 Hazards to the general public and the associated controls in construction activities, including public highways

Four chapters (8, 16, 17 and 18) deal in detail with the protection required for members of the public in many different construction scenarios. These hazards may be conveniently divided between those which are present inside the site and those which may occur outside the site. Inside the site, the public may be visitors (authorised or unauthorised) or joint occupiers of the site if it is a refurbishment project. The hazards from the construction work presented to the public outside the site could include materials falling from working platforms, and the operation of cranes and other lifting equipment.

The control measures begin with the posting of various warning signs and the provision of protected thoroughfares. Separate entrances for people and vehicles are another good safeguard as is the provision of 2 m high perimeter fencing, with no clearance at ground level. This fence will provide good security even for short duration work. Security guards and/or cameras may be required on sites which are deemed particularly vulnerable to trespassers. Such trespassers are deterred from sites which are well lit and are surrounded by a high security fence. Site security is addressed earlier in this chapter.

The protection of the public is a very important issue when the following groups of the public are concerned.

People with disabilities must be considered particularly when pedestrian routes are affected by cable installation work or scaffolds on pavements. During the site assessment, pedestrian movements should be monitored to ascertain whether there is regular usage of the pavement by disabled people, the elderly or children in prams.

Children – this topic is covered in detail earlier in this chapter. Local schools should be made aware of the proposed construction work and pupils warned of the hazards present on the site particularly when it is not being worked. Work on school premises requires extra care and 2 m plywood panels may need to be fitted around the base of scaffolds to prevent access by children. If possible, such work should be done during weekends or school holiday periods. Deliveries should not occur at school start, finish or break times. Tunnels and fans (see Figure 16.13) may be required near playgrounds and entrances to prevent pupils, staff and parents being struck by falling objects.

Occupied premises – again this issue is covered earlier in this chapter. Where work is to be undertaken in or near occupied premises, such as hospitals, schools, office blocks, factories, shops or houses, it may be necessary to completely or partially evacuate the premises either during a hazardous part of the construction work or an emergency. In such cases, a detailed evacuation plan will be needed. For elderly residents or hospital patients, evacuation may present greater risks than the hazardous construction work itself.

Refurbishment of residential properties – the use of hazardous substances during damp course work or timber treatment may require the evacuation of local residents for a short period of time. A plan needs to be in place prior to the start of the work to deal with this eventuality. Good communication between the contractor, clients and residents is essential if risks are to be successfully controlled. Advice should be sought from the emergency services before work begins.

Refurbishment of commercial properties – this is also covered in earlier in this chapter. The principal contractor will need to cooperate closely with the occupiers to manage the risks created by sharing the workplace as well as managing the risks created by his/her work. When work is to be done in areas which remain occupied, the risk assessment must clearly define the perimeter around the construction work area

and detail the methods used to maintain the security of the site and the safety of members of the public passing around, under or over the site.

Healthcare premises – these include hospitals, clinics, surgeries and psychiatric hospitals. It is important that the principal contractor is informed of the nature of the work undertaken in these premises and whether his/her normal mode of operation will need to be modified. Unnecessary obstructions should be removed and adequate signage used to indicate hazards which are present. The accidental release of fungal spores or other organisms could adversely affect patients who are recovering from major surgery. It was decided to use a mobile elevating work platform at a site in a psychiatric hospital due to fears that patients might climb a traditional scaffold out of hours. Construction site vehicles and pedestrians should be separated from hospital traffic as much as possible during the contract.

House building – additional hazards are presented when the building programme takes place beside an established occupied housing estate. Site security, safe vehicular access through the estate to the site and trespass by children are all problems which must be addressed. An additional problem is created when a show house and sales office are located on the site. Prospective buyers must be protected from the hazards on the site.

Street works present particular hazards to pedestrians and many of these hazards are discussed in chapters 8, 16, 17 and 18. Work on a pavement or road is hazardous for both members of the public and the construction workers. Pavements should be kept clear of tripping hazards, such as trailing cables. The site must be well lit at night. Road traffic past the site may also need to be controlled to protect the workforce. Members of the public and traffic vehicles must also be protected from the elbows of loaders, excavators and cranes which may swing into their path. More detailed advice is available from the Code of Practice '*Safety at street works and road works*' related to the New Roads and Street Works Act. The following points for the protection of pedestrians should be considered when work in streets or similar areas is being planned:

▶ temporary traffic controls;
▶ signs for pedestrians and traffic;
▶ cones or other barriers to designate the safety zone;
▶ barriers to protect and guide the public around the site;
▶ temporary walking surfaces for the public;
▶ temporary lighting;
▶ secure and safe storage of working materials;
▶ arrangements for the movement of vehicles to and from the site;
▶ provision of high visibility clothing for workers on the site;

Figure 8.18 Pedestrians separated from the work and traffic

▶ arrangements for other hazards associated with the work at the site, such as noise, dust and buried services.

Construction organisations that undertake street works in England must carry out their works safely and having particular regard to the needs of people with disabilities, as set out in Section 65 of the New Roads and Street Works Act. The statutory 'Code of practice 2013' is a revision to the existing 'Code of practice 2001'. Known informally as the 'safety code' or the 'red book', it came into force on 1 October 2014 and provides updated guidance on safe working at street works and road works sites.

The code has been designed to make it easier to follow and the site layout diagrams have been redrawn to make them easier to understand. It encourages safer working practices and there is more emphasis on risk assessment, provision for pedestrians and other vulnerable road users, and site-specific design. Additional guidance is given on mobile and short duration works and working near tramways and railways, and advice on high visibility clothing and the signing and visibility requirements for works vehicles has been updated. The code also encourages operatives to consider minimising inconvenience to road users.

8.7 Further information

The Construction (Design and Management) Regulations 2015

The Health and Safety (First-aid) Regulations 1981 (as amended)

The Health and Safety (Safety Signs and Signals) Regulations 1996

The Personal Protective Equipment at Work Regulations 1992

The Provision and Use of Work Equipment Regulations 1998

The Workplace (Health, Safety and Welfare) Regulations 1992

The Work at Height Regulations 2005

The health and safety toolbox: How to control risks at work, HSE Books, ISBN 978-0-7176-6587-7 http://www.hse.gov.uk/pubns/priced/hsg268.pdf

Health and safety in construction (Guidance) (HSG150), HSE Books, ISBN 0-7176-6182-2 http://www.hse.gov.uk/pubns/priced/hsg150.pdf

Managing Health and Safety in Construction ACoP (L144), HSE Books (to be replaced by L153 during 2015), ISBN 978-0-7176-6223-4 http://www.hse.gov.uk/pubns/priced/l144.pdf

Workplace, health, safety and welfare, the Workplace (Health, Safety and Welfare) Regulations 1992, ACoP and Guidance, L24, HSE Books, ISBN 978-0-7176-6583-9 http://www.hse.gov.uk/pubns/priced/l24.pdf

Safe use of work equipment (ACoP) (L22), HSE Books, ISBN 978-0-7176-6295-1 http://www.hse.gov.uk/pubns/priced/l22.pdf

Work at Height Regulations 2005 – A Brief guide, HSE Books, INDG 401, rev 2 01/14, http://www.hse.gov.uk/pubns/indg401.pdf

Driving at work, Managing work-related road safety, INDG 382, HSE Books, http://www.hse.gov.uk/pubns/indg382.pdf

8.8 Practice revision questions

1. A new office block is to be constructed on a business park on the outskirts of a busy town.

   (a) **Outline** the issues that should be addressed when undertaking the initial site assessment.

   (b) **Outline** the measures that should be taken during the establishment of the site.

2. The preparation of a construction site requires the removal of topsoil from land known to be contaminated with heavy metals. An overhead power line crosses the site and buried cables may be present.

   (a) **Outline** the specific requirements for this work in relation to:
      (i) safe working practices concerning the overhead power line and buried cables;
      (ii) personal protective equipment; and
      (iii) welfare facilities.

   (b) **Outline** the precautions required in relation to the transportation from the site and disposal of the contaminated soil.

3. Extensive maintenance and refurbishment work is to be undertaken within a working warehouse.

   (a) **Outline** the main hazards faced by:
      (i) construction workers; and
      (ii) the non-construction workers in the warehouse.

   (b) **Identify TEN** items that should be covered at a site induction briefing of the employees of the maintenance contractor.

4. A new golf course is under construction near a residential estate. Several residents have complained about excessive noise levels from the site, dust on their property and mud on residential streets.

   (a) **Outline** the actions that could be taken by the construction company to mitigate such nuisances.

   (b) **Describe** the means by which the site managers could make contractors and operatives aware of, and adhere to, any new procedures adopted.

5. A construction site is where an estate of houses is being built. The site has access to visitors, who wish to view the show house, and is patrolled outside normal working hours by security personnel.

   (a) **Identify** the topics that should be included in a short induction for site visitors.

   (b) **Outline** the issues that should be considered to ensure the health and safety of the security personnel outside of normal working hours.

   (c) **Outline EIGHT** precautions that should be considered to prevent accidents to children who might be tempted to gain access to the site.

6. A project involves the construction of a large warehouse, including the installation of drains, on a brownfield site. Up to 50 people are likely to be involved in the work at any one time. **Outline** the sanitary and washing facilities, drinking water supply and other welfare facilities, including first-aid, that will be required at the site.

7. Extremes of temperature on a construction site may have an adverse effect on the health of workers.

   (a) **Identify** the possible effects on health that may be caused by working in a hot and cold environment, particularly strong sunlight and inclement weather.

   (b) **Outline** the controls required when workers are exposed to:
      (i) a hot environment
      (ii) a cold environment.

8. (a) **Outline** the practical measures that might be taken to reduce the risk of violence to

employees who deal with members of the public as part of their work.

(b) **Outline** an action plan that can be developed to protect employees from the risk of violence in the workplace.

(c) **Identify FOUR** types of security equipment that could be employed to protect vulnerable workers from the risk of violence when dealing with members of the public.

9. (a) **Outline FIVE** possible indications that an employee may be suffering from alcohol, solvent or drug abuse.

(b) **Identify** the risks to health from the misuse of these substances.

(c) **Outline** the control measures that could reduce these risks to health.

10. Slips or trips are the single largest causes of injuries on construction sites.

(a) **Identify EIGHT** types of hazard that may cause such slips or trips on site.

(b) **Identify** ways of controlling slips and trips on a construction site.

11. **Identify EIGHT** design features and/or safe practices intended to reduce the risk of accidents on staircases that are used as internal pedestrian routes during the construction of residential houses.

12. Both pedestrians and vehicles use the same traffic routes on a busy construction site. **Identify** control measures that should be in place to reduce the risks to pedestrians on the site.

13. Following the development of dangerous pot holes during a very wet winter, significant repairs and resurfacing to a road section is to be undertaken. Adjacent to the road is a primary school, a large warehouse and residential properties.

(a) **Identify** the environmental factors that should be taken into account and **outline** how any detrimental environmental effects might be mitigated.

(b) **Describe** the means by which the safe passage of vehicular and pedestrian traffic might be achieved.

8

APPENDIX 8.1 A typical set of site safety rules

1. All contractors are required to comply with the guidance contained in the publication L153 Managing Health and Safety in construction and HSG150 Health and Safety in Construction as the basis upon which their construction activities are organised, planned and implemented. A copy of each will be provided on site for reference by contractors and employees.

2. Contractors must ensure that their employees are made aware of these site rules and fire, first-aid and emergency procedures.

3. Any person who is in doubt about safe working practices and procedures must contact their supervisor.

4. Any person becoming aware of any unsafe act, condition or faulty piece of equipment, which they cannot put right themselves, must where possible warn others, isolate the hazard and immediately inform the person in charge. Where necessary, work should be stopped and, if appropriate, the area evacuated.

5. All persons must report to the site manager when initially arriving on site.

6. Sub-contractors must advise the site manager of any employees under the age of 18 and provide a copy of their assessment of the risks specific to young persons.

7. Young persons on 'work experience' will not be permitted on site without the prior agreement of the site manager.

8. No person who is known to be under the influence of alcohol or illegal substances will be permitted to work.

9. The site manager has the authority to stop work if he/she is not satisfied with the provisions or arrangements made for health and safety whether in respect of unsafe practices, faulty plant and equipment or the conduct of employees.

10. All plant and equipment brought onto site must be in good condition and properly maintained.

11. Operators of plant and equipment must produce 'certificates of competence or training achievement' when requested or a letter from their employer confirming competence and authorisation to operate the plant/equipment.

12. Plant and machinery must be immobilised and secured when the site is unattended.

13. The site manager must be advised before any 'hot work' is undertaken and an appropriate hot work permit form completed.

14. Entrance gates/fencing must be secured and locked when the site is unattended.

15. Contractors must supply details of risk assessment and method statements as required by the site manager.

16. Only 110V electrical tools and equipment are to be used on site.

17. Hard hats must be worn in the designated areas.

18. Safety footwear must be worn at all times.

19. Debris and waste must be cleared away regularly, and dust (and noise) kept to a minimum. There must be no burning on site.

20. All accidents and incidents must be reported immediately to the site manager and entered in the accident book.

21. The statutory reporting of injuries is the responsibility of the individual employer of contracted workers. The site manager is to be provided with a copy of Form F2508.

22. Sub-contractors and employees must cooperate with the site manager in the investigation of any incident or accident.

APPENDIX 8.2 Smoke-free workplaces

1. Introduction

Second-hand tobacco smoke is a major cause of heart disease and lung cancer amongst non-smokers who work with people who smoke. It is estimated that around 700 workers a year die as a direct result of second-hand tobacco smoke in their workplace.

Second-hand smoke is also responsible for many thousands of episodes of illness. For example, Asthma UK reports that it is the second most common asthma trigger in the workplace. Eighty-two per cent of people with asthma say that other people's smoke worsens their asthma and one in five people with asthma feel excluded from parts of their workplace where people smoke.

In 2007, legal restrictions on smoking in workplaces and public places were introduced in England. Similar legislation had been implemented in Wales and Scotland already. The English legislation comes about through the Health Act 2006, which effectively bans smoking in all enclosed workplaces and public places, with some exemptions. The following information comes from the Department of Health website.

2. A quick guide to the smoke-free law is as follows:

▶ England became smoke-free on Sunday, 1 July 2007. The new law was introduced to protect employees and the public from the harmful effects of second-hand smoke.

▶ From 1 July 2007 it is against the law to smoke in virtually all enclosed public places, workplaces and public and work vehicles. There will be very few exemptions from the law.

▶ Indoor smoking rooms in virtually all public places and workplaces are no longer allowed.

▶ Managers of smoke-free premises and vehicles have legal responsibilities to prevent people from smoking.

▶ The new law requires no-smoking signs to be displayed in all smoke-free premises and vehicles.

▶ The new law applies to anything that can be smoked. This includes cigarettes, pipes (including water pipes such as shisha and hookah pipes), cigars and herbal cigarettes.

Failure to comply with the new law is a criminal offence.

Penalties and fines for smoke-free offences are set out below (there are some discounted amounts for quick payment):

▶ **Smoking in smoke-free premises or work vehicles**: a fixed penalty notice of £50 imposed on the person smoking. Or a maximum fine of £200 if prosecuted and convicted by a court.

▶ **Failure to display no-smoking signs**: a fixed penalty notice of £200 imposed on whoever manages or occupies the smoke-free premises or vehicle. Or a maximum fine of £1,000 if prosecuted and convicted by a court.

▶ **Failing to prevent smoking in a smoke-free place**: a maximum fine of £2,500 imposed on whoever manages or controls the smoke-free premises or vehicle if prosecuted and convicted by a court.

Local councils will be responsible for enforcing the law. They will offer information and support to help businesses meet their legal obligations under the law.

You can find out more information on the law on the Smoke-free England website at **smokefreeengland. co.uk**. You can also contact your local council for information.

CHAPTER 9

Vehicle and plant movement – hazards and risk control

This chapter covers the following NEBOSH learning objectives:

1. Explain the hazards and control measures for the safe movement of vehicles/plant within a construction environment, including when using public highways as a workplace
2. Outline the factors associated with driving at work that increase the risk of an incident and the control measures to reduce work-related driving risks

9.1 Safe movement of vehicles and plant within a construction environment

9.1.1 Introduction

The safe movement of vehicles in the workplace is essential if accidents are to be avoided. The more serious accidents between pedestrians and vehicles can often be traced back to excessive speed or other unsafe vehicle practices, such as lack of driver training. More and more workers drive vehicles on public highways as part of their jobs and some are involved in road accidents that may be classed as work – many of the risks associated with these hazards can be significantly reduced by an effective management system.

9.1.2 Hazards from workplace transport operations

Many different kinds of vehicle are used in construction, including dumper trucks, heavy goods vehicles, all terrain vehicles and, perhaps the most common, the fork-lift truck. Approximately 15 persons are killed annually following vehicle accidents on construction sites. There are also over 1,000 major accidents (involving serious fractures, head injuries and amputations) caused by:

► collisions between pedestrians and vehicles;
► people falling from vehicles;
► people being struck by objects falling from vehicles;
► people being struck by an overturning vehicle;
► people injured while inside a vehicle during a collision;
► communication problems between vehicle drivers and employees or members of the public.

Workplace transport means any vehicle that is used in a work setting, such as fork-lift trucks, compact dumpers, tractors and mobile cranes. It can also include cars, vans and large goods vehicles when these are operating at the works premises. A goods vehicle that is loading or unloading on the public highway outside a works premises is considered as workplace transport. The Regulations that cover workplace traffic movements are the Workplace (Health, Safety and Welfare) Regulations. The ACoP to the Workplace Regulations gives specific advice and guidance on how to comply with them.

The main types of accident associated with workplace transport are people being struck by moving vehicles, falling from vehicles (jumping out of cabs or from the tops of high-sided trailers) and being hit by objects falling from vehicles (usually part of a load).

A key cause of these accidents is the lack of competent and documented driver training. HSE investigations, for example, have shown that in over 30% of dumper truck accidents on construction sites, the drivers had little experience and no training. Common forms of these accidents include driving into excavations, overturning while driving up steep inclines and runaway vehicles which have been left unattended with the engine running. Each year, the driving of tractors and other agricultural vehicles, such as all-terrain vehicles, causes several fatalities in farming and other agriculture-related activities.

Risks of injuries to employees and members of the public involving vehicles could arise due to the following occurrences:

► collision with pedestrians;
► collision with other vehicles;
► overloading of vehicles;
► overturning of vehicles;
► general vehicle movements and parking;
► dangerous occurrences or other emergency incidents (including fire);
► access and egress from the buildings and the site.

There are several other more general hazardous situations involving pedestrians and vehicles. These include the following:

► reversing of vehicles, especially inside buildings;
► poor road surfaces and/or poorly drained road surfaces;
► roadways too narrow with insufficient safe parking areas;
► roadways poorly marked out and inappropriate or unfamiliar signs used;
► too few pedestrian crossing points;
► the non-separation of pedestrians and vehicles;
► lack of barriers along roadways;
► lack of directional and other signs;
► poor environmental factors, such as lighting, dust and noise;
► ill-defined speed limits and/or speed limits which are not enforced;
► poor or no regular maintenance checks;
► vehicles used by untrained and/or unauthorised personnel;
► poor training or lack of refresher training.

Vehicle operations need to be carefully planned so that the possibility of accidents is minimised.

9.1.3 Hazards from mobile work equipment

Mobile work equipment is used extensively throughout industry – in factories, warehouses and construction sites. As mentioned in the previous section, the most common is the fork-lift truck.

Accidents, possibly causing injuries to people, often arise from one or more of the following events:

- poor maintenance with defective brakes, tyres and steering;
- poor visibility because of dirty mirrors and windows or loads which obstruct the driver's view; good visibility is essential at all times for mobile plant operators. The HSE advise that operators of mobile construction plant must ensure they regularly clean their windows so they can safely see all around. This should be combined with constant use of mirrors and a banksman where appropriate. Lights on all vehicles should be cleaned regularly to ensure vehicles are visible at all times, and vehicle depots should be well lit to avoid slip and trip hazards;
- operating on rough ground or steep gradients which causes the mobile equipment to turn on its side 90° plus or rollover 180° or more;
- carrying of passengers without the proper accommodation for them;
- people being flung out as the vehicle overturns and being crushed by it;
- being crushed under wheels as the vehicle moves;
- being struck by a vehicle or an attachment;
- lack of driver training or experience;
- underlying causes of poor management procedures and controls, safe working practices, information, instruction, training and supervision;
- collision with other vehicles;
- overloading of vehicles;
- general vehicle movements and parking;
- dangerous occurrences or other emergency incidents (including fire);
- access and egress from the buildings and the site.

The machines most at risk of rollover according to the HSE are:

- compact dumpers frequently used in construction sites;
- agricultural tractors;
- variable reach rough terrain trucks (telehandlers).

9.1.4 Control measures for safe workplace transport operations

Any control strategy involving vehicle operations will involve a risk assessment to ascertain where, on traffic routes, accidents are most likely to happen. It is important that the risk assessment examines both

Figure 9.1 Telescopic materials handler

Figure 9.2 Various construction plant with driver protection

internal and external traffic routes, particularly when goods are loaded and unloaded from lorries. It should also assess whether designated traffic routes are suitable for the purpose and sufficient for the volume of traffic.

The following need to be addressed:

- Traffic routes, loading and storage areas need to be well designed with enforced speed limits, good visibility and the separation of vehicles and pedestrians whenever reasonably practicable.
- Environmental considerations, such as visibility, road surface conditions, road gradients and changes in road level, must also be taken into account.
- The use of one-way systems and separate site access gates for vehicles and pedestrians may be required.
- The safety of members of the public must be considered, particularly where vehicles cross public footpaths.

▶ All external roadways must be appropriately marked, particularly where there could be doubt on right of way, and suitable direction and speed limit signs erected along the roadways. While there may well be a difference between internal and external speed limits, it is important that all speed limits are observed.

▶ Induction training for all new employees must include the location and designation of pedestrian walkways and crossings and the location of areas in the factory where pedestrians and fork-lift trucks use the same roadways.

▶ The identification of recognised and prohibited parking areas around the site should also be given during these training sessions.

Many industries have vehicles designed and used for specific workplace activities. The safe system of work for those activities should include:

▶ details of the work area (e.g. vehicle routes, provision for pedestrians, signage);

▶ details of vehicles (e.g. type, safety features and checks, maintenance requirements);

▶ information and training for employees (e.g. driver training, traffic hazard briefing);

▶ type of vehicle activities (e.g. loading and unloading, refuelling or recharging, reversing, tipping).

To summarise, there are three aspects to a control strategy for safe workplace transport operations:

1. **The design of the site (safe site) —** involves managers in:

    ▶ planning routes to separate pedestrians from vehicles whenever possible;

    ▶ reducing the need to reverse by using one-way systems;

    ▶ avoiding steep gradients and overhead cables and providing traffic routes on firm ground, minimising sharp and blind corners;

    ▶ marking out parking areas for vehicles;

    ▶ providing speed limit signs and traffic warning signs;

    ▶ ensuring a well-lit environment; and

    ▶ maintaining good housekeeping and a tidy site.

2. **Vehicle selection and maintenance (safe vehicle)** requires the provision of:

    ▶ vehicles with suitable and effective headlights, brakes, bumpers and horns, sufficient mirrors to reduce blind spots and seat belts for drivers and passengers;

    ▶ some additional vehicle features such as rear lens or radar sensors to provide extra safety when reversing and speed governors; and

    ▶ a regular and documented inspection and maintenance regime.

3. **Systems of work for system operatives (safe driver)** to include:

▶ effective supervision of everyone who is in areas where vehicles operate, including the provision of banksmen, when required;

▶ adequate training and refresher training for all drivers;

▶ relevant information for all drivers, including speed limits and parking areas;

▶ regular health checks on the suitability of employees for driving roles;

▶ the provision of high visibility clothing, appropriate protective clothing (such as steel toe capped boots and hard hats);

▶ the control of vehicle movements at times of day when there are more people moving around. Access to vehicle areas should be restricted to those that need to be there.

Mobile work equipment legislation

(a) General

The main purpose of the mobile work equipment Provision and Use of Work Equipment Regulations (PUWER) Part III, Regulations 25 to 30 is to require additional precautions relating to work equipment while it is travelling from one location to another or where it does work while moving. All appropriate sections of PUWER 98 will also apply to mobile equipment as it does to all work equipment; for example dangerous moving parts of the engine would be covered by Part II Regulations 10, 11 and 12 of PUWER. If the equipment is designed primarily for travel on public roads, the Road Vehicles (Construction and Use) Regulations will normally be sufficient to comply with PUWER.

Mobile equipment would normally move on wheels, tracks, rollers, skids, etc. Mobile equipment may be self-propelled, towed or remote controlled and may incorporate attachments. Pedestrian-controlled work equipment, such as lawn mowers, is not covered by Part III.

(b) Employees carried on mobile work equipment – Regulation 25

No employee may be carried on mobile work equipment:

▶ unless it is suitable for carrying persons;

▶ unless it incorporates features to reduce risks as low as is reasonably practicable, including risks from wheels and tracks.

Where there is significant risk of falling materials, falling-object protective structures (FOPS) must be fitted.

(c) Rolling over of mobile work equipment

Where there is a risk of overturning it must be minimised by:

▶ stabilising the equipment;

▶ fitting a structure so that it only falls on its side;

▶ fitting a structure which gives sufficient clearance for anyone being carried if it turns over further – rollover protective structure (ROPS);

▶ a device giving comparable protection;

▶ fitting a suitable restraining system for people if there is a risk of being crushed by rolling over.

(d) Self-propelled work equipment

Where self-propelled work equipment may involve risks while in motion it shall have:

▶ facilities to prevent unauthorised starting;

▶ facilities to minimise the consequences of collision (with multiple rail-mounted equipment);

▶ a device for braking and stopping;

▶ emergency facilities for braking and stopping, in the event of failure of the main facility, which have readily accessible or automatic controls (where safety constraints so require);

▶ devices fitted to improve vision (where the driver's vision is inadequate);

▶ appropriate lighting fitted or otherwise it shall be made sufficiently safe for its use (if used at night or in dark places);

▶ if there is anything carried or towed that constitutes a fire hazard liable to endanger employees (particularly, if escape is difficult such as from a tower crane), appropriate fire-fighting equipment carried, unless it is sufficiently close by.

(e) Rollover and falling-object protection (ROPS and FOPS)

Rollover protective structures (ROPS) are now becoming much more affordable and available for most types of mobile equipment where there is a high risk of turning over. Their use is spreading across most developed countries and even some less well-developed areas. A ROPS is a cab or frame that provides a safe zone for the vehicle operator in the event of a rollover.

The ROPS frame must pass a series of static and dynamic crush tests. These tests examine the ability of the ROPS to withstand various loads to see if the protective zone around the operator remains intact in an overturn. A home-made bar attached to a tractor axle or simple shelter from the sun or rain cannot protect the operator if the equipment overturns.

The ROPS must meet International Standards such as ISO 3471:1994. All mobile equipment safeguards should comply with the essential health and safety requirements of the Supply of Machinery (Safety) Regulations but need not carry a CE marking.

ROPS must also be correctly installed strictly following the manufacturers' instructions and using the correct strength bolts and fixings. They should never be modified by drilling, cutting, welding or other means as this may seriously weaken the structure.

Falling-object protective structures (FOPS) are required where there is a significant risk of objects falling on the equipment operator or other authorised person using the mobile equipment. Canopies that protect against falling objects (FOPS) must be properly designed and certified for that purpose. Front loaders work in woods or construction sites near scaffolding or buildings under construction and high bay storage areas, these all being locations where there is a risk of falling objects. Purchasers of equipment should check that any canopies fitted are FOPS. ROPS should never be modified by the user to fit a canopy without consultation with the manufacturers.

ROPS provide some safety during overturning but operatives must be confined to the protective zone of the ROPS. So where ROPS are fitted, a suitable restraining system must be provided for all seats. The use of seat restraints could avoid accidents where drivers are thrown from machines, thrown through windows or doors or thrown around inside the cab. In agriculture and forestry, 50% of overturning accidents occur on slopes of less than 10° and 25% on slopes of 5° or less. This means that seat restraints should be used most of the time that the vehicle is being operated.

Safe driving of mobile work equipment

Drivers have an important role to play in the safe use of mobile equipment. They should include the following in their safe working practice checklist:

▶ Make sure they understand fully the operating procedures and controls on the equipment being used.

▶ Only operate equipment for which they are trained and authorised.

▶ Never drive if abilities are impaired by, for example, alcohol, poor vision or hearing, ill-health or drugs, whether prescribed or not.

▶ Use the seat restraints where provided.

▶ Know the site rules and signals.

▶ Know the safe operating limits relating to the terrain and loads being carried.

▶ Keep vehicles in a suitably clean and tidy condition with particular attention to mirrors and windows or loose items which could interfere with the controls.

▶ Drive at suitable speeds and following site rules and routes at all times.

▶ Allow passengers only when there are safe seats provided on the equipment.

▶ Park vehicles on suitable flat ground with the engine switched off and the parking brakes applied; use wheel chocks if necessary.

▶ Make use of visibility aids or a signaller when vision is restricted.

▶ Get off the vehicle during loading operations unless adequate protection is provided.

▶ Ensure that the load is safe to move.

▶ Do not get off the vehicle until it is stationary, engine stopped and parking brake applied.

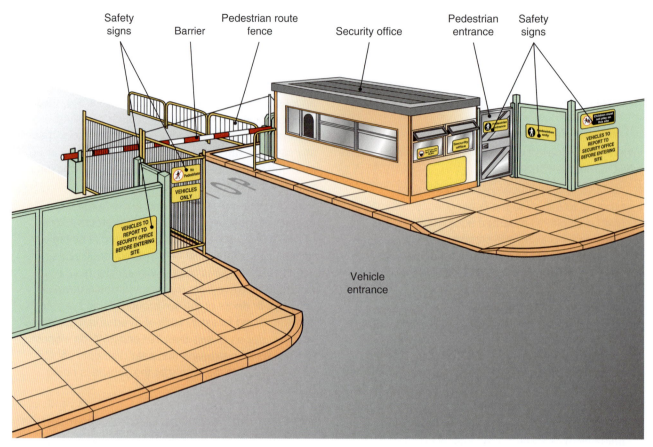

Figure 9.3 Site entrance to large construction site

▶ Where practicable, remove the operating key when getting off the vehicle.
▶ Take the correct precautions such as not smoking and switching off the engine when refuelling.
▶ Report any defects immediately.

The management of vehicle movements

The movement of vehicles should be properly managed, as should vehicle maintenance and driver training. The development of an agreed code of practice for drivers, to which all drivers should sign up, and the enforcement of site rules covering all vehicular movements are essential for effective vehicle management.

All vehicles should be subject to appropriate regular preventative maintenance programmes with appropriate records kept and all vehicle maintenance procedures properly documented. Many vehicles, such as mobile cranes, require regular inspection by a competent person and test certificates.

Certain vehicle movements, such as reversing, are more hazardous than others and particular safe systems should be set up. The reversing of lorries, for example, must be kept to a minimum (and then restricted to particular areas). Vehicles should be fitted with reversing warning systems as well as being able to give warning of approach. Refuges, where pedestrians can stand to avoid reversing vehicles, are a useful safety measure. Banksmen, who direct reversing vehicles, should also

Figure 9.4 Dumper truck with rollover protection (ROP)

be alert to the possibility of pedestrians crossing in the path of the vehicle. Where there are many vehicle movements, consideration should be given to the provision of high visibility clothing. Pedestrians must keep to designated walkways and crossing points, observe safety signs and use doors that are separate to those used by vehicles. Visitors who are unfamiliar with the site and access points should be escorted through the workplace.

Fire is often a hazard which is associated with many vehicular activities, such as battery charging and the storage of warehouse pallets. All batteries should be recharged in a separate well-ventilated area.

As mentioned earlier, driver training, given by competent people, is essential. Only trained drivers should be allowed to drive vehicles and the training should be relevant to the particular vehicle (fork-lift truck, dumper truck, lorry, etc.). All drivers must receive specific training and instruction before they are permitted to drive vehicles. They must also be given refresher training and medical examinations at regular intervals. This involves a management system for ensuring driver competence, which must include detailed records of all drivers with appropriate training dates and certification in the form of a driving licence or authorisation. Competence and its definition are discussed in Chapter 3.

Where large vehicles are routinely stopping to load or unload at loading bays, a certain amount of reversing is probably inevitable. When large vehicles need to reverse in the workplace, the following precautions should be taken:

▶ Undertake a risk assessment and develop a written safe system of work.
▶ Restrict reversing to places where it can be carried out safely.
▶ Keep people on foot or in wheelchairs away from the area.
▶ Provide suitable high visibility clothing for those people who are permitted in the area.
▶ Fit reversing alarms to alert, or a detection device to warn, the driver of an obstruction, or automatically apply the brakes.
▶ Employ banksmen to supervise the safe movement of vehicles.

Vehicular access to the site should be along clearly marked and signed traffic routes. Ideally, the site should have a designated and marked one-way system of traffic flow with convex mirrors sited on any bends or corners. Pedestrian routes should be marked and highlighted and totally separate from the designated traffic routes.

The design features that may need to be considered to minimise risks associated with movement of vehicles in the workplace include:

▶ providing traffic routes with smooth and stable surfaces and with the right width and headroom for the types of vehicles that will use them;
▶ eliminating sharp bends, blind corners and steep gradients, and siting convex mirrors on those corners that are unavoidably blind;
▶ installing a one-way system, to minimise the need for reversing;
▶ including passing places for vehicles;
▶ introducing speed limits and providing speed retarders;
▶ providing a good standard of lighting for the traffic routes, and particularly at the transition areas between the inside and outside of buildings;
▶ segregating vehicles and pedestrians, including separate access and egress, and providing clearly marked crossing places (zebra crossings).

The procedural arrangements that should accompany these design features are:

▶ selecting and training competent drivers;
▶ implementing a regular health screening programme for all drivers;
▶ providing information on site rules for visitors, such as delivery drivers;
▶ procedures for the regular maintenance of the traffic routes and the in-house vehicles, including a system for the reporting of defects and near-miss accidents;
▶ rigorously enforcing speed limits, with the possibility of a points system on drivers' licences or permits.

The HSE publications *Workplace Transport Safety. Guidance for Employers* HSG136, and *Managing Vehicle Safety at the Workplace* INDG199 (revised) provide useful checklists of relevant safety requirements that should be in place when vehicles are used in a workplace.

9.1.5 Hazards and controls of vehicles on construction sites

General

Many of the hazards and controls required for vehicle movements on construction sites have been covered earlier in this chapter. Common vehicles found on construction sites include site dumper trucks, fork-lift trucks with telescopic handlers, all-terrain and rough terrain vehicles, excavators and tipper lorries. On larger construction sites various types of earth-moving equipment may also be used.

The most common hazards are those which cause loss of control of the vehicle, overturning and collision with structures on the site, pedestrians or other site vehicles. Such hazards often arise due to problems of site layout, such as the non-segregation of pedestrians and vehicles, and uneven road surfaces. Attempts to drive vehicles up inclines which are too steep have caused many fatalities on sites and the need for rollover protection and driver restraint systems cannot be overemphasised. The methods of construction can also present hazards, for example excavations, scaffolds and falsework, all of which should have additional protection fitted when sited near a roadway. Additional hazards from falling materials, noise, dust and poor maintenance also need to be addressed.

The key controls, therefore, are the provision of safe routes, the separation of pedestrians and vehicles, the restriction of the carriage of people on site and the prevention of falls into excavations by the erection of strong barriers. Other precautions involve the introduction of one-way systems, speed limits, adequate lighting at road junctions, separate entrances and exits to the site with adequate turning room, and the use of clear signs. All vehicles should be well maintained, fitted with flashing beacons or reversing alarms, not be overloaded and sheeted when loaded.

9

Unintended vehicle movements must be prevented by the proper application of braking systems when the vehicle is stationary.

Clear routes should be set out across the site. These routes should avoid sharp bends, blind corners (unless suitably placed mirrors are located at the corner), narrow gaps, low head room and adverse cambers or steep gradients. Road surfaces should be inspected regularly and repaired using hardcore if necessary.

All vehicle drivers should be properly trained on each of the vehicles which they are expected to drive. Banksmen, wearing high visibility clothing, should be used to direct movement of lorries and excavators. Site rules should cover the use of vehicles on site and site management systems should include details of each driver and renewal dates for refresher training to ensure continued driver competency. A vehicle code of practice should be issued to all drivers. This code should include site safety rules, essential vehicle safety checks and reference to the hazards of driving near overhead power lines.

A level parking area should be provided for vehicles and although they should be drained of fuel overnight this must not be done while the engine is still hot or running. Finally, mud should be cleaned from the wheels of a vehicle before it enters a public highway.

Road works signs and controls

There are more than 4,000 road workers — approximately one for every mile of the Highways Agency's network — working on roads in the UK every day. Between 2003 and 2008, 11 road workers were killed and 104 were seriously injured. The control of vehicle movements during road works is governed by the Department for Transport Code of Practice 'Safety at street works and road works'. This gives detailed information on the temporary signs required and the order of these signs along the approach road. It also covers the way in which traffic cones should be set out with lead-in and exit tapers and it gives an indication of a safety zone between the cones and the work area. Means of controlling traffic, either by the use of traffic lights or stop/go boards, is also important, as is the provision of barriers for pedestrian safety.

The Code of Practice covers all situations from a footpath closure to high speed motorway lane closures and mobile work carried out from a vehicle like lane marking. The following gives an example of signage for advanced signs at road works. See Appendix 9.1 for examples of road works in the Code of Practice.

It does not matter whether the works are small or large, on the ground or overhead; all street works require warning and information. In emergencies as much warning must be given as the circumstances permit, and full signing must be provided as quickly as possible.

High visibility clothing must be worn at all times.

Jackets with sleeves must be worn on dual carriageway roads with a speed limit of 50mph or above, unless operatives stay within the working space at all times.

Signs should be placed where they will be clearly seen, and cause minimum inconvenience to drivers, cyclists, pedestrians and other road users alike, and where there is a minimum risk of their being hit or knocked over by traffic. Where there is a grass verge the signs should normally be placed there; the placing of signs in the footway is permitted but in no circumstances must the footway width be reduced below 1 m. If there are already vehicles parked in the carriageway, place the advance signs so that they are not obscured.

(a) Road Works Ahead

The **Road Works Ahead** sign is the first sign to be seen by the driver, so place it well before the works. Its size, the minimum distance from the start of the lead-in taper, and clear visibility distance will vary according to the type of road and its speed limit. The range of distances is given to allow the sign to be placed in the most convenient position bearing in mind available space and visibility for drivers. Do not simply choose the minimum distance — assess each site carefully.

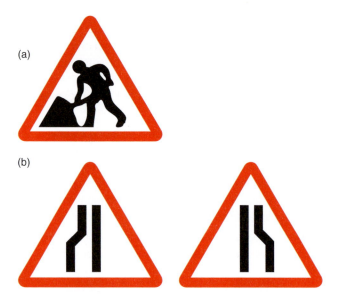

Figure 9.5 (a) Road Works Ahead; (b) Road Narrows

(b) Road Narrows

A **Road Narrows Ahead** sign warns the driver which side of the carriageway is obstructed. Place it midway between the Road Works Ahead sign and the beginning of the lead-in taper.

On roads with speed limits of 50mph or more, all advance signs should have plates giving the distance to the works in yards or miles, not in metres.

(c) Keep Right and Keep Left signs

Place **Keep Right** or, as appropriate, **Keep Left** signs at the beginning and end of the lead-in taper of cones. These signs must be the same size as the Road Works Ahead sign. **Make sure that the signs point in the**

correct direction. Do not turn the sign frame on its side to make it point in the correct direction. These signs must NOT be used for directing pedestrians.

Figure 9.6 Signs for Keep Right and Keep Left

Figure 9.7 Cone and road danger lamp

(d) Cones and lamps

Place a line of traffic cones to guide traffic past the works and add road danger lamps in poor daytime visibility and bad weather. Where the traffic is faster the length of taper must be longer.

Road danger lamps must be used at night on roads with a speed limit of 40mph or above. On roads with a lower speed limit, judgement may be used as to whether road danger lamps are needed, depending on the standard of street lighting.

Road danger lamps must not be higher than 1.5 m above the road or 1.2 m where the speed limit is more than 40mph.

The type of lamp to be used is as follows:

Type of road danger lamp	Conditions of use
Flashing lamp 55 to 150 flashes per minute	Only when ALL of the following conditions apply: — the speed limit is 40mph or less — the road danger lamp is within 50 m of a street lamp, and — the street lamp is illuminated
Steady lamp	On any road with or without street lighting

(e) Barriers

Barriers may comprise separate portable post and plank systems, gate frames linked together or semi-permanent constructions built to enclose the site.

There are several different requirements for the barrier planks associated with post and plank systems. The following explains the requirements and how they may be met using barrier planks which are red and white and manufactured in fully retro-reflective materials. (Note: retro-reflective means that at night the material reflects light back to the light source.)

Figure 9.8 Red and white barrier rail

Barrier planks are required to carry out three functions:

1. As a **traffic barrier**. When a traffic lane is closed for works to take place, the regulations require this to be done with a retro-reflective red and white barrier plank placed across the lane. This is illustrated as a Traffic Barrier Lane Closed sign.

2. As a **pedestrian barrier**. Pedestrians must be separated from the works by barriers which are conspicuous and mounted as part of a portable fencing system. Pedestrian barrier planks may be of several different contrasting colours; yellow, white or orange colours are best detected by partially sighted people, but red and white is one of the acceptable combinations.

3. As a **tapping rail** for blind and partially sighted people. Tapping rails are placed as the bottom rail in a pedestrian fencing system. A red and white barrier plank may be used.

All barriers facing vehicular traffic should be of the fully retro-reflective red and white form. Red and white barrier planks do not have to be used for pedestrian barriers or tapping rails but, if they are, they must be retro-reflective. Other planks used for these purposes do not need to be retro-reflective.

(f) Information board

An information board must be displayed at every site, except for mobile works and minor works which do not include excavation (see (i) Mobile works and minor works carried out from a vehicle). This board should be placed so that it does not obstruct footways or carriageways but can be read mainly by pedestrians, and possibly by drivers who have stopped.

Such boards are too detailed to be read easily by passing traffic. A board must give the name of the organisation for which the works are being carried out, and a telephone number which can be contacted in emergencies. It may also contain other information that will be helpful in explaining to the public why the work is being done, who is doing it and how long it will take. Such additional information is to be encouraged where practical and could include some or all of the following: a brief description of the works, the name of the contractor and a message apologising for inconvenience or delays. A completion date should normally be included if the works are expected to continue for more than a month.

Figure 9.9 Road works sign for footpath closure. Could be improved with walkway for pedestrians beside track. But this is only minor road, and pedestrians can cross to a good pavement opposite

Riveraford Water Company PLC

WATER MAIN RENEWAL
Completion expected
September 2015

Contractor - N E Samson Ltd

Emergency Telephone
020 7123 4567

Sorry for any inconvenience

Figure 9.10 Typical information sign

(g) End sign

The End sign indicates not only the end of works but also the end of any temporary restrictions, including temporary speed limits, associated with the works.

If the permanent speed limit changes within the length of road covered by a temporary speed restriction, signs indicating the new speed limit must be provided on each side of the carriageway at the end of the works, in addition to the End sign.

An End sign must be placed beyond works that are 50 m or more in length, measured between the end of the lead-in taper and the beginning of the exit taper and beyond two or more adjacent sites.

But an End sign is not necessary on a road where ALL of the following conditions are met:

▶ there are no temporary speed limits or other traffic restrictions;
▶ the speed limit is 30mph or under;
▶ there is a total two-way traffic flow of fewer than 20 vehicles counted over 3 minutes (400 veh/h);
▶ fewer than 20 heavy goods vehicles pass the works site per hour.

Figure 9.11 Road works End sign

(h) Site layout

The layout of the work site (which must be marked off with cones) should include a working area, working space and a safety zone. The safety zone must not be used as a working area or for storing plant or materials.

(i) Mobile works and minor works carried out from a vehicle

These include continuous **mobile operations**, as well as those which involve movement with periodic stops and short duration static works. They also include **minor works** which do not include excavations, involving the use of a single vehicle or a small number of vehicles.

Works in this category may omit the use of cones and a traffic barrier Lane Closed sign, provided that safe working methods are used. Single vehicle works must not be carried out on dual carriageways to which the national speed limit applies, unless they can be done at prevailing traffic speeds.

(Consult the supervisor if work is to take place in the centre of the carriageway with traffic passing on both sides.)

The Department for Transport – 'Safety at street works and road works' basic requirements:

- The vehicle must be conspicuously coloured.
- The vehicle must have one or more roof-mounted beacons operating.
- A Keep Right/Left sign must be displayed for drivers approaching on the same side of the carriageway, showing which side to pass. Vehicle-mounted Keep Right/Left signs must be covered when the vehicle is travelling to and from the site. Do NOT simply turn the sign to point up or down.

Additional static signs – will be required when ANY of the following conditions apply:

- the works vehicle cannot be seen clearly because of hills, bends in the road, etc.;
- stationary traffic may tail back;
- there is not enough space for two-way traffic to pass the works vehicle;
- the vehicle is slow moving or is required to make periodic stops.

See the Department for Transport Code of Practice for details. The work must be done by skilled and competent people.

Checklist

The Department for Transport Code of Practice – 'Safety at street works and road works' provides the following checklist.

Before starting – General

- Is high visibility clothing being worn by everyone on the site?
- Are all signs, barriers, cones and lighting correctly placed?
- Are signs obscured by bends, hills or dips in the road?
- Are advanced signs needed?
- Will the site be safe at night or in wind, fog, snow or rain?
- Are parked vehicles, trees or street furniture obscuring signs?
- Is there enough road width remaining for two-way traffic?
- Is traffic control with shuttle lane working required?
- Are there any site-specific risks requiring special guarding?
- Has allowance been made for delivery and removal of materials?
- Is the contact number displayed on the information board?

Before starting – Pedestrians

- Are pedestrians given protected routes which are wide enough?
- Are pedestrian routes clearly indicated?
- If the footway is closed, is there an alternative route? If so, is it clearly marked?
- Are there any special hazards for disabled pedestrians? If so, how can they be made safe?
- If a temporary footway in the road is to be used, are ramps to the kerb provided where necessary?

Before starting – Traffic

- Is type of traffic control right for work, traffic and speed?
- Have any misleading permanent signs and road markings been covered?
- Is there safe access to adjacent premises?
- Have you a copy of portable traffic signals site approval?
- Have you considered the needs of cyclists and horse riders?

When work is in progress

- Does signing and guarding meet changing conditions?
- Are signs, cones and lamps kept clean?
- Can traffic control arrangements be improved to reduce traffic delays as conditions change?
- Are the carriageway and footway being kept clear of mud and surplus equipment?
- Are materials that are left on verges or lay-bys being properly guarded and lit?

When work is suspended

- Will checks be made on signing, lighting and guarding?
- Has the arrangement been changed to reflect conditions?

When work is finished

- Have all signs, cones, barriers, and lamps been removed?
- Have any covered permanent signs been restored?
- Have the authorities been told the work is complete?

9.2 Driving at work

9.2.1 Introduction

The first fatal car accident in the UK occurred in 1896 when Mrs Bridget Driscoll was run over by a Roger-Benz which had a maximum speed of 8mph. Since then, the Royal Society for the Prevention of Accidents (RoSPA) has estimated that more than 550,000 people have been killed on UK roads. It has been estimated that up to a third of all road traffic accidents involve somebody who is at work at the time – accounting for over 20 fatalities and 250 serious injuries every week. Based on recent annual statistics, this means around 800–1,060 deaths a year on the road, compared with 241 fatal injuries to workers in the 'traditional workplace'. Accident rates are 30–40% higher for business drivers than for private drivers.

Some employers believe, incorrectly, that if they comply with certain road traffic law requirements, so that company vehicles have a valid MOT test certificate, and drivers hold a valid licence, this is enough to ensure the safety of their employees, and others, when they are on the road. However, health and safety law applies to on-the-road work activities as it does to all work activities, and the risks should be managed effectively within a health and safety management system.

These requirements are in addition to the duties employers have under road traffic law, for example the Road Traffic Act and Road Vehicle (Construction and Use) Regulations, which are administered by the police and other agencies such as the Vehicle and Operator Services Agency.

Just under half of all those who drive while at work suffer from a road-rage incident at least once a year, and 11% are assaulted. Furthermore, pain in the lower back, neck and shoulder is linked to prolonged driving, as is poor mental health.

Health and safety law does not apply to commuting, unless the employee is travelling from their home to a location which is not their usual place of work.

9.2.2 Benefits of managing work-related road safety

The true costs of accidents to organisations are nearly always higher than just the costs of repairs and insurance claims. The benefits of managing work-related road safety can be considerable, no matter what the size of the organisations. There will be benefits in the area of:

- control: costs, such as wear and tear and fuel, insurance premiums and claims can be better controlled;
- driver training and vehicle purchase: better informed decisions can be made;
- lost time: fewer days will be lost due to injury, ill-health and work rescheduling;
- vehicles: fewer will need to be off the road for repair;
- orders: fewer orders will be missed;
- key employees: there is likely to be a reduction in driving bans.

9.2.3 Managing occupational road risks

Where work-related road safety is integrated into the arrangements for managing health and safety at work, it can be managed effectively. The main areas to be addressed are policy, responsibility, organisation, systems and monitoring. Employees should be encouraged to report all work-related road incidents and be assured that punitive action will not be taken against them.

Figure 9.12 Occupational road risk increases when construction work is undertaken – cranes like this have to be driven to their operating location and operated when on site

The risk assessment should:

- consider the use, for example, of air or rail transport as a partial alternative to driving;
- consider the factors that might increase the risk of becoming involved in a road traffic incident – distance, driving hours, work schedules, stress due to traffic and road conditions and weather conditions;
- attempt to avoid situations where employees feel under pressure;
- make sure that maintenance work is organised to reduce the risk of vehicle failure. This is particularly important when pool cars are used because pool car users often assume another user is checking on maintenance and the MOT. The safety critical systems that need to be properly maintained are the brakes, steering and tyres. Similarly, if the car is leased and serviced by the leasing company, a system should be in place to confirm that servicing is being done to a reasonable standard;
- insist that drivers and passengers are adequately protected in the event of an incident. Crash helmets and protective clothing for those who ride motorcycles and other two-wheeled vehicles should be of the appropriate colour and standard;
- ensure that company policy covers the important aspects of the Highway Code.

9.2.4 Evaluating the risks

A study by RoSPA has found that a new driving for work qualification would be welcomed by employers because the current system of learner training and testing does not adequately prepare young people to drive as part of their jobs. The European Parliament and Council have introduced the Driving CPC (Certificate of Professional Competence) – under EU Directive 2003/59. There are two types of qualification – one for goods vehicles and one for drivers of passenger vehicles – and the Driver CPC will only apply to operators of vehicles over three-and-a-half tonnes or with more than eight passenger seats.

The following actions should be considered by employers to reduce the risks to employees who drive as part of their work:

▶ Journeys should be planned to reduce driving time.
▶ Rest breaks should be included in journey times.
▶ Encourage drivers to remain physically fit to reduce chronic fatigue.
▶ Ensure that drivers have hands-free mobile equipment.
▶ Monitor related working procedures to ensure safety.
▶ Ensure that drivers have necessary communication equipment (mobile phones, GPS, personal alarms).
▶ Keep a record of the location of driver destinations.
▶ Encourage drivers to regularly report back to colleagues in the base office.

The following considerations can be used to check on work-related road safety management.

The driver

(a) Competency

▶ Is the driver competent, experienced and capable of doing the work safely?
▶ Is his or her licence valid for the type of vehicle to be driven?
▶ Is the vehicle suitable for the task or is it restricted by the driver's licence?
▶ Does the recruitment procedure include appropriate pre-appointment checks?
▶ Is the driving licence checked for validity on recruitment and periodically thereafter?
▶ When the driver is at work, is he or she aware of company policy on work-related road safety?
▶ Are written instructions and guidance available?
▶ Has the company specified and monitored the standards of skill and expertise required for the circumstances for the job?

(b) Training

▶ Are drivers properly trained?
▶ Do drivers need additional training to carry out their duties safely?

▶ Does the company provide induction training for drivers?
▶ Are those drivers whose work exposes them to the highest risk given priority in training?
▶ Do drivers need to know how to carry out routine safety checks such as those on lights, tyres and wheel fixings?
▶ Do drivers know how to adjust safety equipment correctly, for example seat belts and head restraints?
▶ Is the headrest 3.8 cm (1.5 in.) behind the driver's head?
▶ Is the front of the seat higher than the back and are the legs 45° to the floor?
▶ Is the steering wheel adjustable and set low to avoid shoulder stress?
▶ Are drivers able to use anti-lock brakes (ABS) properly?
▶ Do drivers have the expertise to ensure safe load distribution?
▶ If the vehicle breaks down, do drivers know what to do to ensure their own safety?
▶ Is there a handbook for drivers?
▶ Are drivers aware of the dangers of fatigue?
▶ Do drivers know the height of their vehicle, both laden and empty?

(c) Fitness and health

▶ The driver's level of health and fitness should be sufficient for safe driving.
▶ Drivers of Heavy Goods Vehicles (HGVs) must have the appropriate medical certificate.
▶ Drivers who are most at risk should also undergo regular medicals. Staff should not drive, or undertake other duties, while taking a course of medicine that might impair their judgement.
▶ All drivers should have regular (every 2 years) eyesight tests. Research has shown that one in four motorists have a level of eyesight below the legal standard for driving which is to be able to read a car

Figure 9.13 Concrete delivery by road

9

Figure 9.14 Must have a valid licence for each type of vehicle

number plate from a distance of 20.5 m.

▶ Drivers should rest their eyes by taking a break of at least 15 minutes every 2 hours.

New offences under the Road Safety Act allow courts to imprison drivers who cause deaths by not paying due care and attention to the road or to other road users. Avoidable distractions which courts will consider when sentencing motorists who have killed include:

▶ using a mobile phone (for either calling or texting);
▶ drinking and eating;
▶ applying make-up;
▶ anything else which takes their attention away from the road and which a court judges to have been an avoidable distraction.

Every year, over 87,000 motorists are disqualified for drink-driving or driving while under the influence of drugs and up to 20% of drink-drivers are caught the morning after drinking. The Department for Transport have calculated that 5% of drivers who failed a breath test after a crash were driving for work at the time.

The vehicle

(a) Suitability

All vehicles should be fit for the purpose for which they are used. When purchasing new or replacement vehicles, management should select vehicles that are suitable for both driving and the health and safety of the public. The fleet should be suitable for the job in hand. Where privately owned vehicles are used for work, they should be insured for business use and have an appropriate MOT certificate test (for vehicles over 3 years old, in the UK).

(b) Condition and safety equipment

Are vehicles maintained in a safe and fit condition? There will need to be:

▶ maintenance arrangements to acceptable standards;
▶ basic safety checks for drivers;
▶ a method of ensuring that the vehicle does not

exceed its maximum load weight;
▶ reliable methods to secure goods and equipment in transit;
▶ checks to make sure that safety equipment is in good working order;
▶ checks on seat belts and head restraints (Are they fitted correctly and functioning properly?);
▶ a defect reporting system for drivers to use if they consider their vehicle is unsafe.

Figure 9.15 Fork-lift truck loading timber trusses onto a trailer. Loading vehicle correctly and evenly is most important for road stability en route

(c) Ergonomic considerations

The health of the drivers, and possibly also their safety, may be put at risk from an inappropriate seating position or driving posture. Ergonomic considerations should therefore be considered before purchasing or leasing new vehicles. Information may need to be provided to drivers about good posture and, where appropriate, on how to set their seat correctly.

(d) The load

For any lorry driving, most of the topics covered in this section are relevant. However, the load being carried is an additional issue. If the load is hazardous, emergency procedures (and possibly equipment) must be in place and the driver trained in those procedures. The load should be stacked safely in the lorry so that it cannot move during the journey. There must also be satisfactory arrangements for handling the load at either end of the journey.

The journey

(a) Routes

Route planning is crucial. Safe routes should be chosen which are appropriate for the type of vehicle undertaking the journey wherever practicable. Motorways are the safest roads. Minor roads are suitable for cars, but they are less safe and could present difficulties for larger vehicles. Overhead restrictions, for example bridges, tunnels and other hazards such as level crossings, may present dangers for long and/or high vehicles, so route planning should take particular account of these.

(b) Scheduling

There are danger periods during the day and night when people are most likely, on average, to feel sleepy. These are between 2 a.m. and 6 a.m. and between 2 p.m. and 4 p.m. Schedules need to take sufficient account of these periods. Where tachographs are carried, they should be checked regularly to make sure that drivers are not putting themselves and others at risk by driving for long periods without a break. Periods of peak traffic flow should be avoided if possible and new drivers should be given extra support while training.

(c) Time

Has enough time been allowed to complete the driving job safely? A realistic schedule would take into account the type and condition of the road and allow the driver rest breaks. A non-vocational driver should not be expected to drive and work for longer than a professional driver. The recommendation of the Highway Code is for a 15-minute break every 2 hours.

▶ Are drivers put under pressure by the policy of the company? Are they encouraged to take unnecessary risks, for example exceeding safe speeds because of agreed arrival times?

▶ Is it possible for the driver to make an overnight stay? This may be preferable to having to complete a long road journey at the end of the working day.

▶ Are staff aware that working irregular hours can add to the dangers of driving? They need to be advised of the dangers of driving home from work when they are excessively tired. In such circumstances they may wish to consider an alternative, such as a taxi.

(d) Distance

Managers need to satisfy themselves that drivers will not be put at risk from fatigue caused by driving excessive distances without appropriate breaks. Combining driving with other methods of transport may make it possible for long road journeys to be eliminated or reduced. Employees should not be asked to work an exceptionally long day.

(e) Weather conditions

When planning journeys, sufficient consideration will need to be given to adverse weather conditions, such as snow, ice, heavy rain and high winds. Routes should be rescheduled and journey times adapted to take adverse weather conditions into consideration. Where

poor weather conditions are likely to be encountered, vehicles should be properly equipped to operate, with, for example, ABS.

Where there are ways of reducing risk, for example when driving a high-sided vehicle in strong winds with a light load, drivers should have the expertise to deal with the situation. In addition, they should not feel pressurised to complete journeys where weather conditions are exceptionally difficult and this should be made clear by management.

9.2.5 Typical health and safety rules for drivers of cars on company business

At least 25% of all road accidents are work-related accidents involving people who are using the vehicle on company business. Drivers are expected to understand and comply with the relevant requirements of the current edition of the Highway Code. The following rules have been produced to reduce accidents at work. Any breach of these rules will be treated as a disciplinary offence.

The following example shows typical rules that have been prepared for use by car drivers.

▶ All drivers must have a current and valid driving licence.

▶ All vehicles must carry comprehensive insurance for use at work.

▶ Plan the journey in advance to avoid, where possible, dangerous roads or traffic delays.

▶ Use headlights in poor weather conditions and fog lights in foggy conditions (visibility 100 m).

▶ Use hazard warning lights if an accident or severe traffic congestion is approached (particularly on motorways).

▶ All speed limits must be observed but speeds should always be safe for the conditions encountered.

▶ Drivers must not drive continuously for more than 2 hours without a break of at least 15 minutes.

▶ Mobile phones, including hands-free equipment, must not be used whilst driving. They must be turned off during the journey and only used during the rest periods or when the vehicle is safely parked and the handbrake on.

▶ No alcohol must be consumed during the day of the journey until the journey is completed. Only minimal amounts of alcohol should be consumed on the day before a journey is to be made.

▶ No recreational drugs should be taken on the day of a journey. Some prescribed and over-the-counter drugs and medicines can also affect driver awareness and speed of reaction. Always check with a doctor or pharmacist to ensure that it is safe to drive.

9

9.2.6 Further information on occupational road safety

The Highway Code, The Stationery Office, 2001, ISBN 0 11 552290 5. Can also be viewed on www.highwaycode.gov.uk

Managing Occupational Road Risk, Royal Society for the Prevention of Accidents, available from http://www.rospa.com

9.3 Further information

The Construction (Design and Management) Regulations 2015

The Health and Safety (Safety Signs and Signals) Regulations 1996

The Road Traffic Act 1988 and 1991

The Work at Height Regulations 2005

The Provision and Use of Work Equipment Regulations 1998 – Part III in particular Road Traffic Act 1988, 1991

Safe use of work equipment (ACoP) (L22), HSE Books, ISBN 978-0-7176-6295-1 http://www.hse.gov.uk/pubns/priced/l22.pdf

Lighting at Work (HSG38), HSE Books, ISBN 978-0-7176-1232-1 http://www.hse.gov.uk/pubns/priced/hsg38.pdf

Safe use of vehicles on construction sites (HSG144), HSE Books, ISBN 978-0-7176-6291-3 http://www.hse.gov.uk/pubns/priced/hsg144.pdf

Workplace Transport Safety – Guidance for Employers (HSG136), HSE Books, ISBN 0-7176-6154-1 www.hse.gov.uk/pubns/priced/hsg136.pdf

The Traffic Signs Manual (Chapter 8: The Traffic Safety Measures and Signs for Roadworks and Temporary Situations) Department for Transport:

Part 1: Design. The current edition is dated 2009. ISBN 9780115530517

Part 2: Operations. The current edition is dated 2009. ISBN 9780115530524

Safety at street works and roadworks: http://www.dft.gov.uk/

Safety at streetworks and roadworks (Code of Practice), HMSO, ISBN 0-1155-1958-0

Driving at work, Managing work-related road safety, INDG 382, HSE Books, http://www.hse.gov.uk/pubns/indg382.pdf

9.4 Practice revision questions

1. A significant number of accidents on construction sites are associated with plant and vehicle movements.
   - (a) **Outline** the possible reasons for these accidents.
   - (b) **Describe** control measures that should be adopted to prevent such accidents occurring.
   - (c) **Describe** the features of a well-designed and maintained traffic management system on the construction site.

2. (a) **Outline** the possible causes of a dumper truck overturn on a construction site.
   - (b) **Identify** the design features of a dumper truck intended to minimise the risk of, or severity of injury from, an overturn.
   - (c) **Identify** the important contents of a training programme for dumper truck drivers so that the risk of accidents to themselves and other employees may be reduced.

3. A vehicle traffic route is to be established on a construction site.
   - (a) **Outline** control measures to consider for a suitable and sufficient traffic route.
   - (b) **Outline** suitable measures to minimise the risks from reversing vehicles on the site.

4. As part of a mains drainage project on a public highway with a speed limit of 40mph, it will be necessary to leave the trench open for several days. Temporary traffic lights will be used during these works. The depth of the excavation will not exceed 1.2 metres. Permission for the works has been obtained from the Highways Agency.

   Outline the traffic control measures that would be required in order to ensure compliance with the requirements of the New Roads and Street Works Act 1991.

5. (a) **Identify SIX** ways in which people may be injured by vehicles that are operating in a workplace.
   - (b) **Identify** the important contents of a training programme for vehicle drivers so that the risk of accidents to themselves and other employees may be reduced.
   - (c) **Outline** the actions required by both drivers and pedestrians to improve the safety of pedestrians in vehicle manoeuvring areas.
   - (d) **Outline** the hazards and controls associated with reversing vehicles in a warehouse.

6. **Describe** the rollover (ROPS) and falling-object (FOPS) protection techniques that are used to protect drivers of mobile work equipment. **Outline** the types of unsafe incidents from which drivers may be protected by these techniques.

7. Managing work-related road risk is an important part of a health and safety management system. **Outline** the possible risks and controls associated with:

(a) The driver
(b) The vehicle
(c) The journey.

8. (a) **Identify SIX** health and safety rules for drivers on company business.

(b) **Identify** the particular risks associated with long distance lorry driving.

9

APPENDIX 9.1 Safety at street works and road works

Figure 9.16 Traffic control by portable traffic signals

Figure 9.17 Works on footway with temporary footway in carriage

CHAPTER 10

Musculoskeletal hazards and risk control

This chapter covers the following NEBOSH learning objectives:

1. Explain work processes and practices that may contribute to musculoskeletal disorders, work-related upper limb disorders and the appropriate control measures
2. Explain the hazards and control measures which should be considered when assessing risks from manual handling activities
3. Explain the hazards and control measures to reduce the risk in the use of lifting and moving equipment with specific reference to manual and mechanically operated load moving equipment

10.1 Musculoskeletal disorders and work-related upper limb disorders

10.1.1 Introduction

Until a few years ago, accidents caused by the manual handling of loads were the largest single cause of over 3-day accidents reported to the Health and Safety Executive (HSE). The Manual Handling Operations Regulations recognised this fact and helped to reduce the number of these accidents. However, accidents due to poor manual handling technique still account for over 25% of all reported accidents and in some occupational sectors, such as the health service, the figure rises above 50%. An understanding of the factors causing some of these accidents is essential if they are to be further reduced. Mechanical handling methods should always be used whenever possible, but they are not without their hazards, many of which are outlined in Chapter 9 and later in this chapter. Much mechanical handling involves the use of lifting equipment, such as cranes and lifts, which present specific hazards to both the users and bystanders. The risks from these hazards are reduced by thorough examinations and inspections as required by the Lifting Operations and Lifting Equipment Regulations (LOLER).

Figure 10.1 Loading pipes onto a barge using a teleporter lift truck

10.1.2 The principles and scope of ergonomics

Ergonomics is the study of the interaction between workers and their work in the broadest sense, in that it encompasses the whole system surrounding the work process. It is, therefore, as concerned with the work organisation, process and design of the workplace and work methods as it is with work equipment. The common definitions of ergonomics, the 'man–machine interface' or 'fitting the man to the machine rather than vice versa' are far too narrow. It is concerned about the physical and mental capabilities of an individual as well as their understanding of the job under consideration. Ergonomics includes the limitations of the worker in terms of skill level, perception and other personal factors in the overall design of the job and the system supporting and surrounding it. It is the study of the relationship between the worker, the machine and the environment in which it operates and attempts to optimise the whole work system, including the job, to the capabilities of the worker so that maximum output is achieved for minimum effort and discomfort by the worker. Cars, buses and lorries are all ergonomically designed so that all the important controls, such as the steering wheel, brakes, gear stick and instrument panel, are easily accessed by most drivers within a wide range of sizes. Ergonomics is sometimes described as human engineering and as working practices become more and more automated, the need for good ergonomic design becomes essential.

The scope of ergonomics and an ergonomic assessment is very wide, incorporating the following areas of study:

▶ personal factors of the worker, in particular physical, mental and intellectual abilities, body dimensions and competence in the task required;
▶ the machine and associated equipment under examination;
▶ the interface between the worker and the machine – controls, instrument panel or gauges and any aids including seating arrangements and hand tools;
▶ environmental issues affecting the work process such as lighting, temperature, humidity, noise and atmospheric pollutants;

- the interaction between the worker and the task, such as the production rate, posture and system of working;
- the task or job itself – the design of a safe system of work, checking that the job is not too strenuous or repetitive and the development of suitable training packages;
- the organisation of the work, such as shift work, breaks and supervision.

The reduction of the possibility of human error is one of the major aims of ergonomics and an ergonomic assessment. An important part of an ergonomic study is to design the workstation or equipment to fit the worker. For this to be successful, the physical measurement of the human body and an understanding of the variations in these measurements between people are essential. Such a study is known as anthropometry, which is defined as the scientific measurement of the human body and its movement. Since there are considerable variations in, for example, the heights of people, it is common for some part of the workstation to be variable (e.g. an adjustable seat).

10.1.3 The ill-health effects of poor ergonomics

Ergonomic hazards are those hazards to health resulting from poor ergonomic design. They generally fall within the physical hazard category and include the manual handling and lifting of loads, pulling and pushing loads, prolonged periods of repetitive activities and work with vibrating tools. The condition of the working environment, such as low lighting levels, can present health hazards to the eyes. It is also possible for psychological conditions, such as occupational stress, to result from ergonomic hazards.

The common ill-health effects of ergonomic hazards are musculoskeletal disorders, and work-related upper limb disorders (WRULDs) including repetitive strain injury (RSI) and deteriorating eyesight.

Figure 10.2 A tilted worktable. The distance between the operator and the work can be reduced by putting the table at a more vertical angle. The table is adjustable in height and angle to suit the particular job

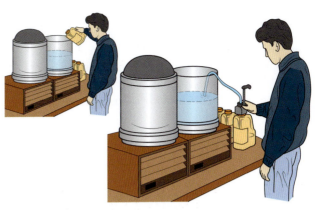

Figure 10.3 Pump liquid from a bulk container to a dispenser to save awkward handling

10.1.4 Work-related upper limb disorders

WRULDs describe a group of conditions which can affect the neck, shoulders, arms, elbows, wrists, hands and fingers. **Tenosynovitis** (affecting the tendons), **carpal tunnel syndrome** (affecting the tendons which pass through the carpal bone in the hand) and **frozen shoulder** are all examples of WRULDs which differ in the manifestation and site of the illness. The term **RSI** is commonly used to describe WRULDs.

WRULDs are caused by repetitive movements of the fingers, hands or arms which involve pulling, pushing, reaching, twisting, lifting, squeezing or hammering. These disorders can occur to workers in offices as well as in factories or on construction sites. Typical occupational groups at risk include painters and decorators, riveters and pneumatic drill operators and desktop computer users. During construction, structural engineers and others can suffer from RSI due to the large torques that they exert on spanners when bolt fixing and tightening nuts and screws.

The main symptoms of WRULDs are aching pain to the back, neck and shoulders, swollen joints and muscle fatigue accompanied by tingling, soft tissue swelling, similar to bruising, and a restriction in joint movement. The sense of touch and movement of fingers may be affected. The condition is normally a chronic one in that it gets worse with time and may lead eventually to permanent damage. The injury occurs to muscle, tendons and/or nerves. If the injury is allowed to heal before being exposed to the repetitive work again, no long-term damage should result. However, if the work is repeated again and again, healing cannot take place and permanent damage can result leading to a restricted blood flow to the arms, hands and fingers.

The risk factors, which can lead to the onset of WRULDs, are repetitive actions of lengthy duration, the application of significant force and unnatural postures, possibly involving twisting and over-reaching and the use of vibrating tools. Cold working environments, work organisation and worker perception of the work organisation have all been shown in studies to be risk factors, as is the involvement of vulnerable workers such as those with pre-existing ill-health conditions and pregnant women.

10.1.5 Display screen equipment

Display screen equipment (DSE), which includes visual display units, is a good example of a common work activity which relies on an understanding of ergonomics and the ill-health conditions which can be associated with poor ergonomic design. A survey of safety representatives by the TUC found that injuries/illnesses caused by the poor use of DSE and repetitive strain injuries together with stress or overwork were among their major concerns.

Legislation governing DSE is covered by the Health and Safety (Display Screen Equipment) Regulations and a detailed summary of them is given in Chapter 19. The Regulations apply to a user or operator of DSE and define a user or operator. The definition is not as tight as many employers would like it to be, but usage in excess of approximately one hour continuously each day would define a user. The definition is important as users are entitled to free eye tests and, if required, a pair of spectacles specifically for DSE use. The basic requirements of the Regulations are:

▶ a suitable and sufficient risk assessment of the workstation, including the software in use, trip and electrical hazards from trailing cables and the surrounding environment;
▶ workstation compliance with the minimum specifications laid down in the schedules appended to the Regulations;
▶ a plan of the work programme to ensure that there are adequate breaks in the work pattern of workers;
▶ the provision of free eye sight tests and, if required, spectacles to users of DSE;
▶ a suitable programme of training and sufficient information given to all users.

The risk assessment of a DSE workstation needs to consider the following factors, many of which are shown in Figure 10.4:

▶ the height and adjustability of the monitor;
▶ the adjustability of the keyboard, the suitability of the mouse and the provision of wrist support;

Seating and posture for typical office tasks

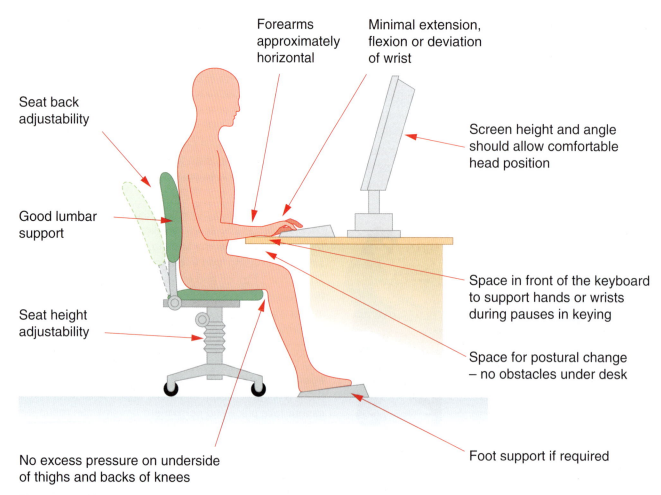

Forearms approximately horizontal

Minimal extension, flexion or deviation of wrist

Seat back adjustability

Screen height and angle should allow comfortable head position

Good lumbar support

Seat height adjustability

Space in front of the keyboard to support hands or wrists during pauses in keying

Space for postural change – no obstacles under desk

No excess pressure on underside of thighs and backs of knees

Foot support if required

Figure 10.4 Workstation design

- the stability and adjustability of the DSE user's chair;
- the provision of ample foot room and suitable foot support;
- the effect of any lighting and window glare at the workstation;
- the storage of materials around the workstation;
- the safety of trailing cables, plugs and sockets;
- environmental issues – noise, heating, humidity and draughts.

Chapter 23 (Form H3) gives an example of a checklist that can be used for a DSE workstation assessment by the user. There are three basic ill-health hazards associated with DSE. These are:

- musculoskeletal problems;
- visual problems;
- psychological problems.

A fourth hazard, of radiation, has been shown from several studies to be very small and is now no longer normally considered in the risk assessment.

Similarly, in the past, there have been suggestions that DSE could cause epilepsy and there were concerns about adverse health effects on pregnant women and their unborn children. All these risks have been shown in various studies to be very low.

The provision of DSE training and risk assessments online has become more common. The risk assessment, however, still has to be managed, made appropriate to the particular workplace setting and reviewed from time to time. Various studies have shown that users of any e-learning package lose concentration after 30–40 minutes.

Musculoskeletal problems

Tenosynovitis is the most common and well-known problem which affects the wrist of the user. The symptoms and effects of this condition have already been covered. Suffice it to say that if the condition is ignored, then the tendon and tendon sheath around the wrist will become permanently injured. Tenosynovitis is caused by the continual use of a keyboard and can be relieved by the use of wrist supports. Other WRULDs are caused by poor posture and can produce pains in the back, shoulders, neck or arms. Less commonly, pain may also be experienced in the thighs, calves and ankles. These problems can be mitigated by the application of ergonomic principles in the selection of working desks, chairs, foot rests and document holders. It is also important to ensure that the desk is at the correct height and the computer screen is tilted at the correct angle to avoid putting too much strain on the neck. (Ideally the user's eyes should be at the same height as the top of the screen.)

The keyboard should be detachable so that it can be positioned anywhere on the desktop and a correct posture adopted while working at the keyboard. The chair should be adjustable in height, stable and have an adjustable backrest. If the knees of the user are lower than the hips when seated, then a footrest should be provided. The surface of the desk should be non-reflecting and uncluttered but ancillary equipment (e.g. telephone and printer) should be easily accessible.

Visual problems

There does not appear to be much medical evidence that DSE causes deterioration in eye sight, but users may suffer from visual fatigue which results in eye strain, sore eyes and headaches. Less common ailments are skin rashes and nausea.

The use of DSE may indicate that reading spectacles are needed and the Regulations make provision for this. It is possible that any prescribed lenses may only be suitable for DSE work as they will be designed to give optimum clarity at the normal distance at which screens are viewed (50–60 cm).

Eye strain is a particular problem for people who spend a large proportion of their working day using DSE. A survey has indicated that up to 90% of DSE users complain of eye fatigue. Eye strain can be reduced by the following steps additional to those already identified in this section:

- Train staff in the correct use of the equipment.
- Ensure that a font size of at least 12 is used on the screen.
- Ensure that users take regular breaks away from the screen (up to 10 minutes every hour).

Sore eyes are a common complaint associated with low humidity. Various professional building research organisations have agreed that an appropriate range of relative humidity in an office environment is 40–60%. Air conditioning systems should be properly maintained to ensure that they maintain a reasonable temperature and humidity level in the workplace. Sore eyes are often a product of dry eyes which may be controlled by:

- restraining from rubbing the eyes;
- focusing on an object in the near distance (preferably green in colour);
- blinking can encourage the moistening of the eye;
- drinking water regularly;
- looking away from the screen periodically to rest the eyes.

The screen should be adjustable in tilt angle and screen brightness and contrast. Finally, the lighting around the workstation is important. It should be bright enough to allow documents to be read easily but not too bright such that either headaches are caused or there are reflective glares on the computer screen.

Psychological problems

These are generally stress-related problems. They may have environmental causes, such as noise, heat, humidity or poor lighting, but they are usually due to high-speed working, lack of breaks, poor training and

10

poor workstation design. One of the most common problems is the lack of understanding of all or some of the software packages being used.

There are several other processes and activities where ergonomic considerations are important. These include the assembly of small components (microelectronics assembly lines) and continually moving assembly lines (car assembly plants).

10.2 Manual handling hazards and control measures

The term 'manual handling' is defined as the movement of a load by human effort alone. This effort may be applied directly or indirectly using a rope or a lever. Manual handling may involve the transportation of the load or the direct support of the load including pushing, pulling, carrying, moving using bodily force and, of course, straightforward lifting. Back injury due to the lifting of heavy loads is very common and several million working days are lost each year as a result of such injuries.

Typical hazards of manual handling include:

▶ lifting a load which is too heavy or too cumbersome, resulting in back injury;
▶ poor posture during lifting or poor lifting technique, resulting in back injury;
▶ dropping a load, resulting in foot injury;
▶ lifting sharp-edged or hot loads resulting in hand injuries.

Figure 10.5 Manual handling: there are many potential hazards

10.2.1 Injuries caused by manual handling

Manual handling operations can cause a wide range of acute and chronic injuries to workers. Acute injuries normally lead to sickness leave from work and a period of rest during which time the damage heals. Chronic injuries build up over a long period of time and are usually irreversible, producing illnesses such as arthritic and spinal disorders. There is considerable evidence to suggest that modern lifestyles, such as a lack of exercise and regular physical effort, have contributed to the long-term serious effects of these injuries.

The most common injuries associated with poor manual handling techniques are all musculoskeletal in nature and are:

▶ muscular sprains and strains – caused when a muscular tissue (or ligament or tendon) is stretched beyond its normal capability leading to a weakening, bruising and painful inflammation of the area affected. Such injuries normally occur in the back or in the arms and wrists;
▶ back injuries – include injuries to the discs situated between the spinal vertebrae (i.e. bones) and can lead to a very painful prolapsed disc lesion (commonly known as a slipped disc). This type of injury can lead to other conditions known as lumbago and sciatica (where pain travels down the leg);
▶ trapped nerve – usually occurring in the back as a result of another injury but aggravated by manual handling;
▶ hernia – this is a rupture of the body cavity wall in the lower abdomen, causing a protrusion of part of the intestine. This condition eventually requires surgery to repair the damage;
▶ cuts, bruising and abrasions – caused by handling loads with unprotected sharp corners or edges;
▶ fractures – normally of the feet due to the dropping of a load. Fractures of the hand also occur but are less common;
▶ work-related upper limb disorders (WRULDs);
▶ rheumatism – this is a chronic disorder involving severe pain in the joints. It has many causes, one of which is believed to be the muscular strains induced by poor manual handling lifting technique.

The sites on the body of injuries caused by manual handling accidents are shown in Figure 10.6. Nearly 11 million working days were lost in 2008/09 due to musculoskeletal disorders, which are thought to cost the UK more than £7 billion each year and affect 6.5 million people.

Musculoskeletal problems are the most common cause of absence, followed by viral infections and stress-related illnesses. These findings are based on an analysis of sickness management records for 11,000 individual employees across a range of private sector organisations. A recent study has found that musculoskeletal disorders (MSDs) account for nearly half (49%) of all absences from work and 60% of permanent work incapacity in the European Union. It is estimated that this costs the UK economy £7 billion every year and costs Europe £240 billion

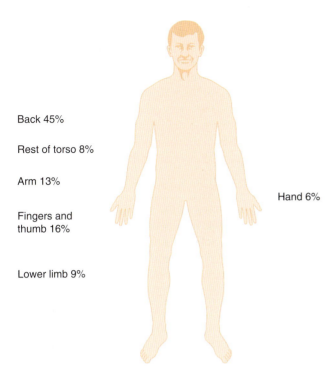

Back 45%

Rest of torso 8%

Arm 13%

Hand 6%

Fingers and
thumb 16%

Lower limb 9%

Figure 10.6 Main injury sites caused by manual handling accidents

each year. The study, conducted across 25 European countries, found that 100 million Europeans suffer from chronic musculoskeletal pain – over 40 million of whom are workers – with up to 40% having to give up work due to their condition. In the UK alone, 9.5 million working days were lost in one year due to musculoskeletal problems.

In general, pulling a load is much easier for the body than pushing one. If a load can only be pushed, then pushing backwards using the back is less stressful on body muscles. Lifting a load from a surface at waist level is easier than lifting from floor level and most injuries during lifting are caused by lifting and twisting at the same time. If a load has to be carried, it is easier to carry it at waist level and close to the body trunk. A firm grip is essential when moving any type of load.

10.2.2 Hierarchy of measures for manual handling operations

With the introduction of the Manual Handling Operations Regulations, the emphasis during the assessment of lifting operations changed from a simple reliance on safe lifting techniques to an analysis, using risk assessment, of the need for manual handling. The Regulations established a clear hierarchy of measures to be taken when an employer is confronted with a manual handling operation:

▶ Avoid manual handling operations so far as is reasonably practicable by either redesigning the task to avoid moving the load or by automating or mechanising the operations.

▶ If manual handling cannot be avoided, then a suitable and sufficient risk assessment should be made.

▶ Reduce the risk of injury from those operations so far as is reasonably practicable, either by the use of mechanical handling or making improvements to the task, the load and the working environment.

The guidance given to the Manual Handling Operations Regulations (available in the HSE Legal series – L23) is a very useful document. It gives very helpful advice on manual handling assessments and manual handling training. The advice is applicable to all occupational sectors. In Chapter 23, Forms H5 and H6 are examples of manual handling assessment forms.

In the construction industry, the typical manual handling tasks that occur on most sites are lifting building blocks, cement bags, doors, windows and lintels. Another common example occurs in road and street works – the lifting and manoeuvring of kerb stones and paving slabs. These examples can lead not only to a strained back and pulled muscles but also to trapped fingers and toes. This indicates the need for strong gloves and safety footwear. The handling of kerb stones is discussed in more detail in 10.2.5.

10.2.3 Manual handling assessments

The Regulations specify four main factors which must be taken into account during the assessment. These are the task, the load, the working environment and the capability of the individual who is expected to do the lifting.

The task should be analysed in detail so that all aspects of manual handling are covered including the use of mechanical assistance. The number of people involved and the cost of the task should also be considered. Some or all of the following questions are relevant to most manual handling tasks:

▶ Is the load held or manipulated at a distance from the trunk? The further from the trunk, the more difficult it is to control the load and the stress imposed on the back is greater.

▶ Is a satisfactory body posture being adopted? Feet should be firmly on the ground and slightly apart and there should be no stooping or twisting of the trunk. It should not be necessary to reach upwards, as this will place additional stresses on the arms, back and shoulders. The effect of these risk factors is significantly increased if several are present while the task is being performed.

▶ Are there excessive distances to carry or lift the load? Over distances greater than 10 m, the physical demands of carrying the load will dominate the operation. The frequency of lifting hand the vertical and horizontal distances the load needs to be carried (particularly if it has to be lifted from the ground and/or placed on a high shelf) are very important considerations.

▶ Is there excessive pulling and pushing of the load? The state of floor surfaces and the footwear of the individual should be noted so that slips and trips may be avoided.

▶ Is there a risk of a sudden movement of the load? The load may be restricted or jammed in some way.

▶ Is frequent or prolonged physical effort required? Frequent and prolonged tasks can lead to fatigue and a greater risk of injury.

▶ Are there sufficient rest or recovery periods? Breaks and/or the changing of tasks enables the body to recover more easily from strenuous activity.

▶ Is there an imposed rate of work on the task? This is a particular problem with some automated production lines and can be addressed by spells on other operations away from the line.

▶ Are the loads being handled while the individual is seated? In these cases, the legs are not used during the lifting processes and stress is placed on the arms and back.

▶ Does the handling involve two or more people? The handling capability of an individual reduces when he/she becomes a member of a team (e.g. for a three-person team, the capability is half the sum of the individual capabilities). Visibility, obstructions and the roughness of the ground must all be considered when team handling takes place.

The load must be carefully considered during the assessment and the following questions asked:

▶ Is the load too heavy? The maximum load that an individual can lift will depend on the capability of the individual and the position of the load relative to the body. There is therefore no safe load. Figure 10.7 is reproduced from the HSE guidance, which does give some advice on loading levels. It recommends that loads in excess of 25 kg should not be lifted or carried by a man (and this is only permissible when the load is at the level of and adjacent to the thighs). For women, the guideline figures should be reduced by about one-third.

▶ Is the load too bulky or unwieldy? In general, if any dimension of the load exceeds 0.75 m (c. 2 ft.), its handling is likely to pose a risk of injury. Visibility around the load is important. It may hit obstructions or become unstable in windy conditions. The position of the centre of gravity is very important for stable lifting – it should be as close to the body as possible.

▶ Is the load difficult to grasp? Grip difficulties will be caused by slippery surfaces, rounded corners or a lack of foot room.

▶ Are the contents of the load likely to shift? This is a particular problem when the load is a container full of smaller items, such as a sack full of nuts and bolts. The movements of people (in a nursing home) or animals (in a veterinary surgery) are loads which fall into this category.

▶ Is the load sharp, hot or cold? Personal protective equipment may be required.

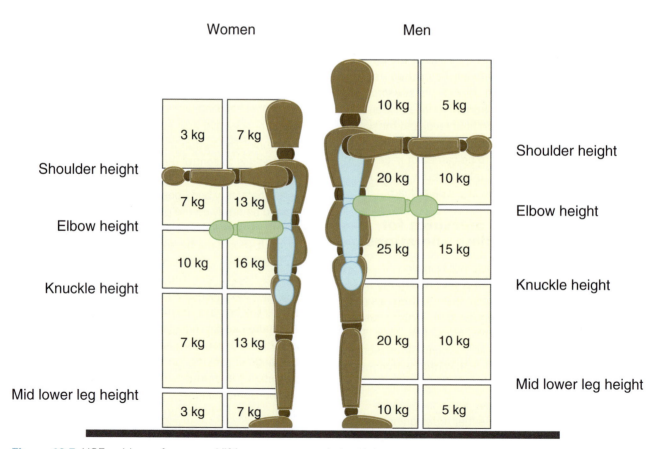

Figure 10.7 HSE guidance for manual lifting – recommended weights

Figure 10.8 Moving bricks or paving blocks using a specially designed barrow

The working environment in which the manual handling operation is to take place must be considered during the assessment. The following areas will need to be assessed:

- any space constraints which might inhibit good posture. Such constraints include lack of headroom, narrow walkways and items of furniture;
- slippery, uneven or unstable floors;
- variations in levels of floors or work surfaces, possibly requiring the use of ladders;
- extremes of temperature and humidity. These effects are discussed in detail in Chapter 8;
- ventilation problems or gusts of wind;
- poor lighting conditions.

Finally, the capability of the **individual** to lift or carry the load must be assessed. The following questions will need to be asked:

- Does the task require unusual characteristics of the individual (e.g. strength or height)? It is important to remember that strength and general manual handling ability depends on age, gender, state of health and fitness.
- Are employees who might reasonably be considered to be pregnant or to have a health problem put at risk by the task? Particular care should be taken to

protect pregnant women or those who have recently given birth from handling loads. Allowance should also be given to any employee who has a health problem, which could be exacerbated by manual handling.

The assessment must be reviewed if there is reason to suspect that it is no longer valid or there has been a significant change to the manual handling operations to which it relates.

10.2.4 Reducing the risk of injury

This involves the introduction of control measures resulting from the manual handling risk assessment. The guidance to the Regulations (L23) and the HSE publication *Manual Handling – Solutions You Can Handle* (HSG115) contain many ideas to reduce the risk of injury from manual handling operations. An ergonomic approach is generally required to design and develop the manual handling operation as a whole. The control measures can be grouped under five headings. However, the first consideration, when it is reasonably practicable, is **mechanical assistance** (see 10.3).

The **task** can be improved by changing the layout of the workstation by, for example, storing frequently used loads at waist level. The removal of obstacles and the use of a better lifting technique that relies on the leg rather than back muscles should be encouraged. When pushing, the hands should be positioned correctly. The work routine should also be examined to see whether job rotation is being used as effectively as it could be. Special attention should be paid to seated manual handlers to ensure that loads are not lifted from the floor while they are seated. Employees should be encouraged to seek help if a difficult load is to be moved so that a team of people can move the load. Adequate and suitable personal protective equipment should be provided where there is a risk of loss of grip or injury. Care must be taken to ensure that the clothing does not become a hazard in itself (e.g. the snagging of fasteners and pockets).

The **load** should be examined to see whether it could be made lighter, smaller or easier to grasp or manage. This could be achieved by splitting the load, the positioning of handholds or a sling, or ensuring that the centre of gravity is brought closer to the handler's body. Attempts should be made to make the load more stable and any surface hazards, such as slippery deposits or sharp edges, should be removed. It is very important to ensure that any improvements do not, inadvertently, lead to the creation of additional hazards.

The **working environment** can be improved in many ways. Space constraints should be removed or reduced. Floors should be regularly cleaned and repaired when damaged. Adequate lighting is essential and working at more than one level should be minimised so that hazardous ladder work is avoided. Attention should be given to the need for suitable temperatures and ventilation in the working area.

The **capability of the individual** is the fifth area where control measures can be applied to reduce the risk of injury. The state of health of the employee and his/her medical record will provide the first indication as to whether the individual is capable of undertaking the task. A period of sick leave or a change of job can make an individual vulnerable to manual handling injury. The Regulations require that the employee be given information and training. The information includes the provision, where it is reasonably practicable to do so, of precise information on the weight of each load and the heaviest side of any load whose centre of gravity is not centrally positioned. In a more detailed risk assessment, other factors will need to be considered such as the effect of personal protective equipment and psychosocial factors in the work organisation. The following points may need to be assessed:

▶ Does protective clothing hinder movement or posture?

▶ Is the correct personal protective equipment being worn?

▶ Is proper consideration given to the planning and scheduling of rest breaks?

▶ Is there good communication between managers and employees during risk assessment or workstation design?

▶ Is there a mechanism in place to deal with sudden changes in the volume of workload?

▶ Have employees been given sufficient training and information?

▶ Does the worker have any learning disabilities and, if so, has this been taken into account in the assessment?

An example of a manual handling assessment in construction would be the moving of a rolled steel joist (RSJ) beam. The assessment would need to consider the weight of the beam, the distance that it needed to be carried and the vertical height of the required lift. Other factors would include the condition of the floor and the space available to manoeuvre the beam into its final position.

Recent amendments to the Manual Handling Operations Regulations have emphasised that a worker may be at risk if he/she:

▶ is physically unsuited to carry out the tasks in question;

▶ is wearing unsuitable clothing, footwear or other personal effects;

▶ does not have adequate or appropriate knowledge or training.

The HSE have developed a Manual Handling Assessment Chart (MAC) tool to help the user identify high-risk workplace manual handling activities and the assessment of common risks associated with lifting, carrying and handling. The tool can be used to assess the risks posed by lifting, carrying and team manual handling activities. It incorporates a numerical and a colour coding score system to highlight high-risk manual handling tasks. There are three types of assessment that can be carried out with the MAC:

1. lifting operations;
2. carrying operations;
3. team handling operations.

The MAC is available on the HSE website.
The training requirements are given in section 10.2.6.

10.2.5 Handling kerb stones

The handling of kerb stones is one of the main causes of back problems and other musculoskeletal disorders. Standard kerb stones weigh approximately 67 kg and are made of pre-cast concrete. Thus the main hazards associated with the manual handling of kerb stones are the weight of the stones, unsuitable body posture and the frequency and repetitive nature of the work. The risk, therefore, of injury to workers who lay kerb stones by hand is high and employers must address all three hazards. The HSE recommends a hierarchy of control measures in their information sheet, CIS No. 57, as follows:

1. Eliminate manual lifting at the design stage perhaps by dispensing with the need for a kerb.
2. Use total mechanisation by ensuring that kerb stones are always lifted and laid mechanically (e.g. vacuum devices or a vehicle-mounted crane and mechanical grab).
3. Use partial mechanisation by ensuring that the maximum amount of kerb handling is done mechanically (e.g. off-loading using a hoist).
4. Use manual handling when it is not possible to use any of the above methods; the risk of injury can be reduced if the workers are well trained, lighter kerb stones are used or devices which allow two people to share the lift are provided.

There are alternatives to the traditional laying of kerbs: using plastic kerbs, hollow kerbs, or slip-form kerbs. Kerb laying may also involve other health risks including hand–arm vibration, dust and noise.

The laying of heavy blocks can also lead to musculoskeletal problems. Previous advice suggested the use of two workers to lay blocks over 20 kg but this has not proved to be workable and is no longer recommended. There is a risk of immediate injury or of long-term physical disability from single-handed repetitive manual handling of heavy masonry units, particularly those weighing above 20 kg. When designing or specifying blocks, the lightest block that fulfils the performance criteria should be chosen. In the majority of cases blocks weighing under 20 kg are available that will fulfil the required design specification. It is not just the weight that matters though; a safe system of work will include suitable precautions to

avoid or minimise all hazards. This includes the weight, the repetitive nature of the work, poor posture, stooping and reaching as well as the risks associated with slips/trips/falls, sharp or rough edges and skin hazards caused by contact with mortar.

Many of the manual handling problems can be reduced at the design and manufacturing stages. Contractors should provide their workers with training in good handling techniques and the safe use of mechanical lifting equipment.

10.2.6 Manual handling training

Training alone will not reduce manual handling injuries – there still need to be safe systems of work in place and the full implementation of the control measures highlighted in the manual handling assessment. The following topics should be addressed in a manual handling training session:

▶ types of injuries associated with manual handling activities;
▶ the findings of the manual handling assessment;
▶ the recognition of potentially hazardous manual handling operations;
▶ the correct use of mechanical handling aids;
▶ the correct use of personal protective equipment;
▶ features of the working environment which aid safety in manual handling operations;
▶ good housekeeping issues;
▶ factors which affect the capability of the individual;
▶ good lifting or manual handling technique as shown in Figure 10.9.

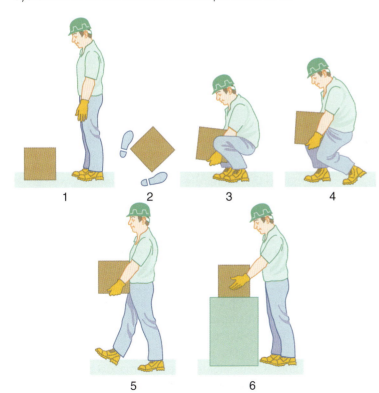

1. Check suitable clothing and assess load. Heaviest side to body.

2. Place feet apart – bend knees.

3. Firm grip – close to body Slight bending of back, hips and knees at start.

4. Lift smoothly to knee level and then waist level. No further bending of back.

5. With clear visibility move forward without twisting. Keep load close to the waist. Turn by moving feet. Keep head up. Do not look at load.

6. Set load down at waist level or to knee level and then on the floor.

Figure 10.9 The main elements of a good lifting technique

The 'good handling technique' given in the Guidance on the Manual Handling Operations Regulations (L23) advises:

▶ think before you lift;
▶ keep the load close to your waist;
▶ adopt a stable position;
▶ ensure a good hold on the load;
▶ at the start of the lift, moderate flexion (slight bending) of the back, hips and knees is preferable to fully flexing the back (stooping) or the hips and knees (squatting);
▶ don't flex your spine any further as you lift;
▶ avoid twisting the trunk or leaning sideways, especially while the back is bent;
▶ keep your head up when handling;
▶ move smoothly;
▶ don't lift more than you can easily manage; and
▶ put down then adjust.

Finally, it needs to be stressed that if injuries involving manual handling operations are to be avoided, planning, control and effective supervision are essential.

10.3 Lifting and moving equipment

10.3.1 Types of manually operated load handling equipment

As mentioned earlier (10.2.4), the first consideration to reduce the risk of injury due to manual handling is, when it is reasonably practicable, to use mechanical assistance in the form of:

▶ simple tools;
▶ wheelbarrows;
▶ trucks and trolleys;
▶ roller tracks and chutes;
▶ pallet trucks;
▶ conveyors; and
▶ various types of hoists that can be used to lift people as well as other loads.

The aids vary from simple, manually-operated tools to power-assisted trucks and lifting devices. All of them 'lighten the load' and reduce the risk of injury.

All kinds of **simple tools** can aid the manual handling of loads. **Lifting hooks** can be used to lift sheets of steel or glass, timber boards and large awkward loads. **Log tongs** will help lift logs and other devices can be used for cylindrical loads. These tools help to grip the load and reduce the need for bending.

Trucks and trolleys allow one person to transport loads between different locations. They can be inexpensive and are available in various sizes to suit the load and type of workplace. **Sack trucks,** like **wheelbarrows**, move loads by balancing them on the truck axle. Some trucks have lifting mechanisms that allow loads to be raised and lowered and others are fitted with special wheels for climbing and descending stairs.

General purpose trucks can be used to support larger loads than sack trucks. They can be flat-topped or fitted with a variety of sides and wheels to suit different uses. The wheels can be fitted with swivels to improve manoeuvrability. Some **platform trucks** are fitted with detachable tug units. This helps to reduce congestion and obstacles in busy production areas. The platform can be designed to be raised or lowered – this further reduces the manual handling because it reduces the need for bending when loading and unloading. Trucks can have big wheels for use on rough ground. When fitted with removable sides they can be used for bulk loads like sand and gravel. **Balance trucks** have a central axle with swivel wheels at each end so they can spin around their centres. This makes them highly manoeuvrable so they can be used in restricted spaces.

There is no clear distinction between general purpose trucks and trolleys. **Trolleys** tend to be of lighter weight construction and designed for more specific applications. Container trolleys allow mixed loads to be carried. A supermarket trolley is a simple example of this. Shelf trolleys can have fixed or removable shelves. Drum trolleys are useful for transporting drums. Brakes can be fitted to trucks and trolleys to keep them stationary and may be necessary where the ground slopes. Garment rails can be used in clothes factories, shops and theatres.

Roller tracks and chutes allow heavy and bulky loads to be moved manually or by gravity under their own weight. They can be portable or set into the floor. Carefully designed work areas with appropriate tracks and chutes can reduce the number of manual handling tasks. When using tracks, the potential for creating a tripping hazard must be kept as low as possible. Using gentle gradients can ensure that the load is moved with very little effort. Chutes are normally used instead of tracks when there are significant changes of level, for example movement between floors. Spiral chutes are typically used for sacks but almost any kind of load can be handled on a straight chute.

Pallet trucks are moved by pedestrians. Manual effort is required to transfer the load but hydraulic power is normally used to raise and lower the load. They can be used in fairly congested and confined areas and are designed to move different types of loads.

Portable conveyors are used to transport loads between places at the same level or different heights. Different types can transport a wide variety of loads including bulk materials like sand and grain. Using portable conveyors can significantly reduce (and often avoid) manual handling. The height of the conveyor at loading and unloading points is important. Recommended heights to reduce manual handling efforts are:

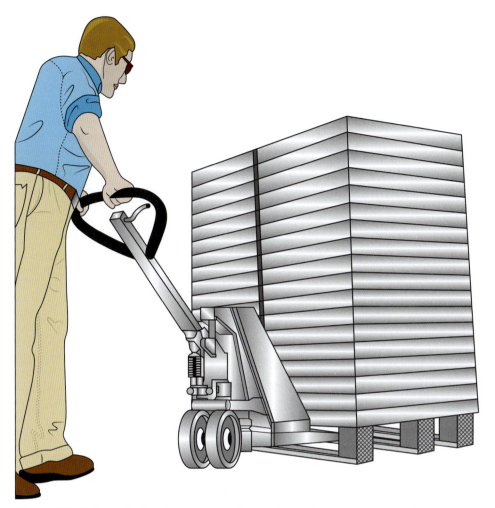

Figure 10.10 Use of a hand-operated pallet truck to raise and move goods

- around 0.9 m above floor level when light loads which can normally be lifted with one hand are to be lifted;
- around 0.75 m above floor level when heavier loads, normally lifted with two hands, are loaded; and
- around floor level when heavy loads, like drums, are loaded onto conveyors, either by rolling or by lifting devices or lift trucks.

Portable conveyors for occasional or semi-permanent occasions can be used in different work areas. Some conveyors are fitted with wheels to increase their mobility. They can be used to move loads at the same level or at different levels. Some types can be inclined by hydraulic rams, to adjust the height of transfer to suit a particular elevation. Non-powered roller conveyors on a slight incline allow loads to move under gravity. Powered conveyers and elevators are covered in 10.3.5.

Mechanical assistance, therefore, involves the use of mechanical aids to assist the manual handling operation such as wheelbarrows, hand-powered hydraulic hoists, specially adapted trolleys, hoists for lifting patients, roller conveyors and automated systems using robots. Some examples of manually operated load handling equipment are shown in Appendix 10.3.

10.3.2 Hazards associated with manually operated load handling equipment

The principal hazards with the use of load handling equipment are due to its incorrect use, such as overloading or attempting to carry unstable loads. Pushing or pulling a truck or trolley is still a manual handling operation and can create different kinds of risks. It may be necessary, for example, to segregate pedestrians from vehicles.

Lack of maintenance can also create additional hazards. This is particularly important for certain lifting machines and tackle where there may be a legal requirement for a regular examination by a competent person (see 10.3.6).

10.3.3 Precautions with the use of manually operated load handling equipment

When handling aids are being selected, the subsequent user should be consulted whenever it is possible. Moving and handling tasks are often made easier by good design. For example, in the health and social care sector, workers are often required to care for service

263

users in bed and the provision of adjustable height beds can prevent the risks of back injuries. Raising the height of laundry equipment, such as washing machines or driers, by placing it on a platform can also reduce bending and stooping. Further measures to reduce the amount of handling by care staff are:

▶ raising the height of beds and chairs using wooden blocks;
▶ handrails at strategic heights adjacent to the bath and toilet;
▶ the use of bath hoists and overhead hoists;
▶ walk-in showers with seats; and
▶ cutting slots in bath panels to allow for mobile hoist wheels to fit underneath a bath fixed against a wall.

All operators of mechanical handling equipment must be properly trained in its use and supervised while they are using it. These precautions are particularly important when people are being lifted or aided using a manual hoist. Other precautions include:

▶ a safe system of work in place;
▶ sufficient room available to easily manoeuvre the equipment;
▶ adequate visibility and lighting available;
▶ the floors in a stable condition;
▶ regular safety checks to identify any faults with the equipment; and
▶ a regular maintenance schedule for the equipment.

Other important considerations are that:

▶ the proposed use will be within the safe working load of the equipment;
▶ it is suitable for the area in which it will operate (e.g. Is there enough room to manoeuvre and enough headroom?);
▶ it suits the terrain in terms of stability and ground surface;
▶ the lifting equipment is CE marked; and
▶ advice is sought from the suppliers/hirers on its suitability for the proposed task and any maintenance requirements.

A useful precaution is to request the equipment on a trial basis, if possible, to check that it is suitable for the required task and to involve the employees who will be expected to use it in this appraisal exercise. Consideration should also be given to other risks associated with the introduction of the lifting aid, such as site safety and the lack of communication with fork-lift truck drivers.

10.3.4 Safety in the use of lifting and moving equipment

An amendment to the Lifting Operations and Lifting Equipment Regulations (LOLER) included the positioning and installation of lifting equipment and the organisation of lifting operations.

Positioning and installation of lifting equipment

Lifting equipment must be positioned and installed so as to reduce the risks, so far as is reasonably practicable, from:

▶ equipment or a load striking a person;
▶ a load drifting, falling freely or being released unintentionally.

Lifting equipment should be positioned and installed to minimise the need to lift loads over people and to prevent crushing in extreme positions. It should be designed to stop safely in the event of a power failure and not release its load. Lifting equipment, which follows a fixed path, should be enclosed with suitable and substantial interlocked gates and any necessary protection in the event of power failure.

The organisation of lifting operations

Every lifting operation, that is lifting or lowering of a load, shall be:

▶ properly planned by a competent person;
▶ appropriately supervised;
▶ carried out in a safe manner.

The person planning the operation should have adequate practical and theoretical knowledge and experience of planning lifting operations. The plan needs to address the risks identified by the risk assessment and identify the resources, the procedures and the responsibilities required so that any lifting operation is carried out safely. For routine simple lifts, a plan will normally be the responsibility of the people using the lifting equipment. For complex lifting operations, for example, where two cranes are used to lift one load, a written plan may need to be produced each time.

The planning should include: the need to avoid suspending loads over occupied areas, visibility, the attaching/detaching and securing of loads, the environment, the location, the possibility of overturning, the proximity to other objects, any lifting of people and the pre-use checks required for the equipment.

In 2010, the Arthritis and Musculoskeletal Alliance launched a charter to help people with MSDs to remain in work. The charter calls for early diagnosis and treatment of MSDs and training of line managers so that they can identify potential problem areas.

Summary of the requirements for lifting operations

There are four general requirements for all lifting operations:

▶ use strong, stable and suitable lifting equipment;
▶ the equipment should be positioned and installed correctly;

- the equipment should be visibly marked with the safe working load (SWL); and
- lifting operations must be planned, supervised and performed in a safe manner by competent people.

10.3.5 Types of mechanical handling and lifting equipment

There are four elements to mechanical handling, each of which can present hazards. These are handling equipment, the load, the workplace and the employees involved.

The **mechanical handling equipment** must be capable of lifting and/or moving the load. It must be fault-free, well maintained and inspected on a regular basis. The hazards related to such equipment include collisions between people and the equipment and personal injury from being trapped in moving parts of the equipment (such as belt and screw conveyors).

The **load** should be prepared for transportation in such a way as to minimise the possibility of accidents. The hazards will be related to the nature of the load (e.g. substances which are flammable or hazardous to health) or the security and stability of the load (e.g. collapse of bales or incorrectly stacked pallets).

The **workplace** should be designed so that, whenever possible, workers and the load are kept apart. If, for example, an overhead crane is to be used, then people should be segregated away or barred from the path of the load.

The **employees** and others and any other people who are to use the equipment must be properly trained and competent in its safe use.

Conveyors and elevators

Conveyors transport loads along a given level which may not be completely horizontal, whereas elevators move loads from one level or floor to another.

There are three common forms of **conveyor** – belt, roller and screw conveyors. The most common hazards and preventative measures are:

- the in-running nip, where a hand is trapped between the rotating rollers and the belt. Protection from this hazard can be provided by nip guards and trip devices;
- entanglement with the power drive requiring the fitting of fixed guards and the restriction of loose clothing which could become caught in the drive;
- loads falling from the conveyor. This can be avoided by edge guards and barriers;
- impact against overhead systems. Protection against this hazard may be given by the use of bump caps, warning signs and restricted access;
- contact hazards prevented by the removal of sharp edges, conveyor edge protection and restricted access;
- manual handling hazards;
- noise and vibration hazards.

Screw conveyors, often used to move very viscous substances, must be provided with either fixed guards or covers to prevent accidental access. People should be prohibited from riding on belt conveyors, and emergency trip wires or stop buttons must be fitted and be operational at all times.

Elevators are used to transport goods between floors, such as the transportation of building bricks to upper

10

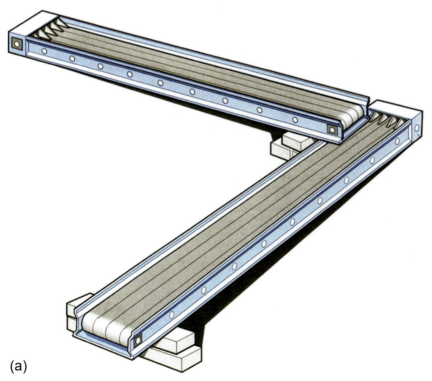

(a)

Figure 10.11 Conveyor systems: (a) belt conveyors; (b) a method of safely moving roofing sheets along a roof valley; (c) a suggested method for manually lifting trusses to eaves level

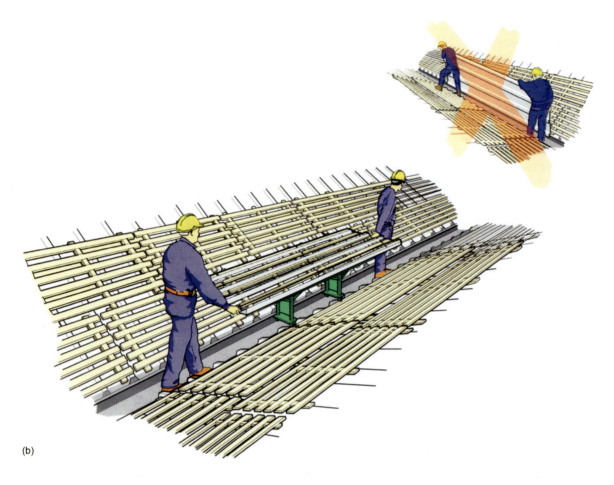

(b)

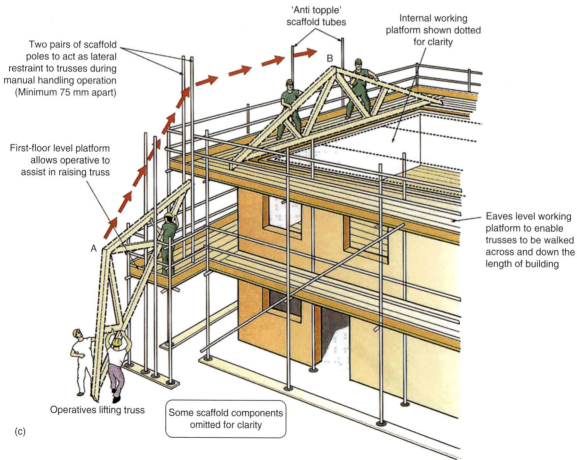

Two pairs of scaffold poles to act as lateral restraint to trusses during manual handling operation (Minimum 75 mm apart)

'Anti topple' scaffold tubes

Internal working platform shown dotted for clarity

B

First-floor level platform allows operative to assist in raising truss

Eaves level working platform to enable trusses to be walked across and down the length of building

A

Operatives lifting truss

Some scaffold components omitted for clarity

(c)

Figure 10.11 (*Continued*)

storeys during the construction of a building or the transportation of grain sacks into the loft of a barn (Figure 10.12). Guards should be fitted at either end of the elevator and around the power drive. The most common hazard is injury due to loads falling from elevators. There are also potential manual handling problems at both the feed and discharge ends of the elevator.

Figure 10.12 A brick elevator

Fork-lift trucks

The most common form of mobile handling equipment is the fork-lift truck (shown in Figure 10.13).

It comes from the group of vehicles known as lift trucks, and can be used in factories, on construction sites and on farms. The term fork-lift truck is normally applied to the counterbalanced lift truck, where the load on the forks is counterbalanced by the weight of the vehicle over the rear wheels. The reach truck is designed to operate in narrower aisles in warehouses and enables the load to be retracted within the wheelbase. The very narrow aisle (VNA) truck does not turn within the aisle to deposit or retrieve a load. It is often guided by guides or rails on the floor. Other forms of lift truck include the pallet truck and the pallet stacker truck, both of which may be pedestrian or rider controlled. Around 400 workers are seriously injured in fork-lift truck accidents every year and some of these accidents cause about

10 fatalities annually. Accidents frequently occur when fork-lift truck drivers or pedestrians in warehouses are distracted, or simply assume that they have been seen by the other party.

Figure 10.13 Rough terrain counterbalanced lift truck

On construction sites, fork-lift trucks of various sizes (including rough terrain, see Figure 10.13) are used to lift loads on to scaffolding working platforms. The loading bay for such working platforms must be constructed to a higher specification than an ordinary scaffold with standards fitted closer together than is normal with extra bracing provided for greater rigidity. There are many hazards associated with the use of fork-lift trucks. These include:

▶ overturning – manoeuvring at too high a speed (particularly cornering); wheels hitting an obstruction such as a kerb; sudden braking; poor tyre condition leading to skidding; driving forwards down a ramp; movement of the load; insecure, excessive or uneven loading; incorrect tilt or driving along a ramp;

▶ overloading – exceeding the rated capacity of the machine;

▶ collisions – particularly with warehouse racking which can lead to a collapse of the whole racking system;

▶ silent operation of the electrically powered fork-lift truck – can make pedestrians unaware of its presence;

▶ uneven road surface – due perhaps to potholes following a heavy storm – can cause the vehicle to overturn and/or cause musculoskeletal problems for the driver;

▶ overhead obstructions – a particular problem for inexperienced drivers (overhead power lines pose a particular problem);

▶ loss of load – shrink wrapping or sheeting will reduce this hazard;

▶ inadequate maintenance leading to mechanical failure;

▶ use as a work platform;

- speeding – strict enforcement of speed limits is essential;
- poor vision around the load;
- pedestrians – particularly when pedestrians and vehicles use the same roadways. Warning signs, indicating the presence of fork-lift trucks, should be posted at regular intervals;
- dangerous stacking or de-stacking technique – this can destabilise a complete racking column;
- carrying passengers – this should be a disciplinary offence;
- battery charging – presents an explosion and fire risk;
- fire – often caused by poor maintenance resulting in fuel leakages or engine/motor burnout, or through using an unsuitable fork-lift truck in areas where flammable liquids or gases are used and stored;
- lack of driver training.

If fork-lift trucks are to be used outside, visibility and lighting, weather conditions and the movement of other vehicles become additional hazards.

There are also the following physical hazards:

- noise – caused by poor silencing of the power unit;
- exhaust fumes – should only be a problem when the maintenance regime is poor;
- vibrations – often caused by a rough road surface or wide expansion joints. Badly inflated tyres will exacerbate this problem;
- manual handling – resulting from manoeuvring the load by hand or lifting batteries or gas cylinders;
- ergonomic – musculoskeletal injuries caused by soft tyres and/or undulating road surface or holes or cracks in the road surface (e.g. expansion joints).

Regular and documented maintenance by competent mechanics is essential. However, the driver should undertake the following checks at the beginning of each shift:

- condition of tyres and correct tyre pressures;
- effectiveness of all brakes;
- audible reversing horn and light working properly;
- lights, if fitted, working correctly;
- mirrors, if fitted, in good working order and properly set;
- secure and properly adjusted seat;
- correct fluid levels, when appropriate;
- fully charged batteries, when appropriate;
- correct working of all lifting and tilting systems.

A more detailed inspection should be undertaken by a competent person within the organisation on a weekly basis to include the mast and the steering gear. Driver training is essential and should be given by a competent trainer. The training session must include the site rules covering items such as the fork-lift truck driver code of practice for the organisation, speed limits, stacking procedures and reversing rules. Refresher training should be provided at regular intervals and a detailed record kept of all training received. Table 10.1 illustrates

some key requirements of fork-lift truck drivers and the points listed should be included in most codes of practice. Appendix 10.4 gives advice from the HSE on the safe use of fork-lift trucks.

Finally, care must be taken with the selection of drivers, including relevant health checks and previous experience. Drivers should be at least 18 years of age and their fitness to drive should be reassessed regularly (every five years after the age of 40 and every year after 65 (HSG6)).

Table 10.1 Safe driving of lift trucks

Drivers must:
▶ drive at a speed to suit road conditions and visibility
▶ use the horn when necessary (at blind corners and doorways)
▶ always be aware of pedestrians and other vehicles
▶ take special care when reversing (do not rely on mirrors)
▶ take special care when handling loads which restrict visibility
▶ travel with the forks (or other equipment fitted to the mast) lowered
▶ use the prescribed lanes
▶ obey the speed limits
▶ take special care on wet and uneven surfaces
▶ use the handbrake, tilt and other controls correctly
▶ take special care on ramps
▶ always leave the truck in a state which is safe and discourages unauthorised use (brake on, motor off, forks down, key out).
Drivers must not:
▶ operate in conditions in which it is not possible to drive and handle loads safely (e.g. partially blocked aisles)
▶ travel with the forks raised
▶ use the forks to raise or lower persons unless a purpose-built working cage is used
▶ carry passengers
▶ park in an unsafe place (e.g. obstructing emergency exits)
▶ turn round on ramps
▶ drive into areas where the truck would cause a hazard (flammable substance store)
▶ allow unauthorised use.

Other forms of lifting equipment

In construction work, there are many different types of vehicles and plant used to lift and move loads. These include excavators, telehandlers, rough terrain vehicles, dumper trucks and hoists, all of which are covered in more detail elsewhere in this book (Chapters 8, 11, 16 and 17).

(a) Excavators

Excavators that have a lifting eye and are used for lifting operations require a periodic thorough examination. However, if the only job they do is conventional 'earth moving' then the requirements of the Provision of Work Equipment Regulations (PUWER) apply. Under PUWER there must be an appropriate maintenance regime and an inspection of the machine at intervals specified by a competent person. But should the machine be used for lifting (even if only occasionally) then it must be thoroughly examined and additional safety devices must be fitted, including a boom lowering device (usually check valves) and an acoustic warning device.

(b) 360° wheeled excavator

A weekly inspection of a 360° wheeled excavator would include the condition of the tyres, wheels, steering controls, lights, brakes and hydraulic systems. Periodically, the seat rigidity and the proper functioning of seat restraints, instrumentation and controls, bucket condition and other attachments need to be inspected. Appendix 10.2 gives an example of a typical risk assessment used for an excavator.

(c) Excavator and crane fly jib quick hitches

The HSE has issued a safety alert to users of excavators fitted with quick hitch devices. There have been a number of serious incidents in the construction industry recently, including four fatalities.

A quick hitch (sometimes called a quick coupler) on an excavator is a latching device that enables attachments to be connected to the dipper arm of the plant and enables the operator to change buckets or attachments rapidly. Quick hitches are in common use throughout the construction industry and, when properly designed, maintained and used can save time when working with excavators. They allow operators to quickly use plant for a wider range of tools other than just buckets, such as mechanical diggers and piling drills. All quick hitches should be designed to the standard (EN 474-1 Annex b) and carry a CE mark. There are many types and sizes of quick hitch – manual, semi-automatic and automatic. Most of the fatal accidents have happened with semi-automatic hitches.

Four fatal incidents involved semi-automatic quick hitches, but there have also been other incidents involving both manual and automatic types. It has been estimated there are approximately 100,000 quick hitch attachments currently in use in the UK.

(b)

(c)

Figure 10.14 (a) Attaching a quick hitch fly jib to a telescopic mobile crane (hard hats missing); (b) crane in use installing a yacht's mast; (c) two excavators with quick hitch bucket couplings

The common cause of all these fatal incidents was a missing retaining pin or bar. The pin holds the attachment in place against the quick hitch and insures against accidental release. A quick hitch may operate for some time without the retaining pin in place and then suddenly, without warning, swing open and fall off. If this happens while the excavator is lifting a load close to a person then a fatal accident may result.

(a)

The HSE Safety Alert, available on the HSE website, lists several precautions that should be identified by a comprehensive risk assessment. These include:

▶ excavator operators should be adequately trained on the use of quick hitches in general;

▶ the manufacturer-specified retaining pin must be available on the machine;

▶ there should be a system for checking that the pin is in place on the hitch before starting the work and every time a different attachment is fitted;

▶ those in control of sites should undertake random checks to ensure the precautions are being implemented.

The Construction Plant-hire Association has published on behalf of the Strategic Forum for Construction a Best Practice Guide 'Safe Use of Quick Hitches on Excavators'.

(d) Compact dumper trucks

Another of these pieces of plant will be discussed in a little more detail here (see Figure 9.4). The small dumper truck is widely used on all sizes of construction site. Compact dumper trucks are involved in about 30% of construction transport accidents. The three main causes of such accidents are:

▶ overturning on slopes and at the edges of excavations;

▶ poorly maintained braking systems;

▶ driver error due to lack of training and/or inexperience.

Some of the hazards associated with these vehicles are collisions with pedestrians, other vehicles or structures, such as scaffolding. They can be struck by falling materials and tools or be overloaded. The person driving the truck can be thrown from the vehicle, come into contact with moving parts on the truck, suffer the effects of whole-body vibrations due to driving over potholes in the roadway and suffer from the effects of noise and dust.

The precautions that can be taken to address these hazards include the use of authorised, trained, competent and supervised drivers only. As with so many other construction operations, risks should be assessed, safe systems of work followed and drivers forbidden from taking short cuts. The following site controls should also be in place:

▶ designated traffic routes and signs;

▶ speed limits;

▶ stop blocks used when the vehicle is stationary;

▶ proper inspection and maintenance procedures;

▶ procedures for starting, loading and unloading the vehicle;

▶ provision of rollover protective structures (ROPS) and seat restraints;

▶ provision of falling-object protective structures (FOPS) when there is a risk of being hit by falling materials;

▶ visual and audible warning of approach;

▶ where necessary, hearing protection.

For other forms of mobile construction equipment, the risk to people from the overturning of the equipment must always be safeguarded. This can usually be achieved by the avoidance of working on steep slopes, the provision of stabilisers and ensuring that the load carried does not affect the stability of the equipment/vehicle.

The other types of lifting equipment to be considered are cranes (mobile overhead and jib), lifts and hoists and lifting tackle. A sample risk assessment for the use of lifting equipment is given in Appendix 10.1.

The lifting operation should be properly prepared and planned. This involves the selection of a suitable crane having up-to-date test certificates and examination reports that have been checked. A risk assessment of the task will be needed which would ascertain the weight, size and shape of the load and its final resting place. A written plan for completing the lift should be drafted and a competent person appointed to supervise the operation.

Lift plans

A lift plan identifies the ways in which the risks involved in a lifting operation can be eliminated or controlled. The degree of planning for a lifting operation should be proportional to the risk and will vary considerably depending upon the complexity of the lifting operation. The complexity will depend on the load to be lifted, the equipment to be used and the environment in which the operation is to be undertaken.

Well-planned lifting operations are a combination of two parts:

1. initial planning to ensure that the lifting equipment provided is suitable for the range of tasks that it will have to undertake; and

2. the planning of individual lifting operations so that they can be performed safely with the lifting equipment provided.

The balance between the two parts of the planning process will vary depending upon the lifting equipment and the particular lifting operation.

The person planning the operation should have adequate practical and theoretical knowledge and experience of planning lifting operations. The plan will need to address the risks identified by the risk assessment and identify the resources, the procedures and the responsibilities required so that any lifting operation is carried out safely. For routine simple lifts a plan will normally be left to the people using the lifting equipment. For complex lifting operations, for example where two cranes are used to lift one load, a written plan may need to be produced each time.

The planning should take account of avoiding suspending loads over occupied areas, visibility, attaching/detaching and securing loads, the environment, location, overturning, proximity to other objects, lifting of people and pre-use checks of the equipment.

Cranes

Cranes may be either a **jib crane** or an **overhead gantry travelling crane**. The safety requirements are similar for each type. All cranes need to be properly designed, constructed, installed and maintained. They must also be operated in accordance with a safe system of work. They should only be driven by authorised persons who are fit and trained. Each crane is issued with a certificate by its manufacturer giving details of the **safe working load (SWL)**. The SWL must **never** be exceeded and should be marked on the crane structure. If the SWL is variable, as with a jib crane (the SWL decreases as the operating radius increases), an SWL indicator should be fitted. Care should be taken to avoid sudden shock loading, as this will impose very high stresses on the crane structure. It is also very important that the load is properly shackled and all eyebolts tightened. Safe slinging should be included in any training programme. All controls should be clearly marked and be of the 'hold-to-run' type.

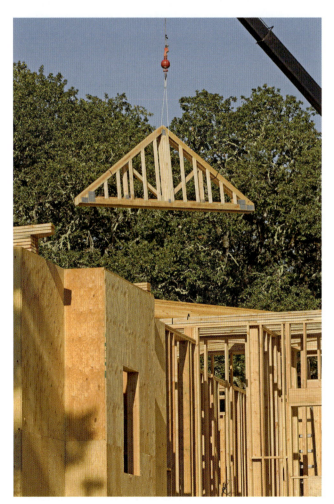

Figure 10.15 Lifting roof trusses

Large cranes, which incorporate a driving cab, often work in conjunction with a banksman, who will direct the lifting operation from the ground. Banksmen are operatives trained to direct vehicle movement on or around site. They are often called traffic marshals Banksmen should only be used in circumstances where other control measures are not possible. It is important that banksmen are trained so that they understand recognised crane signals. Signallers are operatives who are trained to direct crane drivers during lifting operations. It is a hazardous job and measures will help to keep the signaller safe:

▶ Ensure that they are trained and competent to direct lifting operations.
▶ Provide a protected position from which they can work in safety and where they can be seen at all times by the crane driver.
▶ Provide distinctive 'hi viz' clothing for identification.
▶ Tell drivers that if they cannot see the signaller they should stop immediately.
▶ Agree on the use of standard signals.

The lifting operation should be properly prepared and planned. This involves the selection of a suitable crane having up-to-date test certificates and examination reports that have been checked. A risk assessment of the task will be needed which would ascertain the weight, size and shape of the load and its final resting place. A written plan for completing the lift should be drafted and a competent person appointed to supervise the operation. During the planning of a lifting operation close to a public highway, there should be liaison with the police and local authority over any road closures and the lift should be timed to take place when there will be a minimum number of people about. Arrangements may be needed to give advanced warning to motorists and other road users and appropriate signs and traffic controls used.

The basic principles for the safe operation of cranes are as follows. **For all cranes,** the driver must:

▶ undertake a brief inspection of the crane and associated lifting tackle each time before it is used;
▶ check that all lifting accessory statutory inspections are in place and up to date;
▶ check that tyre pressures, where appropriate, are correct;
▶ ensure that loads are not left suspended when the crane is not in use;
▶ before a lift is made, ensure that nobody can be struck by the crane or the load;
▶ ensure that loads are never carried over people;
▶ ensure good visibility and communications;
▶ lift loads vertically – cranes must not be used to drag a load;
▶ travel with the load as close to the ground as possible;

10

271

▶ switch off power to the crane when it is left unattended and lock the cab to prevent unauthorised access.

For **mobile jib cranes**, the following points should be considered:

▶ Each lift must be properly planned, with the maximum load and radius of operation known.
▶ Overhead obstructions or hazards must be identified; it may be necessary to protect the crane from overhead power lines by using goal posts and bunting to mark the safe headroom.
▶ The ground on which the crane is to stand should be assessed for its load-bearing capacity.
▶ If fitted, outriggers should be used with appropriate outrigger mats where ground conditions dictate.

The principal reasons for crane failure, including loss of load, are:

▶ overloading;
▶ poor slinging of load;
▶ insecure or unbalanced load;
▶ overturning;
▶ collision with another structure or overhead power lines;
▶ foundation failure;
▶ structural failure of the crane;

▶ operator error;
▶ lack of maintenance and/or regular inspections;
▶ no signaller used when driver's view is obscured;
▶ incorrect signals given or a misunderstanding of a signal.

Tower cranes

Typical causes of recent serious incidents with tower cranes include:

▶ mechanical failure of the brake or lifting ram;
▶ overturn of the crane;
▶ jib collapse;
▶ a load or dropped load striking a worker; and
▶ sling failure.

(b)

Figure 10.16 (b) Typical saddle jib (horizontal) tower crane (see also Figure 7.6b)

The reasons for some of these incidents were:

▶ poor site induction training – not dealing with site-specific risks and lasting too long (20–30 minutes maximum is sufficient time);
▶ problems with crane maintenance and thorough examinations;
▶ operators working long hours without a break;
▶ poor operator cabin design and too high a climbing distance;
▶ operator health problems;
▶ problems in communicating health and safety issues by crane operators on site.

During lifting operations using cranes, it must be ensured that:

▶ the driver has good visibility;
▶ there are no pedestrians below the load by using barriers, if necessary;
▶ there is adequate space for the installation, manoeuvring and operation of the crane;
▶ all pedestrians are re-routed during the lifting operation;
▶ everyone within the lifting area is wearing a safety helmet;

(a)

Figure 10.16 (a) Typical luffing jib tower cranes operating on a large construction site

▶ an audible warning is given prior to the lifting operation.

If lifting takes place in windy conditions, tag lines may need to be attached to the load to control its movement.

Tower cranes are being used more and more often on construction sites across Europe. Approximately 1,800 tower cranes have operated in the UK. Since 2001 in the UK, there have been 61 accidents involving tower cranes – 9 people have died, and 25 have been seriously injured (e.g. in Canary Wharf, Worthing, Battersea and Liverpool). These accidents have led to public concern over tower crane safety and to the introduction by the HSE of the Notification of Conventional Tower Cranes Regulations in 2010, since revoked. The Regulations required employers who had primary responsibility for the safety of cranes to notify certain information to the HSE. This included the name and address of the crane owner and the site address, as well as sufficient information to identify the crane and the date of its last thorough examination.

The Löfstedt report into health and safety in 2011 noted that, 'when these Regulations were put into place, the Impact Assessment (IA) carried out then did not predict that they would have "direct health and safety benefits (i.e. reductions in injury or ill-health)", but that the main benefit of a tower crane register would be "an increase in public assurance". It was not clear that a statutory requirement to register tower cranes is the most appropriate way to provide public assurance and suggested that non-regulatory methods should be explored.'

The findings of the IA suggested that the revocation of these Regulations would:

▶ have little effect on public perception of risk to their safety arising from the use of tower cranes;
▶ not reduce safety standards required when erecting, using or dismantling tower cranes on construction sites; and
▶ remove a source of cost and burden on the industry.

The key aspects that apply to all cranes including tower cranes are:

1. The planning of safe lifting operations by a competent person following a detailed risk assessment;
2. Safe systems of work for the installation, operation and dismantling stages. The main elements of the safe system or method statement are:

   ▶ planning – including site preparation, crane erection and dismantling;
   ▶ selection, provision and use of a suitable crane and work equipment, including safe slinging and signalling arrangements;
   ▶ maintenance and examination of the crane and equipment;

   ▶ provision of properly trained and competent personnel;
   ▶ supervision of operations by personnel having the necessary authority;
   ▶ thorough examinations, reports and other documents;
   ▶ preventing unauthorised movement or use of the crane; and
   ▶ measures to secure safety of persons not involved in the lifting.

3. Supervision of all lifting operations; and
4. The thorough examination of the crane and, when required, the notification of the HSE.

Recommendations were made by a Construction Industry Forum that site-specific induction training should be given to everybody involved with the tower crane erection, operation and dismantling. A specific risk assessment and a regular inspection of the crane should be undertaken and only competent and trained operators should be involved with tower crane activities. The Health and Safety (Safety Signs and Signals) Regulations and BS 7121 give examples of signals to be used when guiding cranes. The signalman should wear one or more appropriate distinctive items (a jacket, helmet, sleeves or armbands or carry bats). The distinctive items should be all the same colour and for the exclusive use of the signalman.

The basic signals are as follows:

(a) General signals:
   ▶ start operations
   ▶ stop interruption or end of movement
   ▶ end of the operation.

(b) Vertical movements:
   ▶ raise
   ▶ lower
   ▶ vertical distance.

(c) Horizontal movements:
   ▶ move forwards
   ▶ move backwards
   ▶ right
   ▶ left
   ▶ horizontal distance.

(d) Danger:
   ▶ danger emergency stop
   ▶ quick – all movements faster
   ▶ slow – all movements slower.

Finally, if a crane is hired to lift a load, the following points should be covered before the lift takes place:

▶ the suitability of the crane, in particular its load carrying capacity;
▶ the validity of its documentation relating to inspection (test certificate) and maintenance;
▶ an assessment of the area where the lifting is to take place including the ground conditions and the proximity of overhead power lines;
▶ the competence of the driver, slinger and banksman;

10

Figure 10.17 Mobile self-erecting tower crane

▶ the coordination of the lifting operation;

▶ consultation with others affected by the work.

A **lift** or **hoist** incorporates a platform or cage and is restricted in its movement by guides. Hoists are generally used in industrial settings (e.g. construction sites and garages), whereas lifts are normally used inside buildings. Lifts and hoists may be designed to carry passengers and/or goods alone. They should be of sound mechanical construction and have interlocking doors or gates, which must be completely closed before the lift or hoist moves. The hoistway should be properly enclosed so that the moving parts of the hoist are guarded. The loads should be secured on the hoist platform so that they cannot fall and the operator should have a clear view of the landing levels. There

should be no unauthorised use of the hoist by untrained personnel.

Other **items of lifting tackle**, usually used with cranes, include chain slings and hooks, wire and fibre rope slings, eyebolts and shackles (Figure 10.18). Special care should be taken, when slings are used, to ensure that the load is properly secured and balanced. Lifting hooks should be checked for signs of wear and any distortion of the hook. Shackles and eyebolts must be correctly tightened. Slings should always be checked for any damage before they are used and only competent people should use them. Training and instruction in the use of lifting tackle is essential and should include regular inspections of the tackle, in addition to the mandatory thorough examinations. Finally, care should be taken when these items are being stored between use.

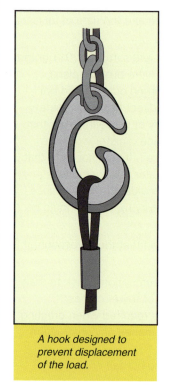

A hook designed to prevent displacement of the load.

A hook with a safety catch.

Figure 10.18 Specially designed safety hooks

Passenger lifts

The Lifting Operations and Lifting Equipment Regulations (LOLER) and the ACoP L113 set out precise requirements for inspection and use of passenger lifts. These are:

▶ Regulation 9 requires passenger lifts to be thoroughly examined for defects before first use and every six months thereafter (12 months for goods-only lifts). Further examinations may be required in accordance with a written scheme, or if an incident occurs that may affect the safe operation of the equipment. This regulation also requires equipment to be inspected at suitable intervals between thorough examinations.

▶ Regulation 10 requires thorough-examiners to notify defects that are a danger to persons to both the employer (duty-holder) and the enforcing authority, and make a report of their inspection in writing as soon as is practicable. This same regulation requires the employer to ensure equipment is, as identified in the report, either not used until a defect is repaired, or is repaired within a specified time frame.

▶ Regulation 11 requires thorough examination reports to be retained for at least two years.

Passenger-carrying lifts must be fitted with an automatic braking system to prevent overrunning, at least two suspension ropes, each capable alone of supporting the maximum working load, and a safety device which could support the lift in the event of suspension rope failure. Maintenance procedures must be rigorous, recorded and only undertaken by competent persons. It is very important that a safe system of work is employed during maintenance operations to protect others, such as members of the public, from falling down the lift shaft and other hazards.

The following issues are relevant to the safe operation of passenger lifts:

▶ An emergency mushroom stop button should be fitted in the lift.

▶ A direct telephone line is required if the internal alarm sounder/siren cannot be heard by rescue staff in the building.

▶ A risk assessment for possible injuries to the lift passengers should be undertaken on the design, sensitivity and performance of the door edges. The lift inspector should have the relevant standard to help with this.

▶ A lift stop switch should be available in the pit as well as the motor room at the top of the lift. PUWER requires that this switch must be 'readily available'.

▶ The lift beam should be inspected by the lift inspector – the safe working load should be stated on it to prevent overload.

▶ If there are any moving or revolving mechanisms accessible by lift technicians performing their maintenance tasks, then they should be protected from inadvertent contact by suitable guards.

▶ It is advisable to locate a fire extinguisher inside the access door of the motor room although this is not a statutory requirement.

▶ Emergency door release devices must not be routinely operated during non-emergency situations. The emergency keys are intended to allow emergency access to the lifting platform in the event of people becoming trapped and should be under strict control.

▶ Emergency unlocking should be undertaken only in exceptional circumstances and by suitably trained and authorised people.

▶ Safe working procedures and arrangements should be in place setting out what to do in the event of an emergency or failure. For example, how to deal with trapped people and the arrangements for repairing faults.

▶ All lifts and lifting platforms must be inspected, serviced and maintained. Where the lift is used by work purposes, it must be thoroughly examined by a competent person.

10.3.6 Requirements for the statutory examination of lifting equipment

The LOLER specify examinations required for lifting equipment by using two terms – an inspection and a thorough examination. Both terms are defined by the HSE in guidance accompanying various Regulations. Lifting equipment includes any equipment used at work to lift or lower loads including any anchoring, fixing or supporting attachments.

An **inspection** is used to identify whether the equipment can be operated, adjusted and maintained safely so that any defect, damage or wear can be detected before it results in unacceptable risks. It is normally performed by a competent person appointed by the employer (often an employee).

A **thorough examination** is a detailed examination, which may involve a visual check, a disassembly and testing of components and/or an equipment test under operating conditions. Such an examination must normally be carried out by a competent person who is independent of the employer. The examination is usually carried out according to a written scheme and a written report is submitted to the employer.

The Code of Practice for the Safe Use of Lifting Equipment defines a thorough examination as a visual examination carried out by a competent person carefully and critically and, where appropriate, supplemented by other means, such as measurement and testing, in order to check whether the equipment is safe to use.

A detailed summary of the thorough examination and inspection requirements of the LOLER is given in Chapter 19.

A thorough examination of lifting equipment should be undertaken at the following times:

▶ before the equipment is used for the first time;
▶ after it has been assembled at a new location;
▶ at least every 6 months for equipment used for lifting persons or as a lifting accessory;
▶ at least every 12 months for all other lifting equipment including the lifting of loads over people;

▶ in accordance with a particular examination scheme drawn up by an independent competent person;

▶ each time that exceptional circumstances, which are likely to jeopardise the safety of the lifting equipment, have occurred (such as severe weather).

The person making the thorough examination of lifting equipment must:

▶ notify the employer forthwith of any defect which, in their opinion, is or could become dangerous;

▶ as soon as practicable (within 28 days) write an authenticated report to the employer and any person who leased or hired the equipment.

The Regulations specify the information that should be included in the report.

The initial report should be kept for as long as the lifting equipment is used (except for a lifting accessory which need only be kept for 2 years). For all other examinations, a copy of the report should be kept until the next thorough examination is made or for 2 years (whichever is the longer). If the report shows that a defect exists that could lead to an existing or imminent risk of serious personal injury, a copy of the report must be sent, by the person making the thorough examination, to the appropriate enforcing authority.

The equipment should be inspected at suitable intervals between thorough examinations. The frequency and the extent of the inspections are determined by the level of risk presented by the lifting equipment. A report or record should be made of the inspection which should be kept until the next inspection. Unless stated otherwise, lifts and hoists should be inspected every week.

It is important to stress that thorough examinations must be accompanied by meticulous in-service inspection that will detect any damage every time the equipment is used. For most workplaces, it is better to store every item of portable lifting equipment in a central store where these inspections can take place and records of the equipment be kept. Users must also be encouraged to report any defects in equipment that they have used.

10.4 Further information

The Health and Safety (Display Screen Equipment) Regulations 1992 (as amended)

The Manual Handling Operations Regulations 1992

The Provision and Use of Work Equipment Regulations 1998

The Lifting Operations and Lifting Equipment Regulations 1998

Work with display screen equipment: Health and Safety (Display Screen Equipment) Regulations 1992 as amended by the Health and Safety (Miscellaneous Amendments) Regulations 2002, (L26) Guidance on Regulations, HSE Books, ISBN 978-0-7176-2582-6 http://www.hse.gov.uk/pubns/priced/l26.pdf

Safe use of work equipment (ACoP) (L22), HSE Books, ISBN 978-0-7176-6295-1 http://www.hse.gov.uk/pubns/priced/l22.pdf

Manual Handling Operations Regulations 1992 (as amended), Guidance on Regulations L23, HSE Books, ISBN 978-0-7176-2823-0 http://www.hse.gov.uk/pubns/priced/l23.pdf

Safe use of lifting equipment (L113), HSE Books, ISBN 978-0-7176-1628-2 http://www.hse.gov.uk/pubns/priced/l113.pdf

Manual handling at work: a brief guide (INDG143 rev3), http://www.hse.gov.uk/pubns/indg143.pdf

Rider-operated lift trucks. Operator training. (ACoP and Guidance) (L117), HSE Books, ISBN 978-0-7176-6441-2 http://www.hse.gov.uk/pubns/priced/l117.pdf

10.5 Practice revision questions

1. (a) **Define** the term 'ergonomics'.
   (b) **Identify SIX** factors that need to be considered during an ergonomic assessment.

2. Work-related upper limb disorders (WRULDs) are responsible for many cases of work-related ill-health:
   (a) **Identify TWO** examples of WRULDs.
   (b) **Identify THREE** work activities that may cause WRULDs and the typical symptoms that might be experienced by affected individuals.
   (c) **Describe** the measures that should be taken to minimise the risk to these individuals.

3. Office employees who work for a construction company are required to use display screen equipment (computers) for eight hours a day.
   (a) **Identify** the risks to the health and safety of these employees.
   (b) **Outline** the factors to consider when making an assessment of a display screen equipment (DSE) workstation.
   (c) **Outline** the control measures that may be taken to minimise the risks to these employees.

(d) **Identify** the features of a suitable seat for use at a DSE workstation.

4. (a) **Identify FOUR** types of injury that may be caused by the incorrect manual handling of loads.

(b) **Outline** a good lifting technique that should be adopted by a person required to lift a load from the ground to a work bench.

(c) **Give TWO** examples of how a manual handling task might be avoided.

5. **Outline** the factors that may affect the risk from manual handling activities in relation to:

(a) The task
(b) The individual
(c) The load
(d) The environment.

6. A rolled steel joist (RSJ) is to be used in the support of the second floor of a town house that is being refurbished. The layout of the house and the arrangement of the supports are such that mechanical means cannot be used either to transport the RSJ from the ground floor or to lift it into its final position.
Outline the factors that would need to be considered when undertaking a manual handling assessment of the task.

7. The handling of kerb stones is one of the main causes of back problems and other musculoskeletal disorders in the construction industry.

(a) **Identify** the main hazards to health and safety associated with kerb laying.

(b) **Outline** a hierarchy of control measures that may address some of the hazards associated with kerb laying.

8. In order to minimise the risk of injury when undertaking a manual handling operation:

(a) **Identify FOUR** types of manually operated load handling equipment that can be used to assist the manual handling operation.

(b) **Outline FOUR** hazards and the corresponding precautions to be taken when using conveyor systems for moving materials within a workplace.

9. Fork-lift trucks are used extensively in busy warehouses for the storage and supply of building materials.

(a) **Identify EIGHT** general and **FOUR** physical hazards associated with their use.

(b) **Outline** the checks that the driver should undertake at the beginning of a shift.

(c) **Identify FOUR** rules that the driver should follow when a fork-lift truck is left unattended.

(d) **Outline** the principal requirements for lifting operations.

10. A pedestrian has been injured after a collision with a fork-lift truck in a supply warehouse.
Outline the possible **immediate** causes of the accident associated with:

(a) the pedestrian
(b) the way in which the vehicle was driven
(c) the workplace
(d) the vehicle.

11. (a) **Identify THREE** types of crane used for lifting operations.

(b) **Outline** the key issues for the safe management of all crane operations.

(c) **Identify EIGHT** reasons for accidents during the operation of cranes.

(d) **Identify EIGHT** requirements of crane drivers to ensure the safe operation of the crane.

12. **Identify** the weekly maintenance inspection requirements for the following items of lifting equipment:

(a) 360° wheeled excavators
(b) safety hooks
(c) dumper trucks
(d) fork-lift trucks.

13. (a) **Outline** a procedure for the safe lifting of a load by a crane, having ensured that the crane has been correctly selected and positioned for the job.

(b) **Outline** issues that could be included in a lifting plan.

14. A factory, that builds timber-frame structures for the construction industry, uses an overhead gantry crane to transport heavy items between machines.

(a) **Identify FOUR** reasons why loads may fall from the crane.

(b) **Outline** the precautions to be taken so that accidents to those working at ground level are prevented when gantry cranes are in use.

15. A mechanical hoist with specially designed lifting tackle is used to raise and lower roofing materials.

(a) **Outline** the precautions to be taken to reduce the risk of injury to employees and others during the lifting operation.

(b) **Identify** the two types of safety inspection required for this equipment.

(c) **Describe FOUR** defects that might be identified in a wire rope sling during routine inspection prior to use.

16. The Lifting Operations and Lifting Equipment Regulations 1998 require the statutory inspection of lifting equipment.

(a) **Identify** the occasions when a thorough examination of a fork-lift truck is required under these regulations.

(b) **Outline** the differences between an inspection and a thorough examination of lifting equipment when applied to these regulations.

(c) **Identify** the duties of the person making a thorough examination of lifting equipment.

APPENDIX 10.1 A typical risk assessment for the use of lifting equipment

INITIAL RISK ASSESSMENT	Use of Lifting Equipment			
SIGNIFICANT HAZARDS		Low	Medium	High
1. Unintentional release of load				✓
2. Unplanned movement of load		✓		
3. Damage to equipment		✓		
4. Crush injuries to personnel			✓	
5.				

ACTION ALREADY TAKEN TO REDUCE THE RISKS:

Compliance with:

Lifting Operations and Lifting Equipment Regulations (LOLER)

Safety Signs and Signals Regulations (SSSR)

Provision and Use of Work Equipment Regulations (PUWER)

British Standard – Specification for flat woven webbing slings

BS – Guide to selection and use of lifting slings for multi-purposes

Planning:

Copies of statutory thorough examinations of lifting equipment will be kept on site. Before selection of lifting equipment, the above standards will be considered as well as the weight, size, shape and centre of gravity of the load. Lifting equipment is subject to the planned maintenance programme.

Physical:

All items of lifting equipment will be identified individually and stored so as to prevent physical damage or deterioration. Safe working loads of lifting equipment will be established before use. Packing will be used to protect slings from sharp edges on the load. All items of lifting equipment will be visually examined for signs of damage before use. Ensuring the eyes of strops are directly below the appliance hook and that tail ropes are fitted to larger loads will check swinging of the load. Banksmen will be used where the lifting equipment operator's vision is obstructed. Approved hand signals will be used.

Managerial/Supervisory:

Only lifting equipment that is in date for statutory examination will be used. Manufacturer's instructions will be checked to ensure that methods of sling attachment and slinging arrangements generally are correct.

Training:

Personnel involved in the slinging of loads and use of lifting equipment will be required to be trained to CITB or equivalent standard. Supervisors will be trained in the supervision of lifting operations.

Date of Assessment...........................Assessment made by..

Risk Re-Assessment Date...................Site Manager's Comments:.................................

APPENDIX 10.2 A typical risk assessment for an excavator to be used for lifting

INITIAL RISK ASSESSMENT	Use of Lifting Equipment			
SIGNIFICANT HAZARDS		Low	Medium	High
1. Unintentional release of load			✓	
2. Unplanned movement of load			✓	
3. Damage to equipment				✓
4. Crush injuries to personnel			✓	
5.				

ACTION ALREADY TAKEN TO REDUCE THE RISKS:

Compliance with:

Lifting Operations and Lifting Equipment Regulations (LOLER)

Provision and Use of Work Equipment Regulations (PUWER)

Health and Safety Executive Guidance Note – Excavators used as chances

British Stadards – Guide to selection and use of lifting slings for Multipurposes

BS – Code of Practice for safe use of wire rope slings for general purposes CDM Regulations

Planning:

Excavators to be used as cranes must be designed for the purpose, and be fitted with sling attachments and check valves. A current test certificate and thorough examination must be available for the excavator. If any exemption certificates apply they should also be available. Excavator safe working load will be greater than the foreseeable weight of loads to be lifted

Physical:

No persons are allowed to stand or work within operating radius without the operator's permission. Loads must not slued over personnel, vehicle cabins or huts. A banksman is to be used where driver's vision is impaired or if operating in congested areas. SWL will be clearly marked on excavator, and a table of SWLs will be clearly visible to the driver; it will not be exceeded. The machine will be on a firm, level base, and the lifts will be carried out with the boom parallel to machine tracks or wheels.

Managerial/Supervisory:

Certification of drivers must be checked; drivers must be over 18. Lifting operations will be supervised to ensure the stability of the machine and the load. Lifting operations will be restricted to those directly associated with excavations. All operatives working within the boom's radius will wear head protection.

Training:

Driver training to CITB standard is required. Operator training for earth-moving machinery. Excavator driving by non-certificate holding operatives is not permitted; this also applies to sub-contractors and the self-employed.

Date of Assessment..	Assessment made by...
Risk Re-Assessment Date.....................................	Site Manager's Comments:...................................

APPENDIX 10.3 Examples of manually operated load handling equipment

Lifting hooks

Paving slab and general purpose handler

Platform truck

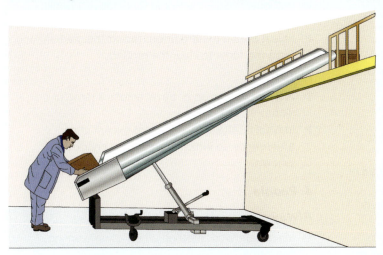

A hydraulically-lifted mobile belt conveyor

A chute being used to transfer sacks from a high to low level

A sack truck with star truck wheels for climbing and descending stairs

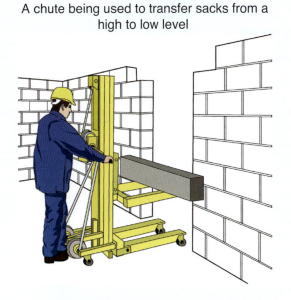

Use of hand-operated lift truck to place lintels

10

APPENDIX 10.4 Safe use of fork-lift trucks (based on an HSE document)

1. Operating

Always:

- carry out a pre-shift check of the lift truck;
- wear operator restraints, where fitted;
- look all around before moving off;
- look in the direction of travel;
- travel at a speed suitable for the location and the load you are carrying;
- travel with the forks lowered, but clear of the ground;
- watch out for pedestrians;
- watch out for obstructions;
- avoid sudden stops and violent braking;
- slow down at corners, doorways, and at every danger spot and sound the horn, where necessary;
- apply the parking brake when leaving the lift truck;
- face the lift truck and use steps and handholds when getting in or out.

Never:

- operate a lift truck unless you are trained and authorised to do so;
- use a lift truck or equipment you know is not working properly;
- operate controls from outside the cab, unless it is designed so you can do this;
- stand on or near the controls to reach the load or anything outside the cab;
- travel on uneven ground unless the lift truck is suitable for this;
- run over unprotected cables or flexible pipes;
- try to carry out repairs – leave this to a qualified maintenance engineer;
- operate a lift truck when under the influence of alcohol or drugs;
- use mobile phones or other hand-held devices while operating or travelling.

2. Loads

Always:

- observe floor loading limits – find out the weight of the laden lift truck;
- ensure there is adequate clearance for the lift truck and load, including overhead;
- lower loads at a safe speed;
- use suitable attachments for lifting unusual or wide loads and follow the manufacturer's instructions;
- ensure you are properly trained, competent and authorised to operate the lift truck with the attachment being used.

Never:

- lift loads greater than the capacity of the lift truck;
- move a load that appears unsuitable or unstable (including on a damaged pallet);
- travel with a bulky load which blocks your view;
- travel with a raised load, unless the lift truck is designed for this;
- use an attachment unless a competent person, an authorised dealer or manufacturer has derated the lift truck (reduced the actual capacity).

3. Slopes

Always:

- travel slowly when going down slopes;
- when carrying a load, ensure the forks face uphill when travelling up or down slopes;
- without a load, ensure the forks face downhill when travelling up or down slopes;
- adjust the tilt (where fitted) to suit the gradient and raise the forks to clear the ground.

Never:

- attempt to turn on or travel across a ramp or a slope;
- leave a lift truck on a gradient except in an emergency, in which case always chock the wheels.

4. People

Always:

- use a safe work method when using working platforms.

Never:

- lift a person on the forks, or on a pallet, or similar, balanced on the forks;
- carry passengers, unless the lift truck is designed for this;
- allow people to walk under raised forks or loads;
- pick up a load if someone is standing close to it.

5. When you have finished working
Always:

- park the lift truck on level ground, never on a slope;
- leave the lift truck with the mast tilted forwards and the forks fully lowered, with the tips on the floor;
- apply the parking brake, select neutral, switch off the engine and remove the key;
- return keys or other activating devices to their place of safe-keeping.

Work equipment hazards and risk control

> **This chapter covers the following NEBOSH learning objectives:**
> 1. Outline general requirements for work equipment
> 2. Outline the hazards and control measures for hand-held tools, both powered and non-powered
> 3. Describe the main mechanical and non-mechanical hazards of machinery
> 4. Explain the main control measures for reducing risks from common construction machinery hazards

11.1 General requirements for work equipment

11.1.1 Introduction

This chapter covers the scope and main requirements for work equipment as covered by Parts II and III of the Provision and Use of Work Equipment Regulations (PUWER). The requirements for the supply of new machinery are also included. Summaries of PUWER and The Supply of Machinery (Safety) Regulations are given in Chapter 19. The safe use of hand-held tools, hand-held power tools and the proper safeguarding of a small range of machinery used in industry and commerce are included.

Any equipment used by an employee at work is generally covered by the term 'work equipment'. The scope is extremely wide and includes hand tools, power tools, ladders, photocopiers, laboratory apparatus, lifting equipment, fork-lift trucks and motor vehicles (which are not privately owned). Virtually anything used to do a job of work, including employees' own equipment, is covered. The uses covered include starting or stopping the equipment, repairing, modifying, maintaining, servicing, cleaning and transporting.

Employers and the self-employed (if on the prescribed list) must ensure that work equipment is suitable, maintained, inspected if necessary, provided with adequate information and instruction and only used by people who have received sufficient training.

Many serious accidents at work involve machinery. Hair or clothing can become entangled in moving parts, people can be struck by moving parts of machinery, parts of the body can be drawn into or trapped in machinery, and/or parts of the machinery or work tool can be ejected.

Many circumstances can increase the risks, including:

▶ not using the right equipment for the task, e.g. ladders instead of access towers for an extended task at high level;
▶ not fitting adequate controls on machines, or fitting the wrong type of controls, so that equipment cannot be stopped quickly and safely, or it starts accidentally;
▶ not guarding machines properly, leading to accidents caused by entanglement, shearing, crushing, trapping or cutting;
▶ not properly maintaining guards and other safety devices;
▶ not providing the right information, instruction and training;

▶ not fitting rollover protective structures (ROPS) and seat belts on mobile work equipment where there is a risk of rollover (*excluding* quad bikes);
▶ not maintaining work equipment or doing the regular inspections and thorough examinations;
▶ not providing (free) adequate personal protective equipment to use.

When identifying the risks, think about:

▶ the work being done during normal use of the equipment and also during setting-up, maintenance, cleaning and clearing blockages;
▶ which workers will use the equipment, including those who are inexperienced, have changed jobs or those who may have particular difficulties, e.g. those with language problems or impaired hearing;
▶ people who may act stupidly or carelessly or make mistakes;
▶ guards or safety devices that may be badly designed and difficult to use or are easy to defeat;
▶ other features of the equipment which could cause risks like vibration, electricity, wet or cold conditions.

Consider the following:

▶ Is the equipment suitable for the task?
▶ Are all the necessary safety devices fitted and in working order?
▶ Are there proper instructions for the equipment?
▶ Is the area around the machine safe and level with no obstructions?
▶ Has suitable lighting been provided?
▶ Has extraction ventilation been provided where required, e.g. on grinding and woodworking machinery?
▶ Has a risk assessment been done to establish a person's competence or training requirements to control particular machinery – this is very important for everyone;
▶ Are machine operators trained and do they have enough information, instruction, training?
▶ Are people adequately supervised?
▶ Are safety instructions and procedures being used and followed?
▶ Are machine operators using appropriate work clothing without loose sleeves, open jackets, dangling jewellery or sandals?
▶ Has the employer supplied all necessary special personal protective equipment (PPE)?
▶ Are safety guards or devices being used properly?

- Is maintenance carried out correctly and in a safe way?
- Are hand tools being used correctly and properly maintained and only used by people who have received sufficient training?

ISO 12100:2010 requires that protective measures to achieve the necessary risk reduction with work equipment is undertaken in the following three-step sequence:

1. Inherently safe design measures which eliminate hazards or reduce risks for people using the machinery;

2. The provision of safeguarding and/or complementary protective measures which take into account the use and reasonably foreseeable misuse of the equipment;

3. The provison of information for use which covers, operating procedures, recommended safe working practices, warning of residual risks and other information for the different phases of the life of the equipment; the description of any PPE required.

11.1.2 Suitability of work equipment and CE marking

Standards and requirements

When work equipment is provided it has to conform to standards which cover its supply as a new or second-hand piece of equipment and its use in the workplace. This involves:

- its initial integrity;
- the place where it will be used;
- the purpose for which it will be used.

There are two groups of law that deal with the provision of work equipment:

- One deals with what manufacturers and suppliers have to do. This can be called the **'supply'** law. One of the most common is the Supply of Machinery (Safety) Regulations 2008 as amended, which requires manufacturers and suppliers to ensure that machinery is safe when supplied and has CE marking. Its primary purpose is to prevent barriers to trade across the EU by ensuring uniform standards, rather than protecting people at work.

- The other deals with what the users of machinery and other work equipment have to do. This can be called the **'user'** law, PUWER 98, and applies to most pieces of work equipment. Its primary purpose is to protect people at work.

Under 'user' law employers have to provide safe equipment of the correct type, ensure that it is correctly used and maintain it in a safe condition. When buying new equipment, the 'user' has to check that the equipment complies with all the 'supply' law that is relevant. Figure 11.1(b) shows the division of responsibility between designers, manufacturers and suppliers, and those of employers. However, whatever hazards have or have not been controlled effectively **the user must check that the machine is safe before it is used**.

Most new equipment, including machinery in particular, should have 'CE' marking when purchased (Figure 11.1(a)). 'CE' marking is only a claim by the manufacturer that the equipment is safe and that they have met relevant supply law. If this is done properly manufacturers will have to:

- find out about the health and safety hazards (trapping, noise, crushing, electrical shock, dust, vibration, etc.) that are likely to be present when the machine is used;
- assess the likely risks;
- design out the hazards that result in risks; or, if that is not possible
  - provide safeguards (e.g. guarding dangerous parts of the machine, providing noise enclosures for noisy parts); or, if that is not possible
  - use warning signs on the machine to warn of hazards that cannot be designed out or safeguarded (e.g. 'noisy machine' signs).

Figure 11.1 (a) This is the CE marking

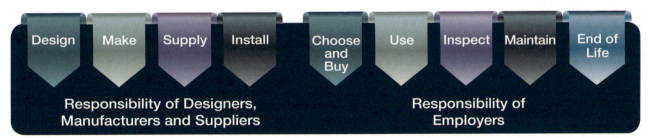

Figure 11.1 (b) Division of responsibility for the safety of machinery

Manufacturers also have to:

► keep information, explaining what they have done and why, in a technical file.

► fix CE marking to the machine where necessary, to show that they have complied with all the relevant supply laws;

► issue a 'Declaration of Conformity' for the machine (see Figure 11.2). This is a statement that the machine complies with the relevant essential health and safety requirements or with the example that underwent type-examination. A declaration of conformity must:

  ▷ state the name and address of the manufacturer or importer into the EU;

  ▷ contain a description of the machine, and its make, type and serial number;

  ▷ indicate all relevant European Directives with which the machinery complies;

  ▷ state details of any notified body that has been involved;

  ▷ specify which standards have been used in the manufacture (if any);

  ▷ be signed by a person with authority to do so;

  ▷ provide the buyer with instructions to explain how to install, use and maintain the machinery safely.

EC-Declaration of Conformity

MACHINERY DIRECTIVE

$C \in$

DC720, DC721, DC722, DC725, DC727, DC730, DC731, DC732, DC735, DC737, DC742, DC743, DC745

DEWALT declares that these products described under "technical data" are in compliance with: 98/37/EC (until Dec. 28, 2009), 2006/42/EC (from Dec. 29, 2009), EN 60745-1, EN 60745-2-1, EN 60745-2-2.

These products also comply with Directive 2004/108/EC. For more information, please contact DEWALT at the following address or refer to the back of the manual.

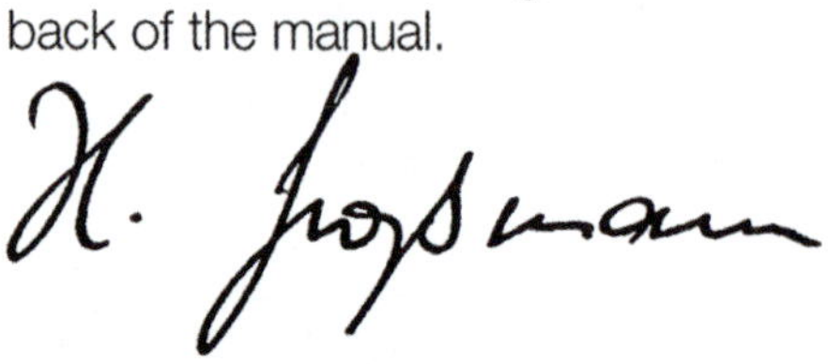

Horst Grossmann
Vice President Engineering and Product Development
DEWALT, Richard-Klinger-Strase 11,
D-65510, Idstein, Germany
30.07.2009

Figure 11.2 Typical Certificate of Conformity

This can help in deciding which equipment may be suitable, particularly if buying a standard piece of equipment 'off the shelf'.

If buying a more complex or custom-built machine the buyer should discuss their requirements with potential suppliers. For a custom-built piece of equipment, there is the opportunity to work with the supplier to design out the causes of injury and ill health. Time spent now on agreeing the necessary safeguards, to control health and safety risks, could save time and money later.

Note: Sometimes equipment is supplied via another organisation, for example an importer, rather than direct from the manufacturer, so this other organisation is referred to as the supplier. It is important to realise that the supplier may not be the manufacturer.

When the equipment has been supplied the buyer should look for CE marking, check for a copy of the Declaration of Conformity and that there is a set of instructions in English on how the machine should be used and, most important of all, **check to see if they think that it is safe**.

Before buying new equipment the buyer will need to think about:

► where and how it will be used;
► what it will be used for;
► who will use it (skilled employees, trainees);
► what risks to health and safety might result;
► how well health and safety risks are controlled by different manufacturers.

Limitations of CE marking

CE marking is not a guarantee that the machine is safe. It is a claim by the manufacturer that the machinery complies with the law. CE marking has many advantages if done properly, for example:

► it allows a common standard across Europe;
► it provides a means of selling to all European Union (EU) member states without barriers to trade;
► it ensures that instructions and safety information are supplied in a fairly standard way in most languages in the EU;
► it has encouraged the use of diagrams and pictorials which are common to all languages;
► it allows for independent type-examination for some machinery like woodworking machinery which has not been made to an EU-harmonised standard – identified by an EN marking before the standard number (e.g. BS EN…).

Clearly there are disadvantages as well, for example:

► Instruction manuals have become very long. Sometimes two volumes are provided, because of the number of languages required.
► Translations can be very poor and disguise the proper meaning of the instruction.

- Manufacturers can fraudulently put on the CE marking.
- Manufacturers might make mistakes in claiming conformity with safety laws.

11.1.3 Prevention of access to dangerous parts of machinery

Regulation 11(1) of PUWER requires employers to take effective measures to prevent access to dangerous parts of machinery or stop their movement before any part of a person enters a danger zone. This regulation also applies to contact with a rotating stock-bar which projects beyond the headstock of a lathe.

The term 'dangerous part' has been established in health and safety law through judicial decisions. In practice, this means that if a piece of work equipment could cause injury if it is being used in a foreseeable way, it can be considered a dangerous part.

There are many HSE publications which are specific to a machine or industry, for example *Safe Use of Woodworking Machinery Approved Code of Practice L114*. They describe the measures that can be taken to protect against risks associated with dangerous parts of machinery. Current national, European and international standards may also be used for guidance, such as ISO 12100:2010 Safety of Machinery.

The risk assessment carried out under Regulation 3 of the Management Regulations should identify hazards presented by machinery. The risk assessment should evaluate the nature of the injury, its severity and likelihood of occurrence for each hazard identified. This will enable employers to decide whether the level of risk is acceptable or if risk reduction measures are needed. In most cases the objective of risk reduction measures is to prevent contact of part of the body or clothing with any dangerous part of the machine, for example, by providing safeguards.

Regulation 11(2) of PUWER specifies the measures which should be taken to prevent access to the dangerous parts of the machinery and achieve compliance with Regulation 11(1). The measures are ranked in the order they should be implemented, where practicable, to achieve an adequate level of protection. The levels of protection are:

(a) fixed enclosing guards;
(b) other guards or protection devices such as interlocked guards and pressure mats;
(c) protection appliances such as jigs, holders and push sticks, etc; and
(d) the provision of information, instruction, training and supervision.

The purpose of the risk assessment is to identify measures that you can take to reduce the risks that the hazards present. When selecting measures employers should consider each level of protection from the first level of the scale listed above and they should use measures from that level so far as it is practicable to

do so, provided that they contribute to the reduction of risk. It may be necessary to select a combination of measures. The selection process should continue down the scale until the combined measures are effective in reducing the risks to an acceptable level thus meeting the requirements of Regulation 11(1). In selecting the appropriate combination employers will need to take account of:

- the requirements of the work;
- evaluation of the risks; and
- the technical features of possible safeguarding solutions.

Most machinery will present more than one mechanical hazard, and employers will need to deal with the risks associated with all of these. For example, at belt conveyors there is a risk of entanglement with the rotating shafts and of being trapped by the intake between drum and moving belt – so appropriate safety measures should be adopted.

Any risk assessment carried out under Regulation 3 of the Management Regulations should not just deal with the machine when it is operating normally, but must also cover activities such as setting, maintenance, cleaning or repair. The assessment may indicate that these activities require a different combination of protective measures from those appropriate to the machine doing its normal work. In particular, parts of machinery that are not dangerous in normal use because they are not then accessible may become accessible and therefore dangerous while this type of work is being carried out.

Certain setting or adjustment operations which may have to be done with the machine running may require a greater reliance on the provision of information, instruction, training and supervision than for normal use.

Regulation 11(3) of PUWER sets out various requirements for guards and protection devices which are covered later in this chapter. These are largely detailed in relevant national, European and international standards such as ISO/EN 12100:2010 Safety of Machinery.

Chapter 19 has a full summary of PUWER.

11.1.4 Use and maintenance of equipment with specific risks

Some pieces of work equipment involve specific risks to health and safety where it is not possible to control adequately the hazards by physical measures alone, for example the use of a bench-mounted circular saw or an abrasive wheel (Figure 11.3). In all cases the hierarchy of controls should be adopted to reduce the risks by:

- eliminating the risks; or, if this is not possible
- taking physical measures to control the risks such as guards; but if the risks cannot be adequately controlled

11

▶ taking appropriate software measures, such as a safe system of work.

PUWER 98 Regulation 7 restricts the use of such equipment to the persons designated to use it. These people need to have received sufficient information, instruction and training so that they can carry out the work using the equipment safely.

Figure 11.3 Using a bench-mounted abrasive wheel

Repairs, modifications, maintenance or servicing is also restricted to designated persons. A designated person may be the operator if he/she has the necessary skills and has received specific instruction and training. Another person specifically trained to carry out a particular maintenance task, for example dressing an abrasive wheel, may not be the operator but may be designated to do this type of servicing task on a range of machines.

Protection against other hazards associated with machinery is dealt with in Regulations 12 and 13. However, the measures required by this regulation may also protect or help to protect against those other hazards such as ejected particles and heat.

11.1.5 Information, instruction and training for specific risks

People using and maintaining work equipment, where there are residual risks that cannot be sufficiently reduced by physical means, require enough information, instruction and training to operate safely. Managers and supervisors will also need to be provided with sufficient information about a piece of equipment to enable them to fulfil their responsibilities towards people using and maintaining equipment.

The information and instructions are likely to come from the manufacturer in the form of operating and maintenance manuals. It is up to the employer to ensure that what is provided is easily understood, and set out logically with illustrations and standard symbols where appropriate. The information should normally

be in good plain English but other languages may be necessary in some cases.

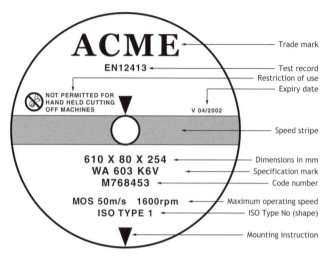

Figure 11.4 British Standard system for specifying abrasive wheels from BS EN 12413:1999 and BS ISO 525:1999

The extent of the information and instructions will depend on the complexity of the equipment and the specific risks associated with its use. For general training requirements see Section 11.6.4 (e) and (f). Examples of specific requirements for woodworking and abrasive wheels are given in the boxes.

Employers should ensure that information and instructions provided on the use of woodworking machinery includes, where relevant:

(a) the speed, range, type and dimensions of tools suitable for the machine;

(b) any limitation on the cutting speeds of the machine, particular operations or size and material of any work-piece;

(c) procedures relating to the repair or replacement of any guard or protection device;

(d) the availability, suitability and use of any additional protection device or protection appliance;

(e) the correct procedures to be followed for setting and adjusting operations;

(f) safe methods of handling tools;

(g) correct procedures for start-up and shutdown, isolation and how to discharge any residual energy;

(h) procedures for cleaning saw blades by hand (which should be carried out with the machine isolated and the blade stopped);

(i) procedures for adjusting any guard, tool, clamp or other part of a machine (which should not be carried out while any part of the machine is in motion, unless they can be done safely).

Training of someone to use a grinding machine should cover the proper methods of mounting an abrasive wheel, see *Safety in the use of abrasive wheels* HSG17, 3rd Edition, HSE 2000. They are summarised in the box.

> **Summary of mounting procedures for an abrasive wheel:**
>
> 1. Wheel mounting should be carried out only by an appropriately trained person. A wheel should be mounted only on the machine for which it was intended. Before mounting, all wheels should be closely inspected to ensure that they have not been damaged in storage or transit.
> 2. The speed marked on the machine should not exceed the speed marked on the wheel, blotter or identification label.
> 3. The bush, if any, should not project beyond the sides of the wheel and blotters. The wheel should fit freely but not loosely on the spindle.
> 4. Flanges should not be smaller than their specified minimum diameter, and their bearing surfaces should be true and free from burrs.
> 5. With the exception of the single flange used with threaded-hole wheels, all flanges should be properly recessed or undercut.
> 6. Flanges should be of equal diameter and have equal bearing surfaces. Protection flanges should have the same degree of taper as the wheel. Blotters, slightly larger than the flanges, should be used with all abrasive wheels except those listed below in a–g. Wrinkles in blotters should be avoided.
>
>   Blotters should not be used with the following types of wheels:
>   (a) mounted wheels and points;
>   (b) abrasive discs (inserted nut discs and cylinders);
>   (c) plate-mounted wheels;
>   (d) cylinder wheels mounted in chucks;
>   (e) rubber-bonded cutting-off wheels 0.5 mm or less in thickness;
>   (f) taper-sided wheels;
>   (g) wheels with threaded inserts.
> 7. Wheels, blotters and flanges should be free from foreign matter. Clamping nuts should be tightened only sufficiently to hold the wheel firmly. When the flanges are clamped by a series of screws they should be tightened uniformly in a criss-cross sequence. Screws for inserted nut mounting of discs, cylinders and cones should be long enough to engage a sufficient length of thread, but not so long that they contact the abrasive.
> 8. When mounting the wheels and points, the overhang appropriate to the speed, diameter of the mandrel and size of the wheel should not be exceeded, and there should be sufficient length of mandrel in the collet or chuck.

The training and supervision of young persons is particularly important because of their relative immaturity, unfamiliarity with a working environment and lack of awareness of existing or potential risks. Some Approved Codes of Practice, for example on the Safe Use of Woodworking Machinery, L114, 2nd Edition, HSE 2014, restricts the use of high-risk machinery. Only young persons with sufficient maturity and competence who have finished their training may use the equipment unsupervised.

11.1.6 Maintenance and inspection

Maintenance

Work equipment needs to be properly maintained so that it continues to operate safely and in the way it was designed to perform. The amount of maintenance will be stipulated in the manufacturer's instructions and will depend on the amount of use, the working environment and the type of equipment. High-speed, high-risk machines, which are heavily used in an adverse environment like salt water, may require very frequent maintenance, whereas a simple hand tool, like a shovel, may require very little.

Maintenance management schemes can be based around a number of techniques designed to focus on those parts which deteriorate and need to be maintained to prevent health and safety risks. These techniques include the following:

▶ **Preventative planned maintenance** – this involves replacing parts and consumables or making necessary adjustments at preset intervals normally set by the manufacturer, so that there are no hazards created by component deterioration or failure. Vehicles are normally maintained on this basis.
▶ **Condition based maintenance** – this involves monitoring the condition of critical parts and carrying out maintenance whenever necessary to avoid hazards which could otherwise occur.
▶ **Breakdown based maintenance** – here maintenance is only carried out when faults or failures have occurred. This is only acceptable if the failure does not present an immediate hazard and can be corrected before the risk is increased. If, for example, a bearing overheating can be detected by a monitoring device, it is acceptable to wait for the overheating to occur as long as the equipment can be stopped and repairs carried out before the fault becomes dangerous to persons employed.

In the context of health and safety, maintenance is not concerned with operational efficiency but only with avoiding risks to people. It is essential to ensure that maintenance work can be carried out safely. This will involve the following:

11

▶ competent well-trained maintenance people;

▶ the equipment being made safe for the maintenance work to be carried out. In many cases, the normal safeguards for operating the equipment may not be sufficient as maintenance sometimes involves going inside guards to observe and subsequently adjust, lubricate or repair the equipment. Careful design allowing adjustments, lubrication and observation from outside the guards, for example, can often eliminate the hazard. Making equipment safe will usually involve disconnecting the power supply and then preventing anything moving, falling or starting during the work. It may also involve waiting for equipment to cool or warm up to room temperature;

▶ a safe system of work being used to carry out the necessary procedures to make and keep the equipment safe and perform the maintenance tasks. This can often involve a formal 'permit-to-work' scheme to ensure that the correct sequence of safety critical tasks has been performed and all necessary precautions taken;

▶ correct tools and safety equipment being available to perform the maintenance work without risks to people. For example special lighting or ventilation may be required.

Inspection under PUWER

Complex equipment and/or high-risk equipment will probably need a maintenance log and may require a more rigid inspection regime to ensure continued safe operation. This is covered by PUWER 98, Regulation 6.

PUWER requires, where safety is dependent on the installation conditions and/or the work equipment is exposed to conditions causing deterioration, which may result in a significant risk and a dangerous situation developing, that the equipment is inspected by a competent person. In this case the competent person would normally be an employee, but there might be circumstances where an outside competent person would be used.

The inspection must be done:

▶ after installation for the first time;

▶ after assembly at a new site or in a new location and thereafter;

▶ at suitable intervals;

▶ each time exceptional circumstances occur which could affect safety.

The inspection under PUWER will vary from a simple visual inspection to a detailed comprehensive inspection, which may include some dismantling and/or testing. The level of inspection required would normally be less rigorous and intrusive than thorough examinations under the Lifting Operations and Lifting Equipment Regulations 1998 (LOLER) for certain lifting equipment (see chapters 7 and 16). In the case of boilers and air receivers the inspection is covered by the Pressure Systems Safety Regulations 2000, which involves a thorough examination (see the summary in Chapter 19). The inspection under PUWER would only be needed in these cases if the examinations did not cover all the significant health and safety risks that are likely to arise.

Examination of boilers and air receivers

Under the Pressure Systems Safety Regulations 2000, a wide range of pressure vessels and systems require thorough examination by a competent person to an agreed specifically written scheme. This includes steam boilers, pressurised hot water plants and air receivers, although these are no longer covered in the Certificate syllabus, but often encountered in small and medium-sized workplaces (Figure 11.5).

The Regulations place duties on designers and manufacturers but this section is only concerned with the duty on users to have the vessels examined. An employer who operates a steam boiler and/or a

(a)

(b)

Figure 11.5 (a) Typical diesel-powered compressor with air receiver and pneumatic chisel; (b) typical electrically powered compressor with air receiver

pressurised hot water plant and/or an air receiver must ensure:

- that it is supplied with the correct written information and markings;
- that the equipment is properly installed;
- that it is used within its operating limits;
- there is a written scheme for periodic examination of the equipment certified by a competent person (in the case of standard steam boilers and air receivers the scheme is likely to be provided by the manufacturer);
- that the equipment is examined in accordance with the written scheme by a competent person within the specified period;
- that a report of the periodic examination is held on file giving the required particulars;
- that the actions required by the report are carried out;
- that any other safety critical maintenance work is carried out, whether or not covered by the report.

In these cases the competent person is usually a specialist inspector from an external inspection organisation. Many of these organisations are linked to the insurance companies that cover the financial risks of use of the pressure vessel.

11.1.7 Operation and working environment

To operate work equipment safely it must be fitted with easily reached and operated controls, kept stable, properly lit, kept clear and provided with adequate markings and warning signs. These are covered by PUWER 98, which applies to all types of work equipment.

Controls

Equipment should be provided with efficient means of:

- starting or making a significant change in operating conditions;
- stopping in normal circumstances;
- emergency stopping as necessary to prevent danger.

All controls should be well positioned, clearly visible and identifiable, so that it is easy for the operator to know what each control does. Markings should be clearly visible and remain so under the conditions met at the workplace. See Figure 11.6 for more information on controls.

Start controls

It should only be possible to start the work equipment by using the designed start control. Equipment may well have a start sequence which is electronically controlled to meet certain conditions before starting can be achieved, for example preheating a diesel engine, or purge cycle for gas-fed equipment. Restarting after

Equipment controls should:
- be easily reached from the operating positions
- not permit accidental starting of equipment
- move in the same direction as the motion being controlled
- vary on mode, shape and direction of movement to prevent inadvertent operation of the wrong control
- incorporate adequate red stop buttons
- incorporate adequate red emergency stop buttons, of the mushroom headed type with lock-off
- have shrouded or sunk green start buttons to prevent accidental starting of the equipment
- be clearly marked to show what they do.

Figure 11.6 Equipment controls – design features

a stoppage will require the same sequence to be performed.

Stoppage may have been deliberate or as a result of opening an interlocked guard or tripping a switch accidentally. In most cases it should not be possible to restart the equipment simply by shutting the guard or resetting the trip. Operation of the start control should be required.

Any other change to the operating conditions such as speed, pressure or temperature should only be done by using a control designed for the purpose.

Stop controls

The action of normal stopping controls should bring the equipment to a safe condition in a safe manner. In some cases immediate stopping may cause other risks to occur. The stop controls do not have to be instantaneous and can bring the equipment to rest in a safe sequence or at the end of an operating cycle. It is only the parts necessary for safety, that is accessible dangerous parts, which have to be stopped. So, for example, suitably guarded cooling fans may need to run continuously and be left on.

In some cases where there is, for example, stored energy in hydraulic systems, it may be necessary to insert physical scotches to prevent movement and/or to exhaust residual hydraulic pressure. These should be incorporated into the stopping cycle, which should be designed to dissipate or isolate all stored energy to prevent danger.

It should not be possible to reach dangerous parts of the equipment until it has come to a safe condition, e.g. stopped, cooled, electrically safe.

Emergency stop controls

Emergency stop must be provided where the other safeguards in place are not sufficient to prevent danger to operatives and any other persons who may be affected. Where appropriate, there should be an emergency stop at each control point and at other locations around the equipment so that action can be taken quickly. Emergency stops should bring the equipment to a halt rapidly but this should be controlled where necessary so as not to create any additional hazards. Crash shutdowns of complex systems have to be carefully designed to optimise safety without causing additional risks.

Emergency stops are not a substitute for effective guarding of dangerous parts of equipment and should not be used for normal stopping of the equipment.

Emergency stop buttons should be easily identified, reached and operated. Common types are mushroom-headed buttons, bars, levers, kick-plates or pressure-sensitive cables. They are normally red and should need to be reset after use. With stop buttons this is either by twisting or a security key (Figure 11.7.)

Mobile work equipment is normally provided with effective means of stopping the engine or power source. In some cases large equipment may need emergency stop controls away from the operator's position.

Figure 11.7 Emergency stop button

Isolation of equipment

Equipment should be provided with efficient means of isolating it from all sources of energy. The purpose is to make the equipment safe under particular conditions, for example when maintenance is to be carried out or where adverse weather conditions may make it unsafe to use. On static equipment isolation will usually be for mains electrical energy; however, in some cases there may be additional or alternative sources of energy. The isolation should cover all sources of energy such as diesel and petrol engines, LPG, steam, compressed air, hydraulics, batteries or heat. In some cases special consideration is necessary where, for example, hydraulic pumps are switched off, so as not to allow heavy pieces of equipment to fall under gravity. An example would be the loading shovel on an excavator.

Stability

Stability is important and is normally achieved by bolting equipment in place or, if this is not possible, by using clamps. Some equipment can be tied down, counterbalanced or weighted, so that it remains stable under all operating conditions. If portable equipment is weighted or counterbalanced, it should be reappraised when the equipment is moved to another position. If outriggers are needed for stability in certain conditions, for example to stabilise mobile access towers, they should be employed whenever conditions warrant the additional support. In severe weather conditions it may be necessary to stop using the equipment or reappraise the situation to ensure stability is maintained.

The quality of general and local lighting will need to be considered to ensure the safe operation of the equipment. The level of lighting and its position relative to the working area are often critical to the safe use of work equipment. Poor levels of lighting, glare and shadows can be dangerous when operating equipment. Some types of lighting, for example sodium lights, can change the colour of equipment, which may increase the level of risk. This is particularly important if the colour coding of pipe work or cables is essential for safety.

Markings

Markings on equipment must be clearly visible and durable. They should follow international conventions for some hazards like radiation and lasers and, as far as possible, conform to the Health and Safety (Safety Signs and Signals) Regulations (see Chapter 19 for a summary). The contents, or the hazards of the contents, as well as controls, will need to be marked on some equipment. Warnings or warning devices are required in some cases to alert operators or people nearby to any dangers, for example 'wear hard hats', a flashing light on an airport vehicle or a reversing horn on a truck.

11.1.8 User responsibilities

The responsibilities for users of work equipment are covered in Section 7 of the Health and Safety at Work Act (HSW Act) and Regulation 14 of the Management Regulations. Section 7 requires employees to:

▶ take reasonable care for themselves and others who may be affected; and
▶ cooperate with the employer.

In addition Section 8 of the HSW Act requires employees not to misuse or interfere with safety provisions.

Regulation 14 of the MHSW Regulations requires employees to use equipment properly in accordance with instructions and training. They must also inform employers of dangerous situations and shortcomings in protection arrangements. The self-employed who are users have similar responsibilities to employers and employees combined.

Where employees provide their own tools the employer still has a responsibility to ensure that work equipment is suitable.

It is not always possible to eliminate every hazard or design safeguards that protect people against all machinery hazards, particularly during commissioning, setting, adjustment, cleaning and maintenance. Safe working practices, in some cases even permits to work, need to be adopted and followed by machinery operators or maintenance staff. These should be considered at the machine design and installation stages as the production of special jigs, fixtures, controls and isolation facilities may be needed.

Where mechanical hazards cannot be avoided there are many precautions which should be observed by properly trained and supervised people, these include:

► keeping the area around the machine clear and free of obstructions,
► wearing suitable clothing and footwear that does not have loose ends which could become entangled;
► avoiding the use of neckties, rings, necklaces and other jewellery;
► appropriate eye protection where there is a risk of particles being ejected;
► using precautions against kickback of work-pieces such as riving knives on circular saws, work rests on abrasive wheels, cutting speeds which are correct for the task and cutting tool;
► taking precautions against bursting of abrasive wheels by ensuring they are not damaged and are running within their design speed;
► limiting closeness of approach to machines like overhead cranes, or removing work from the rear of circular saws;
► using manual handling devices such as tongs for hot steel handling or push sticks for pushing timber through a circular saw;
► using jigs or holders for work-pieces.

11.2 Hazards and controls for hand-held tools

11.2.1 Introduction

Work equipment includes hand-held tools whether powered or not. This section deals with non-powered hand tools. These tools need to be correct for the task, well maintained and properly used by trained people.

Five basic safety rules can help prevent hazards associated with the use of hand-held tools:

► Keep all tools in good condition with regular maintenance.
► Use the right tool for the job.
► Examine each tool for damage before use and do not use damaged tools.

► Use tools according to the manufacturer's instructions.
► Provide and use properly the right personal protective equipment (PPE).

11.2.2 Hazards of non-powered hand tools

Hazards from the misuse or poor maintenance of non-powered hand tools (Figure 11.8) include:

► broken handles on files/chisels/screwdrivers/hammers which can cause cut hands or hammer heads to fly off;
► incorrect use of knives, saws and chisels with hands getting injured in the path of the cutting edge;
► tools that slip causing stab wounds;
► poor-quality uncomfortable handles that damage hands;
► splayed spanners that slip and damage hands or faces;
► chipped or loose hammer heads that fly off or slip;
► incorrectly sharpened or blunt chisels or scissors that slip and cut hands. Dull tools can cause more hazards than sharp ones. Cracked saw blades must be removed from service;
► flying particles that damage eyes from breaking up stone or concrete;
► electrocution or burns by using incorrect or damaged tools for electrical work;
► use of poorly insulated tools for hot work in the catering or food industry;
► use of pipes or similar equipment as extension handles for a spanner which is likely to slip causing hand or face injury;
► mushroom-headed chisels or drifts which can damage hands or cause hammers (not suitable for chisels) and mallets to slip;
► use of spark-producing or percussion tools in flammable atmospheres;
► painful wrists and arms (upper limb disorders) from the frequent twisting from using screwdrivers;
► when using saw blades, knives or other tools, they should be directed away from aisle areas and away from other people working in close proximity.

Figure 11.8 (a) broken and dangerous wood chisel handle

Figure 11.8 (b) range of non-powered hand tools

11.2.3 Non-powered hand tools safety considerations

Use of non-powered hand tools should be properly controlled including those tools owned by employees. The following controls are important:

(a) **Suitability** – All tools should be suitable for the purpose and location in which they are to be used. Using the correct tool for the job is the first step in safe hand tool use. Tools are designed for specific needs. That is why screwdrivers have various lengths and tip styles and pliers have different head shapes. Using any tool inappropriately is a step in the wrong direction. To avoid personal injury and tool damage, select the proper tool to do the job well and safely.

(b) **Inspection** – All tools should be maintained in a safe and proper condition. This can be achieved through:
   ▷ the regular inspection of hand tools;
   ▷ discarding or prompt repair of defective tools;
   ▷ taking time to keep tools in the proper condition and ready for use;
   ▷ proper storage to prevent damage and corrosion;
   ▷ locking tools away when not in use to prevent them being used by unauthorised people.

(c) **Training** – All users of non-powered hand tools should be properly trained in their use. This may well have been done through apprenticeships and similar training. This will be particularly important with specialist working conditions or work involving young people.

(d) **Use well-designed, high-quality tools** – Finally, investing in high-quality tools makes the professional's job safer and easier:

   ▷ If extra leverage is needed, use high-leverage pliers, which give more cutting and gripping power than standard pliers. This helps, in particular, when making repetitive cuts or twisting numerous wire pairs.
   ▷ Serrated jaws provide sure gripping action when pulling or twisting wires.
   ▷ Some side-cutting and diagonal-cutting pliers are designed for heavy-duty cutting. When cutting screws, nails and hardened wire, only use pliers that are recommended for that use.
   ▷ Pliers with hot riveting at the joint ensure smooth movement across the full action range of the pliers, which reduces handle wobble, resulting in a positive cut. The knives align perfectly every time.
   ▷ Induction hardening on the cutting knives adds to long life, so the pliers cut cleanly day after day.
   ▷ Sharp cutting knives and tempered handles also contribute to cutting ease.
   ▷ Some pliers are designed to perform special functions. For example some high-leverage pliers have features that allow crimping connectors and pulling fish tapes.
   ▷ Tool handles with dual-moulded material allow for a softer, more comfortable grip on the outer surface and a harder, more durable grip on the inner surface and handle ends.
   ▷ Well-designed tools often include a contoured thumb area for a firmer grip or colour-coded handles for easy tool identification.
   ▷ Insulated tools reduce the chance of injury where the tool may make contact with an energised source.

Well-designed tools are a pleasure to use. They save time, give professional results and help to do the job more safely.

11.2.4 Hand-held portable power tools

Introduction

The electrical hazards of portable hand-held tools and portable appliance testing (PAT) are covered in more detail in Chapter 12. This section deals mainly with other physical hazards and safeguards relating to hand-held power tools (Figure 11.9).

The section covers, in particular, electric drills, sanders and chainsaws, which are commonly used in the workplace.

General hazards of hand-held power tools

The general hazards involve:

▶ mechanical entanglement in rotating spindles or sanding discs;
▶ waste material flying out of the cutting area;

Figure 11.9 Range of hand-held portable power tools

- coming into contact with the cutting blades or drill bits;
- risk of hitting electrical, gas or water services when drilling into building surfaces;
- electrocution/electric shock from poorly maintained equipment and cables or cutting the electrical cable;
- manual handling problem with a risk of injury if the tool is heavy or very powerful;
- hand–arm vibration, especially with pneumatic drills and chainsaws, disc-cutters and petrol-driven units;
- tripping hazard from trailing cables, hoses or power supplies;
- eye hazard from flying particles;
- injury from poorly secured or clamped work-pieces;
- fire and explosion hazard with petrol-driven tools or when used near flammable liquids, explosive dusts or gases;
- high noise levels with pneumatic chisels, planes and saws in particular (see Chapter 15);
- levels of dust and fumes given off during the use of the tools (but see Chapter 14).

Typical safety controls and instructions

Guarding

The exposed moving parts of hand-held power tools need to be safeguarded. Belts, gears, shafts, pulleys, sprockets, spindles, drums, flywheels, chains or other reciprocating, rotating or moving parts of equipment must be guarded. Machine guards, as appropriate, must be provided to protect the operator and others from the following:

- point of operation;
- in-running nip points;
- rotating parts;
- flying chips and sparks.

Safety guards must never be removed when a tool is being used. For example portable circular saws must be equipped at all times with guards. An upper guard must cover the entire blade of the saw. A retractable lower guard must cover the teeth of the saw, except where it makes contact with the work material. The lower guard must automatically return to the covering position when the tool is withdrawn from the work material.

Operating controls and switches

Most hand-held power tools should be equipped with a constant-pressure switch or control that shuts off the power when pressure is released. On/off switches should be easily accessible without removing hands from the equipment.

Handles should be designed to protect operators from excessive vibration and keep their hands away from danger areas. In some cases handles are also designed to activate a brake of the cutting chain or blade, for example in a chainsaw. Equipment should be designed to reduce lifting and manual handling problems, with special harnesses being used as necessary, for example when using large strimmers or brush cutters.

Means of starting engines and holding equipment should be designed to minimise any musculoskeletal problems.

Safe operations/instructions

When using power tools, the following basic safety measures should be observed to protect against electrical shock, personal injury, ill-health and risk of fire. See also more detailed electrical precautions in Chapter 12. Operators should read these instructions before using the equipment. They must:

- maintain a clean and tidy working area that is well lit and clear of obstructions;
- never expose power tools to rain. Do not use power tools in damp or wet surroundings unless especially designed and protected for this purpose;
- not use power tools in the vicinity of combustible fluids, dusts or gases unless they are specially protected and certified for use in these areas;
- protect against electric shock (if tools are electrically powered) by avoiding body contact with grounded objects such as pipes, scaffolds and metal ladders (see Chapter 12 for more electrical safety precautions);
- keep children away;
- not let other persons handle the tool or the cable. Keep them away from the working area;
- store tools in a safe place when not in use where they are in a dry, locked area which is inaccessible to children;
- not overload tools as they operate better and safer in the performance range for which they were intended;
- use the right tool. Do not use small tools or attachments for heavy work. Do not use tools for purposes and tasks for which they were not intended; for example, do not use a hand-held circular saw to cut down trees or cut off branches;
- wear suitable work clothes. Do not wear loose-fitting clothing or jewellery. They can get entangled in moving parts. For outdoor work, rubber gloves and non-skid footwear are recommended. Long hair should be protected with a hair net;

11

- use safety glasses;
- also use a suitable filtering respirator mask or preferably carry out the work in a suitable work area fitted with exhaust ventilation for work that generates dust;
- not abuse the power cable;
- not carry the tool by the power cable and do not use the cable to pull the plug out of the power socket. Protect the cable from heat, oil and sharp edges;
- secure the work-piece. Use clamps or a vice to hold the work-piece. It is safer than using hands and it frees both hands for operating the tool;
- not over-reach the work area. Avoid abnormal body postures. Maintain a safe stance and maintain a proper balance at all times;
- maintain tools with care. Keep your tools clean and sharp for efficient and safe work. Follow the maintenance Regulations and instructions for the changing of tools. Check the plug and cable regularly and, in the case of damage, have them repaired by a qualified service engineer. Also inspect extension cables regularly and replace if damaged;
- keep the handle dry and free of oil or grease;
- disconnect the power plug when not in use, before servicing and when changing the tool parts, that is blade, bits, cutter, sanding disc, etc.;
- not forget to remove key. Check before switching on that the key and any tools for adjustment are removed;
- avoid unintentional switch-on. Do not carry tools that are connected to power with your finger on the power switch. Check that the switch is turned off before connecting the power cable;
- outdoors, use extension cables. When working outdoors, use only extension cables which are intended for such use and marked accordingly;
- stay alert, keep eyes on the work. Use common sense. Do not operate tools when there are significant distractions;
- check the equipment for damage. Before further use of a tool, check carefully the protection devices or lightly damaged parts for proper operation and performance of their intended functions. Check movable parts for proper function, for whether there is binding or for damaged parts. All parts must be correctly mounted and meet all conditions necessary to ensure proper operation of the equipment;
- ensure damaged protection devices and parts are repaired or replaced by a competent service centre unless otherwise stated in the operating instructions. Damaged switches must be replaced by a competent service centre. Do not use any tool which cannot be turned on and off with the switch;
- only use accessories and attachments that are described in the operating instructions or are provided or recommended by the tool manufacturer. The use of tools other than those described in the operating instructions or in the catalogue of

recommended tool inserts or accessories can result in a risk of personal injury;
- use engine-driven power tools in well-ventilated areas. Store petrol in a safe place in approved storage cans. Stop and let engines cool before refuelling.

Specific hazards and control measures for specified hand-held power tools

The following hand-held power tools have been put in the Construction Certificate syllabus: pneumatic drills/chisels, electric drills, disc-cutters/cut-off saws, sanders, cartridge and pneumatic nail guns and chainsaws, all of which are commonly used on construction sites. The hazards and safety control measures in addition to the general ones covered earlier are set out for each type of equipment.

(a) Pneumatic drills/chisels

Hazards include (Figure 11.10):

- entanglement, particularly of loose clothing or long hair in rotating drill bits;
- accidental disconnection of the air line;
- fracture of air line, particularly when internal diameter exceeds 12 mm;
- accidental expulsion of attachments, particularly on chisels;
- high noise levels from the machine or air compressor;
- eye injury from flying particles and chips, particularly from chisels;
- injury from poorly secured work-pieces;
- electric shock from drilling into a live hidden cable;
- hand–arm vibration hazard in hammer mode;
- dust given off from material being worked on;
- tripping hazard from trailing air lines;
- upper limb disorder from powerful machines with a strong torque, particularly if they jam and kick back;
- foot injury hazards from dropping heavy units onto unprotected feet;
- manual handling hazards, particularly with heavy machines and intensive use;
- hazards associated with a portable air supply compressor such as refuelling, noise, fumes, and dangerous moving parts of the engine, e.g. the fan and belts;
- explosion risk if sparks are given off from chisels near flammable liquids and explosive dusts and gases;
- using equipment in poor weather conditions with wet, slippery surfaces, poor visibility and cold conditions.

Specific control measures include the following:

- Pneumatic tools must be checked to see that the tools are fastened securely to the air hose to prevent them from becoming disconnected. A short wire or positive locking device attaching the air hose to the

Figure 11.10 Pneumatic hammer/chisel

tool must also be used and will serve as an added safeguard.

▶ If an air hose is more than 12.5 mm (1/2 inch) in diameter, a safety excess flow valve must be installed at the source of the air supply to reduce pressure in case of hose failure.

▶ In general, the same precautions should be taken with an air hose that are recommended for electric cables, because the hose is subject to the same kind of damage or accidental striking, and because it also presents tripping hazards.

▶ When using pneumatic tools, a safety clip or retainer must be installed to prevent attachments such as chisels on a chipping hammer from being ejected during tool operation.

▶ Eye protection is required, and head and face protection is recommended, for employees working with pneumatic tools.

▶ Screens must also be set up to protect nearby workers from being struck by flying fragments around chippers, chisels or air drills.

▶ Use of heavy jackhammers (pneumatic chisels) can cause fatigue and strains. Heavy rubber grips reduce these effects by providing a secure handhold.

▶ Workers operating a jackhammer must wear safety glasses and safety shoes that protect them against injury if the jackhammer slips or falls. A face shield also should be used.

▶ Noise is another hazard associated with pneumatic tools. Working with noisy tools such as jackhammers

(heavy pneumatic chisels) requires proper, effective use of appropriate hearing protection.

(b) Electric drills

Hazards are (Figure 11.11):

▶ entanglement, particularly of loose clothing or long hair in rotating drill bits;

▶ high noise levels from the drill or attachment;

▶ eye injury from flying particles and chips, particularly from chisels;

▶ injury from poorly secured work-pieces;

▶ electrocution/electric shock from poorly maintained equipment;

▶ electric shock from drilling into a live hidden cable;

▶ hand–arm vibration hazard in hammer mode;

▶ dust given off from material being worked on;

▶ tripping hazard from trailing cables;

▶ upper limb disorder from powerful machines with a strong torque, particularly if they jam and kick back;

▶ foot injury hazards from dropping heavy units onto unprotected feet;

▶ manual handling hazards, particularly with heavy machines and intensive use or using at awkward heights and/or reaches;

▶ fire and explosion hazard when used near flammable liquids, explosive dusts or gases;

▶ using equipment in poor weather conditions with wet, slippery surfaces, poor visibility and cold conditions.

Figure 11.11 Electric drill with percussion hammer action to drill holes in masonry

Specific control measures include the following:

▶ Use double-insulated tools or earthed reduced voltage tools with a residual current device (see Chapter 12 for electrical safeguards in detail).

▶ Use a pilot hole or punch to start holes whenever possible.

▶ Select the correct drill bit for the material being drilled.

▶ Secure small pieces to be drilled to prevent spinning.

▶ Protect against damage or injury on the far side if the bit is long enough to pass through the material or there are buried services in, say, plaster walls.

▶ Take care to prevent loose sleeves or long hair from being wound around the drill bit; for example, wear short or close-fitting sleeves.

▶ Wear suitable eye protection.

(c) Disc-cutters/cut-off saws

Hazards include:

▶ cutting hazard from high-speed rotating discs;

▶ high noise levels during cutting operations;

▶ overheating of discs due to lack of lubricant such as water;

▶ eye injury from flying hot particles;

▶ eye and puncture injuries from disintegrating, badly mounted discs;

▶ injury from poorly secured work-pieces;

▶ electrocution/electric shock from poorly maintained equipment if electrical;

▶ electric shock from cutting a live hidden cable;

▶ hand–arm vibration hazard;

▶ dust given off from material being worked on;

▶ tripping hazard from trailing cables;

▶ upper limb disorder from powerful machines with a strong torque, particularly if they jam and kick back;

▶ foot injury hazards from dropping heavy units onto unprotected feet;

▶ manual handling hazards, particularly with heavy machines and intensive use or using at awkward heights and/or reaches;

▶ fire and explosion hazard with petrol-fuelled tools;

▶ fire and explosion hazard when using engine or electricity driven tools near flammable liquids, or explosive dusts or gases;

▶ using equipment in poor weather conditions with wet, slippery surfaces, poor visibility and cold conditions.

Figure 11.12 Disc-cutter/cut-off saw

Specific control measures include:

▶ The blade or disc must be fitted with a close-fitting hood guard adjusted to give the maximum protection for the task.

▶ Inspect wheels regularly and discard any cracked wheels immediately before using the cutter again.

▶ Use a wheel with the proper rpm (revolutions per minute) rating and cutting characteristics for the task and material.

▶ Hold the tool firmly with both hands to avoid recoil caused by jamming and wedging.

▶ Wear eye protection.

▶ Do not apply excessive force as this will overheat the work-piece and stress the wheel or disc.

▶ If conditions are dusty suitable filtering face masks should be worn.

▶ Operators should be adequately trained to fit new cutting discs.

▶ If water is used for a cooling lubricant the machine system should be used with the correct fittings to ensure that the water is correctly directed at the cutting edge of the blade.

▶ Operators should wear suitable clothing avoiding loose garments, long hair and jewellery which could catch in the equipment. Protective gloves and footwear are recommended.

(d) Sanders

There is a large range of hand-held sanders on the market, from rotating discs (Figure 11.15), random orbital, rectangular orbital finishing (Figure 11.14) belt sanders and heavy-duty floor sanders of both the rotating drum (Figure 11.13), and, recently, the orbital floor sander. The high-speed rotating discs and drum types are the most hazardous but they all need using with care.

Hazards include:

▶ high noise levels from the sander in operation;

▶ injury from poorly secured work-pieces;

▶ electrocution/electric shock from poorly maintained electrical equipment;

▶ potential of entanglement with rotating disc and drum sanders;

▶ sanding attachments becoming loose in the chuck that can fling off;

▶ injury from contact with abrasive surfaces, particularly with coarse abrasives and high-speed rotating sanding discs and drums;

▶ hand–arm vibration hazard, particularly from reciprocating equipment;

▶ health hazards from extensive dust given off from material being worked on;

▶ fire and health hazards from overheating of abraded surfaces, particularly if plastics are being sanded;

▶ tripping hazard from trailing cables;

▶ large powerful sanders suddenly gripping the surface and pulling the operative off their feet;

▶ foot injury hazards from dropping heavy units onto unprotected feet;

- manual handling hazards, particularly with heavy machines and intensive use or using at awkward heights and/or reaches;
- fire and explosion hazard when used near flammable liquids, explosive dusts or gases;
- using equipment in poor weather conditions with wet, slippery surfaces, poor visibility and cold conditions.

Figure 11.13 Rotary drum floor sander

Figure 11.14 Orbital finishing sander

Figure 11.15 Disc sander

Specific control measures include the following:

- The work-pieces must be securely clamped or held in position during sanding. In some cases a jig will be necessary. The direction of spin of disc sanders (normally anti-clockwise) is important to ensure that a small work-piece is pushed towards a stop or fence, which is normally at the left side of the work-piece, particularly when clamping is impossible.
- Abrasive sanding belts, discs and sheets should be properly and firmly attached to the machine without any torn parts or debris underneath. The manufacturer's instructions and fixing accessories should be used to ensure correct attachment of the abrasive. Operators should be trained, competent and registered to fit abrasive discs.
- Old nails and fixings should be sunk below the surface or removed to prevent the sander snagging.
- Always hold the equipment by the proper handles and particularly on large disc and floor sanders always use both hands (Figure 11.14). Excessive pressure should not be used as the surface will be rutted and the machine may malfunction.
- Ensure that the dust extraction is working properly and has been emptied when about one-third full. Some extraction systems draw dust and air through the sanding sheet, which must have correctly pre-punched holes to allow the passage of air.
- Operators should wear suitable dust respirators, eye protection and, where necessary, hearing protection.
- Operators should wear suitable clothing avoiding loose garments, long hair and jewellery, which could catch in the equipment.
- Protective gloves and footwear are recommended.

(e) Cartridge and pneumatic nail guns

These are used for driving nails either by an explosive powder charge in a cartridge, or compressed air or gas propulsion (Figures 11.16 and 11.17). They can be extremely dangerous if used incorrectly. Although the use of a powder cartridge does not require a firearms licence, correct training of operatives is essential.

Hazards include:

▶ flying debris or particles discharged from the work surface that the nail enters;

▶ using too heavy a charge for the material resulting in the nail being shot right through the material/surface being fixed;

▶ nails ricocheting if the tool is not held properly or the surface material is too hard;

▶ fire and explosion hazard near flammable liquids and explosive dusts or gases;

▶ using tool explosive charges in firearms or vice versa;

▶ high noise levels, particularly from cartridge nail guns;

▶ electric shock from driving into a live hidden cable;

▶ dust given off from material being worked on and fumes from spent cartridges;

▶ tripping hazard from trailing air lines;

▶ upper limb disorder from powerful machines with a strong kickback;

▶ foot injury hazards from dropping heavy units onto unprotected feet;

▶ manual handling hazards, particularly with heavy machines and intensive use;

▶ hazards associated with a portable air supply compressor such as refuelling, noise, fumes, dangerous moving parts of the engine, e.g. the fan and belts;

▶ using equipment in poor weather conditions with wet, slippery surfaces, poor visibility and cold conditions.

Figure 11.16 Cartridge-powered nail gun

Figure 11.17 Pneumatic-powered nail gun

Specific control measures for cartridge-powered nail guns include:

▶ Cartridge powder-actuated tools operate like a loaded gun and must be treated with extreme caution. In fact, they are so dangerous that they must be operated only by specially trained employees.

▶ When using cartridge-actuated tools, an employee must wear suitable ear, eye and face protection. The user must select a cartridge level – high or low velocity – that is appropriate for the cartridge-actuated tool and necessary to do the work without excessive force.

▶ The muzzle end of the tool must have a protective shield or guard centred perpendicular to and concentric with the barrel to confine any fragments or particles that are projected when the tool is fired. A tool containing a high-velocity load must be designed not to fire unless it has this kind of safety device.

▶ To prevent the tool from firing accidentally, two separate motions are required for firing. The first motion is to bring the tool into the firing position, and the second motion is to pull the trigger. The tool must not be able to operate until it is pressed against the work surface with a force at least 2.2 kg greater than the total weight of the tool.

▶ If a powder-actuated tool misfires, the user must hold the tool in the operating position for at least

30 seconds before trying to fire it again. If it still will not fire, the user must hold the tool in the operating position for another 30 seconds and then carefully remove the load in accordance with the manufacturer's instructions. This procedure will make the faulty cartridge less likely to explode. The bad cartridge should then be put in water immediately after removal. If the tool develops a defect during use, it should be tagged and must be taken out of service immediately until it is properly repaired.

▶ Always keep powder-actuated tools, studs and cartridges in a safe place when not in use, preferably under lock and key. Don't leave tools or accessories unattended, even for a short period of time.

Safety precautions that must be followed when using powder-actuated tools include the following:

▶ Do not use a tool in an explosive or flammable atmosphere.
▶ Inspect the tool before using it to determine that it is clean, that all moving parts operate freely, and that the barrel is free from obstructions and has the proper shield, guard and attachments recommended by the manufacturer.
▶ Test the tool each day before loading to see that the safety devices are working. Follow the manufacturer's test methods.
▶ Do not load the tool unless it is to be used immediately.
▶ Do not leave a loaded tool unattended, especially where it would be available to unauthorised persons.
▶ Keep hands clear of the barrel end.
▶ Never carry a loaded tool from one job to another.
▶ Do not rest the tool against the body when loading or making adjustments.
▶ Never point the tool at anyone.
▶ Do not let bystanders stand too close to the operator. Clear workers from the area, particularly from the other side of any partitions being worked on. Screens must also be set up to protect nearby workers from being struck by flying fragments around nail guns, riveting guns or staplers.
▶ Hold the tool perpendicular to the work surface.

When using powder-actuated tools to apply fasteners, several additional procedures must be followed:

▶ Make a thorough study of each job. Know the types of materials you'll be driving into, so that you can select the proper stud and cartridge.
▶ Also know what is on the other side of a wall as well as what is inside it, such as electric wires and gas or water pipes.
▶ Do not fire fasteners into material that would allow the fasteners to pass through to the other side.
▶ Do not drive fasteners into very hard or brittle material that might chip or splatter or make the fasteners ricochet.

▶ Always use an alignment guide when shooting fasteners into existing holes.
▶ When using a high-velocity tool, do not drive fasteners less than 75 mm (3 inches) from an unsupported edge or corner of material such as brick or concrete.
▶ When using a high-velocity tool, do not place fasteners in steel any closer than 12.5 mm (1/2 inch) from an unsupported corner edge unless a special guard, fixture or jig is used.
▶ Never point the tool at anyone.
▶ Do not fire studs into cast iron, high carbon or tempered steel, armour plate, rock, glazed brick, tile or glass.

Specific control measures for pneumatic-powered nail guns include:

▶ Pneumatic nail guns must be checked to see that the tools are fastened securely to the air hose to prevent them from becoming disconnected. A short wire or positive locking device attaching the air hose to the tool must also be used and will serve as an added safeguard.
▶ If an air hose is more than 12.5 mm (1/2 inch) in diameter, a safety excess flow valve must be installed at the source of the air supply to reduce pressure in case of hose failure.
▶ In general, the same precautions should be taken with an air hose that are recommended for electric cables, because the hose is subject to the same kind of damage or accidental striking, and because it also presents tripping hazards.
▶ Pneumatic tools that shoot nails, rivets, staples or similar fasteners and operate at pressures more than 6890 kPa (100 pounds per square inch) must be equipped with a special device to keep fasteners from being ejected, unless the muzzle is pressed against the work surface.
▶ Eye protection is required, and head and face protection is recommended, for employees working with pneumatic tools.
▶ Screens must also be set up to protect nearby workers from being struck by flying fragments around nail guns, riveting guns or staplers.
▶ Compressed-air guns should never be pointed towards anyone. Workers should never 'dead-end' them against themselves or anyone else. A chip guard must be used when compressed air is used for cleaning.
▶ Noise is another hazard associated with pneumatic tools. Appropriate hearing protection should be provided.

(f) Chainsaws

Hazards include:

▶ very serious cutting by contact with the high-speed cutting chain;

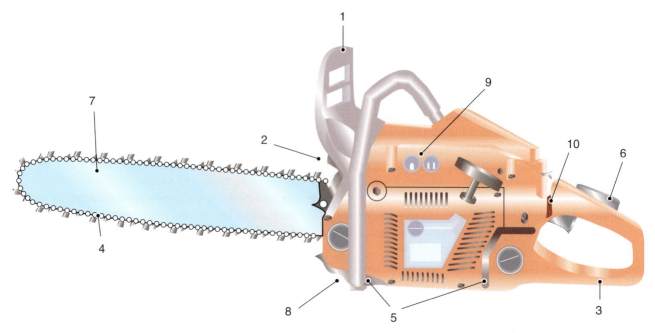

Figure 11.18 Typical chainsaw with rearguard. 1 – hand guard with integral chain brake; 2 – exhaust outlet directed to the right-hand side away from the operator; 3 – chain breakage guard at bottom of rear handle; 4 – chain designed to have low-kickback tendency; 5 – rubber anti-vibration mountings; 6 – lockout for the throttle trigger; 7 – guide bar (should be protected when transporting chainsaw); 8 – bottom chain catcher; 9 – PPE hand/eye/ear defender signs; 10 – on/off switch.

- ▶ kickback due to being caught on the wood being cut or contact with the top front corner of the chain in motion, with the saw chain being kicked upwards – towards the face in particular;
- ▶ pull-in when the chain is caught and the saw is pulled forward;
- ▶ push-back when the chain on the top of the saw bar is suddenly pinched and the saw is driven straight back towards the operator;
- ▶ burns from hot parts of the engine;
- ▶ high noise levels;
- ▶ hand–arm vibration causing white finger and other problems;
- ▶ fire and explosion hazards from the use of highly flammable petrol as a fuel;
- ▶ eye and face puncture wounds from ejected particles;
- ▶ back strain while supporting the weight of the chainsaw when operating it;
- ▶ fire and explosion hazard when using engine or electrical driven tools near flammable liquids, explosive dusts or gases;
- ▶ electric shock if electrically powered;
- ▶ trailing cables on electrically powered equipment;
- ▶ falls from height if using the chainsaw in trees and the like;
- ▶ lone working and the risk of serious injury;
- ▶ contact with overhead power lines if felling trees;
- ▶ being hit by falling branches or whole trees while felling;
- ▶ possible health hazards from cutting due to wood dust, particularly if wood has been seasoned;

- ▶ using the chainsaw in poor weather conditions with wet, slippery surfaces, poor visibility and cold conditions;
- ▶ health hazards from engine fumes (carbon dioxide and carbon monoxide), particularly if used inside a shed or other building.

Competence assessment

Everyone working with chainsaws on or in trees should hold such a certificate or award unless:

- ▶ the work is being done as part of agricultural operations (e.g. hedging, clearing fallen branches, pruning trees to maintain clearance for machines);
- ▶ it is being done by the occupier or their employees;
- ▶ they have used a chainsaw before 5 December 1998.

In any case, operators using chainsaws for any task in agriculture or any other industry must be competent under PUWER 98.

Training should be carried out by specialist instructors at organised training courses. Advice on suitable training courses (and competence assessment where appropriate) is available from the industry Sector Skills Council and

Lantra Awards, Stoneleigh Park, Kenilworth, Warwickshire CV8 2LG, Tel: 02476 419703, Fax: 02476 411655.

NPTC, Stoneleigh Park, Kenilworth, Warwickshire CV8 2LG.

Where training is being consolidated through workplace-based experience, the trainee should be supervised by a person competent in the use of a chainsaw for the work being done by the trainee and who holds the relevant competence certificate or award.

It is recommended that all chainsaw operators have regular refresher/update training to ensure they work to industry best practice and maintain their levels of competence. The suggested intervals for such training are:

occasional users – every two to three years;
full-time users – every five years.

Specific control measures include:

▶ Chainsaws may only be operated by fully trained, fit and competent people.
▶ Avoid working alone with a chainsaw. Where this is not possible, establish procedures to raise the alarm if something goes wrong.
▶ Moving engine parts should be enclosed.
▶ Electrical units should be double insulated and cables fitted with residual current devices.
▶ The saw must be fitted with a top handle and effective brake mechanism.
▶ Chainsaws expose operators to high levels of noise and hand–arm vibration which can lead to hearing loss and conditions such as vibration white finger. These risks can be controlled by good management practice including:
  ▷ purchasing policies for low-noise/low-vibration chainsaws (e.g. with anti-vibration mounts and heated handles);
  ▷ providing suitable hearing protection;
  ▷ proper maintenance schedules for chainsaws and protective equipment;
  ▷ giving information and training to operators on the health risks associated with chainsaws and use of PPE, etc.
▶ Operators need to be trained in the correct chain-sharpening techniques and chain and guide bar maintenance to keep the saw in safe working condition.
▶ Make sure petrol containers are in good condition and clearly labelled, with securely fitting caps. Use containers which are specially designed for chainsaw fuelling and lubrication.
▶ Do not allow operators to use discarded engine oil as a chain lubricant – it is a very poor lubricant and may cause cancer if it is in regular contact with an operator's skin.
▶ When starting the saw, operators should maintain a safe working distance from other people and ensure the saw chain is clear of obstructions.

▶ Kickback is the sudden uncontrolled upward and backward movement of the chain and guide bar towards the operator. This can happen when the saw chain at the nose of the guide bar hits an object. Kickback is responsible for a significant proportion of chainsaw accidents, many of which are to the face and parts of the upper body where it is difficult to provide protection. A properly maintained chain brake and use of low-kickback chains (safety chains) reduces the effect, but cannot entirely prevent it.
▶ Training in good manual handling techniques and using handling aids/tools should reduce the risk of back injuries.
▶ Suitable PPE should always be worn, no matter how small the job. European standards for chainsaw PPE are published as part of EN 381 'Protective clothing for users of hand-held chainsaws'. Protective clothing complying with this standard should provide a consistent level of resistance to chainsaw cut-through. Other clothing worn with the PPE should be close fitting and non-snagging. See Figure 11.19.

NB. No protective equipment can ensure 100% protection against cutting by a hand-held chainsaw.

▶ If conditions are dusty suitable filtering face masks should be worn.

Figure 11.19 *Kevlar gloves, overtrousers and overshoes providing protection against chainsaw cuts. Helmet and face shield protect the head. Apprentice under training – first felling*

11.3 Mechanical and non-mechanical hazards of machinery

11.3.1 Hazard identification

Most machinery has the potential to cause injury to people, and machinery accidents figure prominently in official accident statistics. These injuries may range in severity from a minor cut or bruise, through various degrees of wounding and disabling mutilation, to crushing, decapitation or other fatal injury. It is not solely powered machinery that is hazardous, for many manually operated machines (e.g. hand-operated guillotines and fly presses) can still cause injury if not properly safeguarded.

In the risk assessment of machinery (see ISO 12100:2010) it is essential to consider both permanent hazards and those which can appear unexpectedly. Hazardous situations and events during all phases of the machine life cycle need to be covered. These include: transport, assembly and installation; commissioning; use; maintenance; dismantling, disabling and scrapping.

The human interaction with the machine involving numerous tasks throughout the life cycle of the machine should be considered. They include:

setting; testing; teaching/programming; process/ tool changeover; start-up; all modes of operation; feeding the machine; removal of product from the machine; stopping the machine; stopping the machine in case of emergency; recovery of operation from jam or blockage; restart after unscheduled stop; fault-finding/trouble-shooting (operator intervention); cleaning and housekeeping; preventative maintenance; corrective maintenance.

The possible states of the machine should be taken into account which are:

1. the machine performs the intended function (the machine operates normally);

2. the machine does not perform the intended function (i.e. it malfunctions) due to a variety of reasons, including:
   ▷ variation of a property or of a dimension of the processed material or of the work-piece;
   ▷ failure of one or more of its component parts or services;
   ▷ external disturbances (for example, shocks, vibration, electromagnetic interference);
   ▷ design error or deficiency (for example, software errors);
   ▷ disturbance of its power supply; and
   ▷ surrounding conditions (for example, damaged floor surfaces).

Finally the unintended behaviour of the operator or reasonably foreseeable misuse of the machine must be considered. Examples include:

- loss of control of the machine by the operator (especially for hand-held or mobile machines);
- reflex behaviour of a person in case of malfunction, incident or failure during the use of the machine;
- behaviour resulting from lack of concentration or carelessness,
- behaviour resulting from taking the 'line of least resistance' in carrying out a task;
- behaviour resulting from pressures to keep the machine running in all circumstances; and
- behaviour of certain persons (for example, children, disabled persons).

11.3.2 Mechanical machinery hazards

The hazards of machinery are set out in ISO 12100:2010, which covers the classification of machinery hazards and how harm may occur. The following mechanical machinery hazards follow this standard (Figure 11.20 shows a number of these hazards).

A person may be injured at machinery as a result of:

- **acceleration, deceleration** – potential for impact, being thrown, run-over, slipping, tripping or falling, for example from a piece of mobile plant or an overhead travelling crane
- **angular parts** – potential consequences are impact, crushing or shearing when a person hits a stationary part of the machine or a moving part impacts on an operator particularly in mobile plant;
- **approach of a moving element to a fixed part** – potential consequences are crushing or impact through being trapped between a moving part of a machine and a fixed structure, such as a wall or any material in a machine;
- **cutting parts** – potential consequences are cutting or severing through contact with a cutting edge, such as a band saw or rotating cutting disc;
- **elastic elements** – consequences are crushing or impact when an elastic part is deformed and suddenly straightens or returns to its original dimensions or explodes when an airbag used for lifting fails;
- **falling objects** – potential consequences are crushing or impact, for example if a load fell from an overhead hoist or rail system;
- **gravity** – potential consequences are crushing or trapping, for example if a vehicle lift failed while a person was underneath;
- **height from the ground** – potential consequences are crushing, impact, slipping, tripping or falling, for example getting out of a truck or large piece of mobile plant or gaining access to a tower crane;
- **high pressure** – potential consequences are injection, stabbing or puncture, impact, for example from a high pressure hydraulic system leak;
- **instability** – potential for being thrown, crushing or impact when a machine breaks its holding down

bolts and falls over or a tower crane is not left free to slew to the wide direction and is blown over;

- **kinetic energy –** potential for impact, puncture or severing, for example when a kinetic recovery rope is used to tow a vehicle or ship and it snaps the whiplash can be very hazardous;
- **machinery mobility –** potential consequences are being run over, impact, crushing when, for example, a large machine is being manoeuvred in a small space;
- **moving elements** – potential consequences are crushing, friction, abrasion, impact, shearing, severing, drawing-in, for example on grinding wheels or sanding machines, or when a moving part directly strikes a person, such as with the accidental movement of a robot's working arm when maintenance is taking place. A further example is when persons are drawn-in between in-running gear wheels or rollers or between belts and pulley drives;
- **rotating elements** – potential consequences are severing or entanglement, for example an exposed rotating shaft which grips loose clothing, hair or working material, such as emery paper. The smaller the diameter of the revolving part the easier it is to get a wrap or entanglement. A further example is a shearing which occurs when part of the body, typically a hand or fingers, is trapped between rotating and fixed parts of the machine;
- **rough, slippery surface** - potential consequences are friction, abrasion, slipping, tripping and falling, impact when, for example, the operating area round a machine is very slippery from product spillages and the operator falls over and/or into the machinery hazard zone;
- **sharp edges** - potential consequences are severing, impact, shearing, stabbing or puncture;
- **stored energy** – potential consequences are crushing, impact, puncture, suffocation when the stored energy is released suddenly, for example when the valve on a welding gas bottle (other hazards such as explosion or fire may also be present) is broken off or a high pressure air hose breaks;
- **vacuum** – potential consequences are crushing, drawing-in suffocation for example the sudden failure of a large vacuum vessel.

11.3.3 Non-mechanical machinery hazards

Non-mechanical hazards are also set out in ISO 12100:2010 and include:

Electrical hazards from:

- arc;
- electromagnetic phenomena;
- electrostatic phenomena;
- live parts;
- not enough distance to live parts under high voltage;
- overload;

- parts which have become live under fault conditions;
- short-circuit;
- thermal radiation.

Thermal hazards from:

- explosion;
- flame;
- objects or materials with a high or low temperature;
- radiation from heat sources.

Noise hazards from:

- cavitation phenomena;
- exhausting system;
- gas leaking at high speed;
- manufacturing process;
- moving parts;
- scraping surfaces;
- unbalanced rotating parts;
- whistling pneumatics
- worn parts.

Vibration hazards from:

- cavitation phenomena;
- misalignment of moving parts;
- mobile equipment;
- scraping surfaces;
- unbalanced rotating parts;
- vibrating equipment;
- worn parts.

Radiation hazards from:

- ionising radiation sources;
- low frequency electromagnetic radiation;
- optical radiation (infrared, visible and ultraviolet), including laser;
- radio frequency electromagnetic radiation.

Material/substance hazards from:

- aerosol;
- biological and microbiological (viral and bacteria) agent;
- combustible;
- dust;
- explosive;
- fibre;
- flammable;
- fluid;
- fume;
- gas;
- mist;
- oxidiser.

Ergonomic hazards from:

- access;
- design or location of indicators and visual display units;
- design, location or identification of control devices;
- effort;
- flicker, dazzling, shadow, stroboscopic effect;
- local lighting;
- mental overload/underload;
- posture;

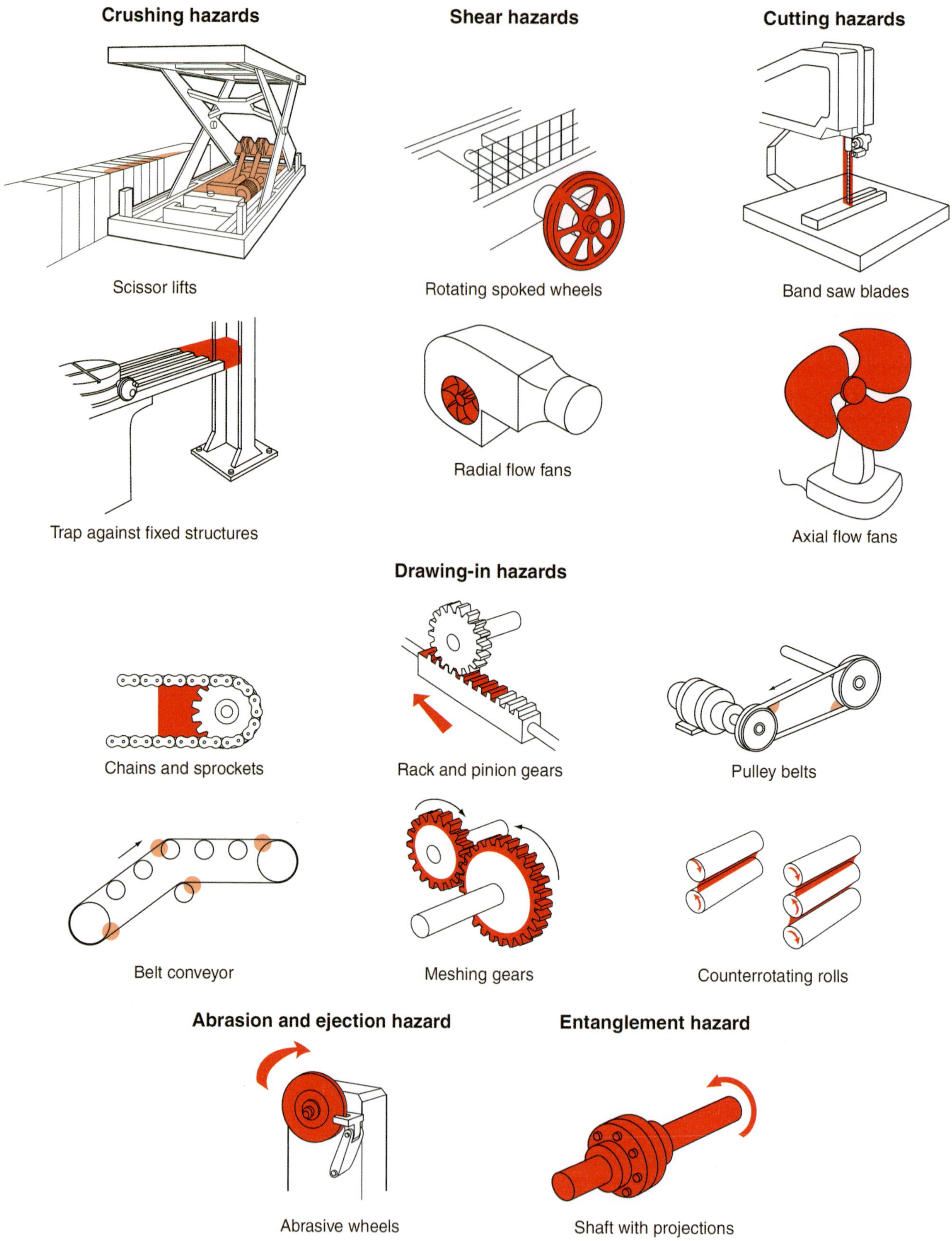

Crushing hazards

Scissor lifts

Trap against fixed structures

Shear hazards

Rotating spoked wheels

Radial flow fans

Cutting hazards

Band saw blades

Axial flow fans

Drawing-in hazards

Chains and sprockets

Rack and pinion gears

Pulley belts

Belt conveyor

Meshing gears

Counterrotating rolls

Abrasion and ejection hazard

Abrasive wheels

Entanglement hazard

Shaft with projections

Figure 11.20 Range of mechanical hazards

▶ repetitive activity;
▶ visibility.

Hazards associated with the environment in which the machine is used from:

▶ dust and fog;
▶ electromagnetic disturbance;

▶ lightning;
▶ moisture;
▶ pollution;
▶ snow;
▶ temperature;
▶ water;

- wind;
- lack of oxygen.

Combination of hazards from:

- for example, repetitive activity + effort + high environmental temperature (or repetitive hard work in a hot climate).

In many cases it will be practicable to install safeguards which protect the operator from both mechanical and non-mechanical hazards.

For example a guard may prevent access to hot or electrically live parts as well as to moving ones. The use of guards which reduce noise levels at the same time is also common.

As a matter of policy, machinery hazards should be dealt with in this integrated way instead of dealing with each hazard in isolation.

11.3.4 Examples of machinery hazards

The following examples are given to demonstrate a small range of machines found in industry and commerce, which are included in the Certificate syllabus. Examples are shown in the application of safeguards in Section 11.4.

Office – printer/photocopier

The hazards are:

- contact (cutting or abrasion) with moving parts when clearing a jam;
- electrical – when clearing a jam or maintaining the machine or through poorly maintained plug and wiring;
- heat through contact with hot parts when clearing a jam;
- health hazard from ozone or lack of ventilation in the area or very rarely an allergy to paper dust.

Office – document shredder

The hazards are:

- drawing-in between the rotating cutters when feeding paper into the shredder;
- contact (cutting or severing) with the rotating cutters when emptying the waste container or clearing a jam;
- electrical through faulty plug and wiring or during maintenance;
- possible noise from the cutting action of the machine;
- possible dust from the cutting action.

Workshop machinery – bench-top grinding machine

The hazards are:

- contact (friction or abrasion) with the rotating wheel;
- drawing-in between the rotating wheel and a badly adjusted tool rest;
- bursting of the wheel, ejecting fragments which puncture the operator;
- electrical through faulty wiring and/or earthing or during maintenance;
- fragments given off during the grinding process causing eye injury;
- hot fragments given off which could cause a fire or burns;
- noise produced during the grinding process;
- possible health hazard from dust/particles/fumes given off during grinding.

Workshop machinery – pedestal drill

The hazards are:

- entanglement around the rotating spindle and chuck;
- contact (cutting, puncturing, severing) with the cutting drill or work-piece;
- being struck (impact) by the work-piece if it rotates;
- being cut or punctured by fragments ejected from the rotating spindle and cutting device;
- drawing-in to the rotating drive belt and pulley;
- contact or entanglement with the rotating motor;
- electrical from faulty wiring and/or earthing or during maintenance;
- possible health hazard from cutting fluids or dust given off during the process.

Workshop machinery – bench-mounted circular saw

The hazards include:

- contact (cutting or severing) with the cutting blade above and below the bench;
- ejection of the work-piece or timber as it closes after passing the cutting blade;
- drawing in between chain and sprocket, V-belt drives or powered feed rollers;
- contact (impact, drawing-in and entanglement) with moving parts of the drive motor;
- likely noise hazards from the cutting action and motor;
- health hazards from wood dust given off during cutting;
- electric shock from faulty wiring and/or earthing or during maintenance.

Workshop machinery – hand-fed power planer

The hazards include:

- contact (cutting, severing, drawing-in) with the cutting cylinder above and below the bench;
- contact (cutting or severing) with the cutting blades under the bridge guard when hands are placed on the work-piece to push it past the planing blades;
- ejection of the work-piece, timber or incorrectly fitted blades;

- drawing-in between chain and sprocket, V-belt drives or powered feed rollers;
- contact (impact or drawing-in) and entanglement with moving parts of the drive motor;
- likely noise hazards from the cutting action and motor;
- health hazards from wood dust given off during cutting;
- electric shock from faulty wiring and/or earthing or during maintenance;
- see also Section 11.4.8.

Workshop machinery – spindle moulding machine

The hazards include:

- contact (cutting or severing) with the cutting blade above the bench;
- contact (cutting or severing) with the cutting blades when hands are placed on the work-piece to push it past the blades and no holding jigs are used;
- ejection of the work-piece, timber or incorrectly fitted blades;
- drawing-in between chain and sprocket or V-belt drives;
- contact (impact/drawing-in) and entanglement with moving parts of the drive motor;
- likely noise hazards from the cutting action and motor;
- health hazards from wood dust given off during cutting;
- electric shock from faulty wiring and/or earthing or during maintenance.

Site machinery – compressor

The hazards include:

- contact (impact/drawing-in) and entanglement with moving parts of the drive motor;
- flying debris and blast from an exploding air receiver;
- drawing-in between chain and sprocket drives;
- electrical, if electrically powered;
- burns from hot parts of petrol or diesel engine;
- fire if highly flammable liquids used as fuel;
- possible noise hazards from the motor and tools;
- contamination of respirable air if used for a supplied air respirator;
- possible health hazards from leaking oil;
- possible hand–arm vibration causing white finger and other problems if used for road or concrete breaking;
- eye puncture wounds from ejected chips from concrete or similar when using a pneumatic chisel.

Site machinery – cement/concrete mixer

The hazards include:

- contact (impact, drawing-in, entanglement) with moving parts of the drive motor;
- crushing between loading hopper (if fitted) and drum;
- drawing-in between chain and sprocket drives;
- electrical, if electrically powered;
- burns from hot parts of the engine;
- fire if highly flammable liquids used as fuel;
- possible noise hazards from the motor and dry mixing of aggregates;
- eye injury from splashing cement slurry;
- possible health hazard from cement dust and cement slurry while handling.

Site machinery – plate compactor

The hazards include:

- noise hazards from the action of the plate and a petrol-driven engine;
- vibration from the action of the plate;
- contact (impact) with foot or lower leg;
- flying splinters/chippings from the material being compacted;
- manual handing injuries as they are generally handled by a single person;
- dust from dry material being compacted.

Site machinery – ground consolidation equipment

The hazards include:

- noise hazards from the action of the vibrating roller and a petrol- or diesel-driven engine;
- contact (impact, drawing-in, cutting) with moving parts of the engine;
- vibration from the action of the roller;
- operative or person standing by being run over or crushed by a self-propelled roller;
- manual handing injuries, as it is generally handled by a single person;
- dust from dry material being consolidated;
- possible health hazard from lubricants or fuels in the engine;
- possible health hazards from exhaust;
- contact with hot parts of the engine;
- fire if highly flammable liquids are used as fuel.

Site machinery – road marking equipment

There is a variety of road and line marking equipment from the simple roller unit used in car parks or sports fields to walk-behind line markers to lorry-mounted specialist units for large highway projects. The hazards here relate to a lorry-mounted unit with a driver and single operative.

The hazards include:

- the danger of being struck (impact, running-over, crushing) by vehicles using the existing road unless it is new and not yet opened to the public;
- injury to fellow workers, who may be hit (impact, running-over, crushing) by the vehicle while travelling

or manoeuvring, or fall off (falling) the side of the vehicle;

▶ contact with dangerous moving parts in the vehicle or equipment engine(s);

▶ fire hazard from burning off existing markings;

▶ fire and explosion hazard if LPG is used as a fuel for burning;

▶ health hazard from any toxic fumes, aerosols or dust given off from the heated products or the line marking sprayed paint;

▶ health hazard from exposure to the UV from the sun;

▶ hazards caused by cold or hot atmospheric conditions because of working outside;

▶ potential burns from heated materials;

▶ injury from damaged air lines.

Site machinery – electrical generator

The hazards are:

▶ contact (impact, cutting and entanglement) with moving parts of the drive motor;

▶ high levels of carbon monoxide rapidly building up in a confined space;

▶ electrocution from contact with medium voltage electrical current;

▶ electric shock from faulty wiring and/or earthing or during use;

▶ if generators are connected into fixed mains installations a very serious electrical hazard may exist for line repair electricians and neighbours who may experience high voltage 'backfeed' into the system;

▶ burns from hot parts of petrol or diesel engines;

▶ fire if highly flammable liquids used as fuel;

▶ possible noise hazards from the motor;

▶ manual handling problems if the equipment is moved frequently.

11.4 Control measures for reducing risks from machinery hazards

11.4.1 Introduction

As discussed in 11.1.1 and 11.1.3, PUWER requires that access to dangerous parts of machinery should be prevented in a preferred order or hierarchy of control methods. The standard required is a 'practicable' one, so the only acceptable reason for non-compliance is that there is no technical solution. Cost is not a factor. (See Chapter 1 for more details on standards of compliance.)

As the mechanical hazard of machinery arises principally from someone coming into contact or entanglement with dangerous components, risk reduction is based on preventing this contact occurring.

This may be by means of:

▶ a physical barrier between the individual and the component (e.g. a fixed enclosing guard);

▶ a device which allows access only when the component is in a safe state (e.g. an interlocked guard which prevents the machine starting unless a guard is closed and acts to stop the machine if the guard is opened);

▶ a device which detects that the individual is entering a risk area and then stops the machine (e.g. certain photoelectric guards and pressure-sensitive mats).

The best method should, ideally, be chosen by the designer as early in the life of the machine as possible. It is often found that safeguards which are 'bolted on' instead of 'built in' are not only less effective in reducing risk, but are also more likely to inhibit the normal operation of the machine. In addition, they may in themselves create hazards and are likely to be difficult, and hence expensive, to maintain.

11.4.2 Fixed guards

Fixed guards have the advantage of being simple, always in position, difficult to remove and almost maintenance-free. Their disadvantage is that they do not always properly prevent access, they are often left off by maintenance staff and they can create difficulties for the operation of the machine.

A fixed guard has no moving parts and should, by its design, prevent access to the dangerous parts of the machinery. It must be of robust construction and sufficient to withstand the stresses of the process and environmental conditions. If visibility or free air flow (e.g. for cooling) is necessary, this must be allowed for in the design and construction of the guard. If the guard can be opened or removed, this must only be possible with the aid of a tool.

An alternative fixed guard is the distance fixed guard, which does not completely enclose a hazard, but which reduces access by virtue of its dimensions and its distance from the hazard. Where perimeter fence guards are used, the guard must follow the contours of the machinery as far as possible, thus minimising space between the guard and the machinery. With this type of guard, it is important that the safety devices and operating systems prevent the machinery being operated with the guards closed and someone inside the guard – that is, in the danger area. Figure 11.21 shows a range of fixed guards for some of the examples shown in Figure 11.20.

11.4.3 Adjustable guards

(a) User-adjusted guard

These are fixed or movable guards, which are adjustable for a particular operation during which they remain fixed. They are particularly used with machine tools where some access to the dangerous part is required (e.g. drills, circular saws, milling machines; Figure 11.22) and where the clearance required will vary (e.g. with the size of the cutter in use on a horizontal milling machine or

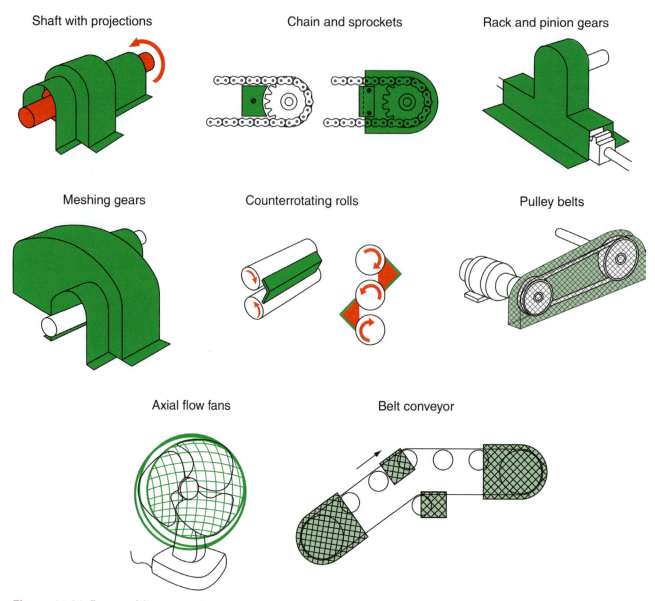

Shaft with projections

Chain and sprockets

Rack and pinion gears

Meshing gears

Counterrotating rolls

Pulley belts

Axial flow fans

Belt conveyor

Figure 11.21 Range of fixed guards

with the size of the timber being sawn on a circular saw bench; Figure 11.23).

Adjustable guards may be the only option with cutting tools, which are otherwise very difficult to guard, but they have the disadvantage of requiring frequent readjustment. By the nature of the machines on which they are most frequently used, there will still be some access to the dangerous parts, so these machines must only be used by suitably trained operators. Jigs, push sticks and false tables must be used wherever possible to minimise hazards during the feeding of the work-piece. The working area should be well lit and kept free of anything which might cause the operator to slip or trip.

(b) Self-adjusting guard

A self-adjusting guard is one which adjusts itself to accommodate, for example, the passage of material.

A good example is the spring-loaded guard fitted to many portable circular saws (Figure 11.23).

As with adjustable guards (see Figure 11.22), they only provide a partial solution in that they may well still allow access to the dangerous part of the machinery. They require careful maintenance to ensure they work to the best advantage.

11.4.4 Interlocking guard

The advantages of interlocked guards are that they allow safe access to operate and maintain the machine without dismantling the safety devices. Their disadvantage stems from the constant need to ensure that they are operating correctly and designed to be fail-safe. Maintenance and inspection procedures must be very strict.

This is a guard which is movable (or which has a movable part), whose movement is connected with the power or control system of the machine.

An interlocking guard must be connected to the machine controls such that:

Figure 11.22 Adjustable guard for a rotating drill bit on a pedestal drill

Figure 11.23 Self-adjusting guard on a circular wood saw

► until the guard is closed the interlock prevents the machinery from operating by interrupting the power medium;

► either the guard remains locked until the risk of injury from the hazard has passed or opening the guard causes the hazard to be eliminated before access is possible.

A passenger lift or hoist is a good illustration of these principles: the lift will not move unless the doors are closed, and the doors remain closed and locked until the lift is stationary and in such a position that it is safe for the doors to open.

Special care is needed with systems which have stored energy. This might be the momentum of a heavy moving part, stored pressure in a hydraulic or pneumatic system, or even the simple fact of a part being able to move under gravity even though the power is disconnected. In these situations, dangerous movement may continue or be possible with the guard open, and these factors need to be considered in the overall design. Braking devices (to arrest movement when the guard is opened) or delay devices (to prevent the guard opening until the machinery is safe) may be needed. All interlocking systems must be designed to minimise the risk of failure-to-danger and should not be easy to defeat (Figure 11.24).

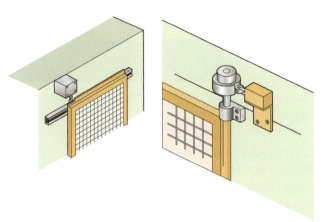

Figure 11.24 Typical sliding and hinged interlocking guards

11.4.5 Other protective devices

(a) Trip devices

A trip device does not physically keep people away but detects when a person approaches close to a danger point. It should be designed to stop the machine before injury occurs. A trip device depends on the ability of the machine to stop quickly and in some cases a brake may need to be fitted. Trip devices can be:

► mechanical in the form of a bar or barrier;
► electrical in the form of a trip switch on an actuator rod, wire or other mechanism;
► photoelectric or other type of presence-sensing device;
► a pressure-sensitive mat.

They should be designed to be self-resetting so that the machine must be restarted using the normal procedure (Figure 11.25).

(b) Two-handed control devices

These are devices which require the operator to have both hands in a safe place (the location of the controls) before the machine can be operated. They are an option on machinery that is otherwise very difficult to guard

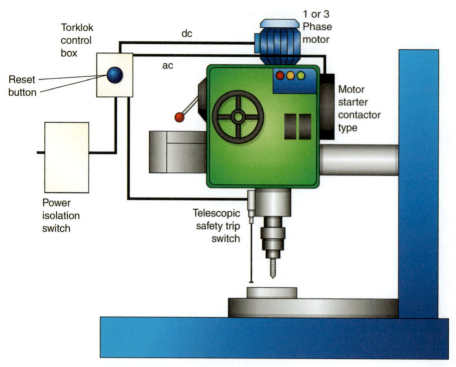

Figure 11.25 Schematic diagram of a telescopic trip device fitted to a radial drill

but they have the drawback that they only protect the operator's hands. It is therefore essential that the design does not allow any other part of the operator's body to enter the danger zone during operation. More significantly, they give no protection to anyone other than the operator.

Where two-handed controls are used, the following principles must be followed:

▶ The controls should be so placed, separated and protected as to prevent spanning with one hand only, being operated with one hand and another part of the body, or being readily bridged.

▶ It should not be possible to set the dangerous parts in motion unless the controls are operated within approximately 0.5 seconds of each other. Having set the dangerous parts in motion, it should not be possible to do so again until both controls have been returned to their off position.

▶ Movement of the dangerous parts should be arrested immediately or, where appropriate, arrested and reversed if one or both controls are released while there is still danger from movement of the parts.

▶ The hand controls should be situated at such a distance from the danger point that, on releasing the controls, it is not possible for the operator to reach the danger point before the motion of the dangerous parts has been arrested or, where appropriate, arrested and reversed (Figure 11.26).

(c) Hold-to-run control

This is a control which allows movement of the machinery only as long as the control is held in a set

Figure 11.26 Two-handed control device

position. The control must return automatically to the stop position when released. Where the machinery runs at crawl speed, this speed should be kept as low as practicable.

Hold-to-run controls give even less protection to the operator than two-handed controls and have the same main drawback in that they give no protection to anyone other than the operator.

However, along with limited movement devices (systems which permit only a limited amount of machine movement on each occasion that the control is operated and are often called 'inching devices'), they are extremely relevant to operations such as setting, where access may well be necessary and safeguarding by any other means is difficult to achieve.

(d) Jigs, holders and push sticks

There are a number of devices to limit the approach of people to mechanical hazards while using machinery. These include manual handling devices such as push sticks for circular saws and spindle moulding machines, or push blocks for wood planing machines. With certain machines jigs and other holding devices may be necessary to hold the work-piece near to the cutter of the machine without exposing the operator's hands to the dangerous part of the machine.

There is a logical progression from push sticks through jigs and holders, through more sophisticated clamping devices and moving work tables for circular saws, to power-driven feeds and automatic feeding arrangements. Emergency stops must always be immediately available.

11.4.6 Information, instruction, training and supervision

(a) Information and instructions

PUWER Regulation 8 requires the provision of information and written instructions for work equipment. This should cover:

(a) the conditions in which the work equipment can be used;
(b) the way in which the work equipment can be used;
(c) any foreseeable difficulties that could arise, and instructions on how to deal with them;
(d) any conclusions drawn from experience using the work equipment, which should be either recorded or steps taken to make sure that all appropriate members of the workforce are aware of them.

Written instructions include all the information provided by manufacturers or suppliers such as manuals, instruction sheets, instruction placards, warning labels and training manuals. In addition there may be in-house instructions and information from training sessions. Manufacturers have a duty to provide sufficient information, including drawings, to enable the correct installation, safe operation and maintenance of the equipment. These are often available to download from the manufacturer's website.

Employers must ensure that written instructions are available to everyone using the equipment. This includes temporary workers and others like maintenance staff

who will need information like maintenance instructions and manuals.

Supervisors and managers should also have access to the information and written instructions. The amount of detailed health and safety information they will need to have immediately available for day-to-day operations will vary, but it is important that they know what information is available and where it can be found.

Information should be provided in writing or verbally where that is considered sufficient. Managers should make a decision based on the unusual nature or complexity of the equipment or task. They should consider:

▶ the skill levels of the workers involved;
▶ their experience and existing training;
▶ the level of supervision required;
▶ the complexity and length of the task.

Information and instructions whether verbal or written should be easy to understand and must be provided in an appropriate language for the user/purchaser. They should be logical, illustrated where appropriate and use standard symbols. Particular attention should be paid where people's first language differs from the local official language, or they have language problems or a disability which makes receiving or understanding information or instructions difficult.

(b) Training and supervision

PUWER Regulation 9 requires that employers ensure that all persons, including supervisors and managers, who use or supervise work equipment, have received adequate training for purposes of health and safety.

Employers should:

▶ consider the existing levels of competence and skills of all users of work equipment including temporary and agency workers;
▶ evaluate the level of management or supervision needed;
▶ carry out a gap analysis to determine the training required to enhance workers' level of competence, taking into consideration, for example, whether they work alone or under close supervision;
▶ carry out the training particularly on recruitment but also when:
  ▷ risks change;
  ▷ tasks change;
  ▷ new technology or equipment is introduced;
  ▷ the system of work changes.

Refresher training needs to be provided as necessary as skills will decline particularly if not used regularly, such as after lengthy absence. Supervisors and managers also require adequate training to carry out their function, particularly if they only supervise a particular task occasionally.

The training and supervision of young persons (under 18) is particularly important because of their relative immaturity, unfamiliarity with a working environment and lack of awareness of existing or potential risks. Some codes of practice, for example on the UK's 'Safe Use of Woodworking Machinery', restrict the use of high-risk machinery, so that only young persons with sufficient maturity and competence who have finished their training may use the equipment unsupervised. ISO 12100:2010 states that the unintended behaviour of the operator or foreseeable misuse of the machinery should be taken into consideration when assessing the risks.

11.4.7 Personal Protective Equipment

This is covered in detail in Chapter 14 and also for chainsaws in Section 11.2.4. The use of PPE for users of work equipment is the last issue to be considered in the hierarchy of protection (see 11.1) after inherently safe design, safeguards, instructions and training. However, there are many tasks and pieces of work equipment where PPE is necessary for general protection such as the use of suitable gloves which fit properly and do not restrict the operation.

Some pieces of equipment such as chainsaws are so inherently dangerous that despite all the safeguards and training specialised PPE is absolutely essential. These include Kevlar-based footwear and overtrousers to prevent chain cuts on the feet and legs; special face screen and helmets to prevent flying chips and saw kickback damaging the face, head and eyes. Harnesses are necessary when working at height in trees.

It is important that the risk assessments identify the equipment which should be used to protect against specific risks as well as general standards required for the use of work equipment. It is becoming increasingly common for all people in process plants to wear eye protection, gloves, high-visibility jackets or overalls and hard hats. These should not be seen as a substitute for well-designed and properly safeguarded equipment.

11.4.8 Application of safeguards to the range of machines

The application of the safeguards to the Certificate range of machines is as follows.

Office – printer/photocopier

Application of safeguards (Figure 11.27):

▶ The machines are provided with an all-enclosing case which prevents access to the internal moving, hot or electrical parts.
▶ The access doors are interlocked so that the machine is automatically switched off when gaining access to clear jams or maintain the machine. It is good practice to switch off when opening the machine.
▶ Internal electrics are insulated and protected to prevent contact.

▶ Regular inspection and maintenance should be carried out.
▶ The machine should be on the PAT schedule.
▶ Good ventilation in the machine room should be maintained.
▶ When changing cartridges or toner use gloves and place in a plastic bag immediately after use.

Figure 11.27 Typical multifunction printer/photocopier

Office – document shredder

Application of safeguards (Figure 11.28):

▶ Enclosed fixed guards surround the cutters with restricted access for paper only, which prevents fingers reaching the dangerous parts.
▶ Interlocks are fitted to the cutter head so that the machine is switched off when the waste bin is emptied.
▶ A trip device is used to start the machine automatically when paper is fed in.
▶ The machine should be on PAT schedule and regularly checked.
▶ General ventilation will cover most dust problems, except for very large machines where dust extraction may be necessary.
▶ Noise levels should be checked and the equipment perhaps placed on a rubber mat if standing on a hard reflective floor.

Workshop machinery – bench-top grinder

Application of safeguards (Figure 11.29):

▶ Wheel should be enclosed as much as possible in a strong casing capable of containing a burst wheel.
▶ Grinder should be bolted down to prevent movement.

Figure 11.28 Typical office shredder

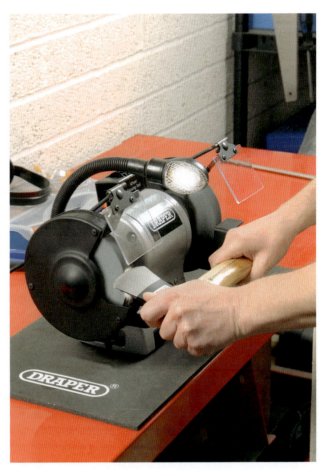

Figure 11.29 Typical bench-mounted grinder

▶ An adjustable tool rest should be adjusted as close as possible to the wheel.

▶ An adjustable screen should be fitted over the wheel to protect the eyes of the operator. Goggles should also be worn.

▶ Only properly trained, competent and registered people should mount an abrasive wheel. See 11.1.5 for a suitable mounting procedure.

▶ The maximum speed should be marked on the machine so that the abrasive wheel can be matched to the machine speed to ensure that the wheel permitted speed exceeds or equals the machine max speed.

▶ Noise levels should be checked and attenuating screens used if necessary.

▶ The machine should be on the PAT schedule and regularly checked.

▶ If necessary extract ventilation should be fitted to the wheel encasing to remove dust at source.

Workshop machinery – pedestal drill

Application of safeguards (Figure 11.30):

▶ Motor and drive should be fitted with fixed guard.

▶ The machine should be bolted down to prevent movement.

▶ The spindle should be guarded by an adjustable guard, which is fixed in position during the work.

▶ A clamp should be available on the pedestal base to secure work-pieces.

Figure 11.30 Typical pedestal drill

▶ The machine should be on the PAT schedule and regularly checked.

▶ Cutting fluid, if used, should be contained and not allowed to get onto clothing or skin. A splash guard may be required but is unlikely.

▶ Goggles should be worn by the operator.

Workshop machinery – bench-mounted circular saw

Application of safeguards (Figure 11.31):

▶ A fixed guard should be fitted to the blade below the bench.

▶ Fixed guards should be fitted to the motor and drives.

▶ An adjustable top guard should be fitted to the blade above the bench which encloses as much of the blade as possible. An adjustable front section should also be fitted.

▶ A riving knife should be fitted behind the blade to keep the cut timber apart and prevent ejection.

▶ A push stick should be used on short work-pieces (under 300 mm) or for the last 300 mm of longer cuts.

▶ Blades should be kept properly sharpened and set with the diameter of the smallest blade marked on the machine.

▶ Noise attenuation should be applied to the machine, for example damping, special quiet saw blades, with, if necessary, fitting in an enclosure. Hearing protection may have to be used.

▶ Protection against wet weather should be provided.

▶ The electrical parts should be regularly checked in addition to all the mechanical guards.

▶ Extraction ventilation will be required for the wood dust and shavings.

▶ Suitable dust masks should be worn.

▶ Suitable warm or cool clothing will be needed when used in hot or cold locations.

▶ Space around the machine should be kept clear.

▶ The machine should be bolted down to prevent movement.

Workshop machinery – hand-fed power planer

Application of controls (Figure 11.32):

▶ Fixed guards should be fitted to the motor and drives.

▶ An adjustable bridge guard should be fitted to the working side of the fence above the bench. It should be strong and rigid, and wide enough to cover the cutter block and, on larger tables, telescopic.

▶ A further bridge guard should be attached to the fence on the non-working side to cover the cutter block as the fence is moved across the table.

▶ Machines should be fitted with a braking device to bring the cutter block to rest within 10 seconds.

▶ Only cylindrical (or 'round form') cutter blocks should be used with maximum cutter projection of 10.1 mm. Keep gaps between the two sides of the table as small as possible.

▶ A push block with well-designed handles should be used when planing short pieces. In some cases a push stick will be required.

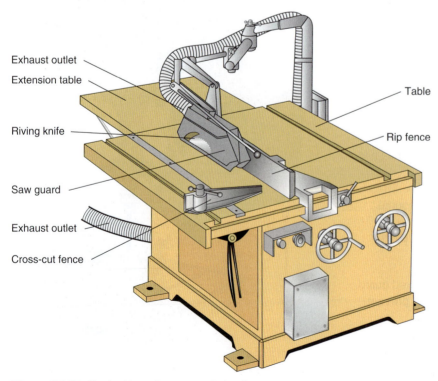

Exhaust outlet
Extension table
Riving knife
Saw guard
Exhaust outlet
Cross-cut fence
Table
Rip fence

Figure 11.31 Typical bench-mounted circular saw

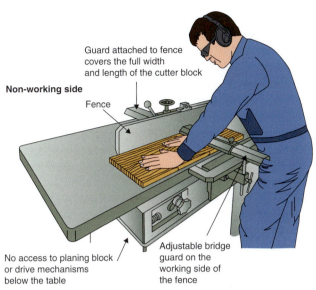

- Guard attached to fence covers the full width and length of the cutter block
- **Non-working side**
- Fence
- No access to planing block or drive mechanisms below the table
- Adjustable bridge guard on the working side of the fence

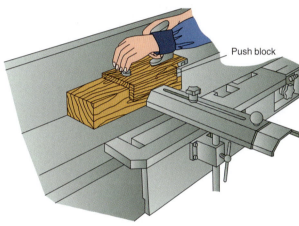

- Push block

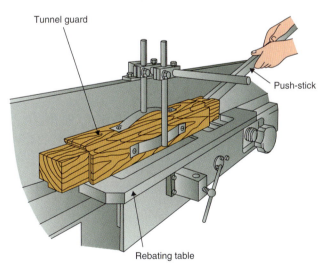

- Tunnel guard
- Push-stick
- Rebating table

Figure 11.32 Hand-fed planing machine safeguards

- ▶ Work-pieces should be supported properly. Tunnel/shaw guards should be used where appropriate.
- ▶ Noise attenuation should be applied to the machine, for example damping, use of special quiet saw blades and, if necessary, fitting in an enclosure. Hearing protection may have to be used.
- ▶ Protection against wet weather should be provided.
- ▶ The electrical parts should be checked regularly plus all the mechanical guards.

- ▶ Extraction ventilation will be required for the wood dust and shavings.
- ▶ Suitable dust masks should be worn.
- ▶ Areas around the machine should be kept clear.
- ▶ Use vacuum for cleaning.
- ▶ The machine should be bolted down to prevent movement.

Workshop machinery – spindle moulding machine

Application of controls (Figure 11.33):

- ▶ Machines should be fitted with a braking device that brings the cutter block to rest within 10 seconds.

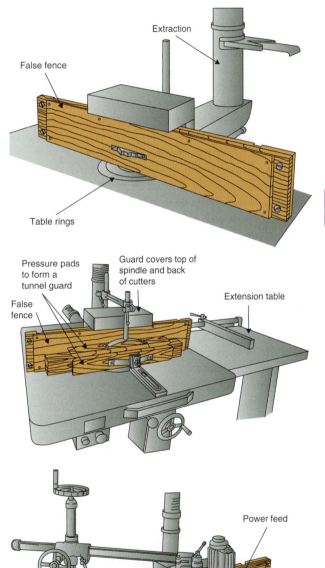

- Extraction
- False fence
- Table rings
- Pressure pads to form a tunnel guard
- Guard covers top of spindle and back of cutters
- False fence
- Extension table
- Power feed
- Shaw guard

Figure 11.33 Spindle moulding machine with various forms of safeguard

317

▶ Extraction ventilation will be required for the wood dust and shavings.
▶ Suitable dust masks should be worn.
▶ Areas around the machine should be kept clear.
▶ Use vacuum for cleaning.
▶ The machine should be bolted down to prevent movement.

Workshop machinery – spindle moulding machine

Application of controls (Figure 11.33):

▶ Machines should be fitted with a braking device that brings the cutter block to rest within 10 seconds.
▶ Table rings should be fitted to reduce the gap between the spindle and the table to a minimum.
▶ For most work the cutters should be effectively guarded. Where this is not possible use jigs or work holders and stops.
▶ Wherever possible, only feed the work-piece to the tool against the direction of spindle rotation.
▶ Only tools marked 'MAN' should be used, meaning hand-free.
▶ False fences should be used for straight work.
▶ Adequate work-piece support should be used.
▶ A demountable power feed should be used for straight cuts where possible. If not, the work area should be enclosed with pressure pads in the form of a tunnel.
▶ Jigs should be used where appropriate with, for example, stopped cuts.
▶ A ring guard or fence should be used for curved cuts.
▶ Dust and chippings should be removed by efficient extraction ventilation and use vacuum for cleaning.
▶ Noise attenuation should be applied to the machine, for example damping, use of special quiet saw blades and, if necessary, fitting in an enclosure. Hearing protection may have to be used.
▶ Protection against wet weather should be provided.
▶ The electrical parts should be regularly checked plus all the mechanical guards.
▶ Suitable dust masks should be worn.
▶ Areas around the machine should be kept clear.
▶ The machine should be bolted down to prevent movement.

Site machinery – compressor

Portable air compressors are very common pieces of machinery on construction sites. For an example see Figure 11.5(a) and (b).

Application of controls:

▶ The moving, dangerous parts of the engine should be fenced locally to prevent access. This will probably be achieved through total enclosure within the engine casing for most components and use of a separate guard for the cooling fan and drive belts.

▶ For electrically driven compressors, components should be properly constructed and designed to prevent contact with dangerous live currents. RCDs should be used and, where possible, reduced voltages at 110V. See Chapter 12 for more electrical hazards information. The equipment should be kept dry and not used in very damp or wet conditions. Extension cables should be fully unwound to avoid overheating and short circuits due to the current flow required for many units.
▶ Compressor air receivers/tanks should be regularly drained of any residual water and the oil levels in the compressor checked.
▶ If using for spray painting or similar keep the compressor at least 3 m away from the work area to reduce the risk of fire and explosion.
▶ The air receiver and associated pressure pipe-work should be thoroughly examined to ensure their integrity.
▶ Machines and tools with reduced noise and/ or vibration characteristics should be used. This will involve use of acoustic enclosures around compressors and anti-vibration engine mountings. For larger units, mounting on pneumatic tyres reduces vibration considerably.
▶ All hoses should be used carefully and regularly checked. Compressors should never be moved by pulling them by a hose. They should be supplied with claw-type quick action couplings or similar to prevent accidental decoupling. Air lines should always be connected to an air tool if they are being supplied with compressed air. A mechanism to cut-off air if a hose becomes disconnected or fails should be fitted.
▶ Compressors are fitted with air pressure gauges so that the pressure in the unit can be monitored and matched to the rating of a particular air tool.
▶ Protective clothing including eye and ear protection should be worn by people using air tools.
▶ Except for small units, moving compressors should be done by site vehicle/tractor. Only use on the road if properly equipped for road use with lights, tow bar and brakes (in most cases).

Site machinery – cement mixer

Application of controls (Figure 11.34(a) and (b)):

▶ Operating position for the hopper hoist should be designed so that anyone in the trapping area is visible to the operator. The use of the machine should be restricted to designated operators only. As far as possible the trapping point should be designed out. The hoist operating location should be fenced off just allowing access for barrows, etc., to the unloading area.

(a)

Figure 11.35 Plate compactor

Site machinery – plate compactor

Application of controls (Figure 11.35):

▶ Select reduced/anti-vibration models.
▶ Have hold-to-run controls.
▶ Always push away from body when operating and wear steel-toe-capped boots.
▶ Engines must only be run in the open air.
▶ Drives and rotating parts of the engine should be enclosed.
▶ If petrol-driven, care is required with flammable liquids and refuelling.
▶ Protection from the sun should be provided in the form of PPE for the operator.
▶ Noise levels should be checked and noise attenuation, for example silencers and damping, fitted if necessary.
▶ Suitable eye protection should be worn.
▶ Respiratory protection to be worn where dusty ground is being compacted.
▶ Manual handling assessments should be carried out. Minimise handling and use two people for lifting or mechanised equipment.

(b)

Figure 11.34 (a) Typical small cement mixer with a petrol engine or electric motor; (b) diesel-powered concrete mixer fitted with a hydraulic loading hopper

▶ Drives and rotating parts of the engine should be enclosed.
▶ The drum gearing should be enclosed and persons kept away from the rotating drum, which is normally fairly high on large machines.
▶ No one should be allowed to stand on the machine while it is in motion.
▶ Goggles should be worn to prevent cement splashes.
▶ If petrol-driven, care is required with flammable liquids and refuelling.
▶ Engines must only be run in the open air.
▶ Electric machines should be checked regularly and be on the PAT schedule.
▶ Noise levels should be checked and noise attenuation, for example silencers and damping, fitted if necessary.

Site machinery – ground consolidation equipment

Application of controls (Figure 11.36):

▶ Select reduced/anti-vibration models.
▶ Engines must only be run in the open air.
▶ Drives and rotating parts of engine should be enclosed.
▶ Operator seat restraints should be fitted and worn.
▶ Hold-to-run controls should be fitted.
▶ Suitable access and egress steps and handholds are required.

11

Figure 11.36 (a) Ground consolidating rider-mounted vibrating roller

Figure 11.36 (b) Ground consolidating roller using remote control which eliminates the whole-body vibration exposure of the operator

▶ Rollover or turnover protection (ROPS) should be fitted to protect the driver.
▶ If petrol-driven, care is required with flammable liquids and refuelling.
▶ Noise levels should be checked and noise attenuation, for example silencers and damping, fitted if necessary.
▶ Respiratory protection to be worn where dusty ground is being consolidated, unless the cab is enclosed with air filtration/air conditioning.
▶ Protection from the sun: should be fitted with a canopy or cab.
▶ Suitable foot protection to be worn.

Site machinery – road marking equipment

Application of controls (Figures 11.37 and 11.38):

▶ Protective clothing must include high visibility overalls or jackets with reflective strips for all

personnel. In addition where appropriate and risk assessments require it:

▷ goggles;
▷ helmets – when working in hazardous conditions, when workers are overhead, where there is overhanging debris or where working under anything that could strike from overhead;
▷ suitable respirators when working in dusty conditions or when fumes and vapours are present.

▶ Roadworks may not commence until sufficient road warning signs as required by the Department for Transport Code of Practice 'Safety at street works and roadworks' (ISBN 0115523103) are in position, and, where appropriate, footpaths have been provided for pedestrians. All road signs must be removed at the completion of the work. Many road marking schemes will use moving signs attached to vehicles as the work proceeds. These should all be in accordance with the Code of Practice. See Section 9.1.5 for more details.
▶ Yellow warning lights with amber lenses must be placed around hazards left overnight. Check that lamps are recharged/refilled and lit before dark.
▶ Operatives of mechanically propelled road marking equipment must be over 18, hold a current driving licence for the category of vehicle, and have been trained and authorised to drive the equipment.
▶ All moving parts of the engine must be properly safeguarded. Engines are usually totally enclosed except for cooling fans, alternators and belts which need to be guarded.
▶ Equipment, flexible pipes, tools and pressure vessels should all be checked for leaks or other defects before using.
▶ No air guns or paint sprays should be used on people even in jest, to prevent compressed air entering the bloodstream.
▶ No smoking or naked flames should be used with flammable road marking paints.

Figure 11.37 Highway line-marking lorry-mounted equipment

Figure 11.38 Walk-behind line-marking equipment

▶ Heat sources should be carefully controlled with thermostatic controls on electrical systems. LPG cylinders should be fixed in an upright position with effective on/off valves, and hoses in date and in good order.

Site machinery – electric generator

Application of controls (see Figures 11.39 and 11.40):

▶ Do not connect portable generators directly to fixed installed wiring installations. Power from generators can back-feed along power lines and electrocute anyone coming in contact with them, including line-workers making repairs.

▶ If the generator is permanently connected into the fixed installation as a standby power source, the system must be properly protected. One good way to avoid back-feeding is to install a double-pole, double-throw transfer-switch gear. The generator must be installed by a competent electrician to make sure it meets local electrical codes. A qualified electrical contractor can install this transfer switch so that dangerous back-feed can be prevented. Since transfer switches can be expensive, another way to install a generator is to have a sub-panel with main breakers and power from the main panel or generator. The main panel breaker and generator breaker in the sub-panel would have handles interlocked to prevent both from being opened and closed at the same time. This prevents back-feed to commercial power when the generator is in use.

▶ The generator must be properly grounded to earth.

▶ The generator must be kept dry and operated under a canopy or open-sided shelter during rain or snow.

▶ Do not use a portable generator in a flooded basement. That could be a dangerous combination. In addition, make sure hands are dry, that the operator is standing in a dry place, and the generator is properly grounded whenever it is operated.

▶ Make sure extension cords used with generators are rated for the load, are free of cuts and worn insulation, and have three-pronged plugs.

▶ Do not overload the generator. A portable generator should be used only when necessary, and only to power essential equipment or appliances.

▶ Never operate the generator in enclosed or partially enclosed spaces. Generators can produce high levels of carbon monoxide very quickly, which can be deadly. Make certain, too, that the generator has enough air to breath and that its exhaust is vented properly.

▶ Use a residual current device (RCD) to help prevent electrocutions and electrical shock injuries.

▶ Make sure fuel for the generator is stored safely, away from living areas, in properly labelled containers, and away from fuel-burning appliances. Before re-fuelling, always turn the generator off and let it cool down.

▶ Turn off all appliances powered by the generator before shutting down the generator.

▶ Keep children away from portable generators at all times.

▶ Make sure the moving parts are correctly guarded.

▶ If noise levels are a problem ensure that the generator doors are kept shut and it is operated away from people. The area should be marked as a noise control zone.

▶ Always follow the instructions which are contained in the generator's operating manual.

11.4.9 Basic requirements for guards and safety devices

The design and construction of guards must be appropriate to the risks identified and the mode

Figure 11.39 Typical portable petrol engine, electric generator

321

Figure 11.40 Larger transportable diesel-powered electric generator in an enclosure

of operation of the machinery in question. ISO 12100:2010 gives the following general requirements for guards and protective devices which should protect against danger, including risks from moving parts. They should:

(a) be of robust construction;

(b) be securely held in place;

(c) not give rise to any additional hazard;

(d) not be easy to bypass or render non-operational, or be easily defeated;

(e) be located at an adequate distance from the danger zone;

(f) cause minimum obstruction of the view of the production process; and

(g) enable essential work to be carried out on the installation and replacement of tools and for maintenance purposes by restricting access exclusively to the area where the work has to be done, if possible without the guard having to be removed or the protective device having to be disabled.

In addition, guards should protect against the ejection or falling of materials or objects and against emissions generated by the machinery.

These general requirements involve consideration of the following factors:

► strength – guards and protection devices must be of good construction, sound material and adequate strength. They must be capable of doing the job they are intended to do. Several factors can be considered:

▷ material of construction (metal, plastic, laminated glass, etc.);

▷ form of the material (sheet, open mesh, bars, etc.);

▷ method of fixing.

They must be adequate for the purpose, able to resist the forces and vibration involved and able to withstand impact (where applicable). Good construction involves design and layout as well as the mechanical nature and

quality of the construction. Foreseeable use and misuse should be taken into account. Guards and protection devices must be designed and installed so that they cannot be easily bypassed or disabled. This refers to accidental or deliberate action that removes the protection offered;

► weight and size – in relation to the need to remove and replace the guard during maintenance;

► compatibility with materials being processed and lubricants, etc.;

► hygiene and the need to comply with food safety Regulations;

► visibility – may be necessary to see through the guard for both operational and safety reasons. Guards and protection devices must not unduly restrict the view of the operating cycle of the machinery, where such a view is necessary. It is not usually necessary to be able to see all the machine – the part that needs to be seen is normally that which is acting directly on material or a work-piece. Operations for which it is necessary to provide a view include those where the operator controls and feeds a loose work-piece at a machine. Examples include manually fed woodworking machines and food slicers. Many of these operations involve the use of protection appliances. Where an operation protected by guards needs to be seen, the guard should be provided with viewing slits or properly constructed panels, perhaps backed up by internal lighting, enabling the operator to see the operation;

► noise attenuation – guards are often utilised to reduce the noise levels produced by a machine. Conversely, the resonance of large undamped panels may exacerbate the noise problem;

► enabling a free flow of air – where necessary (e.g. for ventilation);

► avoidance of additional hazards – for example free of sharp edges. Guards must be constructed so that they are not themselves dangerous parts. If a guard is power-operated or assisted, the closing or opening action might create a potentially dangerous trap which needs secondary protection, for example a leading-edge trip bar or pressure-sensitive strip;

► ease of maintenance and cleanliness – it should be necessary to use a tool such as a spanner or screwdriver to remove a fixed guard. Guards should be interlocked with the main drive if they need to be removed frequently for maintenance or cleaning. Movable panels in guards giving access to dangerous parts, or movable guards themselves, will often need to be fitted with an interlocking device. This device must be designed and installed so that it is difficult or impossible to bypass or defeat. Guards and protection devices must be maintained in an efficient state, in efficient working order and in good repair. This is an important requirement as many accidents have occurred when guards have not been maintained. It is a particular example of the general requirement under Regulation 5 of PUWER to maintain equipment. Compliance can be achieved by the use of an effective check procedure for guards and protection devices, together with any necessary follow-up action. In the case of protection devices or interlocks, some form of functional check or test is desirable;

▶ openings – the size of openings and their distance from the dangerous parts should not allow anyone to be able to reach into a danger zone. These values can be determined by experiment or by reference to standard tables. If doing so by experiment, it is essential that the machine is first stopped and made safe (e.g. by isolation). The detailed information on openings is contained in EN 294:1992, EN 349:1993 and EN 811:1997. Safeguarding is normally attached to the machine, but the regulation does not preclude the use of free-standing guards or protection devices. In such cases, the guards or protection devices must be fixed in an appropriate position relative to the machine.

11.5 Further information

Safe Use of Work Equipment Provision and Use of Work Equipment Regulations 1998 Approved Code of Practice and guidance L22 (4th edition), HSE Books, 2014, ISBN 978 0 7176 6619 5. http://www.hse.gov.uk/pubns/books/l22.htm

Personal Protective Equipment at Work Regulations 1992 (as amended). Guidance on Regulations L25. HSE Books, ISBN 978 0 7176 6139 8 http://www.hse.gov.uk/pubns/books/l25.htm

Safety of machinery – General principles for design – Risk assessment and risk reduction, ISO 12100:2010, ISBN 9780 580 74262 0

The Supply of Machinery (Safety) Regulations 2008 Scope and application, and relationship to CE marking. http://www.hse.gov.uk/pubns/indg270.pdf

Also see HSE micro site on New Machinery. http://www.hse.gov.uk/work-equipment-machinery/new-machinery.htm

The Provision and Use of Work Equipment Regulations 1998 – Part II. http://www.legislation.gov.uk/uksi/1998/2306/contents/made

The Personal Protective Equipment at Work Regulations 1992. http://www.legislation.gov.uk/uksi/1992/2966/contents/made

11.6 Practice revision questions

1. A woodworking company has decided to purchase a new machine.
   (a) **Identify** the questions that must be answered before a purchase is made.
   (b) **Outline** the issues that will need to be considered before the machine is put into use.
   (c) **Identify** the main issues when assessing the suitability of controls (including emergency controls) of the new machine.

2. (a) **Describe** the contents of a 'Declaration of Conformity' that should be issued with a new machine.
   (b) **Outline** the advantages and limitations of CE marking on a new machine.

3. **Outline** a hierarchy of control measures that may be used to prevent contact with dangerous parts of machinery. Give an example of the level of protection required at each stage of the hierarchy.

4. An operator suffers a serious injury after coming into contact with a dangerous part of a machine, **describe**:
   (a) Possible immediate causes.
   (b) Possible root (underlying) causes.

5. Employers are required to provide adequate information, instruction and training to ensure the safe use of work equipment.
   (a) **Identify THREE** categories of employees that should receive information, instruction and training on the safe use of work equipment.
   (b) **Outline** the issues that could be included in such information, instruction and training.

6. (a) **Describe THREE** preventative maintenance schemes that may be used on machinery to prevent health and safety risks.
   (b) **Outline** the measures to be taken to reduce the risk of accidents associated with the routine maintenance of machinery.
   (c) **Identify FOUR** non-mechanical hazards that could lead to injury and/or ill-health when undertaking maintenance work on machinery in the workplace.

7. A bricklayer uses a hammer and chisel to split a building block.
   (a) **Identify FOUR** unsafe conditions, associated with the tools, which could affect the safety of the bricklayer.

(b) **Outline** suitable control measures for minimising the risk of injury to the bricklayer when using the tools.

8. A carpenter regularly uses a hand-held electric sander for the preparation of wooden doors before painting them.

    (a) **Outline** the checks that should be made to ensure the electrical safety of the sander.

    (b) Other than electricity, **identify EIGHT** hazards associated with the use of the sander.

9. Hand-held electric drills are commonly used on construction sites.

    (a) **Outline** the checks that should be carried out by the user of a drill to reduce the likelihood of electric shock.

    (b) Other than electricity, **identify FOUR** hazards associated with the use of hand-held electric drills.

10. **Outline** the ways in which accidents can occur from the use of cartridge-operated nail guns during building work.

11. **Outline** the control measures that should be adopted when cutting paving slabs with a petrol-driven disc-cutter.

12. A petrol-driven chainsaw is to be used to remove several branches from a large tree.

    (a) **Identify EIGHT** hazards associated with the use of the chainsaw for this work.

    (b) **Outline EIGHT** safeguards that should be in place to ensure the safe use of chainsaws.

    (c) **Identify** the items of personal protective equipment that should be used by the chainsaw operative.

13. The preparation of a new construction site requires the complete removal of some trees and the de-branching of others. An overhead power line crosses the site and buried cables may be present. **Describe** a safe system of work for:

    (a) the removal of the timber using chain saws

    (b) the removal of tree roots by means of a wheeled excavator.

14. (a) **Identify FOUR** non-mechanical hazards that may be encountered on woodworking machines and **outline** the possible health and safety effects from exposure in **EACH** case.

    (b) **Identify EIGHT** mechanical hazards associated with moving parts of machinery.

15. **Identify THREE** mechanical and **TWO** non-mechanical hazards, and the associated control measures with the use of the following items of equipment:

    (a) a bench-top grinder

    (b) a document shredder

    (c) a photocopier

    (d) a bench-mounted circular saw.

16. (a) **Identify TWO** types of equipment used in the construction industry which are fitted with an abrasive wheel and **describe** the hazards associated with their use.

    (b) **Outline** the correct method of mounting an abrasive wheel.

    (c) **Prepare** a **10-point** checklist that can be used to monitor safe working practices when abrasive wheels are being used.

17. Work is taking place on a construction site adjacent to a public highway.

    (a) **Outline** practical ways of reducing the risk to employees when using an electrical powered cement mixer on the site.

    (b) **Describe** the hazards associated with the use of a pedestrian-operated plate compactor when used on the public highway.

18. A new pedestal (pillar) drill is to be used in a maintenance workshop.

    (a) **Identify** the issues that should be addressed before it is first used so that the risk of injury to operators of the machine is reduced.

    (b) **Identify FOUR** mechanical hazards presented by the machine and **outline** in each case how injury may occur and the appropriate control measures.

19. A bench-mounted circular saw is used in a carpentry workshop on a regular basis.

    (a) **Identify FOUR** risks to the health **AND FIVE** risks to the safety of the circular saw operators.

    (b) **Outline** the control measures that can be implemented to minimise the health and safety risks to these operators.

20. A local authority uses a rider-operated petrol-powered motor-mower to cut the grass on roadside verges.

    (a) **Identify FIVE** mechanical and **FOUR** non-mechanical hazards that the operator could be exposed to while using the mower.

    (b) **Outline** the precautions that should be taken to address the hazards identified in part (a).

21. Concrete floors inside an empty factory are to be renewed and the old ones are to be broken up using pneumatic jackhammers powered by compressors.

(a) **Outline** the hazards when using this type of equipment.

(b) **Outline** the control measures that must be taken to reduce the risks to site operatives when using this equipment.

22. A large portable diesel air compressor is to be used on a construction site for the supply of air for a variety of tasks and operations.

(a) **Outline** the health, safety and environmental issues directly associated with the running of the diesel compressor.

(b) **Outline** the possible controls for the health, safety and environmental issues identified.

23. **Outline** the principles of operation, advantages and limitations of the following guards and safeguarding devices:

(a) Fixed guards
(b) Interlocked guards
(c) Adjustable guards
(d) Trip devices
(e) Two-handed control devices.

24. Other than contact with dangerous parts, **identify** **FOUR** types of hazard against which fixed guards on machines may provide protection.

11

CHAPTER 12

Electrical safety

> **This chapter covers the following NEBOSH learning objectives:**
> 1. Outline the principles, hazards and risks associated with the use of electricity in the workplace
> 2. Outline the control measures that should be taken when working with electrical systems or using electrical equipment
> 3. Outline the control measures to be taken when working near or underneath overhead power lines

12.1 Hazards and risks associated with the use of electricity in the workplace

12.1.1 Introduction

Electricity is a widely used, efficient and convenient, but potentially hazardous, method of transmitting and using energy. It is in use in every factory, workshop, laboratory and office in the country. Any use of electricity has the potential to be very hazardous with possible fatal results. Legislation has been in place for many years to control and regulate the use of electrical energy and the activities associated with its use. Such legislation provides a framework for the standards required in the design, installation, maintenance and use of electrical equipment and systems and the supervision of these activities to minimise the risk of injury. Electrical work from the largest to the smallest installation must be carried out by people known to be competent to undertake such work. New installations always require expert advice at all appropriate levels to cover design aspects of both the system and its associated equipment. Electrical systems and equipment must be properly selected, installed, used and maintained.

In the UK approximately 2% of all fatalities at work are caused by electric shock. Over the last few years, there have been 378 electrical accidents each year and 25 people died of their injuries. The majority of the fatalities occur in the agricultural, extractive and utility supply and service industries, whereas the majority of the major accidents happen in the manufacturing, construction and service industries.

Only voltages up to and including **mains voltage** (220/240V) and the three principal electrical hazards – electric shock, electric burns and electrical fires and explosions – are considered in detail in this chapter.

12.1.2 Basic principles of electricity

In simple terms, electricity is the flow or movement of electrons through a substance which allows the transfer of electrical energy from one position to another. The substance through which the electricity flows is called a **conductor**. This flow or movement of electrons is known as the **electric current**. There are two forms of electric current – direct and alternating. **Direct current (dc)** involves the flow of electrons along a conductor from one end to the other. This type of current is mainly restricted to batteries, dynamos and similar devices.

Alternating current (ac) is produced by a rotating magnetic alternator that produces a change in polarity every half turn causing an oscillation of the electrons rather than a flow of electrons so that energy is passed from one electron to the adjacent one and so on through the length of the conductor.

It is sometimes easier to understand the basic principles of electricity by comparing its movement to that of water in a pipe flowing downhill. The flow of water through the pipe (measured in litres per second) is similar to the current flowing through the conductor which is measured in amperes, normally abbreviated to **amps (A)**. Sometimes very small currents are used and these are measured in milliamps (mA).

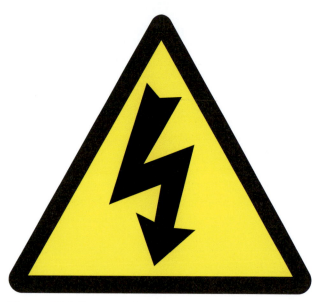

Figure 12.1 Beware of electricity – typical sign

The higher the pressure drop is along the pipeline, the greater will be the flow rate of water and, in a similar way, the higher the electrical 'pressure difference' along the conductor, the higher the current will be. This electrical 'pressure difference', or potential difference, is measured in volts (V).

The flow rate through the pipe will also vary for a fixed pressure drop as the roughness on the inside surface of the pipe varies – the rougher the surface, the slower the flow and the higher the resistance to flow becomes. Similarly, for electricity, the poorer the conductor, the higher the resistance is to electrical current and the lower the current becomes. Electrical resistance is measured in **ohms**.

The voltage (V), the current (I) and the resistance (R) are related by the following formula, known as Ohm's law:

$$V = I \times R \text{ (volts)}$$

and, electrical power (P) is given by:

$$P = V \times I \text{ (watts)}$$

These basic formulae enable simple calculations to be made so that, for example, the correct size of fuse may be ascertained for a particular piece of electrical equipment.

Conductors and insulators

Conductors are nearly always metals, copper being a particularly good conductor, and are usually in wire form but they can be gases or liquids, water being a particularly good conductor of electricity. Superconductor is a term given to certain metals which have a very low resistance to electricity at low temperatures.

Very poor conductors are known as **insulators** and include materials such as rubber, timber and plastics. Insulating material is used to protect people from some of the hazards associated with electricity.

Short circuit

Electrical equipment components and an electrical power supply (normally the mains or a battery) are joined together by a conductor to form a **circuit**. If the circuit is broken in some way producing a fault so that the current flows directly to earth rather than to a piece of equipment, a **short circuit** is made. As the resistance is greatly reduced but the voltage remains the same, a rapid increase in current occurs which could cause significant problems if suitable protection were not available. A short circuit is thus a live to earth contact without resistance.

Earthing

The principle of **earthing** is that the Earth is basically a huge conductor which accepts any electrical current that is conducted to it. Any metal in an electrical device with which a person might come into contact (for example the metal casing of a food mixer) should be connected to earth so that if, due to a fault, the casing becomes live, the electricity will flow away to earth – an easier path than through the body. In the USA or where OSHA standards are adopted, this process is called grounding.

The electricity supply company has one of its conductors solidly connected to earth and every circuit supplied by the company must have one of its conductors connected to earth. This means that if there is a fault, such as a break in the circuit, the current, known as the fault current, will return directly to earth, which forms the circuit of least resistance, thus maintaining the supply circuit. This process is known as **earthing**. Other devices, such as fuses and residual current devices (RCDs), which will be described later, will also be needed within the circuit to interrupt the current flow to earth so as to protect people from electric shock and equipment from overheating. Good and effective earthing is absolutely essential and must be connected and checked by a competent person. Where a direct contact with earth is not possible, for example in a motor car, a common voltage reference point is used, such as the vehicle chassis.

Where other potential metallic conductors exist near to electrical conductors in a building, they must be connected to the main earth terminal to ensure **equipotential bonding** of all conductors to earth. This applies to gas, water and central heating pipes and other devices such as lightning protection systems. **Supplementary bonding** is required in bathrooms and kitchens where, for example, metal sinks and other metallic equipment surfaces are present. This involves the connection of a conductor from the sink to a water supply pipe which has been earthed by equipotential bonding. There have been several fatalities due to electric shocks from 'live' service pipes or kitchen sinks.

Some definitions

Certain terms are frequently used with reference to electricity and the more common ones are defined here.

Low voltage – This is a voltage normally not exceeding 1000V dc or 600V ac between conductors and earth or 1000V ac between phases. Mains voltage falls into this category.

High voltage – This is defined in national and international standards as a voltage exceeding 1000V dc or 600V ac between conductors and earth or 1000V ac between phases.

Mains voltage – The common voltage available in domestic premises and many workplaces and normally taken from three-pin socket points. In the UK, it is distributed by the national grid and is usually supplied between 220 and 240V ac and at 50 cycles/s.

Maintenance – A combination of any actions carried out to retain an item of electrical equipment in, or restore it to, an acceptable and safe condition.

Testing – A measurement carried out to monitor the conditions of an item of electrical equipment without physically altering the construction of the item or the electrical system to which it is connected.

Inspection – A maintenance action involving the careful scrutiny of an item of electrical equipment, using, if necessary, all the senses to detect any failure to meet an acceptable and safe condition. An inspection does not include any dismantling of the item of equipment.

Examination – An inspection together with the possible partial dismantling of an item of electrical equipment, including measurement and non-destructive testing as required, in order to arrive at a reliable conclusion as to its condition and safety.

Isolation – Involves cutting off the electrical supply from all or a discrete section of the installation by separating the installation or section from every source of electrical energy. This is the normal practice so as to ensure the safety of persons working on or in the vicinity of electrical components which are normally live and where there is a risk of direct contact with live electricity.

Competent electrical person – A person possessing sufficient electrical knowledge and experience to avoid the risks to health and safety associated with electrical equipment and electricity in general.

12.1.3 Hazards, risks and danger of electricity

Electricity is a safe, clean and quiet method of transmitting energy. However, this apparently benign source of energy when accidentally brought into contact with conducting material, such as people, animals or metals, permits dangerous releases of energy which may result in serious damage or loss of life. Constant awareness is necessary to avoid and prevent danger from accidental releases of electrical energy.

The principal hazards associated with electricity are:

▶ electric shock
▶ electric burns
▶ electrical fires and explosions
▶ arcing
▶ secondary hazards.

The use of portable electrical equipment can lead to a higher likelihood of these hazards occurring.

Electric shock and burns

There are 1,000 workplace accidents and 30 fatalities involving electric shock and burns reported to the HSE each year.

Electric shock is the convulsive reaction by the human body to the flow of electric current through it. This sense of shock is accompanied by pain and, in more severe cases, by burning. The shock can be produced by low voltages, high voltages or lightning. Most incidents of electric shock occur when the person becomes the route to earth for a live conductor. The effect of electric shock and the resultant severity of injury depend upon the size of the electric current passing through the body which, in turn, depends on the voltage and the electrical resistance of the

skin and body. If a person comes into contact with a voltage above about 50 volts, they can receive a range of injuries, including those directly resulting from electrical shock (such as problems with breathing and heart function), and indirect effects resulting from loss of control (such as falling from height or coming into contact with moving machinery). The chance of being injured by an electric shock increases where it is damp or where there is a lot of metalwork.

If the skin is wet, a shock from mains voltage (220/240V) could well be fatal. The effect of shock is very dependent on conditions at the time but it is always dangerous and must be avoided. Electric burns are usually more severe than those caused by heat, since they can penetrate deep into the tissues of the body.

The effect of electric current on the human body depends on its pathway through the body (e.g. hand to hand or hand to foot), the frequency of the current, the length of time of the shock and the size of the current. Current size is dependent on the duration of contact and the electrical resistance of body tissue. The electrical resistance of the body is greatest in the skin and is approximately 100,000 ohm; however, this may be reduced by a factor of 100 when the skin is wet. The body beneath the skin offers very little resistance to electricity due to its very high water content and, while the overall body resistance varies considerably between people and during the lifetime of each person, it averages at 1,000 ohm. Skin that is wounded, bruised or damaged will considerably reduce human electrical resistance and work should not be undertaken on electrical equipment if damaged skin is unprotected.

An electric current of 1mA is detectable by touch and one of 10mA will cause muscle contraction which may prevent the person from being able to release the conductor, and if the chest is in the current path, respiratory movement may be prevented, causing asphyxia. Current passing through the chest may also cause fibrillation of the heart (vibration of the heart muscle) and disrupt the normal rhythm of the heart, though this is likely only within a particular range of currents. The shock can also cause the heart to stop completely (cardiac arrest) and this will lead to the cessation of breathing. Current passing through the respiratory centre of the brain may cause respiratory arrest that does not quickly respond to the breaking of the electrical contact. These effects on the heart and respiratory system can be caused by currents as low as 25mA. It is not possible to be precise on the threshold current because it is dependent on the environmental conditions at the time, as well as the age, sex, body weight and health of the person.

Alternating current (AC) and direct current (DC) have slightly different effects on the human body, but both are dangerous above a certain voltage. The risk of injury changes according to the frequency of the AC, and it is common for DC to have an AC component (called ripple). The effect on a particular person is very difficult to predict as it depends upon a large number of factors, as mentioned above.

Burns of the skin occur at the point of electrical contact due to the high resistance of skin. These burns may be deep, slow to heal and often leave permanent scars. Burns may also occur inside the body along the path of the electric current, causing damage to muscle tissue and blood cells. These burns result from the direct contact with an electrical source. Burns can also be caused by indirect contact with an electrical source when such a source results in a fire. Successive annual UK fire statistics show that faulty appliances and leads continue to be the single most common cause of accidental fires in 'other buildings' (non-dwellings).

Burns associated with radiation and microwaves are dealt with in Chapter 15.

Treatment of electric shock and burns

There are many excellent posters available which illustrate a first-aid procedure for treating electric shock and such posters should be positioned close to electrical junction boxes or isolation switches (Figure 12.2). The recommended procedure for treating an unconscious person who has received a **low-voltage** electric shock is as follows:

1. On finding a person suffering from electric shock, raise the alarm by calling for help from colleagues (including a trained first aider).
2. Switch off the power if it is possible and/or the position of the emergency isolation switch is known.

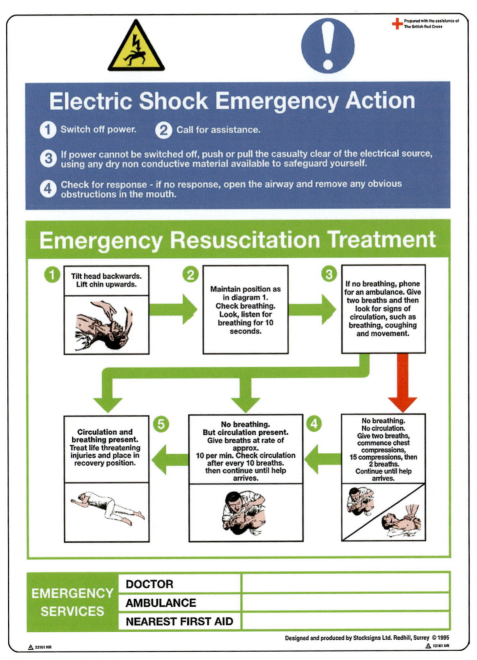

Figure 12.2 Typical electric shock poster

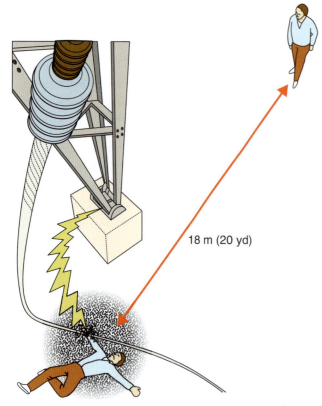

18 m (20 yd)

Figure 12.3 Keep 18 m clear of high-voltage lines

3. Call for an ambulance.
4. If it is not possible to switch off the power, then push or pull the person away from the conductor using an object made from a good insulator, such as a wooden chair or broom. Remember to stand on dry insulating material, for example a wooden pallet, rubber mat or wooden box. If these precautions are not taken, then the rescuer will also be electrocuted.
5. If the person is breathing, place him/her in the recovery position so that an open airway is maintained and the mouth can drain if necessary.
6. If the person is not breathing, apply mouth-to-mouth resuscitation and, in the absence of a pulse, chest compressions. When the person is breathing normally place them in the recovery position.
7. Treat any burns by placing a sterile dressing over the burn and secure with a bandage. Any loose skin or blisters should not be touched nor any lotions or ointments applied to the burn wound.
8. If the person regains consciousness, treat for normal shock.
9. Remain with the person until they are taken to a hospital or local surgery.

It is important to note that electrocution by high-voltage electricity is normally instantly fatal. On discovering a person who has been electrocuted by high-voltage electricity, the police and electricity supply company should be informed. If the person remains in contact with or within 18 m of the supply, then he/she should not be approached to within 18 m by others until the supply has been switched off and clearance has been given by the emergency services. High-voltage electricity can 'arc' over distances less than 18 m, thus electrocuting the would-be rescuer (Figure 12.3).

Electrical fires and explosions

Over 25% of all fires have a cause linked to a malfunction of either a piece of electrical equipment or wiring or both. Electrical fires are often caused by a lack of reasonable care in the maintenance and use of electrical installations and equipment. The electricity that provides heat and light and drives electric motors is capable of igniting insulating or other combustible material if the equipment is misused, is not adequate to carry the electrical load or is not properly installed and maintained. The most common causes of fire in electrical installations are short circuits, overheating of cables and equipment, the ignition of flammable gases and vapours and the ignition of combustible substances by static electrical discharges.

Short circuits happen, as mentioned earlier, if insulation becomes faulty, and an unintended flow of current between two conductors or between one conductor and earth occurs. The amount of the current depends, among other things, upon the voltage, the condition of the insulating material and the distance between the conductors. At first the current flow will be low, but as the fault develops the current will increase and the area surrounding the fault will heat up. In time, if the fault persists, a total breakdown of insulation will result and excessive current will flow through the fault. If the fuse fails to operate or is in excess of the recommended fuse rating, overheating will occur and a fire will result. A fire can also be caused if combustible material is in close proximity to the heated wire or hot sparks are ejected. Short circuits are most likely to occur where electrical equipment or cables are susceptible to damage by water leaks or mechanical damage. Twisted or bent cables can also cause breakdowns in insulation materials.

Inspection covers and cable boxes are particular problem areas. Effective steps should be taken to prevent the entry of moisture as this will reduce or eliminate the risk. Covers can themselves be a problem especially in dusty areas where the dust can accumulate on flat insulating surfaces, resulting in tracking between conductors at different voltages and a subsequent insulation failure. The interior of inspection panels should be kept clean and dust-free by using a suitable vacuum cleaner.

Overheating of cables and equipment will occur if they become overloaded. Electrical equipment and circuits are normally rated to carry a given safe current which will keep the temperature rise of the conductors in the circuit or appliance within permissible limits and avoid the possibility of fire. These safe currents define the

Figure 12.4 Electrical faults through overloading or damaged cables cause a large number of fires on construction sites (see Chapter 13)

(a)

(b)

Figure 12.5 (a) Typical transformer; (b) typical RCD device

12

maximum size of the fuse (the fuse rating) required for the appliance. A common cause of circuit overloading is the use of equipment and cables which are too small for the imposed electrical load. This is often caused by the addition of more and more equipment to the circuit, thus taking it beyond its original design specification. In offices, the overuse of multi-socket incorrectly fused outlet adaptors can create overload problems (sometimes known as the Christmas tree effect). The more modern multi-plugs are much safer as they lead to one fused plug and cannot be easily overloaded. Another cause of overloading is mechanical breakdown or wear of an electric motor and the driven machinery. Motors must be maintained in good condition with particular attention paid to bearing surfaces. Fuses do not always provide total protection against the overloading of motors and, in some cases, severe heating may occur without the fuses being activated.

Loose cable connections are one of the most common causes of overheating and may be readily detected (as well as overloaded cables) by a thermal imaging survey (a technique which indicates the presence of hot spots). The bunching of cables together can also cause excessive heat to be developed within the inner cable, leading to a fire risk. This can happen with cable extension reels, which have only been partially unwound, used for high-energy appliances like an electric heater.

Ventilation is necessary to maintain safe temperatures in most electrical equipment and overheating is liable to occur if ventilation is in any way obstructed or reduced. All electrical equipment must be kept free of any obstructions that restrict the free supply of air to the equipment and, in particular, to the ventilation apertures.

Most electrical equipment either sparks in normal operation or is liable to spark under fault conditions. Some electrical appliances, such as electric heaters,

are specifically designed to produce high temperatures. These circumstances create fire and explosion hazards, which demand very careful assessment in locations where processes capable of producing flammable concentrations of gas or vapour are used, or where flammable liquids are stored.

It is likely that many fires are caused by static electrical discharges. Static electricity can, in general, be eliminated by the careful design and selection of materials used in equipment and plant, and the materials used in products being manufactured. When it is impractical to avoid the generation of static electricity, a means of control must be devised. Where flammable materials are present, especially if they are gases or dusts, then there is a great danger of fire and explosion, even if there is only a small discharge of static electricity. The control and prevention of static electricity is considered in more detail later.

The use of electrical equipment in potentially flammable atmospheres should be avoided as far as possible. However, there will be many cases where electrical equipment must be used and, in these cases, the standards for the construction of the equipment should comply with the Equipment and Protective

Systems Intended for Use in Potentially Explosive Atmospheres Regulations, known as ATEX. Details on the classification or zoning of areas are contained in the Dangerous Substances and Explosive Atmospheres Regulations and ACoPs.

Before electrical equipment is installed in any location where flammable vapours or gases may be present, the area must be zoned in accordance with the Dangerous Substances and Explosive Atmosphere Regulations, and records of the zoned areas must be marked on building drawings and revised when any zoned area is changed. The installation and maintenance of electrical equipment in potentially flammable atmospheres is a specialised task. It must only be undertaken by electricians or instrument mechanics who are trained to ATEX standards.

In the case of a fire involving electrical equipment, the first action must be the isolation of the power supply so that the circuit is no longer live. This is achieved by switching off the power supply at the mains isolation switch or at another appropriate point in the system. Where it is not possible to switch off the current, the fire must be attacked in a way which will not cause additional danger. The use of a non-conducting extinguishing medium, such as carbon dioxide or powder, is necessary. After extinguishing such a fire, careful watch should be kept for renewed outbreaks until the fault has been rectified. Re-ignition is a particular problem when carbon dioxide extinguishers are used, although less equipment may be damaged than is the case when powder is used.

Finally, the chances of electrical fires occurring are considerably reduced if the original installation was undertaken by competent electricians working to recognised standards, such as the Institution of Electrical Engineers' Code of Practice. It is also important to have a system of regular testing and inspection in place so that any remedial maintenance can take place.

Electric arcing

A person who is standing on earth too close to a high-voltage conductor may suffer flash burns as a result of arc formation. Such burns may be extensive and lower the resistance of the skin so that electric shock may add to the ill-effects. Electric arc faults can cause temporary blindness by burning the retina of the eye and this may lead to additional secondary hazards. The quantity of electrical energy is as important as the size of the voltage since the voltage will determine the distance over which the arc will travel. The risk of arcing can be reduced by the insulation of live conductors.

Strong electromagnetic fields induce surface charges on people. If these charges accumulate, skin sensation is affected and spark discharges to earth may cause localised pain or bruising. Whether prolonged exposure

to strong fields has any other significant effects on health has not been proved. However, the action of an implanted cardiac pacemaker may be disturbed by the close proximity of its wearer to a powerful electromagnetic field. The health effects of arcing and other non-ionising radiation are covered in Chapter 15.

Static electricity

Static electricity is produced by the build-up of electrons on weak electrical conductors or insulating materials. These materials may be gaseous, liquid or solid and may include flammable liquids, powders, plastic films and granules. Plastics have a high resistance that enables them to retain static charges for long periods of time. The generation of static may be caused by the rapid separation of highly insulated materials by friction or by transfer from one highly charged material to another in an electric field by induction (see Figure 12.6).

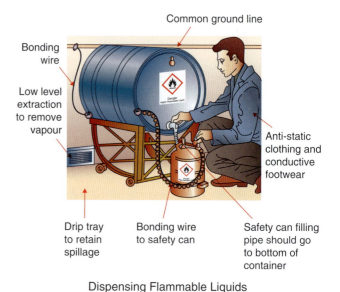

Dispensing Flammable Liquids

Figure 12.6 Prevention of static discharge; container connected to earthed drum

A static electric shock, perhaps caused by closing a door with a metallic handle, can produce a voltage greater than 10,000V. Since the current flows for a very short period of time, there is seldom any serious harm to an individual. However, discharges of static electricity may be sufficient to cause serious electric shock and are always a potential source of ignition when flammable liquids, dusts or powders are present. This is a particular problem in the parts of the printing industry where solvent-based inks are used on high-speed web presses. Flour dust in a mill has also been ignited by static electricity.

Static electricity may build up on both materials and people. When a charged person approaches flammable gases or vapours and a spark ignites the substance, the resulting explosion or fire often causes serious injury. In these situations, effective static control systems must be used.

Lightning strikes are a natural form of static electricity and result in large amounts of electrical energy being dissipated in a short time in a limited space with a varying degree of damage. The current produced in the vast majority of strikes exceeds 3,000 amps over a short period of time. Before a strike, the electrical potential between the cloud and earth might be about 100 million volts and the energy released at its peak might be about 100 million watts per metre of strike.

The need to provide lightning protection depends on a number of factors, which include:

▶ the risk of a strike occurring;
▶ the number of people likely to be affected;
▶ the location of the structure and the nearness of other tall structures in the vicinity;
▶ the type of construction, including the materials used;
▶ the contents of the structure or building (including any flammable substances);
▶ the value of the building and its contents.

Expert advice will be required from a specialist company in lightning protection, especially when flammable substances are involved. Lightning strikes can also cause complete destruction and/or significant disruption of electronic equipment.

Portable electrical equipment

Portable and transportable electrical equipment is defined by the Health and Safety Executive as 'not part of a fixed installation but may be connected to a fixed installation by means of a flexible cable and either a socket and plug or a spur box or similar means'. It may be hand-held or hand-operated while connected to the supply, or is intended or likely to be moved while connected to the supply. The auxiliary equipment, such as extension leads, plugs and sockets, used with portable tools, is also classified as portable equipment. The term 'portable' means both portable and transportable (Figure 12.7).

Almost 25% of all reportable electrical accidents involve portable electrical equipment (known as portable appliances). While most of these accidents are caused by electric shock, over 2,000 fires each year are started by faulty cables used by portable appliances, caused by a lack of effective maintenance. Portable electrical tools often present a high risk of injury, which is frequently caused by the conditions under which they are used. These conditions include the use of defective or unsuitable equipment and, indeed, the misuse of equipment. There must be a system to record the inspection, maintenance and repair of these tools.

Where plugs and sockets are used for portable tools, sufficient sockets must be provided for all the equipment and adaptors should not be used. Many

Figure 12.7 Portable hand-held electric power tools

accidents are caused by faulty flexible cables, extension leads, plugs and sockets, particularly when these items become damp or worn. Accidents often occur when contact is made with some part of the tool which has become live (probably at mains voltage), while the user is standing on, or in contact with, an earthed conducting surface. If the electricity supply is at more than 50V ac, then the electric shock that a person may receive from such defective equipment is potentially lethal. In adverse environmental conditions, such as humid or damp atmospheres, even lower voltages can be dangerous. Portable electrical equipment should not be used in flammable atmospheres if it can be avoided and it must also comply with any standard relevant to the particular environment. Air-operated equipment should also be used as an alternative whenever it is practicable.

Some portable equipment requires substantial power to operate and may require voltages higher than those usually used for portable tools, so that the current is kept down to reasonable levels. In these cases, power leads with a separate earth conductor and earth screen must be used. Earth leakage relays and earth monitoring equipment must also be used, together with substantial plugs and sockets designed for this type of system. Electrical equipment is safe when properly selected, used and maintained. It is important, however, that the environmental conditions are always carefully considered. The hazards associated with portable appliances increase with the frequency of use and the harshness of the environment (construction sites are often particularly hazardous in this respect). These factors must be considered when inspection, testing and maintenance procedures are being developed.

Secondary hazards

It is important to note that there are other hazards associated with portable electrical appliances, such as abrasion and impact, noise and vibration. Trailing leads

12

used for portable equipment and raised socket points offer serious trip hazards and both should be used with great care near pedestrian walkways. Power drives from electric motors should always be guarded against entanglement hazards.

Secondary hazards are those additional hazards which present themselves as a result of an electrical hazard. It is very important that these hazards are considered during a risk assessment. An electric shock could lead to a fall from height if the shock occurred on a scaffold or it could lead to a collision with a vehicle if the victim collapsed on to a roadway.

Similarly, an electrical fire could lead to all the associated fire hazards outlined in Chapter 13 (e.g. suffocation, burns and structural collapse) and electrical burns can easily lead to infections.

High risks associated with the use of electricity

There are several sources of high risks associated with the use of electricity. These include:

▶ working with poorly maintained electrical equipment;
▶ using electrical equipment in adverse or hazardous environments such as wet, flammable or explosive atmospheres;
▶ working on mains electricity supplies;
▶ contact with underground cables during excavation work; and
▶ contact with live overhead power lines.

12.2 Control measures

12.2.1 Safe systems of work and competence

The principal control measures for electrical hazards are contained in the statutory precautionary requirements covered by the Electricity at Work Regulations, the main provisions of which are outlined in Chapter 19. They are applicable to all electrical equipment and systems found at the workplace and impose duties on employers, employees and the self-employed.

The Regulations cover the following topics:

▶ the design, construction and maintenance of electrical systems, work activities and protective equipment;
▶ the strength and capability of electrical equipment;
▶ the protection of equipment against adverse and hazardous environments;
▶ the insulation, protection and placing of electrical conductors;
▶ the earthing of conductors and other suitable precautions;
▶ the integrity of referenced conductors;
▶ the suitability of joints and connections used in electrical systems;

▶ means for protection from excess current;
▶ means for cutting off the supply and for isolation;
▶ the precautions to be taken for work on equipment made dead;
▶ working on or near live conductors;
▶ adequate working space, access and lighting;
▶ the competence requirements for persons working on electrical equipment to prevent danger and injury.

Detailed safety standards for designers and installers of electrical systems and equipment are given a code of practice, published by the former Institution of Electrical Engineers. While these Regulations are not legally binding, they are recognised as a code of good practice and widely used as an industry standard.

BS 7671: Requirements for Electrical Installation – more commonly known as the IEE Wiring Regulations – is the national standard to which all domestic and industrial wiring has to conform. The recent edition includes substantial changes to harmonise with EU requirements. Among the changes to these Regulations is the inclusion of four new regulations for the protection of people and livestock against voltage disturbances and electromagnetic influences. There is also a specific requirement for appropriate documentation for all installations. Seven new special locations are covered to address the risk associated with certain environments or facilities, including exhibition areas, mobile units and temporary installations.

The standard wiring colours in the UK, shown in Table 12.1, are the same as elsewhere in Europe, Australia, and New Zealand and follow the international standard IEC 60445 which defines basic safety principles for identifying electrical conductors by colours in electricity distribution wiring.

Table 12.1 Standard wiring colours

	Colour
Protective earth (PE)	Green-and-yellow
Neutral (N)	Blue
Single phase: Line (L) Three phase: L1	Brown
Three phase: L2	Black
Three phase: L3	Grey

The new edition of the Regulations requires that inspection and testing must be carried out by a 'competent person' to check that the electrical work meets required standards. It defines a competent person as someone 'who possesses sufficient technical knowledge and experience for the nature of the electrical work undertaken and is able at all times to prevent danger, and where appropriate, injury to themselves and others'. After the initial testing, the Regulations recommend that every electrical installation is subject to periodic inspection and testing by a **competent person**.

Regulation 16 of the Electricity at Work Regulations states that

'No person shall be engaged in any work activity where technical knowledge or experience is necessary to prevent danger or, where appropriate, injury, unless he possesses such knowledge or experience, or is under such degree of supervision as may be appropriate having regard to the nature of the work.'

The Memorandum of Guidance on the Electricity at Work Regulations 1989 revised by the HSE in 2007 says that the scope of 'technical knowledge or experience' may include:

▶ adequate knowledge of electricity;
▶ adequate experience of electrical work;
▶ adequate understanding of the system to be worked on and practical experience of that class of system;
▶ understanding of the hazards which may arise during the work and the precautions which need to be taken;
▶ ability to recognise at all times whether it is safe for work to continue.

The risk of injury and damage inherent in the use of electricity can only be controlled effectively by the introduction of employee training, safe operating procedures (safe systems of work) and guidance to cover specific tasks.

Training is required at all levels of the organisation ranging from simple on-the-job instruction to apprenticeship for electrical technicians and supervisory courses for experienced electrical engineers. First-aid training related to the need for cardiovascular resuscitation and treatment of electric burns should be available to all people working on electrical equipment and their supervisors.

A **management system** should be in place to ensure that the electrical systems are installed, operated and maintained in a safe manner. All managers should be responsible for the provision of adequate resources of people, material and advice to ensure that the safety of electrical systems under their control is satisfactory and that **safe systems of work** are in place for all electrical equipment. For certain types of electrical work, such as working on high-voltage equipment or other work on electrical equipment which may give rise to danger, a permit to work will be required. The following items should be included in electrical permits to work:

▶ the permit issue number;
▶ the details of the work and its location;
▶ the significant hazards and risks involved in the work and the precautions to be taken, including any personal protective equipment to be worn;
▶ details of the electrical items to be isolated and the points at which such isolations are to be made;
▶ the test procedures to be followed to confirm that circuits are dead;

▶ details of any special work tools required;
▶ details of the safety warning signs required;
▶ the emergency procedures required;
▶ the date and time of issue of the permit and the length of its duration and the cross-referencing with other permits issued;
▶ acceptance of the permit by the person carrying out the work; and
▶ on completion of the work, a signature of the person who authorised the work to ensure cancellation of the permit.

A permit should not be issued on equipment that is live. It is never absolutely safe to work on live electrical equipment. There are, however, a few circumstances where it is necessary to work live, but this must only be done after it has been determined that it is unreasonable for the work to be done dead. Even if working live can be justified, many precautions are needed to make sure that the risk is reduced 'so far as is reasonably practicable'. (Chapter 4 gives more information on both safe systems of work and permits to work.)

For small factories and office or shop premises where the system voltages are normally at mains voltage, it may be necessary for an external competent person to be available to offer the necessary advice. Managers must set up a high-voltage permit-to-work system for all work at and above 600V. The system should be appropriate to the extent of the electrical system involved. Consideration should also be given to the introduction of a permit system for voltages under 600V.

The additional control measures that should be taken when working with electricity or using electrical equipment are summarised by the following topics:

▶ the selection of suitable equipment;
▶ the use of protective systems;
▶ inspection and maintenance strategies.

These three groups of measures will be discussed in detail.

12.2.2 The selection and suitability of equipment

Many factors which affect the selection of suitable electrical equipment, such as flammable, explosive and damp atmospheres and adverse weather conditions, have already been considered. All conductors in a system that might be dangerous must be covered with an insulating material. Other issues include high or low temperatures, dirty or corrosive processes or problems associated with vegetation or animals (e.g. tree roots touching and displacing underground power cables, farm animals urinating near power supply lines and rats gnawing through cables). Temperature extremes will affect, for example, the lubrication of motor bearings and corrosive atmospheres can lead to the breakdown

12

of insulating materials. The equipment selected must be suitable for the task demanded or either it will become overloaded or running costs will be too high.

The equipment should be installed to a recognised standard and capable of being isolated in the event of an emergency. It is also important that the equipment is effectively and safely earthed. Electricity supply failures may affect process plant and equipment. These are certain to happen at some time and the design of the installation should be such that a safe shutdown can be achieved in the event of a total mains failure. This may require the use of a battery-backed shutdown system or emergency standby electricity generators (assuming that this is cost-effective).

Finally, it is important to stress that electrical equipment must only be used within the rating performance given by the manufacturer and any accompanying instructions from the manufacturer or supplier must be carefully followed.

12.2.3 The planning and installation of a progressively extending electrical system on site

The site distribution system is the cabling system and electrical equipment installed to supply electricity to various points across the site during the construction phase. The supervision, maintenance and development of the system must be the responsibility of a qualified and competent electrician. Any extensions to the system must be agreed beforehand with the electrician.

It is a temporary system, which will be removed at the end of the contract and replaced with the permanent site distribution system. Even though the system is temporary, it must be robust enough to withstand the harsh conditions on site and be adequately protected against damage and contamination. Switchgear and metering equipment should be positioned away from traffic routes in secure and sheltered accommodation. All wiring, no matter how temporary, must be installed to appropriate and recognised standards. Distribution cables must be located where they are least likely to be damaged by site activity. They should also be kept clear of walkways, ladders and other services. If they need to cross a roadway or walkway, they should be placed in ducts with a marker at each end of the duct. If the roadway is used by vehicles, the duct should be buried at least 0.5 m below the surface. A record must be kept of the location of all buried cables to avoid damage as the work progresses. Alternatively, the cables may be carried overhead using a goalpost-type arrangement.

All fixed distribution cables carrying 400 or 230V must be protected with a metal sheath and/or armour, which are continuous, effectively earthed and protected against corrosion. The HSE strongly recommend that any existing or new permanent fixed supply should not be used to supply contractors' equipment during the construction work. Any movable plant such as lifts or hoists should be supplied by armoured cable.

Finally, any work on electrical installations must be subject to a permit to work which must be signed on and off at either end of the work. The detailed requirements of permits to work are given in Chapter 4.

12.2.4 Other construction issues

Approximately three workers are electrocuted each year during refurbishment work on commercial and domestic buildings. These incidents involve both qualified electricians and other trades people. They usually occur when people are working on or close to equipment that is either:

▶ assumed to be dead but is in fact live; or
▶ known to be live but inadequate precautions have been taken.

There are many more incidents which damage equipment and thousands of 'near misses', any of which could have had fatal results. It is very important that all relevant parts of the electrical system have been isolated during refurbishment work. Certain activities, such as labouring, joinery or plumbing, are liable to disturb or damage the existing electrical system and expose people to significant electrical danger. All cabling and bulbs should be protected against breakage. If a bulb breaks, the exposed filament may present a hazard. There should be a system in place for the regular checking of all light bulbs so that electrical safety is maintained and the site is well lit.

12.2.5 The advantages and limitations of protective systems

There are several different types of protective system and technique that may be used to protect people, plant and premises from electrical hazards, some of which, for example earthing, have already been considered earlier in this chapter. However, only the more common types of protection will be considered here.

Fuse

A fuse will provide protection against faults and continuous marginal current overloads. It is basically a thin strip of conducting wire which will melt when an excess of the rated current passes through it, thus breaking the circuit. A fuse rated at 13A will melt when a current in excess of 13A passes through the fuse thus stopping the flow of current. A **circuit breaker** throws a switch off when excess current passes and is similar in action to a fuse. Protection against overload is provided by fuses which detect a continuous marginal excess flow of current and energy. This overcurrent protection is arranged to operate before damage occurs, either to the supply system or to the load, which may be a motor or heater. When providing protection against overload, consideration needs to be made as to whether tripping the circuit could give rise to an even more dangerous situation, such as with fire-fighting equipment.

The prime objective of a fuse is to protect equipment or an installation from overheating and becoming a fire hazard. It is not an effective protection against electric shock due to the time that it takes to cut the current flow.

The examination of fuses is a vital part of an inspection programme to ensure that the correct size or rating is fitted at all times.

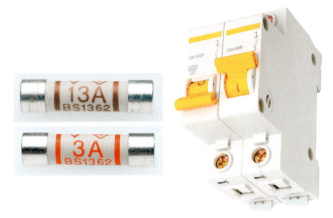

Figure 12.8 Typical 240 volt fuses and mini circuit breaker

Insulation

Insulation is used to protect people from electric shock, the short circuiting of live conductors and the dangers associated with fire and explosions. Insulation is achieved by covering the conductor with an insulating material. Insulation is often accompanied by the enclosure of the live conductors so that they are out of reach to people. A breakdown in insulation can cause electric shock, fire, explosion or instrument damage.

Isolation

The isolation of an electrical circuit involves more than 'switching off' the current in that the circuit is made dead and cannot be accidentally re-energised. It, therefore, creates a barrier between the equipment and the electrical supply which only an authorised person should be able to remove. When it is intended to carry out work, such as mechanical maintenance or a cleaning operation on plant or machinery, isolation of electrical equipment will be required to ensure safety during the work process. Isolators should always be locked off when work is to be done on electrical equipment.

Before working on an isolated circuit, checks must be made to ensure that the circuit is dead and that the isolation switch is 'locked off' and clearly labelled.

Reduced low-voltage systems

When the working conditions are relatively severe, either due to wet conditions or heavy and frequent usage of equipment, reduced voltage systems should be used.

All portable tools used on construction sites, in vehicle washing stations or near swimming pools should operate on 110V or less, preferably with a centre tapped to earth transformer at 55V. This means that while the full 110V are available to power the tool, only 55V are available should the worker suffer an electric shock. At this level of voltage, the effect of any electric shock should not be severe.

Safety Extra Low Voltage (SELV) – a voltage less than 50 volts – is used in low power tools, hand lights or soldering irons. Another way to utilise low-voltage equipment is to use **battery (cordless) operated hand tools**.

Residual current devices

If electrical equipment must operate at mains voltage, the best form of protection against electric shock is the residual current device (RCD). RCDs monitor and compare the current flowing in the live and neutral conductors supplying the protected equipment. Such devices are very sensitive to differences of current between the live and neutral power lines and will cut the supply to the equipment in a very short period of time when a difference of only a few milliamperes occurs. It is the speed of the reaction which offers the protection against electric shock. For protection against electric shock, the RCD must have a rated residual current of 30mA or less and an operating time of 40 milliseconds or less at a residual current of 250mA. RCDs rated above 30mA provide very limited protection against harm from an electric shock.

The protected equipment must be properly protected by insulation and enclosure in addition to the RCD. The RCD will not prevent shock or limit the current resulting from an accidental contact with live electricity, but it will ensure that the duration of the shock is limited to the time taken for the RCD to operate. Electrical circuits must be securely isolated before any work is done on them. The RCD has a test button which should be tested frequently to ensure that it is working properly. RCDs will also protect installations against fire, as they will interrupt the electrical supply before sufficient energy to start a fire has accumulated. However, this is not their prime purpose. Therefore the advantages of a RCD over a fuse are:

▶ a fuse is slower to act than a RCD and protects equipment rather than people;

▶ a blown fuse may be replaced by a fuse having an incorrect value; and

▶ a RCD, with its built-in test button, is easier to test and re-set.

It is advisable to use a residual current device (RCD) whenever possible but particularly in wet or damp locations, such as outdoors.

On construction sites the size and type of the RCD will depend on the particular application and it should never be used in place of reduced low-voltage equipment. If mains voltage has to be used, then RCDs must also be used. In this event, they should be installed in a dust-proof and weather-proof enclosure or be specifically designed to work under these conditions. They should be protected against mechanical damage and vibration and checked daily by operating the test button. It is best practice to use an RCD that is incorporated into the switchboard of the electrical installation so that all circuits fed from that RCD are protected by it. A RCD that is incorporated into an ordinary mains socket, or plugged into it, will protect anything attached to that socket, but it is possible that equipment may have been accidentally plugged into another unprotected socket.

They should be inspected weekly together with the equipment being supplied with power and given an electrical test every three months by a competent electrician.

Double insulation

To remove the need for earthing on some portable power tools, double insulation is used. Double insulation employs two independent layers of insulation over the live conductors, each layer alone being adequate to insulate the electrical equipment safely. This prevents the exposure of live parts to exposed conductive parts such as the outer metal casing of the equipment. Consequently, an internal fault condition cannot make any part of the casing live. If the appliance has an insulating (such as plastic) cover or case and no exposed metal parts, it cannot become live so cannot cause an electric shock. Thus the appliance is insulated in the normal way to avoid electric shock, then a second independent layer (such the insulating plastic housing on a hand drill) is added. If part of the insulation system fails, the user is protected by the second insulation layer.

Double insulation is used on Class II appliances such as hand-held portable appliances and non-hand held portable appliances such as desk fans and desk lamps. Many heating appliances – fan heaters, hair dryers or kettles – are double insulated. These devices are regularly handled and the cases must never become live. For a hair dryer this is especially important since these are sometimes used in a bathroom or with wet hands.

Double insulated tools and equipment must be inspected and maintained regularly and discarded if damaged.

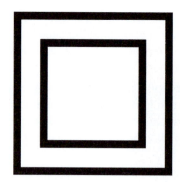

Figure 12.9 Double insulation sign

12.2.6 Protection against contact with buried power lines

The presence of buried power lines is a common problem. Often official knowledge of the exact location of such services is less than accurate, particularly if the power lines have been in place for many years.

When underground cables are damaged, people can be injured or killed by electric shock or electrical arcs. Electric arcs can cause explosions and flames resulting in severe burns to the hands, face and body, even if protective clothing is being worn. Such damage can be caused when a cable is:

► cut through by a sharp object such as the point of a tool; or
► crushed by a heavy object or powerful machine severely enough to cause internal contact between the conductors or between metallic sheathing and one or more conductors.

Cables that have been previously damaged but left unreported and unrepaired can also cause serious incidents. Symbols on electricity cable plans may vary between utilities and advice should be sought from the issuing office. High-voltage cables may be shown on separate plans from low-voltage cables.

The following precautions are suggested when there is uncertainty about the location of underground services in an area to be excavated:

► Check for any obvious signs of underground services, for example valve covers or patching of the road surface.
► Ensure that the excavation supervisor has the necessary service plans and is competent to use them to locate underground services.
► Ensure that all excavation workers are trained in safe digging practices and emergency procedures.
► Use locators to trace any services and mark the ground accordingly. A series of trial holes should be dug by hand to confirm the position of the pipes or cables. This is particularly important in the case of plastic pipes, which cannot be detected by normal locating equipment.
► In areas where underground services may be present, only hand digging should be used with

Figure 12.10 Checking for underground cables with a cable detector

insulated tools. Spades and shovels should be used rather than picks and forks, which are more likely to pierce cables.

▶ Assume that all cables are 'live' unless it is known otherwise.

▶ Hand-held power tools should not be used within 0.5 m of the marked position of an electricity cable. Collars should be fitted to the tools so that initial penetration of the surface is restricted.

▶ Any suspected damage to cables must be reported to the service providers and the health and safety enforcement authority.

▶ All exposed cables should be backfilled with fine material such as dry sand or small gravel.

▶ The service plans must be updated when the new cables have been laid.

Excellent guidance is available in the HSE publication *Avoiding Danger from Underground Services* HSG47.

12.2.7 Inspection and maintenance strategies

Inspection strategies

Regular inspection of electrical equipment is an essential component of any preventative maintenance programme and, therefore, regular inspection is required under Regulation 4(2) of the Electricity at Work Regulations, which has been quoted previously. Any strategy for the inspection of electrical equipment,

particularly portable appliances, should involve the following considerations:

▶ a means of identifying the equipment to be tested;

▶ the number and type of appliances to be tested;

▶ the competence of those who will undertake the testing (whether in-house or brought in);

▶ the legal requirements for portable appliance testing (PAT) and other electrical equipment testing and the guidance available;

▶ organisational duties of those with responsibilities for PAT and other electrical equipment testing;

▶ test equipment selection and re-calibration;

▶ the development of a recording, monitoring and review system;

▶ the development of any training requirements resulting from the test programme.

Maintenance strategies

Regulation 4(2) of the Electricity at Work Regulations requires that 'as may be necessary to prevent danger, all systems shall be maintained so as to prevent so far as is reasonably practicable, such danger'. Regular maintenance is, therefore, required to ensure that a serious risk of injury or fire does not result from installed electrical equipment. Maintenance standards should be set as high as possible so that a more reliable and safe electrical system will result. Inspection and maintenance periods should be determined by reference to the recommendations of the manufacturer, and consideration of the operating conditions and the environment in which equipment is located. The importance of equipment within the plant, from the plant safety and operational viewpoint, will also have a bearing on inspection and maintenance periods. The mechanical safety of driven machinery is vital and the electrical maintenance and isolation of the electrically powered drives is an essential part of that safety.

The particular areas of interest for inspection and maintenance are:

▶ the cleanliness of insulator and conductor surfaces;

▶ the mechanical and electrical integrity of all joints and connections;

▶ the integrity of mechanical mechanisms, such as switches and relays;

▶ the calibration, condition and operation of all protection equipment, such as circuit breakers, RCDs and switches.

Safe operating procedures for the isolation of plant and machinery during both electrical and mechanical maintenance must be prepared and followed. All electrical isolators must, wherever possible, be fitted with mechanisms which can be locked in the 'open/off' position and there must be a procedure to allow fuse withdrawal wherever isolators are not fitted.

12

Working on live equipment with voltages in excess of 110V must not be permitted except where fault-finding or testing measurements cannot be done in any other way. Reasons such as the inconvenience of halting production are not acceptable.

Part of the maintenance process should include an appropriate system of visual inspection. By concentrating on a simple, inexpensive system of looking for visible signs of damage or faults, many of the electrical risks can be controlled, although more systematic testing may be necessary at a later stage.

All fixed electrical installations should be inspected and tested periodically by a competent person, such as a member of the National Inspection Council for Electrical Installation Contracting (NICEIC).

Portable electrical appliances testing

Portable appliances should be subject to three levels of inspection – a user check, a formal visual inspection and a combined inspection and test.

User checks

When any portable electrical hand tool, appliance, extension lead or similar item of equipment is taken into use, at least once each week or, in the case of heavy work, before each shift, the following visual check and associated questions should be asked:

▶ Is there a recent PAT label attached to the equipment?
▶ Are any bare wires visible?
▶ Is the cable covering undamaged and free from cuts and abrasions (apart from light scuffing)?
▶ Is the cable too long or too short? (Does it present a trip hazard?)

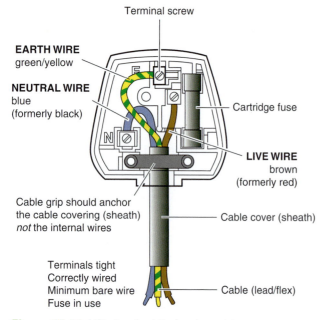

Terminal screw

EARTH WIRE green/yellow

NEUTRAL WIRE blue (formerly black)

Cartridge fuse

LIVE WIRE brown (formerly red)

Cable grip should anchor the cable covering (sheath) *not* the internal wires

Cable cover (sheath)

Terminals tight
Correctly wired
Minimum bare wire
Fuse in use

Cable (lead/flex)

Figure 12.11 UK standard 3-pin plug wiring

▶ Is the plug in good condition (for example the casing is not cracked and the pins are not bent)?
▶ Are there no taped or other non-standard joints in the cable?
▶ Is the outer covering (sheath) of the cable gripped where it enters the plug or the equipment? (The coloured insulation of the internal wires should not be visible.)
▶ Is the outer case of the equipment undamaged or loose and are all screws in place?
▶ Are there any overheating or burn marks on the plug, cable, sockets or the equipment?
▶ Are the trip devices (RCDs) working effectively (by pressing the 'test' button)?

Formal visual inspections and tests

There should be a **formal visual inspection** routinely carried out on all portable electrical appliances. Faulty equipment should be taken out of service as soon as the damage is noticed. At this inspection the plug cover (if not moulded) should be removed to check that the correct fuse is included, but the equipment itself should not be taken apart. This work can normally be carried out by a trained person who has sufficient information and knowledge.

Some faults, such as the loss of earth continuity due to wires breaking or loosening within the equipment, the breakdown of insulation and internal contamination (e.g. dust containing metal particles may cause short circuiting if it gets inside the tool) will not be spotted by visual inspections. To identify these problems, a programme of testing and inspection will be necessary.

This formal **combined testing and inspection** should be carried out by a competent person when there is reason to suspect the equipment may be faulty, damaged or contaminated, but this cannot be confirmed by visual inspection or after any repair, modification or similar work to the equipment, which could have affected its electrical safety. The competent person could be a person who has been specifically trained to carry out the testing of portable appliances using a simple 'pass/fail' type of tester. When more sophisticated tests are required, a competent person with the necessary technical electrical knowledge and experience would be needed. The inspection and testing should normally include the following checks:

▶ that the polarity is correct;
▶ that the correct fuses are being used;
▶ that all cables and cores are effectively terminated;
▶ that the equipment is suitable for its environment.

Testing need not be expensive in many low-risk premises like shops and offices, if an employee is trained to perform the tests and appropriate equipment is purchased.

Frequency of inspection and testing

The law requires an employer to ensure that their electrical equipment is maintained in order to prevent danger. It does not indicate how or how often this should be done.

Electrical equipment should be visually checked to spot early signs of damage or deterioration. The frequency of inspection and testing should be based on a risk assessment which is related to the usage, type and operational environment of the equipment. The harsher the working environment is, the more frequent the period of inspection. Thus tools used on a construction site should be tested much more frequently than a visual display unit which is never moved from a desk. Manufacturers or suppliers may recommend a suitable testing period. Table 12.2 shows the suggested intervals for inspection and testing derived from HSE publications *Maintaining Portable and Transportable Electrical Equipment* (HSG107 and INDG236 and 237).

It is very important to stress that there is no 'correct' interval for testing – it depends on the frequency of usage, type of equipment, and how and where it is used. A few years ago, a young trainee was badly scalded by a boiling kettle of water which exploded while in use. On investigation, an inspection report indicated that the kettle had been checked by a competent person and passed just a few weeks before the accident. Further investigation showed that this kettle was the only method of boiling water on the premises and was in use continuously for 24 hours each day. It was therefore unsuitable for the purpose and a plumbed-in continuous-use hot water heater would have been far more suitable.

The HSE recommend that suppliers, who loan equipment, should formally inspect and test the equipment before each hire to ensure that it is safe to use. The person hiring the equipment should also take appropriate steps to ensure it remains safe to use throughout the hire period.

Records of inspection and testing

Schedules which give details of the inspection and maintenance periods and the respective programmes must be kept together with records of the inspection findings and the work done during maintenance. Records must include both individual items of equipment and a description of the complete system or section of the system. They should always be kept up to date and with an audit procedure in place to monitor the records and any required actions. The records do not have to be paper based but could be stored electronically on a computer. It is good practice to label the piece of equipment with the date of the last combined test and inspection.

The effectiveness of the equipment maintenance programme may be monitored and reviewed if a record

Table 12.2 Suggested intervals for portable appliance inspection and testing

Type of business/ equipment	User checks	Formal visual inspection	Combined inspection and electrical tests
Equipment hire	Yes	Before issue and after return	Before issue
Construction	Yes	Before initial use and then every month	3 months
Industrial	Yes	Before initial use and then every 3 months	6–12 months
Hotels and offices, low-risk environments			
Battery operated (less than 20V)	No	No	No
Extra low voltage (less than 50V ac), e.g. telephone equipment, low-voltage desk lights	No	No	No
Desk-top computers, VDU screens	No	Yes 2–4 years	No if double insulated, otherwise up to 5 years
Photocopiers/fax machines/printers. Not hand-held; rarely moved	No	Yes 2–4 years	No if double insulated, otherwise up to 5 years
Double-insulated equipment: not hand-held. Moved occasionally, e.g. fans, table lamps, slide projectors	No	Yes 2–4 years	No
Double-insulated equipment: hand-held, e.g. some floor cleaners	Yes	Yes 6 months–1 year	No
Earthed equipment (class 1): e.g. electric kettles, some floor cleaners, portable electric heaters, some kitchen equipment and irons	Yes	Yes 6 months–1 year	Yes; 1–2 years
Cables (leads) and plug connected to the above. Extension leads (mains voltage)	Yes	Yes 6 months–4 years depending on the type of equipment it is connected to	Yes 1–5 years depending on the type of equipment it is connected to

Source: Derived from HSE.

Note: Operational experience may demonstrate that the above intervals can be reviewed.

of tests is kept. It can also be used as an inventory of portable appliances and help to regulate the use of unauthorised appliances. The record will enable any adverse trends to be monitored and to check that suitable equipment has been selected. It may also give an indication as to whether the equipment is being used correctly.

Code of practice for portable appliance testing

The Institution of Engineering and Technology (IET) has published a revised edition of its 'Code of Practice for In-service Inspection and Testing of Electrical Equipment'. This guidance on portable appliance testing (PAT) suggests that UK companies are wasting more than £30 million annually by over-maintaining electrical appliances. It emphasises that it is not a legal requirement for companies to carry out PAT on every electrical appliance annually. It advises businesses to take a proportionate, risk-based approach to testing, which should consider the usage, type of the equipment under question and the environment in which it is used.

Advantages and limitations of PAT

The advantages of PAT include:

▶ an earlier recognition of potentially serious equipment faults, such as poor earthing, frayed and damaged cables and cracked plugs;
▶ discovery of incorrect or inappropriate electrical supply and/or equipment;
▶ discovery of incorrect fuses being used;
▶ a reduction in the number of electrical accidents;
▶ monitoring the misuse of portable appliances;
▶ equipment selection procedures checkable;
▶ an increased awareness of the hazards associated with electricity;
▶ a more regular maintenance regime should result.

The limitations of PAT include:

▶ some fixed equipment is tested too often leading to excessive costs;
▶ some unauthorised portable equipment, such as personal kettles, are never tested as there is no record of them;
▶ equipment may be misused or overused between tests due to a lack of understanding of the meaning of the test results;
▶ all faults, including trivial ones, are included on the action list, so the list becomes very long and the more significant faults are forgotten or overlooked;
▶ the level of competence of the tester can be too low;
▶ the testing equipment has not been properly calibrated and/or checked before testing takes place.

Most of the limitations may be addressed and the reduction in electrical accidents and injuries enables the advantages of PAT to greatly outweigh the limitations.

12.2.8 Emergency procedures following an electrical incident

General emergency procedures were covered in Chapter 4. However, even with the best of safety procedures in place, electrical incidents will occur so that specific and effective emergency procedures are essential. If someone is found unconscious as a result of electric shock or suffering from an electric burn, then the treatment procedure listed earlier in this chapter should be followed.

The emergency procedures should ensure that, for a serious electrical accident, the emergency services and key personnel in the organisation are notified and the emergency plan is activated. Such a plan should include the following points:

▶ The isolation of the electrical device that has caused the emergency by switching off and disconnecting the supply.
▶ If electric shock is the main emergency, the victim must not be touched until there is no longer a possibility of contact with the electrical current.
▶ If there is a fire as a result of electrical equipment malfunction, the fire procedures should also be activated.
▶ If the emergency could affect hazardous equipment or processes, essential actions, such as emergency plant shutdown, isolation or making processes safe, should be put into action. It is important that important items such as shut-off valves and electrical isolators are easily identifiable.
▶ Work must not resume after an emergency until a competent person has checked and declared that all electrical equipment affected by the emergency is safe and the electrical supply has been switched back on. If there are any doubts, then advice should be sought from the emergency services.
▶ Following the incident, an investigation should take place, a report produced and, if appropriate, a RIDDOR report submitted to the HSE.

12.3 Control measures for working near overhead power lines

A detailed account of this topic is given in the HSE publication *Avoidance of danger from overhead electric power lines* GS6, HSE Books, and only a summary will be given here.

As a general rule, if the power lines are live, then vehicles, plant or equipment should be brought no closer than:

▶ 15 m of overhead lines suspended from steel towers;
▶ 9 m of overhead lines supported on wooden poles.

Where a closer approach is likely either the lines should be made dead or barriers erected to prevent an approach. Permits to work are likely to be required if work close to the lines is necessary.

Overhead electric lines can carry a voltage as high as 400,000V or as low as 230V. The overhead lines consist usually of uninsulated conductors supported on insulators by wooden poles or metal towers and structures – those on wooden poles can be mistaken for telephone wires. Electric current can flow through wood or plastic (if damp or dirty) and cause fatal shocks. If a crane jib, tipper lorry, scaffold pole, ladder or similar object makes contact with or approaches near to these lines, a significant electric current can flow risking a fatal or severe shock and burns to anybody in the immediate vicinity. A close approach to the line conductors may allow a 'flashover' or arc to take place – the risk increases as the line voltage increases.

Hierarchy of risk control

For work at or near overhead power lines, application of the hierarchy of risk control (see Chapter 4) recommends the following actions:

1. find out if the work has to be carried out under or near overhead lines (can it be avoided altogether) or, if this cannot be done
2. divert all overhead lines clear of the work area or, if it is not reasonable for this to be done
3. make lines dead while the work is in progress or, if this cannot be done
4. work around the live overhead lines using the precautions outlined below.

In some cases it may be necessary to use suitable combinations of these measures, particularly where overhead lines pass over permanent work areas.

Pre-planning work near overhead lines

At tender or negotiation stage, the existence of any overhead lines across the site must be noted and contact made with both the power supply company and the local authority. Information should be available from the local electricity supplier (the service provider) about the location of their lines which should be assumed to be live unless or until this has been proved otherwise. At the site assessment stage, arrangements must be made for any possible line diversions with the electricity company or to confirm with them any safe distances, clearances or other precautions that they may require to be made. If the power line cannot be diverted, the principal contractor and all sub-contractors must be informed of the presence of live overhead power lines.

Where possible all work likely to lead to contact with overhead power lines should be done in an area well clear of the line itself. It may be possible to alter the work and eliminate or reduce the risk. The proposed plan of work near the lines should be discussed with the service provider before work starts. Sufficient time should be allowed for the line to be diverted or

made dead, or for other precautions to be taken as described below. If the lines can only be made dead for short periods, then the passage of tall plant and, as far as is possible, other work around the lines should be scheduled at these times. Liaison between the persons responsible for the work and the owner(s) of the lines should be continued until the work has been completed.

Precautions

The precautions depend on the nature of the work at the site. It is strongly recommended they are taken even when work near the line is of short duration. The HSE recommend the following general precautions for all work near to power lines:

- ▶ **clearance** – the safe clearance required beneath the overhead lines should be ascertained from the owner of the line;
- ▶ **exclusion** – plant, equipment or hand tools that could reach beyond the safe clearance limit should not be taken under the line;
- ▶ **modifications** – plant such as cranes and excavators should be modified by the addition of suitable physical restraints so that it cannot reach beyond the safe clearance limit;
- ▶ **additions** – cranes with telescopic or fly jibs may need additional restraining devices to prevent alteration in length of jib or angle of fly jib; and
- ▶ **supervision** – access for plant and materials and the working of plant should be under the direct supervision of a suitable person appointed to ensure that safety precautions are observed.

There are three broad categories of work near overhead power lines:

(a) Work areas where there will be no work or passage of plant under the lines and barriers can prevent any close approach.
(b) Work areas where plant will pass under the lines. In this case, defined passageways should be made.
(c) Work areas where work will be carried out beneath the lines. Here further precautions must be taken in addition to the erection of barriers with passageways.

For work areas where there will be no work or passage of plant under the lines, danger can be reduced by erecting ground level barriers parallel to the overhead line to prevent any part of the mobile plant approaching too close to the line. The recommended minimum distance from a barrier to the position, projected vertically onto the ground, of the conductor nearest to that barrier is 6 metres. Where this minimum distance could be encroached upon by parts of mobile plant (such as crane jibs and excavator buckets), it is recommended that the plant operators are given an additional indication of the position of the barriers. A line of coloured plastic flags or 'bunting',

12

mounted at a height of 3 to 6 metres above ground level immediately over the barriers, could be used. Fences, posts and oil drums should be made as visible as possible, for example by being painted with red and white stripes. Alternate red and white plastic warning flags may be hung on or immediately above any fence line to improve their visibility. It is strongly recommended that the storage of materials in the area between the overhead line(s) and any ground level barrier is prohibited.

On many construction sites, vehicles will need to move beneath power lines. In these cases, the roadway should be covered by goalposts covered with warning tape. Bunting should be suspended, level with the top of the goalposts and just above ground level (often using empty oil drums), between poles across the site along the length of the power line. This will ensure that vehicles can only pass under the power lines by passing through the goalposts. Suitable warning signs should be placed on either side of the roadway on each side of the power line to warn of the overhead power line. It is recommended that the danger area should be made as small as possible. This should be achieved by restricting the width of the passageway to the minimum needed for the safe crossing of plant. It is safest if the passageway crosses the route of the line at right angles (see Figure 12.12).

Where work is necessary directly beneath the lines, or blasting or other unusual activity has to be done adjacent to the lines, the lines may need to be made dead and a permit-to-work system operated. In certain situations, induced AC voltages and even arcing can be created in fences and pipelines which run parallel to overhead lines. When there are concerns about nearby structures the advice of the safety adviser and/or the electrical supply company must be sought. Risk assessments should

also be made and suitably recorded, the information being made available to all workers who are likely to be working near the power lines.

12.4 Further information

The Electricity at Work Regulations 1989

The Health and Safety (First-Aid) Regulations 1981 (as amended)

The Reporting of Injuries, Diseases and Dangerous Occurrences Regulations 2013

Electricity at Work – safe working practices (HSG85), HSE Books, ISBN 978-0-7176-6581-5 http://www.hse.gov.uk/pubns/priced/hsg85.pdf

Maintaining portable and transportable electrical equipment (HSG107), HSE Books, ISBN 9780717628056 `http://www.hse.gov.uk/pubns/priced/hsg107.pdf

Memorandum of guidance on the Electricity at Work Regulations 1989 (HSR25), HSE Books, ISBN 978-0-7176-6228-9 http://www.hse.gov.uk/pubns/priced/hsr25.pdf

Electrical safety and you; a brief guide, INDG231, HSE Books, http://www.hse.gov.uk/pubns/indg231.pdf

Guidance on Safe Isolation procedures: http://www.select.org.uk/downloads/Electrical%20Safety%20First%20Best%20Practice%20Guide%20No.%202%20(Issue%202).pdf

Avoidance of danger from overhead electrical lines GS6, HSE Books, http://www.hse.gov.uk/pubns/gs6.pdf

Avoiding danger from underground services HSG47 (Third edition, 2014), HSE Books, ISBN 978-0-7176-6584-6 http://www.hse.gov.uk/pubns/priced/hsg47.pdf

DANGER! OVERHEAD ELECTRICAL WIRING

(a)

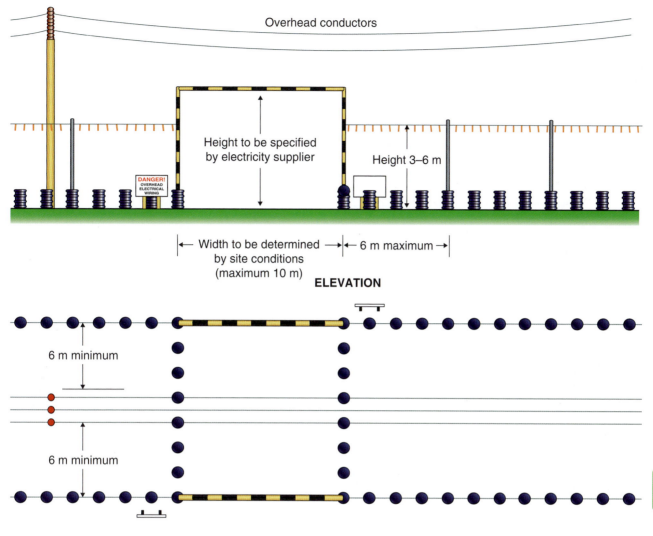

Figure 12.12 (a) Precautions for overhead lines: 'goalpost' crossing points beneath lines to avoid contact by plant; (b) diagram showing normal dimensions for 'goalpost' crossing points and barriers (Reproduced from HSG185 *Health and Safety in Excavations*)

12.5 Practice revision questions

1. (a) **Explain** the relationship between voltage, current and resistance referring to a basic electric circuit.
   (b) **Outline** the main hazards associated with the use of electricity.
   (c) **Identify** the control measures that will reduce the risk of fire from electrical equipment.

2. (a) **Outline** the effects on the human body from a severe electric shock.
   (b) **Outline** the emergency action to take if a person suffers a severe electric shock.

3. **Explain** the following terms used in electrical work:
   (a) 'isolation'
   (b) 'earthing'

   (c) 'conductors and insulators'
   (d) 'overcurrent protection'
   (e) 'electric arcing'
   (f) 'static electricity'.

4. Describe how the following protective measures reduce the risk of electric shock and, in EACH case, give an example of their application.
   (a) fuse
   (b) reduced low voltage
   (c) residual current devices
   (d) double insulation.

5. Electrical plugs and cables can cause accidents in the workplace.
   (a) **Identify FOUR** examples of faults and bad practices that could lead to such accidents.

(b) **Outline** the corresponding control measures that should be taken for **EACH** of the examples identified in (a) above.

6. A construction worker is working on a scaffold on the outside of a building. While using a portable 240V electrical drill to drill into the wall of the building, he makes direct contact with the electrical supply cable installed in the wall.

(a) **Identify FOUR** possible outcomes arising from this incident.

(b) **Identify FOUR** protective devices and procedures that could have reduced the risk of injury to this worker.

(c) **Describe** the types of inspection and/or test that should have been made on the drill.

7. Hand-held electric drills are commonly used on construction sites.

(a) **Outline** the checks that should be carried out by the user of a drill to reduce the likelihood of electric shock.

(b) Other than electricity, **outline FOUR** hazards associated with the use of such equipment.

8. (a) **Outline** the **THREE** levels of inspection that should be included in a maintenance and inspection strategy for portable electrical appliances.

(b) **Identify** the reasons for keeping records of the results of portable appliance testing within an organisation.

(c) **Outline** the issues to be considered when determining the frequency for the inspection and testing of a portable electrical tool.

9. (a) **Identify** the items that should be included on a checklist for the routine visual inspection of portable electrical appliances.

(b) **Identify EIGHT** examples of faults and bad practices that could contribute to accidents following the use of portable electrical appliances.

(c) **Identify** the advantages and limitations of portable appliance testing (PAT).

10. New electrical supplies are being installed into a new extension to an existing building. The work will require an electrical permit to work in order to complete the installation safely. **Outline** the items that should be included in the electrical permit to **work**.

11. Many people are killed or badly injured on construction sites each year from striking overhead or underground electricity supply cables.

(a) **Identify** the precautions and management strategy to be taken to prevent contact with overhead lines.

(b) **Identify** the precautions to be taken to prevent contact with underground cables.

CHAPTER 13

Fire safety

This chapter covers the following NEBOSH learning objectives:

1. Describe the principles of fire initiation, classification and spread and the additional fire risks caused by construction activities in an existing workplace
2. Outline the principles of fire risk assessment
3. Outline the principles of fire prevention and the prevention of fire spread in construction workplaces
4. Identify the appropriate fire detection and fire alarm systems and fire-fighting equipment for construction activities
5. Outline the requirements for an adequate and properly maintained means of escape in the construction workplace
6. Outline the factors which should be considered when implementing a successful evacuation of a construction workplace in the event of a fire

13.1 Principles of fire initiation, classification, spread and fire risks caused by construction activities and legal requirements

13.1.1 Introduction

This chapter covers fire prevention on construction sites and how to ensure that people are properly protected if fire does occur. Each year, UK fire and rescue services attend over 35,000 fires at work in which about 30 people are killed and over 2,500 are injured. Fire and explosions at work account for about 2% of the major injuries reported under Reporting of Injuries, Diseases and Dangerous Occurrences Regulations (RIDDOR 2013).

Every year many construction site workers are killed or injured as a result of their work; others suffer ill-health, such as dermatitis, occupational deafness or asbestosis. However, the hazards are not restricted to those working on sites. Children and other members of the public are also killed or injured because construction activities have not been adequately controlled. Property adjacent to construction sites can also be damaged and occupants put at risk, for example if there is a site fire and it is large enough to spread off-site. The construction industry's performance has improved over the past decade, but the rates of death, serious injury and ill-health are still too high.

There are over 4,000 construction fires annually and about 100 of them cause over £50,000 of damage and can result in complete dislocation of project schedules. Some, like the National Westminster Tower fire in 1996, affect large numbers of people. In this case about 600 workers were at risk when the fire broke out.

The financial costs associated with serious fires are very high including, in many cases (believed to be over 40%), the failure to start up business again. Never underestimate the potential of any fire. What may appear to be a small fire in a waste bin, if not dealt with can quickly spread through a building or structure. The Bradford City Football ground in 1985 or King's Cross Underground station in 1987 are examples of where small fires quickly became raging infernos, resulting in many deaths and serious injuries. Some construction

site fires have caused many millions of pounds in losses.

The Buncefield fuel storage depot fire in December 2005 and the Deepwater Horizon oil drilling platform fire in 2010 are further recent examples of the destruction which a large fire and explosion can cause.

Since the introduction of the Fire Services Act 1947, the fire authorities have had the responsibility for fighting fires in all types of premises. In 1971, the Fire Precautions Act gave the fire authorities control over

(a)

(b)

Figure 13.1 Fire is still a significant risk in many workplaces: (a) multi-storey building on fire during refurbishment; (b) single-storey farm building on fire in the UK

certain fire procedures, means of escape and basic fire protection equipment through the drawing up and issuing of Fire Certificates in certain categories of building. The Fire Certification was mainly introduced to combat a number of serious industrial fires that had occurred, with a needless loss of life, where simple well-planned protection would have allowed people to escape unhurt.

Following a government review in the 1990s, the Fire Precautions (Workplace) Regulations 1999 came into force in December of that year. These Regulations were made under the Health and Safety at Work (HSW) Act. They were amended in 1999 so as to apply to a wider range of premises including those already subject to the Fire Precautions Act 1971. In many ways these Regulations established the principles of fire risk assessment which would underpin a reformed legislative framework for fire safety. There remained a difference of opinion within government, as to the right home for fire legislation. The Home Office believed that process fire issues should remain with the HSW Act, while general fire safety should have a different legislative vehicle.

In 2000 the Fire Safety Advisory Board was established to reform the fire legislation to simplify, rationalise and consolidate existing legislation. It would provide for a risk-based approach to fire safety allowing more efficient, effective enforcement by the fire and rescue service and other enforcing authorities.

The Regulatory Reform (Fire Safety) Order 2005 (RRFSO) SI No 2005 1541 was made on 7 June 2005 and came into force on the delayed date of 1 October 2006. A summary of the Order is provided in the next section. The Fire Precautions Act 1971 was repealed and the Fire Precautions (Workplace) Regulations 1997 were revoked by the Order. Since the CDM 2007 Regulations were introduced the RRFSO has been amended, as the fire sections within the Construction (Health, Safety and Welfare) Regulations 1997 (now revoked) have been transferred to the CDM 2007. This is retained in CDM 2015 which came into force in April 2015.

The RRFSO reforms the law relating to fire safety in non-domestic premises. The main emphasis of the changes is to move towards fire prevention. Fire Certificates under the Fire Precautions Act 1971 were abolished by the Order and cease to have legal status. The RRFSO replaces fire certification with a general duty to ensure, so far as is reasonably practicable, the safety of employees.

There is a duty to carry out fire risk assessments that should consider the safety of both employees and non-employees.

The Order imposes a number of specific duties in relation to the fire precautions to be taken. The Order provides for the enforcement of the Order, appeals,

offences and connected matters. The Order also gives effect in England and Wales to parts of a number of EC Directives including the Framework, Workplace, Chemical Agents and Explosive Atmospheres Directives.

13.1.2 The Regulatory Reform (Fire Safety) Order (RRFSO) – requirements

Outline of RRFSO

Part 1 General – The RRFSO applies to all non-domestic premises other than those listed in Article 6. The main duty-holder is the 'responsible person' in relation to the premises, defined in Article 3. The duties on the responsible person are extended to any person who has, to any extent, control of the premises (Article 5).

Part 2 imposes duties on the responsible person in relation to fire safety in premises.

Part 3 provides for enforcement.

Part 4 provides for offences and appeals.

Part 5 provides for miscellaneous matters including fire-fighters' switches for luminous tube signs, maintenance measures provided to ensure the safety of fire-fighters, civil liability for breach of statutory duty by an employer, special requirements for licensed premises and consultation by other authorities.

Schedule 1 sets out the matters to be taken into account in carrying out a risk assessment (Parts 1 and 2), the general principles to be applied in implementing fire safety measures (Part 3) and the special measures to be taken in relation to dangerous substances (Part 4).

The Order is mainly enforced by the fire and rescue authorities.

Part 1 General

Meanings

The Order defines a responsible person as the person who is in control of the premises – this may be the owner or somebody else.

The meaning of general fire precautions is set out in the Order, which covers:

▶ reduction of fire risks and fire spread;
▶ means of escape;
▶ keeping means of escape available for use;
▶ fire-fighting;
▶ fire detection and fire warning;
▶ action to be taken in the event of fire;
▶ instruction and training of employees.

But it does not cover **process-related fire precautions**. These include the use of plant or machinery or the use or storage of any dangerous substances. Process-related fire precautions still come under either the Health and Safety Executive (HSE) or local authority, and are covered by the general duties imposed by the HSW Act 1974 or other specific Regulations like the Dangerous Substances and Explosive Atmosphere Regulations (DSEAR).

Duties

Duties are placed on a 'responsible person' who is:

▶ the employer in a workplace, to the extent they have control;
▶ any other person who has control of the premises; or
▶ the owner of the premises.

The obligations of a particular responsible person relate to matters within their control. It is therefore advisable that arrangements between responsible persons, where there may be more than one (such as contracts or tenancy agreements) should clarify the division of responsibilities.

Premises covered

The RRFSO does not cover domestic premises, offshore installations, a ship (normal shipboard activities under a master), remote fields, woods or other land forming part of an agricultural or forestry operation (it does cover the buildings), means of transport, a mine (it does cover the buildings at the surface) or a borehole site.

Other alternative provisions cover premises such as sports grounds.

Part 2 Fire safety duties

The responsible person must take appropriate general fire precautions to protect both employees and persons who are not employees to ensure that the premises are safe. Details are given in Chapter 19.

Risk assessment and arrangements

The responsible person must:

▶ make a 'suitable and sufficient' risk assessment to identify the general fire precautions required (usually known as a fire risk assessment) if a dangerous substance is or is liable to be present – this assessment must include the special provisions of Part 1 of Schedule 1 to the Order;
▶ review the risk assessment regularly or when there have been significant changes; with regard to young persons, take into account Part 2 of Schedule 1;
▶ record the significant findings where five or more people are employed or where there is a licence or an Alterations Notice is in place for the premises;
▶ apply the principles of prevention Part 3 of Schedule 1;
▶ set out appropriate fire safety arrangements for planning, organisation, control, monitoring and review;
▶ eliminate or reduce risks from dangerous substances in accordance with Part 4 of Schedule 1.

Fire-fighting and fire detection

The responsible person must ensure that the premises are provided with appropriate:

▶ fire-fighting equipment (FFE);
▶ fire detectors and alarms;
▶ measures for fire-fighting which are adapted to the size and type of undertaking;
▶ trained and equipped competent persons to implement fire-fighting measures;
▶ contacts with external emergency services, particularly as regards fire-fighting, rescue work, first-aid and emergency medical care.

Emergency routes, exits and emergency procedures

The responsible person must ensure that routes to emergency exits and the exits themselves are kept clear and ready for use. See Chapter 19 for specific requirements for means of escape.

The responsible person must establish suitable and appropriate emergency procedures and appoint a sufficient number of competent persons to implement the procedures. This includes arrangements for undertaking fire drills and the provision of information and action regarding exposure to serious, imminent and unavoidable dangers. Where dangerous substances are used and/or stored, additional emergency measures covering the hazards must be set up. There should be appropriate visual or audible warnings and other communications systems to effect a prompt and safe exit from the endangered area.

Where necessary to protect persons' safety, all premises and facilities must be properly maintained and subject to a suitable system of maintenance (see Form F1 in Chapter 23).

Safety assistance

The responsible person must appoint (except in the case of the self-employed or partnerships where a person has the competence themselves) one or more people to assist in undertaking the preventative and protective measures. Competent persons must be given the time and means to fulfil their responsibilities. Competent persons must be kept informed of anything relevant to their role and have access to information on any dangerous substances present on the premises.

Provision of information

The responsible person must provide:

▶ their own employees; and
▶ the employer of any employees of an outside undertaking

with comprehensible and relevant information on the risks, precautions taken, persons appointed for fire-fighting and fire drills, and appointment of competent persons. Additional information is also required for dangerous substances used and/or stored on the premises.

Before a child (16–18 years old) is employed, information on special risks to children must be given to the child's parent or guardian.

Capabilities, training and cooperation

The responsible person must ensure that adequate training is provided when people are first employed or exposed to new or increased risks. This may occur with new equipment, changes of responsibilities, new technologies, new systems of work and new substances used. The training has to be done in working hours and repeated periodically as appropriate.

Where two or more responsible persons share duties or premises, they are required to coordinate their activity and cooperate with each other, including keeping each other informed of risks.

Duties of employees

These are covered in Chapter 19 and are similar to requirements under the HSW Act. Employees must take care of themselves and other relevant persons. They must cooperate with the employer and inform them of any situation which they would reasonably consider to present a serious and immediate danger, or a shortcoming in the protection arrangements.

Part 3 Enforcement

Enforcing authorities

The fire and rescue authority in the area local to the premises is normally the enforcing authority. However, in stand-alone construction sites, premises which require a Nuclear licence, and where a ship is under repair, the HSE is the enforcing authority, except for the general means of escape and fire emergency procedures under CDM 2015.

Fire and rescue authorities officers appointed under the RRFSO have similar powers to those under the HSW Act. These include the power to enter premises, make enquiries, require information, require facilities and assistance, take samples, require an article or substance to be dismantled or subjected to a process or test and the power to issue Enforcement Notices.

In addition to fire safety legislation, health and safety at work legislation also covers the elimination or minimisation of fire risks. As well as the particular and main general duties under the HSW Act, fire risks are covered by specific rules, such as for dangerous substances and explosive atmospheres, work equipment, electricity and other hazards. Thus, environmental health officers or HSE inspectors may enforce health and safety standards for the assessment

and removal or control of process-related fire risks, where it is necessary, for the protection of workers and others.

Alterations and Enforcement Notices

Enforcing authorities can issue (see Chapter 19 for details):

- **Alterations Notices** – where the premises constitute a serious risk to people due either to the features of the premises, or hazards present on the premises. The Notices require the responsible person to notify changes and may require them to record the significant findings of the risk assessment and safety arrangements and to supply a copy of the risk assessment before making changes.
- **Enforcement Notices** – where there is a failure to comply with the requirements of the RRFSO. The Notice must specify the provisions concerned and MAY include directions on remedial action.
- **Prohibition Notices** – where the risks are so serious that the use of the premises should be prohibited or restricted. The Notice MAY include directions on remedial action.

Part 4 Offences

Cases can be tried on summary conviction in a Magistrate's Court or on indictment in the Crown Court. The responsible person can be liable:

- ▶ on summary conviction to a fine not exceeding the statutory maximum;
- ▶ on indictment to an unlimited fine and/or imprisonment for up to 2 years for failure to comply with fire safety duties where there is a risk of death or serious injury and for failure to comply with an Alterations, Enforcement or Prohibition Notice.

Any person who fails to comply with their duties under the Order as regards fire risks can be prosecuted alongside or instead of a responsible person. Fines are limited to the statutory maximum (or levels 3 or 5 on the standard scale) on summary conviction but on indictment, the fine is unlimited.

Employees' Capabilities

The Fire Safety (Employees' Capabilities) (England) Regulations 2010 was introduced on 6 April 2010 to ensure that employers must take into account employees' capabilities as regards fire safety in entrusting tasks to them.

The making of these regulations was a technical exercise to ensure that the EU legislation was transposed into English fire safety law. Not to have done so may have resulted in EU infraction proceedings. Employers already had to comply with the regulation under wider health and safety legislation and the requirements of the 2010 regulations were already implicit in the RRFSO.

13

13.1.3 Construction (Design and Management) Regulations 2015

Details of CDM 2015 are given in Chapter 19. Regulation 36 makes the Fire and Rescue authority responsible for the non-process fire-related sections of CDM 2015 on construction sites where other people are working within or as part of the site – for example, the work is being done in an occupied factory or office building.

The Regulations concerned are as follows.

Regulation 30: Emergency procedures

In so far as they concern fire emergency procedures and involve suitable procedures for evacuation, familiarisation of workers and testing the arrangements.

Regulation 31: Emergency routes and exits

In so far as they concern fire emergency routes and exits and involve the following:

- a sufficient number of emergency routes and exits;
- must lead directly as possible to an identified safe place;
- must be kept clear and free of obstruction;
- provided with emergency lights as necessary;
- suitably signed.

Regulation 32: Fire detection and fire-fighting

This involves the following:

- suitably located fire-fighting equipment (FFE);
- suitably located fire detection and alarm systems;

- proper maintenance, examination and testing;
- ease of access unless automatically activated;
- training for every person on the construction site in the operation of fire equipment that they are likely to use;
- where there is a particular risk of fire, people being instructed before starting the work;
- equipment suitably signed.

See Table 13.1 for the various enforcement authorities and their responsibilities on construction sites.

13.1.4 Basic principles of fire

Fire triangle

Fire cannot take place unless three things are present. These are shown in Figure 13.2.

The absence of any one of these elements will prevent a fire from starting. Prevention depends on avoiding

Fuel
Flammable gases, liquids, solids

Ignition source
Hot surfaces
Electrical equipment
Static electricity
Smoking materials
Naked flame

Oxygen
From the air
Oxidizing substances

Figure 13.2 Fire triangle

Table 13.1 Enforcement in respect of fire on construction sites

Type of premises fire issue		Fire and Rescue Authorities under RRFSO and CDM 2015	HSE under HSW Act, CDM 2015 and RRFSO	Local Authorities, Inspectors under HSW Act, and RRFSO	Defence fire service	A fire inspector or person appointed by the Secretary of State
Fire emergency procedures; escape routes and exits; fire detection and fire-fighting; and training issues on:	Stand-alone construction sites		✓			
	Construction sites where there are other activities (e.g. an occupied factory or office)	✓				
Process fire and explosion risks, where there are, e.g., tar boilers/dangerous substances on:	Stand-alone sites		✓	✓		
	Sites where there are other activities (e.g. an occupied factory or office)					
Licensed nuclear sites and ships under construction or repair		✓				
Defence and Crown armed forces premises other than a ship					✓	✓
Sports grounds and sports stands				✓		
Crown premises except nuclear; UK Atomic Energy premises except nuclear						✓

these three coming together. Fire extinguishing depends on removing one of the elements from an existing fire, and is particularly difficult if an oxidising substance is present.

Once a fire starts, it can spread very quickly from fuel to fuel as the heat increases.

Sources of ignition

Workplaces have numerous sources of ignition, some of which are obvious but others may be hidden inside machinery. Most of the sources may cause an accidental fire from sources inside but, in the case of arson (about 13% of industrial fires), the source of ignition may be brought from outside the workplace and will be deliberately used. The following are potential sources of ignition in the typical construction site:

Naked flames – from smoking materials, cooking appliances, heating appliances and process equipment.

External sparks – from grinding metals, welding, impact tools, electrical switch gear.

Internal sparking – from electrical equipment (faulty and normal), machinery, lighting.

Hot surfaces – from lighting, cooking, heating appliances, process equipment, poorly ventilated equipment, faulty and/or badly lubricated equipment, hot bearings and drive belts.

Static electricity – causing significant high-voltage sparks from the separation of materials such as unwinding plastic, pouring highly flammable liquids, walking across insulated floors or removing synthetic overalls.

Sources of fuel

If it will burn it can be fuel for a fire. The things which will burn easily are the most likely to be the initial fuel, which then burns quickly and spreads the fire to other fuels. The most common things that will burn in a typical workplace are:

Solids – these include: wood, paper, cardboard, wrapping materials, plastics, rubber, foam (e.g. polystyrene tiles and furniture upholstery), textiles (e.g. furnishings and clothing), wallpaper, hardboard and chipboard used as building materials, waste materials (e.g. wood shavings, dust, paper), hair (see Figure 13.3(a) and (b)).

Liquids – these include: paint, varnish, thinners, adhesives, petrol, white spirit, methylated spirits, paraffin, toluene, acetone and other chemicals. Most flammable liquids give off vapours which are heavier than air so they will fall to the lowest levels. A flash flame or an explosion can

(a)

(b)

Danger
Flammable solid

Figure 13.3 (a) Transport flammable solid sign; (b) GHS – packaging sign

occur if the vapour catches fire in the correct concentrations of vapour and air (see Figure 13.4(a) and (b)).

Gases – flammable gases include: LPG (liquefied petroleum gas in cylinders, usually butane or propane), acetylene (used for welding) and hydrogen. An explosion can occur if the air/gas mixture is within the explosive range (see Figure 13.5(a) and (b)).

Oxygen

Oxygen is of course provided by the air all around but this can be enhanced by wind, or by natural or powered ventilation systems which will provide additional oxygen to continue burning.

Cylinders providing oxygen for medical purposes or welding can also provide an additional, very rich source of oxygen.

In addition, some chemicals such as nitrates, chlorates, chromates and peroxides can release oxygen as they burn and therefore need no external source of air.

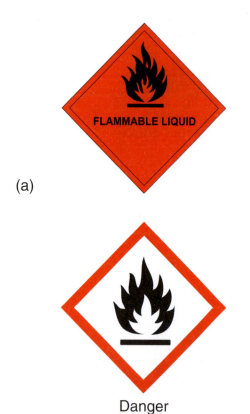

(a)

(b)   Highly Flammable Liquid

Figure 13.4 (a) Transport flammable liquid sign;
(b) GHS – packaging sign

(a)

Danger
May cause or intensify fire
(b)   Oxidiser

Figure 13.6 (a) Transport oxidising agent sign;
(b) GHS – packaging sign

13.1.5 Classification of fire

Fires are classified in accordance with British Standard EN 2:1992 *Classification of Fires*. For all practical purposes there are FIVE main classes of fire: A, B, C, D and F; plus fires involving electrical equipment. BS 7937:2000 *The Specification of Portable Fire Extinguishers for Use on Cooking Oil Fires* introduced the new class F. The categories based on fuel and the means of extinguishing are as follows:

Class A – Fires which involve solid materials such as wood, paper, cardboard, textiles, furniture and plastics where there are normally glowing embers during combustion. Such fires are extinguished by cooling, which is achieved using water.

Class B – Fires which involve liquids or liquefied solids such as paints, oils or fats. These can be further subdivided into:

Class B1 – fires which involve liquids that are soluble in water such as methanol. They can be extinguished by carbon dioxide, dry powder, water spray, light water and vaporising liquid;

Class B2 – fires which involve liquids not soluble in water, such as petrol and oil. They can be extinguished by using foam,

(a)

Danger
Extremely Flammable Gas

Figure 13.5 (a) Transport flammable gas sign;
(b) GHS – packaging sign

carbon dioxide, dry powder, light water and vaporising liquid.

Class C – Fires which involve gases such as natural gas, or liquefied gases such as butane or propane. They can be extinguished using foam or dry powder in conjunction with water to cool any containers involved or nearby.

Class D – Fires which involve metals such as aluminium or magnesium. Special dry powder extinguishers are required to extinguish these fires, which may contain powdered graphite or talc.

Class F – Fires which involve high-temperature cooking oils or fats in large catering establishments or restaurants.

Electrical fires – Fires involving electrical equipment or circuitry do not constitute a fire class on their own, as electricity is a source of ignition that will feed a fire until switched off or isolated. But there are some pieces of equipment that can store, within capacitors, lethal voltages even when isolated. Extinguishers specifically designed for electrical use like carbon dioxide or dry powder units should always be used for this type of fire hazard.

Fire extinguishers are usually designed to tackle one or more classes of fire. This is discussed in Section 13.4

13.1.6 Principles of heat transmission and fire spread

Fire transmits heat in several ways, which need to be understood in order to prevent, plan escape from, and fight fires. Heat can be transmitted by convection, conduction, radiation and direct burning (Figure 13.7).

Convection

Hot air becomes less dense and rises, drawing in cold new air to fuel the fire with more oxygen. The heat is transmitted upwards at sufficient intensity to ignite combustible materials in the path of the very hot products of combustion and flames. This is particularly important inside buildings or other structures where the shape may effectively form a chimney for the fire.

Conduction

This is the transmission of heat through a material with sufficient intensity to melt or destroy the material and ignite combustible materials which come into contact with or close to a hot section. Metals like copper, steel and aluminium are very effective or good conductors of heat. Other materials like concrete, brickwork and insulation materials are very ineffective or poor conductors of heat.

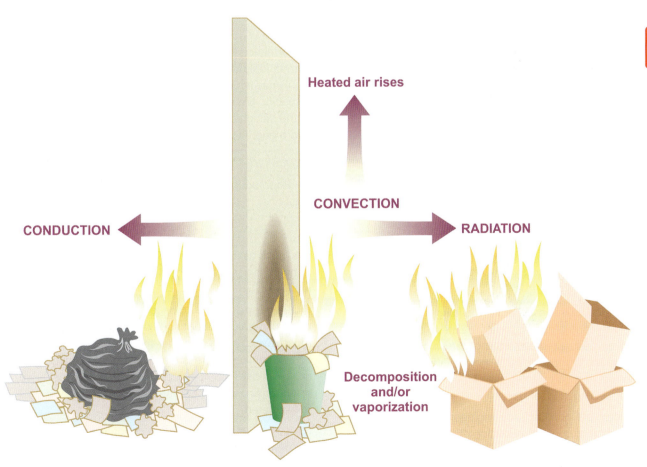

Figure 13.7 Principles of heat transmission

Poor conductors or good insulators are used in fire protection arrangements. When a poor conductor is also incombustible, it is ideal for fire protection. Care is necessary to ensure that there are no other issues, such as health risks, with these materials. Asbestos is a very poor conductor of heat and is incombustible. However, it has very severe health effects, which now outweigh its value as a fire protection material and it is banned in the United Kingdom. Although still found in many buildings where it was used extensively for fire protection, it now has to be managed under the Control of Asbestos Regulations 2012.

Radiation

Often in a fire, the direct transmission of heat through the emission of heat waves from a surface can be so intense that adjacent materials are heated sufficiently to ignite. A metal surface glowing red-hot would be typical of a severe radiation hazard in a fire.

Direct burning

This is the effect of combustible materials catching fire through direct contact with flames which causes fire to spread, in the same way that lighting an open fire, with a range of readily combustible fuels, results in its spread within a grate.

Fire and smoke spread in buildings

Where fire is not contained and people can move away to a safe location, there is little immediate risk to those people. However, where fire is confined inside buildings the fire behaves differently (Figure 13.8).

The smoke rising from the fire gets trapped inside the space by the immediate ceiling, then spreads horizontally across the space deepening all the time until the entire space is filled. The smoke will also pass through any holes or gaps in the walls, ceiling or floor and get into other parts of the building. It moves rapidly up staircases or lift wells and into any areas that are left open, or rooms which have open doors connecting to the staircase corridors. The heat from the building gets trapped inside, raising the temperature very rapidly. The toxic smoke and gases are an added danger to people inside the building, who must be able to escape quickly to a safe location.

The problems in multi-storey buildings under construction or refurbishment relate to the process of building. For example, many of the safety features that make the completed building safe in fire terms, such as correct compartmentation or fire-engineered solutions, are missing or incomplete.

Incomplete or absent fire-engineered solutions or incomplete compartmentation of the structure may lead to a very rapid spread of smoke and flames. Fire-engineered elements of design may differ from what site staff perceive as the 'norm' and their

Figure 13.8 Smoke spread in buildings

performance in the overall fire strategy depends on their correct installation. If incomplete, their performance in a fire may not be as expected. In these circumstances, a specific fire risk assessment should be undertaken at the design stage (and included in pre-construction information) to develop appropriate provisions, primarily to ensure that the building can progress while protecting workers or others. It may be necessary to put in temporary compartmentation or other fire-engineering solutions.

Sites are often in built-up areas and within close proximity to other structures. The risks associated with this, such as access and risk to other properties, need to be considered from the design stage onwards.

It is becoming more common for completed floors of multi-storey buildings to become occupied by the client while construction continues on the other levels. Clients must very carefully consider the feasibility of partial occupation and address it at all stages, including the design phase. If partial occupation is to be considered, the fire risk assessment must be completely re-evaluated.

It may be that the occupier does not allow those involved with the construction process access to occupied parts of the building, in particular stairwells and escape routes. If this is the case, it will be necessary to ensure that adequate means of escape can still be provided for construction personnel and that the fire alarm systems for the occupied parts of the building, and those still under construction, are coordinated. Remember, an emergency in the occupied parts of the building could affect the construction site and vice versa.

As the building progresses, designers and principal contractors need to be aware and work with the fire and rescue service to address this, for example by early commissioning of rising fire mains and fire-fighting shafts as the building is constructed and before work commences inside.

13.1.7 Common causes of fire and consequences

Causes

The Communities and Local Government fire statistics show that the causes of fires in buildings, excluding dwellings, in recent years were as shown in Figure 13.9. The total shows a trend downwards in the last ten years. The majority of fires occurred in:

- private garages and sheds (22%) – 6,700 fires
- retail distribution (14%) – 4,200 fires
- restaurants, cafés, public houses, etc. (9%) – 2,600 fires
- industrial premises (other than construction) (8%) – 2,400 fires
- recreational and other cultural services (6%) – 1,700 fires.

Further information on fire statistics is available at: http://www.communities.gov.uk/publications/corporate/statistics/monitorq1q420091 and *Fire Statistics Scotland* see www.scotland.gov.uk/Topics/Statistics/Search/Forthcoming

About 35 people die in buildings other than dwellings each year. Also in fires in other buildings, 1,300 injuries are sustained each year. These figures represent around 8% of all fire deaths and 10% of non-fatal casualties.

The most common causes of fire on construction sites are:

- the changing flammability of combustible material being cut, planed, sanded, ground or filed into a finely divided form which will readily catch fire;
- easy ignition of loose packaging and other materials left on site;
- poor storage of highly flammable gases and other materials;
- dismantling of tanks with flammable residues;
- coming across buried gas or electric lines during demolition or excavation;
- inadequate arrangements for rubbish disposal;
- discarding of smoking materials;
- overheating of poorly maintained plant and machinery;
- improper use or poor maintenance of welding and cutting equipment;
- overloaded or poorly maintained temporary electrical equipment;
- damaged cables and improper fuses or failure of safety devices;

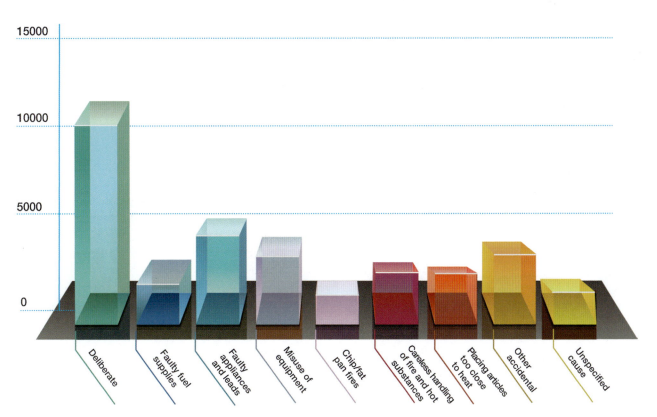

Figure 13.9 Causes of fire in recent years

359

- accumulation of rubbish against electrical equipment, causing overheating;
- overheating of coiled cables used for heavy current usage, such as on heating;
- bonfires getting out of control;
- use of petrol or other accelerants to brighten bonfires;
- arson through intruders or even disgruntled or bored site employees.

The sources of ignition are shown in Figure 13.10. Out of the 19,000 accidental fires each year it shows that cooking appliances and electrical equipment account for over 60% of the total.

Consequences

The main consequences of fire are:

- Death – although this is a very real risk, relatively few people died in building fires that are not in dwellings. Each year about 35 (8%) people died out of a total of about 440 in all fires. The main causes of all deaths are:
  - overcome by gas or smoke – 43%
  - burns – 26%
  - burns and overcome by gas or smoke – 20%
  - other – 2%
  - unspecified – 9%.
  Clearly gas and smoke are the main risks.
- Personal injury – some 1,282 people were injured (10%) of total injuries in all fires.
- Building damage – can be very significant, particularly if the building materials have poor

resistance to fire and there is little or no built-in fire protection.

- Flora and fauna damage – can be significant, particularly in a hot drought or forest fire.
- Loss of business and jobs – it is estimated that about 40% of businesses do not start up again after a significant fire. Many are under- or not insured and small companies often cannot afford the time and expense of setting up again when they probably still have old debts to service.
- Transport disruption – rail routes, roads and even airports are sometimes closed because of a serious fire. The worst cases were of course 11 September 2001 in the USA and the Icelandic ash cloud over Europe in 2010, when airports around the world were disrupted.
- Environmental damage from the fire and/or fighting the fire – fire-fighting water, the products of combustion and exploding building materials, such as asbestos cement roofs, can contaminate significant areas around the fire site.

13.1.8 Additional risk of fire in construction

It is important to note that there will be different things to consider for new builds compared with the refurbishment of an existing building. While completed buildings have the standards of fire protection required by Building Regulations, during construction and before final fire protection is in place the building will be more

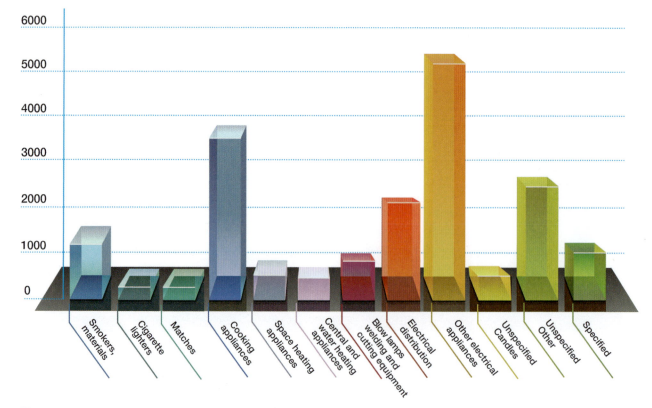

Figure 13.10 Accidental fires – sources of ignition in recent years

vulnerable to fire. This vulnerability can often lead to the whole structure being involved in fire with resultant on- and off-site fire spread issues, e.g. the building could be timber framed and more vulnerable to fire before the external finishes are in place. This vulnerability needs to be taken into account early in the design process. In some situations the additional costs entailed in providing adequate controls might make it more cost effective to specify alternative methods or materials from the outset. For a refurbishment project it will be important to take into account, among other things, the age and construction of the premises, e.g. the building could have a relatively heavy fire load due to lath and plaster ceilings and walls, wooden panelling and floors. There may also have been changes to the fabric of the building that could have significant consequences in a fire.

During construction work many processes such as welding, heating bitumen for roofing, cutting and grinding metals and concrete, and discarding enormous quantities of packaging materials all involve risks which again make the construction site more vulnerable to fire. In addition there will be numerous sub-contractors occupying the building at the same time, some of them extremely temporary for only a few days. This involves careful control and cooperation between the principal contractors and their sub-contractors.

13.2 Fire risk assessment

13.2.1 General

Fire risk assessments come under the RRFSO and not under the Management of Health and Safety at Work Regulations 1999.

A fire risk assessment will indicate what fire precautions are needed. There are numerous ways of carrying out a fire risk assessment: the one described below is based on the method contained within Fire Safety Guides published by the Department of Communities and Local Government (see Appendix 13.1 and Form F2 in Chapter 23). A systematic approach, considered in five simple stages, is generally the best practical method. The RRFSO guides are downloadable on www.firesafetyguides. communities.gov.uk.

13.2.2 Stage 1 – identify fire hazards

There are five main hazards produced by fire that should be considered when assessing the level of risk:

▶ oxygen depletion;
▶ flames and heat;
▶ smoke;
▶ gaseous combustion products;
▶ structural failure of buildings.

Of these, smoke and other gaseous combustion products are the most common cause of death in fires.

For a fire to occur, it needs sources of heat and fuel. If these hazards can be kept apart, removed or reduced, then the risks to people and businesses are minimised. **Identifying fire hazards** in the workplace is the first stage as follows.

Identify any combustibles on construction sites

Most worksites contain combustible materials. Usually, the presence of normal stock in trade should not cause concern, provided the materials are used safely and stored away from sources of ignition. Good standards of housekeeping are essential to minimise the risk of a fire starting or spreading quickly.

The amount of combustible materials to hand at a worksite should be kept as low as is reasonably practicable. Limit materials to half a day's supply or a single shift and return unused materials to the stores. Always choose the least flammable materials and keep site stocks down as far as possible.

Materials should not be stored in gangways, corridors or stairways or scaffolds where they may obstruct exit doors and routes. Fires often start and are assisted to spread by combustible waste or packaging. Such waste should be collected frequently and removed from the site.

Construction work such as cutting, grinding or sanding often alters the flammability of building materials. Finely devised dust or crumbs can often be ignited easily.

Highly combustible materials, such as flammable liquids, paints or plastic foams, ignite very easily and quickly produce large quantities of heat and/or dense toxic smoke. Such materials should be stored outside the buildings under construction in secure storage areas.

If combustible materials are stored inside buildings they should be kept where the means of escape in case of fire would not be affected. In some cases, stores will need to be constructed of fire-resistant materials giving at least 30 minutes' fire resistance and separated from working areas.

Identify any sources of heat or ignition on construction sites

All work sites will contain heat/ignition sources. These may be heaters, boilers, engines, bonfires, smoking materials or heat from processes, or electrical circuits and equipment, whether in normal use, or through misuse or accidental failure.

Where possible, sources of ignition should be removed from the site or replaced with safer forms. Where this cannot be done the ignition source should be kept well away from combustible materials or made the subject of management controls.

13

Particular care should be taken in areas where portable heaters are used or where smoking is permitted. Where heat is used as part of a process, it should be used carefully to reduce the chance of a fire as much as possible. Good security both inside and outside the site will help to combat the risk of arson.

Refuelling of vehicles (especially with petrol) should take place in the open air or in well-ventilated spaces away from sources of ignition.

Lights should be securely fastened to a solid back or tripods made secure so that they cannot be easily dislodged. Make sure electrical equipment cannot be inadvertently covered or that combustible materials are not close to halogen lights (in particular, as they get very hot) or heaters.

Plant should be properly maintained to avoid overheating, particularly in dusty conditions when filters become easily blocked.

Electrical systems, especially temporary ones, should be of sufficient capacity, and inspected and maintained by competent people. For more information see Chapter 12.

Bonfires should not normally be allowed on construction sites. There should be alternative arrangements for the proper disposal of rubbish and other waste materials. If under exceptional circumstances a bonfire is needed it should be separated from combustible materials by at least 10 m, confined in incinerators, never left unattended and checked for dangerous materials. Extinguishers should be available nearby. Bonfires should not be lit on windy days and should be avoided (see Figure 13.11).

Smoking should be discouraged on all sites and where there are high risks it should be banned altogether. Under smoke-free legislation, smoking is not permitted in enclosed or significantly enclosed areas. Outside, designated safe areas should be provided for those who still require to smoke. The smoking rules should be rigorously enforced.

Demolition work can involve a high risk of fire and explosion. In particular:

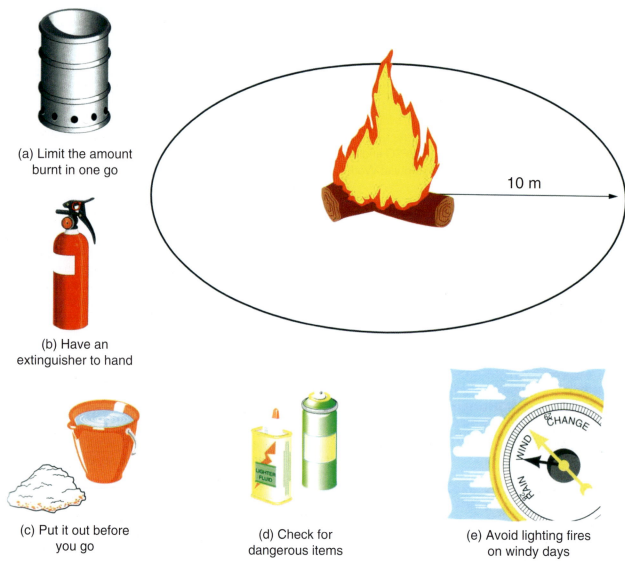

(a) Limit the amount burnt in one go

(b) Have an extinguisher to hand

(c) Put it out before you go

(d) Check for dangerous items

(e) Avoid lighting fires on windy days

10 m

Figure 13.11 Avoid lighting bonfires unless essential. If essential, do make sure points (a)–(e) are followed

- dismantling tank structures can cause the ignition of flammable residues. This is especially dangerous if methods requiring heating are used to dismantle tanks before residues are thoroughly cleaned out. The work should only be done by specialists;
- disruption and ignition of buried gas and electrical services is a common problem. It should always be assumed that buried services are present unless it is positively confirmed that the area is clear. A survey using service detection equipment must be carried out by a competent person to identify any services. The services should then be marked, competently purged or made dead, before any further work is done. A permit to excavate or dig is the normal formal procedure to cover buried services;
- arson is a high risk on construction sites, particularly where trespassers can enter the site fairly easily. Measures should be in place to prevent unauthorised access, especially by children. Flammables should be securely stored or if necessary removed from the site. In some particularly vulnerable sites, security patrols, flood lighting, liaison with local police or even closed-circuit television (CCTV) may be needed.

Identifying sources of fuel on construction sites

Anything that burns is fuel for a fire. Many materials which can burn have to be used during construction work. Reducing the quantity of material on site reduces the chances of fire occurring and limits the extent of any fire which might start. Stocks of high fire hazard material should be managed to balance production needs with the need to reduce the risk of fire. Limit the material present at worksites to what is needed for half a day or a single shift and return unused material to the stores when the work is finished. Where combustible or flammable materials have to be used, select the least combustible alternatives.

It is important to look for the things that will burn and are in enough quantity to provide fuel for a fire or cause it to spread to another fuel source. Some of the most common fuels found on site include:

- stored/in-use building products such as composite panels and timber;
- rubbish;
- flammable liquids such as paints and varnishes;
- protective coverings;
- scaffold sheeting;
- volatile flammable substances such as paints, thinners;
- fuel for portable equipment;
- liquefied petroleum gas (LPG), e.g. bitumen boilers, in temporary site accommodation and similar areas;
- acetylene;
- packaging materials;
- petrol disc-cutters and other portable equipment;
- fall-arrest bags.

Identify sources of oxygen on construction sites

The main source of oxygen for a fire is in the air around us. On construction sites this will be natural airflow through doors, windows and other openings. Wind or the 'chimney effect' can also cause increased oxygen to feed the fire.

Additional sources of oxygen can sometimes be found in site processes or in materials used or stored on site such as oxidising agents; they can provide a fire with additional oxygen and so help it burn. These chemicals should have identification marked on their container (and COSHH/Registration, Evaluation, authorisation and restriction of Chemicals (REACH)/Classification, Labelling and Packaging (CLP) Regulation data sheet) by the manufacturer or supplier, who can advise as to their safe use and storage, Examples include:

- oxygen used in welding processes; and
- oxidising agents.

13.2.3 Stage 2 – identify persons who are at significant risk

Consider the risk to any people who may be present. In many instances and particularly for most small sites the risk(s) identified will not be significant, and specific measures for persons in this category will not be required. There will, however, be some occasions when certain people may be especially at risk from the fire, because of their specific role, disability, sleeping location or workplace activity (see Section 13.17 for more information).

Because fire is a dynamic event, which, if unchecked, may spread throughout a building under construction, all people present will eventually be at risk if fire occurs. Where people are at risk, adequate means of escape from fire should be provided together with arrangements for detecting and giving warning of fire. Fire-fighting equipment suitable for the hazards throughout the site should be provided. See later for control measures.

People may come into the site from outside, such as visitors, the public or sub-contractors. The assessor must decide whether the current arrangements are satisfactory or if changes are needed.

Some people may be at significant risk because they work in areas where fire is more likely or where rapid fire growth can be anticipated. Where possible, the hazards creating the high level of risk should be reduced. Specific steps should be taken to ensure that people affected are made aware of the danger and the action they should take to ensure their safety and the safety of others.

All of the people who use or could be affected by the site should be considered, but particular attention is required for people who may be especially at risk such as:

13

- those who work alone, e.g. security staff;
- people who are in isolated areas, e.g. maintenance staff, or staff on cranes and reach trucks;
- people who are unfamiliar with the site, e.g. new sub-contractors or visitors;
- people with language difficulties;
- people who are using sleeping accommodation on or near the site;
- young people;
- pregnant women;
- people who are physically, visually or mentally challenged;
- people who are unable to react quickly;
- other people in the vicinity of the premises; and
- those occupying adjacent buildings who may be at risk from radiated heat/fire spread.

If a fire does break out, it is most likely that the local fire and rescue service will attend. In implementing the findings of the fire risk assessment, and preparing the emergency plan, consider the risks that fire-fighters would face on site in the event of a fire or other emergency. Fires on construction sites can also cause the closure of roads and railways.

13.2.4 Stage 3 – evaluate and reduce the risks

Having identified the hazards and the persons at risk, the next stage is to reduce the chance of a fire occurring and spreading, thereby minimising the chance of harm to persons on the construction site. The principles of prevention laid down in the RRFSO should be followed at this stage (see Chapter 19). These are based on EC Directive requirements and are therefore the same as those used in the Management Regulations.

Evaluate the risks

Attempt to classify each area as 'high', 'normal' or 'low risk'. If 'high risk', it may be necessary to reconsider the principles of prevention, otherwise additional compensatory measures will be required.

Low risk – Areas where there is minimal risk to persons' lives; where the risk of fire occurring is low; or the potential for fire, heat and smoke spreading is negligible and people would have plenty of time to react to an alert of fire.

Normal risk – Such areas will account for nearly all parts of most worksites, where an outbreak of fire is likely to remain confined or spread slowly, with an effective fire-warning allowing persons to escape to a place of safety.

High risk – Areas where the available time needed to evacuate the area is reduced by the speed of development of a fire, for example where highly flammable or explosive materials are stored or used (other than small quantities under controlled conditions); also where the reaction time to the

fire alarm is slower because of the type of person present or the activity in the workplace, for example the infirm and elderly or persons sleeping on the premises.

Temporary workplaces such as construction sites, temporary buildings, festivals and fêtes all have requirements for fire precautions and means of escape in case of fire. The scale of the temporary workplace will dictate the requirements, which will depend on:

- the number of persons working or visiting the site at any one time;
- the nature of the materials being used to construct the workplace or being used in the workplace. Are they flammable or highly flammable?
- height above or below the ground floor and how far it is to a place of safety;
- the location, whether in a remote area or close to water supplies and/or fire and rescue services;
- the size of the premises and whether audible warnings can be heard.

Determine if the existing arrangements are adequate, or need improvement. Matters that will have to be considered are:

- Means for detecting and giving warning in case of fire – can it be heard by all occupants and operated without putting people at risk? (See Section 13.4.)
- Means of escape – are they adequate in size, number, location, well lit, unobstructed, safe to use, etc.? (See Section 13.5.)
- Signs – for exits, fire routines, etc. (See Section 13.5.)
- Fire-fighting equipment – wall-mounted or in a cradle on fire exit routes, suitable types for hazards present and sufficient in number? (See Section 13.4.)

13.2.5 Stage 4 – record the findings

The findings of the assessment and the actions (including maintenance) arising from it should be recorded. If five or more people are employed, or an Alterations Notice is required, a formal record of the significant findings and any measures proposed to deal with them must be recorded. (See Forms F1 and F2 in Chapter 23 and a summary of the Order in Section 13.1 and Chapter 19.) The record should indicate:

- the date the assessment was made;
- the hazards identified;
- any staff and other people especially at risk;
- what action needs to be taken, and by when (**action plan**);
- the conclusions arising.

Each part of the construction site must be looked at and a decision made on how quickly persons would react to an alert of fire in each area. Adequate safety measures will be required if persons are identified as being at risk. Where maximum travel distances (see 13.5.4 and Table

13.3) cannot be achieved, extra fire safety precautions will be needed.

Where persons are at risk or an unacceptable hazard still exists, additional fire safety precautions will be required to compensate for this, or alternatively repeat previous stages to manage risk to an acceptable level.

13.2.6 Stage 5 – monitor and review on a regular basis

The fire risk assessment is not a one-off procedure. Construction sites are continually changing as work progresses. The site should be continually monitored to ensure that the existing fire safety arrangements and fire risk assessment remain realistic. The assessment should be reviewed frequently as significant changes could affect its validity.

Reasons for review could include:

▶ changes to work activities or the way that you organise them, including the introduction of new equipment;
▶ progression through the various stages of construction, e.g. alterations to the building, including the internal layout;
▶ the introduction, change of use or increase in the storage of hazardous substances;
▶ the failure of fire precautions, e.g. fire detection and alarm systems;
▶ significant changes to types and quantities and/or methods of storage of goods; and
▶ a significant increase in the number of people present.

The assessment should not be amended for a trivial change, but if a change introduces new hazards, it should be considered and, if significant, whatever is necessary to keep the risks under control should be done. In any case, the assessment should be kept under review to make sure that the precautions are still working effectively.

Guidance has been given in Fire Safety in Construction, 2nd Edition, HSG 168, HSE Books, downloadable at http://www.hse.gov.uk/pubns/books/hsg168.htm

13.2.7 Fire emergency plans

The site specific **emergency plan** should be based on the outcome of the fire risk assessment and be available for all workers, contractors, sub-contractors and their representatives (where appointed) and the enforcing authority. It should be produced before the work begins and any control measures identified should be in place from the start of the work. An emergency fire plan should be prepared for every site.

Small- and low-risk sites only require very simple plans, but higher risk sites will need more careful and detailed consideration, including:

▶ an emergency plan, which should be available before work starts;
▶ a responsible person to ensure that fire precautions are in place;
▶ an up-to-date plan that is appropriate for the circumstances and that makes clear who does what during a fire;
▶ staff who know what they need to do if there is a fire;
▶ managers who need to make sure that everyone (including visitors) on their sites knows what to do. On larger high-risk sites, fire drills may be appropriate;
▶ fire drills, which are an important check for the principal contractor on whether induction and fire safety plans work, and training for site workers;
▶ fire action notices, which should be clearly displayed where everyone on site will see them, for example at fire points, site entrances or canteen areas;
▶ on larger or more complex sites, fire wardens to ensure process and general fire precautions (GFP) controls are in order;
▶ arrangements to liaise with fire services; and
▶ arrangements to ensure instruction, information and training is given to all involved with work on the site.

This chapter concentrates on fire. However, there may be other problems, such as flooding in excavations, tunnels, work near the sea or rivers, waterworks, etc. or risk from asphyxiation or toxic gases. These should be integrated within fire procedures. See Chapter 3 for more details.

The purpose of an emergency plan is to make sure that the physical measures will work effectively if they are ever needed and to ensure that the people (including those whose first language is not English or who have poor reading skills) on the site know what to do if there is a fire and that the premises can be safely evacuated.

Some emergencies may only require partial evacuation, e.g. where a series of separate structures are present on the site. Careful thought needs to be given to ensuring that the emergency plans are appropriate and capable of achieving the desired goal. Where whole-structure involvement is a possibility then the emergency plans should include a consideration of the occupants of adjacent premises.

On existing occupied sites, liaise and agree emergency procedures with the other occupiers. Ensure that the means are in place to let each other know straight away if an emergency does arise. If simultaneous evacuation is needed, make sure the escape routes are of sufficient capacity to achieve this.

Site occupiers must give clear and relevant information and appropriate instructions to people on the site, such as sub-contractors (and their employees) and visitors, about how to prevent fires and what they should do if there is a fire. If relevant, contractors must also

13

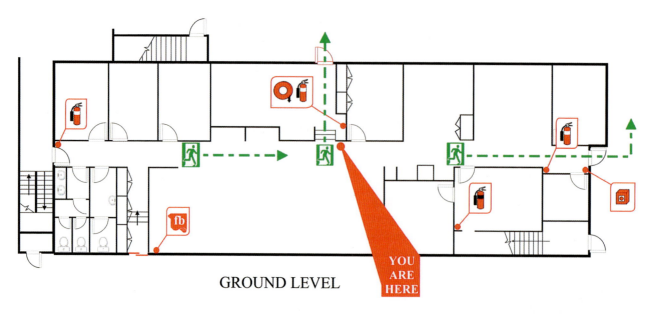

GROUND LEVEL

YOU ARE HERE

LEGEND

- CO2 Fire Extinguisher
- Dry Powder Fire Extinguisher
- Fire Hose Reel
- Manual Call Point
- **fb** Fire Blanket
- Fire Door
- Sliding Fire Door
- Egress Route

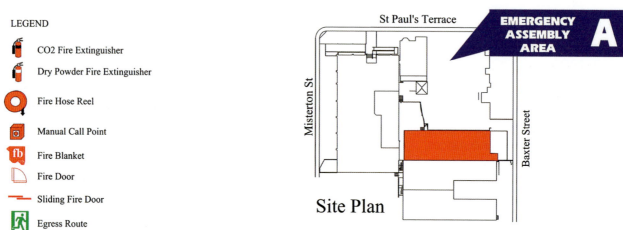

St Paul's Terrace

EMERGENCY ASSEMBLY AREA A

Misterton St

Baxter Street

Site Plan

Figure 13.12 Fire evacuation diagram suitable for refurbishment or when the main structure has been formed

cooperate and coordinate with other responsible people who use any part of the site/premises.

The site emergency fire plans should be attached to the fire risk assessment. A copy should be posted in the workplace. For larger or high risk sites a single-line plan of the area or floor should be produced or an existing plan should be used which needs to show:

- ▶ escape routes, numbers of exits, number of stairs, fire-resisting doors, fire-resisting walls and partitions, places of safety, and the like;
- ▶ fire safety signs and notices including pictorial fire exit signs and fire action notices;
- ▶ the location of fire warning call points and sounders or rotary gongs;
- ▶ the location of emergency lights;
- ▶ the location and type of fire-fighting equipment.

See Section 13.6 for more details on site evacuation.

13.3 Fire prevention and prevention of fire spread

This section provides further information on control measures to prevent fires on construction sites and

reduce the chance of a fire spreading. Sections 13.4 and 13.5 will deal with general fire precautions designed to mitigate the effects of fire once it has started. The following will be covered:

- ▶ control of combustible materials;
- ▶ good housekeeping and waste disposal;
- ▶ control of ignition sources;
- ▶ control of highly flammable liquids and gases (dangerous substances);
- ▶ systems of work;
- ▶ temporary accommodation;
- ▶ structural measures to prevent fires spreading;
- ▶ use of electrical equipment in flammable atmospheres.

13.3.1 Control of combustible materials

Many materials which can burn have to be used during construction work. Combustible materials are not just those generally regarded as highly combustible, such as polystyrene, but all materials that will readily catch fire. Even non-combustible materials may present a fire hazard when packed in combustible materials.

The risk of fire decreases as such material is reduced and the smaller any fire will be. There has to be enough material at hand to do the work, but this needs to be balanced against the need to reduce the risk of fire. Limit the material present at worksites to what is needed for half a day or a single shift and return unused material to the stores when the work is finished. Where combustible or flammable materials have to be used, select the least flammable alternatives. Where the structure itself is flammable special considerations may be necessary to reduce the combustible material likely to be involved. This can include, for example, compartmentation into smaller volumes and fire-retardant treatment.

The amount of material kept on site, which can burn, should be minimised. The need to store such material varies greatly during the life of a site, but try to avoid stockpiling it unless it really is necessary. This can significantly reduce the fire loading and ease congestion on the site.

Construction work can alter the flammability of substances, including nominally flame-retardant ones. For instance, when worked on, solid materials (even nominally fire-resisting ones) produce dust, crumbs or other fine material which are always more easily ignited than the bulk material. Remember this when planning construction fire precautions, especially when hot work is used.

Poor organisation of storage may lead to a concealed fire, and restriction of access to the fire, fire extinguishers, alarm points and escape routes. Discarded packaging materials, e.g. polystyrene and cardboard, and even piles of wooden pallets, can introduce severe fire hazards. Poorly managed storage areas often become overstocked or dumping areas for unwanted material. Do not pile combustible material against electrical equipment or heaters, even if they are turned off for the summer.

Ideally, combustible materials need to be stored outside buildings under construction, especially volatile flammable materials such as LPG. If combustible materials are stored inside buildings, they need to be kept in an area where the safety of people (on and adjacent to the site) is not threatened in the case of a fire. For example, do not put paint stores next to emergency exits or under any means of escape, e.g. steps/staircases.

Access to stores should be controlled so that material does not become dispersed haphazardly around the site.

If storage outside the structure is not possible, internal stores need to be arranged to limit the spread of fire. Internal stores, especially in more enclosed buildings, may need to be separated from the rest of the structure by a partition providing at least 30 minutes' fire resistance. Good quality plasterboard will usually achieve this and can be very useful for constructing small internal stores. Doors should be fire-resisting and self-closing.

13.3.2 Good housekeeping and waste disposal

Good housekeeping will lower the chances of a fire starting, so the accumulation of combustible materials on all construction sites should be monitored carefully. Good housekeeping is essential to reduce the chances of escape routes and fire doors being blocked or obstructed – untidy sites are usually unsafe sites.

Flammable materials such as timber become a lot more vulnerable if waste materials, such as timber shavings, paper and flammable materials, are left lying around. Regular disposal of rubbish from the active areas of construction will help to prevent an accidental fire starting or stop an arsonist. Keep rubbish away from temporary accommodation units (TAUs), buildings being constructed and any escape routes. Organise regular removal off site. It is good practice to use metal or fire-resisting rubbish containers.

All construction sites, especially in the latter stages such as fit-out, can generate large amounts of mostly combustible and easily ignitable rubbish. Implementing simple site rules can prevent the accumulation of rubbish.

The following should be considered.

▶ Set site rules and ensure that they are followed, e.g. contractors must clear rubbish daily or more often.
▶ Provide facilities for storage of rubbish, e.g. skips.
▶ Keep flammable rubbish, such as contaminated rags, in a closed-top, fire-resisting container, e.g. a metal dustbin.
▶ Situate rubbish skips outside (placing the skip so if it does catch fire it does not put at risk the site or other properties nearby).
▶ Store empty bulk fibre bags, sacks and wooden pallets in a safe place until they can be removed from site.
▶ If a skip is less than 3 m away from other structures, precautions to prevent skip fires spreading to the structure include:
  ▷ situating the skip against a fire-resisting wall that is high enough to prevent fire from reaching other flammable parts of the structure, e.g. brick;
  ▷ avoid placing skips beneath canopies or overhanging eaves.

13.3.3 Control of ignition sources

There are many ways to reduce the risks from ignition sources on construction sites, for example:

▶ Wherever possible, replace a potential source with a safer alternative, e.g. replace naked flame and radiant heaters with fixed convector heaters or other types of heaters with no red element.

367

Figure 13.13 Controlling waste on site – waste chute and covered skip

▶ Ignition hazards and combustibles materials should be separated by space, or by incombustible and/or fire-resistant materials, e.g. ensure sufficient clear space between lights and combustibles, and consider building fire-resistant enclosures for hot work processes.

▶ Control and minimise the use of portable heating appliances and guard them when they are unavoidable.

▶ Replace hot work with alternative methods such as cold cutting. Conduct unavoidable routine hot works, e.g. steel cutting, in a designated protected area away from combustible material and the main structure.

▶ Strictly control hot processes/hot work by operating permit-to-work schemes, e.g. when tar boiling for roof work.

▶ Check all areas where hot work (e.g. welding) has been carried out at regular intervals and for at least 60 minutes after work has finished to ensure that

no ignition has taken place and no smouldering or hot materials remain that may cause a fire. In some cases a fire watcher may be necessary and precautions like fire blankets used to protect combustible materials during the hot process, e.g. in a process factory or a warehouse.

▶ Operate a safe smoking policy in compliance with smoke-free legislation – allow smoking only in designated smoking areas outside enclosed spaces and prohibit smoking elsewhere.

▶ Control, inspect and monitor ignition hazards, e.g. temporary lighting, halogen lamps, temporary heating, overloading of power supplies or lights too close to combustibles.

▶ Ensure electrical, mechanical and gas equipment is installed, used, maintained and protected in accordance with the manufacturer's instructions, including any equipment located in temporary accommodation.

▶ Ensure that people carrying out work on gas fittings, or other equipment such as fuel tanks or pipework that has contained or contains flammable materials, follow a safe system of work and do not use any source of ignition such as blow-lamps or hot-air guns.

▶ Turn off equipment when it is not attended or being used.

▶ Take action to avoid any parts of the site, and in particular storage areas, being vulnerable to arson or vandalism, e.g. always lock away gas cylinders after use and/or when the site is closed down.

▶ Dispose of waste quickly and safely and do not permit bonfires on site.

Plant and machinery

Plant and equipment should be appropriate for the task and consideration should be given to the area where it is sited (e.g. it may be acceptable to use a small generator in an open, well-ventilated building constructed of non-combustible materials; however, this would not be appropriate in a basement or enclosed space or an unprotected, timber-framed construction).

Consideration should also be given to the storage of plant and equipment in relation to fire risk. Select plant, both electrical and engine driven, to match the demands placed upon it to prevent overheating during use, especially in dusty conditions. See later information for the use of electrical equipment in flammable atmospheres.

Maintain all plant properly and, in particular, air filters and intakes should be regularly cleaned in dusty conditions. Ensure that air intakes are positioned so that air is free from flammable gases and vapours.

Operating and refuelling (especially with petrol) should not take place within a confined space. Do not refuel on scaffolding or escape routes; this should be done in the open air or in well-ventilated spaces away from ignition sources. Bulk flammable liquid storage tanks should be bunded to current standards.

The use of portable petrol-fuelled generators indoors, or in partially enclosed areas to provide a power source for heating, lighting and other equipment, can put operators at risk of serious illness and death from carbon monoxide (CO) poisoning. This hazard is present from the exhaust fumes of any internal combustion engine and care needs to be taken to avoid the use of other equipment such as disc-cutters, chainsaws, floor polishers and pressure washers. See Chapter 11 for detailed control measures.

Employers, operators, consumers and others should be aware of the hazard and take the following precautionary measures:

▶ Do not operate petrol-fuelled generators or tools indoors or in poorly ventilated areas. Even apparently well-ventilated locations such as partially open temporary enclosures can allow the accumulation of potentially lethal concentrations.
▶ Advise users to read and adhere to the safety instructions supplied with the equipment.
▶ Consider substituting other types of equipment powered by mains electricity, battery or compressed air if they are available and can be used safely.
▶ Recognise potential sources and symptoms of carbon monoxide poisoning.
▶ Design and label equipment for safe operation.

Recharging arrangements for vehicles or plant should be considered; for example, charging batteries in close proximity to ignition sources can cause fire.

Temporary lights can easily become an ignition source if broken or abused. Ideally, lamps should be securely fastened to a solid backing. If they are mounted on tripods, make sure that the tripod cannot be dislodged or overturned. Make sure that electrical equipment is not inadvertently covered and that due care is taken in positioning, especially halogen lamps and heaters, to ensure that they cannot ignite any combustible material nearby, or the structure itself.

Protect plant and equipment when used in areas where a potentially flammable atmosphere may occur, such as in LPG or highly flammable liquids (HFLs) storage areas and/or when paint spraying or floor laying with HFLs.

Smoking

Carelessly discarded cigarettes and other smoking materials are still a major cause of fire. A cigarette can smoulder for several hours, especially when surrounded by combustible material. Many fires are started several hours after the smoking materials have been emptied into waste bags and left for future disposal.

With the smoke-free legislation concerning enclosed premises this has greatly reduced the risk of fire inside premises. However, the risks can still exist in outside areas and designated smoking shelters. Display the smoke-free signs throughout the premises. In those areas where smoking is permitted, provide deep and substantial metal ashtrays filled with sand to help prevent unsuitable containers being used. Empty all ashtrays daily into a metal waste bin and keep it outside.

For very high risk sites the potential for a fire to start from smoking is significantly increased; therefore the controls and policing should reflect the risk.

Bring the smoking rules to the attention of all workers and visitors to the site. Display the appropriate signs, particularly in high-risk or communal areas such as canteens and site access points.

Electrical safety

Ensure electrical installations, especially temporary ones, are of sufficient capacity for the intended use and designed, installed, inspected and maintained by competent personnel. The installation should meet BS 7671: 2008 requirements for electrical installations, which includes a special section on construction sites. Do not allow ad hoc additions or alterations to the electrical installation by personnel who are not competent.

Some common electrical faults posing fire risks include:

▶ overheating cables and equipment, e.g. due to overloading circuits, bunched or coiled cables, or impaired cooling fans;
▶ incorrect installation or use of equipment;
▶ little or no maintenance and testing of equipment;
▶ incorrect fuse ratings;
▶ damaged or inadequate insulation on cables or wiring;
▶ combustible materials being placed too close to electrical equipment and distribution boards which may give off heat even when operating normally or may become hot due to a fault;
▶ cable laid in or near combustible material, frequently in roof and ceiling voids;
▶ arcing or sparking by electrical equipment;
▶ embrittlement and cracking of cable sheathing in cold environments;
▶ use of flat twin and earth cable as extension leads instead of suitable flexible cable;
▶ intentional defeating of safety devices, such as fuses or circuit breakers;
▶ mechanical damage to cables, often as a result of inappropriate routing of cables;
▶ makeshift cable joints made without correct proprietary connectors; and/or
▶ use of non heat-resistant glass or a broken glass cover over a halogen lamp (poor heat-resisting glass covers have been known to ignite flammable vapours being emitted from a freshly applied solvent-based covering laid on to floors).

The proper use of electrical safety devices, such as residual current devices (RCDs), can reduce the risks of

13

fire arising from electrical faults. However, they do not substitute for properly designed, installed, inspected and maintained electrical installations under the supervision of a competent person.

Electrical systems need to be periodically checked to ensure that they remain safe and free from damage or deterioration. They should also be checked before any addition, extension or modification is carried out. On most sites, and particularly larger ones, this will require some form of systematic electrical inspection and maintenance regime.

Electrical equipment should meet standards that reflect the adverse conditions on most construction sites. Guidance regarding site electrical safety and periodic checks can be found in HSG141 *Electrical safety on construction sites.*

All electrical equipment should be installed and maintained in a safe manner by a competent person. If portable electrical equipment is used, including items brought into a workplace by staff, then your fire risk assessment should ensure that it visually inspected and undergoes portable appliance testing (PAT) at intervals suitable for the type of equipment and its frequency of use (refer to HSE guidance 10). If you have any doubt about the safety of your electrical installation then you should consult a competent electrician.

Issues to consider include:

- overloading of equipment;
- correct fuse ratings;
- PAT testing and testing of fixed installations;
- protection against overloading of installation;
- protection against short circuit;
- insulation, earthing and electrical isolation requirements;
- frequency of electrical inspection and test;
- temperature rating and mechanical strength of flexible cables;
- portable electrical equipment;
- physical environment in which the equipment is used (e.g. wet or dusty atmospheres); and
- suitable use and maintenance of personal protective equipment.

13.3.4 Dangerous substances – flammable liquids and gases

Introduction

The Dangerous Substances and Explosive Atmospheres Regulations (DSEAR) apply at most workplaces where a dangerous substance is present or could be present. This is a fire and explosion rather than a health risk which is covered by COSHH (see Chapter 14).

The employer must:

- carry out a risk assessment of any work activities involving dangerous substances;

Figure 13.14 Electrical cabling can often get damaged and overloaded during a construction project. Other poor features here include: lack of head protection; confined space entry; edge protection and trench supports

- provide a way of eliminating or reducing risks as far as is reasonably practicable;
- provide procedures and equipment to deal with accidents and emergencies;
- provide training and information for employees;
- classify places where explosive atmospheres may occur into zones and mark the zones where necessary.

The NEBOSH Construction Certificate covers the storage of small quantities of highly flammable (up to 50 litres) and flammable (up to 250 litres) substances. DSEAR and the approved Code of Practice L138 2nd edition are given as reference documents. Some of the information provided goes beyond the storage of small quantities but is given to provide some basic information to managers in process industries.

The Regulations give a detailed definition of 'dangerous substance', which should be referred to for more information. They include any substance or preparation which, because of its properties or the way it is used, could be harmful because of fires and explosions. The list includes petrol, LPG, paints, varnishes, solvents and some dusts. These are dusts which, when mixed with air, can cause an explosive atmosphere. Dusts from milling and sanding operations are examples of this. Most workplaces contain a certain amount of dangerous substances.

An explosive atmosphere is an accumulation of gas, mist, dust or vapour, mixed with air, which has the potential to catch fire or explode. Although an explosive atmosphere does not always result in an explosion (detonation), if it catches fire, flames can quickly travel through the workplace. In a confined space (e.g. in plant or equipment) the rapid spread of the flame front or rise in pressure can itself cause an explosion and rupture of the plant and/or building.

Risk assessment

This is the process of identifying and carefully examining the dangerous substances present or likely to be present on the construction site, the work activities involving them and how they might fail and cause fire, explosion and similar events that could harm employees and the public. The purpose of a risk assessment is to enable the employer to decide what needs to be done to eliminate or reduce the safety risks from dangerous substances as far as is reasonably practicable. It should take account of the following:

▶ what hazardous properties the substances have;
▶ the way they are used and stored;
▶ the possibility of hazardous explosive atmospheres occurring;
▶ any potential ignition sources.

Regardless of the quantity of dangerous substance present, the employer must carry out a risk assessment. This will enable them to decide whether existing measures are sufficient or whether they need to make any additional controls or precautions. Non-routine activities need to be assessed as well as the normal activities within the workplace. For example, in maintenance work there is often a higher potential for fire and explosion incidents to occur.

Unless the employer has already carried out a detailed assessment under the Management Regulations, there must be an assessment of the risks from fire, explosion and other events arising from dangerous substances, including addressing requirements specified by DSEAR. This is separate from the Fire Risk assessment made under the RRFSO as Dangerous Substances are considered to be process risks covered by the HSW Act.

Employers are required to ensure that the safety risks from dangerous substances are eliminated or, when this is not reasonably practicable, to take measures to control risks **and** to reduce the harmful effects of any fire, explosion or similar events, so far as is reasonably practicable.

Substitution

Substitution is the best solution. It is much better to replace a dangerous substance with a substance or process that totally eliminates the risk. In practice this is difficult to achieve; so it is more likely that the dangerous substance will be replaced with one that is less hazardous (e.g. by replacing a low-flashpoint solvent with a high-flashpoint one).

Designing the process so that it is less dangerous is an alternative solution. For example a change could be made from a batch production to a continuous production process, or the manner or sequence in which the dangerous substance is added could be altered. However, care must be taken when carrying out these steps to make sure that no other new safety or health risks are created or increased, as this would outweigh the improvements implemented as a result of DSEAR.

The fact is that where a dangerous substance is handled or stored for use as a fuel, there is often no scope to eliminate it and very little chance to reduce the quantities handled.

Where risk cannot be entirely eliminated, control and mitigation measures should be applied. This should reduce risk as follows.

Control measures

Control measures should be applied in the following order of priority:

▶ Reduce the amount of dangerous substances to a minimum.
▶ Avoid or minimise releases.
▶ Control releases at source.
▶ Prevent the formation of an explosive atmosphere.
▶ Use a method such as ventilation to collect, contain and remove any releases to a safe place.
▶ Avoid ignition sources.
▶ Avoid adverse conditions (e.g. exceeding the limits of temperature or other control settings) that could lead to danger.
▶ Keep incompatible substances apart.

Mitigation measures

Choose mitigation measures which are consistent with the risk assessment and appropriate to the nature of the activity or operation. These can include:

▶ preventing fires and explosions from spreading to other plant and equipment or to other parts of the workplace;
▶ making sure that a minimum number of employees is exposed;
▶ in the case of a process plant, providing plant and equipment that can safely contain or suppress an explosion, or vent it to a safe place.

In workplaces where explosive atmospheres may occur, employers should ensure that:

▶ areas where hazardous explosive atmospheres may occur are classified into zones based on their likelihood and persistence;
▶ areas classified into zones are protected from sources of ignition by selecting suitable special equipment and protective systems;
▶ where necessary, areas classified into zones are marked with a specified 'EX' sign at their points of entry;
▶ employees working in zoned areas are provided with appropriate clothing that does not create a risk of an electrostatic discharge igniting the explosive atmosphere;
▶ before coming into operation for the first time, areas where a hazardous explosive atmosphere may be present are confirmed as being safe (verified) by a person (or organisation) competent in the field of explosion protection. The person carrying out the verification must be competent to consider the particular risks at the workplace and

13

the adequacy of control and other measures put in place.

Storage

Dangerous substances should be kept in a safe place in a separate building or the open air. Only small quantities of dangerous substances should be kept in a workroom or area as follows.

For flammable liquids that have a flashpoint above the maximum ambient temperature (normally taken as 32°C), the small quantity that may be stored in the workroom is considered to be an amount up to 250 litres.

For extremely and highly flammable liquids and those flammable liquids with a flashpoint below the maximum ambient temperature, the small quantity is considered to be up to 50 litres and this should be held in a special metal cupboard or container (Figure 13.15(a)).

Any larger amounts which cannot be kept outside in a safe area should be kept in a special fire-resisting store (Figure 13.15(b)), which should be:

▶ properly ventilated;
▶ provided with spillage retaining arrangements such as sills;
▶ free of sources of ignition, such as unprotected electrical equipment, sources of static electrical sparks, naked flames or smoking materials;
▶ arranged so that incompatible chemicals do not become mixed together either in normal use or in a fire situation;
▶ of fire-resisting construction;
▶ used for empty as well as full containers – all containers must be kept closed;
▶ kept clear of combustible materials such as cardboard or foam plastic packaging materials.

Flammable gases

Flammable gas cylinders also need to be stored and used safely. The following guidance should be adopted:

▶ Both full and empty cylinders should be stored outside. They should be kept in a separate secure compound at ground level with sufficient ventilation. Open mesh is preferable.
▶ Valves should be uppermost during storage to retain them in the vapour phase of the LPG.
▶ Gas cylinders should be secured in an upright position. Cylinders must be protected from mechanical damage. Unstable cylinders should be together, for example, and cylinders must be protected from the heat of the summer sun.
▶ Hose length should be kept to a minimum. This reduces the likelihood of damage and should help to ensure that the hose is not damaged by the hot work.

(a)

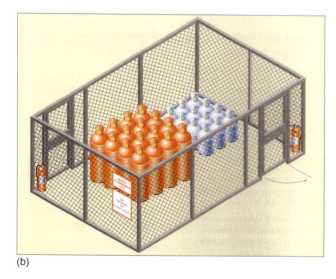

(b)

Figure 13.15 (a) Storage arrangements for highly flammable liquids; (b) LPG outside storage compound

▶ Regulators and hoses should be of a recognised standard, e.g. BS EN ISO 2503:2009 and BS EN 559:1994, respectively.
▶ Flame failure devices are necessary to shut off the gas supply in the event of flame failure.
▶ Use proprietary hoses with properly made hose end connections. Worm drive fasteners are not recommended.
▶ To avoid confusion hoses should be colour coded as:
   ▷ blue – oxygen;
   ▷ red – acetylene; and
   ▷ orange – propane.
▶ Non-return valves at the torch (blowpipe) inlet and flashback arresters at the pressure outlet from the gas cylinders should be provided on both gas lines. All such devices should be to an appropriate standard, such as BS EN 730:2002.
▶ Both high and low ventilation must be maintained where LPG applications are being used.
▶ Make sure that oil or grease do not contaminate the oxygen supply. Only use components that have been specially cleaned and supplied for oxygen use.
▶ Always check equipment visually for damage before use, especially the hoses. Any badly damaged

or suspect hoses should be discarded from use. After assembling the equipment, always check for leaks by applying a soap solution around joints and watching for bubbles.

▶ Nominally empty drums should not be used as supports for hot work activities.

▶ Gas valves must be turned off after use at the end of the shift.

▶ Cylinders must be changed in a well-ventilated area remote from any sources of ignition.

▶ Flammable material must be removed or protected before welding or similar work.

13.3.5 Systems of work

Safe systems of work have an important part to play in preventing fires. This is particularly important in high hazard areas where dangerous substances are being used. Systems of work may involve securing a hot work permit when, for example, welding, cutting or burning are taking place and/or a method statement from a contractor stating how hot roof work using heated bitumen will take place. See Chapter 4 for more details.

Precautions to be taken and reflected in the permit to work before, during and after the work include:

▶ clearing the surrounding area of all loose combustible material;

▶ checking for combustible material on both sides of a wall or partition, where work takes place only on one side;

▶ having suitable extinguishers at hand and a careful watch maintained for fire during the work and following completion;

▶ protecting combustible material which cannot be cleared;

▶ examining the hot work area thoroughly for some time after the work has finished (typically this will be at least an hour, but ignition can sometimes occur much later than this – inform the night security guards where hot work has been going on and ask them to check these areas); and

▶ in view of the potential risk, it is a sensible precaution for all hot work to stop by a safe period before the end of the day.

There may not be a need for a fully-documented permit-to-work system where the risks arising from hot work are low. However, precautions such as having a fire extinguisher are still required. Site rules are an effective means of making these precautions clear to those carrying out such work.

13.3.6 Temporary accommodation units (TAUs)

TAUs should be located away from the construction work (no less than 6 m) in the open air. If TAUs have to be located closer, the risk of a TAU fire spreading can be reduced if either the TAU or the part of the building adjacent to it is fire-resisting. Where TAUs are vertically stacked, the roof/floor assembly and the supporting members should be protected to achieve a minimum of 30 minutes' fire resistance (integrity, insulation and load-bearing capacity). TAUs should be included in emergency plans.

Litter skirts can be installed to stop extraneous material, such as combustibles, from accumulating under the units. The skirts also stop ignition sources, such as cigarettes, from blowing under the units.

Preventing fire is the primary aim, but being able to deal with it is also important. In simple cases, such as a single site hut located on an open site, little beyond basic precautionary measures are necessary, such as:

▶ keeping a tidy office;

▶ providing appropriate fire extinguishers;

▶ enforcing smoking rules;

▶ correct installation and careful use of heaters and cooking equipment; and

▶ properly installed and maintained electrical services.

Individual heating appliances require particular care if they are to be used safely. Heaters should preferably be secured in position when in use and fitted with a fire guard if appropriate.

As a general rule, convector or fan heaters should be preferred to radiant heaters because they present a lower risk of fire and injury.

The following rules should be observed:

▶ All heaters should be kept well clear of combustible materials and where they do not cause an obstruction.

▶ Heaters which burn a fuel should be sited away from draughts.

▶ Portable fuel-burning heaters (including bottled gas (LPG)) should only be used in exceptional circumstances and if shown to be acceptable in the risk assessment.

Figure 13.16 Multiple temporary accommodation units (TAUs) with external staircase

▶ All gas heating appliances should be used only in accordance with manufacturer's instructions and should be serviced annually by a competent person.

In general, staff should be discouraged from bringing their own portable heaters and other electrical equipment (e.g. kettles) onto the premises.

13.3.7 Principles of fire protection in buildings

General

The design of all new buildings and the design of extensions or modifications to existing buildings must be approved by the local planning authority.

Design data for new and modified buildings must be retained throughout the life of the structure.

Buildings' legal standards are concerned mainly with safety of life. Therefore, it is necessary to consider the early stage of fire and how it affects the means of escape, and, also, aim to prevent eventual spread to other buildings.

Asset protection requires extra precautions that will have an effect at both early and later stages of the fire growth by controlling fire spread through and between buildings and preventing structural collapse. However, this extra fire protection will also improve life safety not only for those escaping at the early stages of the fire but also for fire-fighters who will subsequently enter.

If a building is carefully designed and suitable materials are used to build it and maintain it, then the risk of injury or damage from fire can be substantially reduced. Three objectives must be met:

▶ It must be possible for everyone to leave the building quickly and safely.
▶ The building must remain standing for as long as possible.
▶ The spread of fire and smoke must be reduced.

These objectives can be met through the selection of materials and design of buildings.

Fire loading

The fire load of a building is used to classify types of building use. It may be calculated simply by multiplying the weight of all combustible materials by their energy values and dividing by the floor area under consideration. The higher the fire load, the more the effort needed to offset this by building to higher standards of fire resistance.

Surface spread of fire

Combustible materials, when present in a building as large continuous areas, such as for lining walls and ceilings, readily ignite and contribute to spread of fire over their surfaces. This can represent a risk to life in buildings, particularly where walls of fire-escape routes and stairways are lined with materials of this nature.

Materials are tested by insurance bodies and fire research establishments. The purpose of the testing is to classify materials according to the tendency for flame to spread over their surfaces. As with all standardised test methods, care must be taken when applying test results to real applications.

In the UK, a material is classified as having a surface in one of the following categories:

Class 1 – Surface of very low flame spread;
Class 2 – Surface of low flame spread;
Class 3 – Surface of medium flame spread;
Class 4 – Surface of rapid flame spread.
The test shows how a material would behave in the initial stages of a fire.

As all materials tested are combustible, in a serious fire they would burn or be consumed. Therefore, there is an additional Class 0 of materials which are non-combustible throughout or, under specified conditions, non-combustible on one face and combustible on the other. The spread of flame rating of the combined Class 0 product must not be worse than that for Class 1.

The Building Regulations use three of these classes in their Approved Document for lining materials as follows:

Class 0: Materials suitable for circulation spaces and escape routes

Such materials include brickwork, blockwork, concrete, ceramic tiles, plaster finishes (including rendering on wood or metal lathes), wood-wool cement slabs and mineral fibre tiles or sheets with cement or resin binding.

Note: Additional finishes to these surfaces may be detrimental to the fire performance of the surface and if there is any doubt about this then consult the manufacturer of the finish.

Class 1: Materials suitable for use in all rooms but not on escape routes

Such materials include all the Class 0 materials referred to above. Additionally, timber, hardboard, blockboard, particle board, heavy flock wallpapers and thermosetting plastics will be suitable if flame-retardant treated to achieve a Class 1 standard.

Class 3: Materials suitable for use in rooms of less than 30 m²

Such materials include all those referred to in Class 1, including those that have not been flame-retardant treated and certain dense timber or plywood and standard glass-reinforced polyesters.

The equivalent European classification standard will also be acceptable. Further details about internal linings and classifications are available in *The Building Regulations 2000 (as amended 2001): Approved document B Fire Safety* (see Chapter 19), The Stationery Office, ISBN 0 11 753911-2. Appropriate testing procedures are detailed in BS 476-7 and where appropriate BS EN 13501-1.

Fire resistance of structural elements

If structural elements such as walls, floors, beams, columns and doors are to provide effective barriers to fire spread and to contribute to the stability of a building, they should be of a required standard of fire resistance.

In the UK, tests for fire resistance are made on elements of structure, full size if possible, or on a representative portion having minimum dimensions of 3 m long for columns and beams and 1 m² for walls and floors. All elements are exposed to the same standard fire provided by furnaces in which the temperature increases with time at a set rate. The conditions of exposure are appropriate to the element tested. Free-standing columns are subjected to heat all round, and walls and floors are exposed to heat on one side only. Elements of structure are graded by the length of time they continue to meet three criteria:

▶ The element must not collapse.
▶ The element must not develop cracks through which flames or hot gases can pass.
▶ The element must have enough resistance to the passage of heat so the temperature of the unexposed face does not rise by more than a prescribed amount.

The term fire resistance has a precise meaning. It should not be applied to such properties of materials as resistance to ignition or resistance to flame propagation. For example, steel has a high resistance to ignition and flame propagation but will distort quickly in a fire and allow the structure to collapse – it therefore has poor 'fire resistance' (Figure 13.17). It must be insulated to provide good fire protection. This is normally done by encasing steel frames in concrete.

Figure 13.17 Steel structures can collapse in the heat of a fire

In the past, asbestos has been made into a paste and plastered onto steel frames, giving excellent fire protection, but asbestos causes major health problems and its use in new work is banned.

Building materials with high fire resistance are, for example, brick, stone, concrete, very heavy timbers (the outside chars and insulates the inside of the timber), and some specially made composite materials used for fire doors.

Insulating materials

Building materials used for thermal or sound insulation could contribute to the spread of fire. Only approved fire-resisting materials should be used. Many buildings have insulated core panels (Figure 13.18) as exterior cladding or for internal structures and partitions. The food industry, in particular, uses insulated core panels because they are easy to clean and facilitate consistent temperature control within the premises. The simple construction of these panels enables alterations and for additional internal partitions to be erected with minimum disruption to business.

Figure 13.18 Insulated core panels

They normally consist of a central insulated core, sandwiched between an inner and outer metal skin. There is no air gap. The external surface is then normally coated with a PVC covering to improve weather resistance or the aesthetic appeal of the panel. The central core can be made of various insulating materials, ranging from virtually non-combustible through to highly combustible. Differing fire hazards are associated with common types of insulation, when the panels are subjected to certain temperatures. Typical examples are:

▶ Mineral rock/modified phenolic will produce surface char and little smoke or gaseous combustion products, at temperatures above 230°C.
▶ Polyisocyanurate (PIR)/polyurethane (PUR) will char and will generate smoke and gaseous combustion products, at temperatures above 430°C PIR and 300°C PUR.

▶ Expanded polystyrene (EPS) will melt and will generate smoke and gaseous combustion products, at temperatures above 430°C.

Insulation charring can lead to panel delamination/collapse, and the gaseous combustion products can fill areas with the toxic gases carbon monoxide and styrene. A number of fires in buildings where insulated core panels have been used extensively in the fabric of the building have highlighted the particular dangers that may be associated with this form of construction, i.e. where the fabric of the building can contribute to the fire hazard. For more information see: **http://www.communities.gov.uk/fire/firesafety/firesafetylaw/aboutguides/**

Fire compartmentation

A compartment is a part of a building that is separated from all other parts by walls and floors, and is designed to contain a fire for a specified time.

The principal objective is to limit the effect of both direct fire damage and consequential business interruption caused by not only fire spread but also smoke and water damage in the same floor and other storeys.

Buildings are classified into purpose groups, according to their size. To control the spread of fire, any building whose size exceeds that specified for its purpose group must be divided into compartments that do not exceed the prescribed limits of volume and floor area. Otherwise, they must be provided with special fire protection. In the UK, the normal limit for the size of a compartment is 7,000 m³. Compartments must be separated by walls and floors of sufficient fire resistance. Any openings needed in these walls or floors must be protected by fire-resisting doors to ensure proper fire-tight separation.

The early installation and completion of compartments can also provide protection during the construction phase. It should be given priority when planning GFPs but, in practice, there will be limits on how early compartmentation can be installed. Any openings need to be protected to an equivalent standard of fire resistance to the rest of the compartment. Work activities also need to be carefully monitored to ensure that any holes or gaps remaining after services are installed are correctly filled in.

It is essential to maintain the integrity of compartments. Compromised compartments (e.g. with unprotected openings) do not work either during construction or in completed buildings and can undermine fire precautions catastrophically. Compartments can be **horizontal** to assist evacuation where compartments are large and exceed standards of fire resistance, or **vertical** to provide a degree of protection above or below floor/roof level.

Figure 13.19 Plasterboard partitions such as this can form effective compartmentation. It is important that all gaps are filled in. In this case, there are holes in the top of the partition and service ducts in the side rooms that need to be sealed

To be effective, the compartment must be complete, without any voids and/or holes passing through it. The following are examples where compartments may be breached:

▶ refurbishment – consider concealed spaces such as behind panelling and cavity walls;
▶ raised floors for computer suites;
▶ holes requiring patching;
▶ voids/openings for services to pass through; and
▶ damage from site vehicles.

Ventilation and heating ducts must be fitted with fire dampers where they pass through compartment walls and floors. Firebreak walls must extend completely across a building from inside wall to outside wall. They must be stable; they must be able to stand even when the part of the building on one side or the other is destroyed.

No portion of the wall should be supported on unprotected steelwork nor should it have the ends of unprotected steel members embedded in it. The wall must extend up to the underside of a non-combustible roof surface, and sometimes above it. Any openings must be protected to the required minimum grade of fire resistance. If an external wall joins a firebreak wall and has an opening near the join, the firebreak wall may need to extend beyond the external wall.

An important function of external walls is to contain a fire within a building, or to prevent fire spreading from outside.

The fire resistance of external walls should be related to the:

▶ purpose for which the building is used;
▶ height, floor area and volume of the building;
▶ distance of the building from relevant boundaries and other buildings;
▶ extent of doors, windows and other openings in the wall.

A wall which separates properties from each other should have no doors or other openings in it.

13.3.8 Electrical and other equipment in potentially flammable atmospheres

Most electrical equipment either sparks in normal operation or is liable to spark under fault conditions. Some electrical appliances such as electric heaters are specifically designed to produce high temperatures. These circumstances create fire and explosion hazards, which demand very careful assessment in locations where processes capable of producing flammable concentrations of gas or vapour are used, or where flammable liquids are stored.

It is likely that many fires are caused by static electrical discharges. Static electricity can, in general, be eliminated by the careful design and selection of materials used in equipment and plant, and the materials used in products being manufactured. When it is impractical to avoid the generation of static electricity, a means of control must be devised. Where flammable materials are present, especially if they are gases or dusts, then there is a great danger of fire and explosion, even if there is only a small discharge of static electricity. The control and prevention of static electricity is considered in more detail later.

The use of electrical equipment in potentially flammable atmospheres should be avoided as far as possible. However, there will be many cases where electrical equipment must be used and, in these cases, the standards for the construction of the equipment should comply with the Equipment and Protective Systems Intended for Use in Potentially Explosive Atmospheres Regulations, known as ATEX. Details on the classification or zoning of areas are contained in the Dangerous Substances and Explosive Atmospheres Regulations and ACoPs.

Before electrical equipment is installed in any location where flammable vapours or gases may be present, the area must be zoned in accordance with the Dangerous Substances and Explosive Atmosphere Regulations, and records of the zoned areas must be marked on building drawings and revised when any zoned area is changed. The installation and maintenance of electrical equipment in potentially flammable atmospheres is a specialised task. It must only be undertaken by electricians or instrument mechanics who are trained to ATEX standards.

Static electricity is produced by the build-up of electrons on weak electrical conductors or insulating materials. These materials may be gaseous, liquid or solid and may include flammable liquids, powders, plastic films and granules. Plastics have a high resistance that enables them to retain static charges for long periods of time. The generation of static may be caused by the rapid separation of highly insulated materials by friction or by transfer from one highly

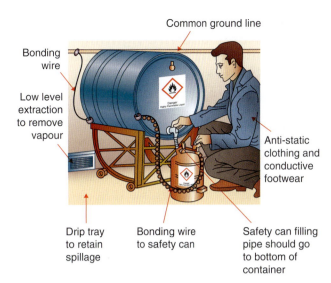

Dispensing Flammable Liquids

Figure 13.20 Safe dispensing of flammable liquids

charged material to another in an electric field by induction (see Figure 13.20).

A static electric shock, perhaps caused by closing a door with a metallic handle, can produce a voltage greater than 10,000V. Since the current flows for a very short period of time, there is seldom any serious harm to an individual. However, discharges of static electricity may be sufficient to cause serious electric shock and are always a potential source of ignition when flammable liquid, dusts or powders are present.

Static electricity may build up on both materials and people. When a charged person approaches flammable gases or vapours and a spark ignites the substance, the resulting explosion or fire often causes serious injury. In these situations, effective static control systems must be used which includes:

► bonding or earthing continuity between pieces of equipment, particularly portable equipment such as containers for carrying highly flammable substances, piping, filling funnels, drip trays and the like;

► not wearing outer clothing which generates static charges. In practice this usually means avoiding man-made fibres and using cotton only;

► using conductive footwear to leak static charges to ground (this must not be used by electricians as it conducts electricity and will not protect them against electric shock);

► avoiding the free fall of highly flammable liquids from one container to another unless anti-static additives have been put into the liquid or its natural properties will not hold static charges. Filling funnels should reach down to near the bottom of the container being filled.

For more information see Chapter 12 on electrical safety.

13.4 Fire detection, fire alarm systems and fire-fighting equipment for construction activities

13.4.1 Introduction

The term general fire precaution (GFP) is used to describe the structural features and equipment needed to ensure everyone reaches safety if there is a fire. GFP includes such things as:

- escape routes and fire exits (see 13.5);
- fire-fighting equipment;
- raising the alarm;
- making emergency plans; and
- limiting the spread of fire (compartmentation, see 13.3.7).

The GFPs needed will vary from site to site. Sometimes they will be very simple and other times much more complicated, depending on the risks involved at each stage of the construction process. But they all need to take account of the size of the site, the number of people present and the nature of the work being done. Individual elements of GFPs must be considered as part of the overall package and not in isolation.

13.4.2 Fire detection and alarm systems

In the event of fire it is vital that everyone in the workplace is alerted as soon as possible. The earlier the fire is discovered, the more likely it is that people will be able to escape before the fire takes hold and before it blocks escape routes or makes escape difficult.

Every site should have detection and warning arrangements. Often the people who work there will detect the fire and in many worksites nothing further will be needed.

It is important to consider how long a fire is likely to burn for before it is discovered. Fires are likely to be discovered quickly if they occur in places that are frequently visited by employees, or in occupied areas of a site. For example, employees are likely to smell burning or see smoke if a fire breaks out in an office or site hut.

The aim of any fire warning system is to ensure that people on the site are alerted to make their escape before a fire becomes life-threatening. The essential requirements of the fire warning signal are that it is distinctive, clearly audible above any other noise and is recognised by all the people on site.

The extent or sophistication of the fire warning will vary from site to site. For example:

- On a small open-air site, or those involving small buildings and structures, 'word of mouth' may well be adequate.

- On larger open-air sites, or those involving buildings and structures with a limited number of rooms, such that a shout of 'fire' might not be heard or could be misunderstood, a klaxon, whistle, gong or small self-contained proprietary fire alarm unit may well be needed.
- On sites for complex multi-storey buildings, it is likely that a wired-in system of call points and sounders will be required to provide an effective fire warning system that meets the relevant British Standard.

The operation and effectiveness of the fire alarm should be:

- regularly checked and tested; and
- periodically serviced.

Where there is concern that fire may break out in an unoccupied part of the premises, for example in a basement, some form of automatic fire detection should be fitted. Commercially available heat or smoke detection systems can be used. In small premises a series of interlinked domestic smoke alarms that can be heard by everyone present will be sufficient. In many cases, staff can be relied upon to detect a fire.

There is not normally any need for automatic fire detectors to be fitted during construction work. However, on high-risk sites or in temporary accommodation units (TAUs) such as site offices, if there are locations where a fire might occur and develop unnoticed until it threatens people's means of escape, detectors may be appropriate. Domestic-type smoke detectors are not considered appropriate on complex multi-storey sites; however, on small lower risk sites or small TAUs they may be acceptable.

Fire alarm systems will often be fitted as part of the construction work. Alternatively, buildings may have a wired-in fire alarm system already installed. Try to plan the work to install the fire alarm system as early as possible and, where a system is already installed, keep it in working order for as long as possible. Where they are relied on during the construction phase, it is vital that existing systems are not inadvertently disabled, for instance during work on electrical systems in refurbishment work. If they are disabled for any reason, alternative arrangements need to be provided.

A fire alarm system will typically include the following:

- manual call points (break-glass call points) at storey exit and final exit doors;
- electronic sirens or bells; and
- a simple control and indicator panel.

In some cases a heat or smoke detector will also be fitted in the system.

13.4.3 Extinguishing media

There are four main methods of extinguishing fires, which are explained as follows:

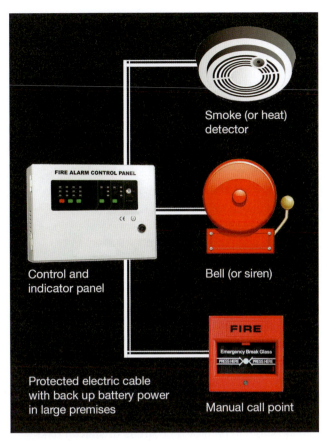

Figure 13.21 Simple electrical fire alarm system components

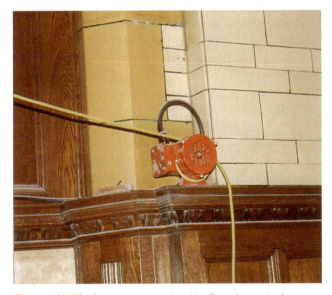

Figure 13.22 A temporary wired-in fire alarm during major renovation of a large and multi-storey complex building

- ▶ **Cooling** – reducing the ignition temperature by taking the heat out of the fire – using water to limit or reduce the temperature.
- ▶ **Smothering** – limiting the oxygen available by smothering and preventing the mixture of oxygen and flammable vapour – by the use of foam or a fire blanket.
- ▶ **Starving** – limiting the fuel supply – by removing the source of fuel by switching off electrical power,

isolating the flow of flammable liquids or removing wood and textiles, etc.

- ▶ **Chemical reaction** – by interrupting the chain of combustion and combining the hydrogen atoms with chlorine atoms in the hydrocarbon chain, for example with Halon extinguishers. (Halons have generally been withdrawn because of their detrimental effect on the environment, as ozone depleting agents.)

Advantages and limitations of the main extinguishing media

Fire extinguishers are all red with 5% of the cylindrical area taken up with the colour code. The colour code (band) denotes on which class of fire the extinguisher can be used.

Water extinguishers (red band)

This type of extinguisher can only be used on Class A fires. They allow the user to direct water onto a fire from a considerable distance.

A 9-litre water extinguisher can be quite heavy and some water extinguishers with additives can achieve the same rating, although they are smaller and therefore considerably lighter. This type of extinguisher is not suitable for use on live electrical equipment, liquid or metal fires.

Water extinguishers with additives (red band)

This type of extinguisher is suitable for Class A fires. They can also be suitable for use on Class B fires and, where appropriate, this will be indicated on the extinguisher. They are generally more efficient than conventional water extinguishers.

Foam extinguishers (cream band)

This type of extinguisher can be used on Class A or B fires and is particularly suited to extinguishing liquid fires such as petrol and diesel. They should not be used on free-flowing liquid fires unless the operator has been specially trained, as they have the potential to rapidly spread the fire to adjacent material. This type of extinguisher is not suitable for deep-fat fryers or chip pans. They should not be used on electrical or metal fires.

Powder extinguishers (blue band)

This type of extinguisher can be used on most classes of fire and achieve a good 'knock down' of the fire. They can be used on fires involving electrical equipment but will almost certainly render that equipment useless.

Because they do not cool the fire appreciably, it can re-ignite. Powder extinguishers can create a loss of visibility and may affect people who have breathing problems and are not generally suitable for confined spaces. They should not be used on metal fires.

13

Carbon dioxide extinguishers (black band)

This type of extinguisher is particularly suitable for fires involving electrical equipment as they will extinguish a fire without causing any further damage (except in the case of some electronic equipment, for example computers). As with all fires involving electrical equipment, the power should be disconnected if possible. These extinguishers should not be used on metal fires.

Wet chemical – class 'F' extinguishers

This type of extinguisher is particularly suitable for commercial catering establishments with deep-fat fryers. The intense heat in the fluid generated by fat fires means that when standard foam or carbon dioxide extinguishers stop discharging, re-ignition tends to occur.

Wet chemical extinguishers starve the fire of oxygen by sealing the burning fluid, which prevents flammable vapour reaching the atmosphere.

Figure 13.23 Fire point on large construction site

13.4.4 Portable fire-fighting equipment

Since 1 April 2006, employers or those who have control of non-domestic premises have had a statutory duty under fire safety and health and safety legislation to ensure that there are appropriate means of fighting fires. The employer or controller of non-domestic premises is known as the 'responsible person'. The responsible person also has to provide suitably trained people to operate non-automatic fire-fighting equipment.

As Fire Certificates issued by fire authorities under the Fire Precautions Act 1971 have been abolished, instead it is up to the responsible person to identify the fire-fighting requirement for their site. If fire breaks out in the workplace and trained staff can safely extinguish it using suitable fire-fighting equipment, the risk to others will be removed. Therefore, all workplaces where people are at risk from fire should be provided with suitable fire-fighting equipment.

The most useful form of fire-fighting equipment for general fire risks is the water-type portable extinguisher or suitable alternative. The number and type of extinguishers present depends on the fire hazard. For a typical spread of fire hazards, the following is considered to provide a reasonable level of cover per 200 m² of floor area, with no fewer than two each of (a) and (b) on each floor:

(a) one 9-litre water or foam; and

(b) one CO_2 extinguisher (at least 1.1 kg).

If each floor has a hose reel which is known to be in working order and of sufficient length for the floor it serves, there may be no need for water-type extinguishers to be provided. **Areas of special risks involving the use of oil, fats or electrical equipment may need carbon dioxide, dry powder or other types of extinguisher (Figure 13.24).**

Fire extinguishers should be sited on exit routes, preferably near to exit doors or, where they are provided for specific risks, near to the hazards they protect. Notices indicating the location of fire-fighting equipment should be displayed where the location of the equipment is not obvious or in areas of high fire risk where the notice will assist in reducing the risk to people in the workplace.

All Halon fire extinguishers should have been decommissioned as from December 2003 and disposed of safely.

Those carrying out hot work should have appropriate fire extinguishers with them and know how to use them.

The primary purpose of fire extinguishers is to tackle fires at a very early stage to enable people to make their escape. Putting out larger fires is the role of the fire and rescue services.

Extinguishers should conform to a recognised standard, BS EN 3:7 'Portable Fire Extinguishers. Characteristics, performance requirements and test methods'.

13.4.5 Fixed fire-fighting equipment – sprinklers

Larger and more complex structures, such as multi-storey buildings, may have fixed fire-fighting systems installed. These may range from dry and wet risers to automatic sprinkler systems. Dry and wet risers are

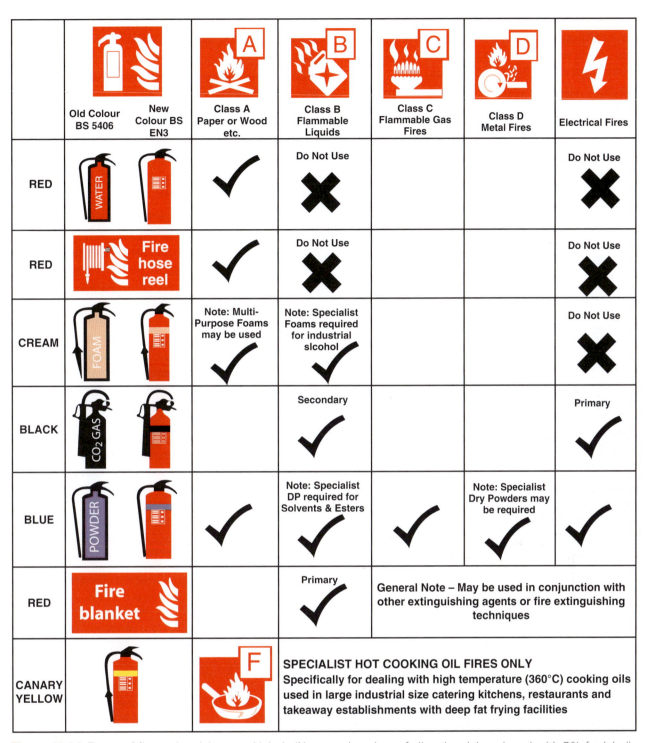

	Old Colour BS 5406 / New Colour BS EN3	Class A Paper or Wood etc.	Class B Flammable Liquids	Class C Flammable Gas Fires	Class D Metal Fires	Electrical Fires
RED	WATER	✓	Do Not Use ✗			Do Not Use ✗
RED	Fire hose reel	✓	Do Not Use ✗			Do Not Use ✗
CREAM	FOAM	Note: Multi-Purpose Foams may be used ✓	Note: Specialist Foams required for industrial slcohol ✓			Do Not Use ✗
BLACK	CO₂ GAS		Secondary ✓			Primary ✓
BLUE	POWDER	✓	Note: Specialist DP required for Solvents & Esters ✓	✓	Note: Specialist Dry Powders may be required ✓	✓
RED	Fire blanket		Primary ✓	General Note – May be used in conjunction with other extinguishing agents or fire extinguishing techniques		
CANARY YELLOW		F	SPECIALIST HOT COOKING OIL FIRES ONLY Specifically for dealing with high temperature (360°C) cooking oils used in large industrial size catering kitchens, restaurants and takeaway establishments with deep fat frying facilities			

Figure 13.24 Types of fire extinguishers and labels (Note: main colour of all extinguishers is red with 5% for label)

provided for the fire service to tackle a fire quickly. The continued availability of these in existing buildings, and their early commissioning in new buildings, is therefore recommended. Similarly with sprinkler systems, it is worth planning the work so that these are available for as much of the construction phase as possible. Where risers are provided, liaison should be established with the fire service and the access points should be reviewed periodically.

Recognition should be given that sprinkler provision may have allowed for reduced fire resistance or extended travel distances. At the construction stage this should be considered and be incorporated into any fire evacuation planning. Sprinklers should be considered as merely one component part of a total fire safety strategy, which is tailored to the existing and projected needs of a building. They have significant benefits to offer in suppressing fires until those best trained to deal with major incidents are on the scene to extinguish them.

Sprinkler systems can be very effective in controlling fires. They can be designed to protect life and/or property and may be regarded as a cost-effective solution for reducing the risks created by fire. If a building has a sprinkler installation, it may have been installed as a result of a business decision, for example for the protection of business assets, or it may have been installed as a requirement, for example imposed under a local Act, or as an integral part of the building design.

Sprinkler systems should normally extend to the entire building. In a well-designed system, only those heads in the immediate vicinity of the fire will actually operate. Sprinkler installations typically comprise a water supply (preferably a stored water supply incorporating tanks), pumps, pipe work and sprinkler heads. There are different types of sprinkler design; sprinklers can be operated to discharge water at roof or ceiling level or within storage racks. Other design types such as ESFR (early suppression fast response) and dry pipe may also be appropriate. In all cases, a competent person/contractor should be used to provide guidance.

The installation should be designed for the fire hazard, taking into account the building occupancy, the fire load and its burning characteristics and the sprinkler control characteristics. For each hazard the sprinkler installation design should take account of specific matters such as storage height, storage layout, ceiling clearance and sprinkler type (e.g. sprinkler orifice, sprinkler sensitivity).

There are some hazardous locations where sprinklers should not be fitted, such as over salt baths and metal melt pans, because water will possibly cause an explosive reaction.

If any significant changes are being made to the premises, for example changing storage arrangements or material stored, the sprinkler installation should be checked to see that it is still appropriate and expert advice sought as necessary.

Sprinkler protection could give additional benefits, such as a reduction in the amount of portable fire-fighting equipment necessary, and the relaxation of restrictions in the design of buildings.

Guidance on the design and installation of new sprinkler systems and the maintenance of all systems is given in the Loss Prevention Council (LPC) Rules, BS EN 12845 or BS 5306, and should be carried out only by a competent person.

Routine maintenance by on-site personnel may include checking of pressure gauges, alarm systems, water supplies, any anti-freezing devices and automatic booster pump(s). For example diesel fire pumps should be given a test run for 30 minutes each week. See Form F1 in Chapter 23 for a maintenance checklist.

13.4.6 Maintenance

It is important that equipment is fit for its purpose and is properly maintained and tested. One way in which this can be achieved is through companies that specialise in the testing and maintenance of fire-fighting equipment.

All equipment provided to assist escape from the premises, such as fire detection and warning systems and emergency lighting, and all equipment provided to assist with fighting fire, should be regularly checked and maintained by a suitably competent person in accordance with the manufacturer's recommendations. The alarm system should be tested every week, while the premises are in normal use. The test should be carried out by activating a different call point each week, at a fixed time. Table 13.2 gives guidance on the frequency of testing and maintenance and provides a simple guide to good practice. A fire maintenance checklist Form F1 is shown in Chapter 23.

Table 13.2 Maintenance and testing of fire equipment

Equipment	Period	Action
Fire detection and fire-warning systems including self-contained smoke alarms and manually operated devices	Weekly	► Check all systems for state of repair and operation ► Repair or replace defective units ► Test operation of systems, self-contained alarms and manually operated devices
	Annually	► Full check and test of system by competent service engineer ► Clean self-contained smoke alarms and change batteries
Emergency lighting including self-contained units and torches	Weekly	► Operate torches and replace batteries as required ► Repair or replace any defective unit
	Monthly	► Check all systems, units and torches for state of repair and apparent function
	Annually	► Full check and test of systems and units by competent service engineer ► Replace batteries in torches
Fire-fighting equipment installation including hose reels	Weekly	► Check all extinguishers including hose reels for correct apparent function
	Annually	► Full check and test by competent service engineer

Figure 13.25 Various sprinkler heads designed to fit into a high-level water pipe system and spray water at different angles onto a fire below

13.5 Requirements for an adequate and properly maintained means of escape in the construction workplace

13.5.1 General

It is essential to ensure that everyone can escape quickly from the site if there is a fire. On open-air sites and unenclosed single-storey structures, escape routes may be obvious and plentiful. In more complex structures which have workplaces above and/or below ground, more detailed provisions will have to be made.

Proper provision is required for everyone on site wherever they are and however temporary their stay on site (e.g. workers on roofs, in cellars or plant rooms near the roof).

Escape routes will change during the construction project and may become unusable. Replacement routes must be identified early on when changes occur. Building designs usually incorporate fire escape routes for the finished building and its occupants. In new buildings these should be installed as early as possible in the contract. For refurbished buildings existing routes should be maintained as far as possible.

In an emergency, escape via a scaffold can be very difficult particularly if there are large quantities of smoke or the weather is poor or it is dark. Where possible provide good access to escape routes inside the main building.

The construction site may contain features that could promote the rapid spread of fire, heat or smoke and affect escape routes. These features may include ducts or flues, openings in floors or walls, or combustible wall or ceiling linings. Where people are put at risk from these features, appropriate steps should be taken to reduce the potential for rapid fire spread by, for example, fitting non-combustible temporary partitions and/or providing an early warning of fire so that people can leave the workplace before their escape routes become unusable.

Combustible wall or ceiling linings should not be used on escape routes and large areas should be removed wherever they are found. Other holes in fire-resisting floors, walls or ceilings should be filled in with fire-resisting material to prevent the passage of smoke, heat and flames.

There should normally be at least two escape routes in different directions. They need to be clear, uncomplicated passageways leading to a place of safety. They must be properly maintained and kept free of obstructions.

A person confronted by a fire or its effects needs to be able to turn away from it or be able to pass the area and reach a place of safety unaided. Where this is not possible it is important to ensure that the risk of being trapped in a dead-end situation is minimised. This can be achieved by, for example, keeping combustible materials out of the area and travel distances to a minimum.

Tower crane operators (or people in similar vulnerable or difficult areas) must be considered and the necessary controls implemented.

13.5.2 Doors

Some doors may need to open in the direction of travel, such as:

▶ doors from a high-risk area, such as a paint spraying room or large kitchen;
▶ doors that may be used by more than 50 persons;
▶ doors at the foot of stairways where there may be a danger of people being crushed.

Some sliding doors may be suitable for escape purposes provided that they: do not put people using them at additional risk; slide easily; and are marked with the direction of opening.

Doors which only revolve and do not have hinged segments are not suitable as escape doors.

13.5.3 Stairways, maintenance and refurbishment

Careful consideration should be given to means of escape from above or below ground level. Stairways and ladders must be located or protected so that any fire will not prevent people using them.

Protected stairways will be a feature in many buildings so it makes sense to install these and make them available early in the construction project, before fire risks increase when finishes are installed. Flame-retardant surfaces should be used in stairways. In buildings over four storeys high, alternative protected stairways will be required.

Doors giving access to protected stairways should be 30-minute fire resistant and fitted with effective self-closing devices. They should open in the direction of escape. Revolving or sliding doors are not suitable.

If internal stairways are impracticable, external temporary stairways should be provided. These can be constructed from scaffolding using wooden treads and platforms.

Sources of heat or combustible materials may be introduced into the workplace during periods of maintenance or refurbishment. Where the work involves the introduction of heat, such as welding, this should be carefully controlled by a safe system of work, for example a hot work permit. All materials brought into the workplace in connection with the work being carried out should be stored away from sources of heat and not obstruct exit routes. See Figure 13.16 for an escape staircase from TAUs.

13

13.5.4 Escape routes and travel distances

Escape routes should meet the following criteria:

▶ Where two or more escape routes are needed they should lead in different directions to places of safety.

▶ Escape routes need to be short and to lead people directly to a place of safety, such as the open air or an area of the workplace where there is no immediate danger.

▶ It should be possible for people to reach the open air without returning to the area of the fire. They should then be able to move well away from the building.

▶ Escape routes should be wide enough for the volume of people using them. A 750-mm door will allow up to 40 people to escape in 1 minute, so most doors and corridors will be wide enough. If the routes are likely to be used by people in wheelchairs, the minimum width will need to be 800 mm.

▶ While the workplace is in use, it must be possible to open all doors easily and immediately from the inside, without using a key or similar device. Doors must be readily opened in the direction of escape. Fire doors should be self-closing (fire doors to cupboards or lockers can be simply latched or locked).

▶ Make sure that there are no obstructions on escape routes, especially on corridors and stairways where people who are escaping could dislodge stored items or be caused to trip. Any fire hazards must be removed from exit routes as a fire on an exit route could have very serious consequences.

▶ Escape routes need regular checks to make sure that they are not obstructed and that exit doors are

Figure 13.27 External access/fire escape in a large scaffold during building construction

not locked. Self-closing fire-resisting doors should be checked to ensure doors close fully, including those fitted with automatic release mechanisms.

The maximum advisable travel distances from any area in a workplace to a fire exit door leading out to a relative place of safety should be in accordance with Table 13.3.

Table 13.3 Maximum travel distances

Fire hazard			
	Lower	*Normal*	*Higher*
Enclosed structures:			
Alternative	60 m	45 m	25 m
Dead-end	18 m	18 m	12 m
Semi-open structures:			
Alternative	200 m	100 m	60 m
Dead-end	25 m	18 m	12 m

Source: HSE.

13.5.5 Lighting

Escape routes must be well lit. If the route has only artificial lighting or if it is used during the hours of darkness, alternative sources of lighting should be considered in case the power fails during a fire. Check the routes when it is dark as, for example, there may be street lighting outside that provides sufficient illumination. In small workplaces it may be enough to

Figure 13.26 Fire escape route clearly signed and free from obstructions

provide the staff with torches that they can use if the power fails. However, it may be necessary to provide battery-operated emergency lights so that if the mains lighting fails the lights will operate automatically. Candles, matches and cigarette lighters are not adequate forms of emergency lighting.

13.5.6 Signs

Exit signs on doors or indicating exit routes should be provided where they will help people to find a safe escape route. Signs on exit routes should have directional arrows, 'up' for straight on and 'left', 'right' or 'down' according to the route to be taken. Advice on the use of all signs including exit signs can be found in chapters 4 and 19.

Figure 13.28 Fire exit sign

13.5.7 Escape times

Everyone on the site should be able to get to the nearest place of safety in between 2 and 3 minutes. This means that escape routes should be kept short. Where there is only one means of escape, or where the risk of fire is high, people should be able to reach a place of safety, or a place where there is more than one route available, in 1 minute.

The way to check this is to pace out the routes, walking slowly and noting the time. Start from where people work and walk to the nearest place of safety. Remember that the more people there are using the route, the longer they will take. People take longer to negotiate stairs and they are also likely to take longer if they have a disability.

Where fire drills are held, check how long it takes to evacuate each floor in the workplace. This can be used as a basis for assessment. If escape times are too long, it may be worth re-arranging the workplace so that people are closer to the nearest place of safety, rather than undertake expensive alterations to provide additional escape routes.

Reaction time needs to be considered. This is the amount of time people will need for preparation before they escape. It may involve, for example, closing down machinery, issues of security or helping visitors or members of the public out of the premises. Reaction time needs to be as short as possible to reduce risk to staff. Assessment of escape routes should include this.

If reaction times are too long, additional routes may need to be provided. It is important that people know what to do in case of fire as this can lessen the time needed to evacuate the premises.

13.6 Evacuation of a construction workplace in the event of a fire

13.6.1 Introduction

Each workplace should have an emergency plan. The plan should include the action to be taken by staff in the event of fire, the evacuation procedure and the arrangements for calling the Fire and Rescue Authority.

For small sites, this could take the form of a simple fire action notice posted in positions where staff can read it and become familiar with it.

High-fire-risk or larger sites will need more detailed plans, which take account of the findings of the risk assessment, for example the staff significantly at risk and their location. For large sites, notices giving clear and concise instructions of the routine to be followed in case of fire should be prominently displayed. The notice should include the method of raising an alarm in the case of fire and the location of an assembly point to which staff escaping from the workplace should report.

As the nature of the workforce changes it is important that any procedures are understood. If there are personnel on site who do not speak English, it is imperative that any instructions or procedures are made clear and they understand what is needed in the event of an emergency.

13.6.2 Fire emergency plan

Site managers must make sure that all employees are familiar with the means of escape in case of fire and their use, and with the routine to be followed in the event of fire.

To achieve this, routine procedures must be set up and made known to all employees, generally outlining the action to be taken in case of fire and specifically laying down the duties of certain nominated persons. Notices should be posted throughout the premises.

While the need in individual premises may vary, there are a number of basic components which should be considered when designing any fire routine procedures:

- ▶ Where will workers gather after evacuation from the site?
- ▶ Who will be in charge of the situation and what will be their role?
- ▶ What information and/or training will that person need to carry out those functions? Fire wardens may need to be appointed to assist the person in charge.
- ▶ The action to be taken on discovering a fire.
- ▶ The method of operating the fire alarm.
- ▶ First-stage fire-fighting by employees.

13

▶ How will the people in charge communicate with each other? (Radios, etc.)

▶ How will you check that everyone has reached the assembly point, e.g. head counts or checking off against site security logs brought to the assembly point? (Possible use of sweep techniques.)

▶ Who will contact the emergency services and how?

▶ Who will meet the emergency services when they arrive and provide them with information? They will need to know of any particular risks, such as the location of LPG cylinders and the likely whereabouts of anyone unaccounted for who may still be on site. Is the fire service aware and are up-to-date details available?

▶ Evacuation of the premises.

▶ Assembly of staff, customers and visitors, and carrying out a roll-call to account for everyone on the premises.

▶ Consider adjacent premises may need evacuation and how this might be done.

▶ The fire and rescue service should be informed of any items in the risk assessment (or changes through the building process) that could affect fire-fighting or emergency operations, e.g. changes to access or water supplies.

▶ If the fire and rescue service is called to a fire when the site is unoccupied, or only occupied by security staff, how will the fire and rescue service obtain relevant information to enable them to work safely and effectively? Even if the site is unoccupied, they will still need to know of any particular risks, such as the location of LPG cylinders.

▶ Consider workers whose first language may not be English.

▶ The procedures must take account of those people who may have difficulty in escaping quickly from a building because of their location or a disability.

▶ Insurance companies and other responsible people may need to be consulted where special procedures are necessary to protect buildings and plant during or after the evacuation of people.

13.6.3 Fire marshals

A member of the staff should be nominated to supervise all fire and emergency arrangements. This person should be in a senior position or at least have direct access to a senior manager. The number of people involved in managing the emergency response should be kept to a necessary minimum. This will reduce the scope for confusion between different parties carrying out different tasks during the emergency. Nominate and train deputies to cover for key personnel when they are absent, e.g. for sickness or holidays.

On larger sites or higher fire risk sites, the appointment of fire wardens (or marshals) may be appropriate. It is important that when such people are appointed they are trained and given the necessary authority to carry out their tasks.

In the event of fire or other emergency, their duties would be, while it remains safe to do so, to ensure that:

▶ the alarm has been raised;

▶ the whole department, including toilets and small rooms, has been evacuated;

▶ the fire and rescue service has been called;

▶ liaising with the fire service is done if there is a fire and provide information on access, people trapped and any special hazards, etc.;

▶ fire doors (if any) are closed to prevent fire spread to adjoining compartments and to protect escape routes;

▶ plant and machinery are shut down wherever possible and any other actions required to safeguard the premises are taken where they do not expose people to undue risks;

▶ a roll-call is carried out at the assembly point and the result reported to whoever is in control of the evacuation.

Under normal conditions, fire marshals should check that good standards of housekeeping and preventative maintenance exist in their areas, that exits and escape routes are kept free from obstruction, that all fire-fighting appliances are available for use and fire points are not obstructed, that smoking is rigidly controlled, and that all members of staff under their control are familiar with the emergency procedure and know how to use the fire alarm and fire-fighting equipment.

13.6.4 Assembly and roll-call

Assembly points should be established for use in the event of evacuation. They should be in positions, preferably under cover, which are unlikely to be affected at the time of fire. In some cases, it may be necessary to make mutual arrangements with the occupiers of nearby premises.

In the case of small premises, a complete list of the names of all staff should be maintained so that a roll-call can be made if evacuation becomes necessary.

In those premises where the number of staff would make a single roll-call difficult, each departmental fire warden should maintain a list of the names of staff in their area. Roll-call lists must be updated regularly.

13.6.5 Fire notices

Printed instructions for the action to be taken in the event of fire should be displayed throughout the premises. The information contained in the instructions should be stated briefly and clearly. The staff and their deputies to whom specific duties are allocated should be identified.

Instruction for the immediate calling of the fire brigade in case of fire should be displayed at telephone switchboards, exchange telephone instruments and security lodges.

A typical fire notice is given in Appendix 13.2.

13.6.6 Fire drills

Fire drills, in which the entire workforce evacuates the site, are a useful means of checking that the GFP routines are effective. However, it is recognised these can often be impracticable and of limited use due to the continually changing nature of sites and the workforce on them. But, as the risks of, and from, fire increase and the number of people on site rises, the need for a drill increases (often when the main structure of the building is complete) in order to check the training and understanding of all site staff and visitors, and also for problems such as 'bottlenecks'.

It is important to check that those on site really do know what to do if there is a fire. Asking individual workers: 'What is the fire alarm?' and 'What would you do?' is a useful way of checking that the instructions and information given have been adequate. Once a fire routine has been established, it must be tested at regular intervals in order to ensure that all staff are familiar with the action to be taken in an emergency.

13.6.7 Fire procedures and people with a disability

People with a disability (including members of the public) need special consideration when planning for emergencies. People with a disability seldom work on construction sites but where necessary their needs must be considered. Under the Equalities Act if people with a disability could realistically expect to use premises, then employers or those in charge of premises must anticipate any reasonable adjustments that would make it easier for that right to be exercised. Employers should:

▶ identify everyone who may need special help to get out;
▶ allocate responsibility to specific staff to help people with a disability in emergency situations;
▶ consider possible escape routes;
▶ enable the safe use of lifts;
▶ enable people with a disability to summon help in emergencies;
▶ train staff to be able to help their colleagues;
▶ consider safe havens.

Advice on the needs of people with a disability, including sensory impairment, is available from the organisations which represent various groups. (Names and addresses can be found in an internet search.)

People with impaired vision must be encouraged to familiarise themselves with escape routes, particularly those not in regular use. A 'buddy' system would be helpful. But, to take account of absences, more than one employee working near anyone with impaired vision should be taught how to help them.

Staff with impaired hearing may not hear alarms in the same way as those with normal hearing but may still be able to recognise the sound. This may be tested during the weekly alarm audibility test. There are alternative means of signalling, such as lights or other visual signs, vibrating devices or specially selected sound signals. Action on Hearing Loss (http://www.actiononhearingloss.org.uk) can advise. Ask the fire brigade before installing alternative signals.

Wheelchair users or others with impaired mobility may need help to negotiate stairs, etc. Anyone selected to provide this help should be trained in the correct methods. Advice on the lifting and carrying of people can be obtained from the Fire Service, Ambulance Service, British Red Cross Society, St John Ambulance Brigade or certain disability organisations.

Employees with learning difficulties may also require special provision. Management should ensure that the colleagues of any employee with a learning difficulty know how to reassure them and lead them to safety.

13.7 Further information

The Management of Health and Safety at Work Regulations 1999

The Health and Safety (Safety Signs and Signals) Regulations 1996

Regulatory Reform (Fire Safety) Order 2005

Fire (Scotland) Act 2005

Fire safety (Scotland) Regulations 2006

Controlling fire and explosion risks in the workplace INDG370(rev1), 2013, ISBN 97807176 64856 http://www.hse.gov.uk/pubns/indg370.htm

The Dangerous Substances and Explosive Atmospheres Regulations 2002 SI 2002 No 2776, ISBN 9780 11 042957 5.

Dangerous substances and explosive atmospheres, L138 2nd Edition, 2013. Dangerous Substances and Explosive Atmospheres Regulations 2002. Approved Code of Practice, 2003, HSE Books, ISBN 978 0 7176 66164 http://www.hse.gov.uk/pubns/books/l138.htm

Fire Safety in Construction, 2nd edition, HSG168, HSE Books, 2010, ISBN 9780 7176 6345 3. http://www.hse.gov.uk/pubns/books/hsg168.htm

13

13.8 Practice revision questions

1. (a) **Outline** the differences between the three types of enforcement notice which may be issued under the Regulatory Reform (Fire Safety) Order.
   (b) **Identify** the principal fire safety duties of the 'responsible person' as defined by the Regulatory Reform (Fire Safety) Order.

2. (a) **Describe** the three sides of a 'fire triangle'.
   (b) **Outline THREE** methods of extinguishing fires.
   (c) **Identify FOUR** different types of ignition source that may cause a fire to occur, and give a typical workplace example of **EACH** type.
   (d) **Identify FIVE** classes of fire, the associated fuel sources and the appropriate fire extinguisher to use for each class.

3. **Outline THREE** methods of heat transfer (other than direct burning) and **explain** how each can cause the spread of fire in buildings.

4. (a) **Identify SIX** causes of fire in a workplace.
   (b) **Identify** the possible consequences on people of fires in a workplace.
   (c) **Outline** the main control measures to minimise the risk of fire in a workplace.

5. (a) **Identify** the five stages of a fire risk assessment in a workplace.
   (b) **Outline** the particular issues that should be considered in the fire risk assessment of temporary workplaces.
   (c) **Identify** additional factors to be taken into account during building refurbishment work.

6. The popularity of timber-framed construction for structures has raised the risk of fires on construction sites.
   (a) **Outline** the issues to be considered when preparing fire risk assessments in order to reduce the risks.
   (b) **Outline** the contents of a typical fire plan that should be attached to the fire risk assessment.

7. (a) **Identify TWO** *flammable* gases contained in cylinders that might be found on a construction site.
   (b) **Describe** the precautions that should be taken to prevent fires and explosions during the transport and use of flammable gases contained in cylinders.

8. (a) **Identify TWO** ways in which electricity can cause a fire on a construction site.
   (b) **Outline** the various ways in which workers might be physically harmed by a fire on a construction site.

9. (a) **Explain** the dangers associated with liquefied petroleum gas (LPG).
   (b) **Describe** the precautions needed for the storage, use and transportation of LPG in cylinders on a construction site.

10. (a) **Outline** effective control measures for the safe storage and handling of small containers containing flammable solvents.
    (b) **Identify EIGHT** types of unsatisfactory working practices using flammable gas cylinders that could increase the risk of a fire or explosion.

11. A major hazard on a refurbishment project is fire.
    (a) **Identify THREE** activities that represent an increased fire risk in such a situation.
    (b) **Outline** the precautions that may be taken to prevent a fire from occurring.

12. Arson on a construction site is a common cause of fire.
    (a) **Outline** reasons why some construction sites may be vulnerable to arson attacks.
    (b) **Identify** ways of reducing the risk of arson on a construction site.

13. Fire protection is an important feature in the design of a new building.
    (a) **Identify SIX** structural measures that can help to prevent the spread of fire and smoke.
    (b) **Outline** the issues involved in the fire compartmentation of a building.

14. The failure of electrical equipment is one of the main causes of workplace fires.
    (a) **Outline** how fires could be caused by the failure of electrical cables and equipment.
    (b) **Identify TWO** types of extinguisher that can be used safely on fires involving electrical equipment and **identify** those extinguishers that should not be used in such fires.
    (c) **Outline** suitable control measures that should be taken to minimise the risk of fire from electrical equipment when used in a flammable atmosphere.

15. (a) **Identify TWO** types of emergency warning systems that can be installed in the building

to ensure that all occupants can be made aware of the need to evacuate the building.

(b) **Outline** the main factors to be considered in the siting of fire extinguishers.

(c) **Outline** suitable arrangements for the inspection and maintenance of fire extinguishers in the workplace.

16. (a) **Outline ONE** advantage and **ONE** limitation of each of the main types of fire extinguishers.

(b) **Outline TWO** advantages and **TWO** disadvantages of using hose reels as a means of extinguishing fires.

(c) **Identify** the issues associated with the access for fire and rescue services to a workplace.

17. Fires on large construction sites are a major hazard with the main concern being the safe evacuation of personnel from the workplace. **Outline** the issues to be considered when planning escape routes and emergency evacuation procedures for a site.

18. (a) **Outline** the principal requirements for a safe means of escape from a building in the event of a fire.

(b) **Outline** the role of fire marshals in the event of a fire emergency.

(c) **Outline** reasons that may delay the safe evacuation of occupants from a workplace during a fire.

19. (a) **Identify** the issues to consider for the location of an assembly point for use in an evacuation of a construction site.

(b) **Outline** the reasons for undertaking regular fire drills in the workplace.

(c) **Outline** the contents of a training programme for employees on the emergency action to take in the event of fire.

13

APPENDIX 13.1 Fire risk assessment checklist as recommended in Fire Safety Guides published by the Department for Communities and Local Government in 2006

Checklist

Follow the five key steps – Fill in the checklist – Assess your fire risk and plan fire safety

1. Fire safety risk assessment

Fire starts when heat (source of ignition) comes into contact with fuel (anything that burns) and oxygen (air).

You need to keep sources of ignition and fuel apart.

How could a fire start?

Think about heaters, lighting, naked flames, electrical equipment, hot processes such as welding or grinding, cigarettes, matches and anything else that gets very hot or causes sparks.

What could burn?

Packaging, rubbish and furniture could all burn just like the more obvious fuels such as petrol, paint, varnish and white spirit. Also think about wood, paper, plastic, rubber and foam. Do the walls or ceilings have hardboard, chipboard or polystyrene? Check outside, too.

Check:

- ▶ Have you found anything that could start a fire?
- ▶ Make a note of it.
- ▶ Have you found anything that could burn?
- ▶ Make a note of it.

2. People at risk

Everyone is at risk if there is a fire. Think whether the risk is greater for some because of when or where they work, such as night staff, or because they're not familiar with the premises, such as visitors or customers.

Children, the elderly or disabled people are especially vulnerable.

Check:

Have you identified:
- ▶ Who could be at risk?
- ▶ Who could be especially at risk?

Make a note of what you have found.

3. Evaluate and act

Evaluate

First, think about what you have found in Steps 1 and 2: what are the risks of a fire starting, and what are the risks to people in the building and nearby?

Remove and reduce risk

How can you avoid accidental fires? Could a source of heat or sparks fall, or be knocked or pushed into something that would burn? Could that happen the other way round?

Protect

Take action to protect your premises and people from fire.

Check:

- ▶ Have you assessed the risks of fire in your workplace?
- ▶ Have you assessed the risk to staff and visitors?
- ▶ Have you kept any source of fuel and heat/sparks apart?
- ▶ If someone wanted to start a fire deliberately, is there anything around they could use?
- ▶ Have you removed or secured any fuel an arsonist could use?
- ▶ Have you protected your premises from accidental fire or arson?

How can you make sure everyone is safe in case of fire?

- ▶ Will you know there is a fire?
- ▶ Do you have a plan to warn others?
- ▶ Who will make sure everyone gets out?
- ▶ Who will call the fire service?
- ▶ Could you put out a small fire quickly and stop it spreading?

How will everyone escape?

- ▶ Have you planned escape routes?
- ▶ Have you made sure people will be able to safely find their way out, even at night if necessary?
- ▶ Does all your safety equipment work?
- ▶ Will people know what to do and how to use equipment?

Make a note of what you have found.

4. Record, plan and train

Record

Keep a record of any fire hazards and what you have done to reduce or remove them. If your premises are small, a record is a good idea. If you have five or more staff or have a licence, then you must keep a record of what you have found and what you have done.

Plan

You must have a clear plan of how to prevent fire and how you will keep people safe in case of fire. If you share a building with others, you need to coordinate your plan with them.

Train

You need to make sure your staff know what to do in case of fire, and, if necessary, are trained for their roles.

Check:

▶ Have you made a record of what you have found, and action you have taken?
▶ Have you planned what everyone will do if there is a fire?
▶ Have you discussed the plan with all staff?

Have you:

▶ informed and trained people (practised a fire drill and recorded how it went)?
▶ nominated staff to put in place your fire prevention measures, and trained them?
▶ made sure everyone can fulfil their role?
▶ informed temporary staff?
▶ consulted others who share a building with you, and included them in your plan?

5. Review

Keep your risk assessment under regular review. Over time, the risks may change.

If you identify significant changes in risk or make any significant changes to your plan, you must tell others who share the premises and, where appropriate, retrain staff.

Have you:

▶ made any changes to the building inside or out?
▶ had a fire or near miss?
▶ changed work practices?
▶ begun to store chemicals or dangerous substances?
▶ significantly changed your stock or stock levels?
▶ planned your next fire drill?

The checklist above can help you with the fire risk assessment but you may need additional information, especially if you have large or complex premises.

Source: Department for Community and Local Government.

APPENDIX 13.2 Typical fire notice

CHAPTER 14

Chemical and biological health hazards and risk control

This chapter covers the following NEBOSH learning objectives:

1. Outline the forms of, the classification of, and the health risks from exposure to, hazardous substances
2. Explain the factors to be considered when undertaking an assessment of the health risks from substances commonly encountered in construction workplaces
3. Explain the use and limitations of Workplace Exposure Limits (WELs) including the purpose of long-term and short-term exposure limits
4. Outline control measures that should be used to reduce the risk of ill-health from exposure to hazardous substances
5. Outline the hazards, risks and controls associated with specific agents
6. Outline the basic requirements related to the safe handling and storage of waste on construction sites

14.1 Forms and classification of, and the health risks from exposure to, hazardous substances

14.1.1 Introduction

Occupational health is as important as occupational safety but generally receives less attention from managers. Every year twice as many people suffer ill-health caused or exacerbated by the workplace than suffer workplace injury. Although these illnesses do not usually kill people, they can lead to many years of discomfort and pain. Such illnesses include respiratory disease, hearing problems, asthmatic conditions and back pain. Furthermore, it has been estimated in the UK that 30% of all cancers probably have an occupational link – that linkage is known for certain in 8% of cancer cases. Occupational exposures to fumes, dusts and chemicals in various forms account for around 4,000 deaths a year and some 38,000 individuals suffer breathing or lung problems possibly caused by their work.

Work in the field of occupational health has been taking place for the last four centuries and possibly longer. The main reason for the relatively low profile for occupational health over the years has been the difficulty in linking the ill-health effect with the workplace cause. Many illnesses, such as asthma or back pain, can have a workplace cause but can also have other causes. Many of the advances in occupational health have been as a result of statistical and epidemiological studies (one well-known such study linked the incidence of lung cancer to cigarette smoking). While such studies are invaluable in the assessment of health risk, there is always an element of doubt when trying to link cause and effect. The measurement of gas and dust concentrations is also subject to doubt when a correlation is made between a measured sample and the workplace environment from which it was taken. Occupational health, unlike occupational safety, is generally more concerned with probabilities than certainties.

In this chapter, chemical and biological health hazards will be considered – other forms of health hazard will be covered in Chapter 15.

The chemical and biological health hazards described in this chapter are covered by the following health and safety regulations:

▶ Control of Substances Hazardous to Health Regulations;
▶ Control of Lead at Work Regulations;
▶ Control of Asbestos at Work Regulations.

Risk assessment is not only concerned with injuries in the workplace but also needs to consider the possibility of occupational ill-health. Health risks fall into the following four categories:

▶ chemical (e.g. paint solvents, exhaust fumes);
▶ biological (e.g. bacteria, pathogens);
▶ physical (e.g. noise, vibrations);
▶ psychological (e.g. occupational stress).

14.1.2 Forms of chemical agent

Chemicals can be transported by a variety of agents and in a variety of forms. They are normally defined in the following ways.

Dusts are solid particles slightly heavier than air but often suspended in it for a period of time. The size of the particles ranges from about 0.4 μm (fine) to 10 μm (coarse). Dusts are created either by mechanical processes (e.g. grinding or pulverising) or construction processes (e.g. concrete laying, demolition or sanding), or by specific tasks (e.g. furnace ash removal). The fine dust is much more hazardous because it penetrates deep into the lungs and remains there – known as **respirable dust**. In rare cases, respirable dust enters the bloodstream directly causing damage to other organs. Examples of such fine dust are cement, granulated plastic materials and silica dust produced from stone or concrete dust. Repeated exposure may lead to permanent lung disease. Any dusts which are capable of entering the nose and mouth during breathing are known as **inhalable dusts**.

Fibres are threads or filaments that can occur naturally (e.g. asbestos) or be man-made such as glass-fibre, nylon and polyester. Man-made fibres are commonly used in insulation boards, blankets for the purpose of heat treatment, electrical insulation and in the reinforcement of plastic and cement. Fibres have a very high length to width ratio of at least 100 and many fibres are in the respirable range causing concern about the effects of exposure to many fibres – fibrosis of the lung and various cancers. Fibres with diameters in excess of 4 μm can cause irritation of the skin and eyes. Exposure to large concentration of such fibres may also cause irritation in the upper respiratory tract. Although the hazards and risks of asbestos and glass-fibre have been well documented, there are many other types of fibre in use. These include synthetic and semi-synthetic fibres (e.g. cellulosic fibres), a range of other non-organic fibres and whiskers and specialist technical fibres including various types of carbon fibres and nanofibres.

Gases are any substances at a temperature above their boiling point. Steam is the gaseous form of water. Common gases include carbon monoxide, carbon dioxide, nitrogen and oxygen. Gases are absorbed into the bloodstream where they may be beneficial (oxygen) or harmful (carbon monoxide).

Vapours are substances which are at or very close to their boiling temperatures. They are gaseous in form. Many solvents, such as cleaning fluids, fall into this category. The vapours, if inhaled, enter the bloodstream and some can cause short-term effects (dizziness) and others long-term effects (brain damage).

Liquids are substances which normally exist at a temperature between freezing (solid) and boiling (vapours and gases). They are sometimes referred to as fluids in health and safety legislation.

Mists are similar to vapours in that they exist at or near their boiling temperature but are closer to the liquid phase. This means that there are suspended, very small liquid droplets present in the vapour. A mist is produced during a spraying process (such as paint spraying). Many industrially produced mists can be very damaging if inhaled, producing similar effects to vapours. It is possible for some mists to enter the body through the skin or by ingestion with food.

Fume is a collection of very small metallic particles (less than 1 μm) which have condensed from the gaseous state. They are most commonly generated by the welding process. The particles tend to be within the respirable range (approximately 0.4–1.0 μm) and can lead to long-term permanent lung damage. The exact nature of any harm depends on the metals used in the welding process and the duration of the exposure.

14.1.3 Forms of biological agent

As with chemicals, biological hazards may be transported by any of the following forms of agent.

Fungi are very small organisms, sometimes consisting of a single cell, and can appear as plants (e.g. mushrooms and yeast). Unlike other plants, they cannot produce their own food but either live on dead organic matter or on living animals or plants as parasites. Fungi reproduce by producing spores, which can cause allergic reactions when inhaled. The infections produced by fungi in humans may be mild, such as athlete's foot, or severe, such as ringworm. Many fungal infections can be treated with antibiotics.

Moulds are a particular group of very small fungi which, under damp conditions, will grow on surfaces such as walls, bread, cheese, leather and canvas. They can be beneficial (penicillin) or cause allergic reactions (asthma). Asthma attacks, athlete's foot and farmer's lung are all examples of fungal infections.

Bacteria are very small single-celled organisms which are much smaller than cells within the human body. They can live outside the body and be controlled and destroyed by antibiotic drugs. There is evidence that bacteria are becoming resistant to most antibiotics. This has been caused by the widespread misuse of antibiotics. It is important to note that not all bacteria are harmful to humans. Bacteria aid the digestion of food, and babies would not survive without their aid to break down the milk in their digestive system. Legionellosis, tuberculosis and tetanus are all bacterial diseases.

Viruses are minute non-cellular organisms which can only reproduce within a host cell. They are very much smaller than bacteria and cannot be controlled by antibiotics. They appear in various shapes and are continually developing new strains. They are usually only defeated by the defence and healing mechanisms of the body. Drugs can be used to relieve the symptoms of a viral attack but cannot cure it. The common cold is a viral infection as are hepatitis, AIDS (HIV) and influenza.

A number of diseases, including bovine spongiform encephalopathy (BSE) in cattle and Creutzfeldt–Jakob disease (CJD) in humans, are caused by another biological agent known as a prion. A **prion** is an infectious agent that is composed primarily of protein. Such agents induce existing substances, called polypeptides, in the host organism to take on a rogue form. All known prion diseases affect the structure of the brain or other neural tissue, are currently untreatable and are always fatal.

14.1.4 Classification of hazardous substances and their associated health risks

A hazardous substance is one which can cause ill-health to people at work. Such substances may include those used directly in the work processes (glues and paints), those produced by work activities (welding fumes) or those which occur naturally (dust). Hazardous substances are classified according to the severity and type of hazard which they may present to people who may come into contact with them. The contact may occur while working or transporting the substances or might occur during a fire or accidental spillage.

14

Until 2015, the supply and hazard classification of substances was regulated in the UK by the Chemicals (Hazard Information and Packaging for Supply) 2009 Regulations (CHIP 4). CHIP 4 required suppliers to identify the hazards of the chemicals that they supplied, provide information regarding those hazards and package their products safely. From June 2015 CHIP was revoked and fully superseded by the Classification, Labelling and Packaging of Substances and Mixtures Regulations (EC) (known as CLP).

A hazard pictogram is an image on a label that includes a warning symbol and specific colours intended to provide information about the damage a particular substance or mixture can cause to health or the environment (see Figure 14.1). The CLP Regulation has introduced a new classification and labelling system for hazardous chemicals in the European Union. The pictograms have also changed and are in line with the United Nations Globally Harmonised System (GHS). The new pictograms are in the shape of a red diamond with a white background, and replace the old orange square symbols which applied under CHIP 4. Since 1 December 2010, some substances and mixtures have already been labelled according to the new legislation, but the old pictogram can still be on the market until 1 June 2017.

There are several classifications but here only the most common under each classification will be described.

CHIP 4 classifications

Irritant is a non-corrosive substance which can cause skin (dermatitic) or lung (bronchial) inflammation after repeated contact. People who react in this way to a particular substance are **sensitised** or **allergic** to that substance. In most cases, it is likely that the concentration of the irritant may be more significant than the exposure time. Many household substances, such as wood preservatives, bleaches and glues, are irritants. Many chemicals used as solvents are also irritants (white spirit, toluene and acetone). Formaldehyde and ozone are other examples of irritants.

Corrosive substances are ones that may destroy living tissue on contact – usually by burning the skin. Usually strong acids or alkalis, examples include sulphuric acid and caustic soda. Many tough cleaning substances, such as kitchen oven cleaners, are corrosive as are many dishwasher crystals.

Harmful is the most commonly used classification and describes a substance which, if swallowed, inhaled or it penetrates the skin, *may* pose limited health risks. These risks can usually be minimised or removed by following the instruction provided with the substance (e.g. by using personal protective equipment). There are many household substances which fall into this category including bitumen-based paints and paintbrush restorers. Many chemical cleansers are categorised as harmful. It is very common for substances labelled harmful also to be categorised as irritant.

Toxic substances are poisonous and impede or prevent the function of one or more organs within the body, such as the kidney, liver and heart. A toxic substance is, therefore, a poisonous one. Lead, mercury, pesticides and the gas carbon monoxide are toxic substances. The effect on the health of a person exposed to a toxic substance depends on the concentration and toxicity of the substance, the frequency of the exposure and the effectiveness of the control measures in place. The state of health and age of the person and the route of entry into the body have influence on the effect of the toxic substance.

Very toxic substances are very hazardous to health when inhaled, swallowed or when they come in contact with the skin and could be fatal. Examples of such substances include prussic acid, nicotine and carbon monoxide.

GHS classifications

The **health hazard** classification combines the CHIP 4 'harmful' and 'irritant' classifications. Such a substance will be harmful if swallowed, inhaled or contacts the skin. It may cause respiratory irritation, drowsiness or dizziness, an allergic skin reaction, a serious eye irritation or skin irritation. It could also be harmful to the environment by destroying ozone in the upper atmosphere. Typical substances include washing detergents, toilet cleaner and coolant fluid.

A **serious health hazard** substance may be fatal if swallowed and/or enters the airways and cause damage to organs. It may damage human fertility or the unborn child and may cause genetic defects. It may also cause allergy or asthma symptoms or breathing difficulties if inhaled or may cause cancer. Turpentine and petrol carry a serious health hazard warning. It is similar to the toxic CHIP 4 classification.

The **corrosive** classification is very similar to the CHIP 4 corrosive classification. A corrosive substance will probably be corrosive to metals and can cause both severe skin burns and eye damage. Strong drain cleaners, acetic acid and hydrochloric acid are all examples of corrosive substances.

Acute toxicity is similar to the very toxic CHIP 4 classification. Such a substance will always be toxic but can also be fatal if swallowed, inhaled or comes into contact with the skin. Biocides, methanol and many pesticides can be acutely toxic.

Other types of hazardous substances

Carcinogenic substances are ones which are known for, or suspected of, promoting abnormal development of body cells to become cancers. Asbestos, hardwood dust, creosote and some mineral oils are carcinogenic. It is very important that the health and safety rules accompanying the substance are strictly followed.

Mutagenic substances are those which damage genetic material within cells, causing abnormal changes that can be passed from one generation to another. A **reproductive toxin** is a chemical which affects adversely the reproductive process including mutations and other effects on the unborn child. One group of such toxins are teratogens that can cause congenital abnormalities, which are also called birth defects.

Each of the classifications may be identified by a symbol and a symbolic letter – the most common of these are shown in Figure 14.1, which shows the relationship between the CHIP 4 and GHS classifications but also note that these are changing to the new Global Harmonisation Scheme (GHS).

The effects on health of hazardous substances may be either acute or chronic.

Acute effects are of short duration and appear fairly rapidly, usually during or after a single or short-term exposure to a hazardous substance. Such effects may be severe and require hospital treatment but are usually reversible. Examples include asthma-type attacks, nausea and fainting.

Chronic effects develop over a period of time which may extend to many years. The word 'chronic' means 'with time' and should not be confused with 'severe'

as its use in everyday speech often implies. Chronic health effects are produced from prolonged or repeated exposures to hazardous substances resulting in a gradual, latent and often irreversible illness, which may remain undiagnosed for many years. Many cancers and mental diseases fall into the chronic category. During the development stage of a chronic disease, the individual may experience no symptoms.

(a)

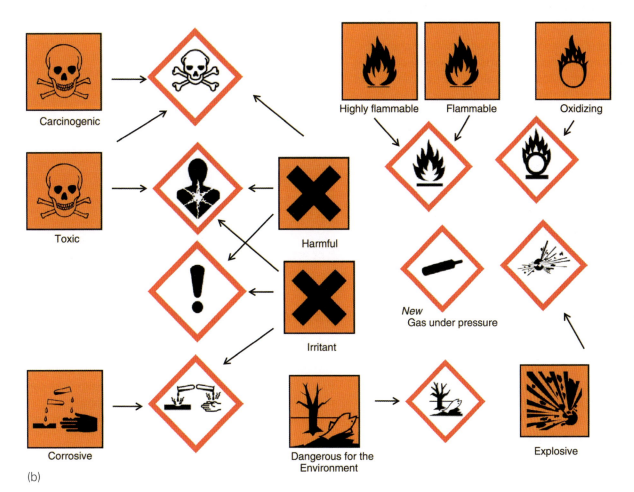

(b)

Figure 14.1 (a) Use of the GHS symbols on site; (b) how the European packaging symbols relate to the new GHS labels

397

14.1.5 Hazard warnings and precautionary statements

The European Regulation on classification, labelling and packaging of substances and mixtures became law across the EU in 2009 and is known as either the CLP Regulation or simply as 'CLP'. It adopted the United Nations Globally Harmonised System on the classification and labelling of chemicals (GHS) across all EU states.

Worldwide, there are many different laws to identify hazardous chemicals (classification) and to communicate this information to users. This is often confusing as the same chemical can have different hazard descriptions in different countries. The UN brought together experts from different countries to create the GHS.

The aim of GHS is to have, throughout the world, the same:

▶ criteria for classifying chemicals according to their health, environmental and physical hazards;
▶ hazard communication requirements for labelling and safety data sheets.

GHS provides a basis for globally uniform physical, environmental, health and safety information on hazardous chemicals through the harmonisation of the criteria for their classification and labelling. It was developed at UN level with the aim of overcoming differing labelling information requirements on physical, health and environmental hazards for the same chemicals around the world. Moreover, it also aims to lower barriers to trade caused by the fact that every time a product was exported, it mostly had to be classified and labelled differently because of differing criteria.

Since GHS is a voluntary agreement, it has to be adopted through a suitable national or regional legal mechanism to ensure it becomes legally binding. That is achieved by the CLP Regulation. As GHS was heavily influenced by the old EU system, the CLP Regulation is very similar in many ways. The duties on suppliers are broadly the same: classification, labelling and packaging. The rules for the classification of hazardous substances have changed and a new set of hazard pictograms (quite similar to the old ones) are used (see Figure 14.1).

The EU legislation on classification, labelling and packaging consists of three acts: The Dangerous Substances Directive (Directive 67/548/EEC, DSD), the Dangerous Preparations Directive (Directive 1999/45/EC, DPD) and the Regulation on Classification, Labelling and Packaging of Substances and Mixtures, Regulation (EC) No 1272/2008 (CLP Regulation or CLP) which applies directly in all member states.

Further information on the stage of implementation of the UN GHS in different countries is available on the UN ECE website, see: http://unece.org/trans/danger/publi/ghs/implementation_e.html.

Under the CLP Regulation there are:

▶ new scientific criteria to assess hazardous properties of chemicals;
▶ two new harmonised hazard warning symbols for labels (known as 'pictograms');
▶ a new design for existing symbols;
▶ new harmonised **hazard warning (H)** and **precautionary statements (P)** for labels; and
▶ two new signal words – 'Danger' and 'Warning'.

A **hazard statement** is a phrase that describes the nature of the hazard in the substance or mixture. A hazard statement will be determined by the application of the classification criteria. It replaces the 'risk or R-phrase' used previously. A list of some of these is given in Appendix 14.1.

A **precautionary statement** is a phrase that describes recommended measure(s) to minimise or prevent adverse effects resulting from exposure to a hazardous substance or mixture due to its use or disposal. Suppliers determine the appropriate precautionary statements (usually no more than six) based on the required hazard statements. It replaces the 'safety or S-phrase' used previously.

As mentioned above, the CLP Regulation also introduces two new **signal words**: 'Danger' and 'Warning'. If the chemical has a more severe hazard, the label includes the signal word 'Danger'; in case of less severe hazards, the signal word is 'Warning'.

Examples of the new hazard statements and the new precautionary statements for labels are shown in Table 14.1.

Table 14.1 Examples of the new hazard warning (H) and precautionary statements (P)

Hazard warning statement (H)	Precautionary statement (P)
H240 – Heating may cause an explosion	P102 – Keep out of reach of children
H320 – Causes eye irritation	P271 – Use only outdoors or in well-ventilated area
H401 – Toxic to aquatic life	P410 – Protect from sunlight

14.2 Assessment of health risks

14.2.1 Types of health risk

The principles of control for health risks are the same as those for safety. However, the nature of health risks can make the link between work activities and employee ill-health less obvious than in the case of injury from an accident (Figure 14.2).

The COSHH Amendment Regulations 2004 set out the principles of good practice for the control of exposure.

Figure 14.3 shows a route map for achieving adequate control.

Figure 14.2 Paint spraying – risk of sensitising particularly if isocyanate based paint used and inadequate local exhaust ventilation

Unlike safety risks, which can lead to immediate injury, the result of daily exposure to health risks may not manifest itself for months, years and, in some cases, decades. Irreversible health damage may occur before any symptoms are apparent. It is, therefore, essential to develop a preventative strategy to identify and control risks before anyone is exposed to them.

Risks to health from work activities include:

▶ skin contact with irritant substances, leading to dermatitis, etc.;

▶ inhalation of respiratory sensitisers, triggering immune responses such as asthma;

▶ badly designed workstations requiring awkward body postures or repetitive movements, resulting in upper limb disorders, repetitive strain injury and other musculoskeletal conditions;

▶ noise levels which are too high, causing deafness and conditions such as tinnitus;

▶ too much vibration, for example from hand-held tools, leading to hand–arm vibration syndrome and circulatory problems;

▶ exposure to ionising and non-ionising radiation, including ultraviolet in the sun's rays, causing burns, sickness and skin cancer;

▶ infections ranging from minor sickness to life-threatening conditions, caused by inhaling or being contaminated with microbiological organisms;

▶ stress causing mental and physical disorders.

Some illnesses or conditions, such as asthma and back pain, have both occupational and non-occupational causes and it may be difficult to establish a definite causal link with a person's work activity or their exposure to particular agents or substances. But, if there is evidence that shows the illness or condition is prevalent among the type of workers to which the person belongs or among workers exposed to similar agents or substances, it is likely that their work and exposure has contributed in some way.

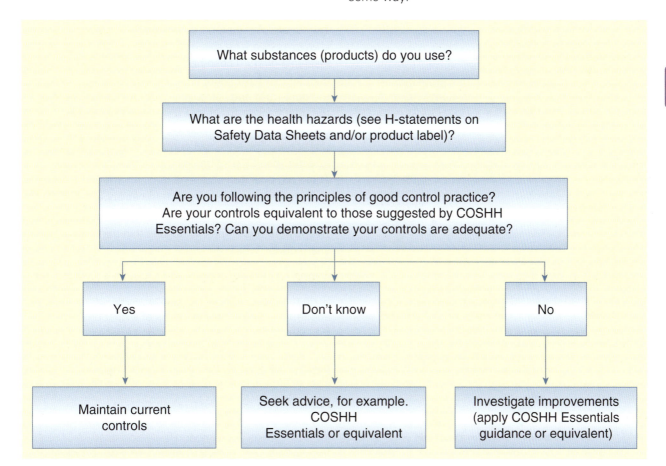

Figure 14.3 Route map for adequate control for SMEs non-experts

14.2.2 Routes of entry of hazardous substances into the body

There are three principal routes of entry of hazardous substances into the human body:

- ▶ **inhalation** – breathing in the substance with normal air intake. This is the main route of contaminants into the body. These contaminants may be chemical (e.g. solvents or welding fume) or biological (e.g. bacteria or fungi) and become airborne by a variety of modes, such as sweeping, spraying, grinding and bagging. They enter the lungs where they have access to the bloodstream and many other organs;

- ▶ **absorption through the skin** – the substance comes into contact with the skin and enters through either the pores or a wound. Tetanus can enter in this way as can toluene, benzene and various phenols;

- ▶ **ingestion** – through the mouth and swallowed into the stomach and the digestive system. This is not a significant route of entry to the body. The most common occurrences are due to airborne dust or poor personal hygiene (not washing hands before eating food).

Another very rare entry route is by **injection**. The abuse of compressed air lines by shooting high pressure air at the skin can lead to air bubbles entering the bloodstream. Accidents involving hypodermic syringes in a health or veterinary service setting are rare but illustrate this form of entry route.

Body cells have a natural defence system against illness caused by harmful invasive foreign particles and bacteria. White blood cells, known as phagocytes, give protection by destroying these harmful foreign particles, bacteria and dead cells. This protection is basically a scavenging action at the level of cells. Other defence mechanisms include the secretion of defensive substances, the prevention of excessive blood loss and the repair of damaged tissue.

The most effective control measures which can reduce the risk of infection from biological organisms are disinfection, proper disposal of clinical waste (including syringes), good personal hygiene and, where appropriate, personal protective equipment. Other measures include vermin control, water treatment and immunisation.

There are five major functional systems within the human body – respiratory, nervous, cardiovascular (blood), urinary and the skin.

The respiratory system

This comprises the lungs and associated organs (e.g. the nose). Air is breathed in through the nose, and passes through the trachea (windpipe) and the bronchi into the two lungs. Within the lungs, the air enters many smaller passageways (bronchioles) and thence to one of 300,000 terminal sacs called alveoli. The alveoli are approximately 0.1 mm across, although the entrance is much smaller. On arrival in the alveoli, there is a diffusion of oxygen into the bloodstream through blood capillaries and an effusion of carbon dioxide from the bloodstream. While soluble dust which enters the alveoli will be absorbed into the bloodstream, insoluble dust (respirable dust) will remain permanently, leading to possible chronic illness (Figure 14.5).

The whole of the bronchial system is lined with hairs, known as cilia. The cilia offer some protection against insoluble dusts. These hairs will arrest all non-respirable dust (above 5 μm) and, with the aid of mucus, pass the dust from one hair to a higher one and thus bring the dust back to the throat (this is known as the ciliary escalator). It has been shown that smoking damages this action. The nose will normally trap large particles (greater than 20 μm) before they enter the trachea. There are over 40 conditions that can affect the lungs and/or airways and impinge on the ability of a person to breathe normally.

Respirable dust tends to be long thin particles with sharp edges which puncture the alveoli walls. The puncture heals producing scar tissues which are less flexible than the original walls – this can lead to fibrosis. Such dusts include asbestos, coal, silica, some plastics and talc. The possible indicators of a dust problem in the workplace are fine deposits on surfaces, people and products, or blocked filters on extraction equipment. Ill-health reports or complaints from the workforce could also indicate a dust problem.

Acute effects on the respiratory system include bronchitis and asthma and chronic effects include fibrosis and cancer. Hardwood dust, for example, can produce asthma attacks and nasal cancer.

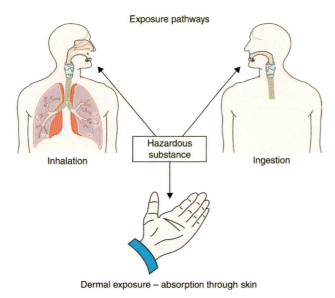

Exposure pathways

Inhalation

Ingestion

Hazardous substance

Dermal exposure – absorption through skin

Figure 14.4 Hazardous substances – principal routes of entry into the human body

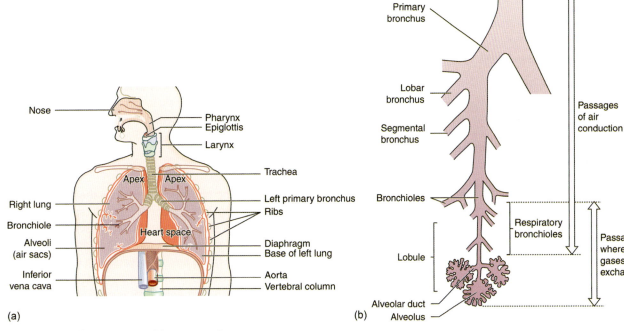

(a)

(b)

Figure 14.5 The upper and lower respiratory system

Finally, asphyxiation, due to a lack of oxygen, is a problem in confined spaces particularly when MIG (metal inert gas) welding is taking place.

The nervous system

The nervous system consists primarily of the brain, the spinal cord and nerves extending throughout the body. Any muscle movement or sensation (e.g. hot and cold) is controlled or sensed by the brain through small electrical impulses transmitted through the spinal cord and nervous system. The effectiveness of the nervous system can be reduced by neurotoxins and lead to changes in mental ability (loss of memory and anxiety), epilepsy and narcosis (dizziness and loss of consciousness). Organic solvents (trichloroethylene) and heavy metals (mercury) are well-known neurotoxins. The expression 'mad hatters' originated from the mental deterioration of top hat polishers in the 19th century who used mercury to produce a shiny finish on the top hats.

The cardiovascular system

The blood system uses the heart to pump blood around the body through arteries, veins and capillaries. Blood is produced in the bone marrow and consists of a plasma within which are red cells, white cells and platelets. The system has three basic objectives:

▶ to transport oxygen to vital organs, tissues and the brain and carbon dioxide back to the lungs (red cell function);

▶ to attack foreign organisms and build up a defence system (white cell function);

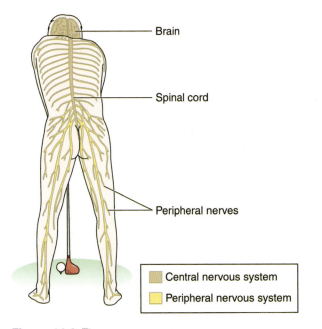

Figure 14.6 The nervous system

▶ to aid the healing of damaged tissue and prevent excessive bleeding by clotting (platelets).

There are several ways in which hazardous substances can interfere with the cardiovascular system. Benzene can affect the bone marrow by reducing the number of blood cells produced. Carbon monoxide prevents the red cells from absorbing sufficient oxygen and the effects depend on its concentration. Symptoms begin with headaches and end with unconsciousness and possibly death.

14

401

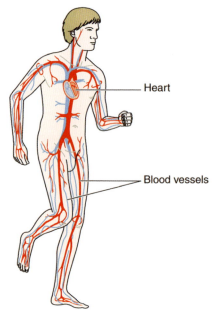

Figure 14.7 The cardiovascular system

The urinary system

The urinary system extracts waste and other products from the blood. The two most important organs are the liver (normally considered part of the digestive system) and kidney, both of which can be affected by hazardous substances within the bloodstream.

The liver removes toxins from the blood, maintains the levels of blood sugars and produces protein for the blood plasma. Hazardous substances can cause the liver

to be too active or inactive (e.g. xylene), lead to liver enlargement (e.g. cirrhosis caused by alcohol) or liver cancer (e.g. vinyl chloride).

The kidneys filter waste products from the blood as urine, regulate blood pressure and liquid volume in the body and produce hormones for making red blood cells. Heavy metals (e.g. cadmium and lead) and organic solvents (e.g. glycol ethers used in screen printing) can restrict the operation of the kidneys, possibly leading to failure.

The skin

The skin holds the body together and is the first line of defence against infection. It regulates body temperature, is a sensing mechanism, provides an emergency food store (in the form of fat) and helps to conserve water. There are two layers – an outer layer called the epidermis (0.2 mm) and an inner layer called the dermis (4 mm). The epidermis is a tough protective layer and the dermis contains the sweat glands, nerve endings and hairs.

The most common industrial disease of the skin is **dermatitis** (non-infective dermatitis). It begins with a mild irritation on the skin and develops into blisters which can peel and weep becoming septic. It can be caused by various chemicals, mineral oils and solvents. There are two types:

irritant contact dermatitis – occurs soon after contact with the substance and the condition reverses after contact ceases (detergents and weak acids);

allergic contact dermatitis – caused by a skin sensitiser such as turpentine, epoxy resin, solder flux and formaldehyde that exerts its effects via the immune system. Once sensitised to a substance, a severe dermatitis may occur following a small exposure to the same substance at a later date.

Dermatitis is on the rise and is costing business more than £20 million a year, even though the cost of control measures to prevent the disease is minimal. Workers in the hotel and catering industry are particularly vulnerable to this debilitating disease as are print workers who have contracted dermatitis after coming into contact with UV curable lacquers used to produce a glossy finish to magazine covers.

For many years, dermatitis was seen as a 'nervous' disease which was psychological in nature. Nowadays, it is recognised as an industrial disease which can be controlled by good personal hygiene, personal protective equipment, use of barrier creams and health screening of employees. Dermatitis can appear anywhere on the body but it is normally found on the hands. Therefore, gloves should always be worn when there is a risk of dermatitis.

The risk of dermatitis increases with the presence of skin cuts or abrasions that allow chemicals to be more

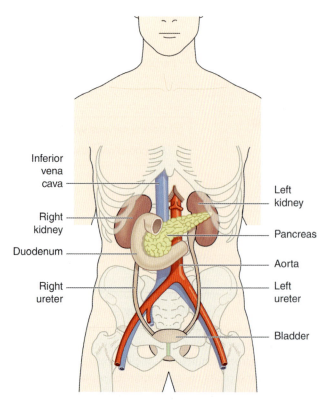

Figure 14.8 Parts of the urinary system

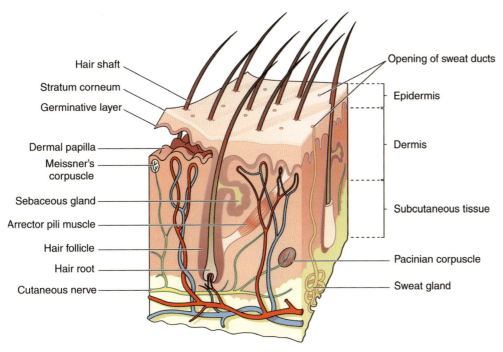

Figure 14.9 The skin – main structures of the dermis

easily absorbed. This risk also depends on the type, sensitivity and existing condition of the skin.

14.2.3 The role of COSHH

The Control of Substances Hazardous to Health Regulations (COSHH) 1988 were the most comprehensive and significant piece of health and safety legislation to be introduced since the Health and Safety at Work Act 1974. They were enlarged to cover biological agents in 1994 and further amended in 1997, 1999, 2002 and 2004. A detailed summary of these Regulations appears in Chapter 19. The Regulations impose duties on employers to protect employees and others who may be exposed to substances hazardous to their health and require employers to control exposure to such substances.

The COSHH Regulations offer a framework for employers to build a management system to assess health risks and to implement and monitor effective controls. Adherence to these Regulations will provide the following benefits to the employer and employee:

▶ improved productivity due to lower levels of ill-health and more effective use of materials;
▶ improved employee morale;
▶ lower numbers of civil court claims;
▶ better understanding of health and safety legal requirements.

The Approved Code of Practice (ACoP) to the COSHH Regulations (COSHH) covers all substances to which the Regulations apply. It outlines the preferred or recommended methods that can be used to comply with the Regulations and the accompanying guidance also provides advice on achieving compliance, such as

the control of carcinogenic substances or those causing occupational asthma, monitoring control measures and conducting health surveillance. The latest edition of this ACoP and guidance takes account of regulatory changes following the introduction of the EU Regulations for REACH (Registration, Evaluation, Authorisation and restriction of Chemicals) and CLP on classification, labelling and packaging of substances and mixtures.

Organisations which ignore COSHH requirements will be liable for enforcement action, including prosecution, under the Regulations.

14.2.4 Requirements of the COSHH Regulations

When hazardous substances are considered for use (or in use) at a place of work, the COSHH Regulations impose certain duties on employers and require employees to cooperate with the employer by following any measures taken to fulfil those duties. The principal requirements are as follows.

1. Employers must undertake a suitable and sufficient assessment of the health risks created by work which is liable to expose their employees to substances hazardous to health and of the steps that need to be taken by employers to meet the requirements of these Regulations (Regulation 6).

2. Employers must prevent, or where this is not reasonably practicable, adequately control the exposure of their employees to substances hazardous to health. WELs, which should not be exceeded, are specified by the HSE for certain substances. As far as inhalation is concerned, control should be achieved by means other than personal protective equipment. If, however, respiratory

14

protective equipment, for example, is used, then the equipment must conform to HSE standards (Regulation 7). The control of exposure can only be treated as adequate if the principles of good practice (discussed under 14.4.1) are applied (Schedule 2A).

3. Employers and employees must make proper use of any control measures provided (Regulation 8).

4. Employers must maintain any installed control measures on a regular basis, keep suitable records (Regulation 9) and review systems of work.

5. Monitoring must be undertaken of any employee exposed to items listed in Schedule 5 of the Regulations or in any other case where monitoring is required for the maintenance of adequate control or the protection of employees. Records of this monitoring must be kept for at least 5 years, or 40 years where employees can be identified (Regulation 10).

6. Health surveillance must be provided to any employees who are exposed to any substances listed in Schedule 6. Records of such surveillance must be kept for at least 40 years after the last entry (Regulation 11).

7. Emergency plans and procedures must be prepared to deal with accidents or incidents involving exposure to hazardous substances beyond normal day-to-day risks. This will involve warnings and communication systems to give appropriate response immediately after any incident occurs.

8. Employees who may be exposed to substances hazardous to their health must be given information, instruction and training sufficient for them to know the health risks created by the exposure and the precautions which should be taken (Regulation 12).

14.2.5 Details of a COSHH assessment

Not all hazardous substances are covered by the COSHH Regulations. If there is no warning symbol on the substance container or it is a biological agent which is not directly used in the workplace (such as an influenza virus), then no COSHH assessment is required. The COSHH Regulations do not apply to those hazardous substances which are subject to their own individual Regulations (asbestos, lead or radioactive substances). The COSHH Regulations do apply to the following substances:

▶ substances having occupational exposure limits as listed in the HSE publication EH40 (Occupational Exposure Limits);

▶ substances or combinations of substances listed in the CLP Regulations;

▶ biological agents connected with the workplace;

▶ substantial quantities of airborne dust (more than 10 mg/m^3 of total inhalable dust or 4 mg/m^3 of respirable dust, both 8-hour time-weighted average (TWA), when there is no indication of a lower value);

▶ any substance creating a comparable hazard which for technical reasons may not be documented.

The following factors should be considered when assessing health risks:

▶ the health risk assessment should be suitable and sufficient (see Chapter 4);

▶ the exposure of employees to hazardous substances are adequately controlled;

▶ the control measures provided are properly used;

▶ the control measures and any associated equipment are properly maintained;

▶ monitor employees exposed to hazardous substances;

▶ there is adequate health surveillance; and

▶ information, instruction and training.

A COSHH assessment is very similar to a risk assessment but is applied specifically to hazardous substances. The HSE has suggested five steps to COSHH assessments but within these steps there are a number of sub-sections. The steps are as follows.

Step 1

Gather information about the substances, the work and working practices. Assessors should:

▶ identify the hazardous substances present or likely to be present in the workplace, the categories and numbers of persons (e.g. employees and visitors);

▶ gather information about the hazardous substances including the quantity of the substances used;

▶ identify the hazards from these substances by reviewing labels, material safety data sheets, HSE guidance and published literature;

▶ decide who could be affected by the hazardous substances and the possible routes of entry to people exposed (i.e. inhalation, ingestion or absorption). There is a need to look at both the substances and the activities where people could be exposed to hazardous substances.

Step 2

Evaluate the risks to health either individually or collectively. Assessors should:

▶ evaluate the risks to health including the duration and frequency of the exposure of those persons to the substances;

▶ evaluate the level of exposure, for example the concentration and length of exposure to any airborne dusts, gases, fumes or vapours;

▶ consider any WELs (see 14.3);

▶ decide if existing and potential exposure presents any insignificant risks to health or poses a significant risk to health.

Step 3

Decide what needs to be done to control the exposure to hazardous substances. Assessors should:

- evaluate the existing control measures, including any PPE and respiratory protective equipment (RPE), for their effectiveness (using any available records of environmental monitoring) and compliance with relevant legislation;
- decide on additional control measures, if any are required (see 14.4);
- decide what maintenance and supervision of the use of the control measures are needed;
- plan what to do in an emergency;
- set out how exposure should be monitored;
- decide what, if any, health surveillance is necessary;
- decide what information, instruction and training is required.

Step 4

Record the assessment. Assessors should:

- decide if a record is required (five or more employed, and significant findings);
- decide on the format of the record;
- decide on storage and how to make records available to employees, safety representatives, etc.

Step 5

Review the assessment. Assessors should:

- decide when a review is necessary (e.g. changes in substances used, processes or people exposed);
- decide what needs to be reviewed.

An example of a typical form that can be used for a COSHH assessment is given in Chapter 23.

It is important that the assessment is conducted by somebody who is competent to undertake it. Such competence will require some training, the extent of which will depend on the complexities of the workplace. For large organisations with many high-risk operations, a team of competent assessors will be needed. If the assessment is simple and easily repeated, a written record is not necessary. In other cases, a concise and dated record of the assessment together with recommended control measures should be made available to all those likely to be affected by the hazardous substances. The assessment should be reviewed on a regular basis, particularly when there are changes in work process or substances or when adverse ill-health is reported.

14.2.6 Assessing exposure and health surveillance

Some aspects of health exposure will need input from specialist or professional advisers, such as occupational health hygienists, nurses and doctors. However, considerable progress can be made by taking straightforward measures such as:

- consulting the workforce on the design of workplaces;

- talking to manufacturers and suppliers of substances and work equipment about minimising exposure;
- enclosing machinery to cut down dust, fumes and noise;
- researching the use of less hazardous substances;
- ensuring that employees are given appropriate information and are trained in the safe handling of all the substances and materials to which they may be exposed.

To assess health risks and to make sure that control measures are working properly, it may be necessary, for example, to measure the concentration of substances in the air to make sure that exposures remain within the assigned WELs. Sometimes health surveillance of workers who may be exposed will be needed. This will enable data to be collected to check control measures and for early detection of any adverse changes to health. Health surveillance procedures available include biological monitoring for bodily uptake of substances, examination for symptoms and medical surveillance – which may entail clinical examinations and physiological or psychological measurements by occupationally qualified registered medical practitioners. The procedure chosen should be suitable for the case concerned. Sometimes a method of surveillance is specified for a particular substance, for example, in the COSHH ACoP. Whenever surveillance is undertaken, a health record has to be kept for the person concerned.

Health surveillance should be supervised by a registered medical practitioner or, where appropriate, it should be done by a suitably qualified person (e.g. an occupational nurse). In the case of inspections for easily detectable symptoms like chrome ulceration or early signs of dermatitis, health surveillance should be done by a suitably trained responsible person. If workers could be exposed to substances listed in Schedule 6 of the COSHH Regulations, medical surveillance, under the supervision of an HSE employment medical adviser or a doctor appointed by the HSE, is required. There is more on health surveillance later in this chapter.

14.2.7 Sources of information

There are other important sources of information available for a COSHH assessment in addition to the HSE Guidance Note EH40.

Product labels include details of the hazards associated with the substances contained in the product and any precautions recommended. They may also bear one or more of the CHIP hazard classification symbols.

Material safety data sheets are another very useful source of information for hazard identification and associated advice. Manufacturers of hazardous substances are obliged to supply such sheets to users, giving details of the name, chemical composition and properties of the substance. Information on the nature of the health hazards and any relevant exposure standard (WEL) should also be given together with recommended exposure control measures and personal protective equipment. The sheets contain useful additional information on first-aid and fire-fighting measures and handling, storage, transport and disposal information. The data sheets should be stored in a readily accessible and known place for use in the event of an emergency, such as an accidental release.

(a)

Product J
(abc chemical)

Danger

Fatal if swallowed
Causes skin irritation

Precautions:
Wear protective gloves.
Take off contaminated clothing and wash before reuse.
Wash hands thoroughly after handling.
Do not eat, drink or smoke when using this product.

Store locked up.
Dispose of contents/containers in accordance with local regulations.

IF ON SKIN: Rinse skin with water/shower.
IF IN EYES: Rinse cautiously with water.
IF SWALLOWED: Immediately call a Poison Center or doctor/physician. Do not induce vomiting.

ABC Chemical Co., 123 Anywhere St., (123) 456-7890
See the SDS for more information

(b)

Figure 14.10 (a) Typical symbols and (b) product label on containers

Other sources of information include trade association publications, industrial codes of practice and specialist reference manuals.

14.2.8 Survey techniques for health risks

An essential part of the COSHH assessment is the measurement of the quantity of the hazardous substance in the atmosphere surrounding the workplace. This is known as air sampling. There are four common types of air-sampling technique used for the measurement of air quality:

1. **Stain tube detectors** use direct reading glass indicator tubes filled with chemical crystals which change colour when a particular hazardous substance passes through them. The method of operation is very similar to the breathalyser used by the police to check alcohol levels in motorists. The glass tube is opened at each end and fitted into a pumping device (either hand or electrically operated). A specific quantity of contaminated air containing the hazardous substance is drawn through the tube and the crystals in the tube change colour in the direction of the air flow. The tube is calibrated such that the extent of the colour change along the tube indicates the concentration of the hazardous substance within the air sample.

This method can only be effective if there are no leakages within the instrument and the correct volume of sampled air is used. The instrument should be held within 30 cm of the nose of the person whose atmosphere is being tested. A large range of different tubes is available. This technique of sampling is known as **grab** or **spot sampling** since it is taken at one point.

The advantages of the technique are that it is quick, relatively simple to use and inexpensive. There are, however, several disadvantages:

▶ The instrument cannot be used to measure concentrations of dust or fume.
▶ The accuracy of the reading is approximately 25%; it will yield a false reading if other contaminants present react with the crystals.
▶ The instrument can only give an instantaneous reading, not an average reading over the working period (TWA).
▶ The tubes are very fragile with a limited shelf life.
▶ Some disposal issues.

2. **Passive sampling** is measured over a full working period by the worker wearing a badge containing absorbent material. The material will absorb the contaminant gas and, at the end of the measuring period, the sample is sent to a laboratory for analysis. The advantages of this method over the stain tube are that there is less possibility of instrument errors and it gives a TWA reading.

3. **Sampling pumps and heads** can be used to measure gases and dusts. The worker, whose breathing zone is being monitored, wears a

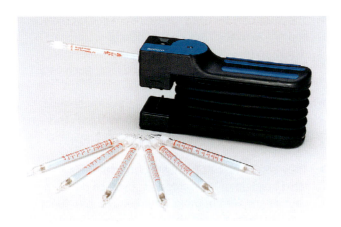

Figure 14.11 Hand pump and stain detector tubes

collection head as a badge and a battery-operated pump on his/her back at waist level. The pump draws air continuously through a filter, fitted in the head, which will either absorb the contaminant gas or trap hazardous dust particles. If this filter is used for dust measurement, it is sometimes called a static dust sampler. Before sampling takes place, the filter is weighed and the pump flow rate preset. After the designated testing period, it is sent to a specialist laboratory for analysis. The quantity of dust present would be determined either by measuring the weight change of the filter or by chemical analysis (e.g. for lead) or by using microscopes to count the number of fibres (e.g. for asbestos). This system is more accurate than stain tubes and gives a TWA result but can be uncomfortable to wear over long periods. Such equipment can only be used by trained personnel.

4. **Direct reading instruments** are available in the form of sophisticated analysers which can only be used by trained and experienced operatives. Infrared gas analysers are the most common but other types of analysers are also available. They are very accurate and give continuous or TWA readings. They tend to be very expensive and are normally hired or used by specialist consultants.

Other common monitoring instruments include **vane anemometers**, used for measuring air flow speeds, and hygrometers, which are used for measuring air humidity.

Qualitative monitoring techniques include smoke tubes and the dust observation lamp. **Smoke tubes** generate a white smoke which may be used to indicate the direction of flow of air – this is particularly useful when the air speed is very low or when testing the effectiveness of ventilation ducting. A **dust observation lamp** enables dust particles which are normally invisible to the human eye to be observed in the light beam. This dust is usually in the respirable range and, although the lamp does not enable any measurements of the dust to be made, it will illustrate the operation of a ventilation system and the presence of such dust.

Regulation 10 requires routine sampling or monitoring of exposure where there could be serious health effects if the controls fail, exposure limits might be exceeded or the control measures may not be working properly. Air monitoring should also be undertaken for any hazardous substances listed in Schedule 5 of the COSHH Regulations. Records of this monitoring should be kept for 5 years unless an employee is identifiable in the records, in which case they should be kept for 40 years.

14.3 Workplace exposure limits

One of the main purposes of a COSHH assessment is to adequately control the exposure of employees and others to hazardous substances. This means that such substances should be reduced to levels which do not pose a health threat to those exposed to them day after day at work. Under the 2004 amendments to the COSHH Regulations 2002, the HSE has assigned WELs to a large number of hazardous substances and publishes any updates in a publication called 'Workplace Exposure limits' EH40. The WEL is related to the concentration of airborne hazardous substances that people breathe over a specified period of time – known as 'time-weighted average'. Before the introduction of WELs, there were two types of exposure limit published: the maximum exposure limit (MEL) and the occupational exposure standard (OES).

The COSHH (Amendment) Regulations 2004 replaced the OES/MEL system with a single WEL. This removed the concern of the HSE that the OES was seen as a 'safe' limit rather than a 'likely safe' limit. Hence, the WEL must not be exceeded. Hazardous substances which have been assigned a WEL fall into two groups.

1. Substances which are carcinogenic or mutagenic or could cause occupational asthma (or listed in section C of the HSE publication '*Asthmagen? Critical assessment for the agents implicated in occupational asthma*' as updated from time to time) or are listed in Schedule 1 of the COSHH Regulations. These are substances which were assigned a MEL before 2005. The level of exposure to these substances should be reduced as far as is reasonably practicable.

2. All other hazardous substances which have been assigned a WEL. Exposure to these substances by inhalation must be controlled adequately to ensure that the WEL is not exceeded. These substances were previously assigned an OES before 2005. For these substances, employers should achieve adequate control of exposure by inhalation by applying the principles of good practice outlined in the ACoP and listed in 14.4.1. The implication of these principles is discussed later in this chapter.

The WELs are subject to time-weighted averaging. There are two such time-weighted averages (TWA): the long-term exposure limit (LTEL) or 8-hour reference

14

period and the short-term exposure limit (STEL) or 15-minute reference period. The 8-hour TWA is the maximum exposure allowed over an 8-hour period so that if the exposure period was less than 8 hours the WEL is increased accordingly with the proviso that exposure above the LTEL value continues for no longer than 1 hour. Table 14.2 shows some typical WELs for various hazardous substances.

For example, if a person was exposed to a hazardous substance with a WEL of 100 mg/m³ (8-hour TWA) for 4 hours, no action would be required until an exposure level of 200 mg/m³ was reached (further exposure at levels between 100 and 200 mg/m³ should be restricted to 1 hour).

Table 14.2 Examples of workplace exposure limits (WELs)

Group 1 WELs	LTEL (8 h TWA)	STEL (15 min)
All isocyanates	0.02 mg/m³	0.07 mg/m³
Styrene	430 mg/m³	1080 mg/m³
Group 2 WELs		
Ammonia	18 mg/m³	25 mg/m³
Toluene	191 mg/m³	574 mg/m³

If, however, the substance has an STEL of 150 mg/m³, then action would be required when the exposure level rose above 150 mg/m³ for more than 15 minutes.

The STEL always takes precedence over the LTEL. When a STEL is not given, it should be assumed that it is three times the LTEL value.

The publication EH40 is a valuable document for the health and safety professional as it contains much additional advice on hazardous substances for use during the assessment of health risks, particularly where new medical information has been made public. The HSE is constantly revising WELs and introducing new ones and it is important to refer to the latest publication of EH40.

It is important to stress that if a WEL from Group 1 is exceeded, the process and use of the substance should cease immediately and employees should be removed from the immediate area until it can be made safe. In the longer term the process and the control and monitoring measures should be reviewed and health surveillance of the affected employees considered.

The overriding requirement for any hazardous substance which has a WEL from Group 1, is to reduce exposure to as low as is reasonably practicable.

Finally, there are certain limitations on the use of the published WELs:

▶ They are specifically quoted for an 8-hour period (with an additional STEL for many hazardous substances). Adjustments must be made when exposure occurs over a continuous period longer than 8 hours.

▶ They can only be used for exposure in a workplace and not to evaluate or control non-occupational exposure (e.g. to evaluate exposure levels in a neighbourhood close to the workplace, such as a playground).

▶ WELs are only approved where the atmospheric pressure varies from 900 to 1100 millibars. This could exclude their use in mining and tunnelling operations.

▶ They should also not be used when there is a rapid build-up of a hazardous substance due to a serious accident or other emergency. Emergency arrangements should cover these eventualities.

The fact that a substance has not been allocated a WEL does not mean that it is safe. The exposure to these substances should be controlled to a level which nearly all of the working population could experience all the time without any adverse effects to their health. Detailed guidance and references are given for such substances in the HSE ACoP and guidance to the COSHH Regulations (L5).

The HSE publishes a revised ACoP and EH40 to include these changes. More information on the COSHH Regulations is given in Chapter 19.

14.4 Control measures

14.4.1 The principles of good practice for the control of exposure to substances hazardous to health

The objective of the COSHH Regulations is to prevent ill-health due to the exposure to hazardous substances. Employers are expected to develop suitable and sufficient control measures by:

1. identifying hazards and potentially significant risks;
2. taking action to reduce and control risks;
3. keeping control measures under regular review.

In order to assist employers with these duties, the HSE has produced the following eight principles of good practice:

(a) Design and operate processes and activities to minimise the emission, release and spread of substances hazardous to health.

(b) Take into account all relevant routes of exposure – inhalation, skin absorption and ingestion – when developing control measures.

(c) Control exposure by measures that are proportionate to the health risk.

(d) Choose the most effective and reliable control options which minimise the escape and spread of substances hazardous to health.

(e) When adequate control of exposure cannot be achieved by other means, provide, in combination with other control measures, suitable personal protective equipment.

(f) Check and review regularly all elements of control measures for their continuing effectiveness.

(g) Inform and train all employees on the hazards and risks from the substances with which they work and the use of control measures developed to minimise the risks.

(h) Ensure that the introduction of control measures does not increase the overall risk to health and safety.

All these principles are embodied in the following sections on COSHH control measures.

The frequency and type of future monitoring of exposure levels will depend on the exposure found in relation to recognised exposure limits. If the exposure level is very much lower than the limit and there is no change in process or other reason, then repeat measurement may only be needed occasionally.

If the exposure level is relatively high, then measurement may be needed several times between assessment reviews, to ensure that these levels have not been altered by some unidentified factor.

14.4.2 Hierarchy of control measures

The COSHH Regulations require the prevention or adequate control of exposure by measures other than personal protective equipment, so far as is reasonably practicable, taking into account the degree of exposure and current knowledge of the health risks and associated technical remedies. The hierarchy of control measures is as follows:

- elimination;
- substitution;
- provision of engineering controls;
- provision of supervisory (people) controls;
- provision of personal protective equipment.

Examples where engineering controls are not reasonably practicable include emergency and maintenance work, short-term and infrequent exposure and where such controls are not technically feasible.

Measures for preventing or controlling exposure to hazardous substances include one or a combination of the following:

- elimination of the substance;
- substitution of the substance (or the reduction in the quantity used);
- total or partial enclosure of the process;
- local exhaust ventilation;
- dilution or general ventilation;
- reduction of the number of employees exposed to a strict minimum;
- reduced time exposure by task rotation and the provision of adequate breaks;
- good housekeeping;
- training and information on the risks involved;
- effective supervision to ensure that the control measures are being followed;

- personal protective equipment (such as clothing, gloves and masks);
- welfare (including first-aid);
- medical records;
- health surveillance.

14.4.3 Preventative control measures

Prevention is the safest and most effective of the control measures and is achieved either by changing the process completely or by substituting for a less hazardous substance (the change from oil-based to water-based paints is an example of this). It may be possible to use a substance in a safer form, such as a brush paint rather than a spray.

The EU has introduced chemical safety regime REACH (Registration, Evaluation, Authorisation and restriction of Chemicals) Regulations, which restrict the use of high-risk substances or substances of very high concern and require that safer substitutes must be used. Manufacturers and importers of chemicals are responsible for understanding and managing the risks associated with their products. Authorisation seeks to ensure that risks from these substances of very high concern (SVHCs) are properly controlled. These are substances that are carcinogenic, mutagenic or toxic for reproduction or substances that are persistent, bio-accumulative and toxic. The objective is that all such substances should be replaced where possible with less dangerous alternatives.

REACH does not replace the COSHH Regulations which place duties on employers to control the exposure of employees to hazardous substances in the workplace.

The Regulations apply to many common items, such as glues, paints, solvents, detergents, plastics, additives, polishes, pens and computers. The three main types of REACH duty-holder are:

- **Manufacturers/importers** – businesses that manufacture or import (from outside the EU) 1 tonne or more of any given substance each year are responsible for registering a dossier of information about that substance with the European Chemicals Agency. If substances are not registered, then the data on them will not be available and it will no longer be legal to manufacture or supply them within the EU. Suppliers will be obliged to carry out an inventory and identify where in the supply chain the chemicals come from. Under the REACH system, industry will also have to prepare risk assessments and provide control measures for safe use of the substance by downstream users and get community-wide authorisations for the use of any substances considered to be of high concern.

- **Downstream users** – downstream users include any businesses using chemicals, which probably includes most businesses in some way. Companies

14

that use chemicals have a duty to use them in a safe way, and according to the information on risk management measures that should be passed down the supply chain.

▶ **Other users in the supply chain** – however, in order for suppliers to be able to assess these risks they need information from the downstream users about how they are used. REACH provides a framework in which information can be passed both up and down supply chains by using the safety data sheet. This should accompany materials down through the supply chain so that users are provided with the information that they need to ensure chemicals are safely managed. It is envisaged that, in the future, the safety data sheets will include information on safe handling and use. Some users may also be importers and have a duty to register.

An important aspect of chemical safety is the need for clear information about any hazardous chemical properties. The classification of different chemicals according to their characteristics (for example, those that are corrosive, or toxic to fish) currently follows an established system, which is reflected in REACH. (Chapter 19 contains more information on REACH.)

14.4.4 Engineering controls

The simplest and most efficient engineering control is the segregation of people from the process; a chemical fume cupboard is an example of this as is the handling of toxic substances in a glove box. Modification of the process is another effective control to reduce human contact with hazardous substances.

More common methods, however, involve the use of forced ventilation – local exhaust ventilation and dilution ventilation.

Local exhaust ventilation

Local exhaust ventilation (LEV) removes the hazardous gas, vapour or fume at its source before it can contaminate the surrounding atmosphere and harm people working in the vicinity. Such systems are commonly used for the extraction of welding fumes and dust from woodworking machines. All exhaust ventilation systems have the following five basic components (Figure 14.12(a)).

1. **A collection hood and intake** – sometimes this is a nozzle-shaped point which is nearest to the work piece, while at other times it is simply a hood placed over the workstation. The speed of the air entering the intake nozzle is important; if it is too low then hazardous fume may not be removed (air speeds of up to 1 m/s are normally required). A suitable airflow indicator should make it easy to see whether airflow is adequate. New LEV systems are likely to be fitted with airflow indicators as standard.

LEV hood design is very important and poor design is the main reason that LEV fails to control exposure. It should be as close as possible to the work-piece and ergonomic principles used in the designing of LEV hoods. LEV suppliers tend to concentrate on the fan and filter system and assume that hoods are standard for all applications.

The most commonly used hood – the capturing hood – has a particular limitation as the process to be controlled takes place outside the hood and the LEV system needs to generate sufficient air flow at and around the process to 'capture' and draw in the airborne contaminant cloud. One company replaced a capturing hood with an enclosing hood fitted with a glass panel so that soldering fumes were separated from the operative's breathing zone.

2. **Ventilation ducting** – this normally acts as a conduit for the contaminated air and transports it to a filter and settling section. It is very important that this section is inspected regularly and any dust deposits removed. It has been known for ventilation ducting attached to a workshop ceiling to collapse under the added weight of metal dust deposits. It has also been known for the deposit to catch fire.

3. **Filter or other air cleaning device** – normally located between the hood and the fan, the filter removes the contaminant from the air stream. The filter requires regular attention to remove the contaminant and to ensure that it continues to work effectively.

4. **Fan** – this moves the air through the system. It is crucial that the correct type and size of fan is fitted to a given system and it should only be selected by a competent person. It should also be positioned so that it can easily be maintained but does not create a noise hazard to nearby workers.

5. **Exhaust duct** – this exhausts the air to the outside of the building. The duct should be positioned on the outside wall of the building such that the air is not discharged into a public area or close to an air inlet for an air conditioning system of any building. It should be checked regularly to ensure that the correct volume of air is leaving the system and that there are no leakages. The exhaust duct should also be checked to ensure that there is no corrosion due to adverse weather conditions.

Controlling Airborne Contaminants at Work, HSG258, HSE Books, is a very useful document on ventilation systems.

The COSHH Regulations require that such ventilation systems should undergo a periodic thorough examination and test at least every 14 months by a competent person to ensure that they are still operating effectively. A record of the examination should be kept for at least 5 years.

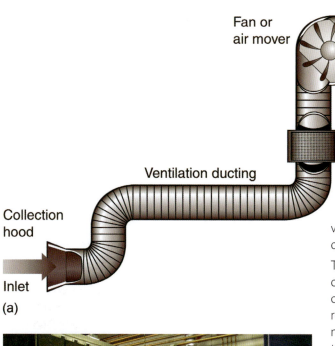

(a)

(b)

Figure 14.12 (a) Common elements of a simple LEV system; (b) welding with an adjustable LEV system to remove dust and fumes

The effectiveness of a ventilation system will be reduced by damaged ducting, blocked or defective filters and poor fan performance. More common problems include the unauthorised extension of the system, poor initial design, poor maintenance, incorrect adjustments and a lack of inspection or testing.

Routine maintenance should include repair of any damaged ducting, checking filters, examination of the fan blades to ensure that there has been no dust accumulation, tightening all drive belts and a general lubrication of moving parts. LEV systems will degrade if they are not regularly checked and maintained. Dirty ventilation systems can seriously affect staff well-being and working environments, and bacteria, MRSA and Clostridium difficile have been found in duct systems. A 'responsible person' should be appointed who follows the instructions in the user manual, arranges monitoring and maintenance, and keeps records. The employer

will also need to check that operators are using the LEV correctly.

The local exhaust ventilation system will have an effect on the outside environment in the form of noise and odour. Both these problems can be reduced by regular routine maintenance of the fan and filter. The waste material from the filter may be hazardous and require the special disposal arrangements described later in this chapter.

Wood dust can cause serious health problems as discussed later in this chapter. It can cause asthma, which carpenters and joiners are four times more likely to get compared with other workers in the UK. The COSHH Regulations require that workers are protected from the hazards of wood dust. Hardwood dust can cause cancer, particularly of the nose. Settled dust contains the fine particles that are most likely to damage the lungs. Both hardwood and softwood dusts have a WEL of 5 mg/m³, which must not be exceeded. These are limits placed on the amount of dust in the air, averaged over an 8-hour working day, but exposure to wood dust must be reduced to as low as 'reasonably practicable'. Dust extraction (local exhaust ventilation – LEV) should be provided at woodworking machines to capture and remove dust before it can spread.

The design of the extraction system should consider the number and type of machines to be connected to it and the required air flow and extraction vacuum for each extraction connection. Much of this information should be obtained from the machine manufacturer and values for common machines are given in Table 14.3.

Table 14.3 Typical airflow rates for various woodworking machines

Type of machine	Air flow (m³/hour)
Moulder	10,000
Saw	3,200
Planer	6,000

Source: HSE and BS EN 12779:2004 Safety of woodworking machines.

Neither brooms nor compressed air lines should be used to sweep up dust from the floor as this will disturb the dust and allow it to become inhaled. A suitable industrial standard vacuum cleaner or a vacuum attached to the extraction system should always be used. For very dusty jobs such as sanding, additional protection such as a suitable face mask should also be worn.

Since wood dust can cause asthma, health surveillance is important. For most woods, low level health surveillance is sufficient. A questionnaire should be completed by each worker at induction, again after six weeks and then annually. Sample questionnaires are available on the HSE website.

A higher level of health surveillance, including lung function testing, is needed for exposures to woods such as western red cedar, which is a known asthmagen.

Dilution (or general) ventilation

Dilution (or general) ventilation uses either natural ventilation (doors and windows) or a fan-assisted forced ventilation system to ventilate the whole working room by inducing a flow of clean air, using extraction fans fitted into the walls and the roof, sometimes assisted by inlet fans. It operates by either removing the contaminant or reducing its concentration to an acceptable level. It is used when airborne contaminants are of low toxicity, low concentration and low vapour

density or contamination occurs uniformly across the workroom. Paint-spraying operations often use this form of ventilation as does the glass-reinforced plastics (GRP) boat-building industry – these being instances where there are no discrete points of release of the hazardous substances. It is also widely used in kitchens and bathrooms. It is not suitable for dust extraction or where it is reasonably practicable to reduce levels by other means.

There are limitations to the use of dilution ventilation. Certain areas of the workroom (e.g. corners and beside cupboards) will not receive the ventilated air and a build-up of hazardous substances occurs. These areas are known as 'dead areas'. The flow patterns are also significantly affected by doors and windows being opened or the rearrangement of furniture or equipment.

14.4.5 Supervisory or people controls

Many of the supervisory controls required for COSHH purposes are part of a good safety culture and were discussed in detail in Chapter 3. These include items such as systems of work, arrangements and procedures, effective communications and training. Additional controls when hazardous substances are involved are as follows:

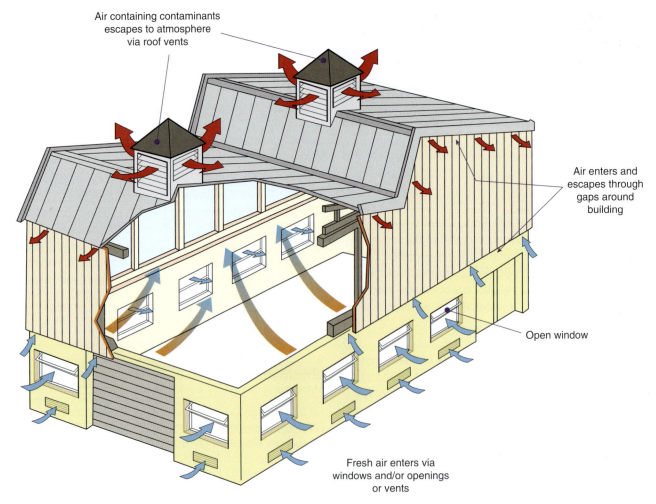

Air containing contaminants escapes to atmosphere via roof vents

Air enters and escapes through gaps around building

Open window

Fresh air enters via windows and/or openings or vents

Figure 14.13 Natural ventilation in a building. Source: HSE

Reduced time exposure – thus ensuring that workers have breaks in their exposure periods. The use of this method of control depends very much on the nature of the hazardous substance and its STEL.

Reduced number of workers exposed – only persons essential to the process should be allowed in the vicinity of the hazardous substance. Walkways and other traffic routes should avoid any area where hazardous substances are in use.

Eating, drinking and smoking must be prohibited in areas where hazardous substances are in use.

Any special rules, such as the use of personal protective equipment, must be strictly enforced.

14.4.6 Personal protective equipment

Personal protective equipment (PPE) is to be used as a control measure only as a last resort. It does not eliminate the hazard and will present the wearer with the maximum health risk if the equipment fails. Successful use of personal protective equipment relies on good user training, the availability of the correct equipment at all times and good supervision and enforcement.

The 'last resort' rule applies in particular to RPE within the context of hazardous substances. There are some working conditions when RPE may be necessary:

during maintenance operations;

as a result of a new assessment, perhaps following the introduction of a new substance;

during emergency situations, such as fire or plant breakdown;

where alternatives are not technically feasible.

The Personal Protective Equipment at Work Regulations state in Regulation 4 'Every employer shall ensure that

Figure 14.14 Personal protective equipment

suitable personal protective equipment is provided to his employees who may be exposed to a risk to their health or safety while at work except where and to the extent that such risk has been adequately controlled by other means which are equally or more effective'. The accompanying guidance recommends that employers should, therefore, provide appropriate PPE and training in its usage to their employees wherever there is a risk to health and safety that cannot be adequately controlled by other means. Employers must do more than simply have the equipment on the premises – it must be readily available. These Regulations cover all PPE except respiratory protective equipment (which is covered by specific Regulations such as COSHH and Lead).

The principal requirements of these Regulations are as follows:

PPE which is suitable for the wearer and the task;

compatibility and effectiveness of the use of multiple personal protective equipment;

a risk assessment to determine the need and suitability of proposed PPE;

a suitable maintenance programme for the personal protective equipment;

suitable accommodation for the storage of the PPE when not in use;

information, instruction and training for the user of PPE including a demonstration on how to use the equipment properly;

the supervision of the use of PPE by employees and a reporting system for defects.

A summary of these Regulations is given in Chapter 19.

Types of personal protective equipment

14

There are several types of personal protective equipment, such as footwear, hearing protectors and hard hats, which are not primarily concerned with protection from hazardous substances; those which are used for such protection include:

respiratory protection PPE;

hand and skin protection PPE;

eye protection PPE;

protective clothing.

Table 14.4 shows the types of PPE recommended by the HSE for various parts of the body.

For all types of PPE, there are some basic standards that should be reached. The PPE should fit well, be comfortable to wear and not interfere with other equipment being worn or present the user with additional hazards (e.g. impaired vision due to scratched eye goggles). Training in the use of particular PPE is essential, so that it is not only used correctly, but the user knows when to change an air filter or to change a type of glove. Supervision is essential, with disciplinary procedures invoked for non-compliance with PPE rules.

Table 14.4 The hazards and types of PPE for various parts of the body

	Hazards	PPE
Eyes	chemical or metal splash, dust, projectiles, gas and vapour, radiation.	safety spectacles, goggles, face-shields, visors.
Head	impact from falling or flying objects, risk of head bumping, hair.	a range of helmets and bump caps.
Respiratory system	dust, vapour, gas, oxygen-deficient atmospheres.	disposable filtering face piece or respirator, half- or full-face respirators, air-fed helmets, breathing apparatus.
Hand and arms	abrasion, temperature extremes, cuts and punctures, impact, chemicals, electric shock, skin infection, disease or contamination.	gloves, gauntlets, mitts, wrist-cuffs, armlets.
Feet and legs	wet, electrostatic build-up, slipping, cuts and punctures, falling objects, metal and chemical splash, abrasion.	safety boots and shoes with protective toe caps and penetration-resistant mid-sole, gaiters, leggings, spats.
Whole body	temperature extremes, adverse weather, chemical or metal splash, spray from pressure leaks or spray guns, impact or penetration, contaminated dust, excessive wear or entanglement of own clothing.	conventional or disposable overalls, boiler suits, specialist protective clothing, e.g. chain-mail aprons, high visibility clothing.

Source: HSE.

It is also essential that everyone who enters the proscribed area, particularly senior managers, wears the specified PPE.

Respiratory protective equipment

Respiratory protective equipment (RPE) can be subdivided into two categories – respirators (or face masks), which filter and clean the air, and breathing apparatus, which supplies breathable air.

Respirators should not be worn in air which is dangerous to health, including oxygen-deficient atmospheres. They are available in several different forms but the common ones are:

▶ a **filtering half-mask** often called disposable respirator – made of the filtering material. It covers the nose and mouth and removes respirable size dust particles. It is normally replaced after 8–10 hours of use. It offers protection against some vapours and gases;

▶ a **half-mask respirator** – made of rubber or plastic and covering the nose and mouth. Air is drawn through a replaceable filter cartridge. It can be used for vapours, gases or dusts but it is very important that the correct filter is used (a dust filter will not filter vapours);

▶ a **full-face mask respirator** – similar to the half-mask type but covers the eyes with a visor;

▶ a **powered respirator** – a battery-operated fan delivering air through a filter to the face mask, hood, helmet or visor.

Breathing apparatus is used in one of three forms (Figure 14.15):

▶ **self-contained breathing apparatus** – where air is supplied from compressed air in a cylinder and forms a completely sealed system;

▶ **fresh air hose apparatus** – fresh air is delivered through a hose to a sealed face mask from an uncontaminated source. The air may be delivered by the wearer, by natural breathing or mechanically by a fan;

▶ **compressed air-line apparatus** – air is delivered through a hose from a compressed air line. This can be either continuous flow or on demand. The air must be properly filtered to remove oil, excess water and other contaminants and the air pressure must be reduced. Special compressors are normally used.

The selection of appropriate RPE and correct filters for particular hazardous substances is best done by a competent specialist person.

There are several important technical standards which must be considered during the selection process. RPE must either be CE marked or HSE approved (HSE approval ceased in 1995 but such approved equipment may still be used). Other standards include the minimum protection required (MPR) and the assigned protection factor (APF). The CE mark does not indicate that the equipment is suitable for a particular hazard. The following information will be needed before a selection of suitable RPE can be made:

▶ details of the hazardous substance, in particular whether it is a gas, vapour or dust or a combination of all three;

▶ presence of a beard or other facial hair which could prevent a good leak-free fit (a simple test to see whether the fit is tight or not is to close off the air supply, breathe in and hold the breath. The respirator should collapse onto the face. It should then be possible to check to see if there is a leak);

▶ the size and shape of the face of the wearer and physical fitness;

▶ compatibility with other personal protective equipment, such as ear defenders;

▶ the nature of the work and agility and mobility required.

Filters and masks should be replaced at the intervals recommended by the supplier or when taste or smell is detected by the wearer.

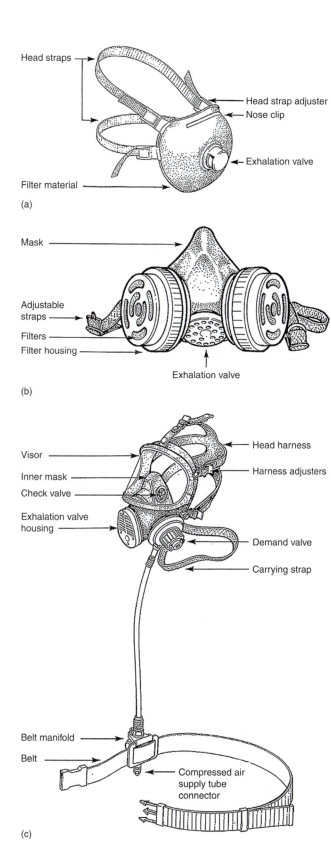

Head straps

Head strap adjuster
Nose clip

Exhalation valve

Filter material

(a)

Mask

Adjustable
straps

Filters
Filter housing

Exhalation valve

(b)

Visor

Head harness

Harness adjusters

Inner mask
Check valve

Exhalation valve
housing

Demand valve

Carrying strap

Belt manifold

Belt

Compressed air
supply tube
connector

(c)

Figure 14.15 Types of respiratory protective
equipment: (a) filtering half-mask; (b) half-mask –
re-usable with filters; (c) compressed air-line breathing
apparatus with full face fitted with demand valve

The performance of RPE with a tight-fitting face piece
depends on a good contact between the wearer's skin
and the face seal of the mask. However, research by
the British Safety Industry Federation indicated that up

to 50% of all RPE used does not offer the wearer the
level of protection assumed, usually because it was not
fitted correctly.

The COSHH Approved Code of Practice states that
'Employers should ensure that the selected face piece
is of the right size and can correctly fit each wearer.
For a tight-fitting face piece, the initial selection
should include fit testing to ensure the wearer has
the correct device. Also, employers must ensure
that whoever carries out the fit testing is competent
to do so'.

The British Safety Industry Federation (working with
the HSE) has developed 'Fit2Fit', an accreditation
scheme for people performing face piece fit testing.
An HSE document 'Fit testing of respiratory protective
equipment face pieces' provides more detail on fit
testing methods. The HSE uses the term 'face piece'
and defines this to mean a full-face mask, a half-face
mask, or a filtering face piece/disposable mask. Fit2Fit
standards are based on this HSE document.

Fit testing is needed when RPE is used as a control
measure not when it is worn by choice for comfort.
Fit testing should be repeated if the shape of the
face of the wearer changes for any reason, such as
weight loss or gain. Re-testing is recommended to
check that the face piece remains suitable and that
the wearer is taking care to wear and remove the
mask correctly. A two-year cycle has been suggested,
and a one year re-test is suggested for work with
asbestos.

Fit testing can be qualitative or quantitative.
Qualitative testing involves a simple pass or fail
based on whether the user can taste or smell
a harmless aerosol or odour through the RPE.
Qualitative testing is only appropriate for disposable
or half-face masks. A quantitative fit test may use a
laboratory test chamber or a portable fit test device,
and produces a numerical 'fit factor' measure. The
British Safety Industry Federation 'Fit2Fit' website
provides details of accredited Fit2Fit training providers.
The results of fit tests should be recorded. Fit test
records must be made available to the HSE on
request, and to the employees who wear RPE. RPE
should be checked every time that it is used to make
sure it has not been damaged and is being worn
correctly.

A filtering face piece (FFP3) device is a mask which is
certified to the PPE Directive. It provides a high level
of filtering capability and face fit. It can be supplied
with an exhale valve so that it can be worn comfortably
over a fairly long period of time. It will provide an
effective barrier to both droplets and fine aerosols and
is the type recommended particularly for people in the
healthcare sector dealing with symptomatic patients
undergoing treatment where aerosols are likely to be
generated.

14

All RPE should be examined at least once a month except for disposable respirators. A record of the inspection should be kept for at least 5 years. There should be a routine cleaning system in place and proper storage arrangements.

Respiratory Protective Equipment – A Practical Guide for Users, HSG53, HSE Books, contains comprehensive advice and guidance on RPE selection, use, storage, maintenance and training, and should be consulted for more information.

Hand and skin protection

Hand and skin protection is mainly provided by gloves (arm shields are also available). A wide range of safety gloves is available for protection from chemicals, sharp objects, rough working and temperature extremes. Many health and safety catalogues give helpful guidance for the selection of gloves. For protection from chemicals, including paints and solvents, impervious gloves are recommended. These may be made of PVC, nitrile or neoprene. For sharp objects, such as trimming knives, a Kevlar-based glove is the most effective. Gloves should be regularly inspected for tears or holes since this will obviously allow skin contact to take place.

Another effective form of skin protection is the use of barrier creams and these come in two forms – pre-work and after-work. Pre-work creams are designed to provide a barrier between the hazardous substance and the skin. After-work creams are general purpose moisteners which replace the natural skin oils removed either by solvents or by washing.

Eye protection

Eye protection comes in three forms – spectacles (safety glasses), goggles and face visors. Eyes may be damaged by chemical and solvent splashes or vapours, flying particles, molten metals or plastics, non-ionising radiation (arc welding and lasers) and dust. Spectacles are suitable for low-risk hazards (low-speed particles such as machine swarf). Some protection against scratching of the lenses can be provided but this is the most common reason for replacement. Prescription lenses are also available for people who normally wear spectacles.

Goggles are best to protect the eyes from dust or solvent vapours because they fit tightly around the eyes. Visors offer protection to the face as well as the eyes and do not steam up so readily in hot and humid environments. For protection against very bright lights, special light filtering lenses are used (e.g. in arc welding). Maintenance and regular cleaning are essential for the efficient operation of eye protection.

When selecting eye protection, several factors need to be considered. These include the nature of the hazard (the severity of the hazard and its associated risk will

Figure 14.16 Variety of eye protection goggles

determine the quality of protection required), comfort and user acceptability, compatibility with other personal protective equipment, training and maintenance requirements and costs.

Protective clothing

Protective clothing includes aprons, boots and headgear (hard hats and bump caps). Head protection is covered in Chapter 16. The law requires head protection on all construction sites where there is a danger of head injury except for turban-wearing Sikhs.

Shorts are often worn on construction sites but it must be remembered that clothing needs to protect against hazards on site. The main reason for protecting the lower legs is to help guard against cuts, grazes and splinters in an environment where any skin damage can lead to infection. Some trades, such as welders, need to keep skin covered for protection. During cold weather it is important to keep warm, especially when working at height where the cold can distract and lead to loss of concentration.

Aprons are normally made of PVC and protect against spillages but can become uncomfortable to wear in hot environments. Other lighter fabrics are available for use in these circumstances. Safety footwear protects against falling objects, collision with hard or sharp objects, hot or molten materials, slippery surfaces and chemical spills. It has metal toe caps and comes in the form of shoes, ankle boots or knee-length boots and is made of a variety of materials dependent on the particular hazard (e.g. thermally insulated against cold environments). It must be used with care near live, unprotected-against electricity. Specialist advice is needed for use with flammable liquids.

Appropriate selection of safety footwear involves the matching of the workplace hazards to the performance requirements of the footwear. The key issues are:

▶ the type of hazard (e.g. physical, chemical or thermal);
▶ the type of environment (e.g. indoors or out of doors);
▶ the ergonomics of the job (e.g. standing up, constant movement).

The footwear must have the correct grip for the environment, a hard wearing sole unit and, possibly, a good shock-absorbing capability. Construction workers are expected to wear protective footwear whilst on site and doing heavy work. The bones in the foot are quite delicate and easily damaged and any muscle or tendon damage can prevent normal movement for several months. Steel toe caps (or equivalent) protect against dropped objects. Mid-sole protection (usually a steel plate) protects against puncture or penetration by a nail

It is important to note that appropriate personal protective equipment should be made available to work-related visitors and other members of the public visiting workplaces where hazardous substances are being used. It is also important to stress that managers and supervisors must lead by example, particularly if there is a legal requirement to wear particular personal protective equipment. Refusals by employees to wear mandatory personal protective equipment must lead to some form of disciplinary action.

In 2002, some extra requirements were added to the Personal Protective Equipment Regulations.

▶ The personal protective equipment must satisfy the basic health and safety requirements which are applicable to that class or type of personal protective equipment.
▶ The appropriate conformity assessment procedures must be carried out.
▶ CE marking must be correctly affixed.
▶ The personal protective equipment must not compromise the safety of individuals, domestic animals or property when properly maintained and used.

Finally, high visibility clothing should be worn on all construction sites where vehicles or plant are operating. This includes drivers when they leave their vehicle. For routine site use it is often sufficient for a tabard (sleeveless top) to be worn.

14.4.7 Health surveillance and personal hygiene

The need for health surveillance in assessing exposure to hazardous substances has already been introduced in 14.2.6. It is required when as a result of reviewing sickness records or when a substance listed in Schedule 6 under Regulation 11 of the COSHH Regulations is being used there appears to be a reasonable chance that ill-health effects are occurring in a particular workplace. Schedule 6 lists the substances and the processes in which they are used. There is a limited number of such substances and surveillance includes medical surveillance by an employment medical adviser or appointed doctor at intervals not exceeding 12 months. The need for health surveillance is not common and further advice on the necessary procedures is available from the Employment Medical Advisory Service.

Health surveillance enables the identification of those employees most at risk from occupational ill-health. It should not be confused with health monitoring procedures such as pre-employment health checks or drugs and alcohol testing, but it covers a wide range of situations, from a responsible person looking for skin damage on hands to medical surveillance by a medical doctor. Health surveillance allows for early identification of ill-health and helps identify any corrective action needed. It may be required by law if employees are exposed to noise or vibration, solvents, isocyanates or other respiratory sensitisers, fumes, dusts, biological agents and asbestos, lead, work in compressed air or with ionising radiation. Health surveillance is a system of ongoing health checks and is important for:

▶ detecting ill-health effects at an early stage, so employers can introduce better controls to prevent them getting worse;
▶ providing data to help employers evaluate health risks;
▶ enabling employees to raise concerns about how work affects their health;
▶ highlighting lapses in workplace control measures, therefore providing invaluable feedback to the risk assessment;
▶ providing an opportunity to reinforce training and education of employees (e.g. on the impact of health effects and the use of protective equipment).

When initiating a health surveillance programme, it is important to avoid blanket coverage for all employees as this can produce misleading results and waste money.

The risk assessment for the organisation should indicate when and where health surveillance is required. In its simplest form, health surveillance could involve employees checking themselves for signs or symptoms of ill-health following a training session – for example, soreness, redness and itching on their hands and arms if they work with substances that can irritate or damage the skin. A responsible person, who has been trained by a competent medical practitioner, can also make routine basic checks, such as skin inspections or signs of rashes.

For more complicated assessments, an occupational health nurse or an occupational health doctor can ask about symptoms or carry out periodic examinations. Statutory medical surveillance involves a medical examination and possibly tests by a doctor with appropriate training and experience.

A health record must be kept for all employees under health surveillance. Health records, or a copy, should be kept in a suitable form for at least 40 years from the date of last entry because often there is a long period between exposure and onset of ill-health. Recorded details of each health surveillance check should include:

▶ the date they were carried out and by whom;
▶ the outcome of the test/check;

14

▶ the decision made by the occupational health professional in terms of fitness for task and any restrictions required. This should be factual and only relate to the employee's functional ability and fitness for specific work, with any advised restrictions.

Health records are different to medical records in that they should not contain confidential medical information.

Personal hygiene has already been covered under supervisory controls. It is very important for workers exposed to hazardous substances to wash their hands thoroughly before eating, drinking or smoking. Protection against biological hazards can be increased significantly by vaccination (e.g. tetanus). Finally, contaminated clothing and overalls need to be removed and cleaned on a regular basis.

For plumbers and roofers who use lead in their work, medical surveillance by a doctor will be required if exposure is significant and the surveillance will include blood lead tests. Similarly workers who come into contact with asbestos will require medical surveillance as described later in this chapter.

An example of a health questionnaire for persons working with respiratory sensitisers is shown in Appendix 14.2.

14.4.8 Maintenance and emergency controls

Engineering control measures will only remain effective if there is a programme of preventative maintenance available. Indeed the COSHH Regulations require that systems of adequate control are in place and the term 'adequate control' spans normal operations, emergencies and maintenance. Maintenance will involve the cleaning, testing and, possibly, the dismantling of equipment. It could involve the changing of filters in extraction plant or entering confined spaces. It will almost certainly require hazardous substances to be handled and waste material to be safely disposed of. It may also require a permit-to-work procedure to be in place since the control equipment will be inoperative during the maintenance operations. Records of maintenance should be kept for at least 5 years.

Emergencies can range from fairly trivial spillages to major fires involving serious air pollution incidents. The following points should be considered when emergency procedures are being developed:

▶ the possible results of a loss of control (e.g. lack of ventilation);
▶ dealing with spillages and leakages (availability of effective absorbent materials);
▶ raising the alarm for more serious emergencies;
▶ evacuation procedures, including the alerting of neighbours;
▶ fire-fighting procedures and organisation;
▶ availability of respiratory protective equipment;
▶ information and training.

The Emergency Services should be informed of the final emergency procedures and, in the case of the Fire Service, consulted for advice during the planning of the procedures. (See Chapter 4 for more details on emergency procedures.)

14.4.9 The transport of hazardous substances by road

Although this topic is not in the NEBOSH National Construction Certificate syllabus, a brief mention of the main precautions required to safeguard the health and safety of those directly involved in the transport of hazardous substances and of general members of the public is important.

Data sheets from the manufacturer of the hazardous substance should indicate the safest method of handling it and will give information on emergency procedures (e.g. for spillages and fire). These sheets should be available to all concerned with the transportation of the substance, in particular those responsible for loading/unloading, as well as the driver. The hazardous substance should be loaded correctly on the vehicle in suitable containers and segregated from incompatible materials. There must be adequate emergency information with the substance containers and attached to the vehicle. Drivers of the vehicles must receive special training which covers issues such as emergency procedures and route planning. There should also be emergency provisions for first-aid and personal protective equipment on the vehicle. Finally, the Transport of Dangerous Goods (Safety Adviser) Regulations require the appointment of a trained and competent safety adviser to ensure the safe loading, transportation and unloading of the hazardous substance. The HSE has produced several guidance publications which offer more detailed advice on this topic.

14.4.10 An illustrative example using COSHH controls

Organic solvents are widely used throughout industry and commerce in paints, inks, glues and adhesives. The COSHH hierarchy discussed under 14.4.2 should be applied to minimise the health risks from the use of these solvents. The top of this hierarchy is to eliminate or substitute the use of the organic solvent by using a less volatile or water-based alternative. If this is not possible, then some form of engineering control should be applied, such as dilution or local exhaust ventilation. Alternatively, the workplace where the solvents are being used could be enclosed or isolated from the main work activities. Other engineering type controls could include the use of properly labelled anti-spill containers, the use of covered disposal units for any used cloths and the transfer of large quantities using pumping/pipe arrangements rather than simply pouring the solvent.

Supervisory controls include the reduction in the length of time any employee is exposed to the solvent and the provision of good housekeeping, such as ensuring that containers are kept closed when not in use and any spills are quickly removed. The provision of barrier creams and personal protective equipment (eye protection, gloves and aprons) and RPE may also be required. Welfare issues will include first-aid provision, washing facilities and the encouragement of high levels of personal hygiene. Smoking and the consumption of food and drink should be prohibited where there might be contamination from organic solvents. All these supervisory items should be reinforced in training sessions and employees given appropriate information on the risks associated with the solvents. Finally, some form of health surveillance will be needed so that employees who show allergies to the solvents can be treated and, possibly, assigned other duties.

14.5 Specific agents

14.5.1 Health risks and controls associated with asbestos

It is estimated that 1.8 million workers remain at risk every day from the possible disturbance of this carcinogenic material. Those most at risk are workers such as electricians, plumbers, joiners and IT installers. Asbestos is responsible for over 4,500 deaths every year and 4 plumbers, 20 tradesmen, 6 electricians and 8 joiners die every week. Younger people, if routinely exposed to asbestos fibres over time, are at greater risk of developing asbestos-related disease than older workers. This is due to the time it takes for the body to develop symptoms after exposure to asbestos (latency). Exposure to asbestos can cause four main diseases:

▶ Mesothelioma (a cancer of the lining of the lungs; it is always fatal and is almost exclusively caused by exposure to asbestos).
▶ Asbestos-related lung cancer (which is almost always fatal).
▶ Asbestosis (a scarring of the lungs which is not always fatal but can be a very debilitating disease, greatly affecting quality of life).
▶ Diffuse pleural thickening (a thickening of the membrane surrounding the lungs which can restrict lung expansion leading to breathlessness).

The number of mesothelioma deaths is more than 2,000 each year and increasing year on year. Most mesothelioma deaths occurring now are a legacy of past occupational exposures to asbestos when it was widely used in the building industry.

Asbestos appears in three main forms – crocidolite (blue), amosite (brown) and chrysotile (white). The blue and brown asbestos are considered to be the most dangerous and may be found in older buildings where they were used as heat insulators around boilers and hot water pipes and as fire protection of structures. White asbestos has been used in asbestos cement products and brake linings. It is difficult to identify an asbestos product by its colour alone – laboratory identification is usually required. Many asbestos-containing materials (ACMs) are difficult to distinguish from other materials. It is easy to drill or cut ACMs unwittingly and release large quantities of airborne fibres that could cause long-term health problems to the operator.

Asbestos produces a fine fibrous dust of respirable dust size which can become lodged in the lungs. The fibres can be very sharp and hard causing damage to the lining of the lungs over a period of many years.

The condition of the ACM will need to be monitored regularly and a record kept. The time between these monitoring checks should not exceed 12 months and may need to be more frequent. Monitoring involves visual inspection to see whether there has been any deterioration at the surface of the ACM. This may be remedied by resealing the surface or removing a section of the ACM.

If asbestos is discovered during the performance of a contract, work should cease immediately and the employer informed. Typical sites of asbestos include ceiling tiles, asbestos cement roof and wall sheets, sprayed asbestos coatings on structural members, loft insulation and asbestos gaskets. Asbestos has its own Regulations (Control of Asbestos Regulations) and a summary of these is given in Chapter 19. These cover the need for a risk assessment, a method statement covering removal and disposal, air-monitoring procedures and the control measures (including personal protective equipment and training) to be used. A summary of these controls is given later in this chapter.

Training is required for the majority of workers involved in maintenance, refurbishment and demolition. The Health and Safety Executive (HSE) has estimated that approximately 50% of buildings still contain some form of asbestos and about 1.5 million workers require asbestos training.

The most recent Asbestos Regulations have removed textured coatings (decorative products such as Artex, Wondertex and Pebblecoat) from the asbestos licensing regime. Until 1992, these products contained white asbestos. This amendment by the HSE followed research work by the Health and Safety Laboratory. The new Regulations introduce an additional training requirement for asbestos awareness training. Such training should include:

▶ the health risks caused by exposure to asbestos;
▶ the materials that are likely to contain asbestos and where they are likely to be found;
▶ the methods to reduce asbestos risks during work;
▶ the action to take in an emergency, such as an uncontrolled release of asbestos dust.

14

Asbestos is the single biggest workplace killer. According to HSE statistics, there are 15 times as many deaths from asbestos as there are deaths from workplace accidents. Asbestos is responsible for at least 4,000 deaths in the UK each year, and the HSE felt that there was a need to increase awareness amongst the workforce of the risks associated with this material.

14.5.2 Managing asbestos in buildings

A recent study published by the *British Medical Journal* has found that there are over 1,800 mesothelioma deaths each year in Britain. Since this disease can take between 15 and 60 years to develop, the peak of the epidemic has still to be reached. In the construction industry, those at risk are asbestos removal workers and those, such as electricians, plumbers and carpenters, who are involved in refurbishment, maintenance or repair of buildings.

It was used widely as a building material until the mid-1980s. Although much asbestos has been removed from buildings, it has been estimated that over half a million non-domestic buildings still have asbestos in them amounting to many thousands of tons.

The strategy of the HSE is to ensure that those involved in the repair, removal or disturbance of asbestos-containing materials (ACMs), such as insulation, coatings or insulation boards, are licensed, competent and working to the strict requirements of the Control of Asbestos Regulations. This requires the identification of ACMs and the planning of any subsequent work. This should prevent inadvertent exposure to asbestos and minimise the risks to those who have to work with it.

The Regulations bring together the three previous sets of Regulations covering the prohibition of asbestos, the control of asbestos at work and asbestos licensing. The duty to manage asbestos is a legal requirement under the Control of Asbestos Regulations (Regulation 4). It applies to the owners and occupiers of commercial premises (such as shops, offices, industrial units, etc.) who have responsibility for maintenance and repair activities. In addition to these responsibilities, they also have a duty to assess the presence and condition of any asbestos-containing materials. If asbestos is present, or is presumed to be present, then it must be managed appropriately.

The Regulations also prohibit the importation, supply and use of all forms of asbestos. They continue the ban introduced for blue and brown asbestos in 1985 and for white asbestos in 1999. They also continue the ban on the second-hand use of asbestos products such as asbestos cement sheets and asbestos boards and tiles, including panels which have been covered with paint or textured plaster containing asbestos. The ban applies to new uses of asbestos. If existing ACMs are in good condition, they may be left in place and their condition monitored and managed to ensure that they are not disturbed. The Regulations only cover the safe management of asbestos in industrial and commercial premises. Domestic premises are covered by the Defective Premises Act and the Civic Government (Scotland) Act, although the Control of Asbestos Regulations apply to the shared parts of some domestic premises.

The Regulations also include the duty to manage asbestos in non-domestic premises. Guidance on the duty to manage asbestos can be found in the HSE ACoP – *Managing and working with asbestos*, L143.

The Regulations require mandatory training for anyone liable to be exposed to asbestos fibres at work, including maintenance workers and others who may come into contact with or who may disturb asbestos (e.g. cable installers) as well as those involved in asbestos removal work.

When work with asbestos or which may disturb asbestos is being carried out, the Regulations require employers and the self-employed to prevent exposure to asbestos fibres. Where this is not reasonably practicable, they must make sure that exposure is kept as low as reasonably practicable by measures other than the use of respiratory protective equipment. The spread of asbestos must be prevented. The Regulations specify the work methods and controls that should be used to prevent exposure and spread.

Worker exposure must be below the airborne exposure limit (the Control Limit) of 0.1 fibres per cm^3 for all types of asbestos. The Control Limit is the maximum concentration of asbestos fibres in the air (averaged over any continuous 4-hour period) and must not be exceeded. Short-term exposures must be strictly controlled and worker exposure should not exceed 0.6 fibres per cm^3 of air averaged over any continuous 10-minute period using respiratory protective equipment if exposure cannot be reduced sufficiently using other means. Respiratory protective equipment is an important part of the control regime but it must not be the sole measure used to reduce exposure and should only be used to supplement other measures.

Most asbestos removal work must be undertaken by a licensed contractor but any decision on whether particular work needs to be licensed is based on the level of risk. Details are given in Chapter 19 of the six possible reasons for work with asbestos being exempt from licensing.

The Control of Asbestos Regulations requires those in control of premises and the duty-holders to:

▶ take reasonable steps to determine the location and condition of materials likely to contain asbestos;
▶ presume materials contain asbestos unless there is strong evidence that they do not;
▶ make and keep an up-to-date record of the location and condition of the ACMs or presumed ACMs in the premises;

- assess the risk of the likelihood of anyone being exposed to fibres from these materials;
- prepare a plan setting out how the risks from the materials are to be managed;
- take the necessary steps to put the plan into action;
- review and monitor the plan periodically;
- provide such information and asbestos awareness training to anyone who is liable to work on these materials or otherwise disturb them.

In addition, the Regulations include the following main provisions:

1. a single tighter control limit for all types of asbestos;
2. specific training requirements for those working with asbestos;
3. a clear hierarchy of controls to be used to reduce exposure.

The HSE has produced a document – *HSG 264 Asbestos: The survey guide* – that gives useful guidance on asbestos surveys. There are **two** types of survey, known in the UK as the **management survey** and the **refurbishment and demolition survey.**

A **management survey** is the standard survey. Its purpose is to locate, as far as reasonably practicable, the presence and extent of any suspect ACMs in the building which could be damaged or disturbed during normal occupancy, including foreseeable maintenance and installation, and to assess their condition. A management survey is required during the normal use of the building to ensure continued management of the ACMs. It should include an assessment of the condition of the various ACMs and their ability to release fibres into the air should they be disturbed in some way. It will often involve minor intrusive work and some disturbance.

Management surveys can involve a combination of sampling to confirm asbestos is present or presuming asbestos to be present. All ACMs should be identified as far as is reasonably practicable. The areas inspected should include under floor coverings, above false ceilings (ceiling voids), cladding and partitions, under carpets, tiles and floors, lofts, inside risers, service ducts and lift shafts and basements, cellars or underground rooms. In these situations, controls should be put in place to prevent the spread of debris, which may include asbestos.

A **refurbishment or demolition survey** is used to locate and describe, as far as reasonably practicable, all ACMs in the area where the refurbishment work will take place or in the whole building if demolition is planned. It is necessary when the building (or part of it) is to be upgraded, refurbished or demolished. It is required for all work which disturbs the fabric of the building in areas particularly where the management survey has not been intrusive. A refurbishment and demolition survey is needed before any refurbishment or demolition work is carried out.

Refurbishment and demolition surveys should only be conducted in unoccupied areas to minimise risks to the public or employees on the premises. Ideally, the building should not be in service and all furnishings removed.

For minor refurbishment, this would only apply to the room involved or even part of the room where the work is small and the room large. In these situations, there should be effective isolation of the survey area (e.g. full floor to ceiling partition), and furnishings should be removed as far as possible or protected using sheeting.

The person who undertakes any of these surveys must be suitably trained and experienced in such work.

Some types of work, of an intermittent and low intensity nature, will not have to be done by a licensed contractor – Artex work is an example of this.

There are many issues which need to be addressed when asbestos is possibly present in a workplace. Several publications are available from HSE Books which cover all these issues in some detail and the reader should refer to them for more information. Here a brief summary of the principal issues will be given.

Identification of the presence of asbestos is the first action. Asbestos is commonly found as boiler and pipe lagging, insulation panels around pillars and ducting for fire protection and heat insulation, ceiling tiles and asbestos cement products, including asbestos cement sheets. The main duty-holder is required to ensure that a written plan is prepared that shows where the ACM is located and how it will be managed to prevent exposure to asbestos, including to contractors and other workers who may undertake work on the fabric of the building that could disturb the ACM. This plan then needs to be put into action and communicated to those affected. The duty-holder should ensure that the plan is reviewed regularly and updated as circumstances change, in consultation with all those who may be affected. The plan, including drawings, should be available on site for the entire life of the premises and should be kept up to date.

Initial investigations will involve the examination of building plans, the determination of the age of the building and a thorough examination of the building. Advice is available from a number of reputable specialist consultants and details may be obtained from the local authority, who often offers such a service. If the specialist is in any doubt, a sample of the suspect material will be sent to a specialist laboratory for analysis. It is important for a specialist to take the sample because the operation is likely to expose loose fibres. When asbestos has been identified, a record, possibly electronic, of its location must be made so that it is available should any future maintenance be necessary.

The **asbestos risk register** is a key component of the asbestos management. The management plan must

14

Figure 14.17 Damaged asbestos lagging on pipework

contain current information about the presence and condition of any asbestos in the building. The asbestos risk register will therefore need to be updated on a regular basis (at least once a year). This will involve:

▶ regular inspections to check the current condition of asbestos materials;

▶ deletions to the register when any asbestos is removed;

▶ additions to the register when new areas are surveyed and asbestos is located;

▶ changes to the register (at any time asbestos-containing materials are found to have deteriorated).

The register can be kept as a paper or electronic record and it is very important that this is kept up to date and easily accessible should any future maintenance be necessary. Paper copies may be easier to pass on to visiting maintenance workers, who will need them to know the location and condition of any asbestos before they start work. Electronic copies are easier to update and are probably better suited for people responsible for large numbers of properties or bigger premises.

An example of an asbestos register is available on the HSE website.

Assessment is an evaluation or asbestos survey to determine whether the location or the condition could lead to the asbestos being disturbed. If it is in good condition, undamaged and not likely to be disturbed, then it is usually safer to leave it in place and monitor it. However, if it is in poor condition, it may need to be repaired, sealed, enclosed or removed. If there is doubt, then specialist advice should be sought. The condition of the ACM will need to be monitored regularly and a record kept. The time between these monitoring checks should not exceed 12 months and may need to be more frequent. Monitoring involves visual inspection to see whether there has been any deterioration at the surface of the ACM. This may be remedied by resealing the surface or removing a section of the ACM.

The asbestos survey should provide enough information so that an asbestos register, a risk assessment and a management plan can then be prepared. The survey will usually involve sampling and analysis to determine the presence of asbestos so asbestos surveys should only be carried out by competent surveyors who can clearly demonstrate they have the necessary skills, experience and qualifications.

Removal must only be done by a licensed contractor. A detailed plan of work is essential before work begins. The plan should give details of any equipment to be used for the protection and decontamination of employees and others. This process will also require an assessment to be made to ensure that people within the building and neighbours are properly protected. At the planning stage, generally, the HSE or the local authority must be notified of the intention to remove asbestos (at least 14 days' notice) and again when the work begins. This is particularly important when the exposure to asbestos is likely to exceed the action level. The assessment should include details of the type and location of the asbestos, the number of people who could be affected, the controls to be used to prevent or control exposure, the nature of the work, the removal methods, the procedures for the provision of personal and respiratory protective equipment and details of emergency procedures. If asbestos cement sheeting is to be removed the following procedure is recommended:

▶ Where reasonably practicable, remove the asbestos sheets before any other operation, such as demolition.

▶ Avoid any breaking of the sheets.

▶ Dampen the sheets while working on them.

▶ Lower the sheets on to a clean hard surface.

▶ Remove all waste and debris from the site as soon as possible to prevent its spread around the site.

▶ Do not bulldoze broken asbestos cement or sheets into piles.

▶ Do not dry sweep asbestos cement debris.

▶ Dispose of the waste and debris safely, separate from general waste as hazardous waste.

Control measures during the removal of asbestos include the provision of personal and RPE including overalls, good ventilation arrangements in and the segregation or sealing of the working area, suitable method statements and air-monitoring procedures. A decontamination unit should also be provided. The sealed area will need to be tested for leaks. Good supervision and induction of the workforce are also essential. A high level of personal hygiene must be expected for all workers with the provision of welfare amenity arrangements, particularly washing and catering facilities and the separation of working and personal clothing. Suitable warning signs must be displayed and extra controls provided if the work is taking place at height. After the work

Figure 14.18 Asbestos removal enclosure

is completed, the area must be thoroughly cleaned and a clean air certificate provided after a successful air test.

Work with high-risk asbestos products, such as asbestos insulating board, sprayed fire or insulation coatings or lagging, must only be carried out by an asbestos licence holder who works within the conditions of their licence.

Medical surveillance in the form of a regular medical examination by an appointed doctor should be given to any employee who has been exposed to asbestos at levels above the action level. The first medical examination should take place within 2 years of exposure and further examinations at intervals of not more than 2 years. A health record of such surveillance should be kept for a period of at least 40 years after the last entry.

Awareness training is an important feature of the Control of Asbestos Regulations and they require that adequate information, instruction and training are given to those employees:

(a) who are or who are liable to be exposed to asbestos, or who supervise such employees; and/or

(b) who carry out work in connection with the employer's duties under these Regulations, so that they can carry out that work effectively.

Such training should cover the following topics:

▶ the properties of asbestos and its effects on health, including its interaction with smoking;
▶ the types of products or materials likely to contain asbestos;
▶ the uses and location of ACMs in buildings and plant;
▶ the operations which could result in asbestos exposure and the importance of preventative controls to minimise exposure;
▶ the presence of other hazards such as working at height;
▶ the requirements of the Control of Asbestos Regulations;
▶ safe work practices, control measures, and protective equipment;
▶ the purpose, choice, limitations, proper use and maintenance of RPE;
▶ emergency procedures;
▶ hygiene requirements;
▶ decontamination procedures;
▶ waste handling procedures;
▶ medical examination requirements;
▶ the control limit and the need for air monitoring.

Refresher training should be given at regular intervals (at least annually) and adapted to take account of significant changes in the type of work carried out or methods of work used by the employer. It should be provided in a manner appropriate to the nature and degree of exposure identified by the risk assessment, and so that the employees are aware of:

(i) the significant findings of the risk assessment; and

(ii) the results of any air monitoring carried out with an explanation of the findings.

Disposal of asbestos waste is subject to the Hazardous Waste Regulations which require it to be consigned to an authorised asbestos waste site only. Asbestos waste describes any asbestos products or materials that are ready to be disposed. This includes any contaminated building materials, dust, rubble, used tools that cannot be decontaminated, disposable PPE and damp rags that have been used for cleaning. Asbestos waste must be placed in suitable packaging to prevent any fibres being released. This should be double wrapped and appropriately labelled. Standard practice is to use a red inner bag marked up with asbestos warning labels and a clear outer bag with appropriate hazard markings. Intact asbestos cement sheets and textured coatings that are firmly attached to a board should not be broken up into smaller pieces. These should instead be carefully double wrapped in suitable polythene sheeting (1000 gauge) and labelled.

The waste container must be strong enough to securely contain the waste and not become punctured; it must be easily decontaminated, kept securely on the site until required and properly labelled. The waste must only be carried by a licensed carrier.

14

Accidental exposure to ACMs can occur even when all reasonable precautions have been taken. If an ACM has been worked on by a worker who did not realise that it was an ACM or has accidentally damaged an ACM, then all work must stop immediately and nobody should be allowed to enter the area in question. The work supervisor must be informed and a sample of the material sent for analysis. If the result is positive, then a specialist licensed contractor should be employed unless the work is exempt from the need for a licence. Any contaminated clothing must be removed and placed in a plastic bag. The affected worker should shower as soon as possible or wash thoroughly.

More information on the legislative requirements is given in Chapter 19.

14.5.3 Health risks and controls associated with other specific agents

The health hazards associated with hazardous substances can vary from very mild (momentary dizziness or a skin irritation) to very serious, such as a cancer.

Cancer is a serious body cell disorder in which the cells develop into tumours. There are two types of tumour – benign and malignant. Benign tumours do not spread but remain localised within the body and grow slowly. Malignant tumours are called cancers and often grow rapidly, spreading to other organs using the bloodstream and lymphatic system. Survival rates have improved dramatically in recent years as detection methods have improved and the tumours can be found in their early stages of development.

Workplace exposures leading to occupational illness and work-related cancers account for 8,000 deaths, of which half are asbestos related (including mesothelioma). Five of the key causes of occupational cancer registrations and deaths are:

▶ Diesel engine exhaust emissions – related deaths in the UK are estimated at 650 a year to lung or bladder cancer.
▶ Solar radiation – at over 1,500 a year.
▶ Asbestos – as covered earlier.
▶ Shift work – causes an estimated 2,000 cases of breast cancer a year.
▶ Silica – over 600 cancer registrations.

A minority of cancers are believed to be occupational in origin.

Occupational asthma has approximately 4 million sufferers in the UK and it is estimated that 13 million working days are lost each year as a result of it. It is mainly caused by breathing in respiratory sensitisers, such as wood dusts, organic solvents, solder flux fumes or animal hair. The symptoms are coughing, wheezing, tightness of the chest and breathlessness due to a constriction of the airways. It can be a mild attack or a serious one that requires hospitalisation. There is some evidence that stress can trigger an attack.

The following common agents of health hazards will be described together with the circumstances in which they may be found (those marked * are **not** in the NEBOSH National Construction Certificate syllabus).

Ammonia* is a colourless gas with a distinctive odour which, even in small concentrations, causes the eyes to smart and run and a tightening of the chest. It is a corrosive substance which can burn the skin, burn and seriously damage the eye, cause soreness and ulceration of the throat and severe bronchitis and oedema (excess of fluid) of the lungs. Good eye and respiratory protective equipment (RPE) is essential when maintaining equipment containing ammonia. Any such equipment should be tested regularly for leaks and repaired promptly if required. Ammonia is also used in the production of fertilisers and synthetic fibres. Most work on ammonia plant should require a permit-to-work procedure.

Chlorine* is a greenish, toxic gas with a pungent smell which is highly irritant to the respiratory system, producing severe bronchitis and oedema of the lungs and may also cause abdominal pain, nausea and vomiting. It is used as a disinfectant for drinking water and swimming pool water and in the manufacture of chemicals.

Organic solvents are used widely in industry as cleansing and degreasing agents. There are two main groups – the hydrocarbons (includes the aromatic and aliphatic hydrocarbons, such as toluene and white spirit) and the non-hydrocarbons (such as trichloroethylene and carbon tetrachloride). All organic solvents are heavier than air and most are sensitisers and irritants. Some are narcotics, while others can cause dermatitis and after long exposure periods liver and kidney failure. It is very important that the hazard data sheet accompanying the particular solvent is read and the recommended personal protective equipment is worn at all times.

Solvents are used extensively in a wide variety of industries as varnishes, paints, adhesives, glue strippers, printing inks and thinners. They are highest risk when used as sprays. One of the most hazardous is dichloromethane (DCM), also known as methylene chloride. It is used as a paint stripper, normally as a gel. It can produce narcotic effects and has been classified as a Category 3 carcinogen in the European Union. The minimum personal protective equipment requirements are impermeable overalls, apron, footwear, long gloves and gauntlet and chemically resistant goggles or visor. Respiratory protective equipment is also required if it cannot be demonstrated that exposure is below the appropriate workplace exposure limit (WEL).

Carbon dioxide is a colourless and odourless gas which is heavier than air. It represses the respiratory system, eventually causing death by asphyxiation. At low concentrations it will cause headaches and sweating followed by a loss of consciousness. The greatest hazard occurs in confined spaces, particularly where the gas is produced as a by-product.

Nitrogen is a colourless, tasteless and odourless gas but is much more benign than many other gases. The Earth's atmosphere, air, comprises 78% of nitrogen by volume. It is an essential constituent of all plants and animals and is used as an ingredient in the manufacture of agricultural fertiliser. It neither burns nor supports combustion – it is an inert gas. Due to these properties, it is used to pressurise systems, particularly hot water systems. In liquid form, it is used both as a refrigerant and freezing agent (see Chapter 17 for an example in excavation work). Since it has a melting point of −196°C, it must be used very carefully to avoid frostbite.

Carbon monoxide is a colourless, tasteless and odourless gas which makes it impossible to detect without special measuring equipment. As explained earlier, carbon monoxide enters the blood (red cells) more readily than oxygen and restricts the supply of oxygen to vital organs. At low concentrations in the blood (less than 5%), headaches and breathlessness will occur, while at higher concentrations unconsciousness and death will result. The most common occurrence of carbon monoxide is as an exhaust gas either from a vehicle or a heating system. In either case, it results from inefficient combustion and, possibly, poor maintenance.

Carbon monoxide exposure may be prevented by ensuring that any work carried out in relation to gas appliances in domestic or commercial premises is undertaken by a **Gas Safe Registered** engineer, competent in that area of work, and that there is enough fresh air in the room containing your gas appliance. In addition the HSE recommends the use of audible carbon monoxide (CO) alarms as a useful back-up precaution.

All gas workers must be signed up to the **Gas Safe Register**. This register replaces the requirement for all gas workers to be registered with CORGI and the scheme is overseen by the HSE. Consumers and landlords should check whether gas workers are Gas Safe Registered since the failure to check that an installer is registered under the **Gas Safe Scheme** may result in a breach of the Building Regulations.

Employers and landlords should ensure their appliances are serviced regularly by an approved and competent engineer. Registered engineers can be identified by the display of a new yellow Gas Safe Register logo, and are required to carry a Gas Safe Register identification card with their own unique licence number.

There have been cases where gas leaks, electricity and water faults have developed in buildings constructed with timber frames as a result of the frame shrinking, thereby placing excessive weight on these services.

Shrinkage usually only takes place with new wood, and within its first year of use.

Diesel engine exhaust emissions* have been classified as carcinogenic by experts from the World Health Organisation (WHO). The International Agency for Research on Cancer (IARC) found sufficient evidence that exposure to diesel exhausts is associated with an increased risk of lung cancer. They also noted limited evidence that exposure could lead to an increased risk of bladder cancer. The classification follows the findings in a US study of occupational exposure to such emissions in underground miners, which showed an increased risk of death from lung cancer in exposed workers. Exposure to these particulates may also aggravate other respiratory diseases such as asthma.

Hydrogen sulphide is a gas with the odour of rotten eggs, although increasing concentrations reduces the sense of smell. It is commonly found in sewers, sewage treatment plants (hydrogen sulphide is often called sewer gas), coal mines and chemical processes. At low concentrations, irritation of the eyes will be experienced. At higher concentrations, headaches and dizziness may result leading to unconsciousness. Pulmonary oedema resulting from irritation of the lungs may also occur. It is a very poisonous gas at concentrations above 0.1% and is also flammable. The health effects of hydrogen sulphide are dependent on the concentration of the gas (see Table 14.5).

Industrial sources of the gas include petroleum and natural gas extraction and refining, pulp and paper manufacturing, rayon textile production, leather tanning, chemical manufacturing and waste disposal. It is heavier than air and will concentrate at and below ground level particularly in confined spaces such as manholes, tanks, pits and chemical reaction vessels.

Hydrogen sulphide has a workplace exposure limit (WEL) of 5 ppm (LTEL) and 10 ppm (STEL) and exposure should be minimised using adequate engineering controls, such as adequate ventilation, and safe working practices. If the engineering controls are not adequate, then, as a last resort, personal protective equipment, such as supplied-air respirators, may be used.

Table 14.5 The health effects of hydrogen sulphide

LOW	0–10 ppm	Irritation of the eyes, nose and throat
MODERATE	10–50 ppm	Headache, dizziness, nausea and vomiting
		Coughing and breathing difficulty
HIGH	50–200 ppm	Severe respiratory tract irritation
		Eye irritation/acute conjunctivitis
		Shock
		Convulsions
		Coma
		Death in severe cases

Isocyanates are volatile organic compounds widely used in industry for products such as printing inks, adhesives and two-pack paints (particularly in vehicle body shops) and in the manufacture of plastics (polyurethane products). They are irritants and sensitisers. Inflammation of the nasal passages and the throat and bronchitis are typical reactions to many isocyanates. When a person becomes sensitised to an isocyanate, very small amounts of the substance often provoke a serious reaction similar to an extreme asthma attack. Isocyanates also present a health hazard to fire-fighters. They are subject to a workplace exposure limit (WEL) and respiratory protective equipment (RPE) should be worn. Two-pack polyurethane paints and varnishes are used in construction as surface coatings.

Lead is a heavy, soft and easily worked metal. It is used in many industries but is most commonly associated with plumbing and roofing work. Over 300,000 tonnes of lead are used annually in the UK of which 80% is used in the manufacture of lead-acid batteries and 6% is extruded into sheets for use in construction. The toxicity of lead affects nerve conduction and red blood cell production. Symptoms include abdominal pains, muscular weakness and tiredness. Inhalation is the most common means for the entry of lead into the body. From the lungs, lead enters the bloodstream and is distributed throughout the body.

Lead enters the body normally by inhalation but can also enter by ingestion and skin contact. The main targets for lead are the spinal cord and the brain, the blood and blood production. The effects are normally chronic and develop as the quantity of lead builds up. Headaches and nausea are the early symptoms followed by anaemia, muscle weakening and (eventually) coma. Regular blood tests are a legal and sensible requirement as are good ventilation and the use of appropriate personal protective equipment. High personal hygiene standards and adequate welfare (washing) facilities are essential and must be used before smoking or food is consumed. The reduction in the use of leaded petrol was an acknowledgement of the health hazard represented by lead in the air. Lead is covered by its own set of regulations – the Control of Lead at Work Regulations (summarised in Chapter 19). These Regulations require risk assessments to be undertaken and engineering controls to be in place. They also recognise that lead can be transferred to an unborn child through the placenta and, therefore, offer additional protection to women of reproductive capacity. Medical surveillance, in the form of a blood test of all employees who come into contact with lead operations, is required by the Regulations. Such tests should take place at least once a year. Regular health surveillance of lead workers by a medical practitioner is required by these Regulations. Lead is used in construction by roofers as a roofing and guttering material and it is also used (to a lesser extent these days) by plumbers.

Silica is the main component of most rocks and is a crystalline substance made of silicon and oxygen. It occurs in quartz (found in granite), sand and flint which are present in a wide variety of construction materials. Harm is caused by the inhalation of silica dust, which can lead to silicosis (acute and chronic), fibrosis and pneumoconiosis.

The dust which causes the most harm is respirable dust which becomes trapped in the alveoli. This type of dust is sharp and very hard and, probably, causes wounding and scarring of lung tissue. As silicosis develops, breathing becomes more and more difficult and eventually, as it reaches its advanced stage, lung and heart failure occur. It has also been noted that silicosis can result in the development of tuberculosis as a further complication. Hard rock miners, quarrymen, and stone and pottery workers are most at risk. Health surveillance is recommended for workers in these occupations at initial employment and at subsequent regular intervals. Prevention is best achieved by the use of good dust extraction systems and respiratory protective equipment. Over 100 enforcement notices were served in 2009/10 on silica-dust related issues compared to 13 in 2008/09.

Silica is commonly produced during construction activities. Such activities which can expose workers and members of the public to silica dust include:

- cutting building blocks and other stone masonry work;
- cutting and/or drilling paving slabs and concrete paths;
- demolition work;
- sand blasting of buildings;
- tunnelling.

In general, the use of power tools to cut or dress stone and other silica-containing materials will lead to very high exposure levels while the work is occurring. In most cases, exposure levels are in excess of workplace exposure limits (WELs) by factors greater than 2 and in some cases as high as 13. In response to the growing evidence of ill-health effects of silica inhalation, the HSE has recently revised the WEL down from 0.3 mg/m^3 to 0.1 mg/m^3.

The use of compressed air to remove dust is not recommended. There is a risk of compressed air entering the operator's bloodstream, which can result in death. Eye injury including blindness can also occur if dust particles bounce back at the operator. The inhalation of very fine silica dust can lead to the development of silicosis. Many construction tasks create dust. High dust levels are caused by one of more the following:

- Equipment – using high energy tools, such as cut-off saws, grinders, wall chasers and grit blasters, produces a lot of dust in a very short time.
- Work method – dry sweeping can make a lot of dust when compared to vacuuming or wet brushing.

▶ Work area – the more enclosed a space, the more the dust will build up.

▶ Time – the longer you work the more dust there will be.

Dust levels are best controlled either by the use of water or an industrial vacuum unit. The HSE has produced a detailed information sheet on silica – CIS No 36 (Rev 1).

In addition to silica, there are three hazardous substances that are particularly relevant to construction activities – cement dust and wet cement, wood dust and the biological hazard tetanus. Respiratory protective equipment (RPE), usually in the form of a mask, will normally be needed for high-risk tasks such as:

▶ using a cut-off saw, grinder or wall chaser on material containing silica;

▶ using powered cross-cut saws and sanders on hardwood, red cedar or MDF;

▶ sanding softwood in an enclosed space.

A filtering face piece FFP3 is the most advisable type of mask to use when high dust levels, silica or wood dust is produced. There are also filtering face pieces FFP1 and FFP2 available but these provide less respiratory protection than a properly fitting FFP3 device.

Cement dust and wet cement is important in construction and is also a hazardous substance. Contact with wet cement can cause serious burns or ulcers which will take several months to heal and may need a skin graft. Dermatitis, both irritant and allergic, can be caused by skin contact with either wet cement or cement powder. Allergic dermatitis is caused by an allergic reaction to hexavalent chromium (chromate) which is present in cement. Cement powder can also cause inflammation and irritation of the eye, irritation of the nose and throat, and, possibly, chronic lung problems. Research has shown that between 5% and 10% of construction workers are probably allergic to cement. And plasterers, concreters and bricklayers or masons are particularly at risk. A plasterer who knelt in wet cement for 5 hours while working required skin grafts to his legs.

Figure 14.19 Dermatitis from wet cement or concrete

Manual handling of wet cement or cement bags can lead to musculoskeletal health problems and cement bags weighing more than 25 kg should not be carried by a single worker. PPE in the form of gloves, overalls with long sleeves and full-length trousers and waterproof boots must be worn on all occasions. If the atmosphere is dusty, goggles and respiratory protection equipment must be worn. An important factor in the possibility of dermatitis occurring is the sensitivity of the worker to the chromate in the cement and the existing condition of the skin including cuts and abrasions. Finally, adequate welfare facilities are essential so that workers can wash their hands at the end of the job and before eating, drinking or using the toilet. If cement is left on the skin for long periods without being washed off, the risk of an allergic reaction to hexavalent chromium will increase.

A 2005 amendment to the COSHH Regulations prohibits the supply of cement which has a concentration of more than 2 parts per million of chromium VI. This measure is designed to prevent allergic contact dermatitis when wet cement comes into contact with the skin. However, since the strong alkalinity of cement will remain, there is still the potential for skin burns.

The Approved Code of Practice (ACoP) gives useful advice on possible control measures by offering two routes for employers to comply with the amended COSHH Regulations. They can either use the generic advice given in COSHH Essentials HSG193 or design a solution themselves with the help of competent advice. In any event, the controls should be proportionate to the health risk. The Regulations stress the need to provide adequate washing, changing, eating and drinking facilities. Details of these welfare facilities are given in Chapter 8.

14

Wood dust can be hazardous, particularly when it is hardwood dust, which is known, in rare cases, to lead to nasal cancer. Composite boards, such as **medium-density fibreboard (MDF)**, are hazardous due to the resin bonding material used, which can also be carcinogenic. There are three types of wood-based boards available: laminated board, particle board and fibreboard. The resins used to bond the fibreboard together contain formaldehyde (usually urea formaldehyde). It is generally recognised that formaldehyde is 'probably carcinogenic to humans' and is subject to a WEL. At low exposure levels, it can cause irritation to the eyes, nose and throat and can lead to dermatitis, asthma and rhinitis. The main problems are most likely to occur when the MDF is being machined and dust is produced. A suitable risk assessment should be made and gloves and appropriate masks should always be worn when machining MDF. However, it is important to stress that safer materials are available which do not contain formaldehyde and these should be considered for use in the first instance.

Wood dust is produced whenever wood materials are machined, particularly sawed, sanded, or bagged as dust from dust extraction units or during cleaning operations, especially if compressed air is used. The main hazards associated with all wood dusts are skin disorders, nasal problems, such as rhinitis, and asthma. There is also a hazard from fire and explosion. A COSHH assessment is essential to show whether the particular wood dust is hazardous. When the wood dust is created inside a woodworking shop, a well-designed extraction system is essential. PPE in the form of gloves, suitable respiratory protective equipment, overalls and eye protection may also be necessary as a result of the assessment. Other substances associated with woodworking which present hazards include:

▶ wood paints, varnishes, stains and preservatives;
▶ stripping chemicals;
▶ various solvents; and
▶ adhesives.

Finally, good washing and welfare facilities are also essential.

Tetanus is a serious, sometimes fatal, disease caused by a bacterium that lives in the soil. It usually enters the human body through a wound from an infected object, such as a nail, wood splinter or thorn. On entering the wound, it produces a powerful toxin which attacks the nerves that supply muscle tissue. It is commonly known as lockjaw because, after an incubation period of approximately a week, stiffness around the jaw area occurs. Later the disease spreads to other muscles including the breathing system and this can be fatal. The disease has been well controlled with anti-tetanus immunisation and it is important that all construction workers are so immunised. Booster shots should be obtained every few years. Any flesh wound should be thoroughly cleaned immediately, an anti-septic cream applied and the wound covered.

Leptospirosis and Weil's disease are caused by a bacterium found in the urine of rats. In humans, the kidneys and liver are attacked causing high temperatures and headaches followed by jaundice and, in up to 20% of cases, it can be fatal. It enters the body either through the skin or by ingestion. The most common source is contaminated water in a river, sewer or ditch and workers, such as canal or sewer workers, are most at risk. Leptospirosis is always a risk where rats are present, particularly if the associated environment is damp. Good, impervious protective clothing, particularly wellington boots, is essential in these situations and the covering of any skin wounds. For workers who are frequently in high-risk environments (sewer workers), immunisation with a vaccine may be the best protection. Weil's disease is, strictly, a severe form of leptospirosis. The symptoms of leptospirosis are similar to influenza but those for Weil's disease are anaemia, nose bleeds and jaundice. While the most common source of infection is from the urine of rats, Weil's

disease has been found in other animals, such as cattle; therefore, farm and veterinary workers may also be at risk.

Legionella is an airborne bacterium and is found in a variety of water sources. It produces a form of pneumonia caused by the bacteria penetrating to the alveoli in the lungs. This disease is known as Legionnaires' disease, named after the first documented outbreak at a State Convention of the American Legion held at Pennsylvania in 1976. During this outbreak, 200 men were affected, of whom 29 died. That outbreak and many subsequent ones were attributed to air-conditioning systems. Legionnaires' disease is a potentially fatal form of pneumonia and everyone is susceptible to infection. However, some people are at higher risk, including:

▶ people over 45 years of age;
▶ smokers and heavy drinkers;
▶ people suffering from chronic respiratory or kidney disease;
▶ anyone with an impaired immune system.

The symptoms are a high temperature, fever and chills, cough, muscle pains and headache. In a severe case, there may also be pneumonia, and occasionally diarrhoea, as well as signs of mental confusion. Legionnaires' disease is not known to spread from person to person.

The legionella bacterium cannot survive at temperatures above 60°C but grows between 20°C and 45°C, being most virulent at 37°C. It also requires food in the form of algae and other bacteria. Control of the bacteria involves the avoidance of water temperatures between 20°C and 45°C, avoidance of water stagnation and the build-up of algae and sediments and the use of suitable water-treatment chemicals. This work is often done by a specialist contractor.

The most common systems at risk from the bacterium are:

▶ water systems incorporating a cooling tower;
▶ water systems incorporating an evaporative condenser;
▶ hot and cold water systems and other plant where the water temperature may exceed 20°C.

An Approved Code of Practice (*Legionnaires' disease – The control of legionella bacteria in water systems –* L8) is available from the HSE. The latest edition of the ACoP and guidance contains revisions to simplify and clarify the text. The main changes are removing Part 2, the technical guidance, which is published separately in HSG274, and giving the following issues ACoP status:

▶ risk assessment;
▶ the specific role of an appointed competent person, known as the 'responsible person';
▶ the control scheme;
▶ review of control measures;

▶ duties and responsibilities of those involved in the supply of water systems.

Where plant at risk of the development of legionella exists, the following is required:

▶ a written 'suitable and sufficient' risk assessment;
▶ the preparation and implementation of a written control scheme involving the treatment, cleaning and maintenance of the system;
▶ appointment of a named person with responsibility for the management of the control scheme;
▶ the monitoring of the system by a competent person;
▶ record keeping and the review of procedures developed within the control scheme.

The code of practice also covers the design and construction of hot and cold water systems and cleaning and disinfection guidance. There have been several cases of members of the public becoming infected from a contaminated cooling tower situated on the roof of a building. It is required that all cooling towers are registered with the local authority. People are more susceptible to the disease if they are older or weakened by some other illness. It is, therefore, important that residential and nursing homes and hospitals are particularly vigilant. The most common source of isolated outbreaks of legionella is showerheads, particularly when they remain unused for a period of time. Showerheads should be cleaned and descaled at least every three months.

All warm water systems should be checked annually by a competent person. However, there are some simple checks that can be made as follows:

▶ Each month, check that the temperature of water from hot water taps reaches 50°C after the water has run for 1 minute.
▶ Each month, check that the temperature of water from cold water taps is below 20°C after the water has run for 2 minutes.
▶ Each quarter, clean and descale shower heads and hoses.
▶ Each week, purge little-used water outlets.

Hepatitis is a disease of the liver and can cause high temperatures, nausea and jaundice. It can be caused by hazardous substances (some organic solvents) or by a virus. The virus can be transmitted from infected faeces (hepatitis A) or by infected blood (hepatitis B and C). The normal precautions include good personal hygiene, particularly when handling food and in the use of blood products. Hospital workers who come into contact with blood products are at risk of hepatitis as are drug addicts who share needles. It is also important that workers at risk regularly wash their hands and wear protective disposable gloves.

Hepatitis is an example of a **blood-borne virus**. Another example is human immunodeficiency virus (HIV) which causes acquired immune deficiency

syndrome (AIDS), affecting the immune system of the body. Some people carry a blood-borne virus in their blood which may cause disease in some people and few or no symptoms in others. The virus can spread to another person, whether the carrier of the virus is ill or not.

In the workplace, direct exposure can happen through accidental contamination by a sharp instrument or through contamination of open wounds or skin abrasions or through splashes to the eyes, nose or mouth. In occupations where there is a risk of exposure to blood-borne viruses, the following measures to prevent or control risks are recommended by the HSE:

▶ Prohibit eating, drinking, smoking and the application of cosmetics in working areas where there is a risk of contamination.
▶ Prevent puncture wounds, cuts and abrasions, especially in the presence of blood and body fluids.
▶ When possible avoid use of, or exposure to, sharps such as needles, glass or metal or if unavoidable take care in handling and disposal.
▶ Consider the use of devices incorporating safety features, such as safer needle devices and blunt-ended scissors.
▶ Cover all breaks in exposed skin by using waterproof dressings and suitable gloves.
▶ Protect the eyes and mouth by using a visor/goggles/safety spectacles and a mask, where splashing is possible.
▶ Avoid contamination by using water resistant protective clothing.
▶ Wear rubber boots or plastic disposable overshoes when the floor or ground is likely to be contaminated.
▶ Use good basic hygiene practices, such as hand washing.
▶ Control contamination of surfaces by containment and using appropriate decontamination procedures.
▶ Dispose of contaminated waste safely.

The risk of first-aiders being infected with a blood-borne virus while carrying out their duties is small. There has been no recorded case of HIV or hepatitis B being passed on during mouth-to-mouth resuscitation. The following precautions are recommended by the HSE to reduce the risk of infection:

▶ Cover any cuts or grazes on your skin with a waterproof dressing.
▶ Wear suitable disposable gloves when dealing with blood or any other body fluids.
▶ Use suitable eye protection and a disposable plastic apron where splashing is possible.
▶ Use devices such as face shields when giving mouth-to-mouth resuscitation, but only if training in their use has been received.
▶ Wash hands after each procedure.

14

14.6 Safe handling and storage of waste

Work-related deaths in the waste management industry are ten times the national average. Accident rates in the industry are four times the national average. The statutory duty of care for the management of waste derives from Part 2 of the Environmental Protection Act (EPA). The principal requirements are as follows:

▶ to handle waste so as to prevent any unauthorised escape into the environment;
▶ to pass waste only to an authorised person as defined by EPA;
▶ to ensure that a written description accompanies all waste. The Environmental Protection (Duty of Care) Regulations 1991 requires holders or producers of waste to complete a 'Transfer Note' giving full details of the type and quantity of waste for collection and disposal. Copies of the note should be kept for at least 2 years;
▶ to ensure that no person commits an offence under the Act.

The EPA is concerned with controlled waste. Controlled waste comprises household, industrial or commercial waste. It is a criminal offence to deposit controlled waste without an environmental permit (see Chapter 19) and/or in a manner likely to cause environmental pollution or harm to human health.

The EPA also covers **hazardous waste and spillage.** Hazardous waste can only be disposed of using special arrangements. There are two categories of hazardous waste: absolute hazardous waste, such as hydrochloric acid, fuel, oil or diesel; or mirror hazardous waste, such as sawdust, shavings, cuttings, wood particle board and veneer containing dangerous substances. If waste does fall into these categories, it needs to be handled, stored, transported and disposed of appropriately. The EPA imposes a legal duty of care to ensure that waste is not illegally disposed of, or dealt with without an Environmental Permit or Registered Exemption (see Chapter 19 for more information on these topics).

The waste must be stored securely and safely in sealed waterproof containers with clear labelling and written instructions for storage and disposal to ensure that the environment is protected. It must be regularly checked for leaks, deteriorating containers, or other potential risks. It is essential to carry out a risk assessment to identify the risks associated with handling and storing hazardous waste.

These substances are sometimes life threatening (toxic, corrosive or carcinogenic) or highly flammable. Clinical waste falls within this category. A consignment note system accompanies this waste at all the stages to its final destination. Before hazardous waste is removed from the originating premises, a contract should be in place with a licensed carrier.

Figure 14.20 Removing waste from a roadside excavation by lorry-mounted loading grab

The Hazardous Waste Regulations, with the exclusion of Scotland, replace the Special Waste Regulations and cover many more substances; for example, computer monitors, fluorescent tubes, end-of-life vehicles and television sets. Hazardous waste is waste which can cause damage to the environment or to human health. (The hazardous properties of waste covered by these Regulations are listed in Appendix 14.3.) Producers of such waste may need to notify the Environment Agency. The Regulations seek to ensure that hazardous waste is safely managed and its movement is documented. The following points are important for construction sites:

▶ Sites that produce more than 200 kg of hazardous waste each year for removal, treatment or disposal must register with the Environment Agency.
▶ Different types of hazardous waste must not be mixed.
▶ Producers must maintain registers of their hazardous wastes.

14.6.1 Waste disposal

The collection and removal of waste from a workplace is normally accomplished using a skip. Every year, activities involving the movement of skips and containers cause death and serious injury. The hazards include:

▶ being struck by vehicles;
▶ falling and slipping;
▶ failures of lifting equipment;
▶ striking overhead cables/obstructions;
▶ vehicle overturns;
▶ runaway vehicles.

The skip should be located on firm, level ground away from the main construction work, particularly excavation work. This will allow clear access to the skip for filling and removal from site. On arrival at site, the integrity of the skip should be checked. It should be filled either by chute or by mechanical means unless items can be placed in by hand. Skips should not be overfilled and should be netted or sheeted over when they are full. Any hazardous waste should be segregated from controlled waste.

Figure 14.21 A designated waste collection area with two types of skip commonly used for waste collection. Heavy materials would be transported in the smaller skip. Sizes of skip range from about 4 cu metres (small skip shown) to about 35 cu metres (large skip shown)

Waste skip selection should be made during the site planning process as discussed in Chapter 8. The selected skip must be suitable for the particular job. The following points should be considered:

- sufficient strength to cope with its load;
- stability while being filled;
- a reasonable uniform load distribution within the skip at all times;
- the immediate removal of any damaged skip from service and the skip inspected after repair before it is used again;
- sufficient space around the skip to work safely at all times;
- the skip should be resting on firm level ground;
- the skip should never be overloaded or overfilled;
- there must be sufficient headroom for the safe removal of the skip when it is filled.

There are hazards present during the movement of a loaded skip from the ground to the back of a skip loader vehicle. Entanglement with the vehicle lifting mechanisms, such as the hydraulic arms and lifting chains, is a major hazard. Other hazards include contact by the skip with overhead obstructions, movement of the skip contents and skip overload leading to mechanical or structural failure. Slip hazards may be present due to spillages from the skip and the skip contents could be contaminated with biological material, asbestos or syringes. Passing traffic during the loading operation may also present a hazard.

The control measures for these hazards include the use of outriggers to increase the stability of the loader vehicle and the provision of steps for the driver to alight from the cab or the vehicle flatbed. The contents of the skip should be secured using netting or tarpaulin. Adhering to the safe working loads of the skip and lifting equipment and the use of a banksman during the lifting process are additional controls. The area around the vehicle may need to be cordoned off to protect passing pedestrians and road traffic. All workers concerned with the operation should wear suitable PPE, such as high visibility jackets, gloves and suitable footwear. Finally, all lifting equipment, including chains and shackles, must be subject to a periodic statutory examination.

14.6.2 Waste management

Some form of training may be required to ensure that employees segregate hazardous and non-hazardous wastes on site and fully understand the risks and necessary safety precautions which must be taken. Personal protective equipment, including overalls, gloves and eye protection, must be provided and used. The storage site should be protected against trespassers, fire and adverse weather conditions. If flammable or combustible wastes are being stored, adequate fire protection systems must be in place. Finally, in the case of liquid wastes, any drains must be protected and bunds used to restrict spreading of the substance as a result of spills.

A hierarchy for the management of waste streams has been recommended by the Environment Agency.

1. **Prevention** – by changing the process so that the waste is not produced (e.g. substitution of a particular material).
2. **Reduction** – by improving the efficiency of the process (e.g. better machine maintenance).
3. **Reuse** – by recycling the waste back into the process (e.g. using reground waste plastic products as a feed for new products).
4. **Recovery** – by releasing energy through the combustion, recycling or composting of waste (e.g. the incineration of combustible waste to heat a building).
5. **Responsible disposal** – in accordance with regulatory requirements.

14.6.3 Other waste issues

In 1998, land disposal accounted for approximately 58% of waste disposal; 26% was recycled and the remainder was incinerated with some of the energy recovered as heat. The Producer Responsibility Obligations (Packaging Waste) Regulations 1997 placed legal obligations on employers to reduce their packaging waste by either recycling or recovery as energy (normally as heat from an incinerator attached to a district heating system). A series of targets have been stipulated which will reduce the amount of waste progressively over the years. These Regulations are

14

enforced by the Environment Agency, which has powers of prosecution in the event of non-compliance. The UK recycled almost two-thirds of all packaging produced in 2008.

The Landfill (England and Wales) Regulations 2002 are part of the government's drive to encourage recycling and reduce the amount of rubbish sent to landfill sites, and this means that companies now have to either recycle or treat their waste before it is taken to landfill. The rules have already come into force in Scotland.

The main changes are that liquid wastes are now banned from landfill and other waste must be treated before it can be passed into landfill. Businesses now need to demonstrate that their waste has been treated in either a physical, thermal, chemical or biological process. One major environmental problem is that of contaminated land. Contaminated land is defined in the Environmental Protection Act and is produced by leakage, accidental spillage and uncontrolled waste disposal. The contaminator of the land has the primary responsibility and liability to clean up the contamination. If the contaminator is unknown, the current owner of the land has the responsibility. If no current owner can be found then the local authority becomes responsible for the remediation.

There is more information on the EPA, the Integrated Pollution Prevention and Control (IPPC) and waste disposal in Chapter 19. The Environment Agency has similar powers following the introduction of the Waste Electrical and Electronic Equipment (WEEE) Directive, the Restrictions of the use of certain Hazardous Substances in electrical and electronic equipment (RoHS) and the End of Life Vehicle (ELV) Directive. The aim of the WEEE Directive is to minimise the environmental impact of electrical and electronic equipment both during their lifetime and when they are discarded. All electrical and electronic equipment must be returned to the retailer from its end user and reused or reprocessed by the manufacturer. Manufacturers must register with the Environment Agency, who will advise on these obligations. The WEEE Regulations have been amended so that producer-compliance schemes report their activities in a more precise way (by providing evidence of recycling in kilograms instead of tonnes). It is hoped that this will reduce the delays in recycling and lead to the recycling of smaller amounts at more frequent intervals. The European Commission has recently proposed changes to both WEEE and RoHS. On WEEE, a major proposal is to increase the amounts of electrical and electronic waste that are separately collected and recycled, while the proposals on RoHS aim for a higher level of environmental protection by revising the scope of the restrictions and the applicable substances.

There is a useful HSE waste management and recycling website available at www.hse.gov.uk/waste/index.htm

Figure 14.22 Electronic waste under WEEE

14.7 Further information

EC Regulation No. 1272/2008 Classification Labelling and Packaging of Substances and Mixtures (CLP) http://www.hse.gov.uk/chemical-classification/legal/clp-regulation.htm

Globally Harmonised System of Classification and Labelling of Chemicals (GHS) http://www.unece.org/fileadmin/DAM/trans/danger/publi/ghs/ghs_rev04/English/ST-SG-AC10-30-Rev4e.pdf

The Control of Asbestos Regulations 2012

The Control of Substances Hazardous to Health Regulations 2002 (as amended)

The Personal Protective Equipment at Work Regulations 1992

Controlling Airborne Contaminants, HSG258, HSE Books, ISBN: 978-0-7176-6415-3 http://www.hse.gov.uk/pubns/priced/hsg258.pdf

Asbestos essentials task sheets http://www.hse.gov.uk/asbestos/essentials/index.htm

Control of substances hazardous to health, ACoP, L5, HSE Books, ISBN 978-0-7176-6582-2 http://www.hse.gov.uk/pubns/priced/l5.pdf

Personal Protective Equipment at Work (Guidance), L25, HSE Books, ISBN 978-0-7176-6139-3 http://www.hse.gov.uk/pubns/priced/l25.pdf

Respiratory protective equipment at work – A practical guide, HSG53, HSE Books, ISBN 978-0-7176-6454-2 http://www.hse.gov.uk/Pubns/priced/hsg53.pdf

Step by step guide to COSHH assessment, HSG97, HSE Books, ISBN 978-0-7176-2785-1 http://www.hse.gov.uk/pubns/priced/hsg97.pdf

EH40/2005 Workplace exposure limits, HSE Books, ISBN 978-0-7176-6446-7 http://www.hse.gov.uk/pubns/priced/eh40.pdf

Managing asbestos in buildings, a brief guide, INDG223, HSE Books, http://www.hse.gov.uk/pubns/indg223.pdf

Managing and working with asbestos, Control of Asbestos Regulations 2012, ACoP and guidance, L143, HSE Books, ISBN 978-0-7176-6618-8 http://www.hse.gov.uk/pubns/priced/l143.pdf

Asbestos: The Survey Guide, HSG264, HSE Books, ISBN 978-0-7176-6502-0 http://www.hse.gov.uk/pubns/priced/hsg264.pd

14.8 Practice revision questions

1. Health hazards may be present in the workplace through various chemical and biological agents.

    (a) **Describe FOUR** chemical and **THREE** biological hazardous agents and **give** an example of each agent.

    (b) **Explain** the term 'respirable dust' and **outline** its health effect on the human body.

2. For each of the following types of hazardous substance, **give** a typical example and **outline** its primary effect on the human body:

    (a) irritant
    (b) corrosive
    (c) toxic
    (d) carcinogenic
    (e) mutagenic.

3. (a) **Describe** the differences between acute and chronic health effects giving an example in each effect.

    (b) **Identify TWO** types of cellular defence mechanisms that the body has as a natural defence system.

    (c) **Identify** the factors that could affect the level of harm experienced by a worker who has been exposed to a toxic substance.

4. (a) **Identify FOUR** possible routes of entry of a hazardous substance into the body and, in **EACH** case, **give** an example of how an employee might be at risk of such exposure.

    (b) **Define** the term 'target organ' within the context of occupational health.

    (c) **Outline** the personal hygiene practices that should be followed to reduce the risk of ingestion of a hazardous substance.

5. (a) **Identify** possible routes of entry of biological organisms into the body.

    (b) **Outline** control measures that could be used to reduce the risk of infection from biological organisms.

6. (a) **Identify FOUR** respiratory diseases that could be caused by exposure to dust at work.

    (b) **Describe** the respiratory defence mechanisms of the body against atmospheric dust.

7. (a) **Describe** the typical symptoms of occupational dermatitis.

    (b) **Identify TWO** common causative agents.

8. (c) **Outline** control measures which may be taken to prevent the occurrence of occupational dermatitis.

9. (a) **Outline** the issues to be considered when undertaking an assessment of health risks from hazardous substances that may be used in a workplace.

    (b) **Identify** the information that should be included on a manufacturer's safety data sheet supplied with a hazardous substance used on a construction site.

    (c) **Outline FOUR** sources of information that might be consulted when assessing the risk of a hazardous substance in a workplace.

9. (a) **Identify** the main advantages and limitations of chemical indicator (stain detector) tubes.

    (b) **Explain** how a dust lamp can give an indication of airborne dust levels in a workplace.

    (c) **Outline** how a personal dust sampler is used to measure the levels of airborne dust in a workplace.

10. (a) **Outline** the main purposes of occupational exposure limits (OELs).

    (b) **Explain** the meaning of the following terms:
       (i)   the short-term exposure limit (STEL)
       (ii) the long-term exposure limit (LTEL)
       (iii) the maximum permissible concentration.

    (c) **Outline** the limitations of occupational exposure limits.

11. (a) **Explain** the meaning of the term 'workplace exposure limit' (WEL).

    (b) **Identify** the two classifications of hazardous substances that require exposure to be reduced to as low a level as is reasonably practicable below the WEL.

    (c) **Outline FOUR** actions that could be taken when a WEL has been exceeded.

12. (a) **Outline** the Principles of Good Practice that will help to prevent ill-health due to exposure of workers to hazardous substances.

    (b) **Outline** a hierarchy of control measures that could be used to implement the Principles of Good Practice for the control of exposure to hazardous substances.

14

433

13. A local exhaust ventilation (LEV) system is to be installed to extract welding fume from an engineering workshop.

    (a) **Identify** the five basic components of the LEV system.

    (b) **Outline** the issues that might reduce the effectiveness of the LEV system.

    (c) **Outline** the routine maintenance that should be carried out between statutory examinations so that the ventilation system continues to operate effectively.

14. (a) **Explain** the meaning of the term 'dilution ventilation'.

    (b) **Identify THREE** instances when dilution ventilation may be used in a workplace.

    (c) **Identify** the limitations to the use of dilute ventilation.

15. **Identify** the possible health effects on workers and **identify FOUR** control measures to minimise those health effects for the following hazardous substances:

    (a) solvent-based adhesives and paints

    (b) wood dust

    (c) glass fibres

    (d) isocyanates.

16. (a) **Outline** the issues that should be considered when selecting and using personal protective equipment.

    (b) **Identify** the hazards against which the following items of personal protective equipment should offer protection and **outline** the practical limitations of each item:
        (i) safety gloves
        (ii) safety footwear
        (iii) safety goggles
        (iv) safety spectacles.

17. (a) **Outline** the health and safety hazards associated with welding operations.

    (b) **Outline FOUR** types of respiratory protective equipment (RPE) that can be used in construction activities to protect against hazards such as dusts, vapours and gases.

    (c) **Outline** the factors to be considered in the selection of RPE for persons undertaking welding activities.

    (d) **Identify** the difference between breathing apparatus and respirators.

18. **Outline** the types of personal protective equipment required when:

    (a) cutting and shaping medium-density fibreboard using an electric saw in a construction site workshop

    (b) cutting up very old painted metalwork using gas cutting equipment.

19. Health surveillance is used as a check to ensure workplace control measures are working effectively.

    (a) **Identify** the particular health issues that should be addressed during routine health surveillance examinations for construction workers and, for each issue identified, give a relevant hazard.

    (b) **Outline** the purpose of health surveillance of workers who may have come into contact with hazardous substances.

    (c) **Give THREE** examples of health surveillance in the workplace.

20. (a) **Identify TWO** respiratory diseases that may be caused by exposure to asbestos.

    (b) **Identify** the **THREE** types of asbestos commonly found in buildings and **identify** the common sources of asbestos in buildings.

    (c) **Outline** the control measures required before and during the removal of asbestos from a building.

21. The refurbishment of a property requires the renewal of the existing heating system and hot water supply. During this work some unidentified pipe insulation material is found. This material has not been disturbed and appears to be in good condition. **Identify** the actions that should be taken to reduce the risk to the health of the people on the site.

22. For **EACH** of the following agents, **outline** the principal health effects **AND identify** a typical workplace situation in which a person might be exposed:

    (a) carbon monoxide

    (b) legionella bacteria

    (c) leptospira

    (d) blood-borne viruses.

23. **Identify** the possible health hazards to which construction workers may be exposed from the following substances. In each case **give** an example of a likely source and suitable control measures:

    (a) silica dust

    (b) wet cement.

24. A carpenter is employed on a long-term refurbishment project which involves the restoration of hardwood windows, doors and flooring.

    (a) **Identify TWO** possible health effects that could arise through the use of a hand-held circular saw by the carpenter.

(b) **Outline** the issues that the carpenter should consider when carrying out the work.

25. A toxic substance has been transported by lorry to a warehouse in a number of drums. At the warehouse, one of the drums is ruptured during unloading and some of the contents are spilt on the floor.

   (a) **Outline** the main controls that should be in place to ensure the safe transport of the hazardous substance by road.

   (b) **Outline** a procedure that would deal safely with the spillage.

   (c) **Identify THREE** ways in which persons working near the spillage might be harmed.

26. A company produces a range of solid and liquid wastes, both hazardous and non-hazardous.

(a) **Identify** the main hazards associated with the storage of wastes.

(b) **Outline** the required arrangements to ensure the safe storage of the wastes prior to their collection and disposal.

(c) **Outline** the issues that should be addressed by the company when developing a system for the safe collection and disposal of waste.

27. Skips are used to store solid waste materials on a construction site prior to collection and disposal.

   (a) **Identify** the hazards a skip collector could be exposed to when moving a full skip from the ground onto the back of a skip loader vehicle.

   (b) **Identify EIGHT** safe practices to be followed when using a skip for the collection and removal of waste from a construction site.

APPENDIX 14.1 GHS hazard (H) statements (Health only)

H-statement	Phrase
300	**Fatal** if **swallowed**.
301	**Toxic** if **swallowed**.
302	**Harmful** if **swallowed**.
303	**May be harmful** if **swallowed**.
304	**May be fatal** if **swallowed** and **enters airways**.
305	**May be harmful** if **swallowed** and **enters airways**.
310	**Fatal** in contact with skin.
311	**Toxic** in contact with skin.
312	**Harmful** in contact with skin.
313	**May be harmful** in contact with skin.
314	Causes **severe burns** and **eye damage**.
315	Causes skin irritation.
316	Causes mild skin irritation.
317	May cause an allergic skin reaction.
318	Causes serious **eye damage**.
319	Causes serious **eye irritation**.
320	Causes **eye irritation**.
330	**Fatal** if **inhaled**.
331	**Toxic** if **inhaled**.
332	**Harmful** if **inhaled**.
333	**May be harmful** if **inhaled**.
334	May cause allergy or asthma symptoms or breathing difficulties if inhaled.
335	May cause **respiratory irritation**.
336	May cause **dizziness** or **drowsiness**.
340	May cause **genetic defects** *(route if relevant)*.
341	Suspected of causing **genetic defects** *(route if relevant)*.
350	May **cause cancer** *(route if relevant)*.
351	Suspected of **causing cancer** *(route if relevant)*.
360	May **damage fertility** or the **unborn child** *(effect if known, route if relevant)*.
361	Suspected of **damaging fertility** or the unborn child *(effect if known, route if relevant)*.
362	May cause harm to breast-fed children.
370	Causes **damage to organs** *(organ if known, route if relevant)*.
371	May cause **damage to organs** *(organ if known, route if relevant)*.
372	Causes **damage to organs** through prolonged or repeated exposure *(organ if known, route if relevant)*.
373	May cause **damage to organs** through prolonged or repeated exposure *(organ if known, route if relevant)*.
EU66	Repeated exposure may cause skin dryness or cracking.
EU70	**Toxic** by **eye contact**.
EU71	**Corrosive** to the **respiratory tract**.

APPENDIX 14.2 Health questionnaire for ongoing surveillance of persons working with respiratory sensitisers

To be completed by the responsible person

Employee's name: Works no:

The questionnaire should be completed 6 weeks, 6 months and annually after employment commences or as advised by the company occupational health adviser.

Further advice will be required from the company occupational health adviser if any Yes box is ticked.

Since starting your present job have you had any of the following symptoms either at work or at home? (Do not include isolated colds, sore throats or flu.):

(a) Recurring soreness of or watering of eyes	Yes	No
(b) Recurring blocked or running nose	Yes	No
(c) Bouts of coughing	Yes	No
(d) Chest tightness	Yes	No
(e) Wheeze	Yes	No
(f) Breathlessness	Yes	No
(g) Have you consulted your doctor about chest problems since the last questionnaire?	Yes	No

To be completed by the responsible person
(a) No further action required.
(b) Refer to company occupational health adviser.

Signature of responsible person: Date:

I confirm that the responses given by me are correct and that I have received a copy of the completed questionnaire.

Signed: Date:

14

APPENDIX 14.3 Hazardous properties of waste as listed in the Hazardous Waste (England and Wales) Regulations 2005

	Hazard	Description
H1	Explosive	Substances and preparations which may explode under the effect of flame or which are more sensitive to shocks or friction than dinitrobenzene.
H2	Oxidising	Substances and preparations which exhibit highly exothermic reactions when in contact with other substances, particularly flammable substances.
H3-A	Highly flammable	► liquid substances and preparations having a flash point below 21°C (including extremely flammable liquids), or ► substances and preparations which may become hot and finally catch fire in contact with air at ambient temperature without any application of energy, or ► solid substances and preparations which may readily catch fire after brief contact with a source of ignition and which continue to burn or to be consumed after removal of the source of ignition, or ► gaseous substances and preparations which are flammable in air at normal pressure, or ► substances and preparations which, in contact with water or damp air, evolve highly flammable gases in dangerous quantities.
H3-B	Flammable	Liquid substances and preparations having a flash point equal to or greater than 21°C and less than or equal to 55°C.
H4	Irritant	Non-corrosive substances and preparations which, through immediate, prolonged or repeated contact with the skin or mucous membrane, can cause inflammation.
H5	Harmful	Substances and preparations which, if they are inhaled or ingested or if they penetrate the skin, may involve limited health risks.
H6	Toxic	Substances and preparations (including very toxic substances and preparations) which, if they are inhaled or ingested or if they penetrate the skin, may involve serious, acute or chronic health risks and even death.
H7	Carcinogenic	Substances and preparations which, if they are inhaled or ingested or if they penetrate the skin, may induce cancer or increase its incidence.
H8	Corrosive	Substances and preparations which may destroy living tissue on contact.
H9	Infectious	Substances containing viable micro-organisms or their toxins which are known or reliably believed to cause disease in humans or other living organisms.
H10	Teratogenic	Substances and preparations which, if they are inhaled or ingested or if they penetrate the skin, may induce non-hereditary congenital malformations or increase their incidence.
H11	Mutagenic	Substances and preparations which, if they are inhaled or ingested or if they penetrate the skin, may induce hereditary genetic defects or increase their incidence.
H12		Substances and preparations which release toxic or very toxic gases in contact with water, air or an acid.
H13		Substances and preparations capable by any means, after disposal, of yielding another substance, e.g. a leachate, which possesses any of the characteristics listed above.
H14	Ecotoxic	Substances and preparations which present or may present immediate or delayed risks for one or more sectors of the environment.

APPENDIX 14.4 Different types of protective gloves

Remember: **Select your gloves according to the Hazard and Risk Assessment. Keep your gloves in good condition. Seek advice if you are in any doubt.**

GENERAL USE

TYPE 1 Heavy work, General Protection (Rigger's type replacement)

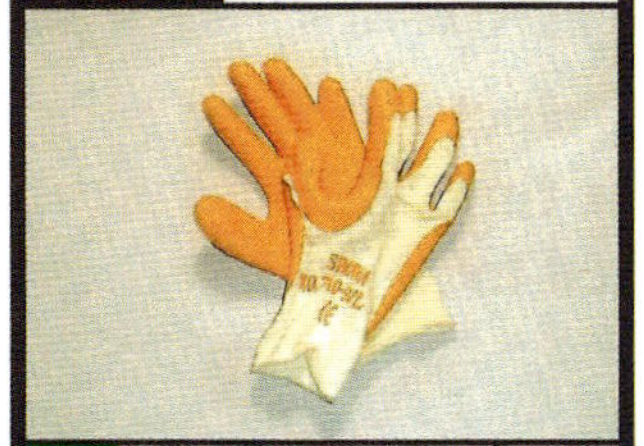

✓ Some Abrasion & Good Tear Resistance
✗ Low Cut & Puncture Protection; No Resistance to Fluids
2142

TYPE 2 Fine, Dextrous Work

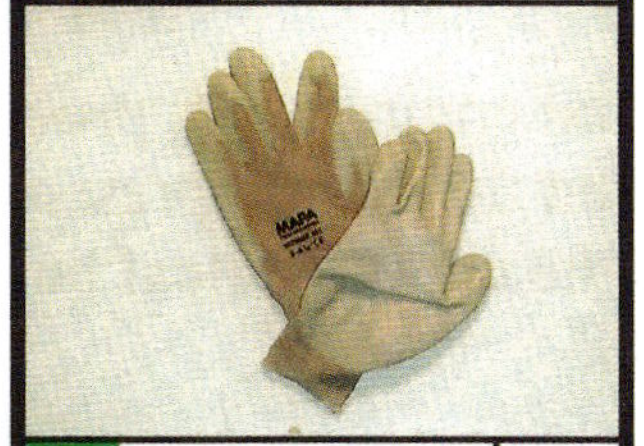

✓ Abrasion & Tear Resistant
✗ Low Cut & Puncture Protection; No Resistance to Fluids
3131

TYPE 3 Work involving Water & Oil

✓ Water & Oil Resistant, Good Abrasion Protection
✗ Low Cut Resistance
3121

TYPE 4 Handling Acid and other Chemicals

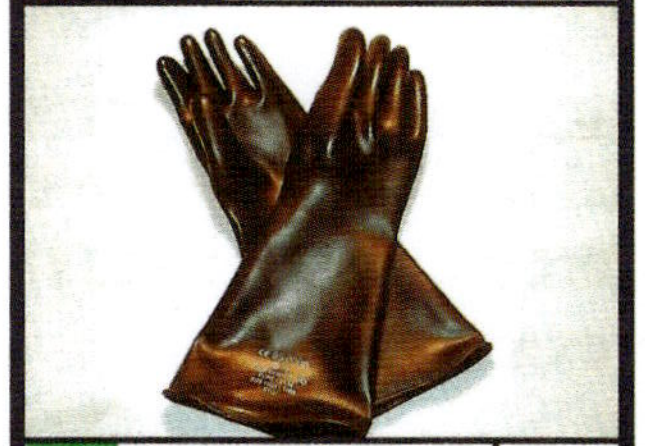

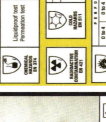

✓ Chemical Resistant, Long to Protect Arm
✗ Minimal Cut & Abrasion Protection
N/A

TYPE 5 Skin Protection from Oils and other Low Hazard Substances

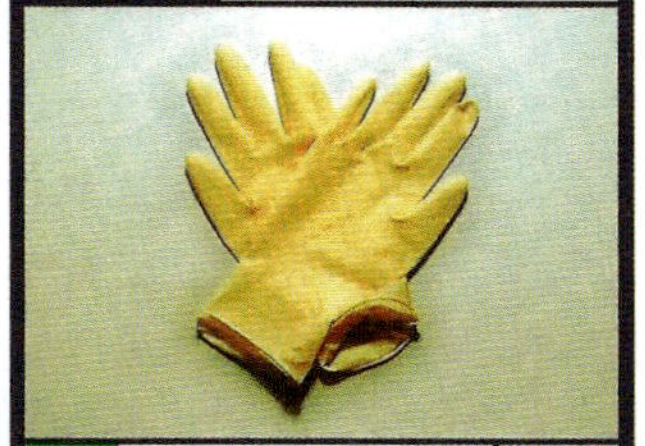

✓ Light and Thin.
✗ No Mechanical Protection at all
N/A

SPECIALIST USE

TYPE 6 Welders Leather Gauntlets – For protection against heat and burns

✓ Very Durable for Heat Protection
✗ Poor Chemical & Cut Resistance
N/A

TYPE 7 High Pressure Water Jetting, Slip Resistant Grip

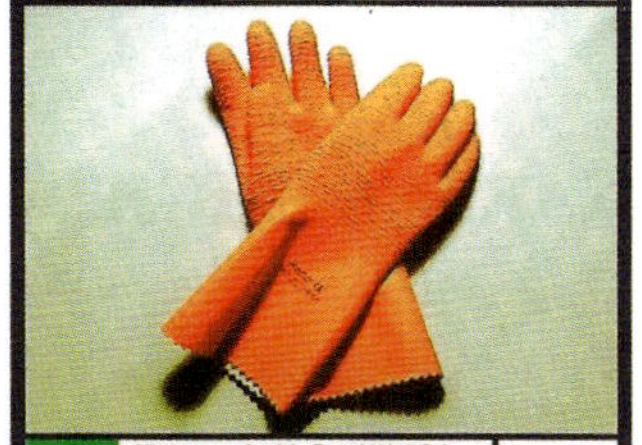

✓ Waterproof with Good Wet Grip, Abrasion & Tear Resistant
✗ Low Cut & Puncture Resistance
4131

TYPE 8 Kevlar, with High Cut Resistance – Use with Sharps, e.g. cent screens, glass

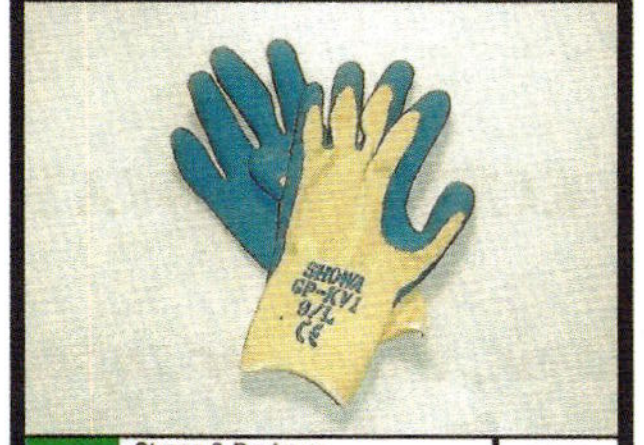

✓ Strong & Dextrous, High Cut Resistance
✗ No Resistance to Fluids
3444

TYPE 9 Heavy Engineering Work – e.g. on Pellet Mills

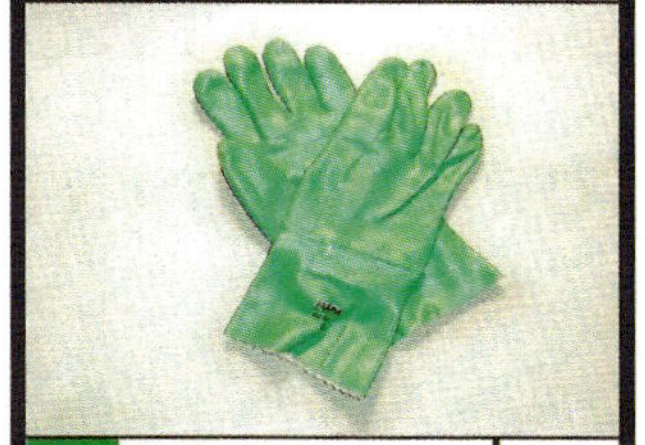

✓ Good Resistance to all Mechanical Hazards, Very Strong
✓ Some Heat and Fluid Protection
4532

Note that types provided may vary from those illustrated above, but will give similar protection levels

Explanation of Protection Ratings

MECHANICAL HAZARDS EN 388	PERFORMANCE LEVELS*			
	0 to 4	0 to 5	0 to 4	0 to 4
	Abrasion resistance	Blade cut resistance	Tear resistance	Puncture resistance

CHEMICAL HAZARDS EN 374 – Liquidproof test, Permeation test

MICRO ORGANISMS EN 374 – Liquidproof according to the air leak test

RADIOACTIVE CONTAMINATION EN 421

COLD HAZARDS EN 511	PERFORMANCE LEVELS*		
	0 to 4	0 to 4	0 to 1
	Convective cold resistance	Contact cold resistance	Water permeability

HEAT AND FIRE EN 407	PERFORMANCE LEVELS*					
	0 to 4	0 to 4	0 to 3	0 to 4	0 to 4	0 to 4
	Burning behaviour	Contact heat resistance	Convective heat resistance	Radiant heat resistance	Resistance to small drops of molten metal	Resistance to large quantities of molten metal

* Level X : The test is not applicable, or the glove is not tested
* Level 0 : The minimum level is not reached
For additional advice regarding CE categories on listed gloves contact the SAFETY DEPARTMENT

ref: CAN-DOC-01 rev.1

CHAPTER 15

Physical and psychological health hazards and risk control

> **This chapter covers the following NEBOSH learning objectives:**
> 1. Outline the health effects associated with exposure to noise and appropriate control measures
> 2. Outline the health effects associated with exposure to vibration and appropriate control measures
> 3. Outline the health effects associated with ionising and non-ionising radiation and outline appropriate control measures
> 4. Outline the causes and effects of work-related stress and appropriate control measures

Introduction

Occupational health is concerned with physical and psychological hazards as well as chemical and biological hazards. The physical occupational hazards have been well known for many years, and recent emphasis has been on the development of lower risk workplace environments. Physical hazards include topics such as electricity, display screen equipment (DSE) and manual handling, which were covered in earlier chapters; and noise, vibration and radiation, which are discussed in this chapter.

However, it is only really in the last 20 years that psychological hazards have been included among the occupational health hazards faced by many workers. This is now the most rapidly expanding area of occupational health, and includes topics such as mental health and workplace stress (as well as violence to staff and substance abuse covered in Chapter 8).

15.1 Noise

15.1.1 Introduction

There was considerable concern for many years over the increasing cases of occupational deafness and this led to the introduction of the Noise at Work Regulations in 1989 and the revised Control of Noise at Work Regulations in 2005. The HSE has estimated that an additional 1.1 million workers will be covered by the revised Regulations. These Regulations, which are summarised in Chapter 19, require the employer to:

▶ assess noise levels and keep records;
▶ reduce the risks from noise exposure by using engineering controls in the first instance and the provision and maintenance of hearing protection as a last resort;
▶ provide employees with information and training;
▶ if a manufacturer or supplier of equipment, provide relevant noise data on that equipment (particularly if any of the three action levels is likely to be reached).

Many occupations have potential noise problems including construction, manufacturing, entertainment, the uniformed services and call centres.

The main purpose of the Noise Regulations is to control noise levels rather than measuring them. This involves the better design of machines, equipment and work processes, and ensuring that personal protective equipment is correctly worn and employees are given adequate training and health surveillance.

Sound is transmitted through the air by sound waves, which are produced by vibrating objects. The vibrations cause a pressure wave which can be detected by a receiver, such as a microphone or the human ear. The ear may detect vibrations which vary from 20 to 20,000 (typically 50–16,000) cycles each second (or Hertz – Hz). Sound travels through air at a finite speed (342 m/s at 20°C and sea level). The existence of this speed is shown by the time lag between lightning and thunder during a thunderstorm. Noise normally describes loud, sudden, harsh or irritating sounds although noise is defined as any audible sound.

Noise may be transmitted directly through the air, by reflection from surrounding walls or buildings or through the structure of a floor or building. In construction work, the noise and vibrations from a pneumatic drill will be transmitted from the drill itself, from the ground being drilled and from the walls of surrounding buildings.

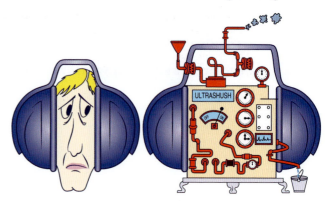

Figure 15.1 Better to control noise at source than wear ear protection

15.1.2 Health effects of noise

The human ear

There are three sections of the ear – the outer (or external) ear, the middle ear and the inner (or internal) ear. The sound pressure wave passes into and through the outer ear and strikes the eardrum, causing it to vibrate. The eardrum is situated approximately 25 mm inside the head. The vibration of the eardrum causes the proportional movement of three interconnected small bones in the middle ear, thus passing the sound to the cochlea, situated in the inner ear.

Within the cochlea the sound is transmitted to a fluid, causing it to vibrate. The motion of the fluid induces a membrane to vibrate which, in turn, causes hair cells attached to the membrane to bend. The movement of the hair cells causes a minute electrical impulse to be transmitted to the brain along the auditory nerve. Those hairs nearest to the middle ear respond to high frequency, while those at the tip of the cochlea respond to lower frequencies. (**Frequency** is the number of occurrences of a repeating event over a given period of time and is measured in Hertz (Hz).)

There are about 30,000 hair cells within the ear and noise-induced hearing loss causes irreversible damage to these cells.

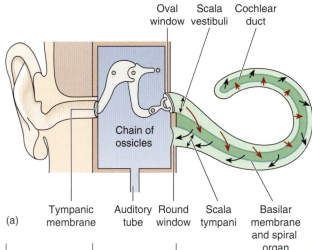

Oval window   Scala vestibuli   Cochlear duct

Chain of ossicles

(a)

Tympanic membrane   Auditory tube   Round window   Scala tympani   Basilar membrane and spiral organ

Outer ear	**Middle ear**	**Inner ear**		
Sound waves in air	Hammer→Anvil→Stirrup Mechanical movement of ossicles	Perilymph Fluid wave	Endolymph Fluid wave stimulates basilar membrane and spiral organ	Perilymph Fluid wave

(b)

Figure 15.2 Passage of sound waves: (a) The ear with cochlea uncoiled; (b) summary of transmission

Ill-health effects of noise

Noise can lead to ear damage on a temporary (acute) or permanent (chronic) basis.

There are three principal **acute** effects:

▶ **temporary threshold shift** – caused by short excessive noise exposures and affects the cochlea by reducing the flow of nerve impulses to the brain. The result is a slight deafness, which is reversible when the noise is removed;

▶ **tinnitus** – a ringing in the ears caused by an intense and sustained high noise level. It is caused by the over-stimulation of the hair cells. The ringing sensation continues for up to 24 hours after the noise has ceased;

▶ **acute acoustic trauma** – caused by a very loud noise such as an explosion. It affects either the

eardrum or the bones in the middle ear and is usually reversible. Severe explosive sounds can permanently damage the eardrum.

Occupational noise can also lead to one of the following three **chronic** hearing effects:

▶ **noise-induced hearing loss** – results from permanent damage to the cochlear hair cells. It affects the ability to hear speech clearly but the ability to hear is not lost completely;

▶ **permanent threshold shift** – this results from prolonged exposure to loud noise and is irreversible due to the permanent reduction in nerve impulses to the brain. This shift is most marked at the 4000 Hz frequency, which can lead to difficulty in hearing certain consonants and some female voices;

▶ **tinnitus** – is the same as the acute form but becomes permanent. It is a very unpleasant condition, which can develop without warning.

It is important to note that, if the level of noise exposure remains unchanged, noise-induced hearing loss will lead to a permanent threshold shift affecting an increasing number of frequencies. The hearing loss caused can be temporary or permanent.

▶ **Temporary deafness** can occur after leaving a noisy place. Hearing usually recovers within a couple of hours. This is a sign that continued exposure to loud noise could permanently damage your hearing;

▶ **Sudden extremely loud noise** can cause instant damage; and

▶ **Repeated exposure** causes gradual hearing loss over time. This is more common and it can take years for a worker to realise just how deaf they have become.

Presbycusis is the term used for hearing loss in older people, which may have been exacerbated by occupational noise earlier in their lives.

15.1.3 Noise assessments

The Control of Noise at Work Regulations specify action levels at which the hearing of employees must be protected. The conclusion as to whether any of those levels has been breached is reached after an assessment of noise levels has been made. However, before noise assessment can be discussed, noise measurement and the statutory action levels must be described.

Noise measurement

Sound intensity is measured by a unit known as a pascal (Pa – N/m²), which is a unit of pressure similar to that used when inflating a tyre. If noise was measured in this way, a large scale of numbers would be required ranging from 1 at one end to 1 million at the other. The sound pressure level (SPL) is a more convenient scale because:

15

▶ it compresses the size of the scale by using a logarithmic scale to the base 10;

▶ it measures the ratio of the measured pressure, p, to a reference standard pressure, p_0, which is the pressure at the threshold of hearing (2×10^{-5} Pa).

The unit is called a decibel (dB) and is defined as:

$$SPL = 20 \log_{10} (p/p_0) \text{ dB}$$

It is important to note that since a logarithmic scale to the base 10 is used, each increase of 3 dB is a doubling in the sound intensity. Thus, if a sound reading changes from 75 dB to 81 dB, the sound intensity or loudness has increased by four times.

Finally, as the human ear tends to distort its sensitivity to the sound it receives by being less sensitive to lower frequencies, the scale used by sound meters is weighted so that readings mimic the ear. This scale is known as the A scale and the readings are known as dB(A). There are also three other scales known as B, C and D.

Originally the A scale was used for sound pressure levels (SPLs) up to 55 dB, the B scale for levels between 55 dB and 85 dB and the C scale for values above 85 dB. However, today the A scale is used for nearly all levels except for very high SPLs when the C scale is used. The B scale is rarely used and the D scale is mainly used to monitor jet aircraft engine noise.

Table 15.1 gives some typical dB readings for common activities and Table 15.2 gives some typical readings for various construction activities.

Table 15.1 Some typical sound pressure levels (SPL) (dB(A) values)

Activity or environment	SPL [dB(A)]
Threshold of pain	140
Pneumatic drill	125
Pop group or disco	110
Heavy lorry	93
Street traffic	85
Conversational speech	65
Business office	60
Living room	40
Bedroom	25
Threshold of hearing	0

Noise is measured using a sound level meter which reads SPLs in dB(A) and the **peak sound pressure** in pascal (Pa), which is the highest noise level reached by the sound. There are two basic types of sound meter – integrated and direct reading meters. Meters which integrate the reading provide an average over a particular time period, which is an essential technique when there are large variations in sound levels. This value is known as the **continuous equivalent noise level (L_{Eq})**, which is normally measured over an 8-hour period.

Table 15.2 Some typical sound pressure levels (dB(A)) for construction processes

Activity or environment	SPL [dB(A)]
Nail gun	130–140
Piling work	1001
Blasting	1001
Shovelling hardcore	94
Sandblasting	85–90
Carpenter	86–96
Excavator driver	85
Dumper truck driver	85
Asphalt paving	85
Bricklayer	81–85

Direct reading devices, which tend to be much cheaper, can be used successfully when the noise levels are continuous at a near constant value.

Another important noise measurement is the **daily personal exposure level** of the worker, $L_{EP,d}$, which is measured over an 8-hour working day. Hence, if a person was exposed to 87 dB(A) over 4 hours, this would equate to an $L_{EP,d}$ of 84 dB(A) since a reduction of 3 dB(A) represents one-half of the noise dose.

The HSE Guidance on the Control of Noise at Work Regulations, L108, offers some very useful advice on the implementation of the Regulations and should be read by anyone who suspects that they have a noise problem at work. The guidance covers 'equipment and procedures for noise surveys' and contains a noise exposure ready reckoner which can be used to evaluate $L_{EP,d}$ when the noise occurs during several short intervals and/or at several different levels during the 8-hour period.

A survey by the Chartered Institute of Environmental Health (CIEH) has found that up to 50% of pub and club owners are failing to protect their staff from excessive noise at work. The findings also show that a majority of responding venue owners had not taken steps to reduce noise exposure since the elements of the Control of Noise at Work Regulations regarding the entertainment sector came into force.

Noise action levels

Regulation 6 of the Control of Noise at Work Regulations places a duty on employers to reduce the risk of damage to the hearing of their employees from exposure to noise to the lowest level reasonably practicable.

The Regulations introduce **exposure action** level values and **exposure limit** values. An **exposure action** value is a level of noise at which certain action must be taken. An **exposure limit** value is a level of noise at the ear above which an employee must not be exposed. Therefore if the workplace noise levels are above this value, any ear protection provided to the employee must reduce the noise level to the limit value at the ear.

These exposure action and limit values are as follows:

1. The lower exposure action levels are:
   (a) a daily or weekly personal noise exposure of **80 dB(A)**;
   (b) a peak sound pressure of **135 dB(C)**.
2. The upper exposure action levels are:
   (a) a daily or weekly personal noise exposure of **85 dB(A)**;
   (b) a peak sound pressure of **137 dB(C)**.
3. The exposure limit values are:
   (a) a daily or weekly personal noise exposure of **87 dB(A)**;
   (b) a peak sound pressure of **140 dB(C)**.

The peak exposure action and limit values are defined because high level peak noise can lead to short-term and long-term hearing loss. Explosives, guns (including nail guns), cartridge tools, hammers and stone chisels can all produce high peak sound pressures.

If the daily noise exposure exceeds the lower exposure action level, then a noise assessment should be carried out and recorded by a competent person. There is a very simple test which can be done in any workplace to determine the need for an assessment. Table 15.3 gives information on the simple test to determine the need for a noise risk assessment and Chapter 23 gives an example of a form that can be used for a noise assessment.

Table 15.3 Simple observations to determine the need for a noise risk assessment

Observation at the workplace	Likely noise level [dB(A)]	A noise risk assessment must be made if this noise level persists for
The noise is noticeable but does not interfere with normal conversation – equivalent to a domestic vacuum cleaner	80	6 hours
People have to shout to be heard if they are more than 2 m apart	85	2 hours
People have to shout to be heard if they are more than 1 m apart	90	45 minutes

If there is a marked variation in noise exposure levels during the working week, then the Regulations allow a weekly rather than daily personal exposure level, $L_{EP,w}$, to be used. It is only likely to be significantly different to the daily exposure level if exposure on one or two days in the working week is 5 dB(A) higher than on the other days, or the working week has three or fewer days of exposure. The weekly exposure rate

is not a simple arithmetic average of the daily rates. If an organisation is considering the use of a weekly exposure level, then the following provisions must be made:

▶ Hearing protection must be provided if there are very high noise levels on any one day.
▶ The employees and their representatives must be consulted on whether weekly averaging is appropriate.
▶ An explanation must be given to the employees of the purpose and possible effects of weekly averaging.

Finally, if the working day is 12 hours, then the action levels must be reduced by 3 dB(A) because the action levels assume an 8-hour working day.

The HSE Guidance document L108 gives detailed advice on noise assessments and surveys. The most important points are that the measurements should be taken at the working stations of the employees closest to the source of the noise and over as long a period as possible, particularly if there is a variation in noise levels during the working day. Other points to be included in a noise assessment are:

▶ details of the noise meter used and the date of its last calibration;
▶ the number of employees using the machine, time period of usage and other work activities;
▶ an indication of the condition of the machine and its maintenance schedule;
▶ the work being done on the machine at the time of the assessment;
▶ a schematic plan of the workplace showing the position of the machine being assessed;
▶ other noise sources, such as ventilation systems, that should be considered in the assessment. The control of these sources may help to reduce overall noise levels;
▶ recommendations for future actions, if any.

Other actions which the employer must undertake when the lower exposure action level is exceeded are to:

▶ inform, instruct and train employees on the hearing risks;
▶ supply hearing protection to those employees requesting it;
▶ ensure that any equipment or arrangements provided under the Regulations are correctly used or implemented.

The additional measures which the employer must take if the upper exposure action level is reached are:

▶ reduce and control exposure to noise by means other than hearing protection;
▶ establish hearing protection zones marked by notices and ensure that anybody entering the zone is wearing hearing protection (Figure 15.3);
▶ supply hearing protection and ensure that it is worn.

15

Figure 15.3 Typical ear protection zone sign

The Control of Noise at Work Regulations place a duty on the employer to undertake health surveillance for employees whose exposure regularly exceeds the upper action level irrespective of whether ear protection was worn. The recommended **health surveillance** is a hearing test at induction, followed by an annual check and review of hearing levels. The checks may be extended to every three years if no adverse effects are found during earlier tests. If exposure continues over a long period, health surveillance of employees is recommended using a more substantial audiometric test. This will indicate whether there has been any deterioration in hearing ability. Where exposure is between the lower and upper exposure action levels or occasionally above the upper action values, health surveillance will only be required if information becomes available, perhaps from medical records, that the employee is particularly sensitive to noise-induced hearing loss.

The Regulations also place the following statutory duties on employees:

▶ For noise levels above the lower exposure action level, they must use any control equipment (other than hearing protection), such as silencers, supplied by the employer and report any defects.
▶ For noise levels above the upper exposure action level, they fulfil the obligations given above and wear the hearing protection provided.
▶ They must take care of any equipment provided under these Regulations and report any defects.
▶ They must see their doctor if they feel that their hearing has become damaged.
▶ Present themselves for health surveillance.

The Management of Health and Safety Regulations prohibits the employment of anyone under the age of 18 years where there is a risk to health from noise.

15.1.4 Noise control techniques

In addition to **reduced time exposure** of employees to the noise source, there is a simple hierarchy of control techniques:

▶ reduction of noise at source;
▶ reduction of noise levels received by the employee (known as attenuation);
▶ personal protective equipment, which should only be used when the above two remedies are insufficient.

Reduction of noise at source

There are several means by which noise could be reduced at source (Figure 15.4):

▶ Change the process or equipment (e.g. replace solid tyres with rubber tyres or replace diesel engines with electric motors).
▶ Change the speed of the machine.
▶ Improve the maintenance regime by regular lubrication of bearings and tightening of belt drives.

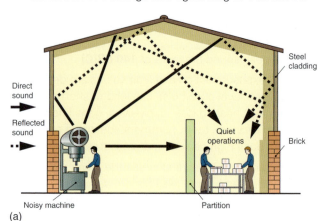

(a)

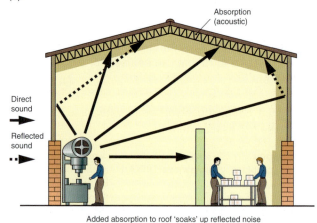

(b)

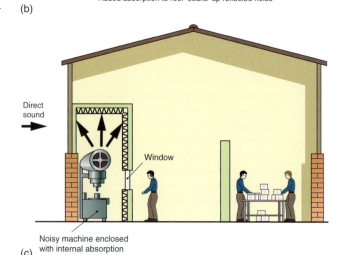

(c)

Figure 15.4 Noise paths found in a workplace: (a) the quiet area is subjected to reflected noise from a machine elsewhere in the building; (b) the correct use of roof absorption will reduce the reflected noise reaching the quiet area; (c) segregation of the noisy operation will benefit the whole workplace

Attenuation of noise levels

There are many methods of attenuating or reducing noise levels and these are covered in detail in the guide to the Regulations. The more common ones will be summarised here.

Orientation or re-location of the equipment – turn the noisy equipment away from the workforce or locate it away in separate and isolated areas.

Enclosure – surrounding the equipment with a good sound-insulating material can reduce sound levels by up to 30 dB(A). Care will need to be taken to ensure that the machine does not become overheated.

Screens or absorption walls – can be used effectively in areas where the sound is reflected from walls. The walls of the rooms housing the noisy equipment are lined with sound absorbent material, such as foam or mineral wool, or sound absorbent (acoustic) screens are placed around the equipment.

Damping – the use of insulating floor mountings to remove or reduce the transmission of noise and vibrations through the structure of the building such as girders, wall panels and flooring.

Lagging – the insulation of pipes and other fluid containers to reduce sound transmission (and, incidentally, heat loss).

Silencers – normally fitted to engines which are exhausting gases to atmosphere. Silencers consist of absorbent material or baffles.

Isolation of the workers – the provision of sound-proofed workrooms or enclosures isolated from noisy equipment (a power station control room is an example of worker isolation).

Immediate noise reduction benefits are often achieved for noise control equipment (e.g. noise enclosures) that is in good repair. Detailed information on established noise control methods for a wide range of woodworking machines is available on the HSE website.

15.1.5 Personal ear protection

The provision of personal hearing protection should only be considered as a last resort. There is usually resistance from the workforce to use it and it is costly to maintain and replace. It interferes with communications, particularly alarm systems, and can present hygiene problems.

The following factors should be considered when selecting personal ear protection:

▶ suitability for the range of sound spectrum of frequencies to be encountered;

▶ noise reduction (attenuation) offered by the ear protection;

▶ pattern of the noise exposure;

▶ acceptability and comfort of the wearer, particularly if there are medical problems;

▶ durability;

▶ hygiene considerations;

▶ compatibility with other personal protective equipment;

▶ ease of communication and ability to hear warning alarms;

▶ maintenance and storage arrangements;

▶ cost.

There are two main types of ear protection – earplugs and ear defenders (earmuffs).

Earplugs are made of sound absorbent material and fit into the ear. They can be reusable or disposable, and are able to fit most people and can easily be used with safety glasses and other personal protective equipment. Their effectiveness depends on the quality of the fit in the ear which, in turn, depends on the level of training given to the wearer. Permanent earplugs come in a range of sizes so that a good fit is obtained. The effectiveness of earplugs decreases with age and they should be replaced at the intervals specified by the supplier. A useful simple rule to ensure that the selected ear plug reduces the noise level at the ear to 87 dB(A) is to choose one with a manufacturer's rating of 83 dB(A). This should compensate for any fitting problems. The main disadvantage of ear plugs is that they do not reduce the sound transmitted through the bone structure which surrounds the ear and they often work loose with time.

Ear defenders (earmuffs) offer a far better reduction of all sound frequencies. They are generally more acceptable to workers because they are more comfortable to wear and they are easy to monitor as they are clearly visible. They also reduce the sound intensity transmitted through the bone structure surrounding the ear. A communication system can be built into earmuffs. However, they may be less effective if the user has long hair or is wearing spectacles or large ear rings. They may also be less effective if worn with helmets or face shields and uncomfortable in warm conditions. Maintenance is an important factor with earmuffs and should include checks for wear and tear and general cleanliness.

Selection of suitable ear protection is very important as it should not just reduce sound intensities below the statutory action levels but also reduce those intensities at particular frequencies. Normally advice should be sought from a competent supplier who will be able to advise on ear protection to suit a given spectrum of noise using 'octave band analysis'.

Finally, it is important to stress that the use of ear protection must be well supervised to ensure that not only is it being worn correctly, but that it is, in fact, actually being worn. This is a particular problem where lone workers are required to wear personal protective equipment, such as ear defenders. The requirement can be emphasised by the use of signs and warning notices on equipment and in company vehicles. Unannounced monitoring visits to the work site also

15

encourage the use of personal protective equipment. If there is a persistent refusal to wear personal protective equipment, then disciplinary procedures must be initiated.

The issue of ear defenders to the workforce is a last resort after measures outlined in 15.1.4 have been considered.

15.1.6 Noise in woodworking machines

Woodworking machines are used widely in the construction industry and the noise levels are usually in excess of the upper exposure action level. There are several Woodworking Information Sheets, published by the HSE, which give a useful indication of some practical solutions to reduce noise levels that do not simply rely on ear defenders. Noise assessments made on typical woodworking machines showed that noise levels ranged from 97 dB for beam panel saws and sanding machines to 105 dB for edge banders and multi-cutter moulding machines. Such machines are often used intermittently during the day so that a worker's daily noise exposure will normally be lower than these values. An electronic spreadsheet on the HSE website (www.hse.gov.uk/noise) can be used to calculate the daily exposure level for a worker who uses any woodworking machine for any period during the day.

Noise in woodworking machines is affected by:

▶ the type, size and moisture content of the timber used;
▶ the condition, speed and balance of the tooling used;
▶ the machine setting; and
▶ the design and effectiveness of the extraction unit used.

Effective action on any of these elements could produce a significant reduction in noise levels. Typical noise levels at various woodworking machines are given in Table 15.4.

Table 15.4 Typical noise levels at woodworking machines

Machine	Noise level (dB)
Beam panel saws and sanding machines	97
Boring machines	98
Band re-saws, panel planers and vertical spindle moulders	100
Portable woodworking tools	101
Bench saws and multiple ripsaws	102
High-speed routers and moulders	103
Thicknessers	104
Edge banders and multi-cutter moulding machines	105
Double-end tenoners	107

15.2 Vibration

15.2.1 Ill-health due to vibration

Hand-held vibrating machinery (such as pneumatic drills, sanders and grinders, powered lawn mowers and strimmers and chainsaws) can produce health risks from hand–arm or whole-body vibration (WBV).

Hand–arm vibration syndrome

Hand–arm vibration syndrome (HAVS) describes a group of diseases caused by the exposure of the hand and arm to external vibration. Some of these have been described under WRULDs, such as carpal tunnel syndrome.

However, the best known disease is **vibration white finger (VWF)**, in which the circulation of the blood, particularly in the hands, is adversely affected by the vibration. The early symptoms are tingling and numbness felt in the fingers, usually some time after the end of the working shift. As exposure continues, the tips of the fingers go white and then the whole hand may become affected. This results in a loss of grip strength and manual dexterity. Attacks can be triggered by damp and/or cold conditions and, on warming, 'pins and needles' are experienced (Figure 15.5). If the condition is allowed to persist, more serious symptoms become apparent including discoloration and enlargement of the fingers. In very advanced cases, gangrene can develop leading to the amputation of the affected hand or finger. VWF was first detailed as an industrial disease in 1911.

The risk of developing HAVS depends on the frequency of vibration, the length of exposure and the tightness of the grip on the machine or tool. Some typical values of vibration measurements for common items of equipment used in industry are given in Table 15.5.

When assessing the risk of HAVS developing among employees, the source of the vibration, such as reciprocating, rotating and vibrating tools and equipment, needs to be considered first together with the age of the equipment, its maintenance record, its suitability for the job and any information or guidance available from the manufacturer. The number of employees using the tooling or equipment, the duration and frequency of its use and any relevant personal factors, such as a pre-existing circulatory problem, all form part of the assessment. Environmental factors, particularly exposure to cold and/or wet weather and the nature of the job itself, are also important factors to be considered during such a risk assessment. Finally, an examination of the existing controls and their effectiveness and the frequency, magnitude and direction of the vibration are important elements of the evaluation. Other issues could include the effectiveness of any personal protective equipment provided and any instruction or training given.

Regular exposure to HAV can cause a range of permanent injuries to hands and arms, collectively known as hand–arm vibration syndrome (HAVS). The injuries can include damage to the:

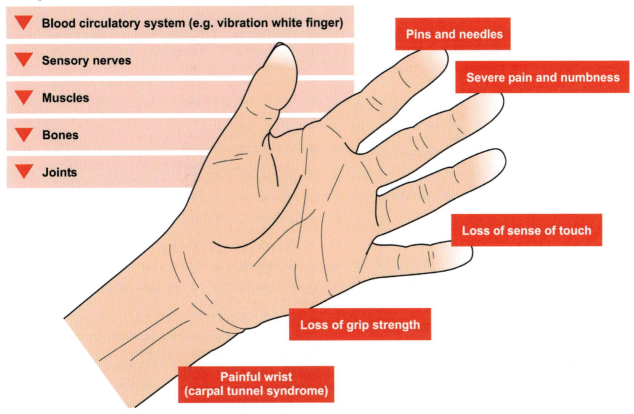

▼ **Blood circulatory system (e.g. vibration white finger)**

▼ **Sensory nerves**

▼ **Muscles**

▼ **Bones**

▼ **Joints**

Pins and needles

Severe pain and numbness

Loss of sense of touch

Loss of grip strength

Painful wrist (carpal tunnel syndrome)

Figure 15.5 Injuries which can be caused by hand–arm vibration. Source: HSE

Table 15.5 Examples of vibration exposure values measured by HSE on work equipment

Equipment	Condition	Vibration reading (m/s²)
Road breakers	Typical	12
	Modern design and trained operator	5
	Worst tools and operating conditions	20
Demolition hammers	Modern tools	8
	Typical	15
	Worst tools	25
Hammer drills	Typical	9
	Best tools and operating conditions	6
	Worst tools and operating conditions	25
Large-angle grinders	Modern vibration-reduced designs	4
	Other types	8
Small-angle grinders	Typical	2–6
Chainsaws	Typical	6
Brush cutter or strimmer	Typical	4
	Best	2
Sanders (orbital)	Typical	7–10

Whole-body vibration

Whole-body vibration (WBV) is caused by vibration from machinery passing into the body either through the feet of standing workers or the buttocks of sitting workers. It is the shaking or jolting of the human body through a supporting surface (usually a seat or the floor), for example when driving or riding on a vehicle along an unmade road, operating earth-moving machines or standing on a structure attached to a large, powerful, fixed machine which is impacting or vibrating.

The most common ill-health effect is severe back pain which, in severe cases, may result in permanent injury. Reasons for back pain in drivers can include:

▶ poor design of controls, making it difficult for the driver to operate the machine or vehicle easily or to see properly without twisting or stretching;
▶ incorrect adjustment by the driver of the seat position and hand and foot controls, so that it is necessary to continually twist, bend, lean and stretch to operate the machine;
▶ sitting for long periods without being able to change position;
▶ poor driver posture;
▶ repeated manual handling and lifting of loads by the driver;

15

▶ excessive exposure to whole-body vibration, particularly to shocks and jolts;

▶ repeatedly climbing into or jumping down from a high cab or one which is difficult to get in and out.

The risk increases where the driver or operator is exposed to two or more of these factors together.

Other acute effects include reduced visual and manual control, and increased heart rate and blood pressure. Chronic or long-term effects include permanent spinal damage, damage to the central nervous system, hearing loss and circulatory and digestive problems.

The most common occupations in which WBV is generated are driving fork-lift trucks, construction vehicles and agricultural or horticultural machinery and vehicles. There is growing concern throughout the European Union about this problem. Control measures include the proper use of the equipment including correct adjustments of air or hydraulic pressures, seating and, in the case of vehicles, correct suspension, tyre pressures and appropriate speeds to suit the terrain. Other control measures include the selection of suitable equipment with low vibration characteristics, work rotation, good maintenance and fault reporting procedures.

The HSE commissioned measurements of WBV on several machines and some of the results are shown in Table 15.6.

Table 15.6 Machines which could produce significant whole-body vibration

Machine	Activity	Vibration reading (m/s²)
3-tonne articulated dumper truck	Removal of spoil	0.78
Dumper truck	Transport of spoil	1.13
25-tonne articulated dumper truck	Transport of spoil	0.91
Bulldozer	Dozing	1.16
4-tonne twin-drum	Finishing tarmac	0.86
80-tonne rigid dumper truck	Transport of spoil	1.03

Preventative and precautionary measures

The control strategy outlined in Chapter 4 can certainly be applied to ergonomic risks. The common measures used to control ergonomic ill-health effects are to:

▶ implement results of task analysis and identification of repetitive actions;

▶ eliminate vibration-related or hazardous tasks by performing the job in a different way;

▶ ensure that the correct equipment (properly adjusted) is always used;

▶ introduce job rotation so that workers have a reduced time exposure to the hazard;

▶ during the design of the job, ensure that poor posture is avoided;

▶ undertake a risk assessment;

▶ take reports from employees and safety reps;

▶ prepare ill-health reports and absence records;

▶ introduce a programme of health surveillance;

▶ ensure that employees are given adequate information on the hazards and develop a suitable training programme;

▶ ensure that a programme of preventative maintenance is introduced and include the regular inspection of items such as vibration isolation mountings;

▶ keep up to date with advice from equipment manufacturers, trade associations and health and safety sources (more and more low vibration equipment is becoming available).

A very useful and extensive checklist for the identification and reduction of WRULDs is given in Appendix 2 of the HSE *Guide to work-related upper limb disorders* HSG60. The HSE has also produced two very useful guides – *Hand–arm vibration, The Control of Vibration at Work Regulations 2005 Guidance on Regulations* L140 and *Whole-body vibration, The Control of Vibration at Work Regulations 2005 Guidance on Regulations* L141. These guidance documents offer detailed advice on the implementation of the Regulations.

15.2.2 The Control of Vibration at Work Regulations

The Control of Vibration at Work Regulations introduced, for both HAV and WBV, a daily exposure limit and action values. These values are as follows:

1. For HAV:
  (a) the daily exposure limit value normalised to an 8-hour reference period is 5 m/s².
  (b) the daily exposure action value normalised to an 8-hour reference period is 2.5 m/s².

2. For WBV:
  (a) the daily exposure limit value normalised to an 8-hour reference period is 1.15 m/s².
  (b) the daily exposure action value normalised to an 8-hour reference period is 0.5 m/s².

An exposure limit value must not be exceeded. If an exposure action value is exceeded, then action must be taken to reduce the value. The term A(8) is added to the exposure limit or action value to denote that it is an average value spread over an 8-hour working day. Thus the daily exposure limit value for HAV is 5 m/s² A(8).

Hand–arm vibration

Many machines and the processes used in industry produce HAV. Typical high-risk processes include:

- grinding, sanding and polishing wood and stone;
- cutting stone, metal and wood;
- riveting, caulking and hammering;
- compacting sand, concrete and aggregate;
- drilling and breaking rock, concrete and road surfaces;
- surface preparation, including de-scaling and paint removal.

(a)

(b)

Figure 15.6 (a) and (b) Powered chisels or breakers mounted on different sizes of excavators to avoid HA vibration

There are several ways to ascertain the size of the vibration generated by equipment and machines. Manufacturers must declare vibration emission values for portable hand-held and hand-guided machines and provide information on risks. Other important sources for vibration information include scientific and technical journals, trade associations and online databases. HSE experience has shown that the vibration level is higher in practice than that quoted by many manufacturers. The reasons for this discrepancy may be that:

- the equipment is not well maintained;
- the equipment is not suitable for the material being worked;

- the tool has not been purchased from a reputable supplier;
- the accessories are not appropriate or are badly fitted;
- the operative is not using the tool properly.

In view of these problems, it is recommended that the declared value should be doubled when comparisons are made with exposure limits. As the exposure limit or action value is averaged over 8 hours, it is possible to work with higher values for a reduced exposure time. Table 15.7 shows the reduction in exposure time as the size of the vibration increases.

Table 15.7 The change in exposure times as vibration increases

Value of vibration (m/s^2)	2.5	3.5	5	7	10	14	20
Exposure time to reach action value (hours)	8	4	2	1	30 min	15 min	8 min
Exposure time to reach limit value (hours)	over 24	16	8	4	2	1	30 min

The Guidance (L140) gives very useful advice on the measurement of vibration, undertaking a suitable and sufficient risk assessment, control measures, health surveillance and training of employees. The following points summarise the important measures which should be taken to reduce the risks associated with HAV:

1. Avoid, whenever possible, the need for vibration equipment.
2. Undertake a risk assessment which includes a soundly based estimate of the employees' exposure to vibration.
3. Develop a good maintenance regime for tools and machinery. This may involve ensuring that tools are regularly sharpened, worn components are replaced or engines are regularly tuned and adjusted.
4. Introduce a work pattern that reduces the time of exposure to vibration.
5. Issue employees with gloves and warm clothing. There is a debate as to whether anti-vibration gloves are really effective but it is agreed that warm clothing helps with blood circulation, which reduces the risk of VWF. Care must be taken so that the tool does not cool the hand of the operator.
6. Introduce a reporting system for employees to use so that concerns and any symptoms can be recorded and investigated.
7. Encourage employees to check regularly for any symptoms of vibration related ill-health and to participate in any health surveillance measures provided.

15

It is important that drill bits and tools are kept sharp and used intact – an angle grinder with a chipped cutting disc will lead to a large increase in vibration as well as being dangerous. Finally, information and training for operators and supervisors is required by the Regulations. A training course for the use of pneumatic drills to break up a concrete platform would include the following advice:

▶ Select the smallest correct tool capable of completing the task.
▶ Hold the drill with a light grip and keep the handles in a horizontal position.
▶ Do not press downwards on the drill but let the weight of the tool do the work.
▶ Stop the drill when lifting it to change position.
▶ Cut the concrete in small pieces so that the drill bit does not jam.

Whole-body vibration

The Regulations require the control of the risks from whole-body vibration. This should be based on an assessment of the risk and exposure. In most cases it is simpler to make a broad assessment of the risk rather than try to assess exposure in detail, concentrating the main effort on introducing controls. The requirements of the Regulations are to:

▶ assess the vibration risk to employees;
▶ decide if they are likely to be exposed above the daily exposure action value (EAV) and if they are:
  ▷ introduce a programme of controls to eliminate or reduce their daily exposure so far as is reasonably practicable;
▶ decide if they are likely to be exposed above the daily exposure limit value (ELV) and if they are:
  ▷ take immediate action to reduce their exposure below the limit value;
▶ provide information and training on health risks and controls to employees at risk;
▶ consult a trade union safety representative or employee representative about the risks and the plans to address these risks;
▶ keep a record of the risk assessment and control actions;
▶ review and update the risk assessment regularly.

If a broad risk assessment has been made and the appropriate and reasonable control actions taken, there is no need to measure the exposure to vibration of the employee. Most machine and vehicle activities in normal use produce daily exposures below the limit value. But some off-road machinery operated for long periods in conditions that generate high levels of vibration or jolting may exceed the exposure limit value.

The measurement of WBV is very difficult and can only be done accurately by a specialist competent person. If the risk assessment has been made and the recommended control actions are in place, there

is no need to measure the exposure of employees to vibration. However, the HSE has suggested that employers can use the following checklist to estimate whether exposure to WBV is high:

1. There is a warning in the machine manufacturer's handbook that there is a risk of WBV.
2. The task is not suitable for the machine or vehicle being used.
3. Operators or drivers are using excessive speeds or operating the machine too aggressively.
4. Operators or drivers are working too many hours on machines or vehicles that are prone to WBV.
5. Road surfaces are too rough and potholed or floors uneven.
6. Drivers are being continually jolted or when going over bumps rising from their seats.
7. Vehicles designed to operate on normal roads are used on rough or poorly repaired roads.
8. Operators or drivers have reported back problems.

If one or more of the above applies, then exposure to WBV may be high.

If checks on exposure levels are required, information in the vehicle manufacturer's handbook or data published by the HSE may be used. An exposure calculator is available on the HSE website (at: www.hse.gov.uk/vibration) for a range of machines and vehicles in different working conditions. However, it is more effective for most employers to direct their efforts towards controlling the risks rather than trying to assess vibration exposures precisely. WBV in industry arises from driving vehicles, such as tractors or fork-lift trucks, over rough terrain or uneven floors. It is highly unlikely that driving vehicles on smooth roads will produce WBV problems. The HSE Guidance document, L141, gives detailed advice to help with the risk assessment and on estimating daily exposure levels. WBV risks are low for exposures around the action value and usually only simple control measures are necessary.

As explained earlier, the most common health problem associated with WBV is back pain. This pain may well have been caused by other activities but WBV will aggravate it.

The actions for controlling the risks from WBV need to ensure that:

▶ the driver's seat is correctly adjusted so that all controls can be reached easily and that the driver weight setting on their suspension seat, if available, is correctly adjusted. The seat should have a back rest with lumbar support;
▶ anti-fatigue mats are used if the operator has to stand for long periods;
▶ the speed of the vehicle is such that excessive jolting is avoided. Speeding is one of the main causes of excessive WBV;
▶ all vehicle controls and attached equipment are operated smoothly;

- only established site roadways are used;
- only suitable vehicles and equipment are selected to undertake the work and cope with the ground conditions;
- the site roadway system is regularly maintained;
- all vehicles are regularly maintained with particular attention being paid to tyre condition and pressures, vehicles suspension systems and the driver's seat;

- work schedules are regularly reviewed so that long periods of exposure on a given day are avoided and drivers have regular breaks;
- prolonged exposure to WBV is avoided for at risk groups (older people, young people, people with a history of back problems and pregnant women);
- employees are aware of the health risks from WBV, the results of the risk assessment and the ill-health reporting system. They should also be trained to drive in such a way that excessive vibration is reduced.

A simple health monitoring system that includes a questionnaire checklist should be agreed with employees or their representatives (available on the HSE website) to be completed once a year by employees at risk. Employees with back problems should be referred to an occupational health specialist. Personal information about the health of individual employees must be treated confidentially.

15.2.3 Role of health surveillance

The Control of Vibration at Work Regulations require the employer to ensure that employees should be given suitable health surveillance if:

- the risk assessment indicates that there is a risk to the health of any employees who are, or are liable to be, exposed to vibration; or
- employees are exposed to vibration at or above an exposure action value.

Records of the health surveillance must be kept, employees given access to their own records and the enforcing authorities provided with copies, as required by them. A range of specified action must be taken if problems are found with the health surveillance results.

More information on the Control of Vibration at Work Regulations is given in Chapter 19.

15.3 Radiation

15.3.1 Ionising radiation

Ionising radiation is emitted from radioactive materials, either in the form of directly ionising alpha and beta particles or indirectly ionising X-rays and gamma rays or neutrons. It has a high energy potential and an ability to penetrate, ionise and damage body tissue and organs.

All matter consists of atoms within each of which is a nucleus containing protons and neutrons, and orbiting electrons. The number of protons within the atom defines the element – hydrogen has 1 proton and lead has 82 protons. Some atoms are unstable and will change into atoms of another element,

(a)

(b)

Figure 15.7 (a) Vibrating roller with risk of whole-body vibration; (b) remote control vibrating plate weighing 1.2 tons with compaction in excess of a 7 ton roller which eliminates the risk of whole-body vibration. The operator is protected from vibrations, noise and dust. The machine can only be operated if line of sight is intact. In case of a loss of control the proximity recognition sensor keeps the operator safe

15

Figure 15.8 Typical ionising sign

emitting ionising radiation in the process. The change is called radioactive decay and the ionising radiations most commonly emitted are alpha and beta particles and gamma rays. X-rays are produced by bombarding a metal target with electrons at very high speeds using very high voltage electrical discharge equipment. Neutrons are released by nuclear fission and are not normally found in manufacturing processes.

Alpha particles consist of two protons and two neutrons and have a positive charge. They have little power to penetrate the skin and can be stopped using very flimsy material, such as paper. Their main route into the body is by ingestion.

Beta particles are high-speed electrons whose power of penetration depends on their speed, but penetration is usually restricted to 2 cm of skin and tissue. They can be stopped using aluminium foil. There are normally two routes of entry into the body – inhalation and ingestion.

Gamma rays, which are similar to **X-rays**, are electromagnetic radiations and have far greater penetrating power than alpha or beta particles. They are produced from nuclear reactions and can pass through the body.

The two most common measures of radiation are the becquerel (Bq), which measures the activity of a radioactive substance per second, and the sievert (Sv), which measures the biological effects of the radiation – normally measured in millisieverts (mSv).

Thus, the becquerel (Bq) measures the amount of radiation in a given environment and the millisievert (mSv) measures the ionising radiation dose received by a person.

Ionising radiation occurs naturally from man-made processes and about 87% of all radiation exposure is from natural sources. The Ionising Radiations Regulations specify a range of dose limits, some of which are given in Table 15.8.

Table 15.8 Typical radiation dose limits

	Dose (mSv)	Area of body
Employees aged 18 years+	20	Whole body per year
Trainees 16–18 years	6	Whole body per year
Any other person	1	Whole body per year
Women employees of child-bearing age	13	The abdomen in any consecutive 3-month period
Pregnant employees	1	During the declared term of the pregnancy

Harmful effects of ionising radiation

Ionising radiation attacks the cells of the body by producing chemical changes in the cell DNA by ionising it (thus producing free radicals), which leads to abnormal cell growth. The effects of these ionising attacks depend on the following factors:

▶ the size of the dose – the higher the dose then the more serious will be the effect;

▶ the area or extent of the exposure of the body – the effects may be far less severe if only a part of the body (e.g. an arm) receives the dose;

▶ the duration of the exposure – a long exposure to a low dose is likely to be more harmful than a short exposure to the same quantity of radiation.

Acute exposure can cause, dependent on the size of the dose, blood cell changes, nausea and vomiting, skin burns and blistering, collapse and death. Chronic exposure can lead to anaemia, leukaemia and other forms of cancer. It is also known that ionising radiation can have an adverse effect on the function of human reproductive organs and processes. Increases in the cases of sterility, stillbirths and malformed foetuses have also been observed.

The health effects of ionising radiation may be summarised into two groups – **somatic effects**, which refer to cell damage in the person exposed to the radiation dose, and **genetic effects**, which refer to the damage done to any future children of the irradiated person.

Sources of ionising radiation

The principal workplaces which could have ionising radiation present are the nuclear industry, medical centres (hospitals and research centres) and educational centres. Radioactive processes are used for the treatment of cancers, and radioactive isotopes are used for many different types of scientific research. X-rays are used extensively in hospitals, but they are also used in industry for non-destructive testing (e.g. crack detection in welds). Smoke detectors, used in most workplaces, also use ionising radiations.

Figure 15.9 X-ray generating unit used for weld testing on site in Russia. The tape states: 'beware of radiation' (shows the value of pictorial signs)

Ionising radiations can also occur naturally – the best example being radon, which is a radioactive gas that occurs mainly at or near granite outcrops where there is a presence of uranium. It is particularly prevalent in Devon and Cornwall. The gas enters buildings normally from the substructure through cracks in flooring or around service inlets.

The Ionising Radiations Regulations have set two action levels above which remedial action, such as fitting sumps and extraction fans, has to be taken to lower the radon level in the building. The first action level is

400 Bq/m^3 in workplaces and 200 Bq/m^3 in domestic properties. At levels above 1000 Bq/m^3, remedial action should be taken within 1 year. The average background radiation in the UK is 2.4 mSv per year. Background radiation levels are much higher elsewhere in the world – 260 mSv have been recorded in Iran.

About half of the background radiation in the UK is caused by radon. Radon gas is responsible for 5% of lung cancer cases. Radon is not generally a problem unless it is able to accumulate in confined spaces such as basements, cellars or lift shafts. An impermeable membrane beneath the floor of the building can be laid to prevent the gas entering the building through the floor. This latter treatment is normally only suitable for new buildings but does cost less than 10% of the cost of a sump and has no running costs. There are now government-defined 'radon-affected areas' throughout the UK, so that checks should be made if underground spaces in buildings are to be used.

About 14% of exposures to ionising radiation are due to medical exposures during diagnostic or treatment processes.

The hazards are categorised as either stochastic or non-stochastic. The principal stochastic hazard is cancer in various forms. The main non-stochastic effects include radiation burns, radiation sickness, cataracts and damage to unborn children.

Personal radiation exposure can be measured using a film badge, which is worn by the employee over a fixed time interval. The badge contains a photographic film which, after the time interval, is developed and an estimate of radiation exposure is made. A similar device, known as a radiation dose meter or detector, can be positioned on a shelf in the workplace for 3 months, so that a mean value of radiation levels may be measured. Instantaneous radiation values can be obtained from portable hand-held instruments, known as Geiger counters, which continuously sample the air for radiation levels. Similar devices are available to measure radon levels.

15.3.2 Non-ionising radiation

Non-ionising radiation includes ultraviolet, visible light (this includes lasers which focus or concentrate visible light), infrared and microwave radiations. As the wavelength is relatively long, the energy present is too low to ionise atoms which make up matter. The action of non-ionising radiation is to heat cells rather than change their chemical composition.

The Control of Artificial Optical Radiation at Work Regulations govern non-ionising radiation (see 15.4.3). The Personal Protective Equipment at Work Regulations are also relevant, since the greatest hazard is tissue burning of the skin or the eyes.

Ultraviolet (UV) radiation occurs with sunlight and with electric arc welding. In both cases, the skin and

Figure 15.10 Radon monitoring equipment

15

the eyes are at risk from the effect of burning. The skin will burn (as in sunburn) and repeated exposure can lead to skin cancer. Skin which is exposed to strong sunlight should be protected either by clothing or sun creams. This problem has become more common with the reduction in the ozone layer (which filters out much ultraviolet light). The eyes can be affected by a form of conjunctivitis which feels like grit in the eye, and is called by a variety of names dependent on the activity causing the problem. Arc welders call it 'arc eye' or 'welder's eye' and skiers 'snow blindness'. Cataracts caused by the action of ultraviolet radiation on the eye lens are another possible outcome of exposure. Most construction work takes place in the open air and therefore protection from strong sunlight is essential although it is seldom witnessed.

The most dangerous form of skin cancer, malignant melanoma, has increased by over 40% over the last 10 years making it the cancer with the fastest rising number of cases in the UK. Outdoor workers that could be at risk include farm or construction workers, market gardeners, outdoor activity workers and some public service workers. Those who are particularly at risk have:

- fair or freckled skin that does not tan, or goes red or burns before it tans;
- red or fair hair and light coloured eyes;
- a large number of moles.

Employers need to be aware of the risks their employees are taking when they work outside without adequate protection from the sun. With growing concern following the rise in skin cancers, the HSE has suggested the following hierarchy of controls for outdoor working:

- Relocate some jobs inside a building or to a shady location.
- Undertake some outdoor work earlier or later in the day.
- Provide personal protection such as:
  - wearing long sleeve shirts or loose clothing with a close weave;
  - wearing hats with a wide brim;
  - using a high factor sunscreen of at least SPF15 on any exposed skin.
- Provide suitable education and training for outdoor workers.
- Provide suitable information and supervision to instigate safe systems of work that protect workers from the sun.

UV radiation has some beneficial effects such as the accumulation of vitamin D, which strengthens bones.

Where UV insectocutors are used, normally in kitchens, to control flying insects, it is important that the correct emitting labels are fitted and not ones that, for example, are used to sterilise surgical instruments as has happened on one occasion.

Lasers use visible light and light from the invisible wavelength spectrum (infrared and ultraviolet). As the word laser implies, they produce 'light amplification by stimulated emission of radiation'. This light is highly concentrated and does not diverge or weaken with distance and the output is directly related to the chemical composition of the medium used within the particular laser. The output beam may be pulsating or continuous – the choice being dependent on the task of the laser. Lasers have a large range of applications including bar code reading at a supermarket checkout, the cutting and welding of metals and accurate measurement of distances and elevations required in land and mine surveying. They are also extensively used in surgery for cataract treatment and the sealing of blood vessels.

The International Electrotechnical Commission (IEC) and the American National Standards Institute (ANSI) have defined seven classes of laser (1, 1M, 2, 2M, 3R, 3B and 4) in ascending size of power output. Classes 1, 2 and 2M are relatively low hazard and only emit light in the visible band. Direct eye contact must be avoided with classes 3R, 3B and 4. These are more hazardous than classes 1 and 2 and the appointment of a laser safety officer is recommended. A Class 4 laser can burn the skin and cause permanent eye damage as a result of direct, diffuse or indirect exposure to the beam. Such Class 4 lasers can ignite combustible materials, and are a fire risk. All lasers should carry information stating their class and any precautions required during use.

The main hazards associated with lasers are eye and skin burns, toxic fumes, electricity and fire. The vast majority of accidents with lasers affect the eyes. Retinal damage is the most common and is irreversible. Cataract development and various forms of conjunctivitis can also result from laser accidents. Skin burning and reddening (erythema) are less common and are reversible.

Infrared radiation is generated by fires and hot substances and can cause eye and skin damage similar to that produced by ultraviolet radiation. It is a particular problem to fire fighters and those who work in foundries or near furnaces. Eye and skin protection are essential.

Microwaves are used extensively in cookers and mobile telephones and there are ongoing concerns about associated health hazards (and several inquiries are currently under way). The severity of any hazard is proportional to the power of the microwaves. The principal hazard is the heating of body cells, particularly those with little or no blood supply to dissipate the heat. This means that tissues such as the eye lens are most at risk from injury. However, it must be stressed that any risks are higher for items such as cookers than for low-powered devices such as mobile phones.

The measurement of non-ionising radiation normally involves the determination of the power output being received by the worker. Such surveys are

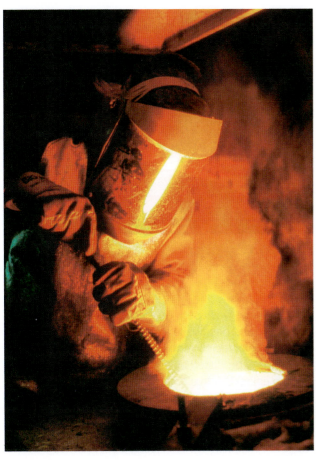

Figure 15.11 Metal furnace – source of infrared heat

best performed by specialists in the field, as the interpretation of the survey results requires considerable technical knowledge.

15.3.3 Radiation protection strategies

Ionising radiation

Protection is obtained by the application of shielding, time and distance either individually or, more commonly, using a mixture of all three.

Shielding is the best method because it is an 'engineered' solution. It involves the placing of a physical shield, such as a layer of lead, steel and concrete, between the worker and the radioactive source. The thicker the shield the more effective it is.

Time involves the use of the reduced time exposure principle and thus reduces the accumulated dose.

Distance works on the principle that the effect of radiation reduces as the distance between the worker and the source increases.

Other measures include the following:

- ▶ effective emergency arrangements;
- ▶ training of employees;
- ▶ the prohibition of eating, drinking and smoking adjacent to exposed areas;
- ▶ a high standard of personal cleanliness and first-aid arrangements;

- ▶ strict adherence to personal protective equipment arrangements, which may include full body protection and respiratory protection equipment;
- ▶ procedures to deal with spillages and other accidents;
- ▶ prominent signs and information regarding the radiation hazards;
- ▶ medical surveillance of employees.

The Ionising Radiations Regulations specify a range of precautions which must be taken, including the appointment of a Radiation Protection Supervisor as the competent person and a Radiation Protection Adviser.

The **Radiation Protection Supervisor** must be appointed by the employer to advise on the necessary measures for compliance with the Regulations and its Approved Code of Practice. The person appointed, who is normally an employee, must be competent to supervise the arrangements in place and have received relevant training.

The **Radiation Protection Adviser** is appointed by the employer to give advice to the Radiation Protection Supervisor and employer on any aspect of working with ionising radiation including the appointment of the Radiation Protection Supervisor. The Radiation Protection Adviser is often an employee of a national organisation with expertise in ionising radiation. A Radiation Protection Adviser is not needed when the only work with ionising radiation is as specified in Schedule 1 of the Regulations.

Non-ionising radiation

For ultraviolet and infrared radiation, eye protection in the form of goggles or a visor is most important, particularly when undertaking arc welding or furnace work. Skin protection is also likely to be necessary for the hands, arms and neck in the form of gloves, sleeves and a collar. For construction and other outdoor workers, protection from sunlight is important, particularly for the head and nose. Sun creams should also be used.

For laser operations, engineering controls such as fixed shielding and the use of non-reflecting surfaces around the workstation are recommended. For lasers in the higher class numbers, special eye protection is recommended. A risk assessment should be undertaken before a laser is used.

Engineering controls are primarily used for protection against microwaves. Typical controls include the enclosure of the whole microwave system in a metal surround and the use of an interlocking device that will not allow the system to operate unless the door is closed.

Intense sources of artificial light in the workplace, particularly from UV radiation and powerful lasers, can harm the eyes and skin of workers and need to be properly managed. The Control of Artificial

15

Figure 15.12 Low level laser beams extensively used for levelling and setting out in construction work

Figure 15.13 Welding shields used to protect against intense ultraviolet radiation which can cause 'arc eye'

Optical Radiation at Work Regulations came into force in 2010 and implement the Physical Agents (Artificial Optical Radiation) Directive (2006/25/EC). The aim of the Regulations is to ensure that standards are set so that all workers are protected from harm arising from exposure to hazardous sources of artificial light. As with the previous Noise and Vibration Regulations, they contain provisions on risk assessment, control of exposure, health surveillance and information, instruction and training. The Regulations are based on the limit values incorporated in the guidelines issued by the International Commission on Non-Ionising Radiation Protection (ICNIRP).

Workers should be protected from hazardous sources of light in the workplace, to ensure that the eyes and skin of workers are properly protected from intense sources of light that can damage the eyes and skin. Such light sources include ultraviolet, visible and infrared radiation produced by artificial sources, such as lasers and welding arcs. Examples of such hazardous light sources include:

▶ welding work (both arc and oxy-fuel) and plasma cutting causing mainly eye damage;
▶ furnaces and foundries causing eye and skin damage;
▶ printing involving the UV curing of inks causing mainly skin damage;
▶ motor vehicle repairs involving the UV curing of paints causing mainly skin damage;
▶ medical and cosmetic treatments involving laser surgery, blue light and UV therapies causing both eye and skin damage;
▶ all use of Class 3B and Class 4 lasers potentially causing permanent eye and skin damage.

More information is given on these Regulations in Chapter 19.

The EMF Directive

Electric, magnetic and electromagnetic fields (EMFs) are a form of non-ionising radiation that arises whenever electrical energy is used. An electric field is generated by electric charges, while a magnetic field occurs around an electric current. Electromagnetic fields are mutually produced by time-varying magnetic and electric fields. When electric fields act on conductive materials (such as the human body), they influence the distribution of electric charges at their surface causing current to flow through the body to the ground. Magnetic fields induce circulating currents within the human body. The strength of these currents depends on the intensity of the outside magnetic field. If sufficiently large, these currents can stimulate nerves and muscles or cause other biological effects such as nausea.

Common sources of electromagnetic fields include work processes such as radiofrequency heating and welding; household electrical wiring and appliances; electrical motors; computer screens; telecommunications; transport and distribution of electricity; broadcasting; and security detection devices.

The EMF Directive covers a frequency spectrum from 0 to 300 GHz, which includes static magnetic fields and low frequency electric and magnetic fields, through to radio frequency and microwave frequencies.

15.3.4 Welding operations

Over 1,000 accidents involving welding work are reported to the HSE each year. There are several different types of welding operation. The most common are:

▶ manual metal arc welding;
▶ metal inert-gas welding (MIG);
▶ tungsten inert-gas welding (TIG);
▶ oxy-acetylene welding.

The non-ionising radiation hazards caused by arc welding are not the only hazards associated with welding operations. The hazards from fume inhalation, trailing cables and pipes and the manual handling of cylinders are also present. There have also been serious injuries resulting from explosions and fires during welding processes. Accidents are often caused by lack of training and faulty equipment and they are often made worse by the lack of complete personal protective equipment. Many of these accidents occur on farms where welding equipment is used to make on-the-spot repairs to agricultural machinery.

15.3.5 The role of monitoring and health surveillance

The monitoring and health surveillance arrangements for exposure to ionising radiation are detailed under Part 5 of the Ionising Radiations Regulations. The Regulations specify that the employer of any individual exposed to ionising radiations as a result of work activities has the responsibility for deciding when such an employee needs to be designated as a classified person in accordance with Regulation 20. Before that person is classified, the employer will need to ensure that the employee has been certified as fit for the intended type of work within the last 12 months. This may require a medical examination. The employer then needs to make arrangements with the appointed doctor (or employment medical adviser) for continuing medical surveillance. Also, the employer will need to arrange for adequate medical surveillance for any employee who has received an overexposure, whether or not that employee has been designated as a classified person.

The Approved Code of Practice further states that the main purpose of medical surveillance is to determine an individual's fitness or continuing fitness for the intended work with ionising radiation. The fitness of the person is not restricted to possible health effects from exposure to ionising radiation. The appointed doctor (or employment medical adviser) will need to take account of specific features of the work with ionising radiation, such as the fitness of the individual:

▶ to wear any personal protective equipment (including respiratory protective equipment) required to restrict exposure;
▶ with a skin disease, to undertake work involving unsealed radioactive materials; and
▶ with serious psychological disorders, to undertake work with radiation sources that involve a special level of responsibility for safety.

The Regulations also require that the employer maintains a health record of any employee receiving health surveillance, and that record (or a copy) is kept until the person to whom the record relates has or would have attained the age of 75 years but in any event for at least 50 years from the date of the last entry made in it.

15.4 Stress

The HSE has estimated that work-related stress costs society around £3.8 billion a year. Approximately 12.8 million working days have been lost due to stress, depression and anxiety in one year alone when 530,000 people reported that they were suffering from work-related stress at a level that was making them ill. Stress and musculoskeletal disorders are the largest causes of work-related ill-health. There are on average over 400,000 workers suffering from stress-related ill-health each year of which 50% have suffered for over one year or longer. Proportionately, cases of work-related stress occurred most commonly in larger organisations, particularly among managerial staff, and in professional occupations (notably public administration, health and social care, and education), and most commonly again among the 35–54 age group. Women suffered higher rates of stress disorders than men. The overall situation seems statistically not to be improving year on year. This has been accompanied by an increase in civil claims resulting from stress at work. Stress and the related issues of workplace bullying and harassment are issues that simply cannot be ignored.

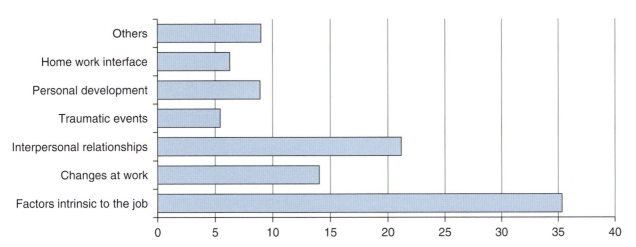

Figure 15.14 Breakdown of mental ill-health cases by type of event which precipitated stress between 2010 and 2012

Work-related stress

In 2001, the HSE defined work-related stress as the reaction people have to excessive pressure or other types of demand placed on them. An important distinction is made in this definition between pressure, which can be a positive state if managed appropriately and a normal reaction to reasonable demands, and stress which can arise in response to intense, continuous or prolonged exposure to excessive pressures and can be detrimental to health. However, a recent report commissioned by the HSE has concluded that there is no simple and universal definition of work-related stress, largely because of the complex nature of work-related stress.

The European Commission defines work-related stress as an 'emotional and psycho-physiological reaction to aversive and noxious aspects of work, work environments and work organisations. It is a state characterised by high levels of arousal and distress and often by feelings of not coping'.

However, the recent HSE report distinguished work-related stress from occupational stress. The former includes cases where work may have aggravated the experience of stress and associated symptoms of ill-health regardless of original cause. Here work may be a contributory factor but not necessarily the sole cause. The latter refers to cases where work is the sole cause of the experience of stress and associated symptoms of ill-health.

Stress is not a disease – it is the natural reaction to excessive pressure. It can be defined as the reaction that people have when they are unable to cope with excessive pressures and demands. Stress can lead to an improved performance but is not generally a good thing as it is likely to lead to both physical and mental ill-health, such as high blood pressure, peptic ulcers, skin disorders and depression.

Symptoms of stress

There are four groups of stress symptoms – physical, behavioural, emotional and cognitive. The physical symptoms include chest pains, nausea and frequent colds. Sufferers may eat less, smoke and drink more and develop nervous habits. They may be unable to relax, become irritable and depressed. They may also have memory problems, be unable to concentrate and become very anxious. Stress victims experience a severe lack of control.

Causes of stress

Most people experience stress at some time during their lives, e.g. during an illness or death of a close relative or friend. However, recovery normally occurs after the particular crisis has passed. The position is, however, often different in the workplace because the underlying causes of the stress, known as work-related stressors, are not relieved but continue to build up the stress levels until the employee can no longer cope.

The basic workplace stressors are:

▶ the job itself – boring or repetitive, unrealistic performance targets or insufficient training, job insecurity or fear of redundancy;
▶ individual responsibility – ill-defined roles and too much responsibility with too little power to influence the job outputs;
▶ working conditions – cramped, dirty and untidy workplace; unsafe practices; lack of privacy or security; inadequate welfare facilities; threat of violence; excessive noise, vibration or heat; poor lighting; lack of flexibility in working hours to meet domestic requirements; and adverse weather conditions for those working outside;
▶ management attitudes – poor communication, consultation or supervision, negative health and safety culture, lack of support in a crisis;
▶ relationships – unhappy relationship between workers, bullying, sexual and racial harassment.

Control measures

Possible solutions to all these stressors have been addressed throughout this book and involve the creation of a positive health and safety culture, effective training and consultation procedures and a set of health and safety arrangements which work on a day-to-day basis. The HSE has produced its own generic stress audit survey tool which is available free of charge on its website and advises the following action plan; that is, to:

▶ identify the problem;
▶ identify the background to the problem and how it was discovered;
▶ identify the remedial action required and give reasons for that action;
▶ identify targets and reasonable target dates;
▶ agree a review date with employees to check that the remedial action is working.

The following additional measures have also been found to be effective by some employers:

▶ take a positive attitude to stress issues by becoming familiar with its causes and controls;
▶ take employees' concerns seriously and develop a counselling system which will allow a frank, honest and confidential discussion of stress-related problems;
▶ develop an effective system of communication and consultation and ensure that periods of uncertainty are kept to a minimum;
▶ set out a simple policy on work-related stress and include stressors in risk assessments;
▶ ensure that employees are given adequate and relevant training and realistic performance targets;

- develop an effective employee appraisal system which includes mutually agreed objectives;
- discourage employees from working excessive hours and/or missing break periods (this may involve a detailed job evaluation);
- introduce job rotation and increase job variety;
- develop clear job descriptions and ensure that the individual is matched to them;
- encourage employees to improve their lifestyle (e.g. many local health authorities provide smoking cessation advice sessions);
- monitor incidents of bullying, sexual and racial harassment and, where necessary, take disciplinary action;
- train supervisors to recognise stress symptoms among the workforce;
- avoid a blame culture over accidents and incidents of ill-health.

The individual can also take action if he/she feels that he/she is becoming over-stressed. Regular exercise, change of job, review of diet and talking to somebody, preferably a trained counsellor, are all possibilities.

Workplace stress has no specific health and safety regulations but is covered by the duties imposed by the Health and Safety at Work Act and the Management of Health and Safety at Work Regulations to:

- ensure so far as is reasonably practicable that workplaces are safe and without risk to health;
- carry out a risk assessment relating to the risks to health;
- introduce and maintain appropriate control measures.

The risk assessment will need to relate to the individual, such as absentee record, production performance and which of the stressors are applicable, and to the organisation, such as training programmes, communication and appraisal procedures. Many stressful conditions can be reduced by an effective 'whistle-blowing' policy that enables individuals to highlight any concerns to senior managers or directors in a confidential manner.

There have been several successful civil actions for compensatory claims resulting from the effects of workplace stress. The Court of Appeal in the case of Sutherland v. Hatton and others in 2002 set out a number of practical propositions for future claims concerning workplace stress:

- Employers are entitled to take what they are told by employees at face value unless they have good reason to think otherwise. They do not have a duty to make searching enquiries about employee's mental health.
- An employer will not be in breach of duty in allowing a willing employee to continue in a stressful job if the only alternative is to dismiss or demote them. The employee must decide whether to risk a

breakdown in their health by staying in the job.

- Indications of impending harm to health at work must be clear enough to show an employer that action should be taken, in order for a duty on an employer to take action to arise.
- The employer is in breach of duty only if he/she fails to take steps which are reasonable, bearing in mind the size of the risk, gravity of harm, the cost of preventing it and any justification for taking the risk.
- No type of work may be regarded as intrinsically dangerous to mental health.
- Employers who offer a confidential counselling advice service, with access to treatment, are unlikely to be found in breach of duty.
- Employees must show that illness has been caused by a breach of duty, not merely occupational stress.
- Compensation will be reduced to take account of pre-existing conditions or the chance that the claimant would have fallen ill in any event.

Their full guidelines should be consulted (and consist of 16 points). However, more recent Court of Appeal decisions have suggested that the provision of a confidential counselling advice service alone may not be enough to discharge an employer's duty where an employee's problems could only be addressed by the intervention of management.

Stress usually occurs when people feel that they are losing control of a situation. In the workplace, this means that the individual no longer feels that they can cope with the demands made on them. Many such problems can be partly solved by listening to rather than talking at people. Many organisations have found that the active participation of union and employee representatives is a critical component in delivering improvements in employee health and well-being.

15.5 Further information

The Control of Noise at Work Regulations 2005

The Control of Vibration at Work Regulations 2005

The Construction (Design and Management) Regulations 2015

The Ionising Radiations Regulations 1999

The Personal Protective Equipment at Work Regulations 1992

The Provision and Use of Work Equipment Regulations 1998

The Control of Artificial Optical Radiation at Work Regulations 2010

Controlling noise at work, L108, HSE Books, ISBN 978-0-7176-6164-4 http://www.hse.gov.uk/pubns/priced/l108.pdf

HSE Stress management standards http://www.hse.gov.uk/stress/standards/

Personal Protective Equipment at Work (Guidance), L25, HSE Books, ISBN 978-0-7176-6139-3 http://www.hse.gov.uk/pubns/priced/l25.pdf

Managing the causes of work-related stress: A step-by-step approach using the Management Standards, HSG218, HSE Books, ISBN 978-0-7176-6273-9 http://www.hse.gov.uk/pubns/priced/hsg218.pdf

Management Standards, HSG218, HSE Books, ISBN 978-0-7176-6273-9 http://www.hse.gov.uk/pubns/priced/hsg218.pdf

Safe use of work equipment (ACoP) (L22), HSE Books, ISBN 978-0-7176-6295-1 http://www.hse.gov.uk/pubns/priced/l22.pdf

Hand–arm vibration, the Control of Vibration at Work Regulations 2005, Guidance on regulations, L140, HSE Books, ISBN 978-0-7176-6125-1 http://www.hse.gov.uk/pubns/priced/l140.pdf

Whole-body vibration – Control of Vibration at Work Regulations 2005, L141, HSE Books, ISBN 978-0-7176-6126-8 http://www.hse.gov.uk/pubns/priced/l141.pdf

Radon in the workplace: http://www.hse.gov.uk/radiation/ionising/radon.htm#testingradon

15.6 Practice revision questions

1. (a) **Outline** the mechanism by which noise is transmitted by the human ear to the brain.
   (b) **Explain** the following terms in relation to noise exposure at work in both the short and long term:
       (i) 'noise-induced hearing loss'
       (ii) 'tinnitus'
       (iii) temporary and permanent threshold shift.
   (c) **Outline** the options that might be considered to reduce the risk of hearing damage to workers who are exposed to high levels of noise.

2. (a) **Give** the meaning of the following terms used in noise measurement:
       (i) frequency;
       (ii) daily personal noise exposure ($L_{EP,d}$);
       (iii) decibel (dB);
       (iv) 'A' weighting.
   (b) Noise action levels are used to limit the exposure of workers to excessive noise levels. **Explain** the difference between a noise exposure value and a noise exposure limit value.
   (c) **Identify TWO** types of noise measurement techniques.

3. (a) **Outline** appropriate control measures to reduce the levels of noise to which the workers are exposed.
   (b) **Outline** the criteria that should be used when selecting suitable hearing protection and **identify** the limitations of such protective equipment.
   (c) **Outline** the contents of a health surveillance programme required to protect workers who are regularly exposed to high noise levels.

4. **Explain** the meaning of the following terms in relation to noise control in the workplace:
   (a) silencing
   (b) absorption

   (c) damping
   (d) isolation
   (e) lagging.

5. During a noise assessment in a woodworking workshop, an employee was observed to be using a sanding machine but wearing no personal hearing protection. An on-the-spot reading indicated that the noise level exceeded 85 dB(A).
   (a) **Outline** the key requirements of health and safety regulations, other than those placed on the employee, that are relevant to this situation.
   (b) **Outline** the recommended actions that should to be taken.

6. (a) **Describe** hand–arm vibration (HAV) and whole-body vibration (WBV).
   (b) **Outline** the ill-health effects and the associated symptoms caused by the exposure of workers to HAV and WBV.

7. A worker operates a hammer drill for long periods of the working day.
   (a) **Outline** the issues to be considered when assessing the risk of hand–arm vibration syndrome (HAVS) being developed by the worker.
   (b) **Outline** the control measures that should be taken to minimise the risk of the worker developing HAVS.

8. A dumper truck driver, working on a construction site, spends much of the working day driving over very rough ground. After several weeks, the driver complains of back pain.
   (a) **Identify** possible reasons for the back pain associated with the truck driving.
   (b) **Outline** actions that could be taken by the employer to control the risks from whole-body vibration.

9. (a) **Identify TWO** types of ionising radiation.
    (b) **Outline** possible means of ensuring that workers are not exposed to unacceptable levels of ionising radiation.

10. (a) For each of the following types of non-ionising radiation, **identify** a source and the possible ill-health effects on exposed individuals:
    (i)   infrared radiation
    (ii) ultraviolet radiation
    (iii) lasers
    (iv) microwaves.
    (b) **Identify** the controls available to protect people against exposure to these non-ionising radiations.

11. Construction workers are often required to work outdoors in a variety of weather conditions. **Outline** the hazards and the associated control measures that should be taken to protect these workers when they are working outdoors:
    (a) in strong sunlight
    (b) in cold temperatures.

12. (a) **Outline** the health and safety risks associated with welding operations on a construction site.
    (b) **Outline** the precautions that should be taken in order that welding may be carried out safely.

13. (a) **Explain** the meaning of 'work-related stress'.
    (b) **Identify** the ill-health effects of excessive stress to workers in the workplace.
    (c) **Identify FOUR** issues that could lead to workplace stress.
    (d) **Outline** the control measures that could be taken to minimise the risk of occupational stress.

15. Stress is responsible for many persons taking time off work from construction sites.
    (a) **Identify** the reasons that are likely to cause stress amongst the workers.
    (b) **Outline** the actions that site management might consider in order to reduce levels of occupational stress amongst workers on site.

15

CHAPTER 16

Working at height – hazards and risk control

16.1 Working at height hazards and control

16.1.1 Introduction

Each year, work at height accounts for about 50–60 deaths – more than any other workplace activity – and 4,000 injuries. Over 200 painters and decorators are seriously injured as a result of falls from height.

In 2010, HSE inspectors visited 93 construction sites in the West Midlands and stopped work on almost one in five of them. Inspectors made unannounced visits to ensure that work at height was being managed safely and that all sites were kept in a reasonable state of cleanliness and good order. The inspection initiative focused on refurbishment or roofing work. They issued 17 prohibition notices for various unsafe situations, such as a lack of edge protection for major roof works, incomplete scaffolds in use, large gaps in scaffold platforms, inadequate toe boards and missing guard rails. During another series of inspections of 2,400 refurbishment projects, the HSE found that one in five sites failed to address work at height risks on the site. More recent inspections have not shown a great improvement.

Attempts have been made to address these concerns by the introduction of the Work at Height Regulations, which apply to all operations carried out at height – not just construction work – so that they are also relevant to, for example, window cleaning, tree surgery, maintenance work at height and the changing of street lamps.

The Work at Height Regulations affect approximately 3 million workers for whom working at height is essential to their work. Among other measures, the height of guard rails on scaffolds is increased by 40mm to 950 mm, thus recognising that the average height of workers has increased over the last 40 years. The design of construction projects can also reduce health and safety hazards and accident rates. A study of accident data on falls from height over the 5-year period by the HSE emphasised the importance of building design in the elimination of work at height hazards. For buildings, factories, warehouses, offices, public buildings, retail premises, etc., sufficient dimensions for guard rails or similar barriers will be achieved by complying with the Building Regulations – which require guard rails to be 1100 mm.

To protect workers at height from serious injury, the Work at Height Regulations gives the following hierarchy of control:

1. Avoid working at height if possible.
2. Use an existing safe place of work.
3. Provide work equipment to prevent falls.
4. Mitigate distance and consequences of a fall.
5. Instruction, training and supervision.

This hierarchy is discussed in more detail later in this chapter.

16.1.2 The Work at Height Regulations

The Work at Height Regulations have no minimum height requirement for work at height. They include all work activities where there is a need to control a risk of falling a distance liable to cause personal injury. This is regardless of the work equipment being used, the duration of the work at height involved or the height at which the work is performed. They include access to and egress from a place of work. They would therefore include:

▶ working on a scaffold or from a mobile elevating work platform (MEWP);
▶ sheeting a lorry or dipping a road tanker;
▶ working on the top of a container in docks or on a ship or storage area;
▶ tree surgery and other forestry work at height;
▶ using cradles or rope for access to a building or other structure like a ship under repair;
▶ climbing permanent structures like a gantry or telephone pole;
▶ working near an excavation area or cellar opening if a person could fall into it and be injured;
▶ painting or pasting and erecting bill posters at height;
▶ work on staging or trestles, for example for filming or events;
▶ using a ladder/stepladder or kick stool for shelf filling, window cleaning and the like;
▶ using man-riding harnesses in ship repair, or offshore or steeple jack work;
▶ working in a mine shaft or chimney;
▶ work carried out at a private house by a person employed for the purpose, for example a painter and decorator (but not if the private individual carries out work on their own home).

However, it would not include:

▶ slips, trips and falls on the same level;
▶ falls on permanent stairs if there is no structural or maintenance work being done;
▶ work in the upper floor of a multi-storey building where there is no risk of falling (except separate activities like using a stepladder).

Figure 16.1 Working at height – mast climbing work platforms

At the centre of the Regulations (Regulation 6), the employer is expected to apply a three-stage hierarchy to all work which is to be carried out at height. The three steps are the avoidance of work at height, the prevention of workers from falling and the mitigation of the effect on workers of falls should they occur.

The Regulations require that:

▶ work is not carried out at height when it is reasonably practicable to carry the work out safely other than at height (e.g. the assembly of components should be done at ground level);

▶ when work is carried out at height, the employer shall take suitable and sufficient measures to prevent, so far as is reasonably practicable, any person falling a distance liable to cause injury (e.g. the use of guard rails);

▶ the employer shall take suitable and sufficient measures to minimise the distance and consequences of a fall (collective measures, for example airbags or safety nets, must take precedence over individual measures, for example safety harnesses).

The risk assessment and action required to control risks from using a kick stool to collect books from a shelf should be simple (not overloading, not overstretching, etc.). However, the action required for a complex construction project would involve significantly greater consideration and assessment of risk.

The Work at Height Regulations were amended in 2007 and now apply to those who work at height providing instruction or leadership to one or more people engaged in caving or climbing by way of sport, recreation, team building or similar activities in Great Britain.

A risk assessment for working at height should first consider whether the work could be avoided. If this is not possible, then the risk assessment should consider the following issues:

▶ the nature and duration of the work;
▶ the competence level of all those involved with the work and any additional training requirements;
▶ the required level of supervision;
▶ use of guard rails, toe boards, working platforms and means of access and egress;
▶ required personal protective equipment (PPE), such as helmets and harnesses;
▶ the presence of fall arrest systems, such as netting or soft landing systems;
▶ the health status of the workers;
▶ the possible weather conditions;
▶ compliance with the relevant legislation, in particular the Work at Height Regulations and the Management of Health and Safety at Work Regulations.

For heights below 2 m, the Regulations require a sensible risk-based approach to preventing falls together with reasonable precautions, such as:

▶ ensuring that handrails to scaffolds and towers are provided and not deliberately removed for work below 2 m;
▶ using edge protection on bandstands (for bricklayers).

Precautions against risks from low falls should only be taken when the scope and duration of the work presents a risk of injury. If the risk is trivial, and it is not reasonably practical to take precautions, then no action needs to be taken apart from training and instruction.

A summary of the Regulations is given in Chapter 19.

16.1.3 Construction hazards and controls from working at height

Safe place of work

Safe access to and egress from the place of work is essential when working at height. All working platforms, ladders, scaffolds, gangways and passenger and materials hoists must be safe for use and be subject to regular inspection. Working platforms should also be kept as clear and clean as possible to avoid slip and trips and the accidental loss of material to the ground below.

Typical work activities and injuries associated with falls from height

Falls from height can result in fractures, serious head injuries and, in some cases, death. It is, therefore, essential that special care is taken to protect workers when they are working at height.

Common construction work activities which are done at height include brick or block laying, roofing, steelwork erection, rendering, cladding, high pressure water-jetting, grit-blasting, concrete repairs, painting and some demolition work. For all these activities safe systems of work are essential. General hazards, such as dust,

16

Figure 16.2 Working platform, pre-fabricated tower scaffolds and bridging unit

hazardous substances, electricity, vibrations and noise can also be present due to the nature of the work being carried out at height.

One particular form of such activity which can be particularly hazardous is the use of false-work. False-work is a temporary structure used to support a non-self-supporting structure during its construction or refurbishment. An example is the wooden structure to support a brickwork arch which is being built. Only competent persons should plan, erect, load or dismantle false-work. The collapse of false-work structures is common and, when occurring at height, can cause serious injury. Most accidents caused by the collapse of false-work result from the lack of liaison between the various trades using it and its suppliers or erectors.

The cleaning of buildings and monuments involves the use of hazardous techniques such as grit-blasting or high pressure water-jets, which is often done from scaffolding or even ladders. With both these techniques, care is needed to protect the workers, the occupiers of the building and passers-by from hazards such as noise, dust, falling debris and possible flooding of walkways. The equipment must be maintained and inspected regularly and operatives provided with training, supervision and suitable personal protective equipment (goggles, ear defenders and gloves) and

waterproof clothing. Other measures will be necessary, such as the boarding up of windows, which may require the provision of temporary lighting inside the building. In the case of water-jetting, the scaffold should be enclosed in tarpaulin sheets and a channel system should be used to enable water to run off.

Finally, stairwells and other holes in floors are another source of falls from height and measures must be taken to address these hazards. Such measures include handrails on stairs, guard rails to stairwells and lift shafts, and ensuring that other holes in floors are similarly protected or covered. However, adequate levels of lighting and good housekeeping are also important as is a high standard of supervision and monitoring control.

Protection against falls

Falls from a height are the most common cause of serious injury or death in the construction industry. The Work at Height Regulations require all duty-holders to ensure that:

▶ all work at height is properly planned and organised;
▶ those involved in work at height are competent;
▶ the risks from work at height are assessed, and appropriate work equipment is selected and used;
▶ the risks of working on or near fragile surfaces are properly managed; and

▶ the equipment used for work at height is properly inspected and maintained.

The regulations require that guard rails on scaffolds are at a minimum of 950 mm and the maximum unprotected gap between the toe and guard rail of a scaffold is 470 mm. This implies the use of an intermediate guard rail although other means, such as additional toe boards or screening, may be used. They also specify requirements for personal suspension equipment and means of arresting falls (such as safety nets).

Figure 16.3 Industrial roof work with safety nets to arrest falls

When working at height, a hierarchy of measures given by the Work at Height Regulations should then be followed, to prevent falls from occurring and to select appropriate access equipment. These measures are:

▶ Eliminate or avoid working at height.
▶ Work from an existing safe workplace provided, such as a properly constructed working platform, complete with toe boards and guard rails.
▶ Ensure that there is sufficient work equipment or other measures to minimise the distance and consequence of a fall by the use of:
  ▷ collective measures rather than individual measures (e.g. a handrail instead of a harness); or where this is not practicable
  ▷ collective fall arrest equipment (airbags or safety nets); or where this is not practicable
  ▷ individual fall restrainers (safety harnesses) should be used.
▶ Provide supervision, training and instruction.
▶ Only when none of the above measures is practicable, should ladders or stepladders be considered.

Fragile roofs and surfaces

Every year several workers are killed or seriously injured by falling through building roofs made of fragile material. Work on or near fragile surfaces is also covered by the Work at Height Regulations (see summary in Chapter 19). Roof work, particularly work on pitched roofs, is

Figure 16.4 Proper precautions should always be taken when working on or near fragile roofs – access system for short-term work

hazardous and requires a specific risk assessment and method statement (see Chapter 4 for a definition of a method statement) prior to the commencement of work. Particular hazards are fragile roofing materials, including those materials which deteriorate and become more brittle with age and exposure to sunlight, exposed edges, unsafe access equipment and falls from girders, ridges or purlins. There must be suitable means of access such as scaffolding, ladders and crawling boards. Suitable barriers, guard rails or covers must be provided where people work near to fragile materials and roof lights. Suitable warning signs, indicating that a roof is fragile, should be on display at ground level.

Where possible, work on a fragile roof should be avoided by doing the following:

▶ work from underneath the roof using a suitable work platform or
▶ where this is not possible, use a mobile elevating work platform that allows people to work from within the basket without having to stand on the roof.

If access onto the fragile roof cannot be avoided, perimeter edge protection should be installed and staging used to spread the load. Unless all the work and access is on staging or platforms that are fitted with guard rails then safety nets should be installed underneath the roof or a harness system used. Where a harness is used, adequate anchorage points will be required.

Fragility can be caused by:

▶ general deterioration of the roof through ageing, neglect and lack of maintenance;
▶ corrosion of cladding and fixings;
▶ poor quality of the original installation and selection of materials;

16

469

- ▶ thermal and impact damage;
- ▶ deterioration of the supporting structure; and
- ▶ weather damage.

Asbestos cement sheets and old roof lights should always be treated as fragile.

There are other hazards associated with roof work – overhead services and obstructions, the presence of asbestos or other hazardous substances, the use of equipment such as gas cylinders and bitumen boilers, and manual handling hazards. All roofs, once fixed, should be treated as fragile until a competent person has confirmed that they are no longer fragile. In particular, the following are likely to be fragile:

- ▶ fibre-cement sheets – non-reinforced sheets irrespective of profile type;
- ▶ liner panels – on built-up sheeted roofs;
- ▶ metal sheets – where corroded;
- ▶ glass – including wired glass;
- ▶ chipboard – or similar material where rotted; and
- ▶ roof lights.

It is essential that only trained and competent persons are allowed to work on roofs and that they wear footwear having a good grip. It is good practice to ensure that a person does not work alone on a roof. Guidance from the Advisory Committee for Roof Work states that any person undertaking roof work needs to be 'both mentally and physically fit, competent to do the work, and be fully aware of all the dangers that exist and the actions necessary to overcome those dangers'.

The HSE has published the third edition of the guidance booklet on safety during roof work: HSG33, *Health and safety in roof work.*

A roof ladder should only be used if more suitable equipment cannot be used and should only be used for low risk, short duration work. Roof ladders should be of an industrial grade, in good condition and secured to prevent movement. The anchorage at the top of the roof ladder should be by some method which does not depend on the ridge capping, as this is liable to break away from the ridge. The anchorage should bear on the opposite slope by a properly designed and manufactured ridge hook or be secured by other means.

Roof work should not be attempted in poor weather conditions such as rain, ice, frost or strong winds (particularly gusting) or if slippery conditions exist on the roof. Winds in excess of 23mph (Force 5) will affect the balance of a roof worker.

Protection against falling objects

This topic is covered in Chapter 8 because it is a general site issue. However, it is particularly relevant when working at height and is also now covered by the Work at Height Regulations. A study by the HSE has indicated that over a 5-year period, 44 construction workers and members of the public (including children) were killed by insecure loads, unsecured equipment and pieces of plant falling on them. Over the same period 59 were fatally injured by the collapse of structures or parts of structures. Whilst 37 of these deaths were caused

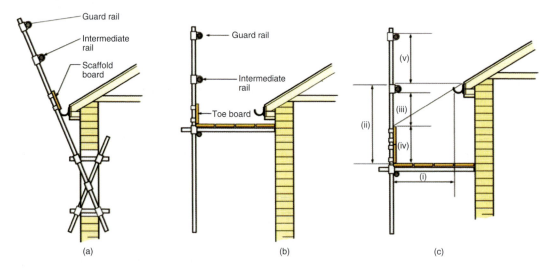

Sloping roof edge protection; typical arrangement in conventional tube and fittings
(a) Supported from window opening
(b) Working platform below the eaves
(c) Top lift of a scaffold. Dimensions should be as follows:
    (i) Working platform minimum width 600 mm
    (ii) Minimum 950 mm
    (iii) Maximum gap 470 mm
    (iv) To rise to the line of the roof slope with a minimum height of 150 mm
    (v) Gap between rails no more than 470 mm

Figure 16.5 Typical sloping roof edge protection: barriers shown in (a) can be useful where space is limited, but they are not capable of sustaining loads so large as (b) and (c) which also provide a working platform

by accidents during demolition work, the remainder occurred during routine construction work.

As mentioned in Chapter 8, workers and members of the public need to be protected from the hazards associated with falling objects by the use of covered walkways or suitable netting to catch falling debris. Waste material should be brought to ground level by the use of chutes or hoists and not thrown from height. Minimal quantities of building materials should be stored on working platforms.

Head protection is essential on site for workers and visitors. Several of the lives lost due to accidents from falling materials could have been saved if hard hats had been worn.

Prevention of falling materials through safe stacking and storage

It is estimated that two million working days are lost each year on account of handling injuries and slip and trip incidents. Such incidents can occur in warehouses and storage facilities for construction materials when palletised goods are stacked higher than two storeys and often weigh several tonnes. Implementing tried and tested methods on safe racking and storage is therefore essential to mitigate the risks of an incident arising.

It is often possible to remove high-level storage from offices and other general workplaces, such as construction sites, and provide storage in warehouses or similar storage facilities. If a storage facility is to be installed for the first time, then the following points should be considered:

- The racking must be erected on and fixed securely to a sound, level floor.
- The storage system must be installed in accordance with manufacturers' instructions.
- If the racking is to be secured to the wall of a building, has this been proved by structural calculations which should be made to ensure that the walls can support the racking and its contents securely.
- Beam-connector locks must be fixed at both ends of the beam.
- Maximum-load notices must be displayed.
- Correct pallets must be used.
- Sufficient protective equipment must be used, such as column guards and rack-end protectors.

Management and employees should familiarise themselves with the racking systems used and ensure they understand the difference between general wear and tear and real damage, in order to help them identify potentially dangerous situations as early as possible.

Storage racking is particularly vulnerable and should be strong and stable enough for the loads it has to carry. Damage from vehicles in a warehouse can easily weaken the structure and cause collapse. Uprights need protection, particularly at corners.

The following action can be taken to keep racking serviceable:

- Inspect them regularly and encourage workers to report any problems/defects.
- Post notices with maximum permissible loads and never exceed the loading.
- Use good pallets and safe stacking methods.
- Band, box or wrap articles to prevent items falling.
- Set limits on the height of stacks and regularly inspect to make sure that limits are being followed.
- Provide instruction and training for staff and special procedures for difficult objects.

Regular visual inspections should be carried out and documented, so that any damage can be quickly resolved. In particular, staff should be trained to act if damage occurs to and affects:

- the cross-sectional profile of a main load beam;
- the straightness of beams, bracing, or uprights; and
- the welds and joints, or bolts and clips.

To guard against back strains and other injuries, shelving and storage solutions can be installed that allow access and retrieval of stock at a comfortable, ergonomic height. Products such as vertical storage machines or pallet pull-out units are possible solutions, as they are designed so that stock can be reached without unnecessary straining.

Generally, storage racks should be examined by a qualified inspector approved by the Storage Equipment Manufacturers' Association (SEMA) at least once or twice a year.

During an inspection, particular attention will be paid to beams, uprights, frame bracing, floor fixings and lock-in clips, as indicated in the SEMA Code of Practice, guideline no. 6. The following will also be subject to general observations:

- pallet locations on beams;
- conditions and types of pallets;
- positioning of loads and types of loads stored on pallets;
- general fork-lift operatives' use of the racking;
- the condition and type of floor on which the racking is fixed;
- general housekeeping of the installation; and possible changes from the original design requirements.

With warehouse managers relying increasingly on temporary and agency workers, who sometimes have little or no prior experience of working in warehouse or storage environments, it is vital that safety training becomes part of the induction process.

All materials used in the construction process on site must be either stacked or stored safely. This will keep the site tidy and reduce slip and trip hazards. It will also help to reduce the risk of fire. Dangerous substances should be kept in a safe place in a separate building or the open air. Only small quantities of

16

hazardous and/or flammable substances should be kept on the site. Any larger amounts, which cannot be kept outside in a safe area, should be kept in a special fire-resisting store which is well ventilated and free of ignition sources. Flammable gas cylinders also need to be stored and used safely. More information on this is given in Chapter 13.

The effect of deteriorating materials and weather

As many materials age, they become brittle, weaker and less stable. This has already been mentioned as a hazard associated with fragile roofs. It is also a problem in other situations found in construction work, such as refurbishment work and the various cladding materials (including asbestos) used inside and outside buildings. The well-publicised concrete 'cancer', where concrete degenerates to a powder leading to serious structural weaknesses, has occurred in several relatively young reinforced concrete structures, causing either very expensive repairs or total rebuilds. Whilst all materials will deteriorate to some extent with age, good design, maintenance, construction methods (using the correct mix ratios for mortars and concrete) and type of cement should significantly reduce the rate of deterioration.

Weather is also a significant factor which can increase the hazards associated with working at height. High winds can affect the stability of scaffolding and ladders as has been mentioned elsewhere in this chapter. They can affect the work process itself by displacing tools and materials and cause injuries to persons working at height or those below the workings. Such winds, if strong enough, can also lift roofing tiles or sheets particularly if the roof is only partially completed. Such problems often occur outside normal working hours causing injuries to members of the public (including trespassers) and damage to neighbouring properties. It

is very important to ensure that the construction work is protected after working hours especially if bad weather is forecast. Lightning is another weather hazard and it is inadvisable for work to continue on roofs or scaffolds when thunderstorms are approaching. The effects of lightning are discussed in more detail in Chapter 12.

Even apparently benign weather can present health and safety hazards. Skin cancer rates due to excessive exposure to sunlight has increased significantly over the last two decades. Even if skin cancer does not occur, serious damage to the skin will result from regular exposure to strong sun over several years. Typical damage is wrinkled skin due to the breakdown of skin elasticity and the formation of red patches of skin. There are also concerns, with the depletion of the ozone layer, that strong sunlight may cause serious deterioration in eye health, leading to possible problems such as arc eye and long-term eye damage. Precautions such as the use of protective creams, sunglasses and reduced exposure by wearing suitable clothing are all essential controls. This topic is covered in more detail in Chapter 15.

Finally, cold weather can also present hazards when working at height particularly when the cold weather is accompanied by wind, which will produce a 'chill factor' and an apparent temperature that is lower than the actual one. Guard rails will become colder and the reactions of workers much slower. Of course, cold weather will have an effect on construction processes, such as the mixing of cement and the laying of bricks. Workers should wear appropriate clothing and, if the temperature is too low, work at height should cease.

Methods of avoiding working at height

Much of the work which is done at height could often be done or partly done at ground level – thus avoiding the hazards of working at height. The partial erection of scaffolding or edge protection at ground level and the use of cranes to lift it into place at height are examples of this. The manufacture of complete window frames in a workshop and then the final installation of a frame into the building is another. By the use of suitable extension equipment high windows can be cleaned from the ground and high walls can be painted from the ground. However, in most construction work at height, the work cannot be done at ground level and suitable control measures to address the hazards of working at height will be required.

Requirements for head protection

The Construction (Head Protection) Regulations 1989 virtually mandated employers to supply head protection (hard hats) to employees whenever there was a risk of head injury from falling objects. (Sikhs wearing turbans were exempted from this requirement.) The employer was also responsible for ensuring that hard

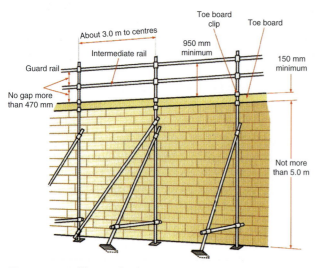

Figure 16.6 Flat roof edge protection supported at ground level. This type of support allows work up to the roof edge without obstruction

About 3.0 m to centres

Toe board clip

Toe board

Intermediate rail

950 mm minimum

Guard rail

150 mm minimum

No gap more than 470 mm

Not more than 5.0 m

hats were properly maintained and replaced when they were damaged in any way. Self-employed workers had to supply and maintain their own head protection. Visitors to construction sites had to be supplied with head protection and mandatory head protection signs displayed around the site.

Following the recommendations of the Löfstedt report in 2011, the Construction (Head Protection) Regulations were revoked since it was felt that they largely replicate the Personal Protective Equipment at Work Regulations which could be relied on to regulate the use of head protection on construction sites. Sikhs wearing turbans are still exempted from the use of head protection.

Under the Personal Protective Equipment at Work Regulations there will still be a requirement for employers to provide, and for workers to wear, head protection where there is a risk of head injury. The HSE will review and update existing guidance (either published or on its website) on compliance with the Regulations to ensure it adequately covers the provision and use of head protection on construction sites.

Therefore employers must ensure, so far as it is reasonably practicable, that each employee (and any other person over whom they have control) wears suitable head protection on site, unless there is no foreseeable risk of injury to their head. The person in control of a site may make rules regulating the wearing of suitable head protection. Such rules should be in writing. An employer may give directions requiring their employees to wear head protection.

Employees who have been provided with suitable head protection must wear it when required to do so by the site rules. They must take reasonable care of head protection, and report any loss or obvious defect to the employer. They must make full and proper use of it and return it to the storage accommodation which should be provided for hard hats when they are not in use.

Some simple hard hat rules

- A hard hat should:
  - be in good condition. If it's damaged, throw it away;
  - fit the person wearing it and be worn properly;
  - not stop you wearing hearing protectors as well (when needed);
  - only be obtained from a reputable supplier – there are fake hard hats on the market.
- Make sure that hard hats are worn:
  - by making it a site rule;
  - by always wearing your hard hat to set an example;
  - by checking others are wearing theirs.

Source: HSE.

Inspections and maintenance

Inspection

Equipment for work at height needs regular inspection to ensure that it is fit for use. A marking system is probably required to show when the next inspection is due. Formal inspections should not be a substitute for any pre-use checks or routine maintenance. Inspection does not necessarily cover the checks that are made during maintenance although there may be some common features. Inspections need to be recorded but checks do not.

Scaffolds must be inspected on a regular basis by a competent person. These inspections should take place before the scaffold is used, after any alteration is made or after adverse weather conditions may have weakened it. In any event an inspection should take place every 7 days and any faults rectified.

Under the Work at Height Regulations, weekly inspections are still required for scaffolding, as previously required by the Construction (Health, Safety and Welfare) Regulations where a person could fall 2 m or more. The reporting requirements for inspection are set out in Schedule 7 to the Regulations as follows:

- the name and address of the person for whom the inspection was carried out;
- the location of the work equipment inspected;
- a description of the work equipment inspected;
- the date and time of the inspection;
- details of any matter identified that could give rise to a risk to the health or safety of any person;
- details of any action taken as a result of any matter identified in the previous point;
- details of any further action considered necessary;
- the name and position of the person making the report.

Whilst there is no longer a specific statutory form to be completed, a record of the inspection should be made on the recommended form for inspections shown in Chapter 23 (Form C1), with a list of typical scaffolding faults (see Appendix 16.2); a possible checklist, which could be used, is given in Appendix 16.3.

Most equipment used while working at height will need regular inspection and, in some cases, a thorough examination. Such equipment includes hoists and mobile elevating work platforms (MEWPs), grit-blasting and water-jetting equipment and electrical appliances.

Maintenance

Inspections and even thorough examinations are not substitutes for properly maintaining equipment. The information gained in the maintenance work, inspections and thorough technical examinations should inform one another. A maintenance log should be kept and be up to date. The whole maintenance

16

system will require proper management systems. The maintenance frequency will depend on the equipment, the conditions in which it is used and the manufacturer's instructions.

16.2 Safe working practices for access equipment and roof work

16.2.1 Hazards and controls associated with working at height

Before any work at height starts, a risk assessment should be undertaken that considers the following hierarchy:

- ▶ Work at height should be avoided where possible.
- ▶ If it is unavoidable, use specific equipment and control measures to prevent falls.
- ▶ If it is impossible to eliminate risk, use measures to reduce the consequences of a fall.

Work at any height involves a risk of falling. The significance of falls from height, for example resulting in fatalities or serious major injuries, has been dealt with earlier in this chapter as have the importance of and legal requirements for head protection. Also covered were the many hazards involved in working at height including fragile roofs/surfaces and the deterioration of materials, unprotected edges and falling materials. Additional hazards include the weather and unstable or poorly maintained access equipment, such as ladders and various types of scaffold. Scaffolding must only be erected or modified by people who are trained and competent to do so. It should be constructed in accordance with recognised standards from components which are of adequate design and strength, and which are inspected at regular intervals. The scaffolding itself should be inspected by a competent person regularly.

The principal means of preventing falls of people or materials include the use of fencing, guard rails, toe boards, working platforms, access boards, ladder hoops, safety nets and safety harnesses. Safety harnesses arrest the fall by restricting it to a given distance due to the fixing of the harness to a point on an adjacent rigid structure. They should only be used when all other possibilities are not practical.

16.2.2 Access equipment

There are many different types of access equipment but only the following four categories will be considered here:

- ▶ ladders
- ▶ fixed scaffolds
- ▶ mobile scaffold towers
- ▶ mobile elevating work platforms (MEWPs).

Ladders

The Work at Height Regulations place a duty on employers to ensure that a ladder is used for work at height only if a risk assessment has demonstrated that the use of more suitable work equipment is not justified because of the low risk and:

- ▶ short duration of use or
- ▶ existing features on the site which cannot be altered.

Falls from ladders caused 13 deaths in 2005 and a third of all major injuries caused by falls at work were due to falls from ladders (1,200). Ten ladder accidents are reported to the HSE every day.

The main cause of accidents involving ladders is ladder movement whilst in use. This occurs when they have not been secured to a fixed point, particularly at the base. Other causes include over-reaching by the worker, slipping on a rung, ladder defects and, in the case of metal ladders, contact with electricity. The main category of ladder accidents is falls.

There are two common materials used in the construction of ladders – aluminium and timber. Aluminium ladders have the advantage of being light but should not be used in high winds or near live electricity. Timber ladders need regular inspection for damage and should not be painted since this could hide cracks and other defects, such as knots.

Every time a ladder is used, a pre-use check should be made. Such a check should be undertaken by the user to check the condition of the feet and rungs at the beginning of the working day or whenever the ladder is dropped or moved from a dirty area to a clean area.

The following factors should be considered when using ladders:

- ▶ Ensure that the use of a ladder is the safest means of access given the work to be done and the height to be climbed.
- ▶ The location itself needs to be checked. The supporting wall and supporting ground surface should be dry and slip free. It should rest on a firm surface, and never be placed on loose bricks or packing. Extra care will be needed if the area is busy with pedestrians or vehicles.
- ▶ The ladder needs to be stable in use. This means that the inclination should be as near the optimum as possible (1 in 4 ratio of distance from the wall to distance up the wall).
- ▶ Wherever possible, a ladder should be tied to prevent it from slipping. This can either be at the top, the bottom or both, making sure both stiles are tied. Never tie a ladder by its rungs.
- ▶ If the ladder cannot be tied, use an 'effective ladder' or one with an 'effective ladder-stability device' that the suppliers or manufacturers can confirm is stable

enough to use unsecured in the worst-case scenario, e.g. a ladder stay with anti-slip device.

▶ If the above three precautions are not possible, then the ladder stiles can be wedged against a wall or other similar heavy object; or, as a last resort, have a second person 'foot' the ladder.

▶ Weather conditions must be suitable (no high winds or heavy rain).

▶ The proximity of live electricity should also be checked. (This last point is important when ladders are to be carried beneath power lines.)

▶ The ladder should be fitted with non-slip feet.

▶ There should be at least 1 m of ladder above the stepping off point.

▶ The work activity must be considered in some detail. Over-reaching must be eliminated and consideration given to the storage of paints or tools which are to be used from the ladder and any loads to be carried up the ladder. The ladder must be matched to work required.

▶ Workers who are to use ladders must be trained in the correct method of use and selection. Such training should include the use of both hands during climbing, clean non-slippery footwear, clean rungs and an undamaged ladder.

▶ Ladders should be inspected (particularly for damaged or missing rungs) and maintained on a regular basis and they should only be repaired by competent persons.

▶ The transportation and storage of ladders is important since much damage can occur at these times. They need to be handled carefully and stored in a dry place.

▶ When a ladder is left secured to a structure during non-working hours, a plank should be tied to the rungs to prevent unauthorised access to the structure.

There are a variety of products which ensure that ladders are stable during use by the provision of anchors for the ladder base or fixing mechanisms to the building. However, certain work should not be attempted using ladders, particularly when the user cannot maintain a safe handhold while carrying a load. This includes work where:

▶ a secure handhold is not available;
▶ the work is at an excessive height;
▶ the ladder cannot be secured or made stable;
▶ the work is of long duration;
▶ the work area is very large;
▶ the equipment or materials to be used are heavy or bulky;
▶ the weather conditions are adverse;
▶ there is no protection from vehicles.

A particular type of timber ladder is the wooden **pole ladder**, which is used by many scaffolders. Pole ladders should be suitable for the job by being of the appropriate length and having complete integrity of

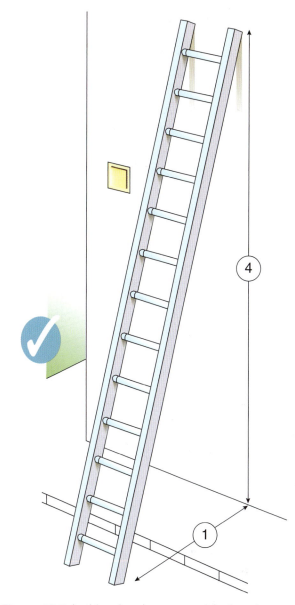

Figure 16.7 Ladder showing correct 1 in 4 angle (means of securing omitted for clarity)

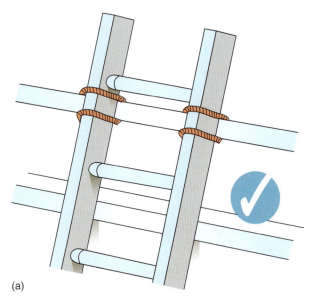

(a)

Figure 16.8 (a) Ladder tied at top stiles (correct for working on, but not for access)

16

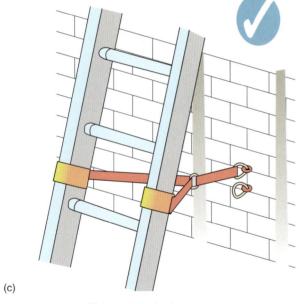

(b)

Figure 16.8 (b) Tying part way down

(c)

Figure 16.8 (c) Tying near the base

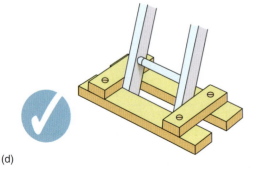

(d)

Figure 16.8 (d) Securing at the base

the steel reinforcement. Prior to use, the ladder should be inspected as outlined earlier in this section and any damage to stiles, rungs or steel reinforcement must be identified, as should any unauthorised repairs.

There have been several rumours that the Work at Height Regulations have banned the use of ladders. This is *not* true. Ladders may be used for access and it is legal to work from ladders. Ladders may be used when a risk assessment shows that the risk of injury is low and the task is of short duration (minutes rather than hours, e.g. replacing a few broken tiles on a roof or adjusting a television aerial) or there are unalterable features of the work site and that it is not reasonably practicable to use potentially safer alternative means of access, such as a MEWP or a mobile tower scaffold. Provided that the ladder is checked before each use, the work is of short duration, only light loads are to be carried, the ladder is stable and secured and users are trained and competent, then ladders may be used. There is no maximum height for using a ladder. However, where a ladder rises 9 m or more above its base, landing areas or rest platforms should be provided at suitable intervals. Ladder guidance from the HSE recommends that users should maintain three points of contact when climbing a ladder and wherever possible at the work position. The three points of contact are a hand and two feet.

More information on ladders and their use within the requirements of the Work at Height Regulations is available from the British Ladder Manufacturers Association. Ladders for industrial work in the UK should be marked to:

- ► Timber BS1129: Kite marked Class 1 Industrial
- ► Aluminium BS2037:1994 Kite marked Class 1 Industrial
- ► Glass fibre BSEN131:1993 Kite marked Industrial
- ► Step stools BS7377:1994.

Stepladders, trestles and staging

Many of the points above for ladders apply to stepladders and trestles where stability and over-reaching are the main hazards. Neither of these types should be used at a workplace above about 2 m in height unless proper edge protection is provided.

All equipment must be checked by the supervisor before use to ensure that there are no defects and must be checked at least weekly whilst in use on site. If a defect is noted, or the equipment is damaged, it must be taken out of use immediately. Any repairs must only be carried out by competent persons.

Supervisors must also check that the equipment is being used correctly and not being used where a safer method should be provided.

Where staging, such as a 'Youngman' staging platform, is being used in roof areas, supervisors must ensure that only experienced operatives are

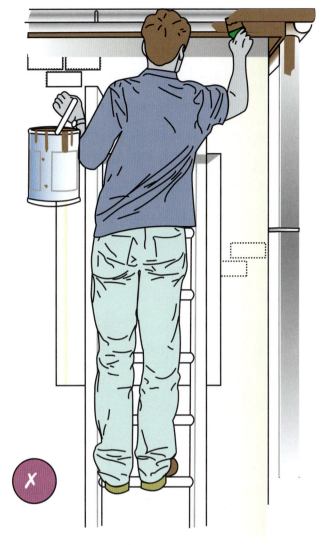

Figure 16.9 Attach paint cans and the like to the ladder

permitted to carry out this work and that all necessary safety harnesses and anchorage points are provided and used.

The main hazards associated with stepladders, trestles and staging are:

- unsuitable base (uneven or loose materials);
- unsafe and incorrect use of equipment (e.g. the use of staging for barrow ramps);
- overloading;
- use of equipment where a safer method should be provided;
- overhang of boards or staging at supports (trap ends);
- use of defective equipment.

Stepladders and trestles must be:

- manufactured to a recognised industrial specification;
- stored and handled with care to prevent damage and deterioration;
- subject to a programme of regular inspection (there should be a marking, coding or tagging system to confirm that the inspection has taken place);

- checked by the user before use;
- taken out of use if damaged – and destroyed or repaired;
- used on a secure surface, and with due regard to ensuring stability at all times;
- kept away from overhead cables and similar hazards.

The small platform fitted at the top of many stepladders is designed to support tools, paint pots and other working materials. It should not be used as a working place unless the stepladder has been constructed with a suitable handhold above the platform. Stepladders must not be used if they are too short for the work being undertaken, or if there is not enough space to open them out fully. If, when on a stepladder, two hands need to be free for a brief period of light work (e.g. to change a light bulb), keep two feet on the same step and the knees and chest supported by the stepladder to maintain three points of contact.

Platforms based on trestles should be fully boarded, adequately supported (at least one support for each 1.5 m of board for standard scaffold boards) and provided with edge protection when the platform is 2 m or higher.

16

477

Fixed scaffolds

It is quicker and easier to use a ladder as a means of access but it is not always the safest. Jobs such as painting, gutter repair, demolition work or window replacement should normally be done using a scaffold. Scaffolds must be capable of supporting building workers, equipment, materials, tools and any accumulated waste. A common cause of scaffold collapse is the 'borrowing' of boards and tubes from the scaffold, thus weakening it. Falls from scaffolds are often caused by badly constructed working platforms, inadequate guard rails or climbing up the outside of a scaffold. Falls also occur during the assembly or dismantling process.

There are two basic types of external scaffold:

1. independent tied – these are scaffolding structures which are independent of the building but tied to it, often using a window or window recess; this is the most common form of scaffolding; and
2. putlog – this form of scaffolding is usually used during the construction of a building; a putlog is a scaffold tube which spans horizontally from the scaffold into the building – the end of the tube is flattened and is usually positioned between two brick courses.

The important components of a scaffold have been defined in a guidance note issued by the HSE as follows:

Standard: an upright tube or pole used as a vertical support in a scaffold.

Ledger: a tube spanning horizontally and tying standards longitudinally.

Transom: a tube spanning across ledgers to tie a scaffold transversely. It may also support a working platform.

Bracing: tubes which span diagonally to strengthen and prevent movement of the scaffold.

Putlog: a tube which is flattened at its end and spans from a ledger to the wall of a building. A scaffold secured to a building in this way is known as a putlog scaffold.

Guard rail: a horizontal tube fitted to standards along working platforms to prevent persons from falling.

Toe boards: these are fitted at the base of working platforms to prevent persons, materials or tools falling from the scaffold.

Figure 16.10 Access ladders should be tied, and extend to at least 1 m above the landing point to provide a secure handhold

✗ Wrong way   Right way ✓

✗ Stepladder too short
✗ Hazard overhead
✗ Over-reaching up and sideways
✗ No grip on ladder
✗ Sideways-on to work
✗ Foot on handrail
✗ Wearing slippers
✗ Loose tools on ladder
✗ Slippery and damaged steps
✗ Uneven soft ground
✗ Damaged stiles
✗ Non-slip rubber foot missing

Steps at right height ✓
No need to over-reach ✓
Good grip on handrail ✓
Working front-on ✓
Wearing good flat shoes ✓
Clean undamaged steps ✓
Firm level base ✓
Undamaged stiles ✓
Rubber non-slip feet all in position ✓
Meets British or European standards ✓

Figure 16.11 Working with stepladders

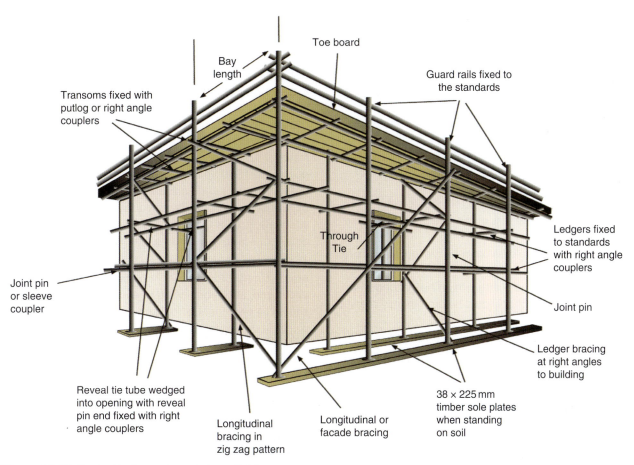

Toe board

Bay length

Guard rails fixed to the standards

Transoms fixed with putlog or right angle couplers

Through Tie

Ledgers fixed to standards with right angle couplers

Joint pin or sleeve coupler

Joint pin

Ledger bracing at right angles to building

Reveal tie tube wedged into opening with reveal pin end fixed with right angle couplers

Longitudinal bracing in zig zag pattern

Longitudinal or facade bracing

38 × 225 mm timber sole plates when standing on soil

Figure 16.12 Typical independent tied scaffold

Raker: an angled or inclined load-bearing tube used to support a cantilevered scaffold working platform. It is fixed to the edge of the working platform at one end and the wall of the building at the other.

Base plate: a square steel plate fitted to the bottom of a standard at ground level.

Sole board: normally a timber plank positioned beneath at least two base plates to provide a more uniform distribution of the scaffold load over the ground.

Fans: specially designed platforms which prevent debris and other materials falling on people passing or working below the scaffold.

Ties: used to secure the scaffold by anchoring it to the building. The scaffold in Figure 16.12 is tied to the building using a through-tie.

Working platform: an important part of the scaffold since it is the platform on which the building workers operate and where building materials are stored prior to use. These are laid on the transoms in various quantities; usually for general purpose, they

should be four boards wide. A working platform can be almost any surface from which work can be undertaken, such as:

▶ a roof
▶ a floor
▶ a platform on a scaffold
▶ mobile elevating work platforms (MEWPs)
▶ the treads of a stepladder.

Other components of a scaffold include access ladders, brick or block guards and chutes to dispose of waste materials.

The following factors must be addressed if a scaffold is to be considered for use for construction purposes:

▶ Scaffolding must only be erected and dismantled by competent people who have attended recognised training courses. Any work carried out on the scaffold must be supervised by a competent person. For complex scaffolds, a written plan may be required. Any changes to the scaffold must be done by a competent person.

▶ In busy town centres, the erection and dismantling of the scaffold should take place during quiet times.

▶ The scaffold must be designed to carry all the required loads and use only sound fittings and materials.

▶ Adequate toe boards, guard rails and intermediate rails must be fitted to prevent people or materials from falling. The top guard rail should be a minimum of 950 mm above the working platform and any gap between the top rail and the intermediate rail should not exceed 470 mm. The toe boards need to be suitable and sufficient to prevent people or materials from falling.

▶ Appropriate bay lengths and widths, bracing and tie-in arrangements should be used together with edge and stop-end protection.

16

479

▶ The scaffold must rest on a stable surface; uprights should have base plates and timber sole boards if necessary.

▶ The scaffold must have safe access and egress.

▶ Work platforms should be fully boarded with no tipping or tripping hazards.

▶ The scaffold must be maintained and cleaned, and have proper arrangements for waste disposal.

▶ The scaffold should be sited away from or protected from traffic routes so that it is not damaged by vehicles.

▶ If the scaffold must be erected near to vehicle roadways then barriers or fenders should be fitted around its base.

▶ The scaffold should be properly braced and secured to the building or structure.

▶ Overloading of the scaffold must be avoided.

▶ Lower level uprights should be prominently marked with red and white stripes.

▶ The public must be protected at all stages of the work and the appropriate highway authority contacted before a scaffold is erected or dismantled beside a public thoroughfare.

▶ Regular inspections of the scaffold must be made and recorded.

There have been many examples of scaffold collapse causing serious injuries to workers and members of the public. Factors which may affect the stability of an independently tied scaffold include:

▶ poor, incompetent erection and/or a lack of regular inspections;

▶ the strength of the supporting ground or foundation;

▶ the proximity to the scaffold of any excavation work;

▶ the effect of excessive surface water in weakening the scaffold foundation;

▶ unauthorised alteration by incompetent persons (perhaps by vandals);

▶ the use of incorrect or damaged fittings during the erection or extension of the scaffold;

▶ the overloading of the scaffold;

▶ adverse weather conditions;

▶ the sheeting in of the scaffold without the use of extra ties;

▶ vehicular impact.

With all scaffolds, brick guards and debris netting should be used to prevent materials and equipment falling to the ground.

For putlog scaffolds, all the previously stated safety requirements apply and some additional ones. The putlog ends, which must be properly flattened, should be fully inserted and securely bedded into well-constructed brickwork. It is recommended that a putlog scaffold should not exceed 45 m in height.

When the work is to take place in a busy area, all the normal safety features of scaffolding apply, but, in addition, special care should be taken with ladder access, debris netting and fans. Appropriate traffic signs (and possibly traffic lights), cones, barriers and lighting arrangements at night will also be necessary.

All scaffolds should have an assembly, use and dismantling plan. Before accepting a scaffold erected by a specialist scaffolding sub-contractor for use, the site supervisor must check the scaffold and obtain a handing-over certificate. A scaffold must either be erected to a recognised standard configuration, such as those in National Access and Scaffolding Confederation (NASC) document TG20 Volume 1 for tube and fitting scaffolds, or the manufacturer's guidance for system scaffolds, or it must be designed by calculation to ensure it has adequate strength and stability. See Appendix 16.5 for details of scaffold structures that need to be designed.

The site supervisor will ensure that all scaffolds are erected in accordance with the standards outlined above and, at the beginning of each week, he/she should inspect the scaffold and ensure that any defect is rectified. A report of the inspection and action taken should be entered in the site inspection register. A similar inspection should also be carried out after high winds or other adverse weather conditions.

Figure 16.13 Fan scaffold to protect people and passing traffic

All materials used for scaffolding must be provided in accordance with the relevant British Standards and checked before use by a scaffolder. Such materials must be properly stored and maintained on the site.

Any scaffold that is being erected, altered or dismantled, or otherwise not suitable for use by employees, must have a warning notice erected stating that it must not be used.

All scaffolds must be checked at the end of each working day to ensure that access to the scaffold by children has been prevented. A checklist of typical scaffolding faults is given in Appendix 16.2.

Requirements for scaffold erectors

All scaffolds must only be erected by scaffolders who are competent and experienced to undertake the work. All the requirements, which were outlined in chapters 3 and 8, apply equally to scaffolders. Only scaffolding firms that have a good reputation and accident record gained from earlier contracts should be employed to erect scaffolds. Scaffolders should be supervised throughout the erection of a scaffold and they should be capable of inspecting scaffolding materials and equipment before it is used. They should be equipped with personal lifelines which must be attached at all times while working at height. All scaffolders must have attended a specialist scaffolding course and be able to provide documentary evidence that they have successfully completed the course.

Design of loading platforms

A loading platform is a particular working platform on a scaffold onto which materials and equipment is loaded to further the construction project. A working platform is part of the scaffold or similar structure on which people work, walk and stand. It must be designed so that it can carry workers, their tools and working materials safely. The platform is built by laying scaffold boards across the transoms and, normally, at least three support points are required for each board. It must be four boards wide. There must be a slight overhang (50–150 mm) across a support at each end of the board. Boards are normally butt jointed. Toe boards and guard rails, at least 950 mm above the working platform, must be fitted inside the standards along the edge of the working platform. Ladders used to access the loading platform must extend at least 1 m above the level of the platform.

All stacking of materials and equipment, such as bricks, must be as close to standards as possible to enable the added weight to be more evenly distributed throughout the scaffold. The boards will need some additional fixing to the structure so that they are protected from lifting during high winds.

The width of the platform will vary with the type of work being undertaken from a minimum of three boards to a maximum of seven (when trestles are to be used on the platform).

The following points are relevant to the design of all working platforms whether loading or not:

▶ They must be wide enough (at least 600 mm) to allow people to pass each other and for any required equipment or materials to be carried and stored.

▶ They must be free of openings and trip hazards.

▶ If members of the public are able to pass beneath, a double-boarded platform with a sheet of polythene placed between the boards may be necessary to prevent debris or tools falling on them.

▶ The principal hazards associated with working platforms are overloading, the fall of materials, tools and equipment and slips, trips and falls of people working on the platform.

▶ Good housekeeping and supervision are essential on all working platforms.

Finally, if it is likely that heavy loads are to be stored on a working platform, the scaffolding firm should be informed so that special support may be designed into the scaffold structure.

Scaffold hoists

Hoists are widely used to transport materials to higher working levels. All hoists must be:

▶ well maintained, inspected regularly and in a good working order;

▶ positioned on firm and stable ground;

▶ erected by trained and experienced people who will follow the manufacturer's instructions and ensure that it is properly secured to the supporting structure;

▶ operated by a driver who is properly trained, and competent;

▶ properly protected at either end; and

▶ marked with its safe working load and its control functions.

The following applies to the use of all hoists whether they are mobile or permanent. The main hazards associated with hoist operations are:

▶ overloading of the hoist;

▶ overloading or incorrect use of lifting gear;

▶ automatic safe load indicator not working;

▶ failure of equipment due to lack of maintenance;

▶ incorrect positioning of lifting appliance;

▶ insecure attachment of load and falls from the platform;

▶ unstable slinging of loads, hook not over centre of gravity;

▶ contact with overhead electricity cables;

▶ lack of operator training;

▶ no banksman used when driver's view is obscured;

▶ incorrect signals.

Only hoists which are suitable for the site and capable of lifting the required loads should be selected and all personnel working with or near lifting appliances must wear hard hats. All hoists must be secured and left in a safe condition at the end of each working period so that any child trespassers are protected.

The following additional precautions should also be taken:

▶ The controls should be set so that the hoist can be operated from one position only (normally ground

16

481

Figure 16.14 Hoist with interlocked gates

level) and that all the landing levels are clearly visible by the operator from the operating position thus preventing people from being struck by the platform or other moving parts.

▶ The hoistway must be enclosed at places where people might be struck, for example working platforms or window openings.

▶ Gates must be provided at all landings and at ground level.

▶ The hoistway must be fenced wherever people could fall down it.

▶ The gates at landings are kept closed except during loading and unloading and they should be secure and not free to swing into the hoistway.

▶ The edges of the hoist platform and the landing should be as close together as possible so that there is no significant gap between for debris to fall through.

▶ Loads, such as wheelbarrows, should be securely chocked and not overfilled.

▶ Loose loads, such as bricks, should be carried in proper containers, or a hoist with an enclosed platform should be used.

▶ The platform must not be overloaded and should be clearly marked with its safe working load.

▶ The hoist must be erected by trained and experienced people who will follow the manufacturer's instructions and ensure that it is properly secured to the supporting structure.

▶ The hoist operator must be properly trained and be competent.

▶ Loads must be evenly distributed on the hoist platform.

▶ The hoist must be thoroughly examined and tested after erection, substantial alteration or repair and at 6-month intervals.

▶ Regular inspections should be carried out at least on a weekly basis and the results recorded.

Finally, no one should be allowed to ride on a goods hoist and a notice, which states this prohibition, should always be posted.

Pre-fabricated mobile scaffold towers

Mobile scaffold towers are frequently used throughout industry. It is essential that the workers are trained in their use since recent research has revealed that 75% of lightweight mobile tower scaffolding is erected, used, moved or dismantled in an unsafe manner.

The following points must be considered when mobile scaffold towers are to be used:

▶ Towers should only be used on level, firm and stable ground.

▶ The selection, erection, dismantling and inspection of mobile scaffold towers must be undertaken by competent and trained persons with maximum height-to-base ratios not being exceeded.

▶ No person must be permitted to erect, alter or dismantle any mobile tower scaffold unless authorised by the site supervisor.

▶ Diagonal bracing and stabilisers should always be used.

▶ Access ladders must be fitted to (not leaning on) the narrowest side of the tower or inside the tower and persons should not climb up the frame of the tower.

▶ All wheels must be locked whilst work is in progress and all persons must vacate the tower before it is moved.

▶ All operatives who are required to use mobile tower scaffolds will be instructed in their safe use and movement.

▶ Outriggers or stabilisers must be extended where applicable.

▶ Tower scaffolds must not be used or moved on sloping, uneven or obstructed surfaces.

▶ Tower scaffolds must only be used in the vertical position.

▶ The tower working platform must be fully boarded, fitted with guard rails and toe boards and not overloaded.

▶ Towers must be tied to a rigid structure if exposed to windy weather or to be used for work such as jet blasting.

▶ Persons working from a tower must not over-reach or use ladders from the work platform.

- Safe distances must be maintained between the tower and overhead power lines both during working operations and when the tower is moved.
- The tower should be inspected on a regular basis and a report made.

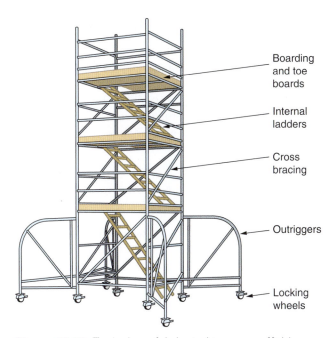

Figure 16.15 Typical pre-fabricated tower scaffold

There are two approved methods for the erection of a pre-fabricated tower scaffold: an advance guard rail system and the 'through the trap' (3T) system. These systems have been developed by the HSE and the Prefabricated Access Suppliers' and Manufacturers' Association and are described in detail on the HSE information sheet *Tower scaffolds* CIS No. 10 (Rev 4).

The main reasons why a mobile tower scaffold may become unstable are:

- brakes defective;
- not erected by a competent person or unauthorised adjustments were made;
- erected on ground that was neither firm nor level;
- the height of the tower exceeding the recommended base to height ratio;
- the failure to fit and/or use outriggers when they were required;
- the safe working load exceeded;
- movement of the tower either with persons and/or materials on the working platform or under windy conditions;
- collision with vehicles or other plant.

The recommended height to base ratio varies from 3:1 to 3.5:1, with no base dimension less than 1 m, dependent on whether the tower is to be used outside or inside the building. The ratio is lower for outside use due to possible wind effects. The ratios quoted should be treated as a guide and the manufacturer's instruction should always be consulted. Anyone erecting a tower scaffold should be competent to do so and should have received training under an industry recognised training scheme, such as the Prefabricated Access Suppliers' and Manufacturers' Association (PASMA), or under a recognised manufacturer or supplier scheme.

Mobile scaffold towers should be inspected regularly as outlined later in this chapter.

Mobile elevating work platforms (MEWPs)

When there is not a fixed structure mobile elevating work platforms are very suitable for many different types of high-level work such as changing light bulbs in a warehouse. Examples of such platforms are scissor lifts and boom-type hydraulic platforms (cherry pickers). The following factors must be considered when using a MEWP:

- The MEWP must only be operated by trained and competent persons and safety harnesses should be available.
- It should be inspected before use.
- The working platform should not be overloaded.
- It must never be moved in the elevated position.
- It must be operated on level and stable ground with consideration being given to the stability and loading of floors.
- The tyres must be properly inflated and the wheels normally immobilised when the platform is in use.
- Outriggers, if fitted, should be fully extended and locked in position.

Figure 16.16 Mobile elevating work platform (MEWP) – scissor lift

16

- Warning signs should be displayed and barriers erected to avoid collisions.
- Due care must be exercised with overhead power supplies and obstructions and adverse weather conditions.
- It should be maintained regularly and procedures should be in place in the event of machine failure.
- Drivers of MEWPs must be instructed in emergency procedures particularly to cover instances of power failure.

Workers using mobile elevating work platforms have been injured by falling from the platform due to a lack of handrails, or inadvertent movement of the equipment because the brakes had not been applied before raising the platform. Injuries have also resulted from the mechanical failure of the lifting mechanism and by workers becoming trapped in the scissor mechanism. All workers on MEWPs should wear safety harnesses.

There is evidence of a growing number of serious and fatal accidents from entrapment accidents between guard rails on MEWPs and adjacent obstructions. There have been several serious accidents caused by the worker becoming trapped by overhead obstructions while the MEWP was in motion preventing him/her from releasing the controls. These accidents often occur during maintenance operations. It is important that risk assessments are undertaken and safe systems of work developed for such work. The Strategic Forum for Construction Plant Safety Group has published guidance on best practice when MEWPs are used in confined overhead spaces. The guidance covers hazards, risk assessment, controls and responsibilities of managers. There have also been accidents caused when a MEWP is reversed into areas where there is poor pedestrian segregation and the driver has limited visibility. During any manoeuvring operation, a dedicated banksman should be used.

Figure 16.17 Mobile elevating work platform (MEWP) – cherry picker with harness and lanyard attached to cradle

If **a MEWP is to be used**, the HSE suggest that the following questions should be considered:

- Height – How high is the job from the ground?
- Application – Do you have the appropriate MEWP for the job? (If you're not sure, check with the hirer or manufacturer.)
- Conditions – What are the ground conditions like? Is there a risk of the MEWP becoming unstable or overturning?
- Operators – Are the people using the MEWP trained, competent and fit to do so?
- Obstructions – Could the MEWP be caught on any protruding features or overhead hazards, such as steelwork, tree branches or power lines?
- Traffic – Is there passing traffic and, if so, what do you need to do to prevent collisions?
- Restraint – Do you need to use either work restraint (to prevent people climbing out of the MEWP) or a fall arrest system (which will stop a person hitting the ground if they fall out)? Allowing people to climb out of the basket is not normally recommended – do you need to do this as part of the job?
- Checks – Has the MEWP been examined, inspected and maintained as required by the manufacturer's instructions and have daily checks been carried out?

Other techniques

Sometimes it is necessary to work from a man-riding skip. It is important that the skip is of sound construction and made from good quality material which has adequate strength. It should be at least 1 m deep and have means to prevent tipping and spinning and persons from falling out. The skip should be properly maintained and be marked with the safe working loads. The crane and lifting tackle used to lift the skip must be regularly inspected and thoroughly examined with appropriate certification being obtained. Only competent persons should undertake this type of work and there must be proper communication with other interested parties, such as the building occupiers, the police and the local authority.

When maintenance operations take place on bridges or similar inaccessible structures, suspended access cradles or platforms may be used. The failure of access cradles has led to serious injuries and fatalities. There are many reasons for such incidents, including:

- poor equipment selection procedures in that the cradle is not suitable for the purpose;
- unsafe access to the cradle;
- overloading of the cradle or non-uniform loading across the platform;

- holes or cracks in the platform floor allowing material to fall below;
- inadequate guard rails and toe boards;
- the structure to which the cradle is attached is not capable of carrying the additional load;
- insecure counterweights and/or braking system;
- the failure of essential components of the cradle structure;
- the failure of the winching and climbing devices;
- poor erection, maintenance and dismantling by incompetent and untrained personnel;
- inadequate emergency procedures.

The safest access to a cradle is at ground level. If access at height is unavoidable, then the cradle must be secured so that it is prevented from swinging away from the structure. It is also important that suitable anchorage points are available for safety harnesses and a secondary rope is fitted. Regular visual inspections should be made at least before each time of use and a weekly detailed inspection should be made by a competent person.

A boatswain's chair is often used for light work of a short-term nature. Whenever it is used, the worker should be protected by a harness and lanyard in case of a possible fall. Abseiling or rope access techniques are also used for inspection work. Devices such as these should only be used when a properly constructed working platform is not practicable.

Equipment, such as cradles, boatswain's chairs and rope access techniques, must be erected either by or under the supervision of a competent person and the worker must be properly trained in the use of the technique. The main rope and safety access rope should always be attached to separate anchorage points. Any tools or equipment used must be secured to the worker to prevent them falling and possibly injuring people below. Whenever possible, the area beneath the work should be fenced off.

Fall arrest equipment

The three most common types of fall arrest equipment are safety harnesses, safety nets and air bags.

Safety harnesses should only be used alone when conventional protection, using guard rails, is no longer practicable. Such conditions occur when it is possible to fall 2 m or more from an open edge. It is important that the following points are considered when safety harnesses are to be used:

1. The length of fall only is reduced by a safety harness. The worker may still be injured due to the shock load applied to them when the fall is arrested. A free fall limit of about 2 m is maintained to reduce this shock loading. Lanyards are often fitted with shock absorbers to reduce the effect of the shock loading.

2. The worker must be attached to a secure anchorage point before they move into an unsafe position. The lanyard should always be attached above the worker, whenever possible.

3. Only specifically trained and competent workers should attach lanyards to anchorage points and work in safety harnesses. Those who wear safety harnesses must be able to undertake safety checks and adjust the harness before it is used.

Three levels of inspection are recommended for safety harnesses as follows:

1. A pre-use check undertaken by the user at the beginning of each shift to check there are no visible or surface defects.

2. A detailed inspection undertaken at least every 6 months. However, for frequently used equipment, this should be increased to at least every three months, particularly when the equipment is used in arduous environments (e.g. demolition, steel erection, scaffolding, steel masts/towers with edges). The results of the inspection should be formally recorded.

3. An interim inspection is also an in-depth, recorded inspection and may be needed between detailed inspections because the associated risk assessment has identified a risk that could result in significant deterioration, affecting the safety of the lanyard before the next detailed inspection is due. The need for and frequency of interim inspections will depend on use. Examples of situations where they may be appropriate include risks from arduous working environments involving paints, chemicals or grit blasting.

Any defects discovered by any of these inspections must be reported to the employer as soon as possible.

Safety nets are widely used to arrest falls of people, tools and materials from height but competent installation is essential. The correct tensioning of the net is important and normally specialist companies are available to fit nets. The popularity of nets has grown since the advocacy of their use by the HSE in 1996. Nets are used for roofing work and for some refurbishment work. Nets, however, have a limited application since they are not suitable for use in low-level construction where there is insufficient clearance below the net to allow it to deflect the required distance after impact. Nets should be positioned so that workers will not fall more than 2 m, in case they hit the ground or other obstructions.

Testing, inspection by a competent person and maintenance are essential activities to ensure that all safety nets are safe for further use on site.

The following points should be considered when safety nets are to be used.

16

Ensure that:

► only nets that comply with the British Standard are used on site;
► all nets are tested annually for UV degradation;
► only competent workers repair nets;
► records of all nets are kept up to date;
► only competent safety net riggers install nets on site;
► a handover certificate for all installed nets is completed by the rigger;
► all nets are inspected every week or after adverse weather conditions;
► damage to the net is reported to the installers; and
► unauthorised alterations to the net are not made.

Air bags are used when it is not either possible or practical to use safety nets. Therefore they are used extensively in domestic house building or when it is difficult to position anchorage points for safety harnesses. When air bags are used, it is important to ensure that the bags are of sufficient strength and the air pressure high enough to ensure that any falling person does not make contact with the ground. Only reputable suppliers should be employed for the provision of air bags. Air bags or bean bags are known as soft landing systems and are used to protect workers from the effects of inward falls. Other possible solutions to this problem of the inward fall are the use of internal scaffolding or lightweight 'crash deck systems'.

Figure 16.18 Airbags to give a safe soft landing

Air bags may be linked together to form an inflated crash deck system. Such a system is suitable when a safety net is not viable. It consists of a series of interlinked air mattresses that are positioned beneath the working area and is suitable for working at height inside a building where safety harnesses would not be practicable. The mattresses are interconnected by secure couplings and inflated in position on site using an air pump. The mattresses are made in various sizes so that any floor area configuration can be covered. A similar form of crash deck can be made from bean bags that are clipped together.

After the completion of the construction project, cable-based fall arrest systems are normally suitable for on-going building maintenance work. They may also be used for the construction of complex roof structures, such as parabolic or dome structures, where a safety net may be more than 2 m below the highest point of the structure.

Emergency procedures (including rescue)

A suitable emergency and rescue procedure needs to be in place for situations that could be expected to occur on the construction site. There is a legal requirement to plan for emergencies and rescue when there is working at height. Such foreseeable situations could include a crane driver trapped in his cab due to a power failure or a serious ill-health problem, or the rescue of a person who has fallen into a safety net. The method of rescue may be simple and straightforward, such as putting a ladder up to a net and allowing the fallen person to descend or the use of a MEWP to lower the person to the ground.

The Work at Height Safety Association (WAHSA) recommends that an appropriate rescue plan should consider:

► the safety of the persons undertaking or assisting with the rescue;
► the anchor points to be used for the rescue equipment;
► the suitability of equipment (anchors, harnesses, attachments and connectors) that has already arrested the fall of the person for use during the rescue;
► the method that will be used to attach the person to the rescue system;
► the direction in which the person needs to be moved to get them to the point of safety; and
► the possible needs of the person after the rescue, including first-aid.

The HSE guidance states that the key is to release the person from the suspended-harness position to the nearest point of safety, in the shortest possible time, and as soon as is safely possible. Otherwise, the person can suffer from **presyncope** (loss of consciousness). The symptoms for this include light-headedness, nausea, sensations of flushing, tingling or numbness of the arms or legs, anxiety, visual disturbance, or a feeling they are about to faint. **Presyncope** can occur within one hour (or sooner for some people) and if the rescuer is unable to immediately release the conscious person, elevation of the legs by the person or rescuer may prolong the tolerance of suspension.

Proprietary rescue systems are available and may well be suitable for many more complex rescues. The system selected needs to be proportionate to the risk and there should not be undue reliance on the emergency services. However, arrangements should be in place to inform the emergency services of any serious incident. It is important that rescue teams are available at all times with a designated leader. Instruction and training are essential for the teams, preferably by simulated rescue exercises.

The term 'suspension trauma' is one that is used to describe the situation of a person falling into suspension in a harness and then becoming unconscious. The emergency or rescue procedures will require the development of an emergency and rescue plan that will get the person down safely in the shortest possible time and before the emergency services have responded. Motionless head up suspension can lead to light-headedness, nausea, sensations of flushing, tingling or numbness of the arms or legs or anxiety.

Figure 16.19 Fall arrest harness and device

16.2.3 **Roof work**

Building owners and occupiers should take the following actions concerning the roofs of their buildings:

▶ Control access to the roofs.
▶ Provide signs warning of fragility at suitable access points.
▶ Check the competence of anyone employed to work on roofs.
▶ Check that the contractors or workers take precautions to prevent persons falling through or off the roof.

Before any roof work is started, a risk assessment should be undertaken. This should include an assessment of the structural integrity of the roof and the methods to be used to repair the roof and a COSHH (Control of Substances Hazardous to Health) assessment of any hazardous substances to be used. The risk assessment should be followed by the provision of a method statement by the sub-contractor. This should be examined to ensure a safe method of working is proposed and which does not conflict with other sub-contractors. Only competent and trained workers should be employed for roof work and any equipment used should be suitable for the job.

The precautions needed when repairing the flat roof of a building include the provision of safe access to the roof (using scaffolding and/or ladders), the use of crawling boards, and the provision of safety harnesses and edge protection, using toe boards, to prevent the fall of persons or materials. Roof battens should be checked to ensure they provide safe hand and footholds. If they do not then crawling ladders or boards will be used. Crawling boards should always be provided when work on fragile roofing materials cannot be avoided.

The perimeter of the roof requires a two-rail scaffold guard rail fitting, together with a toe board, where the gutter would not prevent materials slipping from the roof. Protection of the leading edge must also be provided. Where this is also used as a working platform, it is recommended that the 'Youngman' trolley system or similar fitted with guard rails is used. Suitable safety nets may also need to be provided to prevent people and materials from falling.

Access to the area immediately below the work should be restricted using suitable barriers, netting, safety signs and safety helmets. There should also be no danger to employees from fragile roof lights, voids, overhead obstructions and services.

All walkways for the passage of people and materials should be a minimum of 600 mm wide and protected with double guard rails and toe boards on the exposed edge. Where staging is used care should be taken to ensure that the ends are supported and not cantilevered over the steelwork. Where a gutter is used as a walkway either along the perimeter or along a valley of the roof the need for similar protection still applies. Walkways should be kept clear of materials, tools and debris as far as possible.

Consideration should also be given to the way in which materials are transported to and from the working area,

16

involving lifts, hoists, manual handling and/or using chutes and covered waste skips. When they are to be used on the roof, such materials should be secured in demarcated areas. Materials stacked on the roof should be securely tied down when not required, to avoid the possibility of being blown down from the roof. Where materials are stacked on purlins on the open part of the roof, a suitable form of walkway as described above should be provided around each pack and also for the working area.

Good housekeeping procedures are essential for safe roof work.

All openings left in the roof for lights, vents or similar items should either be protected with guard rails or have some form of temporary covers fitted.

Where an inclined hoist is used to lift materials onto the roof this should have protective barriers around the base, an overhead mesh guard fitted to protect the operator and around the top a suitable walkway with guard rails to provide a safe means of off-loading the materials onto the roof.

Ladders securely tied should be situated as close as possible to the working area on the roof and extend 1 m above the roof level. Where the height to the roof exceeds 9 m, it is normally a requirement that a scaffold tower with ladders be provided such that the height between intermediate platforms does not exceed 9 m.

If asbestos is present, then special precautions will be required including the provision of overalls and

respiratory protection, the need for damping down and the avoidance of breaking asbestos sheets. Safe disposal of waste asbestos is a legal requirement.

Consideration should also be given to the wearing of a safety harness where other forms of protection are not adequate. This applies, for example, when fixing the gutter or the scaffold guard rail at the start of work. Safety helmets should be worn by all persons in accordance with the Construction Head Protection Regulations whilst work is in progress overhead and/or there is a risk of head injury.

During the construction of large industrial roofs, a purlin trolley system is usually used together with a safety harness. Such systems have a double handrail on the leading edge positioned on the side opposite to the working side. As the working side is open to the roof, a safety harness is needed to protect against falls from this edge. A second line of defence has recently been designed into these trolleys by the attachment of a horizontal barrier beneath.

Other issues include the use of netting and protection against adverse weather conditions, by issuing suitable clothing (and possibly sun cream) and limiting exposure. The sensible positioning of bitumen boilers and gas cylinders will also reduce the risk of fire, and electrical risks will be reduced by the use of residual current devices or reduced low voltage for portable electrical equipment.

One final aspect of roof work which must be considered is work that takes place immediately below the roof in the attic or roof space, where access may be gained using MEWPs and 'Youngman' staging platforms. An example of this is the painting of roof girders in a warehouse. As with similar work, continual liaison with the client is essential during the planning stage and after the work begins. At the planning stage, any hazards existing in the roof space must be assessed including any lighting, heating and electrical and other services that may be present. Often other work activities may be taking place in the warehouse while the painting work is in progress. It is essential, therefore, that there is agreement with the client on the sequence of work, working periods and working areas so that there is no accidental contact between painters and high reach pallet fork-lift trucks. The painters will need to be briefed on the operational methods and routes used by the pallet trucks. Similarly, the warehouse workers will require information on the materials and systems of work used by the painters. Other related aspects of the work, such as safe use of the mobile and stationary platforms, the use of harnesses and debris netting, storage arrangements for paints and solvents and the provision of respiratory protective equipment, have been dealt with elsewhere.

Contractors and roof workers should take the following actions:

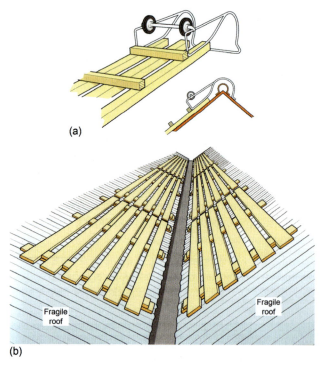

(a)

(b)

Figure 16.20 (a) Roof ladder. The ridge iron should be large enough to be clear of the ridge tile; (b) permanent protection installed at valley gutter (the protection should be supported by at least three rafters beneath the roof sheets)

- Treat any roof as a fragile surface unless there is sound information to prove otherwise.
- Only undertake work for which they are competent.
- Only work on the roof if there is no other reasonable way of tackling the job.
- Take suitable precautions to prevent falls through or off the roof.
- Train and supervise all roof workers properly

The HSE has given the following advice on roof work:

- Almost all domestic roof work needs scaffolding – fit edge protection to stop people and materials from falling from eaves and gable ends.
- On terraced properties make sure scaffolding is provided at the front and back of the property.
- Use scaffolding around chimneys and roof windows.
- Do not throw materials from the roof or the scaffold – use a chute or similar equipment.
- Stop materials falling onto the street, and people – for example, use debris netting sheeting and/or close fitting scaffold boards.
- Keep people away from the area below the roof work.
- Do not go onto fragile roof surfaces such as cement sheets or those with skylights.
- Work from underneath, reach from an access platform or cover fragile areas on the roof.
- Ladders can be used to access the workplace but working from ladders is allowed only as a last resort.
- Make sure your roof workers are properly trained and competent to do the work safely.

Three examples of safe systems of roof work are given in Appendix 16.6.

16.3 Protection of others

Members of the public must be protected at all stages of the construction project and at all times of day and night. Scaffold should be made very visible by painting standards with red and white stripes, masking couplers, and using warning signs, good illumination at night and barriers and other means of safe passage. Safety precautions, such as the boarding of ladders, should be taken during periods when the site is unoccupied. Such protection should be provided to protect anyone, including trespassers, such as children.

Many construction projects take place in occupied premises and important health and safety matters must be discussed with the premises management. This was covered in some detail in Chapter 8 and the following points need to be agreed:

- the location and isolation of the construction working areas;
- the supervision of the work;
- procedures for the evacuation of the premises in the event of an emergency;
- the use of welfare and canteen facilities;
- the protection of the occupier's employees;

- the provision of protected access into the building;
- the safe removal of all waste materials; and
- the procedures for the handing back of areas of the premises when the work has been completed.

The use of sheeting, netting and fans to protect members of the public has been covered earlier in this chapter and in Chapter 8. Similarly demarcation, barriers, tunnels and signs are also covered in Chapter 8. Head protection for visitors is an essential requirement.

16.4 Working over or near water

Where construction work takes place over water, steps should be taken to prevent people falling into the water and rescue equipment should be available at all times. If the work takes place adjacent to a river or canal, special precautions are needed for persons using a towpath. The CDM Regulations cover any work where there is a risk to operatives of drowning. Working on or adjacent to water or when people have to pass near or across water on their way to work gives rise to a risk of drowning. People can also drown in other liquids such as slurries, chemicals and some free-flowing foodstuffs in vats and silos.

At the planning stage of any work over or beside water, the following points need to be considered:

- the provision of suitable fencing to ensure that persons do not fall into the water, i.e. the provision of barriers or fences with toe boards;
- the provision of life jackets, buoyancy aids and grab lines (fast-flowing water such as rivers);
- the provision of rescue equipment, lifebelts, harnesses, safety lines and, in some cases, a rescue boat;
- where transport by water is required, arrangements must be made for safe landing stages and life jackets;
- when boats are used they should be of suitable construction, under the control of a competent person, not overloaded, and properly maintained;
- the provision of rescue teams and procedures;
- specific training for boat operatives, rescue teams and supervisors;
- all precautions must also protect the safety of the public, especially children;
- suitable security measures will be necessary to prevent the theft of rescue equipment and boats outside working hours;
- weather, tides and flooding conditions must be taken into account when planning rescue measures and using boats as transport;
- all openings, drains and vats must be covered, where possible.

Site supervision is required to ensure that all barriers, fencing and rescue equipment is provided before any work commences which could place personnel at risk

Figure 16.21 Working over or near water – large scaffold with protection screens and a small boat moored under the bridge in case rescue is needed

from drowning. The site supervisor will ensure that only authorised personnel alter barriers and operate rescue equipment and boats. The supervisor will also ensure that all rescue equipment is checked regularly and that any defective equipment is repaired or replaced immediately. He/she will ensure that personnel appointed as rescue teams are available during working periods and that replacement personnel are available when necessary.

There are several general hazards associated with working at height over or adjacent to water which are similar to those for any work done at height. These include falling from a height, scaffold stability, falling material, pedestrian safety and possible collision with the scaffolding by boats. Many of these issues can be addressed by the provision of a properly designed scaffold, erected by competent persons and securely tied and provided with guard rails, toe boards and ladder access to working platforms. Other precautions include fall arrest equipment such as safety nets or harnesses, the non-overcrowding of the scaffold, the provision of rescue equipment, such as lifebelts, throw-lines, lights and whistles; and the

availability of a rescue boat under the control of a competent person, should someone fall into the water. Harnesses must be worn by those working on the scaffold in addition to the use of appropriate personal protective equipment. There should be a current register of workers and nobody should work alone on the scaffold.

For pedestrians, using perhaps a towpath, a protected access or safe diversion route must be provided. Signs, lighting and protective barriers will also be required to minimise the risk of boats colliding with the scaffold.

Finally, consideration must be given to health hazards which are associated with working near water. These include biological hazards such as leptospirosis and tetanus and general hazards to health associated with working on any construction site such as noise, vibration, musculoskeletal problems, electricity, mechanical hazards, flying particles, dust and contact with cement.

Appropriate control measures include disinfection, good standards of personal hygiene, the use of appropriate personal protective equipment, adequate washing and first-aid facilities, and the provision of information, instruction, training and supervision.

16.5 Further information

The Construction (Design and Management) Regulations 2015

The Personal Protective Equipment at Work Regulations 1992

The Provision and Use of Work Equipment Regulations 1998

The Work at Height Regulations 2005

Health and safety in roof work (HSG33), HSE Books, ISBN 978-0-7176-6257-3 http://www.hse.gov.uk/pubns/priced/hsg33.pdf

Protecting the public: Your next move (HSG151), HSE Books, ISBN 978-0-7176-6294-4 http://www.hse.gov.uk/pubns/priced/hsg151.pdf

Safe use of work equipment (ACoP) (L22), HSE Books, ISBN 978-0-7176-6295-1 http://www.hse.gov.uk/pubns/priced/l22.pdf

Inspection and reports, HSE Construction Information Sheet No 47 (rev1) Series Code, CIS47REV1 http://www.hse.gov.uk/pubns/cis47.pdf

Personal Protective Equipment at Work (Guidance), L25, HSE Books, ISBN 978-0-7176-6139-3 http://www.hse.gov.uk/pubns/priced/l25.pdf

Working at height, a brief guide, INDG401, HSE Books, http://www.hse.gov.uk/pubns/indg401.pdf

16.6 Practice revision questions

1. (a) **Identify FOUR** hazards associated with work at height above ground level.
   (b) **Outline** a hierarchy of control measures to be considered when a construction worker is at risk of falling while working at height.
   (c) **Outline** the precautions to be taken to ensure the safety of construction workers required to undertake repair work on a fragile roof.

2. (a) Falls from height are the major cause of accidents in the construction industry, particularly falls from ladders and scaffolds. **Outline SIX** possible reasons for such accidents.
   (b) Serious falls are also associated with lift shafts, stairwells and other holes in the floor during construction work in multi-storey buildings. **Outline** the measures to be taken to prevent the occurrence of such falls.

3. The stone façade of a three-storey building within a busy city centre is to be cleaned using a specialist cleaning contractor during a general refurbishment of the building. The work will be undertaken using high pressure water-jetting.

   Describe a safe system and method of work to be adopted by the contractor to ensure the safety of:
   (a) the water-jetting operatives;
   (b) the workers carrying out the internal refurbishment work;
   (c) the public and users of the highway.

4. An upgrade of an existing ventilation system is to be undertaken in a large storage warehouse. The work is to be carried out on an overnight basis by contractors when the only other activities in the warehouse are stock replenishment on the racks and cleaning. The majority of the work will involve access above a suspended ceiling at a height of 5 m above the ground but plant situated on the roof of the store is also to be replaced.

   Outline the main hazards and the associated controls that must be addressed to ensure the health and safety of the contractors and others on site during the work.

5. A wooden pole ladder is to be used as a means of access to a scaffold.
   (a) **Outline** the main issues to be considered when selecting and inspecting the ladder prior to use to ensure its suitability for the job.
   (b) **Describe** the safe practices required when the ladder is used in the workplace.

6. Falls from height account for a large proportion of the fatal and major injury accidents that occur on construction sites, particularly when using scaffolding to reach high areas.
   (a) **Outline** the issues that need to be considered when selecting equipment for working at height.
   (b) **Outline** the items that need to be checked when inspecting a scaffold.
   (c) **Identify** the requirements for inspecting scaffolds.

7. (a) **Outline THREE** causes of a scaffold collapse.
   (b) **Outline** the precautions that should be taken to reduce the risk of injury to members of the public during the erection and use of the scaffold.
   (c) **Identify EIGHT** items to be examined when inspecting a putlog scaffold.

8. Mobile tower scaffolds must be used on stable, level ground.
   (a) **Give** reasons why a mobile tower scaffold may become unstable.
   (b) **Identify SIX** additional points that should be considered to ensure the safe use of a mobile tower scaffold.
   (c) **Identify** measures that should be adopted in order to protect against the dangers of people and/or materials falling from a mobile tower scaffold.

9. Ceiling repairs are to be undertaken in a busy warehouse by construction workers using a mobile elevating work platform (MEWP).
   (a) **Identify** the hazards associated with this work.
   (b) **Outline** the controls required to ensure the safety of the maintenance workers and others who may be affected by the work.

10. A temporary boatswain's chair is to be used to undertake inspection work on the external façade of a four-storey block of flats.

    Outline the factors to be considered in the installation and use of the chair to reduce the risk to the user and others who may be affected by the work.

11. The three most common types of fall arrest equipment are safety harnesses, safety nets and airbags. **Outline** the application, advantages and disadvantages of each of these devices to protect those working at height.

16

12. A flat roof is to be repaired while a building remains occupied. A small platform hoist is to be used for lifting materials to the roof. There is no access to the roof from inside the building.

    (a) **Outline** the issues that need to be addressed to reduce the risks to the workers involved in the roof work and others who may be affected by that work.

    (b) **Outline** the measures required to ensure the safe operation of an inclined hoist used to raise and lower the required roofing materials.

13. A painter has been contracted to paint roof girders in a storage warehouse. The painters will gain access to the roof area from both mobile elevating work platforms and 'Youngman' staging platforms. High reach pallet trucks will continue to be used in the warehouse during the work to remove and replace items from the warehouse racking.

    Outline the hazards and their controls to ensure the health and safety of the painters, warehouse staff and any visitors to the warehouse.

14. A gable end wall of a four-storey building adjacent to a canal needs re-pointing. An independent tied scaffold will be used for access by construction workers. The canal is used by boats of various sizes. The scaffold will obstruct the towpath and will restrict the width of the canal.

    (a) **Outline** the health and safety issues that need to be addressed during the planning of the work.

    (b) **Describe** the specific measures that may be necessary to ensure safety of the construction workers and those who use the towpath and the canal.

APPENDIX 16.1 Inspection timing and frequency chart

For an Inspection report form see Chapter 23 Form C1.

Place of work or work equipment	Timing and frequency of checks, inspection and examinations equipment								
	Inspect before work at the start of every shift (see note 1)	Inspect after any event likely to have affected its strength or stability	Inspect after accidental fall of rock, or other material	Inspect after installation or assembly in any position (see notes 2 and 3)	Inspect at suitable intervals	Inspect after exceptional circumstances which are liable to jeopardise the safety of work equipment	Inspect at intervals not exceeding 7 days (see note 3)	Check on each occasion before use (REPORT NOT REQUIRED)	LOLER Thorough Examination (if work equipment subject to LOLER) (see note 4)
Excavations which are supported to prevent any person being buried or trapped by an accidental collapse or a fall or dislodgement of material	✓	✓	✓						
Cofferdams and caissons	✓	✓							
The surface and every parapet or permanent rail of every existing place of work at height								✓	
Guard rails, toe boards, barriers and similar collective means of fall protection				✓	✓	✓			
Scaffolds and other working platforms (including tower scaffolds and MEWPs) used for construction work and from which a person could fall more than 2 m				✓		✓	✓		✓
All other working platforms				✓	✓	✓			✓
Collective safeguards for arresting falls (e.g. nets, airbags, soft landing systems)				✓	✓	✓			
Personal fall protection systems (including work positioning, rope access, work restraint and fall arrest systems)				✓	✓	✓			✓
Ladders and stepladders					✓	✓		✓	

Source: Timing and frequency chart (reproduced from HSG150).

Notes

1. Although an excavation must be inspected at the start of every shift, only one report is needed in any seven-day period. However, if something happens to affect its strength or stability, and/or an additional inspection is carried out, a report must then be completed. A record of this inspection must be made and retained for three months.

2. 'Installation' means putting into position and 'assembly' means putting together. You are not required to inspect and provide a report every time a ladder, tower scaffold or mobile elevating work platform (MEWP) is moved on site or a personal fall protection system is clipped to a new location.

3. An inspection and a report (see Inspection Report) is required for a tower scaffold or MEWP (used for construction work and from which a person could fall 2 metres) after installation or assembly and every seven days thereafter, providing the equipment is being used on the same site. A record of this inspection must be made and retained for three months. If a tower scaffold is reassembled rather than simply moved, then an additional, pre-use inspection and report is required. It is acceptable for this inspection to be carried out by the person responsible for erecting the tower scaffold, providing they are trained and competent. A visible tag system, which supplements inspection records as it is updated following each pre-use inspection, is a way of recording and keeping the results until the next inspection.

4. All work equipment subject to Lifting Operations and Lifting Equipment Regulations (LOLER) Regulation 9, thorough examination and inspection requirements, will continue to be subject to LOLER Regulation requirements.

APPENDIX 16.2 Checklist of typical scaffolding faults

Footings	Standards	Ledgers	Bracing	Putlogs and transoms	Couplings	Bridles	Ties	Boarding	Guard rails and toe boards	Ladders
Soft and uneven	Not plumb	Not level	Some missing	Wrongly spaced	Wrong fitting	Wrong spacing	Some missing	Bad boards	Wrong height	Damaged
No base plates	Jointed at same height	Joints in same bay	Loose	Loose	Loose	Wrong couplings	Loose	Trap boards	Loose	Insufficient length
No sole plates	Wrong spacing	Loose	Wrong fittings	Wrongly supported	Damaged	No check couplers	Not enough	Incomplete	Some missing	Not tied
Undermined	Damaged	Damaged			No check couplers			Insufficient supports		

APPENDIX 16.3 Checklist for a safety inspection of a scaffold

Explain how the work should be planned and undertaken in order to ensure the health and safety of both painters and warehouse staff.

- Is the scaffold on a firm foundation?
- Are standards (uprights) resting on suitable base plates and (where the scaffolding is not on hard standing) timber sole boards? Bricks, blocks and other building materials should not be used as packing.
- The scaffold should not be undermined by excavations close to its supports.
- Are the uprights vertical and the horizontals horizontal?
- Are the uprights close enough together and is the spacing consistent? (Note: the permissible width of bays will vary with the operations for which the scaffolding is intended. The greater the loading, the closer the uprights will need to be.)
- Are load-bearing couplers used where appropriate?
- Are working platforms properly supported (each board resting on at least three supports – no board should overhang the last support by a distance greater than three times its thickness), fully boarded out (no gaps for people or materials to fall through) and wide enough to allow safe access and the safe movement of materials?
- Has the scaffold been erected by a competent person? Are inspection records in order?
- Are all components in good condition?
- Are guard rails and toe boards fitted to all working platforms to prevent people and materials falling?
- Is additional protection such as wire mesh brick guards used where appropriate?
- Is the scaffold loaded with excessive quantities of materials?
- Are materials safely stacked? (Loads such as piles of bricks should be in line with the standards.)
- Is the scaffold structure adequately stiffened by the use of diagonal bracing?
- Is the scaffold adequately tied to a structure, or suitably buttressed, to prevent collapse?
- Are incomplete sections of the scaffold marked with suitable warning notices?
- Where there is sheeting or some other feature which will increase the windage of the scaffold, has this been allowed for in the design?
- Is there safe access to all the scaffold's working platforms?

APPENDIX 16.4 Scaffold design, inspection, competence and supervision checklist

This guide is intended to clarify when scaffold design is required and what level of training and competence those erecting, dismantling, altering, inspecting and supervising scaffolding operations are expected to have obtained.

Design and inspection issues

- Unless a scaffold is a basic configuration described in recognised guidance, e.g. NASC Technical Guidance TG20 for tube and fitting scaffolds or manufacturer's guidance for system scaffolds, the scaffold should be designed by calculation, by a competent person, to ensure it will have adequate strength and stability.
- All scaffolding should be erected, dismantled and altered in accordance with either NASC guidance document SG4 for tube and fitting scaffolds or the manufacturer's erection guide for system scaffolds.

▶ For scaffolds that fall outside the scope of 'Basic Scaffolds' as described in bullet point 1, the design information should describe the sequence and methods to be adopted when erecting, dismantling and altering the scaffold, if this is not covered by published guidance as detailed in bullet point 2.

▶ Any proposed modifications or alterations outside a generally recognised standard configuration should be designed by a competent person.

▶ Handover certificates should refer to relevant drawings, permitted working platform loadings and any specific restrictions on use.

▶ All scaffolding inspection should be carried out by a competent person whose combination of knowledge, training and experience is appropriate for the type and complexity of the scaffold he/she is inspecting. Competence may have been assessed under the Construction Industry Scaffolders Record Scheme (CISRS) or an individual may be suitably experienced in scaffolding work and have received additional training under a recognised manufacturer/supplier scheme for the specific configuration he/she is inspecting.

▶ A non-scaffolder who has attended a suitable scaffold inspection course and has the necessary background experience would be considered competent to inspect a basic scaffold (i.e. a site manager).

▶ The scaffold inspection report should note any defects and corrective actions taken, even when those actions are taken promptly, as this assists with the identification of any recurring problems.

▶ To prevent use by unauthorised persons of incomplete scaffolds, relevant warning signs identifying the areas where access is not permitted should be displayed at the access points to these areas. In addition, access to the incomplete areas should be prevented by suitable physical means.

Competence and supervision issues

▶ All employees should be competent (or in the case of trainees, supervised by a competent person) for the type of scaffolding work they are undertaking and should have received appropriate training relevant to the type and form of scaffolding they are working on.

▶ Employers must provide appropriate levels of supervision taking into account the complexity of the work and the levels of training and competence of the scaffolders involved.

▶ As a minimum requirement, every scaffold gang should contain an appropriately qualified scaffolder for the type and complexity of the scaffold to be erected, altered or dismantled. This may be an individual who has received training under an industry recognised training scheme, e.g. CISRS, and has been awarded the scaffolder card or someone who has received training under a recognised manufacturer/supplier scheme, to the limit of the configuration(s) involved.

▶ Trainee scaffolders should always work under the direct supervision of a qualified scaffolder (i.e. a working foreman). Scaffolders are classed as 'trainees' until they have completed the approved training and assessment required to be deemed qualified.

▶ Erection, alteration and dismantling of complex designed scaffolding (e.g. suspended scaffolds, shoring, temporary roofs, etc.) should be done under the direct supervision of a competent person. This may be a qualified Advanced scaffolder, a design engineer providing they possess the necessary industry experience or alternatively an individual who has received training under a recognised manufacturer/supplier scheme to the limit of the configuration(s) involved.

APPENDIX 16.5 Scaffold structures that need to be designed

1. Dead shores
2. Flying shores
3. Raking shores
4. Cantilevered scaffolds
5. Truss-out scaffolds
6. Access birdcages
7. Façade retention
8. Access scaffolds with more than the two working lifts allowed with TG20 'Basic Scaffolds'
9. Buttressed free-standing scaffolds
10. Temporary roofs and temporary buildings
11. Support scaffolds
12. Loading bays founded on the ground
13. Mobile and static towers outside base/height limitations
14. Free-standing scaffolds outside base/height limitations
15. Temporary ramps and elevated roadways
16. Staircases and fire escapes
17. Spectator terraces and seating stands
18. Bridge scaffolds
19. Towers requiring guys or ground anchors
20. Offshore scaffolds outside Offshore Contractors Association (OCA) handbook
21. Pedestrian footbridges or walkways
22. Slung and suspended scaffolds
23. Protection fans, nets and pavement frames
24. Marine scaffolds
25. Boiler scaffolds
26. Power line crossings

16

27. Lifting gantries and towers
28. Steeple scaffolds
29. System scaffolds outside users' guide parameters
30. Sign board supports
31. Sealing end structures
32. Temporary storage on site
33. Masts, lighting towers and transmission towers

34. Advertising hoardings/banners
35. Any scaffold structure subject to:
    ▷ Vibration
    ▷ High loading
    ▷ Long-term duration
    ▷ High-risk areas
    ▷ Loading from passenger/goods hoists.

APPENDIX 16.6 Examples of safe systems of work used in roof work

1. Slating work

1. Materials will be delivered as close as possible to the roofing operation.
2. Materials will be loaded onto the scaffold by a mechanical conveyor or ladder.
3. Felt and battens will be laid from the eaves to the ridge in accordance with the specifications.
4. All slating will progress from right to left across the roof and, on completion, the ridge tiles and fixings will be similarly loaded and laid.
5. All waste materials and debris will be cleared from the site daily.
6. The site will be left in a tidy condition on completion.

2. Lead work on roofs

1. The Control of Lead at Work Regulations 2002 and associated Approved Code of Practice will be followed closely during all work involving lead.
2. Appropriate manual handling procedures will be adopted.
3. Work will be confined to clean solid metallic lead and always carried out in the open air.
4. In areas where lead work is being carried out, no food or drink will be consumed and no smoking allowed.
5. Adequate washing facilities will be provided on site. Nail brushes will be provided.
6. All employees working with lead will wear overalls which must remain on site at all times (except when being laundered at a suitably equipped facility).

3. Working on a fragile roof (with a non-asbestos roofing material)

A risk assessment will be provided for each job.

1. The area of roof in question which is to be repaired or renewed will be assessed from the scaffold unless a specific scaffold extension or duckboards have been provided and are in place on the area in question.

2. Fixing bolts will be removed one at a time either manually or with the use of 110 volt portable equipment. No smashing or breaking of sections of sheeting is permitted unless prior authorisation has been obtained from the operations manager.
3. Old sheets and material are to be placed into the skip provided via the installed rubbish chute. In the event of larger sections being removed which will not go into the chute, they will be lowered by hoist or pulley wheel to ground level where they will be received by a colleague.
4. In the event of repair of a fragile roof by walking through the valley gutters the following precautions must be taken:

    (a) The valley will be cleared of debris and water, i.e. leaves and soil. Any areas where the valley is showing dips or deflection should be reported immediately, unless they were briefed on at the start of the work, for a revision of the work method.

    (b) Both elevations of the roof either side of the gutter will be provided with either a continuous 'safe site' supergrip platform to reduce the possibility of slipping and falling through fragile sheeting, or a series of good condition pallets to serve the same purpose.

5. To carry out leak spotting, a safety harness secured to either an internal roof purlin by removing a ridge cap or other secure fixing near the ridge, must be used.
6. To provide a pathway on fragile sheets a series of duckboards should be used with a specially adapted hook assembly on inclined roof areas.
7. Under no circumstances is anyone allowed to walk on sheets, bolts or any other bare roof coverings to gain access to an area requiring attention.
8. If workers are unable to use equipment provided they must report to the operations manager and withdraw from the area until receiving further instructions.

CHAPTER 17

Excavation work and confined spaces – hazards and risk control

> **This chapter covers the following NEBOSH learning objectives:**
> 1. Explain the hazards and risk assessment of excavation work
> 2. Explain the control measures for excavation work
> 3. Explain the hazards associated with confined space working
> 4. Outline the control measures for confined space working

17.1 Excavation work hazards and risk assessment

17.1.1 Introduction

Excavation work is an essential part of the construction process and one of the most hazardous. Building foundations and the installation of drainage, sewage and other services require trenches to be excavated. Because it is a hazardous operation several people are killed or seriously injured each year whilst working in excavations. Many are killed or injured by excavation wall collapses and falls of materials, others by contact with buried services. Excavation work has to be properly planned, managed, supervised and carried out to prevent accidents and ill-health to workers.

A recent survey of accidents has shown that there are on average seven fatalities each year in excavation work. Over a 5-year period, the survey showed that fatal and major injuries to workers who were working in excavations were caused by the following events:

- struck by falling or flying object, including earth (23%);
- struck by a construction vehicle or plant (15%);
- falling into excavations or from ladders or working platforms (14%);
- contact with electricity (12%);
- trapped by a collapse of earth or materials (11%);
- other (25%).

A major factor in such accidents is the type of soil which is being excavated. The soil structure may range from sand to clay and rock. Sands and gravel tend to be more unstable than clays or rock and need more support during the excavation process to protect both workers and plant from the effects of collapse. The stability of soils changes rapidly with water content and many collapses in the past have occurred after storms.

There are several major accidents caused each year by contact with buried services during excavation work. In a recent case, a man was digging a hole with a spike when he struck a high voltage electrical cable causing a massive explosion and serious injuries for him.

The HSE has produced an excellent booklet entitled *Health and Safety in Excavations* HSG185. This offers detailed advice to those involved in the planning and design of excavations.

17.1.2 Hazards

There are about seven deaths each year due to work in excavations and these are often related to the

composition and structure of the soil which forms the walls of the excavation. Many types of soil, such as clays, are self-supporting but others, such as sands and gravel, are not. The walls of excavations can collapse without any warning resulting in death or serious injury. Many such accidents occur in shallow workings. It is important to note that, although most of these accidents affect workers, members of the public can also be injured.

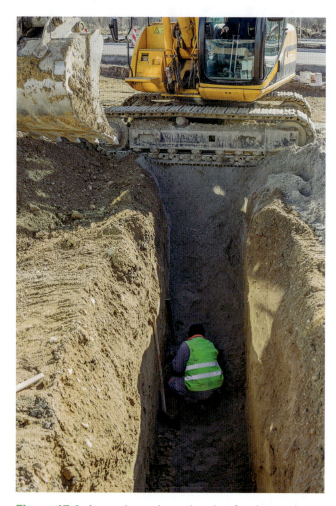

Figure 17.1 A very hazardous situation for the worker with a deep trench, a heavy machine very close to the edge and no trench supports

The specific hazards associated with excavations are as follows:

- collapse of the sides;
- materials falling on workers in the excavation;
- dangers associated with excavation machinery;
- falls of people and/or vehicles into the excavation;

- workers being struck by plant;
- specialist equipment such as pneumatic drills;
- hazardous substances, particularly near the site of current or former industrial processes;
- influx of ground or surface water and entrapment in silt or mud;
- proximity of stored materials, waste materials or plant;
- proximity of adjacent buildings or structures and their stability;
- contact with underground services;
- contact with overhead power lines;
- access to and egress from the excavation;
- contaminated ground; and
- fumes, lack of oxygen and other health hazards (such as Weil's disease).

The build-up of fumes can be particularly hazardous since it may lead to asphyxiation, fire or explosion. When the excavation is taking place in a busy road, there will be hazards associated with passing traffic and the possibility of vehicle collision with the excavation, its workers or passing pedestrians. Clearly, alongside these specific hazards, more general hazards, such as manual handling, electricity, noise and vibrations, will also be present.

If underground electrical cables are damaged, fatalities and injuries may be caused by electric shock, electrical arcs (causing an explosion), and flames. This often results in severe burns to the hands, face and body, even if protective clothing is being worn. Such damage may be caused by the sharp edge of a hand tool or crushing by a heavy machine. Cables that have been previously damaged but left unreported and unrepaired can also cause accidents.

As mentioned earlier, one of the most critical factors to be considered when planning an excavation is the nature of the soil which is to be excavated. All excavations, except the most shallow, need some support using trench sheeting. When excavation work takes place in non-cohesive soil (e.g. sand, gravels and soft clays), close sheeting will be required to prevent ground movement but for more cohesive soils, such as most clays and rock, some support, such as open sheeting, will still be necessary. Since soil cohesiveness decreases as water content increases, the stability of an excavation is considerably reduced during prolonged wet weather. It is, therefore, important that a competent person should assess the likelihood of any changes in ground conditions and adjust the working method accordingly.

If the excavation work takes place below the local groundwater level or water table, then water will flow into the trench. Measures will be required to impede the inflow of water and channel the water in the trench to a sump or pond where it may be pumped to a safe discharge point. These problems will be exacerbated at times of high rainfall, such as late winter or early spring,

Figure 17.2 Undermining of boundary wall

and will be even more hazardous if the soil structure is naturally weak.

Instability may also be caused by additional loading in the area adjacent to the excavation, for example by vehicles, plant or a spoil tip. This is known as **surcharging**.

If liquid nitrogen is used to freeze pipes during an excavation, additional hazards, such as frostbite and breathing difficulties due to nitrogen leakage, are presented. Permits to work should always be used for this type of work.

Finally, low levels of natural light are an additional hazard, which is particularly problematic in the winter and at the beginning and end of the working day.

17.1.3 Planning the work

Before starting any excavations, it is important to plan the work by following a comprehensive planning procedure (as described in HSG185) so that all significant hazards can be addressed. The principal hazards have been outlined in 17.1.2 and the most immediate ones for planning purposes are the following:

- collapse of the sides;
- control of people and vehicles around the surface area of the excavation;
- the proximity of adjacent structures;
- the position of any underground services;
- access to the excavation;
- the protection of members of the public.

A site survey should always be undertaken during the planning process and the following points need to be considered at this stage:

- previous use of site;
- location of existing buildings;
- location of new structures;

17

499

- results of soil investigations;
- ground contamination;
- level of water table and type of soil;
- storage and disposal of excavated material;
- amount of working space and storage required;
- the most suitable method of temporary support of the excavation walls;
- adequate emergency arrangements.

Much excavation work is needed to repair mains (water or gas), piped services or cables (electrical power or telecommunications) which are feeding occupied buildings. It is crucial that there is effective communication between the excavation contractors and the building occupants throughout all stages of the project. The pipe or cable must be isolated before work begins. If the work obstructs the main or emergency exits from the buildings then alternative egress arrangements must be made.

After the site survey has been completed, meetings should be organised with the occupants of surrounding buildings and other properties so that any special arrangements can be communicated. For example, if the neighbouring property is a school, special care will be needed when the children arrive and depart from the school. Continual liaison with neighbours is an important element in the risk control strategy.

Before work begins, there must be available sufficient numbers of trained operatives and competent supervisors and suitable plant and materials for trench support. Relevant monitoring equipment and personnel trained in its use will be required if exposure to toxic substances or lack of oxygen may occur. The location of any existing services must be completed. Most service cables belong to a Distribution Network Operator (DNO) although some belong to other organisations, such as the Highways Authority, Ministry of Defence or Network Rail. Any of these organisations should be contacted if equipment belonging to them is located nearby. In some cases it may be necessary to make the underground electrical services dead. This will need to be planned in advance because electricity companies are required to give five days' notice to customers whose supply is to be disconnected.

Plans or other suitable information about all buried services in the area should be obtained and reviewed before any excavation work starts. If the excavation work is an emergency, and plans and other information cannot be found, the work should be carried out as though there are live buried services in the area. Symbols on electricity cable plans may vary between utilities and advice should be sought from the issuing office. High-voltage cables may be shown on separate plans from low-voltage cables. Plans give only an indication of the location, and number, of underground services at a particular site. It is essential that a competent person traces cables using suitable locating devices as shown in Figure 17.11.

A method statement, which should cover external conditions, such as traffic, weather and existing structures, should also be prepared and, if necessary, Control of Substances Hazardous to Health Regulations (COSHH) assessments made.

During the final stages of planning, it will be necessary to ensure that any special personal protective equipment (PPE) or signs and equipment related to the work are ordered and available on the site. The organisation of the daily inspection of the excavation before each shift and the proper administration of the inspection records will need to be made at this stage.

Arrangements will also be required for the proper induction of any sub-contractors employed on the excavation work so that they are aware of the main hazards and the site rules, including the correct use of personal protective equipment.

Finally, it is important to ensure that all the required equipment, such as trench sheets, props and walings, is available on site before the work starts.

17.1.4 Risk assessment

A risk assessment should be made by the contractor undertaking the work before work begins to identify those hazards that are likely to be encountered and determine the control measures required. The significant findings of an assessment should be recorded, for example in a method statement. Method statements should describe the plant and equipment and the safe methods of working required to control the risks generated. These statements should be relevant to the particular work (rather than generic) and easily understandable by those who need to use them. They should be proportionate to the level of risk and degree of complexity – if the risks and necessary precautions are straightforward, short, simple method statements are all that are needed to convey the necessary information. The documents should identify the hazards associated with the work and any factors which increase the risk of injury, such as:

- the nature of the ground, soil structure and groundwater regime;
- the depth of excavation;
- the nature of the work required to be undertaken within an excavation;
- the location of the work (e.g. readily accessible public place, contaminated ground or heavily serviced urban area).

The method of support selected needs to take account of the work to be done and allowance made for adequate working space. It may be necessary to install support in relatively shallow excavations of less than 1.2 m depth if ground conditions are particularly poor or the nature of the work requires workers to lie or crouch in a trench. The options for the method of support may also be limited by ground conditions or the presence of services crossing the trench line. Support will also be required when plant for excavating or depositing backfill causes an extra loading on the excavation walls.

Support methods are discussed in 17.2, but whatever the support method selected, an essential part of the method statement is a defined safe method of work. No person should ever enter an unsupported excavation to install or remove supports. The use of proprietary support systems which can be installed safely from existing ground level may help should this situation arise. Methods of work need to take into account the following factors to avoid the risk of workers being struck by the plant itself or of the materials being placed where the vision of the plant operator is restricted:

▶ the removal from site of surplus excavated material;
▶ the stockpiling of excavated material which is to be reused;
▶ the provision of storage space for materials and working space for workers and plant;
▶ the type of compaction equipment to be used.

The method statement should also contain the information about the necessary emergency arrangements should an incident occur, unless they are already detailed in the health and safety plan. Details regarding any site-based personnel trained to administer first-aid, and the location of the nearest hospital with accident and emergency facilities, should be included. A method for swift communication should be identified and, if there is no telephone available on site, details given on the location of a public telephone close by; otherwise a mobile telephone should be provided. Thought should be given to the actions to be taken if a person were to be injured within a trench or if any gas monitors in use identify the presence of harmful gas or oxygen deficiency. Employees and other workers should be instructed in all relevant health and safety procedures. The risk assessment also needs to categorise the level of risk. High risks include the collapse of the excavation walls, contact with existing underground or overhead services, plant and materials falling into excavations and the presence of hazardous atmospheres and contaminated soil. The possibility of persons falling into the excavations is normally considered a medium risk whereas the flooding of excavations is usually designated a low risk.

Finally, it is very important to consider the additional risks arising to members of the public, in particular to children, those with impaired mobility and the visually impaired. Work should be planned to ensure that excavations are well lit, securely fenced, and covered or backfilled overnight and at weekends. Plant should be left in a secure compound and any engines isolated. Stored materials should be stacked carefully to prevent displacement and fenced if they can be accessed. The risk assessment should determine the precautions necessary to prevent unauthorised access to any

construction site. Arrangements may be required for traffic management supervised by competent persons, involving traffic controls such as traffic lights, stop/go boards or road closures and protection for pedestrians from the work. Temporary road signs, positioned in the correct order, may be needed either side of the excavation and cones set out with lead-in and exit tapers. There should also be an indication of a safety zone between the cones and the work area. Workers who work on the carriageway must wear high visibility clothing and be trained and competent for such work under the New Road and Street Works Act. The Act also requires that any part of a street that is broken up, is open or is obstructed must be adequately guarded and lit. In addition, any traffic signs must be placed, maintained and operated as reasonably required for the guidance or direction of persons (particularly the disabled) who are using the street.

Figure 17.3 Barriers around excavation by footpath

The New Road and Street Works Act stipulates circumstances under which certain excavation contractors, defined as utilities and other undertakers, must give notice of their planned works. Emergency, urgent and some minor works are excluded from this duty. It also requires that contractors keep up-to-date records of the location of their apparatus and make them available, free of charge, for inspection at all reasonable hours. The exemptions applicable to the Act do not affect contractor liability under the HSW Act.

The New Road and Street Works Act requires that the person who supervises any street works must be qualified under the Street Works (Qualifications of Supervisors and Operatives) Regulations. There must also be an operative who is qualified under the same Regulations on site while work is in progress.

A typical risk assessment for an excavation is given in Appendix 17.2 at the end of this chapter.

17

17.1.5 Use of cofferdams and caissons

Cofferdams and caissons are used in construction for excavation work under or close to water and have many of the hazards associated with confined spaces.

Cofferdams

A cofferdam is an enclosure placed or constructed in waterlogged soil or under water and pumped dry so that construction or repairs can proceed under normal conditions. It is a watertight structure, usually of sheet steel piles driven into the ground to create a walled chamber that encloses an area under water. Seal concrete is placed into the bottom to prevent water from seeping in from underneath the sheet piling, and the water is pumped out. Inside the cofferdam, soil can be excavated to bedrock, or piles can be driven to create foundations for a structure such as a bridge,

Figure 17.4 (a) Cofferdam for building a below-ground shaft and concrete tank – cofferdam removed and backfilled after construction

Figure 17.4 (b) Cofferdam built for repairing bridge piers in a river to be removed after construction completed

pier or building. The cofferdam is removed after the foundations have been constructed. The cofferdam must be tightly and properly constructed to ensure that the water is kept out of the exposed work area. Below a certain depth, however, a caisson is required.

Cofferdams are generally temporary structures that are removed after the construction work is complete. By using a cofferdam, construction engineers can use materials and techniques that cannot be deployed underwater. A cofferdam generally consists of three components:

▶ sheet piles – individual panels with interlocking connections that link the panels together, forming a single barrier;
▶ wales – beams that cover the top edge of the sheet piles around the perimeter of the cofferdam; and
▶ braces – used to maintain the structural integrity of the sheet piles around the circumference of the cofferdam by resisting the pressure of the outside water against the sheet piles.

Cofferdams are generally made of steel to provide the strength needed to hold back water from the construction area. Other materials such as timber, concrete and even earth are sometimes used. The loads imposed on the cofferdam structure by the construction equipment and operations must be considered, both during installation of the cofferdam and during the construction of the structure itself. Removal of the cofferdam must be planned and executed with the same degree of care as its installation, on a stage-by-stage basis. The effect of the removal on the permanent structure must also be considered. For this reason, sheet piles extending below the permanent structure are often cut off and left in place, since their removal may damage the foundation soils adjacent to the structure.

In cofferdam construction, safety is a paramount concern, since workers will be exposed to the hazards of flooding and collapse. Safety considerations require that every cofferdam and its component parts must be of suitable design and construction, of suitable and sound material and of sufficient strength and capacity for the purpose for which it is to be used. The cofferdam must be monitored during its construction and throughout the life of the construction project itself. There must be adequate access, lighting and ventilation and soil stability regularly monitored. Supervision of the workforce is also important to ensure that safe work practices are being followed.

Some of the more common uses for cofferdams include:

▶ bridge construction;
▶ hydroelectric dam construction;
▶ ship repair – cofferdams are used to generate a 'dry dock' condition; and
▶ the recovery of sunken ships in shallow waters.

Some of the advantages of cofferdams are that:

▶ they allow excavation and construction of structures in difficult environments;
▶ they provide a safe workplace;
▶ the steel sheet piles are easily installed and removed; and
▶ many of the materials can be reused.

Caissons

A caisson is a large cylinder or box chamber that is sunk onto the river or sea bed. The excavation and foundation work takes place within the submerged caisson. Some caissons are removed after construction, while others are left in place, filled with concrete, and used as part of a permanent foundation. A caisson can be made of wood, steel, concrete or reinforced concrete and usually forms an air- and watertight chamber in which it is possible for workers to work under air pressure greater than atmospheric pressure where they are able to excavate the material below the water level.

Caissons are used to build foundations for buildings and bridges. They are lowered into water or into dug-out areas in the ground, protecting workers from drowning and wall collapses and providing a shell for some of the foundations. Tall buildings often rest on many small caissons.

Figure 17.5 (b) Steel caisson being used in construction work

The three main types of caisson are box caissons, open caissons and compressed-air or pneumatic caissons.

1. Box caissons are prefabricated concrete boxes with closed sides and bottom and are set down on prepared bases. Once in place they will be filled with concrete to become part of the permanent works, for example the foundations for a bridge or pier. A box caisson is open at the top and closed at the bottom. It is usually constructed on land, then launched, floated to position, and sunk onto a previously prepared foundation, leaving its upper edge above water level. It serves as a suitable shell for a pier, seawall, breakwater, jetty or similar work, remaining permanently in place on the sea bottom.

2. Open caissons are similar to box caissons but are open at the top and at the bottom. They are suitable for use in soft clays (found in some river beds). Open caissons used in soft grounds or high water tables, where open trench excavations are impractical, can also be used to install deep manholes, pump stations and reception/launch pits for micro-tunnelling, pipe jacking and other operations. The open caisson is a cylinder or box, open at the top and bottom, of size and shape to suit the projected foundation and with a cutting edge around the bottom. It is sunk by its own weight and by excavation, then filled with concrete.

3. Pneumatic caissons have an airtight bulkhead high enough above the cutting edge to permit workers to work underneath it. The air in the chamber beneath the bulkhead is kept under pressure great enough to prevent the inflow of water, while shafts through the bulkhead allow the movement of workers, equipment and excavated material between the bottom and the surface. At the top of each shaft is an airlock to prevent variations in the air pressure in the working chamber. As the working chamber moves down, the caisson above the bulkhead and the shafts is filled with concrete,

Figure 17.5 (a) Massive Second World War concrete caissons known as mulberry harbour units – hundreds were constructed in the UK and then towed to Normandy and sunk in position to form a harbour wall

17

and when a sufficient depth or bedrock is reached, the working chamber itself is filled, so that there is a solid block of concrete from base to top. Workers leaving a pneumatic caisson after several hours of working under high air pressure are given special decompression treatment to accustom them to the lower atmospheric pressure and thus to prevent decompression sickness.

Where foundations have to be taken to a considerable depth in water-bearing strata, or deep into the bed of a river, to reach a hard stratum, bottomless caissons sunk by excavating under compressed air are employed. The caisson at the bottom, forming the working chamber, is usually provided with a strong roof, round the top of which, when the caisson is floated into a river, plate-iron sides are erected forming an upper open caisson. Inside this upper caisson, the pier, quay wall or similar structure is built up out of water, on the top of the roof of the lower caisson, as the sinking proceeds. Shafts through the roof up to the open air provide access for workers and materials to the working chamber, through an airlock consisting of a small chamber with an airtight door at each end, enabling locking into and out of the compressed-air portion to be undertaken. When a sufficiently reliable stratum has been reached, the workers leave the chamber and it is filled with concrete through the shafts, the bottomless caisson remaining embedded in the work.

A type of caisson often called a camel is used to raise sunken vessels. It consists of a cylinder filled with water, which is sunk, attached to the vessel, and emptied by pump or compressed air, so that its buoyancy can assist in raising the vessel. Caissons are also sometimes used for closing the entrance to dry docks or as a substitute for gates in canal locks.

Hazards and controls of cofferdams and caissons

The hazards associated with the use of cofferdams and caissons include flooding, collision with shipping, unstable soil, weather and tides, the hoisting of people and materials, and construction work equipment. Additional caisson hazards are oxygen depletion, toxic fumes and fire. There is also a problem of working in the compressed air atmosphere of a caisson – similar to those of a diver when the Compressed Air Regulations apply.

The following general **controls** are given for cofferdams and caissons. These controls do not include the technical controls that should be put into place at the project design stage.

The controls for cofferdams are that:

▶ they must be designed by a qualified engineer and constructed and maintained in accordance with the design;

▶ if the flooding of the cofferdam by high waters is possible, then means must be provided for controlled flooding of the work area;

▶ emergency evacuation procedures of workers must be posted and communicated to the workforce;

▶ there should not be less than two means of egress; and

▶ if they are located close to navigable shipping, then warning devices that are visible to passing ships must be provided.

The controls for caissons are that:

▶ they must be designed by a qualified engineer and constructed and maintained in accordance with the design;

▶ each employee who is required to enter a caisson excavation must have a lanyard attached to a body harness;

▶ all employees shall be removed from the caisson when material is being hoisted from the caisson;

▶ a supervisor must be stationed at the caisson whenever workers are present;

▶ a secondary mechanical means that is capable of removing an employee from the caisson shall be readily available should the crane performing the caisson work break down;

▶ the air quality of a caisson must be tested at regular intervals;

▶ emergency evacuation procedures of workers must be posted and communicated to the workforce;

▶ if they are located close to navigable shipping, then warning devices that are visible to passing ships must be provided;

▶ caissons shall be substantially braced before loading with concrete or other weight;

▶ when two or more caissons are sunk together, employees shall be removed from adjacent caissons during concreting operations;

▶ working in compressed air atmosphere requires specialist competent contractors working under the Compressed Air Regulations.

Legal requirements for cofferdams or caissons

The design and inspection of cofferdams or caissons is included in the CDM Regulations. These require that every cofferdam or caisson shall be:

▶ of suitable design and construction;

▶ appropriately equipped so that workers can gain shelter or escape if water or materials enter it; and

▶ properly maintained.

The Regulations also require that they should be inspected regularly by a competent person:

▶ at the start of the shift in which the work is to be carried out; and

▶ after any event likely to have affected the strength or stability of the cofferdam or caisson; and

▶ the person who carried out the inspection is satisfied that the work can be safely carried out there.

Finally for caissons, the Compressed Air Regulations may well apply.

17.2 Control measures for excavation work

Excavations must be constructed so that they are safe environments in which construction work can take place. This means that they must be fenced and suitable notices posted so that neither people nor vehicles fall into them. However, there are many other precautions and controls that must be taken and several of these should have emerged during the communication and liaison with the occupants of adjacent buildings at the planning stage of the project.

The first form of control is to decide whether the excavation is necessary. There have been many technological advances which have eliminated the need for excavation work. These trenchless technologies include micro-tunnelling, directional drilling, impact moling, auger boring, pipe relining and pipe bursting. More information on these techniques, which are not in the NEBOSH Construction Certificate syllabus, may be found in HSG150.

If excavation work is the only option available, the following precautions and controls should be adopted:

▶ At all stages of the excavation, a competent person must supervise the work. Workers must be given clear instructions on working safely in the excavation.

▶ The walls of the excavation trench must be prevented from collapsing either by digging them at a safe angle (between 5° and 45° depending on soil structure and its dryness) or by shoring them up with timber, sheeting or a proprietary support system. Even shallow trenches may need support if the work involves bending or kneeling in the trench.

▶ Falls of material into the workings can also be prevented by not storing spoil material near the top of the excavation.

▶ Workers must wear hard hats and safety footwear.

▶ Where people are liable to fall into an excavation a substantial barrier, consisting of guard rails and toe boards, should be provided around the surface of the workings. This is particularly important if the work is taking place in a public place where a brightly coloured barrier may be required.

▶ Workers should never enter an unsupported excavation or work ahead of supports.

▶ Vehicles should be kept away as far as possible using warning signs and barriers. Where a vehicle is tipping materials into the excavation, stop blocks should be placed behind its wheels.

▶ Whenever it is possible, workers and moving plant, such as excavators, should be kept separated.

▶ It is very important that the excavation site is well lit at night.

▶ All plant and equipment operators must be competent and non-operators should be kept away from moving plant.

▶ Personal protective equipment (gloves, overalls, boots, ear defenders and hard hats) must be worn by operators of noisy plant.

▶ Nearby structures and buildings may need to be shored up if the excavation may reduce their stability. Scaffolding could also be de-stabilised by adjacent excavation trenches. When there is doubt about this, the views of a structural engineer should be sought.

▶ The influx of water can only be controlled by the use of pumps after the water has been channelled into sumps to enable effective removal.

▶ The presence of hazardous substances or health hazards should become apparent during the original survey work and, when possible, they should be removed or suitable control measures adopted. Any such hazards found after work has started must be reported and noted in the inspection report and remedial measures taken. Exhaust fumes can be dangerous and petrol or diesel plant should not be sited near the top of the excavation.

▶ The presence of buried services is one of the biggest hazards and the position of all underground cables are located, identified and clearly marked using all available service location drawings before work commences. Since these will probably not be accurate, service location equipment, such as cable avoidance tools (CATs), should be used by specifically trained people. Locating devices should always be used in accordance with the manufacturer's instructions, regularly checked and maintained in good working order.

▶ The area around the excavation should be checked for service boxes. If possible, the supply should be isolated. Only hand tools should be used in the vicinity of underground services. Overhead services may also present risks to cranes and other tall equipment. If the supply cannot be isolated then 'goal posts' beneath the overhead supply, together with suitable bunting and signs, must be used.

▶ Safe access by ladders is essential as are crossing points for pedestrians and vehicles. Whenever possible, the workings should be completely covered with steel sheeting outside working hours particularly if there is a possibility of children entering the site.

▶ Finally, care is needed during the filling-in process.

Wells and disused mine shafts are found during construction work and must be treated with caution and in the same way as an excavation. The obvious hazards include falling in and/or drowning and those associated with confined spaces (see Chapter 4 and later in this chapter) – oxygen deficiency, the presence of toxic gases and the possible collapse of the walls. Controls include the fencing off of the well and covering it until

17

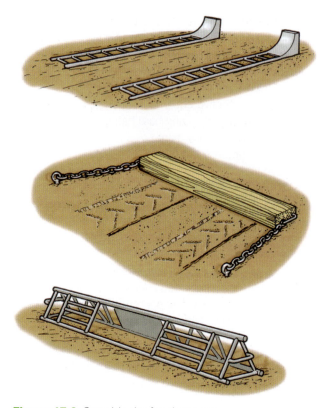

Figure 17.6 Stop blocks for dumpers

the situation has been reviewed by specialists. Shallow wells would normally be drained and filled with hard core whereas deeper ones would be capped.

Some of these precautions will now be discussed in more detail.

A particular hazard, discussed earlier, is **surcharging**. The risk of collapse due to surcharging may be reduced by:

▶ depositing spoil and other waste material at least 1.5 m from the edge of the excavation or a distance equal to the depth of the excavation, whichever is the greatest;

▶ increasing the amount of temporary support to the excavation when there are buildings close by or the excavation is taking place on a slope;

▶ installing suitable barriers and ensuring that all traffic routes are situated at a safe distance from the top of the excavation.

The stability of adjacent buildings and other structures will be affected by the excavation if it weakens or undermines the foundations of the structure. There have been several fatalities caused by the collapse of a retaining wall into a trench. When it is unavoidable that a trench must be dug next to a building, the building should be shored up by suitable supports. Where there is a width restriction on either side of the trench, arrangements will be needed to deposit the spoil material elsewhere at a safe distance.

Most excavations will require some form of **support**. Unsupported excavation is safe only if the sides are

battered back sufficiently and the vertical depth of the trench does not exceed 1.2 m. HSG150 gives details of safe angles of slope for different soils.

There are several methods of excavation support but before they are discussed the basic components of such support and some excavation terminology must be defined.

Sheeting is the general term given to steel sheets or timber boards used to support the walls of an excavation. When vertical or horizontal sheeting is placed close together for added strength, this is known as **close sheeting**. However, **open sheeting** occurs when the sheets are only placed at intervals, normally in clay.

Trench sheets are long, narrow, rolled steel plate sections which either lap-joint or interlock at their edges with adjacent sheets. They are normally installed vertically to support the sides of the trench.

Runner is a vertical member which is driven below the floor of the excavation and supports the side of the trench, and is progressively driven as the excavation proceeds.

Steel sheet piles are interlocked steel sections driven into the ground to resist side pressure on the excavation trench.

Poling boards are sheeting (1 m by 5 m) which supports the sides of the trench. When poling boards are used horizontally across a gap between runners or sheeting, this is known as **cross poling**.

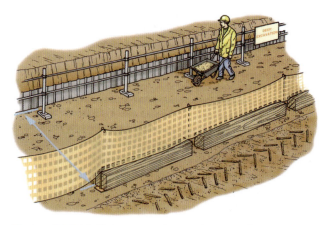

Figure 17.7 Vehicle protection at the top of an excavation

Walings are horizontal beams used to support sheeting or poling boards against pressure from the trench walls.

Props are sometimes called **struts** and provide extra support between pairs of sheeting, and are positioned between the walings.

Puncheons are vertical posts which support a waling or prop from below.

Well point is a long thin tube which is installed vertically into the ground and through which groundwater is pumped. A **well point system** is a system of well points situated around the excavation area and used to

Figure 17.8 Trench sheets with timber walings, screw props, puncheons and sole plates

pump out groundwater and reduce the flooding risk in the excavation trench.

There are six basic excavation support systems in general use:

1. **Battered sides**

   This is the traditional method of trench excavation and relies on the strength of the soil to support the trench walls. There is no safe depth for this technique since all soils vary in cohesiveness and many soils consist of a mixture of sand, gravel and clay. For many soils, the angle of repose of the sides (measured from the horizontal plane) is approximately 30°. This means that such excavation covers a considerable area. It is generally recommended that this type of excavation should not exceed 1.2 m in depth.

2. **Open and close sheeting**

   These are also traditional support systems and HSG150 recommends a safe method statement for each system. Underground services can present problems with these methods of support. Such services should not themselves be used to support any part of the trench support system. It is important that sufficient struts and walings are used to give added strength to the excavation. Regular

supervision is required to ensure that such support is not removed by workers to improve the size of the working space. Care is also required during the dismantling operation.

3. **Hydraulic waling frames**

   These consist of two steel beams braced apart by struts containing integral hydraulic rams. They can be used with open or close sheeting trenching and for supporting close sheeting in very deep trenches.

4. **Manhole shores**

   As the name suggests, these are supporting manhole or sump-type excavations. They are four-sided adjustable frames with integral hydraulic rams. As with waling frames, they must be secured to the trench sheeting (normally by chains).

5. **Trench boxes**

   These consist of solid side panels which are kept apart by integral hydraulic struts which may be adjusted to suit the width of the trench. The bottoms of the side panels have a tapered cutting edge. This enables them to be dug in or lowered by crane into a trench which has already been dug. Trench boxes can be fitted on top of each other when the trench is particularly deep.

6. **Drag boxes**

   These have solid sides with tapered cutting edges to their leading edges. The side sheets are braced apart by specially strengthened struts so that they may be dragged through the advancing excavation by the excavator. This technique is often used for pipe laying. The greatest problem with box systems is that they can only be used where there are no buried services or other obstructions. Their greatest advantage is that they allow trench supports to be

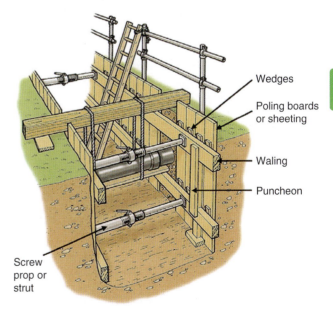

Wedges

Poling boards or sheeting

Waling

Puncheon

Screw prop or strut

Figure 17.9 (a) Timbered excavation with ladder access and supported services (guard removed on one side for clarity)

17

Figure 17.9 (b) A long timbered trench in soft ground

Figure 17.10 Trench box in use

put in place without requiring people to enter the excavation.

The presence of **underground services**, particularly electricity and gas, is a common problem. Often official knowledge of the exact location of such services is less than accurate, particularly if the underground services have been in place for many years.

The following precautions are suggested when there is uncertainty about the location of underground services in an area to be excavated:

▶ Check for any obvious signs of underground services, for example valve covers or patching of the road surface.

▶ Ensure that the excavation supervisor has the necessary service plans and is competent to use them to locate underground services.

▶ Ensure that all excavation workers are trained in safe digging practices and emergency procedures. Once a locating device has been used to determine cable positions and routes, excavation may take place, with trial holes dug using suitable hand tools, to confirm this. Excavation should take place alongside the cable or pipe rather than directly above it. Final exposure of the cable or pipe should take place by horizontal hand digging since the

force applied to hand tools can be controlled more effectively.

▶ Use locators (cable avoidance tools) to trace any services and mark the ground accordingly. A series of trial holes should be dug by hand to confirm the position of the pipes or cables. This is particularly important in the case of plastic pipes which cannot be detected by normal locating equipment.

▶ In areas where underground services may be present, only hand digging should be used with insulated tools. Spades and shovels should be used rather than picks and forks, which are more likely to pierce cables.

▶ Assume that all pipes or cables are 'live' unless it is known otherwise.

▶ Hand-held power tools should not be used within 0.5 m of the marked position of an electricity cable. Collars should be fitted to the tools so that initial penetration of the surface is restricted.

▶ A machine should not be used to excavate within 0.5 m of a gas pipe.

▶ Any suspected damage to services must be reported to the service providers and the health and safety enforcement authority.

▶ All exposed pipes or cables should be backfilled with fine material such as dry sand or small gravel. Backfill which is properly compacted, particularly under cast or rigid pipes, prevents settlement which could cause damage at a later date.

▶ The service plans must be updated when the new services have been laid.

The utility companies use a colour-coding system for buried services:

▶ black – electricity;
▶ red – electricity (some 11kV cables);
▶ blue – water;
▶ yellow – gas;
▶ grey/white – telecommunications; and
▶ green – cable television.

For older services, black cables were generally used for electricity and cast-iron and steel pipes for gas. However, wherever possible all services should be isolated before digging commences.

Excellent guidance is available in the HSE publication *Avoiding Danger from Underground Services* HSG47.

The adoption of safe digging methods is essential if accidents are to be avoided. Several elements of such practice have already been covered and detailed safe digging practice advice is given in Appendix 17.1 at the end of this chapter. Recently a construction worker was seriously burned when his pneumatic drill struck two underground 11kV power cables. He was drilling through

Figure 17.11 Using a cable detector

a concrete pavement when he struck the cables, causing a flash explosion that set light to his clothes.

If an electricity cable is accidentally struck, all workers in the vicinity may receive an electric shock and there is also a fire risk, as there would be if a gas main were damaged. Severe flooding of the trench could result if a water main were intercepted. Any of these events could cause severe problems to the local community, particularly local hospitals. When the purpose of the excavation is to investigate or repair a cable or pipe, it must be isolated while work on it is in progress.

An emergency plan, which includes the requirement to notify the relevant service provider, is essential to deal with damage to any underground cables or pipes. In the case of gas pipe damage, all smoking and naked flames must be banned. An evacuation procedure may be necessary which may include people in nearby properties likely to be affected by any leaks. Suitable signs should be erected to warn of the danger and the relevant health and safety enforcement authority informed.

It is important to note that **overhead power lines** can also cause serious injuries if plant, such as cranes or excavators, comes into contact with them. Although this problem is discussed in more detail in Chapter 12, some obvious precautions will be noted here. If possible, excavation beneath power lines should be avoided. If the work is unavoidable, then highly visible barriers should be erected at least 6 m away from the lines and red and white goalpost-type crossing points should be positioned beneath the lines. When excavation work has to take place beneath power lines, the planning supervisor must, during the planning stage, ascertain whether the line can be diverted or made dead. The use of conductive materials, such as scaffold tubes or aluminium ladders, should be limited in these circumstances.

Access to the excavation is normally provided using strong ladders but consideration must be given to the means of rescue, perhaps using a winch or tripod, for an injured worker who is unable to use a ladder.

Fumes and exhaust gases can be hazardous, particularly if they become concentrated in an excavation. This is a particular problem for some welding gases that are heavier than air. Petrol- or diesel-engined equipment, such as generators or compressors, should not be sited in, or near the edge of, an excavation unless the exhaust gases can be ducted away or the area can be ventilated.

Where a **flooding** risk exists, cofferdams/caissons may need to be installed with pumps of suitable capacity. As mentioned earlier (17.1.5), a cofferdam or caisson is a watertight box-like iron structure which is normally sunk in the bed of a river. It is pumped dry and filled with concrete and used in the construction of underwater foundations for structures such as bridges or quay

17

walls. Cofferdams must be inspected by a competent person on the same day as used by a worker. A permit-to-work system may also have to be used.

Finally, if emergency routes from adjacent buildings could be obstructed by the excavation work, consideration of alternative routes will be necessary. This is a particular problem if the building is a school, surgery or hospital.

17.2.1 Inspection requirements

The duty to inspect and prepare a report only applies to excavations which need to be supported or battered back to prevent accidental fall of material. Only persons with a recognised and relevant competence should carry out the inspection and write the report. Inspections should take place at the following timing and frequency:

▶ after any event likely to affect the strength or stability of the excavation;
▶ before work at the start of every shift;
▶ after an accidental fall of any material.

Although an inspection must be made at the start of every shift, only one report is required of such inspections every 7 days. However, reports must be completed following all other inspections. Any written reports should be completed before the end of the relevant working period and a copy given to the manager responsible for the excavation within 24 hours. Reports must be kept on site until the work is completed and then retained for 3 months at an office of the organisation which carried out the work. If any inspection finds that the excavation is unsafe the manager should be informed and the work stopped.

A suitable form is shown in Appendix 16.1 at the end of Chapter 16.

17.3 Confined space working hazards and risks

The hazards present in confined spaces have caused the deaths of many workers and those who were trying to rescue them. On average 15 people each year are killed in confined spaces. There have been several tragedies in confined spaces due to a lack of oxygen. Several years ago in Cornwall, a young mining trainee was killed whilst attempting to rescue an unconscious colleague from a dead-end underground roadway. The cause of the death was oxygen deficiency.

As with all hazardous situations, work should only take place in a confined space if it is unavoidable. Workers in confined spaces should be skilled and trained or specialist contractors.

The Confined Spaces Regulations require that:

1. no person at work shall enter a confined space for any purpose unless it is not reasonably practicable to achieve that purpose without such entry;

2. other than in an emergency, no person shall enter, carry out work or leave a confined space otherwise than in accordance with a safe system of work, relevant to the specified risks.

17.3.1 Definition of a confined space

A confined space can be defined as any space of an enclosed nature which has limited means of access and egress, restricted natural ventilation and is not intended for continual occupancy by persons. It is, therefore, any space which, by virtue of its enclosed nature, presents a reasonably foreseeable specified risk of serious injury.

Examples of a confined space include manholes, sewers, tunnels, excavations, storage tanks, holds of ships, pits, trenches, ducts, some unventilated areas or rooms within buildings (particularly below ground level), boilers, combustion chambers in furnaces, chambers, vats, silos, pits, trenches, pipes, flues and wells.

17.3.2 Confined space hazards

The principal hazards associated with a confined space are the difficult access and egress, which makes entrapment, escape and rescue more problematic; the accumulation of vapours, gases or fumes; and the lack of ventilation. These hazards may be inherent due to the nature of the confined space (e.g. the leakage of methane from the surrounding strata) or they may be introduced into the confined space due to the nature of the work (e.g. welding work or cleaning work with solvents).

Other hazards associated with confined spaces include:

▶ asphyxiation due to oxygen depletion;
▶ poisoning by toxic substance or fumes;
▶ explosions due to gases, vapours and dust;
▶ fire due to flammable liquids, vapours and oxygen enrichment;
▶ fall of materials leading to possible head injuries;
▶ free-flowing solid such as grain in a silo;
▶ electrocution from unsuitable equipment;
▶ difficulties of rescuing injured personnel;
▶ drowning due to flooding;
▶ fumes from plant or processes entering confined spaces;
▶ excessive heat leading to heat stress;
▶ claustrophobic effects due to restricted space;
▶ diseases from animal wastes, infected materials or micro-organisms, e.g. fungal infections, tetanus, Weil's disease and pigeon droppings.

When oxygen levels fall to 17%, the physical and mental abilities of workers become severely reduced. As the concentration falls below 17%, unconsciousness and death follow very rapidly. In confined spaces, the act of breathing can reduce oxygen levels quickly. Oxygen depletion can be caused by the reaction of some soils to oxygen in the atmosphere, such as the action of precipitated water on chalk. Some operations, like welding and cutting, consume oxygen very rapidly.

Poisonous gases or vapours can be caused by leaching from contaminated surrounding land or by the concentration of gases in sewers or manholes. Some of the more serious hazards are introduced into the confined space. These include welding and flame cutting, painting and coating, and cleaning operations, all of which can produce hazardous gases and vapours.

17.3.3 Risk assessment

Risk assessment is an essential requirement of the Confined Spaces Regulations and must be done so that a safe system of work may be determined. The risk assessment needs to identify the hazards present in the confined space, assess the risks and determine suitable controls to address those risks. The first part of the risk assessment will determine whether the work can only be done in the confined space.

Figure 17.12 Entering a confined space with full breathing apparatus and watcher outside

The risk assessment must be done by a competent person and will form the basis of a safe system of work. This will normally be formalised into a specific permit-to-work procedure which is applicable to a particular task.

The assessment will involve the examination and investigation of the following items:

▶ any previous contents in the confined space;
▶ any residues that have been left in the confined space, for example sludge, rust or scale, and details of gases or vapours which may be generated if these substances are disturbed;
▶ any contamination which may arise from adjacent plant, processes, services, pipes or surrounding land, soil or strata;
▶ any oxygen deficiency and enrichment. There are very high risks if the oxygen content differs significantly from the normal of 20.8%;

▶ the physical dimensions and layout of the space since these can affect air quality;
▶ the use of cleaning chemicals and their direct effect on or interaction with other substances;
▶ any sources of ignition for flammable dusts, gases, vapours, plastics and the like;
▶ the need to isolate the confined space from outside services or from substances held inside such as liquids, gases, steam, water, fire extinguishing media, exhaust gases, raw materials and energy sources;
▶ the requirement for emergency rescue arrangements including trained people and equipment.

On completion of the risk assessment, the site manager, in conjunction with a specialist contractor if one is to be used, will draw up a method statement and a programme of work. This will detail the methods, plant and equipment to be used, general precautions and any special requirements for dealing with health hazards, and the sequencing of work. The details and procedures required for any permit-to-work systems should also be included. This method statement and programme of work will be issued to the supervisor responsible for the work on site prior to the commencement of the work. All this information should also be contained in the on-site construction phase safety plan (as required under CDM). A copy of the risk assessment and method statement should be kept on site for reference purposes. An example of a risk assessment for a confined space is given in Appendix 17.3 at the end of this chapter.

17.4 Control measures for confined space working

The main duty of the employer, so far as is reasonably practicable, is to avoid the need for employees or others to enter a confined space. If entry is unavoidable, he/she should ensure that a safe system of work is available and that adequate emergency rescue arrangements are in place including the provision of appropriate personal protective equipment such as respirators (as described in Chapter 14).

The site supervisor must ensure that all necessary equipment is available on site in accordance with the risk assessment, method statement and any other planned procedures, such as a permit to work, before any person is allowed to enter a confined space. He/she must also ensure that the planned procedures, including any permit-to-work systems, are adhered to strictly and that only authorised trained persons are permitted to enter the confined space. He/she must be notified of any changes in working methods or conditions inside the confined space which were not included in the planning procedures before any changes are implemented.

17

All safety equipment must be regularly checked and maintained. Records should be kept of the checks and any defects in equipment rectified immediately.

The detailed precautions required will depend on the nature of the confined space and the actual work being carried out. The main elements of a safe system of work which may form the basis of a permit to work are as follows:

▶ the type and extent of the supervision of the work;
▶ the competence, training and instruction of the workforce;
▶ the stipulation of a minimum gang size for large confined spaces such as reservoirs;
▶ details of any information required by the workforce;
▶ the methods of communication between people inside, from inside to outside and to summon help in an emergency;
▶ the testing and monitoring of the atmosphere inside the confined space for hazardous gas, fumes, vapour, dust and the concentration of oxygen;
▶ the gas-purging of toxic or flammable substances with air or inert gas such as nitrogen;
▶ the provision of good ventilation, sometimes by mechanical means;
▶ if appropriate, the cleaning of the confined space before the work commences;
▶ the careful removal of residues using appropriate equipment which will not cause additional hazards;
▶ effective protection from gases, liquids and other flowing materials by the removal of redundant piping, the blanking of pipes and the locking of valves;
▶ the effective and complete isolation of electrical and mechanical equipment by the use of a lock-off and a tag system with key security. Checks will need to be made to protect the workforce against hazards such as stored energy in flywheels or the fall under gravity of heavy presses;
▶ the provision of personal and respiratory protective equipment if it is not possible to enter the confined space otherwise. In this event, the supply of gas through pipes and hoses must be carefully controlled;
▶ access to and egress from the confined space so that there is quick and unobstructed access and escape;
▶ the details of fire prevention equipment and procedures;
▶ the lighting arrangements including emergency lighting;
▶ the details of any behavioural rules such as the prohibition of smoking;
▶ the emergency and rescue arrangements and procedures (including first-aid);
▶ the maximum length of any one working period;
▶ the details of any required health surveillance of the workforce;

▶ monitoring and audit arrangements to ensure that the permit-to-work system is working as envisaged.

Any lighting or electrical power tools must be specially protected to the required standard in damp and flammable atmospheres.

Specific, detailed and frequent **training** is necessary for all people concerned with confined spaces, whether they are acting as rescuers, supervisors or those working inside the confined space. The training will need to cover all procedures and the use of equipment under realistic simulated conditions.

Figure 17.13 Training for confined space entry

17.4.1 Monitoring arrangements

Before any work commences in the confined space, the site supervisor must arrange for any necessary environmental surveys and sampling needed to protect the health of the workforce. Specialists and other personnel required to test and monitor the atmosphere must be competent to take the measurements and interpret the results accurately. Continuous gas monitoring using electronic instruments should be used in preference to 'spot' detection devices which only give a measurement at a given time rather than over the complete work period. The workforce should also be given health checks periodically to ensure that they are not suffering any claustrophobic effects or problems with the wearing of breathing apparatus. An occupational health specialist will provide information on ventilation equipment, breathing apparatus, resuscitation apparatus, ropes, harnesses and monitoring equipment, and information on symptoms associated with working in a confined space.

All breathing and resuscitation apparatus, rescue equipment and emergency alarms must be regularly checked and in some cases recalibrated. First-aid provision must be checked regularly, as must the accreditation of first-aiders. The workers must also be monitored and consideration should be given to the

Figure 17.14 Escape breathing apparatus

exclusion from confined spaces of individuals who have any of the following problems:

- heart disorders including high blood pressure;
- fainting, blackouts, fits or loss of balance;
- weak eyesight or hearing;
- asthma or chronic bronchitis;
- poor physical size, poor mobility or back pain;
- claustrophobia or psychological stress.

17.4.2 Emergency arrangements

Before people enter a confined space, suitable and sufficient emergency and rescue arrangements must be in place. These arrangements must reduce the risks to rescuers so far as is reasonably practicable and include the provision and maintenance of suitable resuscitation equipment which must be designed to meet the specific risks associated with the particular confined space.

For the emergency arrangements to be suitable and sufficient, they must incorporate the following:

- contingency plans to deal with an emergency in the confined space. Such plans must include details of rescue teams together with individual responsibilities;
- the contact names and telephone numbers of the local emergency services;

- the details of the communication arrangements from inside the confined space;
- the provision and maintenance of the rescue and resuscitation equipment;
- raising the alarm, alerting the rescue team and maintaining close supervision of the workforce inside the confined space;
- the notification of and consultation with the emergency services;
- the protection and safeguarding of the rescue team;
- the fire safety precautions and procedures;
- the provision and maintenance of fire-fighting equipment;
- the control and possibly the protection of adjacent buildings, plant and equipment;
- the arrangements for dealing with emergencies outside the normal working day;
- the first-aid arrangements including any special training required of first-aiders;
- the training of rescuers and simulations of emergencies, such as fire drills;
- the size of access openings to permit rescue with full breathing apparatus, harnesses, fall arrest gear and lifelines, which is the normal suitable respiratory protection and rescue equipment for confined spaces.

17.5 Further information

The Confined Spaces Regulations 1997

The Construction (Design and Management) Regulations 2015

The Personal Protective Equipment at Work Regulations 1992

The Work at Height Regulations 2005

Avoiding danger from underground services, HSG47, HSE Books, ISBN 978-0-7176-6584-6 http://www.hse.gov.uk/pubns/priced/hsg47.pdf

Safe work in confined spaces, ACoP, regulations and guidance, L101, HSE Books, ISBN 978-0-7176-6233-3 http://www.hse.gov.uk/pubns/priced/l101.pdf

Personal Protective Equipment at Work (Guidance), L25, HSE Books, ISBN 978-0-7176-6139-3 http://www.hse.gov.uk/pubns/priced/l25.pdf

Protecting the public: Your next move (HSG151), HSE Books, ISBN 978-0-7176-6294-4 http://www.hse.gov.uk/pubns/priced/hsg151.pdf

17

17.6 Practice revision questions

1. Excavation work is common on construction sites.

   (a) **Identify** the main hazards associated with excavation work.

   (b) **Outline** the control measures needed to reduce the risks during excavation work.

   (c) **Describe** the elements of a safe system of work to be followed to ensure the safe use and movement of vehicles on a site where excavation work is taking place and where vehicles are entering the site either to deliver materials or to remove surplus material.

2. Excavations in construction can be at risk of flooding.

   (a) **Identify FOUR** ways in which water can enter an excavation.

   (b) **Outline** practical measures that may be taken to prevent water from entering an excavation.

3. Pipe-freezing using a liquid nitrogen jacket is to be carried out in a properly supported, 2-m deep trench.

   (a) **Identify** the specific hazards associated with the pipe-freezing operation.

   (b) **Outline** the precautions that would be appropriate in such circumstances.

4. The construction of a cofferdam requires sheet piles to be driven into a river bed and the water contained by the piles pumped out before the construction work can begin.

   (a) **Identify** the health hazards that may be encountered by workers involved in this operation.

   (b) **Outline** the precautions required to control these hazards.

5. A trench is to be excavated to a depth of 2 m through a recreation park in order to repair the water mains supply to a pavilion.

   (a) **Identify** the possible hazards associated with the work.

   (b) **Outline** the control measures to help minimise the risks to the workers and others who may be affected by the work.

6. A deep trench needs to be excavated and supported next to a four-storey building.

   (a) **Identify** the hazards that may be present in these circumstances.

   (b) **Outline** the control measures required so that the excavation workers, members of the public and the adjacent building are adequately protected.

   (c) **Identify** when inspections of the supported excavation must be carried out by the competent person.

7. A tank that measures 4 m long, 3 m wide and 2 m deep is to be buried in a school sports field to aid surface water drainage.

   (a) **Outline** the principal hazards that should be considered when planning the work.

   (b) **Outline** the contents of a method statement for the excavation and installation operations.

8. Cofferdams and caissons are structures used in construction for excavation work under or close to water.

   (a) **Identify TWO** uses of cofferdams and caissons in construction work.

   (b) **Outline** the hazards associated with these structures.

   (c) **Outline** the precautions required to control these hazards.

9. (a) **Explain** the meaning of the term 'confined space'.

   (b) **Identify FOUR** examples of a confined space that may be encountered on a construction site.

   (c) **Identify FIVE** specific hazards associated with confined spaces.

10. An underground concrete storage chamber requires inspection and possible repair. A risk assessment has identified the possibility of the presence of methane within the chamber.

    (a) **Describe** the items of safety equipment that would be expected for such work, explaining how **EACH** item reduces the risk of harm to the workers in the chamber.

    (b) **Describe** the health risks faced by those repairing the reservoir when pneumatic tools are to be used to remove defective concrete and epoxy resin material used to repair the walls.

    (c) **Outline** the issues that should be considered to ensure that the emergency rescue arrangements are adequate.

11. Construction sites are responsible for the deaths of several workers each year due to the failure to identify the work area as a confined space.

    (a) **Define** the meaning of the term confined space.

    (b) **Outline** the main hazards associated with confined spaces.

(c) **Identify FOUR** typical confined spaces found during construction activities.

(d) **Outline** the control measures for working in a confined space.

12. The water main supplying a school is to be replaced. The work will be carried out in a 1.5-m deep excavation, which will be supported in order to ensure the safety of the employees working in the excavation.

(a) **Identify** when the **THREE** statutory inspections of the supported excavation must be carried out by the competent person.

(b) **Identify** the information that should be recorded in the excavation inspection report.

(c) Other than the provision of supports for the excavation, **outline** additional precautions to be taken during the repair work to reduce the risk of injury to the employees and others who may be affected by the work.

17

APPENDIX 17.1 An example of safe digging practice

Introduction

All pipes and cables must be checked before digging begins. Service plans should be used to see whether the place intended for digging involves working near buried underground services. Proper planning and the execution of a safe system of work should enable contact with services to be avoided.

A safe system of work for digging in an excavation depends upon the use of:

▶ cable or other service plans;
▶ cable and service locators; and
▶ safe digging practices.

Whenever possible, excavations should be kept well away from existing services.

Safe digging practice method statement

Before digging, ensure that:

▶ the person who is going to supervise the digging on site has service plans and is trained in the use of them;
▶ all workers involved in the digging have been trained in safe digging practice and excavation emergency procedures;
▶ all workers are properly supervised;
▶ the locator is used to trace as accurately as possible the actual line of any pipe or cable or to confirm that there are no pipes or cables in the way and the ground has been marked accordingly;
▶ there is an emergency plan to deal with damage to cables or pipes which includes the notification of the service provider in all circumstances. In the event of gas pipe damage, all smoking and naked flames must be banned. If necessary, evacuate all those at risk of injury (this may include people in nearby properties likely to be affected by leaks).

Suitable signs must be erected to warn everyone of the danger.

During the excavation operation the following **safe digging practice** should be used:

▶ Check repeatedly for evidence of pipes or cables during digging using the locator. If unidentified services are found, work must cease until further checks can be made to confirm that it is safe to continue digging.
▶ Trial holes should be dug by hand to confirm the position of the pipes or cables. This is particularly important in the case of plastic pipes which cannot be detected by normal locating equipment.
▶ Always hand dig near buried pipes or cables. Use spades and shovels rather than picks and forks, which are more likely to pierce cables.
▶ Treat all pipes or cables as 'live' unless it is known otherwise. A live cable may be contained in a rusty pipe or conduit. Do not break or cut into any service until its identity is ascertained and it is known that it has been made safe.
▶ Hand-held power tools should not be used within 0.5 m of the marked position of an electricity cable. Fit check collars to the tools so that initial penetration of the surface is restricted.
▶ No machine should be used to excavate within 0.5 m of a gas pipe.
▶ Exposed services must be supported as soon as they are uncovered. This will prevent them from being damaged.
▶ Any suspected damage to services must be reported to the site supervisor immediately.
▶ Pipes or cables should be backfilled with fine material such as dry sand or small gravel. Backfill which is properly compacted, particularly under cast or rigid pipes, prevents settlement which could cause damage to them at a later date.
▶ The service plans are updated when the new services have been laid.

APPENDIX 17.2 Typical excavation work risk assessment

INITIAL RISK ASSESSMENT		Work in and with excavations		
SIGNIFICANT HAZARDS		Low	Medium	High
1.	Collapse of sides			✓
2.	Striking existing services			✓
3.	Persons falling into excavations		✓	
4.	Plant and materials falling into excavations			✓
5.	Flooding of excavations	✓		
6.	Presence of hazardous atmospheres			✓
7.	Presence of contaminated soil			✓
8.				

ACTION ALREADY TAKEN TO REDUCE THE RISKS:

Compliance with:

Provision and Use of Work Equipment Regulations (PUWER)	CDM Regulations
Control of Substances Hazardous to Health Regulations (COSHH)	HSE Avoidance of dangers from underground services
Confined Spaces Regulations	HSE HSG150 Health and Safety in Construction

Planning:

Compliance with British Standards for Earthworks. Sufficient numbers of trained operatives and competent supervision must be available before work starts. Sufficient and suitable plant and materials must be available for trench support before work starts. Suitable monitoring equipment and personnel trained in its use will be required where known exposure to toxic substances or lack of oxygen may occur. Location of existing services must be complete before work starts, also information obtained on ground conditions. A method statement should be prepared before work starts. COSHH assessments to be made as required.

Physical:

Excavations will be supported as appropriate. Where flooding risk exists, cofferdams/caissons will be installed with pumps of suitable capacity. Substantial barriers will be erected around excavations. Where poor ventilation is identified the atmosphere will be continually monitored. A permit to work will be used. Stop barriers will be used to prevent vehicle entry. Spoil and materials will be stacked at least 1.5 m from the edge of excavations. Ladders will be provided for safe access/egress. Cable location devices and local authority drawings will be used to trace buried services prior to commencement of work. Suitable signs and barriers will be provided to warn of the work.

Managerial/Supervisory:

Ensure safe system of work provided, taking account of prevailing conditions including weather, traffic and existing structures. Provide suitable PPE as required and ensure its correct use. Inspect excavations daily or before each shift. Record thorough examination. Ensure personnel selected are capable, fit and experienced unless under direct supervision. COSHH assessments are to be made of substances likely to be found or produced during the work.

Training:

Supervisors must have received training in COSHH appreciation, general site safety, and theory and practice of excavation work. Operatives must have received training in excavation support procedures and use of cable location devices. This applies to sub-contractors as well as direct employees.

Date of Assessment: .... Assessment made by: ....

Risk Re-assessment Date: .... Site Manager's Comments:...........................

17

APPENDIX 17.3 Typical confined spaces risk assessment

INITIAL RISK ASSESSMENT	Work in Confined Spaces		
SIGNIFICANT HAZARDS	Low	Medium	High
1. Poisoning from toxic gases			✓
2. Asphyxiation – lack of oxygen			✓
3. Explosion			✓
4. Fire			✓
5. Excessive heat		✓	
6. Drowning			✓
7.			

ACTION ALREADY TAKEN TO REDUCE THE RISKS:

Compliance with:

Confined Spaces Regulations	HSE The Selection, Use and Maintenance of Respiratory Protective Equipment
HSC ACoP Confined Spaces Regulations	Local Authority/client safety standards, e.g. on sewer entry
HSE Guidance Note – Safe Work in Confined Spaces	
HSE Respiratory Protective Equipment	CDM Regulations

Planning:

Eliminate need for entry or use of hazardous materials by selection of alternative methods of work or materials. Assessment of ventilation available and possible local exhaust ventilation requirements, potential presence of hazardous gases/atmosphere, process by-products, need for improved hygiene/welfare facility. Rescue equipment will be available.

Physical:

Documented entry permit-to-work system will apply. Adequate ventilation will be present or arranged. Detection equipment will be present before entry to check on levels of oxygen and presence of toxic or explosive substances. The area will be tested before entry and continually during the presence of persons in the confined space. Breathing apparatus or airlines will be provided if local ventilation is not possible. When breathing apparatus is assessed as being required, emergency breathing apparatus and rescue harnesses will be provided. Rescue equipment including lifting equipment, resuscitation facilities, safety lines and harnesses will be provided. A communication system with those in the confined space will be established. Air will not be sweetened with pure oxygen. Precautions for safe use of any plant or heavier-than-air gases in the confined space must be established before entry. Necessary PPE and hygiene facilities will be provided for those entering sewers. Electrical equipment will be low voltage only, suitable for the conditions in the confined space. Energy sources will be isolated to acceptable standards with, e.g., lock-off and tags on electrical equipment and valves; blanking or pipe removal is preferred.

Managerial/Supervisory:

The management role is to decide on the nature of the confined space and to put a safe system into operation, including checking the above. Flood potential and isolations must be checked. Rescuers must be in place throughout the operation in the confined spaces.

Training:

Full training required for all entering and managing confined spaces. Rescue surface party to be trained, including first-aid and operation of testing equipment. All operatives must be certified as trained and supervisory staff trained to the same standard.

Date of Assessment:	Assessment made by:.................
Risk Re-assessment Date:	Site Manager's Comments:............

CHAPTER 18

Demolition and deconstruction – hazards and risk control

This chapter covers the following NEBOSH learning objectives:

1. Identify the main hazards of demolition and deconstruction work
2. Outline the control measures for demolition and deconstruction work
3. Identify the purpose and scope of a pre-demolition/refurbishment survey
4. Outline the main control measures that a demolition/refurbishment method statement should include

18.1 Demolition and deconstruction hazards and risks

18.1.1 Introduction

Demolition is one of the most hazardous construction operations and is responsible for more deaths and major injuries than any other activity. Under Construction (Design and Management) Regulations (CDM) 2015 all demolition work requires a written plan to show how danger will be prevented. See Chapter 19 and Chapter 7 for more details. If a demolition project is well planned the risks of injury and death can be minimised. It should be emphasised that the planning and execution of a demolition project should only be done by appropriately competent persons. The work should be supervised by someone with sufficient knowledge of the particular structure being dismantled and an understanding of the demolition method statement. For complex demolition work, expert advice from structural engineers will be necessary.

All demolition work must be carried out so as to minimise, so far as is reasonably practicable, the risks to employees and others who may be affected by the work. CDM 2015 applies to all demolition work. The HSE must be notified before work begins if the construction work, including demolition, is to last for 30 days and has 20 workers working simultaneously or more than 500 person days are involved.

18.1.2 Demolition and deconstruction methods

Both demolition and deconstruction involve the pulling down of a building or structure and are the reverse of the construction process. Whist the two processes are similar in effect, **deconstruction** preserves various components of the structure, such as wood and bricks, for reuse at another building or structure. The **demolition** process reduces the structure to a rubble which is either treated as waste or, in some cases, used as hard core for other construction projects.

Demolition is a simple process and may be done either manually or mechanically using large hydraulic equipment: elevating work platforms, excavators, bulldozers or cranes. Larger structures often require a 'wrecking ball' – a heavy weight on a cable that is swung from a crane into the sides of the structure. Rotational hydraulic shears and rock-breakers attached to excavators are also used to cut or break through wood, steel and concrete.

Deconstruction is a relatively new approach that seeks to minimise the quantity of waste that is sent to landfill sites. The process involves the careful removal of materials and segregating them either for reuse or recycling. A well-designed deconstruction project can result in less than 10% of the structure and its contents being transported to landfill. Deconstruction will take longer than conventional demolition processes but the reduction in landfill fees and cost saving in reused materials can make it a cost-effective process.

There are two forms of demolition:

1. piecemeal – where the demolition is done using hand and mechanical tools such as pneumatic drills, cranes and demolition balls, hydraulic pusher arms or heavy duty grabs; and
2. deliberate controlled collapse – where explosives are used to demolish the structure. This technique should only be used by trained, specialist competent persons. Pre-weakening of the structure, by the removal of several load-bearing elements and their replacement with temporary props, normally precedes the deliberate collapse. This is the most economic form of demolition but it is the most hazardous and everyone must be at a safe distance at the time of the collapse.

Demolition using explosives

If the demolition is to use explosives, an exclusion zone must be established at a distance from and surrounding the structure to be demolished. All persons, with the possible exception of the shot firer, must be outside the exclusion zone at the time of the explosion. The HSE have produced a detailed information sheet (CIS No. 45) on the establishment of exclusion zones when using explosives in demolition.

An exclusion zone comprises four areas:

1. the plan area of the structure to be demolished;
2. the designed drop area where the bulk of the structure is designed to drop;
3. the predicted debris area which is beyond the designed drop area and where the rest of the debris is predicted to drop;
4. the buffer area between the predicted debris area and the boundary of the exclusion zone.

The size of the exclusion zone is not just related to the height of the structure – other issues must be considered such as the effect of ground vibration on buried services.

Figure 18.1 Demolition of old mill in progress

Figure 18.2 High hazard vacuum cleaner to clear up asbestos material

Figure 18.3 Long-reach hydraulic arm for piecemeal demolition

18.1.3 Principal hazards of deconstruction and demolition work

The principal hazards associated with deconstruction and demolition work are:

▶ falls from height or on the same level;
▶ falling debris;
▶ premature collapse of the structure being demolished;
▶ dust and fumes;
▶ the silting up of drainage systems by dust;
▶ the problems arising from spilt fuel oils;
▶ manual handling;
▶ presence of asbestos and other hazardous substances;
▶ noise and vibration from heavy plant and equipment;
▶ electric shock;
▶ fires and explosions from the use of flammable and explosive substances;
▶ smoke from burning waste timber;
▶ pneumatic drills and power tools;
▶ the existence of services, such as electricity, gas and water;
▶ collisions with heavy plant;
▶ plant and vehicles overturning; and
▶ dilapidation of the structure.

The main factors that influence the design of an exclusion zone include:

▶ the designed collapse mechanism;
▶ the materials used in the structure;
▶ the condition of the structure;
▶ the extent of any planned pre-weakening;
▶ the types of explosives to be used;
▶ the topography of the site;
▶ the proximity of surrounding structures;
▶ any adverse effects of ground vibration or the attendant shock wave;
▶ noise; and
▶ dust.

The competent person with prime responsibility for designing the exclusion zone is the explosives engineer. The members of the public must be kept outside the exclusion zone until the all clear is given. Any houses within the exclusion zone must be evacuated and provision for the welfare of evacuees during the evacuation period must also be made. Finally, the local police should be informed of and ideally involved in the planning process.

18.2 Control measures

The control measures for deconstruction and demolition work must ensure the following:

▶ the avoidance of premature collapse;
▶ the protection of workers and members of the public from falls and falling material;
▶ the safe siting and use of plant, vehicles and other equipment;
▶ adequate control of dust and fumes;
▶ adequate protection from noise and vibration;
▶ the general protection of the environment;

- ▶ the competence of the workforce to complete the demolition work;
- ▶ a pre-demolition investigation/survey;
- ▶ a complete knowledge of the type of structure, method of construction, structural condition and the presence of cellars or other underground storage facilities;
- ▶ the identification of services, particularly underground services (gas, electricity, water and telecommunications services). These will need to be isolated or disconnected before demolition work begins. If this is not possible, pipes and cables must be labelled clearly so that they are not disturbed;
- ▶ knowledge of the presence of hazardous substances (particularly asbestos) and relevant control measures; and
- ▶ knowledge of the proximity and condition of other nearby structures, roadways, railways, rivers and canals.

Some of these issues will now be dealt with in more detail.

18.2.1 Specific issues

Premature collapse

Premature collapse of the structure is one of the main causes of serious injury resulting from demolition work. A common reason for this type of incident is the lack of effective planning before the beginning of the demolition. Such planning has already been covered earlier in this chapter – the pre-demolition survey, a detailed site risk assessment and the development of a comprehensive method statement. Premature collapse usually begins with the structural collapse of floors and is often caused by plant operating on floors which are not certified safe and/or back-propped where required. It can also be caused by poor site supervision and a lack of instruction, training and information for the workforce. This will sometimes lead to individual workers or teams being unaware of additional hazards which have been created as the demolition work progresses. A fully competent and trained workforce that has been made

Figure 18.4 Remote-controlled hydraulic arm for pushing, nibbling or hammering

completely aware of the hazards associated with a particular site should lead to a successful conclusion to the demolition.

Protection from falls and falling material

A risk assessment should be undertaken in relation to falls, and fall protection should be provided for all persons exposed to the risk of a fall that could cause an injury. Having regard to the hierarchy of control, the higher order controls of elimination and isolation should be introduced where practicable, to ensure all persons work from solid construction.

Falls at demolition sites can be the result of:

- ▶ falling through fragile roofing material;
- ▶ falling through penetrations or open voids (e.g. skylights and voids for air conditioning and plumbing);
- ▶ falling from open edges (e.g. stairways, landings, fixed platforms, scaffolding and edge of roof);
- ▶ falling out of elevating work platforms (e.g. scissor lifts);
- ▶ failure of plant (e.g. elevating work platforms);
- ▶ falling whilst accessing a roof or other elevated area;
- ▶ falling down lift shafts;
- ▶ collapse of flooring (e.g. concrete slab and wooden decking);
- ▶ collapse of ground above cellars or pits; or
- ▶ tripping over debris on the ground.

Figure 18.5 Controlled collapse

The hazard of being hit, trapped or struck by an object may be present for a number of reasons, which include:

- ▶ falling debris (e.g. down service ducts and lift shafts);
- ▶ accidental or uncontrolled collapse of the structure;
- ▶ use of plant and equipment (e.g. cranes lifting loads); and
- ▶ failure of structural members (e.g. load-bearing steelwork).

Noise

Noise can be a serious problem during the demolition process. However, some of the effect might be mitigated by replacing noisy equipment with quieter machinery and ensuring that plant and equipment are regularly maintained. Noise from the processes may be difficult to control but the location of noisy plant where it would have least effect on the neighbourhood, the erection of noise barriers, the reduction of the time taken by noisy operations and the organisation of the work so that noisy operations are not carried out either early in the morning or during the evening and night, would all help to reduce the risk.

Noise and vibration on demolition sites will normally be caused by the use of plant and equipment, falling debris and/or explosives. Compressors, pneumatic hand-held tools, front-end loaders, excavators and other equipment can create noise levels of more than the lower exposure action value of 80 dB(A) and may at times create peak noise levels of more than 135 dB(C), the lower peak sound pressure action value. A noise assessment will be required and workers will need to wear ear defenders.

Dust

During the demolition process, dust is a considerable hazard which needs to be controlled. Among the possible control measures are damping the process down with water, the sheeting of disposal lorries as they leave the site to prevent the generation of dust and the provision of filters or covers to site drainage systems to prevent the risk of them silting up. A restriction on the speed of vehicles on the site and the regular cleaning of roads will also help to reduce the dust level. Mud on the roads, which will cause dust, can be alleviated by providing 'stoned-up' vehicle routes, ensuring that vehicles pass through a wheel wash before leaving the site, and by setting up procedures for regular road cleaning and sweeping. Designating the vehicle routes to and from the site would also limit the number of affected roads.

Dust and fumes from site machinery and smoke from burning timber waste can further exacerbate the dust problem. As stressed earlier, asbestos dust is a particularly hazardous problem that can only be addressed by specialists.

Fire

Fire is a risk where hot work (using any tools that generate spark, flame or heat) is being done. During structural alteration, the fire plan must be kept up to date as the escape routes and fire points may alter. There must be an effective way to raise the alarm.

Siting and use of machinery

The hazards associated with plant and equipment used on demolition sites are numerous. Only appropriately competent persons should operate plant and equipment

on these sites. Examples of some of the hazards that may result from operating plant and equipment on demolition sites include:

▶ electrocution due to plant or equipment coming into contact with live electricity;
▶ plant failure due to its safe working load being exceeded;
▶ accidental dropping of material due to the incorrect slinging of loads;
▶ plant striking or colliding with persons, particularly on shared access ways;
▶ excessive noise and vibration from machinery, such as front-end loaders and excavators;
▶ flying particles from pneumatic tools such as impact hammers;
▶ welding and cutting operations during maintenance work;
▶ the structural failure of steelwork; and
▶ fire and explosions.

The site supervisor must ensure that all plant and equipment when delivered to site is in good working order and fitted with all necessary safety devices, notices and guards. He/she must also ensure that an ongoing maintenance and defect reporting system is in place. No equipment may be used until all defects have been rectified. The site supervisor will ensure that only authorised licensed operators are permitted to operate any plant and equipment used on site.

Manual handling tasks on demolition sites should be identified, assessed and controlled using the hierarchy of control described in Chapter 10. Wherever possible, plant and other mechanical aids should be used to move debris, materials and equipment. The use of some equipment may also create a manual handling hazard. Manual handling risk assessments should be carried out as necessary. Manual handling hazards on demolition sites may result from operating plant and equipment, manual demolition, lifting materials, loading trucks and bins and tidying the site after demolition.

All wiring, except where temporary installations are required, should be disconnected before demolition work commences. Temporary electrical installations must comply with the Institution of Electrical Engineers Wiring Regulations and legislative requirements. Where there is a possibility of live wires in the structure, areas of danger should be clearly tagged and signs erected warning of the hazard. A detailed plan of these areas should be formulated.

Plant and equipment should be used with care to ensure that no part of them comes into contact with overhead or underground wires or cables. When working near any power lines or cables or other services such as gas mains, any national or local legislative requirements for such work must be followed.

18

Environment

Where any redundant building, structure or plant has contained flammable materials, precautions must be observed to avoid fires or explosions. Specialist advice may be required to identify any residual materials and assess whether any contamination remains. Any residue of flammable materials must be made safe.

Deliberate burning of structures must not be used as a method of demolition. The burning of any waste materials, such as timber, on demolition sites must not be undertaken without the approval of the relevant local authorities.

When welding or cutting at the demolition site, sparks will present a fire hazard. Any flammable materials should be identified and fire risks assessed before this type of work is undertaken.

The escape of explosive gases from accidental damage to pipes during the demolition process is yet another source for fire and explosions. All services, including gas, need to be identified during the pre-demolition survey.

Where services are shown on original plans for the structure and site, but cannot be found, work should cease until a further investigation determines the location of these services.

Arson, especially when the site is unattended, is a further risk of fire. Adequate site security, particularly the boarding of empty window spaces, and good housekeeping of the site, by removing and/or isolating flammable and combustible materials, will assist in reducing the risk of arson at a demolition site.

Portable fire extinguishers and access to local fire hydrants should be maintained throughout the demolition process.

Environmental airborne hazards in the form of pollution can be produced from petrol- or diesel-engine powered vehicles or machinery. This is a particular problem when the engines are poorly maintained or being overworked. Carbon monoxide from petrol- or diesel-driven plant is a serious health problem when such plant is operated in enclosed areas.

Finally, fuel storage tanks should be surrounded by a bund and standing vehicles and vessels should be provided with drip trays beneath them to prevent fuels from contaminating the ground and water courses.

Competence and training

A very important requirement for demolition work is the provision of training for all construction workers involved in the work. Specialist training courses are available for those concerned with the management of the total process. The Construction Industry Training Board (CITB) offers a training course entitled 'A Scheme for the Certification of Competence of Demolition Operatives'.

An induction training which outlines the hazards and the required control measures should be given to all workers before the start of the demolition work. This training should provide information to site workers to make them aware of the hazards that they will face, the demolition and control procedures to be adopted and the site rules. Issues such as first-aid and welfare facilities should also be covered. The raising of environmental issues at progress meetings and the inclusion of the procedures in work specifications and the health and safety plan are further actions that can be used to increase awareness. During the demolition process, communications should be continued using toolbox talks, and posters and signs around the site.

Figure 18.6 Demolition site should be well planned and properly signed and controlled throughout the project

18.2.2 Management of the demolition and general controls

Demolition hazard control involves implementing measures which reduce the risk at the demolition site. Where legislation requires specific methods to control the risk, these must be used.

As discussed in Chapter 4, there is a recognised hierarchy of controls in health and safety ranging from the most to the least effective. This hierarchy can be applied directly to the demolition process. The elements of the hierarchy are:

- ▶ elimination;
- ▶ substitution;
- ▶ isolation;
- ▶ engineering controls;
- ▶ administrative controls;
- ▶ personal protective equipment.

Elimination is the most effective control measure. This involves removing the hazard or hazardous work practice from the site, by the following actions:

- ▶ disconnecting services to the demolition site;
- ▶ ensuring there are no sparks or ignition sources, where a risk of fire or explosion exists;
- ▶ ensuring separation between the public and demolition activities;

excluding unauthorised persons from the site.

Substituting or replacing a hazard or hazardous work practice with a less hazardous one could include the following:

▶ using a controlled collapse technique, in place of people working at heights;
▶ using power shears in place of grinding or oxy-acetylene cutting, where a risk of fire exists.

Isolating or separating the hazard or hazardous work practice from workers and others involved in the demolition area could include the following:

▶ installing screens on plant to protect from dust and noise;
▶ installing barriers and fences;
▶ marking off hazardous areas.

If the hazard cannot be eliminated, substituted or isolated, an engineering control is the next preferred measure, by means including the following:

▶ modification of tools or equipment;
▶ provision of guarding to machinery or equipment;
▶ provision of shatterproof/guarded windscreens on plant;
▶ provision of falling-object protective structures (FOPS) and rollover protective structures (ROPS) on vehicles and plant;
▶ installation of safe working platforms on scaffolds and elevating work platforms;
▶ installation of edge protection to open edges of landings, stairways and fixed platforms;
▶ use of props and bracing to support loads;
▶ installation of static lines and anchor points for fall arrest systems;
▶ water sprays to suppress dust;
▶ use of chutes for dropping debris;
▶ use of cranes for lowering loads;
▶ installation of vehicle buffers where plant is exposed to an open edge;
▶ use of machine-mounted impact hammers, power shears, drills and saws;
▶ provision of flash arresters on gauges and hoses on welding equipment.

Administrative controls, including the introduction of reduced risk work practices, are also important and include the following:

▶ limiting the amount of time a person is exposed to a particular hazard;
▶ implementing and documenting safe working procedures for all hazardous tasks;
▶ training and instructing all personnel;
▶ identifying hazardous substances prior to work commencing;
▶ implementing safe procedures for handling hazardous substances;
▶ implementing procedures for disconnecting services to the site;
▶ using operational observers at the site;

▶ implementing confined space entry procedures;
▶ ensuring that all loads to be lifted are accurately calculated;
▶ consulting with all workers on the site.

Personal protective equipment is the least effective management control because it needs almost constant supervision and enforcement. The advice of a specialist may be required when selecting personal protective equipment. Personal protective equipment should only be considered when other control measures are not practicable or to increase the protection given to a person. Personal protective equipment includes:

▶ safety helmets (required by legislation);
▶ fall arrest harnesses and lanyards;
▶ safety boots;
▶ gloves;
▶ goggles;
▶ respirators;
▶ hearing protectors.

18.3 Purposes and scope of pre-demolition, deconstruction or refurbishment survey

When a building is to be demolished the (non-domestic) client (usually the property owner) has a duty to provide pre-demolition information to the designer and contractor. This will involve a pre-demolition investigation and survey.

Before any work is started, a full site investigation must be made by a competent person to determine the hazards and associated risks which may affect the demolition workers and members of the public who may pass close to the demolition site. The competent person is often a specialist structural engineer who will also advise on the temporary support of adjacent buildings and the correct method of dismantling or demolition.

There are several pre-commencement surveys that may be necessary to ensure the health and safety of all personnel on site. These include:

▶ a services survey to locate any existing services on site;
▶ an asbestos survey to identify asbestos-containing materials on site;
▶ a soil survey to identify any contaminated ground; and
▶ a ground survey to determine the load-carrying capacity of the ground.

The investigation should cover the following topics:

▶ the construction details of the structures or buildings to be demolished (including the materials used, fragile roofs, rot and the identification, significance and extent of any dilapidation of the structure, the presence

18

of cantilevered structures and any general weaknesses) and those of neighbouring structures or buildings;

▶ the identification of key structural elements including pre- and post-tensioned components;

▶ a review of drawings, structural calculations, health and safety file, etc. related to the structure;

▶ the previous use of the premises;

▶ the load-carrying capacity of adjoining land including the presence of underground culverts;

▶ the need for possible temporary support structures for the building being demolished and adjoining buildings;

▶ falls of materials and people;

▶ the location of any dangerous machinery;

▶ the presence of asbestos, lead or other hazardous or radioactive substances and any associated health risks;

▶ environmental issues, such as dust, water pollution and noise;

▶ public safety including the provision of high fencing or hoardings;

▶ manual handling issues;

▶ the location of any underground or overhead services (water, electricity, gas and sewage);

▶ the location of any underground cellars, storage tanks, chimneys, balconies or bunkers particularly if flammable or explosive substances were previously stored;

▶ the means of access to the site;

▶ the removal of waste;

▶ the details of any traffic or pedestrian routes through the site;

▶ the provision of welfare facilities;

▶ the proximity of neighbours;

▶ the location of any public thoroughfares adjacent to the structure or building;

▶ the name of the demolition client.

The structural survey should consider:

▶ the age of the structure;

▶ its previous use;

▶ the type of construction;

▶ nearby buildings or structures; and

▶ the weight of removed material or machinery on floors above ground level.

The details of the structure of the building to be demolished would include whether it was built of brick, pre-stressed concrete, reinforced concrete or steel. There may be certain building regulations which cover the site and the Local Authority Building Department should be contacted to ascertain whether any part of the site is affected by these regulations. All structural alterations carried out on the structure in the past should be reviewed. It is also important to consult with legal advisers to ensure that there are no legal covenants or disputes which could affect operations on the site.

All demolition work requires those in control of the work to produce a written plan showing how danger will be prevented. This will be the responsibility of the CDM coordinator on notifiable projects and the contractor or designer on non-notifiable projects.

The written plan will include a risk assessment of the state and design of the structure to be demolished and the influence of that design on the demolition method proposed. This risk assessment will normally be made by the project designer who will also plan the demolition work. A further risk assessment should then be made by the contractor undertaking the demolition and a written method statement will be required before demolition takes place.

The site manager should arrange for suitable plant and equipment to be provided so that the work can be executed to the standards required by health and safety legislation, in particular the Control of Asbestos Regulations. It may be necessary for the local authority and the police to be consulted about the proposed demolition so that issues of public protection, local traffic management and possible road closures can be addressed. There should be liaison with the occupiers of adjacent properties because, in some cases, they may need to be evacuated.

The provision of temporary access roads, welfare facilities, office accommodation, fuel storage and plant maintenance facilities on site will need to be considered at the planning stage. Effective traffic management systems are essential on site, to avoid putting workers at risk of being hit by vehicles turning, slewing or reversing. Where possible, vision aids and zero tail swing machines should be used.

The presence of hazardous substances and their release during the demolition process must be considered at the planning stage. Most hazardous substances create a hazard for demolition workers by being inhaled, ingested, injected or coming into contact with or being absorbed by the skin. Environmental monitoring may need to be carried out in certain situations. Specialist advice should be obtained from appropriately competent persons. Some of the most common hazardous substances in demolition work include:

▶ **lead** – is most dangerous when it is in the air as a fume or dust (e.g. cutting steelwork coated with lead-based paint or dismantling of tanks containing lead-based petrol);

▶ **asbestos** – where possible it should be removed before any other demolition work starts and must always be removed by a licensed contractor. Asbestos may be found in sprayed coatings, thermal and acoustic insulation materials, fire-resistant walls/partitions, asbestos cement sheets or flooring materials;

▶ **PCBs** – a toxic substance found in electric transformers and capacitors, refrigeration and heating equipment;

- **silica** – occurs in stone, some bricks and concrete aggregate. Any demolition of structures constructed from these materials will give rise to dust containing silica.

Residues of hazardous substances may also create a hazard to demolition workers. These could include:

- acids from industrial processes;
- paints;
- flammable liquids;
- unidentified drums;
- microbiological hazards (especially in old hospital buildings).

Storage tanks, vessels, pipes and other confined spaces may also contain flammable vapours or toxic sludges – especially those which were formerly used in industrial or chemical processes. Plans must be made to dispose of any hazardous or dangerous substance found during the demolition process in a safe way which conforms to legislative requirements.

18.4 Control measures that a method statement should include

18.4.1 Introduction

It is essential that a detailed method statement is in place before the commencement of a demolition, deconstruction or a major refurbishment project. Sometimes, the exterior of a building has to be preserved intact whilst the interior is completely stripped of partitioning, flooring and internal walls ready for refurbishment or redevelopment. This work requires careful control and management and is known as **'soft strip'**. During this process, temporary propping and structural support may well be required.

Any demolition project will require the application of many of the issues covered in previous chapters. These issues include:

- working at height including the use of scaffolding;
- access and egress from site;
- the protection of third parties, such as children, and surrounding structures;
- the isolation of mains gas, water and electricity;
- the installation of temporary services, such as electricity;
- the identification and removal of asbestos;
- the use of plant and equipment;
- effective communication between all workers on the site;
- effective emergency arrangements;
- the management of waste material, both on-site segregation and off-site disposal;
- the effective coordination of all work activities.

18.4.2 Demolition method statement

A risk assessment should be made by the contractor undertaking the demolition – this risk assessment will be used to draw up a method statement for inclusion in the health and safety plan. This statement is a safe system of work that identifies the sequence of work required to prevent accidental collapse of the structure.

A written method statement will be required before demolition takes place. The contents of the method statement will include the following:

- details of the method of demolition to be used including the means of preventing premature collapse or the collapse of adjacent buildings and the safe removal of debris from upper levels to prevent overloading due to the debris;
- establishment of exclusion zones and hard-hat areas, clearly marked and with barriers or hoardings;
- the design and method of temporary supports;
- details of site access and security;
- details of the location of any underground or overhead services;
- details of protection from falling materials arrangements;
- details of equipment, including access equipment, required, and any hazardous substances to be used;
- details and use of high-reach machines and reinforced machine cabs so that drivers are not injured;
- arrangements for the protection of the public and the construction workforce against noise and whether hazardous substances, such as asbestos or other dust, are likely to be released;
- details of the isolation methods for any services which may have been supplied to the site and any temporary services required on the site;
- details of personal protective equipment, such as hard hats, which must be worn by all personnel on site;
- first-aid, emergency and accident arrangements;
- training and supervising site workers;
- welfare arrangements;
- arrangements for the separation and disposal of waste;
- names of site foremen and those with responsibility for health and safety and the monitoring of the work;
- the coordination of all work activities on the site; and
- the expected level of competence of site workers.

Other risk assessments, such as COSHH, personal protective equipment and manual handling, should be appended to the method statement.

The isolation of all services (gas, electricity and water) which feed the site is essential before any demolition takes place. Contact with the appropriate service provider will be necessary since sites often have a complex of feeds for many of their services. The local authority and surrounding properties also need to be informed that services are to be isolated.

Over recent years, more rules and regulations have been introduced concerning the disposal of construction waste and this topic is covered in more depth in

18

Chapter 14. Proper arrangements must be made with a reputable waste disposal contractor for the disposal of demolition waste. If hazardous substances are included in the waste, then specialist waste contractors should be used. Only registered disposal sites should be used and records kept of each load.

The site should be made secure with relevant signs posted to warn members of the public of the dangers.

Asbestos-containing material requires special care and will only be dismantled by a licensed contractor. Indeed, if asbestos is disturbed in any way, a specialist contractor must be consulted. If possible, such work should be done before any other work is started.

Figure 18.7 Clearing up asbestos-containing materials (hazardous waste) after demolition has started is very difficult and expensive

It is important to develop a safe system of work for demolition which ensures that people are distanced as far as possible from the demolition area. A checklist of issues which should be incorporated into a safe system of work is given in Appendix 18.1.

Glass must be removed from all windows before demolition starts. This will avoid hazards from flying broken glass. The window spaces should be boarded up to deter trespassers. Timber can present three hazards: injuries caused by protruding nails or sharp and broken edges, and pollution and fire if the timber is burnt on site.

Young persons who are undergoing training either as an apprentice or work experience trainee require close supervision on demolition sites. No young person (under 18 years old) should be permitted to operate any plant or act as banksman unless being trained under direct supervision of a competent person. Children and unauthorised persons must not be permitted to enter working areas at any time, in particular whilst plant is in use; all necessary control measures must be taken to avoid injuries to such persons on the site both during and after working hours, particularly if it is not possible to fully fence the site. The general public must also be

protected by restricting access to the demolition site and by taking special precautions for walkways and other public areas.

18.4.3 Demolition pre-tender information and a construction phase health and safety plan

Under CDM 2015 pre-tender information and a construction phase health and safety plan must be prepared. They are covered extensively in chapters 7 and 19.

For demolition work, the pre-tender plan will draw heavily on the pre-demolition survey and would take account of conditions that the client imposes, so that the work will not impinge too greatly on his/her business or upon the neighbourhood. It would include:

► the nature of the project (including the name of the client and the principal designer);
► the type and size of the building or structure;
► a description of the work and its scope and timescale;
► the previous use of the building and its proximity to other buildings;
► any likely health risks that might arise (such as dust possibly including asbestos);
► the condition of the structure;
► the position of existing services;
► ground conditions and any possible contamination;
► information relating to residual risks identified by the designers arising from the methods or materials of construction;
► site rules and procedures for continuing liaison with the principal contractor.

Figure 18.8 Type and size of the building is particularly important for unusual demolitions

The construction phase health and safety plan should include the following:

► the coordination of and provision of information to other contractors;
► risk assessments and other statutory assessments such as COSHH and noise;

- emergency procedures;
- health and safety monitoring arrangements;
- site rules;
- welfare arrangements;
- the provision of information and training to the workforce;
- consultation arrangements with the workforce.

The following issues are specific to the demolition work and would include:

- the particular methods of demolition to be employed;
- the arrangements for the control and disposal of waste;
- the provision and use of personal protective equipment;
- the details of permit-to-work systems;
- the arrangements for the exclusion of unauthorised persons.

18.5 Further information

The Construction (Design and Management) Regulations 2015

The Personal Protective Equipment at Work Regulations 1992

The Provision and Use of Work Equipment Regulations 1998

Personal Protective Equipment at Work (Guidance), L25, HSE Books, ISBN 978-0-7176-6139-3 http://www.hse.gov.uk/pubns/priced/l25.pdf

British Standard Code of practice for demolition (BS 6187), British Standards Institution ISBN 0-5803-3206-3

Protecting the public: Your next move (HSG151), HSE Books, ISBN 978-0-7176-6294-4 http://www.hse.gov.uk/pubns/books/hsg151.htm

18.6 Practice revision questions

1. A former factory is to be demolished to make way for a new block of offices on a trading estate.

   (a) **Outline** the criteria for notifying the project to the HSE.

   (b) **Outline** (with reasons) **THREE** types of pre-commencement survey that will be necessary to ensure the health and safety of personnel on site.

2. Major alterations and refurbishment are to be carried out on an occupied office block, including the demolition and rebuilding of partition walls.

   (a) **Outline** the main hazards that may be present during the demolition of a building.

   (b) **Outline** the issues that need to be considered when preparing an overall method statement for the works.

3. **Outline** the health and safety issues to be considered when planning the demolition of a two-storey detached house in a street of occupied houses.

4. **Identify** the key issues to be addressed in a pre-demolition survey of a multi-storey block of flats in a city centre.

5. A large industrial site consisting of several two-storey buildings has been closed. The owners of the site are to have the buildings demolished before selling the site for redevelopment. **Outline** the issues the principal demolition contractor should address as part of the health and safety plan.

6. The owners of a currently disused office block are to have part of the building demolished and a large extension built. With respect to the proposed work, **outline** the areas to be addressed in the health and safety plan:

   (a) as prepared by the client;

   (b) as developed by the principal contractor.

7. A large construction project in a residential area involves demolition and subsequent removal of debris and spoil.

   (a) **Outline** the potential environmental problems that might arise from this work and **explain** how the problems might be mitigated.

   (b) **Describe** the actions that could be taken by site management in order to ensure that contractors and operatives are aware of the procedures adopted.

8. A single-storey farm building is to be demolished using an excavator. **Outline** control measures that the method statement should contain for this activity.

18

APPENDIX 18.1 Checklist for a safe system of work

The key to the development of an effective safe system of work for demolition is to choose a method of working with everyone as far away from the danger area as possible. After the site survey has been completed, the following issues should be addressed:

▶ Does the demolition method to be used, such as a long-reach machine or crane and ball, enable everyone to be distanced from the demolition?

▶ Will the demolition work cause nearby structures to become unstable requiring the use of temporary supports?

▶ Will structural elements of the building, such as intermediate floors, support the weight of demolition machinery and/or falling materials?

▶ Will the HSE be notified using Form 10 or its equivalent before work begins?

▶ Have all electric, gas, water, steam, sewer and other service lines been shut off, capped or otherwise controlled prior to the start of work and have temporary lines to supply power for adequate lighting during demolition work been provided?

▶ Have any hazards that exist from fragmentation of glass been fully controlled?

▶ Have all floor openings, not used as material drops, been covered?

▶ Have worker entrances to sites of multi-storey buildings which are to be demolished been protected through the use of walkway canopies that extend from the face of the building for a minimum of 2 m?

▶ Are materials to be dropped to any point lying outside the exterior walls of the structure into an area which is not effectively protected?

▶ Are material chutes, inclined at an angle of 45° or more to the horizontal, entirely enclosed and is there a substantial gate at the discharge end which is to be operated by a competent worker? Chute openings must be protected by a substantial guard rail approximately 1.6 m above the floor level.

▶ When using material chutes, is the area where the material is dropped enclosed with barricades not less than 2 m around the chute opening? Are warning signs of the hazard of falling material posted at each level?

▶ During mechanical demolition, is a competent person continuously inspecting the structure as work progresses to detect hazards resulting from the weakened floors and walls or from loosened materials?

CHAPTER 19

Summary of the main legal requirements

This chapter contains:

1. An outline of the Health and Safety at Work (HSW) Act, The Environmental Protection Act and associated regulations and relevant parts of The New Roads and Street Works Act 1991
2. Summaries of over 30 sets of regulations mainly made under the HSW Act which are covered by the NEBOSH (National Examination Board in Occupational Safety and Health) Construction Certificate syllabus
3. A summary of the Regulatory Reform (Fire Safety) Order
4. A very brief summary of over 17 other sets of regulations which are of interest to managers and students alike
5. Further information after each piece of legislation

19.1 Introduction

The achievement of an understanding of the basic legal requirements can be a daunting task. However, the health and safety student should not despair, because at NEBOSH National Construction Certificate level they do not need to know the full history of UK health and safety legislation, the complete range of regulations made under the HSW Act 1974 or the details of the appropriate European Directives. These summaries, which are correct up to April 2015, cover those acts and regulations which are required for the Certificate student and will provide the essential foundation of knowledge. Students should check the latest edition of the NEBOSH Construction Certificate guide to ensure that the regulations summarised here are the latest for the course being undertaken.

Many managers and other students will find these summaries a useful quick reference to health and safety legal requirements. However, anyone involved in achieving compliance with legal requirements should ensure they read the regulations themselves, any approved codes of practice and Health and Safety Executive (HSE) guidance.

19.2 The legal framework

19.2.1 General

The HSW Act is the foundation of British health and safety law. It describes the general duties that employers have towards their employees and to members of the public, and also the duties that employees have to themselves and to each other.

The term 'so far as is reasonably practicable' (sfairp) qualifies some of the duties in some sections, notably 2, 3, 4 and 6, of the HSW Act. In other words, the degree of risk in a particular job or workplace needs to be balanced against the time, trouble, cost and physical difficulty of taking measures to avoid or reduce the risk.

The law simply expects employers to behave in a way that demonstrates good management and common sense. They are required to look at what the hazards are and take sensible measures to control/reduce the risks from them.

The Management of Health and Safety at Work Regulations 1999 (MHSWR) (the Management Regulations) give employers specific requirements to manage health and safety under the HSW Act. Like the Act, they apply to every work activity.

The MHSW Regulations require every employer to carry out risk assessments. If there are five or more employees in the workplace, the significant findings of the risk assessments need to be recorded.

In a place like an office, risk assessment should be straightforward; but where there are serious hazards, such as those in a chemical plant, laboratory or an oil rig, it is likely to be more complicated.

In October 2010 Lord Young published his report entitled, *Common Sense Common Safety,* which stated that:

> *'The aim is to free businesses from unnecessary bureaucratic burdens and the fear of having to pay out unjustified damages claims and legal fees. Above all it means applying common sense not just to compensation but to everyday decisions once again.'*

Some effects of this report became evident fairly quickly during 2011 with the introduction of: the Occupational Safety Consultants Register in January 2011; 20-minute risk assessments for low risk offices; a review of the information reported under RIDDOR and a move to reporting to 7 days for 3-day-plus lost time injuries; Police and Fire Services – police and fire-fighters not to be at risk of investigation or prosecution when committing a heroic act.

To read the full report, go to http://news.hse.gov.uk/2010/10/15/lord-young-report/.

Subsequently in March 2011 the then Employment Minister Chris Grayling commissioned an independent review and appointed Professor Ragnar Löfstedt –

Director of the King's Centre for Risk Management at King's College, London – to chair it.

Professor Löfstedt made recommendations in November 2011 aimed at reducing the burden of unnecessary regulation on businesses while maintaining Britain's health and safety performance, which is among the best internationally. The Government accepted his recommendations. As of December 2014 significant work is still being undertaken to complete his recommendations which included changes to: Reporting of Injuries, Diseases and Dangerous Occurrences (RIDDOR) (now RIDDOR 2013); exempting the self-employed whose work imposes no risk to others; amendment, consolidation or withdrawal of 30 Approved Codes of Practice; amendment of Construction (Design and Management) (CDM) Regulations (now CDM 2015); revocation of 14 regulatory measures including Construction (Head Protection) Regulations and Notification of Conventional Tower Cranes Regulations, and the changes to strict liability in Section 47 of the HSW Act 1974.

19.2.2 The relationship between the regulator and industry

In April 2008 the HSC and the HSE merged into a single unified body called the Health and Safety Executive (HSE). The HSE consults widely with those affected by its proposals.

The HSE has many contact points with industry and work through:

▶ the Industry and Subject Advisory Committees, which have members drawn from the areas of work they cover, and focus on health and safety issues in particular industries (such as the textile industry, construction and education or areas such as toxic substances and genetic modification);

▶ intermediaries, such as small firms' organisations providing information and advice to employers and others with responsibilities under the HSW Act;

▶ guidance to enforcers, both the HSE inspectors and those of local authorities (LAs);

▶ the day-to-day contact which inspectors have with people at work.

The HSE consults with small firms through Small Firms' Forums. It also seeks views in detail from representatives of small firms about the impact on them of proposed legislation.

19.2.3 Health and safety in Scotland

Although occupational health and safety is a reserved issue (i.e. not devolved to the Scottish Executive (SE)), there are some significant differences with some issues. The following extracts from the concordat between the HSE and the SE outline some of these issues.

19

Extracts from The Concordat between the Health and Safety Executive and the Scottish Executive

Introduction

1. This concordat is an agreement between the Scottish Executive and the Health and Safety Executive. It is intended to provide the framework to guide the future working relationship between the Health and Safety Executive (HSE) and the Scottish Executive (SE). The objective of the concordat is to ensure that the roles and responsibilities of the HSE and the SE in the new constitutional structure are effectively translated into practical working arrangements between the two organisations. The aim is to promote the establishment of close and harmonious working relationships and good communications at all levels between the HSE and the SE, and in particular to foster constructive cooperation.

   Most day to day liaison between the SE and HSE will, in the first instance, be handled by the HSE's Director for Scotland, based in Edinburgh.

   [*Note*: where the concordat referred to HSC/E this has been changed to HSE meaning the new combined Authority.]

Nature of concordat

3. This concordat is a voluntary arrangement between the HSE and the SE. It is not a binding agreement or contract and so does not create any legally enforceable rights, obligations, or restrictions. It is intended to be binding in honour only. The HSW Act provides for statutory consultation in connection with proposals for regulatory change; this concordat does not create any other right to be consulted or prevent consultation beyond that required by statute. Any failure to follow the terms of the concordat is not to be taken as invalidating decisions taken by the HSE or the Scottish Executive.

Consultation

11. There are a significant number of areas where HSE and Scottish Ministers share or have closely related interests, as set out in the Annex. The HSE and the SE recognise that the extent of these areas makes good communications essential in order to assist the process of policy formation and decision-making in each administration and to meet any consultation or other requirements connected with the exercise of a function. They also recognise that there will be mutual benefit from the exchange of information on scientific, technical and, where appropriate, policy matters.

12. Information obtained by use of the HSE's statutory powers cannot be made available to SE where the release of that information is not for the purposes of the HSE's functions. Subject to that, the HSE and the SE will:

▶ share relevant information, analysis and research;

▶ inform each other of any relevant information which comes to their attention which may require action or have resource consequences for the other party;

▶ seek to involve each other, as and when appropriate, in policy development on all topics where there is a reasonable expectation that a policy initiative might affect the other's responsibilities, or be used or adapted by the other;

▶ inform each other at the earliest practicable stage of any emerging proposal to change primary or secondary legislation which might have an impact upon the other's responsibilities;

▶ inform each other at the earliest practicable stage of substantive new policy announcements which may be relevant to the other's responsibilities; and

▶ coordinate activities where appropriate.

Local authorities

16. The constitution of Scottish local authorities, which includes their establishment, dissolution, assets, liabilities, funding and receipts, is a devolved matter in which the Scottish Ministers have a policy interest as are most functions conferred upon such authorities. The HSE has an interest in the functions of local authorities under health and safety legislation, and under Section 18(4) of HSWA, the Commission may issue guidance to local authorities, as enforcing authorities, which they must follow. The HSE will consult the SE in advance where proposed changes in health and safety legislation, or directions under Section 18(4), might result in new burdens being imposed upon Scottish local authorities.

ANNEX – areas of common and closely related interest – extracts only

This annex lists subjects in which the Scottish Ministers and the HSE share an interest and other areas where there may be a particular need for consultation between the HSE and the Scottish Executive.

Building control

This is a devolved matter, but the HSE has an interest in the related matter of construction safety.

Dangerous substances

The HSE have an interest under the HSW Act 1974 and the European Communities Act 1972 in the control of the keeping, notification, supply and use of explosive or highly flammable or otherwise dangerous substances which have the potential to create a major accident and in the prevention of the unlawful acquisition, possession and use of such substances. The HSE also has an interest in the carriage by road and rail of dangerous and environmentally hazardous goods. The Scottish Ministers, however, have an interest in such matters for the purposes of protection of the environment and the planning system (by virtue of the Planning (Hazardous Substances) (Scotland) Act 1997). The HSE will continue to advise what hazardous substances and in what quantities have significant risk off site. HSE involvement in individual hazardous substances consent/planning applications which are before Scottish Ministers for a decision will be on the basis of established planning and related public inquiry procedures.

Educational facilities; adventure activity centres

The safety of these premises is generally a devolved matter. However, the HSE has interests in the safety of adventure activity centres because of its responsibilities by virtue of the Activity Centres (Young Persons Safety) Act 1995 and subordinate legislation made under that Act. It also has an interest in the health and safety of workers and those affected by the activities of workers in these premises.

Fire safety

General fire safety is a devolved matter and the policy responsibility of the Scottish Ministers. However, the HSE has an interest in the reserved matters of process fire precautions, fire precautions in relation to petroleum and petroleum spirit, fire safety on construction sites, ships under construction or repair by persons other than the master or crew, in mines, on offshore installations, and on any other premises which on 1 July 1999 are of a description specified in Part I of Schedule 1 to the Fire Certificates (Special Premises) Regulations 1976.

Food safety

This is generally a devolved matter and the policy responsibility of the Scottish Ministers. However, the HSE has an interest in the safety of workers in the food processing, manufacturing, cooking and food distribution industries and in related risks to the public.

Local authorities

The Scottish Ministers have an interest in the devolved matter of the constitution of Scottish local authorities, which includes their establishment, dissolution, assets, liabilities, funding and receipts. The HSE has an interest in the functions and performance of local authorities in relation to the enforcement of health and safety legislation.

Pesticides (including biocides and plant protection products)

The Scottish Ministers' interest is in the protection of the environment, public health, animal and plant health and food safety. The HSE's interest is in the protection of workers and those affected by the activities of workers, protection of the environment, and product approvals.

Places of entertainment (cinemas, theatres, casinos, dance-halls, etc.), sports facilities, sports events and zoos

The Scottish Ministers' interest is in public safety, fire safety, structural building standards, sanitation, food standards, pest control, security, etc. The HSE has an interest in the safety of workers and the safety of members of the public and others affected by work activity.

Ports, harbours and inland waterways

Scottish Ministers' interest is in ports policy and communications via the various ferry links between mainland Scotland and the islands. Responsibility for safety enforcement for the ships and their crews rests with the Maritime and Coastguard Agency. The HSE has an interest in the safety of workers and the safety of members of the public and others affected by work activity at inland waterways, the safety aspects of new harbour byelaws and the health and safety of shore side workers loading or unloading from berthed ships or the dock side.

Protection of the environment

This is a devolved matter and the policy responsibility of the Scottish Ministers. Responsibility for enforcement of much environmental legislation in Scotland rests

19

with the Scottish Environment Protection Agency. However, the HSE has enforcement responsibilities for certain environmental legislation under agency arrangements under Section 13 of HSWA.

Public health, occupational health and health promotion

Public health is generally a devolved matter and the policy responsibility of the Scottish Ministers. Occupational Health is a reserved matter for the HSE. General health promotion in Scotland is a devolved matter and the policy responsibility of the Scottish Ministers, but the HSE has an interest in 'lifestyle' issues such as misuse of drugs, alcohol and smoking (see also below), when these impinge on the workplace.

Rail safety

This is a reserved matter. However, the Scottish Ministers have an interest because of their executively devolved functions in relation to a number of aspects of the regulation of the rail industry in Scotland. The HSE has an interest in any matter which could have an impact on its responsibility for the approval of new and altered works on railways, light rail and tramways.

Smoking

The regulation of smoking and passive smoking in workplaces is a reserved matter. However, the Scottish Ministers have an interest. Other aspects of smoking such as the regulation of tobacco advertising are generally devolved and are the policy responsibility of the Scottish Ministers.

Water and sewerage

This is generally a devolved matter and the policy responsibility of the Scottish Ministers. Water and sewerage services in Scotland are provided by three publicly owned water authorities. The HSE's interest is in matters relating to the safety of workers and those affected by the activities of workers.

19.2.4 Health and safety in Northern Ireland

The Health and Safety at Work (Northern Ireland) Order 1978 SI 1978/1039 (NI 9) http://www.legislation.gov.uk/nisi/1978/1039/contents effectively applies the Great Britain, Health and Safety at Work Act 1974 to Northern Ireland. The Health and Safety at Work (Amendment) (Northern Ireland) Order 1998 SI 1998/2795 (NI 18) established the Health and Safety Executive for Northern Ireland.

Much of the subordinate legislation made under the GB HSW Act has been transcribed into NI regulations made under the 1978 Order. For example the Management of Health and Safety at Work Regulations (Northen Ireland) 2000. For a full list as of 1 September 2013 see http://www.hseni.gov.uk/list_of_subordinate_h_s_legislation.pdf.

This also applies to many ACoPs made under the GB HSW Act, which have been approved for use in Northern Ireland. A list as of 18 December 2011 is available at http://www.hseni.gov.uk/resources/legislation.htm. In addition there are a number of NI ACoPs which have been adopted separately from GB but these also follow closely GB legislation. A list up to September 2011 is available at the same website.

19.3 List of Acts, orders and regulations summarised

The following Acts and Regulations are covered in this summary:

The HSW Act 1974
Enterprise and Regulatory Reform Act 2013 (Strict liability issues only)
The Health and Safety at Work etc. Act 1974 (Civil Liability) (Exceptions) Regulations 2013
The Health and Safety (Fees) Regulations 2012
The Environmental Protection Act (EPA) 1990
New Roads and Street Works Act 1991.

Most of the relevant regulations covering health and safety at work have been made under the HSW Act since 1974. Those relevant to the Certificate student are listed here and summarised in this chapter. The first list is alphabetical and some titles have been modified to allow an easier search. The second list is chronological with Statutory Instrument (SI) numbers.

The Classification, Labelling and Packaging of Substances and Mixtures Regulation is an EU regulation which does not need member state regulations to be made. The Regulatory Reform (Fire Safety) Order was made under the Regulatory Reform Act 2001, and the Hazardous Waste Regulations under the EPA.

The regulations/orders covered are as follows.

19.3.1 Alphabetical list of regulations/orders summarised

Control of Artificial Optical Radiation at Work Regulations 2010;
Control of Asbestos Regulations 2012;
Classification, Labelling and Packaging of Substances and Mixtures Regulation (European) adopting into EU UN Globally Harmonised System of Classification and Labelling of Chemicals (GHS);
Confined Spaces Regulations 1997;
Construction (Design and Management) Regulations 2015;

Health and Safety (Consultation with Employees) Regulations 1996;

Control of Substances Hazardous to Health Regulations 2002 and 2004 Amendment;

Dangerous Substances and Explosive Atmospheres Regulations 2002;

Health and Safety (Display Screen Equipment) Regulations 1992;

Electricity at Work Regulations 1989;

Employers' Liability (Compulsory Insurance) Act 1969 and Regulations 1998;

The Regulatory Reform (Fire Safety) Order 2005 including:

Fire Safety (Employees' Capabilities) (England) Regulations 2010;

Fire Safety (Employees' Capabilities) (Wales) Regulations 2012;

Fire (Scotland Act) 2005;

Police and Fire Reform (Scotland) Act 2012;

Health and Safety (First-Aid) Regulations 1981 as amended 2013;

Hazardous Waste (England and Wales) Regulations 2005;

Health and Safety (Information for Employees) Regulations 1989;

Ionising Radiations Regulations 1999;

Lifting Operations and Lifting Equipment Regulations 1998;

Management of Health and Safety at Work Regulations 1999 as amended;

Manual Handling Operations Regulations 1992 as amended;

Control of Noise at Work Regulations 2005;

Personal Protective Equipment at Work Regulations 1992 as amended;

Provision and Use of Work Equipment Regulations 1998 (except Part IV – Power Presses) as amended;

Reporting of Injuries, Diseases and Dangerous Occurrences Regulations 2013;

Safety Representatives and Safety Committees Regulations 1977;

Health and Safety (Safety Signs and Signals) Regulations 1996;

Supply of Machinery (Safety) Regulations 2008 and amendments;

Vibration at Work Regulations 2005;

Waste (England and Wales) Regulations 2011 as amended in 2012;

Workplace (Health, Safety and Welfare) Regulations 1992 as amended;

Work at Height Regulations 2005.

The following acts and regulations have also been included. With a few exceptions like the Road Traffic Acts, they are generally not in the NEBOSH Construction Certificate syllabus. Brief summaries only are given.

Building Regulations 2010 Approved Documents B & M;

Compensation Act 2006;

Corporate Manslaughter and Corporate Homicide Act 2007;

Electrical Equipment (Safety) Regulations 1994;

Equality Act 2010 (Disability) Regulations;

Gas Appliances (Safety) Regulations 1992;

Gas Safety (Installation and Use) Regulations 1998;

Health and Safety (Offences) Act 2008;

Lead at Work Regulations 2002;

Occupiers Liability Acts 1957 and 1984;

Personal Protective Equipment Regulations 2002;

Pesticides Regulations 1986;

Pressure Systems Safety Regulations 2000;

Road Traffic Acts 1988 and 1991;

Sharp Instruments in Healthcare Regulations 2013;

Smoke-free Regulations 2006;

Working Time Regulations 1998, 2001, 2003 and 2007;

Water Resources Act.

19.3.2 Chronological list of regulations/orders summarised

The list below gives the correct titles, year produced and SI number.

Year	SI number	Title
1977	0500	Safety Representatives and Safety Committees Regulations
1981	0917	Health and Safety (First-Aid) Regulations
1989	0635	Electricity at Work Regulations
1989	0682	Health and Safety (Information for Employees) Regulations
1992	2792	Health and Safety (Display Screen Equipment) Regulations
1992	2793	Manual Handling Operations Regulations
1992	2966	Personal Protective Equipment at Work Regulations
1992	3004	Workplace (Health, Safety and Welfare) Regulations
1996	0341	Health and Safety (Safety Signs and Signals) Regulations
1996	1513	Health and Safety (Consultation with Employees) Regulations
1997	1713	Confined Spaces Regulations
1998	2306	Provision and Use of Work Equipment Regulations (except Part IV – Power Presses)
1998	2307	Lifting Operations and Lifting Equipment Regulations
1998	2573	Employers Liability (Compulsory Insurance) Regulations
1999	437	Control of Substances Hazardous to Health Regulations
1999	3232	Ionising Radiations Regulations
1999	3242	Management of Health and Safety at Work Regulations
2002	2677	Control of Substances Hazardous to Health Regulations
2002	2776	Dangerous Substances and Explosive Atmospheres Regulations

19

2004	3386	Control of Substances Hazardous to Health (Amendment) Regulations
2005	asp 5	The Fire Scotland Act
2005	894	The Hazardous Waste (England and Wales) Regulations
2005	735	Work at Height Regulations
2005	1541	The Regulatory Reform (Fire Safety) Order
2005	1093	Control of Vibration at Work Regulations
2008	1597	The Supply of Machinery (Safety) Regulations
2008	EC No 1272	European Regulation on Classification, Labelling and Packaging of Substances and Mixtures (CLP Regulation)
2010	1140	The Control of Artificial Optical Radiation at Work Regulations
2011	988	The Waste (England and Wales) Regulations as amended 2012
2012		The Health and Safety (Fees) Regulations (Regs 23–25)
2012	632	The Control of Asbestos Regulations
2013	1471	The Reporting of Injuries, Diseases and Dangerous Occurrences Regulations
2013	1667	The Health and Safety at Work etc. Act 1974 (Civil Liability) (Exceptions) Regulations
2015		Construction (Design and Management) Regulations

Very brief summaries only for:

Year	SI number	Title
1957	Ch 31	Occupiers Liability Act
1984	Ch 3	Occupiers Liability Act
1986	1510	Control of Pesticides Regulations
1988	Ch52	Road Traffic Act
1991	Ch40	Road Traffic Act
1994	3260	Electrical Equipment (Safety) Regulations
1995	1629	Gas Appliances (Safety) Regulations
1998	2451	Gas Safety (Installation and Use) Regulations
1998	1833	Working Time Regulations
1999	743	The Control of Major Hazards Regulations and amendments
2000	128	Pressure Systems Safety Regulations
2002	1144	Personal Protective Equipment Regulations
2002	2676	The Control of Lead at Work Regulations
2003	1684	Working Time (Amendment) Regulations
2005	1082	The Manufacture and Storage of Explosives Regulations
2006	Ch 29	Compensation Act
2006	3368	Smoke-free (Premises and Enforcement) Regulations
2006	090	Scottish SI – The Prohibition of Smoking in Certain Premises (Scotland) Regulations
2007	W.68	The Smoke-free Premises etc. (Wales) Regulations

2007	Ch 19	Corporate Manslaughter and Corporate Homicide Act
2008	Ch 20	Health and Safety (Offences) Act
2010	471	The Fire Safety (Employees' Capabilities) (England) Regulations
2010	2214	The Building Regulations Approved Documents B and M
2010	Ch 15	Equality Act
2010	2128	The Equality Act 2010 (Disability) Regulations
2012	Asp8	Police and Fire Reform (Scotland) Act
2012	1085 (W133)	The Fire Safety (Employees' Capabilities) (Wales) Regulations
2013	645	The Health and Safety (Sharp Instruments in Healthcare) Regulations
2013	C24	Enterprise and Regulatory Reform Act (Strict Liability Changes only)

19.4 HSW Act 1974 as amended in 2013

The HSW Act was introduced to provide a comprehensive and integrated piece of legislation dealing with the health and safety of people at work and the protection of the public from work activities.

The Act imposes a duty of care on everyone at work related to their roles. This includes employers, employees, owners, occupiers, designers, suppliers and manufacturers of articles and substances for use at work. It also includes self-employed people. The detailed requirements are spelt out in Regulations.

The Act basically consists of four parts.

Part 1 covers:

▶ the health and safety of people at work;
▶ protection of other people affected by work activities;
▶ the control of risks to health and safety from articles and substances used at work.

Part 2 sets up the Employment Medical Advisory Service.

Part 3 makes amendments to the safety aspects of building regulations.

Part 4 consists of general and miscellaneous provisions.

19.4.1 Duties of employers – section 2

The employers' main general duties are to ensure, sfairp, the health, safety and welfare at work of all their employees, in particular:

▶ the provision of safe plant and systems of work;
▶ the safe use, handling, storage and transport of articles and substances;
▶ the provision of any required information, instruction, training and supervision;
▶ a safe place of work including safe access and egress; and
▶ a safe working environment with adequate welfare facilities.

When five or more people (The Employer's Health and Safety Policy Statements (Exception) Regulations 1975 (SI No. 1584) exempted an employer who employs less than 5 people) are employed the employer must:

▶ prepare a written general health and safety policy;
▶ set down the organisation and arrangements for putting that policy into effect;
▶ revise and update the policy as necessary;
▶ bring the policy and arrangements to the notice of all employees.

Employers must also:

▶ consult safety representatives appointed by recognised trade unions;
▶ consult safety representatives elected by employees;
▶ establish a safety committee if requested to do so by recognised safety representatives.

19.4.2 Duties of owners/occupiers – sections 3 and 4

Every employer and self-employed person (a self-employed person must be one who conducts an undertaking of a prescribed description; this can be a type of activity carried out or any other feature or whether a person's health and safety may be affected by the undertaking) is under a duty to conduct their undertaking in such a way as to ensure, sfairp, that persons not in their employment (and themselves for self-employed), who may be affected, are not exposed to risks to their health and safety.

Those in control of non-domestic premises have a duty to ensure, sfairp, that the premises, the means of access and exit, and any plant or substances are safe and without risks to health. The common parts of residential premises are non-domestic.

Note: Section 5 was repealed by the EPA 1990.

19.4.3 Duties of manufacturers/ suppliers – section 6

Persons who design, manufacture, import or supply any article or substance for use at work must ensure, sfairp, that:

▶ it is safe and without risks to health when properly used (i.e. according to manufacturers' instructions);
▶ they carry out such tests or examinations as are necessary for the performance of their duties;
▶ they provide adequate information (including revisions) to perform their duties;
▶ they carry out any necessary research to discover, eliminate or minimise any risks to health or safety;
▶ the installer or erector has done nothing regarding the way in which the article has been installed or erected to make it unsafe or a risk to health.

19.4.4 Duties of employees – section 7

Two main duties are placed on employees:

▶ to take reasonable care for the health and safety of themselves and others who may be affected by their acts or omissions at work;
▶ to cooperate with their employer and others to enable them to fulfil their legal obligations.

19.4.5 Other duties – sections 8 and 9

No person may misuse or interfere with anything provided in the interests of health, safety or welfare in pursuance of any of the relevant statutory provisions.

Employees cannot be charged for anything done, or provided, to comply with the relevant statutory provisions. For example, employees cannot be charged for personal protective equipment (PPE) required by health and safety legislation.

19.4.6 Powers of inspectors – sections 20–25

Inspectors appointed under this Act have the authorisation to enter premises at any reasonable time (or anytime in a dangerous situation), and to:

▶ take a constable with them if necessary;
▶ take with them another authorised person and necessary equipment;
▶ examine and investigate;
▶ require premises or anything in them to remain undisturbed for purposes of examination or investigation;
▶ take measurements, photographs and recordings;
▶ cause an article or substance to be dismantled or subjected to any test;
▶ take possession of or retain anything for examination or legal proceedings;
▶ if practical to do so take samples as long as a comparable sample is left behind;
▶ require any person who can give information to answer questions and sign a statement. Evidence given under this Act cannot be used against that person or their spouse;
▶ require information, facilities, records or assistance;
▶ do anything else necessary to enable them to carry out their duties;
▶ issue an **Improvement Notice(s)**, which is a notice identifying a contravention of the law and specifying a date by which the situation is remedied. If an appeal is to be made, the appeal procedure must be triggered within 21 days. The notice is suspended pending the outcome of the appeal;
▶ issue a **Prohibition Notice(s)**, which is a notice identifying and halting a situation which involves or will involve a risk of serious personal injury to which the relevant statutory provisions apply. A contravention need not have been committed. The notice can have immediate effect or be deferred, for example to allow a process to be shut down safely. Again there is provision to appeal against the notice

19

but the order still stands until altered or rescinded by an employment tribunal;

▶ initiate prosecutions;

▶ seize, destroy or render harmless any article or substance which is a source of imminent danger to health and safety.

19.4.7 Offences – section 33

The Health and Safety Offences Act 2008 increases penalties and provides courts with greater sentencing powers for those who flout health and safety legislation and the range of offences for which an individual can be imprisoned has also been broadened.

Further under the Legal Aid, Sentencing and Punishment of Offenders Act 2012 Section 85(1) from 12 March 2012, Magistrates Courts can issue unlimited fines.

There are strict guidelines which are observed by the regulators in their approach to the prosecution of health and safety offences. The HSE Enforcement Policy Statement makes it clear that prosecutions should be in the public interest and where one or more of a list of circumstances apply. These include where:

▶ death was a result of a breach of the legislation;

▶ there has been reckless disregard of health and safety requirements;

▶ there have been repeated breaches which give rise to significant risk, or persistent and significant poor compliance;

▶ false information has been supplied wilfully, or there has been intent to deceive in relation to a matter which gives rise to significant risk.

Summary of penalties under the HSW Act

Maxima (from 12 March 2015):

Unlimited fines in Magistrates Courts for all summary offences, unlimited fines in higher courts.

Imprisonment for nearly all offences – up to 6 months in Magistrates Courts (until S154 of CJ Act 2003 is enacted, then 12 months) and two years in the Crown Court.

Prosecutions of individuals by health and safety regulators are not undertaken lightly. Any prosecutions of individuals are subject to the same strict considerations set out above and are only taken if warranted, and not in lieu of a case against their employer.

Table 19.1 Summary of maximum penalties under Health and Safety (Offences) Act 2008 for offences committed on or after 16 January 2009

Section and contravention	Magistrates' Court	Crown Court
S.33(1)(a) – S2–S6	Unlimited fine and/or 6 months imprisonment*	Unlimited fine and/or 2 years imprisonment
S.33(1)(a) – S7	Unlimited fine and/or 6 months imprisonment*	Unlimited fine and/or 2 years imprisonment
(b) – S8	Unlimited fine and/or 6 months imprisonment*	Unlimited fine and/or 2 years imprisonment
(b) – S9	Unlimited fine	Unlimited fine
(c) – any H & S regulations	Unlimited fine and/or 6 months imprisonment*	Unlimited fine and/or 2 years imprisonment
(d) – relating to S14 enquiries	Unlimited fine (Level 5)	Summarily only
(e) – any requirement imposed by Inspector under S20 & S25	Unlimited fine and/or 6 months imprisonment*	Unlimited fine and/or 2 years imprisonment
(f) – attempting to/preventing a person speaking to Inspector	Unlimited fine and/or 6 months imprisonment*	Unlimited fine and/or 2 years imprisonment
(g) – contravention of a PN or IN	Unlimited fine and/or 6 months imprisonment*	Unlimited fine and/or 2 years imprisonment
(h) – obstruction	Unlimited fine (Level 5) and/or 6 months imprisonment**	Summarily only
(i) – relating to S27	Unlimited fine	Unlimited fine
(j) – disclosure in contravention to S28 and S27(4)	Unlimited fine and/or 6 months imprisonment*	Unlimited fine and/or 2 years imprisonment
(k) – false statement	Unlimited fine and/or 6 months imprisonment*	Unlimited fine and/or 2 years imprisonment
(l) – false entry in register	Unlimited fine and/or 6 months imprisonment*	Unlimited fine and/or 2 years imprisonment
(m) – use a document with intent to deceive	Unlimited fine and/or 6 months imprisonment*	Unlimited fine and/or 2 years imprisonment
(n) – impersonating an inspector	Unlimited fine (Level 5)	Summarily only
(o) – relating to S42 orders	Unlimited fine and/or 6 months imprisonment*	Unlimited fine and/or 2 years imprisonment
Any other offence under existing statutory provisions	Unlimited fine and/or 6 months imprisonment*	Unlimited fine and/or 2 years imprisonment

* The maximum term of imprisonment that may be imposed by a Magistrates' Court is currently 6 months. When s154(1) of the Criminal Justice Act 2003 is brought into force the maximum term will be increased to 51 weeks in England and Wales or 12 months in Scotland.

** The maximum term of imprisonment that may be imposed by a Magistrates' Court is currently 6 months. When s281(5) of the Criminal Justice Act 2003 is brought into force the maximum term will be increased to 51 weeks in England and Wales or 12 months in Scotland.

If a regulation has been contravened, failure to comply with an Approved Code of Practice is admissible in evidence as failure to comply. Where an offence is committed by a corporate body with the knowledge, connivance or neglect of a responsible person, both that person and the body are guilty of the offence. In proceedings the onus of proving the limits of what is reasonably practicable rests with the accused.

On conviction of directors for indictable offences in connection with the management of a company, the courts may also make a disqualification order as follows:

| Lower court maximum | 5 years disqualification |
| Higher court maximum | 15 years disqualification |

19.4.8 The merged HSE – section 10

The majority of the proposals set out in the original Lord Robens report were adopted in full and formed the basis of the HSW Act. However, contrary to Robens' recommendation, the HSW Act did not provide for a single authority, but two separate Crown Non-Department Public Bodies (NDPBs): the HSC and the HSE.

This recommendation was finally implemented and the HSC and the HSE merged on 1 April 2008 to form a single national regulatory body responsible for promoting the cause of better health and safety at work. The merged body is called the HSE and has provided greater clarity and transparency whilst maintaining its public accountability. Section 10 of the Act has been revised.

19.4.9 Civil liability – section 47

The Enterprise and Regulatory Reform Act 2013 removes the right of civil action against employers for breach of statutory duty in relation to health and safety at work regulations. This will address the potential unfairness that arises where an employer can be found liable to pay compensation to an employee despite having taken reasonable steps to protect them.

This unfairness was identified by Professor Löfstedt in his independent review of health and safety, 'Reclaiming Health and Safety for All' (November 2011) and arises where health and safety at work regulations impose a strict duty on employers, giving them no opportunity to defend themselves on the basis of having done all that was reasonable to protect their employees. The inability of employers to defend themselves in these cases helps fuel the perception of a compensation culture and the fear of being sued is driving businesses to over-comply with regulations, resulting in additional unnecessary costs.

Previously, compensation claims for workplace injuries or illnesses could be brought by two routes: breach of statutory duty in which failure to meet a particular

standard in law has to be proved, or breach of common law duty of care in which negligence has to be proved.

The Act amends the law so that compensation claims can only be made where negligence or fault on the part of the employer can be proved.

This change will help redress the balance of the civil litigation system in respect of health and safety at work legislation. It will help employers' confidence, allowing them to focus on a sensible and practical approach to health and safety and keep costs down by avoiding over-compliance.

Employees will continue to have the same level of protection, as the standards set out in criminal law will not change and they will still be able to claim compensation where an employer has been negligent.

The Health and Safety at Work etc. Act 1974 (Civil Liability) (Exceptions) Regulations 2013 create exceptions from the exclusion of civil liability for breach of statutory duty for pregnant workers and workers who have recently given birth or are breastfeeding ('new and expectant mothers').

19.4.10 Fees for Intervention

Fee for Intervention (FFI) is the HSE's cost recovery regime implemented from 1 October 2012, under regulations 23 to 25 of The Health and Safety (Fees) Regulations 2012.

These Regulations put a duty on the HSE to recover its costs for carrying out its regulatory functions from those found to be in material breach of health and safety law.

Duty-holders who are compliant with the law, or where a breach is not material, will not be charged FFI for any work that the HSE does with them.

A material breach is when, in the opinion of the HSE inspector, there is or has been a contravention of health and safety law that requires them to issue notice in writing of that opinion to the duty-holder.

Written notification from an HSE inspector may be by a notification of contravention, an improvement or prohibition notice, or a prosecution and must include the following information:

▶ the law that the inspector's opinion relates to;
▶ the reasons for their opinion; and
▶ notification that a fee is payable to the HSE.

FFI applies to duty-holders where the HSE is the enforcing authority. This includes employers, self-employed people who put others (including their employees or members of the public) at risk, and some individuals acting in a capacity other than as an employee, e.g. partners. It includes:

1. public and limited companies;
2. general, limited and limited liability partnerships; and
3. Crown and public bodies.

The fee payable by duty-holders found to be in material breach of the law is £124 per hour (December 2014). The total amount to be recovered will be based on the amount of time it takes the HSE to identify and conclude its regulatory action, in relation to the material breach (including associated office work), multiplied by the relevant hourly rate. This will include part hours.

19.4.11 Further information

An Introduction to Health and Safety: Guidance on Health and Safety for Small Firms INDG259 (rev 1), reprinted 2006, HSE Books, ISBN 978 0 7176 2685 4.

Health and Safety made simple: The basics for your business INDG449, HSE Books, 2011 http://www.hse.gov.uk/pubns/indg449.pdf

Health and Safety at Work etc. Act 1974, chapter 37, London, The Stationery Office http://www.legislation.gov.uk/ukpga/1974/37/contents

19.5 Environmental Protection Act 1990

19.5.1 Introduction

The EPA 1990 is still the centrepiece of current UK legislation on environmental protection. It is divided into nine parts, corresponding to the wide range of subjects dealt with by the Act.

Integrated Pollution Control (IPC) was a system established by Part 1 of the Act. Part 1 introduced Part A Processes, which are the most potentially polluting or technologically complex processes. In England and Wales these were enforced by the Environment Agency. In Scotland there was a parallel system enforced by the Scottish Environment Protection Agency.

Less polluting industries were classified as Part B, with only emissions released to air being subject to regulatory control. For such processes LAs were the enforcing body and the system was known as Local Air Pollution Control (LAPC).

Both IPC and LAPC have now been replaced by an Integrated Pollution Prevention and Control (IPPC) regime that implements the requirements of the European Community (EC) Directive 96/61 on IPPC. This was introduced under the Pollution Prevention and Control Act of 1999 (1999, Chapter 24), which repealed Part 1 of the EPA.

The change is outlined in Box 19.1.

Box 19.1 Pollution prevention and control regimes

PPC REGIME

Pollution prevention and control

IPC (regime A)
Integrated pollution control

IPPC (regime A1 and A2)
Integrated pollution prevention and control

LAPC (regime B)
Local air pollution control

LAPPC (regime B)
Local air pollution prevention and control

Under EPA 1990 Part 1

Regime A – This is an integrated permitting regime. Emissions to the air, land and water of the potentially more polluting processes are regulated. The Environment Agency is the regulator.

Regime B – This regime permits processes with a lesser potential for polluting emissions. Only emissions to the air are regulated. The Local Authority is the regulator.

Under Pollution Prevention and Control Act 1999

Regime A1 – This is an integrated permitting regime. Emissions to the air, land and water of potentially more polluting processes are regulated. The Environment Agency is the regulator.

Regime A2 – This is an integrated permitting regime. Emissions to the air, land and water of processes with a lesser potential to pollute are regulated. The Local Authority Agency is the regulator.

Regime B – This is the permitting of processes with a lesser potential to pollute. Only emissions to the air are regulated. The Local Authority is the regulator.

19.5.2 Integrated pollution prevention and control

The system of IPPC applies an integrated environmental approach to the regulation of certain industrial activities. This means that emissions to air, water (including discharges to sewers) and land, plus a range of other environmental effects, must be considered together. It also means that regulators must set permit conditions so as to achieve a high level of protection for the environment as a whole. These conditions are based on the use of the 'Best Available Techniques' (BAT), which balances the costs to the operator against the benefits to the environment (see Box 19.2). IPPC aims to prevent emissions and waste production and, where that is not practicable, reduce them to acceptable levels. IPPC also takes the integrated approach beyond the initial task of permitting, through to the restoration of sites when industrial activities cease.

> ### Box 19.2 Best available techniques (BAT)
>
> This term is defined as:
>
> > The most effective and advanced stage in the development of activities and their methods of operation which indicates the practicable suitability of particular techniques for providing the basis for emission limit values designed to prevent, and where that is not practicable, generally to reduce the emissions and the impact on the environment as a whole.
>
> This definition implies that BAT not only covers the technology used but also the way in which the installation is operated, to ensure a high level of environmental protection as a whole. BAT takes into account the balance between the costs and environmental benefits (i.e. the greater the environmental damage that can be prevented, the greater the cost for the techniques).

19.5.3 Setting the legal framework

The PPC Regulations implement the EC Directive 96/61/EC on IPPC ('the IPPC Directive'), insofar as it relates to installations in England and Wales. Separate regulations apply the IPPC Directive in Scotland and Northern Ireland and to the offshore oil and gas industries.

Prior to the PPC Regulations coming into force, many industrial sectors covered by the IPPC Directive were regulated under Part I of the EPA 1990. This introduced the systems of IPC, which controlled releases to all environmental media, and LAPC, which controlled releases to air only. Other industrial sectors new to integrated permitting, such as the landfill, intensive farming and food and drink sectors, were regulated, where appropriate, by separate waste management licences issued under Part II of the EPA and/or water discharge consents under the Water Resources Act 1991 or Water Industry Act 1991.

The PPC Regulations create a coherent new framework to prevent and control pollution, with two parallel systems similar to the old regimes of IPC and LAPC. The first of these – the 'Part A' regime of IPPC – applies a similar integrated approach to IPC while delivering the additional requirements of the IPPC Directive. 'Part A' extends the issues that regulators must consider alongside emissions into areas such as energy use and site restoration. The main provisions of IPPC apply equally to the ex-IPC processes and the other sectors new to integrated permitting. There are also some further requirements that apply solely to waste management activities falling under IPPC.

The IPPC Directive applies to those landfills receiving more than 10 tonnes per day or with a total capacity exceeding 25,000 tonnes (but excluding landfills taking only inert waste), the landfill Directive applies to all landfills. The PPC Regulations have been amended to include all landfills. For landfills the technical requirements are met through the Landfill Regulations. Department of Environment, Food and Rural Affairs (Defra) issued separate guidance on the Landfill Regulations in 2004.

The Environment Agency regulates Part A(1) installations. Part A(2) installations are regulated by the relevant local authority – usually the district, London or metropolitan borough council in England and the county or borough council in Wales. However, the local authority will always be a statutory consultee where the Environment Agency is the regulator, and vice versa. Moreover, the local authority and the Environment Agency will work together in the permitting process. LAs have expertise in setting standards for noise control, while the Environment Agency will ensure that permit conditions protect water adequately. Annex I describes how IPPC installations are classified into either Part A(1) or Part A(2) installations depending on what activities take place within them.

The second new regime – the 'Part B' regime of Local Air Pollution Prevention and Control (LAPPC) – represents a continuation of the old LAPC regime. LAPPC is similar to IPPC from a procedural perspective, but it still focuses on controlling emissions to air only. Defra provides separate guidance on local authority air pollution control.

19.5.4 Overview of the regulatory process

The basic purpose of the IPPC regime is to introduce a more integrated approach to controlling pollution from industrial sources. It aims to achieve 'a high level of protection of the environment taken as a whole by, in particular, preventing or, where that is not practicable, reducing emissions into the air, water and land'. The main way of doing that is by determining and enforcing permit conditions based on BAT.

The entire regulatory process for IPPC consists of a number of elements. These are outlined below. IPPC applies to specified 'installations' – both 'existing' and 'new' – requiring each 'operator' to obtain a permit (from April 2008 an Environmental Permit) from the regulator – either the Environment Agency or the Local Authority.

19.5.5 Environmental permitting

Introduction

The Environment Agency provides guidance on environmental permitting which aims to provide comprehensive help for those operating, regulating or interested in facilities that are covered by the

Environmental Permitting (England and Wales) Regulations 2010 SI 2010 No. 675 (EP Regulations).

These Regulations revoke and replace the Environmental Permitting (England and Wales) Regulations 2007, the system of waste management licensing in Part II of the EPA 1990 (c. 43) and the Waste Management Licensing Regulations 1994 (SI 1994/1056, as amended), and the system of permitting in the Pollution Prevention and Control (England and Wales) Regulations 2000 (SI 2000/1973, as amended), with a new system of environmental permitting in England and Wales. In July 2009 the 2007 Regulations were amended to bring mining waste operations into the permitting regime. The changes resulted in a local authority requiring the Secretary of State or Welsh Ministers' consent to institute summary proceedings for certain statutory nuisances against waste operations that are permitted or are registered as exempt or mining waste operations that are permitted in the same way as they already were for Part A(1), A(2) and Part B installations and mobile plant.

The Environmental Permitting (England and Wales) Regulations 2010 generally revoked and replaced the 2007 regulations and brought three further regimes into the environmental permitting regime. These included:

▶ water discharge activities and exemptions;
▶ groundwater activities and exemptions; and
▶ radioactive substances activities and exemptions.

What is environmental permitting?

Some activities could harm the environment or human health unless they are controlled. The Environmental Permitting Regime ('the Regime') requires operators to obtain permits for some facilities, the registration of exemptions for other facilities and ongoing supervision by regulators. The Regime was first operated under the EP 2007 Regulations which came into force on 1 April 2008 and now with the 2010 Regulations from 6 April 2010, plus a number of amendment regulations since.

The aim of the Regime is to:

▶ protect the environment so that statutory and Government policy environmental targets and outcomes are achieved;
▶ deliver permitting and compliance effectively and efficiently in a way that provides increased clarity and minimises the administrative burden on both the regulator and the operators of facilities;
▶ encourage regulators to promote best practice in the operation of regulated facilities;
▶ continue to fully implement European legislation.

The Regime covers facilities previously regulated under the PPC Regulations and Waste Management Licensing and exemptions. The Regime extends to England and Wales. It also covers the adjacent sea as far as the seaward boundary of the territorial sea.

The legal framework

The Regulations set out the following:

▶ the facilities that need environmental permits or need to be registered as exempt;
▶ the process for registering exemptions;
▶ how to apply for and determine permit applications;
▶ requirements that environmental permits contain conditions to protect the environment as required by Directives and national policy;
▶ how environmental permits can be changed and ultimately be surrendered;
▶ a simplified permitting system called standard rules;
▶ compliance obligations backed up by enforcement powers and offences;
▶ provisions for public participation in the permitting process;
▶ the powers and functions of regulators, the Secretary of State and the Welsh Assembly Government (WAG);
▶ a simple transition to the new regime;
▶ provisions for appeals against permitting decisions.

19.5.6 What facilities require an Environmental Permit?

The EP Regulations specify which activities and waste operations require an Environmental Permit and provide that some waste operations can be exempt from those requirements. Certain waste operations covered by other legislation are excluded from permitting.

Regulated facilities

The nine classes of regulated facility are:

▶ an installation (where activities listed in Schedule 1 to the Regulations, and any directly associated activities, are carried on – see Installations);
▶ mobile plant (used to carry on either one of the Schedule 1 activities or a waste operation);
▶ a waste operation;
▶ a mining waste operation;
▶ a radioactive substances activity;
▶ a water discharge activity;
▶ a groundwater activity;
▶ a small waste incineration plant;
▶ a solvent emission activity.

There may be more than one regulated facility on the same site. In such cases there are arrangements in the EP Regulations to allow all such facilities to be regulated by the same regulator and to allow, in many cases, for a single permit.

It is an offence under Regulation 12 to operate a regulated facility without a permit.

An Environmental Permit can, however, cover more than one regulated facility (Regulation 17).

When can a single permit be granted?

A single Environmental Permit can only be granted for more than one regulated facility where:

▶ the regulator is the same for each facility;
▶ the operator is the same for each facility;
▶ all the facilities are on the same site (the exceptions to this are set out as follows).

Where the regulator and operator are the same, a single Environmental Permit can be granted to an operator for more than one mobile plant. Mobile plants do not have to be operating on the same site in order to be included in a single permit.

Where the regulator and operator are the same, a single Environmental Permit can be granted to an operator for more than one regulated facility to which standard rules apply. Standard facilities do not have to be on the same site in order to be included in a single permit. However, standard facilities on different sites cannot be combined in a single permit where the IPPC Directive applies to any of the facilities.

Regulated facilities have to be on the same site in order to be covered by the same permit (with the exceptions of mobile plant and standard facilities set out previously). The regulator should consider the following factors in determining whether the facilities are on the same site.

Proximity – there should, however, be no simple 'cut off' distance since some industrial complexes cover very large areas but still can be regarded as one site for permitting purposes.

Coherence of a site – some regulated facilities will be within a single fenced area or may share security or emergency systems.

Management systems – the extent to which the regulated facilities share a common management system is a relevant consideration.

It is expected that a regulator will adopt a common-sense approach to determining when facilities should be regulated under one permit. This consideration should be based on achieving protection of the environment in the most efficient regulatory manner.

19.5.7 Exempt facilities

Some facilities that pose a sufficiently low risk can be exempt from the need to hold a permit. However, this is only where any applicable European Directive allows it. A waste operation, water discharge or groundwater activity must meet certain criteria in order to be exempt from the need for an Environmental Permit. An exempt facility must

▶ meet the criteria of Schedule 2;
▶ be consistent with the need to attain the objectives in Article 24 of the Waste Framework Directive;
▶ be registered.

19.5.8 The regulator

The regulator for each category of regulated facility is identified in Regulation 32.

The Environment Agency regulates:

▶ Part A(1) installations;
▶ waste mobile plant;
▶ waste operations including those carried on at a Part B installation or by Part B mobile plant (unless the waste operation is a Part B activity);
▶ mining waste operations including any carried on at a Part B installation;
▶ radioactive substances activities;
▶ water discharge activities including those carried on at a Part B installation;
▶ groundwater activities including those carried on at a Part B installation.

The relevant local authority regulates:

▶ Part A(2) installations including any waste operations, water discharge activities or groundwater activities carried on as part of the installation or mobile plant;
▶ the 'Part B' installations and Part B mobile plant (except as set out above);
▶ small waste incineration plants;
▶ solvent emission activities.

LAPPC focuses on controlling emissions to air only for Part B installations. Defra and WAG jointly provide guidance on local authority air pollution and control.

19.5.9 Environmental Permit applications

The operator

Only a person who is in control of the facility may obtain or hold an Environmental Permit. This person is the 'operator' (Regulation 7).

The meaning of 'operator' is set out in Box 19.3. In most cases a single operator will have to obtain a single Environmental Permit for each regulated facility.

However, in some circumstances different operators run different parts of a regulated facility. This does not affect the regulator's determination of what actually constitutes the regulated facility.

Where two or more operators run different parts of a single regulated facility, they will each need a separate Environmental Permit and be responsible for complying with their permit conditions. In such cases, there should be no ambiguity over which operator has responsibility for which part of the regulated facility.

The requirement that the permit holder must be the operator of the facility is different from the previous system under waste management licensing. The transitional provisions therefore allow the holder of a waste management licence to be treated as the operator

19

for the purposes of the Regulations (see Regulation 69(2)). However, when permits are transferred over time they must be transferred to the person who will have control of the regulated facility.

Box 19.3 'Operator'

'Operator' is defined in regulation 7 as the person who has control over the operation of a regulated facility.

If a regulated facility has not been put into operation, the person who will have control over it when it is in operation is the operator.

If a regulated facility authorised by an Environmental Permit has ceased to be in operation, the person who holds the permit is the operator.

Legal obligations may be imposed on an operator during the pre- and post-operational phases.

The operator must demonstrably have the authority and ability to ensure the Environmental Permit is complied with.

Pre-application discussions

Pre-application discussions between operators and regulators can help in improving the quality of the formal application and are therefore encouraged. In order for such discussions to make the best use of time, the operator is expected to have read the relevant published guidance. The regulator should not be expected to provide advice that might prejudice its determination of an application.

Operators and regulators may use the discussions to clarify whether a permit is likely to be needed. The regulator may also give operators general advice on how to prepare their applications, focus on the key issues and tell them what additional guidance is available.

Operators should bear in mind that, especially for controversial cases, good engagement with local or national interested parties at the pre-application stage can be beneficial to all sides.

Using existing data

Operators may, where relevant, draw upon or attach other sources of information in their applications. These might include:

▶ environmental impact assessments;
▶ documents relating to an installation's regulation under the Control of Major Accident Hazards (COMAH) Regulations;
▶ prior investigations for compliance with the Groundwater Regulations;
▶ externally certified environmental management systems (EMSs);
▶ site reports prepared for planning purposes.

They should make clear which parts of any attachments are relevant to their Environmental Permit applications and should demonstrate how they relate to the relevant Directive requirements.

Timing of applications

Operators should normally make an application when they have drawn up full designs but before construction work commences (whether on a new regulated facility or when making changes to an existing one). Where facilities are not particularly complex or novel, the operator should usually be able to submit an application at the design stage containing all information the regulator needs. If, in the course of construction or commissioning, the operator wants to make any changes which mean that the permit conditions have to be varied, the operator may apply for this in the normal way.

There is nothing in the regulations to stop an operator from beginning construction before a permit has been issued. However, regulators may not agree with the design and infrastructure put in place. Therefore, to avoid any expensive delays and re-work, it is in the operator's interest to submit applications at the design stages. Any investment or construction work that an operator carries out before they have got an Environmental Permit will be at their own risk and will in no way affect the regulator's decision.

Planning and Environmental Permit applications

If a regulated facility also needs planning permission, it is recommended that the operator should make both applications in parallel whenever possible. This will allow the pollution control regulator to start its formal consideration early on, thus allowing it to have a more informed input to the planning process. For certain waste operations and certain mining waste facilities, where planning permission is required, this must be in force before an Environmental Permit can be granted.

19.5.10 Transitional arrangements

Existing permissions

The EP Regulations provide transitional arrangements for existing permits, licences, authorisations, registrations and consents (existing permissions) so new applications for environmental permits are not required.

The regulator for these permits remains the same and will not change unless there is a subsequent direction from the Secretary of State or the Welsh Ministers.

Except where there is an outstanding application (see below) these existing permissions automatically become environmental permits from the date the Regulations come into force.

Existing permissions with an outstanding application to transfer, surrender, vary or modify

Existing permissions do not become environmental permits on the date the EP Regulations come into force where there is an outstanding, duly-made, application in relation to the existing permission. The relevant applications are those to transfer, surrender, vary or modify the existing permissions.

19.5.11 Application procedures

Applications

The requirements for applications are set out in Schedule 5 to the Regulations. The application must:

▶ be made by the operator (though it may be made by an agent acting on behalf of the operator);
▶ in the case of a transfer application, be made jointly by the current operator and the future operators;
▶ be made on the form provided by the regulator;
▶ include the information required by the application form;
▶ include the relevant fee.

An applicant can withdraw an application at any time before it is determined but the regulator is not obliged to return any of the application fee.

19.5.12 Permit conditions

If the regulator grants a permit it can include any conditions it sees fit. It has a duty to impose conditions in order to secure the objectives that apply to the category of facility.

Where the regulator grants an application for the variation, transfer or partial surrender of an Environmental Permit and there are additional variations needed as a consequence of the application, the regulator should make those necessary variations to the Environmental Permit.

All permit conditions should be both necessary and enforceable. 'Necessary' means that the regulator should be able to justify – at appeal if necessary – the permit conditions it attaches. To be enforceable, conditions should clearly state the objective, standard or desired outcome of the condition so that the operator can understand what is required.

Permit conditions may comprise some or all of the following:

▶ conditions stipulating objectives or outcomes;
▶ standards to mitigate a particular hazard/risk;
▶ conditions addressing particular legislative requirements.

The regulator can include conditions in the permit setting out steps to be taken during, prior to and after the operation of the regulated facility.

19.5.13 Standard rules

The Secretary of State, Welsh Minister and the Environment Agency can make standard rules (Regulation 26).

These rules consist of requirements common to the class of facilities subject to them (standard facilities – Regulation 2) and can be used instead of site-specific permit conditions. Standard rules would be suitable for industry sectors where a number of regulated facilities share similar characteristics in relation to environmental hazards.

The standard rules must achieve the same high level of environmental protection as site-specific conditions.

The rules are the conditions of the standard permit for all purposes other than for appeals. The standard rules cannot be appealed (Regulation 27(3)) as applying for a permit subject to the rules is voluntary.

Developing standard rules

In preparing standard rules, it is necessary to consult widely with those who may be affected by or have an interest in the rules (Regulation 26(2)), including relevant statutory bodies. The standard nature of the facilities for which standard rules will be produced allows a general consideration of the requirements and standards for all such facilities.

It is expected that standard rules will be developed in consultation with the relevant industry.

Assessments of risk can be carried out nationally for common generic activities. This understanding of the hazards and risks posed by these activities would form the basis for the development of standard rules for standard facilities.

19.5.14 Standard permits

It is the operator's decision as to whether they wish to operate under standard rules. Where standard rules have been made, operators of standard facilities can, if they so wish, request that their facility be subject to the relevant rules. This request may be made in an application for a new permit or an application to vary an Environmental Permit.

Operators may apply to operate under a set of standard rules.

The generic assessments of risk for standard facilities should be made available to applicants to assist them in determining whether their activity is within the scope of the standard rules and, if they apply for a standard permit, in the adoption of suitable control measures to meet those rules.

One important difference from other regulated facilities is that any additional site-specific assessment of risk is not necessary for a standard facility. Regulated facilities

19

that require a location-specific assessment of impact and risk are not suitable for standard rules.

Public consultation on applications for individual standard facilities is not required (other than for Part A installations).

19.5.15 Operator competence and management systems

Following an application for the grant or transfer of an Environmental Permit, there is also a specific duty on the regulator not to grant or transfer the permit if it considers that the operator will not operate the facility in accordance with the permit. In making this decision the regulator should consider whether the operator cannot or is unlikely to operate the facility in accordance with the permit. The regulator might doubt whether the operator could or is likely to comply with the permit conditions if, for example:

▶ the operator's management system is inadequate;
▶ the operator's technical competence is inadequate;
▶ the operator has a poor record of compliance with previous regulatory requirements, or;
▶ the operator's financial competence is inadequate.

In order to ensure a high level of environmental protection, operators should have effective management systems in place. The nature of the required management system depends upon the complexity of the regulated facility.

Complex regulated facilities are encouraged to put in place a formal EMS externally certified to the international standard ISO 14001 by the UKAS accredited certification body or other European equivalent and to register for the EU's Eco Management and Audit Scheme (EMAS). These standards require that the management system include safeguards for legal compliance and a commitment to continuous improvement of environmental performance. Additionally EMAS requires organisations to produce an independently validated public report about their environmental performance and progress against targets and objectives. Where relevant the performance should be benchmarked against European legislation, e.g. BAT under IPPC. EMAS and ISO 14001 are also recognised by the Environment Agency's risk rating scheme OPRA (Operator and Pollution Risk Appraisal scheme). OPRA scores are linked to fees and charges. Organisations which have implemented an EMS may achieve a better OPRA score and can pay lower fees and charges.

For simpler regulated facilities, externally certified schemes or a full EMS may be less appropriate but should still be carefully considered by operators and, where appropriate, encouraged by regulators. The stepwise approach provided by BS8555 is particularly appropriate for smaller facilities and can make EMS implementation much simpler. Organisations can

achieve UKAS accredited certification to one or more stages of BS8555 under the IEMA Acorn or BSI Stems schemes. The European Commission has also developed a simplified implementation guide, EMAS 'easy', which aims to help small and medium enterprises (SMEs) achieve registration for EMAS. There is also specific guidance on management systems for some industry sectors on the website of the Institute of Environmental Management and Assessment.

EMSs have relevance to other aspects of regulation, such as determining risk-based inspection frequencies. Recognised quality assurance schemes may also be relevant, and regulators may also take account of non-certified systems where these can be demonstrated to provide an equivalent role in safeguarding compliance and continual improvement of environmental performance.

Operators should be technically competent to operate their facility. The operator's wider management system should contain mechanisms for assessing and maintaining technical competence. The competence of individuals should form part of those management systems.

The development of industry led competence schemes is strongly encouraged by the Environment Agency.

19.5.16 Enforcement

Enforcement notices

Regulation 36 of the EP Regulations allows the regulator to serve an 'enforcement notice' if it believes an operator has contravened, is contravening or is likely to contravene any permit conditions.

Enforcement notices will specify the steps required to remedy the problem and the timescale in which they must be taken. Enforcement notices may include steps to remedy the effects of any harm and to bring a regulated facility back into compliance.

Suspension notices

If the operation of a regulated facility involves a risk of serious pollution, the regulator may serve a 'suspension notice' under regulation 37 of the Regulations. This applies whether or not the operator has breached a permit condition.

The suspension notice must describe the nature of the risk of pollution and the actions necessary to remove that risk. The notice must specify the deadline for taking actions.

When the regulator serves a suspension notice, the permit ceases to authorise the operation of the entire facility or specified activities depending upon what is specified in the notice.

A suspension notice should allow activities to continue unless their cessation is necessary to address the risk of pollution. While the suspension notice is in force, additional restrictions may be necessary for any activities that are allowed to continue. Where this is the case the suspension notice must set out these additional steps.

When the operator has taken the remedial steps required by the notice, the regulator must withdraw the notice.

Prosecutions

If an operator has committed a criminal offence under the Regulations, regulators should consider a prosecution. Conviction in a magistrates' court carries a fine of up to £50,000 and up to 12 months imprisonment for the most serious offences under the Regulations. Conviction in the Crown Court for those offences may lead to an unlimited fine and imprisonment for up to five years.

19.5.17 Wider scope of IPPC

IPPC takes a wider range of environmental impacts into account than IPC which regulated emissions to land, water and air. The IPPC regime additionally takes into account: waste avoidance or minimization, energy efficiency, accident avoidance, and minimization of noise, heat and vibrations. These aims will achieve a higher level of protection as a whole.

IPPC also applies to a wider range of industries than IPC. These industries include all installations that are currently regulated under IPC, some installations currently under LAPC and some installations that are not currently under either regime, such as landfill sites, intensive agriculture, large pig and poultry units, and food and drink manufacturers.

Under IPPC, regulated industries are referred to as 'installations' as opposed to 'processes', which is the term used for IPC. This change in terminology enables a more integrated approach to regulation; a whole installation must be permitted rather than just individual processes within the installation.

Guidelines to establish which techniques are BAT are published by the European Commission's IPPC Bureau. These reference notes are known as BREF notes and provide the basis for national sectoral guidance. The Environment Agency has supplementary guidance to cover many issues; some, for example energy efficiency, site remediation and noise, are new issues under IPPC. Industry sectors not previously regulated under the EPA 1990 such as intensive farming and food and drink installations are also covered by guidance.

19.5.18 Duty of care

Waste and the duty of care

The duty of care is covered in Part II of the EPA 1990. The duty of care applies to anyone who produces or imports, keeps or stores, transports, treats or disposes of waste. It also applies if they act as a broker and arrange these things.

The duty-holder is required to take all reasonable steps to keep waste safe. If they give waste to someone else, the duty-holder must be sure they are authorised to take it and can transport, recycle or dispose of it safely.

The penalty for breach of this law is an unlimited fine.

Waste can be anything owned, or that a business produces, which a duty-holder wants to get rid of. Controlled waste is defined in Box 19.4.

> ### Box 19.4 Definition of controlled waste
>
> Controlled waste means household, commercial or industrial waste. It includes any waste from a house, school, university, hospital, residential or nursing home, shop, office, factory or any other trade or business premises. It is controlled waste whether it is solid or liquid and even if it is not hazardous or toxic.

If the waste comes from a person's own home, the duty of care **does not** apply to them. But if the waste is not from the house they live in – for example, if it is waste from their workplace or waste from someone else's house – the duty of care does apply.

Animal waste collected and transported under the Animal By-Products Order 1992 is not subject to the duty of care.

Duty-holders must take all reasonable steps to fulfil the duty and complete some paperwork. What is reasonable depends on what is done with the waste.

Steps to take if the duty of care applies: when a duty-holder has waste, they must:

► stop it escaping from their control and store it safely and securely. They must prevent it causing pollution or harming anyone;

► keep it in a suitable container. Loose waste in a skip or on a lorry must be covered;

► if the duty-holder gives waste to someone else, check that they have authority to take it. The law says the person to whom they give the waste must be authorised to take it. Box 19.5 shows who is allowed to take waste and how the duty-holder can check;

► describe the waste in writing. The duty-holder must fill in and sign a transfer note for it and keep a copy. To save on paperwork, the description of the waste can be written on the transfer note (see Box 19.6).

When a person takes waste from someone else they must:

► be sure the law allows them to take it. Box 19.5 shows who is allowed to take waste;

► make sure the person giving them the waste describes it in writing. The waste receiver must fill

19

in and sign a transfer note and keep a copy (see Box 19.6).

Box 19.5 Who has authority to take waste?

Council waste collectors

The duty-holder does not have to do any checking, but if they are not a householder, there is some paperwork to complete. This is explained in Box 19.6.

Registered waste carriers

Most carriers of waste have to be registered with the Environment Agency or the Scottish Environment Protection Agency. Look at the carrier's certificate of registration or check with the Agencies.

Exempt waste carriers

The main people who are exempt are charities and voluntary organisations. Most exempt carriers need to register their exemption with the Environment Agency or the Scottish Environment Protection Agency. If someone says they are exempt, ask them why. Check with the Agencies that their exemption is registered.

Holders of environmental permits (from April 2008 – formerly Licences)

Some environmental permits are valid only for certain kinds of waste or certain activities. Ask to see the permit. Check that it covers the kind of waste being consigned.

Businesses exempt from environmental permits

There are exemptions from permitting for certain activities and kinds of waste. Most exempt businesses need to register their exemption with the Environment Agency or the Scottish Environment Protection Agency. Check with the Agencies that their exemption is registered.

Authorised transport purposes

Waste can also be transferred to someone for 'Authorised transport purposes'. This means:
the transfer of controlled waste between different places within the same premises
the transport of controlled waste into Great Britain from outside Great Britain
the transport by air or sea of controlled waste from a place in Great Britain to a place outside Great Britain.

Registered waste brokers

Anyone who arranges the recycling or disposal of waste, on behalf of someone else, must be registered as a waste broker. Check with the Environment Agency or the Scottish Environment Protection Agency that the broker is registered.

Exempt waste brokers

Most exempt waste brokers need to register with the Environment Agency or the Scottish Environment Protection Agency. Those who are exempt are mainly charities and voluntary organisations. If someone tells you they are exempt, ask them why. You can check with the Environment Agencies that their exemption is registered.

Box 19.6 Filling in paperwork

When waste is passed from one person to another, the person taking the waste must have a written description of it. A transfer note must also be filled in and signed by both persons involved in the transfer.

The duty-holder can write the description of the waste on the transfer note. Who provides the transfer note is not important as long as it contains the right information. The Government has published a model transfer note with the Code of Practice which can be used if desired.

Repeated transfers of the same kind of waste between the same parties can be covered by one transfer note for up to a year, for example weekly collections from shops.

The transfer note to be completed and signed by both persons involved in the transfer must include:
- what the waste is and how much there is
- what sort of containers it is in
- the time and date the waste was transferred
- where the transfer took place
- the names and addresses of both persons involved in the transfer
- whether the person transferring the waste is the importer or the producer of the waste
- the details of which category of authorised person each one is. If the waste is passed to someone for authorised transport purposes, you must say which of those purposes applies
- if either or both persons is a registered waste carrier, the certificate number and the name of the Environment Agency which issued it
- if either or both persons have an Environmental Permit (or an old waste management licence), the permit number (or old licence number) and the name of the Environment Agency which issued it

- ▶ the reasons for any exemption from the requirement to register or have an environmental permit
- ▶ where appropriate, the name and address of any broker involved in the transfer of waste.

The written description – The written description must provide as much information as someone else might need to handle the waste safely.

Keeping the papers – Both persons involved in the transfer must keep copies of the transfer note and the description of the waste for 2 years. They may have to prove in court where the waste came from and what they did with it. A copy of the transfer note must also be made available to the Environment Agency or the Scottish Environment Protection Agency if they ask to see it.

19.5.19 Hazardous waste

On 16 July 2005, the Hazardous Waste (England and Wales) Regulations and the List of Wastes (England) Regulations, replacing the Special Waste Regulations, came into force (Scotland has retained The Special Waste Regulations). The Regulations can be found on the Government website at: http://www.legislation.gov.uk/uksi/2005/894/

The regime includes a requirement for most producers of hazardous waste to notify their premises to the Environment Agency. The facility to notify premises has been available since April 2005. Guidance on notification, including the online notification facility, and more general guidance on the regime, can be found on the Environment Agency's hazardous waste pages. See a brief summary of these Regulations in Section 19.21.

19.5.20 Applying for a waste management licence

From April 2008 this has now been replaced by the Environmental Permitting regime. See Section 19.5.5.

19.5.21 Introduction to the waste hierarchy

Article 4 of the revised EU Waste Framework Directive sets out five steps for dealing with waste, ranked according to environmental impact – the 'waste hierarchy'.

Prevention, which offers the best outcomes for the environment, is at the top of the priority order, followed by preparing for reuse, recycling, other recovery and disposal, in descending order of environmental preference.

Table 19.2 The waste hierarchy

Stages	Include
Prevention:	using less material in design and manufacture, keeping products for longer, reuse, using less hazardous materials
Preparing for reuse:	checking, cleaning, repairing, refurbishing, whole items or spare parts
Recycling:	turning waste into a new substance or product, includes composting if it meets quality protocols
Other recovery:	includes anaerobic digestion, incineration with energy recovery, gasification and pyrolysis which produce energy (fuels, heat and power) and materials from waste, some backfilling
Disposal:	landfill and incineration without energy recovery

The waste hierarchy has been transposed into UK law through the The Waste (England and Wales) Regulations 2011. The regulations came into force on 29 March 2011. The provisions relating to the hierarchy (set out in Regulations 12 and 15) came into force on 28 September 2011.

If a business or organisation (including local authorities on behalf of householders) produces or handles waste (this includes importing, producing, carrying, keeping or treating waste; dealers or brokers who have control of waste, and anyone responsible for the transfer of waste), they must take all such measures as are reasonable in the circumstances to:

- ▶ prevent waste
- ▶ apply the waste hierarchy when transferring waste.

19.5.22 Further information

See Environmental Permitting Guidance Core Guidance revised March 2013 https://www.gov.uk/government/uploads/system/uploads/attachment_data/file/211852/pb13897-ep-core-guidance-130220.pdf

Also see new guidance How to comply with your environmental permit June 2013 http://publications.environment-agency.gov.uk

Guidance on applying the waste hierarchy https://www.gov.uk/government/uploads/system/uploads/attachment_data/file/69403/pb13530-waste-hierarchy-guidance.pdf

Small businesses a guide to the Hazardous Waste Regulations, Environment Agency July 2013, HWR01A, http://www.environment-agency.gov.uk/business/topics/waste/32180.aspx

Other sources of information on environmental management systems

- ▶ EMAS: www.iema.net/ems/emas and http://europa.eu.int/comm/environment/emas

19

▶ EMAS Easy: http://ec.europa.eu/environment/emas/toolkit/
▶ ISO 14001: http://www.bsigroup.com/en/
▶ UKAS: www.ukas.com
▶ Acorn: www.iema.net/acorn
▶ BSI Stems: www.bsi-global.com/en/Assessment-and-certification-services/management-systems/Standards-and-Schemes/BS-8555-STEMS/
▶ Green Dragon: www.greendragonems.com/

19.6 New Roads and Street Works Act 1991

19.6.1 Introduction

The prevailing legislation surrounding street works is primarily enshrined in the New Roads and Street Works Act 1991 (NRSWA). This is a detailed Act with 171 sections but it is Part III which applies to street works in England and Wales and Part IV which applies to road works in Scotland that most concern health and safety. A summary of the general requirements only is given.

Prior to this legislation, the Public Utilities and Street Works Act 1950 (PUSWA) gave public utilities the right to dig up roads without needing to obtain prior permission, while local authorities were generally responsible for carrying out the permanent reinstatements. As this was widely believed to be unnecessarily bureaucratic, there was a groundswell in favour of deregulation to create more favourable operating conditions for the private sector. However, the deregulation of street works coincided with the liberalisation of the telecommunications industry, which meant that when the Act came into effect the conditions which had originally spawned it had radically altered and there was soon a plethora of companies empowered to dig up roads. In 1987 there were only two telecommunications companies licensed to dig up roads – there are now nationally over 120 and the scale of coordination required is daunting.

Under the terms of this Act, statutory undertakers (in general, companies or public bodies supplying gas, water, electricity and telecommunications, as well as bodies such as London Underground) have the legal right to dig up roads to either maintain or repair their existing pipes and cables or in order to install additional ones to provide service to new customers.

19.6.2 Application of the Act to street works/road works (Scotland) – Part III Section 48 and Part IV Section 107

A street means:

▶ any highway, road, lane, footway, alley or passage;
▶ any square or court; and
▶ any land laid out as a way whether it is for the time being formed as a way or not.

Where a street passes over a bridge or through a tunnel, references in this Part to the street include that bridge or tunnel.

In Section 107 the definition of 'road' is similar but here it means any way whether or not there is a public right of passage.

Street works or road works (in Scotland) means works for any purpose (other than works for road purposes) executed in a street or road in pursuance of a statutory right or a street licence (permission granted under Section 109 in Scotland). They cover works of any of the following kinds:

▶ placing apparatus; or
▶ inspecting, maintaining, adjusting, repairing, altering or renewing apparatus, changing the position of apparatus or removing it; or
▶ works required for or incidental to any such works (including, in particular, breaking up or opening the street, or any sewer, drain or tunnel under it, or tunnelling or boring under the street).

19.6.3 General requirements

The statutory undertakers are obliged to inform local authorities of the work but the notice required varies dramatically, depending on the nature of the work. The categories are laid down in the Act.

The Act requires local authorities to coordinate the works while obliging undertakers to cooperate with local authorities in the interests of safety, minimising inconvenience to persons using the street and protecting the structure of the street and the integrity of the apparatus within it.

The sheer volume of street works is colossal – with the vast majority giving less than 7 days' notice. Clearly it would be unrealistic and unwieldy to expect local authorities to inform residents and businesses about every last piece of activity.

There is in theory a vested interest for local authorities and statutory undertakers to notify the public in advance (through either mail-drops or signage), especially when large-scale works are carried out, yet there is no legal compulsion to do so.

Other complications are caused by the fact that the records detailing what is located underground are not always 100% accurate. Underground work can pre-date the First World War and unsurprisingly details can be patchy, or non-existent. While sophisticated equipment is available to gauge what might be lurking underground, this is not infallible either.

The introduction of the 'electronic transfer of notices' (ETON) in 1999 established a standard mechanism for undertakers to use to send their notices to highway authorities.

This promised to be far more efficient than previous paper-based systems, as well as dramatically speeding up communication between the undertakers and highway authorities.

While councils have a duty to coordinate the work of statutory undertakers, enforcing this is another matter. Formal coordination meetings are generally held on a regular basis, attended by representatives of the council, the statutory undertakers and the police. However, in central London in particular, the companies carrying out this sort of work are operating in a highly competitive environment and are therefore reluctant to disclose details of their plans to their rivals, beyond their legal obligation to give 1 month's notice of substantial works.

Although councils can try to persuade companies to work together (by laying several cables in one trench or timetabling planned work in close sequence) in order to minimise disruption, they have no authority to do anything more than encouraging them to coordinate their activities.

19.6.4 Code of Practice under sections 65 and 124

This Code of Practice is issued by the Secretary of State for Transport, the Scottish Executive and the National Assembly for Wales under sections 65 and 124 of the NRSWA, and by the Department for Regional Development (Northern Ireland) under article 25 of the Street Works (Northern Ireland) Order 1995. The legislation requires an undertaker, and those working on its behalf, carrying out work under the Act or the Order to do so in a safe manner as regards the signing, lighting and guarding of works.

Failure to comply with this requirement is a criminal offence. Compliance with the Code will be taken as compliance with the legal requirements to which it relates.

Highway authorities in England and Wales and roads authorities in Scotland should comply with this Code for their own works, as recommended by the respective national administrations. The Northern Ireland road authority is legally required to comply with the Code. In the application of this Code to Scotland, all references in the text to highway authorities are to be read as references to roads authorities.

Everyone on site has a personal responsibility to behave safely, to the best of their ability. Under the Health and Safety at Work etc. Act 1974, employers have duties to protect their employees from dangers to their health and safety and to protect others who might be affected by the work activity (e.g. passing pedestrians and motorists). These include proper arrangements for design (including planning and risk assessment) and management (including supervision) of the works. Supervisors qualified under the NRSWA or the Order will know what to do in most situations about which they have to be consulted, and will be able to find out

quickly what to do about the others. It is the employer's responsibility to ensure that these arrangements are properly carried out.

This Code applies to all highways and roads except motorways and dual carriageways with hard shoulders. More detailed advice, and advice on some situations not covered by this Code, can be found in Chapter 8 of the *Traffic Signs Manual* published by the Department for Transport, Local Government and the Regions in conjunction with the Scottish, Welsh and Northern Ireland administrations. This gives authoritative advice, but it does not have the status of a Code of Practice under the Act. In Northern Ireland the use of Chapter 8 is mandatory for undertakers' works on motorways or dual carriageways with hard shoulders, and elsewhere in the United Kingdom undertakers should comply with Chapter 8 when carrying out such works. On all other roads they meet their obligations under section 65 or 124 of the Act, or under article 25 of the Northern Ireland Order, if they comply with this Code, even though further relevant advice may be available in Chapter 8 and other relevant documents.

19.6.5 Changes to legislation

Prolonged occupation

As a result of the widespread perception that street works are often unnecessarily protracted, the Government held consultations before activating Section 74 of the NRSWA under the Street Works (Charges for Unreasonably Prolonged Occupation of the Highway) (England) Regulations 2001.

This provides powers for highway authorities to charge undertakers a daily fee if they fail to complete works by an agreed deadline. The Regulations define a 'prescribed' period for work (i.e. one prescribed in Regulations) and a 'reasonable' period (i.e. the period that the undertaker estimates that the work will take, if not challenged by the highway authority, or, if not agreed, the period determined by arbitration). If the duration of a work exceeds both of these periods, a highway authority may levy a charge. Highway authorities have the power to waive or reduce the level of charges when they believe that circumstances warrant this.

The Government commissioned a report from Halcrow into the effects of the scheme, which was published in 2011.

Lane rental

At the time Section 74 was passed, the Government made it clear that if the legislation failed to lead to a sufficient reduction in disruption, then it would be prepared to consider making lane rental charging powers available to local authorities. The Government decided to undertake a localised test of the proposed

19

new powers and, following the London Borough of Camden's successful application to the Secretary of State to operate a lane rental pilot scheme, this was passed under Section 74A of the NRSWA as the Street Works (Charges for Occupation of the Highway) (London Borough of Camden) Order 2002. The pilot scheme was launched in March 2002 and ran until March 2004 with a similar one running in Middlesborough. Section 74 will not apply while Section 74A is operational.

Under the lane rental scheme, streets in Camden were divided into 'premium routes' and 'ordinary routes' with charges applying per working day from the commencement of works.

The charges for premium routes and ordinary routes differ, depending on whether the jobs are works or remedial works – works in this context consist of the work originally scheduled while remedial works are street works which have been necessary because the local authority was dissatisfied, for instance, with the quality of reinstatement.

19.6.6 Further information

Crossing High-Speed Roads on Foot During Temporary Traffic Management Works. Construction Information Sheet No. 53, 2000, HSE, http://www.hse.gov.uk/pubns/cis53.pdf

New Roads and Street Works Act 1991. Chapter 22, London, http://www.legislation.gov.uk/ukpga/1991/22/contents

Safety at Street Works and Road Works Code of Practice. 2nd impression, 2013 London, https://www.gov.uk/government/uploads/system/uploads/attachment_data/file/321056/safety-at-streetworks.pdf

The Traffic Signs Manual. Chapter 8 (second edition) 2009: The traffic safety measures and signs for roadworks and temporary situations, Department for Transport. https://www.gov.uk/government/uploads/system/uploads/attachment_data/file/203669/traffic-signs-manual-chapter-08-part-01.pdf

19.7 Control of Artificial Optical Radiation at Work Regulations 2010

19.7.1 Introduction

These Regulations implement as respects Great Britain Directive 2006/25/EC of the European Parliament and of the Council (O.J. L114, 27.4.2006, p.38) on the minimum health and safety requirements regarding the exposure of workers to risks arising from physical agents AOR (artificial optical radiation). The Regulations impose duties on employers to protect both employees who may be exposed to risk from exposure to artificial optical radiation at work and other persons at work who

might be affected by that work. AOR includes light emitted from all artificial sources in all its forms such as ultraviolet, infrared and laser beams, but excludes sunlight.

Regulation 1(2) defines exposure limit values as being those set out in Annexes I and II to the Directive, as amended from time to time, and these Annexes provide for exposure limit values for non-coherent radiation and laser radiation respectively.

19.7.2 Duties

The Regulations impose a duty to carry out a specific form of risk assessment where an employer carries out work which could expose its employees to levels of artificial optical radiation (i.e. artificial light) that could create a reasonably foreseeable risk of adverse health effects to the eyes or skin and where those risks have not already been eliminated or controlled (Regulation 3).

Where a risk assessment is necessary the Regulations also impose duties to –

(a) eliminate, or where this is not reasonably practicable, to reduce to as low a level as is reasonably practicable the risk of adverse health effects to the eyes or skin of the employee as a result of exposure to artificial optical radiation where this risk has been identified in the risk assessment (Regulation 4(1));

(b) devise an action plan comprising technical and organisational measures to prevent exposure to artificial optical radiation exceeding the exposure limit values where the risk assessment indicates that employees are exposed to levels of artificial optical radiation that exceed the exposure limit values (Regulation 4(3));

(c) take action in the event that the exposure limit values are exceeded despite the implementation of the action plan and measures to eliminate or reduce so far as is reasonably practicable the risk of exposure (Regulation 4(5));

(d) demarcate, limit access to, and provide for appropriate signs in those areas where levels of artificial optical radiation are indicated in the risk assessment as exceeding the exposure limit values (Regulation 4(6) and (7);

(e) provide information and training if the risk assessment indicates that employees could be exposed to artificial optical radiation which could cause adverse health effects to the eyes or skin of the employee (Regulation 5); and

(f) to provide health surveillance and medical examinations in certain cases (Regulation 6).

19.7.3 Safe light sources

The following information is taken from the HSE Guidance for Employers available at: www.hse.gov.uk/radiation/nonionising/employers-aor.pdf

The majority of light sources are safe, such as those described in List 1 below. If these sources, or similar, are being used workers are not at risk and there is no need to do anything further.

However, when making this decision, employers should consider the following points to ensure that all workers are protected:

▶ If there are any workers whose health is at particular risk (e.g. those with pre-existing medical conditions made worse by light).

▶ If workers use any chemicals (e.g. skin creams) which could react with light to make any health effects worse.

▶ If any workers are exposed to multiple sources of light at the same time.

▶ If exposure to bright light could present unrelated risks (e.g. temporary blindness could lead to mistakes being made in hazardous tasks).

List 1 Safe light sources

▶ All forms of ceiling-mounted lighting used in offices, etc. that have diffusers over bulbs or lamps.

▶ All forms of task lighting including desk lamps and tungsten-halogen lamps fitted with appropriate glass filters to remove unwanted ultraviolet light.

▶ Photocopiers.

▶ Computer or similar display equipment, including personal digital assistants (PDAs).

▶ Light emitting diode (LED) remote control devices.

▶ Photographic flashlamps – when used singly.

▶ Gas-fired overhead heaters.

▶ Vehicle indicator, brake, reversing and fog lamps.

▶ Any exempt or Risk Group 1 lamp or lamp system (including LEDs), as defined in British Standard BS EN 62471: 2008.

▶ Any Class 1 laser light product, as defined in British Standard BS EN 60825-1: 2007, for example laser printers and bar code scanners.

There are also some sources of light that, if used inappropriately, e.g. placed extremely close to the eyes or skin, have the potential to cause harm but which are perfectly safe under normal conditions of use. Examples include:

▶ Ceiling-mounted fluorescent lighting without diffusers over bulbs or lamps.

▶ High-pressure mercury floodlighting.

▶ Desktop projectors.

▶ Vehicle headlights.

▶ Non-laser medical applications such as:
  ▷ operating theatre and task lighting;
  ▷ diagnostic lighting such as foetal/neonatal transilluminators and X-ray light/viewing boxes.

▶ UV insect traps.

▶ Art and entertainment applications such as illumination by spotlights, effect lights and

flashlamps (provided that any ultraviolet emissions have been filtered out).

▶ Multiple photographic flashlamps, for example in a studio.

▶ Any Risk Group 2 lamp or lamp system (including LEDs), as defined in British Standard BS EN 62471: 2008.

▶ Class 1M, 2 or 2M lasers, as defined in British Standard BS EN 60825-1: 2007, for example low-power laser pointers.

The above list is not exhaustive.

19.7.4 Hazardous light sources

Some sources of light can cause a risk of ill-health, such as: burns or reddening (erythema) of the skin or surface of the eye (photokeratitis); burns to the retina of the eye; so-called blue-light damage to the eye (photoretinitis) and damage to the lens of the eye that may bring about the early onset of cataract.

List 2 Hazardous light sources

Examples of hazardous sources of light that present a 'reasonably foreseeable' risk of harming the eyes and skin of workers and where control measures are needed include:

▶ Metal working – welding (both arc and oxy-fuel) and plasma cutting.

▶ Pharmaceutical and research – UV fluorescence and sterilisation systems.

▶ Hot industries – furnaces.

▶ Printing – UV curing of inks.

▶ Motor vehicle repairs – UV curing of paints and welding.

▶ Medical and cosmetic treatments – laser surgery, blue light and UV therapies, Intense Pulsed Light sources (IPLs).

▶ Industry, research and education, for example all use of Class 3B and Class 4 lasers, as defined in British Standard BS EN 60825-1: 2007.

▶ Any Risk Group 3 lamp or lamp system (including LEDs), as defined in British Standard BS EN 62471: 2008, for example search lights, professional projections systems.

Less common hazardous sources are associated with specialist activities – for example lasers exposed during the manufacture or repair of equipment, which would otherwise not be accessible.

19.7.5 Further information

Guidance for Employers on the Control of Artifical Optical Radiation at Work Regulations (AOR) 2010 http://www.hse.gov.uk/radiation/nonionising/employers-aor.pdf

The full text of the Regulations is available from the UK Government Information website at http://www.legislation.gov.uk/uksi/2010/1140/contents/made

19.8 Control of Asbestos Regulations (CAR) 2012

19.8.1 Introduction

The Control of Asbestos Regulations 2012 came into force on 6 April 2012 (Asbestos Regulations – SI 2012/632).

In practice the changes are fairly limited. They mean that some types of non-licensed work with asbestos now have additional requirements, i.e. notification of work, medical surveillance and record keeping. **All other requirements remain unchanged**.

Most work with asbestos needs to be carried out by a **licensed contractor**. This includes work on asbestos coating and asbestos lagging; and work on asbestos insulation or asbestos insulating board (AIB) where the risk assessment demonstrates that the fibre release will be high, e.g. the material is badly damaged, or the work is not short duration work.

'Short duration' means any one person doing this type of work for less than one hour, or more people doing the work for a total of less than two hours, in any seven consecutive days. The total time spent by all workers must not exceed two hours. This includes time spent setting up, cleaning and clearing up. Non-licensed work includes work on asbestos-containing textured coatings and asbestos cement.

From 6 April 2012, some non-licensed work needs to be **notified** to the relevant enforcing authority. Brief written records should be kept of non-licensed work, which has to be notified, e.g. copy of the notification with a list of workers on the job, plus the level of likely exposure of those workers to asbestos. This does not require air monitoring on every job, if an estimate of degree of exposure can be made based on experience of similar past tasks or published guidance.

By April 2015, all workers/self-employed doing **notifiable** non-licensed work with asbestos must be under health surveillance by a Doctor. Workers who are already under health surveillance for licensed work need not have another medical examination for non-licensed work. BUT medicals for notifiable non-licensed work are not acceptable for those doing licensed work.

There has been some modernisation of language and changes to reflect other legislation, e.g. the prohibition section has been removed as the prohibition of supply and use of asbestos is now covered by REACH.

The Asbestos Regulations also include the 'duty to manage asbestos' in non-domestic premises. Guidance on the duty to manage asbestos can be found in the 'Approved Code of Practice. The Management of Asbestos in Non-Domestic Premises', L127. Domestic premises are covered by the Defective Premises Act 1972 or the Civic Government (Scotland) Act.

The Regulations require mandatory training for anyone liable to be exposed to asbestos fibres at work (see Regulation 10). This includes maintenance workers and others who may come into contact with or who may disturb asbestos (e.g. cable installers) as well as those involved in asbestos removal work.

When work with asbestos or which may disturb asbestos is being carried out, the CAR 2012 require employers and the self-employed to prevent exposure to asbestos fibres. Where this is not reasonably practicable, they must make sure that exposure is kept as low as reasonably practicable by measures other than the use of respiratory protective equipment (RPE). The spread of asbestos must be prevented. The Regulations specify the work methods and controls that should be used to prevent exposure and spread.

Worker exposure must be below the airborne exposure limit (Control Limit). The CAR 2012 have a single Control Limit for all types of asbestos of 0.1 fibres/cm³. A Control Limit is a maximum concentration of asbestos fibres in the air (averaged over any continuous 4-hour period) that must not be exceeded.

In addition, short-term exposures must be strictly controlled and worker exposure should not exceed 0.6 fibres/cm³ of air averaged over any continuous 10-minute period using RPE if exposure cannot be reduced sufficiently using other means.

RPE is an important part of the control regime but it must not be the sole measure used to reduce exposure and should only be used to supplement other measures. Work methods that control the release of fibres such as those detailed in the *Asbestos Essentials task sheets* for non-licensed work should be used. RPE must be suitable, must fit properly and must ensure that worker exposure is reduced as low as is reasonably practicable.

Most asbestos removal work must be undertaken by a licensed contractor but any decision on whether particular work is licensable is based on the risk. Work is only exempt from licensing if:

▶ the exposure of employees to asbestos fibres is sporadic and of low intensity (but exposure cannot be considered to be sporadic and of low intensity if the concentration of asbestos in the air is liable to exceed 0.6 fibres/cm³ measured over 10 minutes); and

▶ it is clear from the risk assessment that the exposure of any employee to asbestos will not exceed the control limit; and

▶ the work involves:
  ▷ short, non-continuous maintenance activities. Work can only be considered as short, non-continuous maintenance activities if any one person carries out work with these materials for less than 1 hour in a 7-day period. The total time spent by all workers on the work should not exceed a total of 2 hours.

▷ removal of materials in which the asbestos fibres are firmly linked in a matrix. Such materials include: asbestos cement; textured decorative coatings and paints which contain asbestos; articles of bitumen, plastic, resin or rubber which contain asbestos where their thermal or acoustic properties are incidental to their main purpose (e.g. vinyl floor tiles, electric cables, roofing felt); and other insulation products which may be used at high temperatures but have no insulation purposes (e.g. gaskets, washers, ropes and seals).

▷ encapsulation or sealing of ACMs which are in good condition; or

▷ air monitoring and control, and the collection and analysis of samples to find out if a specific material contains asbestos.

Under the CAR 2012, anyone carrying out work on asbestos insulation, asbestos coating or asbestos-insulating board (AIB) needs a licence issued by the HSE unless they meet one of the exemptions above.

Although work may not need a licence to carry out a particular job, there is still a need to comply with the rest of the requirements of the CAR 2012.

If the work is licensable there are a number of additional duties. Duty-holders need to:

▶ notify the enforcing authority responsible for the site where they are working (for example the HSE or the local authority);

▶ designate the work area (see Regulation 18 for details);

▶ prepare specific asbestos emergency procedures; and

▶ pay for their employees to undergo medical surveillance.

The CAR 2012 require any analysis of the concentration of asbestos in the air to be measured in accordance with the 1997 WHO recommended method.

From 6 April 2007, a clearance certificate for reoccupation may only be issued by a body accredited to do so. At the moment, such accreditation can only be provided by the United Kingdom Accreditation Service (UKAS).

19.8.2 Application Regulation 3 and general note

The Regulations and ACoP apply to all work with asbestos. They apply in particular to work on or which disturbs building materials containing asbestos, asbestos sampling and laboratory analysis with the exception of clearing asbestos-contaminated land which is not specifically covered by this ACoP. An additional ACoP entitled 'The Management of Asbestos in Non-domestic Premises' is aimed at those who have repair and maintenance responsibilities for non-domestic premises.

Most of the duties in the CAR 2012 are placed upon 'an employer', that is the person who employs the workers who are liable to be exposed to asbestos in the course of their work. Although the Regulations always refer to an employer, Regulation 3(1) makes it clear that self-employed people have the same duties towards themselves and others as an employer has towards his or her employees and others.

There is an exemption from certain regulatory requirements for particular, specified types of work with asbestos where any worker exposure will only be sporadic and of low intensity and the exposure level is below the control limit (Regulation 3(2)). Such work will not require a licence. All other work with asbestos will require a licence (Regulation 8); must be notified to the relevant enforcing authority (Regulation 9); must have emergency arrangements in place (Regulation 15(1)); must have designated asbestos areas (Regulation 18); and those working with the asbestos must be subject to medical surveillance and have health records (Regulation 22). Some of the guidance in the ACoP is specifically aimed at this more hazardous work and, for convenience, this work has been referred to as licensable work throughout the ACoP.

If the control limit for asbestos is exceeded in the working area, this triggers particular requirements including:

(a) immediately informing employees and their representatives (Regulation 11(5)(b)(i))

(b) identification of the reasons for the control limit being exceeded and the introduction of appropriate measures to prevent it being exceeded again (Regulation 11(5)(b)(ii))

(c) stopping work until adequate measures have been taken to reduce employees' exposure to below the control limit (Regulation 11(5)(b)(i))

(d) a check of the effectiveness of the measures taken to reduce the levels of asbestos in the air by carrying out immediate air monitoring (Regulation 11(5)(b)(iii))

(e) the designation of respirator zones and

(f) the mandatory provision of RPE (Regulation 11(3)), although such equipment should always be provided if it is reasonably practicable to do so (Regulation 11(2)).

Where work with asbestos forms part of a larger project there will be a particular need to cooperate with other employers, and there may be other Regulations which must be taken into account. However, the responsibility to ensure compliance with the provisions of the Asbestos Regulations remains with the employer or self-employed person.

There are exceptions from some requirements.

Where Regulation 3(2) applies (i.e. non-licensable work):

(a) the work will not need to be notified to the relevant Enforcing Authority;

19

(b) the work will not need to be carried out by holders of a licence to work with asbestos;

(c) the workers will not need to have a current medical and a current health record;

(d) the employer will not need to prepare specific asbestos emergency procedures;

(e) the area around work does not need to be identified as an asbestos area.

Work with the following materials is likely only to produce sporadic and low-intensity worker exposure and can be categorised as complying with Regulation 3(2) as long as 3(2)(b) is fulfilled, that is it is clear from the risk assessment that the control limit will not be exceeded:

(a) asbestos cement;

(b) textured decorative coating which contains asbestos;

(c) any article of bitumen, plastic, resin or rubber which contains asbestos where its thermal or acoustic properties are incidental to its main purpose (e.g. vinyl floor tiles, electric cables, roofing felt); and

(d) asbestos materials such as paper linings, cardboards, felt, textiles, gaskets, washers, and rope where the products have no insulation purposes.

19.8.3 Work with asbestos

'Work with asbestos' includes:

(a) work which consists of the removal, repair or disturbance of asbestos;

(b) work which is ancillary to such work (ancillary work); and

(c) supervising work referred to in sub-paragraphs (a) or (b) above (supervisory work).

'Ancillary work' means work associated with the main work of repair, removal or disturbance of asbestos. Work carried out in an ancillary capacity requires a licence unless the main work (i.e. the removal, repair, disturbance activity) would result in worker exposure which fulfils the conditions for Regulation 3(2) to apply.

'Supervisory work' means work involving direct supervisory control over those removing, repairing or disturbing asbestos. Work carried out in a supervisory capacity requires a licence to work with asbestos unless the work being supervised would result in worker exposure which fulfils the conditions for Regulation 3(2) to apply.

Therefore, compliance with Regulations 8 (licence), 9 (notification), 15(1) (emergency arrangements), 18(1)(a) (designated areas) and 22 (health records and medical surveillance) are not required in such circumstances. Those other Regulations which apply to all work with asbestos must be observed.

19.8.4 Duty to manage asbestos in non-domestic premises and identification of the presence of asbestos – Regulations 4 and 5

Owners and occupiers of premises, who have maintenance and repair responsibilities for those premises, have a duty to assess them for the presence of asbestos and the condition of that asbestos.

The duty to manage asbestos is contained in Regulation 4. It requires the person who has the duty (i.e. the 'duty-holder') to:

▶ take reasonable steps to find out if there are materials containing asbestos in non-domestic premises, and if so, its amount, where it is and what condition it is in;

▶ presume materials contain asbestos unless there is strong evidence that they do not;

▶ make, and keep up to date, a record of the location and condition of the asbestos containing materials – or materials which are presumed to contain asbestos;

▶ assess the risk of anyone being exposed to fibres from the materials identified;

▶ prepare a plan that sets out in detail how the risks from these materials will be managed;

▶ take the necessary steps to put the plan into action;

▶ periodically review and monitor the plan and the arrangements to act on it so that the plan remains relevant and up to date; and

▶ provide information on the location and condition of the materials to anyone who is liable to work on or disturb them.

Other parties have a legal duty to cooperate with the duty-holder.

The employer should not rely on the information of the other duty-holders if they cannot produce reasonable evidence regarding the nature of suspect material (e.g. survey details or analytical reports).

19.8.5 Assessment of work – Regulation 6

If work which is liable to expose employees to asbestos is unavoidable, then before starting the work, employers must make a suitable and sufficient assessment of the risk created by the likely exposure to asbestos of employees and others who may be affected by the work and identify the steps required to be taken by the Asbestos Regulations.

For non-licensable work it is not always necessary to make an assessment before each individual job. Where an employer carries out work which involves very similar jobs on a number of sites on the same type of asbestos material, for example electrical and plumbing jobs, only one assessment for that work may be needed, although the plan of work should always be job specific.

However, for licensable work or where the degree and nature of the work varies significantly from site to site, for example in demolition or refurbishment, or where the type of asbestos material varies, a new assessment and plan of work (see Regulation 7) will be necessary.

19.8.6 Plan of work – Regulation 7

For any work involving asbestos, including maintenance work that may disturb it, the employer of the workers involved must draw up a written plan of how the work is to be carried out before work starts. Employers must make sure that their employees follow the plan of work (sometimes called a method statement) so far as it is reasonably practicable to do so. Where unacceptable risks to health and/or safety are discovered while work is in progress, for example disturbance of hidden, missed or incorrectly identified ACMs, any work affecting the asbestos should be stopped except for that necessary to render suitable control and prevent further spread. Where there is extensive damage to ACMs which causes contamination of the premises, or part of the premises, then the area should be immediately evacuated. Work should not restart until a new plan of work is drawn up or until the existing plan is amended. Some measures, for example, should only be carried out by licensed contractors.

For licensable work in particular, the plan of work should identify procedures to adopt in emergencies and indicate clearly what remedial measures can be undertaken by staff.

19.8.7 Licensing of work with asbestos – Regulation 8

This regulation means that an employer must not carry out work with asbestos (other than that fulfilling the conditions for Regulation 3(2) to apply), including supervisory and ancillary work and work with asbestos in their own premises with their own employees, unless the employer holds a licence issued under this regulation and complies with its terms and conditions. This includes work with asbestos insulation, asbestos coatings (excluding asbestos-containing textured decorative coatings) and AIB.

For supervisory work a licence is needed when directly supervising licensable work but not when the person concerned is:

(a) the client who has engaged a licensed contractor to do the licensable work;

(b) the principal or main contractor on a construction or demolition site if the licensable work is being done by a sub-contractor holding an **asbestos licence**;

(c) an analyst checking that the area is clear of asbestos at the end of a job;

(d) carrying out quality control work such as:

(i) atmospheric monitoring outside enclosures while asbestos removal work is in progress; or

(ii) checking that work has been carried out to a standard which meets the terms of the contract;

(e) a consultant or other preparing the method statement; or

(f) a consultant or other reviewing tender submissions on behalf of the client.

For ancillary work, a licence is needed for:

(a) setting up and taking down enclosures for the asbestos work;

(b) putting up and taking down scaffolding to provide access for licensable work where it is foreseeable that the scaffolding activity is likely to disturb the asbestos;

(c) maintaining negative pressure units;

(d) work done within an asbestos enclosure, such as sealing an electric motor in polythene and installing ducting to the motor to provide cooling air from outside the enclosure; and

(e) cleaning the structure, plant and equipment inside the enclosure.

A licence holder is required to:

(a) notify the work to the appropriate enforcing authority (Regulation 9);

(b) ensure medical surveillance is carried out for their employees and themselves (Regulation 22);

(c) maintain health records for employees and themselves (Regulation 22);

(d) prepare procedures in case of emergencies (Regulation 15(1)); and

(e) demarcate the work areas appropriately (Regulation 18(1)(a)).

All licences issued for work with asbestos are granted by the HSE under the terms of this regulation. Fees are payable for issuing licences, reassessments and changes to licences. These fees are periodically updated by the Health and Safety (Fees) Regulations.

19.8.8 Notification of work with asbestos – Regulation 9

If licensable work is undertaken notification has to be given to the appropriate enforcing authority with details of the proposed work. This gives the enforcing authorities the opportunity to assess your proposals for carrying out work with asbestos and to inspect the site either before or during the work.

Notification will normally be required 14 days before work begins, but the enforcing authority may allow a shorter period, for example in an emergency where there is a serious risk to the health and safety of any person. This shorter period is known as a 'waiver' or dispensation. Each individual job must normally be notified to the enforcing authority.

19

Form FOD ASB5 can be used for notification, available from the HSE website, local HSE offices or the Asbestos Licensing Unit.

19.8.9 Information, instruction and training – Regulation 10

There are three main types of information, instruction and training (simply referred to as training from now on). These are:

1. Asbestos awareness training. This is for those persons who are liable to be exposed to asbestos while carrying out their normal everyday work.
2. Training for non-licensable asbestos work. This is for those who undertake work with asbestos which is not licensable such as a roofer removing a whole asbestos cement sheet in good condition.
3. Training for licensable work with asbestos – for those working with asbestos which is licensable, such as removing asbestos lagging or insulating board.

The Regulations require mandatory training for anyone liable to be exposed to asbestos fibres at work (see Regulation 10). This includes maintenance workers and others who may come into contact with or who may disturb asbestos (e.g. cable installers) as well as those involved in asbestos removal work.

Training must include detailed information on:

▶ recognising asbestos;
▶ how asbestos can affect your health;
▶ the added dangers of smoking;
▶ the uses and likely locations for asbestos in buildings;
▶ what work you are allowed to do by law;
▶ what the law requires you to do;
▶ procedures to protect yourself;
▶ what methods to use;
▶ what equipment you need to do the job properly;
▶ how to choose, use and look after personal protective equipment;
▶ recognising and dealing with other dangers, such as work at height;
▶ decontamination of yourself and work areas;
▶ emergency procedures; and
▶ waste disposal.

Refresher training is needed every year, or more often if:

▶ work methods change;
▶ the type of equipment used changes; or
▶ the type of work changes a lot.

19.8.10 Prevention or reduction of exposure to asbestos – Regulation 11

Work which disturbs ACMs should only be carried out when there is no other reasonably practicable way of doing the work or the alternative method creates a more significant risk. Employers must therefore first decide whether they can prevent the exposure to asbestos sfairp, before considering how they will reduce the exposure to as low as reasonably practicable.

Where it is not reasonably practicable to prevent exposure, it must first be reduced to the lowest level reasonably practicable by means other than the use of RPE.

Airborne levels should be reduced to as low a level as reasonably practicable and exposure should be controlled so that any peak exposure is less than 0.6 fibre/cm^3 averaged over a maximum continuous period of 10 minutes by the use of appropriate RPE if exposure cannot be reduced sufficiently by other means.

Employers must ensure that the number of employees exposed to asbestos is kept as low as reasonably practicable. All unnecessary personnel should be excluded from the working areas if asbestos is being disturbed.

The provision of a sufficient number of suitable viewing panels in enclosures will allow managers to monitor the work of their employees without being unnecessarily exposed.

When it is not reasonably practicable to prevent exposure to asbestos the employer must choose the most effective method, or combination of methods, which minimises fibre release and thereby reduces the exposure to the lowest levels reasonably practicable and document this in the written risk assessment/plan of work.

19.8.11 Use of control measures – Regulation 12

Employers should have procedures in place to make sure that control measures are properly used or applied and are not made less effective by other work practices or other machinery.

These procedures should include:

(a) checks at the start of every shift and at the end of each day; and
(b) prompt action when a problem is identified.

Within the general duties imposed by Regulation 12(2), employees should, in particular:

(a) use any control measures, including RPE, and protective clothing properly and keep it in the places provided;
(b) follow carefully all the procedures set out in the employer's assessment and plan of work, including those for changing and decontamination, and comply with the use of control measures;
(c) keep the workplace clean;
(d) eat, drink and smoke only in the places provided; and
(e) report any defects concerning control measures to their supervisor/manager immediately.

19.8.12 Maintenance of control measures – Regulation 13

When working with asbestos, employers should make sure that maintenance procedures are drawn up for all control measures and for PPE. These should include the equipment used for cleaning, washing and changing facilities, and the controls to prevent the spread of contamination. The procedures should make clear:

(a) which control measures require maintenance;
(b) when and how the maintenance is to be carried out; and
(c) who is responsible for maintenance and for making good any defects.

19.8.13 Provision and cleaning of protective clothing

As part of the assessment, the employer must decide whether or not protective clothing is required for work with asbestos. The assessment should start from the assumption that protective clothing will be necessary unless exposures are extremely slight and infrequent. For work which requires a licence exposure will potentially be significant and employers will always need to provide a full set of PPE.

The protective clothing must be adequate and suitable and include footwear, whenever employees are liable to be exposed to a significant amount of asbestos debris or fibres. It should be appropriate and suitable for the job and must protect the parts of the body likely to be affected. If the assessment has concluded that a risk of contamination exists, disposable overalls (of a suitable standard fitted with a hood) and boots without laces will be required.

To be adequate and suitable and depending on the circumstances, the protective clothing must:

(a) fit the wearer;
(b) be of sufficient size to avoid straining and ripping the joints;
(c) be comfortable and, where appropriate, to allow for the effects of physical strain;
(d) be suitable for cold environments;
(e) prevent penetration by asbestos fibres;
(f) be elasticated at the cuffs and ankles and on the hoods of overalls and designed to ensure a close fit at the wrists, ankles, face and neck;
(g) not have pockets or other attachments which could attract and trap asbestos dust; and
(h) be easily decontaminated or disposable.

Where disposable overalls are used, these should be of a suitable standard.

Non-disposable protective clothing and towels must be effectively washed after every shift. If the employer does not have the facilities and expertise for laundering asbestos-contaminated clothing, it must be sent to a specialist laundry.

Where disposable overalls are used they should be treated as asbestos waste and properly disposed of after every shift.

This may not be necessary for overalls used for occasional sampling where there is a low risk of contamination.

When working in enclosures, clothing for washing should be collected from the airlock and hygiene facility as soon as it has been discarded.

19.8.14 Accidents, incidents and emergencies – Regulation 15

Employers of people removing or repairing ACMs must have prepared procedures which can be put into effect should an accident, incident or emergency occur which could put people at risk because of the presence of asbestos unless, because of the quantity or the condition of the asbestos present at the workplace, there is only a slight risk to the health of employees.

Sufficient information should be made available to the emergency services (e.g. fire and rescue and paramedics), so that when they are attending a relevant incident they can properly protect themselves against the risks from the asbestos.

In any circumstance where there is an accidental uncontrolled release of asbestos into the workplace then measures, including emergency procedures, should be in place to limit exposure and the risks to health. Such procedures should include means to raise the alarm and procedures for evacuation, which should be tested and practised at regular intervals. The cause of the uncontrolled release should be identified, and adequate control regained as soon as possible.

Any people in the work area affected who are not wearing PPE including RPE must leave that area. Where such people have been contaminated with dust or debris then arrangements should be made to decontaminate those affected. Any clothing or PPE should be decontaminated or disposed of as contaminated waste.

19.8.15 Duty to prevent or reduce the spread of asbestos – Regulation 16

Any plant or equipment which has been contaminated with asbestos should be thoroughly decontaminated before it is moved for use in other premises or for disposal. The basic decontamination procedures must be followed every time a person leaves the work area.

Asbestos materials should never be left loose or in a state where they can be trampled, tracked over by plant and machinery or otherwise spread. All asbestos-contaminated waste should be removed at regular intervals in appropriate waste containers.

For non-licensable work where a risk of significant contamination exists, the work area should be enclosed. A full enclosure will be expected where there is large-scale work, e.g. asbestos-containing textured decorative coating removal. A 'mini-enclosure' should be used where the work is minor.

It should be assumed that for most of the work which requires a licence, which is not external/remote, a full enclosure will normally be required.

19.8.16 Cleanliness of premises and plant – Regulation 17

When work with asbestos comes to an end, the work area should be thoroughly cleaned before being handed over for reoccupation or for demolition. All visible traces of asbestos dust and debris should be removed and a thorough visual inspection should be carried out. Where the work is licensable then the 4-stage clearance procedure (which includes air sampling) should be carried out and a certificate of reoccupation issued. Where licensed work is performed out of doors (e.g. soffit removal), then air sampling will not be required. In this situation, the certificate of reoccupation should still be completed but without stage 3 (air monitoring). More information on clearance procedures for non-licensed work is given in Asbestos Essentials.

To aid the process of cleaning and to prevent the spread of asbestos, employers must choose work methods and equipment to prevent the build-up of asbestos waste on floors and surfaces in the working area. Wherever practicable, waste should be transferred direct into waste bags as workers remove the asbestos materials. Employers must make sure that any asbestos dust and debris is cleaned up and removed regularly to prevent it accumulating (and drying out where wet removal techniques have been used), and at least at the end of each shift.

Procedures will need to take account of the necessity for cleaning following an accidental and uncontrolled release of asbestos.

Procedures will need to be set up for cleaning:

(a) working areas including transit and waste routes;
(b) plant and equipment; and
(c) hygiene facilities.

Dustless methods of cleaning should be used including, wherever practicable, a type 'H' (BS 5415: 1986) vacuum cleaner with appropriate tools. Procedures for cleaning should make clear:

(a) the items and areas to be cleaned;
(b) how often they need to be cleaned;
(c) the cleaning methods, which should not create dust; and
(d) any special precautions which need to be taken during cleaning, including the low-dust technique to be used, and the measures to be taken to

reduce the spread of dust. Dry manual brushing, or sweeping or compressed air, must not be used to remove asbestos dust.

Once removal of the asbestos has been completed, the premises must be assessed to determine whether they are thoroughly clean and hence fit to be returned to the client. It is important that this includes the premises, any plant or equipment or parts of the premises where work with asbestos has taken place, and the surrounding areas which may have been contaminated. The areas requiring assessment for site clearance certification for reoccupation include:

(a) the enclosed area including airlocks or the delineated work area where an enclosure has not been used;
(b) the immediate surrounding area (for enclosures this will include the outside of walls and underneath polythene floors; for delineated areas this will include surfaces nearby either where asbestos may have been spread or where the pre-cleaning was not done properly);
(c) the transit route if one has been used; and
(d) the waste route and area around the waste skip.

19.8.17 Designated areas – Regulation 18

All areas where licensable work is being undertaken should be demarcated and identified by suitable warning notices as asbestos areas.

Any area where an employee may be exposed to asbestos to a level which may exceed a control limit, must be designated as a respirator zone. Respirator zones, whether enclosed or not, must be demarcated and identified by suitable warning notices. Notices that RPE must be worn are also necessary.

Only employees who need to do so for their work can enter and remain in asbestos areas and respirator zones.

Only employees who are competent may enter respirator zones or supervise people working in respirator zones. To enter a respirator area, the employee must have received adequate information, instruction and training in accordance with Regulation 10.

Employers should ensure the provision of suitable facilities for employees to eat and drink outside the working area and where appropriate as close as is reasonably practicable to the hygiene facilities. No one should eat, drink or smoke in the enclosure or work area, in the hygiene facilities or in any areas which have been marked as asbestos areas or respirator zones.

Employers should also ensure that toilet facilities are provided, if they are not provided elsewhere on the site.

Where hygiene facilities are not being used, personnel should wash and decontaminate themselves whenever they leave an asbestos area or respirator zone.

19.8.18 Air monitoring – Regulation 19

Air monitoring may be required to protect the health of employees by determining or checking the concentrations of airborne asbestos to which they are exposed and to establish employee exposure records. This should be done at regular intervals for a representative range of jobs and work methods.

Air monitoring should always be done when there are any doubts about the effectiveness of the measures taken to reduce the concentration of asbestos in air (e.g. that engineering controls are working as they should to their design specification and do not need repair), and, in particular, measures taken to reduce that concentration below the control limit or below a peak level measured over 10 minutes of 0.6 fibre/cm^3. Monitoring will also be necessary to confirm that the RPE chosen will provide the appropriate degree of protection where the level of asbestos fibres in air exceeds, or is liable to exceed, the control limit or a peak level measured over 10 minutes of 0.6 fibre/cm^3.

Air monitoring will be appropriate unless:

(a) exposures are known to be low and not likely to approach the control limit or a 10 minute peak of 0.6 f/m^3;

(b) the work is such that it complies with Regulation 3(2) and adequate information is available to enable the appropriate protective equipment to be provided; or

(c) the protective equipment provided is of such a standard that no foreseeable measurement could indicate a need for equipment of a higher standard.

If the employer decides that monitoring is not necessary then he or she should use other sources of information about the likely concentrations of asbestos in air, for instance the guidance issued by the HSE in the *Licensed Contractors' Guide* or exposure data from previous similar work.

Monitoring of employee exposure should be by personal sampling. Static sampling can be used to check that control measures are effective. Analysis must be undertaken using the 1997 WHO recommended method.

19.8.19 Standards for air testing, site clearance certification and analysis – Regulations 20 and 21

Those engaged to carry out air measurements and employee exposure monitoring must demonstrate that they conform with specified requirements in ISO 17025 through accreditation with a recognised accreditation body.

Employers carrying out their own air measurements or employee exposure monitoring should make sure that employees carrying out this work receive similar standards of training, supervision and quality control to those required by ISO 17025.

Those engaged to carry out site clearance certification for reoccupation must demonstrate that they conform with specified requirements in ISO 17020 and ISO 17025 through accreditation with a recognised accreditation body.

Those engaged to analyse samples of material to determine whether or not they contain asbestos must demonstrate that they conform with ISO 17025 by accreditation with a recognised accreditation body.

Employers carrying out their own analysis of samples should make sure that employees carrying out this work receive similar standards of training, supervision and quality control to those required by ISO 17025.

The UKAS is currently the sole recognised accreditation body in Great Britain.

19.8.20 Health records and medical surveillance – Regulation 22

The employer must keep a health record for any employee who undertakes licensable work. The health record must be kept for 40 years in a safe place and should contain at least the following information:

(a) the individual's surname, forenames, sex, date of birth, permanent address, postcode and National Insurance number;

(b) a record of the types of work carried out with asbestos and, where relevant, its location, with start and end dates, with the average duration of exposure in hours per week, exposure levels and details of any RPE used;

(c) a record of any work with asbestos prior to this employment of which the employer has been informed; and

(d) dates of the medical examinations.

Anyone who undertakes licensable work must have been medically examined within the previous 2 years. Employers will need to obtain certificates of examination for any employees who state that they have been examined under these Regulations within the previous 2 years and keep them for 4 years from the date of issue. Employers should check with the previous employer or with the examining doctor that the certificates are genuine.

For work with asbestos that is not licensable and is not exempt by Regulation 3(2) a medical examination must be carried out before 30 April 2015 and after this date not more than 3 years before exposure and at least every 3 years thereafter while such exposure continues (more frequently if the relevant doctor requires it).

Medical examinations should take place during the employee's normal working hours and be paid for by the employer. Employees should cooperate with their employer regarding attendance for medical examinations.

19

Where an employee is diagnosed with a condition related to exposure to asbestos then the employer must review the health of all other current employees similarly exposed, as well as reviewing his/her assessments and methods of work.

If the examination reveals the presence of any potentially limiting health conditions then a decision should be reached on whether a general fitness assessment is required in addition to the asbestos medical examination.

19.8.21 Washing and changing facilities – Regulation 23

The type and extent of washing and changing facilities provided should be determined by the type and amount of exposure as indicated by the risk assessment.

If the work is licensable, separate facilities should be provided for the workers working with asbestos. Employers must ensure that adequate changing and showering facilities are provided so that employees can clean and decontaminate themselves completely each time they leave the work area. This includes providing shampoo, soap or gel and towels. The provision of suitable hygiene facilities (also known as a decontamination unit, DCU), should be on site and fully operational before any work (including ancillary work) commences. Maintenance records for DCUs (or copies of them) should be kept on site. The hygiene facility should not leave the site until the job is complete and the certificate of reoccupation has been issued.

The hygiene facility enables the employer to further comply with their duties to prevent the spread of asbestos and reduce the potential exposure of employees and other people to as low as reasonably practicable. The facilities will need to:

(a) have separate changing rooms for dirty, contaminated work clothing and for clean or personal clothing known as 'dirty' and 'clean' areas respectively. The showers should be located between the two changing rooms so that it is necessary to pass through them when going from one changing facility to the other. All doors between each room and those leading to the outside from the 'dirty end' should be self-closing and provide an airtight seal. The 'clean' and 'dirty' ends should be fitted with adequate seating and be of sufficient size for changing purposes;

(b) be designed so that they can be cleaned easily;

(c) be fitted with air extraction equipment which keeps a flow of air from the clean to the dirty areas. The extracted air should be discharged through a HEPA filter;

(d) be adequately heated, lit (i.e. light switches at both the 'clean' and 'dirty' ends) and have internal vents so that air can pass through the unit;

(e) be of sufficient size, including allowance for sufficient and separate storage for personal clothing and protective clothing and equipment in the 'clean' end and sufficient receptacles for contaminated clothing, towels, filters and so on in the 'dirty' end and shower area;

(f) have an adequate supply of clean running hot and cold or warm water, at a suitable pressure, in the showers, and soap or gel, shampoo, nailbrushes and individual dry towels. If gas heating is provided and the heater is mounted inside the unit, it must be a room-sealed type, and not open-flued. Waste water should be filtered before being discharged to the drains. All filters should be treated as asbestos waste;

(g) have shower areas of sufficient size to allow thorough decontamination and to have means to support the power pack of a full face respirator while it is still required to be worn (the power pack support should be out of the direct line of the shower to avoid contact with water and prevent damage to the batteries);

(h) have a wall-mounted mirror in the clean end of the unit; and

(i) have the electricity supply enter it via a 30mA residual current circuit breaker fitted at the point of entry into the unit, and the unit must be effectively earthed when in use.

19.8.22 Storage, distribution and labelling of raw asbestos and asbestos waste – Regulation 24

Waste should be placed in suitable, labelled containers as it is produced. Where practicable, containers should be sealed and the outside should be cleaned before removal from enclosures or the work area, and they should be taken to a suitable and clearly identified secure storage area if they are not being disposed of at once.

Any friable waste should be placed in UN-approved packaging (available in up to 2 tonnes capacity). The *Licensed Contractors' Guide* provides further advice.

Containers must be designed, constructed and maintained to prevent any of the contents escaping during normal handling. For most waste, double plastic sacks are suitable provided they will not split during normal use. It is important that the inner bag is not overfilled, especially when the debris is wet, and each bag should be capable of being securely tied or sealed. Air should be excluded from the bag as far as possible before sealing. Precautions will need to be taken as the exhaust air may be contaminated. Stronger packages are necessary if the waste contains sharp metal fragments or other materials liable to puncture plastic sacks.

Bags containing asbestos waste should be appropriately labelled and transported to a licensed disposal site in an enclosed vehicle, skip or freight container. The specific requirements of various Hazardous Waste Regulations in

England and Wales and the Special Waste Regulations in Scotland should be adhered to, as appropriate.

Asbestos waste must be labelled:

(a) in accordance with the Carriage of Dangerous Goods and Use of Transportable Pressure Equipment Regulations 2004 where those Regulations apply;

(b) where the Regulations in (a) do not apply, in accordance with Schedule 2 of the Asbestos Regulations 2006.

The *Licensed Contractors' Guide* contains more detailed advice on waste handling.

19.8.23 Prohibitions – Regulations 25, 26, 27, 28

Regulations 25 to 28 deal with the prohibitions of certain exposure to asbestos, and supply and use of asbestos or any product containing asbestos. For details see the Regulations and ACoP.

19.8.24 References

A Comprehensive Guide to Managing Asbestos in Premises, HSG 227, HSE Books, 2002, ISBN 9780717623815. http://www.hse.gov.uk/pubns/books/hsg227.htm

Approved Code of Practice Work with Materials containing Asbestos, L143, HSE Books, 2006, ISBN 978 0 7176 6206 0. http://www.hse.gov.uk/pubns/books/l143.htm

Asbestos Essentials, HSG 210 (third edition), HSE Books, 2012, ISBN 9780717665037. http://www.hse.gov.uk/pubns/books/hsg210.htm

Asbestos Essentials task sheets are available on the Asbestos Essentials area of the HSE website: http://www.hse.gov.uk/asbestos/essentials/index.htm.

Asbestos Kills: A Quick Guide to Protecting Yourself, INDG418, HSE Books, ISBN 978 0 7176 6271 5. http://www.hse.gov.uk/pubns/indg418.pdf

Asbestos: The Analysts' Guide for Sampling, Analysis and Clearance Procedures, HSG 248, ISBN 978 0 7176 2875 9, HSE Books, 2005. http://www.hse.gov.uk/pubns/books/hsg248.htm.

Asbestos: the Licensed Contractors' Guide, HSG 247, HSE Books, 2006, ISBN 9780 717628742. http://books.hse.gov.uk/hse/public/saleproduct.jsf?catalogueCode=9780717628742

Asbestos: The Survey Guide, HSG264, Second Edition, 2012, HSE Books, ISBN 9780717665020. http://www.hse.gov.uk/pubns/books/hsg264.htm

http://www.hse.gov.uk/pubns/books/l127.htm

Managing asbestos in buildings: A brief guide, INDG223(rev5), 2012, HSE Books, ISBN 9780717664870. http://www.hse.gov.uk/pubns/indg223.htm

19.9 Classification, Labelling and Packaging of Substances and Mixtures Regulation (European) adopting into EU UN Globally Harmonised System of Classification and Labelling of Chemicals (GHS)

19.9.1 Introduction

Worldwide, there are many different laws on how to identify hazardous chemicals (classification) and how this information should be communicated to users (via labels and safety data sheets – SDSs).

This is often confusing as the same chemical can have different hazard descriptions in different countries. The UN brought together experts from different countries to create the GHS.

The aim of GHS is to have, throughout the world, the same:

▶ criteria for classifying chemicals according to their health, environmental and physical hazards;

▶ hazard communication requirements for labelling and SDSs.

The GHS is not a law, it's an international agreement. To make the GHS legally apply, each country or bloc of countries must adopt the GHS through legislation. European Union (EU) member states agreed to adopt the GHS across the EU through a direct-acting Regulation, the European Regulation (EC) No 1272/2008 on **C**lassification, **L**abelling and **P**ackaging of substances and mixtures. This is also known as the 'CLP Regulation' or just 'CLP'.

19.9.2 The Classification, Labelling and Packaging (CLP) of Substances and Mixtures Regulation

The CLP Regulation:

▶ is a direct-acting regulation in all EU member states, including the UK, and does not require transposing into national legislation;

▶ adopts the majority of the GHS 'hazard classes' (nature of the hazard) and 'hazard categories' (severity of the hazard);

▶ keeps a few aspects of the existing EU system, where these do not contradict the GHS, to maintain existing EU standards (e.g. harmful to the ozone layer);

▶ replaces the requirements of the Dangerous Substances Directive (67/548/EEC) and the Dangerous Preparations Directive (1999/45/EC) relating to the classification, packaging and labelling of substances and preparations, over a transitional period lasting until 1 June 2015;

▶ introduces newly designed hazard symbols, known as pictograms (see Figure 19.1);

19

▶ introduces new labelling phrases, known as 'hazard statements' and 'precautionary statements';

▶ introduces new provisions for a classification and labelling inventory;

▶ maintains the list of harmonised classifications.

The CLP Regulation became law throughout the EU on 20 January 2009 but it did not apply all at once, to give suppliers and their customers time to adapt to the new system until June 2015.

19.9.3 Requirements of the CLP Regulation

The CLP Regulation requires:

▶ substances to be classified, labelled and packaged according to the CLP Regulation from 1 December 2010 onwards;

▶ mixtures to be classified, labelled and packaged according to the CLP Regulation from 1 June 2015 onwards.

The former system required that suppliers must use certain symbols and warning phrases on the label to inform users about the hazards of chemicals.

The CLP Regulation introduces new hazard pictograms (symbols), and new 'signal' words such as 'warning' and 'danger', together with new phrases for the different hazards.

Although many of the CLP pictograms are similar to the existing EU system, they have been re-designed. The new pictograms are diamond shaped, white and black with a red border.

	Example of hazard statement	Example of precautionary statement
	Heating may cause an explosion	Keep away from heat/sparks/open flames/hot surfaces – no smoking
	Heating may cause a fire	Keep only in original container
	May intensify fire; oxidiser	Take any precaution to avoid mixing with combustibles
	Causes serious eye damage	Wear eye protection
	Toxic if swallowed	Do not eat, drink or smoke when using this product
	Toxic to the aquatic life, with long lasting effects	Avoid release to the environment
	**New pictogram**, reflects serious longer term health hazards such as carcinogenicity and respiratory sensitisation, e.g. may cause allergy or asthma symptoms or breathing difficulties if inhaled	In case of inadequate ventilation, wear respiratory protection
	**New pictogram**, refers to less serious health hazards such as skin irritancy/sensitisation and replaces the CHIP X symbol, e.g. may cause an allergic skin reaction	Contaminated work clothing should not be allowed out of the workplace
	**New pictogram**, used when the containers hold gas under pressure, e.g. may explode when heated	None

Figure 19.1 New GHS pictograms with examples of Hazard statements and Precautionary statements

19.9.4 Chemical manufacture and storage – the European Commission's chemical strategy (REACH)

What is REACH?

REACH (Registration, Evaluation, Authorisation and restriction of Chemicals) is the system for controlling chemicals in the EU. It became law in the UK on 1 June 2007. In October 2006, Defra nominated the HSE to be the UK Competent Authority for REACH, working closely with the Environment Agency and other partners to manage certain key aspects of the REACH system in the UK.

The Competent Authority's responsibilities under REACH are to:

- provide advice to manufacturers, importers, downstream users and other interested parties on their respective responsibilities and obligations under REACH (Competent Authorities' helpdesks);
- conduct substance evaluation of prioritised substances and prepare draft decisions;
- propose harmonised Classification and Labelling for carcinogens, mutagens or reproductive toxicants (CMRs) and respiratory sensitisers;
- identify substances of very high concern for authorisation;
- propose restrictions;
- nominate candidates to membership of ECA committees on Risk Assessment and Socio-economic Analysis;
- appoint members for the Member State Committee to resolve differences of opinion on evaluation decisions;
- appoint a member to the Forum for Information Exchange and meet to discuss enforcement matters;
- provide adequate scientific and technical resources to the members of the Committees that they have nominated;
- work closely with the European Chemical Agency in Helsinki.

Aims

REACH has several aims:

- to provide a high level of protection of human health and the environment from the use of chemicals;
- to allow the free movement of substances on the EU market;
- to make the people who place chemicals on the market (manufacturers and importers) responsible for understanding and managing the risks associated with their use;
- to promote the use of alternative methods for the assessment of the hazardous properties of substances;
- to enhance innovation in and the competitiveness of the EU chemicals industry.

No data, no market

A major part of REACH is the requirement for manufacturers or importers of substances to register them with a central European Chemicals Agency (ECHA). A registration package will be supported by a standard set of data on that substance. The amount of data required is proportionate to the amount of substance manufactured or supplied.

If substances are not registered, then the data on them will not be available and, as a result, the company will no longer be able to manufacture or supply them legally; that is no data, no market.

Scope and exemptions

REACH applies to substances manufactured or imported into the EU in quantities of 1 tonne per year or more. Generally, it applies to all individual chemical substances on their own, in preparations or in articles (if the substance is intended to be released during normal and reasonably foreseeable conditions of use from an article). Some substances, e.g. radioactive substances, are specifically excluded. See REACH for details.

Registration

Registration is a requirement on industry (manufacturers/suppliers/importers) to collect and collate specified sets of information on the properties of those substances they manufacture or supply. This information is used to perform an assessment of the hazards and risks that substance may pose and how those risks can be controlled. This information and its assessment are submitted to the ECHA in Helsinki.

Chemicals already existing (those on EINECS (European Inventory of Existing Commercial Chemical Substances) or manufactured in the EU prior to entry into force of REACH) are known as 'phase-in' substances under REACH. These will be registered in three phases according to their tonnage and/or hazardous properties.

Non-phase-in substances (i.e. those not on EINECS, or those which have never been manufactured previously or not pre-registered) were subject to registration by 1 June 2008.

Evaluation

Registration packages (dossiers) submitted under REACH can be evaluated for:

- **Compliance check:** A quality/accuracy check of the information submitted by industry.
- **Dossier evaluation:** A check that an appropriate testing plan has been proposed for substances registered at the higher tonnage levels ($100 tonnes/annum).

19

▶ **Substance evaluation:** An evaluation of all the available data on a substance, from all registration dossiers. This is done by national Competent Authorities on substances that have been prioritised for potential regulatory action because of concerns about their properties or uses.

Substances of very high concern

Some substances have hazards that have serious consequences; for example they cause cancer, or they have other harmful properties and remain in the environment for a long time and gradually build up in animals. These are 'substances of high concern'. One of the aims of REACH is to control the use of such substances.

Authorisation

Authorisation is a feature of REACH that is new to the area of general chemicals management. As REACH progresses, a list of 'substances of very high concern' will be created. Substances on this list cannot be supplied or used unless an authorisation has been granted. A company wishing to market or use such a substance must apply to the ECHA in Helsinki for an authorisation, which may be granted or refused.

Restrictions

Any substance that poses a particular threat can be restricted. Restrictions take many forms, for example from a total ban to not being allowed to supply it to the general public. Restrictions can be applied to any substance, including those that do not require registration. This part of REACH takes over the provisions of the Marketing & Use Directive.

Classification and labelling

An important part of chemical safety is clear information about any hazardous properties a chemical has. The classification of different chemicals according to their characteristics (e.g. those that are corrosive, or toxic to fish) currently follows an established system, which is reflected in REACH. REACH has been written with GHS in mind (see Section 19.9.1 for more information on GHS).

Information in the supply chain

The passage of information up and down the supply chain is a key feature of REACH. Users should be able to understand what manufacturers and importers know about the dangers involved in using chemicals and how to control risks. However, in order for suppliers to be able to assess these risks they need information from the downstream users about how the chemicals are used. REACH provides a framework in which information can be passed both up and down supply chains.

More detailed information about REACH can be found at the HSE REACH website.

Enforcement

Enforcing this very wide-ranging system presents new challenges to regulators across Europe. REACH places new duties on a range of different businesses. Mostly, the new duties will be on manufacturers and importers of chemicals, but there are also requirements for downstream users of chemicals to share information with their suppliers. The HSE play a key role in enforcing REACH, both as the UK Competent Authority and more generally as the UK occupational health and safety regulator.

19.9.5 Further information

European Chemicals Agency information website for general and specific guidance. http://echa.europa.eu/web/guest/support/guidance-on-reach-and-clp-implementation/guidance-in-a-nutshell

Guidance in a nutshell – Compilation of safety data sheets, European Chemicals Agency 2013. http://echa.europa.eu/documents/10162/13643/sds_nutshell_guidance_en.pdf

HSE Chemical Classification website: http://www.hse.gov.uk/chemical-classification/index.htm

HSE's REACH website: http://www.hse.gov.uk/reach/about.htm

HSE's GHS website: http://www.hse.gov.uk/chemical-classification/index.htm

19.10 Confined Spaces Regulations 1997

19.10.1 Introduction

These Regulations concern any work that is carried on in a place which is substantially (but not always entirely) enclosed, where there is a reasonably foreseeable risk of serious injury from conditions and/or hazardous substances in the space or nearby. Every year about 15 people are killed and a number seriously injured across a wide range of industries (e.g. from simple open top pits to complex chemical plants). Rescuers without proper training and equipment, often become the victims.

19.10.2 Definitions

Confined space – means any place, including any chamber, tank, vat, silo, pit, trench, pipe, sewer, flue, well or similar space, in which, by virtue of its enclosed nature, there arises a reasonably foreseeable specified risk.

Specified risk – means a risk to any person at work of:

▶ serious injury arising from a fire or explosion;

- loss of consciousness arising from an increase in body temperature;
- loss of consciousness or asphyxiation arising from gas, fume, vapour or the lack of oxygen;
- drowning arising from the an increase in the level of liquid;
- asphyxiation arising from a free-flowing solid or because of entrapment by it.

19.10.3 Employers' duties – Regulation 3

Duties are placed on employers to:

- comply regarding any work carried out by employees;
- ensure, sfairp, that other persons (e.g. use competent contractors) comply regarding work in the employer's control.

The self-employed also have duties to comply.

19.10.4 Work in confined space – Regulation 4

1. No person at work shall enter a confined space for any purpose unless it is not reasonably practicable to achieve that purpose without such entry.
2. Other than in an emergency, no person shall enter, carry out work or leave a confined space otherwise than in accordance with a safe system of work, relevant to the specified risks.

19.10.5 Risk assessment

Risk assessment is an essential part of complying with these Regulations and must be done (under the MHSWR 1999) to determine a safe system of work. The risk assessment needs to follow a hierarchy of controls for compliance. This should start with the measures – both in design and procedures – that can be adopted to enable any work to be carried out outside the confined space.

The assessment must be done by a competent person and will form the basis of a safe system of work. This will normally be formalised into a specific permit to work, applicable to a particular task. The assessment will involve the following.

(a) The general conditions to assess what may or may not be present. Consider:
  - what have been the previous contents of the space;
  - residues that have been left in the space, for example sludge, rust, scale, and what may be given off if they are disturbed;
  - contamination which may arise from adjacent plant, processes, services, pipes or surrounding land, soil or strata;
  - oxygen deficiency and enrichment. There are very high risks if the oxygen content differs

significantly from the normal level of 20.8%; if it is above this level increased flammability exists; if it is below then impaired mental ability occurs, with loss of consciousness under 16%;
  - that physical dimensions and layout of the space can affect air quality.

(b) Hazards arising directly from the work being undertaken. Consider:
  - the use of cleaning chemicals and their direct effect or interaction with other substances;
  - sources of ignition for flammable dusts, gases, vapours, plastics and the like.

(c) The need to isolate the confined space from outside services or from substances held inside, such as liquids, gases, steam, water, fire extinguishing media, exhaust gases, raw materials and energy sources.

(d) The requirement for emergency rescue arrangements including trained people and equipment.

19.10.6 Safe system of work

The detailed precautions required will depend on the nature of the confined space and the actual work being carried out. The main elements of a safe system of work which may form the basis of a 'permit-to-work' are:

- the type and extent of supervision;
- competence and training of people doing the work;
- communications between people inside, and from inside to outside and to summon help;
- testing and monitoring the atmosphere for hazardous gas, fume, vapour, dust, etc. and for concentration of oxygen;
- gas purging of toxic or flammable substances with air or inert gas such as nitrogen;
- good ventilation, sometimes by mechanical means;
- careful removal of residues using equipment which does not cause additional hazards;
- effective isolation from gases, liquids and other flowing materials by removal of pieces of pipe, blanked off pipes, locked off valves;
- effective isolation from electrical and mechanical equipment to ensure complete isolation with lock off and a tag system with key security. Need to secure against stored energy or gravity fall of heavy presses etc.;
- if it is not possible to make the confined space safe, the provision of personal and respiratory protective equipment;
- supply of gas via pipes and hoses carefully controlled;
- access and egress to give quick unobstructed and ready access and escape;
- fire prevention;
- lighting, including emergency lighting;
- prohibition of smoking;

19

▶ emergencies and rescue;
▶ limiting of working periods and the suitability of individuals.

19.10.7 Emergency arrangements – Regulation 5

Before people enter a confined space, suitable and sufficient rescue arrangements must be set up. These must:

▶ reduce the risks to rescuers sfairp;
▶ include the provision and maintenance of suitable resuscitation equipment designed to meet the specified risks.

To be suitable and sufficient, arrangements will need to cover:

▶ rescue and resuscitation equipment;
▶ raising the alarm, alerting rescue and watch keeping;
▶ safeguarding the rescuers;
▶ fire safety precautions and procedures;
▶ control of adjacent plant;
▶ first-aid arrangements;
▶ notification and consultation with emergency services;
▶ training of rescuers and simulations of emergencies;
▶ size of access openings to permit rescue with full breathing apparatus, harnesses, fall arrest gear and lifelines, which is the normal suitable respiratory protection and rescue equipment for confined spaces.

19.10.8 Training

Specific, detailed and frequent training is necessary for all people concerned with confined spaces, whether they are acting as rescuers, watchers or those carrying out the actual work inside the confined space. The training will need to cover all procedures and the use of equipment under realistic simulated conditions.

19.10.9 Further information

Guidance on Permit-to-Work Systems: A Guide for the Petroleum, Chemical and Allied Industries, HSG250, HSE Books, 2005, ISBN 978 0 7176 2943 5. http://www.hse.gov.uk/pubns/books/hsg250.htm

Respiratory Protective Equipment at Work: A Practical Guide, HSG53 (Fourth Edition), 2013, HSE Books, ISBN 978 0 7176 6454 2. http://www.hse.gov.uk/pubns/books/hsg53.htm

Safe Work in Confined Spaces. Confined Spaces Regulations 1997. Approved Code of Practice, Regulations and Guidance, L101 (Third Edition), 2014, HSE Books, ISBN 978 0 7176 6622 5. http://www.hse.gov.uk/pubns/books/l101.htm

Safe work in confined spaces, INDG258(rev 1), 2013, HSE Books, ISBN 7176 1442 5. http://www.hse.gov.uk/pubns/indg258.pdf

19.11 Construction (Design and Management) Regulations 2015 (CDM 2015)

19.11.1 Introduction and background

The CDM 2015 Regulations were introduced in April 2015.

The policy objectives behind CDM 2015 are to:

▶ maintain or improve worker protection;
▶ simplify the regulatory package;
▶ improve health and safety standards on small construction sites;
▶ implement the Temporary or Mobile Construction Sites Directive (TMCSD) in a proportionate way;
▶ discourage bureaucracy; and
▶ meet better regulation principles.

The HSE commissioned extensive external research which was published in April 2012 and along with the other elements of the evaluation the broad conclusions were that:

▶ CDM 2007 was viewed more positively by duty-holders than the 1994 version;
▶ its broad structure was fit for purpose;
▶ problems generally arose through mis- and over-interpretation of the Regulations;
▶ significant concerns remained, however, in several areas:
  ▷ the Regulations had not borne down on bureaucracy as hoped;
  ▷ the Regulations had led to an industry approach to competence which was heavy-handed and in many cases burdensome, particularly on SMEs;
  ▷ the coordination function in the pre-construction phase was not in many cases well-embedded.

The HSE considered the implementation of the Temporary or Mobile Construction Sites Directive (TMCSD) and the Government's policy on implementation of EU Directives. The UK remains committed to fully implementing EU Directives and the proposed changes to the Regulations will meet that aim.

The balance of where serious and fatal injuries occur has shifted dramatically in the past 10–15 years. Two-thirds or more of fatalities now occur on small sites – sites where fewer than 15 people work – which is the reverse of the historical picture. The larger, more structured part of the industry has made significant progress in improving its management of health and safety risks over this time frame. Its motivation for achieving higher standards is often one of continuous improvement and innovation leading to best practice, rather than just meeting regulatory requirements.

19.11.2 Main changes and outline of Regulations

The main changes in comparison with the CDM Regulations 2007 include the following:

(a) These Regulations apply to all construction projects in Great Britain, and all clients whether or not a person is acting in the course or furtherance of a business (Regulation 2(1) and 3(1)).

(b) If a client is a domestic client, the duties in Regulations 4(1) to (7) and 6 will be carried out by the contractor, or principal contractor depending on the number of contractors. The client can agree in writing for the principal designer to carry out the duties.

(c) Site surveys and pre-construction archaeological investigations are not included within the scope of the definition of construction work (Regulation 2(1)).

(d) Pre-construction information has now been defined in Regulation 2(1) rather than in a Schedule.

(e) The role of a CDM coordinator has been omitted and instead a new role of a principal designer has been created (Regulation 2(1) and 5).

(f) The client's duty to appoint a principal designer or principal contractor is triggered where there is more than one contractor (Regulation 5), rather than the previous threshold for notification under the CDM Regulations 2007.

(g) The duties of the principal designer are provided for in Regulation 11.

(h) The duties of designers are retained in Regulation 9.

(i) A construction phase plan is required for all projects.

(j) Specific risks like excavation or working at height must have control measures included in the construction phase plan (Regulation 12(2) and Schedule 3).

(k) The principal contractor's duties are set out in Regulations 13 and 14.

(l) The duties of contractors remain largely the same as the CDM Regulations 2007 and are set out in Regulation 15.

(m) The notification requirement has been amended and is now provided for in Regulation 6.

(n) The requirement for the contents of inspection reports are now provided for in Regulation 24 rather than a separate Schedule (previously Schedule 3 of the CDM Regulations 2007 set out the requirements).

(o) The Secretary of State will carry out a review of the Regulations in accordance with the provisions of Regulation 39.

The Regulations are divided into five parts. Part 1 of the Regulations deals with matters of interpretation and application. The Regulations apply to all construction work in Great Britain and its territorial waters, and apply to both employers and self-employed without distinction.

Part 2 covers general client duties which apply to all construction projects, including the requirements for notification.

Part 3 sets out health and safety duties and roles which apply to all construction projects.

Part 4 sets out the general requirements for all construction sites, and covers physical safeguards which need to be provided to prevent danger. Duties to achieve these standards are held by contractors who actually carry out the work, irrespective of whether they are employers or are self-employed. Duties are also held by those who do not do construction work themselves, but control the way in which the work is done. In each case, the extent of the duty is in proportion to the degree of control which the individual or organisation has over the work in question.

This does not mean that everyone involved with design, planning or management of the project legally must ensure that all the specific requirements in Part 4 are complied with. They only have duties if, in practice, they exercise sufficient control over the actual working methods, safeguards and site conditions. If, for example, a client specifies that a particular job is done in a particular way then the client will have a duty to make sure that their instructions comply with the requirements.

Contractors must not allow work to start or continue unless the necessary safeguards are in place, whether they are providing the safeguards, or using something, for example a scaffold, supplied by someone else.

Part 5 of the Regulations covers issues of enforcement in respect of fire; transitional provisions which will apply during the period when the Regulations come into force; review; amendments and revocations of other legislation.

19.11.3 Definition of construction and client – Regulation 2

Construction work

'Construction work' means the carrying out of any building, civil engineering or engineering construction work and includes:

(a) the construction, alteration, conversion, fitting-out, commissioning, renovation, repair, upkeep, redecoration or other maintenance (including cleaning which involves the use of water or an abrasive at high pressure or the use of corrosive or toxic substances), decommissioning, demolition or dismantling of a structure;

(b) the preparation for an intended structure, including site clearance, exploration, investigation (but not site survey) and excavation (but not preconstruction archaeological investigations), and the clearance or preparation of the site or structure for use or occupation at its conclusion;

(c) the assembly on site of prefabricated elements to form a structure or the disassembly on site of pre-fabricated elements which, immediately before such disassembly, formed a structure;

(d) the removal of a structure or of any product or waste resulting from demolition or dismantling of

19

a structure or from disassembly of pre-fabricated elements which immediately before such disassembly formed such a structure;

(e) the installation, commissioning, maintenance, repair or removal of mechanical, electrical, gas, compressed air, hydraulic, telecommunications, computer or similar services which are normally fixed within or to a structure.

This means that there are some things which may take place on a construction site, like tree planting and horticulture work, putting up marquees, positioning lightweight panels in an open plan office, archaeological exploration and surveying, which are not construction work.

Client

'Client' means any person for whom a project is carried out; 'domestic client' means a client for whom a project is being carried out which is not in the course or furtherance of a business of that client.

Note:

Typical operating and owning costs of a building are in the ratio:

- ▶ 1 for construction costs
- ▶ 5 for maintenance and building operating costs
- ▶ 200 for business operating costs.

Source: Royal Academy of Engineering report on 'Long Term costs of Owning and Using Buildings'.

Part 2 – Client duties

19.11.4 Client duties for managing projects – Regulation 4

A client must:

- ▶ make arrangements for managing a project to ensure that construction work is carried out safely sfairp;
- ▶ ensure the welfare requirements in Schedule 2 are complied with;
- ▶ take reasonable steps to ensure that the arrangements are maintained and reviewed throughout the project;
- ▶ provide pre-construction information to every designer and contractor appointed or being considered for appointment;
- ▶ ensure before the construction phase begins that the principal contractor (or contractor if only one) draws up a construction phase plan;
- ▶ ensure that the principal designer prepares an appropriate health and safety file for the project which:
  - ▷ includes information which may be needed for subsequent projects;
  - ▷ is revised from time to time;
  - ▷ is kept available for inspection;

- ▷ is passed on if interest in the project passes to another client. If a client disposes of the client's interest in the structure, the client complies with the duty in paragraph 5(b)(iii) by providing the health and safety file to the person who acquires the client's interest in the structure and ensuring that that person is aware of the nature and purpose of the file;
- ▶ ensure that the principal designer complies with their duties under Regulations 11 and 12;
- ▶ ensure that the principal contractor complies with Regulations 12 to 14.

19.11.5 Appointment of principal designer and contractor – Regulation 5

Where it is foreseeable that more than one contractor will be working on the project at any time the client must appoint in writing:

- ▶ A principal designer for the responsibilities in Regulations 11 and 12.
- ▶ A principal contractor for the responsibilities in Regulations 12 to 14.

If an appointment is not made the Client must fulfil either/or both of these roles. This part of Regulation 5 (3) and (4) does not apply to domestic clients.

19.11.6 Notification – Regulation 6

The client must notify the HSE using the particulars in Schedule 1 (Figure 19.2(a)) if the construction work on a site is scheduled to:

- ▶ last longer than 30 working days and have more than 20 workers working simultaneously at any point in the project; or
- ▶ exceed 500 person days.

The notice must be clearly displayed on the site in a form that can be read by any worker engaged in the construction work. It must be updated if necessary.

19.11.7 Application to domestic clients – Regulation 7

Unless there is a written agreement between the client and the principal designer, the duties in Regulations 4(1) to (7) and 6, must be carried out by:

- ▶ the contractor or principal contractor where there is more than one;
- ▶ the principal designer where there is a written agreement that they will fulfil those duties.

Where no appointment is made:

- ▶ the designer in control of the pre-construction phase is the principal designer; and
- ▶ the contractor in control of the construction phase is the principal contractor.

Regulation 5(3) and (4) does not apply (see 19.11.5).

SCHEDULE 1
PARTICULARS TO BE NOTIFIED TO THE EXECUTIVE

1. Date of forwarding.
2. Exact address of the construction site.
3. The name of the local authority where the site is located.
4. A brief description of the project and the construction work which it includes.
5. Contact details of the client (name, address, telephone number and any e-mail address).
6. Contact details of the CDM co-ordinator (name, address, telephone number and any e-mail address).
7. Contact details of the principal contractor (name, address, telephone number and any e-mail address).
8. Date planned for the start of the construction phase.
9. The time allowed by the client to the principal contractor referred to in Regulation 15(b) for planning and preparation for construction work.
10. Planned duration of the construction phase.
11. Estimated maximum number of people at work on the construction site.
12. Planned number of contractors on the construction site.
13. Name and address of any contractor already appointed.
14. Name and address of any designer already engaged.
15. A declaration signed by or on behalf of the client that he is aware of his duties under these Regulations.

Figure 19.2 (a) CDM 2015 – Schedule 1

Part 3 – Health and safety duties and roles

19.11.8 General duties – Regulation 8

Designers and contractors (including principal) appointed to work on a project must have: skills, knowledge and experience and, if an organisation, the organisational capability to fulfil the role they are given. They must not accept the appointment unless they fulfil these conditions.

Any person responsible for appointing a designer or contractor must take reasonable steps to satisfy themselves that the designer or contractor fulfils the conditions of the paragraph above.

Any duty-holders must cooperate with others on the same or adjoining construction sites.

A person involved in a project has a duty to report anything they are aware of which could endanger them or others.

Information must be comprehensible and provided as soon as is practicable.

19.11.9 Duties of designers – Regulations 9 and 10

A designer must:

▶ before starting work be satisfied that the client is aware of their duties under CDM 2015;
▶ must take into account the general principles of prevention to eliminate risks for any one:
  ▷ working or liable to be affected by construction work;
  ▷ maintaining or cleaning a structure; or

  ▷ using a structure designed as a workplace;
▶ if elimination is not possible the designer must sfairp:
  ▷ take steps to reduce and control the risks through the design process;
  ▷ provide information about the risks to the principal designer;
  ▷ ensure appropriate information is contained in the health and safety file;
▶ provide with the design sufficient information to help clients, other designers and contractors to fulfil their duties under CDM 2015.

Regulation 10 requires that if a design comes from outside Great Britain the person who commissions it or the client must ensure that Regulation 9 is complied with.

19.11.10 Duties of principal designer in relation to health and safety at the pre-construction phase – Regulation 11

A principal designer must plan, manage, monitor and coordinate the pre-construction phase of a project, taking into account the general principles of prevention and the content of the construction phase plan and health and safety file to ensure:

▶ sfairp, the project is carried out without risks to health and safety with regard to design, technical, organisational and time required aspects;
▶ identify and eliminate or control sfairp the foreseeable risks to people:
  ▷ doing or affected by construction work,
  ▷ maintaining or cleaning a structure; or
  ▷ using a structure designed as a workplace;
▶ all designers comply with their duties in Regulation 9;
▶ all people concerned with the pre-construction phase cooperate with the client, the principal designer and each other;
▶ that they assist the client to provide the pre-construction information and provide pre-construction information to every designer and contractor appointed or being considered, in a convenient form;
▶ liaison with the principal contractor for the duration of the project particularly over the preparation of the construction phase plan.

19.11.11 Construction phase plan and health and safety file

The principal contractor must draw up (or make arrangements for) a construction phase plan in the pre-construction phase period. This plan must:

▶ set out health and safety arrangements;
▶ set out site rules taking account of any industrial activity on the site;

19

SCHEDULE 3
WORK INVOLVING PARTICULAR RISKS

1. Work which puts workers at risk of burial under earthfalls, engulfment in swampland or falling from a height, where the risk is particularly aggravated by the nature of the work or processes used or by the environment at the place of work or site;
2. Work which puts workers at risk from chemical or biological substances constituting a particular danger to the safety or health of workers or involving a legal requirement for health monitoring;
3. Work with ionizing radiation requiring the designation of controlled or supervised areas under regulation 16 of the Ionising Radiations Regulations 1999;
4. Work near high voltage power lines;
5. Work exposing workers to the risk of drowning;
6. Work on wells, underground earthworks and tunnels;
7. Work carried out by divers having a system of air supply;
8. Work carried out by workers in caissons with a compressed air atmosphere;
9. Work involving the use of explosives;
10. Work involving the assembly or dismantling of heavy prefabricated components.

Figure 19.2 (b) CDM 2015 – Schedule 3

▶ include specific measures for work within Schedule 3 (Figure 19.2(b)) to the Regulations.

And the:

▶ Principal designer must assist the principal contractor to prepare the construction phase plan by providing all relevant information from the client and designers;
▶ Principal contractor must review, update and revise as necessary the construction phase plan;
▶ Principal designer must prepare a health and safety file during the pre-construction phase;
▶ Principal designer must review, update and revise the health and safety file as necessary;
▶ Principal contractor must provide relevant information to the principal designer for the health and safety file.

And:

▶ If the appointment of the principal designer terminates before the end of the project the health and safety file must be passed on to the principal contractor who must then keep it reviewed, updated and revised as needed;
▶ At the end of the project the health and safety file must be passed to the client.

19.11.12 Duties of the principal contractor in relation to health and safety at the construction phase – Regulation 13

The principal contractor must plan, manage, monitor and coordinate the construction phase taking into account the general principles of prevention to ensure that sfairp:

▶ the project is carried out without risks to health and safety with regard to design, technical, organisational and time required aspects;
▶ cooperation between contractors is achieved;
▶ the implementation of legal requirement by contractors is coordinated;
▶ employers and the self-employed apply general principals of prevention consistently and follow the construction phase plan;
▶ a suitable site induction is provided;
▶ access to unauthorised persons is controlled;
▶ Schedule 2 welfare facilities are provided throughout the construction phase;
▶ there is liaison with the principal designer throughout the project particularly regarding information for the health and safety file and management of the construction phased plan.

19.11.13 Principal contractor's duties to consult and engage with workers – Regulation 14

The principal contractor must:

▶ make and maintain arrangements to enable people on site to cooperate effectively to develop safety measures and check their effectiveness;
▶ consult workers or their representatives in good time on health and safety issues unless already consulted by their own employer;
▶ ensure that workers or their representatives can inspect and take copies of health and safety related information for the project except:
  ▷ if disclosure would be against the interests of national security;
  ▷ if disclosure would contravene a legal prohibition;
  ▷ where the information relates specifically to an individual without their consent;
  ▷ where disclosure would cause substantial damage to the principal contractor's undertaking or other undertaking if they had supplied the information;
  ▷ where the information was obtained by the contractor in connection with legal proceedings.

19.11.14 Duties of contractors – Regulation 15

Contractors must:

▶ not carry out construction work on a project unless satisfied that the client is aware of their duties under CDM 2015;
▶ plan, manage and monitor the way construction work is done to ensure it is safe sfairp taking into account design, technical, organisational and time required aspects;
▶ where there is only one contractor working on the project, ensure that a construction phased plan is drawn up prior to setting up the site which must fulfil

requirements of Regulation 12(2) which includes specific measures of Schedule 3 (see previous box);

▶ not employ or appoint a person unless they have or are obtaining the necessary skills, knowledge, training and experience to carry out the allocated tasks safely;

▶ provide any employees or persons in their control with appropriate supervision, information and instruction to ensure safety of the project including:
  ▷ suitable site induction;
  ▷ emergency procedures;
  ▷ information on risks to their health and safety either identified by their risk assessments or from another contractor's operations;
  ▷ any other information necessary;

▶ not start work until access to unauthorised persons is controlled;

▶ ensure that Schedule 2 welfare facilities are complied with for employees and others under their control.

19.11.15 Part 4 General requirements for all construction sites

The CHSW Regulations 1996 were revoked by CDM 2007 which has been retained by CDM 2015. Without the work at height provisions (Regulations 6, 7 and 8), their requirements form the basis of Part 4 and Schedule 2 (Welfare facilities) of the CDM 2015 Regulations.

Some issues already covered by the Workplace Regulations are particularly important in construction. These include traffic management and lighting. Because these are so important, CDM 2015 duplicates aspects of the Workplace Regulations requirements, although that was not legally necessary.

The issues covered by Part 4 apply to all construction projects and are as follows.

General application of Regulations 17–35 – Regulation 16

Duties to comply with this part are placed on the contractor or other person who controls the way in which construction work is carried on.

Duty placed on workers to report any defect they are aware of which may endanger the health and safety of themselves or another person.

Safe places of construction work – Regulation 17

This regulation requires sfairp:

▶ safe access and egress to places of work;
▶ maintenance of access and egress;
▶ that people are prevented from gaining access to unsafe access or workplaces;

▶ provision of safe places of work with adequate space and suitable for workers.

Good order and site security – Regulation 18

Requires that all parts of a construction be kept, sfairp, in good order and in a reasonable state of cleanliness.

Where necessary in the interests of health and safety the site's perimeter should be identified, signed and fenced off.

No timber or other materials with projecting nails or similar sharp objects shall be used or allowed to remain if they could be a source of danger.

Stability of structures – Regulation 19

All practical steps must be taken to ensure that:

▶ any new or existing structure which may be, or become, weak does not collapse accidentally;
▶ temporary supporting structures are designed, installed and maintained so as to withstand all foreseeable loads;
▶ a structure is not overloaded so as to be a source of danger.

Demolition or dismantling – Regulation 20

Demolition or dismantling must be planned and carried out so as to prevent danger or reduce it to as low a level as is reasonably possible.

Arrangements must be recorded in writing before the work takes place.

Explosives – Regulation 21

Explosives, sfairp, must be stored, transported and used safely and securely.

Explosives may only be fired if suitable and sufficient steps have been taken to ensure no one is exposed to risk of injury.

Excavations – Regulation 22

All practicable steps shall be taken to prevent danger to people including, where necessary, supports or battering to ensure that:

▶ excavations do not collapse accidentally;
▶ no material from side, roof or adjacent ground is dislodged or falls;
▶ a person is prevented from being buried or trapped by a fall of material;
▶ persons, work equipment or any accumulation of material are prevented from falling into the excavation;
▶ any part of an excavation or ground adjacent is not overloaded by work equipment or material.

No construction work may be carried out in an excavation where any supports or battering has been

19

provided unless it has been inspected by a competent person:

- ▶ at the start of the shift (only one report required every 7 days);
- ▶ after any event likely to affect the strength or stability of the excavation;
- ▶ after any material accidentally falls or is dislodged;
- ▶ the person inspecting is satisfied that it is safe.

Where the person who carries out the inspection has informed the person on whose behalf the inspection was carried out, work must cease until the relevant matters have been corrected.

Cofferdams and caissons – Regulation 23

Every cofferdam or caisson must be:

- ▶ suitably designed and constructed;
- ▶ appropriately equipped so that people can gain shelter or escape if water or material enters;
- ▶ properly maintained.

A cofferdam or caisson may be used to carry out construction only if:

- ▶ it and any work equipment and materials which affect its safety have been inspected by a competent person:
  - ▷ at the start of the shift (only one report required every 7 days);
  - ▷ after any event likely to affect its strength or stability;
  - ▷ the person inspecting is satisfied that it is safe.

Where the person who carries out the inspection has informed the person on whose behalf the inspection was carried out, work must cease until the relevant matters have been corrected.

Reports of inspection – Regulation 24

Before the end of the shift in which the report was completed, the person who carries out the inspection under Regulations 22 or 23 must:

- ▶ inform the person for whom they carried out the inspection if they are not satisfied that construction work can be carried out safely;
- ▶ prepare a report with the particulars shown in Figure 19.3.

A copy of the report must be provided within 24 hours of the relevant inspection. If the inspector is an employee or works under someone else's control, that person must ensure that these duties are performed.

Records must be kept at the site where the inspection was carried out until the work is complete and then for three months. Extracts must be provided for an inspector as and when they require. Only one report is required if more than one report is produced under each shift requirement of Regulation 22(4)(a)(i) or 23(2)(a)(i) within 7 days.

Inspection reports which must include—

(i) the name and address of the person on whose behalf the inspection was carried out;
(ii) the location of the place of construction work inspected;
(iii) a description of the place of construction work or part of that place inspected (including any work equipment and materials);
(iv) the date and time of the inspection;
(v) the details of any matter identified that could give rise to a risk to the safety of any person;
(vi) details of any action taken as a result of any matter identified in paragraph (v);
(vii) the details of any further action considered necessary; and
(viii) the name and position of the person making the report.

Figure 19.3 Content of inspection reports

Energy distribution installations – Regulation 25

Energy distribution installations must be suitably located, checked and clearly indicated.

Where there is a risk from electric power cables:

- ▶ they should be routed away from the risk areas; or
- ▶ the power should be cut off; or
- ▶ if the above safety measures are not reasonably practicable suitable warning notices, with barriers to exclude work equipment and suspended protections where vehicles have to pass underneath, should be provided; or
- ▶ something equally as safe must be done.

No construction work which is liable to create a risk to health or safety from an underground service, or from damage to or disturbance of it, shall be carried out unless suitable and sufficient steps have been taken to prevent such risk, sfairp.

Prevention of drowning – Regulation 26

Where persons could fall into water or other liquid, steps must be taken to:

- ▶ prevent, sfairp, people from falling;
- ▶ minimise the risk of drowning if people fall;
- ▶ provide, use and maintain suitable rescue equipment to ensure prompt rescue.

Transport to work by water must be suitable and sufficient and any vessel must not be overcrowded or overloaded.

Traffic routes – Regulation 27

Every construction site shall be organised in such a way that, sfairp, pedestrians and vehicles can move safely.

Traffic routes shall be suitable for the persons or vehicles using them, sufficient in number, in suitable positions and of sufficient size.

Steps must be taken to ensure that:

- ▶ pedestrians or vehicles may use them without causing danger to the health or safety of persons near them;

- any door or gate for pedestrians which leads onto a traffic route is sufficiently separated from it to enable them from a place of safety to see any approaching vehicle;
- there is sufficient separation between vehicles and pedestrians to ensure safety or, where this is not reasonably practicable:
  (i) there are provided other means for the protection of pedestrians;
  (ii) there are effective arrangements for warning any person liable to be crushed or trapped by any vehicle of its approach;
- any loading bay has at least one exit point for the exclusive use of pedestrians;
- where it is unsafe for pedestrians to use a gate intended primarily for vehicles, one or more doors for pedestrians provided in the immediate vicinity of the gate, is clearly marked and is kept free from obstruction.

Every traffic route must be:

- indicated by suitable signs;
- regularly checked;
- properly maintained.

No vehicle shall be driven on a traffic route unless, sfairp, that traffic route is free from obstruction and permits sufficient clearance.

Vehicles – Regulation 28

The unintended movement of any vehicle must be prevented or controlled.

Suitable and sufficient steps shall be taken to ensure that, where a person may be endangered by the movement of any vehicle, the person having effective control of the vehicle shall give suitable and sufficient warning.

Any vehicle being used for the purposes of construction work, shall, when being driven, operated or towed:

- be driven, operated or towed in a safe manner;
- be loaded so that it can be driven, operated or towed safely.

No person shall ride on any vehicle being used for the purposes of construction work except in a safe place provided.

No person shall remain on any vehicle during the loading or unloading of any loose material unless a safe place of work is provided and maintained.

Suitable measures must be taken to prevent any vehicle from falling into any excavation or pit, or into water, or overrunning the edge of any embankment or earthwork.

Prevention of risk from fire, flooding and asphyxiation – Regulation 29

Steps must be taken to prevent injury arising from:

- fires or explosions;
- flooding; or
- substances liable to cause asphyxiation.

Emergency procedures – Regulation 30

Where necessary, emergency procedures must be prepared and, where necessary, implemented to deal with any foreseeable emergency. They must include procedures for any necessary evacuation of the site or any part thereof.

The arrangements must take account of:

- the type of work for which the construction site is being used;
- the characteristics and size of the construction site and the number and location of places of work on that site;
- the work equipment being used;
- the number of persons likely to be present on the site at any one time;
- the physical and chemical properties of any substances or materials on or likely to be on the site.

Steps must be taken to make people on site familiar with the procedures and test the procedures at suitable intervals.

Emergency routes and exits – Regulation 31

Where necessary a sufficient number of suitable emergency routes and exits shall be provided to enable any person to reach a place of safety quickly in the event of danger.

An emergency route or exit (or, as appropriate, traffic route) provided must:

- lead as directly as possible to an identified safe area;
- be kept clear and free from obstruction and, where necessary, provided with emergency lighting so that such emergency route or exit may be used at any time;
- be indicated by suitable signs.

Fire detection and fire-fighting – Regulation 32

Where necessary, duty-holders must provide suitable and sufficient:

- fire-fighting equipment; and
- fire detection and alarm systems which shall be suitably located.

The equipment must be examined and tested at suitable intervals, properly maintained and suitably indicated by signs. If non-automatic, it must be easily accessible.

Every person at work on a construction site shall, sfairp, be instructed in the correct use of any fire-fighting equipment which it may be necessary for them to use.

19

Where a work activity may give rise to a particular risk of fire, a person shall not carry out such work unless suitably instructed.

Fire-fighting equipment must be indicated by suitable signs.

Fresh air – Regulation 33

Steps must be taken to ensure, sfairp, that every place of work or approach has sufficient fresh or purified air to ensure that the place or approach is safe and without risks to health. Any plant used for the provision of fresh air must include, where necessary, an effective device to give visible or audible warning of any failure of the plant.

Temperature and weather protection – Regulation 34

Steps must be taken to ensure that, sfairp, during working hours the temperature at any place of work indoors is reasonable having regard to its purpose.

Every place of work outdoors must, where necessary, sfairp, and having regard to the purpose for which that place is used and any protective clothing or work equipment provided, be so arranged that it provides protection from adverse weather.

Lighting – Regulation 35

Every place of work and approach thereto and every traffic route shall be provided with suitable and sufficient lighting, which shall be, sfairp, by natural light.

The colour of any artificial lighting provided shall not adversely affect or change the perception of any sign or signal provided for the purposes of health and safety.

Suitable and sufficient secondary lighting shall be provided in any place where there would be a risk to the health or safety of any person in the event of failure of primary artificial lighting.

19.11.16 Welfare facilities – Schedule 2

Sanitary conveniences

1. (1)   Suitable and sufficient sanitary conveniences must be provided or made available at readily accessible places.

(2)   So far as is reasonably practicable, rooms containing sanitary conveniences must be adequately ventilated and lit.

(3)   So far as is reasonably practicable, sanitary conveniences and the rooms containing them must be kept in a clean and orderly condition.

(4)   Separate rooms containing sanitary conveniences must be provided for men and women, except where and so far as each convenience is in a separate room, the door

of which is capable of being secured from the inside.

Washing facilities

2. (1)   Suitable and sufficient washing facilities, including showers if required by the nature of the work or for health reasons, must, so far as is reasonably practicable, be provided or made available at readily accessible places.

(2)   Washing facilities must be provided –
(a)   in the immediate vicinity of every sanitary convenience, whether or not also provided elsewhere; and
(b)   in the vicinity of any changing rooms required by paragraph 4, whether or not provided elsewhere.

(3)   Washing facilities must include –
(a)   a supply of clean hot and cold, or warm, water (which must be running water so far as is reasonably practicable);
(b)   soap or other suitable means of cleaning; and
(c)   towels or other suitable means of drying.

(4)   Rooms containing washing facilities must be sufficiently ventilated and lit.

(5)   Washing facilities and the rooms containing them must be kept in a clean and orderly condition.

(6)   Subject to sub-paragraph (7), separate washing facilities must be provided for men and women, except where they are provided in a room the door of which is capable of being secured from inside and the facilities in each room are intended to be used by only one person at a time.

(7)   Sub-paragraph (6) does not apply to facilities which are provided for washing hands, forearms and the face only.

Drinking water

3. (1)   An adequate supply of wholesome drinking water must be provided or made available at readily accessible and suitable places.

(2)   Where necessary for reasons of health or safety, every supply of drinking water must be conspicuously marked by an appropriate sign.

(3)   Where a supply of drinking water is provided, a sufficient number of suitable cups or other drinking vessels must also be provided, unless the supply of drinking water is in a jet from which persons can drink easily.

Changing rooms and lockers

4. (1)   Suitable and sufficient changing rooms must be provided or made available at readily accessible places if a worker –

(a) has to wear special clothing for the purposes of construction work; and

(b) cannot, for reasons of health or propriety, be expected to change elsewhere.

(2) Where necessary for reasons of propriety, there must be separate changing rooms for, or separate use of rooms by, men and women.

(3) Changing rooms must –
(a) be provided with seating; and
(b) include, where necessary, facilities to enable a person to dry any special clothing and any personal clothing or effects.

(4) Suitable and sufficient facilities must, where necessary, be provided or made available at readily accessible places to enable persons to lock away –
(a) any special clothing which is not taken home;
(b) their own clothing which is not worn during working hours; and
(c) their personal effects.

Facilities for rest

5. (1) Suitable and sufficient rest rooms or rest areas must be provided or made available at readily accessible places.

(2) Rest rooms and rest areas must –
(a) be equipped with an adequate number of tables and adequate seating with backs for the number of persons at work likely to use them at any one time;
(b) where necessary, include suitable facilities for any woman at work who is pregnant or who is a nursing mother to rest lying down;
(c) include suitable arrangements to ensure that meals can be prepared and eaten;
(d) include the means for boiling water; and
(e) be maintained at an appropriate temperature.

19.11.17 Part 5 General

Enforcement in respect of fire – Regulation 36

This regulation is given in full.

(1) As regards Regulations 29 and 30 the enforcing authority in respect of a construction site which is contained within, or forms part of, premises which are occupied by persons other than those carrying out construction work or any activity related to this work, in so far as those regulations relate to fire are –
(a) in England and Wales the enforcing authority within the meaning of article 25 of the Regulatory Reform (Fire Safety) Order 2005(a) in respect of premises to which that Order applies; or

(b) in Scotland the enforcing authority within the meaning of Section 61 of the Fire (Scotland) Act 2005(b) in respect of premises to which Part 3 of that Act applies.

19.11.18 Further information

Asbestos Essentials HSG 210(third edition), 2012, ISBN 978 0 7176 6503 7. http://www.hse.gov.uk/pubns/books/hsg210.htm

Asbestos: The licensed contractors' guide HSG247, 2006, ISBN 9780 7176 2874 2. http://www.hse.gov.uk/pubns/books/hsg247.htm

Electricity in Construction HSE website http://www.hse.gov.uk/electricity/information/construction.htm

Fire Safety in Construction Work HSG 168rev, 2010, HSE Books, ISBN 978 0 7176 6345 3. http://www.hse.gov.uk/pubns/books/hsg168.htm.

Health and Safety in Construction HSG150 (rev 3), 2006, HSE Books, ISBN 9780 7176 6182 4 (due for revision). http://www.hse.gov.uk/pubns/priced/hsg150.pdf

Health and Safety in Roof Work (Fourth Edition) HSG33, 2012, HSE Books, ISBN 978 0 7176 6527 3. http://www.hse.gov.uk/pubns/books/hsg33.htm

Safe Use of Vehicles on Construction Sites HSG 144, 2009, HSE Books, ISBN 97 0 7176 6291 3. http://www.hse.gov.uk/pubns/books/hsg144.htm

Protecting the public: Your next move HSG151, 2009, ISBN 978 0 7176 6294 4. http://www.hse.gov.uk/pubns/books/hsg151.htm

Guidance on The Construction (Design and Management) Regulations 2015 L153 http://www.hse.gov.uk/pubns/books/l153.htm

Industry Guidance for clients, contractors, designers, principal designers, principal contractors and workers see http://www.citb.co.uk/health-safety-and-other-topics/health-safety/construction-design-and-management-regulations/

19.12 Health and Safety (Consultation with Employees) Regulations 1996

19.12.1 Application

These Regulations apply to all employers and employees in Great Britain except:

▶ where employees are covered by safety representatives appointed by recognised trade unions under the Safety Representatives and Safety Committees Regulations 1977 (note in workplaces where there is trade union recognition but either the trade union has not appointed a safety representative, or the union safety representative does not cover the whole workforce, the Health and

Safety (Consultation with Employees) Regulations will apply);

▶ domestic staff employed in private households;

▶ crew of a ship under the direction of the master.

19.12.2 Employers' duty to consult – Regulation 3

The employer must consult relevant employees in good time with regard to:

▶ the introduction of any measure which may substantially affect their health and safety;

▶ the employer's arrangements for appointing or nominating competent persons under the Management of Health and Safety at Work Regulations 1999;

▶ any information required to be provided by legislation;

▶ the planning and organisation of any health and safety training required by legislation;

▶ the health and safety consequences to employees of the introduction of new technologies into the workplace.

The guidance emphasises the difference between informing and consulting. Consultation involves listening to employees' views and taking account of what they say before any decision is taken.

19.12.3 Persons to be consulted – Regulation 4

Employers must consult with either:

▶ the employees directly; or

▶ one or more persons from a group of employees, who were elected by employees in that group to represent them under these regulations. They are known as 'Representative(s) of Employee Safety' (ROES).

Where ROES are consulted, all employees represented must be informed of:

▶ the names of the ROES;

▶ the group of employees represented.

Employers shall not consult a ROES if:

▶ the ROES does not wish to represent the group;

▶ the ROES has ceased to be employed in that group;

▶ the election period has expired;

▶ the ROES has become incapacitated from carrying out the necessary functions.

If an employer decides to consult directly with employees, they must inform them and ROES of that fact. Where no ROES is elected, employers will have to consult directly.

19.12.4 Duty to provide information – Regulation 5

Employers must provide enough information to enable ROES to participate fully and carry out their functions. This will include:

▶ what the likely risks and hazards arising from their work may be;

▶ reported accidents and diseases etc. under the Reporting of Injuries, Diseases and Dangerous Occurrences Regulations (RIDDOR) 2013;

▶ the measures in place, or which will be introduced, to eliminate or reduce the risks;

▶ what employees ought to do when encountering risks and hazards.

An employer need not disclose information which:

▶ could endanger national security;

▶ violates a legal prohibition;

▶ relates specifically to an individual without their consent;

▶ could substantially hurt the employer's undertaking or infringe commercial security;

▶ was obtained in connection with legal proceedings.

19.12.5 Functions of ROES – Regulation 6

ROES have the following functions (but no legal duties):

▶ to make representations to the employer on potential hazards and dangerous occurrences related to the group of employees represented;

▶ to make representations to the employer on matters affecting the general health and safety of relevant employees;

▶ to represent the group of employees for which they are the ROES in consultation with inspectors.

19.12.6 Training, time off and facilities – Regulation 7

Where an employer consults ROES, they shall:

▶ ensure that each ROES receives reasonable training at the employer's expense;

▶ allow time off with pay during ROES working hours to perform the duties of a ROES and while the ROES is a candidate for election;

▶ provide such other facilities and assistance that ROES may reasonably require.

19.12.7 Civil liability and complaints – Regulation 9 and Schedule 2

Breach of these Regulations does not confer any right of action in civil proceedings subject to Regulation 7(3) and Schedule 2 relating to complaints to industrial tribunals.

A ROES can complain to an industrial tribunal that:

▶ their employer has failed to permit time off for training or to be a candidate for election;

▶ their employer has failed to pay them as set out in Schedule 1 to the Regulations.

19.12.8 Elections

The guidance lays down some ideas for the elections, although there are no strict rules.

ROES do not need to be confined to consultation related to these Regulations. Some employers have ROES sitting on safety committees and taking part in accident investigation similarly to union Safety Representatives.

19.12.9 Further information

Consulting Workers on Health and Safety. Safety Representatives and Safety Committees Regulations 1977 (as amended) and Health and Safety (Consultation with Employees) Regulations 1996 (as amended). Approved Code of Practice and Guidance L146, HSE Books, 2012, ISBN 978 0 7176 6461 0. http://www.hse.gov.uk/pubns/books/l146.htm

Consulting Employees on Health and Safety. A Brief Guide to the Law INDG 232 (rev 2), HSE Books, 2013, ISBN 978 0 7176 6312 5. http://www.hse.gov.uk/pubns/indg232.pdf

Involving Your Workforce in Health and Safety: Good Practice for All Workplaces HSG263, HSE Books, 2008, ISBN 978 0 7176 6227 2. http://www.hse.gov.uk/pubns/books/hsg263.htm

Involving Your Workers in Health and Safety: A Guide for Small Businesses. Available on the HSE's website www.hse.gov.uk/pubns/web35.pdf

19.13 Control of Substances Hazardous to Health Regulations (COSHH) 2002 and 2004 Amendment

19.13.1 Introduction

The 2002 COSHH Regulations updated the 1999 Regulations with a few changes which include: additional definitions like 'inhalable dust' and 'health surveillance'; clarify and extend the steps required under risk assessment; introduce a duty to deal with accidents and emergencies.

In 2004, the Amendment Regulations were passed, but they were not enforced until 2005. The 2004 Amendment Regulations replace Regulation 7(7) and (8) by substituting new requirements to observe principles of good practice for the control of exposure to substances hazardous to health introduced by Schedule 2A, to ensure that workplace exposure limits are not exceeded, and to ensure in respect of carcinogens and asthmagens that exposure is reduced to as low a level as is reasonably practicable. They also introduce a single new Workplace Exposure Limit for substances hazardous to health which replaces occupational exposure standards and maximum exposure limits. The amendment regulations introduce a duty to review

control measures other than the provision of plant and equipment, including systems of work and supervision, at suitable intervals.

A revised 6th edition of the ACoP was issued in late 2013. The key changes are:

▶ revision of the text to make it easier for duty-holders to understand and comply with their legal duties;
▶ revision of the text to take account of the introduction of the Classification, Labelling and Packaging Regulation (CLP);
▶ moving information on the principles of good control practice from an annex to the guidance associated with Regulation 7;
▶ clarification of the requirements of Regulation 9, relating to thorough examination and test of control measures.

COSHH covers most substances hazardous to health found in workplaces of all types. The substances covered by COSHH include:

▶ substances used directly in work activities (e.g. solvents, paints, adhesives and cleaners);
▶ substances generated during processes or work activities (e.g. dust from sanding and fumes from welding);
▶ naturally occurring substances (e.g. grain dust).

But COSHH does **not** include:

▶ asbestos and lead, which have specific regulations;
▶ substances which are hazardous only because they:
  ▷ are radioactive;
  ▷ are simple asphyxiants;
  ▷ are at high pressure;
  ▷ are at extreme temperatures;
  ▷ have explosive or flammable properties (separate Dangerous Substances and Explosive Atmospheres Regulations, DSEAR, cover these);
  ▷ are biological agents if they are not directly connected with work and are not in the employer's control, such as catching flu from a workmate.

19.13.2 Definition of substance hazardous to health – Regulation 2

The range of substances regarded as hazardous under COSHH are:

▶ substances including preparations (mixtures of two or more substances) which are listed in Table 3.2 of part 3 of Annex VI of the CLP Regulation and for which an indication of danger specified for the substance is very toxic, toxic, harmful, corrosive or irritant;
▶ substances with a workplace exposure limit as listed in EH40 published by the HSE;
▶ biological agents (bacteria and other microorganisms) if they are directly connected with the work;
▶ any kind of dust in a concentration specified in the regulations, that is

19

(i) 10 mg/m³, as a time-weighted average over an 8-hour period, of inhalable dust, or

(ii) 4 mg/m³, as a time-weighted average over an 8-hour period, of respirable dust;

▶ any other substance which has comparable hazards to people's health but which, for technical reasons, is not covered by CLP.

Workplace exposure limit for a substance hazardous to health means the exposure limit approved by the HSC for that substance in relation to the specified reference period when calculated by a method approved by the HSC, as contained in HSE publication 'EH/40 Workplace Exposure Limits 2005' as updated from time to time.

19.13.3 Duties under COSHH – Regulation 3

The duties placed on employers under these Regulations are extended to (except for health surveillance, monitoring and information and training) persons who may be on the premises but not employed whether they are at work or not, for example visitors and contractors.

Table 19.3 The employers' duties – COSHH

Duty of employer relating to:	Duty for the protection of:		
	Employees	Other people on the premises	Other people likely to be affected by the work
Assessment (Regulation 6)	Yes	SFAIRP	SFAIRP
Prevention/control of exposure (Regulation 7)	Yes	SFAIRP	SFAIRP
Use of control measures and maintenance, examination and testing of control measures (Regulations 8 and 9)	Yes	SFAIRP	SFAIRP
Monitoring exposure (Regulation 10)	Yes, where required	SFAIRP	No
Health surveillance (Regulation 11)	Yes, where appropriate	No	No
Information, training, etc. (Regulation 12)	Yes	SFAIRP	No
Arrangements to deal with accidents and emergencies (Regulation 13)	Yes	SFAIRP	No

Source: HSE ACoP 6th Edition.

19.13.4 General requirements

There are eight basic steps to comply with COSHH. They are:

1. Assess the risks to health;
2. Decide what precautions are needed;
3. Prevent or adequately control exposure;
4. Ensure that control measures are used and maintained;
5. Monitor the exposure of employees to hazardous substances;
6. Carry out appropriate health surveillance where necessary;
7. Prepare plans and procedures to deal with accidents, incidents and emergencies;
8. Ensure employees are properly informed, trained and supervised.

19.13.5 Steps 1 and 2: assessment of health risk – Regulation 6

No work may be carried out where employees are liable to be exposed to substances hazardous to health unless:

▶ a suitable and sufficient risk assessment, including the steps needed to meet COSHH, has been made.

The assessment must be reviewed and changes made regularly and immediately if:

▶ it is suspected that it is no longer valid;
▶ there has been a significant change in the work to which it relates;
▶ the results of any monitoring carried out in accordance with Regulation 10 show it to be necessary.

The assessment includes:

▶ the hazardous properties of the substance;
▶ information on health effects provided by the supplier, including information contained in any relevant safety data sheet;
▶ the level, type and duration of exposure;
▶ the circumstances of the work, including the amount of the substance involved;
▶ activities, such as maintenance, where there is the potential for a high level of exposure;
▶ any relevant workplace exposure limit or similar occupational exposure limit;
▶ the effect of preventative and control measures which have been or will be taken in accordance with Regulation 7;
▶ the results of relevant health surveillance;
▶ the results of monitoring of exposure in accordance with Regulation 10;
▶ in circumstances where the work will involve exposure to more than one substance hazardous to health, the risk presented by exposure to such substances in combination;
▶ the approved classification of any biological agent; and
▶ such additional information as the employer may need in order to complete the risk assessment.

Where five or more employees are employed, the significant findings must be recorded and the steps taken to comply with Regulation 7 (this includes any revisions).

19.13.6 Step 3: prevention or control of exposure – Regulation 7

Every employer must ensure that exposure to substances hazardous to health is either:

▶ prevented or, where this is not reasonably practicable;
▶ adequately controlled.

Preference must be given to substituting with a safer substance. Where it is not reasonably practicable to prevent exposure, protection measures must be adopted in the following order of priority:

▶ the design and use of appropriate work processes, systems and engineering controls and the provision and use of suitable work equipment;
▶ the control of exposure at source, including adequate ventilation systems and appropriate organisational measures;
▶ where adequate control of exposure cannot be achieved by other means, the additional provision of suitable PPE.

The measures must include:

▶ safe handling, storage and transport (plus waste);
▶ reducing to a minimum required for the work the number of employees exposed, the level and duration of exposure, and the quantity of substance present at the workplace;
▶ the control of the working environment including general ventilation;
▶ appropriate hygiene measures.

Control of exposure shall only be treated as adequate if:

▶ the principles of good practice as set out in Schedule 2A (see Figure 19.4) are applied;
▶ any workplace exposure limit approved for that substance is not exceeded;
▶ for a carcinogen (in Schedule 1 or with risk phrase R45, R46 or R49, likely to be Hazard Statements 340, 350, 351, 360, 361 under CLP Regulation) or sensitiser (Risk phrases R42 or R42/43, likely to be Hazard Statements 317, 334 under CLP Regulation) or asthmagen (section C of HSE publication on Asthmagens ISBN 0 7176 1465 4) exposure is reduced to as low a level as is reasonably practicable.

PPE must conform to the Personal Protective Equipment (EC Directive) Regulations 2002.

Additionally, respiratory protective equipment must:

▶ be suitable for the purpose;
▶ comply with the EC Directive Regulations; or
▶ if there is no requirement imposed by these regulations, be of a type that shall conform to a standard approved by the HSE.

Principles of good practice for the control of exposure to substances hazardous to health (Schedule 2A)

Design and operate processes and activities to minimize emission, release and spread of substances hazardous to health.

Take into account all relevant routes of exposure-inhalation, skin absorption and ingestion-when developing control measures.

Control exposure by measures that are proportionate to the health risk.

Choose the most effective and reliable control options which minimize the escape and spread of substances hazardous to health.

Where adequate control of exposure cannot be achieved by other means, provide, in combination with other control measures, suitable PPE.

Check and review regularly all elements of control measures for their continuing effectiveness.

Inform and train all employees on the hazards and risks from the substances with which they work and the use of control measures developed to minimize the risks.

Ensure that the introduction of control measures does not increase the overall risk to health and safety.

Figure 19.4 Principles of good practice – COSHH

Where exposure to a carcinogen or biological agent is involved, additional precautions are laid down in Regulation 7.

19.13.7 Step 4: use, maintenance, examination and test of control measures – Regulations 8 and 9

Every employer must take all reasonable steps to:

▶ ensure that control measures, PPE or anything else provided under COSHH, are properly used or applied:
▶ carry out visual checks and observations at appropriate intervals;
▶ see that where more than one item of PPE is being worn, the different items are compatible with each other;
▶ supervise employees to ensure that the defined methods of work are being followed;
▶ monitor systems for the effectiveness of controls and prompt remedial action where necessary.

Employees must make full and proper use of control measures, PPE or anything else provided. Employees must also, so far as is reasonable:

▶ follow the defined methods of work;
▶ return items to accommodation and report defects immediately;
▶ remove any PPE, which could cause contamination, before eating, drinking or smoking;
▶ maintain a high standard of personal hygiene, and make proper use of facilities provided for washing, showering or bathing and for eating and drinking.

19

Employers shall also:

▶ properly maintain and keep clean plant and equipment, other engineering controls and PPE;

▶ in the case of the provision of systems of work and supervision or similar, review at suitable intervals and revise as necessary;

▶ carry out thorough examination and tests on engineering controls:

  ▷ in the case of local exhaust ventilation (LEV) at least once every 14 months (except those in Schedule 4 which covers blasting of castings – monthly; dry grinding, polishing or abrading of metal for more than 12 hours per week – 6 months; and jute manufacture – 6 months);

  ▷ in any other case, at suitable intervals;

▶ carry out thorough examination and tests where appropriate (except on disposable items) on respiratory protective equipment at suitable intervals;

▶ keep a record of examination and tests for at least 5 years;

▶ check, properly store, repair/replace and safely dispose of contaminated PPE as necessary.

19.13.8 Step 5: monitoring exposure – Regulation 10

Employers must monitor exposure:

▶ where this is necessary to ensure maintenance of control measures or the protection of health;

▶ specifically for vinyl chloride monomer or chromium plating as required by Schedule 5; and

▶ keep a record of identifiable personal exposures for 40 years and any other exposures for 5 years.

19.13.9 Step 6: health surveillance – Regulation 11

Employers must ensure employees, where appropriate, who are exposed or liable to be exposed, are under health surveillance. It is considered appropriate when:

▶ an employee is exposed (if significant) to substances or processes in Schedule 6;

▶ an identifiable disease or adverse health effect may be related to the exposure, and there is a reasonable likelihood that disease may occur and there are valid disease indication or effect detection methods.

Records of health surveillance containing approved particulars must be kept for 40 years or offered to the HSE if trading ceases. If a person is exposed to a substance and/or process in Schedule 6, the health surveillance shall include medical surveillance by an employment medical adviser or appointed doctor at intervals not exceeding 12 months.

If a medical adviser certifies that a person should not be engaged in particular work they must not be permitted to carry out that work except under specified conditions.

Health records must be available for the individual employee to see after reasonable notice.

Employees must present themselves, during working hours at the employer's expense, for appropriate health surveillance.

An employment medical adviser or appointed doctor has the power to inspect the workplace or look at records for the purpose of carrying out functions under COSHH.

19.13.10 Step 7: accidents and emergencies – Regulation 13

Prepare plans and procedures to deal with accidents, incidents and emergencies involving hazardous substances beyond the risks of day-to-day use. This involves:

▶ first-aid provision and safety drills;

▶ information on emergency arrangements and specific hazards;

▶ suitable warning and other communications to enable appropriate response;

▶ ensuring suitable PPE and other equipment is available;

▶ making sure that the procedure is put into practice in an incident.

19.13.11 Step 8: information, instruction and training – Regulation 12

Where employees are likely to be exposed to substances hazardous to health, employers must provide:

▶ information, instruction and training on the risks to health and the precautions which should be taken (this duty is extended to anyone who may be affected);

▶ information on any monitoring of exposure (particularly if there is a workplace exposure limit where the employee or their representative must be informed immediately);

▶ information on collective results of health surveillance (designed so that individuals cannot be identified).

19.13.12 Defence – Regulation 21

It is a defence under these regulations for a person to show that they have taken all reasonable precautions and exercised all due diligence to avoid the commission of an offence.

19.13.13 Further information

A Step by Step Guide to COSHH Assessment HSG97, HSE Books, 2004, ISBN 978 0 7176 2785 1. http://www.hse.gov.uk/pubns/books/hsg97.htm

Control of Substances Hazardous to Health Approved Code of Practice and Guidance L5 (sixth edition), 2013,

ISBN 9780717665822. http://www.hse.gov.uk/pubns/books/l5.htm

Working with substances hazardous to health INDG136(rev5), 2012, HSE Books, ISBN 9780717664863. http://www.hse.gov.uk/pubns/indg136.pdf.

COSHH Essentials guidance publications: http://www.hse.gov.uk/pubns/guidance/index.htm

COSHH HSE Micro website: http://www.hse.gov.uk/coshh/index.htm

EH40/2005 Workplace Exposure Limits, HSE Books, 2011, ISBN 978 0 7176 6446 7. http://www.hse.gov.uk/pubns/books/eh40.htm

Respiratory Protective Equipment at Work HSG53 (Fourth Edition), HSE Books, 2013, ISBN 978 0 7176 6454 2. http://www.hse.gov.uk/pubns/books/hsg53.htm

Read the label How to find out if chemicals are dangerous. INDG 352 (rev 1) HSE 2010, ISBN 9780 7176 6414 6. http://www.hse.gov.uk/pubns/indg352.pdf

The Control of Substances Hazardous to Health Regulations 2002, SI 2002, 2677, 2004 Amendment http://www.legislation.gov.uk/uksi/2002/2677/contents/made

19.14 Dangerous Substances and Explosive Atmospheres Regulations (DSEAR) 2002

19.14.1 Introduction

DSEAR deals with the prevention of fires, explosions and similar energy-releasing events arising from dangerous substances.

In summary the DSEAR Regulations require employers and the self-employed to:

▶ carry out an assessment of the fire and explosion risks of any work activities involving dangerous substances;

▶ provide measures to eliminate, or reduce sfairp, the identified fire and explosion risks;

▶ apply measures, sfairp, to control risks and to mitigate the detrimental effects of a fire or explosion;

▶ provide equipment and procedures to deal with accidents and emergencies;

▶ provide employees with information and precautionary training.

Where explosive atmospheres may occur:

▶ the workplaces should be classified into hazardous and non-hazardous places; and any hazardous places classified into zones on the basis of the frequency and duration of an explosive atmosphere, and where necessary marked with a sign;

▶ the equipment in classified zones should be safe so as to satisfy the requirements of the Equipment and

Protective Systems Intended for Use in Potentially Explosive Atmospheres Regulations 1996;

▶ the workplaces should be verified by a competent person, as meeting the requirements of DSEAR.

The revised ACoP (August 2013) consolidates the contents of five existing ACoPs into a single new ACoP as proposed by the first consultation. As a consequence, the following four ACoPs have been withdrawn, L134 (Design), L135 (Storage), L136 (Control) and L137 (Maintenance).

Their contents have been consolidated into a revised and updated version of ACoP L138 – Dangerous Substances and Explosive Atmospheres Regulations 2002.

Changes in this edition include:

▶ Some guidance has been assigned ACoP status and vice versa. For example, previously no ACoP material existed for Regulations 7 and 10. However, these changes were introduced to provide clarification on how to comply and do not introduce any new requirements.

▶ Adjustments in light of European and other legislation introduced, amended or revoked after DSEAR came into force such as:
  ▷ General fire safety legislation
  ▷ Classification for labelling and packaging.

▶ Clarification on the scope of the regulations in relation to incompatible substances being outside the scope of DSEAR but within the scope of the Health and Safety at Work Act (HSWA).

19.14.2 Scope of regulations

Dangerous substance

The Regulations give a detailed interpretation of dangerous substance, which should be consulted. In summary it means:

▶ a substance or preparation that, because of its chemical and sometimes physical properties and the way it is present and/or used at work, creates a fire or explosion risk to people; for example, substances like petrol, LPG, paints, cleaners, solvents and flammable gases;

▶ any dusts which could form an explosive mixture in air (not included in a substance or preparation); for example, many dusts from grinding, milling or sanding.

Explosive atmosphere

The Regulations give the definition as:

'A mixture under atmospheric conditions, of air and one or more dangerous substances in the form of gases, vapours, mists or dusts, in which, after ignition has occurred, combustion spreads to the entire unburned mixture.'

19

19.14.3 Application

The Regulations do not in general apply to:

▶ ships under the control of a master (DSEAR does apply when 'shoreside' workers and the ship's crew work together in a British port);

▶ areas used for medical treatment of patients;

▶ many gas appliances used for cooking, heating, hot water, refrigeration, etc. (except an appliance specifically designed for an industrial process), gas fittings;

▶ manufacture, use, transport of explosives or chemically unstable substances;

▶ mine, quarry or borehole activities;

▶ activity at an offshore installation;

▶ the use of means of transport (but the Regulations do cover means of transport intended for use in a potentially explosive atmosphere).

19.14.4 Risk assessment – Regulation 5

The risk assessment required by Regulation 5 must include:

▶ the hazardous properties of the dangerous substance;

▶ supplier information and safety data sheet;

▶ the circumstances of the work including:

 ▷ work processes and substances used and their possible interactions;

 ▷ the amount of substance involved;

 ▷ risks of substances in combination;

 ▷ arrangements for safe handling, storage and transport and any waste which might contain dangerous substances;

▶ high risk maintenance activities;

▶ likelihood of an explosive atmosphere;

▶ likelihood of ignition sources, including electrostatic discharges, being present;

▶ scale of any possible fire or explosion;

▶ any places connected by openings to areas where there could be an explosive atmosphere;

▶ any additional information which may be needed.

The risk assessment should be reviewed regularly.

19.14.5 Elimination or reduction of risks – Regulation 6

Regulation 6 concerns the reduction of risks and tracks the normal hierarchy as follows:

▶ Substitute a dangerous substance by a substance or process which eliminates or reduces the risk, for example the use of water-based paints, or using a totally enclosed continuous process.

▶ Reduce the quantity of dangerous substance to a minimum; for example, only a half day supply in the workroom.

▶ Avoid release of a dangerous substance or minimise releases, for example keeping them in special closed containers.

▶ Control releases at source.

▶ Prevent the formation of explosive atmospheres, including the provision of sufficient ventilation.

▶ Ensure that any releases are suitably collected, contained and removed; suitable LEV in a paint spray booth is an example.

▶ Avoid ignition sources and adverse conditions, for example keep electrical equipment outside the area.

▶ Segregate incompatible dangerous substances, for example oxidising substances and other flammable substances.

Steps must also be taken to mitigate the detrimental effects of a fire or explosion by:

▶ keeping the number of people exposed to a minimum;

▶ avoidance of fire and explosion propagation;

▶ provision of explosion relief systems;

▶ provision of explosion suppression equipment;

▶ provision of very strong plant which can withstand an explosion;

▶ the provision of suitable PPE.

19.14.6 Classification of workplaces – Regulation 7 and Schedule 2

Where an explosive atmosphere may occur, workplaces must be classified into hazardous and non-hazardous places. Table 19.4 shows the zones specified by Schedule 2 of the Regulations.

19.14.7 Accidents, incidents and emergencies – Regulation 8

In addition to any normal fire prevention requirements employers must under these Regulations:

▶ ensure that procedures and first-aid are in place with tested relevant safety drills;

▶ provide information on emergency arrangements including work hazards and those that are likely to arise at the time of an accident;

▶ provide suitable warning and other communications systems to enable an appropriate response, remedial actions and rescue operations to be made;

▶ where necessary, before any explosion condition is reached, provide visual or audible warnings and withdraw employees;

▶ provide escape facilities where the risk assessment indicates it is necessary.

In the event of an accident, immediate steps must be taken to:

▶ mitigate the effects of the event;

▶ restore the situation to normal;

▶ inform employees who may be affected.

Only essential persons may be permitted in the

Table 19.4 Classification zones

Classification zones	
Zone 0	A place in which an explosive atmosphere consisting of a mixture with air of dangerous substances in the form of gas, vapour or mist is present continuously or for long periods.
Zone 1	A place in which an explosive atmosphere consisting of a mixture with air of dangerous substances in the form of gas, vapour or mist is likely to occur in normal operations occasionally.
Zone 2	A place in which an explosive atmosphere consisting of a mixture with air of dangerous substances in the form of gas, vapour or mist is not likely to occur in normal operations but, if it does, will persist for a short period only.
Zone 20	A place in which an explosive atmosphere in the form of a cloud of combustible dust in air is present continuously, or for long periods.
Zone 21	A place in which an explosive atmosphere in the form of a cloud of combustible dust in air is likely to occur in normal operations occasionally.
Zone 22	A place in which an explosive atmosphere in the form of a cloud of combustible dust in air is not likely to occur in normal operations but, if it does occur, will persist for a short period only.

Figure 19.5 Warning sign for places where explosive atmospheres may occur

affected area. They must be provided with PPE, protective clothing and any necessary specialised safety equipment and plant.

19.14.8 Information, instruction and training – Regulation 9

Under Regulation 9 where a dangerous substance is present, an employer must provide:

▶ suitable and sufficient information, instruction and training on the appropriate precautions and actions;
▶ details of the substances, any relevant data sheets and legal provisions;

▶ the significant findings of the risk assessment.

19.14.9 Contents of containers and pipes – Regulation 10

Regulation 10 requires that for containers and pipes (except where they are marked under legislation contained in Schedule 5 to the Regulations) the content and the nature of those contents and any associated hazards be clearly identified.

19.14.10 Further information

The Dangerous Substances and Explosive Atmosphere Regulations 2002, SI 2002, No. 2776, ISBN 9780 11 042957 5

Dangerous Substances and Explosive Atmospheres, L138 (Second edition). Dangerous Substances and Explosive Atmospheres Regulations 2002. Approved Code of Practice, 2013, HSE Books, ISBN 978 0 7176 2203 2. http://www.hse.gov.uk/pubns/priced/l138.pdf

Fire and Explosions – How Safe is Your Workshop. A Short Guide to DSEAR, HSE INDG370(rev1), 2013, HSE Books, www.hse.gov.uk/pubns/indg370.htm

Safe Handling of Combustible Dusts; Precautions Against Explosion HSG 103, HSE Books, 2003, ISBN 978 0 7176 2726 4. http://www.hse.gov.uk/pubns/books/hsg103.htm

Safe Use and Handling of Flammable Liquids HSG 140, HSE Books, 1996, ISBN 978 0 7176 0967 3. http://www.hse.gov.uk/pubns/books/hsg140.htm

Safe Working with Flammable Substances INDG227, HSE Books, 2011, ISBN 978 0 7176 1154 6. www.hse.gov.uk/pubns/indg227.pdf

19.15 Health and Safety (Display Screen Equipment) Regulations 1992 as amended in 2002

19.15.1 General

These Regulations cover the minimum health and safety requirements for the use of display screen equipment (DSE) and are accompanied by a guidance note. They typically apply to computer equipment with either a cathode ray tube or liquid crystal monitors. But any type of display is covered with some exceptions, for example on board a means of transport, or where the main purpose is for screening a film or for a television. Multimedia equipment would generally be covered.

Equipment not specifically covered by these Regulations or where it is not being used by a defined 'user' is, nevertheless, covered by other requirements under the MHSWR and the Provision and Use of Work Equipment Regulations (PUWER).

19

19.15.2 Definitions – Regulation 1

(a) Display Screen Equipment (DSE) refers to any alphanumeric or graphic display screen, regardless of the display process involved.

(b) A user is an employee and an operator is a self-employed person, both of whom habitually use DSE as a significant part of their normal work. Both would be people to whom most or all of the following apply. A person:
  ▷ who depends on the DSE to do their job;
  ▷ who has no discretion as to use or non-use;
  ▷ who needs particular training and/or skills in the use of DSE to do their job;
  ▷ who uses DSE for a continuous spell of an hour or more at a time;
  ▷ who does so on a more or less daily basis;
  ▷ for whom fast transfer of data is important for the job;
  ▷ of whom a high level of attention and concentration is required, in particular to prevent critical errors.

(c) A workstation is an assembly comprising:
  ▷ DSE with or without a keyboard, software or input device;
  ▷ optional accessories;
  ▷ disk drive, telephone, modem, printer, document holder, chair, desk, work surface, etc.;
  ▷ the immediate working environment.

19.15.3 Exemptions – Regulation 1(4)

Exemptions include DSE used in connection with:
- drivers' cabs or control cabs for vehicles or machinery;
- on board a means of transport;
- mainly intended for public operation;
- portable systems not in prolonged use;
- calculators, cash registers and small displays related to the direct use of this type of equipment;
- window typewriters.

19.15.4 Assessment of risk – Regulation 2

Possible hazards associated with DSE are physical (musculoskeletal) problems, visual fatigue and mental stress. They are not unique to display screen work, nor an inevitable consequence of it, and indeed research shows that the risk to the individual from typical display screen work is low. However, as in other types of work, ill-health can result from poor work organisation, working environment, job design and posture, and from inappropriate working methods.

Employers must carry out a suitable and sufficient analysis of users' (regardless of who has provided them) and operators' (provided by the employer) workstations in order to assess the risks to health and safety. The guidance gives detailed information on workstation minimum standards and possible effects on health.

The assessment should be reviewed when major changes are made to software, hardware, furniture, environment or work requirements.

19.15.5 Workstations – Regulation 3

All workstations must meet the requirements laid down in the schedule to the Regulations. This schedule lays down minimum requirements for display screen workstations, covering the equipment, the working environment and the interface between computer and user/operator.

19.15.6 Daily work routine of users – Regulation 4

The activities of users should be organised so that their daily work on DSE is periodically interrupted by breaks or changes of activity that reduce their workload at the equipment.

In most tasks, natural breaks or pauses occur from time to time during the day. If such breaks do not occur, deliberate breaks or pauses must be introduced. The guidance requires that breaks should be taken before the onset of fatigue and must be included in the working time. Short breaks are better than occasional long ones, for example, a 5–10-minute break after 50 to 60 minutes continuous screen and/or keyboard work is likely to be better than a 15-minute break every 2 hours. If possible, breaks should be taken away from the screen. Informal breaks, with time on other tasks, appear to be more effective in relieving visual fatigue than formal rest breaks.

19.15.7 Eyes and eyesight – Regulation 5

Initially on request, employees have the right to a free eye and eyesight test conducted by a competent person where they are:
- already users (as soon as practicable after the request);
- to become a user (before they become a user).

The employer must provide a further eye and eyesight test at regular intervals thereafter or when a user is experiencing visual difficulties which could be caused by working with DSE.

There is no reliable evidence that work with DSE causes any permanent damage to eyes or eyesight, but it may make users with pre-existing vision defects more aware of them.

An eye and eyesight test means a sight test as defined in the Opticians Act 1989. This should be carried out by a registered ophthalmic optician or medical practitioner (normally only those with an ophthalmic qualification do so).

Employers shall provide special corrective appliances to users where:

▶ normal corrective appliances cannot be used;
▶ the result of the eye and eyesight test shows that such provision is necessary.

The guidance indicates that the liability of the employer extends only to the provision of corrective appliances which are of a style and quality adequate for their function. If an employee chooses a more expensive design or multi-function correction appliances, the employer need only pay a proportion of the cost.

Employers are free to specify that users' eye and eyesight tests and correction appliances are provided by a nominated company or optician.

The confidential clinical information from the tests can only be supplied to the employer with the employee's consent.

Vision screen tests can be used to identify people with defects but they are not a substitute for the full eyesight test, and employees have the right to opt for the full test from the outset.

19.15.8 Training – Regulation 6

Employers shall ensure that adequate health and safety training is provided to users and potential users in the use of any workstation and refresh the training following any reorganisation. The guidance suggests a range of topics to be covered in the training. In summary these involve:

▶ the recognition of hazards and risks, including the absence of desirable features and the presence of undesirable ones;
▶ causes of risk and how harm may occur;
▶ what the user can do to correct them;
▶ how problems can be communicated to management;

▶ information on the regulations;
▶ the user's contribution to assessments.

19.15.9 Information – Regulation 7

Operators and users shall be provided with adequate information on all aspects of health and safety relating to their workstation and what steps the employer has taken to comply with the Regulations (insofar as the action taken relates to that operator or user and their work). Under Regulation 7, specific information should be provided as outlined in Table 19.5.

19.15.10 Further information

Display screen equipment Workstation Checklist, 2013, HSE Books, ISBN 9780 7176 6521 1. http://www.hse.gov.uk/pubns/ck1.htm

Work with Display Screen Equipment, Health and Safety (Display Screen Equipment) Regulations 1992 as Amended by the Health and Safety (Miscellaneous Amendments) Regulations 2002 – Guidance on Regulations L26 (second edition), 2003, HSE Books, ISBN 978 0 7176 2582 6. http://www.hse.gov.uk/pubns/books/l26.htm

Working with Display screen equipment (DSE) A brief guide, INDG36 (rev 4), 2013, HSE Books, ISBN 978 0 7176 6222 7. http://www.hse.gov.uk/pubns/indg36.pdf

19.16 Electricity at Work Regulations 1989

The purpose of these Regulations is to require precautions to be taken against the risk of death or personal injury from electricity in work activities. The Regulations impose duties on persons ('duty-holders') in respect of systems, electrical equipment and

Table 19.5 Provision of information under DSE Regulation 7

		Information on					
		Risks from DSE workstation	Risk assessment and measures to reduce the risks Regs 2 and 3	Breaks and activity changes Reg. 4	Eye and eyesight tests Reg. 5	Initial training Reg. 6(1)	Training when workstation modified Reg. 6(2)
Does employer have to provide information to display screen workers who are:	Users employed by the undertaking	Yes	Yes	Yes	Yes	Yes	Yes
	Users employed by other employer	Yes	Yes	Yes	No	No	Yes
	Operators in the undertaking	Yes	Yes	No	No	No	No

Source: HSE.

19

589

conductors and in respect of work activities on or near electrical equipment. They apply to almost all places of work and electrical systems at all voltages.

Guidance on the Regulations is contained in the Memorandum of Guidance on the Electricity at Work Regulations 1989, published by the HSE.

19.16.1 Definitions

(a) **Electrical equipment** – includes anything used to generate, provide, transmit, rectify, convert, conduct, distribute, control, store, measure or use electrical energy.

(b) **Conductor** – means a conductor of electrical energy. It means any material (solid, liquid or gas) capable of conducting electricity.

(c) **System** – means an electrical system in which all the electrical equipment is, or may be, electrically connected to a common source of electrical energy, and includes the source and equipment. It includes portable generators and systems on vehicles.

(d) **Circuit conductor** – term used in Regulations 8 and 9 only, means a conductor in a system which is intended to carry electric current in normal conditions. It would include a combined neutral and earth conductor, but does not include a conductor provided solely to perform a protective connection to earth or other reference point and energised only during abnormal conditions.

(e) **Danger** – in the context of these Regulations means a risk of injury from any electrical hazard.

Every year about 30 people die from electric shock or electric burns at work. Each year several hundred serious burns are caused by arcing where the heat generated can be very intense. In addition, intense ultraviolet radiation from an electric arc can cause damage to the eyes – known as arc-eye. These hazards are all included in the definition of **Danger**.

19.16.2 Duties – Regulation 3

Duties are imposed on employers, self-employed and employees. The particular duties on employees are intended to emphasise the level of responsibility which many employees in the electrical trades and professions are expected to take on as part of their job. Employees are:

▶ to cooperate with their employer so far as is necessary to enable any duty placed on the employer to be complied with (this reiterates Section 7(b) of the HSW Act);

▶ to comply with the provisions of these Regulations insofar as they relate to matters which are within their control. (This is equivalent to duties placed on employers and self-employed where these matters are within their control.)

19.16.3 Systems, work activities and protective equipment – Regulation 4

Systems must, at all times, be of such construction as to prevent danger. Construction covers the physical condition, arrangement of components and design of the system and equipment.

All systems must be maintained so as to prevent danger.

Every work activity, including operation, use and maintenance or work near a system, shall be carried out in a way which prevents danger.

Protective equipment shall be suitable, suitably maintained and used properly.

19.16.4 Strength and capability of equipment – Regulation 5

No electrical equipment may be put into use where its strength and capability may be exceeded in such a way as may give rise to danger, in normal transient or fault conditions.

19.16.5 Adverse or hazardous environments – Regulation 6

Electrical equipment which may be exposed to:

▶ mechanical damage;
▶ the effects of weather, natural hazards, temperature or pressure;
▶ the effects of wet, dirty, dusty or corrosive conditions;
▶ any flammable or explosive substances including dusts, vapours or gases;

shall be so constructed and protected that it prevents danger.

19.16.6 Insulation, protection and placing of conductors – Regulation 7

All conductors in a system which may give rise to danger shall either be suitably covered with insulating material and further protected as necessary, for example against mechanical damage, using trunking or sheathing; or have precautions taken that will prevent danger, for example, being suitably placed like overhead electric power cables, or by having strictly controlled working practices.

19.16.7 Earthing, integrity and other suitable precautions – Regulations 8 and 9

Precautions shall be taken, either by earthing or by other suitable means, for example double insulation, use of safe voltages and earth-free non-conducting environments, where a conductor, other than a circuit conductor, could become charged as a result of either the use of, or a fault in, a system.

An earth conductor shall be of sufficient strength and capability to discharge electrical energy to earth. The conductive part of equipment, which is not normally live but energised in a fault condition, could be a conductor.

If a circuit conductor is connected to earth or to any other reference point nothing which could break electrical continuity or introduce high impedance, for example a fuse, thyristor or transistor, is allowed in the conductor unless suitable precautions are taken. Permitted devices would include a joint or bolted link, but not a removable link or manually operated knife switch without bonding of all exposed metal work and multiple earthing.

19.16.8 Connections – Regulation 10

Every joint and connection in a system shall be mechanically and electrically suitable for its use. This includes terminals, plugs and sockets.

19.16.9 Excess current protection – Regulation 11

Every part of a system shall be protected from excess current, for example short circuit or overload, by a suitably located efficient means such as a fuse or circuit breaker.

19.16.10 Cutting off supply and isolation – Regulation 12

There should be suitably located and identified means of cutting off (switch) the supply of electricity to any electrical equipment and also isolating any electrical equipment. Although these are separate requirements, they could be affected by a single means. The isolator should be capable of being locked off to allow maintenance to be done safely.

Sources of electrical energy (accumulators, capacitors and generators) are exempt from this requirement, but precautions must be taken to prevent danger.

19.16.11 Work on equipment made dead – Regulation 13

Adequate precautions shall be taken to prevent electrical equipment that has been made dead from becoming live while work is carried out on or near the equipment. This will include means of locking off isolators, tagging equipment, permits to work and removing fuses.

19.16.12 Work on or near live conductors – Regulation 14

No persons shall work near a live conductor, except if it is insulated, unless:

▶ it is unreasonable in all the circumstances for it to be dead;

▶ it is reasonable in **all** the circumstances for them to be at work on or near it while it is live;

▶ suitable precautions (including where necessary the provision of suitable protective equipment) are taken to prevent injury.

19.16.13 Working space access and lighting – Regulation 15

Adequate working space means of access and lighting shall be provided for all electrical equipment at which or near which work is being done in circumstances which may give rise to danger. This covers work of any kind. However, when the work is on live conductors, the access space must be sufficient for a person to fall back out of danger and if needed for persons to pass one another with ease and without hazard.

19.16.14 Competence – Regulation 16

Where technical knowledge or experience is necessary to prevent danger, all persons must possess such knowledge or experience or be under appropriate supervision.

19.16.15 Further information

Electrical Safety and You INDG231(rev1), 2012, HSE Books. http://www.hse.gov.uk/pubns/indg231.pdf

Electrical Switch Gear and Safety. A Concise Guide for Users INDG372(rev1), HSE Books, 2013, http://www.hse.gov.uk/pubns/indg372.pdf

Electricity at Work: Safe Working Practices HSG85, 2013 (third edition), HSE Books, ISBN 978 0 7176 6581 5. www.hse.gov.uk/pubns/books/hsg85.htm.

Guidance on Safe Isolation Procedures, SELECT, http://www.select.org.uk ISBN 978 0 7176 1272 4.

Maintaining Portable and Transportable Electrical Equipment HSG107 (third edition), 2013, HSE Books, ISBN 978 0 7176 6606 5. http://www.hse.gov.uk/pubns/books/hsg107.htm

Maintaining Portable Electrical Equipment in Low-risk Environments INDG236(rev 3), 2013, HSE Books. http://www.hse.gov.uk/pubns/indg236.pdf

Memorandum of Guidance on the Electricity at Work Regulations, Guidance on Regulations HSR25, 2007, HSE Books, ISBN 978 0 7176 6228 9. http://www.hse.gov.uk/pubns/books/hsr25.htm

Safety in Electrical Testing at Work, General Guidance INDG354(rev1), 2013, HSE Books. http://www.hse.gov.uk/pubns/indg354.pdf

Avoidance of danger from overhead powerlines GS6 (Fourth Edition), 2013. http://www.hse.gov.uk/pubns/gs6.htm

19

(a) 'Premises' includes any place and, in particular, includes:

- any workplace;
- any vehicle, vessel, aircraft or hovercraft;
- any installation on land (including the foreshore and other land intermittently covered by water), and any other installation (whether floating, or resting on the seabed or the subsoil thereof, or resting on other land covered with water or the subsoil thereof);
- any tent or movable structure.

(b) 'Risk' means the risks to the safety of persons from fire. 'Safety' means the safety of persons in the event of fire. 'Workplace' means any premises or parts of premises, not being domestic premises, used for the purposes of an employer's undertaking and which are made available to an employee of the employer as a place of work and includes:

- any place within the premises to which such employee has access while at work;
- any room, lobby, corridor, staircase, road or other place:
  - ▷ used as a means of access to or egress from that place of work; or
  - ▷ where facilities are provided for use in connection with that place of work, other than a public road.

(c) 'Responsible person' means:

- in relation to a workplace, the employer, if the workplace is to any extent under their control;
- in relation to any premises not falling within paragraph (a):
  - ▷ the person who has control of the premises (as occupier or otherwise) in connection with the carrying on by them of a trade, business or other undertaking (for profit or not); or
  - ▷ the owner, where the person in control of the premises does not have control in connection with the carrying on by that person of a trade, business or other undertaking.

(d) 'General fire precautions' in relation to premises means:

- measures to reduce the risk of fire and the risk of the spread of fire;
- the means of escape from the premises;
- measures for securing that, at all material times, the means of escape can be safely and effectively used;
- measures in relation to fighting fires;
- the means for detecting fires and giving warning in case of fire;
- action to be taken in the event of fire, including:
  - ▷ instruction and training of employees;
  - ▷ measures to mitigate the effects of the fire.

These issues do not cover process-related fire risks including:

- the use of plant or machinery; or
- the use or storage of any dangerous substance.

Duties are placed on responsible persons in a workplace and in premises which are not workplaces to the extent that they have control over the premises. The order does not apply (Article 8) to:

- domestic premises;
- an offshore installation;
- a ship, in respect of the normal ship-board activities of a ship's crew which are carried out solely by the crew under the direction of the master;
- fields, woods or other land forming part of an agricultural or forestry undertaking but which is not inside a building and is situated away from the undertaking's main buildings;
- an aircraft, locomotive or rolling stock, trailer or semi-trailer used as a means of transport or a vehicle;
- a mine other than any building on the surface at a mine;
- a borehole site.

In addition certain provisions of the Order do not apply to groups of workers such as:

- occasional work which is not harmful to young people in a family undertaking;
- armed forces;
- members of police forces;
- emergency services.

Part 2 Fire Safety Duties

19.18.3 Duty to take general fire precautions – Article 8

The responsible person must:

(a) take such general fire precautions as will ensure, sfairp, the safety of any of his employees; and

(b) in relation to relevant persons who are not their employees, take such general fire precautions as may reasonably be required in the circumstances of the case to ensure that the premises are safe.

19.18.4 Risk assessment and fire safety arrangements – Articles 9 and 11

The responsible person must make a suitable and sufficient assessment of the risks to identify the general fire precautions he/she needs to take.

Where a dangerous substance is or is liable to be present in or on the premises, the risk assessment must include consideration of the matters set out in Part 1 of Schedule 1, reproduced in Figure 19.6(a).

Risk assessments must be reviewed by the responsible person regularly and if it is no longer valid or there have been significant changes. The responsible person must not employ a young person unless risks to young persons have been considered in an assessment covering the following, which is Part 2 of Schedule 1, reproduced in Figure 19.6(b).

The responsible person must make and record (as per risk assessments) arrangements as are appropriate, for the effective planning, organisation, control, monitoring and review of the preventative and protective measures.

Matters to be considered in risk assessment in respect of dangerous substances

The matters are;

(a) the hazardous properties of the substance

(b) information on safety provided by the supplier, including information contained in any relevant safety data sheet

(c) the circumstances of the work including:
   (i) the special, technical and organizational measures and the substances used and their possible interactions
   (ii) the amount of the substance involved
   (iii) where the work will involve more than one dangerous substance, the risk presented by such substances in combination
   (iv) the arrangements for the safe handling, storage and transport of dangerous substances and of waste containing dangerous substances.

(d) activities, such as maintenance, where there is the potential for a high level of risk

(e) the effect of measures which have been or will be taken pursuant to this Order

(f) the likelihood that an explosive atmosphere will occur and its persistance

(g) the likelihood that ignition sources, including electrostatic discharges, will be present and become active and effective

(h) the scale of the anticipated effects

(i) any places which are, or can be connected via openings to, places in which explosive atmospheres may occur

(j) such additional safety information as the responsible person may need in order to complete the assessment.

Matters to be taken into particular account in risk assessment in respect of young persons

The matters are:

(a) The inexperience, lack of awareness of risks and immaturity of young persons;

(b) The fitting-out and layout of the premises;

(c) The nature; degree and duration of exposure to physical and chemical agents;

(d) The form, range and use of work equipment and the way in which it is handled;

(e) The organization of processes and activities;

(f) The extent of the safety training provided or to be provided to young persons;

(g) Risks from agents, processes and work listed in the Annex to Council Directive 94/33/EC(a) on the protection of young people at work.

Figure 19.6 Fire safety order – matters to be considered in risk assessment in respect of: (a) dangerous substances; (b) young persons

19.18.5 Principles of prevention to be applied and fire safety arrangements – Articles 10 and 11

Preventative and protective measures must be implemented on the basis of the principles specified in

Part 3 of Schedule 1 as follows, which are broadly the same as those in the Management Regulations.

Principles of prevention

The principles are:

(a) avoiding risks;

(b) evaluating the risks which cannot be avoided;

(c) combating the risks at source;

(d) adapting to technical progress;

(e) replacing the dangerous substances by the non-dangerous or the less-dangerous substances;

(f) developing a coherent overall prevention policy which covers technology, organisation of work and the influence of factors relating to the working environment;

(g) giving collective protective measures priority over individual protective measures;

(h) giving appropriate instructions to employees.

Fire safety arrangements must be made and put into effect by the responsible person where five or more people are employed; there is a licence for the premises; or an Alterations Notice requires them to be recorded. The arrangements must cover effective planning, organisation, control, monitoring and review of the preventative and protective measures.

19.18.6 Elimination or reduction of risks from dangerous substances – Article 12

Where a dangerous substance is present in or on the premises, the responsible person must ensure that risk of the substance is either eliminated or reduced, sfairp, by first replacing the dangerous substance with a safer alternative.

Where it is not reasonably practicable to eliminate risk the responsible person must, sfairp, apply measures including the measures specified in Part 4 of Schedule 1 to the Order (Figure 19.7).

The responsible person must:

(a) arrange for the safe handling, storage and transport of dangerous substances and waste containing dangerous substances;

(b) ensure that any conditions necessary for ensuring the elimination or reduction of risk are maintained.

19.18.7 Fire-fighting and fire detection – Article 13

The premises must be equipped with appropriate fire-fighting equipment and with fire detectors and alarms; any non-automatic fire-fighting equipment provided must be easily accessible, simple to use and indicated by signs.

Appropriate measures must be taken for fire-fighting in the premises: to nominate and train competent persons to implement the measures; and arrange any necessary

19

Measures to be taken in respect of dangerous substances

1. In applying measures to control risks the responsible person must, in order of priority:
    (a) reduce the quantity of dangerous substances to a minimum
    (b) avoid or minimize the release of a dangerous substance
    (c) control the release of a dangerous substance at source
    (d) prevent the formation of an explosive atmosphere, including the application of appropriate ventilation
    (e) ensure that any release of a dangerous substance which may give rise to risk is suitably collected, safely contained, removed to a safe place, or otherwise rendered safe, as appropriate
    (f) avoid:
        (i) ignition sources including electrostatic discharges
        (ii) such other adverse conditions as could result in harmful physical effects from a dangerous substance
    (g) segregate incompatible dangerous substances.
2. The responsible person must ensure that mitigation measures applied in accordance with Article 12(3)(b) include:
    (a) reducing to a minimum the number of persons exposed
    (b) measures to avoid the propagation of fires or explosions
    (c) providing explosion pressure relief arrangements
    (d) providing explosion suppression equipment
    (e) providing plant which is constructed so as to withstand the pressure likely to be produced by an explosion
    (f) providing suitable personal protective equipment.
3. The responsible person must:
    (a) ensure that the premises are designed, constructed and maintained so as to reduce risk
    (b) ensure that suitable special, technical and organizational measures are designed, constructed, assembled, installed, provided and used so as to reduce risk
    (c) ensure that special, technical and organisational measures are maintained in an efficient state, in efficient working order and in good repair
    (d) ensure that equipment and protective systems meet the following requirements:
        (i) where power failure can give rise to the spread of additional risk, equipment and protective systems must be able to be maintained in a safe state of operation independently of the rest of the plant in the event of power failure
        (ii) means for manual override must be possible, operated by employees competent to do so, for shutting down equipment and protective systems incorporated within automatic processes which deviate from the intended operating conditions, provided that the provision or use of such means does not compromise safety
        (iii) on operation of emergency shutdown, accumulated energy must be dissipated as quickly and as safely as possible or isolated so that it no longer constitutes a hazard
        (iv) necessary measures must be taken to prevent confusion between connecting devices.
    (e) where the work is carried out in hazardous places or involves hazardous activities, ensure that appropriate systems of work are applied including:
        (i) the issuing of written instructions for the carrying out of work
        (ii) system of permits to work, with such permits being issued by a person with responsibility for this function prior to the commencement of the work concerned.

Figure 19.7 Measures to be taken in respect of dangerous substances

---

contacts with external emergency services.

A person is to be regarded as competent where they have sufficient training and experience or knowledge and other qualities to enable them to properly implement the measures referred to.

19.18.8 Emergency routes and exits – Article 14

Where necessary, routes to emergency exits from premises and the exits themselves must be kept clear at all times.

The following requirements must be complied with:

(a) emergency routes and exits must lead as directly as possible to a place of safety;
(b) in the event of danger, it must be possible for persons to evacuate the premises as quickly and as safely as possible;
(c) the number, distribution and dimensions of emergency routes and exits must be adequate having regard to the use, equipment and dimensions of the premises and the maximum number of persons who may be present there at any one time;
(d) emergency doors must open in the direction of escape;
(e) sliding or revolving doors must not be used for exits specifically intended as emergency exits;
(f) emergency doors must not be so locked or fastened that they cannot be easily and immediately opened by any person who may require to use them in an emergency;
(g) emergency routes and exits must be indicated by signs;
(h) emergency routes and exits requiring illumination must be provided with emergency lighting of adequate intensity in the case of failure of their normal lighting.

19.18.9 Procedures for serious and imminent danger and for danger areas – Article 15

The responsible person must:

(a) establish appropriate procedures, including safety drills;
(b) nominate a sufficient number of competent persons to implement evacuation procedures;
(c) provide adequate safety instruction for restricted areas.

Persons who are exposed to serious and imminent danger must be informed of the nature of the hazard and of the steps taken or to be taken to protect them from it. They must be able to stop work and immediately proceed to a place of safety in the event of their being exposed to serious, imminent and unavoidable danger; and procedures must require the persons concerned to be prevented from resuming work in any situation where there is still a serious and imminent danger.

19.18.10 Additional emergency measures in respect of dangerous substances – Article 16

In order to safeguard persons from an accident, incident or emergency related to the presence of a dangerous substance, the responsible person must (unless the risk assessment shows it is unnecessary) ensure that:

(a) information on emergency arrangements is available, including:
 (i) details of relevant work hazards and hazard identification arrangements;
 (ii) specific hazards likely to arise at the time of an accident, incident or emergency;
(b) suitable warning and other communication systems are established to enable an appropriate response, including remedial actions and rescue operations, to be made immediately when such an event occurs;
(c) where necessary, before any explosion conditions are reached, visual or audible warnings are given and relevant persons withdrawn;
(d) where the risk assessment indicates it is necessary, escape facilities are provided and maintained to ensure that, in the event of danger, persons can leave endangered places promptly and safely.

The information required must be:

(a) made available to accident and emergency services;
(b) displayed at the premises, unless the results of the risk assessment make this unnecessary.

In the event of a fire arising from an accident, incident or emergency related to the presence of a dangerous substance in or on the premises, the responsible person must ensure that:

(a) immediate steps are taken to:
 (i) mitigate the effects of the fire;
 (ii) restore the situation to normal;
 (iii) inform those persons who may be affected;
(b) only those persons who are essential for the carrying out of repairs and other necessary work are permitted in the affected area and they are provided with:
 (i) appropriate PPE and protective clothing; and
 (ii) any necessary specialised safety equipment and plant which must be used until the situation is restored to normal.

19.18.11 Maintenance – Article 17

Any facilities, equipment and devices provided must be subject to a suitable system of maintenance and maintained in an efficient state, in efficient working order and in good repair.

19.18.12 Safety assistance – Article 18

The responsible person must (except a competent self-employed person) appoint one or more competent persons to assist them in undertaking the preventative and protective measures. If more than one person is appointed, they must make arrangements for ensuring adequate cooperation between them.

The number of persons appointed, the time available for them to fulfil their functions and the means at their disposal must be adequate having regard to the size of the premises, the risks to which relevant persons are exposed and the distribution of those risks throughout the premises.

19.18.13 Provision of information to employees and others – Articles 19 and 20

The responsible person must provide their employees with comprehensible and relevant information on:

▶ the risks to them identified by the risk assessment;
▶ the preventative and protective measures;
▶ the procedures for fire drills;
▶ the identities of persons nominated for fire-fighting or fire drills;
▶ the risks notified to them regarding shared premises.

Before employing a child, a parent (or guardian) of the child must be provided with comprehensible and relevant information on the risks to that child identified by the risk assessment; the preventative and protective measures; and the risks notified to them regarding shared premises.

Where a dangerous substance is present in or on the premises, additional information must be provided for employees as follows:

(a) the details of any such substance including:
 (i) the name of the substance and the risk which it presents;
 (ii) access to any relevant safety data sheets;
 (iii) legislative provisions which apply to the substance;
(b) the significant findings of the risk assessment.

The responsible person must ensure that the employer of any employees from an outside undertaking who are working in or on the premises is provided with comprehensible and relevant information on:

(a) the risks to those employees;
(b) the preventative and protective measures taken by the responsible person.

The responsible person must ensure that any person working in their undertaking who is not their employee is provided with appropriate instructions and comprehensible and relevant information regarding any risks to that person.

19

19.18.14 Capabilities and training – Article 21

The responsible person must ensure that all employees are provided with adequate safety training:

(a) at the time when they are first employed;

(b) on their being exposed to new or increased risks because of:

   (i) their being transferred or given a change of responsibilities;

   (ii) the introduction of new work equipment or a change respecting work equipment already in use;

   (iii) the introduction of new technology; or

   (iv) the introduction of a new system of work, or a change respecting a system of work already in use.

The training must:

(a) include suitable and sufficient instruction and training on the appropriate precautions and actions to be taken by the employee;

(b) be repeated periodically where appropriate;

(c) be adapted to take account of any new or changed risks to the safety of the employees concerned;

(d) be provided in a manner appropriate to the risk identified by the risk assessment;

(e) take place during working hours.

Under the Fire Safety (Employee's Capabilities) Regulations for both England and Wales, every employer must, in entrusting tasks to employees, take into account their capabilities as regards health and safety, so far as those capabilities relate to fire.

19.18.15 Cooperation and coordination – Article 22

Where two or more responsible persons share, or have duties in respect of, premises (whether on a temporary or a permanent basis) each such person must cooperate with, coordinate safety measures and inform the other responsible person concerned so far as is necessary to enable them to comply with the requirements and prohibitions imposed on them.

Where two or more responsible persons share premises (whether on a temporary or a permanent basis) where an explosive atmosphere may occur, the responsible person who has overall responsibility for the premises must coordinate the implementation of all the measures required.

19.18.16 General duties of employees at work – Article 23

Every employee must, while at work:

(a) take reasonable care for the safety of himself or herself and of other relevant persons who may be affected by acts or omissions at work;

(b) as regards any duty or requirement imposed on an employer by or under any provision of this Order, the employee must cooperate so far as is necessary to enable that duty or requirement to be performed or complied with;

(c) inform their employer or any other employee with specific responsibility for the safety of fellow employees:

   (i) of any work situation which a person with the first-mentioned employee's training and instruction would reasonably consider represented a serious and immediate danger to safety;

   (ii) of any matter which a person with the first-mentioned employee's training and instruction would reasonably consider represented a shortcoming in the employer's protection arrangements for safety.

Part 3

19.18.17 Enforcement – Articles 25–31

Part 3 of the Order is about enforcement and penalties. An enforcing authority may be the fire and rescue authority of the area (most cases), the HSE (nuclear sites, ships and construction sites without other operations) and Fire Inspectors maintained by the Secretary of State (armed forces, UK Atomic Energy Authority and Crown premises).

Fire Inspectors and an officer of the fire brigade maintained by the fire and rescue authority, have similar powers (Articles 27 and 28) to inspectors under the HSW Act.

There are differences between the HSW Act and the Fire order in the notices which can be issued. Appeals can be made against these notices within 21 days and the notices are suspended until the appeal is heard. However, a Prohibition Notice stands until confirmed or altered by the court. Fire Inspectors can issue the following notices.

Alterations Notices – Article 29

These may be issued where the premises concerned constitute a serious risk to relevant people or may constitute a serious risk if any change is made to the premises or its use.

Where an Alterations Notice has been served the responsible person must notify the enforcing authority before making any specified changes, which are as follows:

▶ a change to the premises;

▶ a change to the services, fittings or equipment in or on the premises;

▶ an increase in the quantities of dangerous substances which are present in or on the premises;

▶ a change in the use of the premises.

In addition, the Alterations Notice may also include the requirement for the responsible person to:

▶ record the significant findings of the risk assessment as per Article 9(7) and 9(6);
▶ record the fire safety arrangements as per Article 11(1) and 11(2);
▶ send a copy of the risk assessment to the enforcing authority, before making the above changes, and a summary of the proposed changes to the general fire precautions.

Enforcement Notices – Article 30

Where the enforcing authority is of the opinion that the responsible person has failed to comply with the requirements of this Order or any regulations made under it, they can issue an Enforcement Notice which must:

▶ state that the enforcing authority is of this opinion;
▶ specify the provisions which have not been complied with; and require that person to take steps to remedy the failure within a period (not less than 28 days) from the date of service of the notice.

An Enforcement Notice may include directions on the measures needed to remedy the failures. Choices of remedial action must be left open.

Before issuing an Enforcement Notice the enforcing authority must consult the relevant regulatory enforcing authorities including those under the HSW Act and the Building Regulations.

Prohibition Notices – Article 31

If the enforcing authority is of the opinion that the risks, relating to escape from the premises, are so serious that the use of the premises ought to be prohibited or restricted they may issue a Prohibition Notice. The notice must:

▶ state that the enforcing authority is of this opinion;
▶ specify the provisions which give or may give rise to that risk;
▶ direct that the use to which the notice relates is prohibited or restricted as may be specified until the specified matters have been remedied.

A Prohibition Notice may include directions on the measures needed to remedy the failures. Choices of remedial action must be left open.

A Prohibition Notice takes effect immediately it is served.

Part 4

19.18.18 Offences and Appeals – Articles 32–36

Cases can be tried in a Magistrates' Court or on indictment in the Crown Court.

The responsible person can be liable on conviction on indictment to a fine (not limited), or to imprisonment for a term not exceeding 2 years or both.

Any person can be liable to:

▶ on conviction on indictment to a fine (not limited) for an offence where that failure places one or more relevant people at risk of death or serious injury in the case of fire;
▶ on summary conviction to a fine at standard levels 3 or 5 depending on the particular offence.

In general, where an offence committed by a body corporate is proved to have been committed with the consent or connivance of any director, manager, secretary or other similar officer of the body corporate they, as well as the body corporate, are guilty of that offence – Article 32(9).

19.18.19 The Fire Scotland Act 2005

Fire is a devolved subject. Previous fire safety legislation in Scotland has been replaced by the Fire (Scotland) Act 2005, as amended, and by regulations made under that Act. Domestic premises do not generally fall within the scope of this Act, but there are exceptions which include those requiring a licence under the Houses in Multiple Occupation mandatory licensing scheme, and care home premises. However, the legislation does require any equipment or facilities provided for the protection or use of fire-fighters and located in common areas of private dwellings to be maintained.

The Police and Fire Reform (Scotland) Act 2012 part 2 – Fire Reform sets up the Scottish Fire and Rescue Service (SFRS). The 2005 Act is amended so that references to 'a relevant authority' is replaced by 'SFRS'. The SFRS also has the duty to secure 'best value' when carrying out its functions (see the SFRS website: http://www.firescotland.gov.uk).

Duties imposed by the legislation fall into seven general categories and are similar to those in England and Wales:

1. Carrying out a **fire safety risk assessment** of the premises;
2. Identifying the **fire safety measures** necessary as a result of the fire safety risk assessment outcome;
3. Implementing these fire safety measures using **risk reduction principles**;
4. Putting in place fire safety arrangements for the ongoing **control and review of the fire safety measures**;
5. Complying additionally with the specific requirements of the **fire safety regulations**;
6. Keeping the fire safety risk assessment and outcome under review; and
7. Record keeping.

The following is a summary of the main rights and responsibilities of those covered by the legislation. For a comprehensive list of rights and responsibilities, please refer to the legislation.

19

(a) Employers:
- are **entitled** to the cooperation of their employees in respect of fire safety responsibilities and for employees to take reasonable care for their own safety and others whom their actions could affect.
- **must** carry out (and review regularly) a fire safety risk assessment to identify the potential for fire to occur in the workplace and cause harm to employees and other persons in, or in the immediate vicinity of, the premises.
- **must** ensure, so far as is reasonably practicable, the safety of their employees from harm caused by fire in the workplace and take reasonable measures to ensure their safety and that of others there, or in the immediate vicinity. These measures should be aimed at avoiding or reducing any risks identified.
- **must** provide clear, appropriate information and instruction (and training where necessary) to employees and anyone else working on their premises, e.g. contractors and their employer(s), in relation to any risks identified and fire safety measures provided.
- **must** ensure that a record of a fire safety risk assessment is kept (electronically or paper-based) if they employ five or more employees (whether they are based in the premises or not), have a licence (such as a liquor licence) for the premises, are subject to registration (such as care home registration), or an alterations notice requiring this has been served in relation to the premises by the enforcing authority.
- **must** carry out (or review) a fire safety risk assessment before employing anyone under the age of 18, taking into account their youth and inexperience, and the particular risks they may be exposed to in the workplace. If a child is of school age, their parent must be informed of the risks and the measures taken to avoid or reduce them.
- **must** ensure that the premises are equipped to an appropriate level, with means of escape (ensuring these can be safely used), means of detecting fire and giving warning in the event of fire, means of fighting fires, and arrangements for action to be taken in the event of fire.
- **must** ensure that the premises, and any fire safety facilities, equipment and devices are maintained in good order.
- **must** cooperate with anyone else who has fire safety responsibilities for the same premises (including sharing information about the activities/measures undertaken, where appropriate) and take all reasonable steps to coordinate your fire safety measures regarding the premises.

(b) Employees:
- are **entitled** to the provision of adequate fire safety measures on the premises to ensure, so far as is reasonably practicable, their safety from harm caused by fire.
- are **entitled** to appropriate instruction (and training where necessary) about any risks identified on the premises, fire safety measures provided and what to do in the event of a fire.
- **must** take reasonable care to ensure the workplace is safe from harm caused by fire and do nothing that will place themselves or others at risk.
- **must** inform their employer (or a fellow employee with specific fire safety responsibilities) of anything relating to the premises which could represent a serious and immediate fire safety danger; of anything which they reasonably consider represents a shortcoming in the employer's fire safety protection arrangements; or in the event of fire.
- **must** cooperate with your employer, so far as is necessary, to allow them to comply with their fire safety responsibilities.

(c) Someone who is not an employer but has control of the premises or safety obligations to some extent, such as an owner, tenant, contractor, or occasional user (including voluntary groups):
- is **entitled** to cooperation from others with fire safety responsibilities for the premises.
- **must** carry out (and review regularly as necessary) a fire safety risk assessment of the premises, to the extent of their control, or obligations towards them, identifying any risks to persons in, or in the immediate vicinity of, the premises in respect of harm caused by fire.
- **must** take reasonable fire safety measures on the basis of the results of their fire safety risk assessment to ensure, to the extent of the control or obligations they have, the safety of persons from harm caused by fire in the premises.
- **must** ensure that the premises are equipped to an appropriate level, relative to the extent of their control or obligations, with means of escape (ensuring they can be safely used), means of detecting fire and giving warning in the event of fire, means of fighting fires, and arrangements for action to be taken in the event of fire.
- **must** cooperate (including sharing information where appropriate) with others who have fire safety responsibilities for the premises, take all reasonable steps to coordinate fire safety measures with, taking into account the extent of their control or obligation and others in respect of the premises.

▶ **must** ensure that the premises and any fire safety facilities, equipment and devices are maintained in good order, to the extent of their control or obligation towards the premises.

19.18.20 Fire and Rescue Services (Northern Ireland) to fire prevention.

The law applies to almost all premises which are not private dwellings. Examples include:

▶ Offices and shops;
▶ Premises that provide care, including care homes and hospitals;
▶ Pubs, clubs and restaurants;
▶ Places of worship;
▶ Educational establishments including schools;
▶ Theatres and cinemas;
▶ Sports centres and other community premises;
▶ Hotels and hostels;
▶ Guest houses and B&B accommodation;
▶ Shared areas of properties common to several households;
▶ Houses in multiple occupation;
▶ Factories and warehouses;
▶ Tents and marquees;
▶ Transport premises and facilities;
▶ Animal premises and stables;
▶ Open air events and venues.

It does not apply to people's private homes, including individual flats in a block or house.

The main requirements under the legislation are very similar to other areas of the UK as follows:

▶ Carry out a fire risk assessment for the premises over which you have control;
▶ Identify the fire hazards and risks associated with the premises, materials/substances, activities, etc.;
▶ Identify the people, or groups of people at risk and anyone who may be especially at risk;
▶ Remove and reduce the risks as far as reasonably possible;
▶ Put in place general fire precautions to deal with any remaining risks;
▶ Implement additional preventative and protective measures if flammable or explosive substances are used or stored on the premises;
▶ Develop and implement appropriate emergency procedures in the event of fire;
▶ If you have five or more employees, or require a licence or registration, you must record the significant findings of the risk assessment and any actions you have taken to remove/reduce the risk;
▶ Review the risk assessment periodically or after significant changes in the workplace.

19.18.21 Further information

Regulatory Reform, England and Wales. The Regulatory Reform (Fire Safety) Order 2005, SI 2005 No. 1541.

There are a number of guides produced by the Department of Communities and Local Government including the following subjects:

Regulatory Reform (Fire Safety) Order 2005 – A short guide to making your premises safe from fire. Product code: 05 FRSD 03546

Fire Risk Assessment – Offices and Shops ISBN 978 1 85112 815 0.

Fire Risk Assessment – Sleeping Accommodation ISBN 978 1 85112 817 4.

Fire Risk Assessment – Residential Care Premises ISBN 978 1 85112 818 1.

Fire Risk Assessment – Small and Medium Places of Assembly ISBN 978 1 85112 820 4.

Fire Risk Assessment – Large Places of Assembly ISBN 978 1 85112 821 1.

Fire Risk Assessment – Factories and Warehouses ISBN 978 1 85112 816 7.

Fire Risk Assessment – Theatres, Cinemas and Similar Premises ISBN 978 1 85112 822 8.

Fire Risk Assessment – Educational Premises ISBN 978 1 85112 819 8.

Fire Risk Assessment – Healthcare Premises ISBN 978 1 85112 824 2.

Fire Risk Assessment – Transport Premises and Facilities ISBN 978 1 85112 825 9.

Fire Risk Assessment – Open Air Events and Venues ISBN 978 1 85112 823 5.

Fire Risk Assessment – Animal Premises and Stables ISBN 978 1 85112 884 6.

For free downloads see website: https://www.gov.uk/workplace-fire-safety-your-responsibilities/fire-safety-advice-documents

Fire Safety Scotland see: http://www.firelawscotland.org/v2bbaa.html?pContentID=237

Northern Ireland Fire and Rescue Service see: http://www.nifrs.org/firesafe/

19.19 Health and Safety (First-Aid) Regulations 1981 as amended

19.19.1 Introduction

These Regulations set out employers' duties to provide adequate first-aid facilities. They define first-aid as:

▶ in cases where a person will need help from a medical practitioner or nurse, treatment for the purposes of preserving life and minimising the consequences of injury and illness until such help is obtained;

▶ treatment of minor injuries which would otherwise receive no treatment or which do not need treatment by a medical practitioner or nurse.

From October 2013 the HSE have reissued L74 as Guidance only, taking into account the amendment of regulation 3(2) to remove the requirement for the HSE to approve the training and qualifications of appointed first-aid personnel and other earlier amendments.

19.19.2 Duty of the employer – Regulation 3

An employer shall provide, or ensure that there are provided:

▶ adequate and appropriate facilities and equipment;
▶ such number of suitable persons as is adequate and appropriate for rendering first-aid; they must undergo such training and have such qualifications as may be appropriate in the circumstances;
▶ an appointed person, being someone to take charge of situations as well as first-aid equipment and facilities, where medical aid needs to be summoned. An appointed person will suffice where:
  ▷ the nature of the work is such that there are no specific serious hazards (offices, libraries, etc.); the workforce is small; the location makes further provision unnecessary;
  ▷ there is temporary (not planned holidays) or exceptional absence of the first-aider.

Employers must make an assessment of the first-aid requirements that are appropriate for each workplace. Where an employer provides first-aiders in the workplace, they should ensure they have undertaken suitable training, have an appropriate first-aid qualification and remain competent to perform their role. Typically, first-aiders will hold a valid certificate of competence in either first-aid at work (FAW) or emergency first-aid at work (EFAW). EFAW training enables a first-aider to give emergency first-aid to someone who is injured or becomes ill while at work. FAW training includes EFAW and also equips the first-aider to apply first-aid to a range of specific injuries and illness.

If an employer, after carrying out a needs assessment, decides a first-aider is not required in the workplace, a person should be appointed to take charge of the first-aid arrangements. The role of this appointed person includes looking after the first-aid equipment and facilities and calling the emergency services when required.

There must always be at least an appointed person in every workplace during working hours.

Any first-aid room provided under this Regulation must be easily accessible to stretchers and to any other equipment needed to convey patients to and from the room. It must be sign-posted according to the safety signs and signals regulations.

19.19.3 Employees information – Regulation 4

Employees must be informed of the arrangements for first-aid, including the location of facilities, equipment and people.

19.19.4 Self-employed – Regulation 5

The self-employed shall provide such first-aid equipment as is appropriate to render first-aid to themselves.

19.19.5 Further information

Basic Advice on First-Aid at Work INDG347 (rev2), 2011, ISBN 9780 7176 6435 1. http://www.hse.gov.uk/pubns/indg347.htm

First-Aid at Work INDG214(rev2), HSE Books, 2014, ISBN 978 0 7176 6469 6. www.hse.gov.uk/pubns/indg214.htm

First-Aid at Work. Health and Safety (First-Aid) Regulations 1981. Guidance L74 (Third Edition), HSE Books, 2013, ISBN 978 0 7176 6560 0. http://www.hse.gov.uk/pubns/books/l74.htm

HSE First-Aid Micro Site for General Information, http://www.hse.gov.uk/firstaid/

19.20 Health and Safety (Information for Employees) Regulations 1989

19.20.1 General requirements

These Regulations require that the Approved Poster entitled, 'Health and safety – what you should know', is displayed or the Approved Leaflet is distributed.

This information tells employees in general terms about the requirements of health and safety law.

A modern and easy-to-read version of the poster and pocket cards was introduced on 6 April 2009.

The poster has three sections for completion by the employer. These are 1. Employee representative; 2. Management representative (e.g. appointed competent person to assist managers health and safety appointee); 3. Details of how the Enforcing authority and the Employment Medical Advisory Service (EMAS) can be contacted.

19.20.2 Further information

Health and Safety (Information for Employees) Regulations 1989 (SI No. 682).

Health and Safety Information for Employees (Amendment) Regulations 2009, SI 2009/606.

Health and Safety Law poster A3 size, 2009, ISBN 9780717663699. http://www.hse.gov.uk/pubns/books/lawposter-a3.htm

Health and Safety Law What you need to know, 2009, HSE Books, ISBN 9780717665013. http://www.hse.gov.uk/pubns/lawleaflet.pdf

19.21 Hazardous Waste (England and Wales) Regulations 2005

19.21.1 Introduction

The Hazardous Waste (England and Wales) Regulations 2005 were implemented on 16 July 2005. These are made under the EPA 1990. There were many changes to the previous Special Waste Regulations, but the two key ones are that hazardous waste producers are now required to pre-register before any hazardous waste can be collected from their premises and the Regulations apply the European Waste Catalogue codes of hazardous wastes that will affect a much wider range of producers.

19.21.2 Summary

The list below is not exhaustive, but summarises the main requirements. There is also a range of web links that will help to obtain more detail.

From 16 July 2005, it is an offence for hazardous waste to be collected from a site that has not been registered or is exempt.

All non-exempt sites that produce hazardous waste must be registered even if they are unlikely to have that waste collected for some time. Recent EA (Environment Agency) Guidance has clarified that it is an offence to produce hazardous waste on site and not be registered.

The Regulations implement, through the List of Wastes (England) Regulations 2005, the European Waste Catalogue list of Hazardous wastes for the purposes of collection. This means that things like PC monitors, PC base units, fridges, TVs, oily rags and separately collected fluorescent tubes require collection under the new hazardous waste notification and documentation procedures.

The EA will accept postal registrations and registrations can be made online through the following website: www.environment-agency.gov.uk

Each site producing 500 kg or more of hazardous waste each year has to have a separate registration, although multiple sites can be registered on the same notification. Therefore, a head office could register all its sites centrally, but each site would have a separate unique registration number and require a separate fee. Registration is annual and must be renewed each year.

Domestic waste is excluded from the Hazardous Waste Regulations on collection from the domestic property, but is then subject to the Regulations if it is separately collected or if it consists of asbestos. This includes Prescription Only Medicines (other than cytotoxic and cytostatic medicines) which will be hazardous waste.

Each consignment of Hazardous Waste will require a consignment note (see Figure 19.8) and a fee to be paid to the EA by the consignee with their quarterly returns to the Agency. Clearly, this will be charged back to the collector, but a consignment might well attract more than one consignment fee if, for instance, it goes through a transfer station and the collector would have to ensure that this was considered in the price.

Collection rounds will be possible but again, each site where the waste is collected would have to be left a copy of the consignment note, so the process of tracking and the paper trail will get quite complex especially as each collection will count as a consignment from the fee point of view.

The Regulations ban the mixing of hazardous waste and state that it must be stored separately on site. It prohibits businesses like offices or retail premises from discarding items of hazardous waste in their general waste bins.

Businesses also have a duty of care to store their waste appropriately. The key factors are that they ensure waste is:

▶ secure;
▶ contained so that it cannot escape; and
▶ protected from the weather, vehicles including fork-lift trucks, scavengers and pests.

For certain wastes there are additional storage requirements:

▶ oil – there is further guidance in Pollution Prevention Guidance PPG 2 from the Environment Agency;
▶ chemicals – can have different properties and have the potential to react dangerously if not stored properly.

Registration as a Hazardous Waste producer places a statutory duty on the EA to inspect the site where the hazardous waste arises.

19.21.3 Further information

Small businesses A guide to Hazardous Waste Regulations, Environment Agency 2013, HWR01A

19.22 Ionising Radiations Regulations 1999

19.22.1 Introduction

The Ionising Radiations Regulations 1999 (IRR99) implement the majority of the Basic Safety Standards Directive 96129/Euratom (BSS Directive). From 1 January 2000, they replaced the Ionising Radiations Regulations 1985 (IRR85) (except for Regulation 26 (special hazard assessments)).

The main aim of the Regulations and the supporting ACoP is to establish a framework for ensuring that exposure to ionising radiation arising from work activities, whether from man-made or natural radiation

19

PART A Notification Details

1. Consignment note code: ABC123/AB001

2. The waste described below is to be removed from:
   (name, address, postcode, telephone, e-mail & fax)
   The Green Grocer, High Street , New Town, The Shire, XX12 3YY
   Tel 0123 456789, GG@hotmail.com

3. Premises Code (where applicable) :
   ABC123

4. The waste will be taken to (name address and postcode):
   The Waste Treatment Plant, Low Street, Old Town, The Shire
   YY12 4XX

5. The waste producer was (if different from 2)
   (name, address, postcode, telephone, e-mail & fax) :

PART B Description of waste

1. The process giving rise to the waste(s) was: Grocery shop   2. SIC for the process giving rise to the waste: 52.21
3. WASTE DETAILS (where more than one waste type is collected all of the information given below must be completed for each EWC identified).

Description of Waste	List of Wastes (EWC) code	Qty (kg)	The chemical / biological components of the waste, their concentrations mg/kg or %	Physical Form	Hazard code(s)	Container type, number & size
Part filled tins of gloss paint	20 01 27*	8 kg	di-isobutyl ketone <1% ethyl methyl ketoxime naptha D4 dearomatised 10-30% naptha D6 high flash <1% xylene <1%	liquid	H3-B	3 x 10 litre tins

The information given below is to be completed for each EWC identified

EWC code	Description for Carriage (UN identification number(s), Proper shipping name(s), UN class(es), Packing group(s))	Special Handling requirements
20 01 27*	1263, PAINT, Class 3:Flammable liquids, III	EMS F-E, S-E Emergency Action Code •3YE Hazard No. (ADR) 30 Tunnel Restriction Code (D/E)

PART C Carrier's certificate

(If more than one carrier is used, please attach a schedule for subsequent carriers. If a schedule of carriers is attached tick here) ☐

I certify that I today collected the consignment and that the details in A2, A4 and B3 are correct & I have been advised of any specific handling requirements.

Where this note comprises part of a multiple collection the round number and collection number are:

Carrier driver name (please PRINT) Tony Driver

On behalf of (name, address, postcode, telephone, e-mail & fax)
The Waste Treatment Plant, Low Street, Old Town, The Shire, YY12 4XX
Tel : 0987 654321   e.mail: waste@hotmail.com

Carriers registration no./ reason for exemption reason:
CB/XZ1234AB
Vehicle registration no: AB07 WEE

Signature **T.Driver** Time 16:00 Date 29/12/2012

PART D Consignor / Holders certificate

I certify that the information in A, B & C has been completed and is correct, that the carrier is registered or exempt and was advised of the appropriate precautionary measures. All of the waste is packaged and labelled correctly and the carrier has been advised of any special handling requirements.

I confirm that I have fulfilled my duty to apply the waste hierarchy as required by Regulation 12 of the Waste (England and Wales) Regulations 2011.

1. Consignor name (please PRINT) :   Mr Eric Smith
   On behalf of (name, address, postcode, telephone, e-mail & fax):

   The Green Grocer, High Street, New Town, The Shire, XX12 3YY
   Tel 0123 456789, GG@hotmail.com

Signature **E. Smith**   Time 16:00   Date 29/12/2012

PART E Consignee's Certificate (where more than one waste type is collected all of the information given below must be completed for each EWC)

Individual EWC code(s) received	Quantity of each EWC code received (kg)	EWC code accepted / rejected	Waste Management operation (R or D code)
20 01 27*	8	accepted	R02

1. I received this waste at the address given in A4 on   Date:29/12/2012 Time: 17:00
2. Vehicle registration no (or mode of transport if not by road):   AB07 WEE
3. Where waste is rejected please provide details:

I certify that the waste management licence / permit / authorised exemption no(s).

EPR/XX6598PP/V002

authorises the management of the waste described in B at the address given at A4

Where the consignment forms part of a multiple collection, as identified in Part C,
I certify that the total number of consignments forming the collection are ☐

Name: Peter Harris
On behalf of (name, address, postcode, tel, e-mail & fax):

The Waste Treatment Plant, Low Street, Old Town,
YY12 4XX Tel : 0987 654321   e.mail: waste@hotmail.com

Signature: **P. Harris**

Date: 17:15   Time: 29/02/2008

Figure 19.8 A completed Hazardous Waste Consignment Note

and from external radiation (e.g. X-ray set) or internal radiation (e.g. inhalation of a radioactive substance), is kept as low as reasonably practicable and does not exceed dose limits specified for individuals. IRR99 also:

(a) replaces the Ionising Radiations (Outside Workers) Regulations 1993 (OWR93), which were made to implement the Outside Workers Directive 90/641/Euratom;

(b) implements a part of the Medical Exposures Directive 97143/Euratom in relation to equipment used in connection with medical exposures.

The guidance which accompanies the Regulations and ACoP gives detailed advice about the scope and duties of the requirements imposed by IRR99. It is aimed at employers with duties under the Regulations but should also be useful to others such as radiation protection advisers, health and safety officers, radiation protection supervisors and safety representatives.

19.22.2 Working with ionising radiation

Essentially, work with ionising radiation means:

(a) a practice, which involves the production, processing, handling, use, holding, storage, transport or disposal of artificial radioactive substances and some naturally occurring sources, or the use of electrical equipment emitting ionising radiation at more than 5 kV (see definition of practice in Regulation 2(1));

(b) work in places where the radon gas concentration exceeds the values in Regulation 3(1)(b); or

(c) work with radioactive substances containing naturally occurring radionuclides not covered by the definition of a practice.

19.22.3 Radiation employers

Radiation employers are essentially those employers who work with ionising radiation, that is they carry out:

(a) a practice (see definition in Regulation 2(1)); or

(b) work in places where the radon gas concentration exceeds the values in Regulation 3(1)(b); or

(c) work with radioactive substances containing naturally occurring radionuclides not covered by the definition of a practice.

19.22.4 Duties of self-employed people

A self-employed person who works with ionising radiation will simultaneously have certain duties under these Regulations, both as an employer and as an employee.

For example, self-employed persons may need to take such steps as:

▶ carrying out assessments under Regulation 7;

▶ providing control measures under Regulation 8 to restrict exposure;

▶ designating themselves as classified persons under Regulation 20;

▶ making suitable arrangements under Regulation 21 with one or more approved dosimetry services (ADS) for assessment and recording of doses they receive;

▶ obtaining a radiation passbook and keeping it up to date in accordance with Regulation 2;

▶ if they carry out services as an outside worker, making arrangements for their own training as required by Regulation 14;

▶ ensuring they use properly any dose meters provided by an ADS as required by regulation.

19.22.5 General requirements

Some of the major considerations are:

▶ notification to the HSE of specific work unless specified in Schedule 1 to the Regulations (Regulation 6);

▶ carrying out a prior risk assessment by a radiation employer before commencing a new activity (Regulation 7);

▶ use of PPE (Regulation 9);

▶ maintenance and examination of engineering controls and PPE (Regulation 10);

▶ dose limitations (Regulation 11);

▶ contingency plans for emergencies (Regulation 12);

▶ radiation protection adviser appointment (Regulation 13);

▶ information, instruction and training (Regulation 14).

19.22.6 Prior risk assessment

Where a radiation employer is required to undertake a prior risk assessment, the following matters need to be considered, where they are relevant:

▶ the nature of the sources of ionising radiation to be used, or likely to be present, including accumulation of radon in the working environment;

▶ estimated radiation dose rates to which anyone can be exposed;

▶ the likelihood of contamination arising and being spread;

▶ the results of any previous personal dosimetry or area monitoring relevant to the proposed work;

▶ advice from the manufacturer or supplier of equipment about its safe use and maintenance;

▶ engineering control measures and design features already in place or planned;

▶ any planned systems of work; estimated levels of airborne and surface contamination likely to be encountered;

▶ the effectiveness and the suitability of PPE to be provided;

▶ the extent of unrestricted access to working areas where dose rates or contamination levels are likely to be significant; possible accident situations, their likelihood and potential severity;

▶ the consequences of possible failures of control measures – such as electrical interlocks, ventilation systems and warning devices – or systems of work; steps to prevent identified accident situations, or limit their consequences.

This prior risk assessment should enable the employer to determine:

▶ what action is needed to ensure that the radiation exposure of all persons is kept as low as reasonably practicable (Regulation 8(1));

19

▶ what steps are necessary to achieve this control of exposure by the use of engineering controls, design features, safety devices and warning devices (Regulation 8(2)(a)) and, in addition, by the development of systems of work (Regulation 8(2)(b));

▶ whether it is appropriate to provide PPE and if so what type would be adequate and suitable (Regulation 8(2)(c));

▶ whether it is appropriate to establish any dose constraints for planning or design purposes, and if so what values should be used (Regulation 8(3));

▶ the need to alter the working conditions of any female employee who declares she is pregnant or is breastfeeding (Regulation 8(5));

▶ an appropriate investigation level to check that exposures are being restricted as far as are reasonably practicable (Regulation 8(7));

▶ what maintenance and testing schedules are required for the control measures selected (Regulation 10);

▶ what contingency plans are necessary to address reasonably foreseeable accidents (Regulation 12);

▶ the training needs of classified and non-classified employees (Regulation 14);

▶ the need to designate specific areas as controlled or supervised areas and to specify local rules (Regulations 16 and 17);

▶ the actions needed to ensure restriction of access and other specific measures in controlled or supervised areas (Regulation 18);

▶ the need to designate certain employees as classified persons (Regulation 20);

▶ the content of a suitable programme of dose assessment for employees designated as classified persons and for others who enter controlled areas (Regulations 18 and 21);

▶ the responsibilities of managers for ensuring compliance with these Regulations;

▶ an appropriate programme of monitoring or auditing of arrangements to check that the requirements of these Regulations are being met.

19.22.7 Further information

The HSE have also produced a number of information sheets, which are available free at: http://www.hse.gov.uk/radiation/ionising/publications.htm

Work with ionising radiation, L121, 2000, HSE Books, ISBN 978 0 7176 1746 3. This is a large document and need only be studied by those with a specific need to control IR. http://www.hse.gov.uk/pubns/books/l121.htm

Working safely with ionising radiation: Guidelines for expectant or breastfeeding mothers, INDG334, 2001, http://www.hse.gov.uk/pubns/indg334.pdf

19.23 Lifting Operations and Lifting Equipment Regulations (LOLER) 1998 as amended in 2002

19.23.1 Introduction

This summary gives information about the LOLER 1998 which came into force on 5 December 1998.

The Regulations aim to reduce risks to people's health and safety from lifting equipment provided for use at work. In addition to the requirements of LOLER, lifting equipment is also subject to the requirements of the PUWER 1998.

Generally, the Regulations require that lifting equipment provided for use at work is:

▶ strong and stable enough for the particular use and marked to indicate SWL;

▶ positioned and installed to minimise any risks;

▶ used safely, i.e. the work is planned, organised and performed by competent people;

▶ subject to ongoing thorough examination and, where appropriate, inspection by competent people.

19.23.2 Definition

Lifting equipment includes **any equipment used at work for lifting or lowering loads**, including attachments used for anchoring, fixing or supporting it. The Regulations cover a wide range of equipment including cranes, fork-lift trucks, lifts, hoists, mobile elevating work platforms and vehicle inspection platform hoists. The definition also includes lifting accessories such as chains, slings, eyebolts, etc. LOLER **does not** apply to escalators; these are covered by more specific legislation, i.e. the Workplace (Health, Safety and Welfare) Regulations 1992.

If employees are allowed to provide their own lifting equipment, then this too is covered by the Regulations.

19.23.3 Application

The Regulations apply to an employer or self-employed person providing lifting equipment for use at work, or who has control of the use of lifting equipment. They do not apply to equipment to be used primarily by members of the public, for example lifts in a shopping centre. However, such circumstances are covered by the HSW Act 1974.

LOLER applies to the way lifting equipment is used in industry and commerce. LOLER applies only to work activities, for example:

▶ a crane on hire to a construction site;

▶ a contract lift;

▶ a passenger lift provided for use of workers in an office block;

▶ refuse collecting vehicles lifting on a public road;

▶ patient hoist;

▶ fork-lift truck.

These Regulations add to the requirements of PUWER 98 and should be interpreted with them. For example, when selecting lifting equipment, PUWER Regulation 4, regarding suitability, should be considered in connection with:

▶ ergonomics;
▶ the conditions in which the equipment is to be used;
▶ safe access and egress;
▶ preventing slips, trips and falls;
▶ protecting the operator.

While employees do not have duties under LOLER, they do have general duties under the HSW Act and the MHSWR 1992, for example to take reasonable care of themselves and others who may be affected by their actions and to cooperate with others.

The Regulations cover places where the HSW Act applies – these include factories, offshore installations, agricultural premises, offices, shops, hospitals, hotels, places of entertainment, etc.

19.23.4 Strength and stability – Regulation 4

Lifting equipment shall be of adequate strength and stability for each load, having regard in particular to the stress induced at its mounting or fixing point.

Every part of a load and anything attached to it and used in lifting it shall be of adequate strength.

Account must be taken of the combination of forces to which the lifting equipment will be subjected, as well as the weight of any lifting accessories. The equipment should include an appropriate factor of safety against failure.

Stability needs to take into account the nature, load-bearing strength, stability, adjacent excavations and slope of the surface. For mobile equipment, keeping rails free of obstruction and tyres correctly inflated must be considered.

19.23.5 Lifting equipment for lifting persons – Regulation 5

To ensure safety of people being lifted, there are additional requirements for such equipment. The use of equipment not specifically designed for raising and lowering people should only be used in exceptional circumstances.

The Regulation applies to all lifting equipment used for raising and lowering people and requires that lifting equipment for lifting persons shall:

▶ prevent a person using it from being crushed, trapped or struck, or falling from the carrier;
▶ prevent, sfairp, persons using it while carrying out work from the carrier being crushed, trapped or struck or falling from the carrier;

▶ have suitable devices to prevent the risk of the carrier falling. If a device cannot be fitted, the carrier must have:
  ▷ an enhanced safety coefficient suspension rope or chain;
  ▷ the rope or chain inspected every working day by a competent person;
▶ be such that a person trapped in any carrier is not thereby exposed to any danger and can be freed.

19.23.6 Positioning and installation – Regulation 6

Lifting equipment must be positioned and installed so as to reduce the risks, sfairp, from:

▶ equipment or a load striking another person;
▶ a load drifting, falling freely or being released unintentionally;

and it is otherwise safe.

Lifting equipment should be positioned and installed to minimise the need to lift loads over people and to prevent crushing in extreme positions. It should be designed to stop safely in the event of a power failure and not release its load. Lifting equipment, which follows a fixed path, should be enclosed with suitable and substantial interlocked gates and any necessary protection in the event of power failure.

19.23.7 Marking of lifting equipment – Regulation 7

Machinery and accessories for lifting loads shall be clearly marked to indicate their SWL, and:

▶ where the SWL depends on the configuration of the lifting equipment:
  ▷ the machinery should be clearly marked to indicate its SWL for each configuration;
  ▷ information which clearly indicates its SWL for each configuration should be kept with the machinery.
▶ accessories for lifting (e.g. hooks, slings) are also marked in such a way that it is possible to identify the characteristics necessary for their safe use (e.g. if they are part of an assembly);
▶ lifting equipment which is designed for lifting people is appropriately and clearly marked;
▶ lifting equipment not designed for lifting people, but which might be used in error, should be clearly marked to show it is not for lifting people.

19.23.8 Organisation of lifting operations – Regulation 8

Every lifting operation, that is lifting or lowering of a load, shall be:

▶ properly planned by a competent person;
▶ appropriately supervised;
▶ carried out in a safe manner.

19

The person planning the operation should have adequate practical and theoretical knowledge and experience of planning lifting operations. The plan will need to address the risks identified by the risk assessment and identify the resources, the procedures and the responsibilities required so that any lifting operation is carried out safely. For routine simple lifts a plan will normally be left to the people using the lifting equipment. For complex lifting operations, for example where two cranes are used to lift one load, a written plan may need to be produced each time.

The planning should take account of avoiding suspending loads over occupied areas, visibility, attaching/detaching and securing loads, the environment, location, overturning, proximity to other objects, lifting of people and pre-use checks of the equipment.

19.23.9 Thorough examination and inspection – Regulations 9 and 11

Before using lifting equipment for the first time by an employer, it must be thoroughly examined for any defect unless:

▶ the lifting equipment has not been used before;
▶ an EC declaration of conformity (where one should have been drawn up) has been received or made not more than 12 months before the lifting equipment is put into service;
▶ if it is obtained from another undertaking, it is accompanied by physical evidence of an examination.

A copy of this thorough examination report shall be kept for as long as the lifting equipment is used (or, for a lifting accessory, 2 years after the report is made) (Regulation 11).

Where safety depends on the installation conditions, the equipment shall be thoroughly examined:

▶ after installation and before being put into service;
▶ after assembly and before being put into service at a new site or in a new location to ensure that it has been installed correctly and is safe to operate.

A copy of the thorough examination report shall be kept for as long as the lifting equipment is used at the place it was installed or assembled (Regulation 11).

Lifting equipment which is exposed to conditions causing deterioration that may result in dangerous situations, shall be:

▶ thoroughly examined at least every 6 months (for lifting equipment for lifting persons, or a lifting accessory); at least every 12 months (for other lifting equipment); or in accordance with an examination scheme; and each time that exceptional circumstances, liable to jeopardise the safety of the lifting equipment, have occurred and a copy of the

report kept until the next report is made, or for 2 years (whichever is longer);
▶ inspected, if appropriate, by a competent person at suitable intervals between 'thorough examinations' (and a copy of the record kept until the next record is made).

All lifting equipment shall be accompanied by physical evidence that the last 'thorough examination' has been carried out before it leaves an employer's undertaking (or before it is used after leaving another undertaking).

The user, owner, manufacturer or some other independent party may draw up examination schemes, provided they have the necessary competence. Schemes should specify the intervals at which lifting equipment should be thoroughly examined and, where appropriate, those parts that need to be tested. The scheme should take account, for example, of its condition, the environment in which it is used, the number of lifting operations and the loads lifted.

The 'competent person' carrying out a thorough examination should have appropriate practical and theoretical knowledge and experience of the lifting equipment to be examined to enable them to detect defects or weaknesses and to assess their importance in relation to the safety and continued use of the lifting equipment. They should also determine whether a test is necessary and the most appropriate method for carrying it out.

19.23.10 Reports and defects – Regulation 10

The person making a 'thorough examination' of lifting equipment shall:

▶ notify the employer forthwith of any defect which, in their opinion, is or could become, dangerous;
▶ as soon as is practicable (within 28 days) write an authenticated report to:
  ▷ the employer;
  ▷ any person who hired or leased the lifting equipment, containing the information specified in Schedule 1.
▶ send a copy (as soon as is practicable) to the relevant enforcing authority where there is, in their opinion, a defect with an existing or imminent risk of serious personal injury (this will always be the HSE if the lifting equipment has been hired or leased).

Every employer notified of a defect following a 'thorough examination' of lifting equipment should ensure that it is not used:

▶ before the defect is rectified;
▶ after a time specified in the schedule accompanying the report.

The person making an 'inspection' shall also notify the employer when, in his or her opinion, a defect is, or could become, dangerous and, as soon as is practicable, make a record of the inspection in writing.

19.23.11 Reports – Schedule 1

Schedule 1 lists the information to be contained in a report of a thorough examination. For example, name and address; identity of equipment; date of last thorough examination; SWL; appropriate interval; any dangerous or potentially dangerous defects; repairs required; date of next examination and test; details of the competent person.

19.23.12 Further information

HSE LOLER website: http://www.hse.gov.uk/work-equipment-machinery/loler.htm

Lifting Equipment and Lifting Operations Regulations 1998, SI No., OPSI website: http://www.legislation.gov.uk/uksi/1998/2307/contents/made

Lifting Equipment at work. A brief guide, INDG290(rev1), HSE books, 2013, ISBN 9780 7176 6483 2

LOLER 1998 Lifting Operations and Lifting Equipment Regulations 1998, Open Learning Guidance. 1999, HSE Books, ISBN 978 0 7176 2464 5. http://www.hse.gov.uk/pubns/books/loler.htm

Rider-Operated lift trucks, L117, 2013, HSE Books, ISBN 978 0 7176 6441 2. http://www.hse.gov.uk/pubns/books/l117.htm

Safe Use of Lifting Equipment, Lifting Operations and Lifting Equipment Regulations, 1998, Approved Code of Practice and Guidance, L113 2nd ed., 2014, HSE Books, ISBN 978 0 7176 6586 0. http://www.hse.gov.uk/pubns/books/l113.htm?ebul=hsegen&cr=2/22-dec-14

Thorough Examination and Testing of Lifts. Simple Guidance for Lift Owners INDG339 (rev1), HSE Books, 2008, ISBN 978 0 7176 6255 5. http://www.hse.gov.uk/pubns/indg339.htm

Thorough examination of lifting equipment, A simple guide for employers, INDG422, HSE, 2008, ISBN 97807176 6305 7. http://www.hse.gov.uk/pubns/indg422.pdf

19.24 Management of Health and Safety at Work Regulations 1999 as amended in 2003 and 2006

19.24.1 General

These Regulations give effect to the European Framework Directive on health and safety. They supplement the requirements of the HSW Act 1974 and specify a range of management issues, most of which must be carried out in all workplaces. The aim is to map out the organisation of precautionary measures in a systematic way, and to make sure that all staff are familiar with the measures and their own responsibilities. They were amended slightly in 2003, particularly in relation to civil liability (Regulation 22).

19.24.2 Risk assessment – Regulation 3

Every employer is required to make a 'suitable and sufficient' assessment of risks to employees, and risks to other people who might be affected by the organisation, such as visiting contractors and members of the public.

A systematic investigation of risks involved in all areas and operations is required, together with identification of the persons affected, a description of the controls in place and any further action required to reduce risks.

The risk assessments must take into account risks to new and expectant mothers and young people.

Significant findings from the assessments must be written down (or recorded by other means, such as on a computer) when there are five or more employees. The assessments need to be reviewed regularly and if necessary, when there have been significant changes, they should be modified.

19.24.3 Principles of prevention – Regulation 4

The following principles must be adopted when implementing any preventative and protective measures:

▶ avoiding risks;
▶ evaluating the risks which cannot be avoided;
▶ combating the risks at source;
▶ adapting the work to the individual, especially as regards the design of workplaces, the choice of work equipment and the choice of working and production methods, with a view, in particular, to alleviating monotonous work and work at a predetermined work-rate and to reducing their effect on health;
▶ adapting to technical progress;
▶ replacing the dangerous substances by non-dangerous or less-dangerous substances;
▶ developing a coherent overall prevention policy which covers technology, organisation of work, working conditions, social relationships and the influence of factors relating to the working environment;
▶ giving collective protective measures priority over individual protective measures;
▶ giving appropriate instruction to employees.

19.24.4 Effective arrangements for health and safety – Regulation 5

Formal arrangements must be devised (and recorded) for effective planning, organisation, control, monitoring

19

and review of safety measures. This will involve an effective health and safety management system to implement the policy. Where there are five or more employees the arrangements should be recorded.

Planning involves a systematic approach to risk assessment, the selection of appropriate risk controls and establishing priorities with performance standards.

Organisation involves consultation and communication with employees; employee involvement in risk assessment; the provision of information; and securing competence with suitable instruction and training. Control involves clarifying responsibilities and making sure people adequately fulfil their responsibilities. It involves adequate and appropriate supervision.

Monitoring should include the measurement of how well the policy is being implemented and whether hazards are being controlled properly. It covers inspections of the workplace and management systems; and the investigation of incidents and accidents to ascertain the underlying causes and effect a remedy.

Review is essential to look at the whole of the health and safety management system to ensure that it is effective and achieving the correct standard of risk control.

19.24.5 Health surveillance – Regulation 6

In appropriate circumstances health surveillance of staff may be required – the ACoP describes more fully when this duty will arise. Health surveillance is considered relevant when: there is an identifiable disease or poor health condition; there are techniques to detect the disease; there is a reasonable likelihood that the disease will occur; and surveillance is likely to enhance the protection of the workers concerned. A competent person, who will range from a manager, in some cases, to a fully qualified occupational medical practitioner in others, should assess the extent of the surveillance.

19.24.6 Competent assistance – Regulation 7

Every employer is obliged to appoint one or more 'competent person(s)' to advise and assist in undertaking the necessary measures to comply with the relevant statutory requirements. They may be employees or outside consultants. The purpose is to make sure that all employers have access to health and safety expertise. Preference should be given to an in-house appointee who may be backed up by external expertise.

The competence of the person(s) appointed is to be judged in terms of their training, knowledge and experience of the work involved; it is not necessarily dependent upon particular qualifications. In simple situations, it may involve knowledge of relevant best practice, knowing one's limitations and taking external

advice when necessary. In more complex situations or risks, fully qualified and appropriately experienced practitioners will be required.

Appointed competent persons must be provided with adequate information, time and resources to do their job.

19.24.7 Procedures for serious and imminent danger and contact with external services – Regulations 8 and 9

Procedures must be established for dealing with serious and imminent dangers, including fire evacuation plans and arrangements for other emergencies. A sufficient number of competent persons must be appointed to evacuate the premises in the event of an emergency. The procedures should allow for persons at risk to be informed of the hazards and how and when to evacuate to avoid danger. In shared workplaces employers must cooperate. Access to dangerous areas should be restricted to authorised and properly trained staff. Any necessary contact arrangements with external services for first-aid, emergency medical care and rescue work must be set up.

19.24.8 Information for employees – Regulation 10

Information must be provided to staff on the risk assessment, risk controls, emergency procedures, the identity of the people appointed to assist on health and safety matters and risks notified by others.

The information provided must take into account the level of training, knowledge and experience of the employees. It must take account of language difficulties and be provided in a form that can be understood by everyone. The use of translations, symbols and diagrams should be considered. Where children under school-leaving age are at work, information on the risk assessments and control measures must be provided to the child's parent or guardians before the child starts work. It can be provided verbally or directly to the parent, guardians or school.

19.24.9 Cooperation and coordination – Regulations 11, 12 and 15

Where two or more employers share a workplace, each must:

▶ cooperate with other employers in health and safety matters;
▶ take reasonable steps to coordinate their safety precautions;
▶ inform the other employers of the risks to their employees, i.e. risks to neighbours' employees.

Where people from outside organisations are present to do work, they – and their employers – have to be provided with appropriate information on risks, health and the necessary precautions to be taken.

Temporary staff and staff with fixed-term contracts as well as permanent employees must be supplied with health and safety information before starting work (Regulations 12 and 15).

Regulation 11 does not apply to multi-occupied premises or sites where each unit, under the control of an individual tenant employer or self-employed person, is regarded as a separate workplace. In other cases, common areas may be shared workplaces, such as a reception area or canteen, or they may be under the control of a person to whom Section 4 of the HSW Act applies. Suitable arrangements may need to be put in place for these areas.

19.24.10 Capabilities and training – Regulation 13

When giving tasks to employees, their capabilities with regard to health and safety must be taken into account.

Employees must be provided with adequate health and safety training:

- on recruitment;
- on being exposed to new or increased risks;
- on the introduction of new procedures, systems or technology. Training must be repeated periodically and take place in working hours (or while being paid).

19.24.11 Duties on employees – Regulation 14

Equipment and materials must be used properly in accordance with instructions and training. Obligations on employees are extended to include certain requirements to report serious and immediate dangers and any shortcomings in the employer's protection arrangements.

19.24.12 New or expectant mothers – Regulations 16–18

Where work is of a kind that could present a risk to new or expectant mothers working there or their babies, the risk assessments must include an assessment of such risks. When the risks cannot be avoided the employer must alter a woman's working conditions or hours to avoid the risks; offer suitable alternative work; or suspend from work on full pay. The woman must notify the employer in writing of her pregnancy, that she has given birth within the last 6 months or she is breastfeeding.

19.24.13 Young persons – Regulation 19

Employers must protect young persons at work from risks to their health and safety which are the result of lack of experience, or absence of awareness of existing or potential risks or which arise because they have not yet fully matured. Young persons may not be employed in a variety of situations which pose a significant risk to their health and safety that are a consequence of the following factors:

- physical or psychological capacity;
- pace of work;
- temperature extremes, noise or vibration;
- radiation;
- compressed air and diving;
- hazardous substances;
- lack of training and experience.

The exception to this is young persons over school-leaving age:

- where the work is necessary for their training;
- where they will be supervised by a competent person;
- where the risk will be reduced to the lowest level that is reasonably practicable.

The risk assessment must take the following specific factors into account:

- the fitting-out and layout of the workplace and the particular site where they will work;
- the nature of any physical, biological and chemical agents they will be exposed to, for how long and to what extent;
- what types of work equipment will be used and how this will be handled;
- how the work and processes involved are organised;
- the need to assess and provide health and safety training;
- risks from the particular agents, processes and work.

19.24.14 Provisions as to liability – Regulation 21

This provision is to prevent a defence for an employer by reason of any act or default by an employee or a competent person appointed under Regulation 7. Employees' duties to take reasonable care of their own health and safety and that of others affected by their work activities are unaffected by this provision.

19.24.15 Restriction of civil liability for breach of statutory duty – Regulation 22

Further restrictions have been imposed by the Enterprise and Regulatory Reform Act 2013. It amends the Health and Safety at Work etc. Act 1974 Section 47, to remove the right of civil action against employers and other duty-holders for breach of statutory duty in relation to certain health and safety legislation, other than where such a right is specifically provided for which has been done by The Health and Safety at Work etc. Act 1974 (Civil Liability) (Exceptions) Regulations 2013 which changed Regulation 22 (see below). This

19

will address the potential unfairness that arises where an employer can be found liable to pay compensation to an employee despite having taken reasonable steps to protect them.

Regulation 22 of the Management of Health and Safety at Work Regulations ('the 1999 Regulations') excluded third parties from bringing claims for breach of a duty contained in those Regulations against employers and employees. This exclusion is no longer required because Section 47 of the 1974 Act now automatically excludes liability and express provision is only required for exceptions.

Regulation 22 has been changed to the following:

> **22.** – *(1) Breach of a duty imposed by regulation 16, 16A, 17 or 17A shall, so far as it causes damage, be actionable by the new or expectant mother.*
>
> *(2) Any term of an agreement which purports to exclude or restrict any liability for such a breach is void.*

19.24.16 Further information

Health and Safety made simple: The basics for your business, INDG449, HSE Books. http://www.hse.gov.uk/pubns/indg449.pdf

Risk Assessment HSE INDG163 (rev 4), 2014, HSE Books. http://www.hse.gov.uk/pubns/indg163.pdf

Leading Health and Safety at Work. Action for Directors and Board Members, business owners and organisations of all sizes INDG417(rev1), HSE Books, 2013. http://www.hse.gov.uk/leadership/

The Management of Health and Safety at Work Regulations 1999, ISBN 9780 11 025051 6, OPSI website: http://www.opsi.gov.uk/si/si1999/uksi_19993242_en.pdf

Managing for health and safety HSG65, 2013, ISBN 9780717664566 see: http://www.hse.gov.uk/pubns/books/hsg65.htm

The health and safety toolbox: How to control risks at work HSG268, 2014, ISBN 9780717665877. http://www.hse.gov.uk/toolbox/index.htm

19.25 Manual Handling Operations Regulations (MHO) 1992 as amended in 2002

19.25.1 General

The Regulations apply to the manual handling (any transporting or supporting) of loads, that is by human effort, as opposed to mechanical handling by fork-lift truck, crane, etc. Manual handling includes lifting, putting down, pushing, pulling, carrying or moving. The human effort may be applied directly to the load, or indirectly by pulling on a rope, chain or lever. Introducing mechanical assistance, like a hoist or sack truck, may reduce but not eliminate manual handling, as human effort is still required to move, steady or position the load.

The application of human effort for purposes other than transporting or supporting a load, for example pulling on a rope to lash down a load or moving a machine control, is not a manual handling operation. A load is a discrete movable object, but it does not include an implement, tool or machine while being used.

Injury in the context of these Regulations means to any part of the body. It should take account of the physical features of the load which might affect grip or cause direct injury, for example slipperiness, sharp edges and extremes of temperature. It does not include injury caused by any toxic or corrosive substance which has leaked from a load, is on its surface or is part of the load. See Figure 19.9, which shows a flow chart for the Regulations.

19.25.2 Duties of employers: avoidance of manual handling – Regulation 4(1)(a)

Employers should take steps to avoid the need for employees to carry out MHO which involves a risk of their being injured.

The guidance suggests that a preliminary assessment should be carried out when making a general risk assessment under the MHSWR 1999. Employers should consider whether the operation can be eliminated, automated or mechanised.

19.25.3 Duties of employers assessment of risk – Regulation 4(1)(b)(i)

Where it is not reasonably practicable to avoid MHO, employers must make a suitable and sufficient risk assessment of all such MHO in accordance with the requirements of Schedule 1 to the Regulations (shown in Table 19.6). This duty to assess the risk takes into account the task, the load, the working environment and individual capability.

19.25.4 Duties of employers: reducing the risk of injury – Regulation 4(1)(b)(ii)

Where it is not reasonably practicable to avoid the MHO with which there is a risk of injury, employers must take steps to reduce the risk of injury to the lowest level reasonably practicable.

The structured approach (considering the task, the load, the working environment and the individual capability) is recommended in the guidance. The steps taken will involve ergonomics, changing the load, mechanical assistance, task layout, work routines, PPE, team working and training.

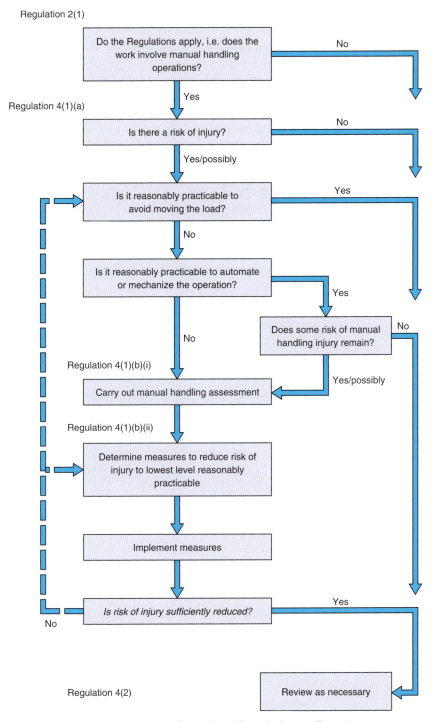

Figure 19.9 Manual Handling Operations Regulations – flow chart

19.25.5 Duties of employers: additional information on the load – Regulation 4(1)(b)(iii)

Employers must take appropriate steps where manual handling cannot be avoided to provide general indications and, where practicable, precise information on:

▶ the weight of each load;
▶ the heaviest side of any load which does not have a central centre of gravity.

The information is probably best marked on the load. Sections 3 and 6 of the HSW Act may place duties on originators of loads, like manufacturers or packers.

19.25.6 Duties of employers: reviewing assessment – Regulation 4(2)

The assessment should be reviewed if there is reason to suspect that it is no longer valid or there have been significant changes in the particular MHO.

19

613

Table 19.6 Schedule 1 to the Manual Handling Operations Regulations

Factors to which the employer must have regard and questions they must consider when making an assessment of manual handling operations	
Factors	**Questions**
1. The tasks	*Do they involve:*
	Holding or manipulating loads at distance from trunk?
	Unsatisfactory bodily movement or posture, especially:
	Twisting the trunk?
	Stooping?
	Reaching upwards?
	Excessive movement of loads especially:
	Excessive lifting or lowering distances?
	Excessive carrying distances?
	Excessive pulling or pushing of loads?
	Risk of sudden movement of loads?
	Frequent or prolonged physical effort?
	Insufficient rest or recovery periods?
	A rate of work imposed by a process?
2. The loads	*Are they:*
	Heavy?
	Bulky or unwieldy?
	Difficult to grasp?
	Unstable, or with contents likely to shift?
	Sharp, hot or otherwise potentially damaging?
3. The working environment	*Are there:*
	Space constraints preventing good posture?
	Uneven, slippery or unstable floors?
	Variations in level of floors or work surfaces?
	Extremes of temperature or humidity?
	Conditions causing ventilation problems or gusts of wind?
	Poor lighting conditions?
4. Individual capability	*Does the job:*
	Require unusual strength, height, etc.?
	Create a hazard to those who might reasonably be considered to be pregnant or have a health problem?
	Require special information or training for its safe performance?
5. Other factors	Is movement or posture hindered by personal protective equipment or by clothing?

19.25.7 Individual capability – Regulation 4(3)(a)–(f)

A further requirement was added in 2002 by Amendment Regulations concerning the individual capabilities of people undertaking manual handling. Regard must be had in particular to:

(a) the physical suitability of the employee to carry out the operation;

(b) the clothing, footwear or other personal effects they are wearing;

(c) their knowledge and training;

(d) results of any relevant risk assessments carried out under Regulation 3 of the Management Regulations 1999;

(e) whether the employee is within a group of employees identified by that assessment as being especially at risk;

(f) the results of any health surveillance provided pursuant to Regulation 6 of the Management Regulations 1999.

19.25.8 Duty of employees – Regulation 5

Each employee, while at work, has to make proper use of any system of work provided for their use. This is in addition to other responsibilities under the HSW Act and the MHSWR.

The provisions do not include well-intentioned improvisation in an emergency, for example rescuing a casualty or fighting a fire.

19.25.9 Further information

Making the Best Use of Lifting and Handling Aids INDG398(rev1), HSE Books, 2013. http://www.hse.gov.uk/pubns/indg398.pdf

Managing upper limb disorders in the workplace, INDG171(rev2), 2013. http://www.hse.gov.uk/pubns/indg171.pdf

Manual Handling Assessment Chart INDG383(rev1), HSE Books, 2014. http://www.hse.gov.uk/pubns/indg383.htm

Manual handling at work. A brief guide, INDG143(rev3), 2012, ISBN 9780 7176 6478 8. http://www.hse.gov.uk/pubns/indg143.htm

Manual Handling, Manual Handling Operations Regulations 1992 (as amended), Guidance on Regulations Revised, L23, HSE Books, 2004, ISBN 978 0 7176 2823 0. http://www.hse.gov.uk/pubns/books/l23.htm

Manual Handling, Solutions You Can Handle HSG 115, 1994, HSE Books, ISBN 978 0 7176 0693 1. http://www.hse.gov.uk/pubns/books/hsg115.htm

Upper limb disorders in the workplace, HSG60, 2002, ISBN 9780 7176 1978 8. http://www.hse.gov.uk/pubns/books/hsg60.htm

19.26 Control of Noise at Work Regulations 2005

19.26.1 Introduction

The Control of Noise at Work Regulations 2005 require employers to prevent or reduce risks to health and safety from exposure to noise at work. Employees also have duties under the Regulations.

The Regulations require employers to:

▶ assess the risks to their employees from noise at work;
▶ take action to reduce the noise exposure that produces those risks;
▶ provide their employees with hearing protection if they cannot reduce the noise exposure enough by using other methods;
▶ make sure the legal limits on noise exposure are not exceeded;
▶ provide their employees with information, instruction and training;
▶ carry out health surveillance where there is a risk to health.

The Regulations do not apply to:

▶ members of the public exposed to noise from their non-work activities, or making an informed choice to go to noisy places;
▶ low-level noise which is a nuisance but causes no risk of hearing damage.

The Control of Noise at Work Regulations came into force for the music and entertainment sectors on 6 April 2008, bringing them in line with all other sectors where the regulations have been in force since April 2006. The Control of Noise at Work Regulations now apply to pubs and clubs, amplified live music events, orchestras and other premises where live music or recorded music is played.

19.26.2 Exposure limit values and action levels – Regulation 4

The Noise at Work Regulations require employers to take specific action at certain action values. These relate to:

▶ the levels of exposure to noise of their employees averaged over a working day or week;
▶ the maximum noise (peak sound pressure) to which employees are exposed in a working day.

The values are:

▶ lower exposure action values:
  ▷ daily or weekly exposure of 80 dB (A-weighted);
  ▷ peak sound pressure of 135 dB (C-weighted).
▶ upper exposure action values:
  ▷ daily or weekly exposure of 85 dB (A-weighted);
  ▷ peak sound pressure of 137 dB (C-weighted).

Figure 19.10 helps decide what needs to be done.

There are also levels of noise exposure which must not be exceeded:

▶ exposure limit values:
  ▷ daily or weekly exposure of 87 dB (A-weighted);
  ▷ peak sound pressure of 140 dB (C-weighted).

These exposure limit values take account of any reduction in exposure provided by hearing protection.

19.26.3 Risk assessment – Regulation 5

If employees are likely to be exposed to noise at or above the lower exposure value, a suitable and sufficient assessment of the risks must be made. Employers need to decide whether any further action is needed, and plan how to do it.

The risk assessment should:

▶ identify where there may be a risk from noise and who is likely to be affected;
▶ contain a reliable estimate of employees' exposures, and compare the exposure with the exposure action values and limit values;
▶ identify what needs to be done to comply with the law, e.g. whether noise-control measures or hearing protection are needed, and, if so, where and what type;
▶ identify any employees who need to be provided with health surveillance and whether any are at particular risk.

The risk assessment should include consideration of:

▶ the level, type and duration of exposure, including any exposure to peak sound pressure;
▶ the effects of exposure to noise on employees or groups of employees whose health is at particular risk;
▶ sfairp, any effects on the health and safety of employees resulting from interaction, e.g. between noise and vibration;
▶ any indirect effects from the interaction between noise and audible warnings;
▶ manufacturers' information;
▶ availability of alternative equipment designed to reduce noise emissions;
▶ any extension to noise exposure due to extended hours or in supervised rest facilities;
▶ information following health surveillance;
▶ availability of personal hearing protectors with adequate attenuation characteristics.

It is essential that employers can show that their estimate of employees' exposure is representative of the work that they do. It needs to take account of:

▶ the work they do or are likely to do;
▶ the ways in which they do the work;
▶ how it might vary from one day to the next.

The estimate must be based on reliable information, for example measurements in their own workplace, information from other similar workplaces or data from suppliers of machinery.

Employers must record the significant findings of their risk assessment. They need to record in an action plan anything identified as being necessary to comply with the law, setting out what they have done and what they are going to do, with a timetable and saying who will be responsible for the work.

19

615

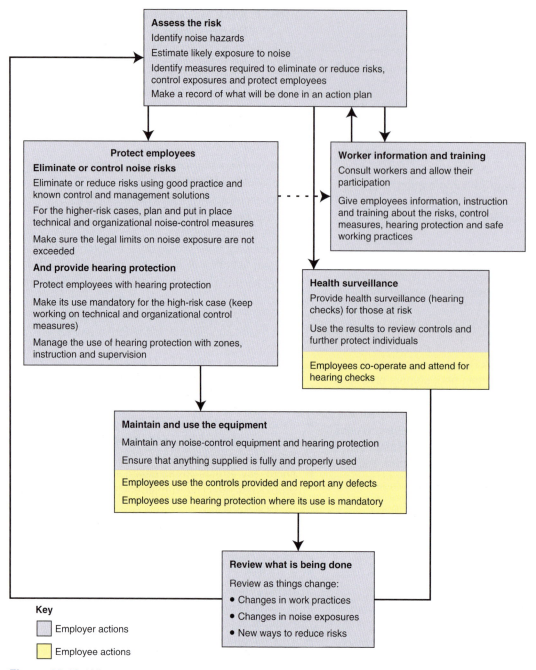

Assess the risk

Identify noise hazards

Estimate likely exposure to noise

Identify measures required to eliminate or reduce risks, control exposures and protect employees

Make a record of what will be done in an action plan

Protect employees

Eliminate or control noise risks

Eliminate or reduce risks using good practice and known control and management solutions

For the higher-risk cases, plan and put in place technical and organizational noise-control measures

Make sure the legal limits on noise exposure are not exceeded

And provide hearing protection

Protect employees with hearing protection

Make its use mandatory for the high-risk case (keep working on technical and organizational control measures)

Manage the use of hearing protection with zones, instruction and supervision

Worker information and training

Consult workers and allow their participation

Give employees information, instruction and training about the risks, control measures, hearing protection and safe working practices

Health surveillance

Provide health surveillance (hearing checks) for those at risk

Use the results to review controls and further protect individuals

Employees co-operate and attend for hearing checks

Maintain and use the equipment

Maintain any noise-control equipment and hearing protection

Ensure that anything supplied is fully and properly used

Employees use the controls provided and report any defects

Employees use hearing protection where its use is mandatory

Review what is being done

Review as things change:
- Changes in work practices
- Changes in noise exposures
- New ways to reduce risks

Key

☐ Employer actions

☐ Employee actions

Figure 19.10 What needs to be done under the Control of Noise at Work Regulations 2005

The risk assessment should be reviewed if circumstances in the workplace change and affect noise exposures. Also it should be reviewed regularly to make sure that the employer continues to do all that is reasonably practicable to control the noise risks. Even if it appears that nothing has changed, employers should not leave it for more than about 2 years without checking whether a review is needed.

19.26.4 Elimination or control of exposure – Regulation 6

The purpose of the Control of Noise at Work Regulations 2005 is to make sure that people do not suffer damage to their hearing – so efforts to controlling noise risks and noise exposure should be concentrated.

Wherever there is noise at work employers should be looking for alternative processes, equipment and/or working methods which would make the work quieter or mean people are exposed for shorter times. They should also be keeping up with what is good practice or the standard for noise control within their industry.

Where there are things that can be done to reduce risks from noise that are reasonably practicable, they should be done. However, where noise exposures are below the lower exposure action values, risks are low and so employers would only be expected to take actions which are relatively inexpensive and simple to carry out.

Where the assessment shows that employees are likely to be exposed at or above the upper exposure action values, employers must put in place a planned programme of noise control.

The risk assessment will have produced information on the risks and an action plan for controlling noise. Employers should use this information to:

▶ tackle the immediate risk, e.g. by providing hearing protection;
▶ identify what is possible to control noise, how much reduction could be achieved and what is reasonably practicable;
▶ establish priorities for action and a timetable (e.g. consider where there could be immediate benefits, what changes may need to be phased in over a longer period of time and the number of people exposed to the noise in each case);
▶ assign responsibilities to people to deliver the various parts of the plan;
▶ ensure the work on noise control is carried out;
▶ check that what has been done has worked.

Actions taken should be based on the general principles set out in the Management Regulations and should include consideration of:

▶ other working methods;
▶ choice of appropriate work equipment emitting the least possible noise;
▶ the design and layout of workplaces, workstations and rest facilities;
▶ suitable and sufficient information and training;
▶ reduction of noise by technical means;
▶ appropriate maintenance programmes;
▶ limitation of the duration and intensity of exposure;
▶ appropriate work schedules with adequate rest periods.

19.26.5 Hearing protection – Regulation 7

Hearing protection should be issued to employees:

▶ where extra protection is needed above what can been achieved using noise control;
▶ as a short-term measure while other methods of controlling noise are being developed.

Hearing protection should not be used as an alternative to controlling noise by technical and organisational means.

Employers should consult with their employees or their representatives on the type of hearing protection to be used.

Employers are required to:

▶ provide employees with hearing protectors if they ask for it and their noise exposure is between the lower and upper exposure action values;
▶ provide employees with hearing protectors and make sure they use them properly when their noise exposure exceeds the upper exposure action values;
▶ identify hearing protection zones, i.e. areas where

the use of hearing protection is compulsory, and mark them with signs if possible. Restrict access to hearing protection zones where this is practicable and the noise exposure justifies it;
▶ provide employees with training and information on how to use and care for the hearing protectors;
▶ ensure that the hearing protectors are properly used and maintained.

19.26.6 Maintenance and use of equipment – Regulation 8

Employers need to make sure that hearing protection works effectively and to:

▶ check that all equipment provided in compliance with the regulations remains in good, clean condition and that there are no unofficial modifications;
▶ check that hearing protection is fully and properly used (except where it is provided for employees who are exposed at or above the lower exposure but below the upper exposure level). This is likely to mean that an employer needs to:
  ▷ put someone in authority in overall charge of issuing it and making sure replacements are readily available;
  ▷ carry out spot checks to see that the rules are being followed and that hearing protection is being used properly;
  ▷ ensure all managers and supervisors set a good example and wear hearing protection at all times when in hearing protection zones;
  ▷ ensure only people who need to be there enter hearing protection zones and do not stay longer than they need to.

19.26.7 Health surveillance – Regulation 9

Employers must provide health surveillance (hearing checks) for all their employees who are likely to be regularly exposed above the upper exposure action values, or are at risk for any reason – for example they already suffer from hearing loss or are particularly sensitive to damage.

The purpose of health surveillance is to:

▶ warn when employees might be suffering from early signs of hearing damage;
▶ give employers an opportunity to do something to prevent the damage getting worse;
▶ check that control measures are working.

The trade union safety representative, or employee representative, and the employees concerned should be consulted before introducing health surveillance. It is important that employees understand that the aim of health surveillance is to protect their hearing. Employers will need their understanding and cooperation if health surveillance is to be effective.

19

617

Health surveillance for hearing damage usually means:

► regular hearing checks in controlled conditions;
► telling employees about the results of their hearing checks (required after reasonable notice by employee);
► keeping health records;
► ensuring employees are examined by a doctor where hearing damage is identified.

If the doctor considers the hearing damage is likely to be the result of exposure, the employer must:

► ensure that the employee is informed by a suitably qualified person;
► review the risk assessment and control measures;
► consider assigning the employee to alternative work;
► ensure continued surveillance.

19.26.8 Information, instruction and training – Regulation 10

Where employees are exposed to noise which is likely to be at or above the lower exposure level, suitable and sufficient information, instruction and training must be provided and kept up to date. This includes the nature of the risks, compliance action taken, significant findings of the risk assessment, hearing protection, how to detect and report signs of hearing damage, entitlement to health surveillance, safe working practices and collective results of health surveillance.

19.26.9 Further information

Control of Noise at Work Regulations 2005 (SI No. 1643) http://www.legislation.gov.uk/uksi/2005/1643/contents/made

Controlling Noise at Work. The Control of Noise At Work Regulations 2005, L108 (second edition), 2005, HSE Books, ISBN 978 0 7176 6164 0. http://www.hse.gov.uk/pubns/books/l108.htm

Noise Don't lose your hearing INDG363 (rev 2) 2012, HSE Books, ISBN 978 0 7176 6510 5. http://www.hse.gov.uk/pubns/indg363.htm

Noise at Work: a brief guide to controlling the risks INDG362 (rev 2), 2012, HSE Books, ISBN 978 0 7176 6482 5. http://www.hse.gov.uk/pubns/indg362.htm

Sound Advice: Control of noise at work in music and entertainment, HSG260, HSE Books, 2008, ISBN 9780 7176 6307. 1 http://www.hse.gov.uk/pubns/priced/hsg260.pdf

19.27 Personal Protective Equipment at Work Regulations 1992 as amended in 2002 and 2013

19.27.1 Introduction

The effect of the PPE at Work Regulations is to ensure that certain basic duties governing the provision and use of PPE apply to all situations where PPE is required. The Regulations follow sound principles for the effective and economical use of PPE, which all employers should follow.

PPE, as defined, includes all equipment (including clothing affording protection against the weather) which is intended to be worn or held by a person at work and which protects them against one or more risks to their health and safety. Waterproof, weatherproof or insulated clothing is covered only if its use is necessary to protect against adverse climatic conditions.

Ordinary working clothes and uniforms, which do not specifically protect against risks to health and safety, and protective equipment worn in sports competitions, are not covered.

Where there is overlap in the duties in these Regulations and those covering lead, ionising radiations, asbestos, hazardous substances (COSHH), and noise then the specific legislative requirements should prevail. Construction head protection is now included within the scope of these Regulations by a 2013 amendment and the revocation of the Construction (Head protection) Regulations.

19.27.2 Provision of PPE – Regulation 4

Every employer shall ensure that suitable PPE is provided to their employees who may be exposed to risks to their health and safety except where it has been adequately or more effectively controlled by other means. (Management Regulations require PPE to be the last choice in the principles of protection.)

PPE shall not be suitable unless:

► it is appropriate for the risks and the conditions of use including the period for which it is worn;
► it takes account of ergonomic requirements and the state of health of the wearer and the characteristics of the workstation of each person;
► it is capable of fitting the wearer correctly, by adjustments if necessary;
► it is, so far as is practicable, able to combat the risks without increasing overall risks;
► it complies with UK legislation on design or manufacture, i.e. it has a CE marking.

Where it is necessary that PPE is hygienic and otherwise free of risk to health, PPE must be provided to a person solely for their individual use.

19.27.3 Compatibility – Regulation 5

Where more than one health and safety risk necessitates the wearing of multiple PPE simultaneously then they shall be compatible and remain effective.

19.27.4 Assessment – Regulation 6

Before choosing any PPE, employers must ensure that an assessment is made to determine whether the PPE is suitable.

The assessment shall include:

▶ assessing risks which have not been avoided by other means;
▶ a definition of the characteristics that PPE must have to be effective, taking into account any risks created by the PPE itself;
▶ a comparison of available PPE with the required characteristics;
▶ information on whether the PPE is compatible with any other PPE which is in use, and which an employee would be required to wear simultaneously.

The assessment should be reviewed if it is no longer valid or there have been significant changes. In simple cases it will not be necessary to record the assessment but, in more complex cases, written records should be made and kept readily available for future reference.

19.27.5 Maintenance – Regulation 7

Every employer (and self-employed person) shall ensure that any PPE provided is maintained, including being replaced and cleaned, in an efficient state, in efficient working order and in good repair.

The guide emphasises the need to set up an effective system of maintenance for PPE. This should be proportionate to the risks and appropriate to the particular PPE. It could include, where appropriate, cleaning, disinfection, examination, replacement, repair and testing. For example, mechanical fall arrestor equipment or sub-aqua breathing apparatus will require planned preventative maintenance with examination, testing and overhaul. Records should be kept of the maintenance work. Gloves may only require periodic inspection by the user as necessary, depending on their use.

Spare parts must be compatible and be the proper part, suitably CE-marked where applicable. Manufacturers' maintenance schedules and instructions should be followed unless alternative schemes are agreed with the manufacturer or agent.

In some cases these requirements can be fulfilled by using disposable PPE which can be discarded after use or when their life has expired. Users should know when to discard and replace disposable PPE.

19.27.6 Accommodation – Regulation 8

When an employer or self-employed person has to provide PPE they must ensure that appropriate accommodation is provided to store it when not in use.

The type of accommodation will vary and may just be suitable hooks for special clothing and small portable cases for goggles. It should be separate from normal outer clothing storage arrangements and protect the PPE from contamination or deterioration.

19.27.7 Information, instruction and training – Regulation 9

Employers shall provide employees with adequate and appropriate information, instruction and training on:

▶ the risks which the PPE will avoid or limit;
▶ the purpose for which and the manner in which PPE should be used;
▶ any action required of the employee to maintain the PPE.

Employers must ensure that the information is kept available. The employer shall, where appropriate and at suitable intervals, organise demonstrations of the wearing of PPE. The guidance suggests the training should include:

▶ an explanation of the risks and why PPE is needed;
▶ the operation, performance and limitations of the equipment;
▶ instructions on the selection, use and storage of PPE;
▶ problems that can affect PPE relating to other equipment, working conditions, defective equipment, hygiene factors and poor fit;
▶ the recognition of defects and how to report problems with PPE;
▶ practice in putting on, wearing and removing PPE;
▶ practice in user cleaning and maintenance;
▶ how to store safely.

19.27.8 Use and reporting of defects – Regulations 10 and 11

Every employer shall take all reasonable steps to ensure that PPE is properly used.

Every employee shall:

▶ use PPE provided in accordance with training and instructions;
▶ return it to the accommodation provided after use;
▶ report any loss or obvious defect.

19.27.9 Further information

A Short Guide to the Personal Protective Equipment at Work Regulations 1992, INDG174(rev2), HSE Books, 2013, ISBN 978 0 7176 6141 1. www.hse.gov.uk/pubns/indg174.pdf

Hard hats: What you need to know as a busy builder. Construction Information Sheet CIS70, HSE Books, 2013. www.hse.gov.uk/pubns/cis70.pdf

Personal Protective Equipment (EC Directive) Regulations 2002. SI 2002 No. 1144, http://www.legislation.gov.uk/uksi/2002/1144/contents/made

19

Personal Protective Equipment at Work. Personal Protective Equipment at Work Regulations 1992. Guidance on Regulations. L25 (second edition), HSE Books, 2005, ISBN 978 0 7176 6139 8. http://www.hse.gov.uk/pubns/books/l25.htm

Selecting Protective Gloves for Work with Chemicals. Guidance for Employers and Health and Safety Specialists INDG330, HSE Books, 2000. www.hse.gov.uk/pubns/indg330.htm

Sun Protection Advice for Employers of Outdoor Workers INDG337, HSE Books, 2001, ISBN 978 0 7176 1982 5. http://www.hse.gov.uk/pubns/indg337.pdf

19.28 Provision and Use of Work Equipment Regulations 1998 (except Part IV) as amended in 2002 and 2013

19.28.1 Introduction

The PUWER Regulations 1998 are made under the HSW Act and their primary aim is to ensure that work equipment is used without risks to health and safety, regardless of its age, condition or origin. The requirements of PUWER that are relevant to woodworking machinery are set out in the Safe Use of Woodworking Machinery Approved Code of Practice. PUWER has specific requirements for risk assessment which are covered under the Health and Safety Management Regulations 1999.

Part IV of PUWER is concerned with power presses and is not part of the Certificate syllabus, and is therefore not covered in this summary.

19.28.2 Definitions

Work equipment means any machinery, appliance, apparatus, tool or installation for use at work.

Use in relation to work equipment means any activity involving work equipment and includes starting, stopping, programming, setting, transporting, repairing, modifying, maintaining, servicing and cleaning.

19.28.3 Duty-holders – Regulation 3

Under PUWER the following groups of people have duties placed on them:

▶ employers;
▶ the self-employed;
▶ people who have control of work equipment, for example plant hire companies.

In addition to all places of work, the Regulations apply to common parts of shared buildings, industrial estates and business parks; to temporary works sites including construction; to home working (but not to domestic work in a private household); to hotels, hostels and sheltered accommodation.

19.28.4 Suitability of work equipment – Regulation 4

Work equipment:

▶ has to be constructed or adapted so that it is suitable for its purpose;
▶ has to be selected with the conditions of use and the users' health and safety in mind;
▶ may only be used for operations for which, and under conditions for which, it is suitable.

This covers all types of use and conditions and must be considered for each particular use or condition. For example, scissors may be safer than knives with unprotected blades and should therefore be used for cutting operations where practicable; risks imposed by wet, hot or cold conditions must be considered.

19.28.5 Maintenance – Regulation 5

The Regulation sets out the general requirement to keep work equipment maintained in:

▶ an efficient state;
▶ efficient working order;
▶ good repair.

Compliance involves all three criteria. In addition, where there are maintenance logs for machinery, they must be kept up to date.

In many cases this will require routine and planned preventative maintenance of work equipment. When checks are made priority must be given to:

▶ safety;
▶ operating efficiency and performance;
▶ the equipment's general condition.

19.28.6 Inspection – Regulation 6

Where the safety of work equipment depends on the installation conditions, it must be inspected:

▶ after installation and before being put into service for the first time;
▶ after assembly at a new site or in a new location;
▶ to ensure that it has been installed correctly and is safe to operate.

Where work equipment is exposed to conditions causing deterioration which is liable to result in dangerous situations, it must be inspected:

▶ at suitable intervals;
▶ when exceptional circumstances occur.

Inspections must be determined and carried out by competent persons. An inspection will vary from a simple visual external inspection to a detailed comprehensive inspection which may include some dismantling and/or testing. However, the level of inspection would normally be less than that required for a thorough examination under, for example, LOLER for certain items of lifting equipment.

Records of inspections must be kept with sufficient information to properly identify the equipment, its normal location, dates, faults found, action taken, to whom faults were reported, who carried out the inspection, when repairs were made, date of the next inspection.

When equipment leaves an employer's undertaking it must be accompanied by physical evidence that the last inspection has been carried out.

19.28.7 Specific risks – Regulation 7

Where the use of work equipment involves specific hazards, its use must be restricted to those persons given the specific task of using it, and repairs, etc. must be restricted to designated persons.

Designated persons must be properly trained to fulfil their designated task.

The ACoP requires that hazards must be controlled using a hierarchy of control measures, starting with elimination where this is possible, then considering hardware measures such as physical barriers and, lastly, software measures such as a safe system of work.

19.28.8 Information, instruction and training – Regulations 8 and 9

Persons who use work equipment must have adequate:

▶ health and safety information;
▶ where appropriate, written instructions about the use of the equipment;
▶ training for health and safety in methods which should be adopted when using the equipment, and for any hazards and precaution which should be taken to reduce risks.

Any persons who supervise the use of work equipment should also receive information, instruction and training. The training of young persons is especially important with the need for special risk assessments under the Management of Health and Safety at Work Regulations.

Health and safety training should take place within working hours.

19.28.9 Conformity with community requirements – Regulation 10

The intention of this Regulation is to require that employers ensure that equipment, provided for use after 31 December 1992, conforms at all times with the relevant essential requirements in various European Directives made under Article 100A (now 95) of the Treaty of Rome. The requirements of PUWER 98 Regulations 11–19 and 22–29 only apply if the essential requirements do not apply to a particular piece of equipment.

However, PUWER Regulations 11–19 and 22–29 will apply if:

▶ they include requirements which were not included in the relevant product legislation;
▶ the relevant product legislation has not been complied with (e.g. the guards fitted on a machine when supplied were not adequate).

Employers using work equipment need to check that any new equipment has been made to the requirements of the relevant Directive, and has a CE marking, suitable instructions and a Certificate of Conformity.

The Machinery Directive was brought into UK law by the Supply of Machinery (Safety) Regulations 1992 as amended, which duplicate PUWER Regulations 11–19 and 22–29. On 29th December 2009 these regulations were replaced by the 2008 Regulations.

The employer still retains the duty to ensure that the equipment is safe to use.

19.28.10 Dangerous parts of machinery – Regulation 11

Measures have to be taken which:

▶ prevent access to any dangerous part of machinery or to any rotating stock-bar;
▶ stop the movement of any dangerous part of machinery or rotating stock-bar before any person enters a danger zone.

The measures required follow the normal hierarchy and should be considered to the extent that they are practicable in each case. They consist of:

▶ the provision of fixed guards enclosing every dangerous part of machinery; then
▶ the provision of other guards or protection devices; then
▶ the provision of jigs, holders, push-sticks or similar protection appliances used in conjunction with the machinery; then
▶ the provision of information, instruction, training and supervision as is necessary.

All guards and protection devices shall:

▶ be suitable for their purpose;
▶ be of good construction, sound material and adequate strength;
▶ be maintained in an efficient state, in efficient working order and in good repair;
▶ not give rise to increased risks to health and safety;
▶ not be easily bypassed or disabled;
▶ be situated at sufficient distance from the danger zone;
▶ not unduly restrict the view of the operating cycle of the machine where this is relevant;
▶ be so constructed or adapted that they allow operations necessary to fit or replace parts and for maintenance work, if possible without having to dismantle the guard or protection device.

19

621

19.28.11 Protection against specified hazards – Regulation 12

Exposure to health and safety risks from the following hazards must be prevented or adequately controlled:

- any article falling or being ejected from work equipment;
- rupture or disintegration of work equipment;
- work equipment catching fire or overheating;
- the unintended or premature discharge of any article, or of any gas, dust, liquid, vapour or other substance which is produced, used or stored in the work equipment;
- the unintended or premature explosion of the work equipment or any article or substance produced, used or stored in it.

19.28.12 High or very low temperature – Regulation 13

Work equipment and any article or substance produced, used or stored in work equipment which is at a high or very low temperature must have protection to prevent injury by burn, scald or sear.

This does not cover risks such as from radiant heat or glare.

Engineering methods of control such as insulation, doors, temperature control guards, should be used where practicable but there are some cases, like cooker hot plates, where this is not possible.

19.28.13 Controls – Regulations 14–18

Where work equipment is provided with (Regulation 14):

- starting controls (including restarting after a stoppage); and
- controls which change speed, pressure or other operating condition which would affect health and safety;

it should not be possible to perform any operation except by a deliberate action on the control. This does not apply to the normal operating cycle of an automatic device.

Where appropriate, one or more readily accessible **Stop controls** shall be provided to bring the work equipment to a safe condition in a safe manner (Regulation 15). They must:

- bring the work equipment to a complete stop where necessary;
- if necessary, switch off all sources of energy after stopping the functioning of the equipment;
- operate in priority to starting or operating controls.

Where appropriate, one or more readily accessible **emergency stop controls** (Regulation 16) must be provided unless it is not necessary:

- by the nature of the hazard;
- by the time taken for the stop controls to bring the equipment to a complete stop.

Emergency stop controls must have priority over stop controls. They should be provided where other safeguards are not adequate to prevent risk when something irregular happens. They should not be used as a substitute for safeguards or the normal method of stopping the equipment.

All controls for work equipment shall (Regulation 17):

- be clearly visible and identifiable including appropriate marking where necessary;
- not expose any person to a risk to their health and safety except where necessary.

Where appropriate, employers shall ensure that:

- controls are located in a safe place;
- systems of work are effective in preventing any person being in a danger zone when equipment is started;
- an audible, visible or other suitable warning is given whenever work equipment is about to start.

Persons in a danger zone as a result of starting or stopping equipment must have sufficient time and suitable means to avoid any risks.

Control systems (Regulation 18) must be safe and chosen so as to allow for failures, faults and constraints. They must:

- not create any increased risk to health and safety;
- not result in additional or increased risks when failure occurs;
- not impede the operation of any stop or emergency stop controls.

19.28.14 Isolation from sources of energy – Regulation 19

Work equipment must be provided with readily accessible and clearly identified means to isolate it from all sources of energy. Re-connection must not expose any person using the equipment to any risks.

The main purpose is to allow equipment to be made safe under particular circumstances, such as maintenance, when unsafe conditions occur, or when adverse conditions such as electrical equipment in a flammable atmosphere or wet conditions occur.

If isolation may cause a risk in itself, special precautions must be taken – for example, a support for a hydraulic press tool which could fall under gravity if the system is isolated.

19.28.15 Stability – Regulation 20

Work equipment must be stabilised by clamping or otherwise as necessary to ensure health and safety.

Most machines used in a fixed position should be bolted or fastened so that they do not move or rock in use.

19.28.16 Lighting – Regulation 21

Suitable and sufficient lighting, taking account of the operations being carried out, must be provided where people use work equipment.

This will involve general lighting and in many cases local lighting, such as on a sewing machine. If access for maintenance is required regularly, permanent lighting should be considered.

19.28.17 Maintenance operations – Regulation 22

So far as is reasonably practicable, work equipment should be constructed or adapted to allow maintenance operations to be:

- conducted while it is shut down;
- undertaken without exposing people to risk;
- carried out after appropriate protection measures have been taken.

19.28.18 Markings and warnings – Regulations 23 and 24

Work equipment should have all appropriate **markings** for reasons of health and safety made in a clearly visible manner – for example, the maximum safe working load, stop and start controls, or the maximum rotation speed of an abrasive wheel.

Work equipment must incorporate **warnings** or **warning devices** as appropriate, which are unambiguous, easily perceived and easily understood.

They may be incorporated in systems of work, a notice, a flashing light or an audible warning. They are an active instruction or warning to take specific precautions or actions when a hazard exists.

19.28.19 Part III: mobile work equipment – Regulations 25–30

The main purpose of this section is to require additional precautions relating to work equipment while it is travelling from one location to another or where it does work while moving. If the equipment is designed primarily for travel on public roads, the Road Vehicles (Construction and Use) Regulations 1986 will normally be sufficient to comply with PUWER 98.

Mobile equipment would normally move on wheels, tracks, rollers, skids, etc. Mobile equipment may be self-propelled, towed or remote controlled and may incorporate attachments. Pedestrian-controlled work equipment such as lawn mowers is not covered by Part III.

Employees carried on mobile work equipment – Regulation 25

No employee may be carried on mobile work equipment unless:

- it is suitable for carrying persons;
- it incorporates features to reduce risks as low as is reasonably practicable, including risks from wheels and tracks.

Rolling over of mobile work equipment – Regulation 26

Where there is a risk of overturning it must be minimised by:

- stabilising the equipment;
- fitting a structure so that it only falls on its side (tip over protection structure such as the vertical mast of a fork-lift truck);
- fitting a structure which gives sufficient clearance for anyone being carried if it turns over further (rollover protective structures – ROPS);
- a device giving comparable protection;
- fitting a suitable restraining system for people if there is a risk of being crushed by rolling over.

This Regulation does not apply:

- to a fork-lift truck fitted with ROPS or vertical mast which prevents it turning on its side;
- where it would increase the overall risks;
- where it would not be reasonably practicable to operate equipment;
- to any equipment provided for use before 5 December 1998.

Overturning of fork-lift trucks – Regulation 27

Fork-lift trucks which carry an employee must be adapted or equipped to reduce the risk to safety from overturning to as low as is reasonably practicable.

Self-propelled work equipment – Regulation 28

Where self-propelled work equipment may involve risks while in motion, it shall have:

- facilities to prevent unauthorised starting;
- (with multiple rail-mounted equipment) facilities to minimise the consequences of collision;
- a device for braking and stopping;
- (where safety constraints so require) emergency facilities for braking and stopping, in the event of failure of the main facility, which have readily accessible or automatic controls;
- (where the driver's vision is inadequate) devices fitted to improve vision;
- (if used at night or in dark places) appropriate lighting fitted or otherwise it shall be made sufficiently safe for its use;

19

▶ if there is anything carried or towed that constitutes a fire hazard liable to endanger employees (particularly if escape is difficult such as from a tower crane), appropriate fire-fighting equipment carried, unless it is sufficiently close by.

Remote-controlled self-propelled work equipment – Regulation 29

Where remote-controlled self-propelled work equipment involves a risk while in motion it shall:

▶ stop automatically once it leaves its control range;
▶ have features or devices to guard against the risk of crushing or impact.

Drive shafts – Regulation 30

Where seizure of the drive shaft between mobile work equipment and its accessories or anything towed is likely to involve a risk to safety:

▶ the equipment must have means to prevent a seizure;
▶ where it cannot be avoided, every possible measure should be taken to avoid risks;
▶ the shaft should be safeguarded from contacting the ground and becoming soiled or damaged.

19.28.20 Part IV – power presses

Regulations 31–35 relate to power presses and are not included here as they are excluded from the NEBOSH Certificate syllabus. Details can be found in the Power Press ACoP.

19.28.21 Further information

Hiring and Leasing Out of Plant: Application of PUWER 98, Regulations 26 and 27, HSE MISC156, 1998, HSE Books. www.hse.gov.uk/pubns/9204.pdf

Providing and using work equipment safely: A brief guide, INDG291(rev1), HSE Books, 2013, ISBN 978 0 7176 2429 4. http://www.hse.gov.uk/pubns/indg291.pdf

Provision and Use of Work Equipment Regulations 1992, SI 1992 No. 2932. http://www.legislation.gov.uk/uksi/1998/2306/regulation/12/made

PUWER 1998, Provision and Use of Work Equipment Regulations 1998: Open Learning Guidance, HSE Books, 2008, ISBN 978 0 7176 6285 2. http://www.hse.gov.uk/pubns/books/puwer.htm

Retrofitting of Roll-over Protective Structures, Restraining Systems and Their Attachment Points to Mobile Work Equipment MISC175, HSE Books, 1999. http://www.hse.gov.uk/pubns/misc175.pdf

Safe Use of Woodworking Machinery. Provision and Use of Work Equipment Regulations 1998 as applied to woodworking machinery. Approved Code of Practice and Guidance L114 (2nd Edition), HSE Books, 2014, ISBN 978 0 7176 6621 8. www.hse.gov.uk/pubns/books/l114.

htm

Safe Use of Work Equipment, Provision and Use of Work Equipment Regulations 1998, Approved Code of Practice and Guidance, HSC L22 (4th Edition), HSE Books, 2014, ISBN 978 0 7176 6619 5. http://www.hse.gov.uk/pubns/books/l22.htm

The selection and management of mobile elevating work platforms, HSE information sheet, 2014, GEIS6. http://www.hse.gov.uk/pubns/geis6.htm

Using Work Equipment Safely INDG229(rev2), 2012, HSE Books, http://www.hse.gov.uk/pubns/indg229.pdf

19.29 The Reporting of Injuries, Diseases and Dangerous Occurrences Regulations 2013

19.29.1 Introduction

RIDDOR is the law that requires employers, and other people who are in control of work premises, to report and keep records of:

▶ work-related deaths
▶ certain serious injuries (reportable injuries);
▶ diagnosed cases of certain industrial diseases; and
▶ certain 'dangerous occurrences' (near-miss incidents).

These new RIDDOR regulations came into force on 1 October 2013.

The main changes are to simplify the reporting requirements in the following areas:

▶ The classification of 'major injuries' to workers is being replaced with a shorter list of 'specified injuries'.
▶ The previous list of 47 types of industrial disease is being replaced with eight categories of reportable work-related illness.
▶ Fewer types of 'dangerous occurrence' require reporting.

There are no significant changes to the reporting requirements for:

▶ Fatal accidents.
▶ Accidents to non-workers (members of the public).
▶ Accidents which result in the incapacitation of a worker for more than seven days.

Recording requirements remain broadly unchanged, including the requirement to record accidents resulting in the incapacitation of a worker for more than three days.

19.29.2 Definitions – Regulation 2

Accident includes an act of non-consensual physical violence done to a person at work.

Dangerous occurrence means an occurrence which arises out of or in connection with work and is of a class specified in Schedule 2 of the regulations.

Reportable incident means an incident giving rise to a notification or reporting requirement under these Regulations.

Disease includes a medical condition.

For the purposes of RIDDOR, an accident is a separate, identifiable, unintended incident that causes physical injury. This specifically includes acts of non-consensual violence to people at work.

Not all accidents need to be reported; a RIDDOR report is required only when:

▶ the accident is **work-related**; and
▶ it results in an injury of a type which is **reportable** (Regulations 4, 5 and 6).

Examples of incidents that do and do not have to be reported are available at www.hse.gov.uk/riddor/do-i-need-to-report.htm

19.29.3 Responsible person

Reports under the RIDDOR regulations should only be submitted by the **responsible person,** such as employers, the self-employed, and those in control of premises where incidents occur as follows:

Employers must report any work-related deaths, and certain work-related injuries, cases of disease, and near misses involving your employees wherever they are working.

Persons in control of premises, must report any work-related deaths, certain injuries to members of the public and self-employed people on their premises, and dangerous occurrences (some near miss incidents) that occur on their premises.

Self-employed if working in someone else's work premises and suffer either a specified injury or an over-seven-day injury, then **the person in control of the premises** will be responsible for reporting. If a self-employed person has a reportable accident while working on their own premises or in domestic premises, or if a doctor tells them that they have a work-related disease or condition, then the **self-employed** person will need to report it.

Employment Agencies are often the legal employer and are responsible for reporting like any other employer. In other cases such as self-employed people, the duty is on the host business. Agents should ensure that responsibilities under RIDDOR are clearly assigned.

There are special cases for mines, quarries, pipelines and wells (other than a water well).

19.29.4 Non-fatal injuries to workers – Regulation 4

Specified injuries

Responsible persons must follow the reporting procedures where as a result of a work-related accident a person at work suffers a specified injury.

The list of 'specified injuries' in RIDDOR 2013 (Regulation 4) includes:

▶ a fracture, other than to fingers, thumbs and toes;
▶ amputation of an arm, hand, finger, thumb, leg, foot or toe;
▶ permanent loss of sight or reduction of sight;
▶ crush injuries leading to internal organ damage;
▶ any burns (covering more than 10% of the body, or damaging the eyes, respiratory system or other vital organs);
▶ scalpings (separation of skin from the head) which require hospital treatment;
▶ unconsciousness caused by head injury or asphyxia;
▶ any other injury arising from working in an enclosed space, which leads to hypothermia, heat-induced illness or requires resuscitation or admittance to hospital for more than 24 hours.

Over seven-day incapacitation of a worker

Accidents must be reported where they result in an employee or self-employed person being away from work, or unable to perform their normal work duties, for more than **seven** consecutive days as the result of their injury. This seven-day period does not include the day of the accident, but does include weekends and rest days. The responsible person must send a report in an approved manner, within 15 days of the accident.

19.29.5 Non-fatal injuries to non-workers – Regulation 5

Accidents to members of the public or others who are not at work must be reported if they result in an **injury** and the person is taken directly from the scene of the accident to hospital for treatment to that injury. Examinations and diagnostic tests do not constitute 'treatment' in such circumstances.

If the accident occurred at a hospital, the report only needs to be made if the injury is a 'specified injury'.

19.29.6 Work-related fatalities – Regulation 6

Where any person dies (up to a year after the date of the accident, regardless of whether it was reported at the time) as a result of a work-related accident or occupational exposure to a biological agent, the responsible person must report the death.

This does not apply to a self-employed person working on their own premises, although other reporting requirements to the police are likely to apply.

19.29.7 Dangerous occurrences

There are 27 categories of dangerous occurrences which must be reported by the responsible person following the procedure.

The list from **Schedule 2 Part 1 – General** is as follows:

19

Lifting machinery, etc.

1. The collapse, overturning, or the failure of any load-bearing part of lifts and lifting equipment.

Pressure systems

2. The failure of any closed vessel or of any associated pipe-work (other than a pipeline), forming part of a pressure system as defined by the Pressure Systems Safety Regulations 2000, where that failure could cause the death of any person.

Overhead electric lines

3. Any plant or equipment unintentionally coming into –
   (a) contact with an uninsulated overhead electric line in which the voltage exceeds 200 volts; or
   (b) close proximity with such an electric line, such that it causes an electrical discharge.

Electrical incidents causing explosion or fire

4. Any explosion or fire caused by an electrical short circuit or overload (including those resulting from accidental damage to the electrical plant) which either –
   (a) results in the stoppage of the plant involved for more than 24 hours; or
   (b) causes a significant risk of death.

Explosives

5. Any unintentional –
   (a) fire, explosion or ignition at a site where the manufacture or storage of explosives requires a licence or registration, as the case may be, under Regulation 9, 10 or 11 of the Manufacture and Storage of Explosives Regulations 2005; or
   (b) explosion or ignition of explosives (unless caused by the unintentional discharge of a weapon, where, apart from that unintentional discharge, the weapon and explosives functioned as they were designed to), except where a fail-safe device or safe system of work prevented any person being endangered as a result of the fire, explosion or ignition.

6. The misfire of explosives (other than at a mine or quarry, inside a well or involving a weapon) except where a fail-safe device or safe system of work prevented any person being endangered as a result of the misfire.

7. Any explosion, discharge or intentional fire or ignition which causes any injury to a person requiring first-aid or medical treatment, other than at a mine or quarry.

8. (1) The projection of material beyond the boundary of the site on which the explosives are being used, or beyond the danger zone of the site, which caused or might have caused injury, except at a quarry.
   (2) In this paragraph, 'danger zone' means the area from which persons have been excluded or

forbidden to enter to avoid being endangered by any explosion or ignition of explosives.

9. The failure of shots to cause the intended extent of collapse or direction of fall of a structure in any demolition operation.

Biological agents

10. Any accident or incident which resulted, or could have resulted, in the release or escape of a biological agent likely to cause severe human infection or illness.

Malfunction of radiation generators and radiography

11. (1) The malfunction of –
    (a) a radiation generator or its ancillary equipment used in fixed or mobile industrial radiography, the irradiation of food or the processing of products by irradiation, which causes it to fail to de-energise at the end of the intended exposure period;
    (b) equipment used in fixed or mobile industrial radiography or gamma irradiation, which causes a radioactive source to fail to return to its safe position by the normal means at the end of the intended exposure period.
    (2) In this paragraph, 'radiation generator' means any electrical equipment emitting ionising radiation and containing components operating at a potential difference of more than 5kV.

Breathing apparatus

12. The malfunction of breathing apparatus –
    (a) where the malfunction causes a significant risk of personal injury to the user; or
    (b) during testing immediately prior to use, where the malfunction would have caused a significant risk to the health and safety of the user had it occurred during use, other than at a mine.

Diving operations

13. The failure, damaging or endangering of –
    (a) any life support equipment, including control panels, hoses and breathing apparatus; or
    (b) the dive platform, or any failure of the dive platform to remain on station, which causes a significant risk of personal injury to a diver.

14. The failure or endangering of any lifting equipment associated with a diving operation.

15. The trapping of a diver.

16. Any explosion in the vicinity of a diver.

17. Any uncontrolled ascent or any omitted decompression which causes a significant risk of personal injury to a diver.

Collapse of scaffolding

18. The complete or partial collapse (including falling, buckling or overturning) of –
    (a) a substantial part of any scaffold more than 5 metres in height;

(b) any supporting part of any slung or suspended scaffold which causes a working platform to fall (whether or not in use); or

(c) any part of any scaffold in circumstances such that there would be a significant risk of drowning to a person falling from the scaffold.

Train collisions

19. The collision of a train with any other train or vehicle, other than a collision reportable under Part 5 of this Schedule, which could have caused the death, or specified injury, of any person.

Wells

20. In relation to a well (other than a well sunk for the purpose of the abstraction of water) –
   (a) a blow-out (which includes any uncontrolled flow of well-fluids from a well);
   (b) the coming into operation of a blow-out prevention or diversion system to control flow of well-fluids where normal control procedures fail;
   (c) the detection of hydrogen sulphide at a well or in samples of well-fluids where the responsible person did not anticipate its presence in the reservoir drawn on by the well;
   (d) the taking of precautionary measures additional to any contained in the original drilling programme where a planned minimum separation distance between adjacent wells was not maintained; or
   (e) the mechanical failure of any part of a well whose purpose is to prevent or limit the effect of the unintentional release of fluids from a well or a reservoir being drawn on by a well, or whose failure would cause or contribute to such a release.

Pipelines or pipeline works

21. In relation to a pipeline or pipeline works –
   (a) any damage to, accidental or uncontrolled release from or inrush of anything into a pipeline;
   (b) the failure of any pipeline isolation device, associated equipment or system; or
   (c) the failure of equipment involved with pipeline works, which could cause personal injury to any person, or which results in the pipeline being shut down for more than 24 hours.

22. The unintentional change in position of a pipeline, or in the subsoil or seabed in the vicinity, which requires immediate attention to safeguard the pipeline's integrity or safety.

Part 2 Dangerous occurences reportable except in relation to an offshore workplace

Structural collapse

23. The unintentional collapse or partial collapse of –
   (a) any structure, which involves a fall of more than 5 tonnes of material; or

   (b) any floor or wall of any place of work, arising from, or in connection with, ongoing construction work (including demolition, refurbishment and maintenance), whether above or below ground.

24. The unintentional collapse or partial collapse of any falsework.

Explosion or fire

25. Any unintentional explosion or fire in any plant or premises which results in the stoppage of that plant, or the suspension of normal work in those premises, for more than 24 hours.

Release of flammable liquids and gases

26. The sudden, unintentional and uncontrolled release –
   (a) inside a building –
      (i) of 100 kilograms or more of a flammable liquid;
      (ii) of 10 kilograms or more of a flammable liquid at a temperature above its normal boiling point;
      (iii) of 10 kilograms or more of a flammable gas; or
   (b) in the open air, of 500 kilograms or more of a flammable liquid or gas.

Hazardous escapes of substances

27. The unintentional release or escape of any substance which could cause personal injury to any person other than through the combustion of flammable liquids or gases.

Additional categories of dangerous occurrences apply to mines, quarries, relevant transport systems and offshore workplaces.

19.29.8 Reporting of cases of disease and occupational exposure – Regulations 8 and 9

Where a medical practitioner notifies the employer's responsible person that an employee suffers from a reportable work-related disease (i.e. receives a diagnosis), a completed disease report form (F2508A) should be sent to the enforcing authority. The full list is as follows:

(a) Carpal Tunnel Syndrome, where the person's work involves regular use of percussive or vibrating tools;

(b) cramp in the hand or forearm, where the person's work involves prolonged periods of repetitive movement of the fingers, hand or arm;

(c) occupational dermatitis, where the person's work involves significant or regular exposure to a known skin sensitiser or irritant;

(d) Hand–Arm Vibration Syndrome, where the person's work involves regular use of percussive or vibrating tools, or the holding of materials which are subject to percussive processes, or processes causing vibration;

19

(e) occupational asthma, where the person's work involves significant or regular exposure to a known respiratory sensitiser; or

(f) tendonitis or tenosynovitis in the hand or forearm, where the person's work is physically demanding and involves frequent, repetitive movements.

For exposure to carcinogens, mutagens and biological agents:

(a) any cancer attributed to an occupational exposure to a known human carcinogen or mutagen (including ionising radiation); or

(b) any disease attributed to an occupational exposure to a biological agent.

Diseases offshore have separate criteria in Schedule 3 to the Regulations.

19.29.9 Gas incidents – Regulation 11

Distributors, fillers, importers and suppliers of flammable gas must report incidents where someone has died, lost consciousness, or been taken to hospital for treatment to an injury arising in connection with that gas. Such incidents should be reported using the online form.

Registered gas engineers (under the Gas Safe Register) must provide details of any gas appliances or fittings that they consider to be dangerous, to such an extent that people could die, lose consciousness or require hospital treatment. The danger could be due to the design, construction, installation, modification or servicing of that appliance or fitting, which could cause:

▶ an accidental leakage of gas;

▶ inadequate combustion of gas; or

▶ inadequate removal of products of the combustion of gas.

Unsafe gas appliances and fittings should be reported using the online form.

19.29.10 Recording and record-keeping – Regulation 12

A record of each incident reported (deaths, specified injuries, over 7-day incapacitations and dangerous occurrences) and each diagnosis received of a reportable disease or exposure must be kept at the place of business.

In addition records must be kept of **over 3-day** incapacitations.

Where an employer must keep an accident book under the Social Security (Claims and Payments) Regulations 1979, that record will be enough for the accident records. Records must be kept for at least 3 years.

19.29.11 Restrictions on reporting – Regulation 14

Reports are not required under RIDDOR in relation to the following:

Accidents during medical or dental treatment , or during any examination carried out or supervised by a doctor or dentist.

Accidents involving the movement of a vehicle on a public road (other than those associated with: loading or unloading operations; work alongside the road such as road maintenance; escapes of substances from the vehicle; and accidents involving trains).

Accidents to members of the armed forces on duty.

Reports are not required under RIDDOR where this would duplicate other similar reporting requirements as listed. These include reports required under:

Nuclear Installations Act 1965;

Merchant Shipping Act 1988;

Ionising Radiations Regulations 1999;

Civil Aviation (Investigation of Military Air Accidents at Civil Aerodromes) Regulations 2005;

Civil Aviation (Investigation of Air Accidents and Incidents) Regulations 1996;

The Electricity Safety, Quality and Continuity Regulations 2002.

19.29.12 Reporting and recording procedures – Schedule 1

For most types of incident, including:

▶ accidents resulting in the **death** of any person;

▶ accidents resulting **in specified injuries** to workers;

▶ non-fatal accidents requiring hospital treatment to **non-workers**; and

▶ **dangerous occurrences**

the responsible person must notify the enforcing authority without delay, in accordance with the reporting procedure (Schedule 1). This is most easily done by reporting online at: http://www.hse.gov.uk/riddor/report.htm#online

Alternatively, **for fatal accidents and specified injuries only**, the responsible person can telephone the Incident Contact Centre on 0845 300 9923 (opening hours Monday to Friday 8.30 a.m. to 5 p.m.).

A report must be received within **10 days** of the incident.

For accidents resulting in the over 7-day incapacitation of a worker, the responsible person must notify the enforcing authority within **15 days** of the incident, using the appropriate online form F2508IE.

Cases of occupational disease, including those associated with exposure to carcinogens, mutagens or biological agents, should be reported as soon as the responsible person receives a diagnosis, using the appropriate online form F2508AE at: https://extranet.hse.gov.uk/lfserver/external/F2508AE

There is no longer a paper form for RIDDOR reporting, since the online system is the preferred reporting mechanism. Should it be essential for the responsible person to submit a report by post, it should be sent to: RIDDOR Reports, Health and Safety Executive, Redgrave Court, Merton Road, Bootle, Merseyside, L20 7HS.

The HSE has an out-of-hours duty officer. Circumstances where the HSE may need to respond out of hours include:

▶ a work-related death or situation where there is a strong likelihood of death following an incident at, or connected with, work;

▶ a serious accident at a workplace so that the HSE can gather details of physical evidence that would be lost with time; and

▶ following a major incident at a workplace where the severity of the incident, or the degree of public concern, requires an immediate public statement from either the HSE or government ministers.

If the incident fits these descriptions ring the duty officer on 0151 922 9235. Further information about contacting the HSE out of hours can be found at www.hse.gov.uk/contact/outofhours.htm

19.29.13 Further information

Accident book BI510(2nd Edition), HSE Books, 2012, ISBN 978 0 7176 6458 0. www.hse.gov.uk/pubns/books/accident-book.htm

HSE website for RIDDOR: http://www.hse.gov.uk/riddor/

Reporting accidents and incidents at work: A brief guide to RIDDOR INDG453(rev1), 2013, HSE Books. http://www.hse.gov.uk/pubns/indg453-rev1.pdf

The Reporting of Injuries, Diseases and Dangerous Occurrences Regulations 2013, SI 2013, No.1471, http://www.legislation.gov.uk/uksi/2013/1471/contents/made

19.30 Safety Representatives and Safety Committees Regulations 1977

These Regulations, made under the HSW ACT Section 2(4), prescribe the cases in which recognised trade unions may appoint safety representatives, specify the functions of such representatives and set out the obligations of employers towards them.

Employers' obligations to consult non-union employees are contained in the Health and Safety (Consultation with Employees) Regulations 1996.

19.30.1 Appointment – Regulation 3

A recognised trade union may appoint safety representatives from among employees in all cases where one or more employees are employed. When the employer is notified in writing the safety representatives have the functions set out in Regulation 4.

A person ceases to be a safety representative when:

▶ the appointment is terminated by the trade union;

▶ they resign;

▶ employment ceases.

A safety representative should have been with the employer for 2 years or have worked in similar employment for at least 2 years.

19.30.2 Functions – Regulation 4

These are functions and not duties. They include:

▶ representing employees in consultation with the employer;

▶ investigating potential hazards and dangerous occurrences;

▶ investigating the causes of accidents (*Note*: re accident book BI510 and Data Protection – if the injured person has ticked the box and signed the form the safety representative may see all details. If not the employer should conceal the injured person's identity and details when giving access to the safety representatives);

▶ investigating employee complaints relating to health, safety and welfare;

▶ making representations to the employer on health, safety and welfare matters;

▶ carrying out inspections;

▶ representing employees at the workplace in consultation with enforcing inspectors;

▶ receiving information;

▶ attending safety committee meetings.

Safety representatives must be afforded time off with pay to fulfil these functions and to undergo training.

19.30.3 Employers' duties – Regulation 4a

Every employer shall consult safety representatives in good time regarding:

▶ the introduction of any measure which may affect health and safety;

▶ the arrangements for appointing or nominating competent person(s) under the Management Regulations;

▶ any health and safety information required for employees;

▶ the planning and organising of any health and safety training for employees;

▶ the health and safety consequences of introducing new technology.

Employers must provide such facilities and assistance as safety representatives may reasonably require to carry out their functions.

19.30.4 Inspections – Regulations 5 and 6

Following reasonable notice in writing, safety representatives may inspect the workplace every quarter (or more frequently by agreement with the employer). They may inspect the workplace at any time, after consultation, when there have been substantial changes in the conditions of work or there is new information on workplace hazards published by the HSE.

Following an injury, disease or dangerous occurrence subject to Reporting of Injuries, Diseases and Dangerous Occurrences Regulations, and after notifying the employer, where it is reasonably practicable to do so, safety representatives may inspect the workplace if it is safe.

Employers must provide reasonable assistance and facilities, including provision for independent investigation and private discussion with employees. The employer may be present in the workplace during inspections.

19.30.5 Information – Regulation 7

Having given reasonable notice to the employer, safety representatives are entitled to inspect and take copies of any relevant statutory documents (except any health record of an identified person).

An exempt document is one:

▶ which could endanger national security;
▶ which could cause substantial commercial injury on the employer;
▶ that contravenes a prohibition;
▶ that relates to an individual without their consent;
▶ which has been obtained specifically for legal proceedings.

19.30.6 Safety committees – Regulation 9

When at least two safety representatives have requested in writing that a safety committee is set up, the employer has 3 months to comply. The employer must consult with the safety representatives and post a notice stating its composition and the workplaces covered by it, in a place where it can be easily read by employees. The guidance gives details on the composition and running of safety committees.

19.30.7 Complaints – Regulation 11

If the employer fails to permit safety representatives time off or fails to pay for time off, complaints can be made to an industrial tribunal within 3 months of the incident.

19.30.8 Further information

Consulting workers on health and safety: Safety representatives and safety Committees Regulations 1977 (as amended) and Health and Safety (Consultation with Employees) Regulations 1996 (as amended); Approved Codes of Practice and guidance. L146 (Second edition), 2014, HSE Books, ISBN 978 0 7176 6461 0. http://www.hse.gov.uk/pubns/priced/l146.pdf

Consulting employees on health and safety. A brief guide to the law. INDG232(rev2), 2013, HSE Books. http://www.hse.gov.uk/pubns/indg232.pdf

Involving your workforce in health and safety: Good practice for all workplaces HSG263, HSE Books, 2008, ISBN 9780 7176 6227 2. http://www.hse.gov.uk/pubns/books/hsg263.htm

19.31 Health and Safety (Safety Signs and Signals) Regulations 1996

19.31.1 Introduction

These Regulations came into force in April 1996. The results of the risk assessment made under the Management of Health and Safety at Work Regulations will have identified situations where there may be a residual risk when warnings or further information are necessary. If there is no significant risk there is no need to provide a sign.

19.31.2 Definitions – Regulation 2

'Safety sign' means a sign referring to a specific object, activity or situation and providing information or instruction about health and safety at work by means of a signboard, a safety colour, an illuminated sign, an acoustic sign, a verbal communication or a hand signal.

'Signboard' means a sign which provides information or instructions by a combination of geometric shape, colour and a symbol or pictogram and which is rendered visible by lighting of sufficient intensity.

'Hand signal' means a movement or position of the arms or hands or a combination, in coded form, for guiding persons who are carrying out manoeuvres which create a risk to the health and safety of people at work.

'Acoustic signal' means a coded sound signal which is released and transmitted by a device designed for that purpose, without the use of a human or artificial voice.

'Verbal communication' means a predetermined spoken message communicated by a human or artificial voice.

19.31.3 Provision and maintenance of safety signs – Regulation 4

The Regulations require employers to use and maintain a safety sign where there is a significant risk to health and safety that has not been avoided or controlled by other means, like engineering controls or safe systems of work, and where the use of a sign can help reduce the risk.

They apply to all workplaces and to all activities where people are employed, but exclude signs used in connection with transport or the supply and marketing of dangerous substances, products and equipment.

The Regulations require, where necessary, the use of road traffic signs in workplaces to regulate road traffic.

19.31.4 Information, instruction and training – Regulation 5

Every employer shall ensure that:

▶ comprehensible and relevant information on the measures to be taken in connection with safety signs is provided to each employee;
▶ each employee receives suitable and sufficient instruction and training in the meaning of safety signs.

19.31.5 Functions of colours, shapes and symbols in safety signs

Safety colours

Red

Red is a safety colour and must be used for any:

▶ prohibition sign concerning dangerous behaviour (e.g. the safety colour on a 'No-Smoking' sign). Prohibition signs must be round, with a black pictogram on a white background with red edging and a red diagonal line (top left, bottom right). The red part must take up at least 35% of the area of the sign;

No fork-lift trucks   No smoking

Figure 19.11 (a) and (b) Prohibition signs

▶ danger alarm concerning stop, shutdown, emergency cut-out devices, evacuate (e.g. the safety colour of an emergency stop button on equipment);
▶ fire-fighting equipment. Rectangular or square shape with white pictogram on a red background (red part must be at least 50% of the area of the sign).

Figure 19.12 (a) and (b) Fire action signs

Red and white alternating stripes may be used for marking surface areas to show obstacles or dangerous locations.

Yellow

Yellow (or amber) is a safety colour and must be used for any warning sign concerning the need to be careful, take precautions, examine or the like (e.g. the safety colour on hazard signs, such as for flammable material, electrical danger). Warning signs must be triangular, with a black pictogram on a yellow (or amber) background with black edging. The yellow (or amber) part must take up at least 50% of the area of the sign. Stripes should be of equal size and at 45 degrees.

Yellow and black alternating stripes may be used for marking surface areas to show obstacles or dangerous locations.

General danger   Explosive

Figure 19.13 (a) and (b) Warning signs

Yellow may be used in continuous lines showing traffic routes.

Blue

Blue is a safety colour and must be used for any mandatory sign requiring specific behaviour or action (e.g. the safety colour on a 'Safety Helmet Must Be Worn' sign or a 'Pedestrians Must Use This Route' sign). Mandatory signs must be round, with a white pictogram on a blue background. The blue part must take up at least 50% of the area of the sign.

Ear protection must be worn   Eye protection must be worn

Figure 19.14 (a) and (b) Mandatory signs

Green

Green is a safety colour and must be used for emergency escape signs (e.g. showing emergency doors, exits and routes) and first-aid signs (e.g. showing location of first-aid equipment and facilities). Escape and first-aid signs must be rectangular or square, with a white pictogram on a green background. The green part must take up at least 50% of the area of the sign. So long as the green takes up at least 50% of the area, it is sometimes permitted to use a green pictogram on a white background, for example where there is a green wall and the reversal provides a more effective sign than one with a green background and white border.

Means of Escape   First Aid

Figure 19.15 (a) and (b) Safety signs

Other colours

White

White is not a safety colour but is used for: pictograms or other symbols on blue and green signs; in alternating red and white stripes to show obstacles or dangerous locations; in continuous lines showing traffic routes.

Black

Black is not a safety colour but is used for: pictograms or other symbols on yellow (or amber) signs and, except for fire signs, red signs; in alternating yellow and black stripes to show obstacles or dangerous locations.

Shapes

- Round signs must be used for any prohibition (red) sign; mandatory (blue) sign.
- Triangular signs must be used for any warning (yellow or amber) sign.
- Square or rectangular signs must be used for any emergency escape sign and any first-aid sign.

Pictograms and other symbols

The meaning of a sign (other than verbal communication) must not rely on words. However, a sign may be supplemented with words to reinforce the message provided the words do not in fact distract from the message or create a danger.

A sign (other than verbal communication, acoustic signals or hand signals) should use a simple pictogram and/or other symbol (such as directional arrows, exclamation mark) to effectively communicate its message and so overcome language barriers.

Pictograms and symbols are included in the Regulations. Employee training is needed to understand the meaning of these as many are not inherently clear, some are meaningless to anyone who has not had their meaning explained and some can even be interpreted with their opposite meaning.

Pictograms and symbols included in the Regulations do not cover all situations for which graphic representation of a hazard or other detail may be needed. Any sign used for a situation not covered in the Regulations should include:

- the international symbol for general danger (exclamation mark!) if the sign is a warning sign and tests show that the sign is effective;
- in any other case a pictogram or symbol which has been tested and shown to be effective.

BS EN 7010:2012

These standards aim to bring about consistency in safety signage internationally. The date for adoption was January 2013, when it became a European Normative. Therefore, at that time, it replaced BS5499-5, the previous British Standard.

British Standards are not law. They are Codes of Practice, generally affecting only new products, as opposed to those previously produced. However, it is possible for standards to be given a type of legal status when they are referred to within legislation or government-issued guidance and Approved Codes of Practice.

The EC Directive 92/58/EEC and UK Health and Safety (Signs and Signals) Regulations 1996 now need to be updated to reflect the BS EN ISO 7010 symbols. However, there is no apparent plan to change them yet to incorporate BS EN ISO 7010.

The guidance on these regulations published in 2009 (still the most current version) stated that businesses could continue to use the BS5499-5 fire escape signage if they preferred. Luckily for many companies, BS5499-5, the old UK standard for fire escape signs, was used as the basis for BS EN ISO 7010.

The text of any words used to supplement a sign must convey the same meaning. For example, a round blue sign with a pictogram showing the white outline of a face with a solid white helmet on the head means 'Safety Helmet Must Be Worn' and so any text used must maintain the obligatory nature of the message.

19.31.6 Further information

The Health and Safety (Safety Signs and Signals) Regulations 1996, SI 1996/341. Stationary Office http://www.legislation.gov.uk/uksi/1996/341/contents/made

Safety Signs and Signals Guidance on the Regulations L64 (2nd Ed), 2009, HSE Books, ISBN 978 0 7176 6359 0. http://www.hse.gov.uk/pubns/books/l64.htm

BS EN ISO 7010: 2012, Graphical symbols. Safety colours and safety signs. Registered safety signs British Standards Institute 2011. http://shop.bsigroup.com/en/ProductDetail/?pid=000000000030254963

19.32 The Supply of Machinery (Safety) Regulations 2008 as amended

19.32.1 Introduction

The Supply of Machinery (Safety) Regulations 2008 and the Supply of Machinery (Safety) (Amendment) Regulations 2011 implement Directives 2006/42/EC and 2009/127/EC on Machinery. Their scope extends to other products such as safety components, lifting tackle and partly completed machinery, but excludes such items as domestic electrical machines and fairground equipment.

The Regulations require that all machinery:

▶ is designed and constructed to be safe, meeting the essential health and safety requirements listed in the Regulations (these are supported by many harmonised standards)
▶ is CE marked
▶ is supplied with instructions in English
▶ has a Declaration of Conformity (or, in the case of partly completed machinery, a Declaration of Incorporation).

These Regulations apply to manufacturers or their authorised representatives. In some cases, they apply to others – such as importers of non-CE-marked equipment from outside the EU, and those who design and construct machinery for their own use. However, the Regulations do not usually apply to intermediate suppliers of CE-marked machinery, which is covered by Section 6 of the HSW Act. (The non-application of the Regulations to intermediate suppliers may change in the next few years, when the Machinery Directive is brought into line with other EU legislation.) Both the HSE and local trading standards enforce the provisions of these Regulations, depending on the field of use of the equipment (the HSE leads where machinery is for use at work).

19.32.2 Main requirements for manufacturers

The Machinery Directives, and therefore the Supply of Machinery Regulations and amendments, require manufacturers of machinery (or their 'authorised representatives') to undertake the following activities before they can place machinery on the European market or put it into service:

▶ carry out a detailed risk assessment by:
  ▷ identifying the health and safety hazards (trapping, crushing, electrical shock, dust or fumes, noise, vibration, etc.) that are likely to be present when the machinery is used. The essential health and safety requirements listed in the Regulations should be used by manufacturers as a guide to ensure all possible hazards have been considered, and that they have identified those relevant to the machine;
  ▷ assessing the likely level of risks involved;
▶ eliminate significant risks or, if that is not possible;
▶ provide safeguards (e.g. guarding dangerous parts of the machinery, providing noise enclosures, etc.) or, if that is not possible;
▶ provide information about any residual risks and place signs on the machinery to warn of risks that cannot be reduced in other ways (e.g. 'noisy machine' signs).

19.32.3 Suppliers requirements

If machinery is manufactured in the European Economic Area (EEA), suppliers should do all they can to ensure it is safe. If importing machinery from outside the EEA and it is CE marked with a Declaration of Conformity, suppliers will still need to follow the same checks as for machinery built inside the EEA. If it is not CE-marked with a Declaration of Conformity, suppliers will need to make sure the machinery meets the requirements of the law in full as if they were the manufacturer. This means ensuring the machinery:

▶ meets the essential health and safety requirements;
▶ has (in certain cases) undergone conformity assessment by a notified body;
▶ has a technical file and Declaration of Conformity; and
▶ is CE marked.

Suppliers who are importing partly completed machinery ensure it is supplied with instructions for its final incorporation with other machinery/equipment and a Declaration of Incorporation. Partly completed machinery should not be CE marked. The manufacturer of partly completed machinery should hold technical documentation in one of the European Community languages, and must be able to produce it upon request to any relevant enforcing authority.

19.32.4 Further information

Buying New Machinery – A Short Guide to the Law INDG271(rev1), 2011, HSE Books, www.hse.gov.uk/pubns/indg271.pdf.

Supplying New Machinery Advice to Suppliers INDG270, HSE ,2011, www.hse.gov.uk/pubns/indg270.pdf

The Supply of Machinery (Safety) Regulations 2008, OPSI: http://www.legislation.gov.uk/uksi/2008/1597/contents/made

19

19.33 Control of Vibration at Work Regulations 2005

19.33.1 Introduction

These Regulations implement European Directive Vibration Directive 2002/44/EC. They came into force in July 2005 with some transitional arrangements for the exposure limits until 2010 (2014 for WBV the whole-body vibration exposure limit value for agriculture and forestry sectors).

The Regulations impose duties on employers to protect employees who may be exposed to risk from vibration at work, and other persons who may be affected by the work.

19.33.2 Interpretation – Regulation 2

'Daily exposure' means the quantity of mechanical vibration to which a worker is exposed during a working day, normalised to an 8-hour reference period, which takes account of the magnitude and duration of the vibration.

'Hand–arm vibration' (HAV) means mechanical vibration which is transmitted into the hands and arms during a work activity.

'Whole-body vibration' (WBV) means mechanical vibration which is transmitted into the body when seated or standing through the supporting surfaces, during a work activity or as described in Regulation 5(3)(f).

19.33.3 Application – Regulation 3

For work equipment first provided to employees for use prior to 6 July 2007 and where compliances with the exposure limit values is not possible, employers had until 2010 to comply and in the case of agriculture and forestry 2014 (for WBV).

However, action must be taken to use the latest technical advances and the organisational measures in accordance with Regulation 6(2).

19.33.4 Exposure limit values and action values – Regulation 4

HAV

(a) 8-hour daily exposure limit value is 5 m/s^2 A(8);
(b) 8-hour daily exposure action value is 2.5 m/s^2 A(8);
(c) daily exposure ascertained as set out in Schedule 2 Part I.

WBV

(a) 8-hour daily exposure limit value is 1.15 m/s^2 A(8);
(b) 8-hour daily exposure action value is 0.5 m/s^2 A(8);
(c) daily exposure ascertained as set out in Schedule 2 Part I.

19.33.5 Assessment of risk to health created by vibration at the workplace – Regulation 5

A suitable and sufficient risk assessment must be made where work liable to expose employees is carried on. The risk assessment must identify the measures which need to be taken to comply with these Regulations.

Assessment of daily exposure should be by means of:

▶ observation of specific working practices;
▶ reference to relevant work equipment vibration data;
▶ if necessary, measurement of the magnitude of vibration to which employees are exposed;
▶ likelihood of exposure at or above an exposure action value or above an exposure limit value.

The risk assessment shall include consideration of:

▶ the magnitude, type and duration of exposure, including any exposure to intermittent vibration or repeated shocks;
▶ the effects of exposure to vibration on employees whose health is at particular risk from such exposure;
▶ any effects of vibration on the workplace and work equipment, including the proper handling of controls, the reading of indicators, the stability of structures and the security of joints;
▶ any information provided by the manufacturers of work equipment;
▶ the availability of replacement equipment designed to reduce exposure to vibration;
▶ any extension of exposure at the workplace to WBV beyond normal working hours, including exposure in rest facilities supervised by the employer;
▶ specific working conditions such as low temperatures;
▶ appropriate information obtained from health surveillance including, where possible, published information.

The risk assessment shall be reviewed regularly and forthwith if:

▶ there is reason to suspect that the risk assessment is no longer valid; or
▶ there has been a significant change in the work to which the assessment relates;
▶ and where, as a result of the review, changes to the risk assessment are required, those changes shall be made.

The employer shall record:

▶ the significant findings of the risk assessment as soon as is practicable after the risk assessment is made or changed;
▶ the measures which they have taken and which they intend to take to meet the requirements of Regulations 6 and 8.

19.33.6 Elimination or control of exposure to vibration at the workplace – Regulation 6

Following the general principles of prevention in the Management Regulations, the employer shall ensure that risk from the exposure of his or her employees to vibration is either eliminated at source or reduced as low as reasonably practicable.

Where this is not reasonably practicable and the risk assessment indicates that an exposure action value is likely to be reached or exceeded, the employer shall reduce exposure to as low as reasonably practicable by establishing and implementing a programme of organisational and technical measures which is appropriate.

Consideration must be given to:

▶ other working methods which eliminate or reduce exposure to vibration;
▶ a choice of work equipment of appropriate ergonomic design which, taking account of the work to be done, produces the least possible vibration;
▶ the provision of auxiliary equipment which reduces the risk of injuries caused by vibration;
▶ appropriate maintenance programmes for work equipment, the workplace and workplace systems;
▶ the design and layout of workplaces, workstations and rest facilities;
▶ suitable and sufficient information and training for employees, such that work equipment may be used correctly and safely, in order to minimise their exposure to vibration;
▶ limitation of the duration and intensity of exposure to vibration;
▶ appropriate work schedules with adequate rest periods;
▶ the provision of clothing to protect employees from cold and damp.

Subject to implementation dates, the employer shall:

▶ ensure that no employees are exposed to vibration above an exposure limit value; or
▶ ensure if an exposure limit value is exceeded, the employer shall forthwith:
  ▷ take action to reduce exposure to vibration below the limit value;
  ▷ identify the reason for that limit being exceeded;
  ▷ modify the organisational and technical measures taken to prevent it being exceeded again.

This shall not apply where the exposure of an employee to vibration is usually below the exposure action value but varies markedly from time to time and may occasionally exceed the exposure limit value, provided that:

▶ any exposure to vibration averaged over 1 week is less than the exposure limit value;
▶ there is evidence to show that the risk from the actual pattern of exposure is less than the corresponding risk from constant exposure at the exposure limit value;
▶ risk is reduced to as low as reasonably practicable, taking into account the special circumstances;
▶ the employees concerned are subject to increased health surveillance, where such surveillance is appropriate.

Account must be taken of any employee whose health is likely to be particularly at risk from vibration.

19.33.7 Health surveillance – Regulation 7

If:

▶ the risk assessment indicates that there is a risk to the health of any employees who are, or are liable to be, exposed to vibration; or
▶ employees are exposed to vibration at or above an exposure action value;

then the employer shall ensure that such employees are under suitable health surveillance. Records must be kept, and employees given access to their own records and the enforcing authorities provided with copies, as may be required. A range of specified action is required if problems are found with the health surveillance results.

19.33.8 Information, instruction and training – Regulation 8

If:

▶ the risk assessment indicates that there is a risk to the health of employees who are, or who are liable to be, exposed to vibration; or
▶ employees are exposed to vibration at or above an exposure action value;

then the employer shall provide those employees and their representatives with suitable and sufficient information, instruction and training.

The information, instruction and training provided shall include:

▶ the organisational and technical measures taken in order to comply with the requirements of Regulation 6;
▶ the exposure limit values and action values;
▶ the significant findings of the risk assessment, including any measurements taken;
▶ the why and how to detect and report signs of injury;
▶ entitlement to appropriate health surveillance;
▶ safe working practices to minimise exposure to vibration;
▶ the collective results of any health surveillance undertaken in accordance with Regulation 7 in a form calculated to prevent those results from being identified as relating to a particular person.

The information, instruction and training required must be adapted to take account of significant changes in the

19

type of work carried out or methods of work used by the employer, and cover all persons who carry out work in connection with the employer's duties under these Regulations.

19.33.9 Further information

Control Back-Pain Risks from Whole-Body Vibration Advice for Employers on the Control of Vibration at Work Regulations 2005 INDG242 (rev 1), HSE Books, 2005, ISBN 978 0 7176 6119 0. http://www.hse.gov.uk/pubns/indg242.pdf

Control of Vibration at Work Regulations 2005 SI No 1093. http://www.legislation.gov.uk/uksi/2005/1093/contents/made

Drive Away Bad Backs Advice for Mobile Machine Operators and Drivers INDG404, 2005, HSE Books, ISBN 9780 0 7176 6120 6. http://www.hse.gov.uk/pubns/indg404.pdf

Hand–Arm Vibration Advice for Employees Pocket Card INDG296 (rev 2), 2014, HSE Books, http://www.hse.gov.uk/pubns/indg296.pdf

Hand–arm vibration at work: A brief guide, INDG175(rev3), HSE Books, 2012, ISBN 9780 7176 6488 7. http://www.hse.gov.uk/pubns/indg175.htm

Hand–Arm Vibration: Control of Vibration at Work Regulations L140, HSE Books, 2005, ISBN 978 0 7176 6125 1. http://www.hse.gov.uk/pubns/books/l140.htm

Vibration Solutions Practical Way to Reduce the Risk of Hand–Arm Vibration HSG170, HSE Books, 1997, ISBN 978 0 7176 0954 3. http://www.hse.gov.uk/pubns/books/hsg170.htm

Whole-Body Vibration: The Control of Vibration at Work Regulations Guidance on Regulations L141, HSE Books, 2005, ISBN 987 0 7176 6126 8. http://www.hse.gov.uk/pubns/priced/l141.pdf

19.34 Workplace (Health, Safety and Welfare) Regulations 1992 as amended in 2002 and 2013

19.34.1 General

These Regulations were made to implement the European Directive on the minimum safety and health requirements for the workplace. A workplace for these purposes is defined very widely, to include any part of non-domestic premises to which people have access while at work, and any room, lobby, corridor, staircase or other means of access to or exit from them. The main exceptions to these rules are constructions sites, means of transport, mines and quarries and other mineral extraction sites. The stability and solidity of workplaces in a building were added in 2002. Ships in a British port not under the control of a master are now covered by the 2013 amendment to these Regulations.

Employers have a duty to ensure that workplaces under their control comply with the requirements of these Regulations. Any workplace and relevant equipment, devices and systems must be properly maintained. A maintenance system must be set up where appropriate, that is where a fault could result in a failure to comply, and/or mechanical ventilation systems provided under Regulation 6.

The main requirements are summarised below. They are expressed in very general terms, and in each case it will be necessary to turn to the Approved Code of Practice associated with these Regulations for clarification of what is necessary to meet the objectives set.

19.34.2 Health: the working environment – Regulations 6–10

Ventilation

Ventilation must be effective in enclosed areas, and any plant used for this purpose must incorporate warning devices to signal breakdowns which might endanger health or safety.

A reasonable temperature

A reasonable temperature must be maintained during working hours and sufficient thermometers must be provided to enable people at work to determine the temperature in any workroom. The temperature should be comfortable without the need for special clothing. Special guidance is available for areas like food processing where it could be very hot or very cold.

Temperature should be at least 16°C, or where strenuous effort is involved, 13°C.

Lighting

Lighting must be suitable and sufficient. This should be natural lighting sfairp.

Emergency lighting

Emergency lighting shall be provided where persons are especially exposed to danger if artificial light fails. Lights should avoid glare and dazzle and should not themselves cause a hazard. They should not be obscured, and be properly maintained.

Workplaces, furniture and fittings

Workplaces, furniture and fittings should be kept sufficiently clean. Surfaces inside buildings shall be capable of being kept sufficiently clean.

Floors

Floors should not be slippery and wall surfaces should not increase fire risks.

Waste materials

Wastes should not be allowed to accumulate, except in suitable receptacles, and should be kept free from offensive waste products and discharges.

Room dimensions

Room dimensions have to allow adequate unoccupied space to work in and to move freely; 11 m³ minimum per person is required, excluding anything over 3 m high and furniture, etc.

Workstations

Workstations shall be suitable for any person in the workplace who is likely to work at that workstation and for any work that is likely to be done there.

Outside workstations

Outside workstations shall provide, sfairp, protection from adverse weather; provide adequate means of escape in emergencies; and ensure that no person is likely to slip or fall.

Seating

Seating shall be provided where work can or must be done sitting and shall be suitable for the person as well as the work. A footrest shall be provided where necessary.

19.34.3 Safety: accident prevention – Regulations 5, 12–19

Maintenance

The workplace and equipment, devices and systems shall be maintained (including cleaned as appropriate) in an efficient state, efficient working order and in good repair, and where appropriate subject to a system of maintenance. This generally means planned rather than breakdown maintenance. Systems include ventilation, emergency lighting, safety fences, window cleaning devices, and moving walkways.

Floors and traffic routes

Floors and the surface of traffic routes shall be suitably constructed for their intended purpose and free of slope and holes (unless fenced). They should not be uneven or slippery. The traffic routes should be of adequate width and height to allow people and vehicles to circulate safely with ease and they should be kept free of obstructions.

Additional precautions are necessary where pedestrians have to cross or share vehicle routes. Open sides of staircases should be fenced with an upper rail 900 mm or higher and a lower rail. Loading bays should have exits or refuges to avoid people getting crushed by vehicles.

Falls and falling objects

Now covered by Work at Height Regulations 2005 (see Section 19.35).

Tanks and pits

Where there is a risk of falling into a tank, pit or structure containing a dangerous substance that is likely to:

▶ scald or burn;
▶ be poisonous or corrosive;
▶ have an asphyxiating gas fume or vapour;
▶ have any granular or free-flowing substance likely to cause harm;

measures must be taken to securely fence or cover the tank, pit or structure.

Ladders and roofs

Now covered by Work at Height Regulations 2005 (see Section 19.35).

Glazing

Windows and transparent doors and partitions must be appropriately marked and protected against breakage.

Windows

Windows and skylights must open and close safely, and be arranged so that people may not fall out of them. They must be capable of being cleaned safely.

Traffic routes

Pedestrians and vehicles must be able to circulate safely. Separation should be provided between vehicle and people at doors, gateways and common routes. Workplaces should have protection from vehicles.

Doors and gates

Doors, gates and moving walkways have to be of sound construction and fitted with appropriate safety devices.

19.34.4 Welfare: provision of facilities – Regulations 20–25

Sanitary conveniences and washing facilities

Suitable and sufficient sanitary conveniences and washing facilities should be provided at readily accessible places. The facilities must be kept clean, adequately ventilated and lit. Washing facilities should have running hot and cold or warm water, soap and clean towels or other method of cleaning or drying. If necessary, showers should be provided. Men and women should have separate facilities unless each facility is in a separate room with a lockable door and is for use by only one person at a time.

19

Drinking water

An adequate supply of wholesome drinking water, with an upward drinking jet or suitable cups, should be provided. Water should only be supplied in refillable enclosed containers where it cannot be obtained directly from a mains supply.

Accommodation for clothing and facilities for changing

Adequate, suitable and secure space should be provided to store workers' own clothing and special clothing. The facilities should allow for drying clothing. Changing facilities should also be provided for workers who change into special work clothing.

Facilities to rest and to eat meals

Suitable and sufficient, readily accessible, rest facilities should be provided. Arrangements should include suitable facilities to eat meals; adequate seats with backrests and tables; and means of heating food (unless hot food is available nearby) and making hot drinks.

Canteens or restaurants

Canteens and restaurants may be used as rest facilities, provided there is no obligation to purchase food.

Suitable rest facilities

Suitable rest facilities should be provided for pregnant women and nursing mothers. They should be near sanitary facilities and, where necessary, include the facility to lie down.

Non-smokers

This is now covered by the smoke-free legislation which bans smoking in largely enclosed or enclosed work or public places. See Section 19.37.17.

19.34.5 Further information

How to Deal with Sick Building Syndrome: Guidance for Employers, Building Owners and Building Managers HSG132, HSE Books, 1995, ISBN 978 0 7176 0861 4. http://www.hse.gov.uk/pubns/books/hsg132.htm

Lighting at Work HSG38, HSE Books, 1997, ISBN 978 0 7176 1232 1. http://www.hse.gov.uk/pubns/books/hsg38.htm

Seating at Work HSG57, HSE Books, 1997, ISBN 978 0 7176 1231 4. http://www.hse.gov.uk/pubns/books/hsg57.htm

The Workplace (Health, Safety and Welfare) Regulations 1992, http://www.legislation.gov.uk/uksi/1992/3004/made/data.pdf

Workplace Health, Safety and Welfare. Workplace (Health, Safety and Welfare) Regulations 1992 Approved Code of Practice and Guidance L24 (second edition), 2013, HSE Books, ISBN 978 0 7176 65839. http://www.hse.gov.uk/pubns/priced/124.pdf.

Workplace Health, Safety and Welfare A Short Guide, INDG244 (rev 2), 2011, HSE Books, ISBN 978 0 7176 6277 7. http://www.hse.gov.uk/pubns/indg244.pdf

19.35 Work at Height Regulations 2005 as amended in 2007

19.35.1 Introduction

These Regulations bring together all current requirements on work at height into one goal-based set of Regulations. They implement the requirements of the 2nd Amending Directive (2001/45/EC) to the Use of Work Equipment Directive (89/955/EEC) which sets out requirements for work at height. The 2nd Amending Directive has become known as the Temporary Work at Height Directive.

The Work at Height (Amendment) Regulations 2007, which came into force on 6 April 2007, apply to those who work at height providing instruction or leadership to one or more people engaged in caving or climbing by way of sport, recreation, team building or similar activities in Great Britain.

The Regulations require a risk assessment for all work conducted at height and arrangements to be put in place for:

▶ eliminating or minimising risks from working at height;

▶ safe systems of work for organising and performing work at height;

▶ safe systems for selecting suitable work equipment to perform work at height;

▶ safe systems for protecting people from the consequences of work at height.

19.35.2 Definitions – Regulation 2

'Work at height' means:

(a) work in any place, including a place at or below ground level;

(b) obtaining access to or egress from such place while at work except by a staircase in a permanent workplace where, if measures required by these Regulations were not taken, a person could fall a distance liable to cause personal injury.

'Working platform':

(a) means any platform used as a place of work or as a means of access to or egress from a place of work;

(b) includes any scaffold, suspended scaffold, cradle, mobile platform, trestle, gangway, gantry and stairway which is so used.

19.35.3 Organisation, planning and competence – Regulations 4 and 5

Work at height must be properly planned, appropriately supervised and carried out in a manner which is, sfairp, safe. The selection of appropriate work equipment is included in the planning. Work must not be carried out if the weather conditions would jeopardise safety or health (this does not apply where members of the police, fire, ambulance or other emergency services are acting in an emergency).

All people involved in work at height activity including planning, organising and supervising must be competent for such work, or if being trained, under competent supervision.

19.35.4 Avoidance of risk – Regulation 6

A risk assessment carried out under the Management Regulations must be taken into account when identifying the measures required by these Regulations. Work at height should be avoided if there are reasonably practicable alternatives.

Where work at height is carried out employers must take suitable and sufficient measures to prevent persons falling a distance liable to cause personal injury. The measures include:

(a) ensuring work is carried out:
   (i) from an existing workplace;
   (ii) using existing means of access and egress that comply with Schedule 1 of the Regulations (assuming it is safe and ergonomic to do so);
(b) where this is not reasonably practicable, providing work equipment (sfairp) for preventing a fall occurring.

Employers must take steps to minimise the distance and the consequences of a fall, if it is not prevented. Where the distance cannot be minimised (sfairp) the consequence of a fall must be minimised and additional training, instruction and other additional suitable and sufficient measures must be adopted to prevent a person falling a distance liable to cause personal injury.

See Figure 19.16, which shows a flow chart for the work at height risk assessment.

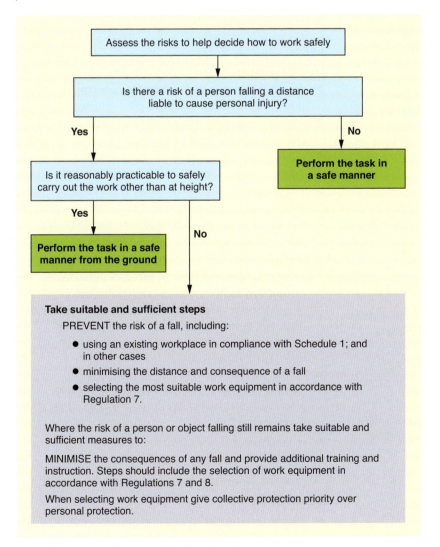

Figure 19.16 Work at height – flowchart

19.35.5 General principles for selection of work equipment – Regulation 7

Collective protection measures must have priority over personal protection measures.

Employers must take account of:

- working conditions and the risks to the safety of persons at the place where the work equipment is to be used;
- the distance to be negotiated for access and egress;
- distance and consequences of a potential fall;
- duration and frequency of use;
- need for evacuation and rescue in an emergency;
- additional risks of using, installing and removing the work equipment used or evacuation and rescue from it;
- other provisions of the Regulations.

Work equipment must:

- be appropriate for the nature of the work and the foreseeable loadings;
- allow passage without risk;
- be the most suitable work equipment having regard in particular to Regulation 6.

19.35.6 Requirements for particular work equipment – Regulation 8

Regulation 8 requires that various pieces of work equipment comply with the schedules to the Regulations as follows:

(a) A guard-rail, toe-board, barrier or similar means of protection comply with Schedule 2.

Schedule 2 in summary:

- they must be suitable, of sufficient dimensions, of sufficient strength and rigidity;
- be so placed, secured and used to prevent accidental displacement;
- prevent fall of persons and materials;
- supporting structure to be of sufficient strength and suitable for the purpose;
- no lateral opening save at point of access to a ladder or stairway where an opening is necessary and:
  - ▷ must be in place except for a time to gain access, perform a particular task and then replaced;
  - ▷ compensatory safety measures must be provided if protection removed temporarily.

(b) A working platform, comply with Schedule 3 Part 1 and, in addition, where scaffolding is provided, Schedule 3 Part 2.

Schedule 3 Part 1 in summary

Supporting structure must:

- be suitable, of sufficient strength and rigidity;
- if wheeled, be prevented, by appropriate device, from moving during work;

- not slip and have secure attachment;
- be stable while being erected, used, modified or dismantled.

Working platform to:

- be suitable and strong enough;
- have no accidental displacement of components;
- remain stable during dismantling;
- be dismantled so as to prevent accidental displacement;
- be of sufficient dimensions for safe passage and use;
- have a suitable surface and be constructed to prevent people falling through;
- have a suitable surface and be constructed to prevent material or objects falling through unless measures have been taken to protect other persons from falling objects;
- be erected, used and maintained so that risk of slipping and tripping is prevented and no person can be caught between the working platform and adjacent structure;
- be not loaded so as to give risk of collapse or deformation.

Schedule 3 Part 2 additional for scaffolds:

- strength and stability calculations required unless they are available already or scaffold is assembled in conformity with a generally recognised standard configuration;
- an assembly, use and dismantling plan must be drawn up. Could be standard plan with supplements;
- a copy of the plan and instructions to be available for persons doing the work;
- dimensions, form and layout to be suitable;
- when not available for use to be marked with warning signs and physical barrier preventing access;
- must be assembled, dismantled or significantly altered under supervision of competent person.

(c) a net, airbag or other collective safeguard for arresting falls which is not part of a personal fall protection system, comply with Schedule 4.

(d) any personal fall protection system to comply with Part 1 of Schedule 5.

- in the case of a work positioning system, comply with Part 2 of Schedule 5;
- in the case of rope access and positioning techniques, comply with Part 3 of Schedule 5;
- in the case of a fall arrest system, comply with Part 4 of Schedule 5;
- in the case of a work restraint system, comply with Part 5 of Schedule 5.

(e) a ladder to comply with Schedule 6.

Schedule 6 in summary:

- only to be used for work at height if the risk assessment demonstrates that the use of more suitable equipment is not justified because of the low risk and the short duration of use or existing features on site that cannot be altered;

- surface on which ladder rests must be stable, firm, of sufficient strength and suitable composition so its rungs remain horizontal;
- must be positioned to ensure stability;
- suspended ladder attached firmly without swing (except flexible ladder);
- portable ladder to be prevented from slipping by securing the stiles near their upper and lower ends and using an anti-slip device or by any other equivalent measures;
- when used for access must protrude above place of landing unless other firm handholds provided;
- interlocking or extension pieces must not move, relative to each other, during use;
- mobile ladder to be prevented from moving before being stepped on;
- where a ladder or run of ladders rises a vertical distance of 9 metres or more above its base, there shall, where reasonably practicable, be provided at suitable intervals sufficient safe landing areas or rest platforms;
- must be used so that a secure foothold and handhold are always available;
- must be used so that user can maintain a secure handhold while carrying a load (exceptions for using stepladders for low-risk, short-duration work).

19.35.7 Fragile surfaces – Regulation 9

Employers must take steps to prevent people falling through any fragile surface. Steps to be taken include:

- avoid, sfairp, passing across or near, or working on or near, a fragile surface;
- where this is not reasonably practicable:
  - ▷ provide suitable and sufficient platforms, covering, guard-rails or other similar means of support or protection and use them;
  - ▷ provide suitable and sufficient guard-rails to prevent persons falling through fragile surfaces;
  - ▷ where a risk of falling remains, take suitable and sufficient measures to minimise the distances and consequences of a fall;
- provide prominent warning signs at approach to the fragile surface, or make people aware of the fragile surface if not reasonably practicable;
- where the risk of falling remains despite the precautions a suitable and sufficient fall arrest system must be provided.

19.35.8 Falling objects and danger areas – Regulations 10 and 11

Suitable and sufficient steps must be taken:

- to prevent the fall of any material or object sfairp;
- if not reasonably practicable to prevent the fall of

any material or object, then prevent persons being struck by falling objects or material (if liable to cause personal injury) by, for example, providing covered walkways or fan scaffolds;

- to prevent material or objects being thrown or tipped from height where it is likely to cause injury;
- to store materials and objects in such a way as to prevent risk to any person arising from collapse, overturning or unintended movement;
- where danger of being struck exists, unauthorised persons must be kept out of the area by suitable devices and the area clearly indicated;
- to store materials and objects so as to prevent them collapsing, overturning or move in a way that could be a risk to people.

19.35.9 Inspection of work equipment – Regulation 12

This Regulation applies only to work equipment to which Regulation 8 and Schedules 2–6 apply. The requirements include the following:

- where work equipment used for work at height depends for safety on how it is installed or assembled, the employer must ensure that it is not used after installation or assembly in any position unless it has been inspected in that place;
- where work equipment is exposed to conditions causing deterioration which is liable to result in dangerous situations, it must also be inspected at suitable intervals and each time that exceptional circumstances which are liable to jeopardise the safety of the work equipment have occurred, for example a severe storm;
- in addition, a working platform used for construction and from which a person could fall more than 2 m has to be inspected in that position (mobile equipment on the site) before use and within the last 7 days. The particulars for this inspection are set out in Schedule 7;
- no work equipment (lifting equipment is covered under LOLER) may either leave the undertaking or be obtained from another undertaking without evidence of in-date inspection;
- results of inspections must be recorded and retained until the next due inspection is recorded. Reports must be provided within 24 hours and kept at the site until construction work is complete and thereafter at the office for 3 months.

19.35.10 Inspection of places of work at height – Regulation 13

The surface and every parapet, permanent rail or other such fall protection measure of every place of work at height must be checked on each occasion before the place is used.

19

19.35.11 Duties of persons at work – Regulation 12

Every person must, where working under the control of another person, report to that person any activity or defect relating to work at height which they know is likely to endanger themselves or others.

Work equipment must be used in accordance with training and instructions.

19.35.12 Further information

A Tool Box Talk on Leaning Ladders and Stepladders Safety INDG403, HSE Books, 2011, ISBN 978 0 7176 6106 0. http://www.hse.gov.uk/pubns/indg403.pdf

Health and Safety in Roof Work HSG 33, 2012, HSE Books, ISBN 978 0 7176 6527 3. http://www.hse.gov.uk/pubns/priced/hsg33.pdf

Height Safe Essential Health and Safety Information for People who Work at Height. HSE website: http://www.hse.gov.uk/pubns/heightsafeleaflet.pdf

Safe Use of Ladders and Stepladders: An Employers' Guide INDG455, HSE 2014, http://www.hse.gov.uk/pubns/indg455.pdf

The Work at Height (Amendment) Regulations 2007, SI 2007 No. 114, http://www.legislation.gov.uk/uksi/2007/114/contents/made

The Work at Height Regulations 2005, SI 2005/735, OPSI, http://www.legislation.gov.uk/uksi/2005/735/contents/made

The Work at Height Regulations 2005 (as amended). A Brief Guide INDG401 (rev 2), 2014, http://www.hse.gov.uk/pubns/indg401.pdf

Tower Scaffolds CIS10, HSE Books, 2005. http://www.hse.gov.uk/pubns/cis10.pdf

The Selection and Management of Mobile Elevating Work Platforms CIS 58, HSE 2008, Web only. http://www.hse.gov.uk/pubns/cis58.pdf

Useful website – HSE's Work at Height: http://www.hse.gov.uk/work-at-height/index.htm

19.36 The Waste (England and Wales) Regulations 2011

19.36.1 Introduction

These regulations implement the revised EU Waste Framework Directive 2008/98 which sets requirements for the collection, transport, recovery and disposal of waste.

In summary the Waste (England and Wales) Regulations 2011 require businesses to confirm that they have applied the waste management hierarchy when transferring waste and include a declaration to this effect on their waste transfer note or consignment note.

They introduce a two-tier system for waste carrier and broker registration, including the new concept of waste dealer.

They make amendments to hazardous waste controls.

They exclude some categories of waste from waste controls.

19.36.2 Who do the Waste (England and Wales) Regulations 2011 affect?

The regulations apply to businesses that:

► produce waste;
► import or export waste;
► carry or transport waste;
► keep or store waste;
► treat waste;
► dispose of waste;
► operate as waste brokers or dealers.

19.36.3 Waste management hierarchy

The hierarchy sets out, in order of priority, the waste management options that should be considered:

► prevention;
► preparing for reuse;
► recycling;
► recovery, e.g. energy recovery;
► disposal.

Since 28 September 2011, whenever waste is passed on to someone else, a declaration has to be made on the waste transfer note, or consignment note for hazardous waste, that the waste hierarchy has been considered.

The waste transfer note must also now include the 2007 Standard Industrial Classification (SIC) code of the person transferring the waste. The 2003 SIC codes should continue to be used on hazardous waste consignment notes.

19.36.4 Hazardous waste controls

Carriers of hazardous waste that operate a multiple collection round will need to include a round number and the number of collection sites on the consignment note.

19.36.5 The Waste (England and Wales) (Amendment) Regulations 2012

These Regulations were laid before Parliament and the Welsh Assembly on 19 July 2012 and came into force on 1 October 2012. The amended regulations relate to the separate collection of waste. They amend the Waste (England and Wales) Regulations 2011 by replacing Regulation 13. Since 1 January 2015, waste collection authorities must collect waste paper, metal, plastic and glass separately. It also imposes a duty on waste collection authorities, from that date, when making arrangements for the collection of such waste, to ensure that those arrangements are by way of separate collection.

These duties apply where separate collection is necessary to ensure that waste undergoes recovery operations in accordance with the directive and to facilitate or improve recovery; and where it is technically, environmentally and economically practicable. The duties apply to waste classified as waste from households and waste that is classified as commercial or industrial waste. The amended regulations also replaced Regulation 14(2) to reflect the changes to Regulation 13 to ensure a consistent approach. Consequential changes are also made to reflect changes in paragraph numbering in the new Regulation 13.

Environmental permitting for waste

The recovery and disposal of waste requires a permit under EU legislation with the principal objective of preventing harm to human health and the environment. This legislation also allows member states to provide for exemptions from the need for a permit, providing general rules are laid down for each type of exempt activity, and the operation is registered with the relevant registration authority. The EU requirements have been effected through the Environmental Permitting (England and Wales) Regulations 2010 (the 2010 regulations). More information is available on the National Archive and on the Environment Agency website.

19.37 Other relevant legislation in brief

The following Regulations are included with brief summaries.

19.37.1 Building Regulations 2010 Approved Documents B and M

These Regulations are not in the Construction Certificate but are included in the National Fire Certificate and could be useful reference for Safety Practitioners.

The Building Regulations 2010 Fire Safety – Approved Document B Volumes 1 and 2

General introduction: Fire safety
Scope:

1. Approved Document B (Fire safety) has been published in two volumes. Volume 1 deals solely with dwelling houses (see Appendix E and Building Regulation 2(1)), while Volume 2 deals with all other types of building covered by the Building Regulations.

Where very large (over 18 m in height) or unusual dwelling houses are proposed some of the guidance in Volume 2 may be needed to supplement that given by Volume 1.

Arrangement of sections

2. The functional requirements B1 to B5 of Schedule 1 of the Building Regulations are dealt with separately in one or more Sections. The requirement is reproduced at the start of the relevant Sections, followed by an introduction to the subject.

3. The provisions set out in this document deal with different aspects of fire safety, with the following aims:

    B1: To ensure satisfactory provision of means of giving an alarm of fire and a satisfactory standard of means of escape for persons in the event of fire in a building.

    B2: To ensure fire spread over the internal linings of buildings is inhibited.

    B3: To ensure the stability of buildings in the event of fire; to ensure that there is a sufficient degree of fire separation within buildings and between adjoining buildings; to provide automatic fire suppression where necessary; and to inhibit the unseen spread of fire and smoke in concealed spaces in buildings.

    B4: To ensure external walls and roofs have adequate resistance to the spread of fire over the external envelope, and that spread of fire from one building to another is restricted.

    B5: To ensure satisfactory access for fire appliances to buildings and the provision of facilities in buildings to assist fire-fighters in the saving of life of people in and around buildings.

4. Whilst guidance appropriate to each of these aspects is set out separately in this document, many of the provisions are closely interlinked. For example, there is a close link between the provisions for means of escape (B1) and those for the control of fire growth (B2), fire containment and/or suppression (B3) and facilities for the fire and rescue service (B5). Similarly there are links between B3 and the provisions for controlling external fire spread (B4), and between B3 and B5. Interaction between these different requirements should be recognised where variations in the standard of provision are being considered. A higher standard under one of the requirements may be of benefit in respect of one or more of the other requirements. The guidance in the document as a whole should be considered as a package aimed at achieving an acceptable standard of fire safety.

Appendices: provisions common to more than one of Part B's requirements

19

5. Guidance on matters that refer to more than one of the Sections is in a series of Appendices, covering the following subjects:
Appendix A – fire performance of materials, products and structures
Appendix B – provisions regarding fire doors
Appendix C – methods of measurement
Appendix D – a classification of purpose groups
Appendix E – definitions
Appendix F – Standards and other publications referred to.

Fire performance of materials, products and structures

6. Much of the guidance throughout this document is given in terms of performance in relation to standard fire test methods. Details are drawn together in Appendix A to which reference is made where appropriate. In the case of fire protection systems, reference is made to standards for system design and installation. Standards referred to are listed in Appendix F.

Fire doors

7. Guidance in respect of fire doors is set out in Appendix B.

Methods of measurement

8. Some form of measurement is an integral part of much of the guidance in this document and methods are set out in Appendix C.

Purpose groups

9. Much of the guidance in this document is related to the use of the building. The use classifications are termed purpose groups, and they are described in Appendix D. This document deals only with buildings in Purpose Groups 1b and 1c.

Definitions

10. The definitions are given in Appendix E.

The Building Regulations 2010 Access to and use of buildings – Approved Document M

The requirements of Part M will be met by making reasonable provision to ensure that buildings are accessible and usable. People, regardless of disability, age or gender, should be able to:

(a) gain access to buildings and to gain access within buildings and use their facilities, both as visitors and as people who live or work in them;
(b) use sanitary conveniences in the principal storey of a new dwelling.

The provisions are expected to enable occupants with disabilities to cope better with reducing mobility and to 'stay put' longer in their own homes. The provisions are not necessarily expected to facilitate fully independent living for all people with disabilities.

Where the requirements apply Application of Part M

1. The requirements apply if:
   (a) a non-domestic building or a dwelling is newly erected;
   (b) an existing non-domestic building is extended, or undergoes a material alteration; or
   (c) an existing building or part of an existing building undergoes a material change of use to a hotel or boarding house, institution, public building or shop.

The terms 'institution', 'public building' and 'shop' are explained in Regulation 2.

It should be noted that, regardless of compliance with Building Regulations, there will be obligations under The Equality Act 2010 for service providers and employers to consider barriers created by physical features in buildings.

Extensions and material alterations: dwellings

2. Under Regulation 4(3), where any building is extended, or undergoes a material alteration, the building work must be carried out so that after it has been completed the building complies with the applicable requirements of Schedule 1, or, where it did not fully comply with any applicable requirement, it is no more unsatisfactory than before.

3. This rule applies to domestic as well as to non-domestic buildings. Under the Limits on Application in Part M, Part M does not apply to an extension of, or a material alteration of, a dwelling. However, an extension of a dwelling, or a material alteration of a dwelling, must not make the building less satisfactory in relation to Part M than it was before.

4. Under Regulation 3, the expression 'material alteration' is defined by reference to a list of 'relevant requirements' in Schedule 1. That list includes Part M. This means that an alteration of a dwelling is a material alteration if the work would result in the dwelling not complying with Part M where previously it did. Alternatively, if the dwelling did not previously comply with Part M, the dwelling should not be more unsatisfactory in relation to Part M after the material alteration.

It is irrelevant whether or not the dwelling was subject to Part M at the time of its construction. Under the general Limits on Application of Part M, the requirements of that Part do not apply to an extension of or a material alteration of a dwelling. This means that the extension or alteration work itself need not comply with Part M. However, a planned alteration to a dwelling that has the potential to reduce the compliance of the dwelling as a whole with Part M must be carried out in such a way that there is no reduction in the extent of Part M compliance. Similarly, an extension of a dwelling need not itself comply with Part M, but it must not result in the dwelling being less compliant with Part M.

Further information

Access to and use of Buildings Approved Document http://www.planningportal.gov.uk/buildingregulations/approveddocuments/partm/approved

Building regulations 2010 SI 2214. http://www.legislation.gov.uk/uksi/2010/2214/pdfs/uksi_20102214_en.pdf

Fire Safety Approved Document B Volume 1 – Dwelling Houses 2006 version with 2010 and 2013 amendments. http://www.planningportal.gov.uk/uploads/br/BR_PDF_AD_B1_2013.pdf

Fire Safety Approved Document B Volume 2 Buildings other than dwellings. http://www.planningportal.gov.uk/uploads/br/BR_PDF_AD_B2_2013.pdf

19.37.2 Compensation Act 2006

The Compensation Act 2006 (c.29) was introduced in response to concerns about a growing compensation culture but conversely to ensure that the public received dependable service from claims management companies.

The Compensation Act contains provisions relating to the law of negligence and breach of statutory duty; damages for mesothelioma; and the regulation of claims management services.

Section 1 of the Act makes statutory provision that, in determining whether the omission of certain steps amounts to a breach of duty, the court MAY consider whether such steps, had they been performed, would prevent some desirable activity. For example, the court must consider whether precautionary and defensive measures might prevent something socially useful. Though this principle had often been observed by the common law (see Miller v. Jackson 1977, Denning's judgement), the Act places it on a statutory footing.

Section 2 stipulates that, in the event of an accident, an apology or offer or redress, such as paying for medical treatment, is not, of itself, an admission of liability.

Sections 1 and 2 came into force on royal assent on 26 July 2006.

Section 3 reverses the common law on allocation of damages in mesothelioma claims arising from unlawful exposure to asbestos. In 2002, the House of Lords had controversially ruled that, where several parties had unlawfully exposed the claimant to asbestos and risk of mesothelioma, all were liable for his injury, even though the claimant could not prove which individual party had provided the critical single fibre believed to cause the disease. However, in *Barker v. Corus UK Ltd* the House of Lords held that the parties who contributed to the risk were severally but not jointly liable. This meant that a single defendant could only be held liable for a fraction of any damages proportional to the exposure for which they were responsible and that a claimant could only

recover all their damages if they succeeded in actions against all such contributors. Section 3 reverses the House of Lords decision by making all such parties jointly and severally liable for the damage so that a claimant could recover the totality of their damages, even from a subgroup of potential defendants.

Section 3 is 'treated as having always had effect' (s.16(3)) (but does not apply to claims settled before 3 May 2006) and also extends to Scotland and Northern Ireland (s.17).

Part 2 of the Act seeks to regulate the provision of claims management services. As of 23 April 2007, an individual or a corporation may not provide claims management services by way of business unless authorised, exempt or otherwise in receipt of a waiver (s.4(1)). The Act creates a Claims Management Services Regulator to authorise and regulate claims management companies and to (s.5):

► Set and monitor standards of competence and professional conduct;
► Promote good practice, in particular as to the provision of information about charges and other matters to users;
► Promote practices likely to facilitate competition;
► Ensure that arrangements are made for the protection of users, including complaints handling.

It is a crime for an unauthorised person to provide or offer claims management services, or to pretend to be authorised. Offenders are punishable, on summary conviction, by a fine of up to level 5 on the standard scale or 51 weeks' imprisonment. If convicted on indictment in the Crown Court, offenders can be sentenced to an unlimited fine or two years' imprisonment (ss.7 and 11). Where a corporate crime is committed, the offender can only be fined and not imprisoned. The Regulator may investigate unauthorised trading and seek an injunction to prevent it or bring a criminal prosecution (s.8). It is a crime to obstruct the Regulator, punishable on summary conviction by a fine of up to level 5 on the standard scale (s.10).

Section 12 creates a Claims Management Services Tribunal to which a person may appeal a decision of the Regulator about authorisation (s.13(1)). There is a further route of appeal to the Court of Appeal (s.13(4)).

Part 2 came progressively into force from 1 December 2006.

19.37.3 Corporate Manslaughter and Corporate Homicide Act 2007

The Act creates a new statutory offence of corporate manslaughter which will replace the common law offence of manslaughter by gross negligence where corporations and similar entities are concerned. In Scotland the new offence will be called 'corporate homicide'. An organisation will have committed the new offence if it:

- owes a duty of care to another person in defined circumstances;
- there is a management failure by its senior managers;
- it amounts to a gross breach of that duty resulting in a person's death.

On conviction the offence will be punishable by an unlimited fine and the courts will be able to make remedial orders requiring organisations to take steps to remedy the management failure concerned. It is important to note that the Act does not create a new individual liability. Individuals may still be charged with the existing offence of manslaughter by gross negligence. Crown immunity will not apply to the offence, although a number of public bodies and functions will be exempt from it (in defined circumstances). The Act came into force on 6 April 2008. Available at: http://www.legislation.gov.uk/ukpga/2007/19/contents

19.37.4 Electrical Equipment (Safety) Regulations 1994

These Regulations came into force in January 1995 and relate to the supply of electrical equipment with a working voltage between 50 and 1000 volts and are made under the Consumer Protection Act 1987. They apply to suppliers, which include both landlords and letting agents.

The Regulations apply to all mains voltage household electrical goods and require them to be safe so that there is no risk of injury or death to humans or pets, or risk of damage to property. They do not apply to fixed electrical wiring and built-in appliances like central heating systems. The Regulations also require that instructions be provided where safety depends on the user being aware of certain issues and equipment should be labelled with the CE marking.

There are other electrical consumer Regulations, like The Plugs and Sockets etc. (Safety) Regulations 1994 and the Low Voltage Electrical Equipment (Safety) Regulations 1989.

Available at: http://www.legislation.gov.uk/uksi/1994/3260/made

19.37.5 The Equality Act 2010 and (Disability) Regulations 2010

The Equality Act 2010 defines a person as disabled if they have a physical or mental impairment that has a substantial and long-term adverse effect on a person's ability to carry out normal day-to-day activities; such as reading, writing, using the telephone, having a conversation and travelling by public transport .

The key thing is not the impairment but its effect. Some people don't realise that impairments such as migraines, dyslexia, asthma and back pain can count as a disability

if the adverse effect on the individual is substantial and long-term. Some conditions automatically count as disabilities for the purposes of The Equality Act 2010, from the point of first diagnosis – these are cancer, HIV and multiple sclerosis (MS).

It is discrimination to treat a disabled person unfavourably because of something connected with their disability (e.g. a tendency to make spelling mistakes arising from dyslexia). This type of discrimination is unlawful where the employer or other person acting for the employer knows, or could reasonably be expected to know, that the person has a disability.

Public authorities and public, private or voluntary organisations carrying out public functions have a new Equality Duty. In summary, those subject to the duty must have due regard to the need to:

- Eliminate unlawful discrimination, harassment and victimisation and other conduct prohibited by the Act.
- Advance equality of opportunity and foster good relations between people who share a protected characteristic and those who do not.

Employers are required to make reasonable adjustments to jobs and workplaces for disabled workers. These may include:

- adjustments to the workplace to improve access or layout;
- giving some of the disabled person's duties to another person;
- transferring the disabled person to fill a vacancy;
- changing the working hours, e.g. flexi-time, job-share, starting later or finishing earlier;
- time off, e.g. for treatment, assessment, rehabilitation;
- training for disabled workers and their colleagues;
- getting new or adapting existing equipment;
- modifying instructions or procedures;
- improving communication, e.g. providing a reader or interpreter;
- providing alternative work (a last resort).

If a disabled person expects an employer to make a reasonable adjustment, they will need to provide the employer with enough information to carry out that adjustment. An employer must not disclose confidential details without explicit consent.

The Equality Act 2010 (Disability) Regulations 2010 came into force in Augist 2010.

These Regulations re-enact with amendments provisions which were previously made under the Disability Discrimination Act 1995 (c.50) and which are revoked by Regulation 15 of these Regulations.

Part 2 contains provisions which supplement those in the Act about when a person is disabled for the purposes of that Act. Regulation 3 excludes from the scope of the definition of disability addictions (other than those medically caused). Regulation 4 excludes certain conditions from being impairments for the purposes of the Act. Regulation 5 provides that severe disfigurements consisting of tattoos and certain body piercings are not to be treated as having a substantial adverse effect on a person's ability to carry out normal day-to-day activities. Regulation 6 contains provision for assessing the ability of a child under six years of age to carry out normal day-to-day activities. Regulation 7 deems a person who is certified by a consultant ophthalmologist as blind, severely sight-impaired, sight-impaired or partially sighted to be a disabled person. The various terms used in this provision reflect the fact that the terminology used in the certificates by consultant ophthalmologists has changed over time.

In Part 3, Regulation 8 sets out things which are to be treated as auxiliary aids or services for the purposes of paragraphs 2 to 4 of Schedule 4 to the Act.

Part 4 contains provisions about reasonable adjustments to physical features of premises. Regulation 9 provides that it is not reasonable for a provider of services, a public authority carrying out its functions or an association to have to remove or alter a physical feature which was provided in or in connection with a building to assist with access to the building or to use facilities and satisfies the design standard. The Schedule to these Regulations provides details of how to determine whether the design standard is satisfied.

Regulation 10 sets out the circumstances in which a relevant landlord is taken to have withheld consent for the purposes of Schedule 21 to the Act. Regulations 11 and 12 set out circumstances in which a relevant landlord is taken to have withheld consent for the purposes of Schedule 21 to the Act unreasonably and reasonably respectively.

Regulation 13 sets out circumstances in which a condition imposed by a landlord to consent to an alteration is reasonable for the purposes of Schedule 21 to the Act.

Regulation 14 provides modifications to Schedule 21 to the Act where the occupier occupies premises under a sub-tenancy.

Act available at: http://www.legislation.gov.uk/ukpga/2010/15/contents

Regulations available: http://www.legislation.gov.uk/uksi/2010/2128/pdfs/uksi_20102128_en.pdf

19.37.6 Gas Appliances (Safety) Regulations 1995

The 1992 Regulations required the gas appliances and fittings to which they applied to conform with the essential requirements detailed in Schedule 3 and to be safe when normally used. Series-manufactured products were also required to undergo EC type-examination together with one of a number of prescribed options for production control, at the choice of the manufacturer, conducted by third-party notified bodies (Regulations 8, 10–18, 20–23). Single appliances or small numbers may be subject to EC unit verification (Regulations 8 and 19). The 1992 Regulations are revoked.

These Regulations re-enact those requirements (with certain minor amendments) (Regulations 4(1)(c) and 4(5), 8 and 23).

Second-hand appliances are not covered by these Regulations but they retain in force, for second-hand gas appliances only, the Gas Cooking Appliances (Safety) Regulations 1989 and the Heating Appliances (Fireguard) (Safety) Regulations 1991 Regulation 1(3).

The Gas Appliance Regulations cover the safety standards on new gas appliances which have to:

▶ satisfy safety and efficiency standards;
▶ undergo type-examination and supervision during manufacture;
▶ carry the CE mark and specified information;
▶ be accompanied by instructions and warnings in the language of destination.

Available at: http://www.legislation.gov.uk/uksi/1995/1629/contents/made

19.37.7 Gas Safety (Installation and Use) Regulations 1998

The Installation and Use Regulations place duties on gas consumers, installers, suppliers and landlords to ensure that:

▶ only competent people work on gas installations;
▶ no one is permitted to use suspect gas appliances;
▶ landlords are responsible, in certain cases, to make sure that fittings and flues are maintained;
▶ with the exception of room-sealed appliances there are restrictions on gas appliances in sleeping accommodation;
▶ instantaneous gas water heaters must be room-sealed or fitted with appropriate safety devices.

Available at: http://www.legislation.gov.uk/uksi/1998/2451/contents/made

From April 2009 a new gas installer registration scheme replaced the former CORGI gas scheme. The new 'Gas Safe Register' is run by the Capita Group Plc on behalf of the HSE. A Gas Safety engineer must be used for any type of gas work, including installation, maintenance and servicing. See: http://www.gassaferegister.co.uk/default.aspx

See: Safety in the installation and use of gas systems and appliances: Gas Safety (Installation and Use) Regulations 1998 Approved Code of Practice and guidance. L56 (Fourth edition), HSE Books, 2013, ISBN, 9780717666171. http://www.hse.gov.uk/pubns/priced/l56.pdf

19

19.37.8 Health and Safety (Offences) Act 2008

The Health and Safety Offences Act 2008 increases penalties and provides courts with greater sentencing powers for those who flout health and safety legislation.

The Act raised the maximum penalties that can be imposed for breaching health and safety regulations in the lower courts from £5,000 to £20,000 and the range of offences for which an individual can be imprisoned has also been broadened.

The Act amended Section 33 of the HSW Act 1974, and raises the maximum penalties available to the courts in respect of certain health and safety offences. It received royal assent on 16 October 2008 and came into force in January 2009.

Available at: http://www.legislation.gov.uk/ukpga/2008/20/contents

19.37.9 Control of Lead at Work Regulations 2002

These Regulations came into force in November 2002 and impose requirements for the protection of employees who might be exposed to lead at work and others who might be affected by the work. The Regulations:

▶ require occupational exposure levels for lead and lead alkyls;

▶ require blood-lead action and suspension levels for women of reproductive capacity and others;

▶ re-impose a prohibition for women of reproductive capacity and young persons in specified activities;

▶ require an employer to carry out a risk assessment;

▶ require employers to restrict areas where exposures are likely to be significant if there is a failure of control measures;

▶ impose requirements for the examination and testing of engineering controls and RPE and the keeping of PPE;

▶ impose new sampling procedures for air monitoring;

▶ impose requirements in relation to medical surveillance;

▶ require information to be given to employees;

▶ require the keeping of records and identification of containers and pipes.

Available at: http://www.legislation.gov.uk/uksi/2002/2676/contents/made

19.37.10 The Manufacture and Storage of Explosives Regulations 2005

The Regulations (SI Number 2005/1082) came into force on 26 April 2005.

These Regulations cover the manufacture, storage and handling of all explosives, including blasting explosives, propellants, detonators and detonating cord, fireworks and other pyrotechnic articles, ammunition, and other explosive articles such as airbags and seat belt pretensioners. The Regulations cover the manufacture of explosives and intermediate products for on-site mixing and storage, and also handling operations that are not in themselves considered to 'manufacture'. These include fusing fireworks, assembling fireworks displays from components, and filling shotgun cartridges and other cartridges for small arms. The main requirements of the Regulations are as follows:

▶ Anyone manufacturing or storing explosives must take appropriate measures to prevent fire or explosion; to limit the extent of any fire or explosion should one occur; and to protect persons in the event of a fire or explosion. These are the key requirements of the Regulations and are backed up by extensive guidance in the ACoP.

▶ In most cases a separation distance must be maintained between the explosives building and neighbouring inhabited buildings. This is intended to ensure that risks to those living or working in the area are kept to an acceptable level. If there is development in this separation zone then the quantity that may be kept must be reduced.

▶ With certain exceptions a licence is required for the manufacture or storage of explosives. The HSE licenses manufacturing activities because of the greater risks involved. The HSE also licenses larger explosives storage facilities. In most cases, stores holding less than 2 tonnes of explosives are either licensed or registered by the local authority or the police.

▶ The HSE may not grant a licence for a manufacturing facility or, in most cases, store until the local authority has given its assent (normally following a public hearing). This is an important safeguard that is retained from the previous system.

Available at: http://www.opsi.gov.uk/si/si2005/uksi_20051082_en.pdf

And the amendment Regulations are available at: http://www.opsi.gov.uk/si/si2007/20072598.htm

Manufacture and storage of explosives guide 3rd edition, 2006, HSE Books. http://www.hse.gov.uk/explosives/forms/exguide.pdf

19.37.11 Occupiers Liability Acts 1957 and 1984 – Civil Law

The 1957 Act concerns the duty that the occupier of premises has towards visitors in relation to the condition of the premises and to things which have or have not been done to them. The Act imposes the following:

▶ there is a duty to take reasonable care to see that a visitor is reasonably safe in using the premises for the purpose for which they were invited or permitted by the occupier to be there;

- the common duty of care will differ depending on the visitor, so a greater duty is owed to children;
- an occupier can expect that a person in the exercise of their trade or profession will appreciate and guard against normal risks, for example a window cleaner;
- no duty of care is owed to someone exercising a public right of way.

The 1984 Act extends the duty of care to persons other than visitors, i.e. trespassers. The occupier has to take reasonable care in all the circumstances to see that non-visitors do not get hurt on the premises because of its condition or the things done or not done to it. The occupier must cover perceived dangers and must have reasonable grounds to know that the trespassers may be in the vicinity.

Available at: http://www.legislation.gov.uk/ukpga/Eliz2/5-6/31/contents

http://www.legislation.gov.uk/ukpga/1984/3/contents

19.37.12 Personal Protective Equipment Regulations 2002

These Regulations relate to approximation of the laws of EU member states. They place duties on persons who place PPE on the market to comply with certain standards. These requirements are that PPE must satisfy the basic health and safety requirements which are applicable to that class or type of PPE, the appropriate conformity assessment procedures must be carried out, CE marking must be correctly affixed and the PPE must not compromise the safety of individuals, domestic animals or property when properly maintained and used.

Available at: http://www.legislation.gov.uk/uksi/2002/1144/contents/made

19.37.13 Control of Pesticides Regulations 1986 as amended

These Regulations made under the Food and Environment Protection Act 1985 all came into force by 1 January 1988. The Regulations apply to any pesticide or any substance used generally for plant control and protection against pests of all types, including anti-fouling paint used on boats. They do not apply to substances covered by other Acts like the Medicines Act 1968, The Food Safety Act 1990, those used in laboratories and a number of other specific applications.

No person may advertise, sell, supply, store, or use a pesticide unless it has received ministerial approval and the conditions of the approval have been complied with. The approval may be experimental, provisional or full, and the Minister has powers to impose conditions.

The Regulations also cover the need for users to be competent and have received adequate information and training. Certificates of competence (or working

under the supervision of a person with a certificate) are required where pesticides approved for agricultural use are used commercially.

Defra issued a consultation document in August 2007 which covers the update of these Regulations as follows.

At present there are four Regulations (plus their amendments) to control and monitor marketing and use of pesticides in England and Wales:

- The Control of Pesticides Regulations 1986 and its amendment in 1997;
- The Plant Protection Products Regulations 2005 and its amendment in 2006;
- The Plant Protection Products (Basic Conditions) Regulations 1997;
- The Plant Protection Products (Fees) Regulations 2007.

Defra's proposal is to consolidate these four Regulations (plus amendments) into one Regulation.

Available at: http://www.legislation.gov.uk/uksi/1986/1510/contents/made

19.37.14 Pressure Systems Safety Regulations 2000 (PSSR)

These Regulations came into effect in February 2000 and replace the Pressure Systems and Transportable Gas Containers Regulations 1989. Transportable gas containers are covered by the Use of Transportable Pressure Receptacles Regulations 1996 (SI 1996, No. 2092).

The aim of PSSR is to prevent serious injury from the hazard associated with stored energy as a result of a pressure system or one of its parts failing. The Regulations cover:

- steam at any pressure;
- gases which exert a pressure in excess of 0.5 bar above atmospheric pressure;
- fluids which may be mixtures of liquids, gases and vapours where the gas or vapour phase may exert a pressure in excess of 0.5 bar above atmospheric pressure.

With the exception of scalding from steam, the Regulations do not consider the effects of the hazardous contents being released following failure. The stored contents are of concern where they can accelerate wear and cause more rapid deterioration and an increased risk of failure.

Available at: http://www.legislation.gov.uk/uksi/2000/128/contents/made

19.37.15 Road Traffic Acts 1988 and 1991

Road rules from Road Traffic Acts 1988 and 1991 are contained in the Highway Code.

19

Many references to legislation can be found abbreviated throughout the Code. The Code contains a table of legislation under the 'Road User and the law' section. It is not intended to be a comprehensive guide, but a guide to some of the important points of law. For the precise wording of the law, please refer to the various Acts and Regulations (as amended) indicated in the Code.

Most of the provisions apply on all roads throughout Great Britain, although there are some exceptions. The definition of a road in England and Wales is 'any highway and any other road to which the public has access and includes bridges over which a road passes' (Road Traffic Act (RTA) 1988 sect 192(1)). In Scotland, there is a similar definition which is extended to include any way over which the public have a right of passage (Road (Scotland) Act (R(S)A) 1984 sect151(1)).

It is important to note that references to 'road' therefore generally include footpaths, bridleways and cycle tracks, and many roadways and driveways on private land (including many car parks). In most cases, the law will apply to them and there may be additional rules for particular paths or ways. Some serious driving offences, including drink-driving offences, also apply to all public places, for example public car parks.

This Highway Code applies to England, Scotland and Wales. The Highway Code is essential reading for everyone who uses the public highways, which is virtually the whole population.

The most vulnerable road users are pedestrians, particularly children, older or disabled people, cyclists, motorcyclists and horse riders. It is important that all road users are aware of the Code and are considerate towards each other. This applies to pedestrians as much as to drivers and riders.

Many of the rules in the Code are legal requirements, and if you disobey these rules you are committing a criminal offence. You may be fined, given penalty points on your licence or be disqualified from driving. In the most serious cases you may be sent to prison. Such rules are identified by the use of the word 'MUST'. In addition, the rule includes an abbreviated reference to the legislation which creates the offence.

Although failure to comply with the other rules of the Code will not, in itself, cause a person to be prosecuted, The Highway Code may be used in evidence in any court proceedings under the Traffic Acts to establish liability. This includes rules which use advisory wording such as 'should/should not' or 'do/do not'. Knowing and applying the rules contained in The Highway Code could significantly reduce road casualties. Cutting the number of deaths and injuries that occur on our roads every day is a responsibility we all share. The Highway Code can help us discharge that responsibility.

The Highway Code covers:

- Rules for pedestrians;
- Rules for animals;
- Rules for cyclists;
- Rules for motorcyclists;
- Rules for drivers and motorcyclists;
- General rules, techniques and advice for all drivers and riders;
- Using the road;
- Road users requiring extra care;
- Driving in adverse weather conditions;
- Waiting and parking;
- Motorways;
- Breakdowns and accidents;
- Road works;
- Railway level crossings;
- Tramways;
- Light signals controlling traffic;
- Signals to other road users;
- Signals by authorised persons;
- Traffic signs;
- Road markings;
- Vehicle markings;
- Annexes
  1. You and your bicycle
  2. Motorcycle licence requirements
  3. Motor vehicle documentation and learner driver requirements
  4. The road user and the law
  5. Penalties
  6. Vehicle maintenance, safety and security
  7. First-aid on the road
  8. Safety code for new drivers;
- Supplementary Notes.

Here we have picked out the legal requirements from the Code applying to drivers of motor vehicles and motorcyclists as an example of the Code. All those who drive for work should be familiar with the latest copy of the code, which is available from all good book stores and at https://www.gov.uk/browse/driving/highway-code

1. **Vehicle condition**. You must ensure your vehicle and trailer comply with the full requirements of the Road Vehicles (Construction and Use) (CUR) Regulations and Road Vehicles Lighting Regulations.

2. **Fitness to drive.** Make sure that you are fit to drive. You must report to the Driver and Vehicle Licensing Agency (DVLA) any health condition likely to affect your driving. Law RTA 1988 sect 94.

3. **Vision**. You must be able to read a vehicle number plate, in good daylight, from a distance of 20 metres (or 20.5 metres where the old style number plate is used). If you need to wear glasses (or contact lenses) to do this, you must wear them at all times while driving. The police have the power to require a driver to undertake an eyesight test. Laws RTA 1988 sect 96 & MV(DL)R reg 40 & sch 8.

4. You must not drive with a breath alcohol level higher than 35 microgrammes/100 millilitres of breath or a blood alcohol level of more than 80

milligrammes/100 millilitres of blood. Law RTA 1988 sects 4, 5 & 11(2).

5. You must not drive under the influence of drugs or medicine. Law RTA 1988 sect 4.

6. **Vehicle towing and loading**. As a driver:
   (a) you must not tow more than your licence permits. If you passed a car test after 1 Jan 1997 you are restricted on the weight of trailer you can tow:
   (b) you must not overload your vehicle or trailer;
   (c) you must secure your load and it must not stick out dangerously.
   Laws CUR reg 100 & MV(DL)R reg 43.

7. **Seat belts and child restraints.** You must wear a seat belt in cars, vans and other goods vehicles if one is fitted. Adults, and children aged 14 years and over, must use a seat belt or child restraint, where fitted, when seated in minibuses, buses and coaches. Laws RTA 1988 sects 14 & 15, Motor Vehicle (Wearing of Seat Belts)[MV(WSB)] Regulations, MV(WSB by Children in Front Seats) Regulations & MV(WSB)(Amendment)R.

8. The driver must ensure that all children under 14 years of age in cars, vans and other goods vehicles wear seat belts or sit in an approved child restraint where required. If a child is under 1.35 metres (approx 4 feet 5 inches) tall, a baby seat, child seat, booster seat or booster cushion must be used suitable for the child's weight and fitted to the manufacturer's instructions.

9. A rear-facing baby seat must not be fitted into a seat protected by an active frontal airbag, as in a crash it can cause serious injury or death to the child. Laws RTA 1988 sects 14 & 15, MV(WSB)R, MV(WSBCFS) R & MV(WSB)(A)R.

19.37.16 The Health and Safety (Sharp Instruments in Healthcare) Regulations 2013

The Health and Safety (Sharp Instruments in Healthcare) Regulations came into force on 11 May 2013. Northern Ireland introduced equivalent regulations on the same date.

The new regulations were introduced to implement a European Directive. They supplement the existing health and safety legislation that already requires employers across all sectors to take effective action to control the risk from sharps injuries.

Employers and contractors working in the healthcare sector are required to:

▶ Have effective arrangements for the safe use and disposal (including using 'safer sharps' where reasonably practicable, restricting the practice of recapping of needles and placing sharps bins close to the point of use);

▶ Provide the necessary information and training to workers;

▶ Investigate and take action in response to work-related sharps injuries.

Guidance for healthcare employers and employees, HSIS7, HSE, 2013 at: www.hse.gov.uk/pubns/hsis7.htm

19.37.17 Smoke-free legislation

Five sets of smoke-free Regulations set out the detail of the smoke-free legislation. Within the UK, smoking law is a devolved issue and therefore similar legislation has been enacted in all parts of the UK.

The smoke-free law has been introduced to protect employees and the public from the harmful effects of secondhand smoke.

Key points are:

▶ It is now against the law to smoke in virtually all 'enclosed' and 'substantially enclosed' public places and workplaces.

▶ Public transport and work vehicles used by more than one person must be smoke free at all times.

▶ No-smoking signs must be displayed in all smoke-free premises and vehicles.

▶ Staff smoking rooms and indoor smoking areas are no longer allowed, so anyone who wants to smoke has to go outside.

▶ Managers of smoke-free premises and vehicles have legal responsibilities to prevent people from smoking.

▶ If anyone is uncertain where they can or cannot smoke, they need to look for the No-smoking signs or ask someone in charge.

Available at: http://www.legislation.gov.uk/all?title=Smoke%20free%20

19.37.18 Working Time Regulations 1998 as amended

These Regulations came into force in October 1998 and, for specified workers, restrict the working week to 48 hours per 7-day period. Individuals can voluntarily agree to disapply the weekly working hours limit. Employers must keep a copy of all such individual agreements. Workers whose working time is not measured or predetermined, or who can themselves determine the duration of their working day, are excepted from weekly working time, night work, rest periods and breaks.

The Regulations were amended, with effect from 1 August 2003, to extend working time measures in full to all non-mobile workers in road, sea, inland waterways and lake transport, to all workers in the railway and offshore sectors, and to all workers in aviation who are not covered by the sectoral Aviation Directive. The Regulations applied to junior doctors from 1 August 2004.

Mobile workers in road transport have more limited protections. Those subject to European Drivers' hours rules 3820/85 are entitled to 4 weeks' paid annual leave

19

and health assessments if night workers, with effect from 1 August 2003. Mobile workers not covered by European drivers' hours rules will be entitled to an average 48 hours per week, 4 weeks paid holiday, health assessments if night workers and adequate rest.

The Regulations were previously amended, with effect from 6 April 2003, to provide enhanced rights for adolescent workers.

The basic rights and protections that the Regulations provide are a:

▶ limit of an average of 48 hours a week which a worker can be required to work (though workers can choose to work more if they want to);
▶ limit of an average of 8 hours work in 24 hours which night workers can be required to work;

▶ right for night workers to receive free health assessments;
▶ right to 11 hours' rest a day;
▶ right to a day off each week;
▶ right to an in-work rest break if the working day is longer than 6 hours;
▶ right to 28 days' paid leave per year from April 2009.

Further amendment in 2009 applied to some doctors in training.

1998 Regulations available at: http://www.legislation.gov.uk/uksi/1998/1833/contents/made

2003 amendment available at: http://www.legislation.gov.uk/uksi/2003/1684/contents/made

See the HSE website for more information at: http://www.hse.gov.uk/foi/internalops/ocs/001-099/1_6.htm

CHAPTER 20

International, environmental and other aspects of health and safety

This chapter covers the following learning objectives:

1. Outline the nature and sources of international health and safety management information
2. Explain the various accident compensation systems that are operated globally
3. Outline the precautions to be taken when travelling abroad
4. Describe the roles and powers of the environmental enforcement agencies
5. Identify the reasons for implementing an Environmental Management System
6. Explain the principles and practice of environmental impact assessment
7. Identify basic health and safety issues in the home
8. Identify the key principles of safe cycling

20.1 Introduction

This chapter covers two topics that are no longer in the syllabus for the NEBOSH National Certificate in Construction Health and Safety but may well be useful to many health and safety managers. They cover the following areas:

▶ International issues;
▶ Environmental considerations.

Additional sections on health and safety in the home and safe cycling have also been included.

20.2 International issues

20.2.1 Introduction

The authors of this book have recently published *International Health and Safety at Work* which covers the syllabus for the NEBOSH International General Certificate and gives a detailed account of the operation of occupational health and safety internationally. A basic introduction is given here to three aspects of the international scene – the International Labour Organisation (ILO), the differences between fault and no fault compensation schemes that exist in various parts of the world and working abroad. Many countries have either fault or no-fault compensation schemes for workers involved in accidents. Knowledge of these schemes is important for those who work in several different countries The hazards of working in unfamiliar countries and/ or climates are important health and safety issues and include snake bites, diseases (such as malaria and yellow fever) and sunstroke.

The International Labour Organisation (ILO), World Health Organisation (WHO) and United Nations (UN) have estimated that there are 270 million occupational accidents and 160 million occupational diseases every year throughout the world – and these are recognised as relatively conservative estimates due to probable under-reporting. The ILO estimates that 2 million women and men die each year as a result of occupational accidents and work-related diseases. Table 20.1 shows the global numbers in more detail.

Table 20.1 Numbers of global work-related adverse events

Event	Average (daily)	Annually
Work-related death	5,000	2,000,000
Work-related deaths to children	60	22,000
Work-related accidents	740,000	270,000,000
Work-related disease	438,000	160,000,000
Hazardous substance deaths	1,205	440,000
Asbestos-related deaths	274	100,000

In the USA in 2002, approximately 2 million workers were victims of workplace violence. In the UK, 1.7% of working adults (357,000 workers) were the victims of one or more incidents of workplace violence.

Ten per cent of all skin cancers are estimated to be attributable to workplace exposure to hazardous substances. In Latin America 37% of miners have silicosis, rising to 50% among miners over 50. In India 54.6% of slate pencil workers and 36.2% of stone cutters have silicosis.

In the course of the 20th century, industrialised countries saw a clear decrease in serious injuries, not least because of real advances in making the workplace healthier and safer. The challenge is to extend the benefits of this experience to the whole working world. However, 1984 saw the worst chemical disaster ever when 2,500 people were killed and over 200,000 injured in the space of a few hours at Bhopal. Not only were the workers affected but their families, their neighbours and whole communities. Twenty years later many people are still affected by the disaster and are dying as a result. The rusting remains of a once magnificent plant remain as a reminder of the disaster.

According to the ILO, deaths due to work-related accidents and illnesses represent 3.9% of all deaths and 15% of the world's population suffers a minor or major occupational accident or work-related disease in any one year. A large number of the unemployed – up to 30% – report that they suffer from an injury or disease dating from the time at which they were employed. The number of fatal occupational accidents, especially in Asia and Latin America, is increasing.

The main preventable factors for accidents are:

▶ lack of a preventative safety and health culture;
▶ poor management systems;
▶ poor supervision and enforcement by the government.

Experience has shown that a strong safety culture is beneficial for workers, employers and governments alike. Various prevention techniques have proved themselves effective, both in avoiding workplace accidents and illnesses and improving business performance. Today's high standards in some countries are a direct result of long-term policies encouraging tripartite social dialogue, collective bargaining between trade unions and employers, and effective health and safety legislation backed by potent labour inspection. The ILO believes that safety management systems like ILO-OSH 2001 provide a powerful tool for developing a sustainable safety and health culture at the enterprise level and mechanisms for the continual improvement of the working environment.

20.2.2 The role and function of the ILO

The ILO is a specialised agency of the United Nations that seeks to promote social justice through establishing and safeguarding internationally recognised human and labour rights. It was founded in 1919 by the Treaty of Versailles at the end of the First World War.

The motivation behind the creation of such an organisation was primarily humanitarian. Working conditions at the time were becoming unacceptable to a civilised society. Long hours, unsafe, unhygienic and dangerous conditions were common in low-paid manufacturing careers. Indeed, in the wake of the Russian Revolution, there was concern that such working conditions could lead to social unrest and even other revolutions. The ILO was created as a tripartite organisation with governments, employers and workers represented on its governing body.

The ILO formulates international labour standards and attempts to establish minimum rights including freedom of association, the right to organise, collective bargaining, abolition of forced labour, equality of opportunity and treatment and other standards that regulate conditions across all work-related activities.

Representatives of all ILO member states meet annually in Geneva for the International Labour Conference, which acts as a forum where social and labour questions of importance to the entire world are discussed. At this conference, labour standards are adopted and decisions made on policy and future programmes of work.

The ILO has 178 Member States but if a country is not a member, the ILO still has influence as a source of guidance when social problems occur.

Figure 20.1 World Cup stadium under construction, Cape Town, 2009

The main principles on which the ILO is based are:

1. labour is not a commodity;
2. freedom of expression and of association are essential to sustained progress;
3. poverty anywhere constitutes a danger to prosperity everywhere;
4. the 'war against want' requires to be carried on with unrelenting vigour within each nation, and by continuous and concerted international effort in which the representatives of workers and employers, enjoying equal status with those of governments, join with them in free discussion and democratic decision with a view to the promotion of the common welfare.

A campaign launched by the ILO has been to seek to eliminate child labour throughout the world. In particular, the ILO is concerned about children who work in hazardous working conditions, bonded child labourers and extremely young working children. It is trying to create a worldwide movement to combat the problem by:

▶ implementing measures which will prevent child labour;
▶ withdrawing children from dangerous working conditions;
▶ providing alternatives;
▶ improving working conditions as a transitional measure towards the elimination of child labour.

ILO Conventions and Recommendations

The international labour standards were developed for four reasons. The main motivation was to improve working conditions with respect to health and safety and career advancement. The second motivation was to reduce the potential for social unrest as industrialisation progressed. Third, the member states want common standards so that no single country has a competitive advantage over another due to poor working conditions.

Finally, the union of these countries creates the possibility of a lasting peace based on social justice.

International labour standards are adopted by the International Labour Conference. They take the form of Conventions and Recommendations. At the present time, there are 187 Conventions and 198 Recommendations, some of which date back to 1920.

International labour standards contain flexible measures to take into account the different conditions and levels of development among member states. However, a government that ratifies a Convention must comply with all its articles. Standards reflect the different cultural and historical backgrounds of the member states as well as their diverse legal systems and levels of economic development.

ILO occupational safety and health standards can be divided into four groups, and an example is given in each case:

1. Guiding policies for action
   The Occupational Safety and Health Convention, 1985 (No. 155) and its accompanying Recommendation (No. 164) emphasize the need for preventative measures and a coherent national policy on occupational safety and health. They also stress employers' responsibilities and the rights and duties of workers.
2. Protection in given branches of economic activity
   The Safety and Health in Construction Convention, 1988 (No. 167) and its accompanying Recommendation (No. 175) stipulate the basic principles and measures to promote safety and health of workers in construction.
3. Protection against specific risks
   The Asbestos Convention, 1986 (No. 162) and its accompanying Recommendation (No. 172) gives managerial, technical and medical measures to protect workers against asbestos dust.
4. Measures of protection
   Migrant Workers (Supplementary Provisions) Convention, 1975 (No. 143) aims to protect the safety and health of migrant workers.

ILO **Conventions** are international treaties signed by ILO member states and each country has an obligation to comply with the standards that the Convention establishes.

In contrast, ILO **Recommendations** are non-binding instruments that often deal with the same topics as Conventions. Recommendations are adopted when the subject, or an aspect of it, is not considered suitable or appropriate at that time for a Convention. Recommendations guide the national policy of member states so that a common international practice may develop and be followed by the adoption of a Convention.

ILO standards are the same for every member state and the ILO has consistently opposed the concept of different standards for different regions of the world or groups of countries.

The standards are modified and modernised as needed. The governing body of the ILO periodically reviews individual standards to ensure their continuing relevance.

Supervision of international labour standards is conducted by requiring the countries that have ratified Conventions to periodically present a report with details of the measures that they have taken, in law and practice, to apply each ratified Convention. In parallel, employers' and workers' organisations can initiate contentious proceedings against a member state for its alleged non-compliance with a convention it has ratified. In addition, any member country can lodge a complaint against another member state which, in its opinion, has not ensured in a satisfactory manner the implementation of a Convention which both of them have ratified. Moreover, a special procedure exists in the field of freedom of association to deal with complaints submitted by governments or by employers' or workers' organisations against a member state whether or not the country concerned has ratified the relevant Conventions. Finally, the ILO has systems in place to examine the enforcement of international labour standards in specific situations.

The ILO also publishes Codes of Practice, guidance and manuals on health and safety matters. These are often used as reference material by either those responsible for drafting detailed Regulations or those who have responsibility for health and safety within an organisation. They are more detailed than either Conventions or Recommendations and suggest practical solutions for the application of ILO standards. Codes of Practice indicate 'what should be done'. They are developed by tripartite meetings of experts and the final publication is approved by the ILO governing body.

For example, the construction industry has a Safety and Health in Construction Convention, 1988 (No. 167) that obliges signatory ILO member states to comply with the construction standards laid out in the Convention – the Convention is a relatively brief statement of those standards. The accompanying Recommendation (No. 175) gives additional information on the Convention statements. The Code of Practice gives more detailed information than the Recommendation. This can best be illustrated by contrasting the coverage of scaffolds and ladders by the three documents shown in Appendix 20.1.

Important ILO Conventions (C) and Recommendations (R) in the field of occupational safety and health include:

▶ C 115 Radiation Protection and (R 114), 1960;
▶ C 120 Hygiene (Commerce and Offices) and (R 120), 1964;
▶ C 139 Occupational Cancer and (R 147), 1974;
▶ C 148 Working Environment (Air, Pollution, Noise and Vibration) and (R 156), 1977;

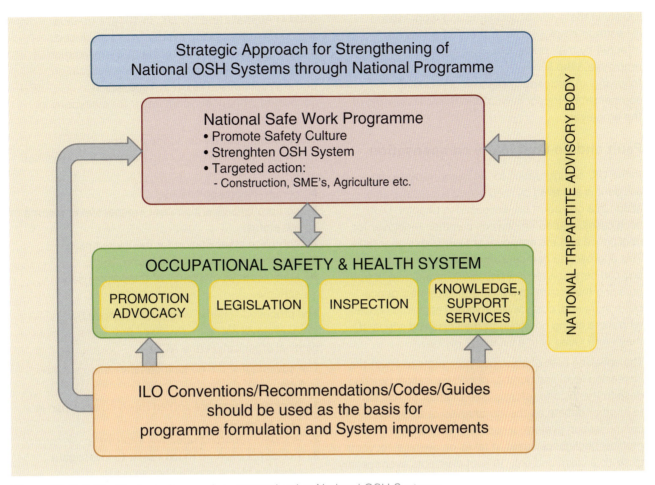

Figure 20.2 ILO's Strategic Approach to strengthening National OSH Systems

Figure 20.3 Excavator at work in France

▶ C 155 Occupational Safety and Health and (R164), 1981;

▶ C 161 Occupational Health Services and (R 171), 1985;

▶ C 162 Asbestos and (R 172), 1986;

▶ C 167 Safety and Health in Construction and (R 175), 1988;

▶ C 170 Chemicals and (R 177), 1990;

▶ C 174 Prevention of Major Industrial Accidents and (R 181), 1993;

▶ C 176 Safety and Health in Mines and (R 176), 1995;

▶ C 184 Safety and Health in Agriculture and (R 192), 2001;

▶ C 187 Promotional Framework for Occupational Safety and Health and (R 197), 2006;

▶ R 97 Protection of Workers' Health Recommendation, 1953;

▶ R 102 Welfare Facilities Recommendation, 1956;

▶ R 31 List of Occupational Diseases Recommendation, 2002.

Copies of Conventions, Recommendations and Codes of Practice can be obtained from the ILO website www. ilo.org/safework.

20.2.3 Compensation and insurance issues

Accidents/incidents arising out of the organisation's activities resulting in injuries to people and incidents resulting in damage to property can lead to compensation claims. There are two facets to this – employers' liability insurance and fault and no-fault compensation.

Employers' liability insurance

In many countries, employers are required to take out employers' liability insurance to cover their liability in the event of accidents and work-related ill-health to employers and others who may be affected by their operations. This ensures that any employee, who successfully sues his/her employer following an accident, is assured of receiving compensation irrespective of the financial position of the employer.

Fault and no-fault injury compensation

In the UK, compensation for an injury following an accident is achieved by means of a successful legal action in a civil court (as discussed in Chapter 1). In such cases, injured employees sue their employer for negligence and the employer is found liable or at fault. This approach to compensation is adversarial, costly and can deter injured individuals of limited means from pursuing their claim. In a recent medical negligence claim in Ireland, costs were awarded against a couple who were acting on behalf of their disabled son, and they were faced with a bill for £3 million.

The spiralling cost of insurance premiums to cover the increasing level and number of compensation awards, despite the Woolfe reforms (see Chapter 5), has led to another debate on the introduction of a no-fault compensation system. It has been estimated that in medical negligence cases, it takes on average six years to settle a claim and only 10% of claimants ever see any compensation.

No-fault compensation systems are available in many parts of the world, in particular New Zealand and several states in the USA. In these systems, amounts of compensation are agreed centrally at a national or state level according to the type and severity of the injury. The compensation is often in the form of a structured continuous award rather than a lump sum and may be awarded in the form of a service, such as nursing care, rather than cash.

The no-fault concept was first examined in the 1930s in the USA to achieve the award of compensation quickly in motor car accident claims without the need for litigation. It was introduced first in the State of Massachusetts in 1971 and is now mandatory in nine other states, although several other states have either repealed or modified no-fault schemes.

In 1978, the Pearson Commission in the UK rejected a no-fault system for dealing with clinical negligence even though it acknowledged that the existing tort system was costly, cumbersome and prone to delay. Its principal reasons for rejection were given as the difficulties in reviewing the existing tort liability system and in determining the causes of injuries.

In New Zealand, there was a general dissatisfaction with the workers' compensation scheme, which was similar to the adversarial fault-based system used in Australia and the UK. In 1974, a no-fault accident compensation system was introduced and administered by the Accident Compensation Corporation (ACC). This followed the publication of the Woodhouse Report in 1966 which advocated 24-hour accident cover for everybody in New Zealand.

The Woodhouse Report suggested the following five principles for any national compensation system:

▶ community responsibility;
▶ comprehensive entitlement irrespective of income or job status;
▶ complete rehabilitation for the injured party;
▶ real compensation for the injured party;
▶ administrative efficiency of the compensation scheme.

The proposed scheme was to be financed by channelling all accident insurance premiums to one national organisation (the ACC).

The advantages of a no-fault compensation scheme include:

1. Accident claims are settled much quicker than in fault schemes.
2. Accident reporting rates will improve.
3. Accidents become much easier to investigate because blame is no longer an issue.
4. Normal disciplinary procedures within an organisation or through a professional body are unaffected and can be used if the accident resulted from negligence on the part of an individual.
5. More funds are available from insurance premiums for the injured party and less used in the judicial and administrative process.

The possible disadvantages of a no-fault compensation system are:

1. There is often an increase in the number of claims, some of which may not be justified.
2. There is a lack of direct accountability of managers and employers for accidents.
3. Mental injury and trauma are often excluded from no-fault schemes because of the difficulty in measuring these conditions.
4. There is more difficulty in defining the causes of many injuries and industrial diseases than in a fault scheme.
5. The monetary value of compensation awards tends to be considerably lower than those in fault schemes (although this can be seen as an advantage).

No-fault compensation schemes exist in many countries including Canada and the Scandinavian countries. However, a recent attempt to introduce such a scheme in New South Wales, Australia, was defeated in the legislative assembly.

(a)

(b)

Figure 20.4 Have to consider different solutions in different countries: (a) safe delivery of furniture in Certaldo, Italy; (b) dangerous access to install overhead low-voltage data lines in Morocco

20.2.4 Working abroad

Introduction

Staff may travel and work abroad in the course of the organisation's business. Most of this work is no more risky than the equivalent activity in their home country and takes place in countries that are considered safe for travellers. Increasingly though, work takes place in countries or parts of countries where the relevant government office advises against travel.

This section outlines the steps needed for safe management of these activities. It applies to work carried out by staff (working alone or with colleagues).

General safety advice

Safety begins when you pack. To help avoid becoming a target, do not dress so as to mark yourself as an affluent tourist. Expensive-looking jewellery, for instance, can draw the wrong attention.

Always try to travel light. You can move more quickly and will be more likely to have a free hand. You will also be less tired and less likely to set your luggage down, leaving it unattended.

Carry the minimum number of valuables, and plan places to conceal them. Your passport, cash and credit cards are most secure when locked in a hotel safe. When you have to carry them on your person, you may wish to put them in various places rather than all in one wallet or pouch. Avoid handbags, fanny packs and outside pockets that are easy targets for thieves. Inside pockets and a sturdy shoulder bag with the strap worn across your chest are somewhat safer. One of the safest places to carry valuables is in a pouch or money belt worn under your clothing.

If you wear glasses, pack an extra pair. Bring them and any medicines you need in your carry-on luggage.

To avoid problems when passing through customs, keep medicines in their original, labelled containers. Bring copies of your prescriptions and the generic names for the drugs. If a medication is unusual or contains narcotics, carry a letter from your doctor attesting to your need to take the drug. If you have any doubt about the legality of carrying a certain drug into a country, consult the embassy or consulate of that country before you travel.

Bring travellers cheques and one or two major credit cards instead of cash.

Pack an extra set of passport photos along with a photocopy of your passport's information page to make replacement of your passport easier in the event it is lost or stolen.

20

Put your name, and organisation's address and telephone numbers inside and outside of each piece of luggage. Use covered luggage tags to avoid casual observation of your identity or nationality. If possible, lock your luggage.

Risks involved in overseas work

Risks to health and safety that arise from overseas work can be:

1. work-related (e.g. physical risks from the fieldwork itself);
2. health-related (e.g. exposure to tropical diseases);
3. related to personal security (e.g. associated with local criminal or political activity, or civil unrest).

The following notes concentrate on the third point above as the health and safety aspects of the work itself are covered throughout this book.

Risk assessment

Risk assessments must be made for all work abroad. This need not necessarily be onerous – the nature and complexity of the assessment should reflect the risks involved in the work. In many cases the work itself is not hazardous but it takes place in hazardous surroundings. In these cases the risk assessment should concentrate on the travel-associated risks. For instance, no written risk assessment would be needed for low risk activities such as attending conferences or visiting colleagues in Australia or the European Union, but similar activities in more dangerous locations would require one.

Travel advice

Many national foreign office departments or consulates will have information about safe travel. The UK's Foreign and Commonwealth Office (FCO) has a website (www.fco.gov.uk/travel) which carries up-to-date travel advice for over 200 countries. It could be used as the basis for informing all those travelling on business or leisure of the risks that they might face. Although the advice pertains to the safety of British travellers, it is relevant to other nationalities as British insurance companies use it to determine the extent or validity of insurance cover.

Health matters

Staff travelling on business should take advice on the potential health risk associated with travel. This should be sought in good time so that any recommended immunisations can be given and/or prophylactic drugs prescribed.

Appendix 20.2 has some useful safe travel recommendations.

(a)

(b)

Figure 20.5 Occupational road risk: (a) unusual and slow-moving large animals mixed with traffic in India; (b) overloaded truck in Morocco

20.2.5 ILO recommendations for road transport drivers

The ILO covers some aspects of driver health and safety in the Hours of Work and Rest Periods (Road Transport) Convention 1979 (No. 153) and its accompanying Hours of Work and Rest Periods (Road Transport) Recommendation 1979 (No. 153).

The Convention states that no driver shall be allowed to drive continuously for more than four hours without a break, although this may be exceeded by no more than one hour under certain circumstances. The maximum total driving time, including overtime, must not exceed nine hours per day or 48 hours per week. The length of the break and, as appropriate, the way in which the break may be split shall be determined by the competent authority or body in each country.

The Convention also states that the daily rest of drivers shall be at least ten consecutive hours during any 24-hour period starting from the beginning of the working day. The daily rest may be calculated as an average over periods to be determined by the competent authority or body in each country. During the daily rest the driver shall not be required to remain in or near the vehicle if he has taken the necessary precautions to ensure the safety of the vehicle and its load.

The Recommendation adds the following detail:

▶ The normal hours of work should not exceed eight per day as an average.
▶ The length of the break after the four-hour driving period and, as appropriate, the way in which the break may be split should be determined by the competent authority or body in each country.
▶ The maximum total driving time, including overtime, should exceed neither nine hours per day nor 48 hours per week (averaged, if necessary, over a maximum period of four weeks).
▶ The daily rest of drivers should be at least 11 consecutive hours during any 24-hour period starting from the beginning of the working day.
▶ The daily rest may be calculated as an average over periods to be determined by the competent authority or body in each country; provided that the daily rest should in no case be less than eight hours (exceptions to the recommended duration of the daily rest periods and the manner of taking such rest periods may be provided in the cases of vehicles having a crew of two drivers and of vehicles using a ferry-boat or a train).
▶ The minimum duration of the weekly rest should be 24 consecutive hours, preceded or followed by the daily rest.
▶ In long-distance transport, it should be possible to accumulate weekly rest over two consecutive weeks. In appropriate cases, the competent authority or body in each country may approve the accumulation of this rest over a longer time.

20.3 Environmental considerations

The current NEBOSH National Certificate in Construction Health and Safety no longer includes general environmental considerations other than waste disposal. However, many health and safety managers also have an 'environmental' responsibility. This section attempts to give some background information on environmental issues. NEBOSH also offers a Certificate in Environmental Management and a Diploma in Environmental Management that covers these issues in much more detail.

Organisations must also be concerned with aspects of the environment. There will be an interaction between the health and safety policy and the environmental policy which many organisations are now developing. Many of these interactions will be concerned with good practice, the reputation of the organisation within the wider community and the establishment of a good health and safety culture. The health and safety data

sheet, used for a COSHH assessment, also contains information of an environmental nature covering ecological information and disposal considerations. An important component in the management of environmental considerations is the Environmental Management System (EMS).

The Environmental Management System should cover all aspects of an organisation's activities, products and services that could interact with the environment. Such aspects or elements include:

▶ the raw materials used and their source;
▶ by-products from processes;
▶ waste material produced;
▶ the energy and water used;
▶ emissions to air and water;
▶ lifespan of products and end-of-life disposal;
▶ packaging and transport issues; and
▶ environmental hazards associated with any materials used.

Environmental assessments (see 15.3.5) need to be reviewed if there are significant changes to some of the elements, such as changes in raw material source or change of product. However, changes in plant or location, the availability of new processes and technologies or recycling opportunities could also be reasons. In addition, new legislation or market and public pressures, evidence of environmental damage or accidents could also prompt a review. Reviews of assessments may also arise as a result of audits or from annual reviews as part of an EMS or striving for continuous improvement.

There are three environmental issues which place statutory duties on employers and are directly related to the health and safety function. These are:

▶ air pollution;
▶ water pollution;
▶ waste disposal.

The statutory duties are contained in the Environmental Protection Act (EPA) 1990 and several of its subsequent Regulations. The Act is enforced by various state agencies (the Environment Agency, local authorities) and these agencies have very similar powers to the HSE (e.g. enforcement, prohibition notices and prosecution). In addition to serving notices, they can enter premises at any reasonable time or at any time if they have reason to believe that there is immediate risk of serious pollution to the environment. They may be accompanied by other authorised persons or a police constable and take with them any equipment or materials required to exercise their powers.

Once there they can make any examinations or inspections necessary and take samples, photographs, measurements and copies of documents. They may cause premises or parts of premises to be left undisturbed and ask for articles to be dismantled, tested or rendered harmless. Articles and substances may be taken and examined or ensured that they are secure from tampering and available as evidence. Any persons involved are required to offer assistance or facilities as necessary and may be required to answer questions and sign a declaration that their statements are true. The end result may be the initiation of legal proceedings.

Environmental managers should have a clear understanding of the harm that can be done to a business when regulatory controls are breached. The potential liabilities arising from non-compliance with environmental regulatory controls can be significant, leading to criminal prosecutions by the enforcing authorities or through private prosecutions. The results of a successful prosecution could result in:

▶ the significant fines that can be given for serious breaches of the law;

▶ the potential for prison sentences and for personal prosecution for directors and managers;

▶ administrative sanctions including serving notices such as revocation notices, suspension notices and enforcement notices;

▶ orders requiring works to be undertaken to prevent pollution;

▶ award of clean-up costs for rectifying environmental damage;

▶ loss of value of property, land or a business; civil liability, resulting in claims for damages or injunctions;

▶ adverse publicity and its effect on relations with key stakeholders, including shareholders, customers, neighbours, insurers and banks; and

▶ insurance issues due to the fact that most standard business policies exclude liability for environmental damage unless sudden and unforeseen.

The Act is divided into nine parts but this chapter will only be concerned with Part 1 (the control of pollution into the air, water or land) and Part 2 (waste disposal).

Pollution is a term that covers more than the effect on the environment of atmospheric emissions, effluent discharges and solid waste disposal from industrial processes. It also includes the effect of noise, vibration, heat and light on the environment. This was recognised in the 1996 EU Directive on Integrated Pollution Prevention and Control, and the Pollution Prevention and Control Act 1999 replaced the Integrated Pollution Control Regulations made under Part 1 of the EPA by extending those powers to cover waste minimisation, energy efficiency, noise and site restoration. Thus the Pollution Prevention and Control Act has now replaced Part 1 of the EPA.

The Solvents Emissions Directive (SED) has produced some tougher rules on the use of solvents in industry. Local Authorities and the Environment Agency are beginning to incorporate SED provisions into Integrated Pollution Prevention and Control Permits (IPPCP). Organisations that use solvents on an industrial scale need an IPPCP to operate.

The Control of Pesticides Regulations have been in place since 1986 and the Department of Food and Rural affairs (Defra) are proposing to consolidate these Regulations with three Plant Protection Products Regulations to form one set of integrated Regulations (see Chapter 19 for more information).

The Environmental Damage (Prevention and Remediation) Regulations came into force in 2009 (in England only) to implement the EU Environmental Liability Directive. They cover environmental damage caused by economic activity of any public or private organisation. Under the Regulations such organisations are obliged to report to the Environment Agency any activity that has or could cause environmental damage. Such damage could be to protected species, natural habitats, surface or groundwater, sites of special scientific interest or land in general. If the Environment Agency considers that such damage has occurred, it can serve a remediation notice on the organisation outlining any remedial action required. If the organisation fails to remedy the problem, the Environment Agency is empowered to undertake the remediation itself and order any associated costs from the organisation.

The Regulatory Enforcement and Sanctions Act has given the Environmental Agency new powers to enforce environmental law using civil sanctions rather than criminal prosecutions. The new powers include the imposition of fixed and variable fines and prohibition notices to prevent the continuation of an activity until compliance has been assured. Other powers include the use of compliance notices that require remedial action to be taken; restoration notices that require the restoration of particular situation and enforcement undertakings. These allow corrective actions to be undertaken before any enforcement is taken. Appeals against any of these measures are possible to an independent tribunal.

Figure 20.6 Example of heavy industrial pollution

20.3.1 Air pollution

The most common airborne pollutants are carbon monoxide, benzene, 1,3-butadiene, sulphur dioxide, nitrogen dioxide and lead. Air pollution is monitored by Integrated Pollution Prevention and Control (IPPC). This is a system which extends the Integrated Pollution Control (IPC) system established by Part 1 of the EPA by introducing three tiers of pollution control.

Regime A1 processes, which are certain large-scale manufacturing processes with a potential to cause serious environmental damage to the air, water or the land. In England and Wales this is enforced by the Environment Agency. In Scotland there is a parallel system enforced by the Scottish Environmental Protection Agency.

Regime A2 processes, which produce emissions to air, water and land with a much smaller potential to pollute than regime A1 processes. The local authority (LA-IPPC) is the regulator for these processes.

Part B processes, which may be classified as those from less polluting industries with only emissions released to air being subject to regulatory control. For such processes local authorities are the enforcing body through Environmental Health Officers. The system is known as Local Air Pollution Control (LAPC) in Northern Ireland and Scotland. In England and Wales, it is known as Local Air Pollution Prevention and Control (LAPPC).

This division has led to some anomalies in that some Part A processes create less pollution than some Part B processes. However, the grouping of three pollution destinations under one arrangement tends to a more holistic approach. The aim of IPPC is to control pollution of the whole environment under a single enforcement system and it offers three principles to prevent and control pollution. These are:

▶ The 'Best Practicable Environmental Option' (BPEO) which considers both the environmental and economic costs and benefits of the possible options available to deal with the pollution problem. BPEO is a legal requirement for Part A processes. It normally requires a technical solution.

▶ The 'Best Available Techniques' are similar to BATNEEC (Best Available Techniques Not Entailing Excessive Cost) introduced by the EPA to minimise the overall environmental impact of a process. Part B processes only need to satisfy the BATNEEC requirement which is not restricted to pollution control technology but can include employee training and competence and building design and maintenance.

▶ 'As low as reasonably practicable' applies the same test to an environmental problem as is applied to a health and safety problem. Any high or unacceptable environmental risk should be reduced to as low as is reasonably practicable.

The EPA proscribes certain listed substances from being released to air, water or land. All proscribed processes must have authorisation. An operator of a proscribed process (such as a vehicle spray booth) must apply to the Environment Agency for prior authorisation

to operate the process. If the application is granted, the operator must monitor emissions and report them to the Environment Agency on a yearly basis. The Agency has the power to revoke the authorisation, enforce the terms of the authorisation or prohibit the operation of the process. Further information on the authorisation process is given in Chapter 19.

The causes, principles of formation and consequences of four common air pollution effects are as follows:

▶ **The greenhouse effect** is caused by the emission of, and increase in, atmospheric levels of specific pollutants such as carbon dioxide, methane and refrigerant gases. Greenhouse gases trap long wave thermal radiation causing a warming effect upon the atmosphere. As concentrations increase, this warming effect increases to the point where there is an imbalance between incoming short wave radiation and outgoing long wave radiation leading to net global warming of the atmosphere. The consequences of this warming include climate change, possible sea level rises and effects on agricultural and natural ecosystems.

▶ **Stratospheric ozone depletion** is caused by emissions of volatile organic compounds containing halogens which have a long half-life in the atmosphere – such as refrigerants, carbon tetrachloride, halons and other chlorinated solvents. The Earth is protected by a natural layer of ozone that absorbs incoming UV radiation. Natural ozone formation is by the process of photo-dissociation of oxygen molecules and reformation as ozone. Chlorine (and other halogen) atoms catalyse the destruction of ozone back to oxygen molecules and are not consumed in the reaction. Therefore, one atom of chlorine can destroy many ozone molecules. The main consequence is an increased level of cell and genetic damage, ultimately leading to skin cancer and crop damage.

▶ **Acid rain** is caused by the emission of gases that form acidic compounds in the presence of water or water vapour, such as sulphur dioxide, hydrogen chloride and nitrogen dioxide. These acid-forming gases dissolve in water in the atmosphere to produce acids such as sulphurous or sulphuric acid. When the acidic water forms raindrops or mist and reaches the ground, it affects the acidity (pH) balance in upland ecosystems, leading to the release of toxic metals into surface run-off with subsequent toxic effects on vegetation, aquatic invertebrates and fish. Acid rain can also damage materials such as limestone in buildings, and cause corrosion to metals such as galvanised steel.

▶ **Photochemical smog** is associated with the emission of volatile organic compounds such as solvents, petrol vapour and other compounds into a warm atmosphere in the presence of sunlight. The emissions react with each other in the presence of ultraviolet radiation and reactive gases such as nitrogen dioxide to create complex mixtures of secondary pollutants and ozone, many of which cause irritation to respiratory systems and degrade materials such as rubber. On a warm day, levels of photochemical air pollutants accumulate in the atmosphere around larger cities to create a brown photochemical haze.

20.3.2 Water pollution

Pollution of rivers and other water courses can produce very serious effects on the health of plants and animals which rely on that water supply. The Environment Agency is responsible for coastal waters, inland freshwater and groundwaters (known as 'controlled waters'). The EC Groundwater Directive seeks to protect groundwater from pollution since this is a source of drinking water. Such sources can become polluted by leakage from industrial soakaways. Discharges to a sewer are controlled by the Water Industry Act that defines trade effluent and those substances which are prohibited from discharge (e.g. petroleum spirit) and the Water Resources Act that covers discharge consent to controlled waters. It is an offence to pollute

(a)

(b)

Figure 20.7 Water pollution from: (a) an oil spillage; (b) plastic and other solid waste

any controlled waters or sewage system. If hazardous substances are being used by the organisation, safety data sheets give advice on the safe disposal of any residues that remain after the particular process has been completed.

The local water company has a right to sample discharges into its sewers because it is required to keep a public trade effluent register. There are two lists of proscribed substances which can only be discharged into a public sewer with the permission of the water company.

Finally, if oil is stored on the premises, a retaining bund wall should surround the oil store. This will not only ensure that any oil leakage is contained but will also stop the contamination of groundwater by fire-fighting foam in the event of a fire. This is a requirement of the Control of Pollution (Oil Storage) Regulations 2001.

20.3.3 Waste management – environmental permits

The UK produces in excess of 330 million tonnes of waste annually – a quarter of which is from households and business. The remainder derives from construction and demolition, sewage sludge, farm waste and spoils from mines and dredging of rivers. In an attempt to manage this waste, the Environmental Permitting (England and Wales) Regulations have been introduced. These Regulations are one-third of the length of the previous legislation and replace over 40 statutory instruments. They have created one single regulatory system by streamlining and integrating Waste Management Licensing and Pollution Prevention and Control. Environmental permits will provide industry, regulators and others with a single permitting and compliance system and could include those systems for:

▶ discharge consenting;
▶ groundwater authorisations;
▶ water abstraction and impoundment;
▶ radioactive substances regulation;
▶ licensing of some waste carriers and brokers.

The Environmental Permitting Regime ('the Regime') requires operators to obtain permits for some facilities, the registration of exemptions for other facilities and ongoing supervision by regulators. The regulator may be the Environment Agency or the Local Authority depending on the type of process involved.

The aim of the Regime is to:

▶ protect the environment;
▶ deliver permitting and compliance effectively and efficiently in a way that provides increased clarity and minimises the administrative burden on both the regulator and the operators of facilities;
▶ encourage regulators to promote best practice in the operation of regulated facilities;
▶ continue to fully implement European legislation.

An Environmental Permit is required for any of the following:

▶ an installation (which carries out the activities listed in Schedule 1 to the Regulations and any activities that are technically linked);
▶ a waste operation; or
▶ a mobile plant (carrying out either one of the Schedule 1 activities or a waste operation).

The collective term used in the EP Regulations for these installations, waste operations and mobile plant is 'regulated facility'. There may be more than one regulated facility on the same site. In such cases there are arrangements in the EP Regulations to allow all such facilities to be regulated by the same regulator and to allow, in many cases, for a single permit. It is an offence under the Regulations to operate a Regulated facility without a permit. More information on environmental permits is available in Chapter 19.

The single environmental permit will combine and streamline the previous waste management licence (WML) and pollution prevention and control (PPC) systems. All existing WML or PPC permits will automatically become environmental permits. All organisations who handle waste must check that they have the correct environmental permits in place to operate legally. The environmental permit outlines the amount and types of waste that the organisation can accept. Recently a company was fined for operating a waste transfer station without the necessary permits in place.

20.3.4 Waste disposal

This has been covered in detail in Chapter 14. Some of the topics covered in Chapter 14 will be covered here in more detail.

Prior to its collection for disposal elsewhere, a producer of waste should take the following measures to reduce the risk associated with its storage:

▶ segregate the waste by separating materials with incompatibility risks;
▶ contain the waste to prevent escape by the choice of suitable containers;
▶ provide bunding arrangements for liquid wastes;
▶ provide secure waste sites and protection from the weather, pests and scavengers;
▶ locate sites away from buildings, watercourses or other potential sites where environmental damage may be caused;
▶ provide proper labelling and site management including record keeping, supervision and site inspections; and
▶ ensure that emergency plans are in place.

The decline in landfill capacity, particularly for hazardous wastes, and the rapid rise of disposal costs is a topical issue for environmental managers.

In addressing the varying reasons why landfill is under pressure, many candidates did include a wide range of relevant pressures, as summarised below. However, some raised only a limited number of issues, preferring instead to address possible strategies for dealing with waste, other than landfill, and hence severely limited the number of marks that were available to them.

Landfill disposal

As a waste disposal option, landfill is coming under increasing pressure and its future as the main disposal method for household, commercial and industrial waste is limited. There are environmental concerns that include the impact of methane emissions on global warming, local air pollution arising from gaseous emissions and potential safety risks from gas migration, nuisance from odours and pests and water pollution, particularly groundwater contamination from leachates. There are also natural resource concerns including the burying of materials that could otherwise be recovered or recycled. Other issues associated with landfill waste disposal are:

▶ the availability of suitable sites and the restriction on development of land due to concerns about health, safety;
▶ environmental issues for buildings on or near landfill sites;
▶ difficulties in obtaining permission for new sites due to the effect of environmental and local protest groups;
▶ economic costs including progressive rises in Landfill Tax making landfill less competitive to its alternatives;
▶ rising operational costs due to tighter regulation under the Pollution Prevention and Control permit system;
▶ tighter restrictions on disposal of some waste types under the Landfill Directive, particularly hazardous wastes, and the shortage of capacity and number of sites capable of dealing with hazardous wastes;
▶ the need for pre-treatment of all wastes prior to landfill disposal.

There are concerns over the liability for future environmental costs arising from landfilled waste and the levels of financial provision required to ensure sites are maintained and restored. There is also a trend in many companies towards greater minimisation and recovery of waste which are reducing requirements for disposal facilities. The effect of such measures as the Packaging Waste Regulations has encouraged higher rates of recovery.

The Waste Electrical and Electronic Equipment Regulations (WEEE)

The WEEE were briefly covered in Chapter 14. The Regulations transpose the EU Directive on waste electrical and electronic equipment into UK law. They apply, with certain exceptions, to the categories of Electrical and Electronic Equipment (EEE) specified in Schedules to the Regulations. The disposal of equipment such as computer monitors and printers are likely to be EEE within the meaning of the Regulations.

The main requirements of the Regulations are:

▶ All producers who put EEE on the market in the UK are responsible for financing the costs of the collection, treatment, recovery and environmentally sound disposal of WEEE from private households that is deposited at designated collection facilities (DCFs) or returned under an in-store take back service.

▶ All producers must join an approved compliance scheme and charges must be paid to the Environment Agency or Scottish Environment Protection Agency (SEPA) for registration of a scheme. A producer must provide a declaration of compliance, together with supporting evidence, to the appropriate Environment Agency.

▶ A producer must mark EEE with the crossed-out wheeled bin symbol, a producer identification mark and a date mark and must also provide information on reuse and environmentally sound treatment for each new type of EEE put on the market by that producer.

▶ Approved compliance schemes must register or notify each producer who is a member of that scheme with the Environment Agency. An operator of a scheme has reporting, compliance and record keeping obligations.

▶ There are special provisions relating to the financing obligation on users of business WEEE that arises from EEE that was put on the market in the United Kingdom before 13 August 2005.

▶ A person who collects or transports WEEE must ensure that reuse and recycling of that equipment, or its components, is optimised and they may refuse to handle contaminated WEEE.

▶ A person who treats WEEE must be an authorised treatment facility (ATF) or an exporter for the purpose of issuing evidence of compliance under these Regulations through 'evidence notes' which may be sold to a producer compliance scheme.

▶ There are exemptions for storing or treating WEEE for the purposes of reuse, recovery or recycling introduced under the Waste Electrical and Electronic Equipment (Waste Management Licensing) Regulations.

Waste incineration

Incineration is a waste treatment technology that involves the combustion of waste at high temperatures and is being used to replace land disposal of waste. However, certain air pollutants that may be released include:

Figure 20.8 Electronic waste under WEEE

▶ acidic gases, for example hydrogen chloride, sulphur oxides, and nitrogen oxides;

▶ metals such as cadmium, mercury, arsenic;

▶ organic substances where combustion has not been complete, e.g. dioxins; oxides of carbon, e.g. carbon monoxide and dioxide; and

▶ particulate matter such as silica.

The advantages of waste incineration include:

▶ an overall reduction in volume of waste; destruction of hazardous components, e.g. incineration of medical waste products;

▶ an end product ash that is sterile and non-hazardous which can be recycled e.g. in road building;

▶ the generation of electricity and steam that can be sold to the regional electric grid and industrial customers and thus replace fossil fuel for energy generation; and

▶ the destruction of organic components of biodegradable waste that may generate landfill gases.

However, there are several disadvantages which include:

▶ the concerns about the health effects of dioxin and furan emissions into the atmosphere;

▶ odour;

▶ public perceptions;

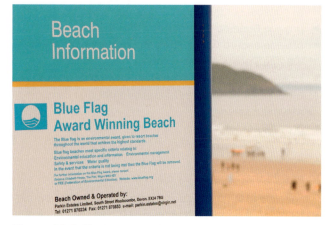

Figure 20.9 Environmental protection commitment

▶ increased road activity;

▶ high capital costs;

▶ stringent monitoring requirements;

▶ high fuel costs on starting the incineration process;

▶ waste inputs need to be carefully controlled since it is not suitable for all wastes (e.g. aqueous);

▶ difficulty in obtaining suitable sites; and

▶ poor maintenance can lead to performance deterioration.

20.3.5 Environmental impact assessments

Environmental impact assessments are a legal requirement in some planning applications (described in schedules to the EPA) or as part of an IPPC application. They are also an essential part of implementing an Environmental Management System (EMS) where they can help identify the impacts from associated aspects. Their use identifies the control measures that may be required to minimise environmental risk or to provide a risk rating for the purpose of prioritisation.

The Town and Country Planning (Environment Impact Assessment) Regulations 1999 require that an environmental assessment should be undertaken by a developer for specified forms of development. These are identified in Schedule 1 and Schedule 11 to the Regulations. Any development listed under Schedule 1 requires an environmental assessment and the submission of an environmental statement to the relevant planning authority. For those developments listed under Schedule 11, the planning authority may require an environmental assessment to be prepared or the developers themselves may opt to submit such an assessment, irrespective of whether the authority requires it. Environmental assessments are also required under applications for authorisations or permits in respect of 'integrated pollution control' or 'pollution prevention and control' and advice is available from planning policy guidance notes.

The types of environmental impacts which should be considered during an environmental impact assessment include:

▶ atmospheric, aquatic and land aspects;

▶ ecosystems;

▶ local communities and loss of amenity;

▶ archaeological or historic features; and

▶ the impact of changes in transport arrangements.

A new industrial development will have an impact on the environment and a pre-development environmental assessment should be made. The environmental criteria that could be in the assessment of a suitable site for a new chemical manufacturing company include:

▶ issues raised by statutory consultees;

▶ the findings of baseline surveys;

▶ characterisation of the development during construction, operation and decommissioning;

▶ the size of the plant and the proposed number of employees;

▶ the location of processes and possible alternatives;

▶ the materials to be used and the arrangements for their storage;

▶ the proximity of the proposed site to sensitive habitats or species, such as nature reserves, areas of outstanding natural beauty and sites of special scientific interest (SSSI);

▶ the presence of protected or vulnerable species on or near the potential site;

▶ the existing topography and hydrogeology of the site;

▶ the proximity to the main sources of raw materials and principal markets;

▶ the environmental consequences of the use of transport;

▶ the availability of appropriate waste disposal, recovery and treatment facilities;

▶ the capacity of sewerage systems to cope with any new discharges;

▶ water quality in nearby rivers, canals and groundwater;

▶ archaeology and visual aspects;

▶ emergency procedures and security;

▶ the sensitivity of neighbouring land uses, such as schools, housing and other commercial and industrial activity;

▶ any relevant statutory plans and designations, such as air quality management plans and planning designation of potential sites allowing industrial development; and

▶ energy emissions (e.g. noise, vibration, heat, light and radiation).

20.4 Health and safety in the home

Why does a book on health and safety at work include a section on home safety? Recent quotes from practitioners in emergency medicine, based on previously unpublished NHS research, show that more accidents occur in the home than anywhere else.

Dr Cliff Mann, president of the College of Emergency Medicine, is quoted in *The Guardian* (12 December 2014), saying that leisure pursuits such as trampolining, horse riding, mountain biking and DIY are more dangerous than previously thought. Because workplace and road safety have improved, home and leisure pursuits now cause more injury and death, averaging 1,000 cases a month. These accidents cost A&E units approximately £1 billion a year.

'The reality is that you are more likely to die sorting out the Christmas lights by taking them out of the loft or trying to fix the faulty plug or flex, than an electrician is on a building site. You are [safer] being at work or driving to and from work than you are being at home or doing a leisure activity with you family and friends. ... People think the kitchen is the

20

most dangerous room in the house but it's actually the living room. Children get scalded from their parents having a hot cup of tea, people get burned by the fire and elderly people can trip on rugs. ...'

'With DIY the use of drills and chainsaws can lead to very severe injuries. You can half-amputate a limb with a chainsaw. Falling off a ladder ... some people die and some never work again. Falls from lofts, typically a drop of 10 ft, can result in broken limbs and bleeding in the brain, and such falls are more common than is generally realised.'

Recent figures from A&E at the John Radcliffe Hospital in Oxford and the Horton General Hospital in Banbury show that over an 18-month period 26,310 patients were treated for injuries: 41% of these were people hurt at home, 25% were hurt during leisure activities, 15% were road accidents, 10% were education related such as PE lessons and only 8% were at work.

Here are some facts and figures taken from RoSPA's research. Unfortunately there are no government statistics after 2002 because the former Department of Trade and Industry (DTI) ceased to collect injury data for its Home Accident Surveillance System (HASS) and Leisure Accident Surveillance System (LASS). (See www.rospa.com)

Home safety facts and figures

▶ More accidents happen at home than anywhere else.
▶ Every year there are approximately 4,000 deaths as the result of a home accident. Around 120 of these are individuals aged below 15 years and 1,300 over 75.
▶ During 2002 there were 2.7 million home accidents requiring hospital treatment, of which 477,500 involved children under 5.
▶ Children under the age of 5 and people over 65 (particularly those over 75) are most likely to have an accident at home.
▶ Falls are the most common accidents, which can cause serious injury at any time of life. Fifty-five per cent of accidental injuries in the home involve falls.
▶ More women than men over the age of 65 die as the result of an accident in the home.
▶ Approximately 1,500 people aged over 75 die annually as the result of a fall.
▶ Every year around 120 children under 14 die as the result of an accident in the home.
▶ Around 25,000 under-fives attend A&E departments each year after being accidentally poisoned.
▶ 26,000 under-fives are burnt or scalded in the home every year. A hot drink can still scald a small child up to 15 minutes after it is made.
▶ More accidents happen in the lounge/living room than anywhere else in the home.
▶ Every year more than 4,200 children are involved in falls on the stairs and 4,000 children under the age of 15 are injured falling from windows.

▶ Boys have more accidents than girls.
▶ The cost to society of UK home accident injuries has been estimated at £45.63 billion (£45,630 million) annually.

Even if you buy a ticket every week in a year you are 20 times more likely to suffer a fatal home accident than you are to win the National Lottery!

Figure 20.10 Gas explosion during the night in Southampton 2015 – two people escaped unhurt

Prevention – a few hints

Fire safety

▶ Keep fires and heaters well guarded. Nursery guards with side clips are essential if there are young children around.
▶ Portable heaters and candles should be positioned so that they cannot be knocked over and away from furniture and curtains.
▶ Clothes should not be dried and aired over or near a fire or cooker.
▶ Never smoke in bed.
▶ Never leave a deep fat fryer or pan of fat unattended on a lit stove.
▶ Keep matches and lighters out of the reach of children.
▶ Fit approved smoke detectors on each floor, either mains operated or with long life batteries.
▶ Plan your escape route.

Electrical safety

▶ If you buy an older property, get the wiring checked straight away.
▶ Worn and damaged flexes should not be used and flexes should not be wired together.
▶ Portable mains operated devices should not be used in the bathroom.
▶ Have electric blankets serviced and checked regularly.
▶ If an appliance seems to be faulty stop using it.

- Residual current devices (RCDs) protect the whole house and make the use of power tools safer.
- Buy electrical equipment with a CE mark, BEAB mark, BS mark or British Standard number.
- Never overload an electrical socket.

Heating and cooking

- Never block or lean anything against air vents or air bricks as they help waste gases to escape. The same applies to outside grilles.
- Be aware that carbon monoxide poisoning causes drowsiness and flu-like symptoms.
- Chimneys in use should be swept annually, more often if burning wood.
- Check the pilot light on a water heater and gas cooker to make sure it is lit.
- Beware of second-hand appliances and cowboy installers when buying gas appliances. Look for the safety mark or seal of approval and only use Gas Safe approved installers.
- Young children and elderly people are especially at risk from fires and heaters.
- If a gas leak is suspected turn off the supply, open the windows, call the gas supplier and avoid operating any switches as they can spark.
- Bioethanol and fuel gel burners should be used with great care as the flame from them is virtually invisible and the fuel is highly flammable.
- Kitchen planning is important – heavy items should not be kept on high shelves. Things in daily use should be easy to reach:
  - ▷ Knives need to be kept sharp and out of the reach of children
  - ▷ Pan handles need to be turned inwards on the stove
  - ▷ Hot liquids need special care if there are children or vulnerable adults around
  - ▷ A cordless kettle is the safest option.

Safety glass

Ordinary glass at low level is dangerous but less so if the area is well lit and people are made aware that it is there. Children should never be allowed to play near it.

Safety glass is much better, either laminated or toughened. In new buildings it is a requirement in areas below 800 mm.

Medicines and cleaners

- Have a lockable medicine cabinet and keep medicines clearly labelled.
- Make sure that chemicals are not kept where children can see or reach them.
- Return old medicines to the pharmacy.

DIY and garden safety

- Gas and electrical work should always be done by a competent qualified person.
- Keep tools clean and in good repair.

- Plan ahead and allow plenty of time. Being unprepared and in a hurry is more likely to result in an accident.
- Tools, paint, chemicals, anything sharp, should be kept out of sight and reach of children.
- Be extra careful with adhesives, especially instant ones.
- Disconnect the electricity before working on electrical appliances and tools.
- Wear strong shoes and appropriate clothing when mowing, strimming, brushcutting. Keep hands and feet away from blades and twine. Children and pets are at risk here too.
- Keep products in original containers.
- Barbecues should be well away from trees, fences and buildings. Never pour petrol on a barbecue or bonfire that is alight.
- A residual current device (RCD) is essential for outdoor electrical equipment unless the property is already wired for this.

Accidents to older people

Higher living standards, better health care and awareness of the need to pay attention to diet and exercise have led to people living and working longer, so we need to think about health and safety in respect of older people. People over 65 years of age are most at risk of an accident in and around the home. Most of these accidents happen on the stairs or in the kitchen. Falls from stairs or steps are the most common, but people also fall from a bed or chair, or trip over a rug or mat. As people age they become less stable for a variety of reasons. Awareness of this fact means that areas of risk can be identified to make the older person's environment safer.

Main areas of risk that should be targeted are:

- Trips and fall from poor eyesight, frailty and balance, negotiating stairs, no or insecure handrails, poor flooring and unsuitable shoes;
- Fire-related accidents from cookers, flammable materials, candles, coal fires, heaters and electric blankets;
- Poisoning accidents from medicines, carbon monoxide and piped gas;
- Burns from radiators, electric fires and cookers; scalds from kettles, hot water bottles and uncontrolled hot water;
- Hypothermia from poor heating, sedentary living, inadequate nutrition and clothes.

Finally a word from Prof. Keith Willett, NHS England's director for acute care:

'The number of people who come through Accident and Emergency with DIY, leisure and gardening injuries is striking. Working as a trauma surgeon for over 30 years, it has been frustrating to see how many people come to A&E with injuries that could

20

have been avoided with a bit more care. However I would much rather people remained active and fit for their general mental and physical well-being than that they avoid sport.'

20.5 Safe cycling

Cycling has gained in popularity over recent years and various initiatives have encouraged people to cycle to work, school and college as well as taking up cycling as a hobby and to improve their fitness. So of course there are many more bikes out there on the road and although local and national authorities are improving cycling conditions by making new cycle routes, there are still a lot of accidents. In fact, according to Sustrans (the sustainable transport charity and organisers of the National Cycle Network), in Britain around 19,000 cyclists are killed or injured every year in reported road accidents.

But there are many ways to cycle safely. Here are some guidelines:

Road cycling

- ▶ follow the Highway Code; cyclists are road users so red traffic lights apply to them and, unless designated, pavements are not for cycling;
- ▶ adjust your speed – wet and icy surfaces alter the way that tyres and brakes behave;
- ▶ claim your space on the road and be decisive. Give parked cars a wide berth in case someone opens a door;
- ▶ wear a helmet;
- ▶ keep your bike in good condition.

Figure 20.11 It is very important to be clearly visible to the driver of large vehicles particularly when they are turning left at junctions. Hi Vis jackets would improve visibility here

Be visible

- ▶ ride where you can be seen especially when passing trucks and buses;
- ▶ use lights in poor visibility and at night and wear bright and reflective clothing;
- ▶ make eye contact with other road users to be sure that they have seen you;

- ▶ signal clearly;
  - ▷ use your bell;
- ▶ be aware that some new LED lights are so bright that they can dazzle oncoming cyclists. There is usually a lower intensity setting to use in town.

Awareness of other vehicles

Many collisions occur when a cyclist is on the inside of a vehicle that is going left. Avoid 'undertaking' large vehicles and never cycle along the inside of trucks and buses. When a large vehicle turns left, it often swings out to the right first to get round the corner and this creates a large empty space. Cyclists are tempted to ride into this space, but the gap will disappear as the vehicle swings round to the left. A lot of accidents have happened this way.

Cycle training

People who are new to cycling or have not cycled for some time will find that cycle training helps to build confidence and develop skills. The CTC (Cyclists' Touring Club) can give advice about courses and trainers (www.ctc.org.uk).

Motorists, be aware!

- ▶ when turning left watch for cyclists coming up the nearside;
- ▶ give cyclists a wide berth when overtaking;
- ▶ dip your headlights at night when approaching cyclists;
- ▶ in wet or icy weather be extra vigilant when passing cyclists.

(a)

(b)

Figure 20.12 (a) and (b) New motorist awareness posters in Southampton

Cyclists and motorists have equal rights to the same road space. On shared pavements the same applies to cyclists and pedestrians. Respect for each other will benefit all of us.

20.6 Further information

The Health and Safety at Work etc. Act 1974

The Management of Health and Safety at Work Regulations 1999 (as amended)

The Environmental Protection Act 1990

The Hazardous Waste (England and Wales) Regulations 2005

ILOLEX (ILO database of International Law). http://www.ilo.org/ilolex/index.htm

Occupational Health and Safety Assessment Series (OHSAS 18000):

Occupational Safety and Health Convention (C155), ILO. http://www.ilo.org/ilolex/cgi-lex/convde.pl?C155

Occupational Safety and Health Recommendation (R164), ILO. http://www.ilo.org/ilolex/cgi-lex/convde.pl?R164

Safety in the Global Village, IOSH Information Sheet, 1999. http://www.iosh.co.uk/index.cfm?go=technical.guidance

APPENDIX 20.1 Scaffolds and ladders

1.1 Convention (Safety and Health in Construction) (167)

Article 14

1. Where work cannot safely be done on or from the ground or from part of a building or other permanent structure, a safe and suitable scaffold shall be provided and maintained, or other equally safe and suitable provision shall be made.
2. In the absence of alternative safe means of access to elevated working places, suitable and sound ladders shall be provided. They shall be properly secured against inadvertent movement.
3. All scaffolds and ladders shall be constructed and used in accordance with national laws and Regulations.
4. Scaffolds shall be inspected by a competent person in such cases and at such times as shall be prescribed by national laws or Regulations.

1.2 Recommendation (Safety and Health in Construction) (175)

Scaffolds

16. Every scaffold and part thereof should be of suitable and sound material and of adequate size and strength for the purpose for which it is used and be maintained in a proper condition.
17. Every scaffold should be properly designed, erected and maintained so as to prevent collapse or accidental displacement when properly used.

18. The working platforms, gangways and stairways of scaffolds should be of such dimensions and so constructed and guarded as to protect persons against falling or being endangered by falling objects.
19. No scaffold should be overloaded or otherwise misused.
20. A scaffold should not be erected, substantially altered or dismantled except by or under the supervision of a competent person.
21. Scaffolds as prescribed by national laws or Regulations should be inspected, and the results recorded, by a competent person:
    (i) before being taken into use;
    (ii) at periodic intervals thereafter;
    (iii) after any alteration, interruption in use, exposure to weather or seismic condition or any other occurrence likely to have affected their strength or stability.

1.3 Code of Practice – Safety and Health in Construction

The Code of Practice covers scaffolds and ladders under the following topics over five pages:

1. general provisions
2. materials
3. design and construction
4. inspection and maintenance
5. lifting appliances on scaffolds
6. prefabricated scaffolds
7. use of scaffolds.

APPENDIX 20.2 International travel tips

1. General International travel tips

▶ Leave copies of your passport, visa and itinerary with your office and home. If you should lose any documents you would have easy access to a copy.
▶ Ask the relevant authorities/travel experts for a country profile if you are travelling to a new destination. They can provide extensive data on weather, local customs, food, political climate and much more.
▶ Your luggage may be searched upon entering a country, so pack accordingly. Know what items are taboo in the country you are visiting.
▶ Be wary of anyone loitering around doors to your room, and notify the desk of your suspicions.
▶ Do not swap or sell any personal effects, such as jewellery or religious items, in countries where it is illegal to do so.
▶ Avoid photographing military installations, border protection points and sensitive government buildings.

▶ Avoid any area of the city which has been the target of a terrorist attack or is a place where they are known to congregate.
▶ Know the location of the embassy of your passport country, and have that phone number with you at all times.
▶ If you go out, let another member of your group know when you expect to return and the general location you are visiting. You may wish to provide a contact number as well.
▶ If you are in a situation where someone starts shooting a gun, drop to the floor or get down as low as possible and don't move until you are sure the danger has passed. If possible, shield yourself behind or under a solid object.

2. Hotel travel tips

▶ Reserve rooms above first floor.
▶ Keep doors and windows locked, and check them before retiring.

▶ Put valuables in the hotel room safe.

▶ Keep curtains/blinds closed after dark.

▶ Verify workmen with the front desk before admitting them into your room.

▶ Locate your nearest emergency exit and fire extinguisher.

▶ Be careful answering the door. Use the peephole, or leave the chain attached when opening the door to talk.

▶ Don't accept packages or items unless you know the source.

▶ Don't answer the phone with your name.

▶ Remember that the doorknob signs tell an outsider whether you are in the room (Do Not Disturb) or out of the room (Please Make Up This Room).

▶ Take your passport and other forms of identification with you when leaving the room.

▶ Do not indiscriminately give out names and room numbers of others in your group.

▶ Know your hotel phone number.

3. Taxi travel tips

▶ If possible, pick your own taxi.

▶ Have the address of your destination and hotel written in the local language.

▶ If you call a taxi, wait for it indoors.

▶ Compare the face of the taxi driver with the posted licence.

▶ When in a metered taxi, make sure the meter is activated.

▶ When boarding a bus or limo, do so promptly.

▶ Keep an eye on the loading and unloading of your baggage, especially if you are sharing a taxi or limo which makes an intermediate stop.

4. Lost luggage travel tips

▶ If your luggage is lost, find an airline representative and report the problem.

▶ You will be given a form to complete. DO THIS IMMEDIATELY, BEFORE YOU LEAVE THE AIRPORT. This form will describe the luggage and its contents.

▶ Obtain a copy of the form before giving it to an airline representative.

▶ If you are asked to surrender your claim checks, make sure you note this on the form.

▶ 98% of lost luggage is found within several hours, and will be put on the next flight to your destination.

▶ If you cannot wait, make sure the airline has your address to forward your luggage.

▶ If you need to buy or rent replacement items, most carriers will front you some cash.

▶ If luggage is never found, you will be asked to estimate the value (depreciated) of its contents. If you claim anything new or costly (i.e. camera), you may be asked to provide receipts and other documentation.

5. Walking travel tips from the travel experts at Safe Harbours

▶ Avoid walking alone or travelling to remote places after dark; avoid poorly lit streets and narrow alleys.

▶ Do not wear flashy jewellery.

▶ Avoid public demonstrations, accidents or civil disturbances.

▶ Do not engage in loud conversations or arguments.

▶ Avoid discussing personal matters with people you do not know.

▶ Keep a phrase book handy.

▶ Know how to use the pay phones in the country you are visiting, and carry exact change in local currency.

▶ Avoid bringing any unnecessary attention to yourself.

20

CHAPTER 21

Study skills

21.1 Introduction

This chapter speaks directly to you, the student. For the NEBOSH certificate you need study skills and exam techniques and this means careful planning from the beginning of the course.

Think in terms of:

▶ clear and realistic goals, both short and long term;
▶ techniques for studying and passing exams;
▶ a well-organised approach;
▶ strong motivation.

People say that genius is 99% application and only 1% inspiration, so good organisation and exam technique are what really matters when it comes to passing exams. Studying is an activity in its own right and there are ways in which you can make your study more effective and give yourself the greatest possible chance of success. The place where you study, the way you plan your study time, how you adapt your study to suit your lifestyle, the way that you take notes, memory and concentration skills, revision techniques, and being clear about the contents of the syllabus and what is expected by the examiner – attention to all these details will give you the best chance of passing the exam.

21.2 Find a place to study

The skill of studying usually has to be learned, so it's worth thinking carefully about the basics. If you can find somewhere quiet and free from distractions, and keep that place just for your studies, it will be easier to get down to work. You need a place with good ventilation to help you to stay alert, and somewhere that has a comfortable temperature. Make sure there is enough light – a reading lamp helps to prevent eyestrain and tiredness.

An upright chair is better than an armchair. Make sure that the workspace is large enough. The size recommended by Warwick University, for example, is a minimum of 60 cm x 1.5 m. If you can, find a place where your study materials do not have to be cleared away – it's easy to put off starting work if they are not instantly accessible.

21.2.1 Make a study plan

Start by making a study timetable. First of all, put aside some time for study so that it does not get squeezed out of your working week. Your timetable will need to include time for carrying out assignments set by tutors, time for going over lecture notes and materials, any further reading rated as 'essential' and, if possible, reading rated as 'desirable'. You should allocate time for revision as the course progresses because the more regularly you revise, the more firmly the information will become fixed in your memory. This will make it much easier to remember when you take the exams.

After every hour of study, take a short break and, if possible, have a change of atmosphere. Physical exercise will help to increase your concentration. Even a short, brisk walk round the block will improve your attention span.

Vary your activities while you are studying. For example, you could spend some time gathering information from books, the internet and so on, then maybe work on a diagram or a graph, followed by writing, reading and so on

21.3 Time management

Finding time to study and take exams when your life is already filled with work, family and social commitments is clearly not going to be easy. The Open University (OU) has been working with adult learners since 1969 and they provide a lot of useful, free information about this at www.open.edu/openlearn/free-course. Here are some ideas from 'manage your time effectively' to help you to plan.

The aim is to develop good work habits and time-management practices. These are often a matter of developing the right attitudes towards your work and towards your time.

Ten tips to help you manage your time:

1. establish goals and targets;
2. work smarter, not harder – value your time;
3. avoid attempting too much;
4. schedule activities, allow time for emergencies, minimise interruption, identify and use your peak energy times, put similar tasks together;
5. stay in control of paper work and electronic work;
6. minimise day dreaming;
7. use a log to track time wasting;
8. put a time limit on some tasks;
9. be decisive, finish things;

10. work at a steady pace and review your progress regularly.

Keep a log of your learning so that you can reflect on how you use your time – are you managing it effectively? Think about how you deal with the unexpected, with emergencies. How do you prioritise your tasks and your time? Can you set deadlines? Can you meet them? If not, how do you deal with the situation?

There is a section on time management in the Guide to the NEBOSH National Certificate in Construction Health and Safety. This subject is also mentioned in 21.12.1 Planning and Revision.

21.4 Blocked thinking

Sometimes your thinking can become blocked. If this happens, try leaving the problem area of study alone for a few days and tackle a different part of the subject, since if you concentrate too hard on something that is too difficult you are likely to lose confidence. Usually, after a break, the difficulty vanishes and the problem clears. This could be because the solution has emerged from another perspective, or because you have learnt something new that has supplied the answer. Psychologists think it's possible that some problems can be solved during sleep when your mind has a chance to wander and think laterally.

21.5 Taking notes

The most efficient way to store notes is in a loose-leaf folder, because you can easily add extra information. Write only on one side of the sheet and use margined paper. The facing pages can be used to make summaries or extra points. If notes are clearly written and well spaced out, they are much more straightforward to work from and more attractive to return to later. Use colour, highlighting and underlining too; notes that look good are more appealing and easier to revise.

During revision it will be important to be able to identify subjects quickly, so use headings, numbering, lettering, bullet points, indentation and so on. It is worth spending a bit of extra time to make revision easier.

Key words and phrases are better than continuous prose when you are taking notes. While you are writing down information it is easy to miss essential points that are being made by the tutor. Try to read through your notes within 36 hours to make sure that they are completely clear, so that when you revise you will not be puzzling over what you have written. At this point you may be able to add more information, while it is fresh in your mind. Reading through notes in this way also helps to fix the information in your memory and the level of recall will be further improved by reading them again a week later. Although it seems time-consuming, this technique will save you a lot of time in the long run.

Many people find that they can understand and remember more effectively by making 'mind maps' (also known as 'pattern notes'). These will be discussed in more detail in the revision section.

21.6 Reading for study

Most people will be working as well as studying, so you will need techniques to help you make the best use of your time. For example, you don't need to read a whole book for study purposes, apart from books written specifically to a syllabus, such as this one. Use the index and be selective. In this book an example would be Chapter 19 on summaries of legislation.

Where a book is not specifically written to a syllabus, time-saving techniques such as skim reading can be used. First of all, flip through the book to see what it contains, looking at the contents, summaries, introduction, tables, diagrams and so on. It should be possible to identify areas that are necessary to the syllabus. These can be marked with removable adhesive strips or 'Post-it' notes.

Second, quickly read through the parts of the book that you have identified as useful. Two ways of doing this are:

(a) run a finger slowly down the centre of a page, watching it as it moves. You will pick up relevant words and phrases as your finger moves down the page;
(b) read in phrases, rather than word by word, which will increase your reading speed.

Both these methods need practice but they save a lot of time and effort. It is better to read through a piece two or three times, quickly, than to read it once slowly; but if there is a really difficult piece of information that is absolutely essential to understand, read it aloud, slowly.

Finally, at the end of each section or chapter, make a few brief notes giving the essential points.

It is worth learning to speed-read if a course involves a lot of reading. Tony Buzan's series The Mind Set explains the technique.

21.7 Free learning resources from the Open University

The OpenLearn section of the Open University is a very useful resource which is free and is available worldwide. There is an extensive study skills section containing 30 units which range from 3 to 50 hours of study. Have a look at 'The Learning Space'. Here are a few examples of what is available:

▶ EAL_1. Am I ready to study in English? (5 hours) This unit provides a series of exercises to help you to reflect on the use of English at an academic (as opposed to everyday) level. It is designed for people who have been educated in a language other than English, or for those who have studied in a country where the conventions are different from those used in the British educational system.

- LDT 101_1. Revision and examinations (6 hours)
This unit is for people who are unsure about exams. Perhaps you have not taken an exam for a long time, or have had a bad experience with exams in the past; or maybe you have never taken an exam? This unit aims to help you to develop techniques for revision and exam taking and to reassure you.
- LDT 101_3. Learning how to learn (6 hours)
This unit looks at learning as an active process and provides activities that will help you to learn more effectively.

You can access these free resources on www.open.edu/openlearn/free-course

21.8 Organising for revision

As the exam approaches, you will need to organise all the information from the course in preparation for revision. Ideally, of course, you should have been doing this throughout, but it is not too late to start. It will just mean extra work. Information from lectures, reading, practical experience and assignments will have to be organised into a form that makes it easier for you to remember. As well as this, you will already possess some knowledge that is relevant to the syllabus. Don't underestimate the importance of this. It is worth spending some time thinking about how this store of information relates to and can be used to expand and back up the new information that you have learnt on the course. Examiners' Reports have, in the past, referred specifically to the advantages to be gained from this. For example:

> *'Some candidates, perhaps from their work experience, showed a good knowledge of a permit-to-work system and produced reasonable answers.'*

But be warned. The examiners also pointed out that:

> *'Answers based solely on "what we do in our organisation" … do not earn high marks'*

21.9 Organising information

You can organise the information in preparation for an exam in several ways:

Pregnant Employees – Risks

Exposure to chemicals	pesticides lead mutagens (inter-cellular change) teratogens (affect embryo)
biological	(e.g. hepatitis)
Physical	(ionizing radiations extremes of temperature)
manual handling	
ergonomic	(prolonged standing awkward body movements)
Stress	
issues relating to	use/wearing of protective equipment

Figure 21.1 Revision notes

- read back through the notes on a regular basis while the course is still in progress; this will help to fix the information in your memory;
- make revision cards: condense the information down so that it fits on a set of postcards. This makes it possible to carry a lot of information around in your pocket. The activity of condensing the information is revision in itself. This way you can keep your revision material available and spend a few spare moments, from time to time, reading it through.

For example: 'Outline the factors that may increase risks to pregnant employees.' On a card, the information could be condensed as shown in Figure 21.1.

Make mind maps (also known as pattern notes): like revision cards, the act of making them is revision in itself. Because they are based on visual images some people find them easier to remember and they can be made as complex or as simple as you like. Use colour and imagery. It will be easier to follow and to recall the information (Buzan). Figure 21.2 shows a mind map based on the report writing section of this book.

Use key words as an aid to memory. For example, a list of key words for 'ways of reducing the risk of a fire

starting in the work place' could be reduced to an eight-letter nonsense word, as follows:

FLICSHEV:

Flammable liquids – provide proper storage facilities
Lubrication – regular lubrication of machinery
Incompatible chemicals – need segregation
Control of hot work
Smoking – of cigarettes and materials, to be controlled
Housekeeping (good) – prevent accumulation of waste
Electrical equipment – needs frequent inspection for damage
Ventilation – outlets should not be obstructed.

The effort involved in making the list and inventing the word helps to fix the information in your memory.

Learning by rote is a popular learning method in many countries and is very useful for specific learning tasks, such as tables and formulae. But it is not appropriate to this type of syllabus. Rote learning can prevent you from applying what you have learnt to a wide range of situations. If you limit yourself in this way you will be in danger of not having sufficient understanding of the depth of the subject, particularly if you learn set answers to exam questions. The NEBOSH examiners'

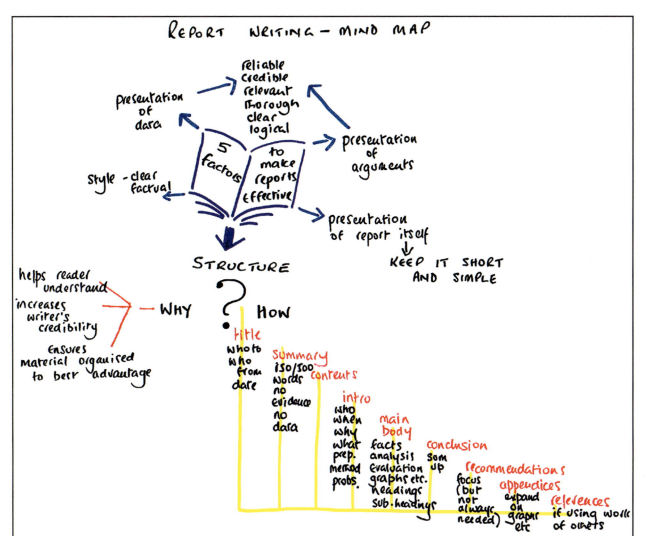

Figure 21.2 Mind map report writing

reports specifically warn against rote learning of set answers. They point out that 'Candidates should prepare themselves for this vocational examination by ensuring their understanding.' Working through old exam papers is very helpful and you can learn a lot by doing revision in this way but don't be tempted to learn a 'perfect' version of an answer as they are all designed to make you think about specific problems that you might encounter in your working life.

It is useful to have a range of techniques for memorising, since what works for one person may not work for another (see 21.10). There are several other memory techniques available, but the ones described in this paragraph are the most commonly used.

21.10 Being aware of your learning style

All of us are different and we learn in different ways, so it makes sense to be able to identify your own preferred style of learning. The main learning styles are **visual, auditory** and **kinaesthetic** and it helps if you can choose ways of learning and remembering that fit in with your own personal style.

For example, if you are a person who learns in a **visual** way, you may find it easier to understand and recall information when it is presented in the form of diagrams, the use of flip charts, handouts, demonstrations and videos. When you are taking notes, and later in revision, you will perhaps find it more effective to:

▶ Use colour
▶ Organise in columns and categories
▶ Use mnemonics in a visual way (see the 'key word' tip in 21.9)
▶ Use mind mapping
▶ Use index cards
▶ Highlight, circle and underline
▶ Make posters and/or use Post-it notes.

If you are a person who leans towards learning in an **auditory** way, you will probably find lectures, discussions, dialogue and music are useful, so here are a few ideas to help you to learn more effectively:

▶ Discussion of new ideas and information
▶ Recording information as well as taking notes
▶ Using word association
▶ Repeating facts with your eyes closed
▶ Watching videos
▶ Using rhyme, rhythm and music to help you to retain information
▶ Using mnemonics aurally (see 21.9).

Maybe you are a person who works best with a **kinaesthetic** approach to learning, and find that touching, feeling and actively doing are the best ways for you to learn. Hands-on activity, role play, simulation and practising skills are likely to be the methods you are most comfortable with. Here are a few possibilities:

▶ Copying notes repeatedly to organise them in your mind
▶ Making on-site visits
▶ Using colour and highlighting
▶ Making posters and models to aid memory
▶ Skim reading, then reading in detail
▶ Taking study breaks
▶ Revise on an exercise bike.

(Arp, 2014)

This is not to say that people only conform to the visual, the auditory or the kinaesthetic model. Most people will be on a spectrum and use more than one type of learning, but you have probably already identified the one that is prevalent in your learning style and you can use that information to make both the learning process and revision a bit easier.

21.11 How does memory work?

Understanding how your memory works will help you to use it more effectively. In common with other scientific subjects, there is still a lot to be discovered about the workings of the mind, and a considerable amount of disagreement about what has been discovered so far; but psychologists generally seem to agree about the way that memory works.

The process of remembering is divided, roughly, into two sections – short-term and long-term memory. Items that go into short-term memory, if no further attention is paid to them, will fade away and be forgotten. If they are rehearsed they will stay in short-term memory for a while. Think, for example, of how you remember a phone number when you make a call. Storing something in long-term memory requires a greater 'depth of processing', that is to say, more mental activity is required so that:

(a) information stays in the memory (storage); and
(b) information can be found when it is needed (retrieval).

If you are studying on the NEBOSH course, it is likely that you are a mature student. This will give you some advantages when it comes to learning. Because your life and work experience will almost always be more comprehensive and of a higher complexity than that of younger students, you are likely to possess more 'schemas' (areas of knowledge) to which new information can be attached, and you will have more experiences already stored in your brain that can provide an explanation for the new pieces of knowledge that you are acquiring. Add to this a very high level of motivation, a stronger level of incentive towards success and more determined application, and it becomes clear that you, as a mature student, have many advantages.

Research shows that, as learners, we take in 10% of what we read; 20% of what we hear; 30% of what we see; 50% of what we see and hear; 70% of what we ourselves say; and 90% of what we ourselves do (Northedge, 2012). You can see from this that people who are simultaneously working at and studying a subject have a marked advantage.

But, it is also true that mature students tend to be very nervous about exams and assignments and often need a good deal of reassurance and support to enable them to realise their full potential.

21.12 How to deal with exams

There are three stages to taking an exam:

▶ planning and revision;
▶ the exam room;
▶ after the exam.

For people who require *special consideration*, there are provisions within NEBOSH to allow extra time or the use of special equipment. If you think you may be eligible for reasonable adjustments/special consideration, you should apply to NEBOSH several weeks before the date of the exam. Have a look at the NEBOSH Guide if you think any of this may apply to you.

21.12.1 Planning and revision

1. It is absolutely essential that you know what you are going to be examined on and what form the exam will take. You should read through the syllabus and if you are concerned about any area of it, this should be raised with your course tutor well before the date of the exam.
2. Read the Guide to the NEBOSH National Certificate in Construction Health and Safety and the examiners' reports. After the exams every year, the examiners highlight the most common mistakes made by students (see Section 21.13 on examiners' reports). They also provide useful information about, for example, pass rates, levels attained (distinction, credit, pass, fail), time management and other hints on exam technique. Also get a copy of the Construction revision guide by Ed Ferrett.
3. You should work through some recent past papers, against the clock, to get used to the 'feel' of the exam. If possible, ask your tutor to set up mock exams and make sure that the papers are marked. Some of the shorter courses will not be able to provide this service, and if this is the case, do try to do at least one or two timed questions as part of your revision. You may find it difficult at first to write an answer as a first draft. Now that everyone is used to the facilities provided by word processing, you may need to practise writing an answer without being able to edit the text. Past papers, examiners' reports and syllabuses are available from NEBOSH. The website is at www.nebosh.org.uk.
4. You will need to know where the exam is to be held and the date and time. If possible, visit the building

beforehand to help build confidence about the location, availability of parking and so on.

5. Make a chart of the time leading up to the exam. Include all activities, work, leisure, and social, as well as the time to be used for revision, so that the schedule is realistic.
6. Try to eat and sleep well and take some exercise.
7. Revision techniques were covered in the sections on revision and memory earlier in this chapter.

Set realistic targets, then achieve them. Make plans and stick to them.

21.12.2 The exam room

▶ Read through the exam paper very carefully.
▶ Check the instructions – how many questions have to be answered? From which section(s)?
▶ Make a time plan.
▶ Underline **command words**, e.g. 'identify' (see Table 21.1). Using the words in the question when writing the answers will help to keep the answer on track.
▶ Stick to the instructions given in the question. A working knowledge of the command verbs is vital to your success. NEBOSH no longer uses the command verb 'sketch', but candidates may use sketches to illustrate their answers where appropriate.
▶ Write clearly. Illegible answers don't get marked.
▶ Look at the number of marks allocated to a question to pick up clues as to how much time should be spent on it.
▶ Mark questions which look possible and identify any that look impossible.
▶ It is rarely necessary to answer exam questions in a particular order. Start with the question that you feel most comfortable with since it will help to boost your confidence. Make sure it is clearly identified by number for the examiner.
▶ Answer the question that is set, not the one you wish was on the paper.
▶ If ideas for other answers spring to mind while you are writing, jot down a reminder on a separate piece of paper. It is easy to forget that bit of information when you are concentrating on something else.
▶ Plan the use of time and plan the answers. Include some time to check over each answer.
▶ Stick to the time plan; stick to the point; make points quickly and clearly.

Early marks in an exam question are easier to pick up than the last one or two, so make sure that all the questions are attempted within the time plan. No marks are given for correct information that is not relevant to the question.

Don't be distracted by the behaviour of other students. Someone who is requesting more paper has not necessarily written a better answer; they may simply have larger handwriting. People who start to scribble

21

Table 21.1 Terminology used in NEBOSH exams

Command word	Definition
Identify	To give reference to an item, which could be its name or title. NB: normally a word or phrase is sufficient provided the reference is clear.
Give	To offer for consideration, acceptance, or use of another. NB: give an example of; give the meaning of.
Outline	To indicate the principal features or different parts of. NB: an exhaustive description is not required. What is sought is a brief summary of the major aspects of whatever is stated in the question.
Describe	To give a detailed written account of the distinctive features of a subject. The account should be factual, without any attempt to explain. When describing a subject (or object) a test of sufficient detail would be that another person would be able to visualise what you are describing.
Explain	To provide an understanding. To make an idea or relationship clear. NB: this command word is testing the candidate's ability to know or understand why or how something happens. Is often associated with the words 'how' or 'why'.

madly as soon as they turn over the question sheet are not in possession of some extra ability – they simply haven't planned their exam paper properly.

Keep calm, plan carefully, don't panic.

21.12.3 After the exam

If there are several exams to be taken, you need to stay calm, relaxed and confident. It is not a good idea to get into discussion about other people's experiences of the exam. After one exam, focus on the next. If something went wrong during the exam (for example, illness or severe family problems), the tutor and the examining board should be alerted immediately.

21.13 The examiners' reports

21.13.1 A few points from the examiners' reports

The examiners' reports provide information to help candidates and tutors in future exams. They aim to be constructive and to help you towards a better understanding of the syllabus and how you are assessed.

The examiners note that many candidates are well prepared and that their answers are relevant and comprehensive and show that they not only have the necessary knowledge, but that they are able to apply it to the workplace situation.

On the other hand, there are some candidates who appear to be unprepared and the examiners emphasise

that good preparation is essential. You will need to study the content of the syllabus, while at the same time understanding how the concepts contained in it apply in the workplace. The examiners also stress the importance of applying the information that you have learned to the question that is asked on the paper. It is especially important to understand that rote-learning will not lead to success in this type of exam.

Here is a quote from the latest examiners' reports:

> 'In order to meet the pass standard for this assessment, acquisition of knowledge and understanding across the syllabus are prerequisites. However candidates need to demonstrate their knowledge and understanding in answering the questions set. Referral of candidates in this unit is invariably because they are unable to write a full, well-informed answer to one or more of the questions asked.'

Some candidates find it difficult to relate their learning to the questions and as a result offer responses reliant on recalled knowledge and conjecture and fail to demonstrate a sufficient degree of knowledge and understanding. Candidates should prepare themselves for this vocational examination by ensuring their understanding, not rote-learning pre-prepared answers.

The section on 'common pitfalls' warns about common mistakes that are made by people taking these exams. Here are seven things which often cause people to fail to reach their full potential:

- ▶ Failure to apply the basic principles of exam technique can make the difference between passing the exam and being referred.
- ▶ Some candidates do not attempt all the required questions or they fail to provide complete answers. Even if your mind goes blank, do make an attempt to answer a question that is compulsory. Applying basic health and safety management principles can generate points that will add to your mark.
- ▶ If you provide information that is relevant to the topic but not relevant to the question that is set, that information cannot gain any marks. Always answer the question that is set.
- ▶ Knowledge of the command words is essential (see Table 21.1). They are the instructions that guide you on the depth of answer that is required. For example, if a question asks you to 'describe' something, you will not get many marks if you 'outline' it.
- ▶ Where a question is divided into sub-sections, you need to indicate clearly which part of the question you are answering. Use the numbering from the question in your answer. If your answer is structured to address the different parts of the question, that will help to draw out the points you need to make in your answer.
- ▶ Time planning is absolutely vital. If you spend a lot of time on one answer and give a lot of unnecessary

information, you may be left with insufficient time to address all the questions.

▶ If you have doubts about the legibility of your handwriting, practise writing at speed. The examiner has to be able to read it to mark it.

The examiners also point out that you do not need to start a new page for each section of a question.

21.13.2 Marks for practice revision questions

We have been asked about the allocation of marks to the practice revision questions shown at the end of each chapter. However, since the marks awarded to each part of a question can vary, only general guidance can be given.

All one- or two-part questions are given 8 marks with a minimum of 2 marks awarded to any part.
Questions with three or more parts are given 20 marks with a minimum of 4 marks for each part.

NEBOSH exam questions have a similar format and allocation of marks.

21.14 Conclusion

Passing health and safety exams and assessments has a lot in common with any other subject being examined. Read the questions carefully and plan your answers. The old carpenter's saying 'Measure twice, think twice, cut once' can be applied to exam technique:

Read twice, think twice and write once.

21.15 Further information

Arp, D. Standards Manager NEBOSH (2014) Presentation to course providers 'Preparing Students for Assessment'.

Buzan, T. A wide range of study skills books by this author can be found at www.thinkbuzan.com

Guide to the NEBOSH National Certificate in Construction Health and Safety.

Health and Safety in Construction Revision Guide by Ed Ferrett, Routledge 2015.

Leicester, Coventry and Nottingham Universities Study Skills booklets.

NEBOSH Examiners' Reports.

Northedge, A. 2012 The Good Study Guide. The Open University. ISBN 978 0 7492 5974 7.

Free learning resources from the Open University www.open.edu/openlearn/free-course

NEBOSH Guidance on command words used in learning outcomes and question papers (March 2013) online.

21

CHAPTER 22

Specimen answers to practice questions

> **This chapter covers:**
> 1. The depth of answer needed for long and short NEBOSH questions
> 2. The different types of questions used by NEBOSH
> 3. The requirements for the practical inspection and the management report

22.1 Introduction

The NEBOSH National Certificate in Construction Health and Safety is assessed by two written examination papers (NGC1 – The management of health and safety; and NCC1 – Managing and controlling hazards in construction activities) and a practical assessment (NCC2 – Construction health and safety practical application). Both examination papers and the practical application must be passed so that the NEBOSH National Certificate in Construction Health and Safety may be awarded. Since the unitisation of the course, students who have recently passed the NEBOSH National General Certificate will not need to re-take NGC1.

NEBOSH is concerned about malpractice problems if published NEBOSH questions are used in publications so the questions used in this book are not used in NEBOSH examinations. In this chapter, specimen answers are given to one long and two short questions for each of the two written papers. A specimen practical assessment and management report is also given for the practical application. It is important to stress that there are no unique answers to these questions but the answers presented should provide a useful guide to the depth and breadth expected by the NEBOSH examiners. Candidates are strongly advised to read past examiners' reports which are published by NEBOSH. These are very useful documents because they indicate some common errors made by candidates.

22.2 The written examinations

The previous chapter on study skills gives some useful advice on tackling examinations and should be read in conjunction with this chapter. NEBOSH has a commendably thorough system for question paper preparation to ensure that no candidates are disadvantaged by question ambiguity. Candidates should pay particular attention to the meaning of action verbs, such as 'outline', used in the questions.

Candidates have always had most difficulty with the action verb '**outline**' and, for this reason, several of the specimen questions chosen use '**outline**' so that some guidance on the depth of answer expected by the examiners can be given.

NEBOSH no longer uses the command word 'sketch; however, candidates may use sketches to illustrate their answers, when appropriate.

It is difficult to give a definitive guide on the exact length of answer to the examination questions because some expected answers will be longer than others and candidates answer in different ways. As a general guide, for the long answer question on the examination paper, it should take about 25 minutes to write about one and a half pages (550–620 words). Each of the 10 short answer questions require about half a page of writing (170–210 words).

Although candidates may not be guaranteed 100% by following the guidance in this chapter, they should obtain a comfortable pass.

22.2.1 NGC1 – The management of health and safety

Paper 1 – Question 1

(a) **Outline** the factors that should be considered when selecting individuals to assist in carrying out risk assessments in the workplace. **(5)**
(b) **Describe** the key stages of a general risk assessment. **(5)**
(c) **Outline** a hierarchy of measures for controlling exposures to hazardous substances. **(10)**

Figure 22.1 Select a competent and experienced person to carry out a risk assessment

Answer:

(a) The most important factor is the competence and experience of the individuals in hazard identification and risk assessment. Some training in these areas should offer evidence of the required competence. They should be experienced in the process or activity under assessment and have technical knowledge of any plant or equipment used. They should have knowledge of any relevant standards, Health and Safety Executive (HSE) guidance and Regulations relating to the activity or process.

They must be keen and committed but also aware of their own limitations. They need good communication skills and to be able to write interesting and accurate reports based on evidence and the detail found in health and safety standards, codes of practice, Regulations and guidance. Some IT skills would also be advantageous. Finally, the views of their immediate supervisor should be sought before they are selected as team members.

(b) There are five key stages to a risk assessment suggested by the HSE as follows:

The first stage is hazard identification, which involves looking at significant hazards which could result in serious harm to people. Trivial hazards should be ignored. This will involve touring the workplace concerned looking for the hazards in consultation with workers themselves and also reviewing any accidents, ill health or incidents that have occurred.

Stage 2 is to identify the persons who could be harmed – this may be employees, visitors, contractors, neighbours and even the general public. Special groups at risk, like young persons, nursing or expectant mothers and people with a disability, should also be identified.

Stage 3 is the evaluation of the risks and deciding if existing precautions or control measures are adequate. The purpose is to reduce all residual risks after controls have been put in to as low as is reasonably practicable. It is usual to have a qualitative approach and rank risks as high, medium or low after looking at the severity of likely harm and the likelihood of it happening. A simple risk matrix can be used to get a level of risk.

The team should then consider whether the existing controls are adequate and meet any guidance or legal standards using the hierarchy of controls and the General Principles of Prevention set out in the Management Regulations.

Stage 4 of the risk assessment is to record the significant findings, which must be done if there are five or more people employed. The findings should include any action that is necessary to reduce risks and improve existing controls – preferably set against a timescale. The information contained in the risk assessment must be disseminated to employees and discussed at the next health and safety committee meeting.

Stage 5 is a timescale set to review and possibly revise the assessment, which must also be done if there are significant changes in the workplace or the equipment and materials being used.

(c) The various stages of the usual hierarchy of risk controls are underlined in **bold** in this answer.

Elimination or **substitution** is the best and most effective way of avoiding a severe hazard and

its associated risks. Elimination occurs when a process or activity is totally abandoned because the associated risk is too high. Substitution describes the use of a less hazardous form of the substance. There are many examples of substitution, such as the use of water-based rather than oil-based paints and the use of asbestos substitutes.

In some cases it is possible to **change the method of working** so that exposures are reduced, such as the use of rods to clear drains instead of strong chemicals. It may be possible to use the substance in a safer form; for example, in liquid or pellets to prevent dust from powders. Sometimes the pattern of work can be changed so that people can do things in a more natural way; for example, by encouraging people in offices to take breaks from computer screens by getting up to photocopy or fetch documents.

Reduced or limited time exposure involves reducing the time that the employee is exposed to the hazardous substance by giving the employee either other work or rest periods.

If the above measures cannot be applied, then the next stage in the hierarchy is the introduction of **engineering controls**, such as isolation (using an enclosure, a barrier or guard), insulation (used on any electrical or temperature hazard) or ventilation (exhausting any hazardous fumes or gases either naturally or by the use of extractor fans and hoods). If ventilation is to be used, it must reduce the exposure level for employees to below the workplace exposure limit (WEL).

Housekeeping is a very cheap and effective means of controlling risks. It involves keeping the workplace clean and tidy at all times and maintaining good storage systems for hazardous substances.

A **safe system of work** is a requirement of the Health and Safety at Work (HSW) Act and describes the safe method of performing the job.

Training and information are important but should not be used in isolation. Information includes such items as signs, posters, systems of work and general health and safety arrangements.

Personal protective equipment (PPE) should only be used as a last resort. There are many reasons for this. It relies on people wearing the equipment at all times and it must be used properly.

Where necessary, **health surveillance** should be introduced to monitor the effects on people and air quality may need to be monitored to check exposure.

Welfare facilities, which include general workplace ventilation, lighting and heating and the provision of drinking water, sanitation and washing facilities, are the next stage in the hierarchy.

22

Figure 22.2 High level of fumes from welding

Figure 22.3 Motivating staff

All risk control measures, including **training and supervision,** must be **monitored** by competent people to check on their continuing effectiveness. Periodically the risk control measures should be **reviewed**. Monitoring and other reports are crucial for the review to be useful. Reviews often take place at safety committee and/or at management meetings. A serious accident or incident should lead to an immediate review of the risk control measures in place.

Finally, special control requirements are needed for carcinogens.

Paper 1 – Question 2

Outline ways in which employers may motivate their employees to comply with health and safety procedures. **(8)**

Answer:

Motivation is the driving force behind the way a person acts or the way in which people are stimulated to act. The best way to motivate employees to comply with health and safety procedures is to improve their understanding of the consequences of not working safely, their knowledge of good safety practices and the promotion of their ownership of health and safety. This can be done by effective training (induction, refresher and continuous) and the provision of information showing the commitment of the organisation to safety and by the encouragement of a positive health and safety culture with good communications systems. Managers should set a good example by encouraging safe behaviour and obeying all the health and safety rules themselves even when there is a difficult conflict between production schedules and health and safety standards. A good working environment and welfare facilities will also encourage motivation.

Involvement in the decision-making process in a meaningful way, such as regular team briefings, the development of risk assessments and safe systems of work, health and safety meetings and effective joint consultation arrangements, will also improve motivation

as will the use of incentive schemes. However, there are other important influences on motivation such as recognition and promotion opportunities, job security and job satisfaction. Self-interest, in all its forms, is a significant motivator.

Although somewhat negative, it is necessary sometimes to resort to disciplinary procedures to get people to behave in a safe way. This is rather like speed cameras on roads with the potential for fines and points on your licence.

Paper 1 – Question 3

(a) **Explain** why young persons may be at a greater risk from accidents at work. **(4)**

(b) **Outline** the measures that could be taken to minimise the risks to young employees. **(4)**

Answer:

(a) Young workers have a lack of experience, knowledge and awareness of risks in the workplace. They tend to be subject to peer pressure and behave in a boisterous manner. They are often willing to work hard and want to please their supervisor and can become over-enthusiastic.

This can lead to the taking of risks without the realisation of the consequences. Some younger workers have underdeveloped communication skills and a limited attention span. Their physical strength and capabilities may not be fully developed and so they may be more vulnerable to injury when manually handling equipment and materials. They are also more susceptible to physical, biological and chemical agents such as temperature extremes, noise, vibration, radiation and hazardous substances.

(b) The Management of Health and Safety at Work Regulations require that a special risk assessment must be made before a young person is employed. This should help to identify the measures which should be taken to minimise the risks to young people.

Measures should include:

- additional supervision to ensure that they are closely looked after, particularly in the early stages of their employment;
- induction and other training to help them understand the hazards and risk at their workplace;
- not allowing them to be exposed to extremes of temperature, noise or vibration;
- not allowing them to be exposed to radiation, or compressed air and diving work;
- carefully controlling levels of exposure to hazardous materials so that exposure to carcinogens is as near zero as possible and other exposure is below the WEL limits which are set for adults;
- not allowing them to use highly dangerous machinery like power presses and circular saws, explosives and mechanical lifting equipment such as fork-lift trucks;
- restricting the weight that young persons lift manually to well below any weights permitted for adults.

There should be clear lines of communication and regular appraisals. A health surveillance programme should also be in place.

22.2.2 NCC1 – managing and controlling hazards in construction activities

Paper 2 – Question 1

It is necessary for maintenance purposes to enter a large cement silo situated on a construction site.

(a) **Identify** the likely hazards. **(5)**

(b) **Outline** the precautions to be taken to reduce the risks to personnel entering the silo. **(15)**

Answer:

(a) There are several hazards in entering and working in this silo.

The main hazards include difficult access and egress, which makes entrapment, escape and rescue more problematic; being engulfed in a free-flowing solid material if the cement has blocked and bridged; the accumulation of vapours, gases or fumes; and the lack of ventilation. Some of these hazards may be introduced into the silo by the nature of the maintenance work (e.g. welding work or cleaning work with solvents) or by incoming services.

Other possible hazards are:

- working at height
- asphyxiation due to oxygen depletion
- poor lighting levels
- claustrophobic effects due to restricted space
- fall of materials leading to possible head injuries

- electrocution from unsuitable equipment
- fire due to flammable liquids and vapours.

Finally, the presence of cement residues is also a hazard. Contact with wet cement can cause serious burns or ulcers which will take several months to heal and may need a skin graft. Dermatitis, both irritant and allergic, can be caused by skin contact with either wet cement or cement powder.

Figure 22.4 Large cement silo on a construction site

(b) The silo is a confined space and the health and safety requirements are covered by the Confined Spaces Regulations. The detailed precautions required will depend on the nature of the silo and the actual maintenance work being carried out.

First a risk assessment must be done so that relevant precautions may be taken and a safe system of work determined. The risk assessment needs to identify the hazards present in the silo, estimate the risks and determine suitable controls to address those risks. It must be done by a competent person and will form the basis of a safe system of work. This will normally be formalised into a specific permit-to-work procedure which will be applicable to the maintenance work. The main elements of the 'permit-to-work' are likely to be as follows:

- the competence, training and instruction of the workforce and the designated rescue team;
- the effective and complete isolation of electrical and mechanical equipment and other services by the use of a lock-off and a tag system with key security;
- the methods of communication between people inside the silo, from inside to outside and to summon help in an emergency;
- the testing and monitoring of the atmosphere inside the silo for hazardous gas, fume, vapour, dust and the concentration of oxygen;

22

689

- ▶ the provision of good ventilation, perhaps by mechanical means, and escape breathing apparatus for all persons inside the silo;
- ▶ harness and attached lanyard for rescue purposes;
- ▶ means of effecting rescue such as a tripod manual winch lifting unit in place during the work;
- ▶ the careful removal of cement residues using appropriate equipment which will not cause additional hazards;
- ▶ the provision of an outside watcher to raise the alarm and summon emergency assistance;
- ▶ the access and egress to the silo so that there is quick and unobstructed access and escape;
- ▶ the details of fire prevention equipment and procedures, such as method of raising the alarm;
- ▶ the lighting arrangements;
- ▶ the maximum length of any one working period.

It is also important that there are monitoring and audit arrangements to ensure that the permit-to-work system is working as envisaged.

The site supervisor must ensure that all necessary equipment is available on site in accordance with the risk assessment, method statement and any other planned procedures, including the permit to work, before any person is allowed to enter the silo. He/she must also ensure that the planned procedures, particularly the permit to work, are adhered to strictly and that only authorised trained persons are permitted to enter the silo. He/she must be notified of any changes in working methods or conditions inside the silo which were not included in the planning procedures before those changes are implemented.

All safety equipment must be regularly checked and maintained. Records should be kept of the checks and any defects in equipment rectified immediately. Appropriate PPE must be provided for the workforce in the silo (helmets, gloves, boots and hearing protection). Any lighting or electrical power tools must be specially protected to the required standard for damp and/or flammable atmospheres.

Specific training is necessary for all people concerned with silo maintenance, whether they are acting as rescuers, supervisors or those working inside it. The training will need to cover all procedures and the use of equipment under realistic simulated conditions.

The workers must also be monitored and consideration should be given to the exclusion from the silo of individuals who suffer from any medical condition which may be exacerbated by working in it (e.g. claustrophobia).

Before people enter the silo, suitable and sufficient emergency, rescue and first-aid arrangements must be in place. These arrangements should reduce the risks to rescuers and include the provision and maintenance of suitable resuscitation equipment.

Paper 2 – Question 2

Outline the precautions that may be needed when carrying out repairs to the flat roof of a building. **(8)**

Answer:

Roof work is hazardous and requires a specific risk assessment and method statement prior to the commencement of work so that the required precautions may be identified. The particular hazards are fragile roofing materials, including those materials which deteriorate and become more brittle with age and exposure to sunlight, exposed edges, unsafe access equipment and falls.

There must be suitable means of access such as scaffolding, ladders and crawling boards. Suitable edge protection will be needed in the form of guard rails to prevent the fall of people or materials and access must

(a)

(b)

Figure 22.5 Flat roof repair with: (a) edge protection; (b) harness and fall arrest device

be restricted to the area below the work using visible barriers. Warning signs indicating that a roof is fragile, should be displayed at ground level. Protection should be provided in the form of covers where people work near to fragile materials and roof lights. The means of transporting materials to and from the roof may require netting under the roof and even weather protection.

Precautions will be required for other hazards associated with roof work, such as overhead services and obstructions, the use of equipment such as gas cylinders and bitumen boilers, and manual handling.

Finally, only trained and competent persons must be allowed to work on roofs and they must wear footwear having a good grip. It is good practice to ensure that a person does not work alone on a roof.

Paper 2 – Question 3

An independent scaffold tied to a 10-storey office block has collapsed into a busy street.

Outline the factors that may have affected the stability of the scaffold. **(8)**

Answer:

The stability of the scaffold will be undermined if its original erection did not follow the intended design or the design was inadequate. Other factors may include one or more of the following:

- ▶ poor, incompetent erection and/or a lack of regular inspections such that the standards were bent or not plumb;
- ▶ poor tying into the building or insecure parts of the building;
- ▶ lack of strength of the supporting ground or foundation;
- ▶ lack of effective sole plates as a means of spreading the load at the base of the scaffold;
- ▶ lack of longitudinal bracing (along the length of the scaffold) or cross-bracing (at right angles to the building);
- ▶ the proximity to the scaffold of any excavation work;
- ▶ the effect of excessive surface water in weakening the scaffold foundation;
- ▶ unauthorised alteration by incompetent persons (perhaps by vandals);
- ▶ the use of incorrect and/or damaged or corroded fittings during the erection or extension of the scaffold;
- ▶ the overloading of the scaffold either with materials or because waste chutes became blocked;
- ▶ adverse weather conditions including high winds;
- ▶ the sheeting-in of the scaffold without the use of extra ties;
- ▶ vehicular impact (either a road vehicle or a load suspended from a crane);
- ▶ the 'borrowing' of boards and tubes from the scaffold, thus weakening it – a common cause of scaffold collapse.

22.3 Unit NCC2 – Construction Health and Safety Practical Application

Guidance and information for accredited course providers and candidates is essential reading before attempting the practical application. The guide contains details on the aims of the application, the rules to be followed, the breadth and depth of observations and management report expected and the mark scheme.

The practical application Guidance v5 (12/13) (under revision as at Aug 2015) is reproduced in Section 22.3.1. A specimen practical application, with associated observations and management report, is given in Appendices 22.1 and 22.2. It is important to emphasise that this specimen practical application (or any parts of it) must not be copied as this is malpractice.

22.3.1 Requirements of the practical application

The aim of the practical application, Unit NCC2, is to assess a candidate's ability to complete successfully two activities:

- ▶ to carry out unaided a safety inspection of a workplace, identifying the more common hazards,

Figure 22.6 Scaffold collapse, Milton Keynes

deciding whether they are adequately controlled and, where necessary, suggesting appropriate and cost-effective control measures;

▶ to prepare a report that persuasively urges management to take appropriate action, explaining why such action is needed (including reference to possible breaches of legislation) and identifying, with due consideration of reasonable practicability, the remedial measures that should be implemented.

This will require candidates to apply the knowledge and understanding gained from their studies of elements of Units NGC1 and NCC1 in a practical environment and to carry out an evaluation of information gathered during the inspection. The time allowed to complete the assessment is not restricted but candidates should aim to complete the inspection and the report in two hours. The practical application may be submitted in the candidate's own handwriting or be word processed.

The submission must include:

▶ completed observation sheets covering a number and range of hazards and good practice, identifying suitable control measures and timescales;
▶ an introduction and executive summary;
▶ main findings of the completed inspection;
▶ conclusions which summarise the main issues identified in the candidate's workplace;
▶ completed recommendations table.

The practical application must be carried out in the candidate's own workplace; it is important that a workplace with construction activities is chosen upon which the practical application will be based. The workplace should be large enough to provide a sufficient range of hazards in the areas covered to provide an adequate range for identification. If the workplace is very large, in order that the practical application is manageable the candidate should limit the area considered.

Where the candidate does not have access to a workplace containing construction activities, the course provider should be consulted to help in making arrangements for the candidate to carry out the practical application at suitable premises. Providers seeking to run the practical in this way should contact NEBOSH for advice and approval.

Management at the premises should be consulted to ensure the candidate can carry out the inspection without endangering their own health and safety. Where confidentiality requires, locations and company names may be omitted from the report, or alternatively guidance should be sought from NEBOSH. Candidates do not require supervision when carrying out the practical application but the candidate must sign a declaration that the submission is their own work.

Date of assessment

The assessment of the practical unit (NGC3) must normally take place **within 10 working days** of (before or after) the date of the NGC1 and/or NCC1 written papers (the 'date of the examination').

Observations and comments should be noted on the observation sheets provided by NEBOSH (see Appendix 22.2).

Completion of observation sheets

Candidates will be supplied with a sufficient number of observation sheets from a course provider which may be photocopied for the purpose. The observation sheets must be completed during the inspection. Only brief details of each hazard are required including where the hazard was located and the nature of the hazard. For example, 'housekeeping could be better' does not give enough information about the particular hazard. Candidates should avoid the tendency to simply identify the tasks being undertaken, or the equipment that is being used, instead of providing an indication of how the tasks or items of equipment pose a risk.

Candidates must demonstrate their understanding of how identified hazards have the potential to cause harm; for example, boxes stored on the floor may cause obstruction of access, egress routes and/or a risk of musculoskeletal injury if lifted. This information must be recorded in the 'Hazards and consequences' column.

The observation sheets should be completed by:

▶ identifying, in the left-hand column, any hazards, unsafe work practices, consequences and examples of good practice observed during the inspection;
▶ commenting in the next column, on the adequacy of existing controls and identifying any immediate, medium-term and long-term remedial actions needed;
▶ stating, in the right-hand column a reasonable timescale for the actions identified.

There should be sufficient information on the observation sheets to enable the candidate to complete a report to management on their findings. Candidates are also advised to make notes on the area inspected, including activities taking place, in order to complete the introduction to their report. Whilst poor spelling and grammar will not be marked or penalised, if the assessor is unable to read or to understand the notes made by candidates during their inspection then invariably fewer marks will be awarded than would otherwise have been the case. The practical application may be submitted in the candidate's own handwriting or word processed.

For most practical assessments, the examiner will be looking for between 20 and 30 observations.

Marking of observation sheets

Candidates are advised to refer to the marking sheet given at Appendix 3 and the marking matrix given at Appendix 6 in the NEBOSH *Guidance and information for accredited course providers and candidates*. This is available on the NEBOSH website.

Range and outline of hazards and consequences (15 marks)

Candidates should identify 20 uncontrolled hazards to gain maximum marks, but are strongly recommended to identify more than 20 (but no more than 30) in case of duplication or inappropriate hazards being identified. The specimen practical application given in Appendices 22.1 and 22.2 contains many more observations than would be expected from candidates. Candidates are expected to identify different types of hazards such as hazardous substances, fire, electrical, work equipment, ergonomic, housekeeping, noise, vibration, transport, manual handling and health hazards and should also consider if there are any welfare and environmental problems. At least five different types of hazard must be included for maximum marks. In addition candidates are expected to comment on situations where there is adequate control of hazards and where good practice is being observed, although candidates should place the emphasis on uncontrolled hazards. A maximum of 1 mark is available for good practice.

Where a candidate repeats a hazard, that hazard will only be credited once (e.g. reference to three items of work equipment requiring portable appliance testing). Candidates should focus on physical conditions and not on poor policies and procedures.

Identification of suitable control measures and timescales (15 marks)

Candidates are expected to give thought to what is required to immediately control the risk from each identified hazard **AND** to identify the need for medium- and long-term actions to control the risk. This requires candidates to distinguish between the symptoms and the root causes of hazards.

For example, the immediate action on a spillage may be 'clean up spillage and inform supervisor' whereas medium-term actions might include appropriate supervisor training, regular inspections and investigation of the source of the leak. A further longer term action may be to modify the work process to tackle the root cause. The proposed control measures must not only remove or control the hazard but must also be realistic in terms of timescales. Candidates should indicate a measure of time, e.g. supervisor training to be completed within three months.

If existing controls are in place and considered adequate, candidates should consider any measures required to maintain this level of control.

Candidates should avoid generic phrases being repeatedly used, e.g. 'monitor' and 'train staff'. Candidates should give appropriate clarification by giving examples of appropriate monitoring and the type of training required.

Candidates should be aware that if unsuitable control measures are suggested full marks cannot be awarded. Short-term measures to improve housekeeping will do little to improve the lack of safety management systems and procedures evidenced by materials and equipment left lying around.

Candidates should also note that where the hazard is not clearly outlined full credit cannot be gained for control measures as assessors will be unable to determine appropriateness of the measures proposed.

Completion of report

Candidates should use the 'Candidate report template' given at Appendix 5 in the NEBOSH *Guidance and information for accredited course providers and candidates,* to structure their report. This Appendix has been used for the specimen practical application given in Section 22.3.2 of this chapter.

The length of the report should be between 700 and 1,000 words and should not simply duplicate the observation sheets. Candidates can consult reference books when preparing the report, but plagiarism will be dealt with as malpractice.

The report should not contain photographs, printed text or any other extraneous material.

The report should be written in such terms that a manager would be able to take reasonable action based on facts. Reports based on unsupported generalities and those that simply reiterate the contents of the observation sheets will be awarded low marks.

Candidates should aim to complete their report in one hour.

If none of the criteria to award marks is met, then candidates should be aware that zero marks will be awarded.

Marking of report

Candidates are advised to refer to the marking sheet given at Appendix 3 and the marking matrix given at Appendix 6 in the *NEBOSH Guidance and information for accredited course providers and candidates*. This is available on the NEBOSH website.

Report – Introduction and executive summary

Introduction providing an overview of the chosen area (5 marks)

Candidates should start with the details of the inspection, stating where and when the inspection

22

took place. A clear and appropriate description of the chosen area and of the activities occurring in the area should be given.

Executive summary (5 marks)

The executive summary should be written after the candidate has completed the rest of the report but it should be inserted at the beginning of the report. The purpose of the executive summary is to provide a **concise** overview of the important points arising from the work and **summarise** the main conclusions and recommendations arising from it.

An executive summary should provide sufficient information to enable a busy manager to make a decision as to whether or not to read the full report and to provide a persuasive case for implementation of recommendations made.

Report – Main findings of the inspection

The main findings of the inspection should form the main body of the report and include the following aspects:

The report should be well structured, the appropriate length and not duplicate observation sheets. The report should be concise, readable and highly selective in terms of action required by management. Candidates should include balanced arguments on why action is needed and explain the effect it would have on the standards of health and safety at the workplace and the possible effects on the business overall.

Quality of interpretation of findings and clear references to strengths and weaknesses (15 marks)

There should be a logical progression from the issues identified on the observation sheets. The key issues should be discussed having clear reference to strengths and weaknesses and should not include any hazards not identified on the observation sheets. Credit will be given for reference to appropriate legislation and standards.

Identification of possible breaches of legislation (5 marks)

Candidates should be able to identify those Acts and Regulations listed in the syllabus that may have been infringed, e.g. the Health and Safety at Work etc. Act, the First-Aid Regulations (full titles normally should be given on first use and may be abbreviated subsequently). For example, Health and Safety at Work Act, First-Aid Regulations, PUWER and the Workplace Regulations are acceptable abbreviations on the observation sheets and in the management report.

Candidates must ensure that any document referred to is relevant to the area/location inspected and must demonstrate a clear understanding of the reasons for

the breaches. Candidates should identify a minimum of five breaches.

The following list of Acts and Regulations would be relevant to all inspections:

▶ Health and Safety at Work etc. Act;
▶ Management of Health and Safety at Work Regulations;
▶ Regulatory Reform (Fire Safety) Order;
▶ Workplace (Health, Safety and Welfare) Regulations.

Most inspections would also include some of the following Regulations:

▶ Manual Handling Operations Regulations;
▶ Provision and Use of Work Equipment Regulations;
▶ Personal Protective Equipment at Work Regulations;
▶ Electricity at Work Regulations;
▶ Control of Substances Hazardous to Health Regulations;
▶ Health and Safety (Display Screen Equipment) Regulations;
▶ Health and Safety (First-Aid) Regulations;
▶ Control of Noise at Work Regulations;
▶ Control of Vibration at Work Regulations;
▶ Health and Safety (Safety Signs and Signals) Regulations;
▶ Work at Height Regulations;
▶ Health and Safety (Consultation with Employees) Regulations.

Persuasiveness/conciseness/technical content (10 marks)

The candidate must convince management to take action on the issues identified from the inspection. This should include clear legal, moral and financial reasons why such action should be included.

Financial benefits, such as increased productivity, may also accrue from making appropriate changes to safety systems. The possible costs of not taking action should be included.

A short list of issues requiring urgent action by management with associated explanations containing convincing arguments why such action is needed should be included. Reference can be made to the list of observations and recommended actions, calling particular attention to any recommendations which could have a high cost in terms of finance, inconvenience or time.

The report should be well structured, the appropriate length and not duplicate observation sheets. It should be concise, readable and highly selective in terms of action required by management. Candidates should include balanced arguments on why action is needed and explain the effect it would have on the standards of health and safety at the workplace and the possible effects on the business overall.

Report – Conclusions and recommendations

Clear and concise conclusions which are clearly related to report findings and are effective in convincing management to take action (15 marks)

This section should provide a concise summary of the findings identified in the main body of the candidate's report. The conclusions should not introduce new issues or additional factors. Relevant and appropriate information should be provided to persuade management to take action.

Recommendations which present realistic actions to improve health and safety in the chosen area (15 marks)

Candidates should include recommendations based upon their conclusions.

Recommendations must be presented using the recommendations table included at Appendix 5 in the NEBOSH guidance documents. The recommendations must be realistic, appropriately prioritised and have appropriate resource implications. Candidates are not expected either to know or to estimate actual costs but should demonstrate that they are aware of cost implications. For example, candidates recommending the complete resurfacing of a site roadway because of a small pothole, without commenting on its overall condition, will receive low marks. If training is recommended as a solution to a problem, candidates should indicate if this is likely to require a few hours of work-based instruction or several days of more costly off-the-job training. It is the assessment of magnitude of the cost that is important, rather than precise figures, e.g. candidates may refer to the number of worker hours as a measure of cost.

Recommendations should be prioritised. The most pressing issues, those which present the highest risk

levels and those that can be done immediately at little or no cost, should be addressed first. Target dates should be included, for example 'plus one week' or 'plus three months'.

If none of the criteria to award marks is met, then zero marks will be awarded.

Important note

The guidelines outlined above and the following specimen practical application are completely valid at the time of publication of this book. However, NEBOSH reviews these guidelines and the format used in the two appendices from time to time so that it is important for course providers and others to consult the NEBOSH website periodically to check whether the guidelines and the report format have been amended. The specimen practical application is given as an illustration only.

22.3.2 Specimen practical application

The specimen practical assessment took place on a construction site, called the Beacon site, and contains more observations than would normally be expected from students but it does attempt to cover most of the major hazards encountered on a construction site.

The observations made may be seen in Appendix 22.2 and Appendix 22.1 is a specimen candidate report. As noted above, Appendix 22.2 contains 42 observations, which is 12 more than the maximum number (30) required by NEBOSH. The authors felt that a list of observations that covered all common construction activities would be useful to students. Course providers will exclude some of these activities during the examined practical assessment.

The following management report (Appendix 22.1) was submitted to the Chief Executive of A B Construction Ltd.

APPENDIX 22.1 The practical application report

NATIONAL CERTIFICATE IN CONSTRUCTION HEALTH AND SAFETY

Candidate report template (2009 specification)

UNIT NCC2 – THE HEALTH AND SAFETY PRACTICAL APPLICATION

Student number: _____ C1416 _____

Location: <u>A B Construction Ltd.</u>

Date of review: _____ 14 / 1 / 15

Introduction including overview of area inspected and activities taking place

This report follows a health and safety inspection of a construction site, called the Beacon site, where 30 domestic houses were being built – 12 erected and being fitted out, 12 half-erected and 6 with just the foundations in place. There were 5 employees, 1 trainee and several sub-contractors at the site on the day of the assessment. The 5 employees were a supervisor, 2 bricklayers, 1 carpenter and 1 building operative. In addition to the 12 houses on the site, there was a site office, welfare facilities, a stores compound and a vehicle compound. The site was surrounded by a security fence.

The inspection took place on 14 January 2015. The report's aim is to highlight good practice and areas of concern requiring urgent action. The observations are listed in Appendix 22.2 attached to this report.

Executive summary

There have been no serious accidents at this site since construction work began six months ago. This good record appears to have led to a relaxation in the monitoring of standards. Several inexpensive improvements in the health and safety arrangements are recommended.

There are some serious concerns involving poor fire precautions, the safeguards during excavation work, missing toe boards on one scaffold, and plant maintenance and control of vehicles on the site, leading to possible breaches of the Health and Safety at Work Act (HSWA), the Management of Health and Safety at Work Regulations (MHSWR), the Construction (Design and Management) (CDM) Regulations, the Work at Height (WAH) Regulations, the Regulatory Reform (Fire Safety) Order (Fire Regs), the Provision and Use of Work Equipment Regulations (PUWER), the Personal Protective Equipment at Work Regulations (PPE Regs) and the Electricity at Work Regulations. Any of these breaches could attract enforcement action from the HSE. All these problems indicate a poor level of supervision and lack of enforcement of basic health and safety rules, such as the wearing of hard hats.

The poor record-keeping could prove costly to the company in the event of a civil claim by workers or others.

The provision of a site induction course for all employees and sub-contractors is a further urgent requirement.

A more detailed description of the topics requiring immediate attention is given below with reference to the observation sheets given in brackets.

Main findings of the inspection

(a) **Health and safety management issues**
Although the strengths outnumbered the weaknesses, many of the observed weaknesses resulted from a lack of health and safety awareness and poor supervision. It is essential that site safety induction training is introduced as a matter of urgency (7) for all workers at the site. Such a course would ensure that everyone is aware of the health and safety plan (6), risk assessments (8), the company health and safety policy (13), the need to wear hard hats (20) and the fire and emergency arrangements (38–41). The

course need not be long (about half an hour) and a simple cheap booklet covering the main points would help to ensure an effective induction for all the site workers.

Improved site supervision and some delegated health and safety responsibilities would significantly reduce risks on the site. This would ensure that the stores were locked at night (4), and the site kept tidy (5) and protected from trespassers (3). Particular attention should be paid to the toilets (15) and the kitchen area (16).

Under the First-Aid Regulations, the first-aider needs to attend a short refresher course at a recognised training centre.

Finally, it was noted that toe boards were missing from one of the scaffolds (21) even though the scaffold had passed an inspection earlier in the week when, presumably, the boards were present. This is a serious health and safety issue and an urgent investigation should be made to discover the reason for the board removal.

The total cost of these recommendations will be small because most of them involve a reorganisation of workloads.

(b) **Documentation**

Most of the documentation was in good order and highly commendable (10, 11, 14).

Risk assessments have been completed except one to cover the activities of the trainee. The Management Regs require a specific risk assessment to be made for persons under the age of 18 years (8). If risk assessments are to be effective, they must be reviewed periodically so that they are relevant to the particular site being worked at the time and that no new risks have arisen. This review could take place at a short site meeting.

There were two further concerns – the absence of any COSHH assessments for the hazardous substances used on the site (36) and the lack of any excavation inspection regime (12) as required by CDM. These are both legal requirements.

Some of the workers' records were incomplete in that courses attended by individuals had not been recorded (9), so their current level of competence was not recorded.

Finally, both the company health and safety policy and the Employers' Liability Insurance certificate should be prominently displayed at the site (13).

The cost of these recommendations will not be large and could be done by existing staff. The

COSHH assessment will be the largest job but the number of hazardous substances on site is limited.

(c) **Site issues**

Several aspects of the site were very commendable (1, 2, 24 and 26). The main concerns were: the excavations (the lack of barriers (29) and the positioning of the spoil material (30)); the lack of an exclusion zone below roof work (27); and the storage of roofing tiles on the scaffold working platform (28). Also, several ladders were not adequately secured (25) or protected from use by trespassers (3). All of these concerns require urgent attention.

The control of traffic on the site also needs urgent attention and a one-way system with pedestrian walkways (34, 35) is recommended.

The total cost of all these recommendations should not be great for the various barriers which can be re-used on future sites.

(d) **Other issues**

The Fire Regs require that fire risk assessments, fire-fighting arrangements, fire information and basic fire training be provided at the site. A simple fire risk assessment should be made as soon as possible – it is a legal requirement.

Other fire precautions that should be in place include the storing of highly flammable substances (38) in a separate store, the designation of a fire assembly point and a regular fire drill (every three months) (40). Details of emergency procedures and the location of the assembly point should be given on a posted emergency procedure notice.

No maintenance records were available for the cement mixer (31) although assurances were given that all plant is regularly serviced. A maintenance schedule for this and any other plant should be recorded and monitored.

Finally, there is a risk of back injury due to the manual handling of kerb stones. It is recommended that a mechanical lifting device should be purchased or hired for this work.

Conclusions

The site was found to be reasonably well run with many health and safety issues being addressed. There were, however, a number of issues which need to be addressed urgently to avoid possible legal action or compensation claims following injuries.

Recommendations – include as a table in the following format:

Recommendation	Likely resource implications	Priority	Review date
No one should be allowed on site without a hard hat on. This topic should be included in the site induction training (20).	This is a legal obligation. Hard hats are not expensive.	High	Immediate
The monitoring of scaffolding (21, 23) and ladders (25) must be improved – toe boards replaced, warning signs on incomplete scaffolds and ladders properly secured.	Low cost for signs and toe boards. Investigation into missing toe board should not take more than one man hour.	High Medium	Immediate 1 month
Restrict to a maximum of 50 litres of highly flammable substances on site and store in a special metal container (38).	A stand-alone secure metal container is the most expensive recommendation but is reusable on future sites.	High	1 week
Complete fire risk assessment as soon as possible (41).	Fire risk assessment should take no longer than two man hours.	High	Immediate
The site is untidy and the stores and ladders (3) must be secured at night.	This is inexpensive and needs to be delegated to an employee as part of a job description.	High	Immediate
Urgent consideration should be given to the introduction of a one-way system on the site (34). Where possible, pedestrians should be separated from vehicles.	This is a low-cost issue that requires some management time in planning the road layout and the purchase of road signs.	High	1 week
Barriers and suitable warning signs must be erected around the excavations (29). Spoil material should be stored further from the excavation and the excavation inspection records saved.	Suitable barriers and warning signs are not expensive and are reusable on future sites.	High	Immediate
Cordon off the area below the roof work and post suitable signs (27). The number of tiles stored on the roof should be reduced to no more than 30 minutes' supply.	Low-cost but reusable items.	High	1 week
Purchase a mechanical lifting device for the kerb stones (38).	Several different products are on the market and the cost is not high. Alternatively hiring the device may be more cost effective.	High	1 week
Site induction training needs to be introduced and a Young Person's risk assessment completed (8).	Induction – 1 hour per group of workers. The risk assessment should be completed in one man hour.	High	1 week
The rest facilities must be cleaned at least twice each day, particularly the kitchen sink and utensils (16).	This should require no more than one man hour each day and should be delegated to a named individual.	High	1 week

APPENDIX 22.2 The practical application observation sheets

NATIONAL CERTIFICATE IN
CONSTRUCTION HEALTH AND
SAFETY

Candidate's observation
sheet

NCC2 – THE CONSTRUCTION HEALTH
AND SAFETY PRACTICAL APPLICATION

Candidate name _____ A Smith _____

Place inspected ____ A B Construction Ltd. _____

Sheet number   1   of   5

Candidate number **C** 1416

Date of inspection 14 / 1 / 2015

Observations	Control measures	Timescale
Hazards and consequences	Immediate and longer-term actions	
General site issues		
1. Site enclosed by a strong security fence – protection against intruders.	Commendable – no immediate actions. Check fence every week.	Every week
2. A comprehensive set of health and safety signs on the fence.	Ongoing monitoring.	Every week
3. All ladders must be fitted with scaffold boards overnight to protect child trespassers.	Delegate responsibility for this to an employee.	Immediate
4. Stores are not secured at night – theft of materials.	Delegate responsibility for this to an employee.	Immediate
5. Site was generally untidy – slip, trip and fire hazards.	Delegate responsibility for this to an employee.	Immediate
Management issues		
6. A health and safety plan and a complete set of method statements were available (CDM) – legal compliance.	Ensure that workers are aware of method statements. Some contractors had not seen them.	Ongoing
7. No site induction training was given to workers – injuries or ill-health to workers.	A simple induction course outlining the company health and safety policy and site hazards should be devised as soon as possible.	1 week
8. All risk assessments completed except for the trainee (Management Regs) – non-compliance with law.	The young worker risk assessment should be completed as soon as possible with his supervisor. All other risk assessments should be reviewed and discussed at site meetings.	1 week 3 months
9. All employees and contractors were competent but some records were incomplete – problems in any future civil actions.	Records for employees and contractors should be kept up to date.	1 month

22

NCC2 – THE CONSTRUCTION HEALTH AND SAFETY PRACTICAL APPLICATION

Candidate's observation sheet

Sheet number ___2___ of ___5___

Candidate name _____A Smith_____

Candidate number **C** 1416

Place inspected ___A B Construction Ltd.___

Date of inspection _14_ / _1_ / _2015_

Observations	Control measures	Timescale
Hazards and consequences	Immediate and longer-term actions	
Documentation		
10. A range of test certificates for plant and equipment were available (e.g. dumper trucks, compressors, cranes, lift trucks).	Good practice.	
11. Register of regular scaffold inspections available.	Good practice.	
12. No register of excavation inspection kept (CDM) – legal non-compliance.	Inspection of excavations is required if it needs to be supported.	Immediate
13. Although a health and safety poster was displayed, no health and safety policy statement or current employers' liability insurance certificate was posted (HSW) – possible HSE enforcement action.	The company health and safety policy is contained in the safety plan but a copy should also be on public display. Employers' liability insurance is a legal requirement and the certificate should be placed on a notice board. The trainee should also be covered by insurance.	1 month

Immediate |
14. Accident book was available and up to date.	Good practice.	
Welfare		
15. Toilet and washing facilities were adequate but the paper towel bins need regular emptying and the facilities more cleaning (CDM) – ill-health problems due to lack of hygiene.	These facilities should be cleaned at least twice a day.	1 week
16. The rest facilities used by workers for meal breaks need attention, particularly the kitchen equipment. (CDM) – hygiene issues leading to ill-health problems.	Regular cleaning at least twice a day especially the kitchen sink and utensils.	1 week
17. First-aid box was fully and properly stocked and appropriate notices displayed.	Commendable	
18. There is a first-aider on site but he must attend a refresher course. (First-Aid Regs)	Arrange for attendance at a refresher course.	Three months

NCC2 – THE CONSTRUCTION HEALTH AND SAFETY PRACTICAL APPLICATION

Candidate's observation sheet

Sheet number ___3___ of ___5___

Candidate name _____A Smith_____

Candidate number **C** 1416

Place inspected ___A B Construction Ltd.___

Date of inspection 14 / 1 / 2015

Observations	Control measures	Timescale
Hazards and consequences	Immediate and longer-term actions	
<u>Personal protective equipment</u>	Commendable.	
19. All relevant personal protective equipment was available and in good condition. (PPE Regs)		
20. Some contractors were not wearing safety helmets. (PPE Regs) – head injuries from falling objects.	No one should be allowed on site without hard hats on. This topic should be included in the site induction training.	Immediate
<u>Scaffolds</u>		
21. All guard rails were adequate but some toe boards were missing. (WAH Regs) – possible falls from height.	The toe boards must be replaced immediately and were present during the weekly inspection. An investigation is recommended.	Immediate 1 month
22. The scaffolding is inspected regularly and records are kept.	Good practice.	
23. No warning signs were fitted to an incomplete scaffold. (WAH Regs) – could be used by mistake leading to serious injury.	Fit signs immediately and monitor in future.	Immediate Monitor daily when scaffold is being erected.
<u>Ladders</u>		
24. All ladders rise to a sufficient height above landing place.	Good practice.	
25. Ladders are not secured to prevent slipping (WAH Regs) – injuries from falls from the ladder.	Secure all ladders at top and bottom.	Immediate

NCC2 – THE CONSTRUCTION HEALTH AND SAFETY PRACTICAL APPLICATION

Candidate's observation sheet

Sheet number ___4___ of ___5___

Candidate name _____A Smith_____

Candidate number **C** 1416

Place inspected ___A B Construction Ltd.___

Date of inspection _14_ / _1_ / _2015_

Observations	Control measures	Timescale
Hazards and consequences	Immediate and longer-term actions	
Roof work		
26. Suitable edge protection is provided.	Commendable.	
27. People are not excluded from the area below the roof work – injuries from falling objects.	Cordon off area below roof work and post suitable signs.	Immediate
28. Too much building material (tiles) stored on the roof – injuries due to tiles falling from the roof, also possible overloading of roof.	Reduce the number of tiles on the roof to 30 minutes' supply.	Immediate
Excavations		
29. No barriers around the drainage excavations (CDM) – injuries from people falling into the excavation.	Erect barriers and suitable warning signs.	Immediate
30. Spoil material stored too close to the excavation – spoil falling on top of excavation workers.	Store the material so that there is no danger of it re-entering the trench.	Immediate
Plant and equipment		
31. No maintenance records were available for the cement mixer. (PUWER) – problem if there is a civil claim or a criminal prosecution.	Either obtain copies or have the mixer serviced.	Immediate
Power and hand tools		
32. Low-voltage equipment used throughout site.	Commendable.	
33. No records of PAT testing of electrical tools. (Electricity Regs) – problem if there is a civil claim following electrical injury.	Arrange for all electrical equipment to be tested on a 3-monthly basis.	1 month

SPECIMEN

NCC2 – THE CONSTRUCTION HEALTH AND SAFETY PRACTICAL APPLICATION

Candidate's observation sheet

Sheet number ___5___ of ___5___

Candidate name _____A Smith_____

Candidate number **C** 1416

Place inspected ___A B Construction Ltd.___

Date of inspection _14_ / _1_ / _2015_

Observations	Control measures	Timescale
Hazards and consequences	Immediate and longer-term actions	
Traffic and vehicles		
34. There have been several near collisions between vehicles on the site. (CDM) – eventually there will be an injury to a person on site.	Urgent consideration should be given to the introduction of a one-way system on the site.	1 month
35. No pedestrian-only walkways have been designated on the site. (CDM) – collision between pedestrians and vehicles.	Where possible pedestrians should be separated from vehicles. (Incorporate into one way system?)	1 month 1 month
Hazardous substances		
36. No COSHH assessments have been done. (COSHH Regs) – possible ill-health problems for workers.	COSHH assessments should be made as soon as possible with particular attention paid to cement.	3 months
Manual handling		
37. Risk of back injury due to manual lifting of kerb stones. (Manual Handling Regs) – musculoskeletal problems for workers.	Use a mechanical lifting device.	Immediately
Fire and emergency procedures		
38. Highly flammable substances not stored separately. (Fire Regs) – serious fire and/or explosion.	Restrict to a maximum of 50 litres on site and store in a special metal container.	1 week
39. Fire extinguishers are adequate and checked regularly.	Good practice	
40. Fire drills do not take place and emergency procedures are not posted – chaos and injuries to workers in an emergency.	Identify a suitable assembly point and mark it. Arrange a fire drill every 3 months and record assembly time. Include details in induction training and in emergency procedures.	1 week
41. No fire risk assessments done. (Fire Regs) – legal non-compliance and increase fire risk.	Complete fire risk assessment as soon as possible.	Immediate

22

CHAPTER 23

International sources of information and guidance

23.1 Introduction

The authors and publishers are not responsible for the contents or reliability of the linked websites and do not necessarily endorse the views expressed within them. Listing should not be taken as endorsement of any kind. We cannot guarantee that these links will work all the time and we have no control over the availability of the linked pages. For this chapter of the book, as in Chapter 21, a less formal style has been chosen, speaking directly to you, the student.

Today using the internet for access to legislation, standards and guidance is universal and of great benefit to the international OH&S practitioner or person seeking information. Clearly there are many organisations that give small amounts of information free in order to gain your membership or custom. As internet users know, documents, guidance, reports, etc. can remain sitting on websites for years. So it is very important to ensure that they are up to date and authoritative. Guidance from good reliable sources is usually kept up to date and often states the date on which the site was last updated.

Fortunately in the OH&S field there are many organisations that freely allow the download of information because they are either governments, government funded or have a strong wish to make good guidance widely available. These include IOSH, RoSPA and many similar OSH organisations around the world; government departments in the USA, UK, Europe and beyond; universities and local authorities. In the UK, the HSE now allows free download of most of their extensive, authoritative and well-produced guidance. This is a significant change of policy in the last few years and provides a wealth of information for the OH&S practitioner and others.

Finding these sources of information and getting the best results from your searches can be a daunting task. The following general guidance may help when searching for occupational, safety, health, fire, chemical and environmental information on the internet. The websites are mainly in English but some are also in several other languages or a local language and English.

There is also a very useful free, UK Open University, 9-hour introductory course, Finding information in information technology and computing,

LIB 5 at: http://openlearn.open.ac.uk/course/view.php?id=2370

23.2 How to search the internet effectively

23.2.1 The key to successful searching

Remember, you are smarter than a computer. Use your intelligence. Search engines are fast, but dumb.

A search engine's ability to understand what you want is very limited. It will obediently look for occurrences of your keywords all over the Web, but it doesn't understand what your keywords mean or why they're important to you. To a search engine, a keyword is just a string of characters. It doesn't know the difference between cancer the crab and cancer the disease … and it doesn't care.

But *you* know what your query means (at least, we hope you do!). Therefore, you must supply the brains. The search engine will supply the raw computing power. While searching, bookmark any other sites of interest. Here are a few tips.

23.2.2 Use unique, specific terms

It is simply amazing how many web pages are returned when performing a search. You might guess that the terms "hard hat" are relatively specialised. A Google search of those terms returned 65,700,000 results! To reduce the number of pages returned, use *unique* terms that are *specific* to the subject you are researching. "Safety hard hat" reduces this to 93,000. Enter a short question, such as "what time is it?"

23.2.3 Use the minus sign (-) to narrow the search

How many times have you searched for a term and had the search engine return something totally unexpected? Terms with multiple meanings can return a lot of unwanted results. The rarely used but powerful minus sign, equivalent to a Boolean NOT, can remove many unwanted results. For example, when searching for the insect caterpillar, references to the company Caterpillar,

Inc. will also be returned. Use Caterpillar -Inc to exclude references to the company or Caterpillar -Inc -Cat to further refine the search.

23.2.4 Use of + sign

Place a **plus sign** (+) prior to each word to view each "word separately" within your search results, such as +writer+grammar+punctuation.

23.2.5 Use quotation marks for exact phrases

I often remember parts of phrases I have seen on a web page or part of a quotation I want to track down. Using quotation marks around a phrase will return only those exact words in that order. It's one of the best ways to limit the pages returned. Example: "Safe as houses". Of course, you must have the phrase exactly right – and if your memory is as good as mine, that can be problematic.

23.2.6 Don't use common words and punctuation

Common terms like *a* and *the* are called stop words and are usually ignored. Punctuation is also typically ignored. But there are exceptions. Common words and punctuation marks *should* be used when searching for a specific phrase inside quotes. There are cases when common words like *the* are significant. For instance, Raven and The Raven return entirely different results.

23.2.7 Capitalisation

Most search engines do not distinguish between uppercase and lowercase, even within quotation marks. The following are all equivalent:

▶ technology
▶ Technology
▶ TECHNOLOGY
▶ "technology"
▶ "Technology"

23.2.8 Drop the suffixes

It's usually best to enter the base word so that you don't exclude relevant pages. For example, *bird* and not *birds*, *walk* and not *walked*. One exception is if you are looking for sites that focus on the act of walking, enter the whole term *walking*.

23.2.9 Maximise AutoComplete

Ordering search terms from general to specific in the search box will display helpful results in a drop-down list and is the most efficient way to use AutoComplete. Selecting the appropriate item as it appears will save time typing.

The standard Google start page will display a drop-down list of suggestions supplied by the Google search engine. This option can be a handy way to discover similar, related searches. For example, typing in *health and safety* will not only bring up the suggestion *health and safety at work* but also *health and safety legislation* and *health and safety jobs*.

23.2.10 Specific file type search

In Google you can search by specific type of file by writing **filetype:<name_of_type>**, for example: "filetype:pdf" and write the subject that you are going to search then you will only get pdf files on that subject in the results.

23.2.11 Authors, institutions and other sources

Do you know the name of any author(s) working in this subject? By using the author's name, you may retrieve other references to similar work on the subject of your choice.

Is there an institution or competent authority(ies) known to have done some work in this area? Again try using the name; you may retrieve even more references.

Are any journals/indexing/abstracting service(s) specialising in the subject which are known to you? Again you can add these to your search.

Are there any information centre(s) specialising in the subject? This is similar to author searches because these information centres may well have produced a publication on the subject.

23.2.12 Don't forget the phone book

Telephone directories (white pages and yellow pages) are widely available now on the Web. This means that if you know someone's name and what town they live in, you can access their address, phone number, even their age. There are also databases ('reverse look-up') that allow you to type in a phone number and get the name and address of the person who owns the phone.

If you know the address of the person you are seeking, you can easily get a map of their town, street, and neighbourhood on the many web map sites. Some maps are precise enough to show the exact location of the person's home.

23.2.13 Privacy issues

The same things that you can find out about other people, other people can find out about you.

Here's a list of some of the databases someone might access when researching you:

▶ Property sales and ownership;
▶ Land Registry;
▶ Postal service change-of-address records;

23

► Telephone books, past and present;
► Automobile registration records;
► Other vehicle registration, i.e. boats, private airplanes;
► Subscription lists for magazines and periodicals;
► National marketing databases;
► National email directories;
► Voter registration lists;
► Bankruptcy filings;
► Court proceedings/judgements;
► Professional licences;
► Tax records;
► Credit bureau records;
► Domain name registration records.

If you are concerned about your privacy, you can ask to have your personal information removed from web databases. It is difficult to remove all trace of yourself, though. Some events and transactions are legitimately matters of public record, and more of these public records are becoming available every year via the web.

23.3 Some useful websites

1. **British Institute for Occupational Hygienists: www.bohs.org/**
   BIOH is the professional body for Occupational Hygienists in the UK and organises examinations and awards in Occupational Hygiene.

2. **CIS: http://www.ilo.org/safework/cis/ WCMS_CON_TXT_SAF_INF_STT_EN/lang--en/ index.htm** International Occupational Safety and Health Information Centres (CIS) and its National and Collaborating Centres which are located all over the world. The goal of this Network is to help its members and the rest of the world to find information from the participating countries on subjects related to occupational safety and health (OSH). The CIS Centres Information Network can be accessed in three languages (English, French and Spanish).

3. **Department for Communities and Local Government (DCLG): https://www.gov.uk/ government/collections/fire-safety-law-and-guidance-documents-for-business**
   DCLG has responsibility for Fire and has published a series of guides on the Regulatory Reform Fire Safety Order 2005 which can be down loaded free from their website.

4. **European Agency for Safety and Health at Work: http://osha.europa.eu/en**
   This Agency was set up by the European Union to promote the exchange of technical, scientific and economic information between member states. Has links to the European Commission legislation and also to each of the member states' legislation. Also contains links to Australia, Canada, Iceland, Switzerland, Norway, USA legislation. Note that the legislation appears in the language of that country.

5. **European Commission EUR-LEX portal: http:// eur-lex.europa.eu/en/index.htm**
   Details of European legislation, publications including the Official Journal of the European Union which is updated daily, links to other Commission sites.

6. **FIRA International Ltd: http://www.fira.co.uk/**
   FIRA has driven the need for higher standards through testing, research and innovation for the furniture and allied industries. A non-government funded organisation, FIRA is supported by all sections of the industry through the Furniture Industry Research Association, ensuring ongoing research programmes which bring benefits to all. With unrivalled support from across the whole industry, FIRA also has the influence and capability to help shape legislation and regulations.

7. **Fire Protection Association: www.thefpa.co.uk/**
   The Fire Protection Association is the UK's national fire safety organisation. The association works to identify and draw attention to the dangers of fire and the necessary fire prevention measures.

8. **Glossary: www.ilo.org/public/english/ protection/safework/cis/products/safetytm/ glossary.htm**
   Glossary of Terms on Chemical Safety from the International Programme on Chemical Safety (IPCS) which is a joint venture between the United Nations Environment Programme (UNEP), the International Labour Organisation (ILO) and the World Health Organisation (WHO). The language of chemical safety is drawn from many sources. These include medicine, toxicology, pharmacology, epidemiology, ecotoxicity and environmental pollution. Its terminology has developed in an unstructured manner with proliferation into multiple terms, some with overlapping, alternative or even ambiguous meanings. To facilitate international communication and comprehension this Glossary of Terms on Chemical Safety has been produced. It does not claim to be an exhaustive compilation nor a definitive list of approved terms, but because language and terminology are not static the compilers welcome suggestions for additions and improvements.

9. **Hazards Forum: http://hazardsforum.org.uk**
   The Hazards Forum was established in 1989 by the four major engineering institutions, the Institutions of Civil, Electrical, Mechanical and Chemical Engineers. The Forum was set up because of concern about several major disasters, both technological and natural. The Forum believes that there is widespread public misunderstanding of the nature of hazards and the risks they pose. The public reaction to the risks from genetically modified foods and those from smoking are two examples of differing attitudes. These attitudes may result from media attention; suspicion of

assurances of safety bodies with a vested interest; or just ignorance. Whatever the reason, the Forum believes there is a need to fill an educated but unbiased role and this the Forum aspires to do.

10. **Health and Safety Executive (HSE): www.hse. gov.uk**

    The Health and Safety Executive's main website containing general information on their areas of work and the latest news.

11. **Health Development Agency (HDA): www.hda-online.org.uk**

    The HDA is a special health authority that aims to improve the health of people in the UK – in particular to reduce inequalities in health between those who are well off and those on low incomes or reliant on state benefits.

12. **Chartered Institute of Waste Management (CIWM): http://www.ciwm.co.uk**

    The IWM is the professional body which represents over 4,000 waste management professionals – predominantly in the UK but also overseas. The IWM sets the professional standards for individuals working in the waste management industry and has various grades of membership.

13. **Institution of Occupational Safety and Health (IOSH): http://www.iosh.org.uk**

    IOSH is the chartered organisation representing occupational safety and health practitioners predominately in the UK but elsewhere also. IOSH has a valuable website with technical services for members, training courses and many local branch meetings for practitioners.

14. **IOSH International Links (IOSH): http://www. iosh.org.uk**

    IOSH has a comprehensive International site. International development is an integral part of IOSH's vision: 'a world of work which is safe, healthy and sustainable'. We need to improve safety and health at work not only within the UK, but also in developing and newly industrialised countries. The site gives access to linked sites like the ILO, WHO and local OSH organisations in many countries on all continents.

15. **International Labour Organisation (ILO): http:// www.ilo.org/global/lang--en/index.htm**

    (IOSH): http://www.iosh.org.uk The ILO is the global body responsible for drawing up and overseeing international labour standards. Working with its member states, the ILO seeks to ensure that labour standards are respected in practice as well as principle. The International Labour Organisation's site contains the Conventions and Recommendations which are continually being developed. Contains the full text of these documents.

    ILO SafeWork Bookshelf: http:// www.ilo.org/safework_bookshelf/ english?d&nd=170000102&nh=0. The SafeWork

Bookshelf is a collection of key OSH documents produced, in whole or in part, by the ILO. It was compiled by CIS, the information arm of the SafeWork Programme of the ILO.

16. **ISO (International Organisation for Standardisation): http://www.iso.org/iso/ home.htm**

    ISO is the world's largest developer and publisher of International Standards.

    ISO is a network of the national standards institutes of 162 countries, one member per country, with a Central Secretariat in Geneva, Switzerland, that coordinates the system.

17. **Legislation – UK: http://www.legislation.gov.uk**

    This website is managed by The National Archives on behalf of HM Government. Publishing all UK legislation is a core part of the remit of Her Majesty's Stationery Office (HMSO), part of The National Archives, and the Office of the Queen's Printer for Scotland. These are freely available in full text. All acts from 1987 and Statutory Instruments from 1988 are available on the website. There are separate pages for Scottish, Northern Ireland and Welsh legislation.

18. **Legislation – Other countries: www.gksoft. com/govt/en**

    Governments around the world on the Web. Lists government websites from over 220 countries. Classified under Worldwide Governments, European Governments, African Governments, American Governments, Asian Governments, Oceanian Governments. There are some multi-government institutions also listed.

19. **National Examination Board in Occupational Safety and Health (NEBOSH): http://www. nebosh.org.uk**

    NEBOSH is the main accredited awarding body for Level 6 and Level 3 health and safety qualifications. Their awards are run by over 400 training organisations.

20. **The National Safety Council USA: http://www. nsc.org/Pages/Home.aspx**

    The National Safety Council saves lives by preventing injuries and deaths at work, in homes and communities, and on the roads, through leadership, research, education and advocacy.

21. **OSH World: www.oshworld.com**

    List of organisations, governments, associations, trade and technical associations, trade unions listed under country and then alphabetically by name. Major organisations listed in each country including the ILO Health and Safety Information Centres.

22. **OSH Links** (formerly known as the Internet Directory): **http://www.ccohs.ca/oshlinks/about. html**

    An extensive resource that can be used to find primarily Canadian websites that cover occupational health and safety subjects.

23

23. **Royal Society for the Prevention of Accidents (RoSPA): http://www.rospa.com/**
The Royal Society for the Prevention of Accidents is a registered charity established more than 90 years ago that aims to campaign for change, influence opinion, contribute to debate, educate and inform – for the good of all. By providing information, advice, resources and training, RoSPA is actively involved in the promotion of safety and the prevention of accidents in all areas of life – at work, in the home, on the roads, in schools, at leisure and on (or near) water. RoSPA's mission is to save lives and reduce injuries.

24. **Safety Health and Environment Intra Industry Benchmarking Association (SHEiiBA): www.sheiiba.com**
The Safety Health and Environment Intra Industry Benchmarking Association offers web-based benchmarking tools designed for intra company knowledge sharing and performance.

25. **Specifications and Standards**
American National Standards Institute: www.ansi.org
British Standards Institution: www.bsi-global.com
European Committee for Standardisation: www.cenorm.be
ILI – Informe London Information: www.ili.co.uk
International Standards Organisation: www.iso.ch

26. **WorkSafe Western Australia:** http://www.commerce.wa.gov.au/WorkSafe/
WorkSafe is a division of the Department of Commerce, the Western Australian State Government agency responsible for the administration of the Occupational Safety and Health Act 1984. The principal objective of the Act is to promote and secure the safety and health of people in the workplace.

27. **World Health Organisation (WHO): http://www.who.int/about/en**
WHO is the directing and coordinating authority for health within the United Nations system. It is responsible for providing leadership on global health matters, shaping the health research agenda, setting norms and standards, articulating evidence-based policy options, providing technical support to countries and monitoring and assessing health trends.

23.4 Health and safety forms

All the forms illustrated are available on the companion website in Microsoft Word format for readers to download and use.

23.4.1 Management

| M1 GENERAL HEALTH & SAFETY RISK ASSESSMENT   EXAMPLE 1 | No. | |

Firm/Company		Department	
Contact name		Nature of Business	
Telephone no.			

Principal Hazards

1.
2.
3.
4.
5.

Persons at Risk

Main Legal Requirements

1.
2.
3.
4.
5.

Significant Risks	**Consequences**
1.	
2.	
3.	
4.	
5.	

Existing Control Measures

1.
2.
3.
4.
5.

Residual risk, i.e. after controls are in place

| Severity | | Likelihood | | Residual Risk = SxL | |

Information - relevant HSE and trade publications

Comments from Line Manager	**Comments from Risk Assessor**
Signed	Date	Signed	Date
Review Date			

M1 – Risk assessment 1

23

M2 RISK ASSESSMENT REPORT FORM EXAMPLE 2

Firm/Company			Date of Assessment				
Assessor			Date of review				

Hazards	Persons Affected	Risks	Initial Risk Level	Existing Controls	Additional Controls	Action by whom?	Action by when?	Done

M2 – Risk assessment 2

M3 CONTRACTORS' RISK ASSESSMENT EXAMPLE FOR CONFINED SPACES

INITIAL RISK ASSESSMENT DATE

	Significant Hazards	Persons at Risk	Low	Medium	High	Residual risk
1	Poisoning from toxic gases	Employees			√	Low
2	Asphyxiation – lack of oxygen	Employees			√	Low
3	Explosion	Employees			√	Low
4	Fire	Employees			√	Low
5	Excessive heat	Employees		√		Medium
6	Drowning	Employees			√	Low
7						
8						

ACTION ALREADY TAKEN TO REDUCE THE RISKS

Compliance with:

HSE Guidance L101 - Entry into Confined Spaces
Local Authority/client safety standards, e.g. on sewer entry.
Entry into Confined Spaces Regulations 1997
The Construction (Design and Management) Regulations 2015
Dangerous Substances and Explosive Atmospheres Regulations 2002

Planning:

Eliminate need for entry or use of hazardous materials by selection of alternative methods of work or materials.
Assessment of ventilation available and possible local exhaust ventilation requirements, potential presence of hazardous gases/atmosphere, process by-products, need for improved hygiene/welfare facility.
Prepare a method statement for the work.

Physical:

Documented entry system will apply, with a Permit to Work and method statement. Adequate ventilation will be present or arranged. Detection equipment will be present before entry to check on levels of oxygen and presence of toxic or explosive substances. The area will be tested before entry and continually during the presence of persons in the confined space. Breathing apparatus or airlines will be provided if local ventilation is not possible. Where no breathing apparatus is assessed as being required, emergency BA and rescue harnesses will be provided. Rescue equipment including lifting equipment, resuscitation facilities, safety lines and harnesses will be provided. A communication system with those in confined space will be established. Air will not be sweetened with pure oxygen. Precautions for safe use of any plant or heavier-than-air gases in the confined space must be established before entry. Necessary PPE and hygiene facilities will be provided for those entering sewers.

Managerial/Supervisory:

The management role is to decide on nature of the confined space and to put a safe system into operation, including checking the above and preparing a method statement. Flood potential and isolations must be checked.

Training:

Full training required for all entering and managing confined spaces. Rescue surface party to be trained, including first aid and operation of testing equipment. All operatives must be certified as trained and supervisory staff trained to the same standard. All persons entering a confined space should be physically fit.

Site Manager's Comments

RISK RE-ASSESSMENT DATE

M3 – Risk assessment – example confined space

M4 CONTRACTORS' RISK ASSESSMENT EXAMPLE FOR WORK ON FRAGILE ROOFS

INITIAL RISK ASSESSMENT

	Significant Hazards	Persons at Risk	Low	Medium	High	Residual risk
1	Falls of persons through materials	Employees and Public below			√	Medium
2	Access across fragile material	Employees		√		Low
3						
4						
5						
6						

ACTION ALREADY TAKEN TO REDUCE THE RISKS

Compliance with:

Lifting Operations & Lifting Equipment Regulations 1998 (LOLER)
Provision and Use of Work Equipment Regulations 1998 (PUWER)
Work at Height Regulations 2005
HSE Guidance HSG 150 – Health and Safety in Construction
HSE Guidance HSG 33 – Safety in Roof work
Construction (Design and Management) Regulations 2015.

Planning:

Fragile materials will be identified before work begins. In each case, a risk assessment will be made to provide a safe system of work taking account of the work to be done, access/egress requirements and protection of the area beneath the work area. A method statement will be prepared.

Physical:

Suitable means of access will be provided, such as roof ladder, crawling boards, scaffolding or mobile elevating work platform. Handrails will be fitted. Where access is possible alongside fragile materials such as roof lights, covers will be provided or the fragile material will be fenced off. Soft landing airbags or catch nets will be provided as appropriate. Barriers and signs will be provided so as to isolate the area below fragile materials while work is in progress. No person is permitted to walk on suspected fragile materials for any purpose, including access and surveying.

Managerial/Supervisory:

The role of management is to prepare a method statement prior to commencement of the work, and to arrange for provision of suitable access equipment and trained personnel as required by the safe system devised. Managers must check risk assessments and method statements supplied by subcontractors and others, including the self-employed, to ensure that the proposed work method is safe.

Training:

All operatives must be given specific instructions on the system of work to be used in each case. Selection may be required of operatives who have experience of the work and are physically fit. MEWP operators must be fully trained with a suitable certificate of competence.

Site Manager's Comments

RISK RE-ASSESSMENT DATE	

M4 – Risk assessment – example fragile roof

M5 WORKPLACE INSPECTION REPORT FORM

Company/Organisation	
Work area covered by this inspection	
Activity carried out in work place	
Person carrying out inspection	Date of inspection

Observations	Priority/risk (H,M,L)	Actions to be taken (if any)	Time Scale
List hazards, unsafe practices and good practices		List all immediate and longer term actions required	Immediate, 1 week etc.

M5 – Workplace Inspection

M6 WORKPLACE INSPECTION CHECKLIST

PREMISES

1	**Work at Height**	Ladders & Step ladders	Right equipment for the job? Level base? Correct angle? Secured at top and bottom? Equipment in good condition?
		Working platforms/ temporary scaffolds	Regularly inspected? Suitable for the task? Properly erected? Maintained and inspected?
		Use of mobile elevating work platforms (MEWPs)	Suitable for task? Operators properly trained? Properly maintained?
2	**Access**	Access ways	Adequate for people, machinery and work in progress? Unobstructed? Properly marked? Stairs in good condition? Handrails provided?
3	**Working Environment**	Housekeeping	Tidy, clean, well organized?
		Flooring	Even and in good condition? Non-slippery? Crowded?
		Comfort /health	Too hot/cold? Ventilation? Humidity? Dusty? Lighting? Slip risk controlled?
		Cleaning	Hygienic conditions? Normal conversation possible?
		Noise	Noise assessment needed/not needed? Noise areas designated? Tasks require uncomfortable postures or actions?
		Ergonomics	Frequent repetitive actions accompanying muscular strain? Workstation assessments needed/not needed?
		Visual display units	Chairs adjustable/comfortable/maintained properly? Cables properly controlled? Lighting OK? No glare? Washing and toilet facilities satisfactory?
4	**Welfare**	Toilets/Washing	Kept clean, with soap and towels/ Adequate changing facilities? Clean and adequate/Means of heating food?
		Eating facilities	For pregnant or nursing mothers?
		Rest room	Kept clean? Suitably placed and provisioned?
		First aid	Appointed person? Trained first aider? Correct signs and notices? Eye wash bottles as necessary? Portable equipment tested?
5	**Services**	Electrical equipment	Leads tidy not damaged? Fixed installation inspected? Equipment serviced annually?
		Gas	Hot and cold water provided?
		Water	Drinking water provided?
6	**Fire Precautions**	Fire extinguishers	In place? Full? Correct type? Maintenance contract?
		Fire instructions	Posted up? Not defaced or damaged?
		Fire alarms	Fitted and tested regularly?
		Means of escape/ Fire exits	Adequate for the numbers involved? Unobstructed? Easily Opened? Properly Signed?
7	**Work Equipment**	Lifting equipment	Thoroughly examined? Properly maintained? Slings etc properly maintained? Operators properly trained? Written schemes for inspection?
		Pressure systems	Safe working pressure marked? Properly maintained?

M6 – Workplace inspection checklist

M6 WORKPLACE INSPECTION CHECKLIST (Cont.)

PREMISES

7	**Work Equipment (cont.)**	Sharps	Safety knives used? Knives/needles/glass properly used/disposed of?
		Vibration	Any vibration problems with hand held machinery or with whole body from vehicle seats etc?
		Tools and equipment	Right tool for the job? In good condition?
8	**Manual & Mechanical Handling**	Manual handling	Moving excessive weight? Assessments carried out? Using correct technique? Could it be eliminated or reduced?
		Mechanical handling	Forklifts and other trucks properly maintained? Drivers authorized and properly trained? Passengers only where specifically intended with suitable seat?
9	**Vehicles**	On site	Speeding limits? Following correct route? Properly serviced? Drivers authorized
		Road risks	Suitable vehicles used? No use of mobile phones when driving? Properly serviced? Schedules managed properly?
10	**Dangerous substances**	Flammable liquids and gases	Stored properly? Used properly/minimum quantities in workplace? Sources of ignition? Correct signs used?
11	**Hazardous substances**	Chemicals	COSHH assessments OK? Exposures adequately controlled? Data sheet information available? Spillage procedure available? Properly stored and separated as necessary? Properly disposed of?
		Exhaust ventilation	Suitable and sufficient? Properly maintained? Inspected regularly?

PROCEDURES

12	**Risk assessments**		Carried out? General and fire? Suitable and sufficient?
13	**Safe systems of Work**		Provided as necessary? Kept up to date? Followed?
14	**Permits to work**		Used for high risk maintenance? Procedure OK? Properly followed?
15	**Personal Protective Equipment**		Correct type? Worn correctly? Good condition?
16	**Contractors**		Is their competence checked thoroughly? Are there control rules and procedures? Are they followed?
17	**Notices, Signs and Posters**	Employers' liability insurance	Notice displayed? In date?
		Health and Safety law poster	Displayed? Latest design?
		Safety Signs	Correct type of sign used? Signs in place and maintained?

PEOPLE

18	**Health surveillance**		Specific surveillance required by law? Stress or fatigue?
19	**People's behaviour**		Are behaviour audits carried out? Is behaviour considered in the safety programme?
20	**Training and supervision**		Suitable and sufficient? Induction training? Refresher training?
21	**Appropriate authorised person**		Is there a system for authorising people for certain special tasks like permits to work, dangerous machinery, entry into confined spaces?
22	**Violence**		Any violence likely in workplace? Is it controlled? Are there policies in place?
23	**Especially at risk categories**	Young persons	Employed? Special risk assessments?
		New or expectant mothers	Employed? Special risk assessments?

23

	M7 JOB SAFETY ANALYSIS				

Job | | **Date** |
Department | | **Carried out by** |

Description of job

Legal requirements and guidance

Task steps	Hazards	Likelihood	Severity	Risk L x S	Controls

Safe systems of work

Job Instruction

Training requirements

Review Date

M7 – Job safety analysis

M8 ESSENTIAL ELEMENTS — PERMIT TO WORK

1. Permit Title		2. Permit No.	

3. Job Title

4. Plant Identification

5. Description of Work

6. Hazard Identification

7. Precautions Necessary	7. Signatures

8. Protective Equipment

9. Authorization	
10. Acceptance	
11. Extension/Shift Handover	
12. Hand back	
13. Cancellation	

NOTES REFER TO BOX NO ON FORM
2. Should have any reference to other relevant permits or isolation certificates.
5. Description of work to be done and its limitations.
6. Hazard identification including residual hazards and hazards introduced by the work.
7. Precautions necessary – person(s) who carry out precautions, e.g. isolations, should sign that precautions have been taken.
8. Protective equipment needed for the task.
9. Authorization signature confirming that isolations have been made and precautions taken, except where these can only be taken during the work. Date and time duration of permit.
10. Acceptance signature confirming understanding of work to be done, hazards involved and precautions required. Also confirming permit information has been explained to all workers involved.
11. Extension/Shift handover signature confirming checks have been made that plant remains safe to be worked on, and new acceptance/workers made fully aware of hazards/precautions. New time expiry given.
12. Hand back signed by acceptor certifying work completed. Signed by issuer certifying work completed and plant ready for testing and re-commissioning.
13. Cancellation certifying work tested and plant satisfactorily recommissioned

Source HSE

M8 – Permit to work

M9 WITNESS STATEMENT FORM

Type of event		Date of occurrence	
Location of event			
Surname			
Forename(s)			
Address			
Telephone No.			
Employer			
Position (if applicable)			

Please provide an account of the event, as you saw or heard it arise, that may be of use in the incident/near miss investigation

Signed			Date	

M9 – Witness statement

M10 ACCIDENT/INCIDENT REPORT

Injured Person		Date		Time	
Position		Place of Accident			
Department		Details of Injury			
Investigation carried out					
Position		Date		Estimated absence	

Brief details of Accident (Diagrams, photographs and any witness statements should be attached where necessary)

Immediate Causes	Underlying or Root causes

Conclusions (How can this kind of incident/accident be prevented?)

Action to be taken		Completion Date	

Please ensure that an accident investigation and report is completed and forwarded to Personnel within 48 hours of the accident occurring. Some Incidents e.g. involving death or major injuries or dangerous occurrences may have to be notified to the Enforcing Authority. (Mandatory in the UK)

Signature of Manager making Report		Date	
Copies to go to	Personnel Manager		Health & Safety Manager		Payroll Controller	

INJURED PERSON

Surname		Forenames	

Home address

Age		Male		Female		Occupation	

By ticking this box I give consent to my employer and Best Practice Co(if not employer) to disclose my personal information and details of the incident which appear on this form to safety representatives and representatives of employee safety for them to carry out the health and safety functions given to them by law. (May not apply outside Europe)

Signature of injured person		Date	

Employee		Agency Temp		Contractor		Visitor		Youth Trainee	

M10 – Accident report 1

23

KIND OF ACCIDENT Indicate what kind of accident led to injury or condition (tick one box)

Contact with moving machinery or material being machined	1	Injured whilst handling or lifting	5	Drowning or Asphyxiation	9	Contact with electricity or an electrical discharge	13
Struck by moving, including flying, or falling object	2	Slip, trip or fall on same level	6	Exposure to or contact with harmful substance	10	Injured by an animal	14
Struck by moving vehicle	3	Fall from height. Indicate approx. distance of fall……mtrs	7	Exposure to fire	11	Violence	15
Struck against something fixed or stationery	4	Trapped by something collapsing or overturning	8	Exposure to an explosion	12	Other kind of accident	16

Detail any machinery, chemicals, tools etc. involved

Accident first reported to:

Position & Dept.		Name	
First Aid/medical attention by	First Aider Name		Dept.
	Doctor Name	Medical Centre	Hospital

WITNESSES

Name	Position and Department	Statement obtained (attach all statements)	
		Yes	No
		Yes	No
		Yes	No
		Yes	No
		Yes	No

For Office Use Only (if relevant)

Date reported to Enforcing Authority	by telephone		On form F2508		by email	
Date reported to Company Insurers		Were the recommendations effective?		

If No say what further action should be taken:

M10 – (*Continued*)

First Aid Treatment and Accident Record

BEST PRACTICE COMPANY logo goes here

Date: _____ FIRST AIDER AND/OR INJURED PERSON **No:** _____

Part A: FIRST AID TREATMENT/EXAMINATION RECORD

1. About the person who was seen/treated

Name _____ Address _____

Postcode _____

Occupation _____ Employer _____

2. About you, the person filling in this record

▼ If the incident **DID NOT** happen to you fill in your name, address and occupation

Name _____ Address _____

Postcode _____

Occupation _____

3. Treatment given (if any)

please provide details _____
_____
_____
_____

Part B: INCIDENT AND INJURY DETAILS (if believed to be a work related incident complete the following)

4. Details of incident

▼ Say when it happened: Date ____ / ____ / ____ Time _____

▼ Say where it happened. State which room or area _____

▼ Say how the incident happened – give the cause if you can _____

▼ If the person who had the incident suffered an injury, say what it was _____

▼ Please sign the record and date it

Signature _____ Date ____ / ____ / ____

5. Further details of injury (if any)

☐ Amputation	☐ Burn - Heat	☐ Electric shock	☐ Puncture
☐ Break/Fracture	☐ Concussion/headache	☐ Foreign body	☐ Scald
☐ Bruises (contusions)	☐ Crush	☐ Graze	☐ Splinters and blisters
☐ Burn - Chemical	☐ Cuts (laceration)	☐ Multiple Injury	☐ Sprain
			☐ Strain

M11 – First aid treatment and accident record 2 revised Jan 2015

23

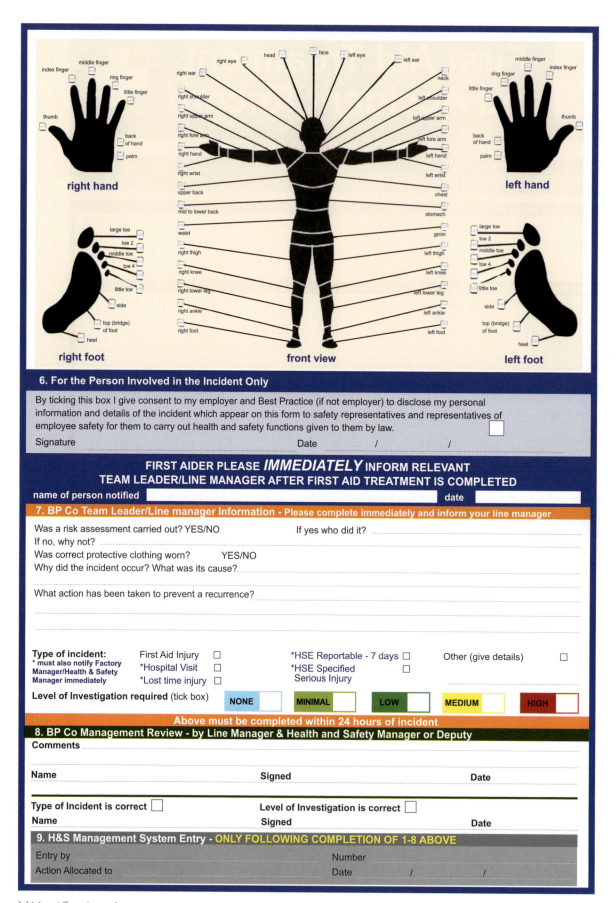

right hand

left hand

right foot

front view

left foot

6. For the Person Involved in the Incident Only

By ticking this box I give consent to my employer and Best Practice (if not employer) to disclose my personal information and details of the incident which appear on this form to safety representatives and representatives of employee safety for them to carry out health and safety functions given to them by law.

Signature   Date   /   /

FIRST AIDER PLEASE *IMMEDIATELY* INFORM RELEVANT TEAM LEADER/LINE MANAGER AFTER FIRST AID TREATMENT IS COMPLETED

name of person notified   date

7. BP Co Team Leader/Line manager Information - Please complete immediately and inform your line manager

Was a risk assessment carried out? YES/NO   If yes who did it?

If no, why not? ....

Was correct protective clothing worn?   YES/NO

Why did the incident occur? What was its cause?.........

What action has been taken to prevent a recurrence?

Type of incident: First Aid Injury ☐   *HSE Reportable - 7 days ☐   Other (give details)   ☐
* must also notify Factory Manager/Health & Safety Manager immediately   *Hospital Visit ☐   *HSE Specified ☐
                                   *Lost time injury ☐   Serious Injury

Level of Investigation required (tick box)   NONE | MINIMAL | LOW | MEDIUM | HIGH

Above must be completed within 24 hours of incident

8. BP Co Management Review - by Line Manager & Health and Safety Manager or Deputy

Comments ....

Name   Signed   Date

Type of Incident is correct ☐   Level of Investigation is correct ☐

Name   Signed   Date

9. H&S Management System Entry - ONLY FOLLOWING COMPLETION OF 1-8 ABOVE

Entry by   Number

Action Allocated to   Date   /   /

M11 – (*Continued*)

23.4.2 Safety

S1 MACHINERY RISK ASSESSMENT							
MACHINE NO	Model		Manufacturer		Risk Assessment No		
	Model no.		Other		Department ID		

HAZARDS	Hazard	**Yes**	**No**	Hazard	**Yes**	**No**
	Trapping/drawing in			Stored energy/vacuum/kinetic		
	Impact/crushing/mobility			Electrical/radiation		
	Cutting/severing/sharp edges			Thermal/noise/vibration		
	Entanglement/ejection			Material/sunstance		
	Rough/slippery surface/falling			Environment/ergonomic		

WHO MIGHT BE HARMED?	Operatives		Cleaners		Maintenance		Visitors		Others	

GUARDING	**Fixed Guards**		**Interlocked Movable Guards**		**Adjustable Guards**		**Fixed Distance Guard**	
	Adequate Enclosure of drives and motors		Fitted to machine		Fitted to machine		Fitted to machine	
			Closed to run		Readily Adjustable		Securely Fixed	
	Securely fixed in position		Design OK with positive switches and robust		Prevent Ejection		Prevents access to danger zone	
	Tools required to remove		Safe to open		Maintained OK		Tool to remove	
			Maintained OK				Maintained OK	

CONTROLS/ WARNINGS/ INSTRUCTIONS/ TRAINING	**Controls**				**Instruction Training**		**Area**	
	Clearly identified		Shrouded start		Warning signs		Lighting OK	
	Function Clear		Warning device		Signs clear		Stability OK	
	Easy to use		Isolator nearby		Safety sheet OK		Ventilation OK	
	Emergency stop		Isolator lockable		SOP available		LEV needed	
	Machine stops OK		Permit required		Training OK		Access operators	
	Safe at stop		Seating needed				Access maintenance	

ACTION REQUIRED

Severity		Frequency		Residual Risk (SxE)	

S1 – Risk assessment machinery revised Jan 2015

ASSOCIATED PERMITS (specify type, if none state NA)		ON THE JOB COPY	BEST PRACTICE COMPANY logo goes here	GENERAL WORK PERMIT	Number GWP 00000000
Type	Prefix & No.				
Type	Prefix & No.				
Type	Prefix & No.				

SECTION 1 ISSUE   PLEASE PRINT DETAILS

Permit Receiver/Competent Person in charge of work	Location of work/equipment to be worked on				
Name of person detailed to carry out work	Details of work to be done				
		Yes	No	NA	Reference
Risk Assessment attached? Yes [ ] No [ ] Reference	Safety method Statement or Safe System of Work attached?	[ ]	[ ]	[ ]	

SECTION 2 ISOLATION of electrical or mechanical plant, liquid or gas pipeline or other energy source – give details

Item	Lock location & reference	Isolated by: Print Name
Attached Isolation sheet? Yes [ ] No [ ] Reference	Location of Keys	

SECTION 3 PREPARATIONS/PRECAUTIONS tick box

	Yes	No	NA		Yes	No	NA
1. Has every source of energy been isolated?				5. Are vessels/pipes free of toxic/flammable gas			
2. Have all isolations been tagged?				6. Any asbestos containing materials present?			
3. Have all isolations been tested?				7. Are the risk assessment control measures in place?			
4a. Does the standard of pipeline isolation meet minimum Corporate standards?				8. Other precautions/control measures required (specify)			
4b. If not, specify additional precautions							

SECTION 4 PERSONAL PROTECTIVE EQUIPMENT tick box

Site Standard		Other specify	
Safety Goggles			
Hearing Protection			
Respiratory Protection	Specify Type		

SECTION 5 TIME LIMITS

From	hrs	On	/	/	To	hrs	On	/	/

Max 24 hours   See overleaf for details of time extensions allowed

SECTION 6 AUTHORISATION Permit Issuer/Authorised Person	SECTION 7 RECEIPT Permit Receiver/Competent Person in charge
I certify that it is safe to work in the area/on the equipment detailed in Section 1 above and that all safety measures detailed in Section 2-4 have been carried out/complied with. ALL OTHER PARTS ARE DANGEROUS	I have read, understood and accept the requirements of this permit. I will ensure that everyone working under my supervision will strictly follow the requirements of this permit. I have checked the isolations.
Print name Date Time	Print Name Date Time

SECTION 8 SUSPENSION OF GENERAL WORK PERMIT this is an exception and must be signed on the front of 'on the job'

I certify that the task for which this Permit was issued has now been suspended. We have agreed and implemented a procedure which complies with the criteria noted in the checklist in Section 13 overleaf

Signed Permit receiver/Competent person in charge	Signed Authorised person
Date Time	Date Time

SECTION 9 CLEARANCE Permit Receiver/Competent person in charge	SECTION 10 CANCELLATION Permit Issuer/Authorised Person
I certify that the work for which the permit was issued is now COMPLETED and that all persons at risk have been WITHDRAWN and WARNED that it is NO LONGER SAFE to work on the plant specified on this permit and that GEAR TOOLS and EQUIPMENT are all CLEAR	This permit to work is hereby CANCELLED, all plant is restored to safe operating conditions, including the replacement of guards.
Print name Date Time	Print name Date Time
Signed	Signed

S2a – General work permit

PERMIT TIME EXTENSION/TRANSFER

Time Extensions

General Work Permits are normally only valid for a maximum of 24 hours.

When a permit is extended beyond 24 hours a Permit issuer/Authorised Person must check ALL the conditions in the Permit. When satisfied there have been no changes and it is safe to continue the work specified in Section 1 Time Extension/Transfer Section 11 must be completed. The permit may be continued beyond a shift and be transferred to another Permit Receiver/Competent Person, who must check that all the precautions are in place and that they understand all of the requirements of the permit. Section 11 must be signed by the new Permit Receiver.

General Note

Except for Section 8 Suspension, the front sheet must not be altered or changed in any way. If there is a need to change the permit conditions (except Section 8 overleaf) then a new permit must be used. Extension sheet Mxxx must be used when Section 11 below is full.

SECTION 11 TIME EXTENSION/TRANSFER OPERATIONAL PERIODS

Permit is extended from	Print Name	Date as FRONT of Permit
_____ hrs to _____ hrs	Permit receiver	Signed
_____ hrs to _____ hrs	Permit receiver	Signed
Permit is extended from		**Date**
_____ hrs to _____ hrs	Authorised Person	Signed
_____ hrs to _____ hrs	Permit receiver	Signed
_____ hrs to _____ hrs	Permit receiver	Signed
_____ hrs to _____ hrs	Permit receiver	Signed
Permit is extended from		**Date**
_____ hrs to _____ hrs	Authorised Person	Signed
_____ hrs to _____ hrs	Permit receiver	Signed
_____ hrs to _____ hrs	Permit receiver	Signed
_____ hrs to _____ hrs	Permit receiver	Signed
Permit is extended from		**Date**
_____ hrs to _____ hrs	Authorised Person	Signed
_____ hrs to _____ hrs	Permit receiver	Signed
_____ hrs to _____ hrs	Permit receiver	Signed
_____ hrs to _____ hrs	Permit receiver	Signed
Permit is extended from		**Date**
_____ hrs to _____ hrs	Authorised Person	Signed
_____ hrs to _____ hrs	Permit receiver	Signed
_____ hrs to _____ hrs	Permit receiver	Signed
_____ hrs to _____ hrs	Permit receiver	Signed
Permit is extended from		**Date**
_____ hrs to _____ hrs	Authorised Person	Signed
_____ hrs to _____ hrs	Permit receiver	Signed
_____ hrs to _____ hrs	Permit receiver	Signed
_____ hrs to _____ hrs	Permit receiver	Signed

SECTION 12 ADDITIONAL PEOPLE WORKING ON PLANT

SECTION 13 SUSPENSION OF GENERAL WORK PERMIT check list

Operation of machinery or plant with safeguards removed must be considered a measure of last resort and may only be carried out where it is not practicable to perform tasks, such as examinations or adjustments, in any other manner with the safeguards in place. Should it be necessary to re-energise part of the machinery to which this permit applies before the task is complete a Permit Issuer/Authorised Person must complete the following checklist & sign Section 8.

CHECK LIST	tick box	Yes	NA	CHECK LIST	tick box	Yes	NA
Is number of people involved as small as possible?				Is area free of obstructions & slipping hazards?			
Have precautions been taken to keep people away?				Are all other parts suitably protected?			
Has loose clothing been avoided?				Are suitable precautions against hazards maintained?			
Is safe access with foothold & handhold provided?				Are people competent, briefed and supervised as necessary?			
Is there a means of stopping equipment?							
Check list completed by: print name				Signed Date			

S2b – General work permit: back of form

23.4.3 Health

	Site	Best Site	Work Area	Laboratory	COSHH Assessment No.	001

Substance / Material: Task / Activity : Hazardous Content: Health Hazard Groups: A / B / C / D / E	FORMALDEHYDE Laboratory Work Formaldehyde* (40%), Methanol (10%). Group "C"

No of Personnel involved in Activity:	One Person
Frequency of Operation:	Once per day
Duration of Exposure:	Up to ½ Hour Daily
Quantity / Amount Exposed To:	Use of small quantities under laboratory conditions.
Specific Work Area:	Inside well ventilated Area

Procedures Reference: Corporate Minimum Standard, H&SMan-002 Section 2 H&SMan-02 Section 3	Routes of Entry to Body:	INHALATION	✓	SKIN	✓
		INGESTION	✓	EYES	✓

Hazard Classification:		Residual Risk: **HIGH HAZARD** LIQUID	(WEL) Workplace Exposure Limits: **Formaldehyde *2.5mg/m3TWA, 2.5mg/m3STEL WEL [15mins]**

Monitoring Arrangements

If using engineering controls, ensure maintenance (Regulation 9.)

Health Surveillance Requirements

Material contains MEL and skin sensitising assigned substance, consider monitoring and skin checks (Regulation 10&11)

Hazards from Activity / Health Risks

AVOID CONTACT WITH SKIN & EYES
CAUSES BURNS
TOXIC IF SWALLOWED, BY INHALATION AND IN CONTACT WITH SKIN
LIMITED EVIDENCE OF A CARCINOGENIC EFFECT
MAY CAUSE SENSITIZATION BY SKIN CONTACT
RISK OF SERIOUS DAMAGE TO EYES
Do not breathe in vapour
When using do not eat, drink or smoke

Control Measures / Storage

Personnel must wear lab coat or protective overalls when working with substance
Chemical Resistant Gloves must be worn [Nitrile]
Eye Protection to BS EN 166 Grade 3
Wear rubber boots
General or local exhaust ventilation should be used to control airborne levels
Store away from oxidising agents, stored at a temperature sufficiently high to avoid the formation of polymers

PPE / Workplace Environment Control Measure Symbols

Chemical Overalls   Glasses   Chemical   Dust Mask

First Aid

INHALATION - REMOVE TO FRESH AIR AND REST
IN THE EVENT OF **SIGNIFICANT EXPOSURE** CALL FOR MEDICAL ASSISTANCE IMMEDIATELY
INGESTION - DO NOT INDUCE VOMITING
INGESTION - GIVE PLENTY OF WATER IN SIPS
INGESTION - GET IMMEDIATE MEDICAL ATTENTION
EYE - IRRIGATE WITH WATER FOR AT LEAST 15 MINUTES
EYE CONTACT - GET IMMEDIATE MEDICAL ATTENTION
SKIN - REMOVE CLOTHING & WASH CONTAMINATED AREA WITH PLENTY OF WATER
SKIN CONTACT - GET IMMEDIATE MEDICAL ATTENTION

Spillage and Disposal

VENTILATE AREA
DO NOT ALLOW SPILLAGE TO ENTER MAINS DRAIN/SEWERS/WATER COURSES
MARK THE AREA AND WARN ALL PERSONNEL
WEAR NITRILE GLOVES
WEAR EYE GOGGLES (GRADE 3)
WEAR RPE WITH ORGANIC FILTER (A)
WEAR PROTECTIVE OVERALLS & CHEMICAL/SAFETY FOOTWEAR
ABSORB IN SAND OR INERT ABSORBENT MATERIAL
COLLECT INTO A CONTAINER, CLOSE LID
DISPOSE OF FOLLOWING SUITABLE SITE PROCEDURE OR SEEK L.A. GUIDANCE

Supplier: **Chemical distribution Ltd** Address: **Bridge Way, Anytown** Post Code: **AN19 1SW** Telephone: **+44 (0) 1206 543210** Emergency Tel: **01234 567891** MSDS Reference: Link to Manufactures Safety Data Sheet

Fire and Emergency

ISOLATED SMALL SCALE FIRE:
DRY POWDER – WATER SPRAY OR MIST – FOAM – CARBON DIOXIDE (CO2)
LARGE FIRE: EVACUATE AREA, CALL FIRE BRIGADE OR FOLLOW SITE PROCEDURE
WEAR SELF-CONTAINED BREATHING APPARATUS AND PROTECTIVE CLOTHING
TOXIC FUMES ARE PRODUCED WHEN SUBSTANCE IS INVOLVED IN A FIRE
KEEP CONTAINERS COOL WITH WATER SPRAY

Assessor Name	A.Another
Date of Assessment	10.01.11
Approved By	A.Another
Date Re-Assessment Due	10.01.13
Assessment Printed Date	10.01.12
Assessment Revision No	01

H1a – COSHH assessment example

	Site		Work Area		COSHH Assessment No.	

Substance / Material:
Task / Activity :
Hazardous Content:

Health Hazard Groups:
A / B / C / D / E

No of Personnel involved in Activity:
Frequency of Operation:
Duration of Exposure:
Quantity / Amount Exposed To:
Specific Work Area:

Procedures Reference:

Routes of Entry to Body:

INHALATION		SKIN	
INGESTION		EYES	

Hazard Classification:

Residual Risk:

(WEL) Workplace Exposure Limits:

Monitoring Arrangements:

Hazards from Activity / Health Risks

Control Measures / Storage

First Aid

PPE / Workplace Environment Control Measure Symbols

Spillage and Disposal

Supplier:
Address:

Post Code:
Telephone:
Emergency Tel:
MSDS Reference:
Link to Manufactures Safety Data Sheet

Fire and Emergency

Assessor Name	
Date of Assessment	
Approved By	
Date Re-Assessment Due	
Assessment Printed Date	
Assessment Revision No	

H1 – COSHH assessment record form example 1

23

COSHH 1 – DETAILS OF SUBSTANCES USED OR STORED

Name of Manager:

Name of Department Area:

SUBSTANCE DETAILS

1. Information from the label

Trade name:	Manufacturer's name:

Name of many chemical constituents listed:

Hazard marking...	corrosive		irritant		harmful		toxic		very toxic	

Risk phrases or Hazard Statements noted on label (eg Harmful in contact with skin)

....

....

....

....

Safety phrases or Precautionary Statements noted on label (eg Avoid contact with skin)

....

....

....

....

PRECAUTIONS noted on label (eg Use in well ventilated area)

....

....

....

....

2. Have you got Safety Data Sheet or MSDS for this product	**YES**		**NO**	

DETAILS OF USE

3. What is it used for?

....

....

....

....

4. By whom?

5. How often?

6. Where?

7. What CONTROL measures (precautions) are used?
  (e.g. local ventilation, goggles, respirator, protective gloves etc)

....

....

....

....

8. Is it ABSOLUTELY ESSENTIAL to keep/use this substance?	**YES**		**NO**	
9. Can it be DISPOSED OF NOW?	**YES**		**NO**	

H2 – COSHH assessment record form example 2, page 1

COSHH 2 – ASSESSMENT OF A SUBSTANCE

1. Name of Substance:

2. The process or description of job where the substance is used:

3. Location of the process where the substance is used:

4. Health & Safety information on substance:

a) Hazards to health:

b) Precautions required:

5. Number of persons exposed:

6. Frequency and duration of exposure:

7. Control measures that are in use:

8. The assessment, an evaluation of the risks to health:

9. Details of steps to be taken to reduce the exposure:

10. Action to be taken (name):	**Date:**

11. Date of next assessment/review:

12. Name and position of person making this statement:

13. Date of assessment:

H2 – COSHH assessment record form example 2, page 2

EXAMPLE OF A WORKSTATION SELF ASSESSMENT CHECKLIST

Name	Department	Date

The completion of this checklist will enable you to carry out a self-assessment of your own workstation. Your views are essential in order to enable us to achieve our objective of ensuring your comfort and safety at work. Please circle the answer that best describes your opinion, for each of the questions listed.
The form should be returned to...as soon as it has been completed.

ENVIRONMENT

1. Lighting

Describe the lighting at your workstation.	About right		Too bright		Too dark	
Do you get distracting reflections on your screen?	Never		Sometimes		Constantly	
What control do you have over local lighting?	Full control		Some control		No control	

2. Temperature and Humidity

At your workstation, is it usually:	Comfortable		Too warm		Too cold	
Is the air around your workstation:	Comfortable		Too dry			

3. Noise

Are you distracted by noise from work equipment?	Never		Occasionally		Constantly	

4. Space

Describe the amount of space around your workstation.	Adequate		Inadequate			

FURNITURE

5. Chair

Can you adjust the height of the seat?		Yes		No
Can you adjust the height and angle of the backrest? Is the chair stable?		Yes		No
Is the chair stable?		Yes		No
Does it allow movement?		Yes		No
Is the chair in a good state of repair?		Yes		No
If your chair has arms, do they get in the way?		Yes		No

6. Desk

Is the desk surface large enough to allow you to place all your equipment where you want it?		Yes		No
Is the height of the desk suitable?	Yes		Too high	Too low
Does the desk have a matt surface (non-reflectant)?		Yes		No

7. Footrest

If you cannot place your feet flat on the floor whilst keying, has a footrest been supplied?		Yes		No

H3 – Workstation self assessment checklist example, page 1

EXAMPLE OF A WORKSTATION SELF ASSESSMENT CHECKLIST (cont.)

8. Document Holder

If it would be of benefit to use a document holder, has one been supplied?	Yes		No	
If you have a document holder, is it adjustable to suit your needs?	Yes		No	

DISPLAY SCREEN EQUIPMENT

9. Display Screen

Can you easily adjust the brightness and the contrast between the characters on screen and the background	Yes		No	
Does the screen tilt and swivel freely?	Yes		No	
Is the screen image stable and free from flicker?	Yes		No	
Is the screen at a height, which is comfortable for you?	Yes		No	

10. Keyboard

Is the keyboard separate from the screen?	Yes		No	
Can you raise and lower the keyboard height?	Yes		No	
Can you easily see the symbols on the keys?	Yes		No	
Is there enough space to rest your hands in front of the keyboard?	Yes		No	

11. Software

Do you understand how to use the software?	Yes		No	

12. Training

Have you been trained in the use of your workstation?	Yes		No	
Have you been trained in the use of software?	Yes		No	
If you were to have a problem relating to display screen work, would you know the correct procedures to follow?	Yes		No	
Do you understand the arrangements for eye and eyesight tests?	Yes		No	

Any other comments

H3 – workstation self assessment checklist example, page 2

23

733

EXAMPLE OF A NOISE ASSESSMENT RECORD FORM

| Name of Department | | | Date of Survey | |

| Lower Exposure Action Level: 80 dBA daily or weekly | Upper Action Level: 85dBA daily or weekly | Peak Pressure: 135 dB(C)/137dB(C) |

Workplace	Number of persons exposed	Noise level (leg) dB (A)	Daily exposure period	$L_{EP,d}$ dB (A)	Peak pressure dB (C)	Comments

General Comments:

| Instrument used | | Date of last calibration | |
| Signature | | Position | | Date | |

H4 – Noise assessment record form example

MANUAL HANDLING OF LOADS: ASSESSMENT CHECKLIST

Section A – Preliminary	Department:	Date:

TASK DESCRIPTION: Factors beyond the limits of the guidelines?	IS AN ASSESSMENT NEEDED? (ie is there a potential risk of injury, and are the factors beyond the limits of the guidelines?) YES/NO

If 'YES 'continue. If 'NO' the assessment need go no further.

Tasks covered by this assessment (detailed description): Locations: People involved: Date of assessment:	Diagrams and other information:

SECTION B – See separate sheet for detailed analysis

SECTION C – Overall assessment of the risk of injury	Low	Med	High

SECTION D – Remedial action needed

Remedial steps that should be taken, in priority order:

a

b

c

d

e

f

g

h

Date by which action should be taken:

Date for reassessment:

Assessor's name:	Signature

H5 – Manual handling of loads assessment record form

23

MANUAL HANDLING RISK ASSESSMENT: EMPLOYEE CHECKLIST

Task Description				Employee's ID no.	
RISK FACTORS					
A. Task Characteristics	**Yes/No**	**Risk Level**			**Current Controls**
1. Loads held away from trunk?		H	M	L	
2. Twisting?					
3. Stooping?					
4. Reaching upwards?					
5. Extensive vertical movements?					
6. Long carrying distances?					
7. Strenuous pushing or pulling?					
8. Unpredictable movements of loads?					
9. Repetitive handling operations?					
10. Insufficient periods of rest/recovery?					
11. High work rate imposed?					
B. Load characteristics					
1. Heavy?					
2. Bulky?					
3. Difficult to grasp?					
4. Unstable/unpredictable?					
5. Harmful (sharp/hot)?					
C. Work environment characteristics					
1. Postural constraints?					
2. Floor suitability?					
3. Even surface?					
4. Thermal/humidity suitability?					
5. Lighting suitability?					
D. Individual characteristics					
1. Unusual capability required?					
2. Hazard to those with health problems?					
3. Hazard to pregnant workers?					
4. Special information/training required?					
Any further action needed?					
Details:					

H6 – Manual handling risk assessment employee checklist

23.4.4 Fire

F1 FIRE SAFETY MAINTENANCE CHECKLIST

A fire safety maintenance checklist can be used as a means of supporting your fire safety policy. This example list is not intended to be comprehensive and should not be used as a substitute for carrying out a fire risk assessment. You can modify the example where necessary, to fit your premises and may need to incorporate the recommendations of manufacturers and installers of the fire safety equipment/ systems that you may have installed in your premises. Any ticks in the grey boxes should result in further investigation and appropriate action as necessary. In larger and more complex premises you may need to seek the assistance of a competent person to carry out some of the checks.

Source: Department for Communities and Local Government Fire Safety Guides

---

Fire Safety Maintenance Checklist	Yes	No	N/A	Comments
DAILY CHECKS (not normally recorded)				
Escape routes				
Can all fire exits be opened immediately and easily?				
Are fire doors clear of obstructions?				
Are escape routes clear?				
Fire warning systems				
Is the indicator panel showing 'normal'?				
Are whistles, gongs or air horns in place?				
Escape lighting				
Are luminaires and exit signs in good condition and undamaged?				
Is emergency lighting and sign lighting working correctly?				
Fire-fighting equipment				
Are all fire extinguishers in place?				
Are fire extinguishers clearly visible?				
Are vehicles blocking fire hydrants or access to them?				
WEEKLY CHECKS				
Escape routes				
Do all emergency fastening devices to fire exits (push bars and pads, etc.) work correctly?				
Are external routes clear and safe?				
Fire warning systems				
Does testing a manual call point send a signal to the indicator panel? (Disconnect the link to the receiving centre or tell them you are doing a test.)				
Did the alarm system work correctly when tested?				
Did staff and other people hear the fire alarm?				
Did any linked fire protection systems operate correctly? (e.g. magnetic door holder released, smoke curtains drop)				
Do all visual alarms and/or vibrating alarms and pagers (as applicable) work?				
Do voice alarm systems work correctly? Was the message understood?				
Escape lighting				
Are charging indicators (if fitted) visible?				
Fire-fighting equipment				
Is all equipment in good condition?				
Additional items from manufacturer's recommendations.				
MONTHLY CHECKS				
Escape routes				
Do all electronic release mechanisms on escape doors work correctly? Do they 'fail safe' in the open position?				

F1 – Fire safety maintenance checklist, page 1

23

Fire Safety Maintenance Checklist	Yes	No	N/A	Comments
Do all automatic opening doors on escape routes 'fail safe' in the open position?				
Are fire door seals and self-closing devices in good condition?				
Do all roller shutters provided for fire compartmentation work correctly?				
Are external escape stairs safe?				
Do all internal self-closing fire doors work correctly?				
Escape lighting				
Do all luminaires and exit signs function correctly when tested?				
Have all emergency generators been tested? (Normally run for one hour.)				
Fire-fighting equipment				
Is the pressure in 'stored pressure' fire extinguishers correct?				
Additional items from manufacturer's recommendations.				
THREE-MONTHLY CHECKS				
General				
Are any emergency water tanks/ponds at their normal capacity?				
Are vehicles blocking fire hydrants or access to them?				
Additional items from manufacturer's recommendations.				
SIX-MONTHLY CHECKS				
General				
Has any fire-fighting or emergency evacuation lift been tested by a competent person?				
Has any sprinkler system been tested by a competent person?				
Have the release and closing mechanisms of any fire-resisting compartment doors and shutters been tested by a competent person?				
Fire warning system				
Has the system been checked by a competent person?				
Escape lighting				
Do all luminaires operate on test for one third of their rated value?				
Additional items from manufacturer's recommendations.				
ANNUAL CHECKS				
Escape routes				
Do all self-closing fire doors fit correctly?				
Is escape route compartmentation in good repair?				
Escape lighting				
Do all luminaires operate on test for their full rated duration?				
Has the system been checked by a competent person?				
Fire-fighting equipment				
Has all fire-fighting equipment been checked by a competent person?				
MISCELLANOUS				
Has any dry/wet rising fire main been tested by a competent person?				
Has the smoke and heat ventilation system been tested by a competent person?				
Has external access for the fire service been checked for ongoing availability?				
Have any firefighters' switches been tested?				
Has the fire hydrant bypass flow valve control been tested by a competent person?				
Are any necessary fire engine direction signs in place?				

F1 – fire safety maintenance checklist, page 2

F2 FIRE RISK ASSESSMENT RECORD – SIGNIFICANT FINDINGS

Risk Assessment for		
Assessment undertaken by		

Company	Date	
Address	Completed by	
	Sign	
Sheet no.	Floor/Area	Use

Step 1 – Identify Fire Hazards

Sources of ignition	Sources of fuel	Sources of oxygen

Step 2 – People at risk

Step 3 – Evaluate, remove, reduce and protect from risk

Evaluate the risk of the fire occurring

Evaluate the risk to people from a fire starting in the premises

Remove and reduce the hazards that may cause a fire

Remove and reduce the risks to people from a fire

Assessment review

Assessment review date	Completed by	Sign

Review outcome (where substantial changes have occurred a new record sheet should be used)

F2 – Fire risk assessment record form

23.4.5 Construction

C1 CONSTRUCTION INSPECTION REPORT

1. Name and address of person for whom inspection was carried out.

2. Site address.	**3. Date and time of inspection.**

4. Location and description of place of work or work equipment inspected.

5. Matters which give rise to any health and safety risks.

6. Can work be carried out safely?	Yes		No	
7. If not, name of person informed.				

8. Details of any other action taken as a result of matters identified in 5 above.

9. Details of any further action considered necessary.

10. Name and position of person making the report.	
11. Date and time report handed over.	
12. Name and position of person receiving report.	

C1 – Construction inspection report form

C2 EXAMPLE RISK ASSESSMENT FOR CONTRACT BRICKLAYERS

Company name:
TVW Contract Bricklayers

Date of risk assessment:
6/3/2006

IMPORTANT REMINDER

This example risk assessment shows the kind of approach a small business might take. Use it as a guide to think through some of the hazards in your business and the steps you need to take to control the risks. Please note that it is not a generic risk assessment that you can just put your company name on and adopt wholesale without any thought. This would not satisfy the law – and would not be effective in protecting people.

Every business is different – you need to think through the hazards and controls required in your business for yourself. Source HSE

What are the hazards?	Who might be harmed and how?	What are you already doing?	What further action is necessary?	Action by who?	Action by when?	Done
Falling from height	Serious injury or even fatal injury could occur if a worker fall	■ **Agree** scaffolding requirements at contract stage, including appropriate load rating and provision of loading bays. ■ **Bricklayers'** supervisor to check with the site manager that the correct scaffold is provided and inspected. ■ **Workers** instructed not to interfere with or misuse scaffold – supervisor to keep an eye out for problems. ■ **Ladders** in good condition, adequately secured (lashed) and placed on firm surface. ■ **Band** stands with handrails to be used for work on internal walls. ■ **Workers** trained to put up bandstands	■ **Scaffold** requirements agreed, including loading bays and appropriate load rating. ■ **Supervisor** to speak regularly to site manager to arrange scaffold alterations and ensure that weekly inspections have been carried out.	TB LG	20/3/06 from 1/5/06	20/3/06
Collapse of scaffold	Operatives on scaffold may incur crush injuries, or worse, if the scaffold collapses on top of them.	■ **Agree** scaffolding requirements at contact stage, including appropriate loading bays. ■ **Bricklayers'** supervisor to check with the site manager that the correct scaffold is provided and inspected.	■ **Supervisor** to keep a check to make sure that scaffold is not overloaded with materials.	LG	from1/5/06	
Falling objects hitting head or body, including feet	Serious head and other injuries to workers, others on site and members of the public.	■ **Brick guards** kept in position on scaffold lifts. ■ **Waste materials** removed from scaffolding and placed in skip. ■ **Safety helmets** and protective footwear (with steel toecaps and mid-soles) supplied and worn at all times.	■ **Supervisor** to monitor use of safety hats and protective footwear.	LG	from1/5/06	
Manual handling	All workers could suffer from back injury and long term pain if regularly lifting/carrying heavy or awkward objects	■ **Bricks** mortar etc. to be transported and lifted to scaffolding using telehandler provided by principal contractor. ■ **Provision** of lifting bay agreed with principal contractor. ■ **Bricks/blocks** to be covered with tarpaulin when stored on site to prevent taking up water. ■ **Spot boards** to be raised with blocks to easy working height. ■ **Trolley** to be used for moving loads of bricks around the scaffold lift. ■ **Check** at tender stage for any blocks or lintels over 20kg and make arrangements.	■ Heaviest blocks are 15kg no special arrangements necessary. ■ Concrete lintels are well over 20kg, to be positioned using telehandler (all are accessible) ■ All workers to be instructed not to carry materials up by hand.	VP VP LG	20/3/06 from1/5/06 from1/5/06	

C2 – Risk assessment bricklayer example, page 1

23

What are the hazards?	Who might be harmed and how?	What are you already doing?	What further action is necessary?	Action by who?	Action by when?	Done
Workers stuck or crushed by moving vehicles on site	Workers could suffer serious or even fatal injuries from vehicles and machines on site – particularly when reversing	■ **Manager** to agree safe route to work area with principal contractor based upon the construction phase health and safety plan. ■ **Induction** to each site to be carried out for all workers on first day.	■ Safe route agreed with principal contractor based.	TB	20/3/06	20/3/06
			■ Supervisor to liaise with site	LG	from 1/5/06	
			■ Manager to ensure safe route stays clear.	LG	from 1/5/06	
			■ Instruct staff that they must never drive vehicles and plant on this site.	LG	from 1/5/06	
			■ High-visibility vests to be provided.	LG	from 1/5/06	
			■ Supervisor to check vests are worn on all sites where the principal contractor requires them.	LG	from 1/5/06	
Slips and trips	All workers may suffer sprains or fractures if they trip over waste including brick bands and pallet debris. Slips at height could result in a serious fall.	■ **Good** housekeeping maintained at all times. ■ **Waste** including brick bands and pallet debris disposed of in skip. ■ **Safety** footwear provided to all workers. ■ **Safe** route to workplace agreed with principal contractor based upon the construction phase health and safety plan.	■ Temporary storage locations to be agreed with site manager	TB	20/2/06	20/3/06
			■ Supervisor to ensure that workers wear safety footwear whenever on site.	LG	from 1/5/06	
Stepping on nails and sharp objects	All workers could suffer foot injuries.	■ **Safety boots** with steel toecaps and mid-soles provided to all workers. ■ **Waste** disposed of in skips.	■ Explain the need to wear safety boots and dispose of waste in skips – repeat annually.	LG	1/5/06	1/5/06
			■ Supervisor to check that safety boots are always worn and waste disposed of properly.	LG	from 1/5/06	
Hazard to eyes, cutting bricks	Bricklayers could suffer eye injury through flying brick fragments.	■ **Safety goggles** (EN 166 B standard) worn when breaking bricks.	■ Use of goggles to be monitored by supervisor.	LG	from 1/5/06	
Hazardous substances, mortar	Direct skin contact with the mortar could also cause bricklayer contact dermatitis and burns.	■ **Risk of dermatitis** or cement burns and precautions explained to all workers. ■ **Use cement** or cement containing products within the use-by date. ■ **Direct skin contact** to be avoided, CE marked PVC gloves used when handling mortar. ■ **Good washing** facilities on site, with hot and cold water, soap and basins large enough to wash forearms. ■ **Principal** contractor's first-aid includes emergency eyewash	■ Training on how to treat exposure to be given to all operatives.	TB	17/4/06	26/4/06
			■ Supervisor to be aware of anyone with early signs of dermatitis.	LG	from 1/5/06	

C2 – Risk assessment bricklayer example, page 2

What are the hazards?	Who might be harmed and how?	What are you already doing?	What further action is necessary?	Action by who?	Action by when?	Done
Dust from cutting bricks	Dust exposure could cause silicosis.	■ **Angle grinders** replaced with block splitter, removing the risk of significant dust exposure. ■ **The use of** a grinder for chasing etc. is not needed on this job.				
Operating cement mixer	Workers could be crushed or cut if the mixer topples or they get caught in moving parts. Damage to electrics could result in a shock.	■ **Cement mixer** located on firm, level ground. ■ **Mixer** is fully guarded and guards in place during operation. ■ **Mixer** is 110 volt and PAT tested every three months.	■ Supervisor to check mixer daily for obvious damage.	LG	from 1/5/06	
Noise from use of equipment, eg angle grinder	Workers using grinders or working near people who may suffer hearing loss.	■ **Angle grinders** replaced with block splitter, removing high noise levels from our work. ■ **Construction** phase plan show other trades using grinders etc. should not be working close enough to cause problems.	■ Supervisor to monitor and talk to site manager if noisy work does start close by.	LG	from 1/5/06	
Vibration from use of equipment such as angle grinder	Exposure to vibration can lead to the development of 'vibration white finger' (VWF).	■ **Angle grinders** replaced with block splitter. No significant vibration left.				
Fire/explosion	All operatives in the vicinity could suffer from smoke inhalation or burns.	■ **Suitable** fire extinguisher kept in site office and welfare block. ■ **Good housekeeping** monitored by supervisor.	■ Supervisor to brief all workers on first day on emergency arrangements agreed with principal contractor.	LG	1/5/06	1/5/06

Welfare/ first-aid	Good facilities help prevent dermatitis etc.	■ **Principal contractor** will have facilities on site by the time bricklaying starts, including: -
		  o flushing toilet;
		  o hot and cold running water, soap, towels and full-size washbasins;
		  o heated canteen with kettle etc;
		  o first-aid equipment;
		  o principal contractor will arrange
		  o clearing and ensure the necessary
		  o electrical and heating safety checks are made;
		  o and site agent is appointed person for first-aid.
			■ Supervisor to brief workers on facilities and keeping them clean.	LG	1/5/06	1/5/06

C2 – Risk assessment bricklayer example, page 3

23

C3 EXAMPLE RISK ASSESSMENT FOR WOODWORK

Company name:
The Woodworking Company

Date of risk assessment:
28/09/07

IMPORTANT REMINDER

This example risk assessment shows the kind of approach a small business might take. Use it as a guide to think through some of the hazards in your business and the steps you need to take to control the risks. Please note that it is not a generic risk assessment that you can just put your company name on and adopt wholesale without any thought. This would not satisfy the law – and would not be effective in protecting people. **Every business is different – you need to think through the hazards and controls required in your business for yourself. Source HSE**

What are the hazards?	Who might be harmed and how?	What are you already doing?	What further action is necessary?	Action by who?	Action by when?	Done
Exposure to wood dust	Staff risk lung diseases, such as asthma, from inhaling wood dust. Hardwood dust can cause cancer, particularly of the nose	■ Local exhaust ventilation (LEV) provided at machines and staff are trained in using it properly. Bricklayers' supervisor to check with the site manager that the correct scaffold is provided and inspected. ■ LEV maintained to keep it in good condition and working effectively. LEV inspected every 14 months by a competent person. ■ Wood dust cleared up using a suitable vacuum cleaner, fitted with an appropriate filter. ■ Suitable respiratory protective equipment (RPE) as well as LEV for very dusty jobs, and staff trained in how to use it. ■ Staff do health surveillance questionnaire before starting, then annually. ■ Any affected staff referred to a medical professional.	■ Remind staff of the risks of wood dust, and why these controls are necessary. ■ Remind staff never to dry sweep wood dust, which just spreads the dust around.	Manager Manager	7/10/07 7/10/07	1/10/07 1/10/07
Machinery	Staff risk serious and possibly fatal cut injuries following contact with moving parts of machinery, particularly saw blades	■ All machines guarded according to manufacturers' instructions. ■ Guards inspected regularly and maintained as necessary to ensure their good condition. ■ Staff have sufficient space at machines to work safely. ■ Staff monitored by manager to ensure guards always used. ■ All staff trained in safe use of machines by a competent person. ■ All machines braked and fitted with necessary safety features, eg chip limited tooling etc.	■ Download information sheets on the safe use of the machines used in the workshop from HSE website and pin them up in mess room.	Manager	4/10/07	1/1/07
Manual Handling	Staff may suffer musculoskeletal disorders, such as back pain, from handling heavy/bulky objects, eg timber boards and machinery parts. Also risk cuts when handling tooling, or splinters when handling pallets.	■ Staff trained in manual handling. ■ Workbenches and machine tables set at a comfortable height. ■ Strong, thick gloves provided for handling tooling and pallets. ■ Panel trolley and lifting hooks available for moving boards. ■ Systems of work in place for the safe and careful handling of assembled furniture.	■ Where possible, store tooling next to the machine to reduce carrying distance. ■ Remind staff to ask for a new set of gloves when old ones show wear and tear, and not to try to lift objects that appear too heavy.	Manager Manager	30/9/07 4/4/07	29/9/07 4/4/07

C3 – Risk assessment woodwork example, page 1

What are the hazards?	Who might be harmed and how?	What are you already doing?	What further action is necessary?	Action by who?	Action by when?	Done
Noise	Staff and others may suffer temporary or permanent hearing damage from exposure to noise from woodworking machinery	■ Noise enclosures used where practicable, and maintained in good condition. ■ Low-noise tooling used where possible. ■ Planned maintenance programme for machinery and LEV systems. ■ Suitable hearing protectors provided for staff and staff trained how to use them. Check and maintain them according to advice given by supplier. ■ Staff trained in risks of noise exposure. ■ Staff trained in systems of work to reduce noise exposure (eg suitable feed rates for certain jobs, timber control etc).	■ Consider if certain machines could be safely mounted on anti-vibration mountings. ■ Include noise emission in specification for new vertical spindle moulder, to be purchased next year.	Manager Manager	30/9/07 30/9/07	29/9/07 29/9/07
Vehicles	Staff may suffer serious, possibly fatal, injuries if struck by a vehicle such as a lift truck or a delivery lorry.	■ Fork-lift truck maintained and inspected as per lease contract. ■ Lift truck operated only by staff who have been trained to use it. ■ Pedestrian walkways marked. ■ Only authorised people allowed in yard for deliveries/dispatch.	■ Ensure drivers get out of their vehicle and stand in a safe area while it is being loaded/unloaded.	Manager and all other staff	31/9/07	15/9/07
Slips, trips and falls	Staff could suffer injuries such as bruising or fractures if they trip over objects, or slip, eg on spillages, and fall.	■ Generally good housekeeping – off-cuts cleared away promptly, dust cleared regularly etc. ■ Staff wear strong safety shoes that have a good grip. ■ Good lighting in all areas.	■ Remind staff to clear up spillages of wax or polish immediately, even very minor spillages.	Manager	31/9/07	15/9/07
Electrical	Staff could get electrical shocks or burns from using faulty electrical equipment, eg machinery, or a faulty installation. Electrical faults can also lead to fires	■ Residual current device (RCD) built into main switchboard. ■ Staff trained to spot and report any defective plugs, discoloured sockets or damaged cable/equipment to manager. ■ No personal electrical appliances, eg toasters or fans, allowed.	■ Ask landlord when the next safety check of the electrical installation will be done. ■ Confirm with landlord the system for making safe any damage to building installation electrics, eg broken light switches or sockets.	Company secretary Company secretary	31/9/07 31/9/07	15/9/07 15/9/07
Work at height	Falls from any height can cause bruising and fractures	■ Strong stepladder, in good condition, provided. ■ Only trained, authorised staff allowed to work at height.	■ Condition of stepladder to be checked periodically.	Manager	31/9/07	as required
Fire	If trapped, staff could suffer fatal injuries from smoke inhalation/burns	■ Fire risk assessment done, see www.communities.gov.uk/fire and necessary action taken.	■ Ensure the actions identified as necessary by the fire risk assessment are completed.	Manager	from now	

C3 – Risk assessment woodwork example, page 2

Index